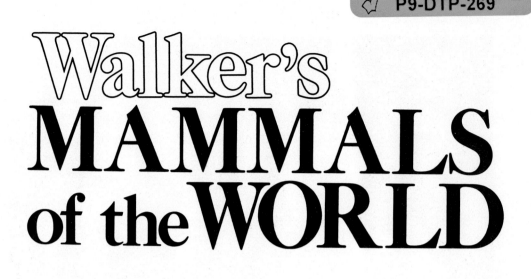

Walker's
MAMMALS
of the WORLD

Ernest P. Walker, 1891–1969

E

RIC AND U. S.

n

GTH

METERS AND FEET   KILOMETERS AND MILES

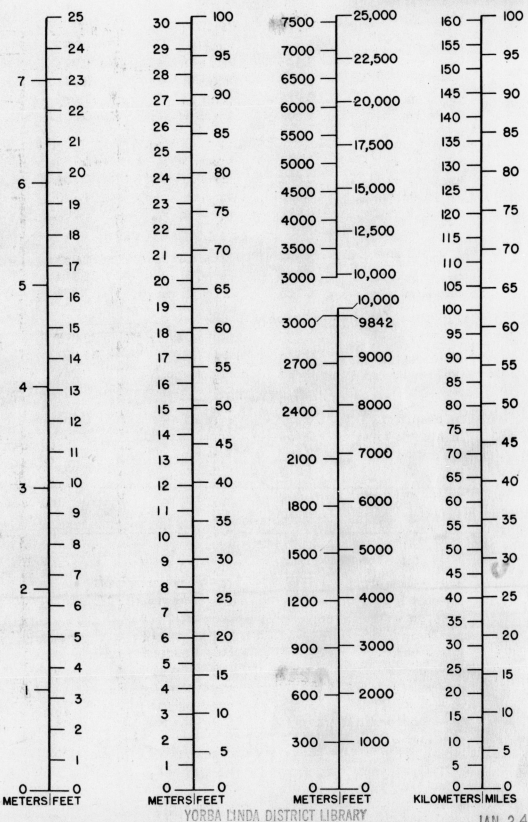

METERS	FEET
25	
24	
23	7
22	
21	
20	6
19	
18	
17	
16	5
15	
14	
13	4
12	
11	
10	3
9	
8	
7	
6	2
5	
4	
3	1
2	
1	
0	0

METERS│FEET

METERS	FEET
30	100
29	95
28	90
27	
26	85
25	
24	80
23	
22	75
21	70
20	65
19	
18	60
17	55
16	
15	50
14	45
13	
12	40
11	35
10	
9	30
8	25
7	
6	20
5	15
4	
3	10
2	5
1	
0	0

METERS│FEET

METERS	FEET
7500	25,000
7000	22,500
6500	
6000	20,000
5500	17,500
5000	
4500	15,000
4000	12,500
3500	
3000	10,000
	10,000
3000	9842
2700	9000
2400	8000
2100	7000
1800	6000
1500	5000
1200	4000
900	3000
600	2000
300	1000
0	0

METERS│FEET

KILOMETERS	MILES
160	100
155	95
150	
145	90
140	
135	85
130	80
125	
120	75
115	70
110	
105	65
100	60
95	
90	55
85	
80	50
75	45
70	
65	40
60	35
55	
50	30
45	
40	25
35	20
30	
25	15
20	10
15	
10	5
5	
0	0

KILOMETERS│MILES

	o Metric	Metric to U.S. Customary	

—— Length ——

To convert	Multiply by	To convert	Multiply by
in. to mm.	25.4	mm. to in.	0.039
in. to cm.	2.54	cm. to in.	0.394
ft. to m.	0.305	m. to ft.	3.281
yd. to m.	0.914	m. to yd.	1.094
mi. to km.	1.609	km. to mi.	0.621

—— Area ——

sq. in. to sq. cm.	6.452	sq. cm. to sq. in.	0.155
sq. ft. to sq. m.	0.093	sq. m. to sq. ft.	10.764
sq. yd. to sq. m.	0.836	sq. m. to sq. yd.	1.196
sq. mi. to ha.	258.999	ha. to sq. mi.	0.004

—— Volume ——

cu. in. to cc.	16.387	cc. to cu. in.	0.061
cu. ft. to cu. m.	0.028	cu. m. to cu. ft.	35.315
cu. yd. to cu. m.	0.765	cu. m. to cu. yd.	1.308

—— Capacity (liquid) ——

fl. oz. to liter	0.03	liter to fl. oz.	33.815
qt. to liter	0.946	liter to qt.	1.057
gal. to liter	3.785	liter to gal.	0.264

—— Mass (weight) ——

oz. avdp. to g.	28.35	g. to oz. avdp.	0.035
lb. avdp. to kg.	0.454	kg. to lb. avdp.	2.205
ton to t.	0.907	t. to ton	1.102
l. t. to t.	1.016	t. to l. t.	0.984

Abbreviations

U.S. Customary	Metric
avdp.—avoirdupois	cc.—cubic centimeter(s)
ft.—foot, feet	cm.—centimeter(s)
gal.—gallon(s)	cu.—cubic
in.—inch(es)	g.—gram(s)
lb.—pound(s)	ha.—hectare(s)
l. t.—long ton(s)	kg.—kilogram(s)
mi.—mile(s)	m.—meter(s)
oz.—ounce(s)	mm.—millimeter(s)
qt.—quart(s)	t.—metric ton(s)
sq.—square	
yd.—yard(s)	

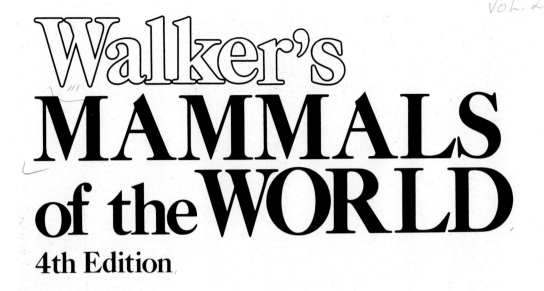

Walker's MAMMALS of the WORLD

4th Edition

Ronald M. Nowak • John L. Paradiso
Volume II

THE JOHNS HOPKINS UNIVERSITY PRESS
Baltimore and London
1983

The Johns Hopkins University Press, Baltimore, Maryland 21218
The Johns Hopkins Press Ltd., London

Originally published, 1964
Second Edition, 1968
Third Edition, 1975

Fourth Edition, 1983
Second printing, 1984

Library of Congress Cataloging in Publication Data

Walker, Ernest P. (Ernest Pillsbury), 1891–1969
Walker's Mammals of the world.

Rev. ed. of: Mammals of the world. 3rd ed. 1975.
Bibliography: pp. 1307–62
Includes index.
1. Mammals   2. Mammals—Bibliography.   I. Nowak,
Ronald M.   II. Paradiso, John L.   III. Title.
IV. Title: Mammals of the world.
QL703.W222   1983   599   82–49056
ISBN 0–8018–2525–3

The first three editions of this work appeared under the title *Mammals of the World* by Ernest P. Walker (senior author) and six coauthors: Florence Warnick, Sybil E. Hamlet, Kenneth I. Lange, Mary A. Davis, Howard E. Uible, and Patricia F. Wright. The second and third editions were revised by John L. Paradiso.

Frontispiece: Photograph by Eugene Maliniak

CONTENTS

Volume I

Volume II

Walker's
MAMMALS
of the WORLD

RODENTIA; *Family MURIDAE*

Rats, Mice, Hamsters, Voles, Lemmings, Gerbils

This family of 228 living genera and 1,063 species is by far the largest mammalian family. Its natural range extends worldwide, except for certain Arctic islands, parts of the West Indies, New Zealand, many oceanic islands, and Antarctica. Introduction through human agency has led to the establishment of several species of murids on many of the islands where the family originally was absent.

There has long been controversy regarding what groups are included in this family and how these groups are interrelated. Many authorities have tended to follow Simpson's (1945) arrangement, shown in the left column of the following table, for those groups assigned in this book to the family Muridae. To the right is the classification proposed by Chaline, Mein, and Petter (1977), which is based in part on recent paleontological evidence.

In this book the actual sequence of the right column has been generally followed. However, all of the families have been made parts of a single family, which in accordance with nomenclatural priority is called the Muridae. Although Simpson (1945) separated the Muridae and Cricetidae, many authorities have considered the differences between the two not to warrant familial distinction. The latter position was taken by Hershkovitz (1962) and Hall (1981), and is followed here. Once the Cricetidae are made part of the

Muridae, it becomes reasonable to do the same with the other families listed by Chaline, Mein, and Petter (1977). Indeed, relatively few authorities have treated these other groups as full families, except for the Spalacidae and Rhizomyidae. While there now seems little doubt that the Spalacidae do not deserve more than subfamilial status within the Muridae (Chaline, Mein, and Petter 1977), there remains considerable support for keeping the Rhizomyidae as a separate family (Lekagul and McNeely 1977; Medway 1978). Chaline, Mein, and Petter's newly proposed subfamilies, Tachyoryctinae and Taterillinae, are not used here. Also, in accordance with Hall (1981) and Corbet (1978), only a single microtine subfamily, with the name Microtinae, is accepted here.

Reig (1980) argued that what are here called the Hesperomyinae should be divided into two subfamilies: the Sigmodontinae, with mainly South American genera; and the Neotominae, with mainly North American genera. Carleton (1980), however, thought it premature to formalize such a division taxonomically. The work of both these authorities suggests the need for some modification of the systematic arrangement accepted here.

The sequence of genera within subfamilies used here basically follows that of Simpson (1945), but has been adjusted in accordance with the arrangements of Hall (1981), Corbet (1978), Cabrera (1961), and Misonne (1969). Changes in the sequence also have been necessitated by the numerous systematic modifications called for by the various authorities cited in the generic accounts. This book's overall sequence of murid subfamilies and genera is shown on the next page.

Simpson (1945)	Chaline, Mein, and Petter (1977)
Family Cricetidae	Family Cricetidae
Subfamily Cricetinae	Subfamily Hesperomyinae (New World Rats and Mice)
Tribe Hesperomyini (New World Rats and Mice)	Subfamily Cricetinae (Hamsters)
Tribe Cricetini (Hamsters)	Subfamily Spalacinae (Blind Mole-rats)
Tribe Myospalacini (East Asian Mole-rats)	Subfamily Myospalacinae (East Asian Mole-rats)
Subfamily Nesomyinae (Malagasy Rodents)	Subfamily Lophiomyinae (Maned Rats)
Subfamily Lophiomyinae (Maned Rats)	Subfamily Platacanthomyinae (Spiny Dormice and Chinese Pygmy Dormice)
Subfamily Microtinae	Family Nesomyidae
Tribe Lemmini (Lemmings)	Subfamily Nesomyinae (Malagasy Rodents)
Tribe Microtini (Voles, Muskrats)	Subfamily Otomyinae (African Swamp Rats)
Tribe Ellobiini (Mole-voles)	Family Rhizomyidae
Subfamily Gerbillinae (Gerbils)	Subfamily Rhizomyinae (Bamboo Rats)
Family Spalacidae (Blind Mole-rats)	Subfamily Tachyoryctinae (African Mole-rats)
Family Rhizomyidae (African Mole-rats and Bamboo Rats)	Family Gerbillidae
Family Muridae	Subfamily Gerbillinae (Pygmy Gerbils)
Subfamily Murinae (Old World Rats and Mice)	Subfamily Taterillinae (Gerbils)
Subfamily Dendromurinae (Climbing Mice)	Family Arvicolidae (Simpson's Microtinae)
Subfamily Otomyinae (African Swamp Rats)	Subfamily Lemminae
Subfamily Phloeomyinae (Cloud Rats)	Subfamily Dicrostonychinae
Subfamily Rhynchomyinae (Shrew-rats)	Subfamily Arvicolinae
Subfamily Hydromyinae (Water Rats)	Family Dendromuridae
Family Platacanthomyidae (Spiny Dormice and Chinese Pygmy Dormice)	Subfamily Dendromurinae (Climbing Mice)
	Subfamily Petromyscinae (African Rock and Swamp Mice)
	Family Cricetomyidae (African Pouched Rats, part of Simpson's Murinae)
	Family Muridae
	Subfamily Murinae (Old World Rats and Mice, including Simpson's Phloeomyinae)
	Subfamily Hydromyinae (Water Rats, including Simpson's Rhynchomyinae)

Subfamily Hesperomyinae

Oryzomys	Akodon	Graomys
Megalomys	Cabreramys	Andinomys
Neacomys	Zygodontomys	Chinchillula
Abrawayaomys	Podoxymys	Irenomys
Scolomys	Lenoxus	Punomys
Nectomys	Juscelinomys	Euneomys
Rhipidomys	Oxymycterus	Neotomys
Thomasomys	Blarinomys	Reithrodon
Megaoryzomys	Notiomys	Holochilus
Phaenomys	Scapteromys	Sigmodon
Chilomys	Kunsia	Neotomodon
Tylomys	Bibimys	Neotoma
Ototylomys	Scotinomys	Xenomys
Nyctomys	Wiedomys	Nelsonia
Otonyctomys	Calomys	Ichthyomys
Rhagomys	Pseudoryzomys	Rheomys
Reithrodontomys	Andagalomys	Anotomys
Peromyscus	Eligmodontia	Neusticomys
Ochrotomys	Phyllotis	Daptomys
Baiomys	Galenomys	
Onychomys	Auliscomys	

Subfamily Cricetinae

Calomyscus	Cricetus	Mesocricetus
Phodopus	Cricetulus	Mystromys

Subfamily Spalacinae

Spalax

Subfamily Myospalacinae

Myospalax

Subfamily Lophiomyinae

Lophiomys

Subfamily Platacanthomyinae

Platacanthomys	Typhlomys

Subfamily Nesomyinae

Macrotarsomys	Eliurus	Hypogeomys
Nesomys	Gymnuromys	Brachyuromys
Brachytarsomys		

Subfamily Otomyinae

Parotomys	Otomys

Subfamily Rhizomyinae

Rhizomys	Cannomys	Tachyoryctes

Subfamily Gerbillinae

Gerbillus	Taterillus	Sekeetamys
Dipodillus	Desmodillus	Meriones
Microdillus	Desmodilliscus	Brachiones
Gerbillurus	Pachyuromys	Psammomys
Tatera	Ammodillus	Rhombomys

Subfamily Microtinae

Clethrionomys	Alticola	Dinaromys
Eothenomys	Hyperacrius	Phenacomys

(continued)

Subfamily Microtinae   (continued)

Arvicola	Ondatra	Dicrostonyx
Microtus	Lemmus	Prometheomys
Lagurus	Myopus	Ellobius
Neofiber	Synaptomys	

Subfamily Dendromurinae

Dendromus	Dendroprionomys	Deomys
Megadendromus	Prionomys	Leimacomys
Malacothrix	Steatomys	

Subfamily Petromyscinae

Petromyscus	Delanymys

Subfamily Cricetomyinae

Beamys	Saccostomus	Cricetomys

Subfamily Murinae

Lenothrix	Mesembriomys	Praomys
Kadarsanomys	Conilurus	Millardia
Diplothrix	Leporillus	Anonymomys
Margaretamys	Zyzomys	Srilankamys
Lenomys	Pseudomys	Chiromyscus
Papagomys	Notomys	Dacnomys
Komodomys	Leggadina	Diomys
Eropeplus	Mastacomys	Maxomys
Chiruromys	Lorentzimys	Niviventer
Pogonomys	Haeromys	Leopoldamys
Pithecheir	Lophuromys	Rattus
Crateromys	Zelotomys	Tryphomys
Mallomys	Colomys	Limnomys
Hyomys	Malacomys	Tarsomys
Batomys	Acomys	Melasmothrix
Carpomys	Uranomys	Tateomys
Tokudaia	Hybomys	Nesoromys
Anisomys	Stochomys	Uromys
Phloeomys	Rhabdomys	Solomys
Apodemus	Lemniscomys	Melomys
Chiropodomys	Pelomys	Pogonomelomys
Vernaya	Aethomys	Xenuromys
Vandeleuria	Arvicanthis	Apomys
Micromys	Dasymys	Mus
Hapalomys	Mylomys	Muriculus
Thallomys	Golunda	Macruromys
Thamnomys	Hadromys	Crunomys
Stenocephalemys	Bandicota	Echiothrix
Oenomys	Nesokia	

Subfamily Hydromyinae

Rhynchomys	Xeromys	Parahydromys
Chrotomys	Hydromys	Crossomys
Celaenomys	Pseudohydromys	Mayermys.
Leptomys	Microhydromys	
Paraleptomys	Neohydromys	

The various groups of murids differ substantially in form and function. Total length ranges from less than 100 mm in *Baiomys* to 800 mm or more in *Phloeomys* and *Cricetomys*. The 16 murid subfamilies can be briefly described as follows:

Hesperomyinae, mostly small animals, generalized in structure;

Cricetinae, mouselike animals with a thickset body, short tail, and cheek pouches;

Spalacinae, molelike animals with a heavyset body, short legs, and no external openings for the eyes;

Myospalacinae, stout animals resembling pocket gophers in appearance;

Lophiomyinae, arboreal, bushy tailed rats of Africa;

Platacanthomyinae, rodents resembling dormice in external form;

Nesomyinae, a highly variable group of ratlike animals found on Madagascar;

Otomyinae, African rodents resembling the microtines, and with large, high-crowned molars;

Rhizomyinae, burrowing rodents resembling pocket gophers in external appearance;

Gerbillinae, African and Asian animals, often resembling jerboas, kangaroo rats, and other genera having a jumping mode of progression;

Microtinae, a Holarctic group with a heavyset body, a tail shorter than the head and body (often shorter than half the head and body length), a bluntly rounded muzzle, and short legs;

Dendromurinae, an African group, sometimes considered arboreal, and usually with the hind feet naked;

Petromyscinae, small mice with a long tail, distinguished mainly by characters of the molar teeth;

Cricetomyinae, African rodents with large internal cheek pouches;

Murinae, the generalized Old World rats and mice, differing from the Hesperomyinae mainly in possessing a functional row of tubercles on the inner side of the upper molars; and

Hydromyinae, shrewlike or muskratlike rodents found in parts of the Australian-Oriental region and characterized chiefly by a reduction in the dentition.

The usual murid dental formula is: (i 1/1, c 0/0, pm 0/0, m 3/3) × 2 = 16. In certain Old World genera, and in *Daptomys* of the New World, the molars are only 2/2, and in *Mayermys*, a member of the Hydromyinae from New Guinea, the molars are 1/1. Murid molars may be rooted or rootless, and either laminate, cuspidate, or prismatic.

Murids are found in a wide variety of habitats. Most species are primarily terrestrial, but some are arboreal, fossorial, or semiaquatic. They shelter in tunnels or crevices, under logs or other suitable objects, in hollow trees or logs, or in nests built on the surface or in bushes or trees. They may be diurnal or nocturnal and are usually active throughout the year. Most genera feed on plant material and invertebrates. Some take small vertebrates, including fish. Seeds and other plant matter are sometimes stored for winter use.

Some members of this family are gregarious, even highly social, while others tend to live alone or in pairs. Breeding may occur throughout the year in warm regions. Females often produce several litters annually. Most individuals probably live less than two years in the wild. Nonetheless, the high reproductive potential of certain species, especially the microtines, sometimes results in remarkable numerical increases. Peak populations are often followed by a sudden crash in numbers, when food supplies in a given area are exhausted. These fluctuations may show a rather cyclic periodicity of around three or four years.

Several diseases, such as plague, tularemia, leptospirosis, and Rocky Mountain spotted fever, can be transmitted from murid rodents to people. Certain murid species are destructive to crops, young trees, and stored human food supplies.

The geological range of this family is Oligocene to Recent in North America, Europe, and Asia; Pliocene to Recent in South America; Recent in Africa; and Pleistocene to Recent in Madagascar and Australia.

RODENTIA; MURIDAE; *Genus ORYZOMYS Baird, 1857*

Rice Rats

There seem to be 7 subgenera and 59 species (Cabrera 1961; Hall 1981; Gardner and Patton 1976; Hershkovitz 1960, 1966a, 1970, 1971; Myers and Carleton 1981; Hutterer and Hirsch 1979; Spitzer and Lazell 1978; Massoia 1973; Massoia and Fornes 1967; Pine 1971; Husson 1978; Handley 1976; Allen 1942; Myers and Wetzel 1979):

subgenus *Oryzomys*

O. palustris, southeastern United States, Mexico to Panama, southern Baja California;

O. argentatus, lower Florida Keys;

O. nelsoni, Maria Madre Island off western Mexico;

O. fulgens, probably central Mexico;

O. dimidiatus, southeastern Nicaragua;

O. gorgasi, northwestern Colombia;

O. capito, eastern Costa Rica to the Guianas and northern Argentina, Trinidad;

O. villosus, northern Colombia;

O. macconnelli, eastern Ecuador and Peru to the Guianas;

O. xantheolus, western Peru, possibly southwestern Ecuador;

O. yunganus, eastern Bolivia;

O. galapagoensis, San Cristobal Island in the Galapagos;

O. bauri, Santa Fe Island in the Galapagos;

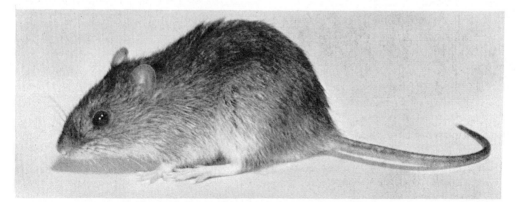

Rice rat (*Oryzomys palustris*), photo by Ernest P. Walker.

O. melanotis, southern Sinaloa and Tamaulipas (Mexico) to El Salvador;

O. caudatus, Oaxaca (southern Mexico);

O. alfaroi, eastern and southern Mexico to northwestern Ecuador;

O. bombycinus, southeastern Nicaragua to northern Ecuador;

O. albigularis, Costa Rica, Panama, the Andes from western Venezuela to northwestern Bolivia;

O. auriventer, the Andes of Ecuador and Peru;

O. nitidus, the Andes of Peru and Bolivia;

O. aphrastus, central Costa Rica;

O. intectus, the Andes of central Colombia;

O. balneator, eastern and southern Ecuador;

O. polius, northern Peru;

O. melanostoma, eastern Peru;

O. subflavus, the Guianas, eastern Brazil;

O. chaparensis, eastern Bolivia;

O. buccinatus, Paraguay, northern Argentina;

O. ratticeps, Paraguay, extreme southern Brazil, northeastern Argentina;

O. hammondi, the Andes of Ecuador;

subgenus *Oligoryzomys*

O. victus, St. Vincent Island in the Lesser Antilles;

O. fulvescens, northeastern and southwestern Mexico to Venezuela;

O. delicatus, Colombia to Surinam and northern Brazil;

O. microtis, Amazonian Brazil;

O. munchiquensis, western Colombia;

O. fornesi, Bolivia, southern Brazil, Paraguay;

O. flavescens, northeastern Argentina, Uruguay;

O. andinus, northwestern Peru;

O. arenalis, northeastern Peru;

O. spodiurus, east slope of the Andes in Ecuador;

O. longicaudatus, Peru, Chile, southern Argentina, Tierra del Fuego;

O. delticola, northeastern Argentina, Uruguay;

O. nigripes, Paraguay, southern Brazil, Uruguay, Argentina;

O. chacoensis, central Brazil, Bolivia, Paraguay, northern Argentina;

O. utiaritensis, central Brazil;

O. mattogrossae, central Brazil;

subgenus *Microryzomys*

O. minutus, the Andes from northwestern Venezuela to Peru;

O. altissimus, the Andes of Ecuador and Peru;

subgenus *Melanomys*

O. caliginosus, Honduras to Ecuador;

O. robustulus, east slope of the Andes in Ecuador;

O. zunigae, west-central Peru;

subgenus *Oecomys*

O. bicolor, tropical zone from Panama to the Guianas and Bolivia;

O. concolor, tropical and subtropical forest zones from Costa Rica to the Guianas and Paraguay, Trinidad;

subgenus *Sigmodontomys*

O. alfari, extreme eastern Honduras to northwestern Venezuela and northern Ecuador;

subgenus *Nesoryzomys*

O. darwini, Santa Cruz Island in the Galapagos;

O. fernandinae, Fernandina Island in the Galapagos;

O. indefessus, Santa Cruz Island in the Galapagos;

O. narboroughi, Fernandina Island in the Galapagos;

O. swarthi, San Salvador Island in the Galapagos.

Certain aspects of the systematics of this genus are indefinite, and the cited sources provide only a partial basis for the sequence of species given above. Gardner and Patton (1976) recommended that both *Oecomys* Thomas, 1906 and *Nesoryzomys* Heller, 1904 be given full generic rank. Gardner and Patton also treated *O. altissimus* as a subspecies of *O. andinus,* stated that *O. albigularis* is actually a composite of several closely related species, and suggested that the name *O. rivularis* might have priority over *O. capito.* The subgenus *Sigmodontomys* was once considered part of the genus *Nectomys,* but is now generally included within *Oryzomys* (Hershkovitz 1970; Gardner and Patton 1976; Hall 1981). The subgenus *Melanomys* Thomas, 1902 has sometimes been treated as a distinct genus. Two other former subgenera of *Oryzomys*—*Micronectomys* and *Macruroryzomys*—were called *nomina nuda* by Hershkovitz (1970). The species *O. dimidiatus* was once put in the genus *Nectomys,* then made the type species of the subgenus *Micronectomys,* and now is considered closely related to *O. palustris* (Hershkovitz 1970; Hall 1981). Another species of *Micronectomys,* formerly designated *Oryzomys borreroi,* actually seems referable to the genus *Zygodontomys* (Gardner and Patton 1976). The only species of *Macruroryzomys, O. hammondi,* is still designated a species of *Oryzomys,* but its precise systematic position is uncertain and it seems closely related to the extinct genus *Megalomys* (Hershkovitz 1966*b,* 1970). Hershkovitz (1966*a,* 1966*b*) suggested that most species of the subgenus *Oligoryzomys* are synonymous with *O. nigripes,* but this view was not followed by Gardner and Patton (1976), Hall (1981), Massoia and Fornes (1967), or Myers and Carleton (1981). Cabrera (1961) listed the species *O. simplex* of eastern Brazil, but noted that its status was doubtful, and Massoia (1980) referred it to the genus *Pseudoryzomys.* Benson and Gehlbach (1979) considered *O. couesi,* found from southern Texas to Central America, to be a species distinct from *O. palustris.* This procedure was followed by Jones, Carter, and Genoways (1979), but not by Hall (1981). Haiduk, Bickham, and Schmidly (1979) found the karyotype of *O. couesi* to be similar to that of *O. palustris texensis.*

Head and body length is 93 to 203 mm, tail length is 75 to 251 mm, and weight is usually 40 to 80 grams. The upper parts are grayish brown to ochraceous tawny, mixed with black; the sides are paler, with less black; the underparts are white to pale buff; and the tail varies from brownish above and whitish below to uniformly dusky. The four longer hind toes in members of the subgenus *Oligoryzomys* bear tufts of silvery bristles that project beyond the ends of the claws. The form in *Oryzomys* is mouselike; the pelage is coarse, but not bristly or spiny; the tail is usually long, with the annulations showing through the sparse hairs; and females have eight mammae. Rice rats may be confused with cotton rats (*Sigmodon*), but the latter have longer, grizzled fur and shorter, stouter tails.

Rice rats live in a variety of habitats including forests, marshes, grassy areas, and brush-covered parts of mountains. In Venezuela, Handley (1976) collected *Oryzomys* under the following conditions: *O. albigularis,* mostly on the ground in moist parts of forests; *O. bicolor,* mostly in trees, but often on the ground, in either moist or dry parts of open country or forest; *O. capito,* mostly in houses near streams or other moist parts of forests or forest openings; *O. concolor* and *O. fulvescens,* in a variety of forested and nonforested areas, mainly in moist places; and *O. macconnelli* and *O. minutus,* mainly on the ground in montane forests.

O. palustris, the best-known species, may be active at any hour of the day throughout the year. It swims and dives readily, on the surface or underwater. Esher, Wolfe, and Layne (1978) reported its swimming speed to be 0.55 meters/second. Its presence in an area can be confirmed by the feeding platforms it constructs by bending vegetation, and by its woven grassy nests, about 457 mm in diameter. The nest is usually placed in a slight depression in the ground in a tangle of vegetation, but may be located 1 meter above ground in areas subject to periodic inundation (Lowery 1974). In drier areas, soil burrows are constructed. The diet includes the succulent parts of grasses and sedges, seeds, fruit, insects, crustaceans, and small fish.

Benson and Gehlbach (1979) reported a population density of 1.5 individuals per 100 sq meters for *O. palustris* (which they called *O. couesi*) in southern Texas. In the Panama Canal Zone, Fleming (1971) found *O. capito* to have a maximum density of 4.3/ha. and an average home range of 1.33 ha. Home ranges overlapped. Breeding in this area occurred throughout the year, with no apparent peaks, and females produced an average of 6.06 litters per year. The mean gestation period was 28.2 days, and litter size averaged 3.29 and ranged from 2 to 5. The average life span was only 4.62 months, with few individuals surviving more than a year.

In coastal Louisiana, according to Lowery (1974), *O. palustris* breeds mainly from February to November. Females are capable of producing seven litters per year, but probably average only five or six, because of their short life span. They have a postpartum estrus and usually mate within 10 hours of giving birth. In this species gestation lasts 25 days, and litter size is one to seven, usually three to five. Newborn *O. palustris* weigh from two to five grams, open their eyes when 6 days old, are usually weaned and independent at 11 to 13 days, and are full grown at 4 months. Females apparently are able to breed when about 7 weeks old. Average longevity is probably much less than 1 year.

The northern limit of the range of *O. palustris* varies with fluctuations in population density. Occurrences in New Jersey, for example, long appeared to be sporadic and the species has actually been considered threatened in that state. Arndt, Rohde, and Bosworth (1978), however, found *O. palustris* to be not uncommon along the shores of Delaware Bay in southern New Jersey.

The silver rice rat (*Oryzomys argentatus*) was discovered in 1973 in a small marshy area on Cudjoe Key, off the south coast of Florida. Although it may eventually be shown to have a wider range, it apparently is extremely rare and on the verge of being eliminated through human modification of its habitat (Layne 1978; Spitzer and Lazell 1978).

A number of other island species or subspecies already seem to have become extinct (Allen 1942; Goodwin and Goodwin 1973). *O. palustris antillarum* of Jamaica apparently disappeared in the 1880s, probably because of predation by the introduced mongoose and Norway rat. *O. victus* of St. Vincent in the Lesser Antilles is known only by a single specimen presented to the British Museum in 1897, and probably also was destroyed by the mongoose. The rice rats of the Galapagos, among the few mammals native to those islands, apparently have suffered through competition with introduced Norway and black rats. *O. galapagoensis* of San Cristobal Island has not been seen since 1835, *O. swarthi* of San Salvador was last collected in 1906, and *O. indefessus* of Santa Cruz disappeared about 1945.

RODENTIA; MURIDAE; Genus MEGALOMYS Trouessart, 1881

West Indian Giant Rice Rats

The following three species apparently survived into historical times (Hall 1981):

M. desmarestii, known only from Martinique;
M. audreyae, known only from Barbuda;
M. luciae, known only from Santa Lucia.

According to Hershkovitz (1966b, 1970), two other fossil species have been reported: *M. curazensis* from Curacao and *M. curioi* from Santa Cruz Island in the Galapagos. The latter species, however, was recently placed in a new genus, *Megaoryzomys* (Lenglet and Coppois 1979). The closest living relative to *Megalomys* appears to be *Oryzomys hammondi* of northwestern Ecuador, and that species may represent an ancestral stock, once widespread in northern South America.

Only a few complete specimens are in existence, including two mounted examples in the National Museum of Natural History in Paris. The largest species, *M. desmarestii*, had a head and body length of about 360 mm and a tail length of about 330 mm. The fur was long and harsh, but not spiny. The upper parts were glossy black or dark reddish brown.

West Indian giant rice rat (*Megalomys* sp.), photo by F. Petter.

The chin, throat, underparts, and base of the tail were white. The well-developed ears were nearly naked. *M. luciae* was smaller and almost all brown. *M. audreyae* is known only from a mandibular fragment and an upper incisor.

Megalomys may once have occupied most of the islands of the Lesser Antilles, but detailed information is available only for *M. desmarestii* on Martinique (Allen 1942). This species apparently occurred in large numbers among the coconut plantations. It is said to have lived in burrows and to have taken to water when driven from shelter. It was commonly eaten by the people, and the European colonists actively tried to exterminate it because of its damage to crops. Its final extinction, however, has been attributed to the great volcanic eruption of Mount Pelee in 1902. *M. luciae* probably was exterminated during the nineteenth century, though one captive individual may have lived from 1849 to 1852 at the London Zoo. *M. audreyae* apparently disappeared soon after the occupation of Barbuda by Europeans and the consequent destruction of cover.

RODENTIA; MURIDAE; **Genus NEACOMYS** Thomas, *1900*

Bristly Mice, Spiny Rice Rats

There are three species (Cabrera 1961; Hall 1981; Handley 1976; Husson 1978):

N. tenuipes, eastern Panama, Colombia, Venezuela, Ecuador;

N. spinosus, Colombia, Ecuador, Peru, Mato Grosso of central Brazil;

N. guianae, southern Venezuela, Guyana, Surinam, probably northern Brazil.

Head and body length is 64 to 100 mm, and tail length is about the same. A male *N. guianae* weighed 20 grams (Husson 1978). The hairy coat is composed of a mixture of tubular bristles, interspersed with slender and fairly soft hairs. The spines are most abundant on the back, fewer on the sides, and almost absent on the belly. The presence of spines is the most conspicuous difference between this genus and *Oryzomys*.

The coloration above ranges from dark rufous, dark fulvous, or orange rufous to bright ochraceous, and is finely lined and darkened with the black tips of the spines. The coloration on the sides becomes lighter and clearer as the numbers and extent of the dark tips of the spines are reduced. In some forms the sides are clear fulvous. The underparts are white, creamy, or fulvous. The tail is scantily haired, brownish above and lighter beneath.

The altitudinal range is from sea level to about 1,100 meters. These rodents have been taken under rocks and logs in dense humid forests, and, in Panama, among grass and bushes along the rocky edge of a sugar cane field. Several pregnant females, with two to four embryos each, were collected in November and December.

Spiny rice rat (*Neacomys* sp.), photo by Bruce J. Hayward. Skull: *N.* sp., photos from American Museum of Natural History.

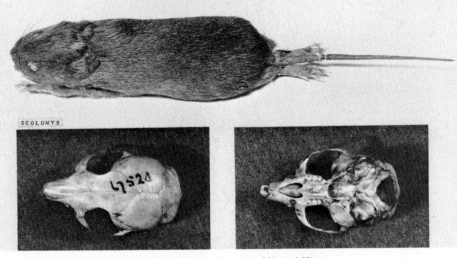

Spiny mouse (*Scolomys melanops*), photos from American Museum of Natural History.

RODENTIA; MURIDAE; **Genus ABRAWAYAOMYS** *Cunha and Cruz, 1979*

The single species, *A. ruschii*, is known only from the type locality in the state of Espirito Santo, southeastern Brazil (Cunha and Cruz 1979).

The type specimen, a young female, has a head and body length of 201 mm, a tail length of 85 mm, and a weight of 46 grams. The general coloration of the upper parts is grayish yellow, the head is darker, and the underparts are yellowish white. The pelage is composed of flattened and grooved spines, mixed with finer hairs. The spines are especially numerous on the back.

Abrawayaomys resembles *Neacomys* in texture of pelage, but is much larger and has longer dorsal hairs, a broader face, a less rounded braincase, a proportionally larger palatal foramen, a more reduced interparietal, and procumbent incisors. The molar teeth resemble those of *Oryzomys* and *Akodon*.

RODENTIA; MURIDAE; **Genus SCOLOMYS** *Anthony, 1924*

Spiny Mouse

The only information available is contained in the original description of the genus and single species, *S. melanops*. Six specimens were taken at Mera in eastern Ecuador at an elevation of about 1,150 meters by G.H.H. Tate in 1924; these specimens are in the American Museum of Natural History. Apparently no others have been collected since.

Head and body length is about 90 mm, and tail length is about 70 mm. The pelage, except for a small patch on the throat, consists of flattened spines mixed with longer, unmodified hairs. The six specimens show great differences in color, but all are some shade of brown above and dull gray below. The upper parts give an impression of sooty black flecked with brown. The black is from the tips of the spines and the brown is from the tips of the unmodified hairs. The color of the head is about the same as that of the back. The color of the sides merges from the gray of the underparts to the black and brown of the upper parts. The hands and feet are gray, and the digits are whitish. The finely haired and annulated tail is the same color as the back above, but is somewhat lighter below.

The thumb has a broad, flat nail, and the proportions of the digits are the same as in *Neacomys*. Aside from the darker coloration, *Scolomys* closely resembles *Neacomys* in all external characteristics. In cranial characters, however, these two genera are different. The skull of *Scolomys*, for example, is short and broad, whereas that of *Neacomys* is long and slender. Female *Scolomys* have six mammae.

RODENTIA; MURIDAE; **Genus NECTOMYS** *Peters, 1861*

Neotropical Water Rats

There are two species (Cabrera 1961; Petter 1979; Goodwin and Greenhall 1961):

N. squamipes, northern and central South America, Trinidad;
N. parvipes, French Guiana.

In *N. squamipes* head and body length is 160 to 255 mm, and tail length is 165 to 250 mm. The body and hind feet of this species seem to grow in length after maturity and into old age. A series of 58 adults weighed 160 to 420 grams. The upper parts are buffy to tawny, with mixtures of brown, and the sides are paler. The underparts are not sharply defined; they are whitish or grayish, with a pale to pronounced wash of reddish orange, at least on the chest and belly. The pelage has long, glossy guard hairs. There is a swimming fringe or keel of short, stiff hairs on the underside of the tail, and a distinct fringe on the hind foot. The hind foot is large and strongly built; it is longer than wide, and the three middle digits are the longest. The webbing of the toes is well developed. Husson (1978) stated that female *N. squamipes* have two pairs of pectoral and two pairs of inguinal mammae.

These rats are usually found in forests, up to about 2,200 meters in elevation. They generally occur near a swamp, lake, or stream. They are good, fast swimmers. Nests are built on the ground under old logs or brush heaps. Linares

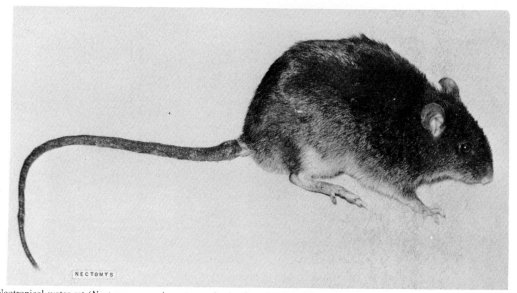

Neotropical water rat (*Nectomys squamipes amazonicus*), photo by Cory T. de Carvalho.

(1969) reported *N. squamipes* to occupy and nest in a cave, through which an underground stream passed.

RODENTIA; MURIDAE; *Genus RHIPIDOMYS* Tschudi, 1844

Climbing Mice

The following seven species (not in systematic order) were listed by Cabrera (1961), Hall (1981), and Hershkovitz (1960):

R. latimanus, Colombia, Venezuela, Ecuador;
R. leucodactylus, the Andes from Ecuador to northwestern Argentina;
R. macconnelli, southeastern Venezuela and probably adjacent parts of Guyana and northern Brazil;
R. maculipes, eastern Brazil;
R. mastacalis, Venezuela, the Guianas, Brazil;
R. scandens, extreme eastern Panama;
R. sclateri, Venezuela, the Guianas, Trinidad.

Handley's (1976) list indicated the following differences: *R. couesi* of northern Venezuela and Trinidad is a species distinct from *R. sclateri*, *R. fulviventer* of Colombia and Venezuela is a species distinct from *R. latimanus*, *R. leucodactylus* occurs in the Orinoco lowlands of Venezuela, *R. venezuelae* of western Venezuela is a species distinct from *R. mastacalis*, and *R. venustus* of northwestern Venezuela is a species distinct from *R. latimanus*.

Head and body length is 80 to 210 mm. The tail is normally longer than the head and body and may be as much as 270 mm in length. The coloration above ranges from pale grayish buffy, clay color, tawny ochraceous, and dull buffy, through fulvous and brown, to dark brown (almost black). The middle of the back is usually the darkest part of the body, because the dark tips of the hairs present in this area are fewer or completely lacking over the rest of the body. The underparts are white, creamy, buffy, fulvous, or grayish. The line of demarcation between the color of the sides and the underparts is usually sharply defined. In some species a lateral line of buffy or yellow color separates the upper and lower surfaces. The digits are generally whitish, yellowish, or buffy, and the upper surfaces of the hands and feet are usually slightly darker than the remainder of the limbs. The tail is slightly darker than the back and is rarely bicolored.

In most species the fur is dense and soft, almost velvety. It

Climbing mouse (*Rhipidomys mastacalis nitela*), photo from Zoological Society of London.

varies from medium to long, though it may be short and dense. The tail is well haired and tufted terminally. The feet and hands are rather large and broad, and are adapted to arboreal life. The fifth toe is quite long, and the claws of both the hands and feet are large and efficiently curved for climbing. Females have six mammae.

Most specimens collected by Handley (1976) in Venezuela were taken in moist parts of forests; some were found on the ground, some in trees, and some in houses. The only known specimen of *R. scandens* was shot in a tree 10.5 meters from the ground. This individual was active at dusk. Husson (1978) reported that all known specimens of *R. mastacalis* from Surinam were taken in buildings. Of the 12 female specimens, 3 were pregnant and 2 had newborn young; litter size was 3 to 5. Females, apparently of this same species, with 2 or 3 young each, were observed near Rio de Janeiro, Brazil, in October and December.

RODENTIA; MURIDAE; *Genus THOMASOMYS* Coues, 1884

Thomas's Paramo Mice

The 25 species (not in systematic order) are as follows (Cabrera 1961; Handley 1976; Pine 1980):

T. aureus, the Andes from western Venezuela to Peru;
T. boeops, western Ecuador;
T. bombycinus, the Andes of Colombia;
T. cinereiventer, the Andes of Colombia and Ecuador;
T. cinereus, southwestern Ecuador, northwestern Peru;
T. daphne, southwestern Peru, central Bolivia;
T. dorsalis, eastern Brazil;
T. gracilis, the Andes of Ecuador and southeastern Peru;
T. hylophilus, northern Colombia, western Venezuela;
T. incanus, the Andes of Peru;
T. ischyurus, the Andes of Ecuador and northwestern Peru;
T. kalinowskii, the Andes of central Peru;
T. ladewi, the Andes of northwestern Bolivia;
T. laniger, the Andes of Colombia and western Venezuela;
T. lugens, the Andes from western Venezuela to Ecuador;
T. notatus, southeastern Peru;
T. oenax, extreme southern Brazil, Uruguay;
T. oreas, the Andes of Bolivia;
T. paramorum, the Andes of Ecuador;
T. pictipes, extreme northeastern Argentina, southeastern Brazil;
T. pyrrhonotus, southern Ecuador, Peru;
T. rhoadsi, the Andes of Ecuador;
T. rosalinda, northwestern Peru;
T. taczanowskii, northwestern Peru;
T. vestitus, the Andes of western Venezuela.

Avila-Pires (1960) placed *T. oenax* in a separate genus, *Wilfredomys.* Barlow (1969) did not recognize this distinction, and Pine (1980) argued that *Wilfredomys* does not warrant generic rank. Gardner and Patton (1976) indicated that *T. monochromos* of northeastern Colombia is a species distinct from *T. laniger.* Gardner and Patton also placed *T. lugens* in a separate genus, *Aepeomys* Thomas, 1898, and recognized the population in Colombia as a distinct species, *Aepeomys fuscatus.*

Head and body length is about 90 to 185 mm, and tail length is 85 to 230 mm. Pine (1980) listed weights of about 35 to 61 grams for *T. oenax.* The tail in *Thomasomys* is normally longer than the head and body. The fur is usually thick and soft. The coloration above varies from olivaceous gray, dull olive fulvous, yellowish rufous, orange rufous, golden brown, reddish brown, and grayish brown to dark brown or almost black. The middorsal region is usually slightly darker than the rest of the body. The sides blend into the underparts, which are silvery grayish, soiled grayish, yellow, buffy, ochraceous buff, dark gray, or dark brownish. The underparts are usually not much paler than the upper parts. The hands and feet are about the same color as the underparts; often the central part of the upper surface is darker, and the fingers and toes lighter, sometimes whitish. The coloration of the tail varies from slightly paler to slightly darker than the back; the tail is normally moderately haired.

The hind foot is usually not modified for arboreal life, but in several species it may be very similar to the hind foot of *Rhipidomys.* It is difficult to differentiate some species of *Thomasomys* from *Rhipidomys.* The two genera can usually be distinguished, however, as follows: *Thomasomys* has an hourglass-shaped interorbital region, and relatively long and robust molar rows; *Rhipidomys* has an interorbital region with a conspicuous shelf or ridge, and relatively short and narrow molar rows. Female *Thomasomys* have six or eight mammae.

Thomas's paramo mouse (*Thomasomys hylophilus*), photo by Norman Peterson.

Thomasomys is common in the forests on the eastern slopes of the Andes, and ranges up to about 4,200 meters in such areas as the moist Urubamba Valley. It has not, however, established itself on the altiplano. In Uruguay, Barlow (1969) found *T. oenax* in subtropical woodland. Apparently, some species are arboreal while others are mainly terrestrial. *T. lugens* is said to nest in trees, though it has small eyes, and most animals with small eyes are terrestrial or fossorial. *T. rhoadsi* is reported to dig burrows similar to those of North American microtine rodents, from 25 to 76 mm below the surface of the ground. *T. hylophilus* has been found living among the natural galleries formed under moss-covered logs, roots, and debris. Barlow (1969) reported the stomachs of three *T. oenax* to contain plant material.

In eastern Brazil, *T. dorsalis* mates from August to at least January or February. There are two litters a year, with from two to four young in each litter.

RODENTIA; MURIDAE; *Genus MEGAORYZOMYS*
Lenglet and Coppois, 1979

Galapagos Giant Rat

The single described species, *M. curioi*, is known only from bones and teeth found in cave deposits on Santa Cruz Island in the Galapagos. This species was originally placed in the genus *Megalomys*, which is otherwise known only from the West Indies. Lenglet and Coppois (1979) erected the new genus *Megaoryzomys* for this species, and suggested affinity to *Oryzomys*. Steadman and Ray (in press) agreed with the need for a new genus, but considered it to be most closely related to *Thomasomys*. They also noted that remains of an undesignated species of *Megaoryzomys* had been found on Isabela Island in the Galapagos.

According to Steadman and Ray, *Megaoryzomys* resembles *Thomasomys* and *Rhipidomys*, but differs in having the following combination of characters: very large size (condylobasal length of skull more than 50 mm), a deep depression along the median suture of the frontals (also present in some species of *Thomasomys*), large and numerous palatal foramina, a very wide zygomatic plate, the zygomatic process of the squamosal joining the braincase at a more obtuse angle, a more rectangular braincase, a straight posterior margin of the interparietal, and planar molars.

The exact geological age of *Megaoryzomys* is not known, but the animal probably became extinct within the last two centuries. It may have survived until about 1900, and then disappeared through predation by introduced dogs, cats, pigs, and *Rattus*. Like many island animals, it had evolved in the absence of mammalian predators, and would not have been wary, when approached by an alien mammal (Steadman and Ray in press).

RODENTIA; MURIDAE; *Genus PHAENOMYS* Thomas, 1917

Rio de Janeiro Rice Rat

The single species, *P. ferrugineus*, is known only from Rio de Janeiro in eastern Brazil. Specimens are in the British Museum (Natural History).

Head and body length is about 150 mm, and tail length is about 190 mm. The upper parts are light red or brilliant rust. The dark tips on the hairs on top of the head and the center of the back give these areas a slightly darker appearance than the remainder of the upper parts. The ears are small, well furred, and rust colored, with a few whitish hairs just behind them. The underparts are white with a slight yellowish tinge.

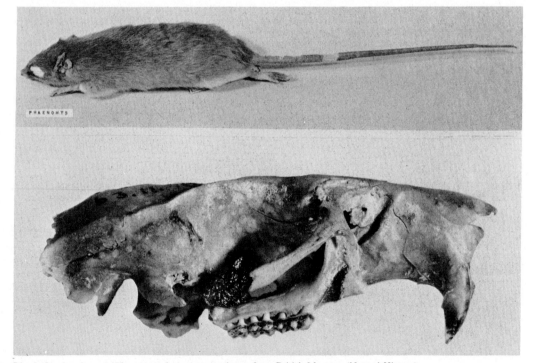

Rio de Janeiro rice rat (*Phaenomys ferrugineus*), photos from British Museum (Natural History).

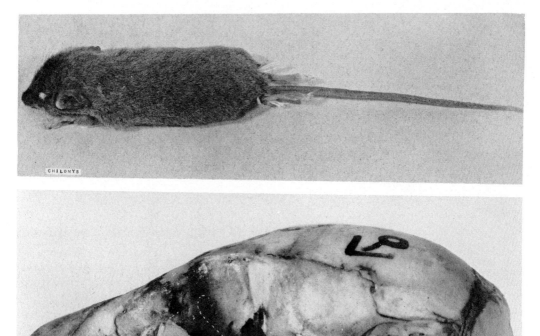

Colombian forest mouse (*Chilomys instans*), photos from British Museum (Natural History).

The hands and feet are pale reddish brown, and the fingers and toes are partly white. The fur is thick and straight.

The skull is slender. The incisor teeth are somewhat heavy. Although the claws are slender and not especially curved, the fifth hind toe is long and the hind foot seems to be slightly modified for arboreal life. The long tail is relatively well haired. Females have eight mammae.

RODENTIA; MURIDAE; *Genus CHILOMYS* Thomas, 1897

Colombian Forest Mouse

The single species, *C. instans,* occurs in Colombia, western Venezuela, and Ecuador (Cabrera 1961; Carleton 1973).

Head and body length is 86 to 99 mm, and tail length is 105 to 130 mm. The coloration above and below is slaty gray, though the face may be blackish from the nose to the eyes. The uniformly brown tail often has a white tip (10 or 20 mm). The hands and feet are brown, but sometimes the digits of all four limbs may be white. Some specimens have a white line from the throat to the middle of the belly, others have a bright buffy pectoral spot. The fur is soft and straight.

The body form is mouselike, not particularly specialized. The skull is delicate, with a large, rounded braincase and a small, slender muzzle. The upper incisor teeth project forward, and the lower incisors are long and slender. The fifth toe reaches to the base of the second phalanx of the fourth digit. This genus is similar to *Oryzomys*, being distinguished from it and other related genera mainly by characters of the skull and dentition.

The Colombian forest mouse inhabits the dark and damp forests, along with *Thomasomys hylophilus*. It is possible that *Chilomys* may be mistaken for the more common *Oryzomys*. In the cloud forests of Venezuela, Handley (1976) collected five specimens of *Chilomys*, either at the base of rotting, moss-covered logs, under moss-covered logs and fallen limbs, or under lichen- and moss-covered tree roots.

RODENTIA; MURIDAE; *Genus TYLOMYS* Peters, 1866

Climbing Rats

There are seven species (Hall 1981; Cabrera 1961):

T. bullaris, known only from the type locality in Chiapas (extreme southern Mexico);

T. mirae, Colombia, northern Ecuador;

T. fulviventer, known only from the type locality in extreme eastern Panama;

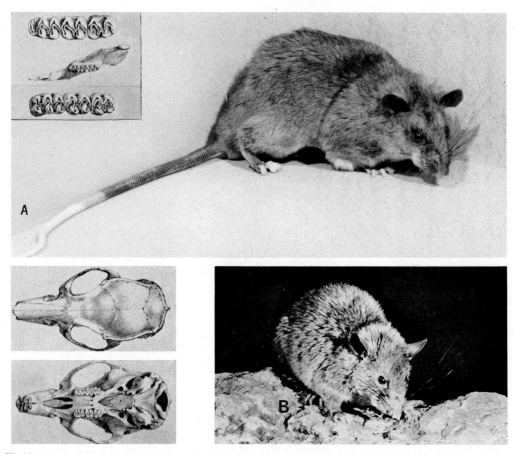

Climbing rat: A. *Tylomys nudicaudus,* photo by Robert Brown through Rollin H. Baker, Michigan State Museum; B. *T. nudicaudus,* photo by O. J. Reichman; Skull: *T. nudicaudus,* photos from *Monatsberichte Königlich. Preuss. Akad. Wissenschaften Berlin.*

T. nudicaudus, southern Mexico to Nicaragua;

T. panamensis, extreme eastern Panama;

T. tumbalensis, known only from the type locality in Chiapas (extreme southern Mexico);

T. watsoni, Costa Rica, Panama.

Cabrera (1961) suggested that *T. fulviventer* might be a subspecies of *T. mirae.*

Head and body length is 170 to 255 mm, and tail length is usually 200 to 250 mm. Helm (1975) listed the average adult weight of *T. nudicaudus* as 280 grams. The upper parts are usually some shade of gray or brown; one species is cinnamon buff with an admixture of brown, and in another the middle of the back is blackish. The underparts are white or fulvous. In the species *T. fulviventer,* a russet median line extends from the pectoral region to the base of the tail. The feet are brown or russet, and the toes are white or brown. The tail is dark brown to blackish, the proximal part being brown, glossy black, yellowish, or whitish. In some species the fur is close, thick, and glossy, and the whiskers are long, smooth, and black. The tail is slender and scantily haired. The ears are large and naked.

These rodents superficially resemble large specimens of *Rattus rattus.* The hind foot is suited to arboreal life, and all the feet are somewhat broad and short.

Heavily forested areas, often around rocky ledges, are preferred habitats. The known elevational range is 175 to 1,375 meters. These animals have been taken among rocks, along logs, on the banks of streams, and at the base of palm trees. Some have been shot as they climbed in palm fronds nine meters above the ground. A captive specimen consumed bird seed, mouse breeders chow, and, occasionally, lettuce and apples (Baker and Petersen 1965). The gestation period of *T. nudicaudus* averages about 40 (36–51) days, and mean litter size is 2.33 young (Helm 1975). A captive individual of this species lived for 5 years and 5 months (Marvin L. Jones, Zoological Society of San Diego, pers. comm.).

RODENTIA; MURIDAE; **Genus OTOTYLOMYS** Merriam, *1901*

Big-eared Climbing Rat

The single species, *O. phyllotis,* is known to occur from the Yucatan Peninsula to Costa Rica. There also is a record from Guerrero, southwestern Mexico, based on remains found in an owl pellet (Hall 1981).

Head and body length is 95 to 190 mm, and tail length is 100 to 190 mm. Two individuals each weighed about 120 grams, though Helm (1975) listed average adult weight as 63 grams. The fur is moderately long, soft, and full. The upper

Big-eared climbing rat (*Ototylomys phyllotis*), photo by Lloyd G. Ingles.

parts are some shade of brown or gray, often intermingled with black hairs. The underparts are white or grayish. The fingers and toes are usually whitish. The insides of the hands and feet are whitish, but the central and outer parts are almost as dark as the outer sides of the arms and legs. The color of the tail ranges from dark, dull, grayish brown to blackish above, and is slightly lighter below, but it may be dull yellowish in some species. The tail is naked except for a few scattered hairs, and the large ears are sparsely haired. *Ototylomys* closely resembles *Tylomys,* but is smaller and has a somewhat shorter tail.

This rat occurs in a wide variety of habitats, but prefers tropical forest with rocks or rocky ledges. Its altitudinal range is sea level to about 2,000 meters. It is nocturnal and mainly arboreal, foraging on creepers or branches, but also is sometimes found on the ground or among rocks (Lawlor 1969; Jones, Genoways, and Lawlor 1974).

Available information indicates that breeding occurs throughout the year in the wild. Litter size is one to four young, averaging about two or three in different areas (Lawlor 1969). In a study of captives, Helm (1975) determined gestation to average 52.8 days and to range from 49 to 69 days. A postpartum estrus was observed for all littering females. The young weighed about 10 grams at birth and most opened their eyes when 6 days old. The earliest recorded mating for a female occurred at 29 days of age.

RODENTIA; MURIDAE; **Genus NYCTOMYS** Saussure, *1860*

Vesper Rat

The single species, *N. sumichrasti,* occurs from southern Mexico to Panama (Hall 1981). Drawing on the morphology of the male reproductive tract, Voss and Linzey (1981) suggested that *Nyctomys* is not closely related to any other South American murid.

Head and body length is 107 to 130 mm, and tail length is 85 to 155 mm. Genoways and Jones (1972) listed weights of 39 to 61 grams for full-grown animals. The fur is fine, is short or long, and has little or no gloss. The short ears are scantily covered with fine hairs. Hairs on the tail hide the scales at all points. Toward the tip of the tail the hairs are slightly longer and thicker, giving the tail a brushy appearance. The upper parts are buff, cinnamon, or tawny, with some dark hairs, especially along the middle of the back. The sides are usually paler, and the underparts are

white or nearly so. The feet are usually white, but in some animals they may be dusky to brownish. The tail is usually brown. There is often a dark ring around each eye, or a dark area between the eye and the base of the whiskers.

The eyes are large and the hind feet are modified for arboreal life. The hallux (big toe) is clawed, as in *Rhipidomys,* and in both *Nyctomys* and *Rhipidomys* the pad representing the pollex (thumb) may be prominent. *Nyctomys* differs from *Rhipidomys* in coloration, and in having a shorter, more fully haired tail. Female *Nyctomys* have four mammae.

The vesper rat is arboreal, usually being found at a considerable height in trees and rarely descending to the ground. It builds outside nests of twigs and fibers, like those of the red squirrel (*Tamiasciurus*). It seems to be mainly, if not entirely, nocturnal. According to Genoways and Jones (1972), the diet of *Nyctomys* consists primarily of seeds, fruits, and other vegetable matter. It has been observed eating wild figs and avocados. Individuals occasionally arrive in the United States in bunches of bananas unloaded from ships.

Birkenholz and Wirth (1965) maintained a captive colony of *Nyctomys.* They reported the most common vocalization to be a faint, rapid, high-pitched musical chirp or trill. The average litter size in this colony was two young. The gestation period was 30 to 38 days. One female gave birth to five litters over a 7-month period. Data summarized by Genoways and Jones (1972) indicated that in the wild the vesper mouse breeds during both the wet and dry seasons and probably is capable of reproduction throughout the year. Litter size evidently varies from one to four, with two probably being the mode. A captive specimen lived for 5 years and 2 months (Marvin L. Jones, Zoological Society of San Diego, pers. comm.).

RODENTIA; MURIDAE; **Genus OTONYCTOMYS** *Anthony, 1932*

Yucatan Vesper Rat

The single species, *O. hatti,* occurs over much of the Yucatan Peninsula and in northern parts of Guatemala and Belize (Hall 1981).

Head and body length is about 104 to 116 mm, and tail length is about 100 to 127 mm. Jones, Genoways, and Lawlor (1974) gave the weight of two individuals as 29.5 and 32.3 grams. The upper parts are nearly uniformly russet to hazel, being darkest on the back. The sides are tawny to ochraceous tawny. The underparts are white with a creamy

Vesper rat (*Nyctomys sumichrasti*), photo from Zoological Society of Philadelphia.

wash. The upper sides of the feet are whitish with a buffy wash, or with tawny tones, and the tail is uniformly brown and heavily haired. The external ear is similar to that of *Nyctomys*.

This rodent is much redder than *Nyctomys sumichrasti*, its pronounced russet coloration making it the showiest among the neotropical climbing rats. It resembles *Nyctomys* in structural features, but its auditory bullae are much larger, its

Yucatan vesper rat (*Otonyctomys hatti*), photo by Robert T. Hatt.

cheek teeth are smaller, and its tarsus is narrower. Females have four mammae.

Jones, Genoways, and Lawlor (1974) collected one specimen on a rafter under the roof of a house and another in a small tree. In both instances, the traps were baited with bananas. Robert T. Hatt trapped two individuals in a thatched hut, on the shelf where the rafters meet the top of the wall. The habits of *Otonyctomys* are probably much like those of *Nyctomys* and *Rhipidomys*. It is possible that *Otonyctomys* is fairly common, and only seems to be rare because of its arboreal habits or because it may have a specialized diet.

RODENTIA; MURIDAE; *Genus RHAGOMYS Thomas, 1917*

Brazilian Arboreal Mouse

The single species, *R. rufescens*, has been reported from Rio de Janeiro in eastern Brazil. Specimens are in the British Museum (Natural History).

Head and body length is about 94 mm, and tail length is about the same. The general color above and below is rich orange rufous. The hairs are slaty blue at the base and rufous at the tip. The underparts are slightly paler than the upper parts. The feet are yellow and the toes are whitish. The tail is scantily covered with brown hairs that form an inconspicuous

tuft at the tip. The ears are short, scarcely projecting above the fur, and are thickly covered with rufous brown hairs.

This rodent is externally modified for an arboreal life. The fifth digit of the hind foot is long, and the hallux (big toe) in the type specimen appears to lack a claw. Females have six mammae.

RODENTIA; MURIDAE; *Genus REITHRODONTOMYS Giglioli, 1874*

American Harvest Mice

There are 2 subgenera and 19 species (Hall 1981; Cabrera 1961):

subgenus *Reithrodontomys*

R. montanus, Great Plains region of the United States, northern Mexico;

R. burti, Sonora and Sinaloa (northwestern Mexico);

R. humulis, southeastern United States;

R. megalotis, southern British Columbia to northeastern Indiana and southern Mexico;

R. raviventris, San Francisco Bay area of California;

R. chrysopsis, mountains of central Mexico;

R. sumichrasti, central Mexico to western Panama;

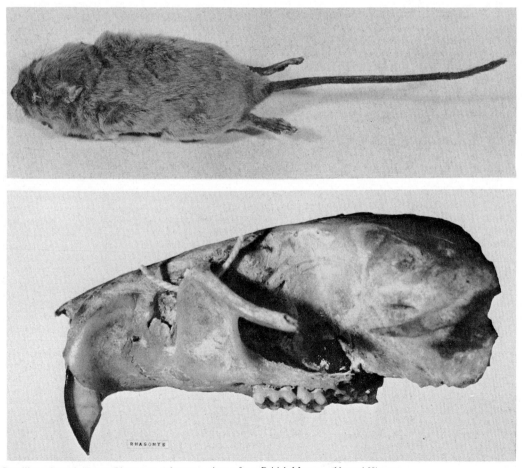

Brazilian arboreal mouse (*Rhagomys rufescens*), photos from British Museum (Natural History).

American harvest mouse (*Reithrodontomys fulvescens*), photo by Ernest P. Walker.

R. fulvescens, southern Arizona and south-central United States to Nicaragua;

R. hirsutus, Nayarit and Jalisco (west-central Mexico);

subgenus *Aporodon*

R. gracilis, southern Mexico to northwestern Costa Rica;

R. darienensis, Panama, probably extreme northwestern Colombia;

R. mexicanus, eastern and southwestern Mexico to western Panama, the Andes of Colombia and Ecuador;

R. spectabilis, Cozumel Island off northeastern Yucatan;

R. brevirostris, north-central Nicaragua to central Costa Rica;

R. paradoxus, southwestern Nicaragua to northwestern Costa Rica;

R. microdon, central Mexico to Guatemala;

R. tenuirostris, central Guatemala;

R. rodriguezi, central Costa Rica;

R. creper, Costa Rica, western Panama.

On the basis of karyological data, Carleton and Myers (1979) suggested that the above subgeneric division might not be valid.

Head and body length is 50 to 145 mm, tail length is 45 to 115 mm, and weight is 6 to 20 grams. Coloration of the upper parts ranges from pale ochraceous gray and pinkish cinnamon through browns to almost black. The sides are paler and usually more ochraceous than the upper parts. The underparts are white or grayish, sometimes tinged with ochraceous buff or pinkish cinnamon. The tail is dark above and light below, or unicolor, and it is slender, scaly, and scantily haired. The juvenile pelage is more or less plumbeous, but the adult pelage is brighter. Adults apparently molt once a year.

American harvest mice resemble house mice (*Mus*), but have more hair on the tail and have grooved upper incisors. The ears of harvest mice are conspicuous and sometimes large. Females have six mammae.

Habitats vary from salt marshes to tropical forests, but American harvest mice are usually associated with stands of short grass. The altitudinal range is from below sea level in places to above the tree line on some Central American mountains. The most noticeable evidence of *Reithrodontomys* is the presence of globular nests of grass about 150 to 175 mm in diameter. The nests are usually constructed above ground in grasses, low shrubs, or small trees. Some winter nests are located in burrows and small crevices. These mice are nocturnal and are active throughout the year. They use the ground runways of other rodents and are nimble climbers. Food consists mainly of seeds and the green shoots of vegetation. The seeds are gleaned from the ground or cut from grass

stems by bending the stem to the ground. Some insects are also eaten.

Reported autumn population densities for *R. megalotis* in southern Wisconsin have varied from only about 0.75/ha. in sandy fields to 45/ha. in an abandoned field with a dense cover of low vegetation (Svendsen 1970; Jackson 1961). O'Farrell (1978) estimated an annual composite home range of 0.95 ha. for *R. megalotis* in a sagebrush desert in Nevada. Packard (1968) reported a population density of about 7.5/ha. for *R. fulvescens* in eastern Texas. He found this species to use a home range of about 0.2 ha. and he observed no marked territorial behavior. There are conflicting reports regarding sociability in *R. megalotis* (Banfield 1974). A high-pitched bugling vocalization has been reported for the genus.

Reproductive activity apparently takes place throughout the year, except during cold winters. In Indiana, for example, *R. megalotis* breeds from March to November (Whitaker and Mumford 1972). Banfield (1974) stated that females of this species are seasonally polyestrous and produce several litters annually, but do not undergo a postpartum estrus, and that the number of young averages 2.6 (1–9). In Nebraska, 75 pregnant female *R. megalotis,* taken from April through November, had an average of 4.3 (2–8) embryos (Jones 1964). In eastern Tennessee, Dunaway (1968) found *R. humulis* to breed mainly from late spring to late fall, but occasionally during the winter, and to have an average litter size of 3.4 (1–8). In Nicaragua, Jones and Genoways (1970) collected pregnant female *R. sumichrasti, R. fulvescens, R. gracilis, R. mexicanus,* and *R. brevirostris* from June to August; the number of embryos found was 3 to 5. Captive female *R. megalotis* produced up to 14 litters and 68 young per year; one gave birth when 2 years and 5 months old (Egoscue, Bittmenn, and Petrovich 1970). Gestation periods of 21 to 24 days have been reported for the genus. The young are born in a woven grass nest and weigh about one gram at birth. In *R. megalotis* the young leave the nest after about 3 weeks and attain adult weight in about 5 weeks. Some females of this species breed when 17 weeks old. Very few *Reithrodontomys* live as long as 1 year, and the known record life span in the wild seems to be about 18 months (Fisler 1971).

American harvest mice are not generally considered to be detrimental to human agriculture. People, however, have had pronounced, and opposite, effects on at least two species of *Reithrodontomys. R. megalotis* apparently was able to extend its range eastward across Illinois and into Indiana in response to the clearing of woodlands and the destruction of tall grass prairies (Ford 1977; Whitaker and Mumford 1972). The salt marsh harvest mouse (*R. raviventris*), one of the few mammals that can drink salt water, is classified as endangered by the IUCN (1972), the USDI (1980), and the

California Department of Fish and Game (1978). It is dependent for survival on the marshes around San Francisco Bay, and its population has been greatly reduced and fragmented through habitat destruction caused by urban and industrial development.

RODENTIA; MURIDAE; *Genus PEROMYSCUS Gloger, 1841*

White-footed Mice, Deer Mice

There are 7 subgenera and 60 species (Hall 1981; Huckaby 1980; Carleton 1977, 1979; Lee and Schmidly 1977):

subgenus *Haplomylomys*

P. eremicus, southwestern United States, northern Mexico, Baja California and several nearby islands;
P. eva, southern Baja California, Carmen Island (Gulf of California);
P. merriami, southern Arizona, northwestern Mexico;
P. guardia, Angel, Granite, and Mejia islands (Gulf of California);
P. interparietalis, Salsipuedes and San Lorenzo islands (Gulf of California);
P. collatus, Turner Island (Gulf of California);
P. dickeyi, Tortuga Island (Gulf of California);
P. pembertoni, San Pedro Nolasco Island (Gulf of California);
P. californicus, central California to northern Baja California;
P. hooperi, Coahuila (northern Mexico);

subgenus *Peromyscus*

P. polionotus, southeastern United States;
P. maniculatus, southeastern Alaska, Queen Charlotte Islands, most of Canada, conterminous United States except parts of Southeast, most of Mexico, Baja California and several nearby islands;
P. sejugis, Santa Cruz and San Diego islands (Gulf of California);
P. slevini, Santa Catalina Island (Gulf of California);
P. sitkensis, several islands off southeastern Alaska, several small islands in the Queen Charlotte group;
P. melanotis, northern and central Mexico;
P. leucopus, southeastern Alberta and Nova Scotia to Arizona and the Yucatan Peninsula;
P. gossypinus (cotton mouse), southeastern United States;
P. crinitus, central Oregon to western Colorado and northern Baja California;
P. caniceps, Monserrate Island (Gulf of California);
P. pseudocrinitus, Coronados Island (Gulf of California);
P. pectoralis, south-central Oklahoma and southeastern New Mexico to central Mexico;
P. boylii, northern California and northwestern Texas to Guatemala;
P. spicilegus, western Mexico;
P. simulus, Sinaloa and Nayarit (western Mexico);
P. madrensis, Tres Marias Islands off western Mexico;
P. attwateri, southeastern Kansas to northern Arkansas and central Texas;
P. polius, Chihuahua (north-central Mexico);
P. stephani, San Esteban Island (Gulf of California);
P. aztecus, central Mexico to Honduras;
P. winkelmanni, Michoacan (western Mexico);
P. truei, southwestern Oregon and Colorado to southern Mexico;
P. comanche, northwestern Texas;

P. difficilis, Colorado to south-central Mexico;
P. bullatus, Veracruz (east-central Mexico);
P. perfulvus, west-central Mexico;
P. melanophrys, Mexico;
P. mekisturus, Puebla (east-central Mexico);
P. ochraventer, southern Tamaulipas and southeastern San Luis Potosi (northeastern Mexico);
P. yucatanicus, Yucatan Peninsula;
P. mexicanus, eastern and southern Mexico to western Panama;
P. gymnotis, extreme southern Mexico, southern Guatemala;
P. mayensis, western Guatemala;
P. stirtoni, Guatemala, southern Honduras, El Salvador;
P. furvus, eastern Mexico;
P. guatemalensis, extreme southern Mexico, western Guatemala;
P. megalops, southwestern Mexico;
P. melanocarpus, north-central Oaxaca (southern Mexico);
P. melanurus, southwestern Oaxaca (southern Mexico);
P. zarhynchus, Chiapas (southern Mexico);
P. grandis, central Guatemala;

subgenus *Megadontomys*

P. thomasi, southern Mexico;

subgenus *Isthmomys*

P. flavidus, western Panama;
P. pirrensis, eastern Panama;

subgenus *Habromys*

P. simulatus, Veracruz (east-central Mexico);
P. chinanteco, northern Oaxaca (southern Mexico);
P. lophurus, extreme southern Mexico to northwestern El Salvador;
P. lepturus, northern Oaxaca (southern Mexico);

subgenus *Podomys*

P. floridanus, Florida;

subgenus *Osgoodomys*

P. banderanus, west-central Mexico.

Hooper (1968) suggested that some of the named species would eventually be shown to be synonyms or subspecies, and that the actual number of biological species is about 40. Of particular interest is the status of the population in northwestern Texas originally called *P. comanche*. It was long recognized as a subspecies of *P. difficilis*, but, on the basis of karyological data, Lee, Schmidly, and Huheey (1972) stated that its affinities actually are with *P. truei*. Subsequently, Schmidly (1973) designated *P. comanche* a subspecies of *P. truei*, and Hall (1981) followed this arrangement. Using an electrophoretic analysis, however, Johnson and Packard (1974) concluded that *P. comanche* is a distinct species, and it was listed as such by Jones, Carter, and Genoways (1979). The following were considered separate genera by Carleton (1980): *Megadontomys* Merriam, 1898; *Isthmomys* Hooper and Musser, 1964; *Habromys* Hooper and Musser, 1964; *Podomys* Osgood, 1909; and *Osgoodomys* Hooper and Musser, 1964.

Head and body length is 70 to 170 mm, and tail length is 40 to 205 mm. Klingener (1968) stated that adults range in body weight from about 15 grams in some small northern species, such as *P. crinitus* and *P. polionotus*, to over 110 grams in a few tropical forms, such as *P. thomasi*. The pelage is usually soft and full. The coloration is variable, but the upper parts may be gray or sandy to golden or dark brown, and the underparts are white or nearly white. Some species, howev-

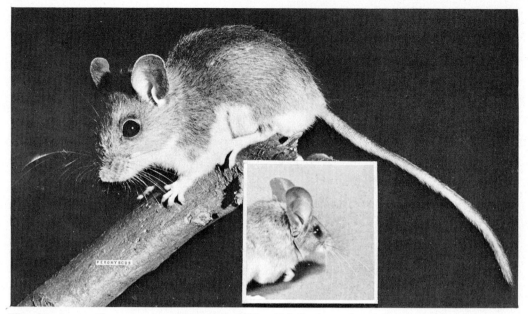

White-footed mouse (*Peromyscus leucopus*), photo by Maslowski & Goodpaster through National Audubon Society. Inset: big-eared cliff mouse (*P. truei*), photo by Ernest P. Walker.

er, are nearly all white and others are nearly black. In general, deer mice inhabiting the cool woods are grayish, whereas those living in open or arid country are pale. Adults molt once a year.

The ears are large in relation to the rest of the body and are covered with fine hairs. The tail is at least one-third the total length of the animal; it is fairly well haired and often tufted. In *P. floridanus* and *P. boylii*, the skin of the distal portion of the tail is known to break readily and freely slip off the vertebrae, apparently serving as a means of escaping predators that seize the tail (Layne 1972). Female *Peromyscus* have four or six mammae.

Deer mice are found in a great variety of habitats and are usually the most abundant mammal in the areas they occupy. *P. maniculatus* is the most widely distributed species, and, indeed, few other mammals can match its tolerance for different conditions—alpine areas, boreal forest,woodlands, grasslands, brushlands, deserts, and arid tropical areas. It does not, however, regularly occur in moist places (Baker 1968; Banfield 1974). In contrast, *P. gossypinus* is restricted to the southeastern United States, where its preferred habitat is bottomland hardwood forest and swamp (Wolfe and Linzey 1977). *P. leucopus* also occurs over a large region, but keeps mainly to deep woodlands or brushy areas (Lowery 1974). *P. attwateri* seems closely associated with cliffs and other rocky areas, where it usually lives in crevices (Schmidly 1974a). Other species build nests in or under logs, in holes in trees, under brush piles, in clumps of vegetation, or in burrows dug by the mice themselves or by other animals. The nests of *P. maniculatus* are spheres of grass, about 100 mm in diameter (Banfield 1974). Nests are lined with down from plants or with shredded materials. A soiled nest is abandoned, so several nests are made during the year. Deer mice are largely, though not exclusively, nocturnal, and are active throughout the year. Periods of torpor, involving a reduction in body temperature for several hours, are known to occur in *P. leucopus* and *P. maniculatus* (Lynch, Vogt, and Smith 1978). The diet includes seeds, nuts, berries, fruits, insects and other small invertebrates, and carrion. One

study of *P. gossypinus* indicated that 68 percent of its diet consisted of animal matter (Wolfe and Linzey 1977). According to Banfield (1974), *P. maniculatus* is primarily a seed eater, and it may store up to three liters of food for winter use.

Normal population density in *P. maniculatus* varies from about 1 to 25 individuals per ha. (Banfield 1974). In Kansas, Hansen and Fleharty (1974) found that the density of this species ranged from 4/ha. in weedy areas during the spring and summer to 19.3/ha. in the favored bluestem grass community during the autumn. In Connecticut, the density of *P. leucopus* ranged from 5.1 to 38.9 per ha. (Miller and Getz 1977). Densities as high as 96.6/ha. have been reported for *P. gossypinus* in wet lowland forests in Tennessee (Wolfe and Linzey 1977).

According to Stickel (1968), home range in *Peromyscus* usually varies from 0.04 to 4.00 ha., is generally smallest in winter and largest in summer, and is generally larger in males than in females. Banfield (1974) stated that the average home range of *P. maniculatus* is about 1.0 ha. in males and 0.6 ha. in females. In a study in Kansas, Hansen and Fleharty (1974) found the home range of this species to average 0.31 ha. in the spring and summer and 0.27 ha. in the autumn and winter. For *P. leucopus* in northern Virginia, average home range was only 0.1 ha. and was about the same for both sexes, but in Quebec the home range of this species averaged 1.26 ha. in males and 0.91 ha. in females (Madison 1977; Mineau and Madison 1977). Home range data on some other species are: *P. attwateri*, average size about 0.2 ha., being about twice as large in males as in females (Schmidly 1974a); *P. gossypinus*, varies from 0.18 to 0.81 ha. (Wolfe and Linzey 1977); *P. californicus*, averages about 0.15 ha. in both sexes (Merritt 1978); and *P. eremicus*, averages about 0.3 ha., there being considerable overlap in males and almost none in females (Veal and Caire 1979).

Banfield (1974) considered *P. maniculatus* to be a sociable rodent, tolerant of conspecifics regardless of age and sex, especially in winter, when up to 13 individuals may huddle together. In this species, and probably some others, the male lives with the family and helps to care for the young. Sadleir (1970) found that captive male *P. maniculatus* formed a single-line hierarchy that remained stable for a month. Studies of *P. leucopus* indicate that males and females form pairs that live together, but that resident females exclude other females from their home ranges (Madison 1977; Mineau and Madison 1977; Harland, Blancher, and Millar 1979). *P. attwateri* is not aggressive, and individuals of all sex and age categories may rest together in captivity (Schmidly 1974a). In *P. californicus*, males are highly aggressive toward one another, but opposite sexes form pairs that cooperate in raising the young (Merritt 1978). Vocalizations of *Peromyscus* include thin squeaks and shrill buzzings. When excited, most species thump rapidly with the front feet, producing a drumming noise.

In some species breeding may continue throughout the year, if the weather does not become too cold or hot. The breeding season of *P. maniculatus* in the northern parts of its range is normally March through October, though some females may mate in winter under favorable conditions. This species is polyestrous and usually produces 3 or 4 litters annually in the wild (Godin 1977; Banfield 1974). In the laboratory, however, female *P. maniculatus* have borne up to 14 litters per year (Egoscue, Bittmenn, and Petrovich 1970). *P. gossypinus* breeds throughout the year in Texas and Florida, though there may be some decline in reproductive activity during the summer (Wolfe and Linzey 1977). In southwestern Missouri, *P. attwateri* has one breeding season in spring and another in autumn (Schmidly 1974a). Year-round reproduction, at least in some areas, has been reported in *P. pectoralis* and *P. eremicus* (Schmidly 1974b; Veal and Caire 1979). In southern Mexico, both *P. mexicanus* and *P. melanocarpus* breed throughout the year, but the latter species shows a definite seasonal peak from March to July (Rickart 1977).

The length of the estrous cycle averages 5.26 days in *P. gossypinus* and about 7 days in *P. californicus* (Merritt 1978). A postpartum estrus may occur in female *Peromyscus*. According to Layne (1968), the gestation period in this genus ranges from 21 to 27 days and averages 23.47 days in nonlactating females, but may be as long as 40 days in lactating females. Somewhat longer gestations in nonlactating females have been reported for several species in southern Mexico (Lackey 1976; Rickart 1977). The overall mean litter size for the genus is 3.4 young (Layne 1968). The range for *P. maniculatus* is 1 to 9 in the wild (Banfield 1974), but up to 11 in the laboratory (Drickamer and Vestal 1973). Apparently, litters average larger in the north: the mean size for *P. leucopus* in central Wisconsin was 4.77 (Long 1973a), while the mean for *P. melanocarpus* in southern Mexico was about 2.0 (Rickart 1977). The overall mean weight for newborn *Peromyscus* is 2.2 grams; the eyes open at around 2 weeks of age. The young of most species are weaned when 3 to 4 weeks old, but may remain with the mother for another month. The average age of first estrus in females is 29.6 days in *P. polionotus*, 39.2 days in *P. eremicus*, and 48.7 days in *P. maniculatus*. These mice may begin to breed while still in subadult pelage, but may continue to grow in weight for as long as 6 months (Layne 1968). Most wild individuals probably live less than 2 years, but record longevity in the laboratory for *P. maniculatus* is 8 years and 4 months (Banfield 1974).

Deer mice are used widely in physiological and genetic studies because they are clean, live well in the laboratory, can be easily fed, and have a high reproductive rate. On the other hand, Banfield (1974) considered *P. maniculatus* to be somewhat of a menace to forest regeneration, through its destruction of tree seeds, especially those of conifers.

A number of southeastern species and subspecies of *Per-*

omyscus are now thought to be endangered or threatened (Layne 1978; Bowen 1968; Dusi 1976). *P. gossypinus allapaticola*, found only in mature tropical hammock forests on Key Largo, Florida, has become restricted to an area of only about 120 to 160 ha., through the commercial and residential development of its habitat. The Florida mouse (*P. floridanus*), which is highly dependent on relatively dry scrub forest, has lost much habitat to real estate projects and orange groves. Several pale-colored races of "beach mice" are in jeopardy because their small ranges are desirable to people for waterfront homes and recreational facilities. These mice include *P. polionotus ammobates*, endemic to the dunes between Mobile Bay and Perdido Bay in southern Alabama; *P. p. trissyllepsis*, found from Perdido Bay to Pensacola Bay, Florida; *P. p. allophrys* of the Gulf coastal dunes of Okaloosa and Bay counties, Florida; and the possibly extinct *P. p. decoloratus*, which occupied a small part of the coast of northeastern Florida.

RODENTIA; MURIDAE; **Genus OCHROTOMYS** Osgood, 1909

Golden Mouse

The single species, *O. nuttalli*, occurs in the south-central and southeastern United States (Hall 1981).

Head and body length is 51 to 115 mm, tail length is 50 to 97 mm, and weight is 15 to 30 grams. This mouse has a striking golden coloration and dense, soft pelage. The upper parts are rich golden brown or bright golden cinnamon, and the feet and belly are white, often washed with orange.

From *Peromyscus*, *Ochrotomys* differs in that the posterior palatine foramina of its skull are nearer to the interpterygoid fossa than to the posterior endings of the anterior palatine foramina; the enamel folds of its molariform teeth are compressed and thick; and its baculum is distinctly capped with a long, cartilaginous cone (Linzey and Packard 1977). Females have six mammae.

The golden mouse inhabits brushy and wooded areas, and is often found in thickets of honeysuckle (*Lonicera*) and greenbrier (*Smilax*). Over most of its range the genus is mainly arboreal and builds nestlike structures in vines, bushes, and trees. At the western limits of its range, however, at least in eastern Texas, it is not arboreal. In Texas, and possibly during the summer in other parts of its range, *Ochrotomys* seems to spend more time on the ground than in trees, and apparently uses nests placed under logs and stumps. Individuals have been trapped in a swamp in Florida in underground runways. The golden mouse is an agile animal, using its tail as a balancing and prehensile organ when climbing.

Typically, two kinds of nestlike structures are used. The nest proper is usually located from 0.4 to 4.5 meters above the ground and has the appearance of a solid mass. It has an outer covering of leaves, grass, and bark, and an inner downy lining of milkweed fibers, feathers, or fur. The outer layer often envelops the abandoned nest of a bird, and the entire structure measures 100 to 200 mm across. The same nest may be used for several generations, and suitable nesting sites may be used year after year. The other nestlike structure, similar in appearance but less bulky and in places 15 meters above the ground, serves as a retreat for feeding. These platforms may be more numerous than the nests proper, and different individuals may use the same feeding platform. The diet consists mainly of plant seeds.

The remainder of this account, dealing with population structure and life history, is taken largely from Linzey and Packard (1977). Reported population densities are 0.5/ha. in

Golden mouse (*Ochrotomys nuttalli*), photo by Ernest P. Walker.

Tennessee, up to 5.4/ha. in early spring on a forested hardwood floodplain in eastern Texas, 0.47 to 6.89 per ha. in Great Smoky Mountains National Park, and 1.7 to 74.1 per ha. in Illinois. Home ranges are relatively small, recorded averages varying from about 0.053 to 0.627 ha. Home ranges apparently overlap extensively and territorial behavior is not evident. *Ochrotomys* is a fairly sociable rodent, and up to 8 individuals have been found in the same nest.

The reproductive period seems to vary considerably, having been reported as throughout the year in Louisiana, mainly from September to spring in Texas, and from March to October in Kentucky and eastern Tennessee. Captives breed throughout the year, though with marked peaks in early spring and late summer. Several litters may be produced annually in the wild, and some captive females have borne as many as 17 litters within 18 months. The gestation period is about 25 to 30 days. Litter size ranges from 1 to 4 young, and averages about 2.7. The young have a mean weight of about 2.7 grams at birth, open their eyes after 11 to 14 days, are weaned when 17 to 21 days old, and reach adult size at an age of 8 to 10 weeks. Captive females have produced young when as old as 6½ years, and have lived as long as 8 years and 5 months.

RODENTIA; MURIDAE; **Genus BAIOMYS** *True, 1894*

Pygmy Mice

There are two living species (Hall 1981):

B. musculus, central Mexico to Nicaragua;
B. taylori, eastern Texas and southeastern Arizona to central Mexico.

A third named species, *B. hummelincki,* is actually referable to *Calomys* (Hershkovitz 1962; Handley 1976).

Head and body length is 50 to 81 mm, tail length is 35 to 55 mm, and weight is 7 or 8 grams. The upper parts are blackish brown to light reddish brown, and the underparts are dark gray to white or buff. This genus differs from *Peromyscus* in its smaller size, in having relatively smaller and more rounded ears, and in the shape of the coronoid process.

Pygmy mice live mainly in dry country, wherever they can find cover in dense grass, shrubs, or cultivated fields. *B. taylori* is probably mainly nocturnal and does not hibernate. Beneath a heavy canopy of matted grasses, it makes a network of small, narrow runways and also uses the runways of other rodents. Tiny piles of green feces are found at the intersections. According to Packard and Montgomery (1978), *B. musculus* apparently lives in underground burrows when occurring in grasslands, but under rocks when occurring in arid rocky areas. This species shows a tendency toward diurnal and crepuscular activity. Although pygmy mice may eat some seeds and insects, the bulk of their food consists of green vegetation. *B. taylori* has a home range less than 30 meters in diameter, and the usual population density seems to be 15 to 20 adults per ha. Pygmy mice are easily kept in captivity and live together peacefully.

Breeding apparently continues throughout the year, but in *B. musculus,* at least, a decline in reproductive activity seems to occur in winter and spring (Packard and Montgomery 1978). Studies of captive *B. taylori* by Quadagno et al. (1970) and Hudson (1974) indicate that the estrous cycle averages 4.9 days, gestation lasts 20 to 25 days, litter size averages about 2.5 young and ranges from 1 to 5, newborn average 1.2 grams in weight, females may conceive young when only 28 days old, and males reach sexual maturity at about 70–80 days of age. One captive female had nine litters

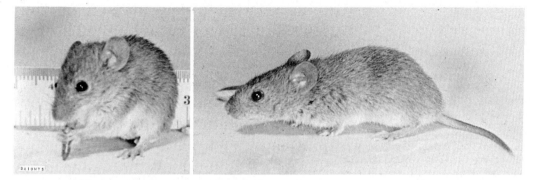

Pygmy mouse (*Baiomys taylori*), photos by Ernest P. Walker.

in 202 days. In the wild the young are born in a surface nest of dry grass, located in a depression in the ground or under a log. Both parents share in caring for the young. The offspring seek their own food when about 18–22 days of age. According to Marvin L. Jones (Zoological Society of San Diego, pers. comm.), a captive *B. taylori* lived for 3 years.

RODENTIA; MURIDAE; **Genus ONYCHOMYS** *Baird, 1857*

Grasshopper Mice

There are three species (Hall 1981; Hinesley 1979):

O. leucogaster, eastern Washington and southern Manitoba to extreme northern Mexico;
O. torridus, southwestern United States, northern Mexico;
O. arenicola, southwestern New Mexico, extreme western Texas, and adjacent parts of north-central Mexico.

Head and body length is 90 to 130 mm, tail length is 30 to 60 mm, and weight is usually 30 to 60 grams. The fur is fine and dense. The upper parts of *O. leucogaster* are brownish to pinkish cinnamon, or buff, while the upper parts of *O. torridus* are grayish or pinkish cinnamon. In both species the basal two-thirds of the tail are colored like the upper parts, but the underside and terminal tip are white. The underparts of the body are white in both species. *O. arenicola* superficially resembles *O. torridus,* but averages smaller in size and differs in karyotype. The genus is distinguished by the stocky body and fairly short, clublike tail. In *O. leucogaster* the tail is usually less than half the length of the head and body, but in *O. torridus* it is usually longer than half the length of the head and body. Female *Onychomys* have six mammae.

Grasshopper mice occur in shortgrass prairies and desert scrub. *O. torridus* prefers relatively xeric areas at lower elevations, while *O. leucogaster* tends to occupy more mesic areas at higher elevations (McCarty 1978). Both species are largely nocturnal and are active throughout the year. They live in practically any shelter they can find at ground level. They are good climbers, but apparently do not climb regularly. The nest may be constructed in a burrow taken over from some other rodent. *O. leucogaster* reportedly excavates a U-shaped burrow, with an average length of 48 cm and an

Grasshopper mouse (*Onychomys leucogaster*), photo by Ernest P. Walker.

average depth beneath the surface of 14 cm. Smaller burrows are constructed for such purposes as emergency retreat and seed storage (McCarty 1978). Although plant material is eaten at times, grasshopper mice are largely carnivorous. The diet includes grasshoppers, beetles, and a variety of other insects. *O. torridus* preys extensively on scorpions. Grasshopper mice also eat small vertebrates, including such rodents as *Peromyscus, Perognathus,* and *Microtus.* The prey is stalked, seized with a rush, and then killed by a bite in the head. While overpowering their prey, grasshopper mice close their eyes and lay back their ears.

Like most predators, grasshopper mice occur at relatively low densities. An average of 1.83 *O. torridus* per ha. has been reported for desert country in Nevada. Usual home range for *Onychomys* is about 2 or 3 ha. Adults are territorial, and captive individuals of the same sex are extremely aggressive toward one another, often fighting to the death (McCarty 1975, 1978). In the wild, according to Hafner and Hafner (1979), grasshopper mice have large, well-defined territories and may associate in male-female pairs all year. Adult vocalizations include a single sharp "chit," given during agonistic encounters; a soft chirp during mating; a quiet, pure tone lasting about 0.8 seconds, given when close to a mate; and a loud, piercing, pure tone, 0.7 to 1.2 seconds long and audible to the human ear up to 100 meters away. When giving this last kind of call, the animal often stands on its hind legs with its nose pointed up. This call appears to be spontaneous, but often is emitted prior to making a kill or upon detecting another *Onychomys* nearby. This shrill call, which is often repeated several times, has been compared to a miniature wolf howl in its qualities of smoothness and prolongation, and in the posture of the animal when the call is given.

The following reproductive information is taken from McCarty (1975, 1978) and Pinter (1970). Grasshopper mice are capable of breeding throughout the year, but most reproductive activity in the wild occurs in the late spring and summer. The estrous cycle of *O. leucogaster* may be about 5 to 7 days in length. This species produces several litters per year (up to 12 in the laboratory). Female *O. torridus* apparently have a high level of sexual activity for just one breeding season; individuals born as early as April may produce two or three litters before the end of the year, while females born late in the season may give birth to as many as six litters during the following breeding season. Reported gestation periods are 26–37 days in nonlactating *O. leucogaster,* 32–47 days in lactating *O. leucogaster,* and 26–35 days in *O. torridus.* Litter size is 1 to 6 young, averaging about 3.6 in *O. leuco-*

gaster and 2.6 in *O. torridus.* The young weigh about 2.6 grams at birth, open their eyes when about 2 weeks old, and are weaned at around 3 weeks of age. Sexual maturity is reached between the ages of 2 and 5 months in *O. leucogaster* and as early as 6 weeks in *O. torridus.* Female *O. leucogaster* remain reproductively active for up to 3 years, but female *O. torridus* seldom breed for more than 2 years.

These rodents are said to make fascinating pets and to become generally quite gentle in captivity. According to Marvin L. Jones (Zoological Society of San Diego, pers. comm.), a captive specimen of *O. torridus* lived for four years and seven months.

RODENTIA; MURIDAE; **Genus AKODON** Meyen, 1833

South American Field Mice, Grass Mice

The 9 subgenera and 41 species (not in systematic order) are as follows (Cabrera 1961; Langguth 1975*a*; Pine 1973*b*, 1976*a*; Pine, Miller, and Schamberger 1979; Ximinez and Langguth 1970; Contreras 1968; Goodwin and Greenhall 1961; De Santis and Justo 1980; Voss and Linzey 1981; Gardner and Patton 1976):

subgenus *Akodon*

A. aerosus, the Andes of Ecuador, Peru, and Bolivia;
A. affinis, western Colombia;
A. albiventer, mountains of southeastern Peru, western Bolivia, northern Chile, and northwestern Argentina;
A. andinus, the Andes from southern Peru to central Chile and western Argentina;
A. arviculoides, eastern Brazil;
A. azarae, extreme southern Bolivia, northern Argentina, southern Brazil, Uruguay;
A. boliviensis, the Andes of southern Peru, Bolivia, and northwestern Argentina;
A. caenosus, mountains of extreme southern Bolivia and northwestern Argentina;
A. cursor, southern Brazil, Paraguay, Uruguay;
A. dolores, central Argentina;
A. iniscatus, central Argentina;
A. llanoi, Tierra del Fuego;
A. markhami, Wellington Island off southern Chile;
A. molinae, eastern Argentina;
A. mollis, the Andes from Ecuador to Bolivia;

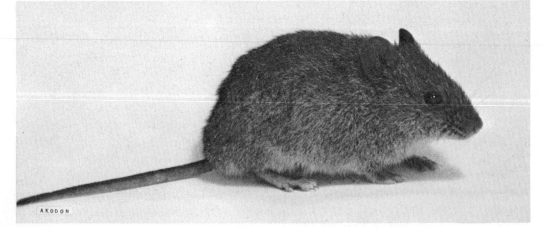

South American field mouse (*Akodon olivaceus*), photo by Luis E. Peña.

South American field mouse (*Akodon varius*), photo by Michael Mares.

A. olivaceus, Chile, western Argentina;

A. orophilus, mountains of Peru;

A. pacificus, mountains of western Bolivia;

A. puer, central and southern Peru, western Bolivia;

A. serrensis, eastern Brazil;

A. surdus, mountains of southeastern Peru;

A. tapirapoanus, central Bolivia, Mato Grosso region of central Brazil;

A. urichi, the Andes from western Venezuela to Bolivia, Trinidad;

A. varius, Bolivia, Paraguay, western Argentina;

A. xanthorhinus, southern Argentina and Chile, including Tierra del Fuego and some nearby islands;

subgenus *Deltamys*

A. kempi, islands of the Parana Delta between Argentina and Uruguay;

subgenus *Hypsimys*

A. budini, mountains of northwestern Argentina;

subgenus *Thalpomys*

A. reinhardti, eastern Brazil;

subgenus *Thaptomys*

A. nigrita, eastern and southern Brazil;

subgenus *Bolomys*

A. amoenus, the Andes of southeastern Peru;

A. berlepschii, the Andes of southeastern Peru, western Bolivia, and northern Chile;

A. lactens, mountains of northwestern Argentina;

A. lasiurus, southern Brazil, Bolivia, Paraguay;

subgenus *Abrothrix*

A. illuteus, mountains of northwestern Argentina;

A. lanosus, southern Argentina and Chile, Tierra del Fuego;

A. longipilis, Chile, Argentina, Tierra del Fuego;

A. mansoensis, south-central Argentina;

subgenus *Microxus*

A. bogotensis, mountains of Colombia and northwestern Venezuela;

A. latebricola, the Andes of Ecuador;

A. mimus, the Andes of southeastern Peru;

subgenus *Chroeomys*

A. jelskii, the Andes from central Peru to extreme northwestern Argentina.

Bianchi et al. (1971) considered the following to be distinct genera: *Hypsimys* Thomas, 1918; *Thalpomys* Thomas, 1916; *Thaptomys* Thomas, 1916; *Bolomys* Thomas, 1916; *Abrothrix* Waterhouse, 1837; *Microxus* Thomas, 1909; and *Chroeomys* Thomas, 1916.

Head and body length is 75 to 140 mm, and tail length is 50 to 100 mm. Dalby (1975) reported that sexually mature *A. azarae* weigh 10 to 45 grams. Animals of the genus *Akodon* have been described as heavy-bodied, short-limbed, short-tailed, volelike mice. The pelage is soft and full, varying above from mouse gray to dark brown. Some species have a red hue to the fur. The underside is white to dark gray, tinged with fulvous. Females have eight mammae.

South American field mice occur in a variety of habitats, including relatively arid country, grasslands, humid forests, and mountain meadows. The altitudinal range is from near sea level to about 5,000 meters. Some species are frequently found in human houses. *A. arviculoides* is very common in the woods and cultivated areas of southeastern Brazil, where it lives in galleries built under humus on the ground. Dalby (1975) reported that burrowing does not seem to occupy

South American field mouse (*Akodon lasiurus*), photo by I. Sazima.

much of the time of *A. azarae,* but that he examined one hole that went straight down for 12–15 cm and ended at a globular nest, and another that leveled off at a depth of 5–6 cm, continued for about 40 cm, and terminated at a globular nest.

Various species of *Akodon* have been reported to be either diurnal, nocturnal, crepuscular, or active at any time. Dalby (1975) stated that *A. azarae* is mainly herbivorous, but takes a substantial amount of animal matter. Barlow (1969) found

Cabreramys obscurus, photos by Abel Fornes and Elio Massoia.

11 stomachs of *A. azarae* to contain 20 percent plant material and 70 percent invertebrates.

On the pampas of northeastern Argentina, Dalby (1975) found that the peak density of *A. azarae* approached 200 individuals per ha., but that density was reduced to approximately 50/ha. by late winter. At a semiarid site in north-central Chile, Fulk (1975) determined the density of *A. olivaceus* to be 30.3/ha. in August and 97.0/ha. in November. The average home range length for this species was 54.0 meters; the ranges of males overlapped considerably with one another and with the ranges of females, but there was no extensive overlap between the home ranges of females. Also in Chile, Greer (1965) found the home range of *A. olivaceus* to be 12.5 sq meters, while that of *A. longipilis* varied from 40 to 241 sq meters.

In general for the genus, the breeding season extends from August to May, there are probably two litters per year, and the number of young is usually 3 or 4, but Pine, Miller, and Schamberger (1979) reported that a female *A. xanthorhinus,* taken in Chile on 2 February, contained 10 embryos. Dalby (1975) reported the reproduction of *A. azarae* to be strongly seasonal, with litters born from November to April. Delayed implantation may occur in this species; gestation averaged 22.7 days and litter size, 4.6 young. The young were weaned at 14–15 days of age and reached sexual maturity when 2 months old, though young born late in the season did not become sexually mature until the following breeding season.

RODENTIA; MURIDAE; *Genus CABRERAMYS*
Massoia and Fornes, 1967

Massoia and Fornes (1967) erected this genus for three species formerly assigned to *Akodon:*

C. obscurus, Uruguay and adjacent parts of Argentina;
C. benefactus, northeastern Argentina;
C. lenguarum, Chaco region of southern Bolivia and
    Paraguay.

Some authorities, such as Ximinez, Langguth, and Praderi (1972) and Reig (1978), have not recognized *Cabreramys* as a full genus.

A specimen of *C. lenguarum* from Paraguay had a head and body length of 107 mm and a tail length of 90 mm. The typical color of the genus is agouti, which is most pronounced on the sides of the body. The back is chestnut to grayish, and the underparts are paler. The texture and length of the fur are about medium for a cricetine rodent. The tail is shorter than the length of the head and body, and is bicolored and tipped with fine hairs. The hands and feet are broad and have large nails like those of fossorial animals.

Cabreramys differs from *Akodon* and *Zygodontomys* in certain characters of the skull, dentition, and digits. The fifth toe of the hind foot of *Cabreramys* is longer than those of the other two genera; its end extends beyond the position of the third postdigital tubercle and its nail extends nearly to the middle of the fourth toe. The nails on the forefeet of *Cabreramys* are normal and of intermediate size for a murid.

Barlow (1969) found *C. obscurus* to be restricted to limnic habitats, and collected it in inundated places with plentiful vegetation. Stomach contents indicated that the species feeds primarily on arthropods. Pregnant females were taken from October to February, and lactating females in December, April, and May. One female *C. obscurus* contained four embryos, another had three embryos, and a lactating female showed five placental scars.

RODENTIA; MURIDAE; *Genus ZYGODONTOMYS*
J. A. Allen, 1897

Cane Mice

There are three species (Cabrera 1961; Hershkovitz 1962; Gardner and Patton 1976; Tranier 1976; Goodwin and Greenhall 1961; Voss and Linzey 1981):

Z. brevicauda, southern Costa Rica to western Ecuador
    and Surinam, Trinidad;
Z. reigi, French Guiana;
Z. borreroi, northern Colombia.

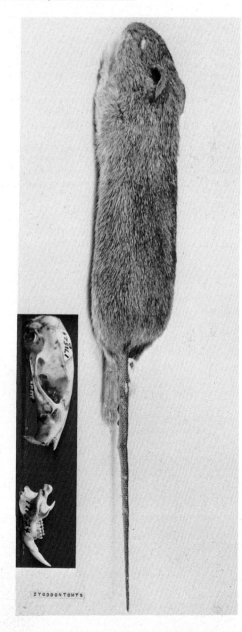

Cane mouse (*Zygodontomys brevicauda*), photos by P. F. Wright of specimen in U.S. National Museum.

Head and body length is 95 to 155 mm, and tail length is 35 to 130 mm. The tail is usually shorter than the head and body. In some species the coat is full, long, and soft, while in others it is short and harsh. The upper pelage is yellowish brown, reddish, or grayish; the sides are grayish to yellowish; and the underparts are grayish white to buffy gray. The tail and ears are brown.

Zygodontomys resembles *Oryzomys*, but may be distinguished by its relatively shorter tail and shorter hind feet. The anterior nails of *Zygodontomys* are smaller than those of *Cabreramys* and *Akodon*. The ears of *Zygodontomys* are small. Females have eight mammae.

These animals live in open country and in areas of low bushes and thick ground cover, usually near water, from sea level to about 600 meters in elevation. They are terrestrial and their habits are much like those of meadow voles (*Microtus*). They make runways through the dense grass and are active at night. Nests are made of grass and the down of flowers. These are built at the ends of short burrows in banks or under tree roots. The diet includes seeds, grasses (including corn and rice), and fruit.

The gestation period is 28 days. Litter size is two to eight young, usually four. The young leave the mother to forage for themselves at the age of 17 to 20 days, and become sexually mature when 3 to 4 months old.

These mice have been used in the laboratory for research on yellow fever. They are easily kept in captivity and feed on a mixture of grains and fresh vegetables. In the wild they are attracted to cultivated fields and may become pests (Husson 1978).

RODENTIA; MURIDAE; Genus *PODOXYMYS* Anthony, 1929

Mount Roraima Mouse

The single species, *P. roraimae*, is known from five specimens taken on Mount Roraima at the point where Guyana, Venezuela, and Brazil come together (Cabrera 1961).

Head and body length of the type specimen is 101 mm, and tail length is 95 mm. The original description (Anthony 1929:1) stated: "Pelage long and lax, 10–11 mm. long on back, blackish slate at base and for most of the length of the hair, only the tip being colored. Above, finely mixed clay-color and blackish, the minute specks of color at the tip of the hairs being insufficient to dominate the black and general impression resulting in a rather dark pelage; there is a tendency (shown by two out of five specimens) for the color pattern to be darkest on the rump, but otherwise the upperparts are fairly uniform; sides of head and underparts slightly lighter in tone than back; hands and feet clove-brown above; tail about half of total length, very sparsely haired, hair brown above and below; ears of fair size but partially hidden in the long pelage; eye rather small; claws of forefeet long (third claw 3 mm. beyond pad), slender, strongly compressed laterally, slightly curved, those of hind feet a trifle shorter."

Anthony also stated that this genus "appears to be somewhat intermediate in character between *Akodon* and *Oxymycterus*," and added: "External appearance that of a dark-colored, long-tailed *Akodon*, with long slender claws (which suggest the generic name); skull with long, slender rostrum, narrow zygomatic plate and general appearance of *Oxymycterus*."

RODENTIA; MURIDAE; Genus *LENOXUS* Thomas, 1909

Andean Rat

The single species, *L. apicalis*, occurs in southeastern Peru and western Bolivia (Cabrera 1961).

Head and body length is 150 to 170 mm, and tail length is 150 to 190 mm. Coloration is grayish black above, paler on the sides, and grayish brown with a buffy wash, or grayish white, below. The tail is entirely brown except for the white terminal part. The large size, black color, and long, white-tipped tail are distinguishing external features. The body form is heavy and ratlike, the ear is large (about 20 mm), and the claws are not enlarged. This genus could be regarded as a

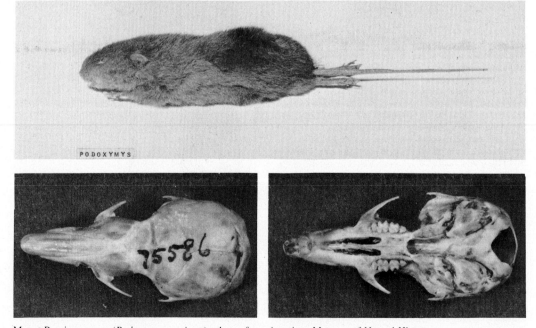

Mount Roraima mouse (*Podoxymys roraimae*), photos from American Museum of Natural History.

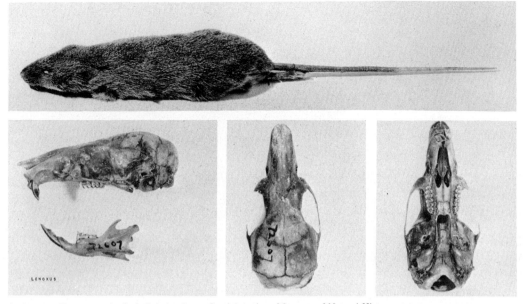

Andean rat (*Lenoxus apicalis boliviae*), photos from American Museum of Natural History.

nonfossorial version of *Oxymycterus*. Some individuals have grooved upper incisors. *Lenoxus* lives in wooded areas between 1,850 and 2,450 meters in elevation.

RODENTIA; MURIDAE; Genus JUSCELINOMYS
Moojen, 1965

The single species, *J. candango*, is known from nine specimens taken on the grounds of the Zoobotanical Foundation in Brasilia, Federal District, Brazil.

This genus differs from *Oxymycterus*, to which it is closely related, in having a shorter muzzle; a much thicker tail, densely covered with hair that completely covers the underlying scales; and a pelage in the ventral region composed of ferrugineous hairs with whitish bases. Cranially, *Juscelinomys* is much more strongly constructed than *Oxymycterus*, with more inflated bullae and a shorter, broader rostrum.

All specimens were taken at an elevation of 1,030 meters in a parklike area with scattered trees under which there were grasses. The entrances to two subterranean nests were recognizable by the tracks that the animals made on the ground during the excavations. The ground around the entrances to the nests was worn as smooth as pavement. The nests were poorly lined with grasses and other fine vegetable matter. Stomach contents included mostly unidentifiable vegetable material, but also some large ants.

RODENTIA; MURIDAE; Genus OXYMYCTERUS
Waterhouse, 1837

Burrowing Mice

The 10 species (not in systematic order) are as follows (Cabrera 1961; Massoia 1963; Ximinez, Langguth, and Praderi 1972):

O. akodontius, northwestern Argentina;
O. angularis, eastern Brazil;
O. delator, Paraguay;
O. hispidus, eastern and southern Brazil, northeastern Argentina;
O. iheringi, northeastern Argentina, extreme southern Brazil;
O. inca, central Peru to western Bolivia;
O. paramensis, southeastern Peru, western and southern Bolivia, northern Argentina;
O. roberti, eastern Brazil;
O. rutilans, eastern and southern Brazil, Uruguay, northeastern Argentina;
O. sanborni, central Chile and adjacent parts of Argentina.

Cabrera (1961) placed *O. iheringi* in the subgenus *Microxus* of Akodon, and *O. sanborni* in the subgenus *Abrothrix* of *Akodon*, but both species were referred to *Oxymycterus* by Massoia (1963). Pine, Miller, and Schamberger (1979) continued to place *O. sanborni* in the genus *Akodon*.

Head and body length is 93 to 170 mm, and tail length is 70 to 145 mm. According to Dalby (1975), a series of sexually mature specimens of *O. rutilans* weighed 46 to 125 grams. The fur of *Oxymycterus* is usually not thick. The upper parts are reddish, yellowish brown, dark brown, or blackish. The underparts are buffy, grayish brown, or grayish white. The tail, usually shorter than the head and body, is moderately or thinly haired. The snout is long and mobile, and the foreclaws are long and prominent. The outer digits of the hind feet are shorter than the central three digits. The dentition is weak in all species. Female *O. rutilans* have eight mammae, and female *O. iheringi* have six mammae.

These rodents have been found in swamps, marshes, grasslands, brushy areas, woodlands, and forests. The following account of *O. rutilans* is based on the work of Barlow (1969) and Dalby (1975) in Uruguay and northeastern Argentina. This species seems to have an ecological role somewhat like that of *Onychomys* in North America. It apparently does not construct its own burrows or runways, but may use those of *Cavia* or *Hydrochaeris* in stands of tall bunch grass. Its long foreclaws and shrewlike, pointed nose function in rooting for subsurface invertebrates. The diet is largely insectivorous, but other small invertebrates and some plant matter are eaten. *O. rutilans* seems to be almost entirely

Burrowing mouse (*Oxymycterus rutilans*), photo by Alfredo Langguth.

diurnal. Population density fluctuates from 5 to 15 individuals per ha. Breeding occurs in all seasons of the year. Known litter size varies from 1 to 6 young, usually being 2 or 3. The young are weaned after 14 days and reach sexual maturity when they are near 3 months old.

RODENTIA; MURIDAE; *Genus BLARINOMYS Thomas, 1896*

Brazilian Shrew-mouse

The single species, *B. breviceps*, has been collected in the highlands of southeastern Brazil in the states of Bahia, Espirito Santo, Rio de Janeiro, and Minas Gerais (Matson and Abravaya 1977).

According to Abravaya and Matson (1977), total length is 129 to 161 mm, and tail length is 30 to 52 mm. The fur is crisp and short. Coloration is a uniform dark slaty gray throughout, with brown tips on the hairs. The back is slightly iridescent with a ruby tinge, at least when the fur is wet. The hands and feet are brown above.

The body form is modified for fossorial life. The short and conical head has extremely reduced eyes, and short ears that are hidden in the fur. The short tail is thinly haired. The hand has four functional digits, and a fifth digit that is strongly reduced; the claws are well developed. The hind foot is broad and has prominent claws. No other South American mammal closely resembles *Blarinomys*. In external appearance this rodent resembles the North American short-tailed shrews of the genus *Blarina*, from which *Blarinomys* received its name.

Specimens have been taken in dense rain forests near the tops of hills at about 800 meters elevation, and at comparable localities in less humid areas. The genus is fossorial. Its burrow is made under the layer of litter on the forest floor. It is dug almost straight down for about 255 mm and then goes into a sloping tunnel that continues downward, but not at such a steep gradient. The diet is unknown, but may consist primarily of insects and worms (Abravaya and Matson 1975). *Blarinomys* reportedly is docile and does not bite when captured. Pregnant females have been found in January, February, and September, and the observed number of embryos has been one or two (Matson and Abravaya 1977).

RODENTIA; MURIDAE; *Genus NOTIOMYS Thomas, 1890*

Long-clawed Mice, Mole Mice

There are six species (Cabrera 1961; Pine, Miller, and Schamberger 1979):

N. angustus, western Argentina;
N. delfini, extreme southern Chile;
N. edwardsii, extreme southern Argentina;
N. macronyx, Chile, Argentina;
N. megalonyx, central and southern Chile;
N. valdivianus, central and southern Chile, and adjacent parts of Argentina.

Head and body length is 75 to 140 mm, and tail length is 25 to 60 mm. The soft fur may be long or short and is molelike in character. The upper parts are yellowish brown to reddish brown, and the lower parts are whitish gray. The tail, covered with short hairs, is pale fawn above and white below. The soles are naked.

These stout-bodied, short-tailed mice are highly modified for a subterranean life. The claws are powerful, sharp, curved, and long (about 7 mm). The tail is about half the length of the head and body. The ears are small. The genus *Notiomys* appears to be based chiefly on the presence of strongly developed foreclaws. Its general appearance is frequently shrewlike, similar to the North American *Blarina* or the Old World *Crocidura*.

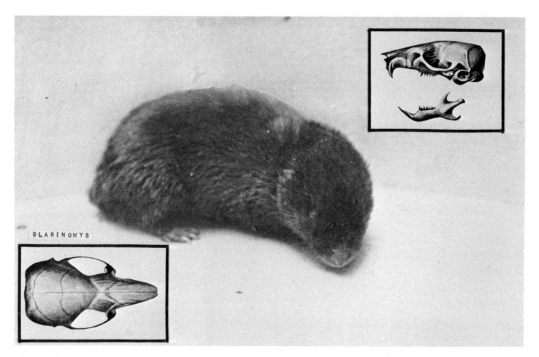

Brazilian shrew-mouse (*Blarinomys breviceps*), photo by Joao Moojen. Insets: photos from *Bol. Mus. Paraense.*

The natural habitat of the various species ranges from dry grassland to humid forest, and extends into the Andes Mountains. They usually construct a network of tunnels through the soil and spend most of their lives underground, though Greer (1965) caught one specimen of *N. macronyx* at the base of a rockpile, where no burrows were evident. A pregnant female *N. valdivianus,* with four embryos, was taken by Pine, Miller, and Schamberger (1979), and a lactating female *N. macronyx* was collected by Greer (1965) in February in central Chile.

Long-clawed mouse (*Notiomys valdivianus*), photo by Richard D. Sage.

Water rat (*Scapteromys tumidus*), photo by Abel Fornes and Elio Massoia.

RODENTIA; MURIDAE; *Genus SCAPTEROMYS* Waterhouse, 1837

Water Rat

The single species, *S. tumidus*, is known from southern Paraguay, extreme southern Brazil, Uruguay, and northeastern Argentina (Hershkovitz 1966a; Myers and Wetzel 1979).

Head and body length is 150 to 200 mm, and tail length is 120 to 170 mm. The upper parts are grayish black and the underparts are grayish white. This rodent is thickset and heavy; it resembles the black rat (*Rattus rattus*), but has a much longer tail. The feet are long, and the middle toes of the hind feet are elongated. All the claws are long. Females have eight mammae.

Barlow (1969) stated that *Scapteromys* is semiaquatic, occupying marshy places and frequently inundated areas. Excellent swimming ability and agility in climbing onto tall plants contribute to its success in such areas. It does not seem to make runways of its own, but follows any open route through the vegetation, and frequently uses the runways of *Cavia*. It digs to obtain food, but apparently does not excavate burrows. It may dig shallow depressions under vegetation, in which it builds nests and bears its young. It is mainly nocturnal and crepuscular, but some specimens were collected during the day. Its diet is largely insectivorous, but other invertebrates and some plant material are also eaten.

Scapteromys has a highly developed sense of hearing and can perceive sounds of extremely high wave lengths. It is easily startled, and seeks safety by plunging into the nearest body of water. Several vocalizations have been reported (Barlow 1969).

Breeding apparently takes place throughout most of the year (Barlow 1969; Hershkovitz 1966a). Males in reproductive condition have been found in every month except January, June, and July. Pregnant females have been taken in January, April, May, October, November, and December.

The number of young varies from two to five and averages about four.

RODENTIA; MURIDAE; *Genus KUNSIA* Hershkovitz, 1966

South American Giant Rats

There are two species (Hershkovitz 1966a; Avila-Pires 1972):

K. fronto, south-central Brazil, northern Argentina;
K. tomentosus, eastern Bolivia to southeastern Brazil.

These are the largest living cricetines. Head and body length is 160 to 287 mm, and tail length is 75 to 160 mm. The pelage is coarse, thick on the upper parts and thinner underneath. The color of the upper parts is dark brown mixed with gray. The underparts are paler, and more or less sharply defined from the upper parts. The tail is uniformly dark brown or black.

The external form is adapted for fossorial and palustrine life. All four feet are large and powerful, and are provided with extremely long claws. The genus differs from *Scapteromys* in its usually larger size, much shorter tail, shorter hind feet, and four-rooted (rather than three-rooted) first lower molar.

Giant rats occur in mixed savannah-forest at elevations from sea level to 1,000 meters. *K. tomentosus* is a burrowing animal. It may live almost entirely underground during the dry season, but mostly above ground during periods of heavy rains or inundations. Perhaps because of the animal's large size and powerful body, it has never been taken in ordinary snap traps. The few preserved specimens have been either seized by hand or captured by dogs.

RODENTIA; MURIDAE; *Genus BIBIMYS* Massoia, 1979

Crimson-nosed Rats

There are three species (Massoia 1979, 1980):

B. torresi, Parana River Delta of eastern Argentina;
B. labiosus, southeastern Brazil;
B. chacoensis, northeastern Argentina.

For three adult specimens of *B. torresi*, head and body length is 84 to 97 mm, tail length is 65 to 78 mm, and weight is 23 to 34 grams. The nose is bulky and of an intense crimson color. The dorsal coloration is dark chestnut, with an almost black medial line; the sides of the body are yellowish chestnut; the underparts and feet are light gray; and the tail is dark gray above and light gray beneath.

This genus is about the same size as certain species of *Akodon*, but has a larger skull and resembles *Scapteromys* and *Kunsia* in skeletal and dental characters. From these two genera, *Bibimys* differs in its much smaller overall size and more complex third lower molar.

South American giant rat (*Kunsia tomentosus*), photos from U.S. National Museum.

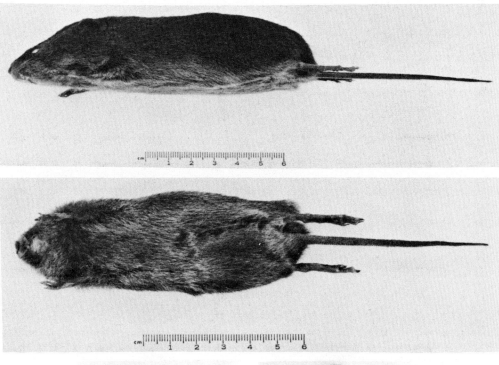

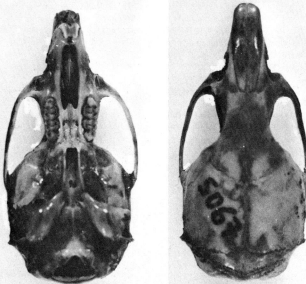

Crimson-nosed rat (*Bibimys torresi*), photos by Carlos Larrañaga.

RODENTIA; MURIDAE; **Genus SCOTINOMYS** Thomas, *1913*

Brown Mice

There are two species (Hall 1981):

S. xerampelinus, Costa Rica, western Panama;
S. teguina, southern Mexico to western Panama.

Head and body length is 80 to 85 mm, and tail length is 48 to 70 mm. Average weight is 12 grams for *S. teguina* and 15 grams for *S. xerampelinus* (Hill and Hooper 1971). The sparsely haired tail is usually about 70 percent of the head and body length. The pelage of *S. teguina* is short and somewhat harsh, while that of *S. xerampelinus* is long and soft. Coloration of the upper parts is dark yellowish brown, reddish brown, or dull dark brown, nearly black. The underparts are reddish brown, buffy brown, or grayish. The ears, hands, feet, and tail are usually blackish. Females may have either two or three pairs of mammae (Hooper 1972).

These mice seem to prefer rough, rocky terrain, but they also inhabit upland savannahs and clearings in forested areas.

Brown mouse (*Scotinomys xerampelinus*), photo by Rexford Lord.

Their altitudinal range is 1,000 to 3,300 meters. They utilize runways through low-growing vegetation, are mainly or exclusively diurnal, and are insectivorous (Hooper 1972).

Hooper and Carleton (1976) provided the following information on social life and reproduction. Both species have a repertoire of vocalizations, the most elaborate of which is the "song," a sequence of pulses that in *S. teguina* lasts 7 to 10 seconds. Males call much more than females. Nests are built mainly or entirely by the female in *S. xerampelinus*, but by both sexes in *S. teguina*. A female with young will tolerate the presence of her mate, and the male may stay with the young while the mother is out. In the laboratory, births occur throughout the year, and many females regularly produce litters at 1-month intervals. In the wild, pregnant females have been found in January, February, April, May, June, July, and August, but adequate collections are lacking for the remainder of the year. Gestation averages 30.8 days in *S.*

teguina and 33.1 days in *S. xerampelinus;* litter size is one to five, usually two to three, in *S. teguina* and two to four in *S. xerampelinus;* the young of *S. teguina* are weaned when 18–21 days old, those of *S. xerampelinus* when 21–24 days old; females attain sexual maturity after an average of 33.8 days of life in *S. teguina* and 51.8 days in *S. xerampelinus;* males of both species are sexually mature when 6 to 8 weeks old.

RODENTIA; MURIDAE; **Genus WIEDOMYS** *Hershkovitz, 1959*

Red-nosed Mouse

The single species, *W. pyrrhorhinos,* is found in eastern Brazil, from the state of Ceara to the state of Rio Grande do

Red-nosed mouse (*Wiedomys pyrrhorhinos*), photo by Michael Mares.

Sul, and possibly in the Mato Grosso and Paraguay (Hershkovitz 1959*b*).

Head and body length is 100 to 128 mm, and tail length is 160 to 205 mm. The coloration of the back is mixed buff and brown, and the underparts are white. In contrast, the nose, eye ring, ears, outer sides of all four limbs, and rump are bright reddish orange. This rodent resembles *Thomasomys oenax* in external appearance, and *Calomys* and *Eligmodontia* in cranial and dental characters.

The red-nosed mouse inhabits scrub forests or caatingas. It is an active climber. It utilizes the abandoned nests of birds for shelter and the rearing of young. Among such nests are those of the thorn bird (*Anabates rufifrons*). These nests are constructed of branches wound around liana vines and may be a few meters long. *Wiedomys* may also inhabit abandoned termite mud nests that were previously used by parrots. It also constructs its own nest of dry leaves and grass or cotton fibers in stone walls, hollow tree trunks, and shrub and palm thickets.

The red-nosed mouse apparently is gregarious: 8 adults and 13 young of varying size were found in one termite nest. Mating usually occurs in August in northeastern Brazil. Litter size is 1 to 6, usually 5, young.

RODENTIA; MURIDAE; **Genus CALOMYS** Waterhouse, *1837*

Vesper Mice

There are seven species (Pearson and Patton 1976; Cabrera 1961; Hershkovitz 1962; Handley 1976; Contreras and Justo 1974):

C. sorellus, the Andes of central and southern Peru;
C. laucha, southern Bolivia and southern Brazil to central Argentina and Uruguay;

Vesper mouse: Top, *Calomys laucha,* photo by R. B. Tesh; Bottom, *C. musculinus,* photo by Abel Fornes and Elio Massoia.

C. fecundus, Bolivia;

C. lepidus, the Andes of southern Peru, western Bolivia, northern Chile, and northwestern Argentina;

C. musculinus, Argentina;

C. callosus, Bolivia and southern Brazil to northern Argentina;

C. hummelincki, northwestern coast of Venezuela and the nearby islands of Aruba and Curacao.

The last species was originally described from Curacao and Aruba as *Baiomys hummelincki,* but was synonymized with *Calomys laucha laucha* by Hershkovitz (1962). He reported the presence of an introduced colony of *C. laucha laucha* in northeastern Venezuela and indicated that the animals on Curacao and Aruba had also been introduced. Handley (1976), however, found a wild population in northwestern Venezuela, which appeared to represent the same distinct species as that on Curacao and Aruba.

Head and body length is about 60 to 125 mm, and tail length is about 30 to 90 mm. The tail is usually shorter than the head and body and is never much longer. A series of adult *C. laucha* from the Mato Grosso of Brazil weighed about 30 to 38 grams. The pelage is usually not very thick, but *C. lepidus* has soft fur. Coloration of the upper parts is buff, tawny, dark brown, grayish buff, or grayish. The underparts are usually grayish or white. Some species have a small white patch behind each ear. The tail is dark above and paler below in some forms, and is all white in others.

The body form is mouselike. The ear is prominent, the hands and feet are narrow, and the tail is moderately haired. The number of mammae is apparently unknown in some species, but has been given as 8, 10, or 14; the number may vary in individual females of the same species. The low-crowned cheek teeth distinguish *Calomys* from related genera. Injuries of the tail, in the form of a missing part or damaged skin, are fairly common in *Calomys* and are probably caused by predators.

Vesper mice occur in a variety of habitats, including montane grasslands, brushy areas, and forest fringes. They may find shelter in bunch grass, in holes in the ground, in rotting tree stumps, or among rocks. They are active mainly at night, and also possibly in the evening and early morning.

Barlow (1969) provided the following information on *C. laucha* in Uruguay. This species is usually found in open grasslands, rocky hillsides, and edge situations, but occasionally is taken in the vicinity of marshes and swamps. It also frequents human dwellings and outbuildings. It is often observed at night in overgrazed rocky pasture. It does not make runways and rarely utilizes those of other species. Nests have been found under boards and rocks, in crevices in the ground, and even high above the ground in trees. The species climbs well, and on the ground it often hops on its hind legs in the manner of *Dipodomys.* The diet is predominantly vegetation, but the stomach of one individual contained beetle fragments. The breeding season apparently extends through much of the austral summer and autumn. Seven pregnant females collected in March and April contained four to eight embryos each, the average being about six. Hershkovitz (1962) added that a captive female *C. laucha* had a gestation period of about 25 days and produced two litters, the first with five young and the second with three.

RODENTIA; MURIDAE; **Genus PSEUDORYZOMYS**
Hershkovitz, 1962

Ratos-do-Mato

There are two species (Hershkovitz 1962; Pine and Wetzel 1975; Wetzel and Lovett 1974; Massoia 1980):

Rato-do-Mato (*Pseudoryzomys wavrini*), photos from Field Museum of Natural History.

P. wavrini, Bolivia, Paraguay, northern Argentina;

P. simplex, southeastern Brazil.

Head and body length is 94 to 140 mm, tail length is 106 to 140 mm, and weight is 30 to 56 grams. The upper parts are brown or yellowish brown, and the underparts are dull whitish with a buffy tinge. The well-haired tail is brown above and white below. This genus is distinguished from other phyllotine rodents by its moderate size, long tail, small ears, large hind feet, fully naked soles, and well-developed webbing between the hind toes.

According to Hershkovitz (1962), *Pseudoryzomys* is semi-aquatic and has adaptations like those of *Oryzomys palustris.*

Wetzel and Lovett (1974), however, found *Pseudoryzomys* to be pastoral, at least during the dry season, and not to be restricted to palustrine habitats.

RODENTIA; MURIDAE; *Genus ANDAGALOMYS* *Williams and Mares, 1978*

There are two species (Williams and Mares 1978):

A. olrogi, northwestern Argentina;
A. pearsoni, Paraguay.

The latter species was originally described as a member of the genus *Graomys* (Myers 1977*b*).

Head and body length is 86 to 115 mm, and tail length is 97 to 130 mm. The upper parts are ochraceous buff or tawny, partly overlain with a suffusion of black hairs. The underparts and feet are white. The tail is brownish above and paler below, and is slightly to moderately haired and penciled. The ears are moderately large and sparsely covered with fine hairs. The soles of the feet are naked.

Andagalomys apparently is closely related to *Eligmodontia* and *Calomys*. *Andagalomys* resembles *Eligmodontia*, but can be distinguished by its longer, more penciled tail; its naked-soled, cushionless feet; and its more ovate molar cusps. *Andagalomys* is most similar cranially to *Calomys*, but has a more slender rostrum and much larger, more globular bullae.

A. olrogi is specialized for desert habitat and *A. pearsoni* occupies dry grasslands. The collection of a subadult male *A. pearsoni* in July suggests that some winter breeding occurs (Myers 1977*b*).

RODENTIA; MURIDAE; *Genus ELIGMODONTIA* *F. Cuvier, 1837*

Highland Desert Mouse

The single species, *E. typus,* occurs from southern Peru and southern Bolivia along the Andes to the Strait of Magellan (Hershkovitz 1962; Pine, Miller, and Schamberger 1979).

Head and body length is 65 to 105 mm, and tail length is 67 to 110 mm (Hershkovitz 1962). The pelage is long and soft. The upper parts are pale brownish yellow. In some forms the underparts are entirely white; in others the white is limited to the throat and upper breast. The tail is well haired. The young are darker and grayer than the parents.

Eligmodontia has some similarity to *Phyllotis;* both genera have slender bodies, long tails, and a silky pelage. *Eligmodontia* is distinguished by the specialization of its hind feet, inflation of the auditory bullae, and brachyodont molars. The hind feet are long, slender, and expanded toward the tips of the toes; the soles are covered with short hairs. In addition, a hairy "cushion" is located on the ball of each foot. The ears of *Eligmodontia* are long and prominent. Females have eight mammae.

This mouse has been reported from dry, gravelly plains; from areas of thorn bushes or scattered grass clumps; and from bare, rocky hillsides up to about 4,575 meters in elevation. *Eligmodontia* can jump and climb with amazing agility, its period of greatest activity being during the night. It is said not to dig its own burrow but to inhabit the deserted dens of *Ctenomys*. It also lives under old logs and bushes. Because it is not seen during the winter, it is assumed to hibernate during this period. Its diet is thought to consist mainly of insects. Two females taken in March in central Chile were

Andagalomys olrogi, drawing by Patricia Barquez.

Highland desert mouse (*Eligmodontia typus*), photo by Michael Mares.

pregnant, one with six embryos and the other with eight (Greer 1965). *Eligmodontia* becomes so abundant at times that it invades human homes and becomes a pest by gnawing furniture and nibbling on unprotected food.

RODENTIA; MURIDAE; **Genus *PHYLLOTIS* Waterhouse, *1837***

Leaf-eared Mice, Pericotes

There are 11 species (Pearson and Patton 1976; Cabrera 1961; Hershkovitz 1962; Pearson 1972; Simonetti and Spotorno 1980):

P. osilae, southern Peru to northwestern Argentina;
P. andium, the Andes from central Ecuador to central Peru;
P. definitus, northwestern Peru;
P. wolffsohni, western Bolivia;
P. magister, western Peru;
P. caprinus, southern Bolivia, northwestern Argentina;
P. darwini, Peru to southern Chile and Argentina;
P. osgoodi, southern Peru, southwestern Bolivia;
P. amicus, northern and central Peru;
P. gerbillus, northwestern Peru;
P. haggardi, the Andes of Ecuador.

Cabrera (1961) placed *P. gerbillus* in a separate genus, *Paralomys* Thomas, 1926, but Pearson and Patton (1976) considered it merely a species of *Phyllotis.* Several additional species of *Phyllotis* were listed by Cabrera (1961) and Hershkovitz (1962), but these are now placed in the genera *Graomys* and *Auliscomys.* The subgeneric name *Loxodon-*

tomys, used by Cabrera, is a *nomen nudum* (Pine, Miller, and Schamberger 1979).

Head and body length is 70 to 150 mm, tail length is 45 to 165 mm, and weight is 20 to 100 grams. The ears are so large (usually 20 to 30 mm long) that they give a leaflike appearance. This condition has given the scientific name to the genus and to a group of related genera, the phyllotines. The upper parts of *Phyllotis* are grayish, brownish, yellowish, orangish, buffy, or cinnamon. The underparts are grayish, whitish, or buffy. In appearance, general habits, and ecology, *Phyllotis* appears to be the Andean equivalent of the North American *Peromyscus.*

Leaf-eared mice occur on savannahs, scrublands, deserts, and humid mountains. The elevational range is sea level to about 5,000 meters. Typical habitat is on bare ground, but

Leaf-eared mouse (*Phyllotis darwini*), photo by Jorge N. Artigas.

Leaf-eared mouse (*Phyllotis magister*), photo by O. P. Pearson.

near vegetation that will furnish food and some sort of shelter. These mice may seek rocky places, stone walls, houses, or burrows constructed by other animals. Some species are nocturnal, some are active occasionally during the day, and others are diurnal. They eat seeds, green plant material, and lichens. At a semiarid site in central Chile, Fulk (1975) found population density of *P. darwini* to be 29 to 46 individuals per ha., and mean home range length to vary from 36.3 meters in February to 41.4 meters in November.

Hershkovitz (1962) stated that peak numbers of pregnant and lactating female *P. darwini* are found from November to February and in August, and that embryo counts average 4.6 (2–7). For *P. osilae* the peak reproductive season is April to June and the average number of embryos is 4.4 (2–6). In central Chile, Greer (1965) collected a pregnant female and a lactating female *P. darwini* in March, and juveniles in March, May, and November.

RODENTIA; MURIDAE; **Genus GALENOMYS** Thomas, *1916*

The single species, *G. garleppi*, is known from five specimens collected at altitudes of 3,800 to 4,500 meters on the altiplano of southern Peru, western Bolivia, and northeastern Chile. Both Cabrera (1961) and Hershkovitz (1962) considered *Galenomys* to be a full genus, while Pearson and Patton (1976) used the name as a subgenus of *Phyllotis,* pending availability of karyotypes.

Head and body lengths of two specimens are 105 and 123 mm, and corresponding tail lengths are 38 and 45 mm. The upper parts are buffy, thinly lined with brown; the head is paler than the trunk; the rump is ochraceous; and the under-

parts and legs are white. The body is stout, the tail is relatively short, and the ears are large. Females have eight mammae. According to Hershkovitz (1962), this genus is distinguished from *Phyllotis* by its vaulted skull; low-slung, anteriorly expanded zygomata; enlarged first upper molars; uniquely procumbent lower incisors; short, stout, thickly haired hind feet with extremely hairy soles; and very short, hairy tail.

RODENTIA; MURIDAE; **Genus AULISCOMYS** Osgood, *1915*

There are four species (Pearson and Patton 1976; Hershkovitz 1962; Simonetti and Spotorno 1980):

A. pictus, the Andes from central Peru to northwestern Bolivia;

A. sublimis, highlands of southern Peru, western Bolivia, northern Chile, and northern Argentina;

A. boliviensis, southern Peru, western Bolivia, northern Chile;

A. micropus, southern parts of Chile and Argentina.

Cabrera (1961) treated *Auliscomys* as a subgenus of *Phyllotis,* and Hershkovitz (1962) did not use it as a genus or subgenus. Based in large part on karyological studies, however, *Auliscomys* was given generic rank by Pearson and Patton (1976), and Gardner and Patton (1976).

The following descriptions are taken from Hershkovitz (1962). In *A. pictus,* head and body length is 100 to 147 mm; tail length is 75 to 118 mm; the head and shoulders are grizzled, contrasting with the brownish back and sides of the trunk; the rump, base of tail, and outer thighs are more tawny

Auliscomys boliviensis, photo by O. P. Pearson.

or ochraceous than the back; and the underparts are gray to dirty white. In *A. sublimis,* head and body length is 96 to 125 mm; tail length is 43 to 62 mm; the upper parts are buffy, finely mixed with black; the sides are paler; the cheeks are whitish; and the underparts are whitish to pale gray. In *A. boliviensis,* head and body length is 105 to 145 mm; tail length is 78 to 105 mm; the upper parts are a mixture of gray, brown, and buff or ochre; the sides are paler; the muzzle and underparts are whitish; and the large ears have prominent ochraceous preauricular tufts.

According to Hershkovitz (1962), the above species are found at high altitudes, from around 3,400 to 5,500 meters. *A. pictus* is active day and night and seems to prefer grassy places, especially near water. *A. sublimis* is nocturnal, shelters among rocks, and possibly remains below ground during much of the wet summer. *A. boliviensis* is diurnal and is found in many kinds of habitat on the altiplano. This species is unusual in its association with the mountain viscacha (*Lagidium*). It suns on rocks with the viscacha, scurries to shelter when *Lagidium* sounds an alarm whistle, and feeds with the larger rodent, "like an elf among grownups."

Hershkovitz (1962) stated that *A. sublimis* seems gregarious, and that nine individuals were reportedly found in a single burrow. In *A. pictus* breeding apparently begins in September or October; a pregnant female with five embryos was taken near Cuzco, Peru, on 13 April. Greer (1965) found two lactating female *A. micropus* in February–March, and two pregnant females in March, one with four embryos and the other with five.

RODENTIA; MURIDAE; **Genus GRAOMYS** Thomas, 1916

There are three species (Pearson and Patton 1976; Hershkovitz 1962; Cabrera 1961):

G. griseoflavus, Bolivia, Paraguay, Argentina, possibly the Mato Grosso of central Brazil;
G. domorum, western and southern Bolivia, northern Argentina;
G. edithae, northwestern Argentina.

A fourth named species, *G. hypogaeus,* was recently shown to be referable to *Eligmodontia typus* (Massoia 1977; Williams and Mares 1978). Cabrera (1961) treated *Graomys* as a subgenus of *Phyllotis.* Hershkovitz (1962) did not use *Graomys* as a genus or subgenus, and also considered *G. domorum* a subspecies of *G. griseoflavus.* Based in large part on karyological studies, however, *Graomys* was given generic rank and *G. domorum* specific rank by Pearson and Patton (1976) and Gardner and Patton (1976).

According to Hershkovitz (1962), head and body length is 106 to 165 mm, and tail length is 120 to 190 mm. The upper parts are buffy to tawny, more or less mixed with gray or black, and the sides and head have less black. The underparts are sharply defined white. The tail is always longer than the head and body, it is well haired, and its terminal part is usually bushy. The skull has a broad supraoccipital region with projecting edges.

Hershkovitz (1962) stated that *Graomys* is scansorial and adapts to most terrestrial and arboreal habitats. It has been found on stream banks, among rocks, in hollow logs, and in brush heaps. It lives in any kind of shelter, including abandoned birds' nests and holes in tree trunks. In summer it moves into human houses and does damage by eating boots and calico, and gnawing on metal teapots. Its natural diet consists mainly of grasses, grains, and the fruits of mesquite. Most pregnant females have been taken in summer (December to March), and none have been found in the winter. The number of young averages about 6 and ranges from 3 to 10. Mares (1977) reported *G. griseoflavus* to be stronger and more aggressive than *Phyllotis darwini,* and to be difficult to handle in the laboratory.

RODENTIA; MURIDAE; **Genus ANDINOMYS** Thomas, 1902

Andean Mouse

The single species, *A. edax,* is known from southern Peru, Bolivia, extreme northern Chile, and northwestern Argentina (Hershkovitz 1962; Pine, Miller, and Schamberger 1979).

Head and body length is 135 to 165 mm, and tail length is 105 to 150 mm. Two males weighed 80 and 91 grams (Pine, Miller, and Schamberger 1979). Coloration is dark buffy above and gray below. The body form is stocky. Females have eight mammae. This genus resembles *Punomys,* but can be distinguished by its longer tail.

The Andean mouse lives on the Altiplano at elevations of 1,675 to 4,850 meters. It has been captured among rocks on the bank of a stream, and in bushy thickets. It is nocturnal. Collectors have remarked that it lives in the branches of trees, where it makes its nest, and in round holes carpeted with very fine straw. Its diet consists of green plants.

Andinomys breeds at the end of the dry season; the young are thus born and raised when plant growth is at its peak. A female, pregnant with three large embryos, was taken on 18 December (Hershkovitz 1962).

Andean mouse (*Andinomys edax*), photo by Michael Mares.

RODENTIA; MURIDAE; *Genus CHINCHILLULA*
Thomas, 1898

Altiplano Chinchilla Mouse

The single species, *C. sahamae*, occurs in the highlands of southern Peru, southwestern Bolivia, northwestern Argentina, and northern Chile (Cabrera 1961).

Head and body length is 122 to 182 mm, and tail length is 90 to 118 mm (Hershkovitz 1962). The color pattern is distinctive: the upper parts are buffy or grayish with black lines, and the underparts are snow white, with the white hips and white rump banded with black. The tail is fully haired. The fur is thick, soft, and silky, resembling the pelage of *Chinchilla* in texture and color. The ears are large and wide. The feet are broad and the digits are normal.

This mouse inhabits the altiplano at elevations of 3,500 to 5,000 meters. In its high, relatively barren habitat, it is associated with such rodent genera as *Phyllotis, Akodon, Lagidium,* and *Punomys*. It lives in rocky places, and is most often caught among boulders and along stone walls. It is active at night, when it feeds on plants. At times its enormous stomach is distended by more than 12 grams of green, finely ground vegetable matter (Hershkovitz 1962). The chinchilla mouse breeds in October and November at the end of the dry

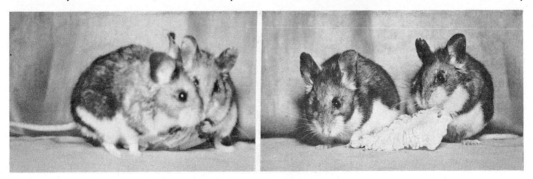

Altiplano chinchilla mice (*Chinchillula sahamae*), photos by Hilda H. Heller.

season, so the young are born and raised when plant growth is at its peak.

Locally, *Chinchillula* is trapped for its warm fur; the skins are used as trimmings or made into robes. A robe made from the pelage of this genus may contain more than 150 skins. In some areas extensive trapping may seriously reduce the numbers of this beautiful rodent.

RODENTIA; MURIDAE; *Genus IRENOMYS* Thomas, *1919*

Chilean Rat

The single species, *I. tarsalis*, occurs in southern Chile, including the islands of Chiloe and Guaitecas, and in the Andes of southern Argentina (Cabrera 1961).

Head and body length is about 100 to 140 mm, and tail length is 134 to 190 mm. Greer (1965) reported the weights of five adults to range from 26.3 to 41.7 grams. The upper parts are grayish red or grayish cinnamon, and the underparts are cinnamon with a buffy or pinkish wash. The ears are a contrasting black color. The hands and feet are mainly whitish, and the tail is usually dark brown throughout. One structural feature, the grooving of the upper incisors, helps to distinguish this genus from related rodents.

The Chilean rat inhabits humid, temperate forests and is thought to be mainly arboreal. Its numbers fluctuate considerably, perhaps in response to the fruiting of various plants, such as bamboos.

RODENTIA; MURIDAE; *Genus PUNOMYS* Osgood, *1943*

Puna Mouse

The single species, *P. lemminus*, inhabits the altiplano of Peru at elevations of 4,450 to 5,200 meters. Several other mammals are found at these elevations, but *Punomys* is the only one that lives entirely at such heights.

Head and body length is 130 to 155 mm, and tail length is usually 50 to 70 mm. The pelage is long, soft, and loose. The upper parts are dull buffy brown or grayish brown, and the belly is whitish or grayish, occasionally with a buffy wash. The hands and feet are usually dusky above and black below. The tail is dusky above and white below, or dusky throughout.

This genus is unique in structure. The body form is stocky and volelike, and the claws are uniformly small. The dentition is peculiar: the upper incisor teeth are fairly stout and heavy, with smooth front surfaces, and the lower incisors are separated by a considerable space. The surfaces of the grinding teeth are unusually complex. The dull color and short tail help to identify this genus.

Specimens of *Punomys* were not collected until 1939, perhaps because of the extraordinary area of its occurrence. The generic name refers to the puna or treeless zone in the higher parts of South America. The puna mouse usually dwells in barren areas, with broken rocks, and where plants of the genus *Senecio* abound. *Punomys* is common within its range. It is active during the daytime, moving about from the shelter of one rock to another. It feeds mainly, if not entirely, on the

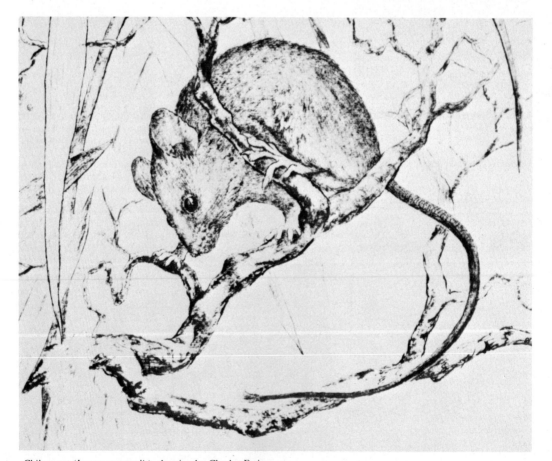

Chilean rat (*Irenomys tarsalis*), drawing by Charles Ewing.

two most offensive-smelling plants in its range: *Senecio adenophylloides*, a fleshy-leaved herb, and *Werneria digitata*, an herb. Twigs, up to 510 mm long, are cut from these plants to be stored under rocks in caches that may contain 30 or more cuttings. These large twigs are manipulated with considerable dexterity as the puna mouse consumes them. The plants are not chewed thoroughly, and so the feces consist of large amounts of undigested material.

Caged individuals show no fear of people; in fact, in the wild, the puna mouse sometimes allows itself to be picked up after its rocky shelter has been removed and it is exposed. The young are apparently born during the warm, wet season

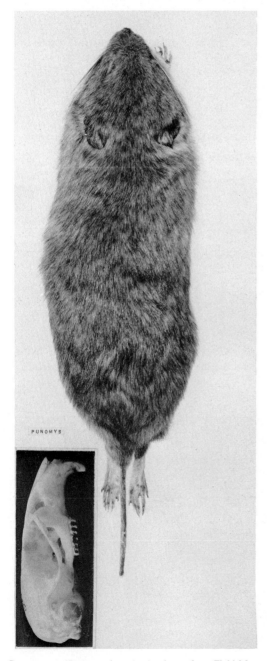

Puna mouse (*Punomys lemminus*), photos from Field Museum of Natural History.

from November to April, when plant growth is at its peak. A pregnant female contained two embryos.

RODENTIA; MURIDAE; *Genus EUNEOMYS Coues, 1874*

Patagonian Chinchilla Mice

Hershkovitz (1962) recognized four living species:

E. chinchilloides, southern Argentina, Tierra del Fuego;
E. noei, central Chile;
E. mordax, west-central Argentina;
E. fossor, northwestern Argentina.

Cabrera (1961) put *E. mordax* and *E. fossor* in a subgenus, *Chelemyscus* Thomas, 1925. Pine, Miller, and Schamberger (1979) suggested that a species of *Euneomys,* in addition to *E. noei,* occurs in central Chile.

Head and body length is 90 to 156 mm, and tail length is 53 to 90 mm (Hershkovitz 1962). The upper parts are cinnamon, reddish, reddish brown, or brownish, and the underparts are buffy or grayish. The hands and feet are usually white. The tail is usually brownish above and whitish below. A white stripe bordering the lips is present in some individuals. The body form is stocky. Females have eight mammae. *E. fossor* is especially modified, externally, for a burrowing life, in that its fur is soft, its ear conch is greatly reduced, and its claws are elongated.

These rodents inhabit brushy and wooded areas. Individuals, not quite fully grown, have been seen in Patagonia in February; this suggests that breeding may take place near the end of the year.

RODENTIA; MURIDAE; *Genus NEOTOMYS Thomas, 1894*

Andean Swamp Rat

The single species, *N. ebriosus,* is known from central and southern Peru, western and southern Bolivia, extreme northern Chile, and northwestern Argentina (Cabrera 1961; Pine, Miller, and Schamberger 1979).

Head and body length is about 120 to 175 mm, and tail length is 62 to 88 mm. Three specimens from Chile weighed 63, 63, and 69 grams (Pine, Miller, and Schamberger 1979). In the subspecies *N. e. ebriosus* the upper parts are grayish brown; the chest is dirty brownish, and this color extends as a line for a short distance down the middle of the belly; the remainder of the belly is whitish; the tail is gray to buffy below and brown above; and the hands and feet are grayish to buffy, sometimes with reddish brown on the ankles. The subspecies *N. e. vulturnus* is slightly paler than *N. e. ebriosus,* both above and below, and the brown sternal band is reduced or lacking.

Neotomys is easily identified by a bright rufous nose, unusually long guard hairs, and broad incisors, the upper ones having a narrow groove at the outer corner. Although it has the usual mouselike body, *Neotomys* is quite stockily built. Females have eight mammae.

Available evidence indicates that this mouse prefers grassy areas with scattered shrubs, along the banks of streams and marshes, at elevations of 3,400 to 4,525 meters. It is not generally present in rocky areas. It is active both day and night. Individuals living at higher elevations are possibly more active during the warmth of the day, whereas those inhabiting somewhat lower elevations are nocturnal. Appar-

Patagonian chinchilla mouse (*Euneomys chinchilloides*), photo from *Mission Scientifique du Cap Horn,* 1882–83, A. Milne-Edwards.

Andean Swamp rat (*Neotomys ebriosus*), photo by B. Elizabeth Horner and Mary Taylor of specimen in Harvard Museum of Comparative Zoology.

Coney rat (*Reithrodon physodes*), photo by Abel Fornes and Elio Massoia.

ently, this genus is not abundant and its range is generally isolated from those of other rodents. Its shelters are usually located under isolated rocks on level ground.

RODENTIA; MURIDAE; *Genus REITHRODON* Waterhouse, 1837

Coney Rat

The single species, *R. physodes*, occurs in Uruguay, Chile, and Argentina, including Tierra del Fuego (Cabrera 1961; Pine, Miller, and Schamberger 1979; Barlow 1969). According to Dalby and Mares (1974), the correct name for this species is *R. auritus*.

Head and body length is 130 to 200 mm, and tail length is 85 to 105 mm. Three females from northwestern Argentina weighed 77, 94, and 95 grams (Dalby and Mares 1974). The fur is thick and soft. The upper parts usually are buffy, often with a mixture of grayish or black hairs. The underparts are whitish or grayish, often with a buffy wash. In addition to the inguinal region and the inner side of the thighs, the hands, feet, and tail are often white. The outer hind toes are reduced, and there is webbing between the middle hind toes. The soles are hairy from the back of the heel to the base of the outer digits. The ears are fairly large. Each upper incisor has a deep median groove. Females have eight mammae.

The coney rat inhabits pampas, cultivated fields, stony hills, and sandy coasts. Barlow (1969) reported that in Uruguay this animal is usually found in overgrazed pasture, among rocky outcrops, or on well-drained slopes with scanty vegetation. It either digs its own burrows or utilizes, with or without modification, the abandoned burrows of *Dasypus* and other fossorial mammals, or natural crevices and holes among the rocks. The burrows constructed by *Reithrodon* were reported to have one or two entrances about 5 cm wide, to descend to about 10 to 25 cm below the surface and then become level, to be about 1 to 2 meters long, and sometimes to contain oval chambers up to 30 cm wide, in one of which was a nest platform composed of fine, dry grass. Barlow found *Reithrodon* to be strictly nocturnal in his study area, but noted that this may simply have reflected a low population level at the time. *Reithrodon* is pastoral, feeding mainly on grasses, other plants with tuberous rhizomes, and roots.

In Barlow's study area the nearness of burrows to each other suggested that *Reithrodon* may be variously gregarious or solitary. Barlow collected males in reproductive condition from January to May, and two pregnant females, one with four embryos in January and one with three embryos in May. Also in Uruguay, a lactating female was taken in January and a female with four nestling young was taken in October. In Chile, Pine, Miller, and Schamberger (1979) collected a pregnant female with five embryos on 3 February. Earlier information indicates that *Reithrodon* breeds throughout the year and that the number of young varies considerably. According to Marvin L. Jones (Zoological Society of San Diego, pers. comm.), a captive individual lived for five years and six months.

RODENTIA; MURIDAE; *Genus HOLOCHILUS* Brandt, 1835

Web-footed or Marsh Rats

There are four species (Massoia 1980, 1981; Cabrera 1961; Handley 1976):

H. brasiliensis, central and eastern Brazil, Paraguay, Uruguay, northeastern Argentina;

H. sciureus, Venezuela, the Guianas, eastern Peru, Brazil, Bolivia;

H. chacarius, northwestern Argentina;

H. magnus, eastern Brazil, Uruguay, northeastern Argentina.

Head and body length is 130 to 220 mm, and tail length is 130 to 230 mm. Twigg (1965) found *H. brasiliensis* to weigh up to 280 grams, but *H. magnus* is the larger species. The upper parts of *Holochilus* are buffy, orangish, or tawny, usually mixed with black; the sides are paler; and the underparts vary from white to orange, except for the white and gray throat and inguinal region. The tail is thinly haired and is uniformly brown, or is somewhat paler beneath. Some animals have a fringe of hairs on the underside of the tail, which may serve as an aid in swimming. The toes are usually webbed. The body form is ratlike. *Holochilus* differs from the genus *Sigmodon* in the conspicuous webbing of its toes and in its soft, not harsh, pelage. According to Husson (1978), female *Holochilus* have eight mammae.

Web-footed rats are semiaquatic. They live in marshes, grassy stream banks, canebrakes, and other moist, non-wooded areas from sea level to about 2,000 meters. They construct nests on the ground or up to 3 meters above the surface in reeds or trees. The nests are made of reeds, cane, and leaves firmly interlaced and woven together. The lower half of the nest is usually packed with plant material, and the upper half contains the living quarters and the entrances. The

Web-footed or marsh rat (*Holochilus brasiliensis*), photo by Joao Moojen.

floor has a bedding of soft plant substances. These nests are 20 to 40 cm in diameter and weigh as much as 100 grams. When disturbed, the occupant either leaps from the nest into the water and swims away or climbs higher in the tree. Marsh rats are active mainly at night, but also may be abroad during the day. They feed primarily on marsh plants and also take mollusks.

In Uruguay, Barlow (1969) found 45 nests of *H. brasiliensis* in an area of about 50 ha. There were as many as 11 nests, some only a meter apart, in a single tree. Barlow found pregnant females in April and May, and suggested that breeding activity was reduced with the onset of winter. In Guyana, Twigg (1965) found reproduction to continue throughout the year, with litter size averaging 3.12 (1–8) young and sexual maturity being attained relatively early in life. In eastern Brazil the period of pregnancy extends from February to April, and litter size is usually 5 to 8, but ranges up to 10.

At certain times marsh rats increase to such numbers that they destroy cultivated crops in fields and warehouses. Some investigators have connected these outbreaks with the fruiting season of the bamboo (*Merostichis*). After depleting the bamboo seeds, the rats begin to feed on cultivated crops. Eventually, however, high mortality rates of uncertain cause bring about a decline in their numbers.

RODENTIA; MURIDAE; **Genus SIGMODON**
Say and Ord, 1825

Cotton Rats

There are eight species (Hall 1981; Cabrera 1961; Husson 1978):

S. hispidus, southern United States to northern Venezuela and northwestern Peru;
S. mascotensis, southern Mexico;
S. arizonae, Arizona, western Mexico;
S. fulviventer, central New Mexico and southeastern Arizona to central Mexico;
S. alleni, western Mexico from southern Sinaloa to southern Oaxaca;
S. leucotis, central Mexico;
S. ochrognathus, north-central Mexico and adjacent parts of Arizona, New Mexico, and Texas;
S. alstoni, northern South America.

Some authorities, such as Handley (1976), place *S. alstoni* in the genus or subgenus *Sigmomys* Thomas, 1901.

Head and body length is 125 to 200 mm, tail length is 75 to 138 mm, and weight is 70 to 211 grams. The upper parts are grayish brown to blackish brown, mixed with buff, and the underparts are grayish or buff. The tail is blackish above and paler below. The fur is often harsh. The body form is stocky,

the ears are small, and the three central digits of each hind foot are larger than the other two. The generic name refers to the S-shaped pattern of the molar cusps.

Cotton rats prefer grassy and shrubby areas. Their presence is indicated by surface trails, shallow burrows beneath small plants, and small nests of grasses and sedges located in cuplike depressions in the ground. At the northern extremity of the range of *S. hispidus,* in Nebraska, nests were found 76.2 cm below the surface in the burrows of other animals (Baar, Fleharty, and Artman 1975). Cotton rats are active day and night throughout the year. They are omnivorous, feeding on vegetation, insects, and other small animals. In some areas they travel along ditches, where they feed on crayfish and fiddler crabs. They destroy both the eggs and chicks of bobwhite quail, and often become pests by eating enormous quantities of sugarcane and sweet potatoes.

These animals are numerous wherever they occur, being among the most abundant rodents in the southeastern United States, as well as in Mexico and Central America. They are, however, subject to population fluctuations. A peak density of 57 individuals per ha. and home ranges averaging about 0.05 ha. have been reported for *S. hispidus* in southern Louisiana (Lowery 1974). In west-central Kansas, the year-round average home range was about 0.4 ha. for males and 0.2 ha. for females; no indication of territoriality was found (Fleharty and Mares 1973). A male *S. ochrognathus* in the Chiricahua Mountains of Arizona had a home range of about 0.3 ha. (Hardin, Barbour, and Davis 1970).

In northeastern Kansas, near the northern extremity of the range of the genus, the breeding season is April to November (McClenaghan and Gaines 1978), but in more southerly areas *Sigmodon* breeds practically throughout the year. In Louisiana, according to Lowery (1974), reproduction occurs all year and litters are born in rapid succession. Females enter estrus every 7 to 9 days, and also immediately after giving birth. The gestation period is 27 days, and litter size is 1 to 12 young, usually 5 to 7. The young weigh 6.5 to 8.0 grams and open their eyes within 24 hours. They may be weaned when only 5 days old, but occasionally remain with the mother for 7 days or more. Sexual maturity is attained at the age of 40 to 60 days, and average life span in the wild is probably less than 6 months. According to Marvin L. Jones (Zoological Society of San Diego, pers. comm.), a captive *S. hispidus* lived for 5 years and 2 months.

RODENTIA; MURIDAE; **Genus NEOTOMODON**
Merriam, 1898

Mexican Volcano Mouse

The single species, *N. alstoni,* inhabits the volcanic mountains of central Mexico (Hall 1981). The original describer thought that the grinding teeth resembled those of *Neotoma,*

Cotton rat (*Sigmodon hispidus hispidus*), photo by Ernest P. Walker.

Mexican volcano mouse (*Neotomodon alstoni*), photo by William B. Davis.

and hence the generic name. Most authorities continue to arrange *Neotomodon* next to *Neotoma* in systematic accounts. Davis and Follansbee (1945:405), however, stated "that on the basis of general shape of skull, dental characters, and habits, *Neotomodon* is much closer to *Peromyscus* than to any other recent genus." Yates, Baker, and Barnett (1979) placed *Neotomodon* in the genus *Peromyscus,* and indicated close relationship to *P. floridanus* and *P. gossypinus.*

Head and body length is about 100 to 130 mm, tail length is about 80 to 105 mm, and adult weight is usually 40 to 60 grams. The fur is soft and dense. The upper parts are grayish to grayish buff, but occasionally, in season, become fulvous brown. The underparts are whitish, often faintly washed with buff on the chest. The tail is bicolored, dusky above and white below; it is relatively well haired. The ears are large and nearly naked. Females have six mammae.

The volcano mouse lives on grassy slopes in forests at an elevation of 2,600 to 4,300 meters. It is nocturnal and is most active shortly after dark. Distinct runways are not usually made. Hall and Dalquest (1963) excavated a burrow that measured 38 mm wide throughout its length of about 1.8 meters. It branched at a depth of about 356 mm, with one branch leading to a nest composed of a hollow ball of dry grass, 127 mm in diameter. The volcano mouse may also use the abandoned burrows of small pocket gophers (*Thomomys*). *Neotomodon* lives in close association with the black-eared mouse (*Peromyscus melanotis*), and has habits much like those of this burrowing, nocturnal species of deer mouse.

Davis and Follansbee (1945) reported that the breeding season of *Neotomodon* extends at least from early June to September; that two or more litters, averaging 3.4 (2–5)

young each, are born each year; and that some individuals reach sexual maturity during the same season in which they are born. Hall and Dalquest (1963) found a pregnant female with 2 embryos in November in Veracruz.

RODENTIA; MURIDAE; **Genus NEOTOMA** Say and Ord, *1825*

Wood Rats, Pack Rats, Trade Rats

There are 4 subgenera and 20 species (Hall 1981):

subgenus *Neotoma*

N. floridana, southern South Dakota and southern New York to Texas and Florida;

N. micropus, New Mexico and southern Kansas to northeastern Mexico;

N. albigula, southwestern United States to central Mexico;

N. nelsoni, known only from the type locality in Veracruz (east-central Mexico);

N. palatina, Jalisco (west-central Mexico);

N. varia, Turner Island (Gulf of California);

N. lepida, central California and southern Idaho to southern Baja California and Arizona;

N. bryanti, Cerros Island off west-central Baja California;

N. anthonyi, Todos Santos Island off northwestern Baja California;

N. martinensis, San Martin Island off northwestern Baja California;

N. bunkeri, Coronados Island off southeastern Baja California;

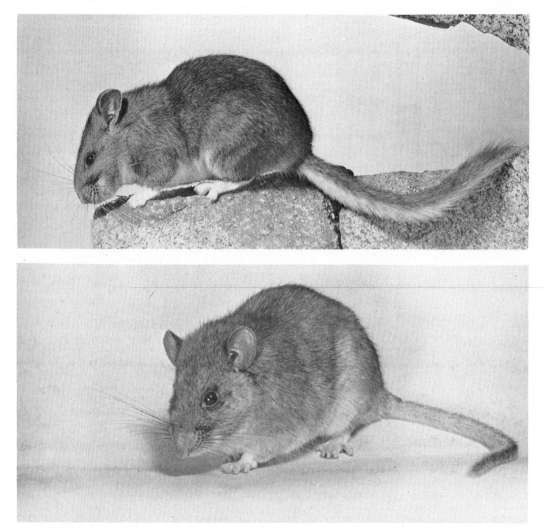

Bushy-tailed wood rat (*Neotoma cinerea*) top, photo by John Wanderer. Allegheny wood rat (*N. floridana*) bottom, photo by Ernest P. Walker.

N. stephensi, extreme southeastern Utah, central and eastern Arizona, western New Mexico;

N. goldmani, north-central Mexico;

N. mexicana, Colorado and southeastern Utah to western Honduras;

N. chrysomelas, central Honduras to western Nicaragua;

N. angustapalata, southern Tamaulipas (northeastern Mexico);

N. fuscipes, western Oregon to northern Baja California;

subgenus *Teonoma*

N. cinerea, northwestern Canada to North Dakota and northern Arizona;

subgenus *Hodomys*

N. alleni, western and central Mexico;

subgenus *Teanopus*

N. phenax, Sonora and Sinaloa (northwestern Mexico).

Some authorities still consider *Hodomys* Merriam, 1894 to be a distinct genus (see Genoways and Birney 1974).

Head and body length is 150 to 230 mm, tail length is 75 to 240 mm, and weight is 199 to 450 grams. The fur is soft to somewhat harsh. The delicate coloration of the upper parts ranges from pale, buffy gray to darker gray and cinnamon buff. The underparts are pure white, pale grayish, or buffy. Coloration is bright rufous in *N. alleni* and *N. chrysomelas.* In some species, the so-called round-tailed wood rats, the tail is very scantily haired and resembles that of *Rattus.* In the bushy-tailed wood rat (*N. cinerea*), however, the tail is fairly well covered and has somewhat the appearance of a short-haired squirrel tail.

Wood rats are found in a great variety of habitats, from low, hot, dry deserts, to humid jungles, to rocky slopes above the timber line. They are generally nocturnal and active throughout the year. Some species build elaborate dens or houses composed of twigs, stems, foliage, bones, rocks, or whatever material is available. These houses often rest on the ground or are placed against rocks or the base of a tree. They may be used and added to by many generations of wood rats, and may eventually grow to over 2 meters in both diameter and height. Bonaccorso and Brown (1972) found that a captive *N. lepida* could construct a complete house, 40 cm high and 100 cm wide, within 7 to 10 nights. Those species inhabiting areas where spiny cactus is available seek this material and build their houses almost entirely over this plant. The dens are so placed that it is almost impossible for an enemy to approach without being impaled, and yet the wood rats dash in and out of their homes unharmed. Several fairly well defined trails lead to these dens, but the paths are often obstructed by cactus or other material. Lowery (1974) wrote that most dens of *N. floridana* have more than one entrance leading to the actual nest. The nest lies in a cavity that measures 127 to 330 mm in diameter and is lined with finely shredded bark, leaves, and grass. The nest contains one alcove for sleeping and another for food storage.

Some species, at least in certain areas, do not build large houses, but utilize crevices among rocky outcrops or high on cliffs; the animals close the openings to the crevices with sticks or other material. There actually appears to be much variation in shelters, even within the same species, depending on habitat conditions and availability of materials. In the Mojave Desert of California, for example, Cameron and Rainey (1972) reported the following situations for *N. lepida:* (1) where the wood rats lived on rocky outcrops, they simply blocked crevices with debris, behind which the nest and food store were in the same chamber; (2) in desert washes, elaborate stick houses were made, averaging 52 cm high and 102 cm wide, each with five to seven entrances and three or four chambers, one for the nest and the others for food storage; and (3) in areas of cactus habitat, houses were built entirely of cholla cactus, used creosote bushes for support, averaged 55 cm high and 160 cm wide, and contained three or four chambers but no food stores or special nest structure.

Wood rats are expert climbers, but do not ordinarily go far up in trees. Both *N. phenax* and *N. fuscipes,* however, regularly nest and move about in trees. Several other species also sometimes build their houses in trees. Dens at heights of about 7.5 meters have been reported for *N. lepida* (Stones and Hayward 1968) and *N. floridana* (Lowery 1974). Some populations that live near lakes and streams build their nests over water in mangrove trees, often 46 to 69 meters from the shore. The animals are not aquatic, however, and regularly forage on land.

Wood rats pick up material for their nests while foraging and carry it to the homesite. If they find a more attractive substance, they will drop the material they are carrying and take the new item. They often select pieces of silverware or other shining objects from camps, and leave the material they had been carrying. This trait has given them the name of "trade rat" or "pack rat."

The diet of wood rats consists almost entirely of such plant tissues as roots, stems, and leaves; seeds; and some invertebrates. They do not drink much water, but during dry seasons they make heavy inroads on the fleshy stems of cacti and other plants that are well filled with water.

Population density of *N. lepida* was listed as 1.2–6.1 individuals per ha. (French, Stoddart, and Bobek 1975), and average home range was determined as 371 sq meters for males and 433 sq meters for females (Bleich and Schwartz 1975). In a radio-tracking study in California, Cranford (1977) found population density of *N. fuscipes* to vary from 14/ha. in winter to 20/ha. in late summer. There was an average of 32 houses per ha. Home range in this study area varied from 590 to 4,049 sq meters, and averaged 2,289 sq meters for adult males, 1,924 sq meters for adult females, and 1,719 sq meters for juveniles. Home ranges overlapped, and they expanded during the reproductive season. *Neotoma* is generally solitary. Dens are seldom occupied by more than one adult (Lowery 1974). In a group of captive *N. floridana,* under study a despotic social organization developed, in

which one individual killed or wounded all the others (Kinsey 1976).

In the southern parts of the range of *Neotoma,* breeding apparently can take place at any time of the year, but the genus is not especially prolific. Lowery (1974) stated that in Louisiana only two or possibly three litters of *N. floridana* are produced annually, usual litter size is two to four young, and sexual maturity is not attained until the age of seven or eight months. Overall range of litter size in this species is one to seven (Goertz 1970). Southern populations of *N. micropus* breed throughout the year, while females in the north apparently begin breeding in December or January, produce at least two and probably three litters before July, and sometimes have one or two additional litters from August to October; litter size is one to four (Birney 1973). The gestation period of *Neotoma* varies from about 30 to 40 days. According to Wiley (1980), the young of *N. floridana* weigh 11.8 to 14.1 grams a few hours after being born, their eyes open in 15 to 21 days, they are weaned in about 4 weeks, and they reach adult weight after 8 months. A captive specimen of *N. albigula* lived for 7 years and 8 months (Marvin L. Jones, Zoological Society of San Diego, pers. comm.).

Wood rats are neat and sanitary animals; they make pleasing pets if their extreme timidity can be overcome. In some parts of their range they live so close to farms that they are considered pests, but in general they are of little economic importance. The subspecies *N. floridana smalli,* found naturally only on Key Largo off the southeast coast of Florida, was considered endangered by Layne (1978). Its maximum population density is only about one individual per 0.8 ha., and, because of human destruction of the mature tropical forest on which it depends, only about 120 to 160 ha. of suitable habitat remain.

RODENTIA; MURIDAE; **Genus XENOMYS** Merriam, *1892*

Magdalena Rat

The single species, *X. nelsoni,* has been recorded from three localities in west-central Mexico: Chamela Bay, Jalisco; Pueblo Juarez, Colima, the type locality; and Armeria, Colima. There are three specimens in the U.S. National Museum of Natural History, one at the University of Michigan, and nine in the Los Angeles County Museum.

Head and body length is 157 to 165 mm, and tail length is 143 to 170 mm. The upper parts are deep tawny red or fulvous, and the underparts are creamy white. There is a white spot over each eye and a white spot below each ear. The upper lips and cheeks are white more than halfway to the eyes. The tail is well haired and the fur is soft. The ears, about half as long as the head, are nearly naked.

This attractive rodent resembles a small wood rat (*Neotoma*) in general external appearance and form, but the coloration is different. The skull and teeth resemble those of *Neotoma* more than those of any other genus. The auditory bullae of *Xenomys* are inflated, and the molar teeth are rooted.

Specimens have been collected from about sea level to approximately 450 meters elevation. *Xenomys* is nocturnal and arboreal. It inhabits thorn forest and tropical deciduous forest in a restricted area of coastal Mexico. It is quite agile in the trees, moving about freely and without apparent hesitation. The collector of the type series stated that *Xenomys* dwells in hollow trees, and that one individual was caught in a low, dense wood near a river. Pregnant females have been taken in August.

Magdalena rat (*Xenomys nelsoni*), photo by Cornelio Sanchez Hernandez.

RODENTIA; MURIDAE; **Genus NELSONIA Merriam,**
1897

Nelson's Wood Rat

The single species, *N. neotomodon*, inhabits the mountains of west-central Mexico (Hall 1981).

Head and body length is 120 to 130 mm, and tail length is 105 to 130 mm. The soft pelage of the upper parts is grayish brown, sometimes with a buffy wash; the sides are usually pale reddish; the underparts are whitish; and the feet are white to dusky. The tail is tufted terminally. Externally, this rodent looks like a large, hairy-tailed species of *Peromyscus*. In cranial and dental characters, however, *Nelsonia* most resembles *Neotoma*.

Nelson's wood rat frequents rocky areas in pine, Douglas fir, or true fir, from 1,810 to 3,010 meters elevation. Crevices in cliffs and in rimrock are used for shelters, these places apparently being cool and moist. *Nelsonia* is mainly nocturnal. It seems to feed mostly on the needles of Douglas fir and juniper. Females collected in March were not pregnant.

RODENTIA; MURIDAE; **Genus ICHTHYOMYS Thomas,**
1893

Fish-eating or Aquatic Rats

There are three species (Cabrera 1961; Handley and Mondolfi 1963):

I. hydrobates, Ecuador, Colombia, northwestern Venezuela;
I. pittieri, north-central Venezuela;
I. stolzmanni, eastern Ecuador, Peru.

Head and body length is 145 to 210 mm, and tail length is 145 to 190 mm. The pelage is thick and not as soft as that of allied genera. The upper parts are dark olive brown, grayish, or blackish, often with a buffy wash, and the underparts are whitish. The tail is fully haired.

These animals, about the size of common rats (*Rattus*), are modified in body form for an aquatic and predaceous life. The head and body are flattened; the skull is fairly smooth, lacking prominent ridging; the eyes and ears are small; the whiskers are long and stout; the skin is very loose, and the fur is thick; the large hind feet are broad and have swimming fringes of hair; the hind toes are partly webbed; and the tail has a bristly undersurface. The outer corner of each incisor projects downward as a sharp point, and the cutting edges of the two upper incisors form an inverted V. Females have six mammae.

Fish-eating rats frequent streams and swampy areas from 600 to 2,800 meters in elevation. Much of their former range is now under cultivation. They are believed to feed mainly, if not entirely, on water animals. Remains of fish having an average length of 150 mm have been found in the intestinal tract of *Ichthyomys*. Presumably these rodents spear fish with their simple, gafflike upper incisors. The cecum is small and not well developed, whereas in vegetarian murids this organ is prominent and well developed. Söderström, who first made known the fish-eating rats, stated that they usually live under rocks during the day.

Nelson's wood rat (*Nelsonia neotomodon*), photo by Howard E. Uible of skin in U.S. National Museum.

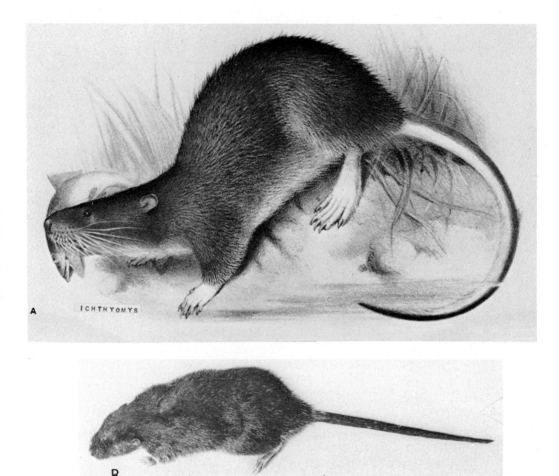

Fish-eating or aquatic rat: A. *Ichthyomys stolzmanni,* photo from *Proc. Zool. Soc. London;* B. *I.* sp., photo by William T. Collins through R. B. Tesh.

RODENTIA; MURIDAE; **Genus RHEOMYS** Thomas, 1906

Central American Water Mice

There are two subgenera and six species (Hall 1981; Cabrera 1961; Handley 1976):

subgenus *Rheomys*

R. thomasi, extreme southern Mexico to El Salvador;
R. hartmanni, Costa Rica, extreme western Panama;
R. underwoodi, Costa Rica, extreme western Panama;
R. raptor, extreme eastern Panama;
R. trichotis, Colombia, western Venezuela;

subgenus *Neorheomys*

R. mexicanus, eastern Oaxaca (southern Mexico).

Handley (1976) placed the species *R. trichotis* in the genus *Anotomys.*

Head and body length is 105 to 188 mm, and tail length is 95 to 150 mm. The fur is short, soft, dense, and glossy. The upper parts are dark brown or mixed black and cinnamon, and the underparts are grayish or whitish. The tail may be tipped with white. Externally, *Rheomys* closely resembles *Ichthyomys.*

Starrett and Fisler (1970) pointed out that *R. underwoodi* is among the most highly hydrodynamically specialized mammals. Its body is streamlined and the forelimbs do not project much, even when walking. The ear pinnae are reduced and do not extend above the surface of the fur. The eyes are also small, and the nostrils open posteriorly behind flaplike valves. The hind feet are strikingly large and have laterally compressed digits interconnected by webbing. The tail flares into the body and apparently serves a stabilizing and supplementary propulsive function.

These rodents live in and near small, swift streams in tropical forests. They are known to feed on aquatic insect larvae and may also eat snails. They are not definitely known to prey on fish, but they are suspected to do so and a captive ate pieces of sardines (Starrett and Fisler 1970).

Central American water mouse (*Rheomys thomasi*), photo from *J. Mamm.*, R. A. Sirton.

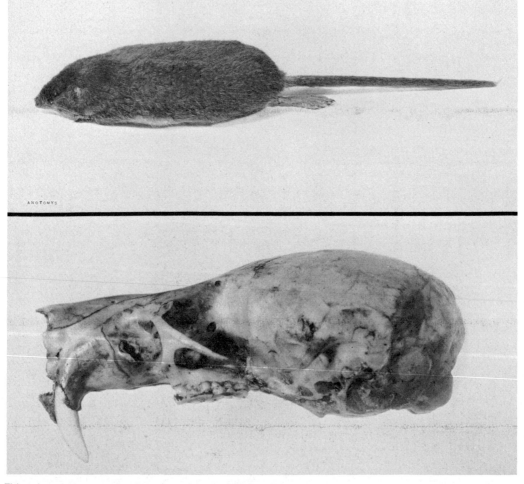

Fish-eating rat (*Anotomys leander*), photos from British Museum (Natural History).

RODENTIA; MURIDAE; *Genus ANOTOMYS Thomas, 1906*

Fish-eating or Aquatic Rat

The single species, *A. leander,* is known only from Mount Pichincha, near Quito, Ecuador, and one locality in central Peru (Cabrera 1961; Gardner 1971a).

Head and body length of one specimen was 128 mm, and tail length was 125 mm. The upper parts are dark slaty, the sides are paler, and the underparts are whitish gray. The whiskers, or vibrissae, are dark on the upper part of the face and white on the lower part. A distinct round white patch lies around each ear opening.

This genus is less highly specialized for catching and eating fish than are some of its close relatives, such as *Ichthyomys* and *Daptomys.* However, it is more specialized for an aquatic existence, as is evidenced by the following characters: the external ear is diminutive and the slitlike ear opening can be closed by muscular contractions to keep out water; the fur is velvety, like the fur of *Hydromys, Neomys, Nectogale,* and other highly specialized aquatic mammals; and the feet are lined by coarse, stiffened hairs that increase the surface of the foot for swimming.

Anotomys was long known only by a few specimens collected in a swift mountain stream at an elevation of 3,600 meters on the slopes of Mount Pichincha. On 15 August 1968, Gardner (1971a) collected a subadult female at an elevation of 2,400 meters on the eastern slope of the Cordillera Carpish in Peru. Subsequent study of this specimen revealed that *Anotomys* has a diploid chromosome number of 92, the highest reported for any mammal.

RODENTIA; MURIDAE; *Genus NEUSTICOMYS Anthony, 1921*

Fish-eating or Aquatic Rat

The single species, *N. monticolus,* inhabits the Andes of northern Ecuador (Cabrera 1961).

Head and body length is about 95 mm, and tail length is about 110 mm. The soft fur is reddish brown above and grayish below. The tail is colored like the back, but has scattered whitish hairs. An external ear is evident, and the feet and toes are weakly fringed with hairs. This genus is similar to the other aquatic rats—*Ichthyomys, Anotomys, Daptomys,* and *Rheomys*—but differs from them in its reduced hallux (first digit of the hind limb) and in dental characters. *Neusticomys* appears to be one of the least specialized of the aquatic rats.

Tate (1931) observed that this rodent was plentiful and could be taken along small "azequias," or irrigation brooks. He said that when trapping, the trap should be placed where the current is rapid and where the banks encroach. *Neusticomys* is frequently taken close to small waterfalls and in places exposed to spray from the water. It is assumed to feed, at least in part, on aquatic life.

RODENTIA; MURIDAE; *Genus DAPTOMYS Anthony, 1929*

Fish-eating or Aquatic Rats

There are three species (Musser and Gardner 1974; Dubost and Petter 1978):

D. venezuelae, known by three specimens from north-eastern and southern Venezuela;

D. peruviensis, known only by a single specimen from east-central Peru;

D. oyapocki, known only by a single specimen from southern French Guiana.

Head and body length is 100 to 131 mm, tail length is 82 to 109 mm, and weight of the single specimen of *D. oyapocki* is 47 grams. The pelage is thick and sleek. The upper parts are blackish brown or buffy brown, the tail is blackish brown, the feet and ears are cream or dark brown, and the underparts are buffy gray to slate. A female specimen of *D. venezuelae* had six mammae.

Daptomys belongs to the fish-eating group of cricetines, typified by *Ichthyomys. Daptomys,* however, is not as highly modified for aquatic life. Its hind feet are not as broad as those of *Ichthyomys.* Its incisors are sharp and the outer surfaces are slightly inclined toward each other. *D. oyapocki* is distinguished by the absence of the upper and lower third molars.

All specimens of *Daptomys* were collected in dense tropical forests along or in streams. The Venezuelan individuals were taken at elevations of 728 and 1,400 meters, and the Peruvian specimen was taken at 300 meters. A female from southern Venezuela, collected on 28 July, was lactating.

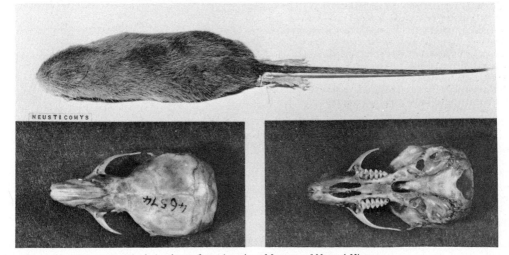

Aquatic rat (*Neusticomys monticolus*), photos from American Museum of Natural History.

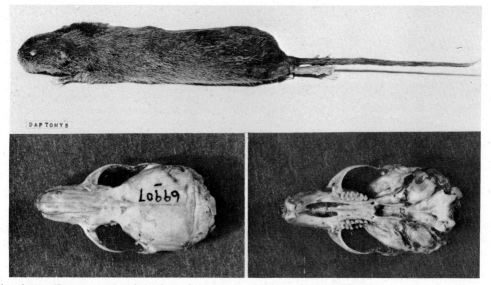

Fish-eating rat (*Daptomys venezuelae*), photos from American Museum of Natural History.

RODENTIA; MURIDAE; **Genus CALOMYSCUS** Thomas, 1905

Mouselike Hamster

The single species, *C. bailwardi,* occurs in southern Turkmen, Iran, Afghanistan, and Pakistan (Corbet 1978). This species was divided into five separate species by Vorontsov, Kartavtseva, and Potapova (1979).

Head and body length is 61 to 98 mm, tail length is 72 to 102 mm, and weight is 15 to 30 grams (Hassinger 1973; Roberts 1977). The pelage is fine and soft in texture. The upper parts are pinkish buff, sandy brown, or grayish brown. The underparts, hands, and feet are white. The tail is dark above and white below. It is thickly haired, tufted, and longer than the head and body. The large ears are conspicuous. There are no cheek pouches.

According to Osgood (1947), this genus is very similar to the New World *Peromyscus,* but most authorities now group it with the Old World hamsters. One feature distinguishing *Calomyscus* from *Peromyscus* is the number of roots of the molar teeth. Females have six mammae.

In Afghanistan, *Calomyscus* is widely distributed between 400 and 3,500 meters, and is found under evergreen oaks in monsoonal areas as well as on barren, scorched hills (Hassinger 1973). In Iran, Lay (1967) found a nest built in a narrow horizontal crevice on a rocky outcrop at 3,300 meters; it consisted of a ball of fine grass and sheep wool. *Calomyscus* is active only at night during the summer, but it is also active by day in the autumn and winter. It feeds mainly on seeds, also consumes flowers and leaves, and readily eats animal matter.

Calomyscus is not highly social, but wild individuals sometimes share favorable shelter sites, and captives sleep huddled together (Roberts 1977). There apparently is a lengthy breeding season, some females have two litters per year, and litter size is three to five young. Lay (1967) found lactating females in August and December, and half-grown males in early August. Hassinger (1973) collected two pregnant females in July, each with three embryos. The newborn are hairless; 13 days after being born, when their eyes open, they have a gray pelage. Adult coloration and size are attained in 6 to 8 months.

Osgood (1947) stated that *Calomyscus* was very limited in distribution and doubtless on the way to extinction. Hassinger (1973), however, found this genus to be abundant in Afghanistan, and Lay (1967) collected a series of 61 specimens in Iran.

Mouselike hamster (*Calomyscus bailwardi*), photo from *Proc. Zool. Soc. London.*

Dwarf hamster (*Phodopus roborovskii*): Top photo from East Berlin Zoo; Bottom photo by P. Rödl.

RODENTIA; MURIDAE; **Genus PHODOPUS** Miller, 1910

Small Desert Hamsters, Dwarf Hamsters

There are two species (Corbet 1978):

P. sungorus, Kazakh, Mongolia and adjacent parts of Siberia, Manchuria;

P. roborovskii, western and southern Mongolia, and adjacent parts of Manchuria and northern China.

Head and body length is 53 to 102 mm, and tail length is 7 to 11 mm. The length of the tail is generally less than one-fifth of the head and body length. The upper parts are grayish to pinkish buff, and the underparts are whitish; the two colors meet to form four reentrants into the dorsal mantle, one before and one behind each leg, leaving a narrow tongue of the mantle extending down on the upper part of each leg. The ears are blackish, with white on the inside. The sides of the muzzle, upper lips, lower cheeks, lower flanks, and limbs, as well as the tail and ventral surfaces, are pure white. Sometimes there is a dark dorsal stripe.

Phodopus has a robust body, prominent ears, and internal cheek pouches. The feet are short, broad, and densely haired throughout. Females have eight mammae.

These hamsters appear to inhabit somewhat arid areas, where there is stiff grass on plains and sand dunes. They seem to be most active in the evening and early morning, but to be active to some extent throughout the night. Figala, Hoffmann, and Goldau (1973) observed daily torpor during the winter in many captive *P. sungorus.* Loukashkin (1940) reported that he observed *P. sungorus* in association with pikas (*Ochotona*) and that the hamsters made use of their paths, tunnels, and burrows, especially in winter. Dwarf hamsters are quite clean. Undoubtedly their feeding habits are somewhat like those of other hamsters, for they eat seeds and any available plant material. Observers report that they will fill their pouches, seemingly almost to the bursting point, with millet or grain seeds, distorting the shape of the body. Then, when teased or disturbed, or upon depositing the food in their burrows, they will push the pouches with their forepaws, thus causing the grain to pour out of their mouths.

These animals are easy to tame, do well in captivity, and make good pets because of their interesting habits and amusing ways. They have a docile disposition and do not attempt to bite or run away.

Figala, Hoffmann, and Goldau (1973) observed reproduction in captive *P. sungorus* only from February to November. The young were born 18 or 19 days after breeding pairs were established. Litter size averaged five and ranged from one to nine young. According to Marvin L. Jones (Zoological Society of San Diego, pers. comm.), a captive specimen of *P. sungorus* lived for 3 years and 2 months.

RODENTIA; MURIDAE; *Genus CRICETUS Leske, 1779*

Common or Black-bellied Hamster

The single species, *C. cricetus,* occurs from Belgium, across central Europe and the Soviet Union, to the Altai region of Siberia (Corbet 1978).

Head and body length is about 200 to 340 mm, tail length is 40 to 60 mm, and weight is 112 to 908 grams. The thick fur is usually light brown above, mostly black below, and white on the sides. Grzimek (1975) noted, however, that in addition to this normal color pattern there is much variation, from albino to melanistic animals. The small tail is almost hairless. Cheek pouches are present, and the feet are broad, usually having well-developed claws. Females have eight mammae.

The common hamster lives on steppes and cultivated land and along riverbanks. In the western parts of its range it is strictly confined to loamy and loess soils. It utilizes burrows throughout the year, hibernating there in the winter and emerging to feed during crepuscular hours at other seasons. It occasionally swims, inflating its cheek pouches with air for greater buoyancy before taking to the water. A food shortage sometimes forces mass population movements, during which large rivers may be crossed (Grzimek 1975).

Studies in Germany (cited in Grzimek 1975) indicate that burrow size is generally dependent on the hamster's age. The rooms and tunnels of the summer and autumn burrows always lie in the same plane, usually at a depth of only 50 cm, but winter burrows are often more than 2 meters below the surface. Each burrow has several oblique and vertical entrance tunnels and several compartments for the nest, food storage, and excrement. The winter burrows contain more space for food, and up to 90 kg of cereal seeds, peas, or potatoes may be stored. Winter hibernation is interrupted every five to seven days, at which time the hamster eats from its cache. The diet consists of grains, beans, lentils, roots, and green parts of plants, as well as insect larvae and frogs. In captivity the hamster does well on dog biscuits, lettuce, and corn.

The common hamster is usually a solitary animal and each burrow has only one adult occupant. In some areas, however, burrows may be crowded together because of limited suitable construction sites, and so give the impression of a colony (Grzimek 1975). In the Kaluga District of the Soviet Union a density of about one burrow per 2 ha. has been reported.

The breeding season extends from early April to August. Males move onto the territory of females and are usually driven out subsequent to mating, but captive investigations

Black-bellied hamsters (*Cricetus cricetus*), photo by Ernest P. Walker.

have indicated that pairs may remain together and raise the young (Grzimek 1975). Females can produce a litter every month throughout the year in captivity, but normally only twice a year in the wild. The gestation period is 18 to 20 days and litter size is 4 to 12 young. The newborn weigh about 7 grams, are weaned when about 3 weeks old, and reach adult size after 8 weeks (Grzimek 1975). Females attain sexual maturity when about 43 days old. The life span is about 2 years.

The common hamster is trapped for its skin in some areas. It was often considered a serious pest, because of its destruction of corn and other crops. The introduction of modern agricultural practices, however, resulted in drastic declines of the species, and it now is considered rare in Europe (Smit and Van Wijngaarden 1976).

RODENTIA; MURIDAE; Genus CRICETULUS
Milne-Edwards, 1867

Ratlike Hamsters

There are eight species (Corbet 1978):

C. migratorius, southeastern Europe to Iran and Mongolia;
C. barabensis, Mongolia, southern Siberia, northern China;
C. longicaudatus, Mongolia and adjacent parts of Siberia and China;
C. kamensis, Tibetan Plateau;
C. alticola, Kashmir, northern Pakistan;
C. eversmanni, northern Kazakh;
C. curtatus, Mongolia and adjacent parts of China;
C. triton, northeastern China, Korea, Ussuri region of Siberia.

Head and body length is 80 to 250 mm, and tail length is 25 to 106 mm. C. triton is the largest species. The fur is quite long (about 15 mm in length on the middle of the back) and is usually mouse gray in color. Sometimes, however, it is reddish or buffy. The underparts are light gray or white. The feet and terminal end of the tail are white. In some species there are other white markings, and in C. barabensis there is a dark brown dorsal stripe. These hamsters have robust bodies, blunt muzzles, fairly short legs and tails, and huge internal cheek pouches. Females have eight mammae.

Ratlike hamsters inhabit open dry country, such as steppes and the borders of deserts. In Afghanistan C. migratorius occurs at 400 to 3,600 meters on rocky slopes and plateaus almost devoid of vegetation (Hassinger 1973). In the spring and summer, ratlike hamsters are active both day and night, but as winter approaches they become more nocturnal. In winter, they do not hibernate continuously, but awaken from time to time to eat stored food. As the weather becomes colder they sleep more heavily, and the intervals between feedings are longer. Their homes are burrows in the ground, which they dig themselves. At least some species burrow almost straight down as far as 1.2 meters. Some burrows have a single entrance, but others have two or three. These burrows contain chambers that serve for food storage, nesting, living quarters, and other domestic uses.

The diet consists of young shoots and seeds. Apparently, the items most preferred are soybeans, peas, and millet seeds. The cheek pouches are so large that it does not take long to store a bushel of beans in the burrow. One author reported that he took 42 soybean seeds from the cheek pouches of an individual, and in this case the head was so large that it comprised one-third of the body. Harrison (1972) stated that the diet of C. migratorius is mainly vegetarian, but that some insects are taken and that captives killed and ate frogs and jerboas.

Ratlike hamsters are reported to be extremely savage rodents. They make a vigorous defense by throwing themselves on their backs and opening their mouths, exposing formidable incisor teeth. In a captive study of C. barabensis, Skirrow and Rysan (1976) found strong intolerance of conspecifics. Females were especially aggressive and were dominant over males.

In C. triton the males are reported to leave their winter burrows in early March and to enter almost all the other burrows they can find, in the search for females. The sexes remain together for about 10 days, and the breeding season continues to the middle of May. In Armenia, C. migratorius breeds throughout the year, when it resides in human habitations, and produces 2 to 11 young per litter. Otherwise, according to Grzimek (1975), the length of the mating season of Cricetulus varies by species, sometimes beginning as early as the end of February and possibly continuing through October. Gestation lasts from 17 to 22 days, and females produce three or four litters a year. Litter size averages about 5 or 6 young, but occasionally is over 10. The young reach adult size in 2 months.

Ratlike hamster (*Cricetulus triton*), photo by Ernest P. Walker.

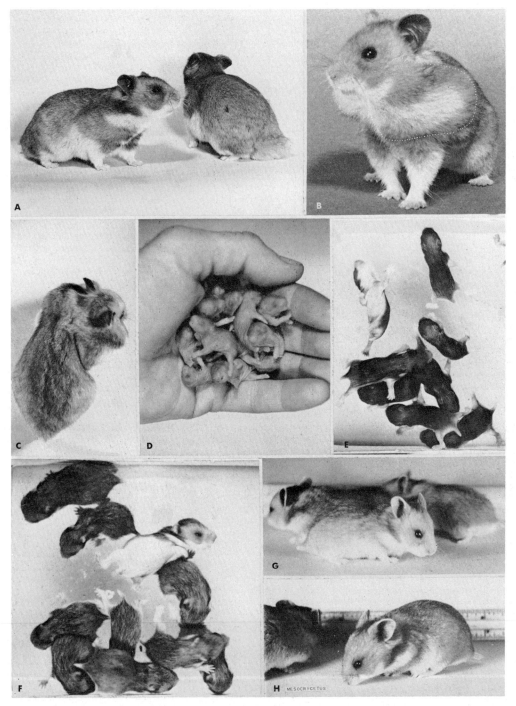

Golden hamsters (*Mesocricetus auratus*): A. Adults; B. A hamster with its cheek pouches filled with food; C. Dorsal view showing distended cheek pouches; D. Two-day-old hamsters; E. Eight days old; F. Thirteen days old; G. Twenty-two days old; H. Twenty-nine days old; Photos by Ernest P. Walker.

RODENTIA; MURIDAE; *Genus MESOCRICETUS*
Nehring, 1898

Golden Hamsters

There are four species (Lyman and O'Brien 1977; Corbet 1978):

M. brandti, Asia Minor, Caucasus, northern Iraq, northwestern Iran, Syria, Palestine;
M. auratus, apparently restricted to the vicinity of Aleppo in northwestern Syria;
M. raddei, region north of the Caucasus between the Black and Caspian seas;
M. newtoni, eastern parts of Rumania and Bulgaria.

Many authorities include *Mesocricetus* in the genus *Cricetus*. Although *M. auratus* has the smallest natural range, it is the best-known species because it has become common in captivity throughout the world. Corbet (1978) listed *M. brandti* as a subspecies of *M. auratus*.

Head and body length is about 170 to 180 mm, and tail length is about 12 mm. Lyman and O'Brien (1977) gave a weight range of 137 to 258 grams for *M. brandti* and 97 to 113 grams for *M. auratus*. Ralls (1976) stated that females are larger than males. The upper parts are generally light reddish brown, and the underparts are white or creamy. One form has a distinct ashy stripe across the breast. The skin of these animals is quite loose, and the enormous cheek pouches, which open inside the lips, extend well back of the shoulders. When filled, they more than double the width of the animal's head and shoulders. In a study of 128 females, Anderson and Sinha (1972) counted 3 animals with 12 mammae each, 30 with 13, 74 with 14, 16 with 15, 4 with 16, and 1 with 17. The greater number of nipples, smaller size, and shorter tail, together with some cranial characters, distinguish *Mesocricetus* from *Cricetus*.

In the wild, golden hamsters live on dry, rocky steppes or brushy slopes. They construct burrows somewhat like those described in the account of *Cricetus*. They are essentially nocturnal, but may occasionally be active by day. They appear to be almost omnivorous, eating many kinds of green vegetation, seeds, fruit, and meat. Hibernation, for uninterrupted periods of up to 28 days, has been experimentally induced through exposure to cold (Lyman and O'Brien 1977).

Adults generally live one to a burrow and will readily fight one another. Lyman and O'Brien (1977) found that litters of *M. brandti* lived amicably together until about the seventh week of life, but then had to be separated. Drickamer, Vandenbergh, and Colby (1973) determined that the social rank of captive males is based on the condition of their flank glands, the hamster with the largest gland being dominant.

Litters have been born in captivity during every month of the year, but with a marked decrease in production during late fall and winter. Lyman and O'Brien (1977) reported that laboratory specimens of *M. brandti*, exposed to natural day length, usually were not in breeding condition between the months of November and March. Successful matings among these animals peaked in May; no female produced more than three litters per season. The reported maximum for wild populations in the Caucasus is two litters a year.

Asdell (1964) stated that in *M. auratus* the estrous cycle is 4 days long, estrus lasts 27.4 hours, and the gestation period is usually 16 and occasionally up to 19 days. Lyman and O'Brien (1977) stated that the estrous cycle of *M. brandti* lasts 4 days, gestation is 15 days, and litter size is 1 to 13, averaging 6 young. Anderson and Sinha (1972) reported the

litter size of *M. auratus* to range from 2 to 16 and average about 9, but Asdell (1964) gave the range as 1–12 and the mode as 6 or 7. Witte (1971) observed that if the young were threatened by danger during their first 3 days of life, up to 12 of them at a time might be placed in the cheek pouches of the mother. The young are born blind and naked, but grow quickly; they are weaned before the 20th day of life and are capable of reproduction when 7 to 8 weeks old. The peak of fecundity is reached at around 1 year of age (Lyman and O'Brien 1977). The life span appears to be 2 to 3 years.

Little was known about these hamsters until 1930, when a female and 12 young *M. auratus* were obtained in Syria and brought to Palestine. Descendants of this small group spread throughout the world as laboratory animals, zoo exhibits, and household pets. In 1938 some of these hamsters were first brought to the United States. In 1965 and 1971, specimens of *M. brandti* were captured in central Turkey, and these animals also formed the basis of a successful laboratory colony (Lyman and O'Brien 1977).

RODENTIA; MURIDAE; *Genus MYSTROMYS Wagner,
1841*

White-tailed Rat

The single species, *M. albicaudatus*, occurs in South Africa and Lesotho (Meester and Setzer 1971).

Head and body length is 136 to 184 mm, and tail length is 50 to 82 mm. The soft, long, rather woolly pelage is buffy grayish to brownish above, suffused with numerous black-tipped hairs, which become fewer on the sides, finally blending into the white underparts. The bases of the hairs on the upper surface are slate gray, whereas the hairs on the lower surface are all white. The sides of the face and the limbs are paler than the back. The hands, feet, and tail are dull white. The ears are dark brown. The tail is covered with short, stiff bristles.

This genus is thick bodied and ratlike. It has large, broad ears, and slender limbs, hands, and feet. The incisor teeth are ungrooved and pale yellow. The short, slightly curved, and sharp-tipped claws are almost concealed by the long, white hairs of the feet. Unlike some of its relatives, *Mystromys* does not have cheek pouches. Females have two pairs of inguinal mammae.

The white-tailed rat inhabits grassy flats and dry sandy areas. It lives in holes in the ground. It has been observed using the burrows of the carnivore *Suricata*. It also shelters in cracks in the soil. It is nocturnal and is said to be especially active and bold during rainy weather. If individuals are evident in a given area, reportedly some can be caught by placing a lighted lantern on the ground, which draws the animals to its light. *Mystromys* eats seeds and other vegetable matter.

Breeding apparently occurs throughout the year. In a study of captive animals, Hallett and Meester (1971) found a mean litter size of 2.9 young and a minimum interval between litters of 36 days. The young became attached to the mother's mammae immediately after being born and remained for about 3 weeks, with the female dragging them about. This process seems an excellent means of protection, and the rate of survival of the young is high for a rodent. After 3 weeks there are increasing periods of separation, and there is no more suckling after about 38 days. The minimum age at which a female produced a litter was 146 days. It has been reported that the white-tailed rat does well in captivity and is tame and playful. Dean (1978) stated that maximum longev-

White-tailed rat (*Mystromys albicaudatus*), photo by H. J. Bohner, National Institutes of Health.

ity in captivity is about 6 years. He also suggested that over-grazing of the high veld by domestic livestock might be a threat to *Mystromys*.

RODENTIA; MURIDAE; **Genus SPALAX** Guldenstaedt, *1770*

Blind Mole-rats

Corbet (1978) recognized three species:

S. microphthalmus, southeastern Europe from Greece to the Volga River;

S. giganteus, plains around the northern Caspian Sea;

S. leucodon, lower Danube Basin and Balkan Peninsula of Europe, Asia Minor and the southern Caucasus to coastal Egypt and Libya.

Most authorities have placed *Spalax* in its own family, the Spalacidae, but Chaline, Mein, and Petter (1977) reduced this group to subfamilial rank, and that procedure is followed here. Topachevskii (1976) divided Spalax into two distinct genera: *Microspalax,* which included *S. leucodon* separated into three species; and *Spalax,* which included *S. giganteus,* and *S. microphthalmus* separated into four species. Corbet (1978) considered Topachevskii's classification to represent excessive splitting, but karyological studies (Nevo and Bar-El 1976; Soldatovic and Savic 1974) have shown the existence of a dozen different chromosomal forms of *Spalax* in Israel and Yugoslavia alone, and have suggested that these forms do indeed represent distinct, though closely related, species.

Head and body length is generally 150 to 300 mm, there is no external tail, and weight is about 133 to 365 grams. The color varies from dark gray to yellowish gray, often with a yellowish or golden sheen. A white central stripe on the snout and one on each cheek may be present. The dense, soft fur is nearly reversible. A line of bristles, apparently serving as tactile aids, extends from each side of the flattened snout along a ridge to about the location of the eye.

These heavy-bodied, short-legged, powerful animals have projecting incisors and small claws. The external form is molelike. *Spalax* can be distinguished from all other rodents by the absence of external openings for the eyes, though small eyes are present beneath the skin. The external ears are reduced to low ridges. Blind mole-rats recognize objects by touching them with the nose. The forefeet and hind feet have five digits. The incisors are broad and heavy, and the cheek teeth are rooted (not ever growing), with Z or S enamel patterns. Females have two pairs of mammae.

Blind mole-rats seem to burrow in any area where the soil is suitable for digging, and they inhabit any type of soil that has more than 100 mm of rainfall annually; they do not live in true desert. They are known from plains below sea level, hilly areas, upland steppes, cultivated fields, and clearings in the mountains up to elevations of 2,600 meters. They are extensive burrowers. Most of their digging is done with the incisors, the whole head acting as a bulldozer blade. Specialized jaws and strong muscles aid the teeth in loosening the soil. Almost all of the orbit, for example, is occupied by muscles for working the teeth. Another aid in burrowing is the broad, padded, horny snout, with which the animal packs excavated ground into the burrow walls. Although the feet are small and quite delicate, the forefeet aid to some extent in breaking the soil, and the hind feet help push the dirt out behind.

Activity goes on even during extremely hot periods or drought, though there is then perhaps less digging and burrows are made deeper. Blind mole-rats are both diurnal and nocturnal, but studies in Yugoslavia by Savic (1973) indicate that *S. leucodon* is more active by day than by night. *Spalax* generally feeds on roots, bulbs, tubers, and various other underground parts of plants. Some food is stored, and up to 18 kg of potatoes and sugar beets have been found in a burrow (Harrison 1972). Although they are truly fossorial, blind mole-rats are occasionally active above ground at night. During such ventures they feed on grasses, seeds, superficial roots, and perhaps insects. They, in turn, are sometimes preyed upon by owls.

The following information on mounds and burrows was taken in part from Grzimek (1975), Nevo (1961), Harrison (1972), and Savic (1973). Except during the breeding sea-

Blind mole-rat (*Spalax leucodon*), photos by P. Rödl.

son, blind mole-rats live mainly in an underground burrow system with many tunnels, nest rooms, storage chambers, and defecation rooms. In at least some areas, during the winter or the hot, dry summer, these burrows are placed at greater depths than at other times. Record depth is 410 cm. Generally, the living and storage rooms are about 20 to 50 cm deep. Tunnels to food sources radiate outward for up to 30 meters or more; they are about 7 cm wide and usually 15 to 25 cm below the surface. The total length of tunnels in a burrow system ranges from about 65 to 195 meters. In certain areas, during the summer, a series of 15 to 20 mounds may be built above the burrow system. One of these, the so-called resting mound, is larger than and is surrounded by the others. It is usually about 100 cm wide and 25 cm tall. In the center is a chamber used for sleeping and resting.

In the Mediterranean region, during the wet fall and winter, female *Spalax* build "breeding" mounds. These are elaborate structures, usually about 160 cm long, 135 cm wide, and 40 cm in height above the surface. They may even reach 250 cm in diameter and 100 cm in height. The mounds are solid domes of earth, through which runs a labyrinth of permanent galleries with hard, smooth walls. The only loose, soft earth is the covering of the dome. In very wet, poorly drained areas, the mounds tend to be much bigger and more elaborate, thereby protecting the occupants from flooding. In the center of each mound is a nest chamber, about 20 cm wide and filled with dry grass, where the young are raised. Beneath this chamber and throughout the mound are rooms for storage and defecation. Each main breeding mound is surrounded by radial rows of smaller mounds, which are built and occupied by the males during the mating season. These smaller mounds are connected to the breeding mound by tunnels.

Home range of *S. leucodon* averages 452 sq meters (Savic 1973). Population densities in the Soviet Union are usually around 1 to 10, but range up to 23 individuals per ha. (Topachevskii 1976). Blind mole-rats are solitary creatures and, except during the breeding season, only a single individual is found per mound or burrow system. Vocalizations include grunting noises, quite unlike the usual rodent squeak, and a hissing threat. In the eastern Mediterranean region the mating season lasts from about November to March, and the single annual litter is born from January to early April (Harrison 1972). In Yugoslavia maximum sexual activity occurs in January and February, and births usually take place in March or April (Savic 1973). The gestation period evidently lasts about a month. Litter size is one to five, usually two to four, young. The young weigh about five or six grams at birth, and are naked, helpless, and pink. After about two weeks they are covered with long, gray fur, which is quite different from the short hair of the adult. At four to six weeks of age they leave the nest. The maximum life span under optimal conditions is four and one-half years (Savic 1973).

Topachevskii (1976) stated that in the Soviet Union *Spalax* does considerable damage to orchards, perennial grasses, forest plantations, and cultivated crops. Grzimek (1975) wrote that the economic significance of *Spalax* varies from region to region. It is considered a serious agricultural pest in the eastern Mediterranean region, but is not regarded as being especially destructive in some other areas. In Libya some people believe that blindness will ensue if one of these mole-rats is touched, and so they will not attempt to catch the animal. Blind mole-rats have sometimes guided archeologists to digging sites by bringing to the surface objects from early civilizations.

RODENTIA; MURIDAE; *Genus MYOSPALAX Laxmann, 1769*

Mole-rats, Zokors

There are two subgenera and four species (Corbet 1978):

subgenus *Eospalax*

M. fontanieri, northern and western China;
M. rothschildi, Kansu and Hupeh (central China);
M. smithi, Kansu (central China);

subgenus *Myospalax*

M. myospalax, southern Siberia, eastern Mongolia, northeastern China.

Head and body length is 150 to 270 mm, tail length is 30 to 70 mm, and weight is 150 to 250 grams (Grzimek 1975). The pelage is composed of long, soft, silky hairs, and lacks coarse guard hairs; there are a few short whiskers. The coloration of the upper parts varies with the species from gray, grayish brown, or light russet to pinkish buff. Usually the individual hairs are bicolored, the tips being lighter than the bases. The lower parts are generally paler than those above. In some species the tail is grayish above and white below. *M. fontanieri* has an all white tail and a white patch on the upper lip and muzzle.

Mole-rats are stout, chunky animals with short, powerful limbs. The forefeet are armed with long, heavy claws for digging. These front claws are doubled under when the animals are walking. The middle or third claw is the largest and heaviest. The claws of the hind feet are much shorter. There are no external ears, and the eyes are so small that they are nearly hidden in the fur. Females have one pectoral and two abdominal pairs of mammae.

Mole-rats frequent cultivated and wooded areas, especially mountain valleys at elevations of 900 to 2,120 meters. They are fossorial and their remarkable digging abilities were described by Grzimek (1975). With its powerful foreclaws one individual dug a tunnel 70 cm long in 12 minutes. It used its incisors to cut through roots and pushed loose earth aside with its head. After it had dug itself completely into the ground, the animal turned around and used its front feet and head to push the accumulated earth to the surface. After a tunnel 3 to 5 meters long had been dug, the mole-rat broke through to the surface to throw out loose soil. In the wild, in this same manner, a series of mounds develops along the route of the mole-rats, thereby allowing their presence to be detected. Their main burrows are located at a depth of about 2 meters, and consist of a nest chamber, a food storage room, and a chamber for defecation. From one to four food tunnels, which may reach a length of 50 to 100 meters, lead gradually from the nest chamber to the surface. These tunnels pass below the food plants and include storage facilities. The diet of *Myospalax* consists of roots and grains.

When frightened or angered, mole-rats utter a peculiar little squeal. Even though their eyes are small, they are sensitive to light, and, if brought to the surface, they always seek a darkened place. They apparently sometimes move about on the surface at night, and are reportedly active throughout the winter. In China, litters of four to six young have been found in March and April, and a pregnant female with two embryos was recorded in May (Grzimek 1975).

Mole-rats are intensively hunted in China, because of their damage to crops. The Chinese farmers say that there will be good weather whenever the mole-rats leave their tunnels open, but that when the tunnels are closed the weather will be bad.

RODENTIA; MURIDAE; *Genus LOPHIOMYS Milne-Edwards, 1867*

Maned or Crested Rat

The single species, *L. imhausi,* occurs in southern Sudan, Ethiopia, Somalia, and Kenya (Meester and Setzer 1971).

Head and body length is 255 to 360 mm, tail length is 140 to 215 mm, and weight is 590 to 920 grams (Kingdon 1974b). Females are usually larger than males. The coloration is usually either black and white or brown and white, in a fairly definite pattern of stripes and spots or blotches. The

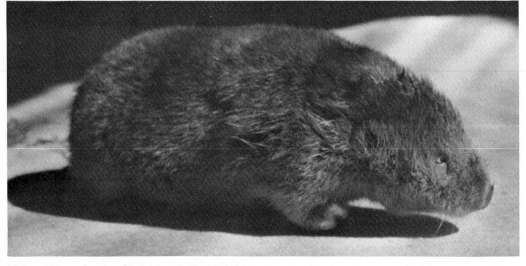

Mole-rat (*Myospalax myospalax*), photo by Paul K. Anderson.

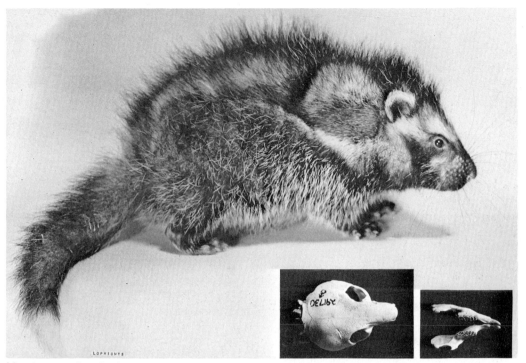

Maned or crested rat (*Lophiomys imhausi*), photo by Ernest P. Walker. Insets: photos by P. F. Wright of specimen in U.S. National Museum.

variation is caused mainly by the intensity or purity of the coloration, and by the relative size of the differently colored areas. Variation in color is also affected by the differing extent of the light and dark parts of those hairs that are banded or tipped with light or dark pigmentation. The most common coloration gives an iron gray effect, but the color ranges from a general overall gray tone (occasioned by a preponderance of white or gray on the tips of the hairs) to almost dark brown or black (brought about by predominance of hairs largely or entirely black).

There is a prominent erectile mane from the top of the head, along the back, and along a quarter of the tail (the tip of the tail is white). On either side of the mane, along the back, the hair is shorter and lighter colored than the surrounding hair, giving the effect of a furrow in the fur and emphasizing the mane. In some animals the mane is scarcely apparent on the head and is reduced on the tail. The fur is generally dense, long, and fine to silky, except along the mane, where it is coarse. The underparts are gray to black, and the hands and feet are jet black.

The tail is densely bushy. The great toe is partially opposable, and the hands and feet have four well-developed digits and are specialized for arboreal life. The ear is small. The skull is unique among rodents in that the temporal fossae are completely roofed by bone. The surfaces of these and some of the other bones of the skull are granulated. The incisors are moderately broad.

This rodent does not resemble a rat, as the head is similar to that of a guinea pig (*Cavia*), and the body, when viewed from a distance, looks like that of a small porcupine (*Erethizon*). When the animal becomes excited or frightened, its crest erects and it exposes a glandular area along its flanks. This trait may be a protective measure to frighten its enemies into mistaking it for a porcupine.

Although the maned rat is generally considered an inhabi-

tant of dense montane forests, Yalden, Largen, and Kock (1976) stated that in Ethiopia the animal has been recorded from sea level to at least 3,300 meters, is clearly tolerant of a wide range of habitats, and is by no means confined to forests. The maned rat is arboreal and quite adept in climbing, even descending head first, but it moves slowly. It is strictly nocturnal, leaving its burrow or hole among rocks at dusk to search for food. The diet consists of leaves and tender shoots. When eating, *Lophiomys* sits on its haunches and grasps its food in its hands. The voice of this rodent is peculiar: one or two hisses or snorts, followed by a growl.

Kingdon (1974*b*) wrote that two or three young are born at once, but Delany (1975) stated that a captive female gave birth to two litters, each with a single young. These young were feeding independently of the mother by the time they were 40 days old. According to Marvin L. Jones (Zoological Society of San Diego, pers. comm.), one captive individual lived for seven years and six months.

RODENTIA; MURIDAE; *Genus PLATACANTHOMYS*
Blyth, 1859

Spiny Dormouse

The single species, *P. lasiurus*, is found in southern India (Ellerman and Morrison-Scott 1966). *Platacanthomys*, along with *Typhlomys*, has sometimes been placed in a separate family, the Platacanthomyidae, and sometimes in a subfamily, the Platacanthomyinae, of the family Gliridae. In accordance with Chaline, Mein, and Petter (1977) and Corbet (1978), the Platacanthomyinae are here considered a subfamily of the family Muridae.

Head and body length is about 130 to 212 mm and tail

Spiny dormouse (*Platacanthomys lasiurus*), photo from *Proc. Zool. Soc. London,* 1865.

length is 75 to 100 mm. The weight of one specimen, an adult female, was 75 grams. The upper parts are densely covered with sharp, flat spines intermixed with thin, delicate underfur, but the underparts have fewer, smaller, and finer spines. The basal half of the tail is sparsely haired and scaly, and the terminal part bears long hairs that form a brush. The general coloration of the upper parts is light rufescent brown. The forehead and crown are more reddish, and the underparts are dull whitish. The tail is somewhat darker than the general body color, becoming paler at the thick, bushy tip. The feet are whitish.

This rodent is much like the Gliridae in form. The muzzle is pointed, the eyes are small, the ears are thin and naked, and the hind feet are broad and elongated. The first toe barely reaches the base of the second toe. The thumb on the forefoot, although short, is well developed. The claws of the digits are slender and compressed. Unlike the Gliridae, *Platacanthomys* has no premolar teeth, and its dental formula is the same as that of the Muridae. The incisors are smooth and compressed. The cheek teeth tend to be high crowned and generally have parallel oblique cross ridges of enamel on the crown. These ridges are broadened, and the depressions tend to become isolated, on the surface of the crown.

The spiny dormouse inhabits rocky hills and forested valleys at elevations of 600 to 900 meters. It lives mainly in the cavities of trunks and branches of trees, and in clefts in rocks. Its nest is constructed mostly of leaves and moss. The long, tufted tail is undoubtedly helpful as a balancing organ as the animal moves about and leaps in trees. The diet consists of fruit, seeds, grain, and roots. A captive specimen was sluggish during the day and, though it allowed itself to be handled without showing fear or attempting to escape, it would inquisitively bite at a finger. Rajagopalan (1968) reported that a wild-caught female was still living after 20 months of captivity.

In some areas the spiny dormouse is so plentiful that it is considered a pest. The native people call it the "pepper rat," because it destroys large quantities of ripe peppers. It also is said frequently to get into "toddy-pots," containers in which palm juice is collected.

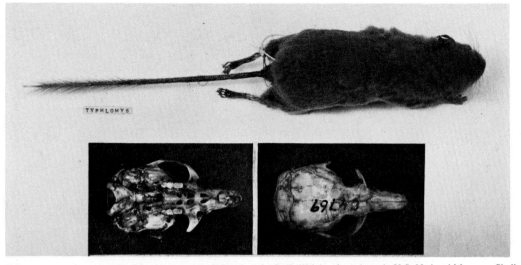

Chinese pygmy dormouse (*Typhlomys cinereus*). Skin: photo by P. F. Wright of specimen in U.S. National Museum. Skull: photos from American Museum of Natural History.

RODENTIA; MURIDAE; **Genus TYPHLOMYS**
Milne-Edwards, 1877

Chinese Pygmy Dormouse

The single species, *T. cinereus*, occurs in southeastern China and northern Viet Nam (Ellerman and Morrison-Scott 1966). *Typhlomys*, along with *Platacanthomys*, has sometimes been placed in a separate family, but the two genera are considered here to represent a subfamily of the Muridae (see the account of *Platacanthomys*).

Head and body length is 70 to 98 mm, and tail length is 95 to 135 mm. The pelage is short, dense, soft, and spineless. The upper parts are uniformly deep mouse gray. The underparts and insides of the limbs are pale grayish, the individual hairs having gray bases and white tips. The hands are white and the feet are dusky. The long gray tail is sparsely haired and scaly on the basal half, but becomes more heavily covered toward the tip with longer hairs that form a distinct terminal brush. The tip is usually white.

Typhlomys is mouselike in external appearance. It has prominent, scantily haired ears; small eyes; and long, slender hind feet. The claws of all the digits are slender and compressed. The dental formula and structure of the molar teeth are the same as those of *Platacanthomys*.

This rodent is found at elevations of 1,200 to 2,100 meters in mountains that are abundantly covered with dwarfed, moss-laden deciduous trees, and an undergrowth of small bamboos. Practically nothing of its natural history has been recorded, but the native people seem to understand its habits and trap it quite readily. The natives claim that cats will not eat this rodent.

RODENTIA; MURIDAE; **Genus MACROTARSOMYS**
Milne-Edwards and Grandidier, 1898

There are two species (Meester and Setzer 1971):

M. bastardi, western Madagascar;
M. ingens, northwestern Madagascar.

In *M. bastardi* head and body length is 80 to 100 mm, and tail length is 100 to 145 mm. This is the smallest of the Malagasy mice. The only other species on the island sometimes under 100 mm in head and body length is *Eliurus minor*. In *M. ingens* the head and body length is about 120 mm and tail length is about 210 mm (Petter 1972). Both species of *Macrotarsomys* have brownish fawn upper parts, whitish underparts, and a long, stiff tail with a terminal tuft.

The body form of *Macrotarsomys* is like that of some Gerbillinae. The ear is large (20 to 25 mm long in *M. bastardi*), the hind foot is long (22 to 28 mm in *M. bastardi*), and the tail is elongated. The three central toes are longer than the first and fifth digits, and the fifth toe is longer than the reduced first toe. The first toe bears a claw that is located fairly low on the foot. The moderate inflation of the bullae gives the skull a unique appearance among the genera of mice from Madagascar.

M. bastardi inhabits dry wooded areas, scrublands, and grassy plains. It lives in burrows often concealed under a rock or small bush. The burrows may be 1.5 meters or more in length, but not very deep, and the outlets are usually kept closed. This species is strictly terrestrial and nocturnal. Its diet consists of berries, fruits, seeds, roots, and plant stems. *M. ingens* digs burrows, which can be located by the soil thrown up, but its nocturnal activity seems largely arboreal (Petter 1972).

M. bastardi lives in pairs (Petter 1972). Litters usually consist of two or three young.

RODENTIA; MURIDAE; **Genus NESOMYS** *Peters, 1870*

The single species, *N. rufus*, is found in the eastern forests and on the western coast of Madagascar (Meester and Setzer 1971).

Head and body length is 186 to 230 mm, and tail length is 160 to 190 mm. The pelage is fairly long, soft, smooth, and shiny. The fine-textured black whiskers on the muzzle extend beyond the tips of the blackish brown ears. The coloration of the upper parts is dark rust brown mixed with brownish yellow, which gives the animal a fawn color. The

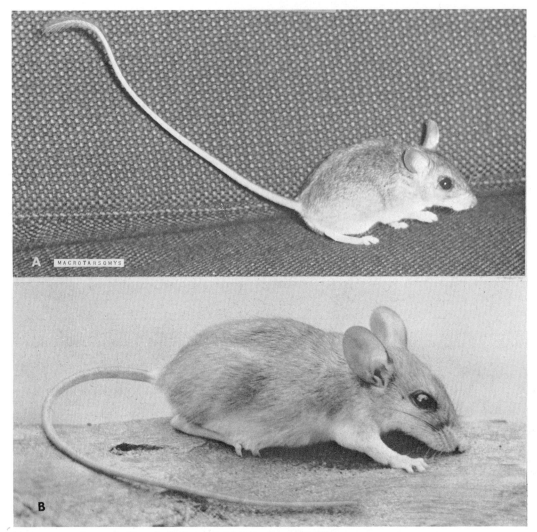

Macrotarsomys ingens: A. Photo by F. Petter; B. Photo by Howard E. Uible.

Nesomys rufus, photo by Constance P. Warner.

basal parts of the hairs are usually slate colored. The sides of the head, body, and limbs are reddish, whereas the lips, throat, breast, middle of the belly, and underside of the tail are white. The upper part of the tail is rust red. A white tuft occurs on the end of the tail in the subspecies *N. r. auderbeti*, and the subspecies *N. r. lambertoni* is remarkable for its hairy tail.

The incisors are smooth and whitish toward the points; the upper teeth are brownish orange and the lower ones are yellowish. The long, well-developed feet help *Nesomys* to leap about freely. The three middle toes are elongated (especially the center one), whereas the two outer toes are much shorter. The claws of the fingers are slightly more than half as big as those of the toes, the latter being very strong. The almost straight, brownish white claws are partially concealed by long, rigid hairs growing from the roots of the nails.

According to Petter (1972), this mouse inhabits wet forests and is terrestrial and nocturnal.

RODENTIA; MURIDAE; *Genus BRACHYTARSOMYS*
Günther, 1875

The single species, *B. albicauda,* is known from the eastern forests of Madagascar (Petter 1972).

Head and body length is 200 to 250 mm, and tail length is about the same (Petter 1972). The tail is prehensile (Meester and Setzer 1971). The soft, dense, woolly pelage is grayish brown on the upper parts, rufous on the sides, and white on the underparts. The head is reddish brown and the nose and lips are blackish. The anterior half of the sparsely haired tail is black, but the posterior half is white.

The appearance of this genus is similar to that of *Nesomys,* but *Brachytarsomys* differs in having shorter, broader hind feet, and a different cranial structure. *Brachytarsomys* has a short snout, small eyes and ears, five rows of mustache hairs, and prominent claws. The hind foot has a long fifth digit. The incisors are not grooved.

Petter (1972) wrote that *Brachytarsomys* is strictly arboreal, lives in holes of trees, and feeds mainly on fruit.

RODENTIA; MURIDAE; *Genus ELIURUS*
Milne-Edwards, 1885

There are two species (Meester and Setzer 1971):

E. myoxinus, Madagascar;
E. minor, eastern Madagascar.

Eliurus minor, photo by Howard E. Uible. Insets: *E. myoxinus,* photos by P. F. Wright of specimen in U.S. National Museum.

Brachytarsomys albicauda, photo by F. Petter.

Head and body length is 80 to 175 mm, tail length is 110 to 200 mm, and weight is 35 to 103 grams. The pelage is fairly soft. The upper parts are uniformly brownish gray or yellowish gray, the latter effect being caused by gray bases and fawn tips of the individual hairs. The feet and lower surfaces are generally light gray. In the subspecies *E. myoxinus myoxinus,* the tail is so well clothed with moderately stiff, deep brown hair that it looks almost bushy. In other subspecies the basal third is almost naked, and the rest of the tail is slightly bushy or penciled. The pencil on the tail of *E. myoxinus tanala* and *E. m. pencillatus* is white. *E. m. tanala* has a dark gray spot in the middle of the back and yellowish white underparts. In *E. myoxinus majori* there is an indistinct ring around the eye. The ears are dark and almost naked, the palms are pink, the mustache is black with long whiskers, and the eyes are large and conspicuous. Females have two pairs of inguinal mammae.

These mice apparently prefer heavily forested areas. *E. myoxinus* is mainly nocturnal and arboreal (Petter 1972), but *E. minor* burrows in the ground (Grzimek 1975). A female *Eliurus* was observed nursing in early November.

RODENTIA; MURIDAE; *Genus* **GYMNUROMYS**
Forsyth Major, 1896

Voalavoanala

The single species, *G. roberti,* is known from eastern Madagascar (Meester and Setzer 1971).

Head and body length is 125 to 160 mm, and tail length is 152 to 175 mm. The upper parts are blackish gray or slaty, and the underparts are white or yellowish white. Long whiskers (from 50 to 60 mm) are present, and the bicolored, scaly tail is scantily haired. The body form is ratlike, and the feet are quite broad with fairly long fifth digits. This genus is best characterized by the pattern of the cheek teeth, which are completely flat crowned, laminate, and tightly compressed.

This genus inhabits forests. Pregnant females have been collected in June and July, each bearing two embryos.

RODENTIA; MURIDAE; *Genus* **HYPOGEOMYS**
Grandidier, 1869

Malagasy Giant Rat

The single species, *H. antimena,* occurs around Morondava on the western coast of Madagascar (Meester and Setzer 1971).

This is the largest rodent on Madagascar. Head and body length is 300 to 350 mm, and tail length is 210 to 250 mm. The ears are large (50 to 60 mm). The pelage is harsh. The upper parts are gray, grayish brown, or reddish; the head is darkest. The limbs, hands, feet, and underparts are white. The dark tail is covered with stiff, short hairs. The hind foot is quite long and has relatively well developed claws.

This giant rat is known only from sandy coastal forests. It is a jumping and running rodent, strictly nocturnal in habit (Petter 1972). It apparently has about the same ecological role as a rabbit. It builds long, deep burrows and feeds mainly on fallen fruit. The usual number of offspring seems to be one. Unfortunately, the numbers and distribution of this unique mammal have declined sharply, because suitable habitat has been reduced through the cutting and burning of virgin forests (Grzimek 1975).

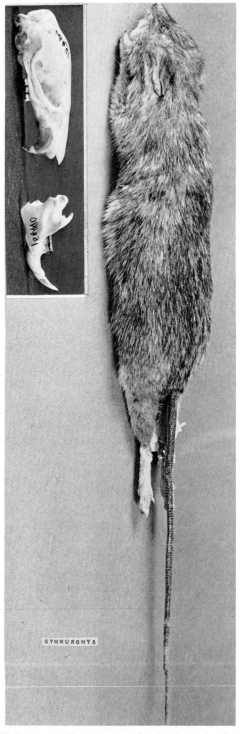

Voalavoanala (*Gymnuromys roberti*), photos by P. F. Wright of specimen in U.S. National Museum.

Malagasy giant rat (*Hypogeomys antimena*), photo by F. Petter.

RODENTIA; MURIDAE; *Genus BRACHYUROMYS* *Forsyth Major, 1896*

There are two species (Meester and Setzer 1971):

B. ramirohitra, eastern Madagascar;
B. betsileonensis, southeastern Madagascar.

Head and body length is 145 to 180 mm, tail length is 60 to 105 mm, and weight is 85 to 105 grams. The upper parts are brown mixed with red and black, and the underparts are reddish. The fur is thick and soft; the individual hairs have gray bases and brown apexes. The external form is somewhat volelike. The tail is short, dark, and somewhat hairy. The skull is broad and massive, and the head is broad and rounded. The ears are well haired and the palms are dark. Females have three pairs of mammae.

Brachyuromys has been trapped in a large, moist meadow on the central plateau of Madagascar. The animals were living in dense, matted grass and reeds, in association with *Microgale, Eliurus,* and *Rattus.* The grass and reed stems were so matted down and densely tangled that no sunlight reached the runways where the mice were traveling. They were active at all hours, during both the night and the day.

RODENTIA; MURIDAE; *Genus PAROTOMYS Thomas, 1918*

Karroo Rats

There are two species (Meester and Setzer 1971; Smithers 1971):

P. brantsi, southern Namibia, southern Botswana, Cape Province of South Africa;
P. littledalei, Namibia, Cape Province of South Africa.

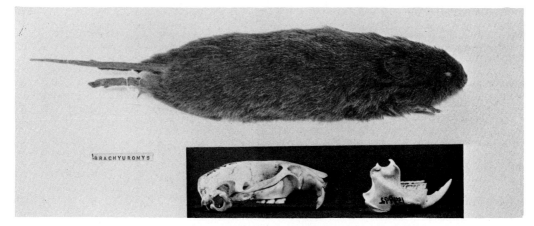

Brachyuromys ramirohitra, photos by P. F. Wright of skin and skull in U.S. National Museum.

Karroo rat (*Parotomys littledalei*), photo by James A. Bateman of mounted specimen.

Head and body length is 135 to 170 mm, and tail length is 75 to 120 mm. Smithers (1971) listed weights of 89 to 155 grams for *P. brantsi*. In *P. brantsi* the general coloration of the upper parts is rusty yellowish, variegated with blackish or brownish to form delicate streaks. The sides of the head, neck, and body, as well as the underparts, are grayish white. In one form the basal half of the tail is reddish orange and the terminal half is brownish red, but in another form the tail is black except for a tawny buffish base. In *P. littledalei* the upper parts vary from tawny to cinnamon buff, and the sides and underparts are buff to whitish buff. The upper portion of the tail is colored like the back, and the underside is paler.

Parotomys has small external ears, but the auditory bullae are much enlarged and spherical. The bullae of *Otomys* are relatively much smaller. *P. brantsi* has definite grooves on the upper incisor teeth, but the incisors of *P. littledalei* lack grooves. Females have four mammae.

Karroo rats inhabit sandy velds, and high and low flats, generally near karroo or salt bushes. They are diurnal and live in burrows with many entrances dug in hard sandy ground. Both species, but especially *P. littledalei*, build conspicuous nests of interwoven sticks and grass, up to about 600 cm in height, among the tangled roots and branches of shrubs above the burrows. Karroo rats seldom venture far from their shelters. They are quite wary, and at the slightest indication of an intruder they quickly seek the safety of their homes. They are vegetarian, feeding on grass, grass seeds, and fresh green shoots of low-growing vegetation (Smithers 1971).

Karroo rats are gregarious. It is said that they breed as often as four times per year, that litters contain up to four young, and that the young become sexually mature when three months old. Of nine pregnant females taken in Botswana in February 1967, one carried a single embryo, six had two embryos each, and two had three embryos. A female has been observed dragging the young attached to the nipples (Smithers 1971).

RODENTIA; MURIDAE; **Genus OTOMYS** F. Cuvier, 1823

African Swamp Rats, Groove-toothed Rats

There are nine species (Meester and Setzer 1971; Kingdon 1974*b*):

O. typus, mountainous areas from southern Sudan and Ethiopia to northeastern Zambia;

O. laminatus, South Africa;

O. anchietae, Angola, southern Tanzania;

O. angoniensis, savannahs from Kenya to Angola and South Africa;

O. saundersiae, Cape Province of South Africa;

O. irroratus, Cameroon, eastern Zaire, Uganda, Kenya, Rhodesia, South Africa;

O. denti, mountainous areas from Uganda to Malawi;

O. sloggetti, eastern South Africa, Lesotho;

O. unisulcatus, Cape Province of South Africa.

The species *O. unisulcatus* has sometimes been placed in a separate genus, *Myotomys* Thomas, 1918. Meester and Setzer (1971) distinguished *O. tropicalis*, found in tropical Af-

African swamp rat (*Otomys irroratus*), photo from South African Union Health Department through D.H.S. Davis.

African swamp rats (*Otomys unisulcatus*), photos by John Visser.

rica, from *O. irroratus*, but noted that the two might be conspecific, as they were treated by Rosevear (1969), Kingdon (1974b), and Delany (1975).

Head and body length is 124 to 217 mm, tail length is 55 to 150 mm, and weight is 60 to 255 grams (Kingdon 1974b; Smithers 1971). The pelage varies in density, texture, and length, but is usually long, thick, and shaggy. The general coloration of the upper parts ranges from pale buffy through shades of brownish to dark brown or bright rusty. The underparts are white, creamy, buffy, brownish, or dull dark grayish, generally paler than the upper parts. The relatively short tail is fairly well haired, but it is not penciled and is usually darker above than below.

African swamp rats are characterized by having at least one conspicuous groove on all four incisor teeth. The upper incisors are strongly curved backward into the mouth. The head of *Otomys* is rounded and rather volelike, and the ears are relatively small. *Dasymys* closely resembles *Otomys* in external appearance, but that genus lacks the grooved incisors and has a longer tail. *Otomys* has a stocky body form. Females have four mammae.

The subfamily Otomyinae, of which *Otomys* is the most widespread living genus, has been called the ecological equivalent of the Microtinae, the voles and lemmings of the Holarctic (Kingdon 1974b). Both groups comprise grass and herb eaters that live mainly in moist, grassy areas and make runways under dense cover. The general habitats of *Otomys* include swamps, grasslands, savannahs, brush country, and alpine meadows. In Ethiopia, *O. typus* is found at elevations of 1,800 to 4,000 meters, where it frequents the edges of streams and marshes (Yalden, Largen, and Kock 1976). In Botswana, *O. angoniensis* usually inhabits the heavily vegetated fringes of rivers and swamps, but during the wet season it may move some distance from waterways (Smithers 1971). *O. sloggetti* seems to prefer rocky areas and *O. unisulcatus* utilizes dry habitat (Meester and Setzer 1971).

Most species usually shelter above ground, where they construct nests of shredded plant material and twigs, often at the base of a small shrub or clump of grass. Others build a nest of fine grass at the bottom of a simple burrow. Still others appear not to make nests at all. A conspicuous system of runways or tunnels is made through the grass to favorite feeding areas. The animals may be either diurnal or nocturnal, but Kingdon (1974b) found them most active around dawn and dusk. They are not primarily aquatic, but they enter water freely to swim from one reed patch to another, and they can dive and stay underwater for a short time to escape their enemies.

In contrast to most other species of *Otomys*, *O. unisulcatus* frequents dry, sandy areas in shrub growth and piles of rock. It is active mainly during the day. Its shelter is usually constructed above ground in shrubs and less commonly in trees or around matted clumps of grass. This species sometimes excavates extensive burrow systems in sandy areas, often under sandstone rocks. In such locations it may not build aboveground homes. The tunneled shelters are made of sticks or of weeds and grasses, the burrows extending below the ground. The female makes a soft nest of dry grass and twigs in a subterranean burrow. Surface paths or runs connect one refuge with another, and when the animals are disturbed, they use these paths to reach another shelter.

Otomys is almost exclusively herbivorous. It feeds mainly on green grasses, semiaquatic plants, and tender shoots. It also takes a variety of grains, seeds, berries, roots, and bark. It apparently does not harm cultivated crops, but has done extensive damage to forest nurseries by gnawing the bark and cambium of young trees (Kingdon 1974*b*).

Population densities of around 16 to 42 individuals per ha. have been reported for *O. irroratus*. Marked individuals of this species have been found to stay within an area of about 20 ha. (Kingdon 1974*b*). Although *Otomys* is sometimes said to occur in pairs, family groups, or colonies, a study of captive *O. irroratus* showed adults to be highly unsocial and aggressive toward one another (Davis 1972). In *O. unisulcatus* a pair of animals usually occupies a single shelter.

According to Kingdon (1974*b*), breeding in East African *Otomys* apparently is continuous, there may be up to five litters annually, and litter size is only 1 or 2 young. In Botswana *O. angoniensis* breeds at least during the warm, wet months from about August to March, and three pregnant females contained 3, 3, and 5 embryos (Smithers 1971). In a study in southern South Africa, Swanepoel (1975) found reproduction in August 1973, January 1974, and June 1974; 10 pregnant female *O. irroratus* contained an average of 2.1 (1–3) embryos, and 4 pregnant female *O. unisulcatus* had 1 to 3 embryos each. Davis (1972) reported the young to cling to the nipples of the female. Kingdon (1974*b*) stated that at birth the young weigh about 12.5 grams, have their eyes open, and are able to run; they reach sexual maturity within three months.

RODENTIA; MURIDAE; **Genus *RHIZOMYS* Gray, 1831**

Bamboo Rats

There are two subgenera and three species (Ellerman and Morrison-Scott 1966):

subgenus *Rhizomys*

R. sinensis, central and southern China, northern Burma;
R. pruinosus, Assam to southeastern China and Malay Peninsula;

subgenus *Nyctocleptes*

R. sumatrensis, Burma, Thailand, Indochina, Malay Peninsula, Sumatra.

Rhizomys, along with *Cannomys* and *Tachyoryctes,* has usually been placed in a distinct family, the Rhizomyidae, but the three genera are considered here to represent only a subfamily of the Muridae, for reasons discussed in the above account of that family.

Head and body length is 230 to 480 mm, tail length is 50 to 200 mm, and weight is 1 to 4 kg. *R. sumatrensis* is the largest species. In northern parts of the range of the genus the pelage is soft, thick, and silky, but in the tropics the fur becomes

harsh and scanty. The coloration of the upper parts ranges from slate through pinkish gray to brownish gray. The underparts are generally paler. Some hairs may be tipped with white.

Bamboo rats resemble American pocket gophers (family Geomyidae), but lack external cheek pouches. They have stout, heavy bodies; short legs; short, naked, and scaleless tails; small eyes and ears; and strong digging claws, the third digit of which has the longest nail. The pads of the feet are granulated; the two posterior sole pads are joined in the subgenus *Nyctocleptes* and distinct in the subgenus *Rhizomys*. The zygomatic arches of the skull are extremely wide and strong. The stout incisor teeth are orange, nearly vertically directed, and not covered by the lips. The dental formula is the same as that of most other Muridae, but there have been suggestions that the first of the three molars, both upper and lower, is actually a premolar. Female *R. sumatrensis* have two pectoral and three abdominal pairs of mammae, and female *R. pruinosus* have one or two pectoral and three abdominal pairs (Lekagul and McNeely 1977).

These rodents generally inhabit bamboo thickets at elevations of 1,200 to 4,000 meters. They spend much of their life underground among the roots of dense stands of bamboo. They dig extensive burrows, using both their teeth and their claws. An individual bamboo rat usually has several burrows, only one or two of which may be in active use. These animals commonly remain underground to cut and eat the bamboo roots that form their staple diet. According to Medway (1978), they sometimes come up at night and roam widely, and may climb bamboo. They straddle the stem, which is gripped between the legs, and with their teeth they cut out sections of stem wall, which are then taken down to the burrow. Other grasses, seeds, and fruits are also eaten. If enough fruit is available, free water will not be drunk.

Although powerfully built, these animals move slowly; their gait is a cumbersome waddle. If they sense that they are going to be overtaken, or if they are cornered, bamboo rats become quite fierce, making short rushes at anything put in front of them and biting savagely. At the same time, a grunting noise is emitted, coupled with a peculiar grinding action of the teeth. Even captives are said to be vicious. If taken young, however, bamboo rats become very tame and tractable in captivity (Medway 1978).

In northern Viet Nam, *R. pruinosus* and *R. sumatrensis* breed from February to April and from August to October, and have litters of one to four young (Tien and Sung 1971). According to Lekagul and McNeely (1977), the gestation period is at least 22 days, and three to five young are born, naked and blind, in an underground nest. Hair begins to grow at around 10–13 days of age, and the eyes open at about 24 days. The young first take solid food when around 1 month old, but still suckle occasionally at 3 months. Longevity is about 4 years.

Wild bamboo rats raid plantations for tapioca and sugar cane roots (Medway 1978). The animals, in turn, are trapped or dug out by native people for use as food.

RODENTIA; MURIDAE; **Genus *CANNOMYS* Thomas, 1915**

Lesser Bamboo Rat

The single species, *C. badius,* is found in Nepal, Assam, Burma, Thailand, Laos, and northern Viet Nam (Ellerman and Morrison-Scott 1966; Sung 1976). *Cannomys,* along with *Rhizomys* and *Tachyoryctes,* has usually been placed in

Bamboo rat (*Rhizomys pruinosus*), photo by Wang Sung.

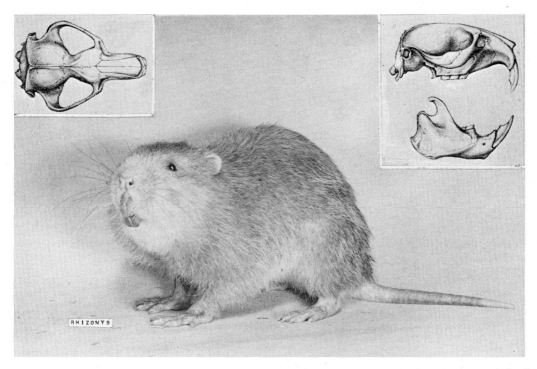

Sumatran bamboo rat (*Rhizomys sumatrensis*), photo by Ernest P. Walker. Insets: *R. pruinosis*, photos from *Anatomical and Zoological Researches: Zoological Results of the Two Expeditions to Western Yunnan in 1868 and 1875*, John Anderson.

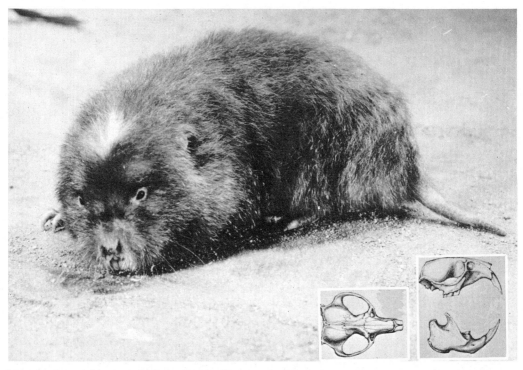

Lesser bamboo rat (*Cannomys badius*), photo by Constance P. Warner. Insets: Photos from *Anatomical and Zoological Researches: Zoological Results of the Two Expeditions to Western Yunnan in 1868 and 1875,* John Anderson.

a distinct family, the Rhizomyidae, but the three genera are here considered to represent only a subfamily of the Muridae, for reasons discussed in the above account of that family.

Head and body length is 147 to 265 mm, tail length is 60 to 75 mm, and weight is 500 to 800 grams (Lekagul and McNeely 1977). The fur is fairly thick on the head and body, and very thin on the tail. The color ranges from reddish cinnamon and chestnut brown to ashy gray and plumbeous. Individuals are almost uniformly colored throughout. There may be a longitudinal white band on the top of the head and a narrow white band from the chin to the throat.

Like *Rhizomys*, *Cannomys* resembles the American pocket gophers, but lacks cheek pouches. *Cannomys* has a thick, heavy body; very wide zygomatic arches in the skull; small eyes and ears; and short legs with long but powerful digging claws. *Cannomys* differs from *Rhizomys* in being smaller, having incisor teeth that protrude forward, and having smooth sole pads. Females have two pectoral and two abdominal pairs of mammae (Lekagul and McNeely 1977).

The lesser bamboo rat constructs burrows in grassy areas, forests, and sometimes gardens. It digs rapidly, using its powerful teeth as well as its claws. The tunnels are often very deep and located in hard, stony ground. Above ground, the lesser bamboo rat moves slowly, though it is said to be fearless when surprised by an enemy. It leaves its burrow in the evening to feed on various plant materials, including shrubs, the young shoots of grasses and cereals, and roots. "Bamboo" rat is something of a misnomer, as the animal consumes all kinds of vegetation.

Captives, maintained by Eisenberg and Maliniak (1973), showed peaks of activity in the morning and evening but slept a great deal during the day. The gestation period in three pregnancies was 40–43 days. Litter size in five deliveries was one or two young. The young developed relatively slow-

ly. One captive lived for 3 years and 3 months (Marvin L. Jones, Zoological Society of San Diego, pers. comm.).

The lesser bamboo rat is sometimes common in tea gardens and has been reported to damage tea plants, but the amount of damage may be exaggerated. This animal is eaten by many of the Burmese hill tribes.

RODENTIA; MURIDAE; **Genus TACHYORYCTES**
Rüppell, 1835

African Mole-rats

There are two species (Meester and Setzer 1971; Yalden 1975):

T. macrocephalus, highlands of Ethiopia;
T. splendens, Ethiopia, Somalia, Uganda, Rwanda, Burundi, Kenya, northern Tanzania.

This fossorial genus, along with *Rhizomys* and *Cannomys,* has usually been placed in a distinct family, the Rhizomyidae, but is here considered part of the Muridae, for reasons discussed in the above account of that family.

In *T. splendens* head and body length is 160 to 260 mm, tail length is 50 to 95 mm, and weight is 160 to 280 grams (Kingdon 1974b). *T. macrocephalus* is a larger species, its head and body length reaching 313 mm and its normal weight range being 330 to 930 grams (Yalden 1975). The short tail of *Tachyoryctes* is about twice the length of the hind foot and is usually well haired. The fur is thick and soft. The coloration is variable: some animals are shining black throughout, while others are brownish, reddish brown, pale gray, or cin-

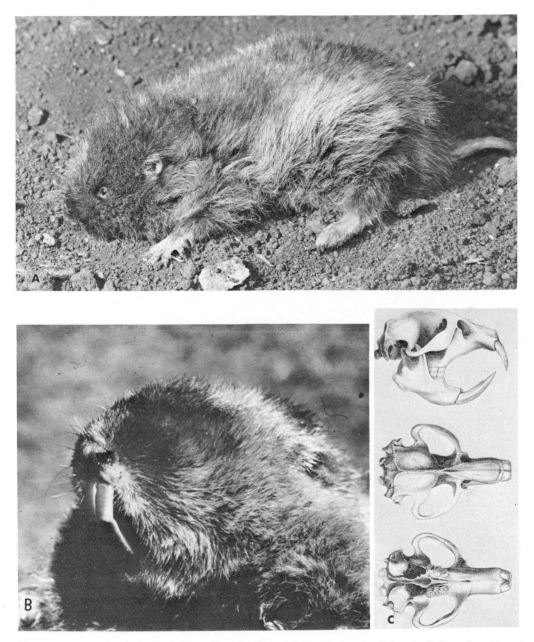

African mole-rat (*Tachyoryctes splendens*): A. Photo by C. A. Spinage; B. Photo by P. Morris; C. Photo from *Museum Senckenbergianum.*

namon buff. Most young are black. Albinos and partial albinos are as common as black adults. The underparts are usually slightly paler than the upper parts, and often have a silvery effect.

The body form is stocky and molelike. The eyes are small, but plainly visible and functional, and the ears are small. There are stiff hairs on the face, which are undoubtedly tactile. The thick, projecting incisors are not grooved and are deep orange in color. Although the claws are not particularly large, the hands and feet are well developed, and the short, powerful legs are suited for digging. African mole-rats resemble American pocket gophers (Geomyidae), but lack ex-

ternal cheek pouches and have somewhat longer and more fluffy fur.

African mole-rats are usually found in areas with more than 500 mm of annual rainfall, and flourish best in wet uplands. They favor open grassland, thinly treed savannah, moorland, and cultivated areas. They can withstand intense cold and are found at elevations of up to 4,150 meters (Kingdon 1974b; Yalden 1975). These animals construct a burrow system consisting of a nest chamber, a nearby bolt hole, and a series of foraging tunnels. The nest chamber contains a nest lined with grasses and herbs, a food store, and a sanitary area. The foraging tunnels may be up to 52 meters long; they

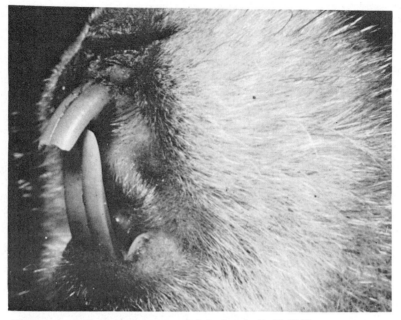

African mole-rat (*Tachyoryctes macrocephalus*), photo by D. W. Yalden.

usually run just below the level of the grass roots on which the animals feed, about 15 to 30 cm deep, but in the dry season *Tachyoryctes* may dig to a depth of 1 meter and become less active. Mounds of earth, about 15 to 40 cm wide and 7 to 15 cm high, develop as the burrow is excavated (Jarvis and Sale 1971; Kingdon 1974b; Delany 1975).

Based on observations of captives, Jarvis and Sale (1971:454–55) described the burrowing procedure of *T. splendens* as follows: "Except where the soil is very soft, all burrowing is done with the incisors. Strong forward and upward sweeps of the lower incisors cut away at the soil face. While biting in this way the mole-rat braces itself by gripping the burrow sides with laterally rotated hind feet. The fore feet scratch up the loosened soil and push it underneath the animal where a pile gradually accumulates. Periodically, when the pile under the abdomen is large enough, the hind feet relax their grip, are rapidly brought forward, collect the soil and kick it vigorously behind the animal. . . . Excavation continues and soil collects until the burrow behind the mole-rat is completely blocked with loose earth. The next phase in burrowing, namely the transportation of excavated soil, follows. The mole-rat first turns round by a lateral 'somersault' in which it curls up and pushes itself round with its fore feet. Then, using one side of its face and one fore foot held close to the chin, it consolidates the earth and pushes it along the burrow."

Kingdon (1974b) reported a captive *T. splendens* to be nocturnal. Through the use of radioactive tagging, however, Jarvis (1973a) found wild individuals to be active mainly from 1000 to 1900 hrs, and to remain in the nest at other times. *T. splendens* sometimes comes to the surface to forage for nesting materials and food, such as grasses and cultivated legumes. Its diet, however, consists mainly of the underground parts of plants—roots, rhizomes, tubers, bulbs, and corms. Some food is stored. Yalden (1975) found *T. macrocephalus,* in contrast to *T. splendens,* to depend largely on aboveground vegetation, mainly grass. Individuals of this species were observed extensively during daylight. They generally foraged by bringing just the head and shoulders out

of the burrow and gathering whatever plants could be reached.

Populations of *T. splendens* sometimes attain high densities—about one individual per 140 sq meters was calculated in one area—but each adult lives alone in its own burrow system. Captives fight savagely with each other for food (Kingdon 1974b). They have a characteristic attitude of defense when cornered, holding the head erect with the mouth wide open, and can give a vicious bite.

Yalden (1975) reported that six adult *T. macrocephalus* occupied an area of approximately 1,100 sq meters on a moor at an elevation of 3,900 meters. No social structure was evident and each animal appeared to be solitary. A single immature individual was also present, which suggests that litter size in the species is small.

A considerable amount of reproductive data on *T. splendens* has now accumulated (Rahm 1969b; Jarvis 1969, 1973b; Kingdon 1974b; Delany 1975). The capability for breeding continues throughout the year, though activity is highest during the rains and lowest in the dry season. Females are polyestrous and generally have 2 litters in fairly rapid succession, the second probably being conceived during a postpartum estrus. In a study in Kenya, Jarvis (1973b) found the average annual number of litters per female to be 2.1, and the gestation period to be 37–40 days. In eastern Zaire, however, Rahm (1969b) determined gestation to last 46–49 days. There may be up to 4 young per litter, but usually there are only 1 or 2. The young are weaned at 4 to 6 weeks of age, leave the mother's burrow about 1 month later, and attain sexual maturity when 6 months old. Average life expectancy is around 1 year, but maximum known longevity in the field is 3.1 years.

African mole-rats are sometimes agricultural pests, because of their raids on sweet potatoes, corn, beans, peas, groundnuts, and root crops. They were formerly regularly used for food by some native tribes. Water was poured into their burrows, thus making them surface and easy to catch. They have become rare in the Buganda area of Uganda, probably because their skins have value as charms (Kingdon 1974b).

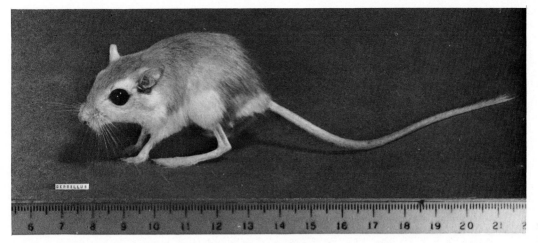

Northern pygmy gerbil (*Gerbillus gerbillus*), photo by Ernest P. Walker.

RODENTIA; MURIDAE; **Genus GERBILLUS Desmarest, 1804**

Northern Pygmy Gerbils

There are 2 subgenera and 27 species (Corbet 1978; Meester and Setzer 1971; Schlitter 1976; Cockrum 1977; Lay and Nadler 1975; Harrison 1972; Misonne 1974; Cockrum, Vaughan, and Vaughan 1976*b*; Lay 1975; Kock 1978*b*; Hubert 1978*b*):

subgenus *Handecapleura*

G. campestris, Atlantic coast of the Sahara to Egypt and Somalia;
G. poecilops, southwestern Arabian Peninsula;
G. famulus, southwestern Arabian Peninsula;
G. nanus, Algeria to Somalia and western India;
G. watersi, Sudan, Somalia, Afars and Issas;
G. dasyurus, Sinai, Syria, Iraq, Arabian Peninsula;
G. mesopotamiae, Tigris-Euphrates Valley in Iraq and western Iran;
G. henleyi, Algeria to Israel and northwestern Arabian Peninsula;
G. syrticus, northeastern Libya;
G. pusillus, southwestern Ethiopia, Kenya;
G. bottai, Sudan, Kenya;
G. muriculus, western Sudan;
G. mauritaniae, known only by a single specimen from central Mauritania;

subgenus *Gerbillus*

G. gerbillus, Algeria and northern Nigeria to Jordan and Sudan;
G. cheesmani, Arabian Peninsula to Afghanistan;
G. gleadowi, Pakistan, western India;
G. andersoni, coastal areas from Tunisia to Israel;
G. pyramidum, throughout northern Africa, Israel;
G. pulvinatus, Somalia, Ethiopia, Afars and Issas, Kenya;
G. perpallidus, northwestern Egypt;
G. riggenbachi, Morocco, Western Sahara;
G. latastei, Morocco to Libya;
G. hoogstraali, known only from the type locality in southwestern Morocco;
G. occiduus, known only from the type locality in southwestern Morocco;

G. rosalinda, central Sudan;
G. nigeriae, southern Mauritania to northern Nigeria;
G. nancillus, central Sudan.

The genera *Dipodillus, Microdillus,* and *Gerbillurus* are sometimes considered only subgenera of *Gerbillus,* but they are treated separately here in accordance with most recent authorities. The species *G. mauritaniae* had been placed in its own genus, *Monodia* Heim de Balsac, 1943, but was based on a single abnormal specimen that actually is referable to the subgenus *Handecapleura* of *Gerbillus* (Schlitter 1976; Meester and Setzer 1971). There now seems general agreement on the above division of *Gerbillus* into two subgenera, but the wealth of recent papers has led to confusion regarding the number of valid species and their relationships, and the arrangement given here is not definite. Meester and Setzer (1971) considered *G. agag* a full species with a range from Niger to Sudan, and with *G. nigeriae* as a subspecies. Schlitter (1976) elevated *G. nigeriae* to specific rank, referring to it part of *G. agag,* and referring the rest of *G. agag* to *G. gerbillus. G. dunni* of Ethiopia, Somalia, and Afars and Issas was treated as a full species by Meester and Setzer (1971), but was included in *G. pulvinatus* by Schlitter (1976). The name *G. aureus* was long applied to a species, or to a subspecies of *G. pyramidum,* inhabiting northwestern Libya, but Cockrum (1977) considered this entity conspecific with the previously named *G. latastei.* He also suggested that *G. latastei* might include *G. dunni, G. perpallidus, G. riggenbachi,* and *G. rosalinda.* Yalden, Largen, and Kock (1976) thought that *G. dunni* and *G. pulvinatus* were inseparable from *G. pyramidum.* Lay and Nadler (1975) considered *G. aquilus* of eastern Iran, southern Afghanistan, and western Pakistan to be a species distinct from *G. cheesmani.*

Head and body length is 50 to 130 mm, tail length is 70 to 150 mm, and weight is 20 to 40 grams. The upper parts range from pale yellowish gray, clay color, or sandy buff, to sandy red, mouse gray, dark fulvous, or brilliant reddish brown. The median line of the back is usually slightly darker. The sides and flanks are paler, blending into the white, creamy, or pale gray of the underparts. The tail is moderately to well furred, generally with a slight tuft near the tip. In some species there is a slight crest on the upper side of the distal half. The underside of the tail is usually light, the upper part near the body is darker, and the terminal portion is darkest (sometimes being dark brown or black).

Gerbillus is characterized by a slender form, a tail that is longer than the head and body, and fairly long ears. The slender upper incisors are grooved, though sometimes indistinctly. Unlike those of *Dipodillus*, the cheek teeth of *Gerbillus* are not hypsodont (Corbet 1978). The claws are fairly long. The hind feet are elongated, usually being over 25 percent of the head and body length. In the subgenus *Handecapleura* the soles of the hind feet are naked, while in the subgenus *Gerbillus* the soles are hairy.

The remainder of this account is based in large part on information compiled by Harrison (1972). Northern pygmy gerbils are found in dry country, sandy or rocky, sometimes with only scarce and coarse vegetation. Some species, however, tend to concentrate in moist, well-vegetated places, and *G. poecilops* seems to prefer cultivated areas and has a tendency to invade farm buildings. All species are primarily nocturnal and some are also crepuscular. They construct burrows, which vary from small, simple holes to fairly elaborate excavations. The burrows of certain species, such as *G. mesopotamiae,* may be two to three meters long and have several entrances, food storage tunnels, and an enlarged chamber with a nest of dry vegetation about a meter below the surface. The burrow entrances are usually kept closed with sand. The diet consists of seeds, roots, nuts, grasses, and insects.

The individuals of most species tend to build their burrows close to one another and thereby give the impression of an extensive colony. The females of most species are polyestrous and breeding has been reported at all seasons of the year. The gestation period is 20–22 days and litter size is one to eight young, usually about four or five. The offspring are born naked and helpless, open their eyes when around 16–20 days old, and are weaned when about 1 month old. According to Marvin L. Jones (Zoological Society of San Diego, pers. comm.), a captive specimen of *G. pyramidum* lived 8 years and 2 months.

RODENTIA; MURIDAE; *Genus DIPODILLUS Lataste, 1881*

Short-tailed Gerbils

There are three species (Corbet 1978; Meester and Setzer 1971; Schlitter and Setzer 1972; Cockrum, Vaughan, and Vaughan 1976*a*):

D. simoni, eastern Morocco to Egypt;
D. zakariai, Kerkennah Islands off eastern Tunisia;
D. maghrebi, known only from the type locality in northern Morocco.

Dipodillus was long considered to be a subgenus of *Gerbillus* and to include the species now placed in *Handecapleura*. Most recent authorities, including all those cited above, have treated *Dipodillus* as a separate genus that contains only a few species. Corbet (1978) listed *D. kaiseri* of the coastal plains of Libya and Egypt as a full species, but suggested that it might be only a subspecies of *D. simoni,* as it was considered by Meester and Setzer (1971).

Head and body length is 69 to 119 mm, tail length is 69 to 111 mm, and weight is 16 to 58 grams. *D. maghrebi* is the largest species. The upper parts are light brownish and the underparts are pale buff or white. *Dipodillus* differs from *Gerbillus* in having a tail that is usually shorter than the head and body, no tuft on the tail, hind feet that are usually less than 25 percent of the head and body length, and slightly hypsodont teeth. The soles of the hind feet are naked.

Ranck (1968) stated that *D. simoni* is found in the coastal plain and littoral deserts of northeastern Libya, where it apparently prefers habitats lacking sand. In northwestern Libya it inhabits narrow, grassy valleys in upland areas. In Tunisia, Harrison (1967) found *D. simoni* on level, sandy ground with sparse vegetation. Its burrows were situated in exposed areas, and specimens were collected at night, several hours after dusk. One of the three females taken by Harrison on 4 October was pregnant and contained four embryos.

RODENTIA; MURIDAE; *Genus MICRODILLUS Thomas, 1910*

The single species, which occurs in Somalia, was described as *Gerbillus peeli* by De Winton (1898). Later, Thomas (1910*a*) stated that the originally described skin was actually referable to the genus *Ammodillus,* but that the original skull, along with subsequently collected material, represented an entirely new genus, which he called *Microdillus*. Although long considered to be only a subgenus of *Gerbillus, Microdillus* was listed as a distinct genus by Meester and Setzer (1971).

According to Funaioli (1971), head and body length is 60 to 80 mm, and tail length is 56 to 62 mm. The general color is pale yellowish brown. The underparts, hands, and feet are whitish. There is a round, white spot behind each ear. The soles are naked.

The skull, unlike that of other gerbils, is peculiarly square and short. It is abnormally bowed, with a strongly convex cranial profile. Like that of *Dipodillus,* the tail is shorter than the head and body, and lacks a terminal tuft. Whereas the upper third molar of *Dipodillus* lacks cusps, this tooth in *Microdillus* contains three or four cusps (Meester and Setzer 1971; Schlitter and Setzer 1972).

Funaioli (1971) stated that the few known specimens have been collected on the predesert steppes of north-central Somalia. *Microdillus* is nocturnal. During the day it apparently takes refuge in an underground burrow.

RODENTIA; MURIDAE; *Genus GERBILLURUS Shortridge, 1942*

Southern Pygmy Gerbils

There are four species (Meester and Setzer 1971; Schlitter 1973):

G. paeba, southwestern Angola, Namibia, Botswana, South Africa;
G. vallinus, southwestern Angola, Namibia;
G. tytonis, southern Namibia;
G. setzeri, Namibia.

Gerbillurus was long considered a subgenus of *Gerbillus*. Meester and Setzer (1971) listed *Gerbillurus* as a separate genus, but with some question as to whether *G. paeba* and *G. vallinus* belonged to the same genus. Schlitter (1976) thought *Gerbillurus* to be more closely related to *Tatera* than to *Gerbillurus*.

Head and body length is about 90 to 120 mm, and tail length is 95 to 156 mm. The weight of *G. paeba* is 20 to 37 grams (Smithers 1971). The upper parts are brownish or gray in *G. paeba,* sandy buff in *G. vallinus,* hazel in *G. tytonis,* and light pinkish cinnamon in *G. setzeri*. The underparts, feet, and ventral surface of the tail are generally white.

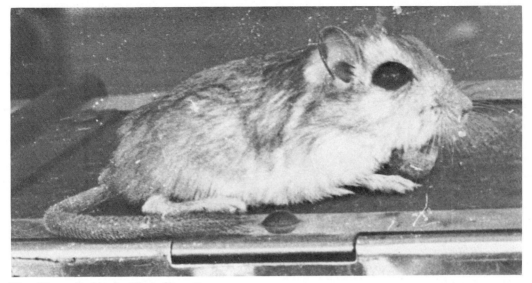

Microdillus peeli, photo by Alberto Simonetta.

Externally, *Gerbillurus* resembles *Gerbillus*. The tail is longer than the head and body and has a tufted tip. Unlike *Gerbillus,* the molar laminae in the upper and lower jaws have no sign of median crest connections. Unlike *Tatera,* the soles of the hind feet of *Gerbillurus* are well haired and the zygomatic plate of the skull does not project far forward (Meester and Setzer 1971).

All species inhabit desert or semidesert regions. Smithers (1971) stated that *G. paeba* is confined to sandy ground or sandy alluvium with a grass, scrub, or light woodland cover. This species is nocturnal and terrestrial, and lives in small warrens with many entrances. The entrances are often concealed under a tuft of grass or at the base of bushes. The diet consists of the seeds of grasses, bushes, and trees. *G. paeba* apparently breeds throughout the year. The number of embryos per pregnant female averages 3.7 and ranges from 2 to 5.

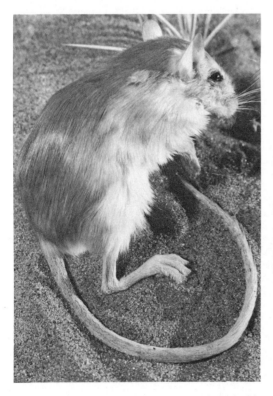

Southern pygmy gerbil (*Gerbillurus tytonis*), photos by Mary Seely.

Southern pygmy gerbil (*Gerbillurus paeba*), photo by John Visser.

RODENTIA; MURIDAE; **Genus TATERA** *Lataste, 1882*

Large Naked-soled Gerbils

There are 2 subgenera and 10 species (Meester and Setzer 1971; Corbet 1978; Hubbard 1970; Harrison 1972; Yalden, Largen, and Kock 1976):

subgenus *Gerbilliscus*

T. boehmi, southwestern Kenya to Angola and Zambia;

subgenus *Tatera*

T. indica, Syria to India, Sri Lanka;
T. leucogaster, Angola and southwestern Tanzania to South Africa;
T. nigricauda, southern Ethiopia to northern Tanzania;
T. robusta, Guinea-Bissau to Somalia and central Tanzania;
T. afra, southwestern South Africa;
T. valida, savannah zones from Senegal to western Ethiopia, and south to Angola and Zambia;
T. brantsii, southern Africa;

T. inclusa, Tanzania to Rhodesia and Mozambique;
T. pringlei, northeastern Tanzania.

Meester and Setzer (1971) provisionally listed *T. pringlei* as a subspecies of *T. inclusa*. Hubert, Adam, and Poulet (1973) recognized *T. guineae* and *T. gambiana* as separate species in Senegal, though these names had been regarded as synonyms of *T. robusta* and *T. valida*, respectively, by Meester and Setzer (1971).

Head and body length is 90 to 200 mm, tail length is 115 to 245 mm, and weight is 30 to 227 grams. The pelage is soft or medium. The coloration above ranges from pale sandy gray or pale buffy gray to dark buffy, blackish cinnamon, sandy brown, or dark gray. The underparts are white or whitish. The hands and feet are light colored. In most forms the tail is slightly darker, both above and below, than the back. The sides of the tail are light, and in most forms the terminal part is white or tufted with long dark hairs. There are usually light-colored or white spots above and behind the eyes and ears, and on the sides of the nose.

The body form is heavy and ratlike. The zygomatic plate of the skull is always very strongly projected forward. In the subgenus *Gerbilliscus* the upper incisors usually have two shallow grooves. In the subgenus *Tatera* the upper incisors are single grooved or plain (Meester and Setzer 1971). The soles of the hind feet are naked. Females have six or eight mammae.

These gerbils inhabit sandy plains, grasslands, savannahs, woodlands, and cultivated areas. They are nocturnal and terrestrial. They usually walk on all four limbs, but when alarmed they flee by means of running bounds of up to 1.5 meters in height. *T. indica* is said to be able to cover 3.5 meters in one leap. Short burrows, the so-called bolt holes, are excavated for sudden retreats, but deeper burrows are used as the living quarters. The animals burrow in sandy soil to a depth of 1 meter or more. Their tunnel systems are practically underground labyrinths in some areas, with numerous entrances. *T. indica* often has the enlarged nest chamber located in the center of the tunnel system. The entrances to the burrows are usually blocked with loose earth. The diet consists of roots, bulbs, seeds, green plant material, and insects. *T. indica* also eats eggs and young birds.

Prakash, Jain, and Rana (1975) found that *T. indica* resided in shallow burrows on the fringes of main streets in an Indian town. In this area the density of the gerbils was 175 to 460 per ha., but in rural grasslands density was only 10/ha. Prakash (1975) reported the home range of *T. indica* in the Rajasthan Desert to be 1,875 sq meters for males and 1,912 sq meters for females. *Tatera* appears to be a gregarious

Large naked-soled gerbils (*Tatera leucogaster*), photo by Ernest P. Walker.

Large naked-soled gerbil (*Tatera afra*), photo by John Visser.

genus, as a number of interconnecting burrows are usually located in the same area, but burrows housing only 1 or 2 adults are sometimes found. Cannibalism is apparently frequent in captive *T. indica*, females consuming their own newborn offspring. Wild individuals of this species have been reported to eat smaller or trapped individuals of their own kind.

Smithers (1971) stated that *T. leucogaster* breeds all year in Botswana, but Sheppe (1973) found reproduction in this species to be heavily concentrated in the rainy season, January–February, in Zambia. Prakash (1975) reported that *T. indica* breeds all year in the Rajasthan Desert of India, with a major peak during the monsoon of July–August and minor peaks in February and November. Females of this species may have three or four litters annually, the estrous cycle averages 4.82 days, and gestation lasts 26 to 30 days. An average gestation period of 22.5 days has been reported for *T. afra*. Overall litter size in the genus is 1 to 13 young, with averages of 3.3 in *T. brantsii* and 4.5 in *T. leucogaster* (Smithers 1971). In *T. indica* the young usually number 4 to 6, they open their eyes at 14 days of age, they are weaned and independent at 21 to 30 days, and they reach sexual maturity when 10 to 16 weeks old (Prakash 1975; Roberts 1977). A

captive *T. indica* lived for 7 years (Marvin L. Jones, Zoological Society of San Diego, pers. comm.).

Gerbils of this genus are suitable hosts for bubonic plague, and are said to be important agents in spreading the disease in southern Africa and the Middle East, especially because of their tendency to enter human habitations. *T. indica* also is considered a serious agricultural pest in some areas, because of its destruction of grains and legumes (Harrison 1972; Roberts 1977).

RODENTIA; MURIDAE; Genus TATERILLUS Thomas, 1910

Small Naked-soled Gerbils

There are apparently seven species (Robbins 1974, 1977; Meester and Setzer 1971):

T. gracilis, Senegal to Nigeria, probably Cameroon;
T. pygargus, southern Mauritania, Senegal, Gambia, southwestern Mali;
T. arenarius, southern parts of Mauritania, Mali, and Chad;
T. lacustris, northeastern Nigeria, northern Cameroon, possibly Niger and Chad;
T. congicus, eastern Cameroon to Sudan;
T. harringtoni, central Sudan and eastern Central African Republic to southern Somalia and northeastern Tanzania;
T. emini, Chad and Central African Republic to northwestern Kenya.

Head and body length is 100 to 140 mm, and tail length is 140 to 176 mm. *T. arenarius* weighs 42 to 52 grams (Robbins 1974). The coloration above ranges from pale yellow, through light buffy, clay color, light reddish brown, and tawny olive, to dark tawny brown. The sides are paler, and the underparts, including the hands and feet, are white or almost white. Most forms have light areas or spots on the sides of the face, above the eyes, and in back of the ears. Some forms have faint dark markings on the face. The tail is usually well tufted and darkest toward the tip. Externally, this genus is similar to *Tatera*, but it is usually smaller and the soles are sometimes partially haired, not completely naked. As in *Tatera*, the upper incisors are grooved.

These gerbils usually inhabit treeless plains or thorny scrub savannahs. Robbins (1973), however, observed that specimens of *T. gracilis* had been taken in a wide spectrum of

Small naked-soled gerbil (*Taterillus gracilis*), photo by Jane Burton.

habitats ranging from moist woodland to very arid Sahel woodland. These gerbils live in underground burrows and excavate the earth in the form of surface mounds.

In studies on a dry thornbush savannah in northern Senegal, Poulet (1972, 1978) found *T. pygargus* to be nocturnal, granivorous, and insectivorous. Population density in this area was calculated to be about 2 to 6 individuals per ha. Following heavy rainfall, however, density increased to as high as 180/ha. Adult home range averaged 1,100 sq meters for males and 300 sq meters for females. The home range of an adult male overlapped those of several females. Births occurred only after rains, in the period from September to March. A female could have a litter every six weeks and she would change her burrow each time. The gestation period was three weeks and there were usually four young per litter. Juveniles were nomadic for a brief period and then settled onto permanent home ranges by the time they were three to five months old.

RODENTIA; MURIDAE; *Genus DESMODILLUS*
Thomas and Schwann, 1904

Cape Short-eared Gerbil

The single species, *D. auricularis,* occurs in Namibia, Botswana, and South Africa (Meester and Setzer 1971).

Head and body length is 90 to 125 mm, tail length is 84 to 99 mm, and weight is 39 to 70 grams. The upper parts vary in color from uniform orangish to tawny brown. There is a white spot behind the ears on the back of the neck. The underparts, hands, and feet are white.

Desmodillus is characterized by short ears with a white spot behind them, enlarged auditory bullae, hind feet with hairy soles, laminate molars, and a tail that is haired but not tufted. The tail is about four-fifths, and the hind foot is about one-fifth, as long as the head and body. The hind feet are not especially modified for jumping.

This gerbil inhabits open sandy plains, dry tablelands, and cultivated areas. It runs like an ordinary rat and is not noticeably saltatorial. Its burrows are recognizable by the small surface heaps of earth thrown up, but are not as complicated as those built by some other genera of gerbil. The tunnels usually do not penetrate more than two meters, and they have blind passages or vertical escape holes. A nest of dried straw or sticks may be located in a chamber at the end of the tunnel. When the burrow is occupied, a heap of prickly seeds of a terrestrial creeping plant lies in front of the entrance on the mound of earth. *Desmodillus* is nocturnal. Its diet consists of seeds, grain, and insects such as locusts and grasshoppers. It

usually takes the food to the entrance of the burrow, where it sits down and eats, leaving husks and wings around the entrance. During a drought this rodent may move to a different area. Laboratory investigations by Pettifer and Nel (1977) have shown *Desmodillus* to store a larder of food in its nestbox.

This gerbil is not sociable. Although there may be many burrows within a short distance of each other, they do not form interconnected warrens (Smithers 1971). In a study of captives, Nel and Stutterheim (1973) found adults apparently solitary, except briefly during mating. Two weeks after two males and two females had been released in an outdoor enclosure, the males had been killed and eaten by the females. Three weeks later one female was killed and eaten by the other.

Available evidence (Smithers 1971; Keogh 1973; Nel and Stutterheim 1973) indicates that females are polyestrous and may bear young at any time of the year. The gestation period is normally 21 days, but a postpartum pregnancy lasted 35 days. Litter size ranges from one to seven young, and usually is about two to four. The young develop slowly, opening their eyes at 22 days of age and being weaned at 33 days.

RODENTIA; MURIDAE; *Genus DESMODILLISCUS*
Wettstein, 1916

The single species, *D. braueri,* is known from northern Senegal, northern Nigeria, and Sudan (Meester and Setzer 1971; Poulet 1972).

Head and body length is 50 to 70 mm, tail length is about three-fourths that of the head and body, and Poulet (1972) stated that adults scarcely exceed 10 grams in weight. The pelage is quite soft, sandy fawn on the upper parts, somewhat paler on the sides, and white on the undersurface. Postauricular white spots are evident. The tail is haired but not tufted.

Desmodilliscus is similar to *Desmodillus* and *Pachyuromys*. The auditory bullae are enlarged. The presence of 3 upper and 2 lower cheek teeth (making a total of 10 cheek teeth) is unique in the family Muridae. The incisors are narrow and furrowed. The skull resembles that of a *Desmodillus* in miniature.

According to Poulet (1972), this tiny gerbil occupies the same kind of habitat as *Taterillus*—dry thornbush savannah. It is nocturnal and granivorous, and constructs more complex burrows than those of *Taterillus*, with many entrances. Poulet (1978) stated that the abundance of *Desmodilliscus* apparently varies inversely with that of all other rodents in its habitat; it is rare during periods of heavy rainfall and becomes more numerous during droughts.

Cape short-eared gerbil (*Desmodillus auricularis*), photo from South African Union Health Department through D.H.S. Davis.

Desmodilliscus braueri, adapted from *Proc. Zool. Soc. London*, St. Leger. Inset: tooth pattern, photo from ''Evolution du dessin de la surface d'usure des molaires des Gerbillidés,'' *Mammalia*, Paris, F. Petter.

RODENTIA; MURIDAE; Genus PACHYUROMYS
Lataste, 1880

Fat-tailed Gerbil

The single species, *P. duprasi,* inhabits the northern part of the Sahara Desert region from western Morocco to Egypt (Corbet 1978).

Head and body length is 105 to 135 mm, and tail length is 45 to 60 mm. The soft pelage of the upper parts ranges from light yellowish gray to buffy brown. The underparts, hands, and feet are white. There is a white spot behind each ear. The tail is bicolored: coloration above is like that of the back, and below it is whitish. Some forms have pinkish buff along the sides and on the rump. The ears are short and buffy white, but the extreme edges are brown.

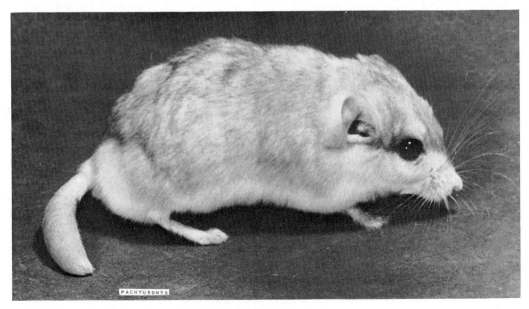

Fat-tailed gerbil (*Pachyuromys duprasi*), photo by Ernest P. Walker.

Pachyuromys is stockily built. It has well-developed claws on the fingers and slightly grooved upper incisors. As in the case of many desert mammals, the auditory bullae and mastoids are so inflated that they extend beyond the foramen magnum. The common name of this genus is derived from the peculiar shape of the tail, which is short, noticeably thickened, and club shaped.

Daly (1979) wrote that *Pachyuromys* is insectivorous and is confined to the type of desert habitat known as the *hamada*—gravelly plains with small patches of bushy perennial vegetation. The gestation period is 19 to 22 days and litter size is three to six young. Captive specimens are used in laboratory studies. One such individual lived for 4 years and 5 months (Marvin L. Jones, Zoological Society of San Diego, pers. comm.).

RODENTIA; MURIDAE; Genus AMMODILLUS
Thomas, 1904

Walo

The single species, *A. imbellis,* occurs in Somalia and southwestern Ethiopia (Meester and Setzer 1971; Yalden, Largen, and Kock 1976).

Head and body length is 85 to 106 mm, and tail length is 134 to 160 mm (Funaioli 1971). The coloration of the upper parts is reddish fawn; the hairs of the back are tipped with black, as are the hairs of the sides, but to a lesser degree, thus being a clearer, lighter fawn. On the eyebrows, cheeks, and fronts of the arms, the hairs are tipped with fawn. Distinct white spots are visible at the bases of the ears and above the eyes. The hands, feet, chin, cheeks, and underparts are white. The tail is darker above than below, and is scantily haired on the basal portion but terminally tufted with brownish hairs 8 to 10 mm in length. The feet and hands are scantily covered. The hairless pads and soles are covered with conspicuous scalelike granulations.

Ammodillus is a desert rodent. Although individuals are noted for fighting among themselves, they have an extraordinary weakness in the lower jaw, as the coronoid process of the mandible is absent. This structural feature led to the choosing of the specific name, *imbellis,* which means "feeble."

RODENTIA; MURIDAE; Genus SEKEETAMYS
Ellerman, 1947

Bushy-tailed Jird

The single species, *S. calurus,* is known from eastern Egypt, Sinai, southern Israel and Jordan, and near Riyadh in central Saudi Arabia (Harrison 1972; Nader 1974). This species was originally described as *Gerbillus calurus* and later was regarded as *Meriones calurus.* In 1947 a new subgenus, *Sekeetamys,* was set up for the retention of this species in *Meriones.* Chromosomal studies, however, have not supported its inclusion in *Meriones* or *Gerbillus.* Since 1956 *Sekeetamys* has generally been considered a full genus (Corbet 1978).

Head and body length is 100 to 125 mm, and tail length is 110 to 160 mm. The upper parts are yellowish or reddish, washed with black. The underparts, hands, and feet are

Walo (*Ammodillus imbellis*), adapted from *Proc. Zool. Soc. London,* St. Leger.

Bushy-tailed jird (*Sekeetamys calurus*), photo by Ernest P. Walker.

whitish. The tail is brownish and bushy; it usually bears a white tuft on the tip, and sometimes has a central black terminal tuft with white tufts on either side. The soles of the hind feet are naked.

The bushy-tailed jird inhabits deserts and rocky slopes that are characterized by broken ground and hard surfaces. It burrows under boulders and rocky ledges. It shows considerable climbing ability in its rocky habitat. The average number of young for 47 litters born in the Giza Zoological Gardens in Cairo was 2.9, and the largest number in 1 litter was 6; births occurred throughout the year. A captive specimen lived for five years and five months (Marvin L. Jones, Zoological Society of San Diego, pers. comm.).

RODENTIA; MURIDAE; Genus *MERIONES* Illiger, 1811

Jirds

There are 4 subgenera and 14 species (Corbet 1978; Meester and Setzer 1971; Ellerman and Morrison-Scott 1966):

subgenus *Parameriones*

M. persicus, eastern Turkey to Pakistan;
M. rex, southwestern Arabian Peninsula;

subgenus *Cheliones*

M. hurrianae, southern Afghanistan, southeastern Iran, Pakistan, western India;

subgenus *Meriones*

M. vinogradovi, eastern Asia Minor, northwestern Iran, Syria;
M. tamariscinus, lower Volga region of Soviet Union to northwestern China;
M. tristrami, Turkey to northwestern Iran and Sinai;

subgenus *Pallasiomys*

M. unguiculatus, Mongolia and adjacent parts of southern Siberia and northern China, Sinkiang, Manchuria;
M. meridianus, Caspian Sea region to Mongolia and northern China;
M. shawi, Morocco to Egypt;
M. libycus, Western Sahara to western Sinkiang;
M. caudatus, northern and western Libya, and probably adjacent parts of Tunisia and Egypt;
M. crassus, northern Africa, southwestern Asia;
M. sacramenti, southern Israel;
M. zarudnyi, southern Turkmen, northeastern Iran, northern Afghanistan.

Meester and Setzer (1971) considered *M. caudatus* a subspecies of *M. libycus.*

Head and body length is 95 to 180 mm, and tail length is 100 to 193 mm. In examining captive individuals of *M. unguiculatus,* Norris and Adams (1972) found weight to average 50 to 55 grams in females and 60 grams in males. Roberts (1977) reported the weight of *M. crassus* to be 29 to 40 grams, and that a male *M. persicus* weighed 100 grams.

Jird (*Meriones unguiculatus*), photo by Ernest P. Walker.

The dense pelage is fairly long and soft in most forms, but is short and harsh in others. The covering on the tail is short near the base and progressively longer toward the tip, so that the tail is slightly bushy in most of the forms and has a slight crest in some of the others. Coloration of the upper parts varies from pale, clear yellowish, through sandy and grayish, to brownish. The sides of the body are generally lighter than the back, because of the absence of black-tipped hairs. The underparts, including the hands and feet, are white, pale yellowish, buffy, or pale gray. There are frequently light areas about the face. Heptner (1975) found the different colors of the subspecies of *M. meridianus* to harmonize with the sand shades found in the habitats of the animals.

Externally, *Meriones* is quite ratlike. It has narrow, well-developed ears; a tail length approximating that of the head and body; upper incisors with a narrow groove on the anterior surface; slightly elongated hind legs for leaping; and strong claws. Harrison (1972) listed the number of mammae as eight in females of *M. persicus, M. tristrami,* and *M. crassus.*

Jirds inhabit clay and sandy deserts, bush country, arid steppes, low plains, cultivated fields, grasslands, and mountain valleys. They are terrestrial and construct burrows in soft soil, where they spend much of their time. The complexity of the burrows varies both within and between species. Young individuals of *M. crassus* may excavate a burrow, which descends at an angle of 15° to 30° to a depth of 0.5 meter, and which has only one entrance. The most complex galleries of this same species often attain a combined shaft length of 30 to 40 meters and have as many as 18 entrances (Koffler 1972). The tunnels of *M. unguiculatus* extend about 0.45 to 0.60 meters underground and are about 4 cm in diameter; a nest and one or two storerooms are in the central part of the system (Gulotta 1971). The burrows of *M. libycus* may be over 1.5 meters deep and radiate outward 3 to 4 meters in a tortuous tunnel system (Harrison 1972). Naumov and Lobachev (1975) classified the burrows of *M. persicus, M. tamariscinus,* and *M. unguiculatus* as simple, and referred to those of *M. vinogradovi, M. meridianus,* and *M. libycus* as complex. The burrows usually include several food storage chambers near the surface and one or more nest chambers at a greater depth. Nests are usually composed of dried vegetation.

Roberts (1977) wrote that *M. persicus* and *M. crassus* are nocturnal, while *M. hurrianae* and *M. libycus* are diurnal. Gulotta (1971) stated that *M. unguiculatus* is active both day and night throughout the year. She reported that this species adapts to a wide range of temperature and humidity, is active on the surface both in subzero winters and on summer days of more than 38° C, and neither hibernates nor estivates. Reports of true hibernation in *Meriones* may have resulted from the tendency of some species to remain underground for long periods in the winter and depend on hoarded food. Roberts (1977) stated that *M. persicus* may undergo intervals of torpidity at such times. Naumov and Lobachev (1975) stated that *M. tristrami* may remain in its burrow for two months during the winter and live entirely off of stored food. The diet of *Meriones* consists of green vegetation, roots, bulbs, seeds, cereals, fruits, and insects. *M. libycus* stores up to 10 kg of seeds in the northern part of its range (Naumov and Lobachev 1975).

Maximum recorded population densities under favorable conditions are: 5 to 10 individuals per ha. for *M. meridianus;* 20–30/ha. for *M. tamariscinus* (Naumov and Lobachev 1975); 11/ha. for *M. crassus* (Koffler 1972); and 477/ha. for *M. hurrianae* (Roberts 1977). The density of *M. vinogradovi* in an area of 401,000 ha. was reported as 37.4/ha. before a poisoning control program and 0.8/ha. afterward (Klimchenko et al. 1975). *M. crassus* has been reported to use a home range of 1,200 to 10,000 sq meters (Koffler 1972), and the individual home range of *M. libycus* varies from 50 to 120 meters in diameter (Naumov and Lobachev 1975). In the Rajasthan Desert of India, Prakash (1975) determined the home range of *M. hurrianae* to average 88.7 sq meters for males and 154.7 sq meters for females. This species usually forages within 20 meters of its burrow (Roberts 1977). In contrast, the daily summer movements of *M. unguiculatus* may cover 1.2 to 1.8 km, and a marked individual moved 50 km (Naumov and Lobachev 1975).

Investigations of *M. libycus* in Algeria (Daly and Daly 1975a; Daly 1979) have revealed that females remain for many months in a single *daya* (a vegetated depression surrounded by desert). In one such *daya,* five adult females occupied home ranges of about 0.6 to 4.0 ha. each. There was little or no overlap between the home ranges of females. Individual adult males, however, occupied relatively large home ranges that covered several *dayas.* Male home ranges overlapped extensively, both with each other and with those of females, but interaction was rare and all adults occupied separate burrows.

Social behavior seems to vary considerably both within and between species of *Meriones.* In contrast to the above information on *M. libycus,* both Roberts (1977) and Naumov and Lobachev (1975) referred to this species as colonial, and noted that many individuals burrowed in the same vicinity. Roberts added that *M. crassus* occurred in smaller colonies than *M. libycus* and that *M. persicus* was not gregarious. He also wrote that adults of *M. hurrianae* were not aggressive and that several sometimes shared a burrow, though probably not the same nest chamber. Naumov and Lobachev stated that 2 or more families of *M. vinogradovi* were sometimes found in the same burrow, and that a burrow of *M. unguiculatus* might contain 3 to 14 animals, probably a mated pair and their most recent litter. *M. unguiculatus* has been studied extensively in the laboratory, the consensus being that adults can be kept together, but that introduction of a stranger may result in a fight to the death (Gulotta 1971). Males scent mark a territory with a ventral sebaceous gland (Yahr 1977). Females are at least as territorial and aggressive as males (Swanson 1974). Monogamous pairs seem to do well in captivity (Norris and Adams 1972), but there is some question about the role of the male. Ahroon and Fidura (1976) suggested that the male disrupts maternal behavior to the extent that many young are lost. Elwood (1975), however, reported that the male shares in caring for the litter, spending considerable time cleaning, grooming, and warming the newborn. Jirds have several vocalizations and also apparently communicate through thumping the hind feet.

Captive *M. unguiculatus* are capable of breeding throughout the year, but in the wild the reproductive season extends from about February to October and up to 3 litters are produced (Gulotta 1971; Naumov and Lobachev 1975). The estrous cycle lasts 4 to 6 days and a postpartum estrus may occur. Gestation periods of 24 to 30 days have been reported in this species, but gestation may actually last only 19 to 21 days. Litter size is 1 to 12, usually 4 to 7, young. Newborn weigh about 2.5 grams each, open their eyes after 16 to 20 days, and are weaned when 20 to 30 days old. Sexual maturity is attained at 65 to 85 days of age. Females are capable of reproduction until they are as much as 20 months old, but average longevity in the wild is only 3 to 4 months.

Reproduction continues all year in *M. hurrianae,* but there are peaks of breeding activity in the late winter, midsummer, and, in the Rajasthan Desert of India, autumn (Prakash 1975; Roberts 1977). Females of this species have an estrous cycle averaging 6.22 days, may experience a postpartum estrus, and give birth to three or four litters per year. The reported gestation period is 28–30 days. Litter size averages 4.4 and ranges from 1 to 9. The young open their eyes at 15–16 days of age, are weaned when 3 weeks old, and reach sexual maturity at 15 weeks.

Data compiled on other species of the Soviet Union and Middle East (Naumov and Lobachev 1975; Roberts 1977; Harrison 1972; Koffler 1972) indicate that reproduction may occur throughout all or most of the year, but usually takes place from late winter to early autumn, and that two or three litters are produced by each female. Gestation periods of 20 to 31 days and litter sizes of 1 to 12 young have been reported. According to Marvin L. Jones (Zoological Society of San Diego, pers. comm.), a captive specimen of *M. crassus* lived for 5 years and 7 months.

Several species of jirds are used in laboratory research. In the early 1950s *M. unguiculatus* was brought to the United States for such purposes, and has been found to be highly adaptable, clean, and in need of little care (Gulotta 1971; Kaplan and Hyland 1972). It has become increasingly popular in medical, physiological, and psychological research, and also is kept now as a pet by many persons (usually being referred to as a ''gerbil''). There is concern, however, that feral populations could become established in North America, and that these might destroy crops and embankments, and even displace native rodents (Fisler 1977). Several species of *Meriones* are sometimes considered pests in the Old World, because they eat cultivated plants, damage irrigation structures by burrowing, and spread disease (Naumov and Lobachev 1975; Klimchenko et al. 1975).

Przewalski's gerbils (*Brachiones przewalskii*): A. Photo from *Mammalia of Central Asia*, E. Büchner; B. Photo by P. F. Wright of specimen in U.S. National Museum.

RODENTIA; MURIDAE; *Genus BRACHIONES Thomas,*
1925

Przewalski's Gerbil

The single species, *B. przewalskii,* occurs in northwestern
China from western Sinkiang to northern Kansu; it is not
found in Mongolia (Corbet 1978).

Head and body length is 80 to 95 mm, and tail length is 70
to 80 mm. The upper parts are pale grayish yellow, pale
yellowish buff, or light sandy gray.The underparts, hands,
and feet are white. The tail is buffy or whitish throughout; it
is slender and tapering, and is not tufted. The soles are
densely haired, whereas the palms are naked.

The thickset body, reduced ears, shortened tail, long foreclaws, and hairy soles indicate that this rodent has developed
burrowing habits. It is found in desert areas.

RODENTIA; MURIDAE; *Genus PSAMMOMYS*
Cretzschmar, 1828

Fat Sand Rat

Both Corbet (1978) and Meester and Setzer (1971) considered that a single species, *P. obesus,* occurs from Algeria to
Palestine, as well as on the coast of Sudan and in parts of
Saudi Arabia. Cockrum, Vaughan, and Vaughan (1977),
however, stated that *P. vexillaris* of Algeria and Libya is a
species distinct from *P. obesus.*

Head and body length is 130 to 185 mm, and tail length is
110 to 150 mm. The upper parts are reddish brown, reddish,
yellowish, or sandy buff. The underparts are yellowish, buffy, or whitish. The tail is fully haired and has a terminal tuft.
The external form is stocky. From *Meriones,* this genus is
distinguished by its nongrooved incisors and short (10 to 15
mm), thick, rounded ears.

The fat sand rat is mainly diurnal and lives in sandy areas
with scant vegetation. It constructs complex burrows with
several entrances and food storage chambers, and a chamber
containing a nest of finely cut vegetation. It sits up on its hind
legs and tail, but quickly retreats into its burrow when alarmed.

Studies in the Sahara Desert (Daly and Daly 1973; Daly
1979) have shown that the preferred food of the fat sand rat is
the leaves and stems of succulent plants of the family Che-
nopodiaceae, which contain much water but also a high proportion of salt. Most mammals could not survive by eating
these plants without also having an abundant source of fresh
water, but the fat sand rat thrives on this diet through the aid
of its extremely powerful kidneys, which produce a highly
salt-concentrated urine. *Psammomys* also feeds on other
kinds of plants and reportedly can be very destructive to
grain; 500 heads of barley were once found stored in a burrow
(Harrison 1972).

During a 5-month study of the fat sand rat in the Algerian
Sahara, Daly and Daly (1975*b*) calculated average home
range length to be 189.6 meters in males and 75.8 meters in
females. Mean weekly range lengths were 67.7 meters in
males and 11.7 meters in females. Females tended to remain
in small areas around the bushes they used for food, and to
move only when edible vegetation was exhausted. Subordinate males also used a small home range, but moved more
frequently. Dominant males ranged over relatively large
areas encompassing the ranges of several females and subordinate males. A conspicuous form of communication observed in this study was audible foot thumping, sometimes
accompanied by a high-pitched squeak. Throughout this
study, which lasted from December to April, females were
pregnant, lactating, or both, and local people said that young
could be found in any month of the year. Associated laboratory work indicated that the gestation period normally is
about 25 days, but is extended to about 36 days after postpartum mating. Litter size in captivity was 2 to 5 young, weaning occurred around 3 weeks of age, and females first conceived when 3 to 3½ months old.

RODENTIA; MURIDAE; *Genus RHOMBOMYS Wagner,*
1841

Great Gerbil

The single species, *R. opimus,* is found in Soviet Central
Asia, Iran, Afghanistan, western Pakistan, Sinkiang, and
southern Mongolia (Corbet 1978).

Fat sand rat (*Psammomys obesus*), photo by Robert E. Kuntz.

Great gerbils (*Rhombomys opimus*): Top, photo from *Mammalia of Central Asia,* E. Büchner; Bottom, photo by Valerij Neronov through X. Misonne.

Head and body length is 150 to 200 mm, and tail length is 130 to 160 mm. The upper parts are sandy yellow, orangish buff, or dark grayish yellow, and the underparts are whitish. The fur is thick and soft, and the tail is hairy, almost bushy. The external form is stocky, the claws are large, and the soles are hairy. Two longitudinal grooves are present on each upper incisor.

The great gerbil usually inhabits sandy deserts, but also occurs in clay deserts and in the foothills of the mountains of Central Asia. According to Naumov and Lobachev (1975), the most important factor in its distribution is proper soil conditions for excavation of its deep and complex burrow. Subsandy soils are best. The burrows are variable in structure, but may contain several levels with nest chambers,

lengthy tunnels, and food storage chambers. The winter nest chambers are 1.5 to 2.5 meters below the surface, where the microclimate is stable all year. The great gerbil is mainly diurnal, being most active at dawn but sometimes remaining on the surface until dusk. It does not hibernate, though its activity apparently is greatly reduced in many areas during the winter.

The winter activity of *Rhombomys* is in direct proportion to the air temperature and in inverse proportion to the depth of the snow. By midwinter, in those parts of the range with a snow cover, only a few of the burrow exits are open and air holes are predominant. In many colonies, where the burrows are completely covered with snow for a long time, the animals come to the surface only rarely.

Little contact is made between individual *Rhombomys* and those living in neighboring colonies, at least in winter when many areas are covered with snow. In a study conducted during the winter in the northern Lake Aral region, the maximum length of a trail in the snow was 20 meters, and 83 percent of the tracks examined were in the immediate vicinity of the burrow exit. Following a harsh winter in Turkmen, Marinina (1971) found a population density of only one individual for every 5 to 10 ha. in areas of open sand ridges, and two per ha. in areas of high ridges and hills.

The diet of *Rhombomys* consists of a variety of desert plants. In some areas the food is stored for winter use (up to 60 kg or more per burrow), but in other regions the animals feed mainly on the surface. The food reserves for winter are located in compartments in the burrow, but in some places the plants are stored in heaps on the surface.

The following information on social life and reproduction is taken mainly from Naumov and Lobachev (1975) and Roberts (1977). The great gerbil is gregarious and many individuals may burrow in the same vicinity to form a large colony. Several adults may share a burrow and work together in its construction. Captives can be kept together without fighting. Females are polyestrous and in captivity can produce six litters in 6 months. In the wild the breeding season lasts from about April to September, and two or three litters are born to each female. The gestation period is 23 to 32 days. Litter size is 1 to 14 young, usually 4 to 7. Some females reach sexual maturity when 3 to 4 months old. Maximum longevity is 3 to 4 years for females and 2 to 3 years for males.

This rodent is considered a pest in the Soviet Union. It is a reservoir for plague, and it damages crops, railway embankments, and the sides of irrigation channels. In some areas it is trapped for its skin.

RODENTIA; MURIDAE; *Genus CLETHRIONOMYS Tilesius, 1850*

Red-backed Mice, Bank Voles

There are seven species (Corbet 1978; Hall 1981):

C. rutilus, northern Scandinavia to the Bering Strait and Manchuria, Sakhalin, Hokkaido, Alaska to Keewatin;
C. gapperi, British Columbia to mainland Newfoundland, northern conterminous United States, Rocky Mountains, Appalachians;
C. californicus, western Oregon, northern California;
C. glareolus, Europe, parts of Asia Minor and Central Asia;
C. rufocanus, Scandinavia to northeastern Siberia and Korea, Hokkaido, Kuril Islands;
C. rex, Hokkaido and nearby Rishiri Island;
C. andersoni, Honshu.

Head and body length is 70 to 112 mm, tail length is 25 to 60 mm, and weight is about 15 to 40 grams. The fur is dense, long, and soft in winter, but shorter and harsher in summer. The general coloration above is dark lead gray with a pronounced reddish wash. The wash becomes less prominent on the sides, which are grayish. The underparts are dark slate gray to almost white. The tail has a slight terminal pencil of hairs. The short thumb is provided with a flat nail. The ears are slightly more conspicuous through the fur in *Clethrionomys* than in *Microtus,* and the eyes are also more prominent. Females have eight mammae.

These rodents inhabit cold, mossy, rocky forests and woodlands in both dry and moist areas. They also inhabit tundra and bogs. They are active night and day, summer and winter, scurrying and climbing about stumps, fallen logs, and rough-barked trees. According to Banfield (1974), *C. gapperi* does not make runways of its own, but often uses those of other small mammals. It does construct spherical nests of grasses, mosses, lichens, or shredded leaves. These nests are usually hidden under the roots of stumps, logs, or brush piles, but may be located in holes or branches of trees high above the ground. During the winter, globular nests of grass may be placed directly on the ground under the snow, with tunnels radiating from the nest under the snow. The diet of *Clethrionomys* consists of tender vegetation, nuts, seeds, bark, lichens, fungus, and insects. Food is often stored in the nest for use when the supply is short.

French, Stoddart, and Bobek (1975) listed population densities for various species in Eurasia, which ranged from less than 1 to nearly 100 individuals per ha. The average home range of *C. glareolus* in Czechoslovakia has been calculated to be 0.82 ha. for males and 0.71 ha. for females. Banfield (1974) wrote that the home range of *C. gapperi* may be as large as 1.4 ha. in the summer and as small as 0.14 ha. in the winter, when foraging is restricted by a blanket of snow. He noted that populations fluctuate widely from year to year, but with no apparent periodicity. Merritt and Merritt (1978) cited population densities ranging from 2 to 74.1 per ha. for *C. gapperi.* In their own study in Colorado, the vole population stabilized at about 18/ha. over the winter, declined to 10/ha. during the spring thaw, and then increased to about 42/ha. in November following the reproductive season.

When disturbed, red-backed mice may utter a chirplike bark that can be heard from one to two meters away, and they

Red-backed mouse (*Clethrionomys glareolus*), photo by P. Morris.

flee or freeze in position, depending on their location and preceding activity. They also gnash or chatter their teeth. Mihok (1976) reported that encounters between captive male and female *C. gapperi* were generally amicable and resulted in female dominance if the female was in breeding condition, but resulted in male dominance and pursuit of the female if she was not in breeding condition. Encounters between animals of the same sex resulted in avoidance behavior and aggression. West (1977) found a lessening of aggressive behavior in *C. rutilus* in Alaska during midwinter, which apparently allowed the animals to huddle and thereby conserve heat.

Breeding may begin as early as late winter and continue to late fall, the gestation period is 17 to 20 days, and overall litter size is 1 to 11 young. Data compiled by Innes (1978) indicate that litter size in *C. gapperi* increases with latitude and elevation. It was found to average 3.33 young in one study in southeastern Kentucky and 7.25 in an investigation in central Alberta. Innes stated that the modal number of offspring is 5 in *C. gapperi* and 7 in *C. rutilus*. Banfield (1974) stated that the reproductive period of *C. gapperi* in Canada is April to early October, and that three or four litters are borne by each female per season. The newborn weigh about 1.9 grams, open their eyes when 9 to 15 days old, and are weaned and independent at 17 to 21 days. The young of the spring litter bear their own first young when they are about 4 months old. According to Marvin L. Jones (Zoological Society of San Diego, pers. comm.), a specimen of *C. glareolus* lived in captivity for 4 years and 11 months.

Red-backed mice destroy quantities of insect larvae and also form a major food source of fur-bearing animals. Mice of this genus have been found to be important in Great Britain, and probably elsewhere, as agents in transporting, burying, and eating tree seeds, and in damaging or killing seedlings (Ashby 1967). The species *C. glareolus* was first recorded in Ireland only in 1964, and was probably introduced there through human agency (Fairley 1971; Corbet 1978).

RODENTIA; MURIDAE; **Genus EOTHENOMYS** Miller, *1896*

Père David's Voles, Pratt's Voles

There are 11 species (Corbet 1978; Lekagul and McNeely 1977):

E. melanogaster, central and southeastern China, northern Burma, northern Thailand;
E. olitor, Yunnan (southern China);
E. proditor, Yunnan and Szechwan (southern China);
E. chinensis, Yunnan and Szechwan (southern China);
E. custos, Yunnan and Szechwan (southern China);
E. smithi, Japan;
E. regulus, Hopei (northeastern China), Korea;
E. shanseius, Shansi and possibly Hopei (northeastern China);
E. inez, northeastern China;
E. eva, central China;
E. lemminus, northeastern Siberia.

Head and body length is 90 to 126 mm, and tail length is 30 to 55 mm. Lekagul and McNeely (1977) gave the weight of *E. melanogaster* as 27 grams. Coloration above is usually dark reddish brown, with a peculiar metallic reflection caused by the burnished tips of a portion of the longer hairs. Some writers refer to this effect as a brassy shine. The underparts are bluish gray. The young are blackish. The pelage is fairly short and smooth. The soles of the feet are hairy behind the pads. The tail is covered with stiff hairs that form a short, thin terminal pencil. Females have four mammae. *Eothenomys* is perhaps intermediate to *Clethrionomys* and *Microtus,* as it possesses some characteristics common to both genera.

These animals frequent banks and slopes that can be easily tunneled, in both wooded growth and mountain meadows. Their elevational range is about 1,800 to at least 4,400

Père David's vole (*Eothenomys smithi*), photo by K. Tsuchiya.

meters. They tend to be active both day and night. Some forms seem to spend more time underground than others. Burrows open to the surface at frequent intervals, and numerous surface runways may be evident. Surface runways of other animals, such as moles, may also be used. The habits of *Eothenomys* are presumably like those of *Clethrionomys* and *Microtus*. Breeding apparently occurs throughout the year, or nearly so, as embryos have been found from February to October, the number per female ranging from one to three.

RODENTIA; MURIDAE; *Genus ALTICOLA Blanford, 1881*

High Mountain Voles

There are four species (Corbet 1978):

A. roylei, mountains of Central Asia from Mongolia to the western Himalayas;

A. stoliczkanus, Altai Mountains of Mongolia, Nan Shan Mountains of north-central China, Tibet, Himalayas;

A. strelzowi, northeastern Kazakh, Altai Mountains of Mongolia;

A. macrotis, mountains of Mongolia and adjacent parts of Siberia.

Head and body length is 80 to 140 mm, and tail length is 15 to 54 mm. Ten specimens, with an average length of 108 mm, weighed 35 to 49 grams. The coloration above is usually some shade of gray or brown; the gray is mixed with ash, silver, buff, or yellow, and the brown, when present, with grays or reds. The underparts are usually white or buffy white. The feet are generally white or grayish white. In most forms the well-haired tail is white or buffy white throughout. The soles are usually haired posteriorly. The fur is thick, soft, and long. It is possible that the winter coat is longer and paler in color than the summer coat.

The external form is not modified for burrowing. The eyes are moderately large, and the ears are of moderate size, though they are often concealed in the long fur. The flattened skull of *A. strelzowi* is probably an adaptation for living under rocks. Female high mountain voles have eight mammae.

Alticola seems to be the Asian equivalent of the snow vole,

Microtus nivalis, of Europe, which has a white tail and lives at high elevations in the mountains. The elevational range of *Alticola* is about 900 meters to at least 5,700 meters. *A. strelzowi* is mainly diurnal, whereas *A. roylei* is partly diurnal but mainly nocturnal. Most, if not all, forms store food for the winter. The stems and leaves of herbaceous plants and shrubs may be dried first in the sun and then piled into heaps or stuffed into stone crevices. *A. roylei* deposits its banana-shaped droppings at the entrances of the burrow, a definite sign of its presence.

A. roylei is said to be fearless, sometimes coming within 4 cm of a person's hand. *A. strelzowi* is known to live in colonies. In the Altai Mountains the breeding season of this species begins in late April or early May and lasts about 2½ months; there are 1 or 2 litters per season, with litter size ranging from 3 to 13 and averaging 7.1 young (Eshelkin 1976). In the mountains of eastern Kazakh, *A. roylei* generally breeds from March or April to September or October, with a peak from April to June, but winter reproduction also has been observed. There are 2 or 3 litters a year and litter size ranges from 2 to 6 and averages 3.6 (Obidina 1972).

RODENTIA; MURIDAE; *Genus HYPERACRIUS Miller, 1896*

Kashmir and Punjab Voles

There are two species (Corbet 1978):

H. wynnei, extreme northern Pakistan;

H. fertilis, Kashmir, northern Pakistan.

Head and body length is 96 to 138 mm, tail length is 24 to 40 mm, and weight is 21.5 to 60 grams (Roberts 1977). The pelage is shorter and more dense than that of *Alticola*. Coloration of *H. fertilis* is deep reddish brown on the upper parts, gradually becoming dull ochraceous below. The feet are dusky. The tail is slightly bicolored, sepia above and dirty white below. *H. wynnei* apparently occurs in two distinct color phases. In the light phase the upper parts are yellowish brown and the underparts are grayish, tinged with wood brown. The feet and hands are grayish brown on the upper surface. The tail is slightly bicolored, dark grayish brown above and paler gray below. In the dark phase the upper parts

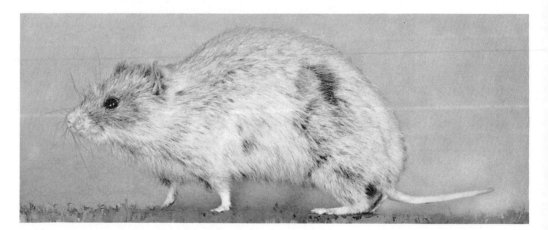

High mountain vole (*Alticola* sp.), photo by Howard E. Uible of mounted specimen in U.S. National Museum.

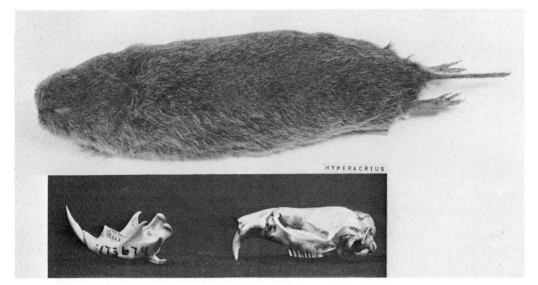

Kashmir vole (*Hyperacrius fertilis*), photos by P. F. Wright of specimen in U.S. National Museum.

vary from a lustrous seal brown to a glossy blackish brown. The underparts are duller and somewhat paler than those above, and are lightened by short whitish- or yellowish-tipped hairs of the ventral surface. The hands and feet are dark gray above. The tail is dusky gray and only slightly bicolored. The whiskers are short, scarcely reaching to the ears, and the feet are well haired. The dark phase of *H. wynnei* is unusual among animals of this kind, and may be a natural result of retaining the darker coat of the immature form, owing to the fact that the animal has adopted a subterranean life away from light. *H. wynnei* differs from *H. fertilis* in being larger in size and in having longer, more lax and luxuriant pelage.

The general appearance of *Hyperacrius* is similar to that of *Microtus* and *Alticola*, the principal means of differentiation being in cranial and dental structure. The well-developed claws on all four limbs are long and slender; the pollex is provided with a flattened, fairly large nail. Females have four inguinal mammae.

According to Roberts (1977), *H. wynnei* occurs in moist temperate forests and meadows at elevations of 1,850 to 3,050 meters, while *H. fertilis* is found in the subalpine scrub zone and alpine meadows at 2,450 to 3,600 meters. *H. wynnei* is highly fossorial, excavating extensive networks of shallow feeding tunnels, about 32 to 75 mm deep, as well as much deeper burrows that descend almost vertically into the ground and that are used for sleeping and breeding. During the winter this species constructs tunnels near or on the surface, through the snow, and lines them with soil and vegetation. It does not hibernate and is active at any time of the day or night. *H. fertilis* is not so fossorial and will forage above the ground. Both species are completely herbivorous, eating grass, stems, and roots. *H. wynnei* reportedly does considerable damage to potato crops.

Roberts (1977) stated that *H. wynnei* is loosely colonial. The female constructs a grass-lined nest at the end of a deep burrow and probably produces two to three litters in a breeding season, with two or three offspring each. Young animals, approximately four to six weeks old, have been found in early June and early October. *H. fertilis* also produces two or three litters, during the spring and summer.

RODENTIA; MURIDAE; **Genus DINAROMYS** Kretzoi, *1955*

Martino's Snow Vole

The single species, *D. bogdanovi*, occurs in the mountains of Yugoslavia and possibly Albania. This species has sometimes been referred to the genus *Dolomys* Nehring, 1898, which is based on a Pleistocene fossil from Hungary, but evidence now indicates that *D. bogdanovi* represents a separate, Recent genus (Corbet 1978).

Head and body length is 99 to 148 mm, and tail length is 74 to 119 mm (Van Den Brink 1968). The tail is usually more than half the length of the head and body. The fur is dense, soft, and moderately long. The upper parts are grayish brown, buffy grayish, or bluish gray, and the underparts are usually paler, often grayish white. The feet are white, and the tail is usually dark brown above and white on the sides and below. The thinly haired tail has a short, stiff, thin pencil line about 5 to 6 mm in length. The palms and soles are naked except near the heels. The large ears are densely haired.

The small thumb bears a small, flattened nail. The other digits have short, sharp claws, about equally long on the hands and feet. Notable cranial characters include the broad and faintly grooved upper incisors and the large, inflated bullae, though the mastoids are comparatively small.

This vole is found in the mountains, generally at elevations of 1,300 to 2,200 meters and rarely below 680 meters. It is nocturnal and active throughout the year. On Mount Trebevic (south of Sarajevo), it finds shelter and builds nests under rocks and isolated stones. Thawing snows and spring rains force it to move to drier ground. It feeds on grasses and stores food for the winter.

Breeding usually occurs twice a year, in March and again in June. The time of the second breeding season varies more than that of the first. A study on Mount Trebevic showed that only one litter was produced during the dry year of 1946. The gestation period is thought to be at least one month. There are usually two or three good-sized young per litter. The young are reported to come out of the burrows in July.

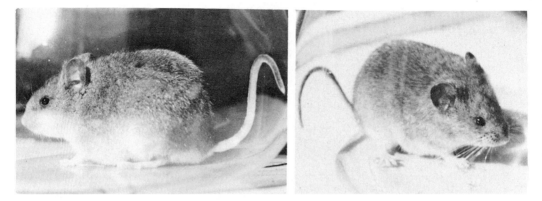

Martino's snow vole (*Dinaromys bogdanovi*), photos by Ivo R. Savić.

RODENTIA; MURIDAE; **Genus PHENACOMYS Merriam, 1889**

Heather Voles, Tree Mice

Hall (1981) listed three species:

P. intermedius, most of subarctic Canada, northern Minnesota, mountains of western conterminous United States;

P. albipes, western Oregon, northwestern California;

P. longicaudus, western Oregon, northwestern California.

Hall provisionally included *P. silvicola* of western Oregon as a subspecies of *P. longicaudus.* He also noted that *P. intermedius* may actually be a composite of two or three allopatric species. Johnson (1973) recommended that certain species of *Phenacomys,* apparently including *P. albipes* and *P. longicaudus,* be placed in the separate genus *Arborimus* Taylor, 1915.

Heather voles: A. *Phenacomys longicaudus,* photo by Alex Walker; B. *P. intermedius,* young are ten days old, photo by J. B. Foster.

Head and body length is about 90 to 120 mm, and weight is 25 to 56 grams. In *P. intermedius* the tail length is 26 to 41 mm, less than half the length of the head and body, while in the other species the tail length is 62 to 87 mm. The pelage is fairly long and fine textured. The upper parts are generally grayish brown in *P. intermedius*, dark brown in *P. albipes*, and reddish brown or cinnamon in *P. longicaudus*. The underparts are whitish, gray, or pinkish buff.

These animals so closely resemble the genera *Microtus* and *Clethrionomys* that positive identification may not be possible until the skull and dentition are carefully examined. Unlike *Microtus*, adult *Phenacomys* have rooted cheek teeth. Unlike those of *Clethrionomys*, the lower molar teeth of *Phenacomys* have the inner reentrant angles much deeper than the outer. There are eight mammae in female *P. intermedius* and four in female *P. longicaudus* (Johnson 1973).

As pointed out by Johnson (1973), the ecology of *P. intermedius* differs markedly from that of *P. longicaudus*, and these differences form part of the case for allotting the two species to separate genera. According to Banfield (1974), *P. intermedius* occurs in a wide range of habitats from sea level to the tree line in the Rocky Mountains. It seems to prefer open coniferous forests with an understory of heaths, or areas of shrubby vegetation on forest borders or in meadows. It is terrestrial and largely nocturnal or crepuscular. It is active throughout the winter, at which time it builds a spherical nest about 150 mm in diameter, composed of twigs, lichens, and grass. The nest is placed on the ground between rocks or under a shrub, and from it tunnels radiate out under the snow. The summer nest is built about 100 to 250 mm underground, and tunnels radiate outward and upward to the surface. *P. intermedius* feeds on a variety of plant material including bark, buds, heaths, forbs, berries, and seeds.

P. longicaudus and *P. albipes* are restricted largely to lower elevations in humid, temperate, coastal forests (Johnson 1973). *P. albipes* is terrestrial and little is known of its biology. *P. longicaudus* is arboreal and has a highly specialized way of life. Unlike most mice, the movements of *P. longicaudus* are slow and cautious, especially along branches, where care is taken to obtain a firm grasp on a new support before leaving a safe foothold. This species is active throughout the year. The males live in burrows on the ground or in piles of debris, but the females live in trees. The males climb the trees to mate and to build small temporary nests. The females also build nests, which are 228 to 305 mm in diameter and 4.5 to 15 meters above the ground. The nests are made of accumulations of twigs and discarded spines of fir and spruce leaves; they are generally placed at a forked branch or over a nest that has been abandoned by a bird or squirrel. The tender terminal twigs of coniferous trees are cut for food. They are carried to the nest, where the needles are eaten and the spines are used to enlarge the nest.

Banfield (1974) stated that female *P. intermedius* are seasonally polyestrous, and that the breeding season extends from May to August. The gestation period is 19 to 24 days, and litter size ranges from two to eight young, averaging about five. The newborn weigh approximately 2.4 grams, open their eyes when 14 days old, and are weaned at 17 to 21 days. Females reach sexual maturity within 4 to 6 weeks, but males do not breed until the following spring. Reproductive aspects of *P. longicaudus* are different (Hamilton 1962). The breeding season lasts from February to September. The estrous cycle apparently averages 5.9 days in length. The gestation period is normally 27 to 28.5 days, but may possibly be extended to as many as 48 days if the female is nursing a previous litter. Litter size is only one to three, and the young are not weaned until they are 30 to 35 days old. The lengthy time required for gestation and weaning are thought to be

physiological adjustments to the difficulties of producing energy from a diet of coniferous needles.

Phenacomys was long considered one of the rarest genera of North American mammals. In the mid-1950s, however, many specimens of *P. intermedius* were collected in Canada (Banfield 1974). *P. longicaudus* also is now known from a large number of specimens, but *P. albipes* apparently is extremely rare throughout its entire range (Olterman and Verts 1972).

RODENTIA; MURIDAE; **Genus ARVICOLA** Lacepede, *1799*

Water Voles, Bank Voles

There are three species (Corbet 1978; Hall 1981):

A. terrestris, most of Europe and Siberia, parts of southwestern Asia;

A. sapidus, France, Spain, Portugal;

A. richardsoni, mountains of southwestern Canada and northwestern conterminous United States.

Hall (1981) considered *Arvicola* a subgenus of *Microtus*, but most other recent authorities (including Banfield 1974; Corbet 1978; Nadler et al. 1978; and Jones, Carter, and Genoways 1979) have recognized *Arvicola* as a separate genus.

Head and body length is 120 to 220 mm, tail length is 65 to 125 mm, and weight is 70 to 250 grams. *A. terrestris* is the largest microtine rodent native to the Old World. The fur is thick. The general coloration ranges from moderately light brown to dark brown above, and from buffy to slate gray below. Black individuals sometimes occur locally. Slight aquatic modifications are developed in some species, such as the small, hairy swimming fringe on the foot. The tail is well haired, and the sole of the hind foot is generally haired. The claws on all four limbs are well developed. Both sexes possess flank glands, and females have eight mammae. Individuals seem to grow throughout life. The cheek teeth are ever growing.

There apparently are both aquatic and nonaquatic populations of these voles. The aquatic populations, which include *A. richardsoni* of western North America and most *A. terrestris* in the western parts of Europe, live along the banks of streams, ponds, and canals, where the water level is constant. They dive freely and swim either on the surface or underwater, sometimes near the bottom of the pool or stream, where they enter burrows that terminate below water level. While *A. terrestris* in Great Britain is found mainly in lowland areas (Corbet and Southern 1977), *A. richardsoni* occurs primarily in alpine meadows and along mountain streams at elevations of about 1,500 to 2,000 meters (Banfield 1974). It constructs large burrow systems in stream banks and beneath the snow in winter. Nests of grass or other plant tissues are built in the burrows, slightly above the water level, or are placed under logs, driftwood, or dense vegetative cover on the surface.

In parts of Europe, fossorial, nonaquatic populations of *A. terrestris* predominate (Meylan 1977). These populations occupy grasslands and cultivated areas, construct extensive burrow systems, and leave conspicuous mounds on the surface. The burrows average 40.2 meters in length, contain one or two nests, and, in autumn and winter, contain storage chambers. Grzimek (1975) noted that *A. terrestris* also freely uses the tunnels of moles.

Bank voles are either nocturnal or diurnal and are active throughout the year. They feed on aquatic plants, herbs,

Water vole (*Arvicola terrestris*), photo from A. Van Wijngaarden.

grass, twigs, buds, roots, bulbs, and fallen fruit. Some food is usually stored for the winter, though the animals continue to forage at this time.

Corbet and Southern (1977) gave the average home range length of *A. terrestris* as 130 meters for males and 77 meters for females. Pelikan and Holisova (1969) reported the average number of *A. terrestris* per 100 meters of stream to vary from 4.0 to 9.3 individuals. Meylan (1977) stated that densities of *A. terrestris* sometimes exceed 1,000 individuals per ha. during peaks of the population cycle, which seem to occur every four to six years. Banfield (1974) noted that populations of *A. richardsoni* also fluctuate dramatically from abundance to scarcity.

Although water voles are sometimes found under crowded conditions, they do not form large social groups. Males mark territories with secretions from flank glands, and may fight fiercely when populations are dense, making high, shrill squeaks (Grzimek 1975). The process of scent marking, which involves raking the hind feet over the flank glands, is peculiar to both *A. terrestris* and *A. richardsoni*, but not to *Microtus* (Jannett and Jannett 1974). Females also fight each other, but not to the same extent as males. Meylan (1977) reported that each burrow of fossorial populations of *A. terrestris* is inhabited by a family group, which includes an adult pair plus two generations of young. These populations usually breed from March to October, but winter reproduction is not unusual.

Other populations of *A. terrestris* also breed from the spring to the early fall (Corbet and Southern 1977; Grzimek 1975). The gestation period is 20 to 22 days. There are one to four litters per female per year, and litter size is 2 to 8 young, usually 4 to 6. The newborn weigh about 5 grams, open their eyes at 8 days, and are weaned at 14 days. In a study of *A. richardsoni* in Montana, Brown (1977) determined that one litter was produced in late June or early July and a second in August. The number of embryos per pregnant female averaged 7.85 and ranged from 6 to 10. Banfield (1974) indicated that the females of this species, if born in the season's first litter, could mature in about 5 weeks and produce small litters (with as few as 2 young) of their own during the summer. Corbet and Southern (1977) stated that mean longevity of *A. terrestris* is 5.4 months under natural conditions, but that captive specimens have lived for up to 5 years.

In the Soviet Union *A. terrestris* is hunted for its fur, but may transmit tularemia to persons handling it. The fossorial populations of *A. terrestris* reportedly sometimes cause ex-

tensive damage to pastures, the root systems of orchard trees, grape vines, tulip bulbs, and vegetable crops (Meylan 1977). *A. richardsoni* is usually not an economic problem, but during population peaks may destroy potatoes, alfalfa, and turnips (Banfield 1974).

RODENTIA; MURIDAE; **Genus MICROTUS** *Schrank,* *1798*

Voles, Meadow Mice

There are 6 subgenera and 62 species (Hall 1981; Corbet 1978; Ellerman and Morrison-Scott 1966; Hassinger 1973; Orlov and Kovalskaya 1978; Kovalskaya and Sokolov 1980; Meier and Yatsenko 1980):

subgenus *Microtus*

M. roberti, Caucasus, northeastern Asia Minor;

M. gud, Caucasus, northeastern Asia Minor;

M. nivalis (snow vole), mountainous areas of southern Europe and southwestern Asia;

M. socialis, southeastern Europe to eastern Kazakh and Iran, northeastern Libya;

M. arvalis, Denmark and northern Spain to central Siberia and Asia Minor;

M. subarvalis, parts of eastern Europe;

M. transcaspicus, southern Turkmen, northern Afghanistan, probably northern Iran;

M. kirgisorum, Tien Shan Mountains of Kirghiz;

M. mongolicus, northeastern Mongolia and adjacent parts of Siberia;

M. cabrerae, central and southeastern Spain;

M. maximowiczii, Lake Baikal to upper Amur Basin in southern Siberia;

M. mujanensis, Muja River Valley in south-central Siberia;

M. evoronensis, coast of Lake Evoron in southeastern Siberia;

M. fortis, southeastern Siberia, Manchuria, Korea, eastern China, possibly Sakhalin;

M. sachalinensis, Sakhalin;

M. kikuchii, Taiwan;

M. clarkei, Yunnan (southern China), northern Burma;

M. millicens, Szechwan (central China);

Vole (*Microtus montebelli*), photo by Ernest P. Walker.

M. montebelli, Japan;

M. agrestis (field vole), Britain and northern Spain to east-central Siberia;

M. pennsylvanicus (meadow vole), Alaska, Canada, the conterminous United States as far south as Georgia and New Mexico, north-central Mexico;

M. breweri, Muskeget Island off Massachusetts;

M. nesophilus, Great Gull and Little Gull islands off eastern Long Island (New York);

M. montanus, south-central British Columbia to east-central California and northern New Mexico;

M. canicaudus, northwestern Oregon;

M. umbrosus, Oaxaca (southern Mexico);

M. oaxacensis, north-central Oaxaca (southern Mexico);

M. californicus, western Oregon, most of California, northern Baja California;

M. guatemalensis, extreme southern Mexico, southern Guatemala;

M. townsendii, Vancouver and nearby islands, extreme southwestern mainland of British Columbia to northwestern California;

M. oeconomus, Scandinavia and the Netherlands to eastern Siberia and north-central China, Alaska, northwestern Canada, many northern Pacific continental islands;

M. middendorffi, northern Siberia;

M. longicaudus, eastern Alaska to California and New Mexico;

M. coronarius, Coronation and nearby islands off southeastern Alaska;

M. oregoni, southwestern British Columbia to northwestern California;

M. chrotorrhinus, southeastern Canada, northeastern Minnesota, northern New York and New England, Appalachians;

M. xanthognathus, central Alaska to northern Manitoba and west-central Alberta;

M. ochrogaster (prairie vole), central Alberta to western West Virginia and northeastern New Mexico, also originally in southeastern Texas and southwestern Louisiana;

M. mexicanus, southwestern United States to southern Mexico;

subgenus *Proedromys*

M. bedfordi, known only from the type locality in Kansu (central China);

subgenus *Lasiopodomys*

M. brandti, Mongolia and adjacent parts of Siberia;

M. mandarinus, north-central Mongolia and adjacent part of Siberia south of Lake Baikal, parts of north-central and northeastern China;

subgenus *Neodon*

M. blythi, Tibetan Plateau and Himalayas;

M. sikimensis, western Nepal to northern Burma and southern China;

M. juldaschi, Tien Shan and Pamir Mountains of Central Asia;

M. afghanus, mountains of southern Soviet Central Asia and Afghanistan;

subgenus *Pitymys*

M. subterraneus, France to western Soviet Union;

M. multiplex, southern side of Alps in Switzerland, Austria, France, and Italy;

M. tatricus, Tatra Mountains between Poland and Czechoslovakia;

M. bavaricus, Bavaria (southern Germany);

M. liechtensteini, northwestern Yugoslavia;

M. schelkovnikovi, Talysh and Elbruz Mountains south of Caspian Sea;

M. majori, Caucasus, Asia Minor;

M. savii, France, northern Spain, Italy, Sicily, possibly Macedonia;

M. lusitanicus, southwestern France, northwestern Spain, Portugal;

M. duodecimcostatus, southeastern France, eastern and southern Spain;

M. thomasi, southern coastal Yugoslavia, Greece, probably Albania;

M. pinetorum (pine vole), extreme southern Ontario, eastern United States;

M. quasiater, east-central Mexico;

subgenus *Stenocranius*

M. gregalis, the Eurasian tundra zone from the White Sea to Bering Strait, southern Siberia, Mongolia, parts of Soviet Central Asia and northern and western China;

M. miurus (singing vole), Alaska, Yukon, western Mackenzie;

M. abbreviatus, Hall and St. Matthew islands (Bering Sea).

Pine vole (*Microtus pinetorum*), photo by Ernest P. Walker.

European authorities, as, for example, Corbet (1978), Ellerman and Morrison-Scott (1966), and Grzimek (1975), have tended to maintain *Pitymys* McMurtie, 1831 as a full genus, sometimes including *Neodon* as a subgenus. Most American authorities, including Hall (1981), Jones, Carter, and Genoways (1979), and Banfield (1974), have considered *Pitymys* to be part of *Microtus*. Ellerman and Morrison-Scott (1966) placed *M. afghanus* in the separate genus *Blanfordimys* Argropulo, 1933, and *M. blythi* (=*Pitymys leucurus*) in the subgenus *Phaiomys*. Corbet (1978), however, indicated that both these species should be joined with *M. sikimensis* and *M. juldaschi* in the subgenus *Neodon*. Jones, Carter, and Genoways (1979) apparently considered *M. miurus* of North America to be conspecific with *M. gregalis* of Eurasia. Morlok (1978) argued that *M. irani* (found in Asia Minor, Iraq, and Iran) and *M. guentheri* (found in Greece, Asia Minor, Syria, Palestine, and Libya) are species distinct from *M. socialis*. Such an arrangement had been followed by Ellerman and Morrison-Scott (1966), but not by Lay (1967), Harrison (1972), or Corbet (1978).

Head and body length is 83 to 175 mm, and tail length is 15 to 95 mm. The tail is always shorter than the head and body, usually being less than half as long. Weight is 17 to 20 grams in *M. oregoni,* a small species; 28 to 70 grams in *M. pennsylvanicus;* and 113 to 170 grams in *M. xanthognathus,* one of the largest species (Burt and Grossenheider 1976). The pelage is usually quite long and loose. The general coloration of the upper parts is grayish brown, the darker forms approaching a sooty black, and the paler forms being tawny, reddish, or yellowish. The underparts range from grayish through pale brown to white.

These voles have short, rounded ears that are nearly concealed by the pelage. The subgenus *Pitymys* has become modified for a semifossorial existence through the reduction of the eyes, external ears, and tail, and in having a close, velvety pelage. The foreclaws are somewhat enlarged for digging. The species *M. afghanus* has enormous bullae, which distinguish it from all other microtines. Female *Micro-*

tus usually have eight mammae, but the species *M. ochrogaster* has only six.

As indicated by the following examples from Banfield (1974) many kinds of habitat are occupied: *M. pinetorum,* dry deciduous forest; *M. ochrogaster,* dry grasslands; *M. miurus,* alpine tundra; *M. pennsylvanicus,* wet meadows; *M. montanus,* arid grasslands; *M. townsendii,* salt marshes; *M. oeconomus,* damp tundra; *M. longicaudus,* grassy areas within forests; *M. chrotorrhinus,* damp talus slopes and rocky outcrops; *M. xanthognathus,* forests and bogs; and *M. oregoni,* humid coniferous forests. *M. pinetorum* is mostly fossorial and seldom comes above ground. Its burrows through thick leafmold and loose soil, however, are normally not more than 100 mm below the surface. It builds a globular nest of shredded leaves, about 150 mm in diameter, either under a log or rock, or in a passage of its burrow. Many other species make both burrows and runways. *M. ochrogaster,* for example, excavates an elaborate system of feeding tunnels, about 50 mm in diameter, 50 to 100 mm deep, and up to 110 meters long (Banfield 1974). Its winter nest is located in a globular chamber, 200 mm wide and 150 to 450 mm below the surface. The burrow also contains a food storage chamber, which may be nearly 1 meter in length. Surface runways are made by cutting and trampling vegetation, and scattering dirt from the connecting underground burrow, to make paths 25 to 50 mm wide (Schwartz and Schwartz 1959). The summer nest may be above ground in a clump of vegetation.

Voles are active throughout the year; they do not hibernate. Although they may move about at any time, most species seem to concentrate their activity at night or in the early morning and late afternoon. *M. chrotorrhinus* has sometimes been called a diurnal species, but Timm, Heaney, and Baird (1977) found it to be largely nocturnal. Voles are terrestrial, but many species swim and dive well. They are strictly vegetarian, and within a 24-hour period may consume their own weight in grasses, leaves, twigs, bulbs, tubers, seeds, nuts, or other plant matter. Food may be stored for the winter.

Home range of *Microtus* generally seems to be about

1,000 sq meters or less (Banfield 1974), though in the Northwest Territories the home range of *M. xanthognathus* was found to vary from 116 to 27,500 sq meters (Douglass 1977). In a study of *M. pennsylvanicus* in Kentucky, utilizing radioisotopes for tracking, Ambrose (1969) determined home range to average only about 300 sq meters in males and 140 sq meters in females.

Population densities of *Microtus* vary considerably and seem to run in cycles, at least in some areas. Gaines and Rose (1976) found a two-year cycle for *M. ochrogaster* in eastern Kansas; in one study area the density fluctuated from about 4 to 115 individuals per ha. Maximum reported density in this species is 358/ha. (Rose and Gaines 1978). Dramatic fluctuations, with peaks every three or four years, have been reported for *M. pennsylvanicus* (Banfield 1974). Normal densities in that species are about 37 to 117 per ha. in old fields and 117 to 370 per ha. in marshes, but peak populations may reach 1,000/ha. In the Humboldt Valley of northwestern Nevada, the density of *M. montanus* reportedly reached about 20,000 to 30,000 per ha. during a "mouse plague" in 1907–8 (Hall 1946). Possibly the highest scientifically documented density of *Microtus* is 1,300/ha., which was attained by *M. oregoni* near Vancouver, British Columbia (Boonstra and Krebs 1978). Some other recorded density figures are: *M. pinetorum,* up to 500/ha. (Banfield 1974); *M. ochrogaster,* up to 35/ha. on the xeric prairie of western Nebraska, but up to 250/ha. on the eastern tall-grass prairie (Meserve 1971); *M. oregoni,* an average of 2 per ha. in spring and 4 per ha. in autumn in Oregon forests (Gashwiler 1972); and *M. agrestis,* 5 to 60 per ha. over a four-year period in Sweden (Hansson 1968).

Myllymäki (1977a, 1977b) reported that Scandinavian populations of *M. agrestis* generally have cycles with peaks every three to four years. Populations are divided into reproductive males, which have fixed home ranges and strict territoriality; reproductive females, which have fixed home ranges but little or no territoriality; nomadic subadults; and juveniles (under 35 days old). Individual home range is 400 to 800 sq meters in males and 200 to 400 sq meters in females. The home ranges of males show little or no overlap, while females share even areas of intensive use. The only social bond is that between the female and her nursing offspring. Young males are forced to disperse by resident males, but young females may remain in the vicinity of their mother's home range.

In general, the social life of *Microtus* is something of an enigma. A number of species are known to live in what appear to be colonies of hundreds of individuals. The animals therein, however, may be totally uncooperative and extremely aggressive toward one another. Banfield (1974) stated that *M. ochrogaster* is a socially tolerant rodent, with up to nine individuals (probably a mated pair plus a late-season litter) sharing the same winter nest and food cache. That this species is highly aggressive, however, is indicated by the presence of wounds in 44 percent of the specimens examined by Rose and Gaines (1976). *M. pennsylvanicus* is also socially aggressive; the males fight viciously among themselves and the females tend to dominate the males (Banfield 1974). There is no evidence of lasting pairs or formal social structure (Getz 1972). Perhaps social tolerance in *Microtus* exists up to a certain point, but aggressiveness increases with density and serves to limit and disperse populations in accordance with available resources.

Compared to *Clethrionomys, Microtus* appears slow moving, docile, and easily trapped and tamed. When upset, these voles may emit a high-pitched squeak, gnash their teeth, and either flee or freeze, depending upon their location and previous activity. Several species are known to have a variety of well-defined sounds. The most vocal vole is *M. miurus,* which produces a series of thin, high, pulsating chirps whenever danger threatens (Banfield 1974).

Female voles are polyestrous and usually give birth several times per year. A postpartum estrus may occur. Breeding generally takes place throughout the year in southern parts of the range of *Microtus,* and from spring to early autumn in the north. Under good conditions, however, some winter reproduction may occur even in the north. Captive female *M. pennsylvanicus* have produced up to 17 litters in one year, but for wild individuals in Canada the annual average is 3.5 (Banfield 1974). *M. pinetorum* gives birth to 4 to 6 litters between January and October in Canada, but breeds all year in Louisiana (Lowery 1974) and Oklahoma (Goertz 1971). In the latter area, peak reproduction occurs between October and May, and from 1 to 4 litters are produced. In eastern Kansas, there is no distinct breeding season for *M. ochrogaster,* though there seems to be less reproduction in summer and winter, and during periods of sharp population decline (Gaines and Rose 1976; Rose and Gaines 1978).

Reported gestation periods in *Microtus* range from 19 to 25 days (Hasler 1975). Gestation was measured precisely as 24 days in *M. pinetorum* by Kirkpatrick and Valentine (1970). A wild female *M. socialis* (reported as *M. guentheri*) gave birth to a record 17 young (Hasler 1975). Usual litter size for *Microtus* is 1–12, and there seems a tendency for the number of young to be larger in the north. Some average (and extreme) litter sizes are: *M. mexicanus* in Mexico, 2.3 (1–4) (Choate and Jones 1970); *M. pinetorum* in Oklahoma, 2.6 (1–5) (Goertz 1971); *M. ochrogaster* in Canada, 3.4 (1–7) (Banfield 1974); and *M. miurus* in northern Alaska, 8.2 (4–12) (Bee and Hall 1956). Innes (1978) found litter size in the genus *Microtus* to be significantly correlated with latitude and elevation, though within particular species litter size was not significantly correlated with either variable. The newborn of *M. pennsylvanicus,* which are typical for the genus, weigh about 2.1 grams, open their eyes at 9 days of age, and are weaned at 12 days. Sexual maturity is reached at 25 days by females and at 45 days by males. One study indicated an average longevity of less than 1 month in this species, though some wild individuals live for up to 1 year (Banfield 1974). A captive specimen of *M. socialis* lived for 3 years and 11 months (Marvin L. Jones, Zoological Society of San Diego, pers. comm.).

When their populations are high, voles may become serious pests, because of their destruction of growing grain, forest plantings, and hay (Jackson 1961). During severe winters, *M. pinetorum* may cause damage to fruit orchards by eating the bark from the roots and lower trunks of trees. Such losses are very heavy when the snow and fallen leaves are mounded around the base of the tree. Moles living within the range of the pine vole are often unjustly accused of damaging bulbs, seeds, and other crops, which are actually taken by *M. pinetorum.*

The species *M. nesophilus,* found only on the Gull Islands off the east coast of Long Island, New York, apparently became extinct when its habitat was destroyed by the construction of fortifications during the Spanish-American War in 1898 (Allen 1942). *M. breweri,* which occurs only on tiny Muskeget Island off southeastern Massachusetts, also has occasionally been thought to be in jeopardy. Both species are sometimes considered only subspecies of *M. pennsylvanicus* of the neighboring mainland. The subspecies *M. californicus scirpensis,* restricted to an isolated area of marshes near Death Valley in southeastern California, was long thought to be extinct, but was recently rediscovered (Bleich 1979). The isolated subspecies *M. ochrogaster ludovicianus* is known by 26 specimens collected on the prairies of southwestern Louisiana in 1899 and by a single specimen taken about the same time in southeastern Texas. This subspecies has not since been collected, despite persistent efforts, and may be extinct (Lowery 1974). Several populations of *M. oeconomus* in Europe are in jeopardy because of human habitat modification (Smit and Van Wijngaarden 1976).

RODENTIA; MURIDAE; *Genus LAGURUS Gloger, 1841*

Sagebrush Voles, Steppe Lemmings

There are three species (Corbet 1978; Hall 1981):

L. luteus, Kazakh to southern Mongolia;
L. lagurus, steppe region from Ukraine to western Mongolia and Sinkiang;
L. curtatus, Great Basin region of western United States, Great Plains region from southern Alberta and Saskatchewan to northern Colorado.

The species *L. luteus* has sometimes been put in its own subgenus, *Eolagurus* Argyropulo, 1946.

Head and body length is 90 to 135 mm, tail length is 10 to 30 mm, and weight is 15 to 38 grams. The pelage is longer and usually softer than in *Microtus*. In the species *L. lagurus* the back is light gray or cinnamon gray, with a black stripe along the spine. *L. luteus* has a sand yellow back. In *L. curtatus* the upper parts are pale buffy gray to ashy gray, the sides are paler, and the underparts are silvery or soiled whitish to buffy. The body is somewhat stocky and lemminglike. The short tail is about the length of the hind foot. Some modifications for burrowing in sandy soil are the small ears, short tail, stout claws, and haired palms and soles. Females have eight mammae.

In the Old World these voles inhabit steppes and semidesert areas, and occasionally pastures and cultivated fields. The ecology of *L. curtatus* of the New World was covered in detail by Carroll and Genoways (1980) and Maser (1974). This species is generally restricted to semiarid prairies, rolling hills, or brushy canyons, which are usually dominated by sagebrush and bunch grasses. Its burrows usually occur in clusters, with 8 to 30 entrances each. The entrances are usually in cover, but are sometimes in the open. There are numerous short tunnels at depths of 80 to 460 mm, and a nest chamber up to 250 mm in diameter. The nest is constructed mainly of leaves and stems. Abandoned tunnels of *Thomomys* are often incorporated into the burrow system. The species also utilizes surface runways, which are usually indistinct, and short adjacent tunnels, which may serve as escape holes. As the food supplies around a given burrow system become exhausted, the animals move to another sys-

tem, and then to still another or back to the original system. *L. curtatus* is active throughout the year and 24 hours a day, but its main periods of activity are from 2 to 3 hours before sunset to 2 to 3 hours after full darkness, and from 1 to 2 hours before sunrise to 1 to 2 hours afterward. The diet consists mainly of the green parts of plants, rather than seeds. Freshly cut vegetation may be heaped before use, and often pulled into burrows, but no long-term storage is evident. Winter food storage has, however, been reported in *L. luteus*.

Populations of *Lagurus* may fluctuate substantially. *L. lagurus* is said to increase dramatically in some years and undergo mass migrations. Shubin (1972) called this species the most abundant rodent in the Kazakh steppe zone, and reported population densities of 30 to 50 individuals per ha. There is no evidence of overall social organization in colonies of *Lagurus,* but females often live together while raising the young, and captive female *L. curtatus* sometimes share nests and suckle each other's offspring. Aggressive behavior sometimes occurs in this species, especially in breeding males (Carroll and Genoways 1980).

L. curtatus appears to breed throughout the year, though the main season may extend only from March to early December in the north. The estrous cycle of this species seems to be about 20 days and a postpartum estrus occurs within 24 hours of birth (Carroll and Genoways 1980). A captive female *L. curtatus* produced 14 litters and 80 young in 1 year (Egoscue, Bittmenn, and Petrovich 1970). *L. lagurus* produces up to 5 litters a year, with reproduction occurring mainly from April to October, but sometimes also in winter; the average number of embryos per female is 8.1 and the maximum is 12 (Shubin 1972). *L. luteus* breeds during the summer in the wild, with females giving birth to at least 3 litters of 4 to 10 young each (Shubin 1974). Sexual maturity in this species is attained at only 3 to 4 weeks of age. According to information compiled by Carroll and Genoways (1980), the gestation period of *L. curtatus* is 24 to 25 days, and the number of embryos averages about 5 and ranges from 1 to 13. The newborn of this species are naked and blind, weigh about 1.5 grams, open their eyes at 11 days of age, and are weaned and independent at 21 days. Sexual maturity is usually reached when the females are about 60 days and the males 60–75 days old.

L. lagurus is considered a serious pest, because of damage to crops and pastures (Shubin 1972). This species was once found in the vicinity of Kiev about 300 km west of its present range, but died out in this area around 1900 because of human disturbance of its habitat (Pidoplichko 1973). *L. luteus* also has declined in range, having disappeared in Kazakh, where it was formerly widespread (Corbet 1978).

Sagebrush vole (*Lagurus curtatus*), photos by Murray L. Johnson.

Round-tailed muskrat (*Neofiber alleni*), photo by James N. Layne.

RODENTIA; MURIDAE; **Genus NEOFIBER** True, 1884

Round-tailed Muskrat, Florida Water Rat

The single species, *N. alleni*, inhabits most of Florida and the Okefenokee Swamp area of southeastern Georgia (Hall 1981).

Head and body length is 163 to 227 mm, tail length is 95 to 170 mm, and weight is 187 to 357 grams. The fairly long pelage is composed of glistening guard hairs and short, soft underfur. The general coloration of the upper parts is brown to blackish brown, whereas the underparts are whitish or tinged with buff. The feet are not modified for aquatic life to the pronounced degree found in *Ondatra,* but they do have small swimming fringes. The scaly, scantily haired tail is round. The ears are small and nearly concealed in the fur. Females have six mammae.

The round-tailed muskrat inhabits fresh-water bogs, swamps, lake margins, stream banks, and brackish waters of river deltas. It seems to choose bogs rather than the aquatic habitat preferred by *Ondatra.* It apparently likes thickly vegetated areas. Although more terrestrial then *Ondatra, Neofiber* is a good swimmer. It is active throughout the year and is primarily nocturnal. Information for this account was taken in large part from Birkenholz (1963, 1972) and Layne (1978).

This rodent constructs a nearly spherical to dome-shaped house about 305 to 457 mm high and 180 to 610 mm in diameter. The house rests on a base of decaying vegetation and is composed of tightly woven grass or emergent aquatic plants. In the middle of the house is a nest chamber, about 100 mm in diameter, and there are usually two exits leading downward in opposite directions from the chamber. These passages are dug through the soft, moisture-soaked ground so they actually become water-filled tunnels. A network of surface trails about 75 mm wide entwines the vegetation near the house and leads into the water. Before giving birth, a female will thicken the walls and raise the floor of her house. Some individuals dig burrows near a marsh, and also burrow into the marsh bottom during a drought. From one to six floating feeding platforms, 100 to 150 mm in diameter and containing one or two plunge holes, are located above the water near the house. Sometimes a protective cover is added to a platform.

The round-tailed muskrat is entirely herbivorous. Its main food is maiden cane (*Panicum hemitomon*). It also eats rushes, sedges, sawgrass, and mangrove bark. Food may be consumed where it is found, carried to a feeding platform, or brought back to the nest.

Population densities of 100 to 300 individuals per ha. have been found in favorable areas. The average space used by one individual is only about 50 sq meters. There is little evidence of territoriality and individuals appear socially tolerant, but there is only one adult occupant per house. Breeding seems to peak in the late fall and early winter, but to continue to some extent throughout the year. Females normally produce four or five litters annually. The estrous cycle lasts 15 days or less, the gestation period is 26 to 29 days, and litter size averages 2.2 young and ranges from 1 to 4. The young weigh about 12 grams at birth, open their eyes by 14 to 18 days of age, and reach sexual maturity when 90 to 100 days old.

The round-tailed muskrat is not usually hunted for its pelt. It may occasionally damage crops, but is generally of little economic significance. Some populations have declined because canals have drained marshes and allowed salt-water intrusion.

RODENTIA; MURIDAE; *Genus ONDATRA Link, 1795*

Muskrat

The single species, *O. zibethicus,* occurs naturally in northern Baja California and in most of Canada and the United States; notable areas of absence include most northern tundra regions, much of California and Texas, the coastal parts of Georgia and South Carolina, and all of Florida (Hall 1981).

Head and body length is 229 to 325 mm, tail length is 180 to 295 mm, and weight is 681 to 1,816 grams. The pelage is composed of a short, soft, dense, fine underfur, interspersed with a thick protective coat of long, coarse, dark, shining guard hairs that produce the dominant color of the upper parts. The general coloration ranges from medium silvery brown to dark brown or almost black. The lower surface is generally somewhat paler than the upper parts. The forefeet, hind feet, and tail are dark brown to black.

The muskrat is morphologically adapted to swimming. The hind foot is partially webbed, and along its edge is a row of closely set, short, stiff hairs commonly called the swimming fringe. The scaly, almost hairless tail is flattened laterally and is used as a rudder. The muskrat gets its name from the pronounced musky odor that comes from the secretions of glands in the perineal area. These glands are especially large in males, and open within the foreskin of the penis, where the secretions mix with the urine and are then deposited along the route of the animal. Females usually have two pairs of inguinal and two pairs of pectoral mammae (Lowery 1974).

Except as noted, the remainder of this account is based primarily on information provided by Lowery (1974) and Banfield (1974). The muskrat is found in a wide range of aquatic habitats—fresh- and salt-water marshes, lakes, ponds, rivers, and sloughs. The water should be deep enough that it does not freeze all the way to the bottom, but not so deep that there is no submerged vegetation. The muskrat spends much of its time in the water and can swim nearly 100 meters underwater. In a stressful situation it can remain submerged for up to 17 minutes, but the usual underwater interval is about 2 or 3 minutes. The muskrat is largely nocturnal and crepuscular, but is sometimes seen abroad by day, especially in the winter.

Two main kinds of shelter are constructed. When living along a flowing stream or canal, the muskrat makes a burrow in the bank or dike. From an entrance located below the lowest depth to which the water freezes, the tunnel extends for up to 10 meters to a dry chamber located above the high water level. Several burrows may be interconnected.

When living in a flat marsh or swamp, the muskrat builds a mound-shaped house of grass, cattails, or other vegetation, and keeps the outside plastered with mud. The house is situated on a stump or clump of vegetation, or directly on the ground. It is usually about one or two meters wide and one meter tall. A dry nest is located inside. One or more tunnels lead from the bottom of the house, pass under the substrate, and sometimes exit underwater.

Houses may be destroyed by spring flooding, and in some areas the muskrat uses a burrow during the spring and builds a new house in the late summer or early fall. The muskrat also may construct fairly well defined channels through vegetation, slides along stream banks, and feeding platforms above the water of a marsh or pond. In the north a characteristic winter structure is the "push-up," a dome of frozen vegetation over a hole in the ice, where the muskrat can come up for air and to eat what has been found underwater. The most important foods are cattails (*Typha*) and bulrushes (*Scirpus*) in North America, and the water lily (*Nymphaea alba*) in Europe (Willner, Chapman, and Goldsberry 1975). The muskrat also takes many other kinds of vegetation and a substantial amount of animal matter. In the delta marshes of Louisiana, studies indicate that 70 percent of the diet is cattails, 15 percent is grass, 10 percent is other vegetation, and 5 percent is crabs, crayfish, mussels, and small fish.

Population density varies from about 7 individuals per ha. in open ponds to 87 per ha. in cattail marshes. Population levels fluctuate periodically and excessive numbers may denude marshes of their vegetation. Shifts of populations sometimes occur, especially in the spring, when there is competition for denning sites and mates. Individuals have been known to travel nearly 13 km. MacArthur (1978), however, found few winter movements to extend beyond 150 meters from the den.

The muskrat is a vicious fighter and utters a whinish growl when disturbed. It appears to be monogamous, but especially quarrelsome, during the breeding season. When the female is suckling young, the male lives in a separate nest on the outside of the house. Later, the mated pair and one litter of young work together to maintain the house and to defend a feeding territory about 60 meters in diameter. When the young reach sexual maturity they are evicted, but may move only about 8 meters away to build houses of their own.

Muskrat (*Ondatra zibethicus*), photo by Ernest P. Walker.

Females are polyestrous, with the estrous cycle lasting an average of 6.1 days. A postpartum estrus may occur. In Canada the breeding season extends from about March to September and 2 litters are normally produced. In the southern United States reproduction is continuous, with peaks in November and March, and there are usually 5 or 6 litters per year. The gestation period is 25 to 30 days. Litter size varies from 1 to 11 young, but averages as low as 2.4 in some parts of the south and as high as 7.1 in the north. The newborn are naked and blind, and weigh about 22 grams. Their eyes open at 14 to 16 days of age, they are weaned at 21 to 28 days, and they are independent when a month old. In the south some females may reach sexual maturity when only 6 to 8 weeks old, but in Canada the young are not able to breed until the year after they are born. Longevity is thought to be about 3 years in the wild and up to 10 years in captivity.

The muskrat yields a thick, glossy, durable fur, but was long not fully appreciated by commercial interests. In the early 1900s muskrat skins sold for only 5 or 10 cents each. Subsequent large-scale exploitation of fur resources in the marshes of Louisiana led to that state's becoming the foremost fur-producing area in North America, and to the muskrat's often surpassing all other fur bearers of the continent in total annual market value. In recent years the muskrat has lost this position to the raccoon, but continues to be of major importance in the fur trade. For the 1976/77 season, 45 states of the United States reported a total harvest of 7,148,370 muskrat skins, with an average value or $4.75. In Canada the take was 2,554,879 skins, with an average value of $4.20 (Deems and Pursley 1978). The muskrat also is sought by some persons for use as food.

In 1905 a few muskrats escaped from captivity in central Europe, and their descendants eventually spread over a wide area. Later, when the value of muskrat fur had increased, deliberate introductions were made in several parts of Europe and Asia. The species now ranges from Sweden and France to eastern Siberia and central Honshu. It is considered a serious pest, especially in the Netherlands, because of its undermining of dikes and dams (Grzimek 1975; Corbet 1978; Danell 1977). Introductions also have been made to California, islands off the coast of British Columbia, and southern South America (Willner et al. 1980).

RODENTIA; MURIDAE; *Genus LEMMUS Link, 1795*

True Lemmings

There are three species (Corbet 1978; Hall 1981):

L. lemmus, Norway, Sweden, Finland, extreme northwestern Soviet Union;

L. sibiricus, arctic parts of Soviet Union from the White Sea to Bering Strait, Alaska to Baffin Island and central British Columbia;

L. amurensis, parts of eastern Siberia.

Head and body length is 100 to 135 mm, tail length is 18 to 26 mm, and weight is 40 to 112 grams. The coloration remains the same throughout the year. The upper parts are grayish or brownish, with a heavy tinge of buffy and dark brown in some species. The underparts are light gray, buffy, or brownish. The subspecies *L. sibiricus nigripes* of St. George Island in the Bering Sea has black feet throughout the year. True lemmings are heavily furred, stockily built, and well adapted to the rigorous conditions of their environment. They have short tails, and ears so small that they are almost concealed by the fur. A claw on the thumb, which is most conspicuous in winter, is long and flat.

True lemmings are found mainly in alpine and tundra areas. They are active both day and night throughout the year. Apparently, in some areas there are regular movements in the spring and autumn to find suitable nesting sites and materials (Grzimek 1975). During the summer the lemmings construct burrows up to about 1 meter in length and about 50 to 300 mm below the surface. A nest, about 150 to 180 mm in diameter and lined with grass and lemming fur, may be located in one chamber of the burrow, and there are other chambers for retreat and defecation. Winter nests are hollow

True lemming (*Lemmus sibiricus*), photo by Ernest P. Walker.

cones or balls of dry vegetation, usually placed on the surface of the ground under the snow and from which subniveal trails radiate to feeding areas (Banfield 1974; Grzimek 1975). All lemmings feed on vegetable matter—sedges, grasses, bark, leaves, berries, lichens, and roots.

In a study of *L. sibiricus* in northern Alaska, Banks, Brooks, and Schnell (1975) found home range to vary from 0.0004 to 2.85 ha., and to average about 1 ha. for males and 0.5 ha. for females. Banfield (1974) wrote that densities vary from about 50 to 325 individuals per ha. in an expanding lemming population. There are regular fluctuations with peak numbers reached every two to five years. Eventually an excess population may overutilize the food supply and force the lemmings to begin a large-scale movement. At such times the animals may swim across rivers and lakes, and appear in human settlements. Many of them die because they are unable to find enough suitable habitat. There have been reliable accounts, especially in northern Europe, of great swarms of lemmings moving over wide areas and eventually plunging into the sea to drown. Such phenomena appear to be unusual and to represent a modification of the typical microtine population cycle, rather than any special behavioral aspect of *Lemmus* (Grzimek 1975).

True lemmings are prolific animals. In a captive colony, Rausch and Rausch (1975) found litters to be delivered regularly throughout the year. Some females became pregnant when they were only 14 days old. One pair produced eight litters in 167 days, after which the male died. The breeding season in the wild usually lasts from spring to autumn, with each female having one to three litters, but under ideal conditions reproduction may continue through the winter. A postpartum estrus may take place. The gestation period is 16 to 23 days, and litter size is 1 to 13, averaging 7.3 young in Canada. The newborn weigh about 3.3 grams, open their eyes when 11 days old, and are weaned at 14 to 16 days. Life expectancy is around 1 to 2 years (Grzimek 1975; Banfield 1974; Hasler 1975).

RODENTIA; MURIDAE; *Genus MYOPUS Miller, 1910*

Wood Lemming

The single species, *M. schisticolor,* occurs in the coniferous forest zone from Norway to eastern Siberia and northern Mongolia (Corbet 1978).

Head and body length is 75 to 110 mm, and tail length is 14 to 18 mm. Weight is 20 to 30 grams (Grzimek 1975). The pelage is soft, dense, and slaty black above, with a definite reddish brown area on the center of the back extending from the shoulders to within about 15 mm of the base of the tail. The remainder of the coat is uniformly a dark slaty gray, slightly paler on the ventral surface. The upper parts have a peculiar metallic luster that is produced by silvery tips on the shorter hairs, with an indistinct showing of black guard hairs. The feet and tail are black, but the hairs on the undersurface of the tail have a silvery luster. The tail is heavily furred. The palms are naked. The hind feet are densely haired behind the pads, but are naked in front of the pads, like the palms of the forefeet. There is no substantial difference in the coloration of the sexes, nor is there much seasonal variation in color. The winter coat is slightly longer than the summer pelage.

This tiny, thickset rodent has ears that are small and project little beyond the fur, but that are well developed, rounded, and well haired. The ears have valves that can regulate the size of the ear openings. The thumb of the hand is small, but bears a large, flattened nail with parallel sides and a notch at the end, somewhat resembling the thumbnail of *Lemmus*, but being smaller. Female *Myopus* have eight mammae.

The wood lemming lives in mossy bogs and coniferous forests at elevations of 600 to 2,450 meters. Its runways have been found in moss, under the roots of trees, and beneath fallen tree trunks. It is known to eat certain mosses (*Dicranum*), stems of red wortleberry, and the bark of juniper. It is apparently rare and is only taken by accident or in years when its populations become unusually large. Its numbers fluctuate and it makes local migrations. Following a study of captives, Ilmen and Lahti (1968) reported that one female delivered 11 litters in 299 days. Average time between litters was 25 days and litter size was one to six young. Females could successfully mate when about 1 month old. Mysterud, Viitala, and Lahti (1972) suggested that breeding during the winter takes place in Norway.

RODENTIA; MURIDAE; *Genus SYNAPTOMYS Baird, 1857*

Bog Lemmings

There are two species (Hall 1981):

S. cooperi, southeastern Canada, northeastern United States;

S. borealis, Alaska and Canada south of the tundra, parts of the extreme northern conterminous United States.

Head and body length is 100 to 130 mm, tail length is 16 to 27 mm, and weight is 21 to 50 grams. The general coloration above is a grizzled, grayish cinnamon brown, composed of a mixture of gray, yellowish brown, and black in varying degrees. The underparts are soiled whitish over the slate-colored bases of the hairs. The tail is brownish above and whitish below. The body form is thickset, with long, loose, and coarse pelage. The incisors are orange. The upper incisors have a shallow longitudinal groove at the outer edge. The nail of the first finger is flat and strap shaped. There are three pairs of mammae in female *S. cooperi* and four pairs in female *S. borealis.*

These animals seem to occupy small, local, isolated areas; their spotty occurrence makes a general map of their range misleading. In the southern parts of their range they are limited mainly to scattered cold bog and cold spring areas. Farther north they have a more general distribution, usually living in moist areas, though in the humid Pacific Northwest they have been found near rocky cliffs. They do not hibernate and are active at any hour of the day or night. They make surface runways and subsurface tunnels, as do voles, though they occasionally occupy the tunnels and use the runways made by other small mammals. Nests about 100 to 150 mm in diameter are concealed in hollows under sphagnum mounds, logs, or stumps, or in grass tussocks, and are made of dry, shredded grasses and sedges (Banfield 1974). The powerful jaws of bog lemmings suggest that they gnaw through tangles of roots, moss, and soil. Apparently they feed largely on plant tissues, mainly green parts of low vegetation, and to a lesser extent on slugs, snails, and other invertebrates.

Population density varies from about 6 to 35 individuals per ha., with regular cycles from a low in the spring to an autumn peak. Individual home range varies from about 0.08 to 0.20 ha., with males usually traveling more extensively than females. The area around the nest is defended (Banfield 1974). It is believed that each sex makes its presence known by secretions from glands in the groin. The voice has been described as a sharp, short squeak.

According to Banfield (1974), the breeding season of *S. cooperi* usually lasts from April to September, but may ex-

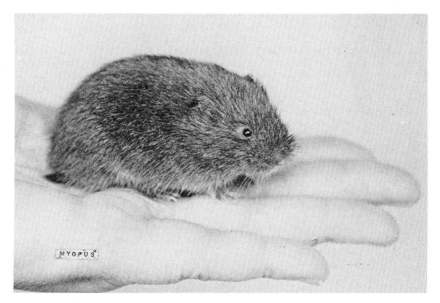

Wood lemming (*Myopus schisticolor*), photo by Erna Mohr.

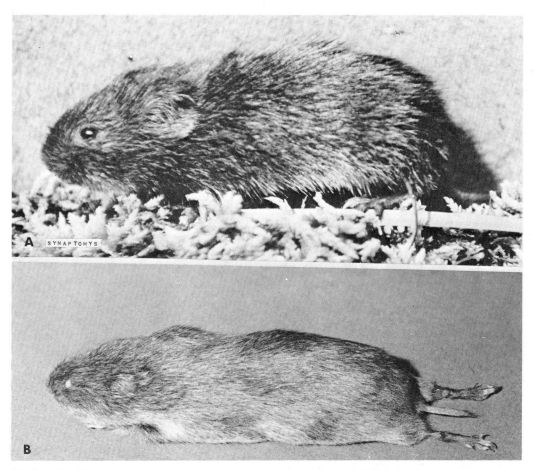

Bog lemming (*Synaptomys cooperi*): A. Photo by Paul F. Connor; B. Photo by Elizabeth Horner and Mary Taylor of specimen in Harvard Museum of Comparative Zoology.

tend through the winter under ideal conditions of food and cover. The breeding season of *S. borealis* lasts from May to August. Females are seasonally polyestrous and may have a postpartum estrus. A captive female bore six litters in 22 weeks, but wild females probably have litters every 67 days during the spring and summer. The gestation period is 23 days. Litter size ranges from one to eight young, averaging three in *S. cooperi* and four in *S. borealis*. The newborn weigh about 3.7 grams, open their eyes after 10 to 12 days, and are weaned at 16 to 21 days. Males reach sexual maturity when 5 weeks old.

Handley (1980*b*) stated that the subspecies *S. cooperi helaletes*, restricted to the Dismal Swamp of Virginia and North Carolina, is at least endangered and possibly extinct. Although seemingly common in the 1890s, it subsequently declined, probably because of habitat destruction and the drying up and overgrowth of sphagnum bogs. Recently, Rose (1981) discovered a living population of this subspecies.

RODENTIA; MURIDAE; **Genus DICROSTONYX** Gloger, *1841*

Collared or Varying Lemmings

There are four species (Hall 1981; Corbet 1978):

D. torquatus, tundra region of northeastern Europe and Siberia;
D. groenlandicus, tundra region of Alaska, northwestern and north-central Canada, and Greenland;
D. exsul, St. Lawrence Island (Bering Sea);
D. hudsonius, northern Quebec, mainland Newfoundland.

Corbet (1978) and Jones, Carter, and Genoways (1979) considered *D. groenlandicus* conspecific with *D. torquatus,* but Hall (1981) did not agree.

Head and body length is 100 to 157 mm, tail length is 10 to 20 mm, and weight is 30 to 112 grams. In summer the general coloration above varies from light grayish buff to dark gray, strongly tinged with buffy to reddish brown. Some *Dicrostonyx* are colored a deep, rich buff, and others are merely grayish tinged with reddish brown. The underparts are whitish, grayish, or buffy. There are definite dark lines from the top of the head along the middle of the back, and on the sides of the head. In winter all species are entirely white. They are the only rodents that turn white in winter. Both the winter and summer coats are thick and heavy.

These lemmings are short, stockily built creatures. The third and fourth claws of their forefeet are unusually large and well adapted to digging. In summer these claws are about normal in appearance, but with the approach of winter they develop a peculiar double effect in a vertical plane, which makes them very strong and well suited for burrowing in the frozen earth or in snow and ice. This remarkable development of the third and fourth claws in winter distinguishes *Dicrostonyx* from all other small rodents. The ears of this genus are so small that they are little more than rims, completely hidden in the fur. Females have eight mammae.

These lemmings occur on the treeless arctic tundra, mainly in dry, sandy, and gravelly areas with some plant cover. They are active throughout the year. During the summer they live in shallow underground burrows or under rocks. The burrows are simple structures, up to 6 meters long, but usually less than 0.5 meters (Brooks and Banks 1973). The tunnels are about 100 to 200 mm deep and 50 to 75 mm wide. They lead to nesting chambers, about 100 to 150 mm in diameter and lined with dry grasses or other soft material. There are also small, bare resting chambers and toilet areas (Banfield 1974). An individual male may use up to 50 burrows, many only for retreat but several with nests (Brooks and Banks 1973). Faint runways on the surface of the ground may also be made, but these paths are not as pronounced as those of some other microtines. In the winter collared lemmings place their grass nests on the surface of the earth under a snowbank, or even in the middle of a snowbank. They dig their tunnels through the snow with their special winter claws (Banfield 1974). They swim readily. In the winter they eat buds, twigs, and bark, and their summer diet consists of fruits, flowers, grasses, and sedges.

Population densities fluctuate dramatically, from 0.6 to

Collared lemming (*Dicrostonyx hudsonius*); Inset: left foreclaws in winter, photos by Ernest P. Walker.

400 individuals per ha. (Banfield 1974), but normal peak densities are 15 to 40 per ha. (Brooks and Banks 1973). At times these lemmings become so plentiful that they are forced to make short migrations in search of food, but they rarely make the lengthy mass movements sometimes observed in *Lemmus*. In a radio-tracking study of *D. groenlandicus* in Manitoba, Brooks and Banks (1971) found home range to average 2.03 ha. for males and only 0.16 ha. for females. Although nesting burrows are defended (Banfield 1974), and captive males fight and form dominance relationships, there is no clear evidence of territoriality (Bowen and Brooks 1978). Several distinctive sounds have been identified, including a "squeal-squawk-grind" complex associated with agonistic behavior (Brooks and Banks 1973).

The breeding season usually extends from early March to early September, but, under ideal winter conditions, may begin in January (Banfield 1974). Females are seasonally polyestrous and have a postpartum estrus. An average estrous cycle of 9.6 days has been reported for *D. groenlandicus* (Hasler 1975). In captivity, females may give birth up to five times from April to September, but in the wild they probably produce only two or three litters per year (Banfield 1974). They are very aggressive after giving birth, often attacking and occasionally killing males that attempt to mate with them. Without such aggressiveness, however, the males may kill the young already born. The normal gestation period is 19 to 21 days, but gestation can be extended to as long as 31 days if the female is nursing a previous litter and is able to protect it from the male (Mallory and Banks 1978). Litter size ranges from 1 to 11 (Brooks and Banks 1973), and averages 3.3 in captivity (Hasler and Brooks 1975) and 4.5 in the wild (Banfield 1974). The young weigh an average of 3.8 grams at birth, open their eyes after 12–14 days of life, and are weaned and disperse at 15 to 20 days (Banfield 1974; Brooks and Banks 1973). In a study of captives, Hasler and Banks (1975) found the youngest female to give birth at an age of 58 days and the oldest at 580 days; the respective figures for males were 85 and 683 days. Captives have lived for as long as 3 years and 3½ months (Banfield 1974).

The beautiful white winter coats of *Dicrostonyx* are used by the Eskimos for garment trimming. The Eskimo children use the little skins to make doll clothes.

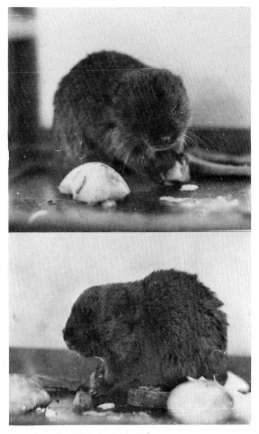

Long-clawed mole-vole (*Prometheomys schaposchnikowi*), photos by R. Zimina.

RODENTIA; MURIDAE; Genus PROMETHEOMYS Satunin, 1901

Long-clawed Mole-vole

The single species, *P. schaposchnikowi*, occurs in the Caucasus and extreme northeastern Asia Minor (Corbet 1978).

Head and body length is 125 to 160 mm, and tail length is 45 to 65 mm. Grzimek (1975) gave the weight as 70 grams. The upper parts are dull grayish brown, suffused with cinnamon, while the underparts are pale plumbeous gray with a cinnamon tint. The fully haired tail is unicolored brownish gray. This stocky rodent has a round head and large claws on the forefeet. The claw of the middle digit of the hand is the longest, being over 5 mm in length. The eyes and ears are reduced. The upper incisors are grooved and the molar teeth are rooted. The tail is swollen at the tip. Females have eight mammae.

Prometheomys is found in alpine and subalpine meadows and open places at elevations of 1,500 to 2,800 meters (Grzimek 1975). It builds complex burrows and throws the excavated dirt on top of the ground in the form of heaps. The

claws, rather than the teeth, are used in digging. A grass-lined nest appears to be the center of the burrow system, as many tunnels radiate from there, gradually approaching the surface. This rodent is active throughout the year. Zimina and Yasny (1977) found maximum activity to be from 1100 to 1300 hrs, and the diet to consist entirely of the green parts of plants found above the ground. *Prometheomys* also is said to feed on subterranean roots.

Dzneladze (1974) reported population densities of 3 to 32 individuals per ha. Grzimek (1975) stated that several older males live together with the females in a single burrow, and that litter size is 2 to 6 young. Zimina and Yasny (1977) reported that females produce two litters during the summer, with an average of 3 young each.

RODENTIA; MURIDAE; Genus ELLOBIUS Fischer, 1814

Mole-voles or Mole-lemmings

There are two species (Corbet 1978):

E. talpinus, steppe region from the Ukraine to Mongolia and northern Afghanistan;

E. fuscocapillus, eastern Turkey, Iraq, Iran, southern Turkmen, Afghanistan, Pakistan.

Head and body length is 100 to 150 mm, and tail length is 5 to 22 mm. The upper parts are dull or bright cinnamon,

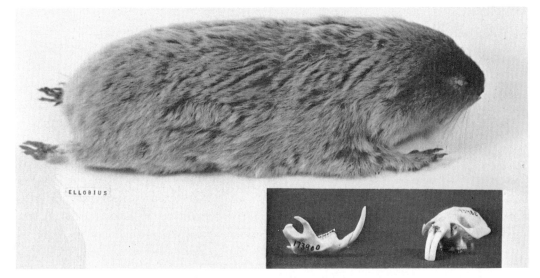

Mole-vole or mole-lemming (*Ellobius fuscocapillus*); Inset: *E. talpinus;* photos by P. F. Wright of specimen in U.S. National Museum.

pinkish buff, buff, or brownish, and the underparts are grayish, brownish, or whitish. A dark face patch may be present. The fur is velvety or plushlike, resembling the pelage of other mammals that spend most of their lives underground.

These burrowing rodents have a round head and reduced eyes and ears. The claws are small. The incisors are large, those of the upper jaw protruding forward and downward; earth is loosened and roots are cut with the incisor teeth. *Ellobius* is probably the most specialized fossorial microtine genus.

According to Grzimek (1975), mole-voles live mainly in steppe and semidesert areas, only rarely moving into woodlands. In Mongolia they prefer the deep, moist soil on the banks of lakes and streams. They construct a system of branching tunnels, including food storage rooms, at a depth of 20 to 30 cm, and a nest chamber at a depth of at least 50 cm. They feed on the underground parts of plants; tulip bulbs and pulpy tubers are especially favored. The gestation period is 26 days. Females produce six or seven litters annually, each with three to five young. The offspring remain in the parents' nest for 2 months before becoming independent, and attain sexual maturity by 96 days of age.

Mole-voles damage crops in some areas. They have some value in the fur trade.

RODENTIA; MURIDAE; **Genus DENDROMUS** A. Smith, *1829*

African Climbing Mice, Tree Mice

There are six species (Meester and Setzer 1971; Dieterlen 1969; Kingdon 1974*b*; Meester 1973):

D. lovati, mountains of Ethiopia;
D. nyikae, central Angola to Malawi, eastern escarpment of Rhodesia and Transvaal;
D. melanotis, Guinea to Ethiopia, and south to South Africa;
D. mesomelas, Cameroon to Ethiopia, and south to South Africa;

D. mystacalis, southeastern Nigeria to Ethiopia, and south to South Africa;
D. kahuziensis, known only by a single specimen from Mount Kahuzi in eastern Zaire.

The following subgeneric terms are sometimes applied: *Chortomys,* for *D. lovati;* and *Poemys,* for *D. nyikae* and *D. melanotis* (see Rosevear 1969).

Head and body length is 50 to 100 mm, tail length is 65 to 132 mm, and weight is 5 to 21 grams. The pelage is soft and woolly. The upper parts are grayish or brownish. There is usually one black dorsal stripe, but *D. lovati* has three stripes on the back. The underparts are whitish or yellowish. The finely scaled and scantily haired tail is either entirely dull brownish or slightly bicolored, in which case the underside is paler. Like many nocturnal tree mammals, these mice usually have well-defined, dusky "spectacles" encircling the eyes. The tail is remarkably long and is semiprehensile. The forefeet have only three well-developed digits, but the hind feet are normal. The upper incisors are grooved. Females have eight mammae.

African climbing mice are found in a variety of habitats at elevations ranging from sea level to more than 4,300 meters. Although these animals are sometimes said to occur mainly in trees, Rosevear (1969) pointed out that positive and reliable recorded evidence that they are arboreal is almost nonexistent. Some populations are known to spend most of their time on the ground or even in subterranean burrows. Kingdon (1974*b*) listed the following ecological conditions: *D. mesomelas,* the driest and most terrestrial habitats whenever other species of *Dendromus* are present, but very wet areas of grass and herbaceous growth in alpine and subalpine zones and in other places where other species of *Dendromus* are absent; *D. nyikae,* savannahs and woodlands; *D. melanotis,* usually relatively open, dry savannahs, often in short grass and scrub growing on sandy plains; and *D. mystacalis,* forests and other densely vegetated areas. In Botswana the species *D. melanotis, D. mesomelas,* and *D. nyikae* have all been found in swampy lowlands (Smithers 1971).

Most species climb with great agility, though Yalden, Largen, and Kock (1976) observed that *D. lovati* lives in grasslands, nests beneath boulders, and shows no inclination to climb. All species seem to be primarily nocturnal and to

African climbing mouse (*Dendromus mesomelas*), photos by John Visser.

spend daylight hours in globular nests of shredded vegetation (Kingdon 1974b; Rosevear 1969). *D. mystacalis*, evidently the species best adapted to climbing, may build its nest 3 meters or more above the ground in banana trees or thatched roofs, though more often nests are placed somewhat lower in thick vegetation. *D. melanotis* and *D. nyikae* may build their nests in underground burrows, apparently to avoid annual fires. The burrows of *D. melanotis* are simple, being about 30 to 60 cm deep and consisting of an open entrance tunnel, a small grass-lined nest chamber, and an emergency exit hole opposite the entrance. African climbing mice also have been known to reline and occupy the nests originally built by sunbirds and weaverbirds. Their diet consists of seeds, berries, insects, small lizards, birds' eggs, and nestlings.

Recorded population densities have varied as follows: *D. mystacalis*, from 0.6 individuals per ha. in savannah to 19.8/ha. in herbaceous growth bordering marsh; and *D. mesomelas*, from 0.5/ha. in herbaceous growth bordering marsh to 6.6/ha. in savannah (Kingdon 1974b). In a laboratory study, Choate (1972) found *D. mesomelas* to be a socially tolerant species, with groups containing individuals of all ages and sexes, including several mature males, living together without antagonism. Rosevear (1969) noted, however, that male and female *D. mystacalis* are solitary and live in separate nests, and Smithers (1971) stated that individuals of *D. melanotis* are territorial and may fight to the death.

Evidence of reproductive activity in *Dendromus* has been collected throughout the year, though apparently breeding is restricted to certain periods in many populations. The following information on seasonality is taken from Kingdon (1974b), Delany (1975), and Smithers (1971): *D. nyikae*, a lactating female collected in August and a litter of young found in June in Tanzania; *D. melanotis*, probably a strictly seasonal breeder where subject to climatic fluctuation and fires, a pregnant female taken in Uganda in July, young observed from December to March in Rhodesia; *D. mesomelas*, probably a seasonal breeder, many young found in February in Tanzania; *D. mystacalis*, near the equator a birth peak from November to January, but scattered records of breeding in other months in eastern Zaire and Uganda. Overall litter size in the genus is two to eight young. In *D. melanotis* the gestation period is 23 to 27 days and the young leave the nest when 30 to 35 days old (Delany 1975). A captive *D. mesomelas* lived for 3 years and 3 months (Marvin L. Jones, Zoological Society of San Diego, pers. comm.).

RODENTIA; MURIDAE; Genus *MEGADENDROMUS* Dieterlen and Rupp, 1978

The single species, *M. nikolausi*, is known from the highlands of eastern Ethiopia (Dieterlen and Rupp 1978).

In three specimens, head and body length is 117 to 129 mm, tail length is 92 to 106 mm, and weight is 49 to 66 grams. The upper parts are brown, except for a black dorsal stripe extending from behind the ears to the rump. The underparts are light brown with some gray showing through. The sides of the head and the shoulders are reddish brown, and there are dark "spectacles" around the eyes. The hands and feet are silvery gray. The tail is whitish gray, except for a thin black stripe on top.

Megadendromus has a striking resemblance to *Dendromus* in color pattern, but differs in being much larger and in having a tail length that is only about 80 percent of the head and body length. There also are important cranial and dental distinctions, including the presence of two lingual cusps on the first upper molar of *Megadendromus*.

This genus inhabits mountainous areas, thickly vegetated with grasses and scrub, at elevations of 3,000 to 4,000 meters. All specimens were collected at night.

Long-eared or gerbil mouse (*Malacothrix typica*), photo by John Visser.

RODENTIA; MURIDAE; *Genus MALACOTHRIX Wagner, 1843*

Gerbil Mouse, Long-eared Mouse

The single species, *M. typica,* occurs in southern Angola, Namibia, southern Botswana, and South Africa (Meester and Setzer 1971; Smithers 1971).

Head and body length is 65 to 95 mm, and tail length is 28 to 42 mm. Smithers (1971) gave the weight as 7 to 13 grams. The pelage is long, dense, and silky. The upper parts range from pale brownish or buffish to reddish brown, and have a more or less distinct dark dorsal stripe and a dark crown spot. There is a sprinkling of black or brown hairs on the back and sides. The underparts, feet, and tail are white, the latter being scantily haired. The hind feet have only four toes each, the soles are hairy, the limbs are slender, and the ears are large. The teeth are similar to those of *Dendromus* and *Steatomys.* Females have eight mammae.

The gerbil mouse is associated with sandy plains and inland grassy velds. It is nocturnal and terrestrial. It constructs a burrow by digging a passageway that slants downward to a depth of 60 to 120 cm, where a chamber is excavated and the soil is used to fill up the original tunnel. A nest is made of grass and often of feathers. From the nest chamber a new tunnel is made to the surface. The soil from this tunnel is also transported to the original passageway beyond the chamber, so that a mound of dirt is not formed near the tunnel exit. The deceiving pile of freshly excavated soil is some distance away. The gerbil mouse may travel up to 4 km from its burrow. It generally follows cattle tracks, foot paths, or sandy roads, and its presence is frequently noted by tiny footprints along these routes. It returns to its burrow before dawn. According to Smithers (1971), the natural diet appears to consist entirely of green vegetable matter.

Smithers (1971) stated that the breeding season in Botswana appears to fall during the warmer, wetter months of the year, from about August to March. More than one litter may be produced during this season. The gestation period evidently can vary from 22 to 35 days, but usually is around 22 to 26 days. Litters consist of two to seven young, which are born naked and weigh about 1.1 gram. They may reach sexual maturity at an age of 51 days.

RODENTIA; MURIDAE; *Genus DENDROPRIONOMYS F. Petter, 1966*

The single species, *D. rousseloti,* is known only by four specimens from the city of Brazzaville, southern Congo (Meester and Setzer 1971).

Dendroprionomys, as implied by its generic name, exhibits certain characters of *Dendromus* and others of *Prionomys,* and appears to be intermediate to these other genera.

The pelage is velvety and molelike, as in *Prionomys,* and the dorsal coloration is brownish, like that of *Dendromus.* The flanks are paler, and the line of demarcation between the dorsal and ventral surfaces is tawny. The venter is entirely white, but the bases of the hairs are gray for half their length. A black area underlines the eyes.

The ears appear to be relatively well developed and are pigmented gray. The tail is scaled and is longer than the head and body. The forefeet are provided with four well-developed toes, as in *Prionomys.* The hind feet have five digits, as in both *Prionomys* and *Dendromus.* The dentition is such as to imply a diet that is at least partially insectivorous. The upper incisors are furrowed longitudinally.

Dendroprionomys rousseloti, photo by F. Petter.

RODENTIA; MURIDAE; *Genus PRIONOMYS Dollman, 1910*

Dollman's Tree Mouse

The single species, *P. batesi,* is known only from Bitye in southern Cameroon, and from around Bangui in the Central African Republic (Meester and Setzer 1971).

Head and body length is about 60 mm, and tail length is about 100 mm. The short, velvety fur resembles that of shrews. The coloration of the upper parts is pinkish chocolate. The sides are paler and blend almost imperceptibly into the grayish pinkish buff underparts. The face is slightly lighter than the remainder of the upper parts. The sides of the face below the eyes are washed with pinkish buff, and the eyes are encircled with narrow black rings. The scantily haired, finely scaled tail is dark brown, almost black. There appears to be no hair on the dorsal tip of the tail, which suggests that it may be prehensile.

The size of *Prionomys* is comparable to that of a large *Dendromus.* The ears are rather small and rounded. The

Dollman's tree mice (*Prionomys batesi*), photo by F. Petter.

forefeet have four well-developed fingers, though the thumb is absent. The second and fourth fingers are moderately elongated and have short claws. The outer finger is only half the length of the two middle fingers and bears a small nail. The hind foot has five toes, all of which are armed with claws. The inner toe is about half as long as the middle digit, and its claw is smaller and more blunt than those of the other toes. The upper incisor teeth are short, slender, ungrooved, and project forward. The slender lower incisors are sharply pointed.

The type specimen was taken at Bitye, Ja River, at an elevation of about 600 meters. Meester and Setzer (1971) referred to *Prionomys* as a burrowing animal.

RODENTIA; MURIDAE; **Genus STEATOMYS** Peters, *1846*

Fat Mice

There are six species (Meester and Setzer 1971; Swanepoel and Schlitter 1978; Anadu 1979*a*):

S. pratensis, Cameroon to southern Sudan, and south to northern Namibia and eastern South Africa;

S. caurinus, Senegal to Nigeria;

S. jacksoni, western Ghana, southwestern Nigeria;

S. parvus, southern Sudan and Somalia to northern Namibia and eastern South Africa;

S. cuppedius, Senegal, southern Niger, northern Nigeria;

S. krebsi, Angola, western Zambia, northern Botswana, South Africa.

Genest-Villard (1979) recognized *S. opimus* of Cameroon, the Central African Republic, and Zaire as a species distinct from *S. pratensis*.

Head and body length is 65 to 145 mm, tail length is 34 to 59 mm, and weight is 5 to 70 grams (Swanepoel and Schlitter 1978; Meester and Setzer 1971). The pelage is thick, short, soft, and silky. The general coloration of the upper parts ranges through various shades of brown and buff, frequently mixed with black. The underparts are white. The dorsal surface of the scantily bristled tail is generally colored like the back, and the ventral surface is white, but in at least one form the tail is all white.

Fat mice (*Steatomys pratensis*), photo by Ernest P. Walker.

These animals have a thick body, rather large and rounded ears, and a thick tail that tapers to a point. The four fingers have slightly curved claws that are suited for burrowing, and there are five toes. The upper incisor teeth are grooved. The females of most species have 8 mammae, but female *S. pratensis* and *S. caurinus* usually have 12 and sometimes have up to 16 mammae (Meester and Setzer 1971).

Fat mice are found in forest clearings, cultivated fields, savannahs, and semidesert areas (Meester and Setzer 1971). They are nocturnal and terrestrial, and construct burrows in sandy soil. The excavated ground is deposited in mounds on the surface. The burrows of fat mice may reach a depth of 1.2 meters and a length of 1.7 meters. A nest, constructed of shredded vegetation, is concealed in a large chamber at the end of the tunnel. According to Meester and Setzer (1971), the burrows of *S. pratensis* are simple, with a main entrance dug at an angle of 30° to the surface, a nest chamber about 30 cm below the ground, and two or more side passages that are normally sealed at ground level and used as escape routes.

The diet of fat mice consists of seeds, grass bulbs, and insects.

Rosevear (1969) declared that the most noted fact about fat mice is their ability to accumulate thick layers of fat, enabling them to sleep, or at least to remain highly inactive, in their underground nests throughout the dead season when food is scarce. In the tropics this season is not associated with wintry conditions, as there is still sunshine and heat, and the process of dormancy is called estivation, not hibernation. In West African tropics the fat mice estivate during the dry season from November to March. In temperate zones farther south, estivation corresponds more to the cooler season, approximately from April to October.

Fat mice evidently live alone in their burrows, except for females with young (Meester and Setzer 1971; Genest-Villard 1979). Numerous burrows, however, are often present in a small area. Although *Steatomys* is well known and frequently collected, reproductive data are limited. Young of *S. pratensis*, *S. caurinus*, and *S. parvus* have been recorded

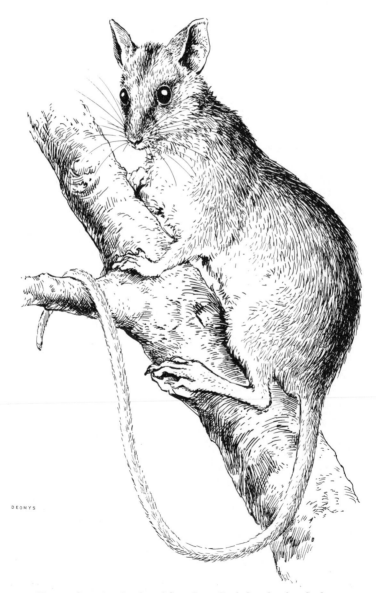

DEOMYS

Congo forest mouse (*Deomys ferrugineus*), adapted from *Proc. Zool. Soc. London*, St. Leger.

from early summer to late autumn in temperate areas (Meester and Setzer 1971). *S. cuppedius* may breed during the dry season, and produce two litters (Rosevear 1969). Pregnant females, one with two embryos and the other with seven, were reported by Smithers (1971), and other available records of litter size fall between these extremes.

Because of the high fat content of its body, *Steatomys* is regarded as a special delicacy by native peoples throughout its range. It is easily dug up, particularly during estivation (Meester and Setzer 1971).

RODENTIA; MURIDAE; *Genus DEOMYS Thomas, 1888*

Congo Forest Mouse

The single species, *D. ferrugineus*, is found from southern Cameroon to Uganda and southwestern Congo (Meester and Setzer 1971).

Head and body length is 120 to 144 mm, tail length is 150 to 215 mm, and weight is 40 to 70 grams (Kingdon 1974*b*). The coloration of the upper parts is pale red or reddish fawn, mixed with black along the middle of the back. The individual hairs on the upper parts have white bases, slate-colored middle portions, and reddish tips. Some of the hairs along the back are completely black. The face and sides of the head are paler and more dull in color than the back, except for a poorly defined blackish ring around each eye. The underparts, the insides of the arms and legs, and the hands are white. The slender, coarsely haired tail is distinctly bicolored in one form, slate gray above and white below, but in another form the terminal 25 to 76 mm of the tail is entirely white.

Deomys has large, rounded ears; long, narrow hind feet; orange upper incisor teeth, each with two minute grooves; and smooth yellow lower incisors. Females have four mammae.

According to Kingdon (1974*b*), *Deomys* occurs in tropical forests, especially in or near seasonally flooded swamps. It constructs leaf and fiber nests in holes or crevices at the base of trees. When frightened it may leap up to 50 cm high using its powerful hind legs, but it normally progresses on all fours across the forest floor in search of prey. It also climbs well. It apparently is crepuscular, but may also be active at night or in the late afternoon. During the rainy season in Uganda, *Deomys* may be found in well-drained areas, but in dry periods the animals concentrate in moist swamps. The diet consists of insects, other small invertebrates, and occasionally vegetable matter.

Kingdon (1974*b*) stated that the main social units are male-female pairs and mothers with young. There seem to be extensive reproductive peaks, and *Deomys* may produce closely spaced litters throughout the year. Litter size is one to three, usually two young.

RODENTIA; MURIDAE; *Genus LEIMACOMYS Matschie, 1893*

Groove-toothed Forest Mouse

The single species, *L. buettneri*, is known only by two specimens collected in 1890 near Yege in central Togo. The specimens are now in the Zoologisches Museum of Humboldt University in Berlin. Although sometimes considered to belong to the subfamily Murinae, *Leimacomys* is now thought to be a dendromurine (Rosevear 1969).

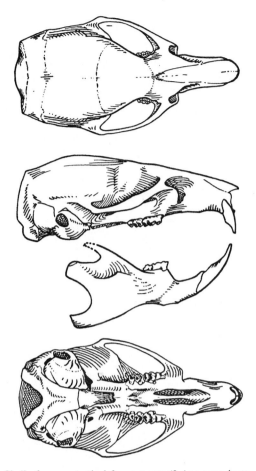

Skull of groove-toothed forest mouse (*Leimacomys buettneri*), from *The Rodents of West Africa*, Rosevear, British Museum (Natural History), 1969.

Head and body length of the type specimen is 118 mm, and tail length is 37 mm. The upper parts are dark brown to grayish brown, the shoulders and flanks are light brown, and the underparts are pale grayish brown. The feet are covered by short brown hairs. The small ears are well haired. The claws on the forefeet are fairly long, but somewhat shorter than those of the hind feet. The tail is extremely short and apparently naked. *Leimacomys* differs from the other dendromurines, *Dendromus* and *Steatomys*, in that the third upper and lower molars are not so reduced, the incisors are only weakly grooved, and the zygomatic plate is broader and well forward of the upper branch, its messeteric knob being relatively poorly developed (Rosevear 1969).

RODENTIA; MURIDAE; *Genus PETROMYSCUS Thomas, 1926*

Rock Mice

There are two species (Meester and Setzer 1971):

P. monticularis, southern Namibia;
P. collinus, southern Angola, Namibia, western South Africa.

Rock mouse (*Petromyscus collinus*), photo by C. T. Stuart.

Head and body length is about 70 to 90 mm, and tail length is 80 to 100 mm. The pelage is fine, straight, soft, silky, and rather thin, and lacks guard hairs. The coloration of the upper parts ranges from various shades of buffy to brownish or drab gray. The individual hairs of the underparts have slaty bases with white tips, giving a grayish appearance. The hands, feet, and tail are uniformly white to buffy in all forms except one, in which the bicolored tail is drab gray above and white below. The tail is moderately well haired, but the hairs are too scanty to hide the coarse scales. In *P. collinus* the tail is longer than the head and body, but in *P. monticularis* the tail is shorter than the head and body. The ears are rather long, especially in *P. collinus,* and the legs are short. The short, broad feet have the normal number of digits. The upper incisor teeth are not grooved. Females usually have six mammae, but occasionally have only four.

Rock mice inhabit dry, barren mountains; they prefer areas where large quantities of loose boulders and rocky outcrops predominate. They are nocturnal, hiding by day under rocks or in crevices, and stealthily creeping between the boulders at night. Reports indicate that rock mice are omnivorous in diet. A female *P. collinus,* collected in September, had two embryos.

RODENTIA; MURIDAE; **Genus DELANYMYS** Hayman, *1962*

Delany's Swamp Mouse

The single species, *D. brooksi,* occurs in extreme southwestern Uganda, and in the mountainous area near Lake Kivu in eastern Zaire (Meester and Setzer 1971).

Head and body length is 50 to 63 mm, tail length is 87 to 111 mm, and weight is 5.2 to 6.5 grams (Delany 1975). The coloration of the upper parts is a warm russet. The chin is whitish and the remaining underparts are a warm buff. The bases of the hairs, both dorsally and ventrally, are grayish. *Delanymys* and *Petromyscus* are placed together in the sub-family Petromyscinae, because of the similar structure of their molar teeth, but the two genera are very different in external appearance and habitat (Kingdon 1974b).

Delanymys has been taken at elevations of 1,700 to 2,625 meters, usually in vegetation associated with marshes within bamboo and montane forests (Kingdon 1974b; Delany 1975). It is nocturnal and a very good climber. It constructs a small, round, grass nest with two entrances. One such nest was located 50 cm above the ground in a bush. The diet consists mainly of seeds. Two pregnant females had three embryos each, and a nest found in June contained four blind young.

RODENTIA; MURIDAE; **Genus BEAMYS** Thomas, *1909*

Long-tailed Pouched Rat

The single species, *B. hindei,* is known from the coast of Kenya, Tanzania, Malawi, and northeastern Zambia (Kingdon 1974b; Ansell and Ansell 1973).

Head and body length is 130 to 187 mm, tail length is 100 to 155 mm, and weight is 55 to 150 grams (Kingdon 1974b). Coloration is grayish or grayish brown above, and white below. *Beamys* is distinguished externally by the naked, white-tipped tail of medium length. The underside of the tail is dark on the proximal quarter and whitish for the rest of its length. The tail is flattened, with sharp edges, and the lower side is wider than the upper, so that in cross section its shape resembles that of a truncated pyramid. The tail is not noticeably scaly, but only the basal centimeter is hairy. Like the other two genera of the subfamily Cricetomyinae, *Cricetomys* and *Saccostomus, Beamys* has cheek pouches.

According to Kingdon (1974b), *Beamys* lives in forest and moist woodland, from sea level to 2,100 meters. It must live near water on soft and sandy soil. It constructs burrows up to 9 meters in length and about 60 cm deep. The burrows consist of a vertical entrance tunnel, passages with enlarged chambers for nesting and food storage, and at least one cul-de-sac, the end of which is used as a latrine. The nest is made of dry

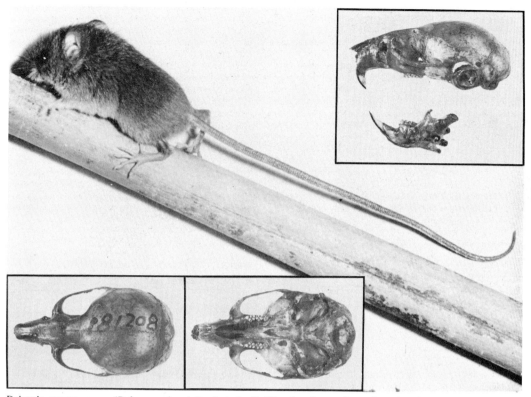

Delany's swamp mouse (*Delanymys brooksi*), photo by F. Dieterlen. Inset: photos from American Museum of Natural History.

Long-tailed pouched rat (*Beamys hindei*), photo from Nyasaland Museum through P. Hanney.

vegetation. *Beamys* is completely nocturnal and mainly ter-restrial, but it is a confident climber and harvests some food that is still attached to plants. Intensive food collection and storage activity continues throughout the year. The diet con-sists mainly of seeds and fruit.

Kingdon (1974*b*) added that burrows are occupied by a single adult, but that burrows of a male and female may be only a few meters apart, and that usually a number of burrows occur in the same vicinity. In Malawi, breeding has been found to coincide with the rainy season. Litter size in the wild is reportedly 4 to 7 young. In a study of captives, Egoscue (1972) found females to be reproductively active throughout the year and to produce up to five litters annually. The gesta-tion period was 22 to 23 days, and litter size averaged 2.8 (1–5) young. The young weighed an average of 3.2 grams at birth, opened their eyes when about 3 weeks old, and were completely weaned at 5 to 6 weeks. The youngest breeding female gave birth to her first litter when 5½ months old, but most females and males did not reach sexual maturity until another 1½ to 3 months had passed. One female produced two litters when she was at least 3 years old.

RODENTIA; MURIDAE; *Genus SACCOSTOMUS* Peters, 1846

African Pouched Rats

Two species are now recognized (Hubert 1978*a*):

S. campestris, Angola, Zambia, Malawi, Mozambique, Rhodesia, Botswana, South Africa;
S. mearnsi, Ethiopia, Somalia, Uganda, Kenya, Tanzania.

Head and body length is 94 to 188 mm, tail length is 30 to 81 mm, and weight is 40 to 85 grams (Kingdon 1974*b*; Hubert 1978*a*; Delany 1975). The pelage is quite long, dense, and fine in texture. The upper parts are gray or brownish gray, being darker on the back than on the sides. The underparts are white in *S. campestris* and gray in *S. mearnsi.* The tail, lightly covered with short hairs, is usually bicolored (dark on top and pale below).

African pouched rats have robust bodies; broad, thick heads; and short, strong legs and toes. The ears are short and rounded, and the eyes are small. The tail is short and thick at the base. *Saccostomus* derives its common name from the cheek pouch that opens inside the lips at each side of the head and extends back to about the shoulders. The incisor teeth are not grooved. Females have 10 or 12 mammae.

African pouched rats inhabit savannahs, scrubby areas, grassy places in open forests, cultivated fields, and sandy plains. They are nocturnal, terrestrial, and rather slow mov-ing. They may utilize holes dug by other animals, or dig their own simple burrows, which often have two entrances and a single enlarged sleeping and storage chamber (Kingdon 1974*b*). At least in the southern portion of their range, they accumulate seeds during the summer for use in the winter. This food is transported from the field to the burrow in the cheek pouches. The diet consists of seeds, berries, grains, acacia nuts, and, occasionally, insects.

According to Kingdon (1974*b*), burrows are often in close proximity, but each contains only a single adult. Breeding evidently continues throughout the year in East Africa. Smithers (1971) stated that in Zambia pregnant females were taken from January to April, and juveniles in September, October, and February; the number of embryos ranged from 5 to 10. According to Delany (1975), there may be as few as 2 young per litter.

RODENTIA; MURIDAE; *Genus CRICETOMYS* Waterhouse, 1840

African Giant Pouched Rats

There are two species (Meester and Setzer 1971):

C. gambianus, Senegal to central Sudan, and south to South Africa;
C. emini, Sierra Leone to Lake Tanganyika, Fernando Poo.

Head and body length is 240 to 450 mm, and tail length is 365 to 460 mm. Males weigh about 1.5 kg, and females weigh up to 1 kg (Ajayi 1975). The fur is short and thin; in some forms it is coarse and harsh, but in others it is relatively fine and sleek. The coloration of the middorsal region ranges from dark grayish brown to medium grayish with a tinge of brown or clear reddish brown. The general coloration be-comes paler on the sides, sides of the face, and flanks, rang-ing from soft gray with a brownish tinge to reddish brown, vinaceous, or buffy. The underparts are soiled white, white, or creamy. The conspicuous ears are practically naked. The tail is almost naked; the dorsal two-thirds are dark grayish and the remainder is soiled white or soiled creamy. Many individuals are mottled with gray, or almost spotted, mainly on the anterior half of the upper parts. This mottling may be inconspicuous, may form irregular small spots or large

African pouched rats (*Saccostomus campestris*), photo by Ernest P. Walker.

Giant pouched rat (*Cricetomys gambianus*). Inset: animal with its cheek pouches well filled; photos from New York Zoological Society.

blotches, or may cover almost all of the upper parts. In some forms there is a fairly definite, almost white, stripe across the back just behind the shoulders.

Cricetomys has a rather long, narrow head; ungrooved incisor teeth; cheek pouches; and a scaly tail. Females have eight mammae.

These rats dwell in forests and thickets. They are nocturnal, but have been seen foraging during the day, at which time they behave as if they were almost blind, sitting on their haunches and sniffing in all directions. They can climb well, using the long tail for balancing, and can swim. For shelter they often use natural crevices and holes, termite mounds, or hollow trees, but they can dig their own burrows, which consist of a long passage with side alleys and several chambers, one for sleeping and the others for storage (Kingdon 1974*b*). The burrows have two to six openings and are frequently located at the base of a tree or among dense vegetation; the entrances are often closed from the inside with leaves. Ajayi (1975) stated that these rats are completely omnivorous, feeding on vegetables, insects, crabs, snails, and other items, but apparently preferring palm fruits and palm kernels. Ewer (1967) stated that the cheek pouches are used to carry food and bedding material, and that there is regular coprophagy. According to Kingdon (1974*b*), these rats store a considerable amount of food and also many nonedible items, such as coins, metal, and bits of cloth.

Cricetomys is generally solitary; a captive male and female may be kept together, but two mature males may fight to the death (Ewer 1967). Reproduction takes place at various times of the year. In a study of captives, Ajayi (1975) observed one female to give birth 5 times in 9 months, and he

thought that probably females could produce 10 litters annually. The gestation period was 27–36, usually 30–32, days. Litters numbered 1 to 5, most commonly 4, young. Sexual maturity was attained at about 20 weeks of age. According to Marvin L. Jones (Zoological Society of San Diego, pers. comm.), a captive specimen lived for 7 years and 10 months.

These animals are timid, but soon become tame in captivity and make delightful pets. They are in great demand as food by the native tribes. Studies have been made of their potential for domestic production of food. In some West African towns, *Cricetomys* has become a sewer rat and is killed along with *Rattus* by the rat catchers.

RODENTIA; MURIDAE; Genus *LENOTHRIX* Miller, 1903

Gray Tree Rat

The single species, *L. canus*, is known from the Malay Peninsula, Sarawak, and Tuangku Island off northwestern Sumatra (Musser 1981*a*). *Lenothrix* was long considered to be only a subgenus of *Rattus*, and to comprise a number of species in addition to *L. canus* (Laurie and Hill 1954). Misonne (1969), however, explained that *Lenothrix* has no close relationship to *Rattus*, is one of the most primitive genera in the subfamily Murinae, and includes only the species *L. canus*. This view was supported by Medway (1977, 1978), Medway and Yong (1976), and Musser and Boeadi (1980). Nonetheless, on the basis of an electrophoretic analysis,

Chan, Dhaliwal, and Yong (1978) supported retention of *L. canus*, as well as other species formerly assigned to the subgenus *Lenothrix*, within *Rattus*.

Head and body length is 165 to 220 mm, tail length is 190 to 265 mm, and weight is 81 to 273 grams. The fur is soft and rather woolly, and lacks spines. The upper parts are gray or brownish gray, and the underparts are white. The tail is pigmented above and below at the base and is distally unpigmented. The only other rat having a tail with this same color pattern is *Rattus bowersi*, and that species is larger than *L. canus*. *Lenothrix* also is distinguished from *Rattus* by dental characters, most notably the presence of a rudimentary extra cusp or posteriorly directed fold of enamel in the midline of the posterior face of the transverse ridges of the first and second upper molars. Female *Lenothrix* have 10 pairs of mammae (Musser 1981*a*).

According to Medway (1978), the gray tree rat is locally common on the Malay Peninsula in plantations and primary and secondary forests. It is arboreal. Litter size averages three and ranges from two to six young. The estimated mean life span in the wild is five months, but record longevity in captivity is three years and nine months.

RODENTIA; MURIDAE; Genus KADARSANOMYS Musser, 1981

The single species, *K. sodyi*, is known by 17 specimens from the southwestern slopes of Gunang Pangrango-Gede, a volcanic massif in western Java. Although *sodyi* was long considered only a subspecies of *Lenothrix canus*, Musser (1981*c*) made it the basis of an entirely new genus, which he considered not to be closely related to any other murid genus.

Head and body length is 182 to 210 mm, and tail length is 263 to 305 mm. The middle of the head and back is dark brown, and this color pales to grayish brown on the sides and cheeks. The underparts are white. The tail, which is longer than the head and body, is pale brown and covered with scales. On each scale are three hairs, which are short near the base of the tail but increase in length toward the tip. The front and hind feet are long and broad. Females have four pairs of mammae.

Kadarsanomys resembles *Lenothrix* in body size and proportion, but has smaller ears, shorter hind feet, a different color pattern, and different proportions of the feet and skull. *Kadarsanomys* has a thicker cranium, a shorter rostrum, a higher braincase, narrower zygomatic plates, longer incisive foramina, a narrower palatal bridge, and larger bullae.

Kadarsanomys is arboreal. All specimens were taken in forest at an elevation of about 1,000 meters. Most were caught inside of bamboo stems, in which they had gnawed holes 3 to 4 cm in diameter. A nest of dry leaves, containing four young, was found in such a stem on 24 January. In addition, juveniles were caught in June.

RODENTIA; MURIDAE; Genus DIPLOTHRIX Thomas, 1916

The single species, *D. legata*, is found on the Ryukyu Islands south of Japan. Thomas (1906) originally placed this species in the genus *Lenothrix*, along with *L. canus*, but subsequently (1916) erected for it the separate genus *Diplothrix*. Ellerman and Morrison-Scott (1966) questionably arranged *Diplothrix* as a subgenus of *Rattus*, but Misonne (1969) indicated that *Diplothrix* is a distinct genus related to *Lenothrix*. Musser and Boeadi (1980) stated that *Diplothrix* is a good genus and is not closely related to *Rattus*.

Head and body length of the type specimen is 230 mm, and tail length is 246 mm. The fur is very long and thick. The general color above approaches clay color, but is more grayish. The underparts are dirty grayish. The upper surfaces of the hands and feet are dark brown. The tail is evenly well haired throughout its length; its color is dark brown on the basal three-fifths and white beyond. The ears are short and thinly haired, with a buffy patch behind their posterior bases. From that of *Lenothrix*, the skull of *Diplothrix* differs mainly in being much broader between the parietal ridges. In addition, the posterior lamina of the third upper molar of *Diplothrix* consist of two elements, an internal and a median cusp, not of a single cusp as is the case in *Lenothrix canus*.

RODENTIA; MURIDAE; Genus MARGARETAMYS Musser, 1981

Margareta's Rats

There are three species (Musser 1981*a*):

M. beccarii, northeastern and central Celebes;
M. elegans, central Celebes;
M. parvus, central Celebes.

Musser (1977*b*) suggested affinity between *M. beccarii* (then known as *Rattus beccarii*) and the genus *Limnomys* of Mindanao, but later (1981*a*) explained that the resemblance is superficial, that *Margaretamys* is not closely related to *Limnomys* or *Rattus*, and that it may actually have phylogenetic ties to *Lenothrix* or *Lenomys*.

In *M. beccarii*, head and body length is 117 to 152 mm; tail length is 150 to 200 mm; the pelage is thick, short, and semispinous; the upper parts are generally grayish brown; the underparts vary from cream through yellow to dark ochraceous buff; there are wide, dark brown rings around the eyes; and the tail is usually brown above and paler below. In *M. elegans*, head and body length is 183 to 197 mm; tail length is 248 to 286 mm; the fur is thick, long, and very soft; the upper parts are brown; the underparts are whitish gray; and the tail is brown on the proximal half to two-thirds and white on the remainder. In *M. parvus*, head and body length is 96 to 114 mm; tail length is 154 to 184 mm; the pelage is dense, short, and soft; the upper parts are reddish brown; the underparts are dark gray; and the tail is monocolored (Musser 1981*a*).

From *Rattus* and *Limnomys*, *Margaretamys* is distinguished in having a prominently penicillate tail, smaller bullae, the squamosal roots of the zygomatic arches originating higher on the sides of the braincase, the posterior margin of the palatal bridge not extending past the molar toothrow, the pterygoid fossa nearly flat and not perforated by a large foramen, and the mesopterygoid fossa at least as wide as the palatal bridge. From *Niviventer*, *Margaretamys* is distinguished in having a more penicillate tail, larger bullae, a more rounded cranium, and a much larger coronoid process relative to the size of the dentary.

Margareta's rats are arboreal forest dwellers. *M. beccarii* occurs from sea level to the upper limits of lowland tropical rain forest. *M. elegans* and *M. parvus* are found in montane forest at elevations of up to 2,300 meters.

RODENTIA; MURIDAE; Genus LENOMYS Thomas, 1898

Trefoil-toothed Giant Rat

The single species, *L. meyeri*, is endemic to Celebes (Musser 1970*c*:18; Groves 1976).

Margareta's rats: Left, *Margaretamys elegans;* Upper right, *M. parvus;* Lower right, *M. beccarii;* drawings by Fran Stiles, courtesy of Guy G. Musser.

Head and body length is 235 to 290 mm, and tail length is 210 to 285 mm. The pelage is thick, soft, and rather woolly, with elongated guard hairs. The general coloration of the upper parts is fuscous to drab olive black, with a sprinkling of whitish; a dorsal line is present and is somewhat darker than the sides. The head is the same color as the body. The under-parts are buffy white, blending gradually into the color of the sides and leaving no conspicuous line of demarcation. The individual hairs of the underparts have gray bases and white tips. The feet and hands are brownish gray. The terminal section (one-half to three-fifths) of the tail is flesh colored. The tail is quite scantily haired.

Trefoil-toothed giant rat (*Lenomys meyeri*), photo by Margareta Becker through Guy G. Musser.

According to Musser (1970c), *Lenomys* differs from *Rattus callitrichus* of Celebes in having a much larger and more massive skull, shallower zygomatic arches, an hourglass-shaped interorbital region, shorter incisive foramina, a longer and arched bony palate, and wider incisor teeth. The molariform teeth of *Lenomys* have three distinct rows of cusps; the lateral and lingual cusps are distinct and well set off from the middle row.

RODENTIA; MURIDAE; *Genus PAPAGOMYS* Sody, 1941

Flores Giant Rat

The single living species, *P. armandvillei*, occurs only on Flores Island in the East Indies. Another specis, *P. theodorverhoeveni*, is known from subfossil material, collected on Flores and estimated to be 3,550 years old. The same deposits contained the remains of one extinct genus, *Spelaeomys*, which seems closely related to *Papagomys*, and a second extinct genus, *Floresomys*, which may have phylogenetic ties to the murid genera native to New Guinea and adjacent islands, and Australia (Musser 1981b).

Head and body length is 410 to 450 mm, and tail length is 330 to 370 mm. According to Musser (1981b), the upper parts are dark brown or tan, the middle of the head and body is darker than the sides, and the underparts are pale gray with a slight tan suffusion. The pelage is dense and harsh, especially on the upper parts, where it consists of flattened, flexible spines mixed with regular hairs. The tail is stout, is shorter than the head and body, and appears naked, but is covered with large scales, each with three short, stiff hairs

emerging from the base. The basal two-thirds of the tail are blackish brown, and the distal third may be either the same color, pale brown, or white. The ears are small, rounded, and covered with fine hairs. The front feet are short and wide, and the hind feet are long and wide.

Papagomys has sometimes been considered to have phylogenetic affinity to *Mallomys*, but the two are not closely related. *Papagomys* is larger, has a tail that is shorter (rather than longer) than the head and body, has short and harsh (rather than long, soft, and woolly) pelage, has a long and slender (rather than short and wide) rostrum, has relatively much larger bullae, and differs in numerous other cranial and dental characters (Musser 1981b).

The body structure of *Papagomys* is like that of a rat adapted for life on the ground and refuge in burrows. The diet may include leaves, buds, fruit, and certain kinds of insects (Musser 1981b).

RODENTIA; MURIDAE; *Genus KOMODOMYS* Musser and Boeadi, 1980

The single species, *K. rintjanus*, is currently found on the islands of Rintja and Padar, between Sumbawa and Flores in the East Indies (Musser and Boeadi 1980). The species is also known from subfossil material, collected on Flores Island and estimated to be 3,550 years old (Musser 1981b). *K. rintjanus* was originally described as a species of *Rattus*, but is now known to represent a distinct genus, not closely related to *Rattus* and apparently with affinity to *Papagomys* and other relictual murid genera of the eastern Sunda Islands.

Head and body length is 130 to 200 mm, and tail length is

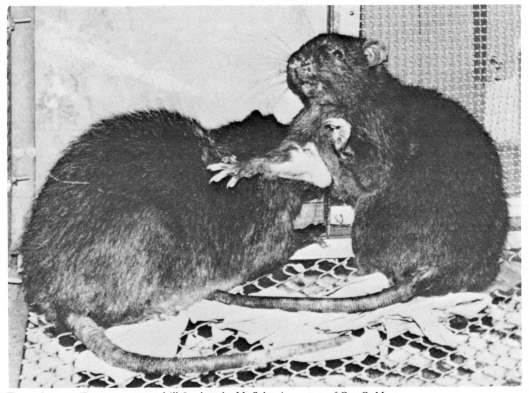

Flores giant rats (*Papagomys armandvillei*), photo by M. Sukaeri, courtesy of Guy G. Musser.

112 to 160 mm. The dorsal pelage is thick, coarse, and spinous. The upper parts are sandy colored, with the middle of the head and body from the nose to the rump being darker and suffused with brown, and the cheeks and sides of the head and body being paler and suffused with gray. The feet are hairy and white. The ears are tan and covered with fine short hairs. The tail is coarsely scaled and covered with long silver hairs that give a hairier appearance than that of *Rattus*. The tail is brown on the top and sides, and is paler below.

Komodomys is a medium-sized rat with a tail much shorter than the head and body. The hind feet are long and narrow, and each has six plantar tubercles. The digits and nails are long. The dorsal profile of the cranium is strongly arched, the rostrum is long and narrow, and the incisive foramina are very long and narrow. The bullae are relatively larger than those of *Rattus*. Also unlike *Rattus*, but like *Papagomys*, the molar teeth of *Komodomys* have very high cusps. *Papagomys* is a much larger rat than *Komodomys*. Female *Komodomys* have 10 mammae.

This rodent may be associated with monsoon forest. Its sandy-colored upper parts, densely haired white feet, moderately large ears, and short, hairy tail suggest that it is a ground dweller in dry scrub or forest. Its cranial and dental specializations may reflect adaptation to a dry, or seasonally dry, tropical forest habitat, where the tall scrub and short, partly deciduous trees provide dense cover above sparse undergrowth at ground level.

RODENTIA; MURIDAE; **Genus EROPEPLUS** *Miller and Hollister, 1921*

Celebes Soft-furred Rat

The single species, *E. canus,* is known by five specimens, all obtained from elevations above 1,500 meters in middle Celebes (Musser 1970c:21).

Head and body length is 195 to 240 mm, and tail length is 210 to 315 mm. The soft pelage is made up of a coat of underfur, which is about 25 to 28 mm in length, and of longer guard hairs, which are 35 to 45 mm long on the back and flanks. The general coloration of the upper parts is brownish gray. Each hair has a pale slate base that terminates in a 3- to 5-mm tip of pale buff color. The underparts are light gray, but there is no sharp line of demarcation. The feet are scantily covered with short black hairs, and the whiskers are also black. The moderately haired tail is long, and the terminal third or half is white.

Eropeplus resembles *Rattus callitrichus* of Celebes in external features, but differs in cranial and dental characters (Musser 1970c). In *Eropeplus*, for example, the skull is smaller, the rostrum is shorter and narrower, the interorbital region is shaped like an hourglass, the incisive foramina are short and do not extend to or beyond the front margins of the toothrows as in *Rattus callitrichus,* and the third upper molars are conspicuously longer than wide.

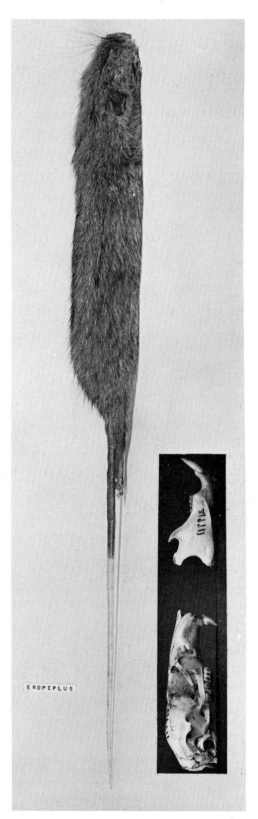

Celebes soft-furred rat (*Eropeplus canus*), photo by Howard E. Uible of specimen in U.S. National Museum. Inset of skull: photo by P. F. Wright of specimen in U.S. National Museum.

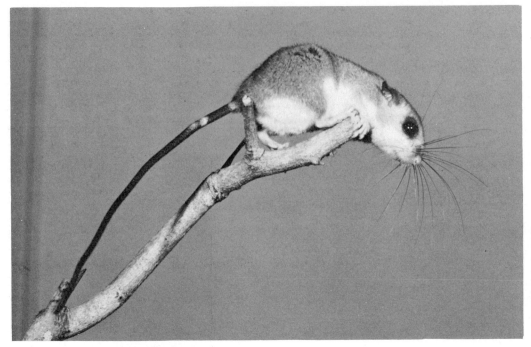

Chiruromys vates, photo by J. I. Menzies.

RODENTIA; MURIDAE; **Genus CHIRUROMYS** Thomas, 1888

There are four species (Dennis and Menzies 1979):

C. forbesi, southeastern New Guinea, D'Entrecasteaux and Louisiade archipelagos;
C. kagi, southeastern New Guinea;
C. lamia, southeastern New Guinea;
C. vates, eastern New Guinea.

Chiruromys was long considered to be only a subgenus of *Pogonomys*, but following a study of morphology and karyology, Dennis and Menzies (1979) concluded that the two taxa are not closely related and should be designated as separate genera.

Head and body length is 115 to 175 mm, and tail length is 160 to 245 mm (Tate 1951*b*). *Chiruromys* resembles *Pogonomys* in external appearance. The pelage is soft and dense. The upper parts vary in color from grayish through fawn and brown to rich reddish brown. The underparts are usually white. Females have six mammae. The tail characteristics are generally the same as in *Pogonomys*, but *Chiruromys* can be distinguished by the rough appearance of its tail; the scales stand out like teeth on a rasp, whereas the tail of *Pogonomys* has a smoother appearance.

These rodents are found in forested areas from sea level to 2,000 meters. Unlike *Pogonomys*, they are completely arboreal, nest in hollow trees, and probably do all their foraging in the forest canopy. The diet and colonial habits are the same as in *Pogonomys*. A pregnant female and a newborn individual were found in December. Litter size is one to three young (Menzies and Dennis 1979).

RODENTIA; MURIDAE; **Genus POGONOMYS** Milne-Edwards, 1877

Prehensile-tailed Rats

There are three species (Dennis and Menzies 1979):

P. macrourus, New Guinea;
P. loriae, highlands of New Guinea, D'Entrecasteaux Archipelago;
P. sylvestris, highlands of New Guinea.

Prehensile-tailed rat (*Pogonomys loriae*), photo by J. I. Menzies and E. S. Dennis.

Chiruromys, long considered a subgenus of *Pogonomys,* was regarded as a separate genus by Dennis and Menzies (1979), and that position is followed here. According to Menzies and Dennis (1979), several *Pogonomys* were recently collected in northern Queensland.

Head and body length is 95 to 148 mm, and tail length is 150 to 214 mm (Tate 1951*b*). Mean adult weight is about 40 grams in *P. sylvestris* and 74 grams in *P. loriae* (Dwyer 1975*b*). The pelage is dense, woolly, and quite soft, with few guard hairs. The numerous whiskers are long, dark, and conspicuous. The general coloration of the upper parts ranges from grayish through rufous and reddish brown to dark brown or almost black. The paler-colored forms have a sprinkling of silver hairs on the upper parts. The underparts are white, buffy, or grayish. The long, scantily haired, prehensile tail is coarsely scaled and mostly brown in color. The terminal 10 to 20 mm of the upper surface of the tail, however, is flesh colored, naked, and smooth. Unlike most prehensile tails, the tail of *Pogonomys* curls backward or up at the tip, rather than downward.

Pogonomys has a short head and large eyes. The hands and feet are short, broad, and modified for climbing. There is a slightly opposable hallux with a fully developed claw. The molar teeth have a complex folded pattern. Females have six mammae.

Prehensile-tailed rats are found in forests from sea level to about 2,700 meters in elevation. Menzies and Dennis (1979)

reported that they are nocturnal and only partly arboreal, foraging in low vegetation or on the ground and nesting in underground burrows. The nests of *P. loriae* are said to have several entrance tunnels that may be many meters in length. The diet consists largely of shoots of grass and bamboo, and young leaves.

Menzies and Dennis (1979) stated that these rats are colonial. A single nest may contain a group of up to 15 individuals of all ages and sexes, though usually there are only 3 to 5 animals per colony. In a study in eastern Papua New Guinea, Dwyer (1975*b*) found that breeding apparently ceases during the dry season from July to September, but is otherwise continuous. In both *P. loriae* and *P. sylvestris* the usual litter size is 2 or 3 young. According to Marvin L. Jones (Zoological Society of San Diego, pers. comm.), a captive *P. macrourus* lived for two years and five months.

RODENTIA; MURIDAE; *Genus PITHECHEIR F. Cuvier, 1842*

Monkey-footed Rat, Malay Tree Rat

The single species, *P. melanurus,* is known from the Malay Peninsula, Sumatra, and Java (Chasen 1940). Muul and Lim (1971) and Lim and Muul (1975) considered the correct generic name to be *Pithechir* Müller, 1840, and *P. parvus* of

Malay tree rat (*Pithecheir melanurus*), photo by Lim Boo Liat. Inset: photo from Museum Zoologicum Bogoriense.

Bushy-tailed cloud rats (*Crateromys schadenbergi*), photo by Ernest P. Walker. Insets: A. Upper molars; B. Lower molars; photos from *Trans. Zool. Soc. London*, "Mammals from the Philippines," Oldfield Thomas.

the Malay Peninsula a species distinct from *P. melanurus* of Sumatra and Java. Medway (1978), however, continued to use the generic name *Pithecheir*, and to use the specific name *P. melanurus* for the Malay population.

Head and body length is 122 to 180 mm, tail length is 157 to 215 mm, and weight is 58 to 146 grams (Muul and Lim 1971; Lim and Muul 1975). The long, soft, and dense pelage is reddish to tawny on the upper parts, but becomes buffy on the sides of the abdomen and rump, and blends into pure white on the underparts. The bases of the individual hairs on the back are mouse gray. The scantily haired tail, ears, hands, and feet are usually reddish, though in some animals the feet are white. The nails are horn brown, and the whiskers are long and numerous.

The hind foot is modified for arboreal life. The first toe on the hind foot is thumblike and opposable, hence the origin of the scientific name, which means "ape hand." The pollex is very small, the tail is prehensile, and the incisor teeth are moderate. Females have four mammae.

The monkey-footed rat is found in dense forests up to about 1,600 meters elevation. It builds globular nests of leaves or moss in the branches or hollows of trees. Two nests found in Java were about 150 mm in diameter and located about 2 and 3.5 meters above the ground. Lim and Muul (1975) reported that captives were active only at night and spent most of their time climbing. The movements of this rodent are quite slow and cautious. The tail is used for grasping as the animal climbs and scampers about limbs and branches. The diet may be mostly green plant material, though a captive lived on plantains and crickets.

Medway (1978) stated that pregnancies have been recorded at all times of the year on the Malay Peninsula and that normally two young are born. In Java, single young have been noted from April to September. Lim and Muul (1975) reported that some captives bred four times per year, that the young are carried about attached to the mother's nipples until they are weaned, and that captive specimens have lived up to three years.

RODENTIA; MURIDAE; *Genus CRATEROMYS Thomas, 1895*

Bushy-tailed and Ilin Island Cloud Rats

There are two species (Taylor 1934; Sanborn 1952; Musser and Gordon 1981):

C. paulus, known only by the holotype from Ilin Island (Philippines);
C. schadenbergi (bushy-tailed cloud rat), northern Luzon.

In the single specimen of *C. paulus*, head and body length is 255 mm, tail length is 215 mm, the pelage is short and coarse, the upper parts are generally dark grizzled brown, and the underparts are cream colored. The tail is densely furred, but not bushy, and is tricolored—the same color as the rump on the basal 45 mm, blackish brown on the next 150 mm, and cream colored on the distal 20 mm (Musser and Gordon 1981).

The remainder of this account applies only to *C. schadenbergi*. Head and body length is 325 to 394 mm, and tail length is 355 to 475 mm. The pelage is quite dense, consisting of woolly underfur and long, straight or wavy, guard hairs. The general coloration is highly variable, but the animal is usually dark brown to black on the upper parts, dark grayish on the sides, and iron gray on the underparts. Some individuals, however, are white or brownish on the anterior part of the body, and occasionally the underparts are irregularly white. The long tail is exceptionally heavily haired and is thickly bushy, a unique characteristic in the family Muridae.

This large rodent has an elongated body, a slender muzzle, and small eyes and ears. The hands and feet each have five digits; the thumb has a well-developed flattened nail, and the remaining fingers and toes have powerful, slender claws. There are strong tufts of hair at the base of each claw. The hind foot is broad.

The bushy-tailed cloud rat appears to be quite common among the high mountains and plateaus of northern Luzon. It prefers a forested habitat, is arboreal, and is most active after sunset. During the day it sleeps in tree cavities or in holes among the roots of trees. It has a strange cry, so shrill that it seems like that of some insects. It is said to feed upon the buds and bark of young pine tree sprouts and on fruit; the latter is eaten while on the tree rather than after it has fallen to the ground.

Igorote natives from Mount Data trap this animal and sell its pelt in the market at Baguio. The woollike pelt is attractive and serviceable. Several bushy-tailed cloud rats have been kept as pets, but no notes were made on their temperament. One captive lived for four years and three months (Marvin L. Jones, Zoological Society of San Diego, pers. comm.).

RODENTIA; MURIDAE; *Genus MALLOMYS Thomas, 1898*

Giant Tree Rat

The single species, *M. rothschildi*, inhabits most of New Guinea between elevations of 1,350 and 2,750 meters (Laurie and Hill 1954).

Head and body length is 345 to 440 mm, and tail length is 365 to 420 mm. Menzies and Dennis (1979) reported the weight to be up to 2 kg. There appears to be considerable variation in both the texture and the color of the pelage. The

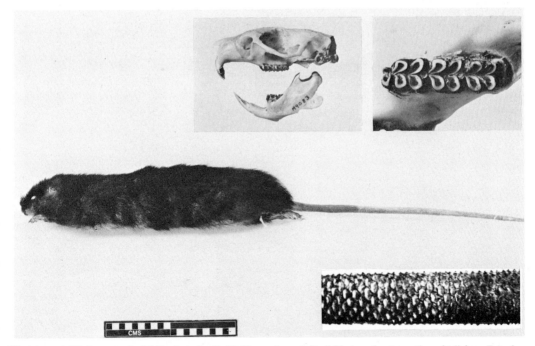

Giant tree rat (*Mallomys rothschildi*), photo by C. W. Turner through Basil Marlow. Insets: section of tail from *Zeitschr. Saugetierk;* photos of skull and right lower molar row by C. W. Turner through Basil Marlow.

fur is quite long and thick, and has a tendency to be somewhat woolly in most specimens. The general coloration of the upper parts is dark brown to fuscous brown or grayish, with the guard hairs being all black or black with white tips. The shades of coloration are dependent upon the abundance and color of the guard hairs. In at least one form the crown is light brown. The underparts are creamy or dull white, and the individual hairs are unicolored. In a few specimens a white band is present, extending across the middle of the underparts and well up onto the sides. The hands, feet, and whiskers are black. The tail is scantily haired and coarsely scaled; the basal part is brownish and approximately the terminal half is white.

The giant tree rat has the usual murid body form and large, broad, heavy feet. The pollex has a short nail, and the other digits have considerably enlarged and slightly curved claws. The muzzle is short, and the skull is heavy and thick. The moderately broad incisors are ungrooved and project slightly forward. Menzies and Dennis (1979) stated that the molar teeth are very distinct, their crowns being divided into an intricate triple series of cusps with a regularity not seen in any other genus. Females have four mammae.

According to Menzies and Dennis (1979), *Mallomys* is evidently common in montane forests throughout New Guinea. It seems to be mainly arboreal. Its lair is usually in a hollow tree, often at a considerable height, but is sometimes on the ground. It is entirely vegetarian, feeding largely on shoots such as those of climbing bamboo. In some areas the local people say there is a second kind of *Mallomys* that is more terrestrial, is found in open country as well as forest, and lives in holes in the ground. The large size of *Mallomys* makes it a desirable game animal. Its teeth are sometimes extracted and used for engraving.

RODENTIA; MURIDAE; *Genus* **HYOMYS** *Thomas, 1903*

White-eared Giant Rat

The single species, *H. goliath,* is found in New Guinea (Laurie and Hill 1954).

Head and body length is 295 to 390 mm, and tail length is 255 to 380 mm. The weight is around 1 kg (Menzies and Dennis 1979). The fur is coarse and harsh. The color of the upper parts is usually mixed gray and fuscous or dark slaty gray. In some individuals the guard hairs are gray with white subterminal bands, and in others they are black with white tips or are white throughout. The underparts are grayish, buffy gray, or dull white. The tail in *Hyomys* is coarsely scaled and almost naked, and the scales appear even rougher and more naked than those of *Mallomys*.

Hyomys is a large rat with a somewhat bulky form. A period of growth apparently continues after maturity is reached. The large toe has a broad nail. *Hyomys* has small ears, well-developed claws, and a tail with large, overlapping scales. The scales are subject to considerable wear. The skull is very massive and the molar teeth are large. Menzies and Dennis (1979) stated that *Hyomys* can be distinguished from *Mallomys* in that the three cusps of each loph of its molar crowns are merged and form transverse ovals, while in *Mallomys* the separate cusps are distinct. Female *Hyomys* have four mammae.

According to Menzies and Dennis (1979), *Hyomys* is fairly common in hill forest from 1,200 to 3,000 meters throughout New Guinea. Although it climbs readily, it seems more terrestrial than *Mallomys*, and is said to make a large nest of leaves between the roots of a tree, between rocks, or in a hollow log. The diet consists mainly of bamboo and other

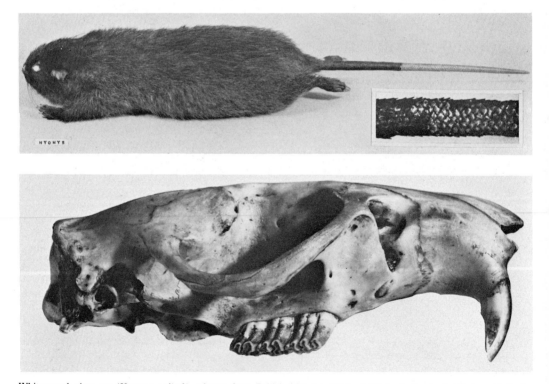

White-eared giant rat (*Hyomys goliath*), photos from British Museum (Natural History). Inset of tail from *Zeitschr. Saugetierk.*

such shoots, but people occasionally complain that it raids gardens. Litters may ordinarily contain only a single young.

RODENTIA; MURIDAE; **Genus BATOMYS** Thomas, 1895

Luzon and Mindanao Forest Rats

There are three species (Taylor 1934; Sanborn 1952, 1953):

B. granti, known only by five specimens from Mount Data in northern Luzon;

B. dentatus, known only by the type specimen collected in Benguet Province in northern Luzon;

B. salomonseni, known only by three specimens from Bukidnon Province in Mindanao.

The specimens of B. salomonseni were collected in 1951 and described by Sanborn (1953) as a new genus, Mindanaomys. Misonne (1969), however, listed Mindanaomys as a synonym of Batomys.

In B. granti the head and body length of the type specimen is 204 mm, and the tail length (probably not perfect) is 121 mm; the upper parts are fulvous and black, becoming rufous toward the rump; the underparts are slaty buff; the hands and feet are brown with whitish digits; the tail is dark brown to black and covered with thick hair; and the eyes are surrounded by a seminaked or finely haired ring. In B. dentatus the head and body length is 195 mm, and the tail length is 185 mm; the upper parts are uniformly light brown; the underparts are ochraceous buff, more buffy than in B. granti; the hands and feet are dull buffy gray; the tail is uniformly blackish brown basally, white toward the tip, and thinly covered with hair; and the area around the eyes is furred normally. In

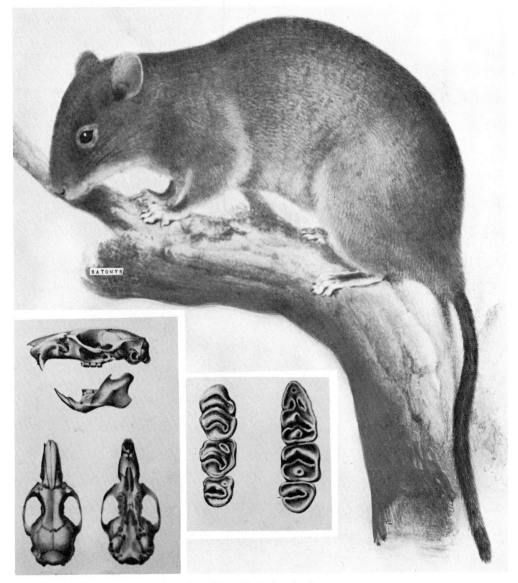

Luzon forest rat (*Batomys granti*), photos from *Trans. Zool. Soc. London.*

B. salomonseni the head and body length of the type specimen is 175 mm, and the tail length is 140 mm; the upper parts are generally dark brown, becoming lighter and more buffy on the sides; the hairs of the underparts have gray bases and long buffy yellow tips; the hands and feet of the type specimen (with the exception of one hand, which is wholly white) have a dark line from the wrist and the ankle to the digits; the remainder of the hands, feet, and digits is white; and the tail is blackish brown above and below, and fairly well haired. The fur of *Batomys* is thick and soft.

These rats resemble *Carpomys* externally, but can be distinguished by their relatively shorter tails, and by cranial and dental characters. In *Batomys,* for example, the skull is more elongate and the bullae are smaller than in *Carpomys.*

Specimens of *Batomys* have been taken in areas of dense vegetation at elevations of about 1,600 to 2,400 meters. Native people, with the aid of their small terriers, captured the specimens of *B. granti.*

RODENTIA; MURIDAE; *Genus CARPOMYS* Thomas, *1895*

Luzon Rats

There are two species (Taylor 1934; Sanborn 1952):

C. melanurus, known only by four specimens from Mount Data in northern Luzon;

C. phaeurus, known only by three specimens from Mount Data and one from Mount Kapilingan in northern Luzon.

In *C. melanurus* head and body length is about 200 mm, and tail length is about 210 mm; the tail is a deep, shining black and is thickly furred for 25 to 50 mm next to the body; and the incisor teeth are quite large. In *C. phaeurus* head and body length is 175 to 195 mm, and tail length is 160 to 180 mm; the tail is dark brown or blackish, but never shining black, and is thickly furred for only a short distance; and the incisors are much smaller than in *C. melanurus.*

Otherwise, both species look much alike. The soft fur is fluffy and thick. The upper parts are deep fulvous, coarsely lined with black, and the underparts are yellowish white. The bases of the hairs are slate or buffy white in *C. melanurus,* but they are not slaty in *C. phaeurus.* The body form in both species is somewhat heavy. The broad hind foot is adapted to arboreal life, but the large toe is clawed and not opposable. The thumb is the only digit with a nail instead of a claw. Females have four mammae.

The known specimens were collected in areas of dense vegetation at elevations of about 2,100 to 2,400 meters. Taylor (1934) stated that both species are arboreal, but the specimens of *C. phaeurus* from Mount Data were obtained by digging them out from among the roots of trees.

Luzon rat (*Carpomys melanurus*), photos from *Trans. Zool. Soc. London.*

Ryukyu spiny rat (*Tokudaia osimensis*), photo by K. Tsuchiya.

RODENTIA; MURIDAE; **Genus TOKUDAIA** Kuroda, 1943

Ryukyu Spiny Rat

The single species, *T. osimensis*, is known from Amami-Oshima and Okinawa in the Ryukyu Islands south of Japan (Ellerman and Morrison-Scott 1966).

Head and body length is 125 to 175 mm, and tail length is 100 to 125 mm. The upper parts are mixed black and orange tawny, and the underparts are grayish white with a faint orange wash. The tail is bicolored throughout. Externally, this rodent looks like a large vole. The body is short and thick, and the pelage is dense. Two kinds of hair have been noted: fine hair, and coarse, grooved spines. The latter are present on all parts of the body except for the region around the mouth and ears, the feet, and the tail. The dorsal spines are black throughout, and the ventral spines are usually white with rufous tips. Females have four mammae.

In the U.S. National Museum of Natural History are 13 specimens that were collected on northern Okinawa in a thick, shrubby forest, about three meters high, with an undergrowth of coarse grasses and brake ferns. The IUCN (1972) classifies *T. osimensis* as indeterminate, and believes the species seriously endangered by the general destruction of natural areas on the Ryukyu Islands.

RODENTIA; MURIDAE; **Genus ANISOMYS** Thomas, 1903

Powerful-toothed Rat

The single species, *A. imitator*, occurs in New Guinea (Laurie and Hill 1954).

Head and body length is 244 to 300 mm, and tail length is 285 to 330 mm. Weight is around 500 to 600 grams (Menzies and Dennis 1979). The coat is composed of short, coarse hairs. The upper parts are blackish fawn, and the underparts are dull buffy white. The head is nearly black. The arms and legs are grayish, and the hands and feet are brown, becoming white on the digits. Hexagonal scales show through the scantily haired tail. *Anisomys* has small ears, strong scansorial feet, and a long tail. The great toe has a broad nail. Females have six mammae.

Anisomys resembles *Uromys* in external appearance, but has highly distinctive dental characters (Menzies and Dennis 1979). The lower incisor teeth are laterally compressed and bladelike, and are less than half the width of the upper incisors. The molar teeth are remarkably small, the length of the toothrow in a given specimen being less than half that of a specimen of *Uromys* with the same external measurements.

Anisomys is apparently distributed throughout the rain forests of New Guinea between elevations of 900 and 2,700 meters (Laurie and Hill 1954). Most specimens have been trapped on the ground, but the broad scansorial feet indicate an ability to climb easily. The small size of the molars may be associated with a diet that includes nuts with hard shells but soft, pulpy interiors. *Anisomys* also eats other vegetable matter and is said to raid gardens (Menzies and Dennis 1979).

RODENTIA; MURIDAE; **Genus PHLOEOMYS** Waterhouse, 1839

Slender-tailed Cloud Rats

There are two species (Taylor 1934):

P. cumingi, southern Luzon, Mindoro, and Marinduque islands (Philippines);
P. pallidus, northern Luzon.

Schauenberg (1978) suggested that *P. pallidus* is only an individual or seasonal variant of *P. cumingi*. He also recognized the establishment of a separate family, the Phloeomyidae, for *Phloeomys*. Taylor (1934) listed another species, *P. elegans*, known only by a single specimen from an unknown locality in the Philippines, but he suspected that it actually belonged to one of the other two named species.

These rats are the largest members of the subfamily

Powerful-toothed rat (*Anisomys imitator*), photos from Museum Zoologicum Bogoriense.

Murinae. Head and body length is 280 to 485 mm, tail length is 200 to 350 mm, and weight is about 1.5 to 2.0 kg. In *P. cumingi* the pelage is rather rough, suberect, and intermixed with long hairs; the general coloration of the upper parts is blackish brown washed with dirty yellowish or reddish yellow, and there may be an irregular reddish brown blotch on the dorsal surface; and the underparts are paler than the upper parts. In *P. pallidus* the pelage is long, dense, and soft in comparison with that of *P. cumingi;* the general coloration of the upper parts results from the longer individual hairs that have brown or reddish brown bases and white tips, and the shorter fur is uniformly brownish; the hairs on the anterior part of the body have gray to brown bases and black tips; the ears are black; the cheeks are gray; and the tail is black to brownish black. In both species the well-haired tail is not bushy.

Phloeomys has broad incisor teeth, a blunt muzzle, and small ears. The feet are large and wide, adapted to arboreal life, and the foreclaws are large. Females have four mammae.

Slender-tailed cloud rats live in forested areas from sea level to high mountains. Information compiled by Schauenberg (1978) suggests that they may shelter in either burrows or hollow tree trunks. They apparently are mainly arboreal and nocturnal. Their movements seem relatively sluggish. The natural diet is not known, but captives have taken a wide variety of vegetable matter.

According to Schauenberg (1978), births in captivity have been recorded in all months except January, March, and May, and a wild pregnant female was taken on 2 August. There is only one young at a time, which the mother carries firmly attached to a nipple. *Phloeomys* thrives in captivity and one specimen lived for 13 years and 7 months.

RODENTIA; MURIDAE; **Genus APODEMUS Kaup, 1829**

Old World Wood and Field Mice

There are 12 species (Corbet 1978; Dolan and Yates 1981):

A. mystacinus, Balkan Peninsula, Asia Minor and adjacent parts of Georgia and Iraq, Palestine, Crete, Rhodes, several Aegean islands;

A. flavicollis, most of Europe, Asia Minor, Palestine;

A. sylvaticus, Europe, parts of central and southwestern Asia, Himalayas, northwestern Africa, British Isles and many nearby islands, Iceland;

A. krkensis, Krk Island off northwestern Yugoslavia;

A. microps, east-central Europe;

A. argenteus, Japan;

A. speciosus, Japan;

A. peninsulae, eastern and southern Siberia, Manchuria, Korea, northeastern and central China, Sakhalin, Hokkaido;

A. draco, southern and eastern China, Assam, Burma;

A. latronum, Szechwan and Yunnan (southern China), northern Burma;

A. semotus, Taiwan;

A. gurkha, Nepal;

A. agrarius, central and eastern Europe, parts of Central Asia and southern Siberia, Manchuria, Korea, eastern and southern China, Taiwan.

Head and body length is 60 to 150 mm, tail length is 70 to 145 mm, and weight is 15 to 50 grams. The fur is usually soft, though it may be bristly in *A. speciosus,* and the tail is moderately haired. The general coloration above is grayish

Slender-tailed cloud rat (*Phloeomys cumingi*), photo by Ernest P. Walker.

buff, grayish brown, brown mixed with yellow or red, light brown, or pale sand color. The underparts are white or grayish, often suffused with yellow, and the hands and feet are usually white. Some forms have a reddish yellow chest patch, and *A. agrarius* has a black middorsal stripe. The tail may be longer or shorter than the head and body. The tail is not prehensile like that of *Micromys,* a genus that resembles *Apodemus* in general appearance. Female *Apodemus* have six or eight mammae.

These mice inhabit grassy fields, cultivated areas, woodlands, and forests. They climb well, though not as well as *Micromys,* and are more active jumpers. They are also good swimmers. Grzimek (1975) wrote that while *A. agrarius* is diurnal, *A. flavicollis* and *A. sylvaticus* are nocturnal or crepuscular. They may move into human habitations in the fall and winter, but generally dig deep burrows, usually with a nest of shredded grass and leaves at the end of the tunnel. *A. flavicollis* sometimes nests between roots, in rocky crevices, or in hollow trees. Jennings (1975) stated that burrows of *A. sylvaticus* are usually about 3 cm wide and 8 to 18 cm below the surface; tunnels of other animals are sometimes incorporated. All burrow systems that he examined consisted of a circular tunnel around the roots of a hazel tree, another tunnel penetrating below the tree to a nest chamber, and other tun-

Yellow-necked mouse (*Apodemus flavicollis*), photo by K. Rudloff through East Berlin Zoo.

Old World field mouse (*Apodemus agrarius*), photos by Ernest P. Walker.

nels leading out to entrances. As has been observed in some other species, the burrows contained stores of food. The diet of *Apodemus* includes roots, grains, seeds, berries, nuts, and insects.

These rodents normally spend their lives within an area about 180 meters in diameter. In a study of *A. sylvaticus* in Ireland, Fairley and Jones (1976) found average range lengths of about 109 meters for males and 64 meters for females. For the same species in England, Crawley (1969) calculated average home ranges of 2,250 sq meters for males and 1,817 sq meters for females. Over a seven-year period in England, Montgomery (1976) found population density to range from 8.7 to 40.1 individuals per ha. in *A. flavicollis* and 2.9 to 18.2 per ha. in *A. sylvaticus*. The latter species may undergo periodic fluctuations in population.

Jennings (1975) observed possible cooperative burrowing in *A. sylvaticus*. In this species, at least, several adults may live in the same nest, though females apparently do not permit the males to enter when young are present. The breeding season of *Apodemus* may vary from year to year, and females produce up to six litters annually. While Larsson, Hansson, and Nyholm (1973) reported winter reproduction for *A. sylvaticus* in north Sweden, Grzimek (1975) stated that both this species and *A. flavicollis* generally breed from March to September. They usually produce four litters annually, averaging 5 young each, after a gestation period of 23 days. The young of these species weigh about 2.5 grams at birth, open their eyes 13 days later, are weaned when 3 weeks old, and reach sexual maturity at 2 months of age. In *A. agrarius* the gestation period is 21 to 23 days and litter size averages 6

young. In a study of *A. microps* in Czechoslovakia, Holisova, Pelikan, and Zejoa (1962) found the reproductive season to last from late February to early October, most pregnancies to occur from March to August, and litter size to average 6.1 and range from 2 to 10 young. Lay (1967) reported a range of 2–10 embryos in pregnant female *A. sylvaticus* in Iran. A captive individual of this species lived for 4 years and 5 months (Marvin L. Jones, Zoological Society of San Diego, pers. comm.), but average life span in wild members of the genus is probably 1 year or less.

In woodland ecosystems of Great Britain, these mice have been found to be important as agents for the transportation and burying of tree seeds, but also for damage and destruction of seedlings. Their influence on regeneration of forests is complex and needs further detailed study (Ashby 1967). *A. speciosus* of Japan is a carrier of scrub typhus, and *A. agrarius* is a possible carrier of hemorrhagic fever.

RODENTIA; MURIDAE; **Genus CHIROPODOMYS** *Peters, 1868*

Pencil-tailed Tree Mice

There are five species (Musser 1979):

C. karlkoopmani, North Pagi Island off western Sumatra;
C. major, northern Borneo;

Pencil-tailed tree mouse (*Chiropodomys gliroides*), photo by Ernest P. Walker.

C. calamianensis, Busuanga, Palawan, and Balabac islands (Philippines);

C. muroides, northeastern Borneo;

C. gliroides, Assam to southeastern China and Malay Peninsula, Sumatra, Java, Bali, Borneo, Natuna Islands.

Head and body length is 66 to 122 mm, tail length is 85 to 171 mm, and weight is 15 to 43 grams (Musser 1979). The pelage is soft, dense, and uniform in length, without conspicuous guard hairs or spines. The general coloration of the upper parts is dull grayish brown, buffy brown, or chestnut, and that of the underparts is white to gray or orange red. The tail is brown, except in *C. karlkoopmani,* in which the basal third of the tail is brown and the remainder is white. The tail is rather thinly covered with short hairs near the base, becoming more or less penciled terminally.

These rodents have short, broad feet and long tails. The first digit of both the hands and the feet is short and stumpy, and has a flat nail. The other digits have short, slightly curved claws. The digit pads are large and the sole of the hind foot is naked. The ears are moderately large, thin, and nearly naked. Females usually have four mammae.

Pencil-tailed tree mice occur in forests and are primarily arboreal. The best-known species is *C. gliroides* (Medway 1978; Musser 1979). It inhabits a variety of forest types and is especially common where there is bamboo. It is most active at night and spends the day in nests in hollow trees or the internodes of bamboo. To reach the internodal space of a standing bamboo stem, the mouse gnaws a circular hole, 25 mm in diameter, in the side of the internode. The opening is then lined with leaves. *C. gliroides* is apparently herbivorous, but little is known about its diet in the wild. Captives have thrived on mixed roots, sweet potatoes, grain, and fruit, occasionally supplemented with fresh bones and raw meat. Individuals are very fierce and not easy to tame, even when born and reared in captivity. Females are polyestrous, with estrous periods of 1 day recurring at intervals of not less than 7 days. One female produced four litters in a 10-month period. Breeding occurs throughout the year in the Malay Peninsula, but apparently peaks from September to March. A gestation period lasted 19 to 21 days. Litter size is 1 to 4, averaging 2.4, young. The newborn cling persistently to their mother's nipples and will be dragged about if she is disturbed. They are partly independent at 17 days of age, completely weaned when 1 month old, and fully mature at 100 days. Mean longevity in the wild has been calculated as 23.8 months, but captives have lived as long as 43 months.

RODENTIA; MURIDAE; **Genus VERNAYA** Anthony, 1941

Vernay's Climbing Mice

There are two species (Musser 1979; Wang, Hu, and Chen 1980):

V. fulva, known only by two specimens from northern Burma and one from Yunnan (southern China);

V. foramena, Szechwan (central China).

It is possible that other specimens of *Vernaya* are in collections under the names *Chiropodomys, Vandeleuria,* or *Micromys,* as these genera are so similar in appearance that they can be distinguished only by careful examination.

In *V. fulva,* head and body length is about 90 mm, and tail length is about 115 mm. This species resembles *Vandeleuria* in the nature of the pelage, the reddish color of the coat, and the long, nontufted tail. In *V. foramena,* the tail is also longer than the head and body, the general color is cinnamon brown, and the dorsal pelage is long, thick, and velvety.

Vernaya differs from *Vandeleuria* and *Chiropodomys* in the development of the digits of the hands and feet. In *Vernaya,* all digits except the pollex, which is vestigial and has an extremely small, flat nail, have characteristic pointed claws. The incisor teeth are not grooved. Female *V. foramena* have six mammae.

The specimens of *V. fulva* were collected at elevations of around 2,100 to 2,700 meters; the Burmese material was taken in areas of low, dense vegetation among low cliffs and rocky outcrops (Musser 1979). The specimens of *V. foramena* were collected in subalpine coniferous forest at elevations of 2,460 to 2,500 meters (Wang, Hu, and Chen 1980).

RODENTIA; MURIDAE; **Genus VANDELEURIA** Gray, *1842*

Long-tailed Climbing Mice

There are apparently two species (Lekagul and McNeely 1977; Musser 1979):

V. oleracea, western India to southern China and Viet Nam, lowlands of Sri Lanka;

V. nolthenii, highlands of Sri Lanka.

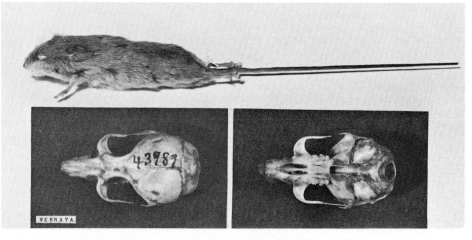

Vernay's climbing mouse (*Vernaya fulva*), photos from American Museum of Natural History.

Long-tailed climbing mouse (*Vandeleuria oleracea*), photo by Rita Maser.

Head and body length is about 55 to 85 mm, and tail length is usually 90 to 130 mm. Weight is approximately 10 grams (Lekagul and McNeely 1977). The upper parts vary from pale buffy through dull brown to dark reddish brown, and the underparts are white or cream color. The fur is full, soft, and almost silky. The tail is fairly well haired, but is not tufted at the tip as in *Chiropodomys*.

This genus of small tree mice is distinguished by the structure of the hands and feet, and by features of the teeth and skull. The first and fifth digits on both the hand and the foot have a flat nail instead of a claw; the claws on the remaining digits are small. The limbs are adapted to grasping and climbing by the opposition of the first digit on both the hand and the foot, the shortness of the foot, and the development of the terminal pads of the fingers and toes. This genus may be distinguished from *Chiropodomys*, which has similar hands and feet, by its mouselike size and nontufted tail. Females of *Vandeleuria* have eight mammae.

These mice are nocturnal and essentially arboreal. They usually spend the day in a nest in the branches of a tree, in a hole in a tree trunk, or, occasionally, on rooftops. The nest is made of grass, sometimes mixed with dried leaves. In Thailand, they may live in tall cane, where they build a globular nest of grass a couple of meters above the ground (Lekagul and McNeely 1977). These mice are quite active as they run among branches and twigs, using the tail as a balancing organ while they run up or down vertical shoots and out to the tips of slender twigs. Sometimes the long, almost prehensile tail is wound loosely around a twig to steady the animal. Long-tailed climbing mice are much less active and are slower on

the ground, where they come only to collect leaves and grasses for their nests.

The diet in the wild is probably fruit and the buds and shoots of trees and shrubs. An individual kept in captivity thrived on bread, milk, and such fruits as plantain and pawpaw. This captive specimen became fairly tame after a few weeks, and spent most of the day asleep with its tail curled around its body.

Apparently, the older males live alone. The young are born and raised in a nest. Usually three or four are born at one time, but litters of six young have been seen.

RODENTIA; MURIDAE; **Genus MICROMYS** Dehne, 1841

Old World Harvest Mouse

The single species, *M. minutus*, occurs from western Europe to east-central Siberia and Korea, in parts of southern China and Assam, in Great Britain and Japan, and on Taiwan (Corbet 1978).

This is among the smallest of rodents. Head and body length is 55 to 75 mm, tail length is 50 to 75 mm, and weight is usually 5 to 7 grams. The pelage is brownish with a yellowish or russet tinge on the upper parts, and is white to buffy beneath. The tail is bicolored. The fur is somewhat longer in winter than in summer. Typical features include the short, rounded head; small, rounded ears; fairly broad feet; and hairless condition of the upper portion of the tip of the tail.

Old World harvest mice (*Micromys minutus*), photo by Eric J. Hosking.

The foot structure facilitates scampering up stems, and the nakedness of the tail suggests that it is semiprehensile.

The remainder of this account is based in large part on the review papers by Trout (1978a, 1978b). The Old World harvest mouse usually is found in tall vegetation, such as hedgerows, weed beds, tall grass, reeds, grain or rice fields, and bamboo thickets. During the spring and summer breeding season, special nests are constructed, one for each litter of young. These nests are globular in shape, about 60 to 130 mm in diameter, and usually suspended between vertical grass stems about 100 to 130 cm above the ground. They are composed of three layers of grass leaves, compactly woven together. The leaves of the inner layer are finely shredded to form a soft lining for the young. There are one or more entrance holes, but these are kept closed by the female during the first week following birth. Such a nest requires 2 to 10 days to build. Nonbreeding animals may construct similar nests, but these are more flimsy and lack the inner lining. *Micromys* sometimes builds its nest on the ground or in a hole, especially during cooler months, and at such times it also may live under haystacks or straw, or in structures made

by people. It does not hibernate and may be active by day or night. Most investigations, however, have found it to be primarily nocturnal, with peaks of activity after dark and before dawn. Its diet consists mainly of seeds, green vegetation, and insects, but it is known sometimes to take the eggs of small birds and to eat meat in captivity. Storage of food has been noted in captivity but not confirmed in the wild.

Micromys normally lives within small, overlapping home ranges. One study found the individual range size to average 400 sq meters for males and 350 sq meters for females. Population levels fluctuate over the years and seasonally, generally reaching a peak in the autumn. Maximum annual densities during a three-year period at three different sites in England varied from 17 to 207 individuals per ha. During the winter, however, large numbers of mice may congregate in a barn or grain storage facility. As many as 5,000 individuals have been found in such structures. Studies have shown that the animals tolerate one another under such conditions, but that aggression increases as density declines and the breeding season approaches. If two captive males are placed together they will fight savagely. Adults of opposite sexes come together only to construct a breeding nest and to mate; the female then drives the male away. Vocalizations include a chatter during courtship and an aggressive squeal.

Reproduction is generally concentrated during the warmer, drier months, starting around April or May and peaking from July to September, but sometimes lasting until late autumn. Females are polyestrous, undergo a postpartum estrus, and, under favorable conditions, can give birth several times in rapid succession. Because of the short natural longevity, a female normally produces only one or two litters in its lifetime, but captives have produced up to nine. The gestation period and minimum interval between litters are about the same, 17–18 days. The number of young per litter ranges from 1 to 13, usually is around 3 to 8, and has averaged about 5 in England and 7 in Bulgaria. The young weigh about a gram at birth, open their eyes and are fully furred at 8–10 days, are weaned and leave the nest at 15–16 days, and reach sexual maturity at as early as 35 days. As many as four generations of mice may breed during one reproductive season. Few individuals live over 6 months; maximum known longevity is 16–18 months in the wild and just under 5 years in captivity.

Although *Micromys* is widely distributed, in areas where modern farm machinery is used this little mouse appears to be on the decline. It food and shelter may be destroyed by reaping machines that leave a much shorter stubble than does a manual scythe.

RODENTIA; MURIDAE; **Genus HAPALOMYS** Blyth, 1859

Asiatic Climbing Rats, Marmoset Rats

There are two species (Musser 1972a):

H. longicaudatus, southeastern Burma, southern Thailand, Malay Peninsula;
H. delacouri, Indochina, Hainan.

Head and body length is 121 to 168 mm, and tail length is 140 to 202 mm. In some forms the tail is considerably longer than the head and body, but in others it is only slightly longer. The fur is thick and soft. The coloration above is buffy, dull reddish gray, or grayish brown. The sides are generally paler than the back. The underparts, including all four legs to below the knee, are white. The limbs are usually buffy gray, and the hands and feet are buffy gray or brownish white. The

Asiatic climbing rat (*Hapalomys longicaudatus*), photos by Jane Burton. Inset: skull, photo by P. F. Wright of specimen in U.S. National Museum.

Acacia rat (*Thallomys paedulcus*), photo from the Nyasaland Museum through P. Hanney.

tail is thinly haired and sometimes penciled terminally. The vibrissae are prominent and slightly longer than the head.

The hind feet are specialized for arboreal life, in that they have enlarged toe pads consisting of two flat plates with a groove between them. The toes are long and slender, the middle three being almost equal in length. The large toe is wide, without a claw, and opposable. There are four fingers; the second, third, and fourth each has a small claw nearly embedded in the pad; the first finger is said to be little more than a slight projection on the inner side of the hand and to have no trace of a nail. The general body form of *Hapalomys* is ratlike. Females have eight mammae.

The incisor teeth are broadened and powerful. *Hapalomys* is the only murid genus with three rows of approximately equally developed cusps on the lower cheek teeth; some authorities consider this a very primitive character.

These mice are arboreal and inhabit tropical forests. According to Medway (1964, 1978), *H. longicaudus* is a skillful climber and appears to be associated strictly with bamboo. By day it retires to a nest in the internodal cavities of dead or living bamboo stems. It gains access to the internodes by gnawing a circular hole, about 35 mm in diameter, in the outer wall. All nests examined have been lined exclusively with bamboo leaves. The natural diet appears to consist of the shoots, flowers, and fruit of bamboo.

RODENTIA; MURIDAE; *Genus THALLOMYS* Thomas, 1920

Acacia Rat

The single species, *T. paedulcus*, occurs from southern Ethiopia and western Somalia to Angola and South Africa (Meester and Setzer 1971; Yalden, Largen, and Kock 1976).

Head and body length is 120 to 162 mm, tail length is 130 to 210 mm, and weight is 63 to 100 grams (Kingdon 1974b; Smithers 1971). The fur is generally quite long. In some animals it is soft and fine, in others it is rather coarse. The coloration of the upper parts ranges from pale buff through yellow fawn to brownish gray. The dorsal surface is usually darker than the sides of the body. The sides of the face are generally grayish. The underparts are white. The tail is generally brown. The hands and feet are pale grayish or white. The scantily haired ears have a reddish tinge. *Thallomys* is characterized by its long tail and short feet. Females have three pairs of mammae.

As the common name suggests, this rodent is usually associated with acacia trees, where it nests in forks or branches, in crevices in the trunk, in hollow limbs, under loose bark, or in holes in the ground at the base of the tree. The nests are built of twigs and other vegetable matter. *Thallomys* is shy, but quite active; it peers from its nest at the slightest strange noise and runs to the dense, thorny foliage to hide from predators. It is an expert climber, but occasionally will drop to the ground in an effort to escape danger. It normally emerges from its retreat in the early evening to search for food. The diet is based principally on acacia trees (Kingdon 1974b). Buds and leaves are staples and are eaten in the canopy, gum is gnawed directly off the bark, and acacia and grass seeds are foraged for on the ground beneath the trees. Berries, roots, and an occasional insect are also taken.

A large tree occasionally harbors a small colony of *Thallomys*. The group probably represents a family, since not more than two pairs of adults and their young have been found together in a single tree. In a study of captives from the Transvaal, Meester and Hallett (1970) learned that no litters

were produced in the winter period from mid-April to mid-July, that the minimum interval between litters was 26 days, that mean litter size was 2.7 young, and that the minimum age at which a female produced a litter was 107 days. For Botswana, Smithers (1971) reported that young apparently are born from October through May, and litter size ranges from 2 to 5 young. In Zambia, Sheppe (1973) found a pregnant female with 7 embryos on 26 March. Kingdon (1974b) stated that in East Africa the breeding season centers on the rains, with pregnant females having been recorded in April and young animals in July and August. According to Marvin L. Jones (Zoological Society of San Diego, pers. comm.), a captive specimen lived for 3 years and 6 months.

RODENTIA; MURIDAE; *Genus THAMNOMYS* Thomas, 1907

Thicket Rats

Meester and Setzer (1971) listed two subgenera and four species:

subgenus *Thamnomys*

T. venustus, eastern Zaire and adjacent parts of Uganda and Rwanda;

subgenus *Grammomys*

T. rutilans, Guinea to Uganda and northwestern Angola;
T. cometes, Kenya to Rhodesia;
T. dolichurus, Guinea and Mali to Sudan, and south to Angola and southeastern South Africa.

The last species also has been recorded from Senegal under the name *T. buntingi* (Hubert, Adam, and Poulet 1973). In addition, Olert, Dieterlen, and Rupp (1978) discussed, but did not name, a new species, related to *T. dolichurus*, from southern Ethiopia.

There has been disagreement regarding the systematics of this genus. *Grammomys* Thomas, 1915 has often been treated as a distinct genus, and the species *T. rutilans* has sometimes been grouped with *T. venustus*. Primarily on the basis of dental characters, Rosevear (1969) considered *Grammomys* a separate genus, but stated that it was very possible that *T. rutilans* was more closely related to *Grammomys* than to *T. venustus*. Most subsequent authorities (Meester and Setzer 1971; Ansell 1978; Ansell and Ansell 1973; Delany 1975; Olert, Dieterlen, and Rupp 1978) have treated *Grammomys* as a subgenus that includes *T. rutilans*.

Head and body length is 85 to 161 mm, tail length is 130 to 222 mm, and weight is 30 to 100 grams (Kingdon 1974b). The pelage is fairly long and soft in some animals, but crisp in others. At higher elevations the fur on the back may be 20 mm long. The coloration of the back ranges from dark brown or reddish brown to orange rufous and clay color. The flanks are usually paler. The underparts are white, buffy, or gray. In *T. venustus* the tail is fairly well haired and slightly tufted at the tip. In the other species the tail is scantily haired, except for the terminal third, which is somewhat bushy. *Thamnomys* is characterized by a long tail and short, curved claws. Females usually have four inguinal mammae, but sometimes have two additional abdominal mammae.

According to Kingdon (1974b), *T. venustus* inhabits montane forests, *T. rutilans* prefers farms or areas of rank secondary growth within forests, *T. dolichurus* is adapted primarily to tall grass and secondary scrub, and *T. cometes* lives in wet forests. All species are arboreal and nocturnal. They con-

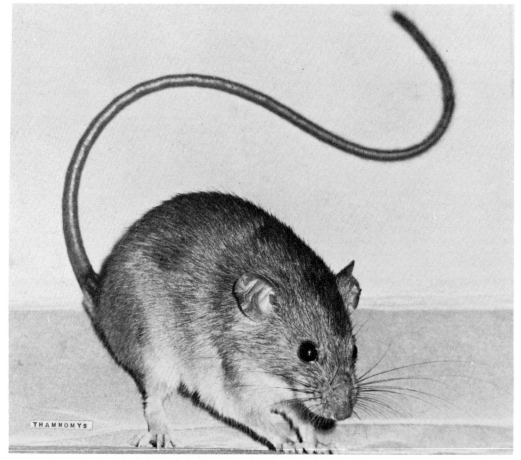

Thicket rat (*Thamnomys venustus*), photo by F. Petter.

struct formless nests of leaves and dried grass in hollow trees, forks, tangles of vegetation, or human buildings. Rosevear (1969) stated that *T. rutilans* normally lives at heights of about 1 to 2 meters above the ground and makes a nest of very fine, threadlike strips of grass, gathered into a kind of bag about 150 mm long by 100 mm wide. The nests of *T. dolichurus* are placed at heights of 0.5 to 4 meters (Delany 1975). The diet of *T. dolichurus* consists mainly of green stems, fruits, nuts, flowers, and other vegetable matter, but also includes insects; other species probably take the same foods (Kingdon 1974*b*).

Thamnomys seems usually to be solitary, though one observer reported seeing the parents and young of two litters in the same nest. Delany (1975) stated that in Uganda female *T. dolichurus* give birth from May to July and in November, apparently at intervals of 5 to 6 weeks. In a study of captive *T. dolichurus*, however, Bland (1973) determined that pregnancy could occur every 28 to 29 days. It was also learned that the gestation period of six litters was 24 days, that litter size ranged from one to seven (usually being two to four) young, and that sexual maturity was attained at an age of 50 to 70 days. Other reproductive data, reported by Delany (1975), are that pregnant female *T. venustus* collected in Zaire contained one or two embryos each, that pregnant female *T. rutilans* taken in Uganda in March had one to three embryos, and that the gestation period of the latter species is about 25 days. Kindgon (1974*b*) observed that the young of *Thamnomys* remain firmly attached to the nipples of the

mother for a period of about 2 weeks. A captive specimen of *T. dolichurus* was still living after 4 years and 5 months (Marvin L. Jones, Zoological Society of San Diego, pers. comm.).

Kingdon (1974*b*) stated that *T. dolichurus* has become an important experimental animal in malarial research.

RODENTIA; MURIDAE; **Genus STENOCEPHALEMYS**
Frick, 1914

Ethiopian Narrow-headed Rats

There are two species (Yalden, Largen, and Kock 1976):

S. albocaudata, northern and central Ethiopia;
S. griseicauda, south-central Ethiopia.

Head and body length is 120 to 195 mm, and tail length is 114 to 165 mm (Van der Straeten 1981). The thick fur is soft and long, and the tail is fairly well haired. The coloration of the back is mottled umber, the mottling caused by intermixed black hairs. The sides are pinkish buff and the underparts are gray. The outer parts of the upper arms and thighs are pinkish buff mixed with gray, and the inner parts are gray. The lower arms and legs, and the hands and feet, are generally white. In *S. albocaudata* the tail is mostly white, while in *S. griseicauda* the tail is usually gray on the dorsal side.

Ethiopian narrow-headed rat (*Stenocephalemys albocaudata*), photos by D. W. Yalden.

Stenocephalemys has large ears and a small first digit on the hand, which has a nail instead of a claw. *Stenocephalemys* is distinguished from *Rattus* by the extremely constricted frontal bones of its skull. *Dasymys* also has narrow frontals, but differs from *Stenocephalemys* in molar pattern and other cranial features. Misonne (1969) considered *Stenocephalemys* an advanced version of *Thamnomys*. Yalden, Largen, and Kock (1976) stated that *Stenocephalemys* is closely related to *Praomys*, but is much larger and has a thicker coat.

According to Yalden, Largen, and Kock (1976), *Stenocephalemys* occurs in scrub and grassland at elevations of 3,000 to 4,000 meters. D. W. Yalden (University of Manchester, pers. comm.) reported that *S. albocaudata* is strictly nocturnal, and often uses burrows and runways that are utilized by *Arvicanthis* during the daytime.

RODENTIA; MURIDAE; **Genus OENOMYS** Thomas, 1904

Rufous-nosed Rat

The single species, *O. hypoxanthus,* occurs from Sierra Leone to central Ethiopia, and south to northern Angola and southwestern Tanzania (Meester and Setzer 1971; Dieterlen and Rupp 1976).

Head and body length is 105 to 220 mm, tail length is 135 to 205 mm, and weight is 50 to 121 grams (Kingdon 1974*b;* Delany 1975; Rosevear 1969). The pelage consists of soft, woolly underfur, about 15 to 17 mm in length, and long, fine guard hairs, which on the back reach a length of 25 to 30 mm. The general coloration of the upper parts ranges from sepia, rufous, and olive to slate, occasionally brushed with black. The rump is generally somewhat reddish, and the flanks are somewhat paler than the rump. The nose, or at least the sides of the face, is reddish, thereby providing the common name. The underparts are white, often tinged with buff, and some forms have a definite buffy line where the pale color of the underparts joins the darker coloration of the sides. The feet and hands are white, brownish gray, or rufous.

The tail is practically naked, so the large scales are exposed (there are about 12 scales per cm). The ears are fairly large, rounded, and hairy. The hind feet have naked soles and large plantar pads, and the fifth digit is slightly longer than the first. Females have four, six, or eight mammae (Kingdon 1974*b*).

The rufous-nosed rat is plentiful in forest clearings at elevations of about 300 to 3,000 meters. It favors moist areas with thick vegetation. Available information (Kingdon 1974*b;* Rosevear 1969; Delany 1975) indicates that *Oenomys* is semiarboreal and a good climber, often being found in vegetation a few meters above the ground. Its nests may be

Rufous-nosed rat (*Oenomys hypoxanthus*), photo by F. Petter.

built underground in tunnels, directly on the surface among grass roots or under piles of debris, or in shrubs or the fork of a tree at heights of 1 to 3 meters. The nests, which are used for both sleeping and breeding, are woven of dry grass and other plant material, and measure about 150 to 200 mm in diameter. Both nocturnal and diurnal activity has been reported. The diet consists mainly of fresh green vegetation, but insects also are taken. *Oenomys* reportedly damages growing rice, millet, and other crops. Recorded population densities are 17.3 individuals per ha. along the borders of a marsh, and 12.5 per ha. in marshland. Adult individuals are usually found alone. Breeding probably occurs throughout the year, at least in Zaire and western Uganda. Litter size is one to six, usually two to four, young.

RODENTIA; MURIDAE; **Genus MESEMBRIOMYS**
Palmer, 1906

Tree Rats

There are two species (Ride 1970):

M. macrurus, northern parts of Western Australia and Northern Territory;

M. gouldii, northeastern Western Australia to Cape York Peninsula of northern Queensland, Melville and Bathurst islands.

In *M. gouldii* head and body length is 240 to 350 mm, and tail length is 270 to 390 mm. Crichton (1969) reported it to be the largest Australian rat, with a weight just under 1 kg. In *M. macrurus* head and body length is 240 to 260 mm, and tail length is 350 to 370 mm. The fur in both species is rather rough. The upper parts of *M. gouldii* are yellowish gray or yellowish brown, with black guard hairs, and the underparts are whitish, creamy, or slaty gray. The hands and feet are black, though the feet are irregularly blotched with cream and brown. The upper parts of *M. macrurus* are usually bright rufous or buffy brown, the sides are buffy gray, and the underparts are creamy white. The hands and feet are white. In both *M. gouldii* and *M. macrurus* the tip of the white tail is tufted, but the tail is not so well haired near its base, where the scales can be traced.

The ears are moderate in size and their outer surfaces are quite densely haired. The foreclaws are fairly large, and the hind foot is somewhat broad, with large claws. The skull is heavily built. Females have four mammae.

Tree rats are found in woodland savannah (Ride 1970). They are nocturnal and arboreal, but may descend to the ground to feed. By day they shelter in hollow trees. At times they live near wooden buildings and climb about the rafters at night. A nest, mainly of leaves and bark, may be located in a tree. One nest, covering part of the main cavity of the trunk, was found in a tree about 3.5 meters above the ground. Studies of fecal samples showed that in Queensland large seeds or nuts are a major food of *M. gouldii*, and that some insects also are taken (Watts 1977). According to Ride

Tree rat (*Mesembriomys gouldii*), photo by A. C. Robinson.

(1970), all naturalists who have handled *M. gouldii* have noted its savage temper and the severity of its bite. When enraged, it is loud voiced, and is said to raise its voice progressively into a sort of whirring, machinelike crescendo.

In Western Australia *M. macrurus* has been reported to breed throughout the year; a pregnant female *M. gouldii* has been recorded in mid-August (Taylor and Horner 1971). In a laboratory study of *M. gouldii*, Crichton (1969) found females to be polyestrous, to have a mean estrous cycle of 26 days, and to undergo postpartum estrus and mating. The gestation period was 43–44 days and litters contained one to three young. The young weighed about 35 grams at birth, opened their eyes after 11 days, clung tenaciously to the teats for the first few weeks of life, and were weaned at 42 days. Females attained sexual maturity when about 2 or 3 months old. According to Marvin L. Jones (Zoological Society of San Diego, pers. comm.), a captive *M. gouldii* lived for 3 years and 11 months.

Crichton (1969) indicated that *M. gouldii* has declined considerably in numbers. Watts (1979) listed both *M. gouldii* and *M. macrurus* as rare, and thought the latter species to be "at risk."

RODENTIA; MURIDAE; **Genus CONILURUS** *Ogilby, 1838*

Tree Rats, Rabbit Rats

There are two species (Ride 1970; Menzies and Dennis 1979):

C. penicillatus, western Papua New Guinea, northeastern Western Australia, northern Northern Territory and nearby Melville Island;
C. albipes, southern Queensland, New South Wales, Victoria, South Australia.

Head and body length is 165 to 200 mm, and tail length is 180 to 215 mm. The pelage of *C. albipes* is close and soft, while that of *C. penicillatus* is rigid and almost spiny. Coloration of the upper parts ranges from blackish brown to grayish and sandy, with tinges of buff. The underparts and the feet are white or buffy, and occasionally there is a rusty tinge on the nape and crown. The tail of *C. albipes* is distinctly bicolored, dark brown above and white below for its entire length, whereas the tail of *C. penicillatus* is wholly black or has a terminal black tip. The tail is usually uniformly haired throughout and tufted at the tip. *Conilurus* has fairly large ears and a moderately long hind foot. Females have four mammae.

Rabbit rats are found in a variety of habitats—beaches along the salt water, swamps, grassy plains, and well-timbered areas. *C. penicillatus* is nocturnal in habit, sleeping by day in hollow trees and limbs, where it constructs a comfortable, well-lined nest of pieces of vegetation. In the evenings it may run along the edge of the surf, apparently feeding on matter that is thrown up on the beach by the waves.

Taylor and Horner (1971) studied a population of *C. penicillatus* on the Cobourg Peninsula of the Northern Territory. Favored habitats were near the beach in either grassy or sand dune areas. The animals were nocturnal, and both arboreal and terrestrial activity was noted. If breeding was seasonal, it apparently extended from May through, or even beyond, August. Three pregnant females each contained two embryos. Cited observations indicate that the young cling tenaciously to the nipples, even when the female scampers about away from the nest.

Although *C. penicillatus* is still common in the Northern Territory, *C. albipes* has not been seen alive in the twentieth century (Ride 1970). Watts (1979) presumed the latter species to be extinct, and suggested that possible causes were habitat disruption by introduced rabbits and livestock, and predation by introduced foxes.

Rabbit rat (*Conilurus penicillatus*), photo by H. and J. Beste.

RODENTIA; MURIDAE; **Genus *LEPORILLUS* Thomas,**
1906

Australian Stick-nest Rats

There are two species (Ride 1970):

L. conditor, South Australia and nearby Franklin Island,
western New South Wales, possibly northwestern
Victoria;
L. apicalis, central and southeastern Australia.

Head and body length is about 140 to 200 mm. In *L.
conditor* the tail is shorter than the head and body, while in *L.
apicalis* the tail is longer and may be as much as 250 mm. The
tail is fairly well haired, with somewhat longer hairs toward
the tip. The pelage is thick and soft. The upper parts are light
yellowish brown, light brown, dull brown, or pale grayish
brown. The underparts in *L. conditor* are usually grayish,
whereas in *L. apicalis* they are white. When resting, these
fluffy-haired animals, with their blunt noses and large ears,
look like small rabbits with ratlike tails. Females have four
mammae.

According to Ride (1970), stick nest rats inhabit scle-
rophyll woodland, shrub and tree heaths, and salt bush plain.
They are nocturnal and build houses of sticks, which presumably enable them to withstand the desiccating heat of the

desert days. Robinson (1975) found that these houses range
from just a few twigs against a bush to elaborate structures up
to 1.5 meters high. Size and construction method depend on
local conditions. Houses are often built around a small bush
or placed against a rock, but in areas where there is little
woody growth *Leporillus* may live in loose heaps of sticks
placed over rabbit warrens that allow instant escape. The
latter type of home often has small stones placed on top and
among the sticks to anchor the structure against strong winds.
The shelters, or "wurlies," contain numerous passageways
with several grass nests. They are shared at times with bandicoots, penguins, and even snakes. Along the beaches the
shelter may consist of some debris or seaweed placed among
the boulders. The species *L. apicalis* is not well known; it
was said to shelter in hollow trees and to occupy the abandoned homes of *L. conditor,* but it probably did build its own
nest (Ride 1970).

The population of *L. conditor* on Franklin Island is independent of fresh water, and the staple diet there was once
reported to consist of the succulent *Tetragona implexicana*
(Ride 1970). More recently, Robinson (1975) found no remains of this plant in stomachs of *L. conditor* collected on
Franklin Island, but did confirm that the diet is entirely
vegetarian.

Stick-nest rats are usually gentle and tame in captivity.
They are gregarious, and some large houses may shelter a
colony of animals, each family occupying one of a number of

Australian stick-nest rat (*Leporillus conditor*), photo from the South Australian Museum.

interconnected cavities (Australian National Parks and Wildlife Service 1978). Robinson (1975) found *L. conditor* to be an opportunistic breeder on Franklin Island, with reproduction apparently possible at any time of the year when conditions are proper. Gestation lasts about 44 days and litters contain one or two young. Ride (1970) stated that females drag their young around attached to their nipples.

Both species of *Leporillus* have declined drastically, and *L. apicalis* is apparently already extinct (Watts 1979; Australian National Parks and Wildlife Service 1978; Robinson 1975). Suggested factors in the decline include competition from introduced sheep and rabbits, habitat destruction by people and livestock, and killing by introduced predators. *L. conditor* may survive in some palces on the mainland, but specimens were last collected there in 1921. The only currently known population is that on the 500-ha. Franklin Island, and estimates of its numbers have ranged from 1,500 to 5,000. *L. conditor* is listed as endangered by the USDI (1980), and is on appendix 1 of the CITES.

rats and mice lose their tails with ease, the hair and flesh strip from the vertebrae of rock rats especially easily; the animals soon amputate the remaining skeletal section to leave a shortened stump (Ride 1970). The feet of *Zyzomys* are short and broad, the ears are fairly small and rounded, and females have four mammae. The skin is quite thin and easily torn.

These rodents occur among rocky outcrops and are thought to be nocturnal. Watts (1977) found the feces of wild *Z. woodwardi* to contain the remains of the fruit and seeds of a dicotyledon, and the feces of *Z. argurus* to contain mostly plant, but also some insect, remains. Begg and Dunlop (1980) identified 11 species of plant from seeds accumulated in crevices by *Z. woodwardi*. Ride (1970) stated that female rock rats drag their young around attached to their nipples.

Watts (1979) considered *Z. argurus* to be generally common, but *Z. woodwardi* and *Z. pedunculatus* to be rare. *Z. pedunculatus* was last collected in 1960; the species is listed as endangered by the USDI (1980) and is on appendix 1 of the CITES.

RODENTIA; MURIDAE; *Genus ZYZOMYS* Thomas, 1909

Thick-tailed Rats, Rock Rats

There are three species (Ride 1970):

Z. argurus, northern Western Australia to northern Queensland,
Z. woodwardi, northeastern Western Australia, northern Northern Territory;
Z. pedunculatus, southern Northern Territory.

Head and body length is 85 to 177 mm, and tail length is 91 to 135 mm. Gordon and Johnson (1973) listed weights of 35 to 65 grams for eight specimens of *Z. argurus* from Queensland. The pelage is crisp, harsh and almost spinous; it is brownish, grayish, or buffy, to reddish sandy on the dorsal surface, and shades into white on the ventral surface. On some animals all ventral hairs are white, whereas on others the white-tipped hairs have grayish bases. The hands and feet are white. The tail is bicolored in some forms and entirely white in others.

These delicately built animals have a characteristic thickening of the tail, which is caused by a proliferation of the skin; it is not primarily a reservoir of fat, as is the case in some mammals. The oldest individuals have the thickest tails, while the youngest specimens show only a slight enlargement. The tails are very fragile. Although some other

RODENTIA; MURIDAE; *Genus PSEUDOMYS* Gray, 1832

Australian Native Mice

There are 20 species (Ride 1970; Mahoney and Posamentier 1975; Fox and Briscoe 1980; Baverstock, Watts, and Cole 1977; Watts 1979; Borsboom 1975; Hocking 1980; Posamentier and Recher 1974; Kitchener 1980):

P. delicatulus, northern Australia;
P. pilligaensis, northern New South Wales;
P. hermannsburgensis, arid parts of Australia;
P. chapmani, Hammersley Range in northwestern Western Australia;
P. novaehollandiae, coastal New South Wales, Mornington Peninsula of Victoria, northeastern Tasmania;
P. albocinereus, southwestern Western Australia and islands of adjacent Shark Bay;
P. apodemoides, southeastern Australia;
P. glaucus, southern Queensland;
P. fumeus, Victoria;
P. occidentalis, southwestern Western Australia;
P. praeconis, Peron Peninsula and islands of Shark Bay in Western Australia;
P. gouldii, southern Western Australia, South Australia, western New South Wales;
P. australis, central and southern Queensland, inland New South Wales, South Australia;

Thick-tailed rat: Top, *Zyzomys argurus,* photo from Queensland Museum; Bottom, *Z. woodwardi,* photo by H. and J. Beste.

Australian native mouse (*Pseudomys nanus*), photo by W. D. L. Ride.

P. fieldi, vicinity of Alice Springs in southern Northern Territory;

P. higginsi, Tasmania;

P. oralis, southeastern Queensland, northeastern New South Wales;

P. shortridgei, southwestern Western Australia, western Victoria;

P. desertor, western and central Australia;

P. gracilicaudatus, northeastern Queensland to coastal New South Wales;

P. nanus, Western Australia, Northern Territory.

Ride (1970) regarded *Leggadina* Thomas, 1910 as part of *Pseudomys,* and thus included the species *L. forresti* within the latter genus. Subsequent authorities have tended to maintain *Leggadina* as a separate genus, with *L. forresti* as one of its species (Mahoney and Posamentier 1975; Watts 1979; Fox and Briscoe 1980). *Leggadina* also is sometimes considered to include the first five species in the above list. *Gyomys* Thomas, 1910 and *Thetomys* Thomas, 1910 are sometimes employed as genera or subgenera for a number of the species listed above. According to Mahoney and Posamentier (1975), however, the characters of these two taxa do not hold up and it now seems best not to use subgenera within *Pseudomys.*

Head and body length is 60 to 160 mm, tail length is 60 to 180 mm, and weight is 12 to 90 grams. The pelage is soft in most species, but is harsh in *P. desertor.* The upper parts of the different species are various shades of gray, brown, or yellow. The underparts are usually white, grayish white, or yellowish white. The head and ears are usually paler than the remainder of the upper parts. The tail is moderately haired and in most species is bicolored, being dark above and white to buffy below. Ride (1970) observed that these mice are rather unspecialized in appearance, some species resembling big fluffy rats and others being as small as very tiny mice. The ears are relatively small in some species, but large in others. Depending on the species, the tail is either shorter than, equal to, or longer than head and body length. Females have two inguinal pairs of mammae.

Habitats of the different species include sandy plains, stony ridges, coastal dunes, grasslands, shrub and tree heaths, swamps, timbered areas, and rain forests. Ecology and behavior are not well known, but evidently these mice are nocturnal and usually spend the day in burrows. *P. australis* sometimes burrows in clay flats and river banks. On the flats its burrows are not more than 150 to 200 mm deep, usually have only one opening, and lack side passages. The burrows of *P. delicatulus* are located some 60 mm below the surface, and, after about 1.5 meters, they terminate in a circular space containing a nest of dried grass.

The burrows of *P. hermannsburgensis* are about 200 mm in depth and 1 meter in length. This species apparently builds low mounds of pebbles over its burrow systems. These pebbles are of a uniform size and cover a large area, often 1 meter in diameter. The pebbles are probably collected both by excavation and from the surface. Some local mammalogists believe these are used as dew traps. Since the air around the pebbles warms more rapidly as the sun rises than the pebbles themselves, dew forms on the pebbles by condensation. As the areas in which these mounds are found are quite dry, except after heavy rain, these dew traps solve the problem of water shortage. Local farmers use the many pebble mounds for mixing concrete. It is believed that the ancient people of the Mediterranean region used a dew trap method comparable to that of *P. hermannsburgensis.*

The species *P. albocinereus* tunnels in the sand to about one meter below the surface. The burrows have several openings, at least one of which is used exclusively for throwing sand to the surface from the tunnel. Many of the entrances have surplus sand piled around the opening. Most of the occupied burrows are difficult to locate, as the mice often close the entrances from the inside. One observer on Bernier Island noted that the mice surrounded the open "escape exits" by a delicate network of small twigs and shredded vegetation. It is believed that this barricade prevents loose sand from filling the openings.

The diet of Australian native mice is thought to consist of a variety of seeds, roots, other vegetable matter, and sand-dwelling insects. Watts (1977) suggested that several species

Australian native mouse (*Pseudomys delicatulus*), photo by Basil Marlow. Inset: pile of pebbles assembled by *P. hermannsburgensis*, photo by Stephen Davis.

of *Pseudomys* are primarily grass eaters, while other species eat very little grass. In captivity some of these mice readily take mealworms, sunflower seeds, and fruit.

Australian native mice reportedly have a relatively mild and gentle disposition. Some species appear to be sociable; the highest number of *P. hermannsburgensis* to be taken from a single burrow is five. Happold (1976a) found that captive groups of *P. albocinereus* lived amicably together, but that intruders might be attacked and killed. In the wild, several individuals of this species are found in the same burrow, generally an adult male, one or two adult females, and the young of one or more generations. In contrast, individuals of *P. desertor* were found to exhibit strong mutual repulsion and to be solitary, except when pairs associate briefly for mating and when a female has dependent young. Watts (1976a) identified 11 basic calls in *P. australis*, including a defensive screech heard in agonistic encounters between adults.

Taylor and Horner (1972) observed that *Pseudomys* appears to have a regular breeding season where it occurs in areas with a regular climate and assured rainfall, but that in locations with erratic precipitation the genus may be opportunistic and breed following adequate rains at a time coincident with the proliferation of grass and herbs. In a summary of reproduction in the genus, Watts (1974) listed gestation periods of 28 to 40 days and litter sizes of 3 to 5 young. Happold (1976b) provided the following data: *P. albocinereus*, reproduction all year in Victoria, but probably a peak season in August after winter rainfall, a normal estrous cycle of 7–10 days, pospartum estrus and mating within 12 hours of parturition, intervals of 38 and 39 days between copulation and birth, and an average litter size of 3.8 (2–5) young; *P. shortridgei*, pregnant females taken only from October to December in Victoria, no postpartum estrus and only one litter per year, and 3 young born in each of six litters; *P. desertor*, probably breeds throughout the year, estrous cycles of 7–9 days, postpartum estrus and mating

within 7 hours of parturition, intervals of 28 to 34 days between successive births, and a mean litter size of 3 (1–4).

In a laboratory study of *P. novaehollandiae,* Kemper (1976a, 1976b) found a mean estrous cycle of 6 days and a postpartum estrus and mating. The gestation period averaged 31.5 (29–33) days in nonlactating females and 33.2 (32–37) days in lactating females. Litter size averaged four and ranged from one to six young. The young weighed about one or two grams at birth, opened their eyes after about 15 days, and were weaned when 3 to 4 weeks old. Females reached sexual maturity when about 13 weeks old, and males when 20 weeks old.

Smith, Watts, and Crichton (1972) reported the mean length of 52 estrous cycles in captive *P. australis* to be 8.5 days. The gestation period was 30–31 days in five cases, and the interval between litters was 30–100 days. Mean size of 140 litters was 3.6 young, with the range being 1–7. The young clung to the nipples of the mother and were dragged about wherever she went. Weaning occurred at 22–30 days and females were capable of giving birth when 12 weeks old. According to Marvin L. Jones (Zoological Society of San Diego, pers. comm.), a captive specimen of *P. australis* lived for 5 years and 7 months.

Several species of *Pseudomys* have been adversely affected by the European colonization of Australia. Ride (1970) noted that the hardest-hit species were those that inhabited areas taken over by people for settlement or agriculture, but that there are still good populations of certain species that occupy offshore islands, inland areas, undeveloped coastal sand heaths, and rain forests. Cockburn (1978) suggested that the decline of some species was associated with a reduction in summer burning of heathlands by aboriginal people, and the resulting increased destructiveness of subsequent fires.

The USDI (1980) lists the following species as endangered: *P. fieldi, P. gouldii, P. novaehollandiae, P. praecornis, P. shortridgei, P. fumeus,* and *P. occidentalis.*

The Australian National Parks and Wildlife Service (1977, 1978) covered the following species in the series "Australian Endangered Species": *P. fumeus*, *P. occidentalis*, *P. praecornis*, and *P. shortridgei*. The CITES covers *P. fumeus* and *P. praecornis* on appendix 1, and *P. shortridgei* on appendix 2. Watts (1979) listed *P. fieldi* and *P. gouldii* as presumed extinct; *P. australis*, *P. desertor*, and *P. gracilicaudatus* as rare; and *P. occidentalis*, *P. oralis*, *P. praecornis*, and *P. shortridgei* as rare and at risk. The range of *P. occidentalis* apparently has declined drastically because of clearing of its habitat for cereal growing, but some remaining populations are secure in national parks. *P. oralis* had been known only from two specimens collected in the first half of the nineteenth century, but a few more specimens were recently taken in southern Queensland. Subfossils of *P. praecornis* have been found in southwestern Australia, and a specimen was collected in 1858 on the Peron Peninsula, but the species now seems to survive only on Bernier Island in Shark Bay. *P. shortridgei* apparently has disappeared in Western Australia, where it was discovered in 1906, but good populations were found in Victoria in 1961.

The New Holland mouse (*P. novaehollandiae*) was long known with certainty only from specimens collected in the early nineteenth century in New South Wales. In 1967, however, the species was rediscovered near Sydney (Mahoney and Marlow 1968). Subsequent investigation indicated a wide if somewhat patchy distribution from the northern coast of New South Wales to the Mornington Peninsula of southern Victoria, and that the species is not in immediate danger of extinction (Posamentier and Recher 1974; Keith and Calaby 1968). Its optimum habitat is dry heath, which has been disturbed by fire and is actively regenerating. *P. novaehollandiae* is known from subfossil remains found on Tasmania and Flinders Island, in addition to its mainland range, and in 1976 a living population was located along the northeastern coast of Tasmania (Hocking 1980).

RODENTIA; MURIDAE; **Genus NOTOMYS** Lesson, 1842

Australian Hopping Mice, Jerboa Mice

There are nine species (Ride 1970; Mahoney 1975, 1977; Watts 1979):

N. mitchelli, southern Australia;
N. mordax, known only by a single specimen from southeastern Queensland;
N. alexis, central Australia;
N. fuscus, southeastern Western Australia to southwestern Queensland;
N. cervinus, south-central Australia;
N. macrotis, known only by two specimens from southwestern Western Australia;
N. longicaudatus, southwestern Western Australia, southern Northern Territory, northwestern New South Wales;
N. amplus, known only by two specimens from the extreme southern part of the Northern Territory;
N. aquilo, northeastern Northern Territory and the nearby island of Groote Eylandt, western Cape York Peninsula of northern Queensland.

Head and body length is 91 to 177 mm, and tail length is 125 to 225 mm. Weight is 20 to 50 grams (Watts 1975a). The general coloration of the upper parts varies from pale sandy brown through yellowish brown to ashy brown or grayish. The underparts are white in all species except *N. mitchelli*, which is whitish gray below. The body covering is fine, close, and soft in most forms. The long hairs near the tip of the tail give the effect of a brush. These rodents are characterized by their strong incisor teeth, long tail, large ears, and extremely lengthened and narrow hind feet, which have only four sole pads. Females have four mammae.

All species that have been examined have a well-devel-

Australian hopping mouse (*Notomys fuscus*), photo from Queensland Museum. Inset: *N. sp.*, photo by Shelley Barker.

oped sebaceous glandular area on the underside of the neck or chest. Watts (1975b) determined this area to be active in all adult males, but in females only during pregnancy and lactation. He suggested that the glands are used for territorial marking and marking of group members, including newborn young.

Australian hopping mice inhabit sand dunes, grasslands, tree and shrub heaths, and lightly wooded areas. They are nocturnal and saltatorial, and are often considered the ecological equivalent of the gerbils (Meriones, Gerbillus), jerboas (Jaculus), and kangaroo rats (Dipodomys) found in the deserts of the Northern Hemisphere (Watts 1975a). Normally, they move about awkwardly on all fours or make short hops, but if they are startled they bound rapidly and gracefully using only the large hind feet. They dig their own burrows, some being of simple construction but others being rather complex. In sandy soil the burrow of N. mitchelli leads downward at an angle of about 40° for 1.2 to 1.5 meters, and then levels off and leads into a nest chamber, from which it leads almost straight upward to the exit. There is often no mound of excavated soil around the opening. Notomys is commonly found in association with Antechinomys, its marsupial counterpart, and sometimes both genera inhabit the same tunnel system.

The diet of Notomys consists of berries, leaves, seeds, and other available vegetation. Ride (1970) stated that although hopping mice will drink water readily when it is available, at least some species do not require water at all. They survive through their ability to excrete waste nitrogen in the form of one of the most highly concentrated urines yet known in a rodent. They also avoid extreme heat and desiccation by remaining underground during the day. They are subject to great fluctuations in numbers; following seasons of good rainfall and plant growth, some localities may be swarming with them.

The species N. alexis and N. cervinus apparently live and move about in groups. In a study of captive N. alexis by Happold (1976a), animals placed together displayed little aggression and soon became amicable. Tolerance was shown toward newborn and juveniles, and females suckled young other than their own. Watts (1975a) identified about eight vocalizations in four species of Notomys. The calls included twittering given during aggressive chases, but none of the sounds was associated with threats or fighting.

The following reproductive information was taken from Crichton (1974); Smith, Watts, and Crichton (1972); Aslin and Watts (1980); and Watts (1974). N. alexis, N. fuscus, and N. cervinus are apparently opportunistic breeders in the wild, giving birth after suitable rainfall. N. mitchelli, however, inhabits areas of regular winter rainfall and breeds during that season. In captivity the females of all four species are polyestrous, with no evidence of seasonality. The estrous cycle lasts about 7–8 days in N. alexis, N. mitchelli, and N. fuscus, but varies from 8 to 38 days in N. cervinus. The gestation period for nonlactating females is 32 days in N. alexis, 34–37 days in N. mitchelli, and 38–43 days in N. cervinus. All three of these species have a postpartum estrus and mating. Lactation seems to delay implantation up to 11 days in N. cervinus and probably in N. mitchelli, but not in N. alexis. In N. fuscus, gestation lasts 32–38 days; a postpartum estrus is not common in this species, but some females entered estrus 14–22 days after giving birth. Usual litter size is 2–4 young in N. fuscus, N. cervinus, and N. mitchelli. In 176 litters of N. alexis, mean litter size was 4 young and range was 1 to 9. The young weigh about 2 to 4 grams at birth and open their eyes when 18 to 28 days old. They cling tenaciously to the nipples of the mother and are dragged about wherever she goes; weaning occurs at around 1 month. Female N. alexis and N. mitchelli can give birth when 3

months old, but female N. cervinus do not bear young until at least 6 months old. Both sexes of N. fuscus reach reproductive maturity at the age of 70 days; one female of this species produced 9 litters in its lifetime and died when 26 months old; males of this species are capable of breeding up to the age of 38 months. According to Marvin L. Jones (Zoological Society of San Diego, pers. comm.), a captive N. alexis lived for 5 years and 2 months.

The species N. macrotis, N. mordax, N. longicaudatus, and N. amplus are known only by specimens collected between 1843 and 1901 (Ride 1970). Watts (1979) presumed all four to be extinct, and also listed N. aquilo and N. fuscus as rare. N. aquilo may have disappeared from the mainland; it is classified as endangered by the USDI (1980) and is on appendix 1 of the CITES. All other species of Notomys are on appendix 2 of the CITES.

RODENTIA; MURIDAE; **Genus LEGGADINA** Thomas, 1910

There are two species (Ride 1970; Watts 1976b; Morton 1974):

L. forresti, arid areas from west-central Western Australia to central Queensland and western New South Wales;
L. lakedownensis, northeastern Queensland.

Ride (1970) included Leggadina within the genus Pseudomys, but Misonne (1969) stated that the two taxa have little to do with one another. The species Pseudomys delicatulus, P. hermannsburgensis, and P. novaehollandiae have sometimes been placed in the genus Leggadina. Mahoney and Posamentier (1975) considered the latter two species part of Pseudomys, but stated that P. delicatulus doubtfully belongs in Leggadina. Fox and Briscoe (1980) placed all three species in Pseudomys, thereby leaving Leggadina only with L. forresti and L. lakedownensis.

Head and body length is 60 to 100 mm, and tail length is 40 to 69 mm (Watts 1976b). The tail is noticeably shorter than the head and body. The pelage is crisp, with a considerable portion of stout guard hairs. L. forresti is yellowish olive brown above and white below. In L. lakedownensis the hairs on the back have buffy brown tips and grade to pale olive buff on the sides; the underparts are white. The head is broad, the muzzle blunt, and the ears short and broad. Females have four mammae.

Morton (1974) indicated that L. forresti frequents grassy plains, and shelters in cracks in the soil. Watts (1979) stated that L. lakedownensis is found in well-grassed areas within or near tropical open woodland. He noted that this species normally occurs in low numbers, and he considered it rare and at risk.

RODENTIA; MURIDAE; **Genus MASTACOMYS** Thomas, 1882

Broad-toothed Rat

The single species, M. fuscus, occurs in southeastern New South Wales, Victoria, and Tasmania. Certain late Pleistocene and subfossil remains, found over a somewhat larger region of southeastern Australia, formerly were considered to represent other species, but all have now been assigned to M. fuscus (Wakefield 1972).

Head and body length is 145 to 195 mm, and tail length is 95 to 135 mm. Green (1968) listed weights of 100 to 196

Leggadina forresti, photo by B. G. Thomson.

Broad-toothed rat (*Mastacomys fuscus*), photo by Ederic Slater.

grams. The dense hair is long and silky, and the coloration throughout is sooty brown to grayish brown. The skin of the tail and feet is dark colored. The broadened molar teeth and narrowed palate are characteristic of the genus. It is volelike in appearance and habits. Females have four mammae.

The remainder of this account is based largely on Ride (1970). The broad-toothed rat appears to be a relict genus, surviving only in isolated colonies in places that provide it with the cold, humid, or alpine conditions that it requires. In the Kosciusko State Park of southeastern New South Wales, this rat lives along small creeks among shrubs and long grasses. The climate is rigorous, with a heavy winter snowfall. Conditions beneath the snow are relatively mild, and *Mastacomys* dwells in tunnels among the shrubs. By contrast, Victorian colonies live at low altitudes, which are wet but less cold. In Tasmania, *Mastacomys* also lives in areas with severe winter climates, but it is very localized in distribution and is found in only one type of habitat. This consists of vegetation on very boggy ground, such as the drainage systems of wet sedgelands in openings in rain forest, or button grass or tussock grass areas. In Tasmania *Mastacomys* is invariably found in association with *Rattus lutreolus* and *Antechinus minimus*.

In the Kosciusko State Park, breeding probably begins in the spring and litters are produced in the summer. In Tasmania, the breeding season is from October to March, and during this period each female produces more than one litter. Captive observations indicate that the female enters estrus soon after one litter is born, and then, subsequent to mating, aggressively keeps the male away until the young are inde-

pendent. Gestation lasts about 5 weeks, and litters of up to three young have been recorded. The young are large and well furred at birth; they attach themselves firmly to the nipples of the mother and are dragged about on their back. They apparently attain independence at 30 days and adult weight at 16 weeks of age.

Green (1968) observed that while the range of *Mastacomys* covers an extensive part of Tasmania, its habitat is specialized and subject to human environmental alterations. He cautioned that the survival of the genus is threatened by hydroelectric development, road construction, and cattle grazing.

RODENTIA; MURIDAE; **Genus LORENTZIMYS** Jentink, *1911*

New Guinea Jumping Mouse

The single species, *L. nouhuysi,* is found throughout forested parts of New Guinea (Menzies and Dennis 1979).

Head and body length is 75 to 85 mm, and tail length is about 103 to 110 mm. In the subspecies *L. n. nouhuysi,* which occurs up to an elevation of about 900 meters, the pelage is relatively short and crisp. In the subspecies *L. n. alticola,* which occurs on mountain slopes from about 600 to 3,000 meters, the pelage is long, soft, and dense. The upper parts of *Lorentzimys* are reddish brown or grayish, and the underparts are buff or grayish white. The feet are white. The

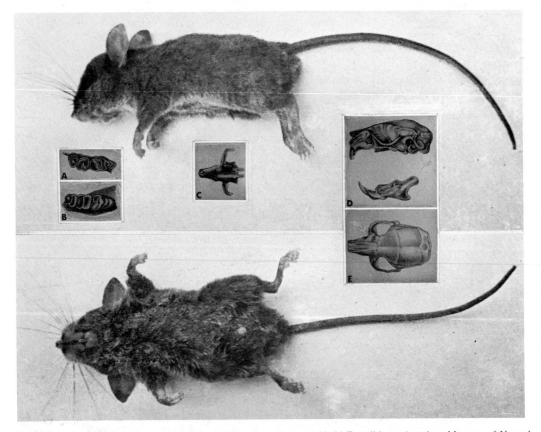

New Guinea jumping mouse (*Lorentzimys nouhuysi*), photos from Archbold Expeditions, American Museum of Natural History. Insets: A. Upper right teeth; B. Lower right teeth; C. Ventral view of skull; D. Lateral view of skull; E. Dorsal view of skull; photos from *Mammals Collected by the Dutch New Guinea Expedition,* F. A. Jentinck.

head is slaty around the eyes and ears, and the underside of the neck is whitish. In one form the hairs of the tail are white, and in the other they are black, but in both forms a pencil of hairs appears at the tip of the tail. The whiskers are numerous and fairly long.

This genus is distinguished by the cranial structure. The head is short and broad, and the upper incisor teeth are inclined slightly forward. The feet are slender, and the ears are narrow and taper to a rounded point. There are five fingers; four are armed with sharp, arched claws, but one is so small that only a flat nail is visible. Females have six mammae.

According to Menzies and Dennis (1979), *Lorentzimys* seems to be completely arboreal, but may prefer open or disturbed forest with abundant low shrubbery, rather than mature forest. It nests in *Pandanus* and similar trees. Its morphology suggests that it runs and jumps through the branches, rather than climbs. Several individuals may live together in one nest, and one pregnant female carried two embryos.

RODENTIA; MURIDAE; **Genus HAEROMYS** *Thomas, 1911*

Pygmy Tree Mice

There are three species (Medway 1977; Laurie and Hill 1954):

H. margarettae, Borneo;
H. pusillus, northern and eastern Borneo;
H. minahassae, northern Celebes.

Haeromys is among the smallest genera of rodents. Head and body length of the type specimen of *H. margarettae* is 76 mm, and tail length is 144 mm. In *H. minahassae* head and body length is 72 mm, and tail length is about 110 mm. The pelage is long and soft. *H. minahassae* is rufous on the upper parts, with a dull back and bright sides. In *H. margarettae* and *H. pusillus* the general coloration of the upper parts is deep chestnut rufous, with a grayish cast resulting from gray hair bases, but the coloration is clearer on the sides, where a rufous lateral band is formed. In *H. minahassae* the tail is brown, and in *H. margarettae* and *H. pusillus* it is greenish gray. In all species the underparts are white, the whiskers are black, and the tail is unicolored and covered with short hairs.

This genus resembles *Chiropodomys* in having a muzzle that is short and slender, and ears that are relatively small, oval, and sparsely haired. The hands and feet are adapted for arboreal life; the great toe is opposable. The thumb has a large nail, whereas the other digits have short, sharp, curved claws. Female *H. margarettae* have six mammae.

Musser (1979:438) wrote: "Species of *Haeromys* inhabit tropical evergreen forests in both lowlands and mountains. They build globular nests in cavities in trees. The animals I observed in Celebes ate only small seeds, mostly from figs."

RODENTIA; MURIDAE; **Genus LOPHUROMYS** *Peters, 1874*

Brush-furred Mice

Dieterlen (1976) recognized two species groups and nine species:

woosnami group

L. woosnami, mountains of northeastern Zaire, southwestern Uganda, and western Rwanda;
L. luteogaster, northeastern Zaire;
L. medicaudatus, mountains around Lake Kivu in northeastern Zaire and western Rwanda;

sikapusi group

L. sikapusi, Sierra Leone to western Kenya and northern Angola;

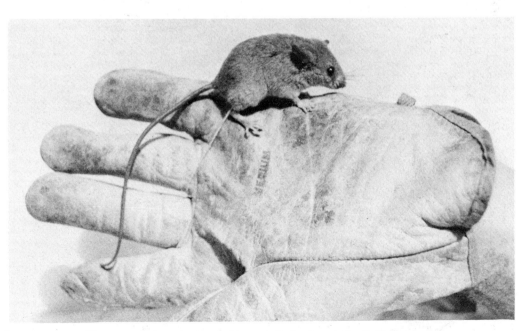

Pygmy tree mouse (*Haeromys minahassae*), photo by Margareta Becker through Guy G. Musser.

Brush-furred mouse (*Lophuromys sikapusi*), photo by D.C.D. Happold.

L. flavopunctatus, Ethiopia to northeastern Angola and northern Mozambique;

L. rahmi, mountains around Lake Kivu in northeastern Zaire and western Rwanda;

L. nudicaudus, southern Cameroon, Equatorial Guinea including Fernando Poo;

L. cinereus, mountains west of Lake Kivu in northeastern Zaire;

L. melanonyx, mountains of south-central Ethiopia.

Head and body length is 84 to 160 mm, tail length is 46 to 148 mm, and weight is 23 to 111 grams (Dieterlen 1976; Kingdon 1974b; Delany 1975). In the *woosnami* group of species the tail is about equal in length to the head and body, while in the *sikapusi* group the tail averages only about half as long as the head and body. The pelage in most forms is fairly long, sleek, and thick, and is unique in structure. The hairs are slightly coarse, and flattened and tapered at either end. The coloration is unusual for a murid. The upper parts range from sandy buff through olive gray to dark brown. In some species there is fine streaking or various degrees of speckling from whitish to orangish. The underparts range from cream through cinnamon to dark orange or wood brown. The upper surfaces of the hands and feet are usually somewhat lighter than the back. The tail is well furred in some forms, but in others it is scantily covered with fine, short hairs. The upper surface of the tail is almost as dark as the back, whereas the lower surface is usually light colored.

In addition to their coloration and peculiarly textured fur, these animals are distinguished by their short legs and chunky shape (Kingdon 1974b). They have five fingers and five toes, all provided with claws. According to Dieterlen (1976), the *woosnami* group is characterized by its long tail, long feet, big ears, short claws, and sausagelike stomach glands, whereas the *sikapusi* group has a short tail, relatively short hind feet and ears, relatively long claws, and packet-shaped stomach glands. The incisor teeth of *Lophuromys* are thin and are occasionally inclined forward. The skin is delicate and easily torn, and it is not unusual for the tip of the tail to be missing.

Brush-furred mice occur in a variety of habitats, from grassy plains and swampy areas to montane forests at elevations above 4,000 meters. They are exclusively terrestrial (Rosevear 1969). According to Kingdon (1974b), moisture and grass are probably essential for these animals, as they are not found in closed canopy forest or in areas subject to periodic drought. They construct nests of dry grass and leaves under rocks, roots, or fallen timber, or in crevices or short, straight burrows that they dig themselves. Activity seems to vary by species, with *L. flavopunctatus* tending to be more diurnal and *L. woosnami* more nocturnal. Kingdon noted that the success of these mice may stem from their specializing in eating insects, especially ants. They also feed on other invertebrates, frogs and other small vertebrates, and vegetation. Dieterlen (1976) stated that the amount of animal food taken by each species of *Lophuromys* varied from 40 to 100 percent.

Happold (1977) reported a mean home range of 2,100 sq meters for *L. sikapusi* in western Nigeria. For *L. flavopunctatus*, Kingdon (1974b) cited a home range of 350 sq meters in Uganda and a density of 14 to 19 individuals per ha. in the Ituri Forest; he thought that densities would be considerably higher in arable areas of Uganda. Rosevear (1969) stated that brush-furred mice are generally solitary and fight extensively among themselves. Collectors have long been impressed by the large proportion of specimens with body wounds, torn ears, or partial or complete loss of the tail.

Breeding cycles have been shown to be highly plastic and responsive to ecological conditions. In tropical areas reproductive activity tends to be continuous, and in southern parts of the range of *Lophuromys* there is a single prolonged breeding season during the rains from October to May (Kingdon 1974b). In studies in Uganda, Delany (1971, 1975) found pregnant female *L. flavopunctatus* in every month of the year, but he indicated that breeding peaks occurred during the rainy seasons, October to December and March to June. The reproductive cycle of *L. sikapusi* in Uganda was considered to be much the same. Happold (1977) found pregnant female *L. sikapusi* in western Nigeria in February, March, April, May, July, and October, and noted that one female had consecutive litters in October, February, and March. Reported gestation periods are 30.5 days for *L. flavopunctatus* (Delany 1971) and 32 days for *L. woosnami* (Dieterlen 1976). Litter sizes are one to four young, averaging about two in *L. flavopunctatus*, two to five in *L. sikapusi*, and one to two in *L. woosnami* (Delany 1975). Young *L. flavopunctatus* weigh five to eight grams at birth and open their eyes after 4 to 7 days (Kingdon 1974b).

Live animals have been reported to be very sensitive and hard to handle without causing injury or death. Nonetheless, Kingdon (1974b) stated that they have been used in the laboratory, and are clean and easy to keep and breed. Captives may survive up to 2 years.

Broad-headed mouse (*Zelotomys* sp.), photo by C. L. Cheeseman.

RODENTIA; MURIDAE; *Genus ZELOTOMYS Osgood, 1910*

Broad-headed Mice

There are two species (Meester and Setzer 1971; Smithers 1971; Germain and Petter 1973; Meester 1976): '

Z. hildegardeae, Central African Republic and Kenya to Angola and Zambia;

Z. woosnami, northern and eastern Namibia, Botswana, northwestern South Africa.

Germain and Petter (1973) recognized *Z. instans* of northern Zaire and the Central African Republic as a species separate from *Z. hildegardeae.*

Head and body length is 104 to 137 mm, tail length is 80 to 118 mm, and weight is 38 to 85 grams (Kingdon 1974b; Delany 1975; Smithers 1971; Rautenbach and Nel 1975). The upper parts range from pinkish buff to dark gray, slate gray, and buffy brown. The flanks are grayish to yellowish; the underparts are light gray, pinkish buff, or whitish; and the hands and feet are pinkish buff to white. The short, stiff hairs that form the sparse tail covering are white to grayish brown.

There are no outstanding external peculiarities except the slightly broadened head, a feature that is reflected in the common name. The generic name, meaning imitator, refers to the external resemblance of *Zelotomys* to the genus *Praomys,* subgenus *Mastomys.* The incisor teeth of *Zelotomys* are rather slender and slightly protruding, whereas in most other rodents the incisors curve back toward the mouth. The ears are of moderate size and rounded. Females have 10 mammae.

According to Kingdon (1974b) and Delany (1975), *Z. hildegardeae* is associated mainly with moist, grassy savannahs and scrub. It is insectivorous and probably forages for its food under the grass cover. Its density apparently is kept low by fires, predators, and the scarcity of insects during the dry season. Pregnant females have been taken in November in Kenya, and in February, March, May, June, and July in Uganda. Litter size has ranged from three to seven young.

In Botswana, Smithers (1971) collected *Z. woosnami* mostly on sandy ground with sparse grass and a thin, open scrub cover. He observed that this species is nocturnal, terrestrial, and graminivorous, but that it may eat some meat. Birkenstock and Nel (1977) reported *Z. woosnami* to live in burrows thought to have been dug by other animals, but that it is capable of digging and making a nest by itself. Individuals were found to be widely spaced and asocial. When two captives were put together, one was usually killed, even if it was of a different sex and age. Birkenstock and Nel also stated that the breeding season of *Z. woosnami* corresponded with the warm, wet months from December to March, that the minimum period between births was 31 days, and that litters numbered four or five young. Meester (1976) reported that three litters of *Z. woosnami* arrived between March and the end of July, and that the number of young was four, five, and five. He considered the species to be rare, but not endangered.

RODENTIA; MURIDAE; *Genus COLOMYS Thomas and Wroughton, 1907*

African Water Rat

The single species, *C. goslingi,* has been recorded from Cameroon to Ethiopia and Angola. The generic name *Nilopegamys* Osgood, 1928 is a synonym of *Colomys,* and the name *Nilopegamys plumbeus* probably represents a subspecies of *Colomys goslingi* (Meester and Setzer 1971).

Head and body length is 117 to 140 mm, tail length is 145 to 180 mm, and weight is 50 to 75 grams (Kingdon 1974b). The pelage is thick, short, and soft, with a velvety texture. The upper parts range in color from cinnamon to wood brown and dark brown, the darkest coloration being just behind the middle of the back. The underparts, including the hands, feet, forearms, and lower legs, are generally white. Occasionally the coloration of the underparts extends in a line onto the upper arms, and sometimes appears as a ring about the

African water rat (*Colomys goslingi*), photo by F. Dieterlen.

ankles. The dark color of the upper parts and the light color of the underparts meet in a distinct line of demarcation. There is generally a white or pale-colored spot just below the ear. The large grayish ears are practically naked. The tail varies from light brown to almost black on the dorsal surface, and is slightly paler beneath. The hairs on the distal part of the tail are white or pale in color, and are slightly longer than those of the basal part. The tail is, however, scaly and poorly haired.

Colomys has a light and slender body form. The feet are elongate and the four fingers are long. Females have eight mammae. The incisor teeth are not grooved, and are flat or slightly concave on the front surface. Kingdon (1974*b*) pointed out that in the related genus *Malacomys* the upper lip slopes sharply away from the nose to allow maximum exposure of the incisors, and that the animal is incapable of closing its mouth. In *Colomys*, however, both the upper lip and chin are swollen and cover the incisors. This modification may be associated with the semiaquatic lifestyle of *Colomys*.

According to Kingdon (1974*b*) and Delany (1975), *Colomys* is found in swamps and along rivers in forested country, at both high and low elevations. It is nocturnal and constructs short tunnels in stream banks. It swims rapidly with powerful thrusts of its long hind legs. It searches for food by wading through watery mud, with its nose held near the water, and pounces on worms, slugs, crustaceans, and aquatic insects. It also eats some vegetable matter. Individuals are probably solitary. Records suggest a breeding season of March to July in western Uganda and eastern Zaire. Pregnant females have contained one to three embryos.

RODENTIA; MURIDAE; **Genus MALACOMYS**
Milne-Edwards, 1877

Big-eared Swamp Rats, Long-eared Marsh Rats

There are four species (Van der Straeten and Verheyen 1979*a*; Rautenbach and Schlitter 1978; Verheyen and Van der Straeten 1977; Ansell 1974):

M. edwardsi, Guinea to southern Nigeria;
M. cansdalei, eastern Liberia to Ghana;

M. longipes, southeastern Nigeria to Uganda and northwestern Zambia;
M. verschureni, known only by a single specimen from northeastern Zaire.

Rautenbach and Schlitter (1978), as well as Cole (1972), classified the forms *cansdalei* and *giganteus,* which occur from eastern Liberia to Ghana, as subspecies of *M. longipes.* Van der Straeten and Verheyen (1979*a*), however, considered *M. cansdalei* to be a full species, more closely related to *M. edwardsi* than to *M. longipes,* and considered *M. l. giganteus* to be a synonym of *M. cansdalei.*

Head and body length is 117 to 190 mm, tail length is 117 to 222 mm, and weight is 50 to 145 grams (Van der Straeten and Verheyen 1979*a*; Kingdon 1974*b*; Cole 1972). The pelage is thick, soft, and velvety, much like that of certain shrews. The texture of the fur is so fine that it has been compared to the bloom of a peach. General coloration of the upper parts is dark brown with a sprinkling of grayish. The sides are ochraceous and the underparts are gray. *M. edwardsi* has an oval, sooty-colored patch around each eye. The hands and feet are plentifully sprinkled with fine-textured short hairs. The whiskers are fine, long, and dark. The tail is so poorly haired that in some specimens it appears almost naked.

These medium-sized, slender-bodied rats have large, naked ears; long, narrow skulls; long, slender hind feet; and long tails. It has been suggested that the structure of the feet permits the paws to spread more than those of most terrestrial rats, as an adaptation for traveling over swampy ground. Females usually have four or six mammae.

These rodents seem to be mainly nocturnal and usually inhabit areas of dense vegetation near water. Kingdon (1974*b*) stated that *M. longipes* is a capable climber, but makes its grass or leaf nests in crevices or among tree roots. It apparently eats animal and vegetable matter in nearly equal proportions, feeding on fallen fruit, seeds, nuts, roots, insects, slugs, snails, and crabs. Misonne and Verschuren (1976) reported that eight stomachs of *M. edwardsi* from Liberia contained fruit and one had a small proportion of insect remains.

Kingdon (1974*b*) wrote that *Malacomys* is generally solitary, has a mild disposition, and has no clearly defined breeding season. In a study of *M. edwardsi* in the rain forests of

Big-eared swamp rat (*Malacomys edwardsi*), photo by D.C.D. Happold.

western Nigeria, Happold (1977) found home range to average 0.42 ha. for males and 0.37 ha. for females. Pregnant females were taken from November to July, which is mainly a dry period, and peak reproduction occurred in January. Females could produce litters every one or two months during this season. Misonne and Verschuren (1976) collected four pregnant female *M. edwardsi* in Liberia in December and January, each with two embryos. Delany (1975) stated that pregnant female *M. longipes*, with three or four embryos each, had been taken in Uganda in May, July, and September, and that the number of embryos reported in specimens from Zaire was one to five. In Zambia, two pregnant females, each with three embryos, were taken in August (Ansell 1974).

RODENTIA; MURIDAE; Genus ACOMYS
I. Geoffroy St.-Hilaire, 1838

Spiny Mice

There are five species (Meester and Setzer 1971; Kingdon 1974*b*; Corbet 1978; Spitzenberger 1978; Yalden, Largen, and Kock 1976):

A. cahirinus, Mauritania to southern Pakistan and Kenya, Crete, Cyprus;

A. cilicius, Asia Minor;

A. russatus, northeastern Egypt, Palestine, Jordan, Arabian Peninsula;

A. spinosissimus, Zambia, Rhodesia, Mozambique, eastern Botswana;

A. subspinosus, Ethiopia to South Africa.

Some authorities (as Meester and Setzer 1971; Kingdon 1974*b*; and Harrison 1972) recognize *M. dimidiatus* of south-western Asia and northeastern Africa as a full species distinct from *A. cahirinus*. Ellerman and Morrison-Scott (1966) considered *A. cahirinus* to be a small commensal form of *A. dimidiatus,* but, since the name *A. cahirinus* had priority, it was applied to the entire species. Neither Rosevear (1969), Corbet (1978), nor Osborn and Helmy (1980) recognized *A. dimidiatus* as a separate species, though Corbet suggested that *A. cahirinus* might actually comprise several closely related species.

Head and body length is 70 to 175 mm, tail length is 42 to 125 mm, and weight is 11 to 90 grams. The brittle tail is easily separated, partly or entirely, from the rest of the animal, and therefore measurements of tail length are often misleading.Coloration above is pale yellowish or reddish brown, reddish, or dark grayish. The underparts are white. *A. cahirinus* and *A. russatus* are sometimes melanistic. The back and tail are covered with coarse, inflexible spines. The tail appears to be practically naked, and the scales are conspicuous. It may be whitish above and below, bicolored, or uniformly dark. The ears are large and erect. Females have four or six mammae.

Spiny mice are found in arid regions. The African forms inhabit rocky country, semidesert areas, and dry woodland or savannah (Kingdon 1974*b*). They shelter in rocky crevices, cracked soil, gerbil or other rodent burrows, and termite mounds. They are terrestrial, but have been taken in trees. In some areas, most notably Egypt, *A. cahirinus* has become a human commensal and lives in and around buildings. Most species are primarily nocturnal, but are sometimes active in the early morning and late afternoon. In the Arabian Peninsula, *A. russatus* is largely diurnal in its extremely rocky habitat, and seems much better adapted to high temperatures than the sympatric *A. cahirinus* (Harrison 1972). Spiny mice are omnivorous, but feed mainly on plant material, especially grains and grasses. Osborn and Helmy (1980:299) wrote that dates are a staple in some areas, and that ''the dried flesh and bone marrow of mummified humans is a

Egyptian spiny mouse (*Acomys cahirinus*) light and dark subspecies, photos by Ernest P. Walker. Animal in lower picture has lost tip of tail.

source of food for *A. cahirinus* in the tombs of Gebel Drunka southwest of Asyut.''

Most knowledge of the reproduction of *Acomys* has been derived from captive studies, mainly of Middle Eastern *A. cahirinus* (Harrison 1972; Kingdon 1974*b*; Rosevear 1969; Grzimek 1975). Spiny mice are gregarious; females that already have given birth will help to clean new mothers and bite their umbilical cords. Reproduction is continuous, there being a postpartum estrus and mating, and sometimes a series of 12 or more litters without cessation. The gestation period is five to six weeks, which is quite long for a murid. There are one to five young, with the larger litters generally being born to older mothers. Nest construction is rudimentary, but the young are remarkably well developed. They weigh up to seven grams when born, their eyes are open either at birth or two or three days later, and they are weaned after about two weeks of life. Sexual maturity is attained at two to three months of age. Average longevity is around three years, but some individuals have lived up to five years.

Under good conditions, wild *Acomys* probably breed over the greater part of the year (Kingdon 1974*b*). Collections from the Arabian Peninsula indicate that reproduction in *A. cahirinus* there continues at least throughout the spring and summer months (Harrison 1972). In Tanzania, Hubbard (1972) collected pregnant female *A. subspinosus*, each with one to four embryos, from October to February. The young of this species were found to travel with the mother during their first night of life, but not to hang from the mammae. In contrast, the young of *A. cahirinus hystrella* in this area were born naked and blind, and did attach themselves firmly to the mammae. Hubbard noted that neither species usually built a nest. In Botswana, however, Smithers (1971) observed *A. spinosissimus* to construct a rough nest of grass and leaves, in which the young were born. Pregnant females, with two to five embryos each, were taken in Botswana in the warm, wet months of December, January, March, and April.

Spiny mice have become popular as pets, and are easy to keep and breed in the laboratory. In Egypt, however, they may be a public health menace, as a significant proportion have been found to carry the causal organism of typhus (Harrison 1972).

RODENTIA; MURIDAE; **Genus URANOMYS** Dollman, *1909*

White-bellied Brush-furred Rat

The single species, *U. ruddi*, has been recorded from Senegal to Nigeria, and in northeastern Zaire, Uganda, Kenya,

White-bellied brush-furred rat (*Uranomys ruddi*), photo by Jane Burton.

Malawi, and Mozambique (Kingdon 1974b; Meester and Setzer 1971).

Head and body length is 84 to 134 mm, tail length is 53 to 79 mm, and weight is 41 to 53 grams (Kingdon 1974b; Delany 1975). The hairs on the back are fairly long, especially on the rump, and are slightly stiffened, thereby giving the coat a crisp or brittle texture. The upper parts vary in color from dark brown to reddish and gray. The underparts vary from grayish to white, with a tinge of buff in some individuals. The tail, which is scantily covered with short hairs, is dark above and slightly paler below.

In external appearance *Uranomys* resembles *Lophuromys*, except that the backs of its hands and feet are covered with fine white hairs. Also, the underparts of *Lophuromys* are never pure white. The skull of *Uranomys* resembles that of *Acomys*, most notably in having the palate extended well behind the molars by a roof to the mesopterygoid fossa (Rosevear 1969). The cheek teeth also closely resemble those of *Acomys*, but are unique in that the anterior cusps of the front lower molars are flattened (Kingdon 1974b). The incisors usually project forward. The hind foot is broad, and the middle three digits are relatively long. Females have 12 mammae.

This genus is generally considered rare, but has been taken in large numbers in the Ivory Coast. It lives mainly on savannahs, and is reportedly nocturnal and insectivorous (Kingdon 1974b). It constructs a burrow about 15 cm deep with two exits, each topped by a mound of soil, and a blind tunnel up to 40 cm long. A grass-lined nest is built within the burrow, and there is no food storage. Each burrow system is occupied by two adults. In the Ivory Coast, breeding occurs throughout the year, but litters are larger from September to December, when they average 4.0 to 5.7 young, than at other times of the year, when they average 2.6 to 3.7 young (Delany 1975).

RODENTIA; MURIDAE; **Genus HYBOMYS** Thomas, 1910

Back-striped Mice

There are two species (Meester and Setzer 1971; Ansell 1978):

H. univittatus, Guinea to Uganda, extreme northeastern Zambia;
H. trivirgatus, Guinea to Nigeria.

Head and body length is 100 to 160 mm, tail length is 77 to 132 mm, and weight is 30 to 70 grams (Rosevear 1969; Kingdon 1974b; Genest-Villard 1978a). The fur is usually soft, though inclined to be coarse in some individuals. The upper parts range from light yellowish brown to quite dark brown, almost black in some individuals. Generally, *H. univittatus* is much darker than *H. trivirgatus*. The former species usually has a dark brown or black middorsal stripe that extends from the nape to the base of the tail. The latter species usually has a dark middorsal stripe, with a lateral stripe close to it on either side. These lines are often faint or entirely lacking in some individuals. The underparts range from tawny or ochraceous to grayish white. The upper surfaces of the hands and feet are usually the same color as the back. The thinly haired tail is about the same color, but in some animals is entirely black.

Although *Hybomys* resembles *Arvicanthis* externally in such features as the dark dorsal stripe, it differs from the latter genus by its more slender form and almost naked tail. The soles of the feet have five well-developed tubercles and a rudimentary sixth, whereas the palms have six tubercles that are all well developed. The thumb is small and has a blunt nail. The hind foot is narrow. Females have two or three pairs of mammae.

These mice inhabit forests, thick brush, and the edges of cultivated areas. They are strictly terrestrial, but seem to require moist places and reportedly are good swimmers (Rosevear 1969). *H. univittatus* is an animal of the forest floor, preferring areas near rivers or swamps, with abundant leaf litter and shade (Kingdon 1974b). Both species have been reported to be active by day and night, but Genest-Villard (1978a) found *H. univittatus* to be entirely diurnal. Individuals spent the night, and resting periods during the day, in burrows, within which they built nests of twigs. Each animal used several burrows, but males, at least, spent most inactive time in one favored lair. *Hybomys* also may shelter in rotting logs, piles of twigs, or termite mounds. *H. univittatus* has been reported to be mainly frugivorous and to be destructive to cassava roots. Its diet also includes other vege-

Back-striped mouse (*Hybomys trivirgatus*), photo by Jane Burton.

table matter and insects (Delany 1975). *H. trivirgatus* feeds mainly on insects (Cole 1975).

In a radio-tracking study of *H. univittatus* in a dense equatorial forest, Genest-Villard (1978a) determined home range to be 4,500 to 6,100 sq meters for males and 1,400 to 1,800 sq meters for females. A male spent its entire life (about 12 months) in one area, and after death was replaced by a younger animal. Home ranges of males did not overlap, and intruders were chased and viciously bitten. The home range of one male could overlap that of one or two females, but individuals were solitary and did not regularly share burrows. Males did occasionally visit the burrows of females.

In captivity, and in cultivated areas where food is abundant, *H. univittatus* is capable of reproduction throughout the year. In the forests of Zaire, however, the main breeding season extends from February to May, the period corresponding with the rains and maximum fruit production (Kingdon 1974b). In Uganda pregnant females of this species have been collected in April, July, and October (Delany 1975). Young *H. trivirgatus* were found in Nigeria during November and December, which is the dry season (Happold 1977), and pregnant females, each with 2 embryos, were collected in Liberia in January and February (Misonne and Verschuren 1976). Laboratory studies of *H. univittatus* show that gestation lasts 29–31 days, litter size averages about 3 (1–4) young, the eyes of the young open when they are 7 to 10 days old, and females attain sexual maturity by the age of 3 months (Rosevear 1969).

RODENTIA; MURIDAE; **Genus STOCHOMYS** Thomas, 1926

Target Rats

There are two species (Meester and Setzer 1971; Kingdon 1974b):

S. longicaudatus, Nigeria to western Uganda;
S. defua, Guinea to Ghana.

These species sometimes have been placed in the genera *Rattus* or *Aethomys*. Both Misonne (1969) and Rosevear (1969), however, used the generic names *Stochomys* for *S.*

longicaudatus and *Dephomys* Thomas, 1926 for *S. defua.* Subsequent authorities (Meester and Setzer 1971; Cole 1975; Misonne and Verschuren 1976) have recognized *Stochomys* as a genus that includes *Dephomys*.

Except as noted, the remainder of this account is based on Rosevear (1969). Head and body length is 115 to 172 mm, and tail length is 177 to 250 mm. Reported weights are 30 to 60 grams for *S. defua* (Misonne and Verschuren 1976), and 60 to 104 grams for *S. longicaudatus* (Kingdon 1974b). The pelage is long and rather coarse. Remarkably long guard hairs, especially in *S. longicaudatus,* project above the general coat contour. The common name of the genus is derived from the fancied resemblance of these bristles to arrows sticking in a target. The upper parts, face, and neck are reddish brown, becoming more gray on the flanks, and the underparts are white. Both the long tail and the small to moderate-sized ears appear naked, but are sparsely covered with short hairs.

Stochomys is readily distinguished from related murids by its large size, its relatively long and seemingly naked tail, and the peculiarities of its pelage. The cusps of the first upper molar tooth are rather well defined in *S. defua* but not in *S. longicaudatus* (Meester and Setzer 1971). The hands and feet each have five digits. The thumb is greatly reduced and bears a flat nail. There are four mammae in female *S. defua* and six in female *S. longicaudatus*.

Target rats are found in forests and plantations. They have been collected in swampy and densely vegetated areas. The long tail suggests arboreal habits. Both species are apparently active at night, and Rahm (1972) found *S. longicaudatus* exclusively nocturnal. Examination of stomach contents indicates that both species feed largely on fruit, but also take some insects (Misonne and Verschuren 1976; Rahm 1972). A female *S. longicaudatus* used a home range of only 0.15 ha. in a two-year period (Happold 1977).

A pregnant female *S. longicaudatus,* carrying two embryos, was taken on 20 July in Cameroon, and young individuals were found there from 24 July to 3 August. In Nigeria, pregnancies in this species were recorded in March and June, and young were found in January, June, and December (Happold 1977). Juvenile *S. defua* have been taken in early October and mid-December, and Misonne and Verschuren (1976) collected four pregnant females, with two or three embryos each, in Liberia in January and February.

Target rat (*Stochomys longicaudatus*), photo by D.C.D. Happold.

RODENTIA; MURIDAE; **Genus RHABDOMYS** Thomas, *1916*

Four-striped Grass Mouse

The single species, *R. pumilio*, is found from Uganda and Kenya to Angola and South Africa (Meester and Setzer 1971).

Head and body length is 90 to 137 mm, tail length is 80 to 135 mm, and weight is 30 to 73 grams (Kingdon 1974*b*; Swanepoel 1975). The fur is coarse. The striping on the dorsal surface of most forms is constant, but the tones of the color vary considerably. The striping consists of a pale mid-dorsal streak that extends from the back of the neck to the base of the tail. This streak is usually bordered on each side by two dark lines. The dark lines range from light golden brownish or clay color to dark chocolate brown or almost black. The pale stripes range from pale yellowish gray to buff. The remainder of the upper parts vary in color from yellowish gray or pale yellowish brown to light grayish brown. The underparts are paler. The tail is scaly and thickly covered with short hairs.

These medium-sized mice have tails that are generally shorter than the head and body. From *Lemniscomys*, *Rhabdomys* is distinguished by its functional fifth digit. The skull is similar to that of *Arvicanthis*. Females have four pairs of mammae.

In East Africa the four-striped grass mouse has a discontinuous range, being strictly limited to grassy uplands, moorlands, and the subalpine zone at elevations of 1,700 to 3,500 meters (Kindgon 1974*b*). Farther to the south it is more widely distributed, occurs at lower elevations, and has been reported in a variety of habitats including dry river beds, bush

Four-striped grass mouse (*Rhabdomys pumilio*), photo by Don Davis.

and scrub country, forest edge, and cultivated areas. It is sometimes found along stone walls and in outbuildings. It is rather common and is primarily diurnal (Smithers 1971; Christian 1977). It seems to enjoy basking in the sun on cool days. Although not classified as arboreal, it has been seen climbing about the low branches of shrubs. It has sometimes been reported to live in burrows with a single slanting entrance that usually is located near a shrub. Its nests, however, are more often located above the ground in a dense shrub, in a bunch of grass, or among tree roots. In the wild the nests are made of grass, leaves, fibers, and moss. In a captive study comparing *Rhabdomys* with *Aethomys* and *Praomys*, *Rhabdomys* built the best nests, these being spherical or cuplike structures of cotton and paper strips (Stiemie and Nel 1973). The diet of *Rhabdomys* is primarily vegetarian, consisting of roots, seeds, berries, and cultivated grains, but also includes snails, insects, and eggs. A home range of about 20 ha. has been calculated (Kingdon 1974b).

Males in the wild are well spaced, and studies by Choate (1972) indicate that they do not tolerate one another in captivity. A mated pair, however, jointly constructs the nest and remains together until the female gives birth, at which time she sometimes excludes the male. The weaned young of one litter may be allowed to remain when a second litter is produced three to four weeks after a postpartum estrus. This process may be responsible for reports of groups of 12 to 30 individuals living in the same nest. Laboratory investigations have shown that females can breed when three months old and produce four litters per year, each consisting of 4 to 12 young (Kingdon 1974b). In the wild in Botswana, breeding apparently continues throughout the year, with embryo counts averaging 5 and ranging from 3 to 9 (Smithers 1971).

RODENTIA; MURIDAE; **Genus LEMNISCOMYS**
Trouessart, 1881

Striped Grass Mice

There are eight species (Meester and Setzer 1971; Kingdon 1974b; Van der Straeten 1975, 1976, 1980a, 1980b; Van der Straeten and Verheyen 1978a, 1979b, 1980):

L. griselda, Angola;

L. linulus, Senegal to Ivory Coast;
L. rosalia, Kenya to eastern South Africa;
L. roseveari, known from two localities in Zambia;
L. striatus, savannah zones from Sierra Leone to Ethiopia, and south to northern Angola and Malawi;
L. macculus, northern and eastern Zaire, southern Sudan, Uganda, Kenya;
L. barbarus, Morocco to Tunisia, Senegal to Sudan and Tanzania;
L. bellieri, Ivory Coast.

Van der Straeten and Verheyen (1980) indicated that the form *mittendorfi* from Cameroon, which was included in *L. striatus* by Meester and Setzer (1971), might actually be a separate species with affinity to *L. macculus* and *L. bellieri.*

Head and body length is 83 to 140 mm, tail length is 75 to 159 mm, and weight is 18 to 68 grams (Kingdon 1974b; Rosevear 1969). The fur is thin, and unusually rough and coarse. This genus shows three main kinds of color pattern. In *L. rosalia* there is a single dark brown or black stripe down the middle of the back, sometimes extending between the eyes and onto the tail. In *L. striatus, L. macculus,* and *L. bellieri* there is again a dark middorsal stripe, but on both sides of it there are about five or six longitudinal rows of pale spots extending from the shoulders to the rump. In some forms the spots are definitely separated, while in others they tend to run together on their long axes. In *L. barbarus* the rows of spots are replaced by solid pale stripes, extending from shoulders to rump, on either side of the dark middorsal line. The general color of the upper parts of *Lemniscomys,* other than the spots and stripes, ranges from medium clay to dull dark brown. The pale spots and stripes range from light buffy to a dull dark clay color. The flanks range from clear buffy through light clay to fairly dark gray tinged with buff. The underparts vary from white or buffy white to medium buffy gray. The tail above is almost as dark as the middorsal stripe, and the remainder of the tail is almost as pale as the pale spots or stripes on the body. Females have two pairs of pectoral and two pairs of inguinal mammae.

According to Kingdon (1974b), *L. rosalia* is the most primitive species and is declining, but remains common along the edge of grassy pans within the woodland zone. *L. striatus* is a more successful species and occupies a wide variety of grassy habitats from sea level to 3,500 meters. It is especially common in forest clearings and also is found along

Striped grass mouse (*Lemniscomys striatus striatus*), photo from New York Zoological Society.

streams, in swampy places, and in cultivated areas. The considerably smaller species *L. macculus* and *L. barbarus* are partly sympatric with *L. striatus,* but generally live in more arid country. *L. barbarus* is typical of the drier savannahs and steppes that border the Sahara. All species construct small, spherical nests of grass and leaves near the ground, and make well-defined surface runways around the nests for foraging. They may sometimes seek refuge in burrows dug by other animals. *Lemniscomys* is terrestrial and has been reported to be active by day, but studies have indicated that *L. striatus* and *L. barbarus* are sometimes crepuscular and nocturnal (Rosevear 1969). The diet includes grass, soft seeds, cultivated crops and, occasionally, insects.

Relatively high population densities have been recorded, but these mice do not seem to be gregarious (Kingdon 1974*b*). Neal (1977*a*) stated that mean life expectancy is very short in the wild and nearly all adults die after each breeding season, resulting in a near total population turnover twice a year.

There is a general consensus that breeding in the wild corresponds mainly with the two annual rainy seasons (Rosevear 1969; Kingdon 1974*b*; Delany 1975; Neal 1977*a*; Gautun 1975; Chidumayo 1977). In Uganda these periods extend from April to June and from September to December. The reported number of young per litter is 1 to 12, most commonly being 4 or 5. Most wild females apparently give birth only once during a wet season, but a captive pair produced 4 litters in 15 weeks. The young weigh about three grams at birth, open their eyes when 6 to 8 days old, and attain full size by 6 months. Recent investigations of captive *L. striatus* (Danny C. Wharton, New York Zoological Society, pers. comm.) have shown that the estrous cycle lasts 3 to 5 days, the precise gestation period is 21 days (some older reports gave it as 28 days), consecutive litters may be produced 21 to 25 days apart when a male and female are kept together, young begin eating solid food at 14 days of age,

males attain sexual maturity by an age of 2½ months, females do not give birth until about 5 months old and produce their last litter when 14 months old, and longevity seldom exceeds 2½ years.

RODENTIA; MURIDAE; Genus *PELOMYS* Peters, 1852

Groove-toothed Creek Rats

There are three subgenera and seven species (Meester and Setzer 1971; Yalden, Largen, and Kock 1976; Dieterlen 1974):

subgenus *Komemys*

P. isseli, islands in northwestern Lake Victoria (Uganda);
P. hopkinsi, southwestern Uganda, Rwanda;

subgenus *Desmomys*

P. harringtoni, highlands of Ethiopia;
P. rex, known only by the type skin from southwestern Ethiopia;

subgenus *Pelomys*

P. minor, northeastern Angola, southern Zaire, northwestern Zambia;
P. fallax, eastern Zaire and southern Kenya to Botswana and northern Mozambique;
P. campanae, southwestern Zaire, western Angola.

Misonne (1969) thought there to be no justification for the use of subgenera in *Pelomys.* Dieterlen (1974) doubted the validity of *P. rex,* but Yalden, Largen, and Kock (1976) accepted it as a species. An additional species listed by Meester and Setzer (1971), *P. dembeensis,* is now referred to the genus *Arvicanthis.*

Groove-toothed creek rat (*Pelomys campanae*), photo by F. Petter.

Head and body length is 100 to about 215 mm, and tail length is usually 100 to 180 mm. The tail may be shorter than, the same length as, or longer than the head and body. Reported weights are 46 to 170 grams (Smithers 1971; Kingdon 1974*b*; Delany 1975). The hairs are usually coarse, slightly stiff, and shiny, sometimes with a greenish brown or olive iridescence. The upper parts are tawny yellowish, yellow and black, dusky and tawny, or buffy clay, and blend into the gray, buff, or whitish of the underparts. Some forms have a single dark middorsal stripe, which is often indistinctly defined from the color of the upper parts. The tail is moderately haired, and is frequently dark above and pale below. The nose may be more brightly colored than the face, and the ears are usually thinly covered with reddish hairs.

The short fifth finger is clawed in the subgenus *Komemys*, but has a nail in the subgenera *Pelomys* and *Desmomys*. The outer toes are short. The genus *Pelomys* has grooved upper incisor teeth, and this character distinguishes it from the closely related genera *Lemniscomys*, *Arvicanthis*, and *Dasymys*. It is difficult, however, to distinguish *Pelomys minor* from *Lemniscomys griselda*, as both species have a narrow dark stripe along the middle of the back and other common external features. Female *Pelomys* have eight mammae.

In Ethiopia, *P. harringtoni* lives on plateaus at elevations of 1,800 to 2,800 meters, and is semiarboreal (Yalden, Largen, and Kock 1976). Otherwise, the genus is found mainly in marshes and wet grasslands, and along streams. *P. fallax* is a good swimmer and in some places has become semiaquatic. It usually makes a nest of grass or leaves at ground level, but has been reported to burrow in some areas. This species is active both by day and by night. Its diet consists of grasses, swamp vegetation, and cultivated crops (Kingdon 1974*b*; Delany 1975).

In Zambia, according to Sheppe (1972), *P. fallax* apparently has no fixed breeding season, and seven pregnant females contained seven to nine embryos each. Earlier reports indicated that litter size in this species is usually only two or three young, though Delany (1975) referred to overall embryo counts of one to nine. Delany (1969) also reported that of four pregnant female *P. isseli* collected on Bugala Island in Lake Victoria in April 1966, two were lactating and one was in estrus.

RODENTIA; MURIDAE; **Genus AETHOMYS** Thomas, 1915

Bush Rats, Rock Rats

There are two subgenera and nine species (Meester and Setzer 1971; Ansell 1978):

subgenus *Aethomys*

A. kaiseri, eastern Angola to southern Kenya and Malawi;
A. thomasi, western and central Angola;
A. hindei, northern Nigeria to southern Sudan and northeastern Tanzania;
A. bocagei, western and central Angola, possibly Zaire and southern Congo;
A. selindensis, known only by two specimens from Mount Chirinda in eastern Rhodesia;
A. chrysophilus, southeastern Kenya to Angola and eastern South Africa;
A. nyikae, southern Zaire, northeastern Angola, Zambia, Malawi, eastern Rhodesia;

Bush rat (*Aethomys namaquensis*), photo by John Visser.

subgenus *Micaelamys*

A. namaquensis, Angola and Zambia to South Africa;
A. granti, the Central Karroo of South Africa.

Aethomys has sometimes been regarded only as a subgenus of *Rattus*, but Misonne (1969) considered it a separate genus, and all authorities cited in this account have taken the same position. *Stochomys* has sometimes been included as a subgenus of *Aethomys*, but was distinguished as a full genus by Meester and Setzer (1971).

The following descriptive data were compiled from Kingdon (1974*b*), Meester and Setzer (1971), Smithers (1971), Ansell (1960, 1974), and Ansell and Ansell (1973). Head and body length is about 100 to 184 mm, tail length is 120 to 202 mm, and weight is 63 to 150 grams. The tail varies in length, according to the species, from about 70 to 150 percent of the head and body length. The fur is short and may be sleek or rough. The upper parts range in color through a variety of brown, gray, red, and yellow shades. The hairs of the upper parts are basally gray in all species. The hairs of the underparts may be either gray based or white based, and either gray or white at the tips.

Aethomys has a ratlike shape, a pointed head, and limbs of equal size. Unlike species of *Rattus* that may occur in the same region, the feet of *Aethomys* are usually white above. The tail may appear naked, but has a light to moderate covering of hairs. Unlike those of *Pelomys*, the upper incisor teeth of *Aethomys* are not grooved. Females have four to six mammae.

The following ecological data were compiled from Smithers (1971), Kingdon (1974*b*), Okia (1976), Hubbard (1972), Ansell (1960), and Ansell and Ansell (1973). These rats inhabit grassland with open scrub, cultivated areas, savannahs, and forest edge. Several species, especially *A. namaquensis*, favor localities with rocky outcrops or kopjes. *A. nyikae* has been collected in montane country at an elevation of 2,120 meters. All species are mainly nocturnal and largely terrestrial, though *A. chrysophilus* and *A. namaquensis*, at least, are partly arboreal. Shelters include burrows excavated under bushes or rocks, crevices among rocks, holes in termite mounds, and hollow trees and logs. Nests may be large accumulations of grass, twigs, and debris. *A. namaquensis* sometimes constructs its nest in the fork or hole of a tree, though usually not more than two meters above the ground. The diet is almost entirely vegetarian and includes grains, grass seeds, roots, nuts, fallen fruit, and cultivated crops.

In Uganda, Okia (1976) found *A. hindei* to use individual home ranges of 30 to 1,600 sq meters. Kingdon (1974*b*) stated that *A. chrysophilus* lives in small social groups, but that *A. kaiseri* appears to be less social and therefore may achieve a more scattered distribution. Hubbard (1972), however, located a large group of *A. kaiseri* concentrated in one area, apparently because of dry surrounding conditions. Ansell (1960) noted that *A. nyikae* is thought to be communal or semicommunal, and Smithers (1971) referred to *A. namaquensis* as colonial. In studies of *A. namaquensis* in Rhodesia, Choate (1972) found each family group to have a minimum exclusive space of 100 sq meters centered on the nest. Captives showed territorial behavior and marked with urine. When placed together in small cages, two males would fight to the death. Individuals of opposite sexes usually lived peacefully together and cooperated in building the nest. Choate reported wild populations of *A. chrysophilus* to consist of widely spaced, apparently territorial pairs. Captives, however, showed less aggression than *A. namaquensis*, and small groups of males could be kept together. Hubbard (1972) described the call of *A. kaiseri* as a medium-pitched, rhythmical "chit, chit, chit."

The following reproductive data were compiled from Smithers (1971), Kingdon (1974*b*), Ansell (1960), Okia (1976), Brooks (1972), Choate (1972), and Hubbard (1972). In captivity, and in the wild in East Africa, reproduction seems to continue throughout the year, sometimes with summer peaks. In Botswana, however, *A. namaquensis* evidently does not breed during the cool, dry period from June to September. There may be a postpartum estrus. The average time between successive litters of captive *A. chrysophilus*

was about 29 days. Gestation periods of 21 to 25 days have been reported. Most litters of *A. hindei* consist of only one or two young; in other species the range is one to seven and the average about three or four. The young attach themselves tenaciously to the nipples of the mother and may be dragged about for 3 weeks. They open their eyes when around 14 days old. Female *A. chrysophilus* produce their first litter at an average age of 138 days.

RODENTIA; MURIDAE; **Genus ARVICANTHIS** Lesson, *1842*

Unstriped Grass Mice, Kusu Rats

The following five species now appear to be recognized (Yalden, Largen, and Kock 1976; Kingdon 1974*b*; Allen 1939*b*; Rosevear 1969; Harrison 1972; Meester and Setzer 1971):

A. niloticus, Nile Delta of Egypt, southwestern Arabian Peninsula;
A. abyssinicus, Senegal to Ethiopia and Zambia;
A. blicki, central Ethiopia;
A. dembeensis, southern Sudan and Ethiopia to Tanzania;
A. somalicus, east-central Ethiopia, Somalia, Kenya.

Meester and Setzer (1971) listed only one species, *A. niloticus,* but acknowledged that others might be distinct. Rosevear (1969) suggested agreement that *A. niloticus* is the only species. Yalden, Largen, and Kock (1976), however, explained that studies in progress indicated that populations in Egypt, the type locality of *A. niloticus,* are distinct and isolated from those farther to the south in Africa. Therefore,

Kusu rat (*Arvicanthis abyssinicus*), photo by Ernest P. Walker.

the proper name for the most widespread African species would be *A. abyssinicus*. Yalden, Largen, and Kock also considered *A. blicki* and *A. somalicus* to be separate species. Various authorities, such as Kingdon (1974*b*), used the name *A. lacernatus* for the population here called *A. dembeensis*. The type of *A. lacernatus,* however, is now known to represent the genus *Meriones* (Yalden, Largen, and Kock 1976). *A. dembeensis* was listed as a species of *Pelomys* by Meester and Setzer (1971), but Dieterlen (1974) showed this species to be referable to *Arvicanthis,* and Yalden, Largen, and Kock used its name for the population formerly called *A. lacernatus.*

Head and body length is 106 to 204 mm, tail length is 90 to 160 mm, and weight is 50 to 183 grams (Kingdon 1974*b*; Delany 1975; Rosevear 1969). The tail is usually shorter than the head and body. The hairs are generally coarse and stiff, so the coat is slightly harsh and somewhat spiny. The upper parts vary from blackish chestnut to tawny olive, grayish olive, and light gray. A definite buffy tinge is present in some animals, but others are quite pale in color. Some forms have a slight dorsal stripe. The tips of the hairs are usually dark or black, but a subterminal band of gray or buffy hair produces a lighter overall tone in some individuals. The underparts are slightly paler than the upper parts—buffy white or grayish in the pale-colored forms, and lead color or brownish dark gray in the darker forms. There is usually no sharp line of demarcation between the upper parts and underparts, though some individuals do have such a line. The ears are usually reddish (occasionally brick red); the hands and feet are gray, yellowish, or almost as dark as the body; and the fairly well haired tail is definitely bicolored.

The fifth finger is considerably reduced, but is functional and bears a claw. The second, third, and fourth toes are rather long, though the first and fifth are short. The ears are fairly large and rounded. The incisor teeth are not grooved. Females have six mammae (Rosevear 1969).

These mice are most common in grasslands and savannahs, but also have been reported from forests and scrubby thickets. In Ethiopia they are found at elevations of up to 3,700 meters (Müller 1977). In Uganda they tend to live under dense herb mats, where they construct long tunnels radiating out from their holes. The latter may be dug in soft earth, but are usually in crevices, under rocks or fallen trees, or in termite mounds (Kingdon 1974*b*). Nests of fine grass are built in burrows or on the surface. Both diurnal and nocturnal activity have been recorded (Rosevear 1969; Harrison 1972). The diet is primarily vegetarian and includes seeds, leaves, grass, and cultivated crops.

Unstriped grass mice are gregarious and often live in colonies. Population density varies and may increase greatly under proper conditions. Hubbard (1972:425) described an October population explosion in the Serengeti of Tanzania, during which the mice became "so numerous, one could hardly avoid stepping on them and many were killed by passing trucks. Survivors were feeding on the dead." Poulet and Poupon (1978) reported that density reached about 100 individuals per ha. in the Sahel of Senegal in February 1976, but collapsed in 1977. In a study in the Simien Mountains National Park of Ethiopia, Müller (1977) found densities of 65 to 250 per ha. Average individual home range during the rainy season in this area was 2,750 sq meters for males and 950 sq meters for females and juveniles. Respective figures for the dry season were 1,400 and 600 sq meters. Adult home ranges overlapped considerably. In this same area breeding occurred in the first half of the dry season. However, Kingdon (1974*b*) indicated that several litters might be born to female *Arvicanthis* during the wet season. Rosevear (1969) thought it likely that this genus breeds throughout the year. Delany (1975) stated that breeding does occur all year in Uganda. He cited a laboratory investigation that showed that the gestation period is approximately 18 days and that litter size averages 5.3 young and ranges from 1 to 11. According to Marvin L. Jones (Zoological Society of San Diego, pers. comm.), a captive individual lived for 6 years and 8 months.

RODENTIA; MURIDAE; **Genus DASYMYS** Peters, 1875

Shaggy Swamp Rat

The single species, *D. incomtus,* occurs from Senegal to Ethiopia, and south to South Africa (Meester and Setzer 1971; Hubert, Adam, and Poulet 1973; Yalden, Largen, and Kock 1976).

Head and body length is 113 to 190 mm, tail length is 97 to 185 mm, and weight is 48 to 150 grams (Kingdon 1974*b*; Sheppe 1973; Smithers 1971). The tail is usually somewhat shorter than the head and body. The texture of the fur varies considerably. In some animals it is long, soft, straight, and silky; in others it is loose and coarse; in still others it is shaggy. The upper parts are olive brown, yellowish brown, brownish mixed with black, or slaty black. The underparts are grayish white, whitish, pale buff, or olive buff. The hands and feet are usually light brownish and only scantily haired. The tail is almost naked, and is dark brown above and below, or slightly paler below.

Dasymys is a thickset, heavy rodent. Both the hands and

Shaggy swamp rat (*Dasymys incomtus rufulus*), photo by Jane Burton.

the feet have five digits. The toes are fairly short, and the claws are not suited to climbing. The ears are small, rounded, and evenly fringed with hair. The incisor teeth are not grooved. Females have six mammae.

Although it is reported to occur in forests and savannahs, *Dasymys* is usually found in wet, grassy areas, such as marshes and reed beds. Kingdon (1974*b*) wrote that this rat is very successful in highland bogs and marshes, occurring up to an elevation of 4,000 meters in the Ruwenzori Mountains. It is primarily nocturnal, but is sometimes active by day. It is basically terrestrial, but swims and dives readily. Smithers (1971) stated that it builds domed nests of cut grass and other vegetation at ground level, with a nearby short refuge tunnel running into the ground. It also establishes distinct runways that radiate from the nest to feeding areas. Delany (1975) added that the burrow may be up to 2 meters long and 30 cm deep. *Dasymys* also has been reported to find shelter in holes along the banks of streams. Its diet apparently consists largely of plants growing in or near the water, but traces of insects have been found in some stomachs.

Reported population densities are 34 per ha. in marshland and 2.6 per ha. in savannah (Kingdon 1974*b*). Breeding records are scattered, but extend over most of the year in eastern and southern Africa (Smithers 1971; Ansell 1960; Delany 1975; Sheppe 1973). Pregnant females were taken in Liberia in December and February (Misonne and Verschuren 1976). Reported litter size in *Dasymys* ranges from one to nine young, but seems usually to be about two to five.

Head and body length is 120 to 194 mm, tail length is 104 to 180 mm, and weight is 46 to 190 grams (Kingdon 1974*b*; Delany 1975). The upper parts have a grizzled appearance much like that of *Arvicanthis,* but the hairs are longer and glossier. The coloration above is yellowish buff lined with black, brown or dull buff, or mixed black and buff. The type specimen of one subspecies (*M. d. richardi* from Zaire) has a pronounced bluish green iridescence on the upper parts. The rump may be ochraceous or rufescent. The underparts are usually white (buff in the subspecies *M. d. richardi*). The upper surfaces of the hands and feet are pale buff, tawny, or black, and the moderately haired tail is dark above and lighter below.

The body form is stocky. Only three digits are well developed on the hands and feet. Each of the upper incisors has a single groove, but the lower incisors are not grooved. *Pelomys* is similar to *Mylomys* in external appearance, but the grooving of its incisor teeth is much less pronounced and the surface structure of its molars is different.

The following data were taken from Kingdon (1974*b*) and Delany (1975). *Mylomys* has been found in a variety of moist grassland habitats, often at high elevations of up to 2,400 meters. It is mainly diurnal, but may also be active by night. It nests on the surface and does not burrow. The diet seems to be entirely herbivorous and consists of grass stems and leaves. In Uganda, pregnant females have been taken during all months of the year except January and February, and have been taken in higher proportions during the wet months.

RODENTIA; MURIDAE; **Genus MYLOMYS** *Thomas, 1906*

African Groove-toothed Rat

The single species, *M. dybowskyi,* is found from Ivory Coast to Kenya and the lower Congo River (Meester and Setzer 1971).

RODENTIA; MURIDAE; **Genus GOLUNDA** *Gray, 1837*

Indian Bush Rat, Coffee Rat

The single species, *G. ellioti,* occurs in Pakistan, India, Nepal, Bhutan, and Sri Lanka (Ellerman and Morrison-Scott 1966).

African groove-toothed rat (*Mylomys dybowskyi*), photo by C. L. Cheeseman.

Indian bush rat (*Golunda ellioti gujerati*), photo by P. J. Deoras.

Head and body length is usually 110 to 155 mm, and tail length is usually 90 to 130 mm. Four specimens weighed 50 to 80 grams (Roberts 1977). The texture of the pelage varies considerably. In some animals the covering is fairly soft with only a few harsh hairs; in others it is coarse, and the longer hairs, some of which are spiny, are flattened and grooved. The coat is generally thin, but the hairs are rather long. The coloration above is grayish, yellowish brown, reddish brown, or fairly dark brown. The darker colors are produced by a fine speckling of fulvous and black hairs. The coloration below is light gray, bluish gray, or white.

Golunda is a thickset, heavy rodent, rather volelike in appearance. The head is short and rounded, and the ears are rounded and hairy. The tail is shorter than the head and body, is stout at the base, tapers toward the tip, and is covered with coarse, short hairs. The upper incisor teeth are grooved, and the pattern of the molars is somewhat unusual. Females have eight mammae.

This rodent usually inhabits swamps, grasslands, the edge of cultivated areas, and (at least in Sri Lanka) jungles. Prakash and Rana (1972) observed that in a grassy area of the Indian Desert the burrows of *Golunda* are invariably found under bushes. Otherwise, this rodent has been reported to burrow only rarely, and to locate its nest on the ground or just above the ground, usually in dense grass. The globular nest is made of plant fibers and is from 150 to 230 mm in diameter. *Golunda* is a good climber, but usually moves about on the ground in definite runways. Some observers have said that *Golunda* is slow in its movements; others have considered it to be quick. It is most active in the early morning and evening. Its food is almost exclusively plant material, normally the roots and stems of such plants as the "dub" or "nariyali grass" (*Cynodon*).

Individuals generally live alone or in family groups. In studies in the Indian Desert, Prakash (1975) found pregnant females from March to August. The number of embryos averaged 6.6 and ranged from 5 to 10. Earlier reports indicated the usual number of young to be 3 or 4.

When coffee was grown in Sri Lanka the bush rat became extremely plentiful and did great damage to the crops by eating the buds and blossoms of the coffee plant. It has been suggested that the animals migrated to the plantations, but it is probable that they simply increased in numbers in response to the abundant food supply. Coffee is no longer planted in Sri Lanka, and bush rat numbers have decreased substantially.

RODENTIA; MURIDAE; **Genus HADROMYS** *Thomas, 1911*

Manipur Bush Rat

The single species, *H. humei*, occurs in Assam at the eastern edge of India (Ellerman and Morrison-Scott 1966).

Head and body length is about 95 to 125 mm, and tail length is about 105 to 140 mm. An adult male weighed 65.5 grams. The fur is thick and soft. The upper parts are blackish gray. The middle area of the back is sometimes darker than the head and shoulders, and the rump has a reddish wash. The belly is yellow or orange, with a slaty tinge. The inner sides of the thighs are rufous, and the hands and feet are yellowish white. The fairly well haired tail is distinctly bicolored—black above and whitish below.

The body has a fairly stout form. The fifth finger and the fifth toe are short. The teeth are large and powerful; the incisors are not grooved, and the molars are broad and heavy. Females have eight mammae. *Hadromys* resembles *Golunda* in general external appearance, but its tail is slightly longer, and its fur is not spiny or less spiny. The upper incisors of *Golunda* are grooved.

Seven specimens were collected for the Zoological Survey of India in connection with a scrub typhus outbreak in 1945. These specimens came mainly from an oak parkland habitat described by Roonwal (1949) as follows: "Fairly thickly covered semi-scrub of moderately tall oak trees (up to 9–12 meters high) . . . on hill-sides at 1,200 meters and probably higher altitudes. . . . Soil loamy with rock beneath and with practically no humus. Usually not associated with streams. . . . Tree canopy very open. Very tall grass, over 3–4 meters high. Hardly any shrubs. Ground canopy also very open."

Leaves of grass that had been cut and eaten were found in the stomach of an adult male. A pregnant female was collected in September.

RODENTIA; MURIDAE; **Genus BANDICOTA** *Gray, 1873*

Bandicoot Rats

There are three species (Ellerman and Morrison-Scott 1966; Lekagul and McNeely 1977; Medway 1978):

Manipur bush rat (*Hadromys humei*), photo from *Proc. Zool. Soc. London*.

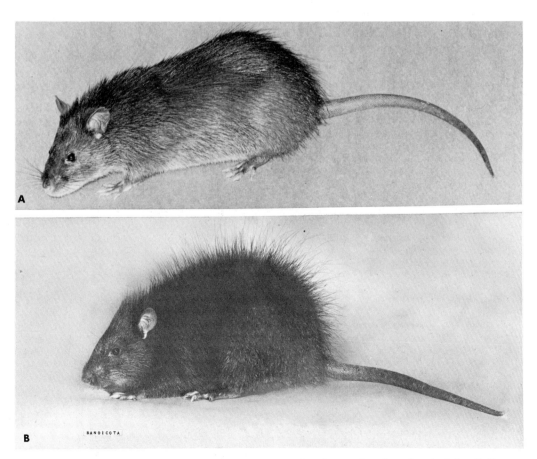

Bandicoot rat: Top, *Bandicota bengalensis*, photo by Ernest P. Walker; Bottom, *B. indica*, photo by Robert E. Kuntz.

B. bengalensis, Pakistan to Burma, Sri Lanka, Penang Island off west coast of Malay Peninsula, Sumatra, Java;

B. indica, western India to southeastern China and Indochina, Sri Lanka, Taiwan, Sumatra, Java;

B. savilei, central Burma to southern Viet Nam.

Gunomys Thomas, 1907 is a synonym of *Bandicota.*

Head and body length is 160 to at least 360 mm, and tail length is 130 to 258 mm. The tail is usually somewhat shorter than the head and body, but may be slightly longer. Lekagul and McNeely (1977) listed average weights of 199 grams for *B. savilei* and 545 grams for *B. indica,* but Grzimek (1975) noted that weight could reach 1,500 grams. The texture of the pelage varies from soft and fairly dense to thin and coarse, and there is much variation in the length of the guard hairs. The general coloration of the upper parts ranges from light grayish to various shades of brown or almost black. The lower parts are a dirty white. The tail is scantily haired. The front claws are large, and the muzzle is short and broad. The incisor teeth are yellow or orange in color. Females have from 12 to 18 mammae.

The natural habitat of *B. bengalensis* has been reported to be evergreen jungle and oak scrub, but this species has become an urban commensal of humans. *B. indica* is also a commensal in some areas and is now especially common in lowland rice fields. Lekagul and McNeely (1977) did not locate *B. indica* in purely natural habitat in Thailand. Bandicoot rats are good diggers and construct elaborate burrows at the edges of fields, in dikes and stream banks, and even in city streets. There are two to six entrances, each marked by a small mound of earth (Medway 1978). The burrow may extend to a depth of 60 cm and contains enlarged chambers used for nesting and food storage. Ordinarily these rodents leave their retreats only under cover of darkness. They are good swimmers and divers. They are omnivorous in diet, but in some areas feed largely on products of cultivation, such as rice, grains, sugar cane, fruit, nuts, and potatoes.

There is usually only a single adult per burrow, but high population densities can occur in favorable areas (Grzimek 1975). Female *B. bengalensis* have an estrous cycle of about 5 days (Sahu and Maiti 1978). This species breeds throughout the year in Pakistan and India (Prakash 1975; Roberts 1977). In the city of Rangoon, Burma, some reproduction occurs at all times of the year, but there are peaks during the dry season from November to April (Walton et al. 1978). Studies in this city indicate that females have an average of about 62 days between litters, whereas earlier investigations in Calcutta determined a wait of only 31 to 34 days, and thus a probable postpartum estrus and mating. In Rangoon litter size averaged 7.4 young and ranged from 1 to 14. In Viet Nam litter size of *B. indica* ranged from 2 to 12 young

(Medway 1978). In southern Pakistan the largest litters of *B. bengalensis,* consisting of 14 to 18 young, are born from September to November, while during the remainder of the year the usual litter size is 5–10. The young weigh an average of 3.5 grams, open their eyes at 14 days, are weaned at 28 days, and reach sexual maturity when around 3 months old (Roberts 1977).

Bandicoot rats are characterized by their ferocious nature. When disturbed, they emit a harsh, nasal barking. Lekagul and McNeely (1977) reported that an effort to breed *B. indica* for laboratory use was not successful, because the animals remained too savage. Lekagul and McNeely also noted that since this species is large, delicious to eat, and generally found as a commensal, it may have been geographically spread through human agency. In 1946 an introduction was made into the states of Kedah and Perlis in mainland Malaysia. The population of *B. indica* on Taiwan reportedly originated from releases made by the Dutch. Several other peculiarities in the range of both *B. indica* and *B. bengalensis* might be interpreted as having come about through introduction. Whatever the case, these animals are now human commensals over large areas, and are often considered serious pests in both agricultural and urban settings (Walton et al. 1978; Grzimek 1975). They raid crops, and eat and contaminate stored foods. Their tunnels and mounds damage irrigation structures, block gutters, and cause the collapse of streets and buildings. They are involved in the spread of plague and other diseases. In return, however, many people hunt bandicoot rats for food, and dig up their burrows to obtain the food stored by the animals.

RODENTIA; MURIDAE; *Genus NESOKIA Gray, 1842*

Pest Rat, Short-tailed Bandicoot Rat

The single species, *N. indica,* occurs from Egypt to Sinkiang and northern India (Corbet 1978).

Head and body length is 140 to about 215 mm, tail length is 88 to 129 mm, and adult weight is approximately 112 to 175 grams. The texture of the pelage varies from coarse, short, harsh, and semispinous to long, fine, and silky. In some forms the coat is dense and in others it is thin, but in all forms the tail is practically naked, and the hands and feet are scantily haired. The general coloration of the upper parts ranges from fawn or yellowish brown to grayish brown, mixed in places with red. The underparts are grayish to whitish. The color of the back gradually merges into that of the sides and flanks without a distinctive line of demarcation.

Nesokia is characterized by a robust form; a short, rounded head; a short, broad muzzle; rounded ears; the relatively

Pest rat (*Nesokia indica*), photo by Harry Hoogstraal.

short, almost naked tail; and broad feet. All the claws, except the rudimentary thumb, are armed with strong, nearly straight nails. The incisor teeth are stout and broad. Females have eight mammae.

The short-tailed bandicoot rat usually lives in moist areas, or along streams and canals, within a variety of general habitats, including deserts, steppes, cultivated areas, and forests. The elevational range is from about 26 meters below to 1,500 meters above sea level. This animal burrows extensively, usually at depths of 15 to 60 cm, and throws the loose soil onto the surface of the ground around the tunnel entrance. All individuals observed by Lay (1967) seemed to confine their activities to their burrow systems, each of which consisted of several to many tunnels, a single enlarged nest chamber lined with finely chewed vegetation, and several enlarged, unlined chambers. In Iraq the holes and prodigious diggings of *Nesokia* are seen frequently along the banks of irrigation canals (Harrison 1972). One burrow system excavated in this area consisted of a nest chamber, about 30 cm in diameter, and a connecting main tunnel that ran for about 4.5 meters, with periodic side tunnels and ventilation holes leading to the surface. Taber et al. (1967) reported that in Pakistan the burrows are as long as 23 meters. In some areas, at least, *Nesokia* also makes well-defined runways along the surface. The diet consists of grass, grains, roots, and cultivated fruits and vegetables. Sometimes food is stored in the burrows. Up to 454 grams of grain have been found in storage chambers.

Lay (1967) reported that usually only a single rat occupies each burrow, and that all individuals captured alive fought fiercely. A captive colony in Teheran bred throughout the year. Al-Robaae (1977) indicated that wild populations in Iraq probably breed all year and that litter size there is 1 to 8 young. Litter sizes listed by Prakash (1971) ranged from 2 to 10 young.

In Iraq this rat does considerable damage to watermelons, sweetmelons, and tomatoes (Al-Robaae 1977). It also causes extensive damage by raiding grain fields and tunneling through irrigation walls. Native people eat *Nesokia*, as well as the grain stored in its burrows. They smoke the tunnels, or pour water into them, until the animal comes to the surface, where it is captured, sometimes with the aid of a dog.

RODENTIA; MURIDAE; *Genus PRAOMYS* Thomas, 1915

African Soft-furred Rats

There are 4 subgenera and 26 species (Meester and Setzer 1971; Robbins, Choate, and Robbins 1980; Van der Straeten and Verheyen 1978*b*; F. Petter 1975, 1977; Green et al.

1980; Gordon 1978; Robbins and Setzer 1979; Hubert, Adam, and Poulet 1973; Ansell and Ansell 1973; Bishop 1979):

subgenus *Mastomys* (multimammate rats)

P. pernanus, southwestern Kenya, northern Tanzania, Rwanda;
P. natalensis, throughout Africa south of the Sahara;
P. coucha, southern Africa;
P. erythroleucus, southern Morocco, Sahel and Sudan zones of West and Central Africa;
P. angolensis, western Zaire, Angola;
P. shortridgei, Namibia, Botswana;

subgenus *Myomyscus*

P. albipes, central Sudan, highlands of Ethiopia;
P. fumatus, southern Sudan, Ethiopia, Somalia, Uganda, Kenya, Yemen;
P. verreauxi, southwestern South Africa;
P. butleri, southwestern Sudan;
P. daltoni, Senegal to Nigeria;
P. derooi, Ghana to Nigeria;

subgenus *Praomys*

P. tullbergi, Senegal to Central African Republic;
P. hartwigi, mountains of Cameroon;
P. morio, Cameroon;
P. jacksoni, Cameroon to southern Sudan and Zambia;
P. lukolelae, Cameroon, Central African Republic, Congo;
P. delectorum, Kenya, Tanzania, northeastern Zambia, Malawi;

subgenus *Hylomyscus*

P. baeri, Ivory Coast, Ghana;
P. aeta, Cameroon and Gabon to Uganda, Fernando Poo;
P. carillus, Angola;
P. denniae, mountains from northeastern Zaire and Uganda to Zambia;
P. parvus, Cameroon, Congo, Gabon;
P. alleni, Guinea to Central African Republic and Gabon, Fernando Poo;
P. stella, Zaire, Uganda, Kenya;
P. fumosus, Cameroon, Gabon, possibly Central African Republic.

Praomys; Mastomys Thomas, 1915; *Myomyscus* Shortridge, 1942; and *Hylomyscus* Thomas, 1926 have sometimes been included within the genus *Rattus,* but Misonne (1969) considered each to be a subgenus of the separate genus *Praomys.* This procedure is now generally followed, as, for example, in Meester and Setzer (1971) and Kingdon (1974*b*). Some authorities, such as Rosevear (1969), and Robbins, Choate, and Robbins (1980), have treated the above subgenera of *Praomys* as full genera. Both Rosevear

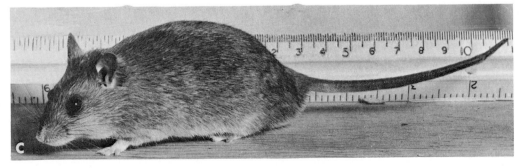

African soft-furred rat (*Praomys natalensis*), photo by H. S. Davis.

(1969), and Van der Straeten and Verheyen (1978b) used the name *Myomys* Thomas, 1915 in place of *Myomyscus,* and treated it as a full genus.

Perhaps the most difficult systematic problem within any of the subgenera of *Praomys* involves the three multimammate rats *P. natalensis, P. coucha,* and *P. erythroleucus.* Many authors, as, for example, Rosevear (1969) and Kingdon (1974b), have dealt with this group under a single specific name, *P. natalensis.* Meester and Setzer (1971) stated that the group included two species, *P. natalensis* and *P. erythroleucus,* but they did not distinguish the two by distribution or morphology. Gordon (1978) showed that in Rhodesia the populations formerly called *P. natalensis* actually comprise two species, one with a diploid chromosome number of 32, and the other with 36. Meanwhile, F. Petter (1977) pointed out that in West Africa there are also two species, one with 32 chromosomes, which he called *P. huberti* (a name included within the *P. natalensis–P. erythroleucus* group by Meester and Setzer), and the other with 38 chromosomes, which he called *P. erythroleucus.* The latter species was assigned a range in the Sahel and Sudan zones of West and Central Africa. In addition, Corbet (1978) reported *P. erythroleucus* to be the only multimammate rat in Morocco. Green et al. (1980) showed that the species with 36 chromosomes is widespread in southern Africa, and that the correct name is actually *P. coucha* (a name included within the *P. natalensis–P. erythroleucus* complex by Meester and Setzer). The species with 32 chromosomes also was found to occur in much of this region, and, on the basis of the karyotype of specimens from the type locality, the name *P. natalensis* was thought to apply to this species. Green et al. thus suggested that the species with 32 chromosomes in West Africa, as discussed by F. Petter (1977), might also represent *P. natalensis.* This suggestion has been adopted in the above list.

The species *P. (Hylomyscus) fumosus* was named by Brosset, Dubost, and Heim De Balsac (1965). Misonne (1969) thought the dentition of this species to differ widely from that of *Hylomyscus,* but to be close to that of *Dephomys* (now considered a part of *Stochomys*). He suggested a new generic name, *Heimyscus,* for the species. Robbins, Choate, and Robbins (1980) indicated that they planned a review of the status of the species.

The following descriptive data were taken primarily from Rosevear (1969) and Kingdon (1974b). Head and body length is 62 to 168 mm, tail length is 52 to 172 mm, and weight is 8 to 75 grams. All species have soft, gray-based fur, which is either long or short. The coloration of the upper parts ranges through a great variety of red, brown, gray, and yellow shades. The underparts are white, gray, or yellowish. Both the tail and the ears appear naked, but actually have a light covering of hairs.

The body is slender, the head is pointed, and the ears are large and rounded. The tail ranges from slightly shorter to about one-third longer than the head and body. *Praomys* is usually considerably smaller than those species of *Rattus* that have been introduced to Africa. Whereas those species have a skull length over 36 mm, a first upper molar with five roots, and harsh fur, *Praomys* has a skull length below 36 mm, a first upper molar with three or four roots, and soft fur. Females of the subgenera *Praomys* and Hylomyscus have 6 to 8 mammae, and female *Myomyscus* have 10 mammae. In *Mastomys,* the multimammate rats, the number of nipples varies from 16 to 24.

African soft-furred rats occur in many types of habitat (Kingdon 1974b; Rosevear 1969). *P. natalensis* may once have been restricted mainly to savannahs, but now can be found throughout sub-Saharan Africa, chiefly in association with people. It occupies cultivated and abandoned fields, buildings, and villages, but not large towns, probably because of competition with *Rattus rattus. M. erythroleucus* inhabits relatively dry but well-vegetated areas around the edges of the Sahara. The subgenus *Myomyscus* occurs in a variety of habitats from dry woodland to lush forest. The subgenera *Praomys* and *Hylomyscus* are found primarily in humid forests. *P. denniae* occupies montane forests at elevations of up to 3,500 meters. All species of the genus *Praomys* are primarily nocturnal.

Smithers (1971) referred to *P. natalensis* as terrestrial, but Delany (1975) noted that it is a good climber and swimmer. Apparently all species are at least partly arboreal, though some may nest underground (Kingdon 1974b; Rosevear 1969). *P. natalensis* uses a crevice or digs its own burrow, which usually consists of a network of galleries without a central chamber. *P. jacksoni* and *P. delectorum* nest in short burrows and also make concealed runways. Other species of the subgenus *Praomys* nest in hollow logs or trees. The subgenus *Hylomyscus* seems largely arboreal and builds nests of leaves or other vegetation in tree holes, forks of branches, and palm or banana leaf axils. This subgenus evidently feeds mainly on fruit and other vegetation, but, like other subgenera, does eat some animal matter. Smithers (1971) stated that under field conditions *P. natalensis* eats mostly grass and other seeds. Kingdon (1974b) observed, however, that in suitable habitat this species may take as much animal matter—mainly insects—as vegetation. In the vicinity of human settlements *P. natalensis* eats nearly everything that people do.

Periodic population outbreaks have been reported for *P. natalensis,* and at times hundreds of individuals may be seen at once. This species seems to have a relatively mild disposition and lacks the aggressiveness of *Rattus rattus.* Several families may live together in one burrow, and strangers are accepted into the group (Kingdon 1974b; Choate 1972). Average home ranges of 0.23 ha. for males and 0.22 ha. for females have been reported in a population of *P. tullbergi* (Happold 1977).

There seems general agreement that *P. natalensis* is capable of breeding throughout the year, but that peak reproduction comes toward the end of the rains and in the early part of the dry season (Kingdon 1974b; Rosevear 1969; Sheppe 1972, 1973; Smithers 1971; Delany 1975). In Uganda these peak periods are centered around May and November (Neal 1977b). Roughly the same pattern has been reported for *P. daltoni* (Anadu 1979b), *P. derooi* (Van der Straeten and Verheyen 1978b), *P. tullbergi* (Happold 1977), and *P. stella* (Kingdon 1974b). *P. jacksoni* is thought to breed throughout the year in Uganda (Delany 1975).

Postpartum estrus and mating have been reported in *P. natalensis* (Rosevear 1969; Kingdon 1974b). Females of this species may have two litters per season (Neal 1977b; Sheppe 1973). The gestation period in *P. natalensis* is 23 days (Delany 1975). In other species that have been investigated, gestation ranges from 21 to 37 days but generally averages just under 1 month (Rosevear 1969; Kingdon 1974b; Delany 1975; Anadu 1979b). Litters contain 1 to 22 young in *P. natalensis,* usually about 10 to 12 (Smithers 1971; Neal 1977b; Kingdon 1974b). Other species evidently have smaller litter sizes: 1–8 in *P. jacksoni,* 3–6 in *P. denniae,* 2–4 in *P. alleni* and *P. fumatus* (Delany 1975), 1–5 in *P. stella* (Kingdon 1974b), 3–10 in *P. daltoni* (Anadu 1979b), and 2–5 in *P. derooi* (Van der Straeten and Verheyen 1978b). Young *P. natalensis* weigh about 1.8 grams at birth, open their eyes when around 16 days old, are weaned and independent at 3 weeks, and are usually able to breed when 3½ months old (Neal 1977b; Delany 1975; Rosevear 1969). Sexual maturity is attained at only 2½ months by *P. jacksoni* (Kingdon 1974*b*), but not until the age of 4½–5½ months by *P. daltoni* (Anadu 1979*b*). In a wild population of *P. tullbergi* in Nigeria most individuals were replaced after only 6 months, but some lived to at least 20 months (Happold 1977). A captive individual of this species lived for 5 years and 2 months (Marvin L. Jones, Zoological Society of San Diego, pers. comm.).

Kingdon (1974*b*) thought it likely that *P. natalensis* was originally restricted to the savannahs of southern Africa, but spread northward as a commensal of people. It is now declining in many areas through competition with *Rattus rattus*, but still is the dominant commensal in more remote villages. It is sometimes a serious pest in human habitations and to agriculture, and is also the principal African indigenous host of human plague.

RODENTIA; MURIDAE; **Genus *MILLARDIA*** Thomas, *1911*

Asian Soft-furred Rats, Cutch Rats

There are two subgenera and seven species (Ellerman and Morrison-Scott 1966; Misonne 1969; Mishra and Dhanda 1975):

subgenus *Millardia*

M. meltada, Pakistan, India, Nepal, Sri Lanka;
M. kondana, southwestern India;
M. kathleenae, central Burma;
M. gleadowi, Pakistan, western India;

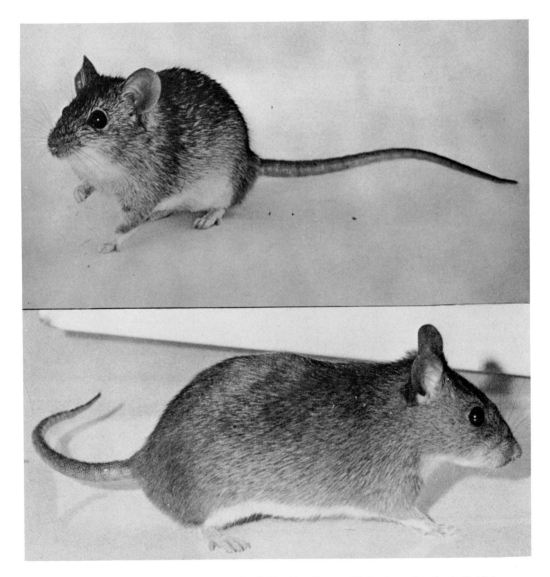

Asian soft-furred rat: Top, *Millardia meltada*, photo by K. Tsuchiya; Bottom, *M. kondana*, photo by A. C. Mishra.

subgenus *Cremnomys*

M. blanfordi, India, Sri Lanka;
M. cutchicus, India;
M. elvira, southern India.

Both *Millardia* and *Cremnomys* are sometimes considered only subgenera of *Rattus,* but Misonne (1969) suggested the arrangement given above. Mishra and Dhanda (1975), while recognizing *Millardia* as a full genus, considered the species of *Cremnomys* to be representative of a separate genus.

The following descriptive data apply only to the subgenus *Millardia* and were taken in part from Mishra and Dhanda (1975). Head and body length is 80 to 200 mm, and tail length is 68 to 186 mm. A specimen of *M. meltada* weighed 70 grams. The fur is soft. In *M. meltada* the upper parts are sandy gray, grayish brown, or whitish buff lined with brown; the underparts are whitish or grayish; and the tail is dark above and light below. In *M. kondana* the upper parts are dark brown, the underparts are grayish white, and the tail is dark above and grayish below. In *M. gleadowi* the upper parts are pale grayish brown or sometimes fawn colored, the underparts are white, and the tail is light brown above and white below. *M. kathleenae* is a pale-colored species with a white-tipped tail.

Millardia resembles *Rattus* in external appearance but has been characterized as follows: total or partial suppression of the pads of the soles of the feet; toes shortened, the fifth scarcely reaching the base of the fourth; tail usually subequal in length to the head and body; and palate long, exceeding half the occipitionasal length. In most species the ears are rather large and the tail is well haired. *M. kondana,* however, has relatively small ears and a poorly haired tail, and also comparatively well developed pads on the feet. Female *M. meltada* and *M. kondana* have eight mammae, female *M. gleadowi* have six, and female *M. kathleenae* have four.

The species *M. meltada* has been found in grasslands, swamps, broken rocky ground, and cultivated fields. It shelters in short burrows, stone heaps, and the abandoned burrows of other rodents. During the hot months it often seeks the cracks formed in drying soil. The diet consists of grain, seeds, and swamp vegetation. At times, M. meltada may increase in numbers to plague proportion, and cause considerable damage to crops. Roberts (1977) stated that both *M. meltada* and *M. gleadowi* are nocturnal and spend the day in shelters. The latter species makes short burrows, usually under the roots of a bush, and extending at an angle of about 45° to a depth of not more than 61 cm. Prakash and Rana (1972) found *M. cutchicus* to be common in rocky parts of the Indian Desert, and to shelter in rock crevices.

Prakash (1975) reported the average home range size for *M. meltada* in the Rajasthan Desert to be 1,217 sq meters. In at least some parts of India this species is said to live in groups of 2 to 6 individuals. In the Rajasthan Desert it apparently is capable of breeding throughout the year, with pregnancies reaching a peak in October, and litter size ranging from 4 to 10 and averaging 5.9 young (Prakash 1971). Farther north, in the Punjab, Bindra and Sagar (1968) found captive *M. meltada* to breed during the hot months from March to May and August to October. Females produced from 2 to 7 litters annually, gestation lasted about 20 days, and litter size ranged from 1 to 8 young, with averages of 3.4 in the laboratory and 6 in the field. Females attained sexual maturity when 3 to 4½ months old.

In the Rajasthan Desert, reproduction of *M. cutchicus* is evidently affected by rainfall (Prakash 1971). During 1968 there was a drought in this area, and only a single pregnant female was found from November 1968 to April 1969. In 1970, however, the rains exceeded the normal level, and pregnant females were taken from March to October, with a peak in August. Litter size averaged four and ranged from two to eight young.

Prakash (1971) observed breeding in *M. gleadowi* only from August to October in the Indian Desert. Litters consisted of two or three young. According to Marvin L. Jones (Zoological Society of San Diego, pers. comm.), a captive *M. gleadowi* lived for four years.

RODENTIA; MURIDAE; **Genus ANONYMOMYS** *Musser, 1981*

Mindoro Rat

The single species, *A. mindorensis,* is known only by three specimens from Ilong Peak, northeastern Mindoro Island, Philippines (Musser 1981*a*).

In an adult male, head and body length is 125 mm, and tail length is 206 mm. The upper parts are bright buffy or tawny brown, and the underparts are cream. The tail is monocolored and tipped by a tuft. The pelage is long and dense, and has numerous spines. The body is stocky, the ears are small and sparsely haired, and the hind feet are short and wide.

From *Rattus* and *Limnomys, Anonymomys* is distinguished in having smaller bullae, the squamosal roots of the zygomatic arches originating higher on the sides of the braincase, the posterior margin of the palatal bridge not extending far past the molar tooth row, the pterygoid fossa nearly flat and not perforated by a large foramen, and the mesopterygoid fossa at least as wide as the palatal bridge. *Anonymomys* resembles *Niviventer cremoriventer* externally, but differs in having paler and less spiny pelage, a shorter and broader rostrum, a more domed braincase, much larger bullae, and relatively larger teeth.

The Mindoro rat is considered to be arboreal. All specimens were taken in a mountainous area at an elevation of about 1,350 meters.

RODENTIA; MURIDAE; **Genus SRILANKAMYS** *Musser, 1981*

Ceylonese Rat

The single species, *S. ohiensis,* occurs in central Sri Lanka. This species was long included in the genus *Rattus,* but Musser (1981*a*) showed that it belongs in its own genus.

In an adult female, head and body length is 145 mm, and tail length is 173 mm. The upper parts are glossy dark gray, slightly suffused with brown. The underparts are pale cream and are sharply demarcated from the dark dorsum. The tail is sharply bicolored, being dark grayish brown above and white below. The upper surfaces of the feet are mostly white. The ears are small and scantily haired. Females have eight mammae.

Srilankamys is distinguished from *Rattus* in having shorter, softer, and finer pelage; smaller bullae; the alisphenoid canal of the skull hidden by a lateral strutlike wall; the squamosal roots of the zygomatic arches originating higher on the sides of the braincase; the posterior margin of the palatal bridge not extending far past the molar tooth row; the pterygoid fossa nearly flat and not perforated by a large foramen; the mesopterygoid fossa at least as wide as the palatal bridge; thinner, more delicately built dentaries, with smaller coronoid processes; relatively smaller teeth; and ivory-colored, rather than bright orange, incisors.

Srilankamys occurs in mountain forests and is considered

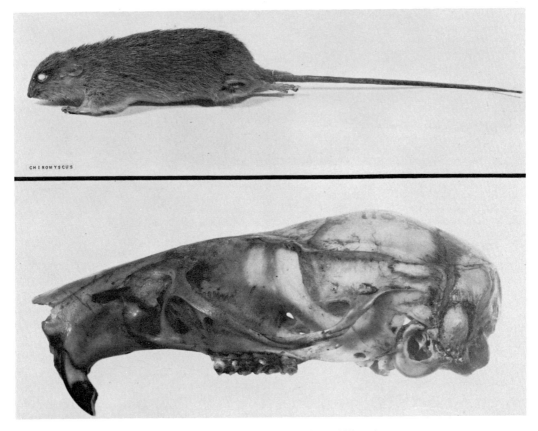

Fea's tree rat (*Chiromyscus chiropus*), photos from British Museum (Natural History).

terrestrial. The type specimen was collected at an elevation of about 1,800 meters.

RODENTIA; MURIDAE; Genus CHIROMYSCUS Thomas, 1925

Fea's Tree Rat

The single species, *C. chiropus*, occurs in eastern Burma, northern Thailand, Laos, and Viet Nam (Lekagul and McNeely 1977). The species was first described by O. Thomas in 1891 as *Mus chiropus*, from a specimen that had been collected in the Karin Hills of Burma and preserved in alcohol. It did not again come to attention until 1924, when H. Stevens, on the Sladen-Godman Expedition to northern Viet Nam, collected a specimen at Bao-Ha at an elevation of 90 meters in the forest country of the Red River. This specimen became the type of the genus established by Thomas.

Head and body length is 127 to 160 mm. Van Peenen, Ryan, and Light (1969) listed tail length as 198 to 252 mm. The pelage is rather bristly. Coloration of the upper parts is "warm-lined buffy" suffused with black. The sides of the face are ochraceous, the color extending up behind the ears, and there is a dark area surrounding each eye. The rump is bright ochraceous, the throat is rusty, and the underparts are creamy. The flesh-colored ears are covered with short hairs. The long, slender tail is also covered with fine, short, light-colored hairs, and is slightly penciled.

This genus is similar to *Niviventer*, but differs in having a nail, rather than a claw, on the hallux (Musser 1981*a*). The hind foot of *Chiromyscus* is specialized for climbing; the great toe is opposable and large digital pads are present. The claws on all the digits are short. Females have eight mammae.

Chiromyscus lives in forests and is probably arboreal. No other information on its natural history has been recorded.

RODENTIA; MURIDAE; Genus DACNOMYS Thomas, 1916

Large-toothed Giant Rat, Millard's Rat

The single species, *D. millardi*, is known from the Darjeeling district of northeastern India, Assam, and Laos (Ellerman and Morrison-Scott 1966). The elevational range is about 1,000 to 1,800 meters.

Head and body length is 228 to 290 mm, and tail length is 308 to 335 mm. The somewhat coarse pelage is short and thin, and the individual hairs on the back are 15 to 16 mm in length. The general coloration of the upper parts is olive brown to grayish or smoky brown, becoming lighter on the sides as it blends into the cinnamon drab or pale brownish of the underparts. The gular, inguinal, and axillary areas are creamy or white. The ears and hands are brown, and the fingers are white. The finely ringed tail (about 10 scales to the centimeter) is scantily sprinkled with fine, short hairs.

Dacnomys is a large, ordinary-looking rat, with no special external distinguishing characters. It derives one of its common names from the relatively large size of the molar teeth. The sole pads are large and rounded, and the fifth toe of the hind foot lacks a claw. Females have eight mammae.

RODENTIA; MURIDAE; *Genus DIOMYS Thomas, 1917*

Manipur Mouse, Crump's Mouse

The single species, *D. crumpi*, was long known only by a single specimen, a skull collected at an elevation of 1,300 meters on Mount Paresnath in east-central India, among rocks. In 1942, 25 years after publication of the description, a series of 30 specimens was secured at Bishenpur, Manipur, Assam in extreme eastern India. In 1978, two skins and four skulls were collected in south-central Nepal at an elevation of 200 meters in moist deciduous forest (Ingles et al. 1980).

Head and body length is 100 to 145 mm, and tail length is 105 to 135 mm. The back is blackish gray, the rump is usually black, and the feet are white. The underparts are whitish gray, and the tail is black above and white below. The fur is thick and soft, but the tail is scantily haired.

Ellerman (1946:205) wrote: "The genus *Diomys* stands on the important character that the incisors are so proodont, that the condylobasal length normally is at least as long as the occipitonasal length, and often exceeds it (three exceptions in twenty measurable skulls). This is very rare in Asiatic Muridae; the only other murid genera from Asia with this character normally present are the very different Bandicoot Rats, *Nesokia* and *Bandicota*."

RODENTIA; MURIDAE; *Genus MAXOMYS Sody, 1936*

Rajah Rats, Spiny Rats

There are 16 species (Musser, Marshall, and Boeadi 1979; Medway 1977, 1978; Chasen 1940):

M. surifer, southern Burma, Thailand, Indochina, Malay Peninsula, Sumatra, Java, many small islands of East Indies;

M. moi, Laos, Viet Nam;

M. pagensis, Mentawi Islands off western Sumatra;

M. panglima, Balabac, Palawan, Busuanga, and Culion islands (Philippines);

M. rajah, Malay Peninsula, Rhio Archipelago, Sumatra, Borneo;

M. hellwaldi, Celebes;

M. dollmani, Celebes;

M. bartelsii, Java;

M. alticola, mountains of northern Borneo;

M. ochraceiventer, Borneo;

M. inas, mountains of Malay Peninsula;

M. hylomyoides, Sumatra;

M. musschenbroekii, Celebes;

M. baeodon, northern Borneo;

M. whiteheadi, Malay Peninsula, Sumatra, Borneo, many small islands of East Indies;

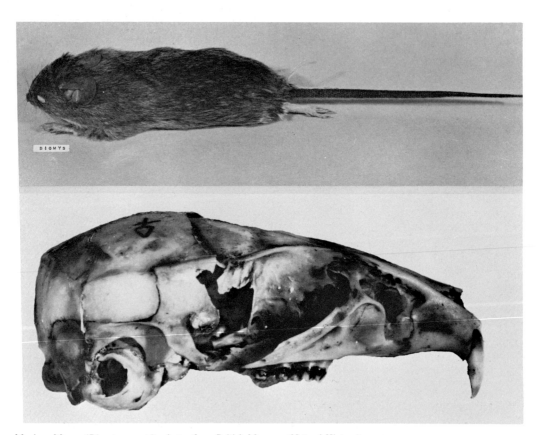

Manipur Mouse (*Diomys crumpi*), photos from British Museum (Natural History).

Rajah rat (*Maxomys surifer*), photo by Lim Boo Liat.

M. inflatus, mountains of Sumatra.

Maxomys was originally proposed as a genus for the species *M. bartelsii*. Subsequently it came to be generally considered a subgenus of *Rattus* and to include a number of species, particularly those now placed in the genus *Niviventer*. Most of the species in the above list were actually put in *Lenothrix*, which also was treated as a subgenus of *Rattus* (Musser, Marshall, and Boeadi 1979; Ellerman and Morrison-Scott 1966; Laurie and Hill 1954). Misonne (1969) reinstated *Maxomys* as a full genus, which included *M. bartelsii*, as well as species now placed in the genus *Niviventer*. Misonne also, however, put the species listed above as *M. baeodon*, *M. hellwaldi*, *M. musschenbroekii*, *M. whiteheadi*, *M. rajah*, and *M. surifer* into the subgenus *Leopoldamys* of *Rattus*. Later authorities (Van Peenen et al. 1974; Medway and Yong 1976; Lekagul and McNeely 1977) indicated that *M. bartelsii* is not closely related to the other species placed in *Maxomys* by Misonne, but is actually associated with most of the species that Misonne put into *Leopoldamys*. Finally, Musser, Marshall, and Boeadi (1979) removed *Rattus niviventer* and its relatives from *Maxomys* and established the contents of the genus as given in the above list.

The following descriptive data were taken from Musser (1969b); Musser, Marshall, and Boeadi (1979); Medway (1977, 1978); and Lekagul and McNeely (1977). Head and body length is 100 to 235 mm, tail length is 88 to 226 mm, and weight is 35 to 284 grams. The dorsal pelage is short, soft, and dense, or is short and spiny. The upper parts range in color from dark brown, bright reddish brown, and iron gray to bright yellowish mixed with black. The underparts are white, grayish, or chestnut. The tail is finely scaled and is usually dark above and white below.

Maxomys comprises small to medium-sized rats of a generalized appearance. Key characters of the genus include a long and slender hind foot, with smooth and naked plantar surfaces, a tail usually shorter or only slightly longer than the head and body, a broad and inflated braincase, broad incisive foramina, and small bullae. Species of *Rattus* generally have a wider hind foot with well-developed plantar pads, a less inflated braincase, narrower incisive foramina, and larger bullae. Female *Maxomys* have six or eight mammae.

Musser, Marshall, and Boeadi (1979) stated that these rats are cursorial animals, adapted for life on the floor of evergreen or semievergreen forests in tropical lowlands and mountains. Lekagul and McNeely (1977) wrote that *Maxomys* is never found in trees. They noted that *M. surifer* occurs in all types of forest and is almost as ubiquitous as *Rattus rattus*, but is not found in association with people. Also, in contrast to *R. rattus*, *M. surifer* was described as being gentle, and naturally tame and friendly.

According to Medway (1978), *M. surifer* lives in burrows in the forest floor, each with one or two entrances and a nest chamber lined with leaves. Its diet consists mainly of vegetation, including roots and fallen fruit, but also includes some invertebrates and small vertebrates. Some food, such as nuts, may be stored in the burrow. Medway listed the following average diameters of effective lifetime ranges: *M. surifer*, 280 meters; *M. rajah*, 263 meters; and *M. whiteheadi*, 476 meters. Medway wrote that *M. whiteheadi* breeds throughout the year in the Malay Peninsula, and has an average of 3 (1–6) young per litter. For *M. surifer* and *M. rajah*, in the same area, litter size averaged 3.3 and ranged from 2 to 5.

RODENTIA; MURIDAE; **Genus NIVIVENTER** Marshall, 1976

White-bellied Rats

There are 15 species (Musser 1981a; Lekagul and McNeely 1977):

N. andersoni, mountains of central and southern China;

N. excelsior, Szechwan (central China);

N. brahma, Assam, northern Burma;

N. eha, Nepal, Sikkim, northeastern India, northern Burma, Yunnan (south-central China);

N. langbianis, Assam, Burma, northeastern Thailand, Laos, Viet Nam;

N. hinpoon, central Thailand;

N. cremoriventer, Malay Peninsula, Sumatra, Java, Borneo, and many small nearby islands;

N. niviventer, China, extreme northeastern Pakistan, northern India, Nepal, Sikkim;

N. confucianus, much of China, northern Burma, northern Thailand, northern Indochina, Taiwan, Hainan;

N. tenaster, mountains of Assam, southern Burma, and Viet Nam;

N. fulvescens, southeastern Tibet, Yunnan (south-central China), northern India, Nepal, Sikkim, Bangladesh, Assam, Burma, Thailand, Indochina, Hainan;

N. coxingi, northern Burma, Taiwan;

N. rapit, known from scattered localities in Malay Peninsula, Sumatra, and Borneo;

N. lepturus, western and central Java;

N. bukit, southern Burma, southern Thailand, Malay Peninsula, Sumatra, Java, Bali.

The above species were long considered part of *Rattus.* Misonne (1969) placed them in the genus *Maxomys.* Musser, Marshall, and Boeadi (1979), however, showed that these species did not belong in *Maxomys.* Lekagul and McNeely (1977) used *Niviventer* as a subgenus of *Rattus.* Finally, Musser (1981a) elevated *Niviventer* to generic rank and established its contents as given in the above list. Lekagul and McNeely (1977) considered *N. langbianis* to be a subspecies of *N. cremoriventer,* but Musser (1973c, 1981a) treated the two as separate species.

Head and body length is 110 to 198 mm, and tail length is 120 to 270 mm (Musser 1970a, 1973c; Musser and Chiu 1979; Medway 1978). Some average weights are: *N. confucianus,* 65 grams; *N. rapit,* 80 grams; *N. bukit,* 73 grams; and *N. hinpoon,* 61 grams (Lekagul and McNeely 1977). The upper parts of the head and body are grayish brown in some species, reddish brown in a few, and either gray, buffy gray, or reddish brown in one. The underparts are white or cream in most species, but dark gray in a few. The tail is usually dark above and light below, but is monocolored in some species. The pelage is dense, and is either soft and woolly or is semispinous. Females have six or eight mammae (Musser 1981a).

From *Rattus,* both *Niviventer* and *Maxomys* are distinguished in having much smaller bullae, the squamosal roots of the zygomatic arches originating higher on the sides of the braincase, the posterior margin of the palatal bridge not extending far past the molar tooth row, the pterygoid fossa nearly flat and not perforated by a large foramen, and the mesopterygoid fossa as wide as or wider than the palatal bridge. From *Maxomys, Niviventer* is distinguished in having the tail usually well longer than the head and body, a relatively shorter rostrum, a less inflated braincase, smaller lacrimals, and relatively narrower incisive foramina.

These rats occur in various kinds of forest in both lowlands and mountains. They are cursorial, scansorial, or arboreal (Musser 1981a; Musser and Chiu 1979; Lekagul and McNeely 1977). Litters of two to five young have been recorded for *N. cremoriventer* (Medway 1978). *N. hinpoon,* which is restricted to a remote area of wooded limestone cliffs, may be threatened with extinction.

RODENTIA; MURIDAE; *Genus LEOPOLDAMYS* *Ellerman, 1947–1948*

Long-tailed Giant Rats

There are four species (Musser 1981a; Lekagul and McNeely 1977; Chasen 1940):

L. neilli, southern Thailand;

L. edwardsi, northeastern India to southeastern China and central Viet Nam, mainland Malaysia, Sumatra;

L. sabanus, Bangladesh, Thailand, Indochina, Malay Peninsula, Sumatra, Java, Borneo, and many small nearby islands;

L. siporanus, Mentawi Islands off western Sumatra.

Leopoldamys was originally described as a subgenus of *Rattus.* Misonne (1969) included in this subgenus the species now known as *Maxomys surifer, M. rajah, M. musschenbroekii, M. whiteheadi, Srilankamys ohiensis,* and *Rattus nativitatis,* as well as *Leopoldamys edwardsi, L. sabanus,* and *L. siporanus.* Subsequent work (Lekagul and McNeely 1977; Musser, Marshall, and Boeadi 1979; Musser 1981a) established *Leopoldamys* as a full genus, with contents as given in the above list.

Head and body length is 180 to 275 mm, tail length is 300 to 420 mm, and weight is 200 to 495 grams (Medway 1978; Yong 1970). The pelage is short and sleek. The upper parts, which are generally brown or buffy, are sharply demarcated from the white underparts. The body is large, and the length of the tail is much greater than that of the head and body. The hind feet are long and slender, and the ears are small and scantily haired. Females have four pairs of mammae (Musser 1981a).

Leopoldamys is distinguished from *Rattus* in having a usually larger body and relatively much longer tail, shorter and sleeker pelage, much smaller bullae, the alisphenoid canal of the skull hidden by a lateral strutlike wall, the squamosal roots of the zygomatic arches originating higher on the sides of the braincase, the posterior margin of the palatal bridge not extending past the molar tooth row, the pterygoid fossa nearly flat and not perforated by a large foramen, the mesopterygoid fossa at least as wide as the palatal bridge, and smaller incisor teeth. *Leopoldamys* superficially resembles *Niviventer,* but is much larger and has relatively smaller bullae and larger incisors (Musser 1981a).

According to Medway (1978), *L. edwardsi* is found in montane forest above 750 meters, and *L. sabanus* occurs in

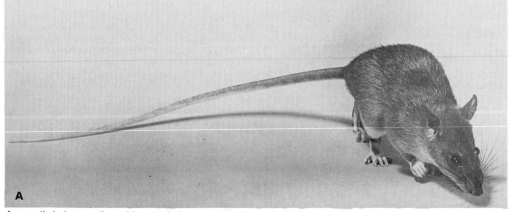

Long-tailed giant rat (*Leopoldamys sabanus*), photo by Ernest P. Walker.

evergreen forest, generally below 750 meters. The latter species is usually found at ground level, but is able to climb freely. The diet includes insects, other invertebrates, and a great variety of vegetable matter. The young are born in short burrows, dug into sloping hillsides, and provided with one or two entrances and a nest chamber lined with leaves. Pregnant females have been taken in all months of the year, but most frequently from July to September and least frequently from January to March. The number of young per litter averages 3.1 and ranges from 1 to 7. Mean longevity in the wild is only 4.1 months, but captives have lived up to 2 years. *L. neilli,* only recently discovered in an area of remote limestone cliffs, may be threatened with extinction.

RODENTIA; MURIDAE; *Genus RATTUS Fischer, 1803*

Rats

The following 4 subgenera and 78 species are recognized here (Misonne 1969, 1979; Musser 1970*a,* 1970*b,* 1970*c,* 1971*a,* 1971*b,* 1971*c,* 1971*d,* 1971*e,* 1972*a,* 1972*b,* 1973*a,* 1973*b,* 1973*c,* 1977*a,* 1977*b,* 1979; Musser and Boeadi 1980; Musser and Chiu 1979; Musser, Marshall, and Boeadi 1979; Taylor and Horner 1973*a;* Lekagul and McNeely 1977; Ellerman and Morrison-Scott 1966; Chasen 1940; Laurie and Hill 1954; Taylor 1934; Medway 1977, 1978; Medway and Yong 1976; Dennis and Menzies 1978; Menzies and Dennis 1979; Sanborn 1952; Agrawal and Ghosal 1969; Barbehenn, Sumangil, and Libay 1972–73; Corbet 1978; Calaby and Taylor 1980):

subgenus *Berylmys*

R. manipulus, Assam, western Burma;
R. berdmorei, southern Burma to Viet Nam;
R. mackenziei, Assam to western Thailand;
R. bowersi, Assam to southeastern China and Malay Peninsula;

subgenus *Bullimus*

R. xanthurus, northern Celebes;
R. bontanus, southern Celebes;
R. marmosurus, northern and central Celebes;
R. celebensis, Celebes;
R. adspersus, Celebes;
R. penitus, central Celebes;
R. andrewsi, Buton Island off southern Celebes;
R. dominator, Celebes;
R. muelleri, Malay Peninsula, Sumatra, Borneo, Palawan, and many nearby islands;
R. culionensis, Culion Island (Philippines);
R. infraluteus, Sumatra, Java, Borneo;
R. baluensis, Mount Kinabalu in northern Borneo;
R. korinchi, Sumatra;
R. chrysocomus, Celebes;
R. callitrichus, Celebes;
R. fratorum, northeastern Celebes;
R. taerae, northern and central Celebes;
R. hamatus, central Celebes;
R. arcuatus, southeastern Celebes;
R. punicans, central Celebes;
R. salocco, southeastern Celebes;
R. foramineus, southern Celebes and nearby Peleng Island;
R. latidens, Luzon;
R. everetti, Philippines;
R. gala, Mindoro Island (Philippines);

R. luzonicus, Luzon;
R. bagobus, Mindanao;
R. rabori, Mindanao;
R. tyrannus, known only by the type specimen from Ticao Island (Philippines);

subgenus *Rattus*

R. rattus (black rat, roof rat, house rat), probably originated in Malaysian region, now found throughout the world as a human commensal;
R. ranjiniae, extreme southern India;
R. montanus, Sri Lanka;
R. nitidus, apparently occurs naturally from Nepal to southeastern China and Viet Nam, now also found as a human commensal in highland areas of Luzon, Celebes, New Guinea, and other islands of the East Indies;
R. rattoides, Soviet Central Asia and northeastern Iran to northeastern India, possibly southern China and Burma;
R. annandalei, Malay Peninsula, Sumatra, and nearby islands;
R. tiomanicus, found partly as a human commensal on Malay Peninsula, Sumatra, Java, Borneo, Palawan, and many small nearby islands;
R. koratensis, Nepal to southeastern China and Viet Nam, Hainan;
R. germaini, Condor Islands off southern Viet Nam;
R. remotus, islands off west coast of peninsular Thailand;
R. rogersi, South Andaman Island (Bay of Bengal);
R. palmarum, Nicobar and Andaman Islands (Bay of Bengal);
R. simalurensis, Simalur and Siumat islands off western Sumatra;
R. babi, Babi Island off western Sumatra;
R. lasiae, Lasia Island off western Sumatra;
R. mentawi, Siberut and Sipora islands off western Sumatra;
R. lugens, Pagi Islands off western Sumatra;
R. enganus, Engano Island off southwestern Sumatra;
R. dammermani, northern and central Celebes;
R. argentiventer (rice field rat), found largely as a human commensal and partly through introduction from southern Thailand and Viet Nam to the Philippines and New Guinea;
R. losea, found partly as a human commensal from southeastern China to peninsular Thailand, and on Taiwan and Hainan;
R. hoogerwerfi, Sumatra;
R. macleari, Christmas Island (Indian Ocean);
R. exulans (Polynesian rat), found as a human commensal and largely through introduction from Burma to Viet Nam, throughout the East Indies, and on the Hawaiian and many other Pacific islands;
R. norvegicus (Norway rat, brown rat), may have originated in northern China, now found throughout the world as a human commensal;
R. mindorensis, Mindoro Island (Philippines);
R. rennelli, Rennell Island (Solomons);
R. hoffmanni, Celebes;
R. elaphinus, Soela Islands east of Celebes;
R. lutreolus, southeastern Australia, Tasmania;
R. fuscipes, eastern and southern coasts of Australia;
R. leucopus, southern New Guinea, northern Queensland;
R. sordidus, south-central New Guinea, central and eastern Australia;
R. gestroi, southeastern Papua New Guinea;
R. bunae, southeastern Papua New Guinea;
R. tunneyi, northern, central, and southwestern Australia;

A. Norway rat (*Rattus norvegicus*), photo from U.S. Fish and Wildlife Service. B. Black rat (*R. rattus*), photo by Ernest P. Walker. C. Black Rat (*R. rattus*), photo from U.S. Fish and Wildlife Service.

subgenus *Stenomys*

R. niobe, mountains of New Guinea;
R. verecundus, New Guinea;
R. owiensis, Owi Island off northwestern New Guinea;
R. omichlodes, mountains of western New Guinea;
R. richardsoni, mountains of western New Guinea;
R. praetor, New Guinea and nearby islands from the
  Moluccas to the Solomons;

R. doboensis, Aru Islands southwest of New Guinea;
R. morotaiensis, Morotai Island in the northern Moluccas;
R. nativitatis, Christmas Island (Indian Ocean).

The last species in the above list was placed by Misonne (1969) in *Leopoldamys*, which he considered a subgenus of *Rattus*. Musser (1981a), who raised *Leopoldamys* to generic level, stated that *R. nativitatis* did not belong therein. Earlier, Chasen (1940) had written that *R. nativitatis* is in some respects a link between *Rattus* and *Uromys*, a genus of New Guinea and Australia.

In number of species, *Rattus* is among the largest of mammalian genera. *Rattus* was once considered to be even more extensive, but many of its former species are now generally assigned to other genera (see accounts of *Lenothrix, Diplothrix, Margaretamys, Stochomys, Aethomys, Praomys, Millardia, Srilankamys, Maxomys, Niviventer, Leopoldamys, Limnomys, Tarsomys,* and *Apomys*). Musser (1981*b*) and Musser and Boeadi (1980) suggested that the limits of *Rattus* be reduced still further by removal of the species in the subgenera *Bullimus* and *Berylmys*. A major turning point in our understanding came with Misonne's (1969) arrangement of the subfamily Murinae, especially in regard to his separation of several African lineages from *Rattus*. While most authorities have followed Misonne in not recognizing the presence of any native *Rattus* in Africa, his description of the situation in Asia has been challenged (see account of *Maxomys*). In the above list, three of the four subgenera of *Rattus* designated by Misonne—*Bullimus, Rattus,* and *Stenomys*—have been utilized, but *Berylmys* has also been employed as a subgenus, distinct from *Bullimus,* in accordance with Lekagul and McNeely (1977). Misonne's fourth subgenus, *Leopoldamys,* is now considered a separate genus and to have a somewhat different content than he indicated (Musser 1981*a*).

Some additional questions and disagreements regarding the classification of this complex genus are as follows. Tiwari, Ghose, and Chakraborty (1971) suggested that *R. rufescens* of India is a species distinct from *R. rattus*. Schlitter and Thonglongya (1971) considered *R. turkestanicus* to be the proper name for the species listed above as *R. rattoides,* but Corbet (1978) saw no need for this change. Medway and Yong (1976) indicated that *R. diardii* of the Malay Peninsula and East Indies is a species distinct from *R. rattus,* rather than only an ecological subspecies as is generally considered. Lekagul and McNeely (1977) suggested that *R. tiomanicus* may not be a species distinct from *R. rattus.* Chromosomal data gathered by Baverstock et al. (1977) may indicate that the three Australian subspecies of *R. sordidus* recognized by Taylor and Horner (1973*a*)—*R. s. sordidus* in the northeast, *R. s. villosissimus* of the central and southeastern regions, and *R. s. colletti* of Arnhem Land—are actually separate species. Misonne (1969) put the Australian species *R. leucopus* in the subgenus *Stenomys,* along with the forms *assimilis* and *greyi.* Dennis and Menzies (1978) also associated *R. leucopus* with species usually placed in *Stenomys.* Taylor and Horner (1973*a*), however, arranged *assimilis* and *greyi* as subspecies of *R. fuscipes,* which Misonne had put in the subgenus *Rattus,* and also indicated a close relationship between *R. leucopus* and *R. fuscipes.* Neither Ellerman and Morrison-Scott (1966) nor Lekagul and McNeely (1977) recognized the subgenus *Bullimus,* and both assigned *R. muelleri* to the subgenus *Stenomys.* Laurie and Hill (1954) also did not recognize *Bullimus,* and divided some of the species assigned to it in the above list between the subgenera *Rattus, Stenomys,* and *Paruromys* (synonymized with *Bullimus* by Misonne). Barbehenn, Sumangil, and Libay (1972–73) noted that *R. gala* is probably conspecific with *R. everetti,* and questioned the distinction of *R. culionensis,* described by Sanborn (1952), from *R. muelleri.* Laurie and Hill (1954) stated that *R. owiensis* is possibly a race of *R. verecundus,* and that *R. doboensis* is possibly a race of *R. praetor.*

Head and body length is 80 to 300 mm. The tail varies in length by species, ranging from substantially shorter to much longer than head and body length. The best-known species, *R. norvegicus,* usually weighs 200 to 400 grams, with a few individuals reaching 500 grams (Grzimek 1975). Most other species are considerably smaller. In Hawaii, average weights for *R. exulans* were only 37 grams in males and 39 grams in females, while respective averages for *R. rattus* were 108 and 77 grams (Tamarin and Malecha 1972). Average weight of 10 species in Thailand ranged from 36 grams in *R. exulans* to 420 grams in *R. bowersi* (Lekagul and McNeely 1977). In some species the pelage is soft, in others it is coarse, and in still others the hairs are enlarged and stiffened into bristles or spines. The coloration of the upper parts ranges through a great variety of black, brown, gray, yellow, orange, and red shades. The underparts are usually white or grayish.

The body form may be stocky or somewhat slender. Certain digits are reduced in some forms, but are long in others. In some species the feet are modified for a terrestrial life, while in others the feet are adapted to an arboreal existence. The number of mammae in females is 4, 6, 8, 10, or 12.

The lifestyle of the several well-known commensal species is not representative of the genus as a whole. Most species of *Rattus* live in natural forest, ranging from tropical lowland to montane type, and tend to avoid human settlements. In Australia, *Rattus* is found in grasslands, heaths, savannahs, and sandy flats, as well as forests (Ride 1970). Of the 12 species in Thailand discussed by Lekagul and McNeely (1977), 6 are fully wild and occur primarily in forests. The remaining 6 Thai species are commensals: *R. rattus,* the most abundant mammal in the nation, occupies cities, villages, cultivated fields, and some natural habitat; *R. nitidus* is found exclusively inside the houses of hill villages; *R. norvegicus* usually occurs in urban areas; *R. argentiventer* is entirely dependent on human rice fields and plantations; *R. losea* is found mainly in gardens and rice fields, but also is known to occur in the wild state in forests; and *R. exulans* inhabits houses and rice fields.

According to Musser (1973*a,* 1977*a*), three related commensal rats of Southeast Asia have divided the available habitat between them. *R. rattus diardii* lives in and around human dwellings; *R. tiomanicus* is found in gardens, plantations, scrub country, and secondary forests; and *R. argentiventer* occurs primarily in rice fields and grasslands. Musser's studies in the Philippines and Celebes indicate that these commensals, along with *R. nitidus, R. norvegicus,* and *R. exulans,* do not occur in primary forest, but are widely distributed in close association with people. In contrast, the many indigenous species of *Rattus* in the Philippines and Celebes have very restricted ranges and occur mainly in primary forest.

Rats shelter in such places as burrows, under rocks, in logs, and in piles of rubbish. Species that are good climbers sometimes build nests in trees or other elevated positions. *R. rattus* is an extremely agile climber and can run along a wire only 1.6 mm in diameter (Ewer 1971). It constructs a loose, spherical nest of shredded vegetation, cloth, or other suitable material (Medway 1978). When occupying buildings, this species is usually found in the dry upper levels. *R. norvegicus* is a ground dweller. It may originally have lived along stream banks in Asia, and then spread rapidly as canals and rice fields were developed. Its underground burrows contain long, branching tunnels, one or more exits, and rooms for nests and food storage. When using buildings, it generally occurs in cellars, basements, and lower floors. It also occupies sewers and rubbish heaps. It swims and dives well (Grzimek 1975). *R. argentiventer* also burrows in the soil, while *R. tiomanicus* constructs a loose, spherical nest of vegetation on the surface of the ground under some shelter (Medway 1978).

Rats are omnivorous, eating a wide variety of plant and animal matter. Seeds, grains, nuts, vegetables, and fruit may be preferred by most species, but insects and other invertebrates sometimes predominate in the diet. Lim (1970) reported that mollusks are eaten by *R. bowersi* and *R. muelleri,* and that the latter species also feeds on crabs. Kingdon (1974*b*) stated that *R. rattus* and *R. norvegicus* eat everything that people eat and much else, such as soap, hides, paper, and beeswax. Food may be carried back to the nest and stored. Grzimek (1975) noted that *R. norvegicus* seems to prefer animal matter, including birds and eggs, and is excellent at catching fish. This species also feeds on mice, poultry, and young lambs and pigs; it may attack larger animals, even humans.

The reported density of rat populations varies widely: for example, 0.38 to 3.04 individuals of *R. fuscipes* per ha. in Australia (Barnett, How, and Humphreys 1977); 70 to 188 *R. exulans* per ha. in Hawaii (Wirtz 1972); and up to 1,200 *R. tiomanicus* per ha. on a Malay oil palm plantation (Medway 1978). In the United States there are an estimated 100 million to 175 million rats, mostly *R. norvegicus* (Jackson 1961; Schwartz and Schwartz 1959; Banfield 1974; Pratt, Bjornson, and Littig 1977). Densities of this species range from about 25 to 150 individuals per block in some U.S. cities, and from 50 to 300 individuals per country farm. As in some other species of *Rattus,* the population density of *R. norvegicus* appears to be cyclic and sometimes increases sharply. Large-scale movements from an area may then seem to occur. Favorable conditions of food and shelter have resulted in the taking of over 200 *R. norvegicus* during a five-day period in a town area of less than 0.2 ha., and of 4,000 rats on about 2.5 ha. of a midwestern farm. The normal home range of *R. norvegicus* is only about 25 to 150 meters in diameter. According to Grzimek (1975), however, individuals have been known to move 3 km from their nests, and return, in a single night. Calculated home range diameters of some other species are: *R. muelleri,* 409 meters; *R. tiomanicus,* 315 to 357 meters; *R. argentiventer,* 273 meters; and *R. exulans,* 280 meters (Medway 1978). A radio-tracking study of *R. exulans* in Hawaiian sugarcane fields showed an average home range of 1,845 sq meters for males and 607 sq meters for females (Nass 1977).

Several forms of social structure have been reported for *R. norvegicus,* but the differences were basically resolved by Calhoun (1963). In a study of wild-caught rats and their descendants in a 0.10-ha. enclosure, he demonstrated the functioning of parallel social systems. In one part of the enclosure, higher-ranking males established individual territories around burrows containing several females. Each male excluded other males and only he mated with the resident females. The females of a colony collectively nurtured their young and, when they were reproductively active, excluded other rats from the burrow. These territorial colonies kept well-organized burrows and carefully maintained nests. Reproduction occurred regularly and successfully. As the young matured, subordinate individuals among them were forced into another part of the enclosure. This downward change in status, involving mostly males, was not reversible; there was no permanent movement of lower-ranking individuals into the area occupied by the dominant rats. No territories were established in the area used by the lower-ranking animals. Large packs were formed, and when a female entered estrus she would be followed by numerous males and mounted several hundred times in a night. Such stressful conditions resulted in poor reproduction; burrows were not well organized and nests were poorly maintained. This process was considered responsible for limiting the size of the overall rat population, so that there were never more than about 200 animals in the pen. Grzimek (1975), however, noted that the size of wild packs may exceed 200 individuals.

In a study of *R. rattus* in Accra, Ghana, Ewer (1971) found social groups to contain a single dominant male and sometimes a linear male hierarchy. There also were two or three

equally ranking top females, which were subordinate to the dominant male but dominant to all other members of the group. Females were much more aggressive than males. A group territory was formed around a feeding place and defended against outsiders, but only the females drove away other females. Attacks were frequent and directed downward, but serious fights were usually avoided through appeasement by the subordinate. Fighting took several forms, including biting, jumping, standing on hind legs and hitting with the front paws, and striking with the rump from the side. Infants had almost complete immunity from aggression and could take food literally from under the nose of dominant adults.

During Ewer's study, young appeared throughout the year. The gestation period was found to be 21–22 days in nonlactating females and 23–29 days in lactating females. Litters most commonly numbered 8 young, and females could give birth when 3 to 5 months old. Medway (1978) reported that in the Malay Peninsula *R. rattus* also breeds throughout the year, has a litter size of 1 to 11 young, and reaches sexual maturity at about 80 days of age. Captive individuals of this species have lived as long as 4 years and 2 months.

In *R. norvegicus,* reproduction may occur all year in captive and some wild populations, but there are usually spring and autumn peaks. Females are polyestrous and each may have from 1 to 12 litters per year. There is a postpartum estrus within 18 hours of birth. The estrous cycle lasts 4 to 6 days, heat 20 hours, and the gestation period 21 to 26 days. Litter size is 2 to 22 young, averaging about 8 or 9. The young are born naked and blind, but they are fully furred and their eyes open at about the age of 15 days. They are weaned and leave the nest when around 22 days old, and attain sexual maturity at 2 to 3 months (Asdell 1964; Grzimek 1975; Banfield 1974; Lowery 1974).

Most species of *Rattus* that have been investigated seem less prolific than *R. rattus* and *R. norvegicus.* Medway (1978) indicated that the other species of the Malay Peninsula breed all year, but that most have litter sizes averaging only around 3 to 6 young. In New Guinea, reproduction of several native species apparently is reduced, or ceases altogether, during the dry season from about June to October, and litters commonly number only about 1 to 3 young (Dwyer 1975*b*; Menzies and Dennis 1979). All native *Rattus* of Australia appear to be seasonal breeders in the wild (Taylor and Horner 1973*b*). The more northerly populations of that continent have only a brief winter lull in reproduction and a peak during the wet summer. More southerly populations undergo a lull of up to 7 months and have their breeding peaks in spring and early summer. Laboratory colonies are capable of reproduction throughout the year. Females of most Australian species are known to be polyestrous and to undergo a postpartum estrus. Their gestation periods vary from 21 to 30 days, and litter sizes from 3 to 14 young.

The commensal species *R. exulans* has been studied extensively on the Pacific islands (Egoscue 1970; Tamarin and Malecha 1972; Wirtz 1972, 1973). Females are polyestrous and capable of breeding all year in the laboratory, with up to 13 litters annually, but wild populations have reproductive seasons centered in the summer months and usually produce only 1 to 3 litters per year. The gestation period is 19 to 30 days, litter size averages about four young, and weaning takes place at 2 to 3 weeks of age. Some young females do not reach sexual maturity until the season following their first winter.

Most species of *Rattus* are not known to be of any direct importance to people. They live in natural habitat and do not enter human habitations or intensively cultivated areas. Many of these species have relatively restricted ranges and habitat tolerances, and some, such as *R. remotus* of Thailand, may be threatened with extinction. Taylor and Horner (1973*a*) observed that *R. tunneyi* had disappeared in southern and southwestern Australia, probably because of excessive grazing by domestic livestock. Both *R. macleari* and *R. nativitatis,* found only on Christmas Island in the Indian Ocean, apparently became extinct by 1908, perhaps because of disease transmitted by introduced *R. rattus* (Harper 1945).

The activities of the few commensal and pest species of *Rattus* have adversely affected the reputation of the entire genus and, indeed, of all mammals that contain the term *rat* in their vernacular names. Most economic and health problems are actually associated with only seven species: *R. rattus* (black rat), found throughout the world and responsible for the spread of disease and immense agricultural losses; *R. nitidus* and *R. tiomanicus,* agricultural pests and residents of human homes in Southeast Asia; *R. argentiventer* and *R. losea,* known mainly for their depredations on rice fields and gardens in Southeast Asia; *R. exulans,* apparently carried over a vast part of Southeast Asia and the Pacific in association with early human migrations, and now sometimes considered a menace to health and agriculture; and *R. norvegicus* (Norway rat), known over much of the world for its destruction of property and stored food, and its threats to the health and safety of people, domestic animals, and wildlife. Under certain conditions, several other species of *Rattus* have been known to act as human commensals and/or become agricultural pests in local areas (Braithwaite 1980; Menzies and Dennis 1979).

The numbers and distributions of all seven major commensal species of *Rattus* have apparently been increased through human agency, either directly by transportation in boats and caravans, or indirectly by habitat modification, poor sanitary practices, and elimination of predators. The most spectacular range expansions have been those of the black and Norway rats. There have been reports that *R. rattus* was present in Europe as early as the Pleistocene (Grzimek 1975), but Kurten (1968) stated that both the black and Norway rats probably entered Europe in postglacial times as human commensals. *R. rattus* may actually have been brought to Europe at the time of the Crusades and then to the Western Hemisphere during the explorations of the sixteenth century. *R. norvegicus* was not known in Europe until about 1553 and did not reach North America until 1775. Although *R. rattus* is by far the more common and widespread of the two species in the tropics, *R. norvegicus* has proved the more adaptable in temperate zones, especially in urban areas. As the latter species spread through North America and Europe, it excluded *R. rattus* from favorable habitat on the ground and in the lower levels of buildings. The black rat has thus become rare in many areas, and has even been designated as endangered in the state of Virginia by Handley (1980*b*).

Norway and black rats consume vast quantities of food stored for people and their livestock, and contaminate much more than they eat. They also gnaw the insulation from electrical wires, sometimes causing fires, and occasionally cut through lead pipe and concrete dams. On a world basis, the direct damage caused each year by these two species amounts to billions of dollars. In the United States alone, direct economic losses to rats have been estimated at $500,000,000 to $1,000,000,000 annually (Pratt, Bjornson, and Littig 1977). Further expenses are incurred in the process of poisoning campaigns and other efforts at control, but rats can be permanently eradicated only through elimination of garbage and carelessly stored food supplies and of environmental conditions that provide suitable shelter for the animals.

Among the many diseases spread by Norway and black rats are bubonic (black) plague, murine typhus, food poisoning (*Salmonella*), leptospirosis, trichinosis, tularemia, and rat-bite fever. Rat-borne diseases are thought to have taken more human lives in the last 10 centuries than all the wars and revolutions ever fought. Bubonic plague, which is transmitted to humans by rat fleas, killed more than one-quarter of the population of Europe from 1347 to 1352, about 11 million people in India from 1892 to 1918, and 60,000 people in Uganda from 1917 to 1942 (Grzimek 1975; Kingdon 1974*b*). Serious outbreaks of plague recurred at San Francisco (1902–41), Galveston (1920–22), and New Orleans (1912–26) (Lowery 1974). Rats make direct attacks on about 14,000 persons annually in the United States, and occasionally inflict mortal wounds (Pratt, Bjornson, and Littig 1977). Rats also kill poultry, domestic livestock, and game birds. Rat predation and competition has contributed to the endangerment or extinction of many species of wildlife.

Lowery (1974) stated that *R. norvegicus,* especially in its white mutant form, has partially redeemed itself through its role in medical and basic research. The species is used by biological laboratories throughout the world and has led to important discoveries in immunology, pathology, epidemiology, genetics, and physiology.

RODENTIA; MURIDAE; **Genus TRYPHOMYS** Miller, *1910*

Mearns' Luzon Rat

The single species, *T. adustus,* is known only by a few specimens from the mountains of northern Luzon. *Tryphomys* was considered a synonym of *Rattus* by Sanborn (1952) and Misonne (1969), but was treated as a full genus by Taylor (1934) and Musser (1981*b*).

Head and body length is 174 mm, and tail length is 150 mm. The fur of the back is coarse and harsh. The tips of the shorter hairs tend to curve toward the head, giving the pelage a peculiarly scorched appearance. The back and sides are a coarse grizzle, tending toward wood brown and black, with the brown especially predominant on the sides. The median dorsal region is abruptly more grizzled than the sides. The bases of the individual hairs are slate gray, streaked by the paler gray of the slender, grooved bristles. The color of the head and face is similar to that of the back, but has a grayish cast. The underparts and cheeks are buffy white, dulled by a slate gray undercolor. The feet are also the color of the upper parts, but they have a brownish tinge. The tail is uniformly dark brown.

Tryphomys is of medium size and robust form. Its tail is coarsely and conspicuously scaled. The claws are well developed, those of the hind foot being larger. The nail of the thumb is small and appressed. The sole of the foot is naked, and has five well-developed tubercles. The teeth are similar to those of *Rattus.* The skull is large and heavily built. Females have five pairs of mammae.

RODENTIA; MURIDAE; **Genus LIMNOMYS** Mearns, *1905*

The single species, *L. sibuanus,* is known by four specimens collected in 1904 and 1906 on the island of Mindanao in the Philippines. The complex history of these specimens, during which they were sometimes thought to represent three separate species and sometimes placed within the genus *Rattus,* was explained by Musser (1977*b*).

The single adult specimen is a female and has a head and

Mearns' Luzon rat (*Tryphomys adustus*), photos from Field Museum of Natural History.

body length of 125 mm, and a tail length of 150 mm. The top and sides of the head and body are covered with tawny, dense, and long pelage. Long, black guard hairs are scattered over the back and rump. The underparts are cream, the ears are dark brown, and the tail is dark brown on all surfaces and is densely haired. The front and hind feet are brownish white, and the hind feet have a dark brown stripe from each ankle to the bases of the digits. The hind feet are short and broad. There are six mammae.

Musser considered *Limnomys* a genus distinct from *Rattus,* because of a combination of characters. In *Limnomys* these are: short body, long and hairy tail, short and broad hind feet, six nipples, short rostrum, narrow zygomatic plates, short bony palatal bridge, large bullae, and details of molar topography.

The adult female specimen of *Limnomys* was collected at an elevation of about 2,000 meters on a wet, mossy growth of vegetation beside a small stream. The date was 30 June and the mammae were probably then functional. The three other specimens were taken on 6 June 1906 in a montane forest at an elevation of about 2,800 meters. The three specimens represent immature animals, all of about the same age, and may have been from the same litter.

RODENTIA; MURIDAE; *Genus TARSOMYS Mearns, 1905*

The single species, *T. apoensis,* is known by five specimens collected in montane forests on the island of Mindanao in the Philippines (Taylor 1934). Some authorities, as, for example, Ellerman and Morrison-Scott (1966) and Misonne (1969), have considered *Tarsomys* a synonym of *Rattus,* but Musser (1977*a,* 1977*b*) treated it as a distinct genus.

According to Taylor's (1934) description, head and body length of the type specimen is 135 mm, and tail length is 120 mm. The pelage is long and rather coarse, but not spiny. The upper parts are brownish slate, the underparts are grizzled yellow brown, and the ears are dark purplish brown. The tail is hairy and is dark purplish throughout. The feet are naked below, but covered with slightly grizzled, drab brown hairs above.

The skull and teeth of *Tarsomys* resemble those of *Batomys* but are relatively broader, and the braincase is more inflated. The rostrum is elongated, similar to that of *Apomys,* but the auditory bullae and anterior palatal foramen are different. The bullae resemble those of *Rattus,* but are more flattened and compressed externally.

RODENTIA; MURIDAE; *Genus MELASMOTHRIX* *Miller and Hollister, 1921*

Celebes Lesser Shrew Rat

The single species, *M. naso,* is known only by the type specimen in the U.S. National Museum of Natural History, which was collected in 1918 at Rano Rano in Middle Celebes. Musser (1969*a*) analyzed this specimen and suggested a morphological relationship with both *Rattus chrysocomus* and *Tateomys rhinogradoides,* but not with *Echiothrix* and *Rhynchomys.*

Head and body length of the type specimen is 124 mm, and tail length is 90 mm. The pelage is dense and has a velvety texture. The general coloration of the upper parts is rich dark bay. The hairs of the underfur have slate-colored bases and golden brown tips. Slightly longer hairs, scattered throughout the pelage, are glossy black for their entire length. The underparts are nearly the same color as the back and sides, but may be somewhat paler and glossier. The hands, feet, ears, and tail are brownish black.

Melasmothrix has an elongated snout and long claws on the forefeet. The thumb is reduced to a tubercle with a flattened nail. The ears are short. The rings of scales on the tail are practically concealed by dense, close hairs.

The specimen was trapped under a rotten, moss-covered

Celebes lesser shrew rat (*Melasmothrix naso*), photo by Howard E. Uible of skin in U.S. National Museum.

log. Musser (1969a:33) speculated: "*Melasmothrix naso* is volelike in form and probably utilizes runways through litter and moss, alongside of fallen tree trunks, and around rocks. It may utilize, and even construct, burrows for nesting sites and refuge. *Melasmothrix naso* also probably digs here and there into the moss, leaf litter, and soft soil of the forest floor to find food. It likely eats a variety of invertebrates, but fruit and seeds may form a greater part of its diet than do soft-bodied arthropods. Morphology of its incisors and maxillary teeth suggests that *M. naso* may be more catholic in food preference than is *T. rhinogradoides,* and it probably shares more of the omnivorous habits of a species like *Rattus rattus.*"

RODENTIA; MURIDAE; *Genus TATEOMYS Musser, 1969*

The single species, *T. rhinogradoides,* is known only by the type specimen in the American Museum of Natural History, which was collected in 1930 in southwestern Celebes. *Tateomys* is morphologically most closely allied to *Melasmothrix* and, through that genus, to *Rattus chrysocomus* and related species of Celebes (Musser 1969a).

Head and body length of the type specimen is 137 mm, and tail length is 168 mm. The pelage is thick, short, velvety, and compact. The upper parts are dark brown with burnished highlights, the sides of the body are slightly paler, and the underparts are mostly buffy gray. The dorsal surface and the entire distal third of the finely haired tail are pale brown, and the proximal two-thirds of the undersurface are unpigmented.

The eyes are tiny, and the muzzle is elongate and shrewlike. The forefeet are stout and broad, and the cylindrical front claws are nearly as long as their respective digits. The hind feet are long and slender, with long digits and claws. The tail is slender and gradually tapered. The skull is elongate and delicately built, with a long and tapered rostrum. The incisors are short, weak, thin, and orthodont. The maxillary teeth are small and low crowned, and their occlusal surfaces are basined and simple in structure.

The external form of *Tateomys* resembles that of a terrestrial species of *Rattus,* but its velvety pelage, elongated facial area, and tiny eyes are shrewlike. *Tateomys* resembles *Melasmothrix,* but can be distinguished by its much larger head and body, relatively smaller eyes, more elongate and delicately built skull, and weaker incisors, and by differences in coloration.

The specimen of *Tateomys* was taken on the ground in a dense virgin cloud forest at an elevation of 2,200 meters. Musser (1969a:32) wrote, based on an analysis of its morphology: "I view *T. rhinogradoides* as a scampering form. Morphologically, it is perfectly capable of constructing burrows through moist and soft soil, and it probably utilizes such burrows, either those made by itself or ready-made shelters it has found, for nesting sites and refuge. Outside of activities related to the nesting area, *T. rhinogradoides* probably spends much of its time scampering over the moist forest floor, stopping here and there to probe with its long nose the leaf litter and moss. Where there is promise of food, the rodent could dig into the moss and litter and probably well into the soft soil. *Tateomys rhinogradoides* is well suited for this activity with its long and strong claws, and its stout and broad front feet with their lateral metacarpal tubercles, which are angular and tough. The tubercles add surface area and strength to the inner surface of each hand, which the animal may use to push dirt and litter away from the depression. Within the leaf litter and moss, and underneath the surface of the soil, the rodent would probably locate what are the main components of its diet: earthworms, grubs, and other soft-bodied arthropods. The short incisors of *T. rhinogradoides,* with their sharp V-shaped cutting tips and smooth enamel-covered sides, are admirably suited to pierce and impale squirming, mucous-covered earthworms, and soft, tugid beetle larvae. The molars are also well adapted for crushing and masticating such items of food. A variety of fungi, as well as soft fruits, are probably also eaten. On occasion, the animal may even seek grubs in tree trunks and limbs scattered over the forest floor, in which it may dig into the rotten wood that has decomposed to a soft and friable point."

RODENTIA; MURIDAE; *Genus NESOROMYS Thomas, 1922*

Ceram Island Rat

The single species, *N. ceramicus,* is known only from the island of Ceram between Celebes and New Guinea (Laurie and Hill 1954). The type specimen was trapped in heavy jungle on Mount Manusela at an elevation of about 1,800 meters. This species originally was placed in the genus *Stenomys,* which now is considered a subgenus of *Rattus.* Misonne (1969) observed that *Nesoromys* seems related to *Stenomys,* but is differentiated enough to be a separate genus. Musser (1981b) included *Nesoromys* in *Rattus,* and indicated close relationship to *R. niobe.*

Head and body length of the type specimen is 135 mm, and tail length is 140 mm. The pelage is fine, soft, and thick. The general coloration of the upper parts is a fine speckling of olive brown. The underparts are somewhat lighter, because of the dull, drab tips of the hairs. The short ears are almost black; the hands and feet are dark brown; and the tail, which is almost naked, is also dark brown. *Nesoromys* is distinguished from *Rattus* by its long, narrow muzzle. The hind foot is also narrow.

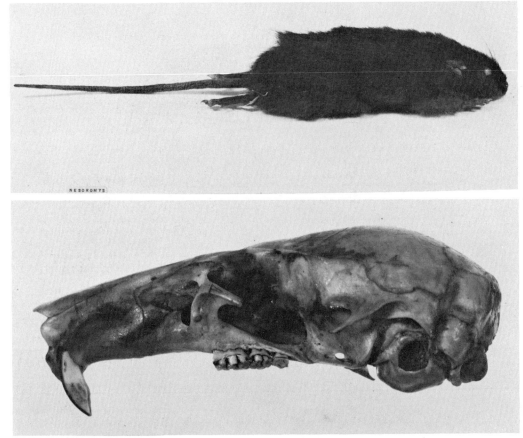

Ceram Island rat (*Nesoromys ceramicus*), photos from British Museum (Natural History).

RODENTIA; MURIDAE; **Genus UROMYS** Peters, 1867

Giant Naked-tailed Rats

There are six species (Laurie and Hill 1954; Menzies and Dennis 1979):

U. caudimaculatus, New Guinea, Kei and Aru Islands, northern Queensland;
U. anak, New Guinea;
U. neobritannicus, New Britain Island;
U. rex, Guadalcanal Island (Solomons);
U. imperator, Guadalcanal Island (Solomons);
U. salamonis, Florida Island (Solomons).

Head and body length is about 200 to 400 mm, and tail length is about 200 to 350 mm. The weight of *U. caudimaculatus* is around 600 to 700 grams, and that of *U. anak* is over 1,000 grams (Menzies and Dennis 1979). The pelage is fairly short, dense, and relatively coarse, except in *U. caudimaculatus*, in which it is rather scant. The coloration of the upper parts ranges from grayish through various shades of brown to gray black. The underparts are white or grayish. In some species, such as *U. anak*, the tail is entirely dark, but in others, such as *U. caudimaculatus*, the tail is white or yellowish toward the tip. The hairs of the tail number only

one per scale. The scales of the tail form a mosaic pattern, whereas in most other genera of the subfamily Murinae the scales of the tail are flat and overlapping.

In *U. neobritannicus*, the largest species, the squamosal crest of the skull is well developed and the ears are relatively large. The other species lack well-developed squamosal crests. The ears are relatively long in *U. caudimaculatus* and *U. anak*, and short in the remaining three species. The teeth of *Uromys* are similar to those of *Melomys*, but much longer. The incisors are large and powerful, with uppers and lowers being about the same size. The hands and feet are broad and well clawed. Females have two pairs of mammae.

Giant naked-tailed rats are arboreal and are excellent climbers. Their tails are not truly prehensile, but do facilitate climbing by curling over limbs and gripping with the rasplike scales. According to Menzies and Dennis (1979), *U. anak* inhabits forests at elevations of 1,000 to 2,500 meters, and *U. caudimaculatus* occurs in both forests and grasslands in the lowlands. Nests are usually made in hollow trees, though *U. caudimaculatus* sometimes seems to raise its young on bare rock in caves or old mine tunnels. The diet consists of coconuts, other nuts, fruit, and flowers. Litters of *U. caudimaculatus* have been recorded in New Guinea during November and December, and may contain one to three young. The young remain with the mother until at least half grown. *U. anak* is often hunted for food by the highland natives of New Guinea.

Giant naked-tailed rat (*Uromys caudimaculatus*), photo by Stanley Breeden.

RODENTIA; MURIDAE; **Genus SOLOMYS** Thomas, *1922*

Naked-tailed Rats

There are two subgenera and three species (Laurie and Hill 1954):

subgenus *Solomys*

S. *sapientis,* Santa Ysabel Island (Solomons);
S. *salebrosus,* Bougainville Island (Solomons);

subgenus *Unicomys*

S. *ponceleti,* Bougainville Island (Solomons).

Some authorities have included *Solomys* in the genus *Melomys,* but Laurie and Hill (1954) did not agree, noting that *Solomys* is much larger and has a larger and much heavier skull with large bullae.

The species S. *sapientis* has a head and body length of about 250 mm, and a tail length of about 250 mm; coloration is cinnamon brown above and on the sides, and pinkish buff below. The species S. *salebrosus* has a head and body length of about 230 mm, and a tail length of about 215 mm; coloration is yellowish brown above, cinnamon buff on the sides, and pinkish buff below. S. *ponceleti* is the largest species; a young adult female measured 330 mm in head and body length, and 340 mm in tail length. The color of this species is brownish black above and below. The hair of S. *ponceleti* is long and fine, without woolly underfur, whereas it is coarser in the other two species.

The tail of *Solomys* appears prehensile and lacks hairs for most of its length. The feet are well padded and are supplied with mobile and strongly clawed digits. The incisor teeth are broad and stout. Females have four mammae.

Naked-tailed rats inhabit thick woods and are apparently arboreal. According to Troughton (1936:347), S. *sapientis* "cracks the ngali (*Canarium*) nuts and gnaws coconuts, and is found in trees felled by the natives." The natives of the Solomon Islands occasionally eat these rats.

RODENTIA; MURIDAE; **Genus MELOMYS** Thomas, *1922*

Mosaic-tailed Rats, Banana Rats

There are 20 species (Laurie and Hill 1954; Tate 1951b; Menzies and Dennis 1979; Knox 1978; Ride 1970; Watts 1979; Baverstock et al. 1980):

M. *albidens,* known only from the type locality at Lake Habbema in western New Guinea;
M. *fellowsi,* mountains of eastern New Guinea;
M. *levipes,* New Guinea, possibly Cape York Peninsula of northern Queensland;
M. *lorentzi,* New Guinea;
M. *aerosus,* Ceram Island between Celebes and New Guinea;
M. *cervinipes,* coastal Queensland, northeastern New South Wales;
M. *capensis,* rain forests of northeastern Queensland;
M. *rubicola,* Bramble Cay off south-central New Guinea;
M. *moncktoni,* New Guinea;
M. *obiensis,* Obi Island between Celebes and New Guinea;
M. *platyops,* New Guinea;
M. *rubex,* New Guinea;
M. *lutillus,* New Guinea, Cape York Peninsula of northern Queensland;
M. *burtoni,* coastal areas from northeastern Western Australia to northeastern New South Wales, Melville and Groote Eylandt islands;
M. *rufescens,* New Guinea, Bismarck Archipelago, Solomon Islands;
M. *leucogaster,* southern New Guinea;

Naked-tailed rat (*Solomys salebrosus*), photo from the Australian Museum, Sydney.

Rümmler's mosaic-tailed rat, right (*Pogonomelomys ruemmleri*), photos from Museum Zoologicum Bogoriense.

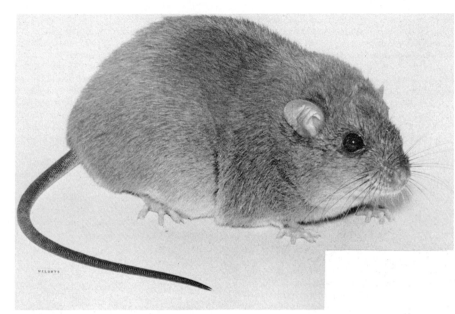

Mosaic-tailed rat (*Melomys burtoni*), photo by Stanley Breeden.

M. fulgens, Ceram and Talaud islands between Celebes and New Guinea;

M. arcium, Rossel Island off eastern New Guinea;

M. porculus, Guadalcanal Island (Solomons);

M. fraterculus, Ceram Island between Celebes and New Guinea.

Head and body length is 90 to 175 mm, tail length is 110 to 170 mm, and weight is 30 to 200 grams. The pelage is usually soft, thick, and woolly, and is sometimes fairly long. The coloration of the upper parts ranges from tawny through several shades of brown to russet. The underparts are white, creamy, or pale brownish or gray. Like *Uromys*, these rodents are easily recognized by their almost naked, filelike tails. The scales of the tail form somewhat of a mosaic pattern, which differs strikingly from the evenly ringed arrangement seen on the tails of the typical rats. In some species the tail is slightly prehensile (Menzies and Dennis 1979). The feet are broad and have well-developed pads. From *Uromys*, *Melomys* is distinguished externally by its smaller size. Females of most species of *Melomys* have two pairs of mammae, but female *M. lorentzi* have only a single pair (Menzies and Dennis 1979).

Most species of *Melomys* are forest dwellers, but *M. lutillus*, *M. rufescens*, and *M. burtoni* are primarily inhabitants of grassland and open scrub country; the elevational range of the genus is sea level to over 3,000 meters. Most species are basically terrestrial, but are capable of climbing. *M. lorentzi* and *M. rubex* live in burrows in the forest floor, while *M. lutillus*, *M. rufescens*, and *M. cervinipes* may nest in vegetation up to 3 meters above the ground (Menzies and Dennis 1979; Taylor and Horner 1970b; Ride 1970; Dwyer 1975b). Nests are also located among roots, in hollow timber, and under rock ledges. Aboveground nests are made of grass or leaves, are spherical, and are usually from 125 to 200 mm in diameter. The diet of *Melomys* is thought to consist of fruits, berries, and other vegetable matter. Wood (1971) found *N. cervinipes* to be extensively arboreal and strictly nocturnal. Barnett, How, and Humphreys (1977) reported

population densities of 0.37 to 0.97 individuals per ha. for this species.

Studies in eastern Australia (Taylor and Horner 1970b; Wood 1971) indicate that *M. cervinipes* and *M. burtoni* breed throughout the year, but have seasonal peaks that differ by locality. Female *M. cervinipes* are known to be polyestrous. In more southerly areas, reproduction occurs at least in late spring and summer, and farther north it extends into the autumn and winter. In New Guinea some species apparently breed all year, but data are limited (Menzies and Dennis 1979; Dwyer 1975b). Overall litter size for the genus is one to five, with two young being most common. The young cling tenaciously to the nipples of the mother and are dragged about almost continuously during their first two weeks of life. Subsequently the young become more independent, but still hold onto the mammae when disturbed. Captives usually produce their first litters at the age of seven months.

Watts (1979) considered a possible newly discovered population of *M. levipes* on the Cape York Peninsula of Queensland to be at risk, but noted that this species is common in New Guinea. *Melomys* is not known to occur in plague proportions in Australia, but *M. burtoni* becomes locally abundant in cane plantations and is regarded there as a major pest (Taylor and Horner 1970b).

RODENTIA; MURIDAE; **Genus POGONOMELOMYS**
Rümmler, 1936

Rümmler's Mosaic-tailed Rats

There are four species (Laurie and Hill 1954; Menzies and Dennis 1979):

P. mayeri, mountains of western New Guinea;

P. bruijni, extreme western and south-central New Guinea;

P. sevia, mountains of northeastern New Guinea;

P. ruemmleri, mountains of New Guinea.

The name *Rattus shawmayeri,* which was associated with *Niviventer eha* by Laurie and Hill (1954) and with the subgenus *Stenomys* of *Rattus* by Misonne (1969), is probably a synonym of *Pogonomelomys ruemmleri* (G. G. George 1979).

Head and body length is 130 to 190 mm, and tail length is about 140 to 210 mm. Two adult male *P. sevia* weighed 57.1 and 64.7 grams (Dwyer 1975*b*). The dorsal pelage of *P. bruijni* is slightly crisp, and that of *P. sevia* and *P. ruemmleri* is long and dense. The upper parts are brownish, usually reddish brown or dark brown, and the underparts are whitish, grayish, or buffy. The hands and feet are yellowish brown to white. The tail is prehensile. The scales on the tail are six sided, and each has one, three, or many hairs. The feet have large claws. Female *P. mayeri, P. bruijni,* and *P. sevia* have four mammae, but female *P. ruemmleri* have six mammae.

One species, *P. bruijni,* occurs in lowland rain forest, and the other three inhabit montane forest or alpine grassland at elevations of up to 3,300 meters. Both *P. bruijni* and *P. sevia* are evidently arboreal, but *P. ruemmleri* is at least partly terrestrial. Limited evidence indicates that the diet of *P. ruemmleri* consists only of vegetable matter, including leaves (Menzies and Dennis 1979). *P. sevia* appears to live off the ground beneath the dead fronds of pandanus palms, and to be solitary. Observations suggest that *P. sevia* begins to breed in early October and that pregnancy may be concurrent with lactation. Two females taken in October were pregnant, each with a single embryo (Dwyer 1975*b*).

RODENTIA; MURIDAE; *Genus XENUROMYS*
Tate and Archbold, 1941

White-tailed New Guinea Rat

The single species, *X. barbatus,* is known only by three specimens from New Guinea (Menzies and Dennis 1979). One was collected in about 1900 in the southeastern part of the island, one was taken in 1939 in a heavy forest at an elevation of 75 meters near the Idenburg River in the northwestern part, and one was taken in the mountains along the upper Sepik River in the north-central part.

Two male specimens measure 275 and 310 mm in head and body length, and 220 and 281 mm in tail length. One male is reddish brown above and buffy below, while the other, the larger specimen, is gray above and white below. The hands and feet are so scantily haired that they appear naked. The tail is white for at least half its length. The tail is coarsely scaled, and the unkeeled scales are arranged annularly instead of spirally.

According to Tate (1951*b*:284–85), *Xenuromys* comprises "large rats having the general appearance of *Uromys,* but with the normal overlapping scales of *Rattus.* . . . Unique or unusual habits may be implied by its rarity, but its feet and tail show no structural feature to indicate whether it is particularly arboreal or fossorial, nor does the pelage suggest aquatic habits."

RODENTIA; MURIDAE; *Genus APOMYS* Mearns, 1905

There are nine species (Taylor 1934; Sanborn 1952; Johnson 1962):

A. hylocoetes, southern Mindanao;

A. datae, northern Luzon;

A. abrae, northern Luzon;

A. sacobianus, known by a single specimen from southwestern Luzon;

A. insignis, Mindanao;

A. petraeus, southern Mindanao;

A. musculus, northern Luzon;

A. microdon, known by a single specimen from the island of Catanduanes (Philippines);

A. littoralis, known by a single specimen from southern Mindanao.

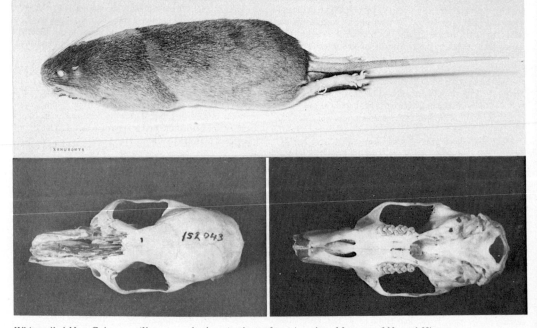

White-tailed New Guinea rat (*Xenuromys barbatus*), photos from American Museum of Natural History.

Apomys was included within *Rattus* by Sanborn (1952) and Ellerman and Morrison-Scott (1966), but more recent authorities have indicated that it is a distinct genus and not closely related to *Rattus* (Misonne 1969; Musser 1977*a*, 1977*b*).

Descriptive data provided by Taylor (1934), Sanborn (1952), and Johnson (1962) indicate that head and body length is about 86 to 143 mm, and tail length is 82 to 176 mm. The tail ranges from slightly shorter to considerably longer than the head and body. The pelage is long, soft, and thick. The upper parts are some shade of brown or buff, the sides are generally paler, and the underparts are grayish, buff, or fawn. The tail is scantily haired and, in most species, is darker above than below. The skull is elongate. Females have two pairs of mammae.

Taylor (1934) stated that the external form of *Apomys* is like that of *Rattus,* but that its eyes are relatively smaller, and that its pollex is less rudimentary and has a rather broad nail, Musser (1977*a*) wrote that the interorbital region of *Apomys* is smooth and the braincase smooth and globular, without the supraorbital ridges that extend back along the margins of the braincase, a characteristic of most species of *Rattus.* In *Apomys* the occlusal surfaces of the first and second upper molars are simple, and the third upper molar is tiny relative to the first and second upper molars.

Most specimens of *Apomys* have been collected in montane forests at elevations of 600 to 2,770 meters, but *A. littoralis* was found at only 15 meters, and Johnson (1962) referred to *A. sacobianus* as a lowland species. Musser (1977*a*) stated that *Apomys* is apparently terrestrial.

RODENTIA; MURIDAE; *Genus MUS* Linnaeus, *1758*

Mice

There are 4 subgenera and 36 species (Marshall 1977, 1979; Meester and Setzer 1971; F. Petter 1978*a*; Ellerman and Morrison-Scott 1966; Lekagul and McNeely 1977; Yalden, Largen, and Kock 1976; Kingdon 1974*b*; Hubert, Adam, and Poulet 1973; Ansell 1978; Vermeiren and Verheyen 1980):

subgenus *Pyromys* (spiny mice)

M. shortridgei, central Burma to Cambodia;
M. saxicola, Pakistan, India;
M. platythrix, India;
M. phillipsi, India;
M. fernandoni, Sri Lanka;

subgenus *Coelomys* (shrew mice)

M. mayori, Sri Lanka;
M. pahari, Sikkim to Indochina;
M. crociduroides, mountains of Sumatra;
M. vulcani, mountains of Java;
M. famulus, southern India;

subgenus *Mus* (house and ricefield mice)

M. caroli, found partly as a human commensal and perhaps partly through introduction in southern China,

A. House mouse (*Mus musculus*), photo by Ernest P. Walker. B. African pygmy mouse (*M. minutoides*), photo by K. B. Newman. C. Laboratory mice (*M. musculus*), photo by Ernest P. Walker.

Thailand, Indochina, Ryukyu Islands, Taiwan, Hainan, Sumatra, Java, Madura, and Flores;

M. cervicolor, found partly as a human commensal and perhaps partly through introduction from Nepal to Indochina, and on Sumatra and Java;

M. cookii, Nepal to Indochina;

M. booduga, India, Burma, Sri Lanka;

M. dunni, India;

M. terricolor, Pakistan, India, Nepal;

M. musculus (house mouse), perhaps naturally distributed from the Mediterranean region to China, now found partly as a human commensal throughout the world;

subgenus *Nannomys* (African pygmy mice)

M. callewaerti, southern Zaire, Angola;

M. setulosus, Guinea to Congo;

M. baoulei, Ivory Coast;

M. pasha, northern Zaire;

M. triton, northwestern Zaire, Uganda, Tanzania, Zambia;

M. bufo, eastern Zaire, western Uganda;

M. tenellus, Sudan and Somalia to Tanzania;

M. haussa, the Sahel from Senegal to Nigeria;

M. mattheyi, savannah zone from Senegal to Ghana;

M. indutus, Botswana, South Africa;

M. setzeri, Zambia, Botswana;

M. gratus, western Uganda;

M. minutoides, throughout Africa south of the Sahara;

M. proconodon, Ethiopia;

M. mahomet, Ethiopia;

M. sorella, western Kenya;

M. goundae, northern Central African Republic;

M. neavei, Rhodesia, Zambia;

M. oubanguii, Central African Republic.

It has long been recognized that *Mus musculus* comprises several fairly distinct kinds of mice, some of which might not regularly interbreed with one another. Certain kinds appear to be wild-living animals, native to the areas they now inhabit. Others are also usually wild, but apparently spread over much of their range through human agency. Still others are primarily commensals of people. Most authorities have treated these various kinds of mice as subspecies of *Mus musculus,* and that procedure has been followed in the above list.

Recently, Marshall and Sage (1981) pointed out that biochemical and morphological analyses indicate that what is listed above as *Mus musculus* actually consists of seven separate species. This report, as augmented by earlier information from Ellerman and Morrison-Scott (1966) and Grzimek (1975), suggests that there are two primarily commensal species, with long tails, and five basically wild-living species, with short tails. Both wild and commensal species may occupy the same area. The commensal species are: *M. domesticus,* found in Europe west of the Elbe River, and from the Mediterranean region to Nepal; and *M. castaneus,* found in the cities of India, China, and Southeast Asia. The wild species are: *M. musculus,* Sweden, Europe east of the Elbe River, Siberia, Mongolia, and China; *M. spretus,* western Mediterranean region; *M. abbotti,* eastern Mediterranean region; *M. hortulanus,* steppe country from Austria to the Caucasus; and *M. molossinus,* Manchuria, Korea, central China, and Japan. Several of these kinds are thought to have been introduced to the Americas, Africa, Australia, and other parts of the world.

Misonne's (1969) classification of *Mus* is substantially different from that given in the above list. He did not recognize subgenera, but divided the genus into three sections of related species: a *booduga* section with the species *M. fernandoni, M. booduga, M. musculus, M. cervicolor, M. triton,* and *M. pasha;* a *minutoides* section with the species *M.*

minutoides, M. musculoides (found in West Africa and considered a subspecies of *M. minutoides* by Meester and Setzer [1971], but also treated as a separate species by Rosevear [1969] and Hubert, Adam, and Poulet [1973]); *M. setulosus, M. gratus, M. birungensis* (found in eastern Zaire and not recognized as a species by Meester and Setzer 1971), *M. neavei, M. sorellus, M. tenellus, M. wamae* (found in Kenya and not recognized as a species by Meester and Setzer 1971), and *M. platythrix;* and a *pahari* section with the species *M. pahari, M. shortridgei, M. crociduroides,* and *M. mayori,* and probably *M. bufo* and *M. callewaerti.* Misonne also did not consider *Muriculus* to be a genus distinct from *Mus,* and placed its one species, *imberbis,* in his *booduga* section.

At one time or another certain of the species in the above list have been put into various additional genera or subgenera. Perhaps most important is *Leggada* Gray, 1837, the type species of which is *M. booduga* of India, but which eventually came to be used as the genus for all native African species of *Mus* (Allen 1939b). *Leggada* was considered a synonym of *Mus* by Ellerman and Morrison-Scott (1966) and Misonne (1969), but was used as a subgenus for African species by Rosevear (1969), Ansell (1978), and Yalden, Largen, and Kock (1976). Marshall (1977, 1979) placed *M. booduga* in the subgenus *Mus,* while for the native African species he used the subgenus *Nannomys* Peters, 1876, the type species of which is *Mus setulosus.* Marshall's arrangement has been followed in the above list.

The generic name *Gatamiya* Deraniyagala, 1965 was established for a single specimen that had been run over by a car in Sri Lanka. The animal appeared to resemble *Mus* closely, but reportedly differed in not possessing a bristly or spiny coat, in having a completely hairy tail, and in being smaller. Eisenberg and McKay (1970:91) did not recognize *Gatamiya* as a genus, and noted: "An examination of the single type of *Gatamiya* indicates no departure in the hairiness of the tail than is found in many immature specimens of the Muridae."

In *Mus musculus,* head and body length is generally 65 to 95 mm, tail length about 60 to 105 mm, and weight usually 12 to 30 grams. In *M. minutoides,* one of the smallest of living mammals, head and body length is 45 to 82 mm, tail length is 28 to 63 mm, and weight is 2.5 to 12 grams (Delany 1975). For the genus as a whole, head and body length is up to 125 mm. The tail ranges from substantially shorter to slightly longer than the head and body. Lekagul and McNeely (1977) listed the following ranges of average measurements for six species in Thailand: head and body length, 74 to 110 mm; tail length, 58 to 86 mm; and weight, 12 to 34 grams.

The fur of *Mus* may be soft, harsh or spiny. The tail appears to be naked, but has a covering of fine hairs. The coloration above ranges from pale buff or pale gray through dull grayish browns and grays to dark lead color or dull brownish gray. The sides may be slightly lighter and the underparts are usually lighter than the upper parts. *Mus musculus* is commonly light brown to black above, and white below, often with a buffy wash, and the tail is lighter below than above. Commensal forms of *M. musculus* tend to have longer tails and darker coats than the wild forms. Domesticated strains of *M. musculus* have been developed, the most common being the albinos. Other such strains have a black and white piebald pattern, and some carry various shades of black or gray.

The "singing," "waltzing," and "shaker" mice are common house mice. Faint but audible twittering sounds, emitted by house mice when in their shelters, have been reported from various parts of the world. The other types mentioned are defective in their balancing apparatus, and "waltz" or "shake" instead of moving about like normal mice.

African pygmy mouse (*Mus minutoides*), photo by John Visser.

Some of the larger species of *Mus* are larger than some of the smaller species of *Rattus,* and the two genera are often confused, though they are not closely related. With respect to those kinds found in the Americas, the two genera can be distinguished as follows. In *Mus,* total length is less than 250 mm, tail length is less than 110 mm, occipitonasal length of the skull is less than 35 mm, and the first upper molar is longer than the combined length of the second and third upper molars. In *Rattus,* total length is more than 250 mm, tail length is more than 110 mm, occipitonasal length is more than 35 mm, and the first upper molar is shorter than the combined length of the second and third upper molars (Hall 1981). The upper incisors of *Mus* are notched, but those of *Rattus* are not. The skull of *Mus* is light and usually rather flat. The hind foot is usually narrow and the outer digits tend to be shortened. Most female *Mus* have 10 or 12 mammae.

The natural habitat of *Mus* includes forests, savannahs, grasslands, and rocky areas. Some species are dependent on human habitations or cultivated fields. As noted in the above systematic discussion, there are a number of wild and commensal kinds of mice grouped under the name *Mus musculus.* Wild forms have been found from the tropics to the Faeroes (62° N) and Macquairie (54° S) islands, from the swamps of Georgia to the deserts of Peru and central Australia, from sea level to high mountains, and even in coal mines at depths of 550 meters (Bronson 1979). Some wild populations exist for only part of the year, disappearing during the winter and being replenished in the spring by stock that had sheltered in buildings. In Britain, three categories of populations exist: those occupying buildings, those inhabiting stacks of grain, and those living free in fields (Berry 1970). There is a net movement into the grain stacks after they are erected in the autumn, and a net movement out into the fields in the spring. Colonies in British buildings, however, tend to be isolated, and there is little interchange with field populations. Some other species of *Mus,* including most of those in the subgenus *Mus,* and *M. minutoides* of Africa, also sometimes act as human commensals.

Wild-living populations of *M. musculus* dwell in cracks in rocks or walls, or make underground burrows, which usually consist of a complex network of tunnels, several chambers for nesting and storage, and three or four exits (Berry 1970). When occupying human habitations, *M. musculus* nests behind rafters, in woodpiles, in storage areas, or in other hidden spots near a source of food. The nest is a loose structure of rags, paper, or other soft substances, lined with finer, shredded material (Jackson 1961). *M. minutoides* may shelter in any sort of cover during the day, and also digs shallow burrows in soft soil. Its natal nest consists of a ball of soft grass or other fibers (Smithers 1971). *M. caroli* and *M. cervicolor* are burrowing species, but *M. shortridgei* and *M. pahari* build grass nests and do not burrow (Lekagul and McNeely 1977).

The commensal forms of *Mus* are active at any hour, but the wild forms seem to be mainly nocturnal. Most species are basically terrestrial, but are good climbers, and *M. musculus* also swims well. The daily movements of a commensal mouse usually cover only a few square or cubic meters, but dispersing feral individuals have been known to wander up to 2 km.

Wild mice eat many kinds of vegetable matter, such as seeds, fleshy roots, leaves, and stems. Insects and some meat may be taken when available. Delany (1975) wrote that *M. triton* is mainly insectivorous. Commensal mice feed on any human food that is accessible, and also on paste, glue, soap, and other household materials. Some mice store food or live within a human storage facility.

Bronson (1979) explained that commensal populations of *M. musculus* are characterized by relative stability, high densities of up to 10 mice per sq meter, individual home ranges of under 10 sq meters, and little potential for sharp increase; and that field populations are characterized by instability, densities of up to one mouse per 100 sq meters, individual home ranges of a few hundred to a few thousand sq meters, and good potential for irruption. Such irruptions have sometimes attained plague proportions. Lowery (1974) reported that densities have reached nearly 1,250 per ha. in Louisiana. During 1926 and 1927, in the dry bed of Buena Vista Lake in Kern County, California, the number of mice in places was estimated at over 205,000 per ha.

There now seems a general consensus that *M. musculus* is both territorial and colonial, when living under commensal or laboratory conditions (Bronson 1979; Lidicker 1976; Berry 1970; Poole and Morgan 1973, 1976; Lloyd 1975; Mackintosh 1973). Territoriality, however, evidently is not pronounced among wild populations, and may break down in overcrowded laboratory colonies. Studies show that when a number of unrelated mice are placed together in an enclosure, considerable fighting will result. A male and a female, or several nonpregnant females, will soon establish an amicable relationship, but two or more males may fight savagely until one establishes dominance. A dominant male sets up a territory with definite boundaries. This territory eventually will include a family group of several females and their young. There also may be one or more subordinate males, though there is some evidence that several males occasionally share a territory on an equal basis. The female members of a family may establish a loose hierarchy among themselves, but they are far less aggressive than the males. Normally, aggression within a family is rare, but the members join in strict defense of the territory against outsiders. Territories are cohesive and long lasting; some have been observed to remain stable for 11 months. As young mice mature they are generally made to disperse through adult aggression, though some, especially females, may remain in the vicinity of the parents. The forced dispersal of young is known to take place in nonterritorial, wild-living populations, as well as in territorial populations. This kind of social arrangement tends to resist genetic modification of established colonies, but to encourage strongly the formation of new groups. A system of pheromonal cues seems to promote successful colonization by allowing females to avoid pregnancy before dispersal, but then to ovulate rapidly once a new home is established. Such a situation, along with the remarkable adaptiveness and reproductive potential of the species, is considered responsible for the success of *M. musculus* in spreading over much of the world.

The following reproductive information on *M. musculus* was taken largely from Berry (1970), Lowery (1974), and Bronson (1979). Breeding continues throughout the year in laboratory, most commensal, and some wild populations. In Britain, however, free-living mice have a reproductive season extending from April to September. The estrous cycle is 4 to 6 days long, with estrus lasting less than a day. Females may experience a postpartum estrus 12 to 18 hours after giving birth. There are usually 5 to 10 litters per year if conditions are suitable, but there may be as many as 14. The gestation period is 19 to 21 days, but may be extended by several days if the female is lactating. Litters consist of 3 to 12, usually 5 or 6, young. They weigh about one gram at birth, and are naked and blind. They are fully furred after 10 days of life, open their eyes at 14 days, are weaned at about 3 weeks, and reach sexual maturity at 5 to 7 weeks. There is usually 60 to 70 percent mortality before independence. Average life span is 2 years in the laboratory, but some captive individuals have lived for 6 years.

Considerably less is known about the reproduction of other species. In the Rajasthan Desert of India, *M. booduga* breeds over much of the year and produces litters of 1 to 13 young, *M. cervicolor* has been known to give birth to 2–6 young in July and December, and *M. platythrix* has a litter size of 3–10 (Prakash 1975). *M. minutoides* is capable of reproduction throughout the year in the laboratory, but breeds mainly during the wet seasons, April–June and September–December, in the wild in East Africa; its recorded litter size is 2–8 young (Delany 1975; Kingdon 1974*b*). A captive specimen of *M. minutoides* lived for three years and one month (Marvin L. Jones, Zoological Society of San Diego, pers. comm.).

Mus musculus is known from Pleistocene fossils in Europe (Kurten 1968) and is evidently a natural inhabitant of the southern parts of that continent, as well as much of Asia. The earliest known association of the house mouse with an urban community was at a neolithic site in Turkey about 8,000 years ago (Brothwell 1981). The species was familiar to the ancient Egyptians and Greeks, and may have reached Britain during Roman times. Its subsequent spread over most of the world was facilitated by human construction of houses and barns, which provided shelter to the mouse, and the development of agriculture, which provided food. It was also inadvertently carried about in ships and caravans. Several other species of the subgenus *Mus* expanded their ranges in Asia and now live partly as human commensals. *M. musculus* and its relatives do not cause such serious health and economic problems as *Rattus norvegicus* and *R. rattus*. Mice are, however, agricultural pests in some areas, and they do consume some stored human food and contaminate far more. They also destroy woodwork, furniture, upholstery, and clothing. They contribute to the spread of such diseases as murine typhus, rickettsial pox, tularemia, food poisoning (*Salmonella*), and bubonic plague. On the other hand, the albino strains of *M. musculus* are used extensively in laboratory research and have added immeasurably to our knowledge of medicine and genetics. In 1965 alone, some 800,000 mice were used at the National Institutes of Health in Bethesda, Maryland (Grzimek 1975; Lowery 1974; Berry 1970).

RODENTIA; MURIDAE; **Genus MURICULUS** Thomas, *1902*

Stripe-backed Mouse

The single species, *M. imberbis*, is known only from the highlands of Ethiopia. Misonne (1969) considered *Muriculus* a synonym of *Mus*, Meester and Setzer (1971) wrote that it probably should be included in *Mus*, and Yalden, Largen, and Kock (1976) stated that it is at least debatable whether *Muriculus* is distinct from *Mus*.

Head and body length is 70 to 95 mm, and tail length is about 45 to 60 mm. The thick pelage is rather crisp, or it may be soft with no suggestion of spines. The hairs of the upper parts, at least in some individuals, have dark blue gray bases and dark yellowish brown tips. A middorsal stripe is usually present from about the middle of the back to the base of the tail, though it may extend forward as a faint line almost to the shoulders. The underparts are tawny ochraceous, creamy, or almost white. Sometimes there is a tawny ochraceous tuft of hairs at the front base of the ears. The closely and finely haired tail is usually dark brown above and yellowish gray below.

Muriculus resembles *Mus* in external appearance, but is easily distinguished by the dark dorsal stripe. This stripe, however, may be poorly defined in some individuals. The head is relatively short, and the ears are of medium size and rounded. The claws are short. The incisor teeth are narrow and have a smooth front surface; they are inclined so far forward that the tips of the upper incisors do not curve backward toward the body.

The stripe-backed mouse inhabits open slopes in grassy or rocky areas extending to the upper limit of forests. It shelters in holes in the ground and is associated with *Arvicanthis abyssinicus*, at least in some areas. Of *Muriculus*, Yalden, Largen, and Kock (1976:32) wrote: "This little-known endemic mouse has been recorded from altitudes between 1900–3400 m on both sides of the Rift Valley. It has not been

Stripe-backed mouse (*Muriculus imberbis*), photo from Museum Senckenbergianum.

found by the most recent collectors and it seems possible that the species has become rarer since 1940, perhaps as a result of the conversion of its habitat to agricultural land.''

RODENTIA; MURIDAE; *Genus **MACRUROMYS** Stein, 1933*

New Guinean Rats

There are two species (Laurie and Hill 1954; Menzies and Dennis 1979):

M. elegans, Weyland Mountains of western New Guinea;
M. major, mountains from west-central to northeastern New Guinea.

In *M. elegans*, head and body length is 150 to 160 mm, tail length is 205 to 220 mm, coloration is grayish brown above and whitish below, and the tail is dark above and white below. In *M. major*, head and body length is 225 to 250 mm, tail length is 315 to 340 mm, coloration is mottled yellowish black above and grayish below, and the terminal two-thirds of the tail is white. *M. major* has coarser pelage than *M. elegans*. Females have two pairs of mammae (Menzies and Dennis 1979).

Macruromys superficially resembles *Uromys caudimaculatus*, but its tail has overlapping scales, like those of *Rattus*, with each scale carrying three hairs. *Macruromys* is also distinguished by its remarkably small and simple molar teeth, a character partly shared with *Anisomys*. The length of the molar tooth row in *Macruromys major* is only about half that of a specimen of *Uromys* of comparable size (Menzies and Dennis 1979).

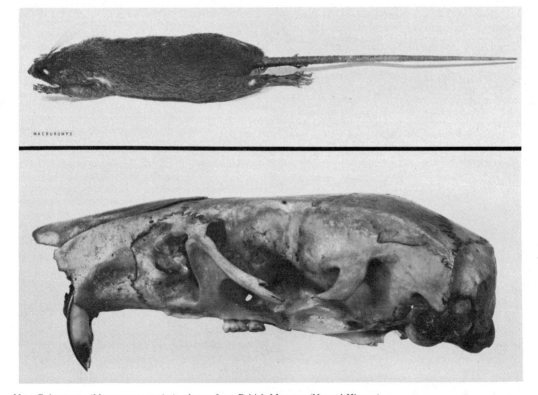

New Guinean rat (*Macruromys major*), photos from British Museum (Natural History).

Both species of *Macruromys* occur in mountains. The elevational range of *M. elegans* is 1,400 to 1,800 meters, and that of *M. major* is 1,200 to 1,500 meters.

RODENTIA; MURIDAE; **Genus CRUNOMYS** Thomas, *1897*

Philippine Swamp Rats

Two species have been described (Taylor 1934):

C. fallax, known by a single specimen from northeastern Luzon;
C. melanius, known by two specimens from Mindanao.

Musser (1977*a*) indicated the existence of an undescribed species of *Crunomys* on Celebes. Misonne (1969) stated that *Crunomys* showed evolutionary trends that associated it with the subfamily Hydromyinae.

The type specimen of *C. fallax* has a head and body length of 105 mm, and a tail length of 79 mm. The pelage consists of close, short fur, which is profusely mixed with flattened spines. The whiskers are long and the tail is uniformly covered with short hair. The upper parts are generally grayish, but are rather yellowish on the back. The underparts are grayish white. The dorsal spines have white bases and black tips. The sides of the nose and the ears are brown, the hands and feet are grayish brown, and the digits are white. The tail is black above and paler on the undersurface.

The two specimens of *C. melanius* have head and body lengths of 98 and 122 mm, and tail lengths of 68 and 79 mm. The fur, which is close and fine, is intermixed with flattened spines. The general coloration of the upper parts is blackish brown, and the limbs, hands, and feet are colored much the same. The tail is uniformly blackish. The underparts are only slightly paler than the upper parts. One specimen, a female, has six mammae.

The type specimen of *C. fallax* was collected in 1894 in a forest at an elevation of approximately 307 meters. It was shot beside a stream, where it was foraging for food. One specimen of *C. melanius* was collected in 1906 at an elevation of about 924 meters on Mount Apo. The other specimen was obtained in 1923 in a forest at sea level (Taylor 1934).

RODENTIA; MURIDAE; **Genus ECHIOTHRIX** Gray, *1867*

Celebes Spiny Rat or Shrew Rat

The single species, *E. leucura,* is known from northern and central Celebes (Laurie and Hill 1954). Misonne (1969)

Philippine swamp rat (*Crunomys fallax*), photos from ''Mammals from the Philippine Islands,'' Thomas, *Trans. Zool. Soc. London.*

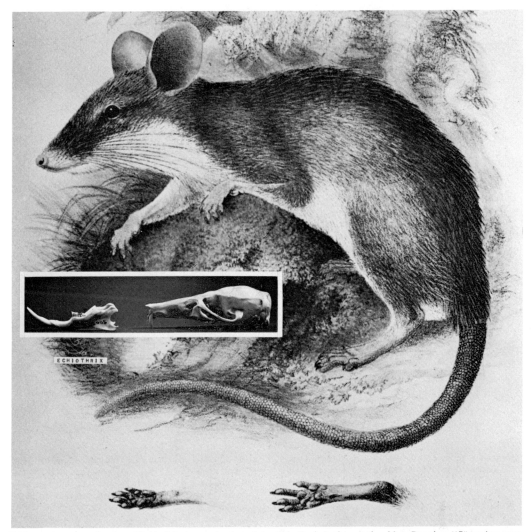

Celebes spiny rat or shrew rat (*Echiothrix leucura*), photo from *Abh. Zool. Anthrop.-ethn. Mus. Dresden*, "Säugetiere vom Celebes- und Philippinen-Archipel," Adolph B. Meyer. Skull: photo by P. F. Wright of specimen in U.S. National Museum.

suggested a distant relationship between *Echiothrix* and the subfamily Hydromyinae. Musser (1969*a*) summarized other views on the affinities of *Echiothrix*, and noted that specimens had been taken only on coastal plains and adjoining foothills.

Head and body length is 200 to 250 mm, and the tail is always shorter. The upper parts are grayish or dark gray brown, with black-tipped hairs on the back and sides. The underparts are whitish, yellowish, creamy buff, or reddish buff. The pelage contains both soft hairs and bristles. The nearly naked, cylindrical tail has rings of square scales. The head and nose are elongated, and give *Echiothrix* a somewhat shrewlike appearance.

The upper incisor teeth are short, and each has two faint longitudinal grooves. The lower incisors are elongated, arched, and widely divergent from each other, and "it is difficult to see how they can function, as they close on either side of the premaxillae, more or less, and evidently do not touch the upper incisors at all" (Ellerman 1941:269).

RODENTIA; MURIDAE; **Genus RHYNCHOMYS** *Thomas, 1895*

Shrewlike Rats

There are two species (Musser and Freeman 1981; Taylor 1934):

R. soricoides, known by seven specimens from Mount Data in northern Luzon;

R. isarogensis, known by a single specimen from Mount Isarog in southeastern Luzon.

In *R. soricoides*, head and body length is 188 to 215 mm, and tail length is 132 to 146 mm. In the type specimen of *R. isarogensis*, head and body length is 187 mm, and tail length is only 105 mm. The pelage is thick, short, and velvety. The upper parts are uniformly dark olivaceous gray. The underparts are dirty gray and not sharply defined from the back in

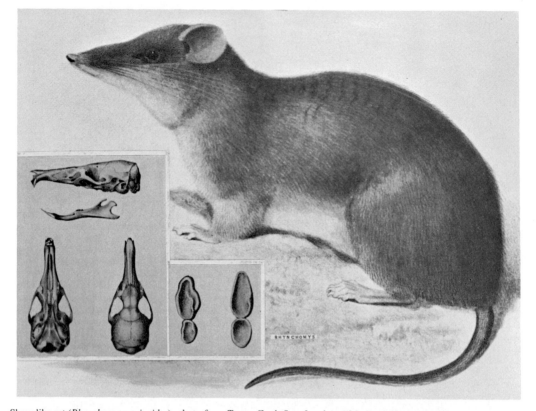

Shrewlike rat (*Rhynchomys soricoides*), photo from *Trans. Zool. Soc. London*, "Mammals from the Philippines," Oldfield Thomas. Insets: skull; left teeth, upper row; right teeth, lower row; photos from *Proc. Zool. Soc. London*.

R. soricoides, but appear whitish gray in *R. isarogensis*. The dorsal surfaces of the feet are dark brown in *R. soricoides*, but white in *R. isarogensis*. The tail is fairly well covered with hair and is not tufted; in *R. soricoides* it is blackish above and slightly paler below, but in *R. isarogensis* it is dark brown above and mostly unpigmented below.

These rodents resemble shrews in having an elongated muzzle and relatively small eyes. The rostrum of the skull is greatly elongated, and the lower jaw is long, slender, and straight. There are no third molar teeth, and the remaining molars are extraordinarily reduced in size and complexity. The upper incisors are white and the lower ones are pale yellow. The feet are similar to those of *Rattus*. The great toe has a small nail.

Shrewlike rats are apparently rare and little is known about their biology. The specimens of *R. soricoides* were taken in thick bushes and mossy forests at elevations of 2,286 and 2,460 meters. *R. isarogensis* was collected in a montane forest at 1,660 meters. Because of the small cheek teeth of these animals, it has been suggested that their diet consists of soft food, mainly insects and worms, rather than plant material.

RODENTIA; MURIDAE; *Genus* **CHROTOMYS** *Thomas, 1895*

Luzon Striped Rat

The single species, *C. whiteheadi*, is known from Luzon and Mindoro islands in the Philippines (Taylor 1934; Temme 1974; Barbehenn, Sumangil, and Libay 1972–73).

Head and body length is 150 to 196 mm, tail length is 90 to 120 mm, and weight is 115 to 160 grams (Temme 1974). The fur is short, soft, and straight. The general coloration above is grayish brown, and some individuals have a rufous tinge. A well-defined bright buff or orange line, bordered on either side by broad, shining black bands, extends from the middle of the face along the middle of the back almost to the tail. The ground color of the back gradually shades lighter on the sides to slaty gray on the underparts. The fingers and toes are white, and the remainder of the hands and feet is shiny gray. The tail is thinly haired, black above and paler beneath, with the tip sometimes white. The eyes are quite small, and the ears are fairly large and covered with short, fine hairs. The pollex has a rounded nail, whereas the other digits have well-developed, slightly curved claws. The incisor teeth are pale yellow and project somewhat forward. Females have two pairs of mammae.

Specimens have been taken in montane forests at elevations of up to 2,600 meters, but also in lowland rice fields and grasslands. The diet is said to include sweet potatoes, grass, and earthworms.

RODENTIA; MURIDAE; *Genus* **CELAENOMYS** *Thomas, 1898*

The single species, *C. silaceus*, is known by six specimens from northern Luzon (Taylor 1934; Sanborn 1952).

Head and body length of the type specimen is 195 mm, and tail length is 110 mm. The pelage is soft, close, and velvety. The upper parts are uniformly gray, and the underparts are paler, but there is no sharply defined line of demarcation.

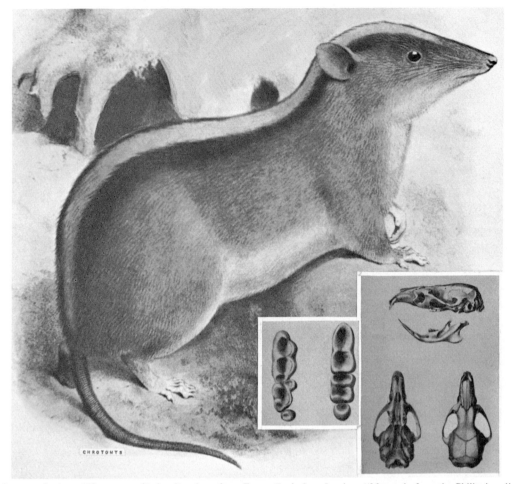

Luzon striped rat (*Chrotomys whiteheadi*), photo from *Trans. Zool. Soc. London,* "Mammals from the Philippines," Oldfield Thomas. Insets: left teeth, upper row; right teeth, lower row; skull; photos from *Trans. Zool. Soc. London.*

Most individual hairs are slaty gray on the basal part and washed with buffy white on the tip. The sides of the muzzle are almost black, the ears are grayish, and the feet are dark gray as far as the digits, but the digits are whitish or flesh colored. The thinly haired tail is white except for the upper basal part, which is brownish.

Celaenomys has a shrewlike body form, and at a glance could be mistaken for *Rhynchomys,* but it is distinguished from that genus by its larger teeth and shorter muzzle. The incisors project forward, like those of *Rhynchomys,* but they are larger and more powerful. The eyes are small and the ears are short. The sharply pointed muzzle is produced by a wedge-shaped skull.

Specimens have been obtained in mountainous areas at elevations of 2,100 to 2,460 meters. Some have been collected in densely vegetated gullies and mossy forests.

RODENTIA; MURIDAE; **Genus LEPTOMYS** Thomas, *1897*

The single species, *L. elegans,* occurs over much of New Guinea (Laurie and Hill 1954; Menzies and Dennis 1979).

The elevational range is sea level to about 3,000 meters.

Head and body length is 144 to 162 mm, and tail length is 150 to 160 mm. The fur is close, soft, and velvety. The general coloration of the upper parts is rufous fawn to brownish. The shoulders, flanks, and hips are bright rufous. The black on the upper side of the muzzle extends backward to form an indistinct ring around the eyes. The cheeks, the inner sides of the arms, and the underparts from the chin to the arms are creamy white, the hairs evenly colored to the bases. One form, however, has a grayish brown abdomen resulting from the individual hairs having gray bases. The upper surfaces of the thinly covered hands and feet are white. The finely scaled tail is brown on the basal dorsal surface, and has a white tip and white undersurface.

This genus does not show any particular modification for aquatic life. The third molar teeth are retained, though they are much reduced in size (Menzies and Dennis 1979). The incisor teeth are broad, flattened in front, and pale yellow with white tips. The eyes are small, and the ears are rather small and naked. The forefeet are normal for a murid, but the hind feet are elongated, suggesting the possibility of leaping habits. The three center toes of the hind feet are considerably larger than the others. There are five sole pads on the front feet and six on the hind feet. Females have four mammae.

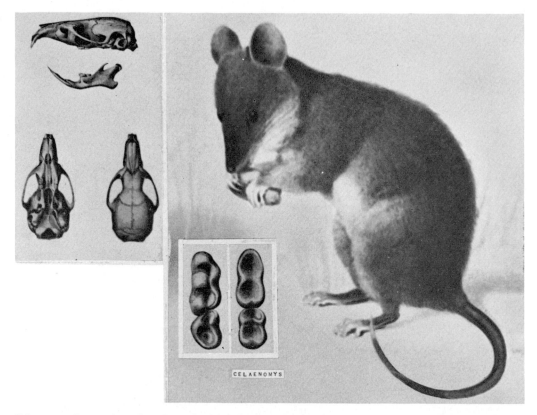

Celaenomys silaceus, photos from *Trans. Zool. Soc. London*, "Mammals from the Philippines," Oldfield Thomas.

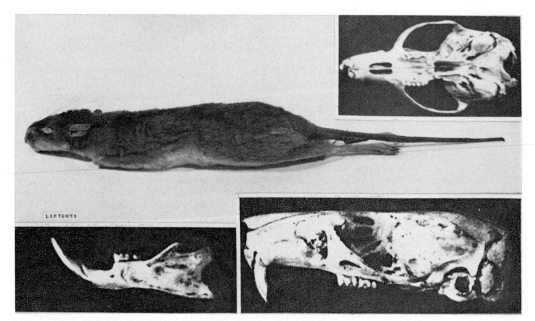

Leptomys elegans, photo of skin from British Museum (Natural History). Insets: skull (*L. elegans*), photos from *Das Aquarium*.

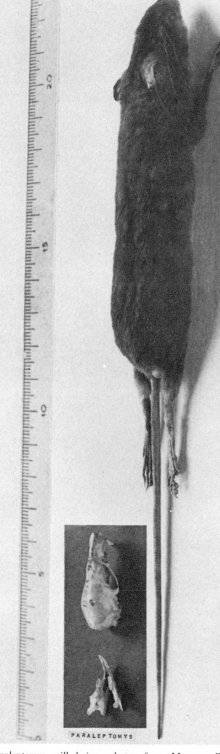

Paraleptomys wilhelmina, photos from Museum Zoologicum Bogoriense.

RODENTIA; MURIDAE; *Genus PARALEPTOMYS*
Tate and Archbold, 1941

There are two species (Laurie and Hill 1954; Menzies and Dennis 1979):

P. wilhelmina, mountains of western and central New Guinea;
P. rufilatus, Cyclops Mountains of north-central New Guinea.

Head and body length is 120 to 140 mm, and tail length is 130 to 140 mm. *P. wilhelmina* is grayish brown above and dusky white below. Its tail is gray above and white below, and has a white tip. *P. rufilatus* is distinguished from *P. wilhelmina* by its white throat, broad reddish brown lateral line, and more reddish head and hind legs. *Paraleptomys* is similar to *Leptomys,* but differs from that genus in having neither an elongated foot nor the third upper and lower molars.

Although *P. wilhelmina* is common at elevations of 1,800 to 2,700 meters, little is known of its biology. *P. rufilatus* is apparently known only by two specimens, taken at 1,450 meters on Mount Dafonsero.

RODENTIA; MURIDAE; *Genus XEROMYS Thomas, 1889*

False Water Rat

The single species, *X. myoides,* is known by six specimens from the Mackay area of southeastern Queensland, five from the coast of the Northern Territory, and three from nearby Melville Island (Watts 1979).

Head and body length is 110 to 114 mm, and tail length is about 88 mm. The general color of the upper parts is dark slaty gray, which gradually blends into the lighter underparts, with no distinctive line of demarcation. The limbs are like the back in color, and the hands, feet, and tail are scantily covered with fine white hairs. Fine scales are conspicuous on the tail.

The body is streamlined, the head is long and like that of *Hydromys,* the ears are short and round, the eyes are small, the tail is much shorter than the head and body, and the coat is water resistant. Nonetheless, the feet are not webbed and there is some question as to whether *Xeromys* is aquatic. There are only two molar teeth on each side of the skull. The ungrooved incisors project slightly outward; the upper incisors are yellow or orange and the lower ones are white. Females have four mammae.

According to the Australian National Parks and Wildlife Service (1978), the false water rat has been recorded in coastal swamps with a variable cover of mangrove forest, in swamps lined with mangroves and paperbark trees, and in reed swamps with tall grass, shrubs, and pandanus palms. it is an agile climber and well adapted to tidal areas. It builds its nest among the roots of mangrove trees. The nest is a mound of leaves and mud, about 60 cm tall, with a single opening near the apex. A narrow tunnel from the opening leads to a nest chamber, from which radiate numerous interconnecting tunnels, some going well below ground level.

Redhead and McKean (1975) found a captive animal to be nocturnal, spending the day in a burrow dug into the bottom of its cage. Although those authors suggested that *Xeromys* is a capable swimmer and adapted for aquatic life, Magnusson, Webb, and Taylor (1976) spent thousands of hours observing the rivers of Arnhem Land and never saw this genus in the water, day or night. It was seen only in small trees lining the

False water rat (*Xeromys myoides*), photo by T. Redhead.

rivers, climbing agilely among the branches. A captive took animal food in preference to vegetable matter. It ate insects, fish, and lizards, and was able to attack and kill crabs larger than itself.

The false water rat was long known only from southeastern Queensland and by a single specimen taken in 1903 in the Northern Territory. Since 1970, additional animals have been found on the coast of the Northern Territory and on Melville Island. Nonetheless, *X. myoides* is considered rare, and its habitat is thought to be jeopardized by agriculture, livestock, and urban and recreational development (Australian National Parks and Wildlife Service 1978; Watts 1979). The species is listed as endangered by the USDI (1980) and is on appendix 1 of the CITES.

Water rat (*Hydromys chrysogaster*), photo by Stanley Breeden.

RODENTIA; MURIDAE; *Genus HYDROMYS*
E. Geoffroy St.-Hilaire, 1805

Water Rats, Beaver Rats

There are three species (Laurie and Hill 1954; Ride 1970; Menzies and Dennis 1979):

H. chrysogaster, New Guinea, Aru and Kei islands,
  Australia, Tasmania;
H. neobritannicus, New Britain Island east of New
  Guinea;
H. habbema, mountains of New Guinea.

The generic name *Baiyankamys* Hinton, 1943 was proposed for the species *B. shawmayeri*. Mahoney (1968), however, showed *B. shawmayeri* to be a junior subjective synonym of *Hydromys habbema*. It was based on a composite specimen, the skin and skull of which represented *Hydromys habbema*, while the mandible belonged to a *Rattus niobe*.

Head and body length is 205 to 350 mm, tail length is 200 to 350 mm, and weight is 400 to approximately 1,300 grams. The pelage is composed of shiny guard hairs and dense, soft underfur. Coloration of the upper parts ranges from dark brown, almost black, to golden brown or dark gray. The underparts range from brownish to yellowish white or bright orange. The well-haired tail is dark except for a white tip.

These large water rats have a sleek, streamlined appearance. Aquatic adaptations include the long and flattened head, forward-thrust nostrils, high-set eyes, small ears, seal-like fur, and broad, partially webbed feet. Females have four mammae.

Beaver rats are aquatic and occur wherever suitable habitat is provided by rivers, swamps, marshes, backwaters, or estuaries. *H. habbema* of New Guinea is found along mountain streams at elevations up to and above 3,000 meters (Menzies and Dennis 1979). Beaver rats live in a nest of shredded weeds in a hollow log or a burrow in a river bank. They hunt along the bottom of streams, but generally carry their prey to a favorite log, stone, or island before eating.

In a study in New South Wales, Woolard, Vestjens, and MacLean (1978) observed *H. chrysogaster* to swim at any time, but especially in the evening. Its eyesight was good and the eyes were kept open underwater. All prey was caught and carried by mouth. Fish (some up to 30 cm long) and large aquatic insects were the most important foods. Other prey included spiders, crustaceans, mussels, frogs, turtles, birds, and bats.

The reproductive season in Australia is spring and summer, when each female produces about three litters at 2-month intervals, Litter size is one to seven, usually three to five, young. Independence is attained at the age of 34 days (Ride 1970; Watts 1974).

Beginning about 1937 in Australia, *H. chrysogaster* was trapped extensively for its beautiful pelt. At that time there was a shortage of muskrat fur, and a single skin of *Hydromys* would bring up to 65 cents. Trapping regulations have since been established, at least in some areas.

RODENTIA; MURIDAE; *Genus PSEUDOHYDROMYS*
Rümmler, 1934

New Guinea False Water Rats

There are two species (Laurie and Hill 1954; Lidicker and Ziegler 1968; Menzies and Dennis 1979):

P. murinus, known by 10 specimens from the mountains
  of Papua New Guinea;
P. occidentalis, known by 5 specimens from the mountains of western and central New Guinea.

Head and body length is 85 to 115 mm, and tail length is 90 to 115 mm. A female taken by Lidicker and Ziegler (1968) weighed 19.9 grams. The pelage is dense, short, and soft. The upper parts are dark gray or gray brown, the underparts are paler, and the tail is brown. The skull is elongate and rather flat, contributing to an overall shrewlike appearance. Females have four mammae (Menzies and Dennis 1979).

These animals appear to be nonaquatic; their feet have pads and soles of a type associated with terrestrial life. They evidently live on the ground in montane forests and feed on insects (Menzies and Dennis 1979). Specimens have been taken at elevations of 2,100 to 3,600 meters. A female *P. murinus*, trapped on 23 October by Lidicker and Ziegler (1968), was pregnant with a single embryo.

RODENTIA; MURIDAE; *Genus MICROHYDROMYS Tate and Archbold, 1941*

The single species, *M. richardsoni*, is known by a very few records from west-central to southeastern New Guinea (Menzies and Dennis 1979).

The type specimen has a head and body length of 80 mm, and a tail length of 92 mm. The upper parts are grayish black, the underparts are slightly paler, and the terminal 10 mm of the tail is white. The genus is distinguished from other members of the subfamily Hydromyinae by its small size, short muzzle, and grooved upper incisors.

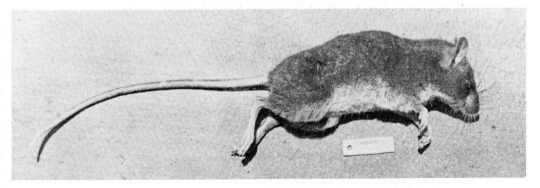

New Guinea false water rat (*Pseudohydromys* sp.), photo from Archbold Expeditions, American Museum of Natural History.

Microhydromys richardsoni, photo from American Museum of Natural History.

Microhydromys evidently lives in hill forests at somewhat lower elevations than related genera. The type specimen was taken in 1939 at 850 meters, and Tate (1951*b*) considered the likely elevational range to be 600 to 900 meters. *Microhydromys* is apparently not aquatic, and probably lives on the ground and feeds on insects (Menzies and Dennis 1979).

is brownish above and below, except for a whitish terminal area of some 15 mm. This small, mouselike rodent is not modified for aquatic life. It is distinguished from other members of the subfamily Hydromyinae by its small molar teeth and fairly long muzzle, as well as other dental and skeletal features. Females have four mammae.

RODENTIA; MURIDAE; *Genus NEOHYDROMYS*
Laurie, 1952

The single species, *N. fuscus,* is known from scattered localities in the montane rain forests of central and eastern New Guinea, at elevations of 2,500 to 3,000 meters (Laurie and Hill 1954; Menzies and Dennis 1979).

Head and body length of the type specimen, an adult female, is 92 mm, and tail length is 78 mm. The upper parts are smoky gray and the underparts are slightly paler. The tail

RODENTIA; MURIDAE; *Genus PARAHYDROMYS*
Poche, 1906

Mountain Water Rat

The single species, *P. asper,* is widely distributed in the mountains of New Guinea (Laurie and Hill 1954; Menzies and Dennis 1979).

Head and body length is about 240 mm, and tail length is about 260 mm. The pelage is firm to bristly, short, and not as

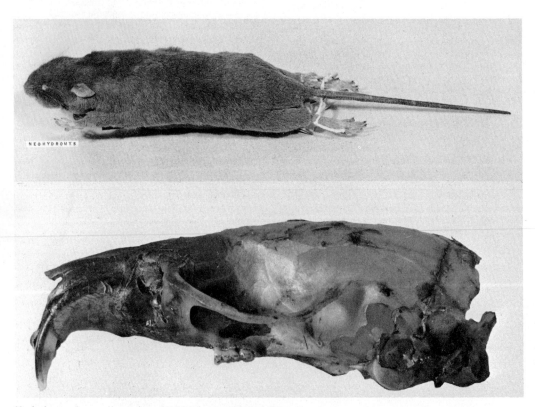

Neohydromys fuscus, photos from British Museum (Natural History).

Mountain water rat (*Parahydromys asper*), photo from Museum Zoologicum Bogoriense.

dense or sleek as the pelage of *Hydromys*. The general coloration of the upper parts is brownish gray; the ends of the longest hairs are black and those of the shorter ones are dull creamy whitish. The underparts are dull white with a buffy wash, and without a sharp demarcation from the color of the sides. The numerous whiskers are stiff; the upper ones are black and the lower ones are white. The finely haired ears are grayish brown; the upper surfaces of the hands and feet are pale brownish; and the well-haired tail is basally brownish black, with a distinctive white, brushed tip.

The muzzle in this genus is exceptionally wide and bears highly developed vibrissae, which may be associated with the detection of food. The hind feet are partially webbed. The soles are smooth and slightly granulated, and the pads are distinct. The incisor teeth grow with their roots wide apart and their tips converging. Females have four mammae.

Parahydromys occurs both in forest and in vegetation along mountain streams, even when these run through cultivated areas. The elevational range is about 600 to 2,700 meters. This rodent does not have marked aquatic adaptations, but may be associated with waterways. It is said to feed on insects and other invertebrates that it digs up (Menzies and Dennis 1979).

RODENTIA; MURIDAE; **Genus CROSSOMYS Thomas, 1907**

Earless Water Rat

The single species, *C. moncktoni,* is known from the highlands of eastern New Guinea, and probably occurs throughout the island (Laurie and Hill 1954; Menzies and Dennis 1979).

Head and body length is about 205 mm, and tail length is about 220 mm. The dorsal pelage is long and soft, with dense, woolly, and glossy underfur. The general coloration of the upper parts is mottled brownish gray with a pale yellowish olivaceous wash. The few scattered guard hairs have black tips and subterminal brownish rings. The soft, cottony hairs of the underparts are pure white to the base. The well-defined line, where the two colors meet, occurs quite high on the sides of the animal. The thick tail is light gray above and white below. Two rows of long white hairs begin on the sides of the body and gradually converge into a single raised crest along the underside of the tail. This development is similar to the swimming fringe in the European water shrew (*Neomys*), and may assist the tail in serving as a rudder.

In addition to this tail arrangement, *Crossomys* has waterproof fur, greatly reduced ears, small eyes, a smoothly rounded head, and very large, webbed hind feet. These adaptations make *Crossomys* even more specialized for aquatic life than *Hydromys*. The hands of *Crossomys* are relatively small, the wrists are slender, and the claws are small, delicate, and strongly curved. The incisor teeth are narrow and beveled on the sides. Females have four mammae.

The elevational range of *Crossomys* is about 600 to 3,000 meters, and this animal apparently occurs only along waterways. The typical habitat is a small, swift mountain stream, but when frogs are spawning, *Crossomys* may move away from the rivers to hunt for tadpoles. It lives in holes in stream banks, and, in addition to tadpoles, feeds on insects, mollusks, and small aquatic vertebrates (Menzies and Dennis 1979).

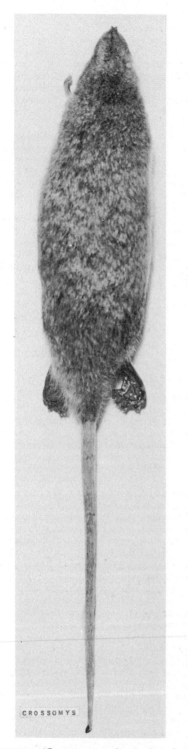

Earless water rat (*Crossomys moncktoni*), photo from Basil Marlow, Australian Museum, Sydney.

RODENTIA; MURIDAE; *Genus MAYERMYS*
Laurie and Hill, 1954

Shaw-Mayer's Mouse

The single species, *M. ellermani,* is known by eight specimens from the mountains of northeastern New Guinea (Laurie and Hill 1954; Lidicker and Ziegler 1968).

Head and body length is 92 to 103 mm, tail length is 97 to 107 mm, and weight is about 17 to 21 grams. The upper parts are predominantly smoky gray. The underparts are pale gray, and some specimens have a small white spot on the middle of the chest. The tail is brownish, both above and below, and may have patches of white or pale gray brown.

Mayermys is distinguished from all other members of the family Muridae, and indeed from all other rodents, in having only one molar tooth on each side of the upper and lower jaws. The incisor teeth are well developed, the uppers slightly curved forward and not grooved. The feet seem adapted for terrestrial life, but, as in *Pseudohydromys* and *Neohydromys,* there is a slight membrane between the fingers and toes.

Specimens have been collected at elevations of 2,100 to 2,700 meters in montane forests. Four were trapped under logs. All reported specimens are males.

RODENTIA; *Family GLIRIDAE*

Dormice

This family consists of 7 living genera and 16 species. One genus, *Graphiurus,* the sole representative of the subfamily Graphiurinae, inhabits Africa south of the Sahara. The other six genera, which compose the subfamily Glirinae, occur in the Palearctic region—Europe, northern Africa, central and southwestern Asia, and Japan. Ellerman and Morrison-Scott (1966), among others, recognized two additional glirid subfamilies: the Platacanthomyinae, which are considered here a subfamily of the Muridae; and the Seleviniinae, which are treated here as a full family. Corbet's (1978) use of the name Gliridae is followed here, but the name of this family is sometimes given as Muscardinidae or Myoxidae. The sequence of genera presented here follows that of Simpson (1945).

Dormice look like squirrels, but are smaller, and one genus (*Glirulus*) resembles chipmunks. Head and body length is 60 to 190 mm, and tail length is 40 to 165 mm. The pelage is soft, and the tail (except in *Myomimus*) is bushy. The eyes are prominent and the ears are rounded. The legs and toes are short, and the short, curved claws are adapted to climbing. The forefeet have four digits and the hind feet have five. The underside of the feet and digits is naked. Female dormice have 6 to 12 mammae.

The skull has well-developed zygomatic bones, no postorbital processes, and relatively large bullae (Ognev 1963; Arata 1967). The dental formula is: (i 1/1, c 0/0, pm 1/1, m 3/3) $\times$ 2 = 20. The cheek teeth are low crowned, and have a series of parallel ridges of enamel across the crown.

Dormice live in wooded areas, hedgerows, gardens, and rocky places. They are generally nocturnal and scansorial, and are squirrellike in some habits. They shelter in hollow trees, on the branches of trees or shrubs, in rocky crevices, in the deserted burrows of other animals, and in the attics of buildings, often in a nest of plant material. During the late summer and early autumn they generally become quite fat, and in the Palearctic from October to April they hibernate in a

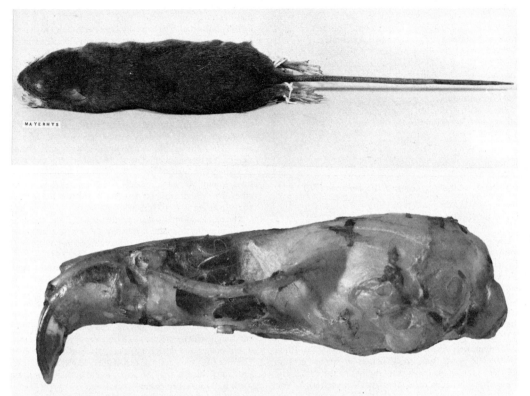

Shaw-Mayer's mouse (*Mayermys ellermani*), photos from British Museum (Natural History).

African dormouse (*Graphiurus murinus*), photo by John Visser.

curled-up, circular position. They may awake from time to time to eat food that they have stored. The diet includes fruits, nuts, insects, eggs, and small vertebrates.

There are usually one or two litters per year, each consisting of 2 to 10 young. Gestation periods of 21 to 30 days have been reported. The young are born in a nest, often lined with moss, in a tree hollow, on a branch, or sometimes in a ground shelter. Longevity is up to about 5½ years in the wild.

The known geological range of the Gliridae is middle Oligocene to Recent in Europe, and Recent in Asia and Africa (Arata 1967).

RODENTIA; GLIRIDAE; *Genus GLIS Brisson, 1762*

Fat or Edible Dormouse

The single species, *Glis glis,* occurs from France and northern Spain to the Volga River and northern Iran, and on the islands of Sardinia, Corsica, Sicily, Crete, and Corfu (Corbet 1978). It was also introduced to England in 1902, and has since become common in parts of that country. There is some support for use of the name *Myoxus* Zimmermann, 1780 in place of *Glis,* but Corbet (1978) thought such a change to be neither necessary nor desirable.

Head and body length is 130 to 190 mm, tail length is 110 to 150 mm, and weight is 70 to 180 grams (Van Den Brink 1968). The short, soft, thick pelage is silvery gray to brownish gray on the upper parts, lighter on the flanks, and white or yellowish on the underparts. This squirrellike animal has large and rounded ears, small eyes, and a long, densely bushy tail. The hands and feet, with their rough pads, are adapted to climbing. Females have 10 or 12 mammae (Ognev 1963).

Except as noted, the information for the remainder of this account was taken from Ognev (1963) and Grzimek (1975). The edible dormouse inhabits deciduous or mixed forests, and fruit orchards, in both lowlands and mountains. Its populations in any given area are partly dependent on the existence of a suitable number of hollow trees, as these are the most common sites for daily shelter, hibernation, and natal nests. The hollows may be heavily lined with grass or other vegetation, especially if being used for hibernation or rearing of young. *Glis* also shelters in crevices between rocks, burrows among tree roots, woodpecker holes, piles of mulch, attics, barns, and artificial nest boxes. Hibernation burrows sometimes are 50 to 100 cm below the surface of the ground.

The edible dormouse is primarily nocturnal and crepuscular, though occasionally it is active by day. It is highly arboreal and its agility in the trees may exceed that of squirrels. Some of its leaps have been reported to cover 7 to 10 meters. It has exceptionally good senses of vision, hearing, smell, and touch (through its vibrissae). Individuals visit many trees each night in the search for food. If a food shortage occurs, there may be movement to a different area. During winter hibernation, *Glis* sleeps on its side or back and curls its body so that the feet touch the muzzle. The hibernation period is very long, with the animals entering torpor from September to November and emerging from early May to early June. The diet consists mainly of seeds, nuts, acorns, berries, and soft fruits. Insects may be important at certain times, and small birds are taken on rare occasions.

Fat Dormouse (*Glis glis*), photo by R. Pucholt.

The population density in the northern Caucasus was calculated at 30/ha. In Moravia, however, Gaisler, Holas, and Homolka (1976) found a minimum density of only 1/ha. and an individual home range diameter of about 200 meters. The edible dormouse is apparently territorial, and marks its space with glandular secretions. It is highly vocal; a variety of chirps, whistles, squeaks, and squeals have been noted. Individuals are quarrelsome, and males are reported to fight savagely during the breeding season. Nonetheless, small groups may hibernate together, and as many as eight individuals have been found in a single tree hollow. In studies of captives, males have been observed to remain in the vicinity of females after mating and to be allowed back into the nest about 16 days following birth, where they help to clean and protect the young; families may stay together through winter hibernation. A wild male, however, probably leaves a female after mating, in order to pursue other estrous females.

The mating season extends from June to early August. In Moravia the young are born in August and early September (Gaisler, Holas, and Homolka 1976). Although two litters per year have been reported in some areas, the very short active season of Glis suggests that a single annual litter per female is the usual case. Gestation periods of 20 to 30 days have been reported. Litter size is 2–10 young, and the average in Gaisler, Holas, and Homolka's (1976) study was 4.5. The young are born naked and blind, open their eyes and are weaned after about 4 weeks, and do not begin to mate until after their first hibernation. Wild individuals have been known to live for more than 4 years, and a captive survived for 8 years and 8 months (Marvin L. Jones, Zoological Society of San Diego, pers. comm.).

In some areas, Glis is considered extremely harmful to the production of fruit and wine. It consumes large amounts of apples, pears, plums, and grapes, and has been reported to destroy one-third of the grape crop in the northern Caucasus. However, it is easily trapped, there is some demand for its luxuriant fur, and it is hunted for use as food and a source of fat. In ancient Rome, Glis was considered a delicacy, and colonies were kept in large enclosures planted with nut-bearing bushes and provided with nesting sites. Prior to a feast, individual animals would be confined to earthen urns and fattened on acorns and chestnuts. The meat of Glis is still a gourmet dish in some parts of Europe.

RODENTIA; GLIRIDAE; *Genus MUSCARDINUS Kaup, 1829*

Common Dormouse, Hazel Mouse

The single species, *M. avellanarius,* occurs from France and southern Sweden to the European part of the Soviet Union and northern Asia Minor, and in southern Britain, Sicily, and Corfu (Corbet 1978).

Head and body length is 60 to 90 mm, tail length is 55 to 75 mm, and weight is 15 to 40 grams (Van Den Brink 1968). The general coloration of the upper parts is rich yellowish brown or yellowish red. The throat and chest are creamy white, and the belly is pinkish buff. The well-haired tail is brownish above and paler below. The snout is blunt, the eyes are large, and the ears are relatively small. Females have eight mammae (Ognev 1963).

The common dormouse dwells in thickets and in forests that have an abundance of secondary growth. It is arboreal and scrambles about with great agility, but does not usually ascend as high into the trees as does Glis. The common dormouse is nocturnal and crepuscular, sleeping by day in a globular nest constructed of shredded bark, leaves, grass, or moss, and located in a bush or the lower branches of a tree. The nest is usually about 1–2 meters above the ground (Grzimek 1975). The nest of an individual animal is 60 to 80 mm in diameter. The nest of a female with young is more extensively lined and measures about 120 mm in diameter.

Winter hibernation is spent in another nest, which is located in a burrow, in a stump, or on the ground beneath moss, leaves, and other debris. This nest is constructed of vegetation and is bound together by a sticky secretion from the

Hazel mouse (*Muscardinus avellanarius*), photo from Zoological Society of London.

salivary glands. The period of dormancy may begin as early as August and extend to May in the Soviet Union (Ognev 1963). In England this period lasts from late October to early April. Apparently, hibernation is triggered when the external temperature falls below 16° C. During deepest dormancy, the blood temperature of *Muscardinus* falls to 0.25–0.50° C, whereas its normal temperature is 34–36° C.

The common dormouse appears to be associated with nut-bearing trees, and hazelnuts are known to be an important part of its diet (Ognev 1963; Grzimek 1975). Nonetheless, one observer considered *Muscardinus* just as incapable as other small rodents of opening hard hazelnuts. *Muscardinus* is able to gnaw a hole in fresh nuts, thereby getting the seed out in little pieces. Its diet also includes other seeds, buds, berries, and insects, and possibly birds' eggs and nestlings. Some food may be stored during the summer and utilized during brief periods of winter arousal.

Van Den Brink (1968) wrote that *Muscardinus* sometimes lives in small colonies. It is not as vocal as *Glis,* but does produce soft chirping and whistling sounds. Ognev (1963) stated that young had been found from June to mid-October, that there are probably at least 2 litters per year, and that litter size is 3–5 young. In a study in Moravia, Gaisler, Holas, and Homolka (1976) learned that there was usually only 1 annual

litter, but that occasionally there were 2; that average litter size was 4.7 young; and that the young normally bred after their first hibernation, but rarely bred in their first summer of life at an age of only 2½ months. According to Corbet and Southern (1977), birth peaks occur in Europe from June to early July and from late July to early August, the gestation period is 22 to 24 days, litters contain two to seven young, the eyes open after 18 days, independence is attained at 40 days of age, and maximum known longevity is 4 years in the wild and 6 years in captivity.

RODENTIA; GLIRIDAE; *Genus ELIOMYS Wagner, 1843*

Garden Dormouse

The single species, *E. quercinus,* occurs from France and Spain to the Ural Mountains, in northern Africa and the lands at the eastern end of the Mediterranean Sea, and on Sicily and most other islands of the western Mediterranean (Corbet 1978). Harrison (1972) considered the populations of *Eliomys* in Asia and Africa to represent a distinct species with the name *E. melanurus.*

Garden dormouse (*Eliomys quercinus*): A. Photo by Eviatar Nevo; B. Individual with extra fatty tissue in preparation for hibernation, photo from Archives of Zoological Garden Berlin-West.

Head and body length is 100 to 175 mm, tail length is 90 to 135 mm, and weight is 45 to 120 grams (Van Den Brink 1968; Grzimek 1975). The pelage is short except at the tip of the tail, where it is long and forms a tuft. The general coloration of the upper parts ranges through several gray and brown shades. The underparts are creamy or white. There are usually some black markings on the face. In European populations of *Eliomys* the tail is distinctly tricolored dorsally, having a cinnamon brown proximal half, a broad black preterminal band, and a white tip (Ognev 1963). In Asian and African populations the proximal third of the tail is brownish white, and the remainder of the tail is usually black (Harrison 1972). Females have eight mammae.

The term *garden dormouse* is not fully appropriate, as *Eliomys* is generally found in extensive forests in Europe (Ognev 1963; Grzimek 1975). It also occurs in a variety of other habitats including swamps, rocky areas, and cultivated fields. In the Middle East it has been found from the steppe desert of Hejaz to the high mountains of Lebanon (Harrison 1972). It shelters in such diverse places as hollow trees, branches of shrubs, crevices among rocks, and foresters' huts. A bird or squirrel nest may be used as a foundation for a shelter, but when *Eliomys* builds a complete nest it is made of leaves and grass, globular in shape, compact, and usually 0.8 to 3.0 meters above the ground. In some areas *Eliomys* is highly arboreal and is reported to be extremely graceful and agile (Ognev 1963), but in other places it occurs where there are no trees at all (Harrison 1972). It is apparently most active by night. In some areas it is known to gain weight in the fall and to become dormant during the coldest part of winter. The diet consists of acorns, nuts, fruit, insects, small rodents, and young birds. Ognev (1963) suggested that *Eliomys* is primarily a predator.

According to Grzimek (1975), large numbers of individuals may live in close proximity and share sleeping and feeding sites. Except during the mating season there is no fighting, even when two unrelated groups come together. *Eliomys* is very noisy and has a variety of calls (Ognev 1963). Females are thought to be polyestrous and to breed from May to October, at least in Europe. The gestation period is about 22–28 days and litter size is two to seven young. The eyes of the newborn open after about 21 days. A captive specimen lived for 5 years and 6 months (Marvin L. Jones, Zoological Society of San Diego, pers. comm.).

RODENTIA; GLIRIDAE; **Genus *DRYOMYS* Thomas, 1906**

Forest Dormice

There are two species (Corbet 1978):

D. nitedula, central Europe to Sinkiang and Iran;
D. laniger, known only from the type locality in southwestern Turkey.

Head and body length is 80 to 130 mm, tail length is 60 to 113 mm, and weight is 18 to 34 grams (Ognev 1963; Van Den Brink 1968; Gaisler, Holas, and Homolka 1976). The general coloration is grayish brown to yellowish brown on the upper parts, and buffy white on the underparts. *Dryomys* is similar to *Eliomys*, but is smaller, has a skull with much smaller tympanic bullae and a more rounded braincase, and has a more uniform tail, which is flattened and moderately bushy. Females have eight mammae.

Forest dormice inhabit dense forests and thickets at elevations of up to 3,500 meters; they sometimes utilize cultivated areas, gardens, and rocky meadows (Grzimek 1975). Their nests are usually located in dense shrubbery or the lower branches of trees. Hibernation sites may be in hollow trees, among tree roots, or in underground burrows. The temporary nest of an individual animal is rather flimsy, but natal nests

Forest dormouse (*Dryomys nitedula*), photo by R. Pucholt.

are solidly constructed (Ognev 1963). Studies in Israel showed that nests are located 1 to 7, usually about 3, meters above the ground in the branches of trees. They are globular, measure 150–250 mm in diameter, have an outer layer of twigs and leaves and an inner lining of bark or moss fragments, and have an entrance hole on the side or top (Harrison 1972).

Forest dormice are nocturnal and arboreal. They climb with great agility, and their leaps from branch to branch cover up to two meters. In Israel they are active all year, even in the high mountains, though they may undergo torpor for some hours during winter days (Harrison 1972). Farther north the data are conflicting. Evidently, hibernation does sometimes occur from October to April in Europe. During this period the animals curl up like a ball while sitting on the hind legs; the tail is wrapped around the body and the hands are pressed onto the cheeks. There may be occasional emergence to eat from stores of food. Observations in the Soviet Union, however, suggest that extensive, deep hibernation does not necessarily occur, and that forest dormice may be active throughout most or all of the winter (Ognev 1963). The diet of *Dryomys* consists of seeds, acorns, buds, fruits, arthropods, eggs, and young birds; animal matter seems to be preferred during the summer (Grzimek 1975).

Nests tend to be clustered in small groups in the same tree or adjacent trees (Harrison 1972). In one area of about 8,000 sq meters, 11 inhabited nests were found (Ognev 1963). *Dryomys* has a variety of vocalizations, most notably a delicate, melodious squeak that serves as an alarm call. The breeding season extends from March to December in Israel, where each female gives birth two or three times annually, and from May to August in Europe, where there usually seems to be just a single litter per year (Harrison 1972; Grzimek 1975; Gaisler, Holas, and Homolka 1976; Ognev 1963). Gestation lasts at least a month. There are usually two to five, and occasionally up to seven, young. The young weigh about two grams at birth, open their eyes at 16 days, and attain independence after 4 or 5 weeks. In Europe they do not mate until after their first winter.

Forest dormice are quite aggressive and never really become tame in captivity. They may be brought to a point at which they will allow themselves to be petted, but they usually bite with their sharp incisors when an attempt is made to hold them. If disturbed while resting, they often lie on their back or side and scratch with their hind legs. If disturbed further, they may suddenly leap high into the air and spit and hiss. Wild populations sometimes cause local damage by raiding fruit orchards and gnawing the bark of coniferous trees.

RODENTIA; GLIRIDAE; *Genus GLIRULUS Thomas, 1906*

Japanese Dormouse

The single species, *G. japonicus,* is found on the Japanese islands of Honshu, Shikoku, and Kyushu (Corbet 1978).

Head and body length is 65 to 80 mm, and tail length is 40 to 55 mm. The coloration is pale olive brown with a dark brown to black dorsal stripe. This stripe varies in width and sometimes is very obscure. The fur is soft and thick. A tuft of long hairs lies in front of the ear, and the tail is flattened from top to bottom.

This dormouse inhabits mountain forests, usually from about 400 to 1,800 meters in elevation. One individual, however, was captured in a cottage at 2,900 meters. *Glirulus* is arboreal and nocturnal. It shelters in a tree hollow or in a nest among branches. The round nest is covered on the outside with lichens, and lined on the inside with bark. Winter hibernation is spent in hollow trees, cottages, or bird houses. A semidormant individual was once captured in July in a snow depression in a ravine of the Japanese Alps, but after several minutes it awakened and escaped from the container where it had been placed. The diet of *Glirulus* consists of seeds, fruits, insects, and birds' eggs. In captivity it does well on rice, peanuts, sweet potatoes, fruit, and insects.

Japanese dormouse (*Glirulus japonicus*), photo by Michi Nomura.

The young are usually born in June or July. Occasional births in October probably represent the second litters of certain females. There may rarely be up to seven young, but observations of captives by Michi Nomura (Oiso-Machi, Kanagawa Pref., Japan, pers. comm.) indicate the usual range to be three to five and the average to be four. She also determined the gestation period to be about a month.

RODENTIA; GLIRIDAE; *Genus MYOMIMUS Ognev, 1924*

Mouselike Dormouse

Corbet (1978) recognized only a single species, *M. personatus*, which occurs in southeastern Bulgaria, Thrace, western Turkey, and extreme southwestern Turkmen near the shores of the Caspian Sea, and which also is known from subfossil material in southern Asia Minor and Palestine. Rossolimo (1976a, 1976b), however, considered the Bulgarian populations to represent a distinct species, *M. bulgaricus*, and described a new species, *M. setzeri*, from northwestern Iran. Ognev (1963) suggested that the range of *M. personatus* may extend as far east as Afghanistan.

Head and body length is 61 to 112 mm, and tail length is 53 to 94 mm (Rossolimo 1976a, 1976b; Van Den Brink 1968). The general coloration of the upper parts is a closely mixed combination of ochraceous and gray. The underparts, insides of the limbs, and the feet are white. There is a sharply defined line of demarcation between the upper and lower parts. Un-like other dormice, which have rather bushy tails, *Myomimus* has a thinly haired, mouselike tail, covered with short, white hairs.

Ognev (1963) indicated that *Myomimus* is the only glirid that is not specialized for arboreal life, and that one specimen was caught among stones scattered amidst small bushes. Van Den Brink (1968) stated that *Myomimus* seems to live on and under the ground.

RODENTIA; GLIRIDAE; *Genus GRAPHIURUS Smuts, 1832*

African Dormice

There are nine species (Genest-Villard 1978b; Robbins and Schlitter 1981; Smithers 1971):

G. parvus, Mali and Sierra Leone to Nigeria, southern Ethiopia and Somalia to Rhodesia and probably Angola;
G. murinus, throughout Africa south of the Sahara;
G. lorraineus, Sierra Leone to Zaire;
G. christyi, southern Cameroon, eastern Zaire;
G. surdus, southern Cameroon, Equatorial Guinea;
G. crassicaudatus, Liberia to Cameroon, Fernando Poo;
G. platyops, southern Zaire, Zambia, Rhodesia, Namibia, Botswana, probably eastern South Africa;
G. hueti, Senegal to Central African Republic and Angola;
G. ocularis, western South Africa.

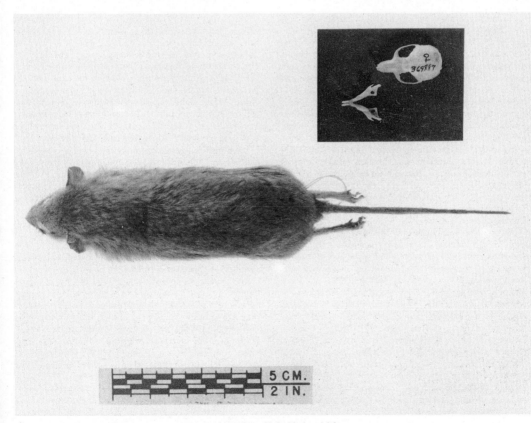

Mouselike dormouse (*Myomimus personatus*), photo from U.S. National Museum.

African dormouse (*Graphiurus murinus*), photo by Ernest P. Walker.

Head and body length is 70 to 165 mm, and tail length is 50 to 135 mm. The weight of *G. murinus* is 18 to 30 grams (Kingdon 1974*b*). The fur differs in texture; it is rather soft and dense in most forms, quite soft in a few, and slightly coarse in others. The general coloration of the upper parts ranges from pale ashy gray to dark slaty gray, and from buffy to reddish brown, tinged with grayish. The underparts are white to grayish, often tinged with buff or reddish brown. Occasionally the fur on the throat and chest is stained by plant or fruit juices. There are black and white markings on the face, variously arranged. The top of the tail is usually black or dark brown, and the bottom is whitish. In most forms the tail is well furred. As is the case with many crevice-dwelling mammals, the skull is flattened. Females have six or eight mammae.

African dormice inhabit forests, and, in southern and eastern Africa, rocky areas in the dry tablelands, generally along waterways. They are arboreal, but may frequently be found on the ground. They often shelter in trees and shrubs, making their globular nests in cavities and among the branches. Some animals use crevices in cliffs or stone fences, and others live in thatched roofs. They are sometimes found in boxes of rubble or in the upholstery of old furniture in houses and storerooms. Their occurrence in human habitations has declined through competition with the introduced *Rattus rattus* (Kingdon 1974*b*). They are principally nocturnal, but in dense, dark forests they are occasionally active by day. There is little precise information on hibernation, but apparently in certain areas *Graphiurus* becomes fat by the end of autumn and retires to a shelter, where it may remain dormant through the winter. The diet includes grains, seeds, nuts, fruits, insects, eggs, and small vertebrates. African dormice occasionally become a nuisance by raiding poultry yards.

According to Kingdon (1974*b*), *G. murinus* shows strong signs of territoriality in captivity, and males have been known to kill and eat one another, but in the wild as many as 11 adults of both sexes have been found in a single nest. Vocalizations include various twittering sounds and a surprisingly loud shriek. There is no conclusive evidence of reproductive seasonality, but there do appear to be breeding peaks. In Uganda many pregnant and lactating females were recorded in March, and one pregnant female was found in October. In Kenya five pregnant females were found in November, and in southwestern Tanzania a pregnant female was taken in January and a juvenile in March. Litters contain one to five young, each of which weighs about 3½ grams at birth. A captive specimen of *G. murinus* lived for five years and nine months (Marvin L. Jones, Zoological Society of San Diego, pers. comm.).

RODENTIA; *Family SELEVINIIDAE; Genus SELEVINIA Belosludov and Bashanov, 1939*

Desert Dormouse

The single known genus and species, *Selevinia betpakdalensis,* occurs in deserts to the west and north of Lake Balkhash in eastern Kazakh (Corbet 1978). The family Seleviniidae has a short but confusing taxonomic and nomenclatural history (see Arata 1967). No fossils referable to this family have been found.

Head and body length is 75 to 95 mm, and tail length is 58 to 77 mm (Ognev 1963). Two pregnant females weighed 21.4 and 24 grams, and a small male weighed 18 grams. The dense fur is grayish above and whitish below. The method of molt is peculiar; instead of a sloughing off of individual hairs, the epidermis becomes detached along with the hairs growing on it. A dense growth of new hair is found already in place in areas where patches of skin have fallen off. The molt begins at the back of the neck (between the ears) and then proceeds along the back and sides; the entire process takes about 1 month. The new hairs come in quickly, growing up to 1 mm in a 24-hour period. In winter the individual hairs attain lengths of 10 mm. The tail is covered with short hairs and the scales are not visible. The palms and soles are naked.

It has been suggested that *Selevinia* is a highly modified and aberrant dormouse. It has a round, stocky body and a long tail. The hand has four digits and the foot has five. The external ears extend beyond the fur. The skull has enormous tympanic bullae and relatively weak zygomatic arches. The dental formula is usually given as: (i 1/1, c 0/0, pm 0/0, m 3/3) $\times$ 2 = 16. However, there are also two upper premolars that are lost early in life (Arata 1967). The upper incisor teeth are massive and have a deep groove on their front surface. The cheek teeth are small and single rooted, and have a relatively simple enamel pattern.

Selevinia has a sporadic distribution in clay and sandy deserts. It occurs among thickets of spirianthus, in growths of wormwood, and among boyalych (*Salsola laricifolia*). It may live in burrows under the bushes, but a captive dug a burrow (28 cm long) only when the temperature was low. At other times this individual sheltered under a leaf or small stone. Some observers have reported *Selevinia* to be diurnal, but most individuals come out at twilight and remain active throughout the night, thus avoiding the heat. The captive mentioned above, when brought out into the sunlight for five minutes at midday in March, had its ear muscles so badly burned that it almost died.

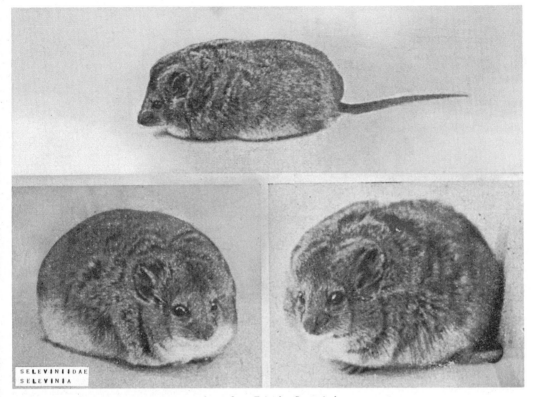

Desert dormouse (*Selevinia betpakdalensis*), photos from *Zeitschr. Saugetierk.*

When undisturbed, the desert dormouse travels at a leisurely, ambling gait, but when alarmed it progresses by small leaps. It does not jump higher than about 20 cm, but is a good climber. It is quite active at moderate temperatures and may then venture far from its shelter. Signs of its presence are not extensive in a given area, and there have been suggestions that it leads a nomadic life in the summer. When temperatures fall below about 5° C, however, *Selevinia* goes into a state of dormancy, during which its breathing rate may decrease from 108 to 25 per minute. Although such behavior has been observed only in captivity, it is reasonable to assume that it takes place in nature. Such a period of lethargy might explain the apparent scarcity of *Selevinia* during the colder parts of the year.

The diet seems to consist solely of invertebrates such as insects and spiders. The intestinal tracts of insects, at least of mealworms, are not eaten. *Selevinia* eats up to three-fourths of its weight in a 24-hour period. About one-fourth of the insect food eaten by a captive was discharged as feces, mostly in the form of indigestible chitinous material. *Selevinia* readily drinks water in captivity.

The desert dormouse is quite gentle in captivity. It does not attempt to bite when caught and held in the hand. Peculiar chirping sounds are emitted when it is feeding or disturbed. Mating takes place in May and possibly again in July. Pregnant females with six or eight embryos have been reported.

The Union of Soviet Socialist Republics Ministry of Agriculture (1978) designated *S. betpakdalensis* as rare.

RODENTIA; *Family ZAPODIDAE*

Birch Mice and Jumping Mice

This family of 4 living genera and 11 species occurs from central Europe to China, and in much of Canada and the United States. There are two subfamilies: Sicistinae, with the single genus *Sicista*; and Zapodinae, with *Zapus, Eozapus,* and *Napaeozapus* (Arata 1967).

The Zapodidae are small, mouselike animals, modified for jumping. In the subfamily Zapodinae the hind limbs and feet are especially long, while in the Sicistinae the hindquarters are only slightly modified for jumping. Head and body length is 50 to 110 mm, tail length is 65 to 165 mm, and weight is normally 6 to 28 grams. Females have four pairs of mammae.

In the Zapodidae the neck vertebrae are not fused, the auditory bullae are relatively small, and the three central bones of the hind feet are not united as they are in most members of the family Dipodidae. The dental formula is: (i 1/1, c 0/0, pm 0–1/0, m 3/3) × 2 = 16 or 18. The upper incisors are grooved in the Zapodinae, but smooth in the Recent Sicistinae. The cheek teeth are low crowned in *Sicista,* but tend to be high crowned and cuspidate in the Zapodinae.

These scampering and jumping rodents inhabit thickets, forests, meadows, swamps, bogs, and sagebrush flats. When startled, members of the Zapodinae can jump as far as two meters. The long tail is used as a balancing organ during leaps. These rodents shelter under logs or in underground burrows, either dug by themselves or abandoned by other animals. Their rather inconspicuous burrows are not marked by a mound of earth. Runways are not generally made. Activity is mainly nocturnal. Zapodids gain weight in the fall and then become dormant for six to eight months. Their diet consists of berries, seeds, fungi, and small invertebrates. Food is not stored. They drink free water.

The geological range of this family is Oligocene to Recent in Europe, lower Pliocene to Recent in Asia, and early Miocene to Recent in North America.

RODENTIA; ZAPODIDAE; *Genus SICISTA* Gray, 1827

Birch Mice

There are six species (Corbet 1978; Sokolov, Baskevich, and Kovalskaya 1981):

S. subtilis, steppes from eastern Austria to Lake Baikal in south-central Siberia;

S. betulina, forests from Norway and Austria to the Ussuri region of southeastern Siberia;

S. napaea, northwestern Altai Mountains in south-central Siberia;

S. pseudonapaea, southern Altai Mountains;

S. concolor, northern Caucasus, Kashmir, Tien Shan Mountains of Central Asia, mountains of Kansu and Szechwan (central China), Sakhalin Island;

S. kluchorica, northern Caucasus.

Head and body length is 50 to 90 mm, tail length is 65 to 110 mm, hind foot length is 14 to 18 mm, and weight is 6 to 14 grams. The upper parts are light or dark brown to brownish yellow, and the underparts are paler, usually a lighter brown. *S. betulina* and *S. subtilis* differ from the other species in having a sharply defined black stripe down the middle of the back. *Sicista* is a small, mouselike genus, with a fairly long and semiprehensile tail, but without special elongation of the legs or feet.

Birch mice inhabit forests, thickets, moors, subalpine meadows, and steppes. They excavate shallow burrows, in which they build oval-shaped nests of dry grass and cut plant stems. They travel on the ground by leaping, and readily climb bushes and shrubs—the outer toes hold on to twigs and the tail curls around other branches for additional support. They are generally active at night. It has been suggested that *S. betulina* spends the summer in wet meadows and migrates to forests in the winter. For at least six months of the year *Sicista* hibernates in an underground burrow. The species *S. napaea* of the Altai Mountains is active only from mid-May to early September (Shubin and Suchkova 1975). There is a considerable weight loss during hibernation; an *S. betulina* weighing about 12 grams in the fall may weigh only about 6 grams in the spring. The diet includes seeds, berries, and insects.

The reproductive season of *S. napaea* in the Altai Mountains is late June to mid-August, gestation lasts about 20 days, and most females produce litters of 1 to 6 young (Shubin and Suchkova 1975). In the Karelian A.S.S.R. of the northwestern Soviet Union, Ivanter (1972) learned that during the summer each adult female *S. betulina* raised a single litter containing 3 to 11 young. Sexual maturity was attained after almost 1 year of life, and a few individuals lived for 3 or 4 years. In another study of *S. betulina,* in the Bialowieza National Park of Poland, females were found to be in estrus from May to about the end of June. There was one litter per

Birch Mouse (*Sicista betulina*), photo by Liselotte Dorfmüller.

year and each female evidently gave birth only twice in her lifetime. The gestation period was estimated to be 4 to 5 weeks, and the duration of parental care was 4 weeks. Sexual maturity was attained following the first winter hibernation, and maximum longevity was estimated at 40 months.

RODENTIA; ZAPODIDAE; *Genus ZAPUS* Coues, 1876

Jumping Mice

There are three species (Hall 1981; Hafner, Petersen, and Yates 1981):

Z. hudsonius, southern Alaska to Labrador and northern Georgia, isolated populations in mountains of Arizona and New Mexico;

Z. princeps, southern Yukon to northeastern South Dakota and the southwestern United States;

Z. trinotatus, southwestern British Columbia to northwest coast of California.

Head and body length is 75 to 110 mm, tail length is 108 to 165 mm, hind foot length is 28 to 35 mm, and summer weight is about 13 to 20 grams. The pelage is coarse. The back is grayish brown, the sides are yellowish brown, and the underparts are white. The posterior part of the body is much heavier than the forepart, and the hind limbs are much larger and more powerful than the forelimbs. Although cheek pouches have sometimes been reported in this genus, none are present. The upper incisor teeth are narrow and grooved in front.

Except as noted, information for the remainder of this account was taken from Whitaker (1972) and Banfield (1974). Jumping mice live in wooded areas, grassy fields, and alpine meadows. They are especially common in the thick vegetation bordering streams, ponds, and marshes. During the summer they build spherical nests of woven grass, about 100 mm in diameter. These are usually placed under a log, board, root, or clump of vegetation. The winter

Jumping mouse (*Zapus hudsonius*), photo by Karl H. Maslowski.

hibernation nest is made of grass and leaves, and is usually located in a burrow about 30 to 90 cm below the ground. These rodents do not normally progress by jumping, but crawl slowly on all fours or make short hops. When startled, however, they often take several leaps of up to 1 meter in length, using their powerful hind legs for propulsion and their long tails for balance. They are capable of climbing in bushes and are excellent swimmers, sometimes diving to depths of over 1 meter.

Jumping mice are primarily nocturnal, but are occasionally seen by day. They are evidently more nomadic than most small rodents, and during the summer may wander over nearly 1 km in search of moist habitat. They are among the most profound of mammalian hibernators, generally entering dormancy by late September or October and emerging in late April or early May. When in torpor they are rolled up in a ball, with the head tucked between the hind feet and the tail curled around the body. The normal body temperature is about 37° C, but drops to as low as 2° C during torpor. Over a period of about two weeks in the fall, prior to hibernation, *Zapus* accumulates fat, and its body weight may double, reaching as much as 37 grams. Seeds are generally the most significant component of the diet, but *Zapus* also eats fungus, nuts, berries, fruits, and insects, the last item being especially important in the spring. There is evidently no storage of food. Captives have been observed to drink water regularly.

Populations of *Z. hudsonius* fluctuate from year to year; there are usually about 5 to 20 individuals per ha., but density has been known to reach 48 per ha. Average individual home range in this species varies from about 0.15 to 1.10 ha., and is usually slightly larger in males. Mean home ranges of *Z. princeps* in the mountains of Colorado have been reported as follows: 0.31 ha. for males and 0.24 ha. for females (Myers 1969), and 0.17 ha. for males and 0.10 ha. for females (Stinson 1977). The latter authority found males to have overlapping home ranges and to be tolerant of one another, but females to have more exclusive territories. In general, jumping mice are considered solitary creatures, but are not antagonistic, even when strangers are placed together. Captives are docile and do not attempt to bite.

The reproductive season extends from about May to September, with most young being born from June to August. In some areas *Z. hudsonius* is known to produce at least two,

and occasionally three, litters per season, but *Z. princeps* and *Z. trinotatus* evidently have only a single annual litter. The gestation period is 17–21 days and litter size is 2–9, usually 4–6, young. The newborn weigh about 0.8 grams each and are naked and blind. During the fourth week of life their eyes open and they become fully furred, and by the end of that week they are weaned. Female *Z. hudsonius* that are born early in the breeding season may be able to produce their own litters in the same season, but most individuals do not mate until after their first hibernation. *Z. princeps* has a relatively long life span for a rodent, with some individuals surviving up to 4 years in the wild (Brown 1970). A captive *Z. hudsonius* lived for 5 years (Marvin L. Jones, Zoological Society of San Diego, pers. comm.).

RODENTIA; ZAPODIDAE; **Genus EOZAPUS** Preble, 1899

Chinese Jumping Mouse

The single species, *E. setchuanus*, occurs in central China, from western Szechwan to southern Kansu and southeastern Chinghai (Corbet 1978). *Eozapus* is sometimes considered only a subgenus of *Zapus*.

Head and body length is 80 to 100 mm, tail length is 100 to 150 mm, and hind foot length is 25 to 33 mm. The upper parts are tawny orangish. The underparts are white, with or without a median tawny or buffy stripe. The tail is dark above and white below. As in *Zapus* and *Napaeozapus*, the hind legs, hind feet, and tail of *Eozapus* are very long. *Eozapus* is distinguished from the American jumping mice by the pattern of its molar teeth, the dark streak down the middle of its breast and belly, and the white tip on its tail.

Only about a dozen specimens are known to have been collected. These are in the Paris Museum of Natural History, the Museum of the Academy of Sciences at Leningrad, the British Museum (Natural History), the U.S. National Museum of Natural History, the American Museum of Natural History, and the Academy of Sciences in Philadelphia. The specimens have come from high mountain elevations, and there is some indication that the preferred habitat of *Eozapus* is beside streams in cool forests.

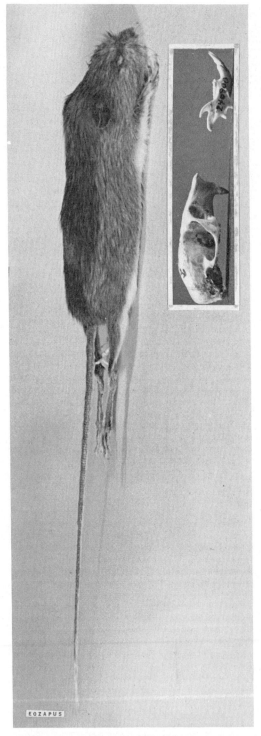

Chinese jumping mouse (*Eozapus setchuanus*), photos by Howard E. Uible (skin) and P. F. Wright (skull) of specimens in U.S. National Museum.

RODENTIA; ZAPODIDAE; *Genus NAPAEOZAPUS*
Preble, 1899

Woodland Jumping Mouse

The single species, *N. insignis*, ranges from southeastern Manitoba to Labrador and Pennsylvania, and south along the Appalachian Mountains to northern Georgia. The information in this account was taken entirely from Wrigley (1972) and Whitaker and Wrigley (1972).

Head and body length is 80 to 100 mm, tail length is 115 to 160 mm, hind foot length is 28 to 34 mm, and weight (without embryos or fat accumulated prior to hibernation) is 17 to 26 grams. The pelage is coarse, because of the stiff guard hairs, and has a tricolor pattern. The back is brown to black, the sides are orange with a yellow or red tint, and the underparts are white. This pattern probably affords camouflage protection by simulating dead vegetation. The tail is distinctly bicolored, grayish brown above and white below, and nearly always has a white tip. *Napaeozapus* differs from *Zapus* in having this white tail tip, and also in having three, rather than four, molariform teeth. As in *Zapus*, the upper incisors are grooved and females have four pairs of mammae.

Napaeozapus inhabits spruce-fir and hemlock-hardwood forests, within which it selects cool, moist places with dense vegetation. It is often found in bogs and swamps, or along streams, but also may occur far from free surface water. It shelters in piles of brush or in underground burrows that it either digs itself or takes over from other small animals. The entrance is kept concealed during the day. A globular nest is constructed from dry leaves and grass. *Napaeozapus* uses a quadrupedal walk when moving slowly, and a quadrupedal hop for greater speed. Such a hop can cover up to 1.8 meters, but normally is about 0.6–0.9 meters long and 0.3–0.6 meters high. *Napaeozapus* climbs well in bushes, but does not ascend trees. It can swim rapidly for short distances. It is mainly nocturnal, but may be active in the late morning or early evening. Most mature individuals enter a state of hibernation by late September, but most young begin in October. Emergence is usually in May, with males appearing about two weeks before females. There is a rapid accumulation of body fat in the several weeks before hibernation, and an adult that weighed 20 grams in the spring may exceed 30 grams by autumn. There is a weight loss of 30 to 35 percent during hibernation. Seeds, fungi, and insects are important components of the diet, and fruits, nuts, and other kinds of vegetation are also eaten. There is evidently little or no food storage in the wild.

Population densities of from 0.64 to 59.0 individuals per ha. have been reported, but the average in good habitat is about 7.5 per ha. Home ranges vary from 0.4 to 3.6 ha., and those of both sexes overlap. Individuals are highly tolerant of one another, and groups can be kept together in captivity without displays of aggression. *Napaeozapus* is usually silent, but may utter a soft clucking sound, or squeal if disturbed.

The reproductive season extends from early May to early September, with pronounced peaks in June and August. Many females give birth twice during this period, especially in more southerly populations, but many others have only a single annual litter. The gestation period is about 23 days. Litter size averages approximately 4.5 young and ranges from 2 to 7. The newborn are naked and blind, and weigh about 0.9 grams each. They are fully furred after 24 days of life, their eyes open at 26 days, and they are weaned at 34 days. Neither sex is able to breed until after the first hibernation. By the fall, about 70 percent of the animals in a given

Woodland jumping mouse (*Napaeozapus insignis*), photo by Ernest P. Walker.

population of *Napaeozapus* are young of the year. Many wild individuals, however, live for 2, 3, or even 4 years.

RODENTIA; *Family DIPODIDAE*

Jerboas

This family of 11 living genera and 29 species occurs throughout the arid part of the southern Palearctic region, from the Sahara, across southwestern and Central Asia, to the Gobi. There are three subfamilies: Dipodinae, with the genera *Dipus, Paradipus, Jaculus, Stylodipus, Allactaga, Alactagulus,* and *Pygeretmus;* Cardiocraniinae, with the genera *Cardiocranius, Salpingotus,* and *Salpingotulus;* and Euchoreutinae, with the single genus *Euchoreutes* (Corbet 1978; Pavlinov 1980).

Head and body length is 36 to 263 mm, tail length is 70 to 308 mm, and the extraordinarily great hind foot length is 18 to 98 mm. The pelage is generally satiny or velvety. The coloration is usually sandy or buffy, resembling the color of the ground in which these animals burrow. The tail is longer than the head and body and usually has a terminal tuft. The eyes, like those of most nocturnal animals, are relatively large. Either the external ears or the tympanic bullae, or both, are enlarged. The dental formula is: (i 1/1, c 0/0, pm 0–1/0, m 3/3) × 2 = 16 or 18. The cheek teeth are rooted, high crowned, and cuspidate.

The Dipodidae are characterized by their remarkable adaptations for jumping. The hind legs are at least four times longer than the front legs. Except in the subfamily Cardiocraniinae, the three central foot bones are fused to form a single cannon bone that gives great strength and support. The number of hind toes is sometimes reduced to three. If five hind digits are present, the first and fifth (the lateral digits) are so small that only the three central digits actually support the foot. In the species *Allactaga tetradactyla* there is only one lateral digit.

Jumping is probably an adaptation for escape from predators in open country. When moving rapidly, jerboas leap and spring with their hind legs, covering up to three meters in a single bound. The long tail serves as a balancing organ. Even when moving slowly, jerboas generally use only their hind legs, bringing them forward alternately in a bipedal walk. Jerboas sometimes, however, hop slowly, rabbitlike, on all four legs.

The Dipodidae are also characterized by their adaptations

for burrowing in the arid desert, semidesert, and steppe regions where they live. Jerboas living in sandy soils have tufts of bristly hairs under the digits and soles of the hind feet. These tufts act as friction pads and tend to support the animal on loose sand; they also aid in kicking the soil backward, after it has been accumulated by the forelimbs in burrowing. Another adaptation of jerboas that live in sandy areas is the tuft of bristly hairs around the opening of the external ear, which guards against the entrance of wind-blown sand. These ear and foot tufts are lacking, or are not so well developed, in jerboas living in loamy soils. At least in the genus *Jaculus*, a thickened fold of skin over the nose can be drawn forward to protect the nostrils, when the animal is pushing

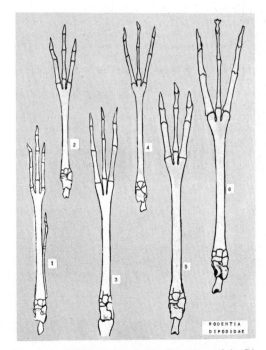

The bones of the specialized feet of members of the Dipodidae: 1. *Allactaga tetradactyla;* 2. *Stylodipus telum;* 3. *Dipus sagitta;* 4. *Jaculus lichtensteini;* 5. *J. orientalis;* 6. *Paradipus ctenodactylus;* photos from *Faune de l'URSS, Mammifères,* B. S. Vinogradov.

earth with its snout. A bony mechanism in front of the orbital cavity on the skull of some jerboas protects the eye, when the head is used in digging.

In the late spring and summer, jerboas often plug their burrows during the day and sometimes after leaving at night. Thus, they keep the heat out and the moisture in, and maintain a suitable microclimate within the burrow during the hottest parts of the year. This is characteristic of many burrowing mammals, and is an essential survival mechanism in arid regions. The permanent burrows may have emergency exits—side tunnels ending at or near the surface—through which the jerboa ''bursts'' when threatened by a predator. Jerboas often lie on their sides, when sleeping in the burrow, so as to accommodate their long legs better. Most species become dormant during the winter. Dipodids feed on seeds, the succulent parts of plants, and insects. They apparently do not require free water in nature, but they drink readily in captivity.

When handled or disturbed, these animals sometimes emit grunting noises or shrill shrieks, but they are usually silent. Tapping sounds, made within the burrow by the hind legs, have been reported for some species. Except for females with dependent young, a burrow is generally inhabited by only one animal. Females often give birth to more than one litter per year.

The known geological range of the Dipodidae is Pliocene to Recent in Asia, and Pleistocene to Recent in Europe and Africa (Arata 1967).

RODENTIA; DIPODIDAE; *Genus DIPUS* Zimmermann, *1780*

Rough-legged or Northern Three-toed Jerboa

The single species, *D. sagitta,* is found from the northern Caucasus and northeastern Iran to Manchuria (Corbet 1978; Brown 1978).

Head and body length is 105 to 157 mm, tail length is 140 to 190 mm, and hind foot length is 60 to 70 mm. Weight is about 70 to 85 grams during the spring and summer, and about 90 to 110 grams in the fall, just before dormancy. The upper parts are apparently bright orangish mixed with black during the winter, and pale sandy buff during the summer. The hip stripe, underparts, and tail tuft are white. There are three digits on each hind foot and a stiff brush of hairs on each toe. The upper incisors are grooved and yellow. Females have eight mammae (Ognev 1963).

The rough-legged jerboa inhabits sandy areas with a covering of shrubs, pines, or other vegetation. It lives in burrows, usually located in the larger sand hillocks. Three main

kinds of burrows have been reported (Ognev 1963). In the spring, following hibernation, *Dipus* excavates a permanent burrow about 40 to 100 cm below the surface. The burrow includes a nest chamber in a side passage, a refuge tunnel extending below the nest, a regular entrance passage, and one or more emergency exits that extend to just below the surface. The entrance is kept sealed during the summer to maintain a suitably cool and moist microclimate, but is left open in the spring and fall. There is a characteristic mound of excavated soil at the entrance. The nest is a loose structure made of dry vegetation. The recorded average length of burrow passageways is 319 cm for males and 480 cm for females. The females use such burrows to rear their young. In addition to this permanent home, individuals dig holes for temporary shelter during surface movements. These tunnels are nearly straight and extend only 50 to 60 cm into the ground at an angle of 10° to 30°. The third kind of burrow is used for winter hibernation; it consists of a sealed tunnel leading to a nest chamber about 80–100 cm below the surface. Instead of leaving a mound at the entrance of its hibernation burrow, *Dipus* scatters and tramples the excavated soil. The rough-legged jerboa is a rapid digger, tunneling through as much as 125 cm of soil in 10 to 15 minutes (Ognev 1963). Naumov and Lobachev (1975) noted that the burrows of *Dipus* may be up to 150 cm deep and contain three to five separate chambers.

Dipus is nocturnal and terrestrial, but is able to climb bushes. Its bipedal hops normally cover 10–15 cm, but extend for 120–40 cm if the animal is disturbed (Ognev 1963). All individuals in a given area tend to leave their burrows at about the same time each evening and travel to the feeding areas, which may be several hundred meters away. On colder nights, activity may last less than an hour. *Dipus* generally hibernates from November to March, but in the Kyzyl Kum Desert of Uzbek the period of dormancy lasts only 36–60 days (Naumov and Lobachev 1975). In certain areas there may be no hibernation at all (Ognev 1963). The diet of *Dipus* includes all parts of vegetation (roots, stems, seeds, flowers, and fruits); insects may be important in some areas. *Dipus* relishes plants with milky juices; it uses its sense of smell in digging for subsurface sprouts and the insect grubs in gallnuts on the underground parts of certain plants.

Population density generally peaks in July or August (Naumov and Lobachev 1975). An average of 7 to 9 and a maximum of 21 individuals per ha. have been reported at that time. Most of these animals die by autumn, however, and spring density is only 2 to 3 per ha. Individuals spend most of their time in a home range of two to three hectares, but may shift their range several times during a summer. According to Ognev (1963), there is only a single adult per burrow and males fight savagely among themselves during the breeding season.

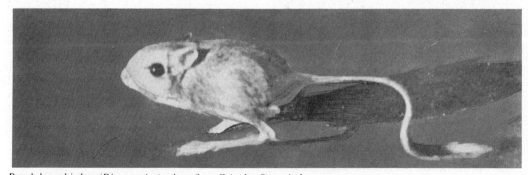

Rough-legged jerboa (*Dipus sagitta*), photo from *Zeitschr. Saugetierk.*

In some areas, reproduction continues from February to October, and up to three litters may be produced by each female, but mating usually does not begin until March. There is generally an intensive breeding peak from March to May and a lesser peak from July to August. The gestation period is 25–30 days. Litters contain one to eight young, the average being about three or four. A female born during the first breeding peak may give birth to its own litter during the second period (Naumov and Lobachev 1975).

RODENTIA; DIPODIDAE; *Genus PARADIPUS* Vinogradov, 1930

Comb-toed Jerboa

The single species, *P. ctenodactylus,* occurs from the Syr Darya River in south-central Kazakh to the shores of the Caspian Sea in southwestern Turkmen (Corbet 1978). The remainder of this account is based largely on works by Ognev (1963), and Naumov and Lobachev (1975).

Head and body length is 110 to 155 mm, tail length is 206 to 221 mm, and hind foot length is 73 to 82 mm. The upper parts are hazel to pinkish cinnamon, and the underparts are white. There is a broad white zone around each eye, and there are rusty yellow patches on each cheek and the chest. The long tail has a large, white terminal tuft. The length of the ears, about 33 to 38 mm, exceeds that of most other jerboas. The upper incisors are white and ungrooved. There are no premolar teeth.

The hind feet of *Paradipus* have three digits, the medial one being noticeably the longest. Along each side of each hind digit is a comb of stiff hairs. These bristles are longest on the middle toe and on the internal edges of the two lateral toes. This arrangement is thought to assist *Paradipus* in moving about on a sandy surface. The forelimbs lack such hairy digits, but have prominent claws, up to 7 mm long, that are employed in burrowing.

The comb-toed jerboa inhabits sandy deserts with a sparse covering of shrubby vegetation. Its relatively simple burrow extends down at nearly a right angle to the slope of a sand dune. The entrance tunnel is about 10 cm in diameter and is not kept sealed. There are usually no side passages or emergency exits. The nest chambers of summer burrows are located about 1.5 to 2.5 meters below the surface. Some summer burrows are occupied for only a day, but others are used for up to 30 days. During winter hibernation the burrow is located on the protected side of a dune, and the nest chamber is lined with vegetation and is often 4 to 5 meters deep.

Paradipus is nocturnal and basically terrestrial, but it can climb bushes and small trees. It is reportedly among the fastest of jerboas, being able to cover up to 3 meters in a single leap and 180 meters per minute. Individuals travel from 7 to 11 km per night, depending on the distance to feeding areas. Hibernation occurs from early December to mid-February, a period when food supplies are low. The diet consists entirely of desert vegetation, and all parts of plants are utilized—shoots, stems, buds, seeds, flowers, and fruits.

Each individual uses a home range with a radius of 1.5 to 2.0 km, but there is considerable overlap of ranges. The observed number of burrows and tracks indicates that *Paradipus* is common in the area it inhabits. A characteristic tapping sound of two quick beats and then a long pause has been reported to occur when an animal is in its burrow. There are apparently two periods of reproduction, April–May and July. The recorded number of embryos per pregnant female is two to six.

RODENTIA; DIPODIDAE; *Genus JACULUS* Erxleben, 1777

Desert Jerboas

There are two subgenera and five species (Corbet 1978; Ellerman and Morrison-Scott 1966; Hassinger 1973; Harrison 1972):

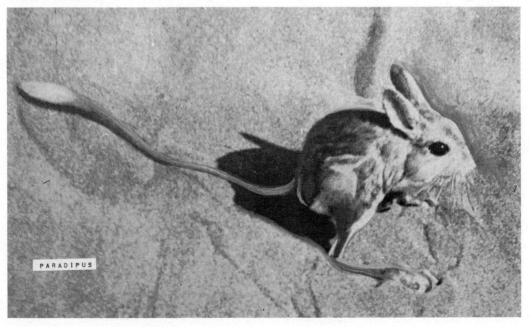

Comb-toed jerboa (*Paradipus ctenodactylus*), photo from *Zeitschr. Saugetierk.*

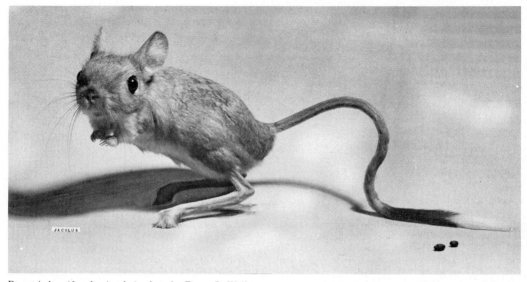

Desert jerboa (*Jaculus jaculus*), photo by Ernest P. Walker.

subgenus *Eremodipus*

J. lichtensteini, southern Soviet Central Asia, from the southeastern shore of the Caspian Sea to the area south of Lake Balkhash;

subgenus *Jaculus*

J. jaculus, desert and semidesert areas from Morocco and Mauritania to southwestern Iran and Somalia;

J. blanfordi, southern and eastern Iran, southern and western Afghanistan, southwestern Pakistan;

J. orientalis, Morocco to southern Israel;

J. turcmenicus, southern Soviet Central Asia, from the southeastern shore of the Caspian Sea to the Kyzyl Kum Desert.

Eremodipus Vinogradov, 1930 was treated as a distinct genus by Naumov and Lobachev (1975). Ranck (1968) considered *J. jaculus* to comprise two sibling species, *J. jaculus* and *J. deserti*, but Harrison (1978) rejected this view.

Head and body length is 95 to 160 mm, tail length is 128 to 250 mm, and hind foot length is 50 to 75 mm. Osborn and Helmy (1980) listed average weights of 55 grams for *J. jaculus*, the smallest species in the genus, and 134 grams for *J. orientalis*, the largest species. The upper parts are pale to dark sandy or buffy, the underparts are whitish, and there is a whitish stripe on the hip. Each hind foot has three toes, and there is a cushion or fringe of hairs under the toes. The eyes and ears are relatively large, the upper incisor teeth are grooved, and there are no premolars. Females have eight mammae (Harrison 1972).

These jerboas are found in a variety of habitats, including both rolling and relatively flat sandy deserts, saline deserts, rocky valleys, and meadows. Osborn and Helmy (1980) stated that the burrows of *J. jaculus* and *J. orientalis* are generally excavated in hard ground, are kept sealed during summer and left open in winter, slant to a depth of about 1 or 2 meters, and include a sleeping chamber with a nest of camel's hair or shredded vegetation, one or two escape tunnels, and, in the case of *J. orientalis*, a food storage chamber. According to Harrison (1972), the burrow of *J. jaculus* descends in a counterclockwise spiral to a depth of about 1.2

meters, where the nest is placed; several additional entrances radiate from the main burrow.

These jerboas progress along the ground with leaps that may measure several meters in length. Captives were observed to jump from a standing position to a height of almost a meter. When standing, the long tail is held curved and the tuft touches the ground, thus providing support. *Jaculus* is exclusively nocturnal and may move over a considerable distance in a single night; the track of one individual was reportedly followed for 14 km (Osborn and Helmy 1980). Wet weather is thought to cause reduced activity, but there is conflicting information on winter dormancy. According to Harrison (1972), *J. jaculus* remains active in Iraq on even the coldest nights. Nonetheless, hibernation has been found for some individual *J. jaculus* in Iraq (Hatt 1959), and for *J. orientalis* in Morocco (El Hilali and Veillat 1975). It is also apparent that *Jaculus* retires to its burrow and estivates during long, hot, dry periods. The diet of *Jaculus* includes roots, sprouts, seeds, grains, and cultivated vegetables (Osborn and Helmy 1980). Sufficient metabolic water is probably derived from vegetation, as captive animals have refused to drink for long periods, even though water was placed in their cages daily.

From 1 to 50 or more individual *J. orientalis* have been counted over a distance of 0.8 km. This species is evidently sociable, nearly always being found in small groups, but *J. jaculus* is strictly solitary (Osborn and Helmy 1980). Two or three desert jerboas have often been found sleeping together in a single nest. Five captive adults were quite gentle and slept together in huddled masses.

Captive female *J. jaculus* have given birth to litters of 3–4 young every 3 months, and have a gestation period of 25 days or less (Harrison 1972). Wild young of this species appear in May in northern Saudi Arabia, and a litter of 4 was found in Iraq in December. Osborn and Helmy (1980) wrote that reproductive periods in *J. jaculus* have lasted from February to September in Egypt and from October to February in Sudan, that litter size in this species is 2–10 young, and that *J. orientalis* evidently has a 5- to 6-month breeding season beginning in February and a mean litter size of 3. A gestation period of about 40 days has been reported for *J. orientalis*.

Naumov and Lobachev (1975) reported that *J. turcmenicus* may have three annual litters of 3–6 young each, in the spring, summer, and autumn, but that *J. lichtensteini* apparently has only one litter per year, the size of which is 2–8. A captive *J. jaculus* lived for 6 years and 3 months (Marvin L. Jones, Zoological Society of San Diego, pers. comm.).

J. orientalis sometimes raids sprouting barley and ripe grain in the fields of Bedouins, and in turn is captured and eaten by these people (Osborn and Helmy 1980).

RODENTIA; DIPODIDAE; *Genus STYLODIPUS G. M. Allen, 1925*

Thick-tailed Three-toed Jerboa

The single species, *S. telum,* is found from the Ukraine to Mongolia (Corbet 1978). The generic name *Scirtopoda* Brandt, 1843 is often used for this species.

Head and body length is 100 to 130 mm, tail length is 132 to 163 mm, and hind foot length is 45 to 60 mm. The upper parts are sandy or buffy, being darkened somewhat by a sprinkling of black-tipped and completely black hairs. The hairs along the sides of the body have white bases and bright buffy tips. The underparts, the backs of the feet, and the hip stripe are white. The tail is about the same color as the back, except that the base may be encircled by white; there is no distinct terminal tuft or white tip.

Each hind foot has three digits, the middle one being the longest. Each of these toes has a stout claw concealed by stiff hairs. The soles of the hind feet are also haired. The ears are relatively short. The incisor teeth are white and grooved. Females have eight mammae (Ognev 1963).

The remainder of this account is based largely on Naumov and Lobachev (1975). *Stylodipus* inhabits deserts and steppes, and occasionally has been reported in cultivated fields and pine forests. It generally requires compact soils. It excavates two kinds of burrows for summer use. Simple temporary holes are dug for one day's rest or for escape during the night. The permanent burrows are more complex, usually having a main entrance, emergency exits, and one or more chambers. Overall length of the passageways is 65 to 270 cm, and depth is 20 to 70 cm below the surface. The entrance is kept sealed by day. *Stylodipus* is generally nocturnal, with peaks of activity from about 2200 to 2400 hrs and at around 0300 hrs. It hibernates from September or October to mid-March. The diet consists of lichens, rhizomes, bulbs, seeds, and wheat.

Population density sometimes reaches 12 to 20 animals per ha. Individual home ranges are only 20 to 45 meters in diameter during the summer and do not overlap. Following its participation in reproductive activity, however, an individual may shift its range once or twice a month. The overall breeding season lasts from March to August, but it is not known whether females give birth more than once. The number of young per litter is 2–8, usually 3–5.

RODENTIA; DIPODIDAE; *Genus ALLACTAGA F. Cuvier, 1836*

Four- and Five-toed Jerboas

There are 2 subgenera and 11 species (Corbet 1978; Ellerman and Morrison-Scott 1966; Womochel 1978; Hassinger 1973; Roberts 1977; Sokolov 1981):

subgenus *Allactaga*

A. sibirica, Caspian Sea to Manchuria;
A. elater, lower Volga River and eastern Asia Minor to Sinkiang and western Pakistan;
A. euphratica, Asia Minor and Jordan to Afghanistan;

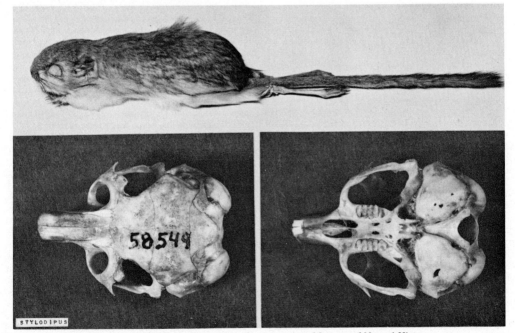

Thick-tailed three-toed jerboa (*Stylodipus telum*), photos from American Museum of Natural History.

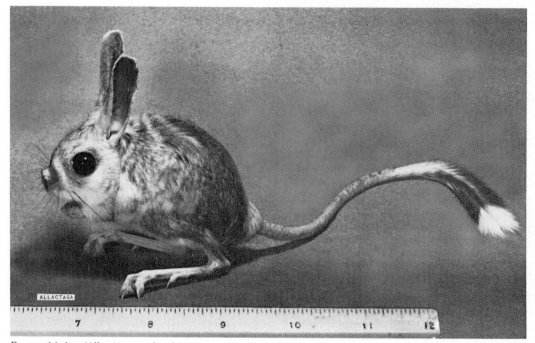

Four-toed jerboa (*Allactaga tetradactyla*), photo by Ernest P. Walker.

A. hotsoni, southern Afghanistan, southeastern Iran, southwestern Pakistan;

A. firouzi, southwestern Iran;

A. nataliae, Mongolia;

A. bullata, western and southern Mongolia and adjacent parts of China;

A. bobrinskii, deserts of Uzbek and Turkmen;

A. severtzovi, Soviet Central Asia;

A. major, Moscow and Ukraine to southwestern Siberia and Tien Shan Mountains;

subgenus *Scarturus*

A. tetradactyla, coastal plains of Libya and Egypt.

Head and body length is 90 to 263 mm, tail length is 142 to 308 mm, and hind foot length is 46 to 98 mm. Ognev (1963) listed weights of 44 to 73 grams for *A. elater,* the smallest species in the genus; 95 to 140 grams for *A. sibirica;* and 280 to 420 grams for *A. major,* the largest of all jerboas. Osborn and Helmy (1980) gave the average weight of *A. tetradactyla* as 52 grams. The coloration of the upper parts is mixed russet and black to sandy and grayish buff. The underparts are whitish, and there is a white stripe on the hip.

In all species except *A. tetradactyla* there are five digits on each hind foot. The two lateral toes are much smaller than the three central digits and do not contribute substantially to the support of the foot. *A. tetradactyla,* the four-toed jerboa, has only one small lateral digit. *A. bobrinskii* differs from other species in the Soviet Union in that the hind digits are covered beneath by a brush of hairs. The sole of the hind foot of *Allactaga* has a tuft of stiff hairs. The eyes are large, and the ears are long and slender, being about the same length as the head. There is one small premolar tooth on each side of the upper jaw. Females have eight mammae.

These jerboas inhabit deserts, semideserts, and steppes. At least some species prefer areas of compact soil with a thick cover of vegetation. *A. tetradactyla* occurs in coastal salt marshes and clay deserts; its burrows are simple, about 60 to

150 cm deep, and occupied only briefly (Osborn and Helmy 1980). *A. major* has been reported to construct four different kinds of burrows (Ognev 1963). Its permanent summer home consists of a main passage up to 3 meters long that leads to a chamber 65–150 cm below the surface. An extra escape hole is sometimes added. Only females actually construct a nest in the summer home. For temporary shelter on summer days, *A. major* mav dig a long, but simple tunnel, up to 13 meters in length and 20–30 cm deep. For emergency refuge at night it digs a blind tunnel, 80–150 cm long, straight into the ground. Finally, for winter hibernation, *A. major* establishes a nest chamber at a depth of 100–250 cm. The other species of *Allactaga* also excavate simple temporary shelters, as well as permanent homes that range from the simple to the complex (Naumov and Lobachev 1975). The burrow of *A. euphratica* is a relatively straight tunnel, about 120–70 cm long and extending 45–70 cm below the surface. The burrow of *A. severtzovi* is much longer and is especially deep and complex. Females of the latter species dig tunnels up to 5.3 meters long, and there may be several chambers, escape exits, and side branches. The small species *A. elater* burrows in hard soil, sometimes making several chambers at considerable depths. Nonetheless, its heaps of excavated soil are not much larger than those left by certain ground beetles. *A. elater* often digs in wagon tracks or other depressions in the ground, and the excavated soil may fill the depression and make it difficult to find the burrow. Unlike those of some other species of *Allactaga,* the occupied burrows of *A. elater* are often left unsealed (Ognev 1963).

Allactaga is primarily nocturnal, but some species are known to emerge from their burrows before sunset to begin foraging (Naumov and Lobachev 1975). When undisturbed, these jerboas move with a slow bipedal walk, but if startled they begin leaping in a zigzag pattern, covering a meter or more per jump (Ognev 1963). *A. elater* has been timed at a speed of 48 km/hr (Roberts 1977). In *A. euphratica,* one of the more southerly species, winter dormancy lasts only from late October to February, and some populations do not hibernate at all. In the other species that occur in the Soviet Union, hibernation generally extends from September or October to about March or April (Naumov and Lobachev 1975). Most species of *Allactaga* are primarily vegetarian. *A. major* is almost exclusively a seed eater, but also takes bulbs and some other plant matter. In contrast, *A. sibirica* is mainly insectivorous (Naumov and Lobachev 1975). Observations in Uzbek indicate that when *A. elater* initially emerges from hibernation it feeds mostly on vegetation, but that in May it begins to concentrate on insects (Pavlenko and Davaletshina 1971).

Naumov and Lobachev (1975) reported population densities of 1/ha. for *A. euphratica* in the Talish Mountains, and up to 7/ha. for *A. elater* in favorable habitat. Pavlenko and Davaletshina (1971) stated that the mean density of *A. elater* was not over 3/ha. in Uzbek, but Mazin (1975) gave the density of this species as 12–15/ha. in southeastern Kazakh. According to Ognev (1963), there is usually only one adult *A. elater* per burrow.

Extensive data compiled by Naumov and Lobachev (1975), Mazin (1975), Bekenov and Mirzabekov (1977), and Lay (1967), indicate that the breeding season of *A. elater* extends from March to as late as November, but that there are peak periods of reproductive activity that vary from place to place. There are generally two or three such peaks, corresponding with spring, summer, and fall. Many females evidently have three annual litters. Most other Soviet species also are thought to have lengthy breeding seasons, with several peak periods, and possibly two or three litters per year. There are usually about three to five young at a time, but overall recorded litter size in the Soviet Union is one to eight. The young of *A. elater* remain with the mother for 30–35 days and attain sexual maturity by 3½ months. Therefore, it is possible for a female of this species that was born early in the reproductive season to give birth to its own litter before the season ends. The young of *A. major* remain with the mother for at least 1½ months and apparently do not mate until the year after they are born.

According to Harrison (1972), a pregnant female *A. euphratica*, with nine embryos, was taken in Syria on 23 April. Several litters per year are apparently the rule in this area. Studies of captives indicate that the young of this species open their eyes after about two weeks, and that adult females will care for young other than their own. One captive *A.*

euphratica lived for four years and two months (Marvin L. Jones, Zoological Society of San Diego, pers. comm.).

Ognev (1963) wrote that *A. major* sometimes destroys melons and pumpkins, as it tries to get at the seeds. It also reportedly does damage to wheat, other grains, and rubber plants. Its skin is sometimes utilized by industry, but is too delicate to be of much value. The unique four-toed African species, *A. tetradactyla*, has a very restricted range and ecological tolerance, and may be threatened with extinction by desert reclamation projects (Osborn and Helmy 1980).

RODENTIA; DIPODIDAE; Genus ALACTAGULUS Nehring, 1897

Lesser Five-toed Jerboa, Little Earth Hare

The single species, *A. pumilio*, is found from the Don River and northeastern Iran to Inner Mongolia (Corbet 1978).

Head and body length is 88 to 125 mm, tail length is 107 to 185 mm, and hind foot length is 34 to 52 mm. Ognev (1963) listed weight as 44 to 61 grams. The back and proximal part of the tail are generally dull buffy mixed with black, the flanks and face are paler, and the underparts are white. The tail has a pronounced tuft that is mostly dark, but white at the tip. Each hind foot has five toes; the first and fifth are extremely short, but the three central digits are long. The toes are fringed with hair on the sides, but the soles are naked. *Alactagulus* differs from *Allactaga* in having shorter ears, no premolar teeth, and a simpler enamel pattern on the molars. Female *Alactagulus* have eight mammae.

The lesser five-toed jerboa inhabits steppes and clay and saline deserts. It is nocturnal and terrestrial. According to Naumov and Lobachev (1975), it builds both permanent and temporary burrows. The latter are used for emergency refuge and are excavated in soft soil with a cover of vegetation. The permanent burrows are usually located in hard, bare soil. The excavated earth is deposited in a characteristic, flat mound above the burrow. The initial passage of the permanent burrow may be up to 6 meters long, but it is sealed and not used after the entire burrow is complete. The initial tunnel usually divides into two branches, one leading downward to a nest

Lesser five-toed jerboa (*Alactagulus pumilio*), photo by F. Petter.

chamber and the other leading first to the side and then up to a regular entrance. There may also be a narrow airhole or vent leading to the surface. The main entrance is kept plugged with earth during the day. The chamber is lined with soft nesting material. The nests of breeding females are about 30–70 cm beneath the surface of the ground. Hibernation chambers contain no bedding material and may be more than 60 cm deep. In the region west of the Aral Sea hibernation lasts from November to February or March. The diet includes succulent vegetation, bulbs, roots, and rhizomes. Some food may be stored in the burrow.

Naumov and Lobachev (1975) reported a population density of about 10 individuals per ha. in the clay desert near the Aral Sea. The reproductive season extends from March to September. There are two peak breeding periods, one in the spring or early summer and one in the late summer or early fall. Some females have more than one annual litter. The number of young ranges from 1 to 7 and usually is 2 to 4. A female born in the spring may produce its own litter in the fall.

RODENTIA; DIPODIDAE; **Genus PYGERETMUS** Gloger, 1841

Fat-tailed Jerboas

There are two species (Corbet 1978):

P. platyurus, western and extreme eastern Kazakh;
P. shitkovi, eastern Kazakh in the area of Lake Balkhash.

In P. platyurus head and body length is 75 to 95 mm, tail length is 78 to 90 mm, and hind foot length is 30 to 35 mm. In P. shitkovi head and body length is 97 to 122 mm, tail length is 94 to 128 mm, and hind foot length is 38 to 45 mm. Coloration is sandy brown above and white below. The tail has no terminal tuft. Considerable subcutaneous fat deposits accumulate in the tail, often causing it to become nearly cylindrical and to be as much as 12 mm in diameter (Ognev 1963). Each hind foot has five digits, the first and fifth of which are extremely short and do not contribute substantially to the support of the foot. The ears, when pressed forward, do not reach the nose. There are no premolar teeth.

Fat-tailed jerboas inhabit clay and saline deserts and semi-deserts. They are nocturnal and terrestrial, and have been reported from areas of both scant and thick vegetation. Compared to most other jerboas, they seem to be poor jumpers and diggers (Ognev 1963). According to Naumov and Lobachev (1975), their burrows are simple, consisting of a passage about 20 cm long and 10–15 cm deep, and a chamber for both daily use and hibernation. Burrows seem to be changed frequently. Hibernation lasts from October to April. The diet includes green vegetation, bulbs, spiders, and insects. There is apparently only one reproductive season, with young having been collected in May and June. Litters of five and six young have been recorded.

RODENTIA; DIPODIDAE; **Genus CARDIOCRANIUS** Satunin, 1903

Five-toed Dwarf Jerboa

The single species, C. paradoxus, has been recorded from the area north of Lake Balkhash in Kazakh, the Tuva Autonomous Region of south-central Siberia, various parts of Mongolia, and the Nan Shan Mountains of northern China (Corbet 1978).

Fat-tailed jerboa (*Pygeretmus platyurus*), photo from *Traité de Zoologie* . . . , Pierre-P. Grasse. Inset: head (*P. shitkovi*), photo from *Faune de l'URSS, Mammifères*, B. S. Vinogradov.

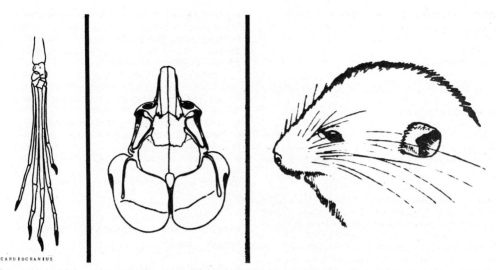

Five-toed dwarf jerboa (*Cardiocranius paradoxus*), photos from *Faune de l'URSS, Mammifères*, B. S. Vinogradov.

Head and body length is 50 to 75 mm, tail length is 70 to 78 mm, and hind foot length is 25 to 30 mm. The upper parts are grayish buff, the underparts are white, and the scantily haired tail is light brown above and white below. The tail is constricted at the base, but suddenly expands and then tapers toward the tip, where there is a small tuft of hairs.

In addition to its small size and unusual tail, *Cardiocranius* differs from most jerboas in several characters. The three central metatarsals of the hind foot are not fused into a single cannon bone. The hind foot has five separate toes. The outermost toe (digit 5) is only about 4 mm shorter than the fourth digit, but the innermost toe (digit 1) is very small and extends only about 8 mm beyond the base of the second digit. The sole of the hind foot has a tuft of bristly hairs. The external ears are remarkably small, but the tympanic bullae are enormously inflated. The skull has a heart-shaped appearance when viewed from above, hence the origin of the generic name. The upper incisor teeth are deeply grooved on the front surface. There is a small upper premolar on each side of the jar. Females have eight mammae (Ognev 1963).

According to Naumov and Lobachev (1975), *Cardiocranius* is a specialized genus, adapted to rocky deserts. It is nocturnal, foraging mainly from 2200 to 0400 hrs. It is a seed eater, and accumulates considerable body fat toward midsummer. Specimens in reproductive condition have been found in Kazakh in July.

Two specimens, taken in the northern Gobi Desert, "were dug out of the burrows which were found among sandhills covered with *Nitaria schoberi*. Both these specimens lived some time in captivity; they refused whatever vegetable food was offered to them. When they were put together into a box they fought very fiercely, this made it necessary to separate them" (Allen 1940:1084).

The Union of Soviet Socialist Republics Ministry of Agriculture (1978) listed *C. paradoxus* as a rare species.

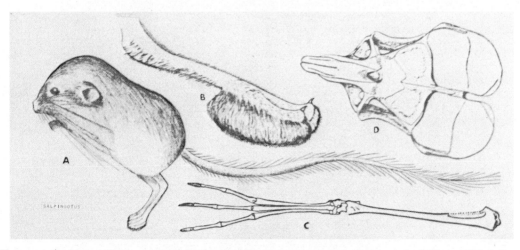

Three-toed dwarf jerboa (*Salpingotus kozlovi*), drawings by B. S. Vinogradov from *The Mammals of China and Mongolia*, Glover M. Allen.

RODENTIA; DIPODIDAE; *Genus SALPINGOTUS*
Vinogradov, 1922

Three-toed Dwarf Jerboas

There are four species (Corbet 1978; Hassinger 1973):

S. kozlovi, Gobi Desert of southern Mongolia;
S. crassicauda, desert areas from Aral Sea to Mongolia;
S. thomasi, known only by the type specimen from either
  Afghanistan or Tibet;
S. heptneri, known only from the type locality south of the
  Aral Sea.

Head and body length is about 41 to 57 mm, tail length is
93 to 126 mm, and hind foot length is 20 to 25 mm (Ognev
1963). The upper parts are sandy or buffy and the underparts
are pale yellowish. There is no terminal tuft on the tail of *S.
crassicauda;* a small, pale-colored tuft on the tail of *S.
kozlovi;* and a dense, black terminal tuft on the tail of *S.
heptneri.* In *S. crassicauda* and *S. thomasi* the proximal third
or half of the tail is swollen from fat accumulation under the
skin, but the tail of *S. kozlovi* is not expanded.

Each hind foot has only three toes, but the metatarsals are
not fused to form a cannon bone. There are tufts of hair
beneath the toes. The external ears are short, but the auditory
bullae are greatly inflated. The upper incisor teeth are not
grooved. There is a small premolar on each side of the upper
jaw. Females have eight mammae.

In Mongolia and the Soviet Union *S. crassicauda* inhabits
sand dunes overgrown with tamarisk, saxaul, and saltwort
(Ognev 1963). Its burrows are up to three meters in length. It
eats either animal (insects and arachnids) or vegetable food.
In the Zaysan Depression in extreme eastern Kazakh, there
are two consecutive breeding peaks in the period from May to
July. Average litter size is about 2.7 young. Sexual maturity
is not attained within the year of birth (Naumov and Lobachev 1975).The Union of Soviet Socialist Republics Ministry of Agriculture (1978) listed *S. crassicauda* as a rare
species.

RODENTIA; DIPODIDAE; *Genus SALPINGOTULUS*
Pavlinov, 1980

Baluchistan Pygmy Jerboa

The single species, *S. michaelis,* occurs in southwestern
Pakistan and possibly the adjacent part of Afghanistan

Baluchistan pygmy jerboa (*Salpingotulus michaelis*), photo by J. FitzGibbon.

(Roberts 1977; Hassinger 1973). This species was not discovered until 1966, and was placed in the genus *Salpingotus* until Pavlinov (1980) erected for it the new genus *Salpingotulus*.

Head and body length is 36 to 47 mm, tail length is 72 to 94 mm, and hind foot length is 18 to 19 mm. The upper parts are pale yellow ochre or sandy buff, the hind legs are pinkish, and the underparts are pure white. The body fur is soft and silky. The long tail is well haired and has a terminal tuft of dark hairs (Roberts 1977).

Salpingotulus resembles *Salpingotus*, but is generally smaller. Each hind foot has three toes and a brush of stiff hairs. The forelimbs have four digits each. The eyes are comparatively large. Females have only six mammae (Roberts 1977). The tail is not generally swollen from fat storage (Corbet 1978); only 2 of 15 specimens had fat accumulation in the tail.

According to Roberts (1977), *Salpingotulus* inhabits sand dunes, barren gravel flats, and sandy plains. It is exclusively nocturnal and resides in extensive, self-excavated burrow systems. Locomotion is by small, rapid hops on the hind limbs, and digging is accomplished with both the fore- and hind limbs. There evidently is no winter hibernation, but captives became torpid each day during the colder months, and could then be picked up and handled. The diet consists of grass seeds and stems and other vegetation. The genus is loosely colonial, there being a number of burrows in the same locality. Up to six adults will sleep huddled together. There seem to be two breeding periods: the first litter is born at the end of June, and the second is produced in August, almost immediately after the first is weaned. The usual litter size is two to four young.

RODENTIA; DIPODIDAE; *Genus EUCHOREUTES Sclater, 1891*

Long-eared Jerboa

The single species, *E. naso,* is known from western Sinkiang, north-central China, and extreme southern Mongolia (Corbet 1978).

Head and body length is 70 to 90 mm, tail length is 150 to 162 mm, and hind foot length is 40 to 46 mm. The upper parts are reddish yellow to pale russet, rather than sandy or buffy as in most jerboas. The underparts are white. The tail is mostly covered with short hairs, but has a terminal tuft. For most of its length the tail is colored like the body, but it becomes white at the beginning of the tuft, then black in the middle of the tuft, and then white again at the tip.

The large ears are about a third longer than the head. Each hind foot has five toes, the two lateral digits being much shorter than the three central digits. The central metatarsals are fused for only a part of their length. The toes are tufted beneath with bristly hairs. The incisor teeth are narrow and white. There is a small premolar on each side of the upper jaw. Females have eight mammae.

The habits and biology of *Euchoreutes* are not known, and there are only a few specimens in study collections. They have been taken in sandy valleys covered with low bushes of *Haloxylon ammodendron,* and sometimes in the huts of nomads. One long-eared jerboa was kept alive in captivity for four days; it had a strong tendency to bite.

Long-eared jerboa (*Euchoreutes naso*), photo from *Proc. Zool. Soc. London.*

RODENTIA; *Family HYSTRICIDAE*

Old World Porcupines

This family of 3 living genera and 11 species occurs in parts of southern Europe (possibly through introduction), all across southern Asia, on many islands of the East Indies, and throughout Africa. The sequence of genera presented here follows that of Van Weers (1979).

Head and body length is 350 to 930 mm, tail length is 25 to 260 mm, and weight is 1.5 to 30 kg. The head and body, and in some species the tail, are covered by spines or quills—thick, stiff, sharp hairs as much as 350 mm in length. The coloration is brownish or blackish, often with white bands on the hairs and spines.

These large rodents have a heavyset body, a tail not more than half as long as the head and body, and short limbs. There are five digits on the forefoot, the thumb is reduced, and there are five digits on the hind foot. The soles of the feet are smooth. The intestinal tract is approximately 790 mm in length, not including the stomach or cecum. The facial part of the skull is inflated by pneumatic cavities. The dental formula is: (i 1/1, c 0/0, pm 1/1, m 3/3) x 2 = 20. The four cheek teeth are flat, with a wavy enamel pattern. They are incompletely rooted, and moderately to strongly high crowned. Females have two or three pairs of mammae, which are located on the sides of the body.

Old World porcupines inhabit deserts, savannahs, and forests. They shelter in caves, crevices, holes dug by other animals, or burrows they have excavated themselves, and construct a nest of plant material within the den. They are mainly nocturnal and terrestrial, walking ponderously on the sole of the foot with the heel touching the ground. They run with a shuffling gait, or gallop clumsily when pursued. Except for *Trichys,* they rattle their spines, when moving about, and may also stamp their feet when alarmed. The diet consists mainly of vegetation, but carrion feeding and bone gnawing have been reported. Temperament, at least in captivity, differs by species and individuals, ranging from shy and nervous to docile. At least some species are moderately gregarious, and up to 10 individuals have been found in a single burrow. Litters usually contain one or two young, which are born with their eyes open and with soft, short quills. After a week or so, when the quills have hardened, the young may leave the nest with the mother. Old World porcupines have a relatively long life span and do well in zoos; captive members of all genera are known to have lived at least 10 years.

The geological range of the Hystricidae is Oligocene to Recent in Europe, middle Pliocene to Recent in Asia, and Pleistocene to Recent in Africa.

RODENTIA; HYSTRICIDAE; *Genus TRICHYS Günther, 1876*

Long-tailed Porcupine

The single species, *T. fasciculata,* is found on the Malay Peninsula, Sumatra, and Borneo (Van Weers 1976).

Head and body length is 350 to 480 mm, tail length is 175 to 230 mm, and weight is 1,750 to 2,250 grams. The body is covered with flattened, flexible, grooved spines of moderate length. Interspersed among these spines are a few longer, cylindrical, spinelike hairs. The head and underparts are hairy, and the underfur is rather woolly. The tail is scaly for most of its length, but its base is clothed like the back, and its tip has a tuft of flat, stiff bristles. The general coloration of the upper parts is brownish; the individual hairs have white bases and brown tips. The coloration becomes paler on the sides and finally blends into the white underparts.

Trichys is the most primitive of the Old World porcupines. With its relatively long, scaly tail and short spines it bears a superficial resemblance to *Rattus* (Grzimek 1975). Many specimens of *Trichys,* however, lack any appreciable tail, thus suggesting that this organ breaks easily from the body. *Trichys* cannot bristle or rattle its quills like most other porcupines. Its broad forefoot has four well-developed digits, each armed with a thick claw. The skull has well-marked postorbital processes (Van Weers 1979).

The long-tailed porcupine is a forest dweller. It is an agile climber and gets its food from the tops of trees and bushes (Grzimek 1975). It is reported to destroy pineapples in some areas. The tail seems to have value to some native people, who therefore sever it from the hide. A captive specimen was still alive after 10 years and 1 month (Marvin L. Jones, Zoological Society of San Diego, pers. comm.).

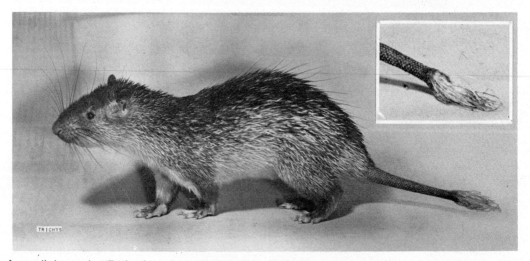

Long-tailed porcupine (*Trichys fasciculata*), photos by Ernest P. Walker.

A. West African brush-·tailed porcupine (*Atherurus africanus*), photos by Ernest P.Walker. B. Asiatic brush-tailed porcupine (*A. macrourus*), photo by Lim Boo Liat.

RODENTIA; HYSTRICIDAE; *Genus ATHERURUS*
F. Cuvier, 1829

Brush-tailed Porcupines

There are two species (Van Weers 1977; Lekagul and McNeely 1977; Chasen 1940; Meester and Setzer 1971):

A. macrourus, south-central China, Assam, Burma, Thailand, Indochina, Malay Peninsula and several small nearby islands, Hainan;
A. africanus, Gambia to western Kenya and southern Zaire.

Head and body length is 365 to 570 mm, tail length is 102 to 260 mm, and weight is 1.5 to 4.0 kg (Van Weers 1977; Kingdon 1974*b*; Medway 1978). The pelage is almost entirely spiny, though the spines are softer on the head, legs, and underparts. The middorsal spines are the longest. Most of the spines are flattened and have a groove on their dorsal surface. Interspersed among the flattened, grooved spines on the lower back are a few round, thick bristles. The basal part of the tail is clothed like the back, the middle part is scaly, and the terminal part has a thick tuft of bristles. The general coloration of the upper parts is grayish brown to blackish brown, with whitish tips on the individual hairs. The sides

are paler than the back, and the underparts are nearly white. The tail tuft is whitish to creamy buff.

Atherurus has a relatively long body; short, stout limbs; and short, rounded ears. The feet are partially webbed and are armed with blunt, straight claws. Postorbital processes are either lacking or very weak in the skull (Van Weers 1979). The tail is about one-fourth to a half as long as the head and body; it breaks easily and is often lost. Females have two pairs of lateral thoracic mammae (Weir 1974*b*).

These porcupines live in forests at elevations of up to 3,000 meters (Grzimek 1975). They are strictly nocturnal, sheltering by day in a hole among tree roots, a rocky crevice, a termite mound, a cave, or an eroded cavity along a stream bank. They evidently do not usually dig their own burrows, but do line the natal chamber with vegetation. They run swiftly and are basically terrestrial, but are capable of climbing, can jump to a height of over 1 meter, and can swim well (Kingdon 1974*b*; Lekagul and McNeely 1977). The diet consists mainly of green vegetation, bark, roots, tubers, and fruit, and may sometimes include cuitivated crops, insects, and carrion.

Except as noted, the information for the remainder of this account was taken from Kingdon (1974*b*) and applies to *A. africanus.* Small colonies of 6 to 8 individuals may share a burrow and a feeding territory. Group territories have been found to measure 0.18 to 5.00 ha., and to be regularly marked in certain spots with dung deposits. Evidently the groups represent a mated pair and their offspring of several litters. A number of such families, comprising up to 20 porcupines, may live in close proximity in a favorable area. Lekagul and McNeely (1977) indicated that *A. macrourus* also may live and move about in small groups.

There seems to be no definite breeding season, up to three litters may be born per year, and the gestation period is 100–110 days. Overall litter size is one to four young, but data compiled by several authorities (Rahm 1962; Asdell 1964; Lekagul and McNeely 1977) suggest that a single offspring is the usual case in both Africa and Asia. The newborn are well developed, weigh 150 grams each, and open their eyes within hours of birth. They are able to eat solid food after 2 or 3 weeks, but lactation continues until they are 2 months old. Adult weight is attained at the age of 2 years. A captive specimen of *A. africanus* lived for 22 years and 11 months (Snyder and Moore 1968).

RODENTIA; HYSTRICIDAE; *Genus HYSTRIX Linnaeus, 1758*

Old World Porcupines

There are three subgenera and eight species (Van Weers 1978, 1979; Meester and Setzer 1971; Corbet 1978; Kingdon 1974b):

subgenus *Thecurus*

H. pumila, Palawan, Busuanga, and Balabac islands (Philippines);
H. sumatrae, Sumatra;
H. crassispinis, Borneo;

subgenus *Acanthion*

H. brachyura, Nepal to eastern China and Malay Peninsula, Hainan, Sumatra, Borneo;

H. javanica, islands from Java to Flores;

subgenus *Hystrix* (crested porcupines)

H. indica, Asia Minor and Arabian Peninsula to Soviet Central Asia and India, Sri Lanka;
H. cristata, Italy, Sicily, possibly the Balkans, Mediterranean coast of Africa to northern Zaire and Tanzania;
H. africaeaustralis, southern half of Africa.

Thecurus Lyon, 1907 has often been treated as a distinct genus. Its head and body length is 420 to 665 mm, its tail length is 25 to 190 mm, and its weight is 3.8 to 5.4 kg (Van Weers 1978; Grzimek 1975). Its body is densely covered with flattened spines, each of which is deeply grooved longitudinally and increases in rigidity toward the tip. The spines are smaller along the tail and are more flexible on the underparts. Coarse, bristlelike hairs cover the feet. The back is dark brown to black, the sides are paler, and the underparts are buffy white. *Thecurus* resembles the other subgenera of *Hystrix* in having a relatively short tail with rattle quills. It resembles the subgenus *Acanthion* in that it lacks a well-developed crest and its quills have only one black band.

In the subgenus *Acanthion,* head and body length is 455 to 735 mm, tail length is 60 to 115 mm, and weight is about 8 kg (Van Weers 1979; Lekagul and McNeely 1977; Medway 1978). Quill structure and arrangement generally resemble those of the subgenus *Thecurus.* In the species *H. brachyura,* according to Lekagul and McNeely (1977), the front half of the body is covered with short, dark brown spines, while the hindquarters have long, pointed, whitish quills, usually with one distinct blackish ring. There is a short whitish crest on the neck and upper back. The short tail has both long, pointed quills and rattle quills.

In the subgenus *Hystrix,* head and body length is 600 to 930 mm, tail length is 80 to 170 mm, and weight is about 10 to 30 kg. The head, neck, shoulders, limbs, and underside of the body are covered with coarse, dark brown or black bristles. There are long quills along the head, nape, and back, and these qills can be raised into a crest. The sides and back half of the body are covered with stout, cylindrical quills, up to 350 mm long and mostly marked with alternating light and

Sumatran porcupine (*Hystrix sumatrae*), photo by Ernest P. Walker.

A. African porcupines (*Hystrix cristata*), mother and young 51 days old, photo by Ernest P. Walker. B. Malayan porcupine (*H. brachyura*), photo by Lim Boo Liat.

dark bands. There may be some longer, more slender, and more flexible quills, which are usually all white. The rattle quills of the short tail are better developed than those in other subgenera.

Hystrix is distinguished externally from the other genera of Old World porcupines by its relatively shorter tail and the presence of rattle quills. These qills are located at the end of the tail. They are slender for most of their length, but are of much greater diameter for about the terminal fifth. The expanded portion is hollow and thin walled, so several quills vibrating together produce a hisslike rattle.

The broad forefoot of *Hystrix* has four well-developed digits, each armed with a thick claw. The hind foot has five digits. The eyes and external ears are very small. The facial region of the skull is inflated by pneumatic cavities, and the nasal bones are enlarged; these characters are especially pronounced in the subgenus *Hystrix*. Female *H. cristata* have two or three pairs of lateral thoracic mammae (Weir 1974b).

These porcupines are highly adaptable, being found in all types of forests, plantations, rocky areas, mountain steppes, and sandhill deserts; the elevational range is sea level to 3,500 meters (Roberts 1977; Kingdon 1974b; Medway 1978). They shelter in caves, rock crevices, aardvark holes, or burrows they dig themselves. These burrows often have several entrances, are sometimes used for many years, and can become quite extensive. One was 18 meters in length, terminated in a chamber 1.5 meters below the surface, and had three escape holes. *Hystrix* is nocturnal and terrestrial. It does not usually climb trees, but does swim well. Its movement was described in Grzimek (1975) as "easy and graceful." Kingdon (1974b), however, wrote that the normal gait is a ponderous, plantigrade walk, with a trot or gallop used in alarm. Paths tend to be followed and a distance of up to 15 km per night may be covered in the search for food. According to Grzimek (1975), European animals sometimes remain in their holes through the winter, but do not truly hibernate. The diet of *Hystrix* includes bark, roots, tubers, rhizomes, bulbs, fallen fruit, and cultivated crops. Insects and small vertebrates are occasionally taken. Carrion feeding has been reported, but is not common. Bones are frequently found in and around burrows, probably having been carried there and gnawed to obtain calcium and to hone the incisor teeth.

Piping calls and a piglike grunt have been reported for *Hystrix*. Kingdon (1974b) stated that there is considerable grunting and quill rattling as the porcupines shuffle around by night. At the least encounter with another animal, they raise and fan their quills, thereby more than doubling their apparent size. If still bothered, they stamp their feet, whirr their quills, and finally charge backward, attempting to drive the thicker, shorter quills of the rump into the enemy. Lions, leopards, hyenas, and even humans are sometimes injured or killed in this manner.

Small family groups commonly share a burrow, but the female may establish a separate den in which to bear its young, and foraging is generally done alone, except when the parents are accompanying the young (Roberts 1977; Kingdon 1974b). Breeding has been reported throughout the year at the London Zoo (Asdell 1964), from March to December in Indian zoos (Prakash 1975), from July to December in Central Africa, and during the summer in South Africa (Kingdon 1974b). There may be two litters per year. Females, at least of the African species, have an estrous cycle of about 35 days and a gestation period of 112 days (Weir 1974b). Overall litter size in the genus is one to four, but there seem usually to be one or two offspring. They are born in a grass-lined chamber within the burrow and are well developed at birth. Old World porcupines have a remarkably long life span, with some individuals probably surviving for

12 to 15 years in the wild. A captive *H. pumila* lived 9 years and 6 months, and a captive *H. brachyura* lived for 27 years and 3 months (Marvin L. Jones, Zoological Society of San Diego, pers. comm.). The latter specimen apparently holds the longevity record for the order Rodentia.

There is controversy regarding the status of *H. cristata* in Europe. Corbet (1978) gave the range there as Italy, Sicily, Albania, and northern Greece, and noted that the species had perhaps been introduced to Europe. Smit and Van Wijngaarden (1976) stated that most authorities think that *H. cristata* was introduced to Italy and Sicily by the Romans, but they concluded that it probably does not occur in the Balkans. They noted that the species is rare and decreasing in numbers in the Mediterranean region, because it is killed by persons who consider it an agricultural pest or who use it for food. Osborn and Helmy (1980) wrote that in Egypt *H. cristata* is probably now found only around the cliffs north of Salum, if it has not already disappeared. Farther south in Africa, and in parts of Asia, *Hystrix* reportedly does much damage by gnawing the bark of trees on rubber plantations, and by eating corn, pumpkins, sweet potatoes, cassava, and young cotton plants. Because of this factor, and because its quills are used as ornaments and talismans, *Hystrix* has been exterminated in heavily settled parts of Uganda (Kingdon 1974b). It may also have disappeared on Singapore (Medway 1978).

RODENTIA; *Family ERETHIZONTIDAE*

New World Porcupines

This family of 4 living genera and 10 species is found from the arctic coast of North America to northern Mexico and the Appalachian Mountains, and from southern Mexico to Ecuador and northern Argentina. There are two subfamilies: Chaetomyinae, with the single genus *Chaetomys;* and Erethizontinae, with the genera *Echinoprocta, Coendou,* and *Erethizon* (Starrett 1967). The sequence of genera presented here is based on that of Cabrera (1961), and on data cited by Woods (1973), which indicate that *Coendou* is ancestral to *Erethizon.*

Head and body length is 300 to 860 mm, and tail length is 75 to 450 mm. Some of the hairs are modified into short, sharp spines with overlapping barbs. The spines vary in distribution, being primarily on the head, neck, and forelimbs in *Chaetomys;* primarily on the back, tail, and hindquarters in *Erethizon;* and over most of the body in *Coendou* and *Echinoprocta* (Starrett 1967). Long, wavy bristles are present on the back. The pelage is marked with blackish to brownish, yellowish, or whitish bands.

Like the Old World porcupines (Hystricidae), the New World porcupines are heavyset, relatively large rodents. In the Erethizontidae, however, the foot is modified for arboreal life: the sole is widened and, in some forms, the first digit on the hind foot is replaced by a broad, movable band. The functional digits have strong, curved claws. The limbs are fairly short. The tail is relatively short in *Echinoprocta* and *Erethizon,* but long in *Chaetomys* and *Coendou.* In *Erethizon* and *Coendou* the axis and third cervical vertebra are fused, as in the genus *Dinomys.* The dental formula is: (i 1/1, c 0/0, pm 1/1, m 3/3) x 2 = 20. The cheek teeth are rooted and have reentrant folds.

The geological range of the Erethizontidae is Oligocene to Recent in South America, and late Pliocene to Recent in North America.

Thin-spined porcupine (*Chaetomys subspinosus*), photo by Joao Moojen.

RODENTIA; ERETHIZONTIDAE; *Genus CHAETOMYS* Gray, 1843

Thin-spined Porcupine

The single species, *C. subspinosus*, is apparently restricted to a small part of southeastern Brazil, in the south of the state of Bahia and in the north of the state of Espirito Santo (Avila-Pires 1967; Coimbra-Filho 1972).

Head and body length is about 430 to 457 mm, and tail length is 255 to 280 mm. The pelage of the back is peculiar, as the hairs are more like bristles than spines. The hairs of the head, neck, and forelimbs, however, are spinelike and less flexible than those on other parts of the body. All the large, coarse hairs are cylindrical and wavy. The general coloration of the upper parts is dull brownish, or sometimes grayish white, and the underparts are slightly rufous brown. The feet and tail are brownish black. The fairly long tail is scaly throughout and only moderately enlarged in the basal part. The underside of the tail is covered with short, stiff hairs.

Although the tip of the tail is naked and resembles that of *Coendou*, most authorities agree that it is not prehensile. The hands and feet of *Chaetomys* each have four digits, all armed with long, curved claws. The skull is unique among rodents, as the orbit is almost ringed by bone, there being a broadened jugal and pronounced postorbital process of the frontal (Starrett 1967). The incisor teeth are narrow.

The thin-spined porcupine inhabits densely vegetated brush country and woodland around open savannahs and cultivated areas. It is known to live in the vicinity of cocoa trees, the nuts of which it eats. It generally moves slowly, but is a quick jumper and climber. It is nocturnal (Coimbra-Filho 1972). It utters hoarse sounds and is docile in captivity. It occurs in a limited area, where natural habitat is being destroyed or modified by deforestation, and is thus classified as rare by the IUCN (1974) and endangered by the USDI (1980).

RODENTIA; ERETHIZONTIDAE; *Genus ECHINOPROCTA* Gray, 1865

Upper Amazonian Porcupine

The single species, *E. rufescens*, is known from Colombia (Cabrera 1961).

The length of this animal, from the tip of its snout to the end of its tail, is about 500 mm. The tail is short, being about as long as the hind foot. The coloration of the back and sides is pale brown to blackish. The chin, throat, and undersides are pale brown, and the feet and tail are dark gray to black. There is a short white streak in the center of the nose, and a slight crest on the nape, formed by a few white spines. The spines gradually become thicker, stronger, and shorter from the head to the rump. On the posterior part of the back, above the tail, the spines are well developed, short, and thick. The tail is hairy and not prehensile. The whiskers are black, and the incisor teeth are slender and yellow.

The habits and biology of this porcupine are not well known. It is arboreal and appears to be fairly common around Bogota at elevations of 800 to 1,200 meters. Specimens are in the Paris Natural History Museum, the British Museum (Natural History), the American Museum of Natural History, and the La Salle Institute in Bogota.

RODENTIA; ERETHIZONTIDAE; *Genus COENDOU* Lacépède, 1799

Prehensile-tailed Porcupines, Coendous

The two subgenera and seven species (not in systematic order) are as follows (Cabrera 1961; Hall 1981; Husson 1978; Handley 1976; Goodwin and Greenhall 1961):

subgenus *Coendou*

C. bicolor, Panama to Bolivia;
C. mexicanus, east-central Mexico to western Panama;
C. pallidus, known only by two specimens purported to be from somewhere in the West Indies;
C. prehensilis, Venezuela, the Guianas, Brazil, Bolivia, Trinidad;

subgenus *Sphiggurus*

C. insidiosus, Surinam, Brazil;
C. spinosus, southeastern Brazil, Paraguay, northeastern Argentina;
C. vestitus, Colombia, western Venezuela.

Husson (1978) considered *Sphiggurus* a full genus and *C. villosus* of southeastern Brazil a species distinct from *C. insidiosus*.

Upper Amazonian porcupine (*Echinoprocta rufescens*), photo by Eugene Maliniak.

Head and body length is about 300 to 600 mm, tail length is 330 to 485 mm, and weight is about 0.9 to 5.0 kg. In the subgenus *Coendou* the body is clothed with short, thick spines. In *Sphiggurus* the spiny covering of the back is mixed with or covered by long, thick fur, and the covering of the chest and underparts is considerably softer than in *Coendou*. In some forms there is a woolly underfur. The general coloration of the upper parts varies considerably from light yellowish through shades of brown to almost black, and some forms are speckled. The underparts are usually grayish.

The long tail is prehensile and lacks spines, in contrast to the short, spine-clad tail of *Erethizon*. The upper surface of the terminal part of the tail in *Coendou* is naked and modified for direct contact in coiling about branches. The tip coils upward and has a callus pad on the hairless upper side near the tip. The hands and feet are highly specialized for climbing. There are four digits on each limb, each of which is armed with a long, curved claw. Alvarez del Toro (1967) reported that female *C. mexicanus* have 12 mammae.

Prehensile-tailed porcupines live mainly in forests, though sometimes they enter cultivated areas. In Peru they occur in both coastal and Amazonian areas, at elevations of 150 to 2,500 meters (Grimwood 1969). They are principally nocturnal and arboreal. All but one of the specimens taken in Veracruz by Hall and Dalquest (1963) were found in tall trees, 18

to 30 meters above the ground, and one was found at a height of 6 meters. Handley (1976) collected 91 percent of his specimens of *C. prehensilis* in trees, and the remainder on the ground. These porcupines are slow in their movements, but are sure-footed climbers, using their tails in conjunction with their hands and feet. They prefer to sleep in tangled vegetation among the treetops, but also shelter in hollow limbs, tree trunks, and shallow burrows. Several specimens of *C. mexicanus,* reported by Jones, Genoways, and Lawlor (1974), were dug from retreats in a rocky cave. The diet includes leaves, tender stems, fruits, blossoms, and roots. The bark of certain trees may be peeled away to reach the cambium layer (Starrett 1967).

In a study of *C. prehensilis* on the llanos of Venezuela, Montgomery and Lubin (1978) found individuals to move up to 700 meters per night, and to rest by day in trees, usually at heights of 6 to 10 meters. The home ranges of three radio-tracked animals were 8, 10, and 38 ha.

These porcupines are said to be pugnacious and to show no fear upon capture; they bite and try to hit an adversary with their spines. They also have been observed to stamp their hind feet when excited and to roll up in a ball if caught in the open (Starrett 1967). They frequently sit on their haunches and shake their spines by moving the skin. Both deep growls and plaintive cries have been reported.

Prehensile-tailed porcupines (*Coendou prehensilis*), photo by Ernest P. Walker.

In the Tuxtla Gutierrez Zoo in southern Mexico, Alvarez del Toro (1967) found that more than one male *C. mexicanus* could not be kept together, because of fighting, but that a male could live peacefully with two or three females. One pair bred three times in six years. The female never made a nest, and the young were born in branches provided in the cage. A single young is probably normal, and it is relatively large, about 250 mm long. It resembles the parents in shape and color, but it is hairier and has soft, short quills. The quills harden in about a week. The young of *C. prehensilis* also are large at birth, weighing from 280 to 426 grams.

An old source cited by Leopold (1959) indicated that the litter size of *C. mexicanus* is about four young, but this does not seem likely, considering the large size of the offspring.

The same source gave the gestation period as 60–70 days. Pregnant female *C. mexicanus*, with only one embryo, have been reported by Hall and Dalquest (1963) and Jones, Genoways, and Lawlor (1974), and a single young also is on record for *C. insidiosus* (Asdell 1964). Births and pregnant females of various species have been reported in January, February, March, May, July, August, and September. A captive *C. prehensilis* lived for 17 years and 4 months (Marvin L. Jones, Zoological Society of San Diego, pers. comm.).

Coendous occasionally raid plantations to feed on guavas, bananas, and corn (Starrett 1967). They are used as food in some parts of Peru, but are not regarded as endangered (Grimwood 1969).

Prehensile-tailed porcupine (*Coendou* sp.), baby, photo by R. B. Tesh.

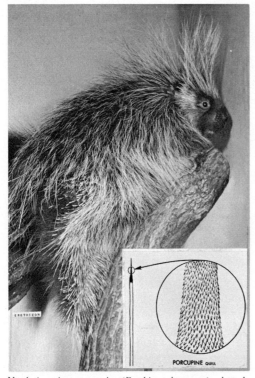

North American porcupine (*Erethizon dorsatum*), photo by Ernest P. Walker. Inset from *The Mammal Guide*, Ralph S. Palmer.

RODENTIA; ERETHIZONTIDAE; **Genus ERETHIZON**
F. Cuvier, 1822

North American Porcupine

The single species, *E. dorsatum*, occurs from northern Alaska to Newfoundland, and south to northern Mexico and Tennessee (Hall 1981).

Head and body length is 645 to 860 mm, tail length is 145 to 300 mm, and weight is usually 3.5 to 7.0 kg, but large males occasionally weigh as much as 18 kg. The upper parts of the body are covered with thick, sharp, barbed quills, which are distributed among longer, stiff guard hairs; the underfur is woolly. The quills are about 2 mm in diameter, are up to 75 mm long, and may exceed 30,000 in number (Woods 1973). The general coloration of the upper parts is usually dark brown or blackish; the individual quills have yellowish white bases and dark tips. The underparts lack the sharp quills and are covered only with stiff, dark hairs.

Erethizon has a robust body, a small head, moderately small ears, short legs, and a short, thick tail that is armed with quills above and stiff bristles below. The feet are heavy, have naked soles, and are modified for arboreal life. There are four toes on the forefoot and five on the hind foot, all with strong, curved claws. The skull is massive and somewhat swollen between the orbits. The incisor teeth are heavy, project anteriorly, and are deep orange or almost ochraceous in color. Females have one pectoral and one abdominal pair of mammae (Jackson 1961).

The preferred habitat of *Erethizon* is forest with mixed hardwood and softwood trees. The genus is highly adaptable and may even be found in open tundra, rangeland, and desert, but it usually stays in vegetated riparian areas, when it is away from forest (Woods 1973). It is primarily terrestrial, but frequently ascends trees to heights of 18 meters in order to obtain food. It climbs surely, though slowly, and has excellent balance. It does not jump in climbing or descending. It swims well, the hollow quills furnishing considerable buoyancy. It dens in a cave, crevice, hollow log or tree, burrow, snow bank, or crude nest in a tree. More than one den is used, and temporary shelters often supplement the main habitation. Brander (1973) found that dens in northern Michigan were in trees, logs, and stumps, and that entrances were usually at ground level, but sometimes over 6 meters high.

This porcupine is thought to have relatively poor vision, but good senses of hearing and smell; it sniffs the air almost continuously. It is mainly nocturnal, but sometimes forages by day. It does not hibernate and is active throughout the winter, but may remain in its den when the weather is very cold or stormy. Seasonal migrations occur in some areas, depending on the availability of food. One individual, recaptured six times, was found to have moved an average of 1,500 meters from its winter den to the point of capture in the summer (Woods 1973). Brander (1973) observed an autumn dispersal in northern Michigan, during which individuals moved an average of 480 meters. In contrast, the mean daily distance between the den and feeding site was found to be only 8 meters in winter and 150 meters in summer. Other reports of such daily movements have ranged up to an average of 130 meters in winter and 1,200 meters for one animal in May. *Erethizon* uses regular runways through the vegetation or snow, and may wear a distinct path from the den to the feeding area.

The winter diet consists mainly of evergreen needles, and the cambium layer and inner bark of trees. During the spring and summer the porcupine eats buds, tender twigs, roots, stems, leaves, flowers, seeds, berries, nuts, and other vegetation. It is fond of salt, and avidly gnaws bones and antlers found on the ground because of their high mineral content (Banfield 1974). Food may be held in the hands. The digestive tract is about 850 cm long, 46 percent of which is small intestine containing bacteria that can decompose cellulose (Woods 1973). Sometimes, 450 grams of food is consumed in a day, and about one-fifth of this weight is excreted in the form of brown, crescent-shaped droppings.

There may be population cycles with peaks 12 to 20 years apart. Reported population densities have varied from about 1 to 37 individuals per sq km, but have usually been about 5–10 per sq km (G. W. Smith 1977; Woods 1973). A winter feeding range was calculated to cover 5.4 ha., and summer ranges of 13.0 and 14.6 ha. were determined during a 30-day radio-tracking study. *Erethizon* evidently does not defend a territory and shows little aggression, but may defend a specific feeding tree (Woods 1973). This genus is generally solitary, with only one adult per den, but pairs may share a single feeding tree, several animals sometimes share a den on a rotating basis, and during the winter a number of individuals may occupy a den simultaneously. Brander (1973) found that during the summer and early autumn about a dozen porcupines might gather at certain nocturnal feeding sites and spend much time vocalizing. A great variety of sounds has been reported for *Erethizon*, including moans, grunts, coughs, wails, whines, and tooth chatters.

This slow-moving, seemingly clumsy creature is said never to attack. It may protect itself by climbing or fleeing, but if cornered it erects the quills, turns the rump toward the source of danger, and rapidly lashes out with the barbed tail. The quills are not thrown or shot, but they are so lightly attached that when they enter the skin of an enemy, they become

North American porcupine (*Erethizon dorsatum*), baby only a few days old, photo by Arthur Ellis, *Washington Post*.

detached from the porcupine. The quills then continue to work their way deeper at a rate of up to one mm or more per hour, and can cause death if they puncture vital organs. Despite this defense mechanism, the porcupine is preyed upon by the great horned owl and many carnivorous mammals. The bobcat, wolverine, and fisher are especially adept at flipping the porcupine on its back to expose the unprotected underparts (Banfield 1974). The fisher (*Martes pennanti*) has even been reintroduced in some areas in an effort to control porcupine numbers. Powell and Brander (1977) noted a 76 percent decline in the porcupine population of a study area in northern Michigan, during a 13-year period subsequent to the establishment of the fisher.

In the fall or early winter, males seek out females for mating. There is an elaborate courtship, involving extensive vocalization and a comical kind of dance (Banfield 1974). The male usually showers the female with urine before mating, and afterward the female repels the male. Females may be polyestrous and will recycle in 25 to 30 days if fertilization does not occur at the time of first ovulation; heat lasts 8–12 hours (Woods 1973). The gestation period is 205–17 days, and the young are born from April to June. Litters generally contain a single offspring, though there are rare records of twins. The newborn is well developed; it weighs 340–640

grams, its eyes are open, it can walk about unsteadily, and it is covered with long, black hair and short, soft quills. It exhibits the typical defense reaction of turning the rump toward danger, and within days it can climb trees. Although nursing has lasted several months in the laboratory, weaning apparently occurs much sooner in the wild, as the young can survive on a diet of vegetation within two weeks of birth. It gains weight at a rate of about 450 grams per day. Some males are capable of mating when only 16 months old, though Banfield (1974) stated that sexual maturity in *Erethizon* is reached at approximately 2½ years of age. The natural life span is evidently long. Both a laboratory captive and a marked wild individual are known to have lived at least 10 years (Brander 1971).

The porcupine sometimes causes the death of timber and ornamental trees by girdling the bark of the trunk or stripping off all the bark above the snow line in winter. There have been extensive debates and analyses regarding the seriousness of such activity (Woods 1973). While Banfield (1974) indicated that the porcupine is one of the most important mammalian forestry pests, Jackson (1961) suggested that damage has been greatly exaggerated. There is general agreement that individual porcupines sometimes become a nuisance by gnawing woodwork, furniture, tools, cooking utensils, saddles, canoe paddles, and other objects that have received salt deposits through contact with human perspiration. The porcupine is considered tasty food by some persons, and its quills have been used in decorative work by Indian tribes. Perhaps in part through human agency, *Erethizon* has vanished, except for occasional wandering animals, from that section of the eastern United States to the south of Minnesota, northern Wisconsin, Michigan, and Pennsylvania (Jackson 1961; Handley 1980*b*). The genus may, however, already have been rare in the southern Appalachians during colonial times, and its disappearance there could be associated with climatic changes.

RODENTIA; *Family CAVIIDAE*

Cavies and Patagonian "Hares"

This family of 5 Recent genera and 16 species occurs over most of South America. The sequence of genera presented

Wild cavies (*Cavia aperea*), photo by J. P. Rood.

here follows that of Cabrera (1961), who recognized two subfamilies: Caviinae, with *Microcavia, Galea, Cavia,* and *Kerodon;* and Dolichotinae, with *Dolichotis.*

Head and body length is 200 to 750 mm, and the tail is vestigial. The pelage in wild forms is fairly coarse or crisp. In the subfamily Caviinae (cavies), the body form is robust, the head is large, and the ears and limbs are short. In the Dolichotinae (Patagonian "hares"), the proportions are rabbitlike, the ears are long, and the limbs are long and thin. In both subfamilies there are only four digits on the forefoot and three on the hind foot. The nails are short and sharp in *Cavia* and *Microcavia,* blunt but well developed in *Kerodon,* and hooflike on the hind foot and clawlike on the forefoot in *Dolichotis.* The soles of the feet are naked in the Caviinae, but mostly haired in the Dolichotinae.

The dental formula is: (i 1/1, c 0/0, pm 1/1, m 3/3) x 2 = 20. The incisors are short. The tooth rows tend to converge anteriorly, and the cheek teeth are rootless (ever growing), with a rather simple pattern of two prisms having sharp folds and angular projections.

Caviids occur in habitats ranging from marshy, tropical floodplains to dry, rocky meadows at elevations of up to 4,000 meters. They live on pampas and savannahs—in grassy, brushy, or rocky areas—but not generally in heavy jungle (Starrett 1967). None of these rodents hibernates, even when living at high altitudes and temperatures are very low. The diet consists of many kinds of plant material. Breeding may continue throughout the year if climatic conditions are favorable. The gestation period, ranging from about 50 to 70 days, is relatively short for the suborder Hystricomorpha. Nonetheless, the young are well developed at birth and reach sexual maturity rather early. The potential life span is relatively long.

The geological range of the Caviidae is middle Miocene to Recent in South America.

RODENTIA; CAVIIDAE; **Genus MICROCAVIA**
Gervais and Ameghino, 1880

Mountain Cavies

There are two subgenera and three species (Cabrera 1961; Pine, Miller, and Schamberger 1979):

subgenus *Microcavia*

M. australis, Argentina, southern Chile;
M. shiptoni, mountains of northwestern Argentina;

subgenus *Monticavia*

M. niata, mountains of Bolivia, possibly southeastern Peru and northern Chile.

The information for the remainder of this account was taken entirely from Rood's (1970b, 1972) papers, which dealt primarily with studies of *M. australis.* Superficially, this species resembles a small, tailless ground squirrel. Head and body length is about 200 to 220 mm, there is no external tail, and weight is about 200 to 500 grams. The upper parts are olive gray agouti and the underparts are pale gray. There is a prominent white ring around the eye.

Microcavia closely resembles *Galea,* but the latter genus has a less prominent eye ring, darker pelage, yellow incisors (white in *Microcavia*), a smaller eye, and a prominent submandibular gland (absent in *Microcavia*). Females of both genera have four mammae.

The species *M. shiptoni* and *M. niata* live in high mountains, but *M. australis* occurs throughout Argentina except for the humid northeastern provinces. It is the most abundant member of the Caviinae in the semiarid thornbush of central Argentina. In such habitat it uses clumps of thornbush for

Mountain cavy (*Microcavia australis*), photo by J. P. Rood.

cover, and makes runways through the open areas between clumps of bushes. Under the bushes it digs a shallow depression, in which it rests and sleeps, and it keeps the surrounding area clear. It also may dig a burrow; the main tunnel of one was 134 cm long, 5 cm wide, and 28 cm below the surface at the deepest point. Most activity is diurnal. *Microcavia* sometimes climbs bushes and low trees to feed. The diet consists mostly of leaves and also includes fruit. Animals were never seen to drink and presumably got sufficient water from succulent vegetation.

Population size in a given area varies substantially, depending on reproductive activity and movements. Average density during one year of study was calculated at 24 individuals per ha. Mean home range was 7,720 sq meters for males and 3,525 sq meters for females. The ranges of males overlap those of both other males and females. A male regularly wanders about and visits several females during a day. He sometimes forms a temporary association with a female and her young and will follow her about closely, nose to rump, especially if she is near estrus. Encounters between males, however, almost always involve aggression. The males in an area establish a linear dominance hierarchy, and subordinates are chased. Sometimes there are fights to the death. Females tend to stay within a small area of bushes and do not often meet members of their own sex. If they do, the reaction can be friendly or hostile. Two females, especially a mother and daughter, may kiss upon meeting and then sit or forage together. Young animals, up to about a month old, are tolerated by adults of both sexes. *Microcavia* apparently has an amicable relationship with the sympatric *Galea,* and cavies of both genera may be seen together in groups. The vocalizations of *Microcavia* include an alarm "tsit," a "twitter" of annoyance, and a "shriek" of fear.

The breeding season in east-central Argentina lasts from August to April, with most births occurring in the spring (September–December). Females are polyestrous, have a postpartum estrus immediately after giving birth, and may be able to mate again within 15 days if fertilization does not occur at the postpartum estrus. When a pregnant female is near term, males gather about it and display much aggression

toward one another. As soon as the offspring are delivered, as many as six males pursue the female. A female can theoretically produce five litters during a season, and one individual wild female is known to have given birth to four. The gestation period averages about 54 days. Litters of one to three young were observed in the wild, but captive females sometimes had litters of four or five. The young weigh around 30 grams at birth, are able to run about and nibble solid food during their first day of life, and are ordinarily weaned at 3 weeks. A young animal sometimes nurses from a female other than its own mother. Young females usually come into estrus at the age of 40–50 days, but do not necessarily conceive at that time. One female is known to have become pregnant when 82 days old. Life expectancy in the wild is apparently short, mainly because of predation by the grison (*Galictis cuja*). Of the 44 cavies marked on Rood's study area in April 1966, only 15 were left in February 1967.

Microcavia is generally considered a pest by the people in the area in which it lives. It supposedly destroys crops and its holes are a hazard to horses. It also is sometimes hunted by people for use as food. Nonetheless, it seems to adapt well to civilization and continues to exist in large numbers.

RODENTIA; CAVIIDAE; *Genus GALEA*
Meyen, 1833

Yellow-toothed Cavies, Cuis

The three species (not in systematic order) are as follows (Cabrera 1961; Wetzel and Lovett 1974):

G. flavidens, Brazil;
G. musteloides, southern Peru, Bolivia, Argentina;
G. spixii, Brazil, eastern Bolivia, Paraguay.

The following descriptive data were provided mostly by Barbara J. Weir (Wellcome Institute, Zoological Society of London, pers. comm.). Head and body length is 150 to 250 mm, there is no external tail, and weight is 300 to 600 grams.

Cui (*Galea musteloides*), photo by T. D. Dennett. Inset: chin gland and view of the yellow-colored incisors, photo by Barbara J. Weir.

The upper parts are agouti colored and the underparts are grayish white. The pelage is paler than that of *Cavia* and does not molt so readily, and is not as long or coarse as that of *Microcavia*. The body form is more compact and stocky than that of *Cavia*. The incisor teeth are yellowish or light orange, rather than white as in *Cavia* and *Microcavia*. There is a prominent submandibular gland in *Galea*, but not in the other two genera. The hind foot has three toes, the forefoot has four toes, and all digits have strong, sharp claws. Females have one pair of inguinal and one pair of lateral thoracic mammae.

Yellow-toothed cavies inhabit grasslands at high or low elevations, as well as rocky and brushy areas. They sometimes excavate their own dens, and extensive colonies reportedly honeycomb the ground with their burrows in the Bolivian highlands (Starrett 1967). They also use the abandoned holes of larger mammals, such as armadillos (*Chaetophractus*), viscachas (*Lagostomus*), and tuco-tucos (*Ctenomys*), as well as crevices in stone walls (Rood 1972). They are primarily terrestrial and diurnal. The diet consists of grasses, forbs, and other kinds of vegetation.

Rood (1972) studied *G. musteloides* both in captivity and the wild. One female had a home range of 4,275 sq meters. Animals lived in groups, but only a single adult male could be kept in an enclosure. Free-living males established a linear dominance hierarchy that was maintained by aggression. The female hierarchy was less stable than that of males. Generally, however, females were dominant to males, and older animals were dominant to younger ones. Subordinates usually retreated, but aggressive displays, fighting, and serious wounds were common. Immature individuals were generally tolerated until they were about one month old, when adults would begin to chase them. Vocalizations included a "churr," indicating sexual arousal; a "tooth chatter," signifying a threat; and a "rusty gate screech" during an attack.

Observations by Rood (1972) and Weir (1974b) indicate that breeding continues throughout the year but that most litters are born during spring and summer in the wild. Females are polyestrous and can produce 7 litters per year; one had 12 in 21 months. The estrous cycle averages about 22 days, receptivity lasts 2 or 3 hours, and there is a postpartum estrus. Females, however, are apparently induced, rather than spontaneous, ovulators, and the presence of a male is needed to stimulate estrus. The submandibular gland, which is more developed in the male, and a behavioral pattern in which the male closely follows the female, chin to rump, seem to be involved in inducing estrus. As soon as females give birth, they are pursued by males. The average gestation period is 53 days and the range is 49–60 days. The mean number of young per litter is three and the range is one to seven. The young weigh about 40 grams each at birth, are well developed, and can move about. They normally nurse for 3 weeks. Sexual maturity is generally attained at the age of 3 months in males and 2 months in females, but some females reach puberty much earlier; one conceived when only 9 days old, a record for mammals. *G. musteloides* is known to have lived over 15 months in the wild and over 22 months in captivity. According to Marvin L. Jones (Zoological Society of San Diego, pers. comm.), a captive *G. spixii* survived for 4 years and 7 months.

RODENTIA; CAVIIDAE; **Genus CAVIA**
Pallas, 1766

Cavies, Guinea Pigs

The seven species (not in systematic order) are as follows (Cabrera 1961; Weir 1974c; Rood 1972; Handley 1976; Pine, Miller, and Schamberger 1979; Husson 1978; Gade 1967):

C. anolaimae, vicinity of Bogota in Colombia;
C. aperea, southern Surinam, eastern and southern Brazil, Paraguay, Uruguay, northern Argentina;
C. fulgida, southeastern Brazil;
C. guianae, southern Venezuela, Guyana, and probably adjacent parts of northern Brazil;
C. nana, western Bolivia;
C. porcellus, about 500 years ago found in domestication from northwestern Venezuela to central Chile;
C. tschudii, Peru, southern Bolivia, northern Chile, northwestern Argentina.

Cabrera (1961) listed the wild species *C. anolaimae* and *C. guianae* as subspecies of the domestic guinea pig, *C. porcellus,* because the original describers of these wild species had suggested strong affinity to the domestic species. Weir (1974c), however, stated that there is no wild form of *C. porcellus* and that this species probably originated within *C. aperea,* *C. fulgida,* or *C. tschudii.* It therefore becomes necessary to treat *C. anolaimae* and *C. guianae* as separate species again.

The following descriptive information was provided in part by Barbara J. Weir (Wellcome Institute, Zoological Society of London, pers. comm.). Head and body length is 200 to 400 mm, there is no external tail, and weight is 500 to 1,500 grams. Wild species have fairly coarse and long pelage, a crest of hairs at the neck, and agouti-banded hairs that produce an overall grayish or brownish color. In the domestic guinea pig the pelage may be smooth and short, smooth and long, or coarse and short, and in some breeds rosettes are formed; there is a wide variety of colors and patterns. In both wild and domestic forms the body hairs are easily shed.

Cavies have a stocky build, fairly short legs, and short, unfurred ears. There are three digits on the hind foot and four on the forefoot, all armed with sharp claws. The incisor teeth are white. Females have a single pair of inguinal mammae (Weir 1974b).

Cavies occur in a wide variety of habitats, including open grasslands, forest edge, swamps, and rocky areas. They are sometimes found at elevations of up to 4,200 meters (Grzimek 1975). They have been reported either to dig their own burrows or to take over the abandoned holes of other mammals. In a study in eastern Argentina, however, Rood (1972) found *C. aperea* not to burrow but to shelter in brush piles and clumps of vegetation. Feeding activity was mainly crepuscular in this area, but occasionally extended throughout the day during cold weather. *Cavia* is terrestrial and seems to follow well-defined tracks from den to feeding area, but has also been reported to swim several kilometers during floods (Rood 1972). The diet consists of many kinds of vegetation.

Cavies generally associate in small groups, usually comprising 5 to 10 individuals, but in favorable habitat a number of such groups may converge and give the impression of a large colony (Barbara J. Weir, Wellcome Institute, Zoological Society of London, pers. comm.). In his study area, Rood (1972) found *C. aperea* to have a maximum population density of about 38 per ha. Home ranges there averaged 1,387 sq meters for males and 1,173 sq meters for females. The ranges were stable and centered around a clump of vegetation that was used for food and shelter. Neither sex maintained an exclusive territory, and ranges frequently coincided. Each sex had its own linear dominance hierarchy, and subordinate animals either retreated or were attacked. If an alpha male eventually lost a fight, it became passive and withdrawn, and sometimes died. Vocalizations included "bubbly squeaks"

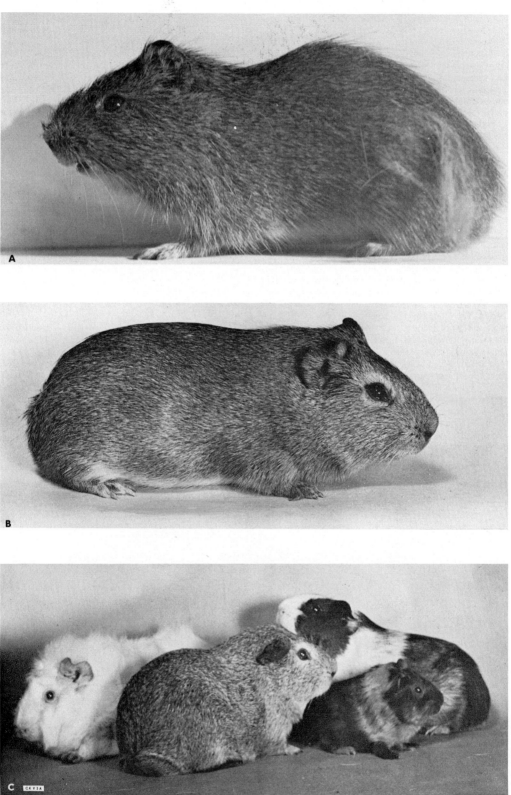

A. Wild cavy (*Cavia tschudii*), photo by Hilda Heller. B & C. Guinea pigs, domestic strain (*C. porcellus*), photos by Ernest P. Walker.

of excitement, "chirps" of anxiety, and a "tooth chatter" threat.

Data compiled by Rood (1972) show that in captivity both *C. aperea* and *C. porcellus* breed throughout the year, with birth peaks in the spring. Females are polyestrous and produce a maximum of five litters annually. Wild *C. aperea* may also reproduce throughout the year under favorable conditions, but did not breed during the severe winter in Rood's study area. The estrous cycle averages about 16½ days in *C. porcellus* (Asdell 1964) and 20½ days in *C. aperea* (Weir 1974b). Females of both the wild and domestic species experience a postpartum estrus immediately after giving birth, and receptivity lasts less than half a day. When a pregnant female is about to give birth, males gather about her. The most dominant male aggressively guards the female from the others, and he usually mates first. If he subsequently can no longer protect the female, several subordinates pursue and mate with her (Rood 1972).

The overall gestation range in the genus is 56 to 74 days, with averages of about 62 days in *C. aperea,* 63 in *C. tschudii,* and 68 in *C. porcellus.* The mean (and extreme) numbers of young per litter are 4.0 (1–13) in *C. porcellus,* 2.3 (1–5) in *C. aperea,* 1.35 in *C. fulgida,* and 1.9 (1–4) in *C. tschudii.* A newborn *C. aperea* weighs about 60 grams, and one of *C. porcellus* weighs about 100 grams (Weir 1974b). The young are well developed at birth, and can run about and eat solid food on their first day of life (Rood 1972). They can survive without further nursing after 5 days, but normally suckle for 3 weeks. Sexual maturity is usually reached after 3 months in males and 2 months in females, but the minimum recorded age of conception in a female is only 21 days (Weir 1974b). Longevity may be as much as 8 years, at least in captivity.

The domestic guinea pig (*C. porcellus*) apparently has been bred for meat production in South America for at least 3,000 years (Gade 1967; Weir 1974c). Its precise origin is unknown, but domestication probably occurred just once. Cross breeding and biochemical studies indicate that *C. por-* *cellus* probably was derived from *C. aperea, C. tschudii,* or *C. fulgida,* but that it is distinct from each of these other species. During the period of the Inca Empire, from about 1200 to 1532, highly selective breeding produced a variety of strains, differentiated by color pattern and flavor. At that time the range of domestication extended from northwestern Venezuela to central Chile. Subsequent to the Spanish conquest there was a disruption of breeding and a contraction of range. The guinea pig is still widely kept as a source of food by the native people of Ecuador, Peru, and Bolivia. In some areas the animals are restricted to huts and are well treated. In other areas they are allowed to run free and scavenge, and under such conditions some feral populations have become established. The guinea pig was brought to Europe in the sixteenth century, and since the mid-1800s has been used by laboratories around the world for research on pathology, nutrition, genetics, toxicology, and development of serums. It also makes an ideal pet.

RODENTIA; CAVIIDAE; Genus **KERODON**
F. Cuvier, 1825

Rock Cavy, Moco

The single species, *K. rupestris,* is found in eastern Brazil, from the state of Piaui to the northern part of Minas Gerais (Cabrera 1961).

Kerodon is about the same size as *Cavia,* or is somewhat larger. The tail is either absent or only a vestigial projection. Adult weight is 900 to 1,000 grams. The upper parts are generally grayish with white and black mottling, the throat is whitish, and the underparts are yellowish brown. Unlike other genera in the subfamily Caviinae, the well-developed nails of the digits are blunt rather than sharp.

The moco inhabits arid and pebbly areas near stony mountains or hills. It seeks shelter under rocks or in the fissures

Moco (*Kerodon rupestris*), photo from Field Museum of Natural History. Inset: left hand and foot, photo from *Proc. Zool. Soc. London.*

between stones, sometimes making a burrow under the stones. It leaves its shelter late in the afternoon or evening, and runs on the ground or climbs trees looking for food, mainly tender leaves. It descends by leaps or by a single jump at the slightest sign of danger. The feces are small, green, and capsule shaped. The voice sounds like a shrill whistle; this is imitated by hunters trying to attract *Kerodon*. Mating has been noted in March near Ceará. There are probably two litters a year. According to Kleiman, Eisenberg, and Maliniak (1979) gestation lasts 77 days, litters contain one or two young, and birth weight is 80 grams. One individual reportedly lived in captivity for 11 years. *Kerodon* is easily domesticated and is suitable as a pet. Persons native to its range are very fond of its flesh and prepare it as food or medicine.

RODENTIA; CAVIIDAE; **Genus *DOLICHOTIS* Desmarest,** *1819*

Patagonian Cavies or "Hares," Maras

There are two species (Cabrera 1961):

D. salinicola, southern Bolivia, Paraguay, northern Argentina;

D. patagonum, central and southern Argentina.

Cabrera placed *D. salinicola* in the separate genus *Pediolagus* Marelli, 1927. This procedure was not followed by Starrett (1967) or Wetzel and Lovett (1974).

Head and body length is 690 to 750 mm in *D. patagonum,* but only about 450 mm in *D. salinicola.* In both species the maximum length of the tail is about 45 mm. Large indi-

viduals of *D. patagonum* weigh 9 to 16 kg. The general coloration of the upper parts is grayish, and that of the underparts is whitish. In *D. salinicola* there is a horizontal band of white or yellowish fur running from the flank to the belly. The pelage is dense. Although the hairs are fine, they have a crisp texture and stand at nearly right angles to the skin.

Dolichotis is modified for a cursorial life. The body form is similar to that of long-legged rabbits and hares. The hind limbs are long, the hind foot has three digits, and each digit has a hooflike claw. The forefoot has four digits, each bearing a sharp claw. Female *D. salinicola* have two pairs of mammae, and female *D. patagonum* have four pairs (Weir 1974*b*).

Maras inhabit arid areas with coarse grass or scattered shrubs. They shelter in burrows of their own construction or in the abandoned holes of other animals, such as *Lagostomus.* They are terrestrial and diurnal. In a study of an introduced colony of *D. patagonum* in Brittany, Dubost and Genest (1974) found that individuals rested at night in wooded areas, and then traveled to open grassland to feed during the day. *Dolichotis* uses a variety of locomotions: walking when undisturbed, hopping like a rabbit or hare, galloping, and stotting—a sort of bounce on all four limbs at once—for covering long distances at high speed. Rood (1972) clocked a mara running beside his car at 45 km per hour over a distance exceeding 1 km. When not moving about, *Dolichotis* stands with its legs straight, sits on its haunches with the forepart of the body resting on fully extended front legs, or reclines like a cat, with the front limbs turned under the chest—an unusual position for a rodent. It spends considerable time basking in the sun, but is ever on the alert for danger. The diet consists of any available vegetation.

Three or four wild individuals are often seen traveling

Patagonian cavy (*Dolichotis patagonum*), photo from East Berlin Zoo.

Patagonian cavies (*Dolichotis* sp.), photo from Zoological Society of London.

together in single file, but occasionally a group of up to 40 may be observed. Studies of an introduced colony of about 70 *D. patagonum* in Brittany (Dubost and Genest 1974; Genest and Dubost 1974) showed that the group dispersed by night and concentrated in feeding areas by day. There was no territoriality and the entire colony shared available space, but males had a dominance hierarchy maintained by aggression. The basic social unit was the mated pair and strict monogamy was practiced throughout the year. The female initiated most movements, and the male always remained close and defended her against other males. Eisenberg (1974) listed a number of vocalizations, including a ''wheet'' for seeking contact and a ''grunt'' used as a threat.

According to Dubost and Genest (1974), the introduced colony of *D. patagonum* in Brittany breeds throughout the year, but most intensively during the European winter, a period corresponding with summer in the natural range of the species. Females are polyestrous, experience a postpartum estrus, and produce three of four litters annually. The interval between litters is about 3 months during the European winter and 3½ months during summer. Litters usually contain two young, but sometimes have one or three. The young are born outside of the burrow, but move inside soon afterward. They are well developed, their eyes are open, and they can move about freely. Weaning occurs by the age of 11 weeks, and 8-month-old females may give birth to their own young. Asdell (1964) stated that a pregnant female *D. salinicola*, with two embryos, was taken in Argentina in August. Weir (1974*b*) listed the gestation period of *D. salinicola* as approximately 77 days. Grzimek (1975) reported that a captive *D. patagonum* lived for almost 14 years, but that most individuals do not survive beyond 10 years.

Rood (1972) noted that *D. patagonum* appears to be declining in numbers in the wild and now is rare in the province of Buenos Aires, Argentina, where it was formerly abundant. The decline is attributable to habitat destruction by people and competition with the introduced and more adaptable European hare (*Lepus capensis*).

RODENTIA; *Family HYDROCHAERIDAE; Genus HYDROCHAERIS Brünnich, 1772*

Capybara

The single living genus and species, *Hydrochaeris hydrochaeris,* occurs in Panama, to the east of the Canal Zone, and on the east side of the Andes in South America, from Colombia and the Guianas to Uruguay and northeastern Argentina (Cabrera 1961; Hall 1981). The use of the above scientific names, rather than Hydrochoeridae, *Hydrochoeris,* and *hydrochoeris,* is in keeping with the explanation of Husson (1978).

This is the largest living rodent. Head and body length is 100 to 130 cm, the tail is vestigial, shoulder height is up to 50 cm, and weight is 27 to 79 kg. Zara (1973) reported the average weight of captive adults to be 50 kg in males and 61 kg in females. The long, coarse pelage is so sparse that the skin is visible. The coloration is generally reddish brown to grayish on the upper parts and yellowish brown on the underparts. Occasionally there is some black on the face, the outer surface of the limbs, and the rump. In the mature male, a bare, raised area on the top of the snout contains greatly enlarged sebaceous glands.

Hydrochaeris resembles *Cavia,* but is much larger and has a proportionally shorter body. The limbs are short, the head is relatively large and broad, the ears are short and rounded, and the small eyes are placed dorsally and relatively far back on the head. The muzzle is heavy and truncate, and the upper

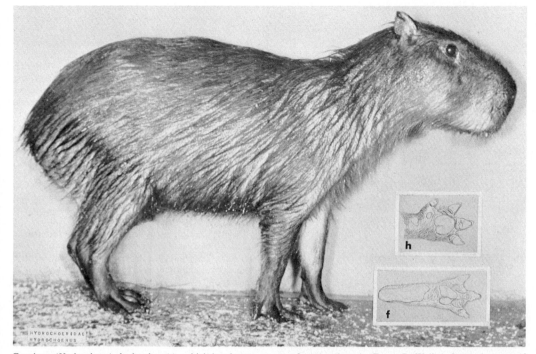

Capybara (*Hydrochaeris hydrochaeris*), which has just come out of water, photo by Ernest P. Walker. Insets: bottoms of hand and foot, photos from *Proc. Zool. Soc. London*.

lip is enlarged. The forefoot has four digits and the hind foot has three. The digits are arranged in a radial pattern, and all are partially webbed and armed with short, strong claws. Females have five pairs of ventral mammae (Weir 1974b).

The dental formula is: (i 1/1, c 0/0, pm 1/1, m 3/3) x 2 = 20. The tooth rows tend to converge anteriorly. The incisor teeth are white and shallowly grooved. The cheek teeth are rootless (ever growing), have an unusually large amount of cement, and have a surface pattern that is similar to, but more complex than, that of the Caviidae.

The capybara inhabits densely vegetated areas around ponds, lakes, rivers, streams, marshes, and swamps. It does not use a den, but shelters in thickets (Schaller 1976). In areas where it is not disturbed it is active in the morning and evening, resting during the heat of the day in a shallow bed in the ground. Like some other mammals, however, it apparently has become nocturnal in areas where molested by people. When alarmed on land, it runs like a horse, and when closely pursued it enters the water, where it swims and dives with ease. While swimming, only the nostrils, eyes, and ears project above the water. The capybara also swims completely underwater, often for a considerable distance, or it may hide in floating vegetation with only the nostrils exposed. Although it thus seems to be an aquatic mammal, water is used primarily as a place of refuge, and most normal activity is on land (Schaller 1976). The diet consists mainly of grasses, and the capybara is sometimes seen grazing with cattle. It also eats aquatic plants, grains, melons, and squashes. Characteristic mounds of elongate fecal pellets are deposited.

Hydrochaeris is commonly found in groups of about 20 individuals. Schaller (1976) once saw 64 together, and old reports told of colonies of 100 or more, but such large congregations are apparently not stable. Observations in captivity (Donaldson, Wirtz, and Hite 1975) indicate that the basic social unit is a family group, and that outsiders are not readily accepted. There is a stable and viciously enforced social hierarchy, with much aggression and fighting, and it is inadvisable to place two adults of the same sex in a small area. Reported vocalizations include low, clicking sounds of contentment; sharp, prolonged whistles; abrupt grunts; and weak barks.

Breeding occurs throughout the year in Venezuela, but mating reaches a peak in April and May, shortly before the rains begin. In Argentina, mating was observed in October, also just before the rainy season (Schaller 1976). Two definite gestation periods, lasting 149 and 156 days, were recorded by Zara (1973). Weir (1974b) listed a single gestation of 104–11 days in the small northern subspecies *H. h. isthmius*. She also reported litter size in the genus to average five (one to eight) young, weight of each newborn to be about 1,500 grams, and normal lactation to last 16 weeks. The young are precocial, and able to follow their mother and eat grass shortly after being born (Schaller 1976). Both sexes reach puberty at the age of 15 months (Zara 1973). Life expectancy in the wild is apparently 8 to 10 years, but some captives have survived over 12 years (Grzimek 1975).

The capybara is sometimes killed by persons who consider it an agricultural pest, and it is also intensively hunted for its meat and hide. Its thick, fatty skin provides a grease used in the pharmaceutical trade, and its incisors are used as ornaments by native people. In Venezuela, capybara populations have declined in the face of heavy market hunting, but efforts have been made to regulate the kill (Ojasti and Padilla 1972). In Peru, this genus has disappeared from many waterways where it was formerly common (Grimwood 1969). Nonetheless, the capybara remains widespread and common in much of South America and has even been raised commercially on some ranches (Schaller 1976). Such ranching is thought to have the advantage of maintaining natural wetlands, rather

Capybara (*Hydrochaeris hydrochaeris*), photo from Riverbanks Zoo.

than draining them, as might be required in the case of intensive cattle raising (Vasey 1979).

The geological range of the Hydrochaeridae is early Pliocene to Recent in South America, Recent in Central America, and middle to late Pleistocene in the southeastern corner of North America (Mones 1973, 1975).

RODENTIA; *Family DINOMYIDAE; Genus DINOMYS* Peters, 1873

Pacarana

The single living genus and species, *Dinomys branickii*, occurs in the highlands from Colombia to western Bolivia (Cabrera 1961).

Head and body length is 730 to 790 mm, tail length is about 200 mm, and weight is 10 to 15 kg. The upper parts are black or brown, with two more or less continuous broad white stripes on each side of the midline of the back, and two shorter rows of white spots on the sides. In older individuals the stripes seem to be broader and more conspicuously white. The underparts are paler than the upper parts and are not

marked. The pelage is rather coarse, scant, and of varied length. The tail is fully haired. The whiskers are about the same length as the head.

"Pacarana" is a Tupí Indian term meaning "false paca," and signifies the resemblance, in size and color pattern, between *Dinomys* and the pacas (*Agouti*). *Dinomys*, however, has a more thickset body than *Agouti*, and also has a stout tail, which is approximately one-fourth the length of the head and body. The overall appearance of *Dinomys* is reminiscent of an immense guinea pig (*Cavia*). Its head is broad, its ears are short and rounded, and its limbs are short. Each of the broad feet bears four digits, all armed with a long, powerful claw. Females have two lateral thoracic and two lateral abdominal pairs of mammae (Weir 1974b).

The dental formula is: (i 1/1, c 0/0, pm 1/1, m 3/3) x 2 = 20. The incisors are broad and heavy. The cheek teeth are probably rootless (ever growing), and are extremely high crowned, each tooth consisting of a series of transverse plates.

The pacarana occurs in valleys and on lower mountain slopes. Grimwood (1969) described it as an animal of the high selva (rain forest) zone, and of the upper, or better drained, parts of the low selva zone. He gave the elevational range as 240 to 2,000 meters. Wild individuals are said to shelter in natural crevices, which they enlarge by digging with their strong claws, but captives have not been seen to dig. Captives have used the claws to climb trees ably and have preferred to sleep in elevated places. Captives are active mainly after dark, and *Dinomys* is thought to be naturally nocturnal (Collins and Eisenberg 1972). When on the ground, the pacarana normally progresses by a waddling gait. When threatened by an agile predator, such as an ocelot or coati, *Dinomys* backs against a cliff or into a hole to protect its vulnerable hind parts. When eating, *Dinomys* sits upon its haunches, and, in a dexterous manner, carefully examines its food before consuming it. The diet seems to consist mainly of fruit, leaves, and tender stems.

Although its generic name means "terrible mouse," the pacarana fights only as a last resort. Captives are generally slow moving, good natured, and peaceful, and do not attempt to bite or escape. They allow themselves to be scratched, occasionally indicating displeasure by a low, guttural growl. Collins and Eisenberg (1972) reported the most common vocalizations, heard in encounters between a male and female pacarana, to be a "hiss," "growl," and "staccato whimper."

Pregnant females, with two embryos each, were taken in the wild in February and May. Another female, which was pregnant when captured alive, gave birth to two young on 8 January 1970. Still another female produced a litter of two young, both dead, on 1 February 1971, following a gestation period of 223–83 days (Collins and Eisenberg 1972). According to Grzimek (1975), of those litters born in captivity in 1974, half contained a single offspring and half contained two. Weir (1974b) listed newborn weight as 900 grams. Although the pacarana tames readily and makes an interesting pet, most captives do not survive for very long. One, however, lived for 9 years and 5 months (Marvin L. Jones, Zoological Society of San Diego, pers. comm.).

Dinomys was not discovered by science until 1873, has always seemed rare, and at several times was feared extinct (Grzimek 1975; Allen 1942). Although the number of captive specimens has increased since the 1940s, wild populations are thought to be declining through habitat destruction and excessive hunting by people for use as food (Grimwood 1969).

The geological range of the Dinomyidae is early Miocene to Recent in South America (Starrett 1967).

Pacarana (*Dinomys branickii*), photo from New York Zoological Society.

RODENTIA; *Family HEPTAXODONTIDAE*

This apparently extinct family is known only from skeletal remains found in cave deposits in the West Indies. Hall (1981) listed two genera—*Elasmodontomys* and *Quemsia*—that evidently still lived when humans occupied the islands, and four other genera—*Clidomys, Spirodontomys, Speoxenus,* and *Amblyrhiza*—that are thought to have disappeared before the arrival of people. He also referred to an additional genus—*Alterodon*—that is related to the others but not definitely a member of the Heptaxodontidae. According to Allen (1942), it is probable that *Amblyrhiza* did live well after Pleistocene times, and therefore an account of it appears below, even though it may not have been a contemporary of people. It is possible that a number of Pliocene genera from South America should be included in the Heptaxodontidae (Starrett 1967). Some authorities, such as Ray (1964), consider the Heptaxodontidae to be only a subfamily of the Dinomyidae.

The skulls of this family are massive, and remains suggest that the bodies were stout. Each of the cheek teeth has four to seven laminae, the crests of which are nearly parallel and are arranged obliquely to the long axis of the skull. All genera were probably terrestrial. The presence of their remains with human refuse and artifacts suggests that some of these animals, at least, were used as food.

RODENTIA; HEPTAXODONTIDAE; *Genus* ELASMODONTOMYS *Anthony, 1916*

The single species, *E. obliquus,* is known only by osseous material from cave deposits at Utado, Morovis, and Ciales in Puerto Rico (Hall 1981). The name *Heptaxodon bidens* was evidently based on young individuals of *E. obliquus,* and is now considered a synonym of the latter (Ray 1964).

Elasmodontomys appears to have been about the size of a paca (*Agouti*) and to have had a heavy body. The skull resembles that of a nutria (*Myocastor*); it is flat topped and has rather large, laterally compressed bullae. The short bones of

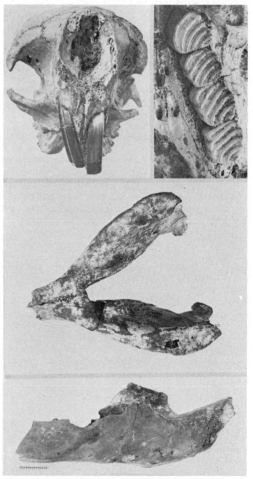

Elasmodontomys obliquus, photos from American Museum of Natural History.

the digits indicate that *Elasmodontomys* was terrestrial rather than arboreal. Available evidence suggests that this genus was distributed throughout the forested parts of Puerto Rico and apparently survived until about the time that European explorers arrived.

RODENTIA; HEPTAXODONTIDAE; *Genus QUEMISIA* Miller, 1929

The single species, *Q. gravis*, is known only by osseous material from cave deposits near St. Michel in Haiti, and Samana Bay in the Dominican Republic (Hall 1981).

Quemisia was apparently about the same size as *Elasmodontomys*. Its cranial peculiarities include a long union of the lower jaws, short lower incisors, and an unusual twisting of the enamel pattern of the cheek teeth.

Allen (1942) summarized the available data on *Quemisia*. The original describer of the genus, G. S. Miller, considered it identical with an animal called the "quemi" by Oviedo in his account of the animals of Hispaniola, published slightly more than 25 years after the Spanish discovery of this island. Oviedo's brief description indicated that the "quemi" resembled the hutias in color but was larger. That *Quemisia* was utilized as food by the native Indians of Hispaniola is suggested by the presence of its limb bones at a depth of about 1.2 meters in a kitchen midden near the entrance of a cave. The genus probably became extinct during the first half of the sixteenth century.

RODENTIA; HEPTAXODONTIDAE; *Genus AMBLYRHIZA* Cope, 1868

The single species, *A. inundata*, is known only by cranial fragments and teeth from cave deposits on the islands of Anguilla and St. Martin at the extreme northern point of the Lesser Antilles (Allen 1942).

Amblyrhiza was related to *Quemisia* and *Elasmodontomys*, but was much larger—nearly the size of an American black bear (*Euarctos*). Its skull is estimated to have been about 400 mm long. Although *Amblyrhiza* probably lived

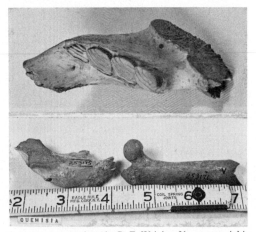

Quemisia gravis, photo by P. F. Wright of bone material in U.S. National Museum.

Amblyrhiza inundata, the last two teeth of upper molar series, photo from *Bull. Amer. Mus. Nat. Hist.*, "New Fossil Rodents from Porto Rico," H. E. Anthony.

well after Pleistocene times, there is no evidence to indicate how it reached the tiny islands where its remains were found or when it became extinct.

RODENTIA; *Family DASYPROCTIDAE*

Agoutis and Pacas

This family of 3 living genera and 15 species occurs from east-central Mexico to southern Brazil, and on the Lesser Antilles. There are two subfamilies: Dasyproctinae (agoutis), with the genera *Dasyprocta* and *Myoprocta;* and Agoutinae (pacas), with the single genus *Agouti*. Cabrera (1961) considered the Agoutinae to be a separate family, and *Stichomys* to be a genus distinct from *Agouti*, but otherwise his nomenclature and sequence are followed here. Some authorities, such as Starrett (1967), have used the name Cuniculinae in place of Agoutinae. It is perhaps unfortunate, and should be remembered, that the animals with the common name "agouti" are not the same as those with the scientific name *Agouti*.

Head and body length is 320 to 795 mm, tail length is 10 to 70 mm, and weight is approximately 1 to 10 kg. The pelage is coarse and thick. Pacas display a spotted pattern on a brownish or blackish ground color, but agoutis are usually uniformly colored above and pale below. Both groups have a piglike body and a rabbitlike head. The external form is modified for running, in that the limbs, especially the hind legs, are lengthened and the lateral toes are reduced in size. The thumb is vestigial. In the pacas the forefoot has four functional digits and the hind foot has five. In the agoutis the forefoot has four functional digits and the hind foot has three. The claws are thick and hooflike. Females have four pairs of ventral mammae.

The dental formula of the Dasyproctidae is: (i 1/1, c 0/0, pm 1/1, m 3/3) x 2 = 20. The incisors are fairly thin, and the cheek teeth are high crowned and semirooted. Part of the zygomatic arch of the skull in *Agouti* is greatly enlarged and contains a large sinus, a condition not found in any other mammal. The surface of this enlargement becomes extremely rugose in the adult. This character is the principal basis on which some authorities place *Agouti* in its own family.

If the extinct subfamily Cephalomyinae is included in the Dasyproctidae, the geological range of the family is Oligocene to Recent in South America (Starrett 1967). Otherwise the family is known only from the Recent.

Agouti (*Dasyprocta* sp.), photo from San Diego Zoological Garden. Inset: right forepaw and hind foot, photo from *Proc. Zool. Soc. London.*

RODENTIA; DASYPROCTIDAE; **Genus DASYPROCTA**
Illiger, 1811

Agoutis

There are 11 species (Cabrera 1961; Ojasti 1972; Hall 1981; Husson 1978):

D. mexicana, southern Mexico, introduced in Cuba;
D. ruatanica, Ruatan Island off northern Honduras;
D. coibae, Coiba Island off southwestern Panama;
D. punctata, southern Mexico to northern Argentina, introduced in Cayman Islands;
D. fuliginosa, Colombia, Venezuela, Surinam, Amazon Basin of Peru and northern Brazil;
D. kalinowskii, southeastern Peru;
D. leporina, Venezuela, Guianas, Brazil, introduced in the Lesser Antilles;
D. azarae, central and southern Brazil, Paraguay, northeastern Argentina;
D. guamara, Orinoco Delta in northeastern Venezuela;
D. cristata, the Guianas;
D. prymnolopha, eastern Brazil.

The use of the name *D. leporina,* instead of *D. aguti,* follows Husson (1978).

Head and body length is 415 to 620 mm, and tail length is 10 to 35 mm. Grzimek (1975) listed weight as 1.3 to 4.0 kg. The pelage is usually quite coarse and glossy, with the longest and thickest hairs on the posterior part of the back. In most forms the general coloration of the upper parts ranges from pale orange through several shades of brown to almost black.

The underparts are generally whitish, yellowish, or buffy. In some forms slight stripes may be present, and in others the rump is of a color contrasting to the remainder of the back. The body form is slender, the ears are short, and the hind foot has three toes with hooflike claws. Females have four pairs of ventral mammae.

Agoutis are found in forests, thick brush, savannahs, and cultivated areas. In Peru, according to Grimwood (1969:34–35), *Dasyprocta* "is confined to the Amazonian region, where it is found in all parts of the low selva [rain forest] zone and many parts of the high selva zone, in which it appears to extend to greater altitudes than *Cuniculus* [=*Agouti*] being regularly reported up to 2,000 m. or more. Agoutis live closely associated with water, but they are to be found on the banks of all types of streams, down to the merest runnel, and are therefore more widespread and common than pacas."

In some areas *Dasyprocta* constructs burrows among limestone boulders, along river banks, or under the roots of trees. In a study of *D. punctata* on Barro Colorado Island, Panama, however, Smythe (1978) found that each animal had several sleeping spots, usually located in hollow logs, among tree roots, or under tangled vegetation. Well-defined paths radiate out from the shelters of *Dasyprocta.*

Agoutis are basically diurnal, but in areas where they have been molested by people they may not leave their shelters until dusk. They are terrestrial and are adapted for a cursorial life. They walk, trot, or gallop on their digits, and also can jump vertically at least two meters from a standing start. They often sit with the body erect and the ankles flat on the ground, a position from which they can dart off at full speed. If danger threatens they may pause, motionless, with one forefoot raised, but if discovered they can travel with remarkable speed and agility. They usually sit erect to eat, holding the food in their hands. The diet consists of fruit, vegetables, and various succulent plants. On Barro Colorado Island, Smythe (1978) learned that *D. punctata* lived mainly on fruit, when it was available, and then buried the fruit seeds at scattered sites, to be eaten when fruit was not in season. Individuals often followed bands of monkeys and picked up the fruit that was dropped from trees. *D. punctata* also sometimes browsed and ate crabs.

Reported population densities of *D. punctata* are about 0.1/ha. in Tikal National Park, Guatemala, and 1.0/ha. on Barro Colorado Island (Cant 1977). In the latter area, Smythe (1978) found the basic social unit to be the mated pair, and that the bond between male and female apparently lasted until death. Each pair occupied a territory of about 1 or 2 ha., which included several sleeping spots and food trees, and a length of creek bed, where the natal nests were located. The pair, especially the male, aggressively drove off intruding agoutis. Territorial defense sometimes resulted in vicious fighting and severe wounds. When behaving aggressively, *D. punctata* sometimes erected the long hairs of its rump. When disturbed, this species may thump the ground with its hind feet. It also has a number of vocalizations, most notably an alarm bark, sounding like that of a small dog, which it makes as it runs away from danger.

In Venezuela, wild *Dasyprocta* apparently breed throughout the year (Ojasti 1972). Both seasonal and continuous reproduction have been reported for captive agoutis (Asdell 1964; Weir 1974*b*). On Barro Colorado Island, Smythe (1978) observed young almost every month, with a maximum from March to July, when fruit was abundant. Some populations of *Dasyprocta* evidently mate twice a year. During the courtship the male sprays the female with urine, causing her to go into a "frenzy dance;" after several sprays she allows the male to approach (Smythe 1978). Females have an average estrous cycle of 34 days, experience a postpartum estrus, and have a gestation period of about 104–20 days (Weir 1974*b*). Litters usually contain one or two young, there are sometimes three, and there is one record of four (Asdell 1964; Ojasti 1972; Weir 1974*b*; Smythe 1978). The newborn are fully furred, have their eyes open, and are able to run in their first hour of life (Smythe 1978). Lactation normally lasts about 20 weeks (Weir 1974*b*). The breakup between parents and offspring is associated with the coming of a new litter, increasing aggression by the adults, and a decline in food supplies (Smythe 1978). On Barro Colorado Island it was found that no young survived unless they were born during the fruiting season, and that even those born in that favorable period suffered 70 percent mortality in the following nonfruiting season. Deaths were caused both by lack of food and by predation from the coati (*Nasua*), males of which become carnivorous in the nonfruiting season. Subsequently the chances for survival greatly increased, and *Dasyprocta* has a potentially long life span. One captive *D. leporina* survived for 17 years and 9 months (Marvin L. Jones, Zoological Society of San Diego, pers. comm.).

Agoutis tame easily and make affectionate pets, but they are intensively hunted throughout their range by people for use as food. In some areas their numbers have declined seriously, because of both hunting and habitat destruction (Smythe 1978; Grimwood 1969). In parts of the Amazon Basin, however, the breaking up of virgin forests seems to have created more favorable habitat for *D. leporina*. There have been suggestions that in this area the species would be superior to beef cattle for commercial food production (N. Smith 1974).

As noted in the above systematic list, several species of *Dasyprocta* were introduced in the West Indies. These introductions apparently were carried out by native Caribbean peoples, before the coming of European explorers, for the purpose of supplying food. Populations of *D. leporina* were established as far north as the Virgin Islands, and were present long enough to develop into several named subspecies (Hall 1981). With the increase of humans on the islands, the clearing of the natural forests, and the introduction of predators, such as the mongoose, agoutis became very rare. They survived on some islands, however, through the early twentieth century (Allen 1942).

RODENTIA; DASYPROCTIDAE; **Genus MYOPROCTA**
Thomas, 1903

Acouchis

There are two species (Cabrera 1961; Husson 1978):

M. exilis (red acouchi), to the east of the Andes in Colombia, southern Venezuela, the Guianas, Ecuador, northern Peru, and the Amazonian basin of Brazil;

M. acouchy (green acouchi), to the east of the Andes in southern Colombia, eastern Ecuador, northern Peru, and the Amazonian basin of Brazil.

Cabrera used the name *M. acouchy* for the red acouchi and *M. pratti* for the green, but Husson pointed out that *M. exilis* is the proper name for the red and that *M. acouchy* applies to the green. Husson also indicated that both forms might be merely color variations of the same species.

The following descriptive data were provided in part by Barbara J. Weir (Wellcome Institute, Zoological Society of London, pers. comm.). Head and body length is 320 to 380 mm, tail length is 45 to 70 mm, and weight is 600 to 1,300 grams. The pelage is coarse. The upper parts are generally reddish to blackish or greenish. The muzzle and sides of the head are often brightly colored with yellow, orange, or red. The underparts are brownish, orangish, yellowish, or whitish. The short-haired tail is white underneath, and perhaps is used as an intraspecific signaling device.

Myoprocta closely resembles *Dasyprocta,* but is usually much smaller and has a more prominent tail. The incisor teeth are pale orange in color. Females have four pairs of ventral mammae.

Myoprocta seems restricted to the tropical forests of the Amazon Basin. Handley (1976) collected all specimens in moist areas of evergreen forest. Grimwood (1969) stated that *Myoprocta* generally occurs in the same areas as *Dasyprocta* in the low selva zone; it is reportedly less dependent on water than are other members of the Dasyproctidae, but is never found far from it. It is basically diurnal and terrestrial, and sometimes lives in holes in river banks. Its diet is thought to be about the same as that of *Dasyprocta*, and it also sits erect to eat and buries seeds in scattered places for future use (Smythe 1978).

Myoprocta apparently is more sociable than *Dasyprocta*. It has been reported to live in colonies. Observations of captives indicate that females tolerate and initiate contact with males to a much greater extent than do female *Dasyprocta*. As in the latter genus, urine spraying is a prominent feature of courtship (Kleiman 1971). Eisenberg (1974) listed about a dozen different vocalizations, including a "sharp twit," when startled; a "tooth chatter" threat; "peeping," when seeking contact; and a "snort" before an attack.

In captivity in London, breeding occurs at any time of the year, but there is a birth peak during the summer. Males are fertile throughout the year, but females may go into anestrus

Acouchi (*Myoprocta acouchy*), photo by Barbara J. Weir.

during the summer. The estrous cycle averages about 42 days. A postpartum estrus may occur, but females usually mate after lactation has ceased, unless the young are lost before that time. The mean gestation period is about 99 days. The average number of young is two and the range is one to three. The newborn weigh about 100 grams each, are fully furred, and have their eyes open. The minimum period of lactation needed for survival is 14 days. Both sexes reach puberty at an age of 8 to 12 months (Weir 1974b; Kleiman 1970). According to Marvin L. Jones (Zoological Society of San Diego, pers. comm.), a captive specimen lived for 10 years and 2 months.

Grimwood (1969) stated that *Myoprocta* is less common in Peru than *Dasyprocta*, but is an important source of food for the native people.

RODENTIA; DASYPROCTIDAE; **Genus AGOUTI**
Lacépède, 1799

Pacas

There are two species (Cabrera 1961; Hall 1981; Grimwood 1969):

A. paca, east-central Mexico to Paraguay;
A. taczanowskii, the Andes of northwestern Venezuela, Colombia, Ecuador, and possibly Peru.

The name *Cuniculus* Brisson, 1762 is often used in place of *Agouti*. Cabrera (1961), along with various other authorities, placed *A. taczanowskii* in a separate genus, *Stichomys* Thomas, 1924. Handley (1976), however, put this species in *Agouti*.

Head and body length is 600 to 795 mm, tail length is 20 to 30 mm, and weight is about 6.3 to 10.0 kg. The pelage of *A. paca* is rather coarse and lacks fine underfur. *A. taczanowskii* has a thicker and softer pelage. The upper parts of *Agouti* are brownish to black, and usually have four longitudinal rows of white spots on each side. The underparts are white or buffy.

The body is robust, the ears are medium sized, the forefoot has four digits, and the hind foot has five digits. Part of the zygomatic arch of the skull in *A. paca* is specialized as a resonating chamber, a feature that is not found in any other mammal. Females have four pairs of mammae (Weir 1974b).

Pacas live in a variety of habitats, but usually seek forested areas near water. *A. taczanowskii* lives on the high Andean paramo. Grimwood (1969) reported that *A. paca* is confined to the Amazonian part of Peru, where it is found throughout the low selva (rain forest) zone, but may occur locally up to elevations of 3,000 meters. It prefers small, swift, tree-shaded streams to open river banks.

Pacas are nocturnal, often spending the day in burrows that they excavate themselves or take over from some other animal. These burrows may be located in banks, on slopes, among tree roots, or under rocks, and usually have one or more escape exits. Leopold (1959) stated that an individual generally has several burrows, or uses such alternative shelters as hollow logs, stumps, and rock piles. In the limestone country of the Yucatan Peninsula, pacas frequent caves or sinkholes, and do not dig burrows. Hall and Dalquest (1963) reported that all the burrows they examined were simple tubes, but that they were mostly about two meters below the surface of the ground; one was excavated for a distance of over six meters, but the end was not reached.

After dusk, pacas emerge to follow well-defined pathways that lead to feeding grounds and water. Although pacas are terrestrial, they enter water freely and swim well. If alarmed, they generally attempt to make their escape in water. Their diet consists of leaves, stems, roots, seeds, and fallen fruit. Their favorite foods appear to be avocados and mangoes.

Leopold (1959) stated that pacas are not social animals, but live alone, each with its own holes and runways. Only about half as many vocalizations have been noted for *Agouti* as for the more sociable *Myoprocta* (Eisenberg 1974). In the Yucatan, pacas reportedly mate in early winter, and the

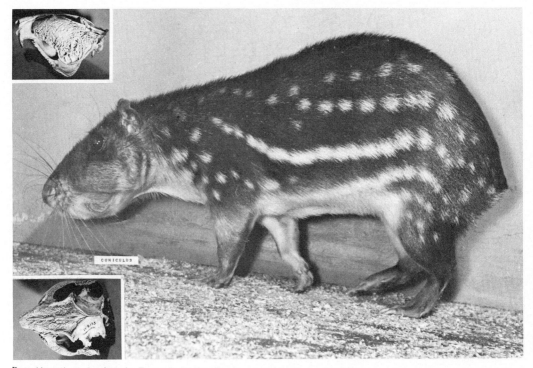

Paca (*Agouti paca*), photo by Ernest P. Walker. Insets: photos by Howard E. Uible.

females produce litters of two young in the dry season (winter–early spring). Other reports indicate that a single young is usually born, that twins are rare, and that there are probably two litters per year. In Rio de Janeiro, one litter was noted in February and another in July. According to Kleiman, Eisenberg, and Maliniak (1979), *A. paca* has a gestation period of 118 days, and produces a single young that weighs 710 grams. A captive specimen in the National Zoological Park in Washington, D.C., lived slightly more than 16 years.

Pacas are sometimes considered agricultural pests, as they may destroy yams, cassava, sugar cane, corn, and other crops. They are killed by people for this reason, and are also intensively hunted throughout their range for their flesh, which is said to be unusually delicious and to have a far higher price than any other meat, domestic or wild. Such hunting, in combination with habitat destruction, has resulted in pacas being exterminated or becoming rare over large areas (Hall and Dalquest 1963; Leopold 1959; Grimwood 1969; Baker 1974).

RODENTIA; *Family CHINCHILLIDAE*

Viscachas and Chinchillas

This family of three Recent genera and six species occurs in western and southern South America. The sequence of genera presented here follows that of Cabrera (1961).

Head and body length is 225 to 660 mm, and tail length is 75 to 400 mm. All genera have thick fur, that of the montane genera, *Lagidium* and *Chinchilla*, being finer than that of *Lagostomus*, which occurs on the pampas. The pelage is coarser on the tail than on the body. The palms and soles of most forms are naked. Coloration varies considerably between the genera.

The body form is slender. The forelimbs are relatively

short, but the hind limbs are long and muscular. The forefoot has four long, flexible digits. The hind foot is elongate. *Lagidium* and *Chinchilla* have four digits on the hind foot, with relatively weak claws. *Lagostomus* has only three hind digits, but they are armed with strong claws. Fleshy pads or pallipes are present on the feet. *Chinchilla* and *Lagostomus* have stiff bristles on the hind digits, which may be used for grooming.

The head, eyes, and ears are relatively large. The bullae are small in *Lagostomus*, intermediate in *Lagidium*, and enormously expanded in *Chinchilla* (Starrett 1967). The dental formula is: (i 1/1, c 0/0, pm 1/1, m 3/3) x 2 = 20. The incisors are fairly narrow. The cheek teeth are ever growing and have a pattern of tightly pressed transverse laminae without cement (Starrett 1967).

Chinchillids live in burrows or rocky crevices. They are basically cursorial, but often jump bipedally, and sit erect while eating, sunbathing, or grooming. All three genera have been observed to dustbathe. They are active throughout the year and are mainly vegetarian. They tend to be gregarious, and are seen in groups ranging from families to colonies of several hundred animals. There is a long gestation period and the young are born fully furred and with their eyes open. All three genera are intensively hunted by people.

The geological range of the Chinchillidae is early Oligocene to Recent in South America (Starrett 1967).

RODENTIA; CHINCHILLIDAE; *Genus LAGOSTOMUS*
Brookes, 1828

Plains Viscacha

The single living species, *L. maximus*, occurs in extreme southern Paraguay, and in northern and central Argentina (Cabrera 1961). Another species, *L. crassus*, is known only

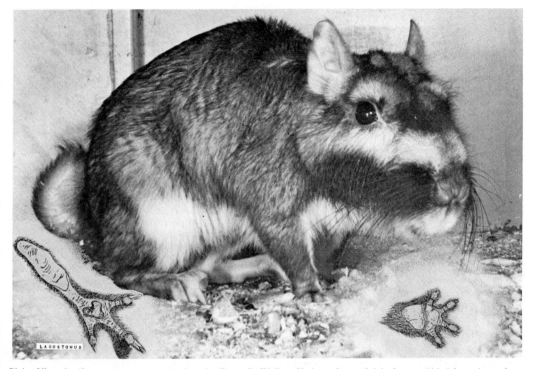

Plains Viscacha (*Lagostomus maximus*), photo by Ernest P. Walker. Undersurfaces of right front and hind feet, photos from *Proc. Zool. Soc. London.*

by a single skull found in southern Peru. Since the specimen was not fossilized, it may have represented an animal that lived in Recent times, but any associated population evidently disappeared long ago.

Head and body length is 470 to 660 mm, and tail length is 150 to 200 mm. Weight is about 2.0 to 4.5 kg in females and 5 to 8 kg in males (Weir 1974a). The guard hairs of the pelage are dark and coarse, but the underfur is very soft. The general coloration of the upper parts varies with the habitat, ranging from light brown in sandy areas, such as southern Buenos Aires province, to dark gray in Entre Rios. The underparts are white. The face is strikingly marked in black and white. The mustache is much thicker and more pronounced in males than in females. The tail is fully furred, the hairs on the dorsal surface being the longest. The base of the underside is usually bare and horny, and is used as a third leg when the animal sits erect on its haunches (Weir 1974a).

The head is large and blunt. The forefoot has four well-developed digits, and the hind foot has three digits armed with stout, sharp claws. The middle hind digit has the largest claw and also has a pad of stiff bristles on the inner side. The forefeet are used mainly for digging, and soil is pushed out with the nose. The rhinarium is furred and intricately folded, thus preventing earth from entering the nostrils. Females have two pairs of mammae, both placed laterally on the thorax (Weir 1974b).

The following ecological information was taken largely from Starrett (1967) and Weir (1974a). *Lagostomus* inhabits relatively barren parts of the pampas at elevations of up to 2,680 meters. It constructs extensive burrow systems, called viscacheras, some of which are used for centuries. Such a system may contain 4 to 30 entrances and cover up to 600 sq meters. Some entrances are so large that a person could stand waist deep in one. A round chamber is generally located about 1.5 meters from the entrance, and tunnels lead off from it in different directions. During construction of a burrow

system, 80 cubic meters of soil may be moved, and much of it deposited in flattened mounds about 50 cm high. The animals carry many inedible items—sticks, stones, bones, dung, and objects dropped by people—from the surrounding area and place them on the ground above the burrows. Various kinds of animals, such as insects, toads, lizards, snakes, burrowing owls, other rodents, and skunks, sometimes share the burrows with *Lagostomus*. Even foxes and boa constrictors may move in and prey on the viscacha.

Lagostomus is crepuscular, emerging from its burrow at dawn and dusk to feed. The surface areas in the vicinity of long-established viscacheras become denuded of food supplies, and the animals travel some distance over well-worn pathways to reach new feeding grounds. If a viscacha is pursued, it can travel at 40 km per hour, alternating running with 3-meter leaps and sharp turns. The diet consists mostly of grass and seeds, but nearly any kind of vegetation may be eaten.

Colonies contain 15 to 30, sometimes up to 50, individuals (Grzimek 1975). The exact composition probably varies from place to place, and in some areas the adult males are thought to live apart from the main colony at certain times of the year, especially when the young are born (Barbara J. Weir, Wellcome Institute, Zoological Society of London, pers. comm.). Although individuals tend to be concentrated in groups, overall population density in eastern Argentina has been estimated at two to five per ha. At the onset of the breeding season, adult males become more aggressive and engage in frequent fights with one another (Starrett 1967). Nearly all males bear scars from such conflicts, and deaths sometimes occur. Males express displeasure or aggression by a fearsome crescendo of notes, culminating in a high whine and accompanied by foot stomping and tail wagging. The most common vocalization of *Lagostomus*, however, is a two-syllable "uh-huh?" made when the animal is investigating (Weir 1974a).

Lagostomus is capable of breeding throughout the year in captivity. In some northern parts of the range, where the climate is favorable, wild females produce two litters annually. In east-central Argentina, however, there is a single breeding season, with mating in March or April and births in July or August. Females have an estrous cycle of 45 days and rarely experience a postpartum estrus, but there is generally a fertile postlactation estrus. The gestation period averages 154 (145–66) days. The mean number of young per litter is two and the range is one to four. The precocial offspring are fully furred and weigh about 200 grams each. The minimum period of lactation is 21 days, but weaning generally occurs by 8 weeks. Both sexes may attain sexual maturity as early as 5½ months of age, but the average is 8½ months for females and 15 months for males (Weir 1974b; Starrett 1967; Asdell 1964). A captive specimen lived for 9 years and 5 months (Barbara J. Weir, Wellcome Institute, Zoological Society of London, pers. comm.).

A hand-reared viscacha makes a delightful pet, but in the wild *Lagostomus* is generally considered a pest. It competes with domestic animals for forage, it sometimes raids cultivated crops, its huge burrows may cause livestock and even humans to stumble and break limbs, and its probably acidic urine devalues the soil for many years. For these reasons, and also because of its valuable flesh and fur, *Lagostomus* has been intensively hunted and poisoned by people. It once occurred in large numbers on the pampas, and it was said to have been possible for a person to ride 800 km without losing sight of the viscacha. In recent decades the range and numbers of *Lagostomus* have been greatly reduced by systematic extermination campaigns, and Weir (1974a) stated that the genus probably would be nearly extinct in about 20 years. Barlow (1969) indicated that the viscacha was once introduced into Uruguay, but had subsequently been extirpated there.

RODENTIA; CHINCHILLIDAE; **Genus *LAGIDIUM* Meyen,** *1833*

Mountain Viscachas

There are three species (Cabrera 1961; Rowlands 1974; Grimwood 1969):

L. peruanum, Peru;
L. viscacia, extreme southern Peru, western and southern Bolivia, northern and central Chile, western Argentina;
L. wolffsohni, southern Chile, southwestern Argentina.

Some authorities, including Rowlands (1974) and Barbara J. Weir, recognize *L. boxi* of northwestern Patagonia as a species distinct from *L. viscacia.*

The remainder of this account is based in part on data provided by Barbara J. Weir (Wellcome Institute, Zoological Society of London, pers. comm.). Head and body length is 300 to 450 mm, and tail length is 200 to 400 mm. The smallest species, *L. peruanum,* weighs 900 to 1,600 grams, but the other species may weigh up to 3,000 grams. The pelage is thick and soft except on the upper part of the tail, where it is coarse. The coloration of the upper parts ranges, as elevation increases, from dark gray to almost chocolate brown. There is a black middorsal stripe, which is more prominent in southerly populations. The underparts are usually white, yellow, or pale gray. The tip of the tail is black to reddish brown.

Mountain viscachas look like long-tailed rabbits, and in Patagonia they are called rock squirrels. The ears are long and covered with hair, and have a fringe of white hairs along the edge. Both the hind foot and the forefoot have four digits. The enamel of the incisor teeth is not colored. Females have a single pair of lateral thoracic mammae.

Mountain viscachas (*Lagidium viscacia*), photo by Ernest P. Walker. Insets: hand and foot, photos from *Proc. Zool. Soc. London.*

These viscachas inhabit dry, rocky, rugged, mountainous country, with sparse vegetation. In Peru they are most commonly found at elevations of 3,000 to 5,000 meters (Grimwood 1969). They are diurnal and do not hibernate. They live in rocky clefts and crevices, and spend much of the day on suitable perches, sunning themselves and dressing their fur. They are poor diggers and are rarely found in earth burrows. They run and leap among the rocks with great agility. Pearson (1948) found that *L. peruanum* never traveled more than about 70 meters from a rocky shelter, and always lived near water, possibly because of the succulent vegetation in the vicinity. *Lagidium* eats almost any kind of plant, including lichens, moss, and grass.

In Peru, *L. peruanum* has a remarkably sparse distribution, with colonies often being separated by 10 to 20 km of seemingly suitable habitat (Grimwood 1969). Pearson (1948) referred to this species as being among the most abundant mammals in the highlands of southern Peru, but gave its overall population density as only about 11 per sq km. He found colonies of 4 to 75 individuals, with the larger assemblies divided into probable families of two to five animals each. There also have been reports of colonies containing several hundred individuals. Although each family maintained its own burrow and sunning rocks, Pearson found intergroup relations to be generally peaceful, and he observed no serious fighting. At the onset of the breeding season, however, females drove males from the burrows, and the males then moved about considerably and became more promiscuous. A variety of vocalizations were heard, most notably a high-pitched, mournful whistle that evidently serves to warn others of danger.

In Peru, mating begins in October and all adult females are pregnant by December (Pearson 1948). In Patagonia, mating occurs in May or June. Females have a mean estrous cycle of 56.7 days (Weir 1974b). Pearson (1948) indicated that there is sometimes a postpartum estrus in *L. peruanum*, and that it might be possible for females to produce two or even three litters per year. This is unlikely, however, as the gestation period of *L. peruanum* is now known to be 140 days, and that of *L. viscacia* has been estimated at 120 to 140 days. Only one young is born at a time. It is fully haired, its eyes are open, and it can eat solid food on its first day of life. Newborn weight is about 180 grams in *L. peruanum* and 260 grams in *L. viscacia*. The young are normally weaned after about 8 weeks and both sexes reach puberty when 1 year old (Weir 1974b). Although few wild viscachas live more than 3 years (Pearson 1948), one captive *L. peruanum* survived for 19 years and 6 months (Marvin L. Jones, Zoological Society of San Diego, pers. comm.).

People hunt mountain viscachas for both meat and fur, and their numbers seem to have declined in some areas. In Chile they are protected by law, except in the extreme north, but illegal hunting is resulting in their becoming increasingly rare (Pine, Miller, and Schamberger 1979).

RODENTIA; CHINCHILLIDAE; *Genus* **CHINCHILLA** *Bennett, 1829*

Chinchillas

Cabrera (1961) recognized two species:

C. brevicaudata, the Andes of Peru, Bolivia, and northwestern Argentina;
C. laniger, the Andes of northern Chile.

Some authorities, as Pine, Miller, and Schamberger (1979) treat *C. brevicaudata* as a subspecies of *C. laniger*.

The remainder of this account is based in large part on data provided by Barbara J. Weir (Wellcome Institute, Zoological Society of London, pers. comm.). Head and body length is 225 to 380 mm, and tail length is 75 to 150 mm. The female is the larger sex and weighs up to 800 grams, while males rarely weigh over 500 grams. The silky pelage is extremely dense and soft, with as many as 60 hairs growing from each hair follicle. The general coloration of the upper parts is bluish, pearl, or brownish gray, and there is usually a black tip to each hair. The underparts are yellowish white. The furry tail

Chinchilla (*Chinchilla laniger*), the well formed feces are clearly shown, photo by Ernest P. Walker. Insets: undersurfaces of right front and right hind foot, photos from *Proc. Zool. Soc. London*.

is covered with coarse hairs on the dorsal surface. Chinchillas of many colors, ranging from white to black, have been bred in captivity.

The head is broad, the external ears are large, and the auditory bullae are relatively enormous. The eyes are large and black, and have a vertical slit pupil. Both the short forefoot and the narrow hind foot have four digits, with stiff bristles surrounding the weak claws. There are vestigial cheek pouches. The enamel of the incisor teeth is usually colored. Females have one pair of inguinal and two pairs of lateral thoracic mammae.

The natural habitat is the relatively barren areas of the Andes Mountains at elevations of 3,000 to 5,000 meters. Chinchillas shelter in crevices and holes among the rocks. They are basically crepuscular and nocturnal, though they have been observed on bright days, sitting in front of their holes and climbing and jumping about the rocks with remarkable agility (Grzimek 1975). To eat, they sit erect and hold the food in their forepaws. The diet consists of any available vegetation.

In former times, colonies of about 100 individuals were frequently seen. Although chinchillas are sometimes referred to as monogamous, there is no substantial evidence for this view. Females are very aggressive toward each other and toward males, even when in estrus, but serious fighting probably is rare in the wild. Threats are expressed by growling, chattering the teeth, and urinating.

The breeding season of *C. laniger* extends from November to May in captive populations in the Northern Hemisphere, and from May to November in the Southern Hemisphere. There are usually two litters during this season. *C. brevicaudata* occasionally produces three litters annually (Grzimek 1975). Females have an estrous cycle averaging 38 days and a postpartum estrus (Weir 1974*b*). The mean gestation period is 111 days. Litters contain one to six, usually two to three, young. The newborn weigh about 35 grams each, are fully furred, and have their eyes open. Lactation normally lasts 6 to 8 weeks. Sexual maturity is attained at an average age of 8 months by both male and female, but may occur as early as the age of 5½ months. The life span is probably about 10 years in the wild, but captives have lived for over 20 years, and some have bred when 15 years old.

Since the time of the ancient Incas, chinchilla fur has been highly prized as human apparel (Allen 1942). The coming of Europeans to the Western Hemisphere resulted in a greatly increased demand for this fur and a corresponding decline in the number and distribution of wild chinchillas. It is said that it was once possible for a person to see thousands of chinchillas in the course of a day's journey. As late as 1900 an estimated 500,000 skins were being exported annually from Chile (IUCN 1972). Shortly thereafter, however, the genus became rare and the price of its skin went higher. Indeed, a chinchilla pelt is the most valuable of any in the world, considering its size and weight. Coats made of wild chinchilla fur have sold for as much as $100,000, and that was years ago. Although chinchillas are now protected by law in their natural habitat, enforcement is difficult in the remote areas involved. Wild populations may actually be extinct, but all are on appendix 1 of the CITES, the species *C. laniger* is classified as vulnerable by the IUCN (1972), and the subspecies *C. brevicaudata boliviana* is listed as endangered by the USDI (1980). Some chinchillas were brought into captivity in the early twentieth century, and today probably hundreds of thousands, if not millions, of their descendants are bred commercially throughout the world (Grzimek 1975). Attempts have been made to introduce chinchillas into the wild in various places, including back into the Andes, but no successes have yet been reported.

RODENTIA; *Family* CAPROMYIDAE

Hutias and Nutria

This family contains 7 Recent genera, only 3 of which still survive, and 32 species. One of the living genera, *Myocastor* (nutria), is native to southern South America. All other Recent genera (hutias) occur only in the West Indies. The sequence of genera presented here follows that of Hall (1981), who recognized three subfamilies: Capromyinae, with *Capromys, Macrocapromys,* and *Hexolobodon;* Plagiodontinae, with *Plagiodontia, Isolobodon,* and *Hyperplagiodontia;* and Myocastorinae, with *Myocastor.* Some authorities, such as Packard (1967), consider the Myocastorinae to be a separate family, and others have classed it as a subfamily of the Echimyidae. Cabrera (1961) listed one additional genus, *Procapromys* Chapman, 1901, with the species *P. geayi,* supposedly from northern Venezuela. The place of origin of the single known specimen is questionable, however, and it may actually represent a young individual of the species *Capromys pilorides* (Packard 1967).

Head and body length is 200 to 635 mm, tail length is 35 to 425 mm, and weight is up to 9 kg. The pelage is harsh in hutias, but soft and thick in the nutria. Coloration is generally brownish or grayish. The head is broad in hutias, and is somewhat triangular in the nutria. All genera have a robust body form, small eyes and ears, short limbs, a reduced thumb, and prominent claws. There are two pairs of mammae in female hutias, and four pairs in the female nutria.

The dental formula is: (i 1/1, c 0/0, pm 1/1, m 3/3) x 2 = 20. In hutias the incisors are narrow, and the rootless cheek teeth are high crowned and grow throughout life. In the nutria the incisors are broad and strong. The cheek teeth are extremely high crowned, are semirooted, and decrease in size as they converge anteriorly.

Hutias are terrestrial and live in forests and plantations. The nutria is modified for an aquatic life, and inhabits marshes, lakes, and sluggish streams. The diet consists of vegetation and small animals. The young are born after a relatively long gestation period and are highly precocious. All genera have been either completely exterminated or reduced in numbers in their native habitat, through human agency. The nutria, however, also has been introduced in various parts of the Northern Hemisphere, where it has become abundant.

The geological range of the subfamilies Capromyinae and Plagiodontinae is Pleistocene to Recent in the West Indies. The geological range of the Myocastorinae is early Miocene to Recent in South America (Packard 1967).

RODENTIA; CAPROMYIDAE; *Genus* CAPROMYS
Desmarest, 1822

Cuban, Bahaman, and Jamaican Hutias

The following 6 subgenera and 17 species apparently lived during Recent times (Hall 1981; Varona 1979; Varona and Arredondo 1979; Kratochvil, Rodriguez, and Barus 1978):

subgenus *Capromys*

C. garridoi, Cayos Maja off south-central Cuba;
C. arboricolus, eastern Cuba;
C. pilorides, Cuba and several small nearby islands;

subgenus *Mysateles*

C. melanurus, eastern Cuba;
C. prehensilis, Cuba and Isle of Pines;

A & B. Cuban hutia (*Capromys pilorides*), photos by Ernest P. Walker. C. *C. melanurus,* photo by Erna Mohr.

subgenus *Mesocapromys*

C. auritus, Cayo Fragoso off north-central Cuba;
C. sanfelipensis, Cayo Juan Garcia off southwestern Cuba;

subgenus *Stenocapromys*

C. gracilis, known only by skeletal remains from cave deposits in western Cuba;

subgenus *Pygmaeocapromys*

C. minimus, known only by skeletal remains from eastern Cuba;
C. beatrizae, known only by skeletal remains from western Cuba;
C. silvai, known only by skeletal remains from south-central Cuba;
C. nana, Cuba;
C. angelcabrerai, Cayos de Ana Maria off south-central Cuba;

subgenus *Geocapromys*

C. brownii, Jamaica, Little Swan Island off northeastern Honduras;
C. columbianus, known only by skeletal remains from Cuba;
C. ingrahami, Bahamas;
C. pleistocenicus, known only by skeletal remains from Cuba.

Geocapromys Chapman, 1901 has often been treated as a separate genus. Unidentified and extinct species of *Geocapromys* are also known from Atwood Cay in the Bahamas and Cayman Brac in the Grand Cayman Islands (Hall 1981; Oliver 1977).

According to Kratochvil, Rodriguez, and Barus (1978), the living hutias of Cuba should be arranged as follows: (1) the genus *Capromys* contains only the single species *C. pilorides;* (2) *Mysateles* Lesson, 1842 is a full genus containing the subgenus *Mysateles* with the species *C. melanurus* and *C. prehensilis,* and the new subgenus *Leptocapromys* with the species *C. garridoi* and *C. arboricolus;* and (3) *Mesocapromys* Varona, 1970 is a full genus containing the subgenus *Mesocapromys* with the species *C. auritus,* and the new subgenus *Paracapromys* with the species *C. nana* and *C. sanfelipensis.*

Head and body length is about 220 to 500 mm. In the subgenus *Geocapromys,* tail length is only about 35 to 37 mm. In the other subgenera, tail length is 150 to 300 mm. Weight is about 0.5 to 7.0 kg. The furry pelage consists of long, coarse guard hairs and moderately dense, softer underfur. The coloration of the upper parts ranges through a variety of buff, yellow, red, gray, brown, and black shades. The underparts are usually paler. The tail is well haired and unicolored. In the subgenus *Mysateles* the tail is known to be prehensile. The feet are broad and have prominent claws. The stomach, at least in the subgenus *Capromys,* has two constrictions that divide it into three compartments, and it is thus one of the most complex stomachs in the Rodentia. Female *C. pilorides* and *C. nana* have two lateral thoracic pairs of mammae (Weir 1974*b*).

These hutias inhabit forests or rocky areas. The living members of all subgenera, except *Geocapromys,* seem to be basically arboreal. Some species are diurnal and sun themselves during the morning on leafy boughs of tall trees, curling up in such a way that from the ground they look like clumps of foilage. *C. melanurus* has been reported to be nocturnal, and to run along branches and leap from tree to tree like a squirrel (Allen 1942). Although some species are said to seek refuge in holes in the ground, Varona (1979) reported *C. angelcabrerai* and *C. auritus* to construct communal nests in mangrove trees. These nests are circular, measure about a meter in diameter, and are composed of

mangrove branches and leaves. The diet of Cuban hutias includes leaves, bark, fruit, lizards, and other small animals.

Clough (1972, 1973, 1974) made a detailed field study of *C. ingrahami* on East Plana Cay, a small semiarid island in the Bahamas. The animals were found to be almost completely nocturnal. By day they stayed underground in caves, crevices, or holes in the limestone. They showed no tendency to dig their own burrows, and probably did not make nests. They normally moved with a slow waddle, but were quicker when alarmed. They were facile climbers, ascended trees to look for food, and descended headfirst. They ran and climbed up steep, rocky cliffs with agility and speed. They stood on their hind limbs to reach into shrubs, and used their forepaws to grasp and manipulate food. Most of the night was spent eating, and the diet consisted mainly of bark, small twigs, and leaves. A supplementary laboratory investigation showed that individuals could survive well without water, and could tolerate water with half the salt concentration of sea water for over a week.

There were calculated to be 30 individual *C. ingrahami* per ha. on East Plana Cay, and this density was thought to have been maintained for many years. There were no natural predators on the island, but the hutias appeared to be in balance with available food supplies. Despite the crowded conditions, there was little social antagonism and no serious fighting. *C. ingrahami* was basically solitary, but there was no territoriality and considerable intraspecific tolerance. Two or more adults were often seen together, and 15 captives were kept in a small enclosure without trouble. In a laboratory investigation of *C. ingrahami,* Howe (1976) also observed a minimum of agonistic behavior, but found that when strangers were placed together there would be some strife and eventually establishment of an amicable dominance hierarchy. Threats between males increased when a female was in estrus. A possibly unique kind of nonagonistic wrestling was observed, generally in combination with mutual grooming. Howe (1974) also learned that individuals scent-marked with urine, not as a territorial measure but as a means of communication, especially of sexual information.

In the wild, *C. ingrahami* probably breeds throughout the year. In captivity, *C. pilorides* also breeds throughout the year but has a birth peak in June. *C. brownii* may produce two or more litters annually in captivity. Scattered records of gestation indicate a usual period of around 110 to 140 days. There is apparently only a single offspring per birth in *C. ingrahami* and *C. nana,* but one to three in *C. brownii* and one to four, usually two, in *C. pilorides.* The newborn are fully furred, have their eyes open, and are capable of moving about freely; those of *C. ingrahami* weigh about 80 grams. In *C. pilorides,* nursing continues for 153 days and sexual maturity is attained by the age of 10 months (Johnson, Taylor, and Winnick 1975; Weir 1974*b*; Clough 1972, 1974; Oliver 1977). A wild female *C. ingrahami* was recaptured over 5 years after being tagged, and within 20 meters of the original site of capture and release (Clough 1974). A captive *C. pilorides* lived for 11 years and 4 months, and a captive *C. brownii* lived for 8 years and 4 months (Marvin L. Jones, Zoological Society of San Diego, pers. comm.).

As indicated in the above systematic list, six species of *Capromys* are known only by skeletal remains, and these species are presumed to be extinct. The remains were often found in association with Indian artifacts and kitchen middens. It is thus evident that hutias have long been used as food by people, and some species may have disappeared before the time of Columbus. The arrival of European colonists accelerated the decline of hutias, through increased demand for meat, clearing of forests, and introduction of predators (Oliver 1977; Clough 1972, 1974, 1976; Allen 1942).

The subspecies *C. ingrahami abaconis* and *C. i. irrectus,* which once occurred throughout the Bahamas, were proba-

Bahaman hutia (*Capromys ingrahami*), photo by Garrett C. Clough.

bly exterminated in early colonial times. The well-studied subspecies *C. i. ingrahami* was not discovered until 1891, and it probably survived because East Plana Cay was never inhabited by people. There are an estimated 6,000 to 12,000 of the animals on the 456-ha. island, and some have also been introduced to a small cay in Exuma National Sea and Land Park. The only other relatively common species of *Capromys* is *C. pilorides* of Cuba, but even it has declined in numbers. It can still be legally hunted for two or three months of the year. All other species are now fully protected by the laws of the countries in which they live, but illegal hunting is jeopardizing some species.

There are serious problems in addition to hunting. *C. nana* was first known only by skeletal fragments from cave deposits, but was found alive in 1917. It formerly occurred throughout Cuba, but is now restricted to the Zapata Peninsula in the south-central part of the island, where it is highly endangered through agricultural development and predation by the introduced mongoose. *C. melanurus* of eastern Cuba is jeopardized mainly by deforestation. The subspecies *C. brownii brownii* of Jamaica has declined drastically in the last 30 years, because of the clearing of forests for agriculture and excessive hunting. *C. brownii thoracatus* of Little Swan Island seemed common in the early 20th century, but apparently disappeared following a severe hurricane in 1955 and subsequent predation by introduced house cats. The IUCN (1973, 1974, 1978) classifies *C. melanurus*, *C. nana*, and *C. ingrahami* as rare, and *C. brownii* as vulnerable. All surviving species, however, are probably in serious danger of extinction, except *C. pilorides* and *C. ingrahami* (Oliver 1977; Clough 1976).

RODENTIA; CAPROMYIDAE; **Genus MACROCAPROMYS** Arredondo, 1958

The single species, *M. acevedoi*, is known only by skeletal remains from cave deposits in western Cuba, and is appar-

ently extinct. The humerus is one-fifth longer than that in the largest species of *Capromys* (Hall 1981).

RODENTIA; CAPROMYIDAE; **Genus HEXOLOBODON** Miller, 1929

There are two species (Hall 1981):

H. poolei, known only by skeletal remains from the type locality in north-central Haiti;
H. phenax, known only by skeletal remains from north-central Haiti.

Hexolobodon was about the size of *Capromys pilorides*, but the skull differs from that of *Capromys* in having a shorter rostrum and a generally more robust form. *Hexolobodon* resembles the subgenus *Geocapromys*, but differs in that the root of the premolar is thrown conspicuously forward and away from that of the first molar; the roots of the cheek teeth are less specialized and appear to close with age, instead of continually growing; and the lower incisor root is extended to the outer side of the mandibular tooth row. The remains of *Hexolobodon* have been found in cave deposits. The genus presumably was exterminated soon after the coming of European colonists.

RODENTIA; CAPROMYIDAE; **Genus PLAGIODONTIA** F. Cuvier, 1836

Hispaniolan Hutias

The following seven species apparently lived during Recent times (Hall 1981):

P. aedium, Hispaniola;
P. hylaeum, Hispaniola;
P. ipnaeum, known only by skeletal remains from Hispaniola;

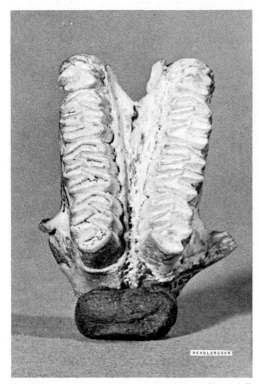

Hexolobodon phenax, upper molars, photo by Howard E. Uible of specimen in U.S. National Museum.

P. spelaeum, known only by skeletal remains from the type locality in Haiti;
P. araeum, known only by a single tooth from the type locality in Haiti;
P. caletensis, known only by skeletal remains from the type locality in the Dominican Republic;
P. velozi, known only by skeletal remains from the type locality in Haiti.

In *P. aedium,* head and body length is about 312 mm, and tail length is about 153 mm. Kleiman, Eisenberg, and Maliniak (1979) listed adult weight as 1,267 grams. In *P. hylaeum,* head and body length is 348 to 405 mm, and tail length is 125 to 145 mm. Judging from the skeletal remains, the largest species in the genus is *P. ipnaeum* and the smallest is *P. spelaeum.* In the living species, the short, dense pelage is brownish or grayish on the upper parts and buffy on the underparts. The tail is scaly and practically naked. Both the forefoot and the hind foot have five digits, all armed with claws except the thumb, which has a short, blunt nail. Females have three pairs of lateral thoracic mammae (Weir 1974*b*).

Hispaniolan hutias inhabit forests. Captives have been observed to be nocturnal and arboreal, and to use nest boxes placed high off the ground (Oliver 1977). Wild *P. aedium* were reported to be active only at night, to hide during the day, to feed principally on roots and fruit, and to live in male-female pairs. Specimens of *P. hylaeum* were caught in December in hollow trees near a lagoon; four pregnant females each contained a single embryo (Allen 1942). According to Johnson, Taylor, and Winnick (1975), captive female *P. aedium* have an estrous cycle of 10 days, a gestation period of 119 days, and apparently a single offspring. Weir (1974*b*) listed gestation as 123–50 days and litter size as one to two young in this species. A captive *P. aedium* was still alive after 9 years and 11 months (Marvin L. Jones, Zoological Society of San Diego, pers. comm.).

Five of the seven species in this genus are known only by skeletal remains, often found in association with human

Hispaniolan hutia (*Plagiodontia hylaeum*), photo from New York Zoological Society.

kitchen middens. These five species probably disappeared by the 17th century, because of excessive hunting by people (Allen 1942; Goodwin and Goodwin 1973; Oliver 1977). *P. aedium* and *P. hylaeum* have been greatly reduced in range and numbers and are now endangered through deforestation, hunting, and predation by the introduced mongoose. Both species are officially classified as rare by the IUCN (1972).

RODENTIA; CAPROMYIDAE; *Genus ISOLOBODON*
J. A. Allen, 1916

There are two subgenera and three species (Hall 1981):

subgenus *Isolobodon*

I. portoricensis, known only by skeletal remains from the Dominican Republic, Puerto Rico, and the Virgin Islands;
I. levir, known only by skeletal remains from Hispaniola;

subgenus *Aphaetreus*

I. montanus, known only by skeletal remains from Hispaniola.

These extinct hutias were about the size of *Plagiodontia* (Allen 1942). They are distinguished mainly by dental characters. According to Hall's (1981) key, in *Plagiodontia* the root of the upper incisor occurs at the anterior margin of the zygomatic process of the maxillary, and the upper premolar has one external reentrant angle. In *Isolobodon* the root of the upper incisor occurs in the infraorbital foramen, and the upper premolar has two external reentrant angles.

Available evidence suggests that *I. portoricensis* was native to Puerto Rico, but was transported to other islands by the early Indians. It may have been domesticated, and its abundant remains in kitchen middens indicate that it formed a staple article of diet for the aborigines. *I. levir* was evidently native to Hispaniola and seems not to have been so heavily utilized as human food. Both species apparently disappeared shortly after the coming of European explorers. *I. montanus* inhabited mountainous areas. Its remains have not been found in association with people and its date of extinction is not known. Skeletal fragments of all three species are present in cave deposits made by the extinct giant barn owl (*Tyto ostolaga*). This owl preyed on the hutias, and perhaps disappeared when their numbers were reduced by excessive human hunting (Allen 1942).

RODENTIA; CAPROMYIDAE; *Genus HYPERPLAGIODONTIA* Rímoli, 1977

The single species, *H. stenocoronalis*, is known only by skeletal remains from the type locality in north-central Haiti. This evidently extinct hutia was larger than *Plagiodontia*, its nearest generic relative. Its molariform teeth are slightly high crowned and have short roots. The folds in the molariform teeth are more compressed anteriorly, and are sharper and more conspicuous than in *Plagiodontia* (Hall 1981).

RODENTIA; CAPROMYIDAE; *Genus MYOCASTOR*
Kerr, 1792

Nutria, Coypu

The single species, *M. coypus*, is native to southern Brazil, Bolivia, Paraguay, Uruguay, Argentina, and Chile (Cabrera 1961).

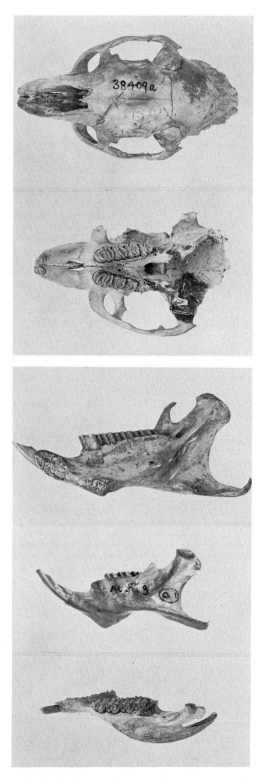

Isolobodon portoricensis, photos from American Museum of Natural History.

Nutria or coypu (*Myocastor coypus*), photo by P. Rödl.

Head and body length is 430 to 635 mm, and tail length is 255 to 425 mm. Weight is usually 5 to 10 kg, but occasionally reaches 17 kg (Grzimek 1975). The pelage of the upper parts contains many long, coarse guard hairs with a general coloration of yellowish brown or reddish brown. These hairs nearly conceal the soft, thick, velvety, dark gray underfur. The belly fur is pale yellow and is not as coarse as that of the upper parts. The tail is scaly and thinly haired, except at the base.

Myocastor looks like a large, robust rat. The eyes and ears are small. The tail is cylindrical, not laterally compressed as in the muskrat (*Ondatra*). The webbed hind foot is much longer than the forefoot, and contains five digits; the first four are connected by skin, and the fifth is free and used for grooming the fur. The forefoot has four long, flexible, unwebbed digits and a vestigial thumb. The claws are sharp and strong. The incisor teeth are large and colored a brilliant orange. Females have four pairs of thoracic mammae that are situated well up on the sides of the body.

The nutria is semiaquatic, inhabiting marshes, the edges of lakes, and sluggish streams. Although it generally prefers fresh water, the population of the Chonos Archipelago in Chile occurs in brackish and salt water. According to Lowery (1974), *Myocastor* is adept on land but at home in the water, and is often seen by day but is most active at night. It commonly makes platforms of vegetation, where it feeds and grooms itself. For shelter the nutria takes over the hole of another animal or constructs its own burrow. The latter may be a simple tunnel or a complex system containing passages that extend 15 meters or more and chambers that hold crude nests of vegetation. The nutria also makes runways through the grass and wanders within a radius of about 180 meters of its den. Much time is spent swimming and browsing on aquatic plants. The diet is strictly vegetarian (Lowery 1974). A study in Maryland determined roots to be the most important food, and population density to range from 2.7 to 16.0 nutria per ha. (Willner, Chapman, and Pursley 1979).

Myocastor lives in pairs, but the presence of many animals in favorable habitat may give the impression of a large colo-

ny. Breeding occurs throughout the year in captivity and in at least some wild populations. Females are polyestrous and may produce two or three litters annually (Grzimek 1975; Brown 1975; Willner, Chapman, and Pursley 1979). Females experience a postpartum estrus within a day or 2 of giving birth, have an estrous cycle of about 24–26 days, and have a period of receptivity lasting 1–4 days (Lowery 1974). Ovulation may be induced. The gestation period is usually 128–30 days. Litter size averages about 5, and ranges from 1 to 13, young. The newborn weigh approximately 225 grams each, are fully furred, and have their eyes open. They are capable of surviving away from the mother after only 5 days of nursing, but they usually remain for 6 to 10 weeks. Sexual maturity may be attained when the young are only 3–4 months old if they were born in the summer, and when 6–7 months old if they were born in the fall (Weir 1974*b*). A captive nutria lived for 6 years and 2 months (Marvin L. Jones, Zoological Society of San Diego, pers. comm.).

A demand for the velvety, plushlike underfur of the nutria developed in the early 19th century and has continued to the present. *Myocastor* also has been hunted by people for use as food. The genus became rare in much of its natural habitat, but by the early 1900s efforts were being made to regulate the harvest and to establish captive breeding farms (Allen 1942). Such farms were begun both in the original range of the nutria and in many other parts of the world. Some animals escaped from captivity, or were deliberately introduced, and feral populations developed in the United States, Canada, England, France, Holland, Scandinavia, Germany, Asia Minor, the Caucasus, Soviet Central Asia, and Japan (Corbet 1978; Van Den Brink 1968; Hall 1981). Although intensively trapped for its fur, the nutria is considered a pest in some of these places because its burrows damage dikes and irrigation facilities, it raids rice and other cultivated crops, and it competes with native fur-bearing animals.

The most extensive populations in the United States are in the south-central part of the country, especially in the marshes of Louisiana. Most of these animals are apparently descended from 20 individuals brought to Louisiana in 1938.

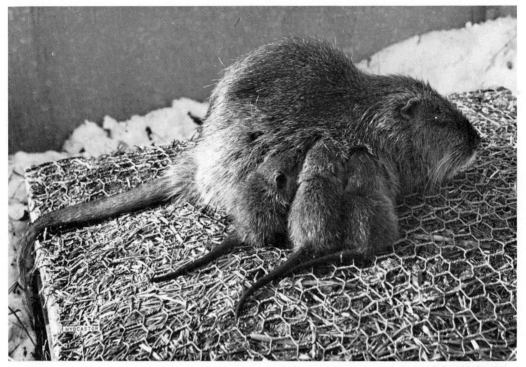

Nutria or coypu (*Myocastor coypus*) with nursing young, photo by H. L. Dozier from U.S. Fish and Wildlife Service.

By the late 1950s there were an estimated 20 million feral nutria in Louisiana, and by 1962 the nutria had replaced the muskrat as the leading fur animal in the state (Lowery 1974). In the 1976/77 season, 1,890,853 nutria pelts from Louisiana were marketed. The total number for the United States was 2,018,815, and the value was $10,598,779 (Deems and Pursley 1978).

RODENTIA; *Family OCTODONTIDAE*

Octodonts

This family of five Recent genera and eight species occurs in Peru, Bolivia, Argentina, and Chile. The sequence of genera presented here follows that of Cabrera (1961)

Head and body length is 125 to 310 mm, and tail length is 40 to 180 mm. In the genera *Octodon*, *Octodontomys*, and *Octomys*, the tail is about as long as the head and body. In *Aconaemys* and *Spalacopus*, the tail is much shorter than the head and body. The fur on the body is usually long, thick, and silky. The coarse hair of the tail is short at the base and becomes longer toward the tip, where it forms a tuft.

The head is large and the nose is pointed. The moderate-sized ears are rounded and covered with short hair. The whiskers are long. The hind foot has four well-developed toes, each with a sharp, curved claw and a row of stiff bristles extending beyond the claw. The thumb is reduced (Packard 1967). The dental formula is: (i 1/1, c 0/0, pm 1/1, m 3/3) x 2 = 20. The grinding surfaces of the cheek teeth are shaped like a figure eight, or sometimes like a kidney. Females have three or four pairs of mammae.

Octodonts inhabit coastal areas, foothills, and the Andes Mountains to elevations of 3,500 meters or more. They live both in cultivated areas and on barren, rocky slopes. They are good diggers, especially *Aconaemys* and *Spalacopus*. They either construct their own burrows or dwell in crevices, rock piles, fences, hedgerows, or the holes of other animals. These agile rodents carry their tails erect while running, and sometimes climb shrubs or trees. They often sit on their haunches. The diet consists of plant matter, such as bulbs, tubers, bark, and cactus. Growing corn stalks are occasionally eaten. Females may give birth to two litters annually, each with 1 to 10 young.

The geological range of this family is early Oligocene to Recent in South America, and Pleistocene in the West Indies (Packard 1967).

RODENTIA; OCTODONTIDAE; *Genus OCTODON*
Bennett, 1832

Degus

There are three species (Cabrera 1961; Pine, Miller, and Schamberger 1979):

O. degus, Atacama to Curico Province in northern and central Chile;
O. bridgesi, base of the Andes in the provinces of O'Higgins, Colchagua, Curico, and Concepcion in central Chile;
O. lunatus, coastal parts of Coquimbo, Aconcagua, and Valpariso provinces in central Chile.

A report of *O. degus* from Peru may have been based on an escaped pet (Woods and Boraker 1975).

Head and body length is 125 to 195 mm, and tail length is 105 to 165 mm. Woods and Boraker (1975) gave the weight

Degu (*Octodon degus*), photo by Luis E. Peña, from *Trans. Zool. Soc. London.*

of *O. degus* as 170 to 300 grams, and noted that the other species are larger. The upper parts are grayish to brownish, often with an orange cast, and the underparts are creamy yellow. There is a black brush at the tip of the tail, which is most prominent in *O. degus.*

When holding a degu by the tail, care must be taken or the animal may spin like a top and leave the person holding only the skin (Woods and Boraker 1975). The degu then bites off the naked vertebrae and tendons, there is little bleeding, and the wound heals. The lost section is not replaced, and many animals are found with a portion of the tail missing. When the degu is familiar with the handler, tail shedding rarely occurs.

On all four limbs the first four digits are well developed and have sharp claws, but the fifth is reduced. There are long, comblike bristles extending over the claws of the hind feet. The auditory bullae are moderate in size and smaller than those of *Octodontomys* and *Octomys.* The anterior upper cheek teeth of *O. degus* are kidney shaped, with only moderate internal indentations. In the other two species, these teeth have deep indentations that nearly touch the opposite side of the teeth (Woods and Boraker 1975). The incisor enamel is pale orange. Females have eight mammae.

The only species for which there is substantial natural history information is *O. degus,* and it has been investigated in both the wild and the laboratory (Woods and Boraker 1975; Fulk 1975, 1976; Boulenge and Fuentes 1978; Meserve, Rodriguez-M., and Martin 1978; Weir 1974b). This species inhabits the west slope of the Andes at elevations of up to 1,200 meters. It is found in relatively open areas near thickets, rocks, or stone walls. It is active throughout the year and is diurnal, with peaks of activity in the morning and late afternoon. It feeds on the ground and will also climb into the branches of small trees and bushes. An elaborate communal burrow system is constructed, with the main section under rocks or shrubs. A complex network of tunnels and surface paths leads out to feeding sites. Piles of sticks, stones, and dung accumulate on the ground above the tunnels, and apparently serve as territorial markers. The diet includes grass, leaves, bark, herbs, seeds, and fruit, and during the dry season may be supplemented with fresh droppings of cattle and horses. The degu sometimes becomes an agricultural pest by raiding orchards, vineyards, and wheat fields. It stores food for winter use.

This species is the most common mammal of central Chile. Reported population densities have ranged from 10 to 259 individuals per ha., with 40 to 80 per ha. possibly being near average. Densities are highest in the spring (September–November), when juveniles are entering the population. *O. degus* lives in small colonies with a strong social organization based on group territoriality. The burrow is the center of the defended territory. Females of the same social group may rear their young in a common burrow. Captive animals exhibit considerable social tolerance, and a group of strangers may be placed together without fighting. Eisenberg (1974) reported a variety of vocalizations.

Captive colonies in the Northern Hemisphere breed throughout the year, and females may have more than one annual litter. In the natural range in central Chile, however, pregnant and lactating females have been found only in September. In northern Chile, pregnant females or juveniles have been taken in February, April, and November. There seems to be no regular estrous cycle, and female *O. degus* may require the presence of a male to induce ovulation. They experience a postpartum estrus, but do not usually mate at this time. The gestation period averages about 90 days. Litters contain 1 to 10 young, with an average of 6.8 in a laboratory colony in Vermont. In this colony the young have been born fully furred and with their eyes open. In another group in London, however, the newborn have been sparsely haired and their eyes have not opened for 2 or 3 days. Each newborn weighs about 14 grams. The offspring must nurse for at least 14 days and normally do so for 4 weeks. Wild parents have been observed to cut fresh grass and bring it into the burrow for the young to eat. The age of sexual maturity has been reported to be as early as 45 days and as late as 20 months, but the average is about 6 months. According to Marvin L. Jones (Zoological Society of San Diego, pers. comm.), a specimen of *O. degus* was still alive after 7 years and 1 month in captivity.

RODENTIA; OCTODONTIDAE; **Genus OCTODONTOMYS**
Palmer, 1903

Chozchoz (plural = chozchoris)

The single species, *O. gliroides,* occurs in the highlands of southwestern Bolivia, northern Chile, and northwestern Argentina (Cabrera 1961; Pine, Miller, and Schamberger

Chozchoz (*Octodontomys gliroides*), photo by Barbara J. Weir.

1979). Except as noted, the information for this account was provided by Barbara J. Weir (Wellcome Institute, Zoological Society of London, pers. comm.).

Head and body length is about 180 mm, tail length is about the same, and weight is 100 to 200 grams. An adult female from Chile had a head and body length of 170 mm, a tail length of 132 mm, and a weight of 158 grams (Pine, Miller, and Schamberger 1979). The fur resembles that of *Chinchilla* in being very dense and soft. Coloration is pearly gray above and white below. The eyes and external ears are relatively large. The auditory bullae and the tail brush are larger than those of *Octodon*. There are no apparent modifications for fossorial life. Females have one inguinal and two lateral pairs of mammae (Weir 1974b).

The chozchoz is found in mountains and on the altiplano. It lives in burrows at the bases of cacti and scrubby acacias, in rock crevices, in caves, and in old Indian tombs. In northern Argentina, at least, this rodent is active only after dusk. It eats acacia pods during the winter and fruits of the cactus during summer. Burrows can be detected by the piles of empty yellow seed pods outside. Eisenberg (1974) stated that the threat vocalization of *Octodontomys* is a "twitter."

On 17 November, Pine, Miller, and Schamberger (1979) collected both a lactating female and a juvenile in northern Chile. Weir (1974b) reported that females have a postpartum estrus, gestation lasts 100 to 109 days, and litters contain one to three young, each weighing 15 grams at birth. The young are born fully furred and with their eyes open. One specimen was still living after 7 years and 4 months in captivity (Marvin L. Jones, Zoological Society of San Diego, pers. comm.).

RODENTIA; OCTODONTIDAE; **Genus OCTOMYS**
Thomas, 1920

Viscacha Rats

There are two species (Cabrera 1961):

O. mimax, mountains of northwestern Argentina;
O. barrerae, plains of eastern Mendoza Province in west-central Argentina.

Cabrera put *O. barrerae* in the separate genus *Tympanoctomys* Yepes, 1940, but Packard (1967) included *Tympanoctomys* in *Octomys*.

Head and body length is 160 to 170 mm, and tail length is 170 to 180 mm. The upper parts are pale buff. The under-

parts, hands, and feet are white. The body form is not modified for fossorial life, and the tail is bushy. This genus resembles *Octodontomys* externally, but the reentrant folds of the upper molars in *Octomys* meet in the middle of the teeth. The auditory bullae of *Octomys* are the largest in the family Octodontidae (Packard 1967).

Viscacha rats appear to be rare and are difficult to trap. There is no evidence that the scarcity has been caused by human agency. These rodents spend the day sleeping in their burrows. Females produce several litters per year.

RODENTIA; OCTODONTIDAE; **Genus SPALACOPUS**
Wagler, 1832

Coruro

The single species, *S. cyanus*, occurs in central Chile, from the coast to the Andean slopes, and from Coquimbo to Maule Province (Reig 1970; Cabrera 1961; Pine, Miller, and Schamberger 1979).

Head and body length is 140 to 160 mm, tail length is 40 to 50 mm, and weight is about 60 to 110 grams. The pelage is soft and glossy. The coloration is brownish black or black throughout, though one specimen had large, irregular white pectoral and pelvic patches. The feet are dark gray.

The body form is stocky. The tail is cylindrical, scaly, almost hairless, and only about as long as the hind foot. The claws are not greatly enlarged, but the forefeet are specialized for digging. The eyes and external ears are small. The cheek teeth are ever growing and have crowns with an 8-shaped pattern, but the reentrant folds do not meet in the middle of the teeth. The incisors are long, broad, and strongly curved forward, being even more procumbent than in *Ctenomys* (Reig 1970).

The coruro is fossorial. It inhabits both coastal areas and mountain slopes, and has an elevational range of sea level to about 3,000 meters. Pine, Miller, and Schamberger (1979) found the genus to be very abundant on wet flats along streams. Reig (1970) studied colonies in a semiarid area with scattered clumps of vegetation. Except as noted, the remainder of this account is based on that study.

Spalacopus maintains a complex communal burrow system. There are long, interconnecting tunnels, about 5 to 7 cm in diameter and generally 10 to 12 cm below the surface. Leading from the main tunnels are short passages, some of which are used to push excavated earth to surface mounds and some of which terminate under clusters of food plants.

Viscacha rat (*Octomys mimax*), photo by Michael Mares.

Entrances are not kept plugged. In one colony there were 350 recently made openings, of which 314 were in an area of 400 sq meters. *Spalacopus* uses its incisors and forefeet to dig in a manner much like that of *Ctenomys*. It is evidently active only during daylight and is rarely seen above ground, except when excavating soil. Colonies are nomadic, using up the food resources of one area in a few days and then tunneling on to new sites. The diet consists primarily, if not ex-

clusively, of the tubers and underground stems of the lily *Leucoryne ixiodes*. Free water evidently is not required. According to reports made prior to Reig's investigation, *Spalacopus* hoards bulbs and tubers in its burrows for the winter, and these stores are raided by people for use as food.

The colonies are small but usually adjoin one another, so *Spalacopus* can sometimes be found almost continuously over an extensive area. One colony consisted of 15 animals,

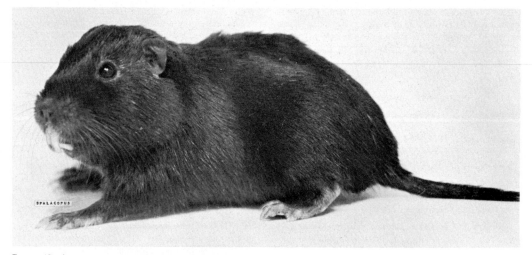

Coruro (*Spalacopus cyanus*), photo by Luis E. Peña.

including at least 3 adult males, 4 nonpregnant adult females, 2 pregnant females, 2 juvenile males, and 2 juvenile females. The pregnant females, which Reig caught in late February, each contained 3 small fetuses. Another source indicated that females have two litters per year.

The call of the coruro, as heard when wild individuals push soil to the surface, is a sequence of three or four trills, each lasting about 5 seconds and separated by about 3 seconds of silence. Captives were active and noisy. Eisenberg (1974) listed several other sounds, including a "tooth chatter," used as a threat. According to Marvin L. Jones (Zoological Society of San Diego, pers. comm.), one captive was still living after 5 years and 10 months.

RODENTIA; OCTODONTIDAE; *Genus ACONAEMYS* Ameghino, 1891

Rock Rat

The single species, *A. fuscus,* occurs in the highlands of west-central Argentina and in that part of Chile from Curico to Osorno Province (Cabrera 1961; Pine, Miller, and Schamberger 1979).

Head and body length is 150 to 185 mm, and tail length is 55 to 75 mm. Greer (1965) listed weights of 118.9 and 135.3 grams, respectively, for a male and a nonpregnant female, and 152.2 grams for a pregnant female. The coloration of the subspecies *A. fuscus fuscus* is usually dark brown throughout. Some specimens have bicolored tails and a suffusion of white or cream-colored hairs on the dorsal surface of the feet (Pine, Miller, and Schamberger 1979). The southerly subspecies, *A. f. porteri,* has a more woolly pelage and a bicolored tail, black above and white below.

This stocky rodent is modified for fossorial life. The tail is much shorter than the head and body. *Aconaemys* differs from *Spalacopus* in that the reentrant folds of the upper molars meet in the middle of the teeth and the upper incisors do not curve forward so markedly.

Aconaemys is found in the high Andes and in coastal mountains. Most of what is known about this genus was recorded by Osgood (1943). He, however, thought that it was restricted to coniferous forests and that it would disappear as the large trees (*Araucaria*) under which it burrowed were cut down by people. Greer (1965) found *Aconaemys* to inhabit open, cleared areas as well as forest, and to burrow under bushes as well as trees. Both authorities agreed that *Aconaemys* makes a complex, though shallow, burrow system and prefers well-drained ground near ledges and boulders. There are a series of entrances that open flush with the ground and are connected by numerous branching tunnels. The entrances are also connected by surface runways, which are either open or partly concealed by vegetation. In the middle of the 19th century, *Aconaemys* was reported to be very common on the east side of the Andes and to honeycomb the ground with its burrows, thus posing a danger to horses and riders. The genus is mainly nocturnal, but occasionally is seen by day. It is active throughout the winter, under the snow. Newborn young and pregnant females were collected in November. One of the females contained two embryos and another had five.

RODENTIA; *Family CTENOMYIDAE; Genus CTENOMYS* De Blainville, 1826

Tuco-tucos

The 2 subgenera and 38 currently recognized species (not in systematic order) of the single Recent genus, *Ctenomys,* are as follows (Cabrera 1961; Roig and Reig 1969; Contreras, Roig, and Suzarte 1977; Ximenez, Langguth, and Praderi 1972; Pine, Miller, and Schamberger 1979):

subgenus *Ctenomys*

C. australis, east-central coast of Argentina;
C. azarae, central Argentina;
C. boliviensis, central and southern Bolivia;
C. brasiliensis, eastern Brazil;
C. colburni, extreme southwestern Argentina;
C. coludo, northwestern Argentina;
C. dorsalis, Paraguay;

Rock rat (*Aconaemys fuscus*), photo by Richard D. Sage.

Tucu-tuco (*Ctenomys* sp.), photos by Ernest P. Walker.

C. emilianus, west-central Argentina;

C. famosus, west-central Argentina;

C. frater, southern Bolivia, northwestern Argentina;

C. fulvus, northern Chile;

C. johanis, west-central Argentina;

C. knighti, northwestern Argentina;

C. latro, northwestern Argentina;

C. leucodon, region of Lake Titicaca in southern Peru and western Bolivia;

C. lewisi, southern Bolivia;

C. magellanicus, extreme southern Chile and Argentina, including Tierra del Fuego;

C. maulinus, central Chile;

C. mendocinus, Argentina;

C. minutus, southern Brazil, Uruguay;

C. nattereri, Mato Grosso of central Brazil;

C. occultus, northwestern Argentina;

C. opimus, southern Peru, southwestern Bolivia, northern Chile, northwestern Argentina;

C. perrensis, northeastern Argentina;

C. peruanus, extreme southern Peru;

C. pontifex, west-central Argentina;

C. porteousi, east-central Argentina;

C. robustus, northern Chile;

C. saltarius, northwestern Argentina;

C. sericeus, extreme southwestern Argentina;

C. steinbachi, eastern Bolivia;

C. talarum, east-central Argentina;

C. torquatus, Uruguay;

C. tuconax, northwestern Argentina;

C. tucumanus, northwestern Argentina;

C. tulduco, west-central Argentina;

C. validus, west-central Argentina;

subgenus *Chacomys*

C. conoveri, Paraguay.

Part of the information used in the remainder of this account was provided by Barbara J. Weir (Wellcome Institue, Zoological Society of London, pers. comm.). Head and body length is 150 to 250 mm, and tail length is 60 to 110 mm. Weight ranges from 100 grams in *C. talarum* to 700 grams in *C. tucumanus.* The skin on the body is loose. The pelage ranges from short to long and is usually thick. The coloration of the upper parts ranges from gray or creamy buff through various shades of brown to almost black. The underparts are usually paler.

In outward appearance these South American burrowing rodents strongly resemble the North American pocket gophers (Geomyidae), and their habits are much the same. Pocket gophers have external cheek pouches, which tuco-tucos lack, and have better-developed fringes of hair on the forefeet. In *Ctenomys,* however, the hair fringes on the hind feet are stiffer and form comblike bristles. These "combs," which are the basis of the familial and generic names, are used to groom dirt from the fur.

The body is robust and cylindrical, the head is large, the tail is short and sparsely haired, and the neck and limbs are short and musuclar. The forelimbs are somewhat shorter than the hind limbs. All the digits have strong claws, those on the forefeet being the longest. The eyes are small, though perhaps somewhat larger than would be expected for fossorial rodents, and the external ears are reduced. The auditory bullae are enlarged. The dental formula is: (i 1/1, c 0/0, pm 1/1, m 3/3) x 2 = 20. The prominent incisor teeth have a bright orange enamel and are large and thick, but are not as procumbent as those of *Spalacopus.* The crown pattern of the molars is kidney shaped, and the last molar is reduced. Females, at least those of *C. opimus, C. peruanus,* and *C. talarum,* have one inguinal and two lateral thoracic pairs of mammae.

Tuco-tucos are found from the tropics to the subantarctic, but seem to prefer coastal areas, grassy plains, forests, and the altiplano. The elevational range is sea level to 4,000 meters (Weir 1974a). *C. torquatus* has been found in cultivated fields, pastures, and vacant lots in the city of Montevideo (Barlow 1969). The main requirement of this fossorial genus seems to be sandy, somewhat dry soil. There is one species, *C. lewisi,* that tunnels in stream banks and may be semiaquatic.

Burrows usually consist of a long main tunnel, either sinuous or nearly straight, and several short passages that either have a blind ending or lead to feeding sites at the surface. The main tunnel may be 14 meters long and is about 5 to 7 cm wide and 30 cm below the surface (Packard 1967; Barlow 1969). There is a grass-lined nest chamber, usually below the level of the main tunnel, and several chambers for food storage. The surface entrances may be marked by heaps of excavated soil and are often plugged if the burrow is occupied.

Tuco-tuco (*Ctenomys knighti*), photo by Michael Mares.

Short lengths of chopped grass are commonly found in the burrows of *C. talarum* and may serve as air filters. Both this species and *C. torquatus* keep burrow temperature at an average of 20° to 22° C by opening and closing the surface entrances according to the direction of wind and sun. A burrow is sometimes shared with other animals, such as lizards, mice, and cuis (*Galea*).

Most digging takes place in the hours of daylight, mainly during early morning and late afternoon. The earth is loosened with the forefeet and then swept away with the hind feet. The incisors are used mainly to cut through portruding roots (Weir 1974*a*). *C. talarum* is known to pack the earth of its tunnel walls firmly by standing on its forefeet, walking the hind feet up the wall, urinating, and then stamping the moistened soil into position as the hind feet are lowered. Tuco-tucos are rarely seen completely out of their burrows, but short surface forays may occur on sunny days. The position of the eyes, almost level with the top of the head, enables an animal to look out of its shelter without appreciably exposing itself. If danger threatens, the animal rapidly retreats backward, its nearly naked tail apparently acting as a sensory organ. *C. peruanus* and *C. opimus* can distinguish a moving human at a distance of about 50 meters. Some food, in the form of roots, stems, and grass, is probably pulled into the burrows from below. *Ctenomys* also obtains the aboveground parts of plants by reaching out from the surface entrance of the burrow (Barlow 1969). The diet is thought to be entirely vegetarian.

Population densities as high as 207 individuals per ha. have been recorded (Pearson et al. 1968), but the figure is usually much lower. Barlow (1969) noted 80 burrow systems

in an area of 7 ha. In most species a burrow is occupied by only one individual, but several female *C. peruanus* may share the same system.

The common name, "tuco-tuco," is an attempt to express in words the vocalization of some species, such as *C. talarum*. This call is made mainly by the males, probably to show territoriality or fear. The actual sound is something like "tloc-tloc-tloc," and it seems to come bubbling up from the ground. A single vocalization lasts from 10 to 20 seconds, starting slowly and then becoming more rapid.

Females are generally monestrous and produce only a single litter per year. They may experience a postpartum estrus, but do not usually mate at that time (Weir 1974*b*). In Peru, *C. opimus* mates in the dry season, and most births occur in the wet season, when plant growth is at its peak. In Uruguay, however, *C. torquatus* mates in the wet season, and pregnant females are found from September to January (Barlow 1969). In Argentina, *C. talarum* mates from May to July. In Chile, Pine, Miller, and Schamberger (1979) found a lactating female *C. opimus* on 15 September and one of *C. robustus* on 26 April. Weir (1974*b*) listed average gestation periods of 120 days for *C. peruanus*, 102 days for *C. talarum*, and 107 days for *C. torquatus*; and litter sizes of one to three young in *C. opimus*, one to five in *C. peruanus*, and one to seven in *C. talarum*. Newborn *C. peruanus* are well developed, and are able to give the same call as adults, to leave the nest, and to feed on green vegetation almost immediately. The young of captive *C. talarum*, however, do not have their eyes open at birth and have only the guard hairs projecting through the skin. They weigh about eight grams each. They grow quickly and can usually fend for themselves when 10 days old, but normally nurse for 5 weeks. Both sexes of *C. talarum* reach puberty at an average age of 8 months, in the next breeding season after they were born. Mean longevity in the wild is probably less than 3 years.

Tuco-tucos are sometimes considered pests, because they damage cultivated crops and compete with livestock for forage, and thus have been intensively persecuted by people. They are now absent or greatly reduced in numbers in some areas in which they were formerly abundant, as, for example, in southern Patagonia to the east of the Andes, an area now largely fenced and devoted to sheep raising. Horseback riding in country frequented by tuco-tucos is sometimes dangerous, as the burrows cave in and cause the horses to break their legs.

The geological range of the family Ctenomyidae is Pliocene to Recent in South America. This family is thought to be a fossorial offshoot from the Octodontidae.

RODENTIA; *Family ABROCOMIDAE; Genus ABROCOMA Waterhouse, 1837*

Chinchilla Rats or Chinchillones

The single Recent genus, *Abrocoma*, contains two living species (Cabrera 1961; Pine, Miller, and Schamberger 1979):

A. bennetti, central Chile;
A. cinerea, southeastern Peru, southern Bolivia, northern Chile, northwestern Argentina.

Remains of an extinct species, *A. oblativa*, have been recovered from burial sites in Peru.

In *A. cinerea*, head and body length is 150 to 201 mm, and tail length is 59 to 144 mm. In the somewhat larger *A. bennetti*, head and body length is 195 to 250 mm, and tail length is 130 to 180 mm. An adult male *A. bennetti* weighed 224 grams, and an adult female weighed 307 grams (Pine, Miller, and Schamberger 1979). The pelage consists of long, soft, dense underfur, and fine, long guard hairs; it resembles that of *Chinchilla*, but is not so woolly. Coloration is silvery gray above and white or yellowish below in *A. cinerea*, and is brownish gray above and brownish below in *A. bennetti*. There may be a white or yellowish patch on the chest, marking a glandular region.

Abrocoma bears some resemblance to *Chinchilla*, but has proportionately longer head and ears, which give a ratlike appearance—hence the common name, "chinchilla rats." The nose is pointed, the eyes are large, and the ears are large and rounded (dishlike). The tail is cylindrical, shorter than the head and body, and covered with fine, short hairs. The limbs are short. The forefoot has four digits and the hind foot has five. The soles of the feet are naked and covered with small tubercles. The small, weak claws are hollow on the underside. Stiff hairs project over the claws of the three middle toes of the hind foot (as in the families Chinchillidae, Octodontidae, and Ctenomyidae), forming combs that are probably used to remove parasites and groom the pelage, or perhaps to aid in removing dirt loosened by digging.

The intestinal tract of *Abrocoma* is long; the small intestine measures 1.5 meters and the large intestine, 1 meter. The voluminous cecum is 205 mm in length. *A. bennetti* has more ribs than any other rodent, 17 pairs. Female *Abrocoma* have four mammae.

The rostrum of the skull is long and narrow, the braincase is rounded, and the bullae are enlarged. The dental formula is: (i 1/1, c 0/0, pm 1/1, m 3/3) x 2 = 20. The incisors are narrow and the cheek teeth continue to grow throughout life. The lower cheek teeth differ from the uppers in having two deep, sharply angular, inner folds (Packard 1967).

The species *A. cinerea* is known only from the altiplano, generally occurring in rocky areas at elevations of 3,700 to 5,000 meters. *A. bennetti* is known from the Andes and coastal hills of Chile up to elevations of 1,200 meters. Chinchillones live in tunnels in the ground and among the crevices of rocks. The entrances to these tunnels are usually located at the base of a bush or under rocks. Fulk (1976) found that *A. bennetti* sometimes shared a burrow with *Octodon degus*, a rodent of comparable size; on two occasions nest burrows were determined to be occupied by mothers and young of both species. *A. bennetti* has been reported to climb bushes and trees. The diet of *Abrocoma* includes many kinds of plant material.

Chinchillones appear to be colonial; about half a dozen *A. cinerea* were noted living within 18 meters of each other. This species is reported to "grunt" before or after an attack, to "squeak" while avoiding intruders, and to "gurgle" when being groomed (Eisenberg 1974). A pregnant female *A. cinerea*, with two embryos, was taken in December in Peru, and a mother with two young, about a week old, was collected in April in northern Chile. Weir (1974b) listed a single gestation period of 115–18 days for this species, and gave litter size as one to two young. Fulk (1976) collected a pregnant female *A. bennetti* near Santiago, Chile, on 25 July, took another 300 km to the north on 1 June, and observed a female carrying two newborn at Santiago on 30 August. He

Chinchilla rat (*Abrocoma bennetti*), photo by Ernest P. Walker.

gave the embryo counts for three pregnant females of this species as four, five, and six. According to Marvin L. Jones (Zoological Society of San Diego, pers. comm.), a captive *A. bennetti* lived 2 years and 4 months.

Chinchillones are said to be rare, but this supposed scarcity may be a factor of the inaccessible habitat. In some areas the native people sell the pelts of these animals to gullible travelers as *Chinchilla*. In parts of Chile the skins are taken to local fur markets, but they do not bring high prices.

The geological range of the family Abrocomidae is late Miocene to Recent in South America.

RODENTIA; *Family ECHIMYIDAE*

Spiny Rats

This family contains 16 Recent genera and 58 species. One genus, *Heteropsomys*, occurred in the West Indies, but apparently became extinct in the nineteenth century. The living genera are found from southern Honduras to Peru, Paraguay, and southeastern Brazil. The sequence of genera presented here is based on those of Cabrera (1961) and Hall (1981), who recognized two subfamilies: Echimyinae, for most genera; and Dactylomyinae, for *Dactylomys, Kannabateomys,* and *Thrinacodus*.

Head and body length is 105 to 480 mm, and tail length is 50 to 430 mm. The tail ranges from less than one-fourth the head and body length (in some *Euryzygomatomys*) to more than the head and body length (Packard 1967). All living genera, except the dactylomyines, and *Trichomys* and *Isothrix*, have a spiny or bristly pelage, consisting of flattened, stiff, sharp-pointed hairs attached to the flesh by narrow basal stalks. The tail varies from well haired to scantily haired. The tail is often lost, because it fractures easily, and this factor may help in escape from predators.

Echimyids are ratlike in general appearance, with pointed or somewhat truncate noses and moderate-sized eyes and ears. The ears are rounded or bluntly pointed, and extend beyond the pelage. The limbs are of moderate length. The first digit of the forefoot (pollex) is vestigial. The other digits are short in most genera; moderately elongate in the semiarboreal *Isothrix, Diplomys, Echimys,* and *Makalata;* and greatly elongate, somewhat united, and modified in the Dactylomyinae, so the first and second toes of the hind foot, and the second and third of the forefoot, are on one side of a limb while the more lateral toes are on the other side (Packard 1967). The dental formula is: (i 1/1, c 0/0, pm 1/1, m 3/3) x 2 = 20. The cheek teeth are flat crowned and rooted.

Spiny rats inhabit forests or clearings, often near water. They are often common and the most abundant mammals in parts of their range, but most genera are little known. They may dig their own burrows, or seek shelter beneath stumps, logs, or rocks. Several genera live in the hollows of trees, and *Euryzygomatomys* is apparently fossorial. *Trichomys* has been noted in captivity leaping on its hind legs, but the other genera are climbing or scampering animals. Echimyids are reported to be good swimmers. They are usually active in the evening or night. Most spiny rats die when exposed to heat and dryness. The herbivorous diet includes grass, sugar cane, bananas, other fruit, and nuts. Some echimyids consume large amounts of water.

These rodents somteimes live in small groups. Various scolding sounds have been noted, but captives are often peaceful. If handled carefully, they do not attempt to bite. Breeding may occur throughout the year in much of the range of the Echimyidae. Litter size is one to seven young. The newborn are well furred and their eyelids are formed. Within a few hours they are active and nimble, and can emit soft whistling sounds. They begin to eat solid food at about 11 days, and leave the mother at about 2 months.

The geological range of this family is late Oligocene to Miocene, and Pleistocene to Recent, in South America, and Pleistocene to Recent in Central America and the West Indies.

RODENTIA; ECHIMYIDAE; *Genus PROECHIMYS;*
J. A. Allen, 1899

Spiny Rats, Casiragua

The 2 subgenera and 19 currently recognized species (not in systematic order) are as follows (Cabrera 1961; Patton and Gardner 1972; Hall 1981; F. Petter 1978*b*; Martin 1970; Husson 1978; Handley 1959, 1976; Goodwin and Greenhall 1961):

Spiny rat (*Proechimys setosus*), photo by Joao Moojen.

subgenus *Proechimys*

P. brevicauda, eastern Peru, northwestern Brazil;
P. canicollis, Colombia to Guyana;
P. cuvieri, French Guiana;
P. goeldii, Amazonian Brazil;
P. guairae, northeastern Venezuela;
P. guyannensis, Colombia, Venezuela, Trinidad, the
    Guianas, Ecuador, northeastern Peru, Brazil, Bolivia;
P. hendeei, eastern Ecuador, northeastern Peru,
    possibly southern Colombia;
P. hoplomyoides, southern Venezuela;
P. longicaudatus, Peru, Brazil, Bolivia, northern Paraguay;
P. oris, northeastern Brazil;
P. quadruplicatus, eastern Ecuador, extreme northern
    Bolivia;
P. semispinosus, southern Honduras to northern Peru
    and northwestern Brazil;
P. steerei, western Brazil;
P. warreni, Guyana, Surinam;

subgenus *Trinomys*

P. albispinus, eastern Brazil;
P. dimidiatus, southeastern Brazil;
P. iheringi, southeastern Brazil;
P. myosuros, eastern Brazil;
P. setosus, eastern Brazil.

Petter (1973) suggested that the generic name *Cercomys* Cuvier, 1829 should replace *Proechimys*. As explained by Patton and Gardner (1972), the systematics of *Proechimys* are highly unstable, and the above list could easily be altered through further investigation of different taxonomic interpretation. Hall (1981) listed *P. corozalus*, known only by a fossilized jawbone of uncertain geologic age from Puerto Rico, but noted that it may not actually represent *Proechimys*.

Head and body length is 160 to 300 mm, and tail length is 109 to 320 mm. The weight of *P. guyannensis trinitatus* is 300 to 380 grams. *P. semispinosus* weighs up to 500 grams (Gliwicz 1973). The fur is spiny, but not as much so as in other genera of the Echimyidae. The general coloration of the upper parts is orangish brown, reddish brown, or light tawny mixed with black. The underparts are usually whitish. The thinly haired tail is brown above and white below. The tail is frequently absent through autotomy—it readily fractures at the fifth caudal vertebra and does not regenerate. This process probably serves as a means of escape from predators that seize the tail. Females have one inguinal and two lateral thoracic pairs of mammae (Weir 1974b).

Casiragua usually inhabit forests, often near coasts and waterways. They occasionally move into human habitations, but generally live in burrows in the ground, beneath tree roots, and among rocks. They are capable of climbing, but rarely do so. They are nocturnal, leaving their dens in the evening to forage on the forest floor. Many kinds of plant material are eaten. In captivity they accept such items as fruit, corn, coconuts, cereals, potatoes, and cabbages. *P. semispinosus* is known to cache food in its burrow (Maliniak and Eisenberg 1971).

In the Panama Canal Zone, Fleming (1971) determined that population density of *P. semispinosus* increased from 1.1 individuals per ha. in June to 3.8 per ha. in October (the wet season), and then declined to 0.6 per ha. in the following June. Gliwicz (1973) found an average density of 8.5 per ha. in the same area. *P. semispinosus* defends its burrow against adults of the same sex, but there is considerable overlap in individual home range (Maliniak and Eisenberg 1971). In a study of *P. guyannensis* on Trinidad, Everard and Tikasingh (1973) found a maximum population density of about 14 per

ha. and an average home range of 0.1 ha. *Proechimys* is generally docile in captivity, but adults may be aggressive to one another and inflict serious wounds. *P. semispinosus* growls and chatters its teeth, when threatening a rival, and has several other vocalizations (Eisenberg 1974).

There is breeding throughout the year by *P. semispinous* in the Panama Canal Zone and by *P. guyannensis* on Trinidad (Fleming 1971; Everard and Tikasingh 1973), and the same is probably true for most other populations of *Proechimys*. Litters may follow in rapid succession, the average annual number per female *P. semispinosus* in Panama being 4.68. Female *P. guairae* have a mean estrous cycle of about 23 days, 50 percent of them experience a fertile postpartum estrus, and 50 percent have a fertile prepartum estrus, 1 or 2 days before giving birth (Weir 1974b). Patton and Gardner (1972) reported collection of pregnant female *P. guyannensis*, *P. hendeei*, and *P. longicaudatus*, each with 1 to 3 embryos, from June to August in northeastern Peru. Everard and Tikasingh (1973) reported a gestation period of 62–64 days and a mean litter size of 2.4 young for *P. guyannensis*. Weir (1974b) listed gestations of 61–64 days for *P. guairae* and 63–66 days for *P. semispinosus*, and the following average (and extreme) numbers of young per litter: *P. dimidiatus*, 3 (1–5); *P. guairae*, 3 (1–7); and *P. semispinosus*, 3 (1–5). She also reported that the newborn are highly precocious and that those of *P. guairae* weigh 22 grams each. The normal period of nursing in this species is 3 weeks; puberty may be reached at a minimum age of 35 days, but usually occurs when females are 2 months old and males 3 months old. Everard and Tikasingh (1973) reported that the young of *P. guyannensis* are weaned at 21–35 days of age and become sexually mature at 146 days, that wild individuals have been recaptured after 20 months, and that captives have been kept 3½ years. Fleming (1971) stated that *P. semispinosus* attains sexual maturity at an age of 6 or 7 months, and that wild individuals over 2 years old are common. According to Marvin L. Jones (Zoological Society of San Diego, pers. comm.), a captive *P. semispinosus* lived 4 years and 10 months.

RODENTIA; ECHIMYIDAE; **Genus HOPLOMYS;**
J. A. Allen, 1908

Armored or Thick-spined Rat

The single species, *H. gymnurus*, is found from southern Honduras to the western side of the Andes in Colombia and northwestern Ecuador (Handley 1959; Hall 1981).

Head and body length is 220 to 320 mm, and tail length is 150 to 255 mm. Weight is about 450 grams (Kleiman, Eisenberg, and Maliniak 1979). Males are heavier than females. The pelage is more spiny than in other echimyid genera. The spines are best developed in the middle of the anterior part of the back, where they are up to 33 mm long and up to 2 mm in diameter. They are white basally and colored toward the tip. The spines on the upper parts are tipped with black; those on the sides are usually tipped with orange or banded with orange and black. The spines come out easily if they are brushed the wrong way. Beneath the spines are soft hairs, usually orange or yellow in color. In one form, these soft hairs are blackened and form a distinct stripe along the middle of the back from the snout to the base of the tail. The underparts are whitish. The scantily haired, scaly tail is usually brownish above and whitish below. The sparsely furred ears are dark brown to black. *Hoplomys* is similar in body form to *Proechimys*. The hind feet are long and narrow, and the tail is shorter than the head and body. The tail is easily lost through autotomy.

Thick-spined rat (*Hoplomys gymnurus*), photo by R. B. Tesh. Inset: photo by P. F. Wright of specimen in U.S. National Museum.

The thick-spined rat has been found in fairly open rain forests, thickets, and grassy clearings, usually near large decaying logs. The maximum recorded elevation is about 780 meters. This rodent is terrestrial and burrows in the banks of creeks. A short, simple horizontal shaft is made, with an enlarged chamber near the entrance and a nest chamber at the end. The nests are dry and kept free from feces. Runways are made in the vegetation above the ground. A wide variety of plant food is eaten, but insects are also taken. One to three precocious young are produced, probably throughout the year. In Panama, Tesh (1970b) took pregnant females from February to July, but did not collect any females from October to January. He found a mean litter size of 2.1. The newborn were furred, open eyed, and ambulatory.

RODENTIA; ECHIMYIDAE; *Genus*
EURYZYGOMATOMYS Goeldi, 1901

Guiara

The single species, *E. spinosus,* occurs in eastern and southern Brazil, Paraguay, and northeastern Argentina (Cabrera 1961).

Guiara (*Euryzygomatomys spinosus*), photo by Ernest P. Walker.

Head and body length is 170 to 270 mm, and tail length is 50 to 55 mm. Males may be somewhat larger than females. As in the other Echimyidae, spines are present on the back. The coloration is drab brown above and whitish below. The hands and feet are brown. Females have one inguinal and two lateral thoracic pairs of mammae (Weir 1974b).

The guiara lives in areas covered with grass or bushes, preferably near water. The well-developed claws and short tail indicate fossorial habits, but very little information on natural history is available. A pregnant female, with three embryos, was taken near Rio de Janeiro in November. Two pregnant females, with a single embryo each, were collected in the Brazilian state of Minas Gerais.

RODENTIA; ECHIMYIDAE; *Genus CLYOMYS*
Thomas, 1916

There are two species (Avila-Pires and Caldatto Wutke 1981):

C. *laticeps,* central and southern Brazil;
C. *bishopi,* southeastern Brazil.

Head and body length is 105 to 230 mm, and tail length is 55 to 87 mm. The coloration above is grayish brown and black, mixed with rufous. The underparts are whitish or buffy. Grayish patches may be present in the throat region and in the middle of the chest and belly. The fur is bristly. *Clyomys* is distinguished from other genera of the family Echimyidae by the well-developed fossorial claws on the forefeet and the enlarged bullae of the skull.

Burrows of *Clyomys* have been found in a natural clearing, and individuals of this genus appear to live in colonies. Joao Moojen (pers. comm. to Ernest P. Walker) described the collection of two specimens as follows: "The traps that caught both specimens alive were set at the mouth of the most elaborate rodent nest I ever saw. A very regular three-spiral helix wound around a large vertical root. The sloping passageway was 80 mm. wide and 90 mm. high and went down 850 mm. vertically to the main nest that was 250 mm. in diameter. Another nest was midway in the spiral and was only 170 mm. in diameter. The lower nest was covered with scant grassy materials, quite worn and old; the midway nest was clean with a few bits of a meaty sweet root (I tasted it) which I was not able to identify."

RODENTIA; ECHIMYIDAE; *Genus CARTERODON*
Waterhouse, 1848

The single species, C. *sulcidens,* occurs in eastern Brazil (Cabrera 1961). The species was first described in 1841 from fossil remains collected at Lagoa Santa in the state of Minas Gerais. In 1851, living animals were found on the open

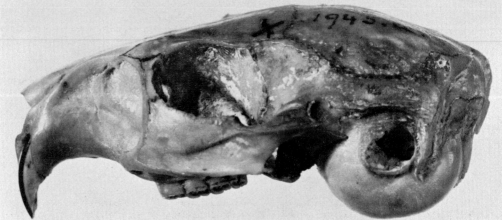

Clyomys laticeps, photos by Joao Moojen. Skull: photo from British Museum (Natural History).

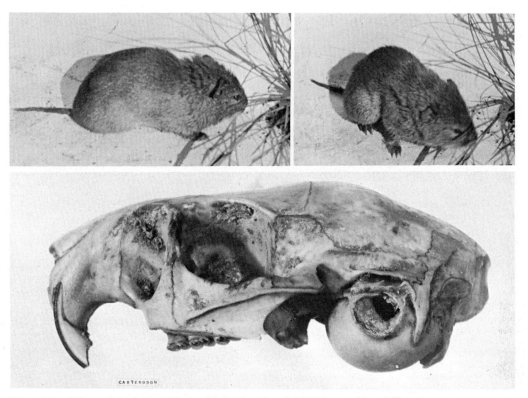

Carterodon sulcidens, photos by Joao Moojen. Skull: photo from British Museum (Natural History).

pampas in Lagoa Santa. Another specimen, collected by Miranda Ribeiro in Campos Novos, Mato Grosso, is now in the Museu Nacional, Rio de Janeiro, Brazil. Two more specimens were taken by Joao Moojen in Brasilia.

Head and body length is about 155 to 200 mm, and tail length is about 68 to 80 mm. The upper parts are yellow brown, shaded with black, the sides are grayish, the lower parts of the neck and throat are reddish, the sides of the belly are yellowish red, and the middle of the belly is white. The well-haired tail is black above and pale yellow below. The back of *Carterodon* is supplied with bristles and spines; the relatively soft spines end in a hairlike point and are long and flexible. The upper incisor teeth are grooved.

Carterodon inhabits the large mesa, savannahlike *cerrado* region, about 400 meters above sea level, where there are two definite seasons, one from October to March, when there are about 2,030 mm of rainfall each year, and the other for the remaining months of the year, when the rains completely stop. The vegetation is grass with scattered trees.

A few notes on the natural history of *Carterodon* were made many years ago by John Reinhardt, who stated: "It inhabits the open *Campos*, overgrown with shrubs and trees, where it digs its residence, consisting of a rather long tube 3 to 4 inches [76 to 102 mm] in diameter, and leading in a slanting direction into a chamber, scarcely beyond a foot [300 mm] from the surface of the ground, which the animal lines with grass and leaves. The stomach of the two specimens which I examined was entirely filled with a yellow pasty substance, evidently of vegetable origin." The specimens analyzed by Reinhardt (1852), apparently taken in the Southern Hemisphere in late fall or early winter, were an immature male and a pregnant female bearing a 25- to 37-mm embryo.

The two specimens collected by Joao Moojen (pers. comm. to Ernest P. Walker) were a young male with only two upper molariform teeth and a female in puberty, about adult size, with three molariform teeth, the third molar just breaking through. These were found in a nest about 200 mm in diameter filled with dry grass and chewed roots at the end of a tube 800 mm in length and 70 mm in diameter. Moojen cited Burmeister, who noted that *Carterodon* remains in its underground galleries during the day but emerges in the late afternoon or night, when it is easily captured by owls (*Tyto alba*). For this reason, the genus was first known from balls vomited by this owl, particularly inside the caves of the Lagoa Santa area.

RODENTIA; ECHIMYIDAE; **Genus TRICHOMYS**
Trouessart, 1880

Punaré

The single species, *T. apereoides,* occurs in eastern Brazil and Paraguay (Cabrera 1961). The generic name *Cercomys* Cuvier, 1829 and the specific name *Cercomys cunicularis* have generally been applied to this animal. Petter (1973), however, found that the type specimen of *Cercomys* and *C. cunicularis* is actually a composite consisting of the skin of a cricetid rodent, the skull of a *Proechimys,* and the mandible of an *Echimys.* He thus concluded that the name *Cercomys cunicularis* is invalid, and that the animals that have been referred to *Cercomys* should be given the later name, *Trichomys apereoides.* He added, however, that the name *Cercomys* is applicable to the genus currently called *Proechimys.*

Punaré (*Trichomys apereoides*), photo by Joao Moojen.

Head and body length is 200 to 290 mm, and tail length is 180 to 220 mm. Weight is about 335 grams (Kleiman, Eisenberg, and Maliniak 1979). The upper parts are dull brown and the underparts are grayish to white. The fur is soft, the tail is hairy, and, unlike most echimyids, there are no spines or bristles.

Trichomys usually inhabits rocky and thickly vegetated areas, but is common even in swamps. It lives in rock crevices and under such plants as cactus. Its straw-lined nest is built in a hollow tree or log, in a stone fence, or under a rock. In captivity the punaré has been seen to leap only on its hind legs. It eats cotton seeds or cotton fruits, and during the dry season it feeds on coconuts or the tender parts of some cacti. Pregnant females have been taken in February and July. According to Kleiman, Eisenberg, and Maliniak (1979), the gestation period is 89 days, the number of young per litter averages 3.4 and ranges from 1 to 5, and weight at birth is 29 grams. A captive specimen lived for 1 year and 9 months (Marvin L. Jones, Zoological Society of San Diego, pers. comm.).

RODENTIA; ECHIMYIDAE; **Genus MESOMYS**
Wagner, 1845

The three species (not in systematic order) are as follows (Cabrera 1961; Husson 1978):

M. didelphoides, probably Brazil;

M. hispidus, southern Venezuela, Surinam, eastern Ecuador, northern Peru, Amazonian Brazil;

M. obscurus, known only by description of a specimen from an indefinite locality in Brazil, and possibly a subspecies or synonym of *M. hispidus.*

Head and body length is 150 to 200 mm, and tail length is 120 to 220 mm. The heavily spined pelage resembles that of *Hoplomys.* The coloration of the upper parts ranges through various shades of brown. Dark brown and pale buffy bands on the spines of some species produce a speckled effect. The

Mesomys hispidus, photo from *Proc. Zool. Soc. London.*

underparts are orangish, pale buffy, or whitish. The tail is usually brown throughout and tufted at the tip.

These rodents live mainly in forests at elevations of 90 to 1,950 meters. The short, rather broad feet suggest that *Mesomys* is arboreal. In the lowland forests of southern Venezuela, Handley (1976) collected most specimens in trees or houses, and near streams or in other moist places. A female *M. hispidus*, with a single embryo, was taken in March near the Madeira River in the center of Amazonian Brazil.

RODENTIA; ECHIMYIDAE; *Genus LONCHOTHRIX*
Thomas, 1920

The single species, *L. emiliae*, is found in Brazil to the south of the Amazon River (Cabrera 1961). Specimens have been taken along the Madeira and Tapajos rivers. They are in the British Museum (Natural History), the American Museum of Natural History, and the Departamento de Biologia, Sao Paulo, Brazil.

Head and body length is about 200 mm, and tail length is about 190 mm. The upper parts are dark brown, with buffy spots in the shoulder region; the sides are tawny; and the underparts are pale tawny. The spines on the body are well developed and even the belly is somewhat spiny. The tail is mostly furless, scaly above and covered with short spines, but tufted at the end with hairs that may be 70 mm long. *Lonchothrix* lives in forests, and the broadened hind feet suggest that it is arboreal in habit.

RODENTIA; ECHIMYIDAE; *Genus ISOTHRIX*
Wagner, 1845

Toros

The three species (not in systematic order) are as follows (Cabrera 1961):

I. bistriata, southern Colombia and Venezuela, Brazil;
I. picta, eastern Brazil;
I. villosa, eastern Peru.

Head and body length is 220 to 290 mm, and tail length is about 250 mm. *I. picta* has a black or dark brown, and white color pattern, but the other species are dull brown. The soft fur lacks spines and bristles, and the long bushy tail is almost squirrellike. The hind feet are of the arboreal type.

Isothrix is said to live in holes about 10 mm from the bases of large trees along river banks, and, during the afternoon, to sit in the entrance to the hole with only its head exposed. Handley (1976) collected all his specimens of *I. bistriata* near streams in evergreen forests, and most in trees. A female *I. bistriata*, with a single embryo, was taken in May near the Cunuconuma River, Amazonia, Brazil.

RODENTIA; ECHIMYIDAE; *Genus DIPLOMYS*
Thomas, 1916

The three species (not in systematic order) are as follows (Cabrera 1961; Hall 1981):

Lonchothrix emiliae, photo from British Museum (Natural History).

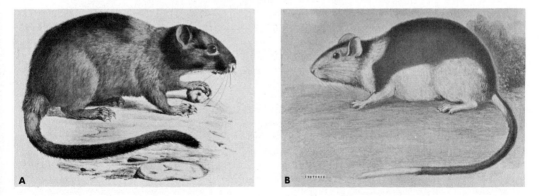

A.   Toro (*Isothrix villosa*), photo from "Animaux Nouveaux ou rares recueillis pendant l'Expédition dans les Parties Centrales de l'Amerique du Sud," *Mammifères,* Paul Gervais. B. Toro (*I. picta*), photo from *Natural History of the Mammalia,* G. R. Waterhouse.

Diplomys sp., photo by R. B. Tesh.

D. caniceps, western Colombia, northern Ecuador;
D. labilis, Panama and nearby San Miguel Island, and
   probably the adjacent part of northwestern Colombia;
D. rufodorsalis, extreme northern Colombia.

Head and body length is 250 to 480 mm, and tail length is
200 to 280 mm. Specimens of *D. labialis* collected by Tesh
(1970*a*) weighed up to 492 grams, all those over 400 grams
being pregnant females. The fur is harsh, but lacks spines.
The upper parts are rusty brown, reddish, or orange buffy
mixed with black. The underparts are buffy, reddish, or reddish white. The tail is mostly brownish, thinly haired, and
slightly tufted with black or white. In *D. labialis* the ears are
short and conspicuously tufted. The feet are of the arboreal
type—short and broad with strongly curved claws.

In central Panama, Tesh (1970*a*) collected 74 specimens
of *D. labialis* in a hilly area covered with patches of moist
tropical forest and scrubby secondary growth. All specimens
were taken from tree holes, usually some distance above the
ground, and in trees near streams. Observations of both wild
and captive animals indicated that activity is strictly nocturnal. They climbed easily and could jump about a meter

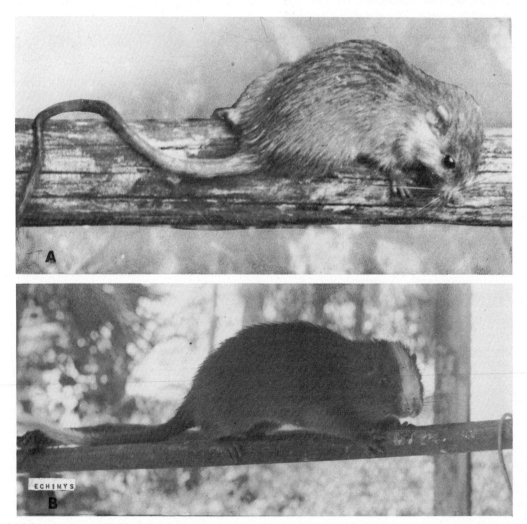

A. Arboreal spiny rat (*Echimys dasythrix*), photo by Joao Moojen. B. Arboreal white-faced spiny rat (*E. chrysurus*),
photographer unknown.

through the air. Captives ate bananas, apples, mangoes, sunflower seeds, almonds, and coconuts. They were difficult to maintain and much more hostile than *Proechimys* and *Hoplomys,* readily attacking objects placed in their cages. Pregnant females were taken in January, February, March, April, May, August, September, and November. Most of these females contained one embryo, and some had two.

RODENTIA; ECHIMYIDAE; **Genus ECHIMYS**
G. Cuvier, 1809

Arboreal Spiny Rats

The 10 species (not in systematic order) are as follows (Cabrera 1961):

E. blainvillei, southeastern Brazil;
E. braziliensis, southeastern Brazil;
E. chrysurus, the Guianas, northeastern Brazil;
E. dasythrix, southeastern Brazil;
E. grandis, northeastern Peru, Amazonian Brazil;
E. macrurus, south of the Amazon in Brazil;
E. nigrispinus, southeastern Brazil;
E. saturnus, eastern Ecuador;
E. semivillosus, northern Colombia, Venezuela, Margarita Island;
E. unicolor, an undetermined part of Brazil.

An additional and widespread species, *E. armatus,* is now assigned to a distinct genus, *Makalata* (Husson 1978).

Head and body length is 170 to 350 mm, and tail length is 150 to 320 mm. An adult female *E. chrysurus* weighed 640 grams (Husson 1978). The fur is bristly or spiny. The upper parts are usually some shade of brown or red. The underparts are usually white or buffy. In *E. saturnus* the upper parts are dark, the middle of the back being black, and the underparts are whitish or coppery brown. Some species have hairy tails, and in a few the tail is slightly tufted or bushy, but in others the tail is scaly. The broad feet have prominent claws and are adapted to arboreal life. Female *E. chrysurus* have one inguinal and three abdominal pairs of mammae (Husson 1978).

Echimys lives in trees, often on river banks or in flooded areas. It is reported to slink along the branches rather than to scamper like a squirrel. In Venezuela, Handley (1976) collected nearly all specimens of *E. semivillosus* in trees, and most near streams or in other moist places within thorn forest. The nests of *Echimys* are built of dry leaves in hollow tree trunks or in holes in trees. The animals remain in their upper holes (usually they have two) during the day and come out on the branches at night. Small groups may live together. At night they give loud calls that sound like: "cró, cró" or "tró, tró." The usual number of young appears to be one or two. A pregnant female *E. blainvillei,* with two embryos, was taken in September, and Husson (1978) collected a pregnant female *E. chrysurus,* with two embryos, on 2 September in Surinam.

RODENTIA; ECHIMYIDAE; **Genus MAKALATA**
Husson, 1978

Husson (1978) erected this new genus for the species previously known as *Echimys armatus* and now designated as *Makalata armata.* It is known to occur in Ecuador, Colombia, Venezuela, Trinidad, the Guianas, and northeastern Brazil. This species also has been reported to be present, through introduction, on the island of Martinique in the Lesser Antilles, but Hall (1981) indicated that the basis of the report is a single specimen that was brought in by a vessel and died without reproducing.

Head and body length is 170 to 202 mm, tail length is 182 mm in each of two specimens, and reported weight is 147 to 317 grams. The upper parts are dark yellowish brown, heavily lined with black. The posterior third of the back is conspicuously speckled with yellow. The spines are pale gray at the base, becoming darker distally, and sometimes, in the speckled area of the body, have a distinct pale yellowish terminal band. The sides of the body are lighter than the back and have fewer spines. The ventral surface is pale yellowish or grayish brown, and has no spines. The tail is relatively short, has no tuft of hairs on the tip, and frequently is lost by autotomy. Compared to *Echimys,* the body is more uniform in color, the spines are smaller, and the folds of the cheek teeth open lingually instead of bucally. One female specimen had two well-developed abdominal mammae, and some other specimens also had a pair of inguinal mammae, perhaps not functional.

Available information, as summarized by Husson (1978), suggests that in Surinam this spiny rat lives in trees, preferably along rivers, and nests between the roots of trees. A female with one embryo was taken in November, and newborn were collected in October. According to Marvin L. Jones (Zoological Society of San Diego, pers. comm.), a captive specimen lived for three years and one month. *Makalata* is reportedly harmful to the cultivation of bananas in Surinam, climbing the trees at night and eating the green fruit.

RODENTIA; ECHIMYIDAE; **Genus DACTYLOMYS**
I. Geoffroy St. Hilaire, 1838

Coro-coros

The three species (not in systematic order) are as follows (Cabrera 1961):

D. boliviensis, southeastern Peru, central Bolivia;
D. dactylinus, eastern Ecuador, Amazonian Brazil, probably Colombia;
D. peruanus, the Andes of southeastern Peru.

The last species has sometimes been placed in a separate genus, *Lachnomys* Thomas, 1916.

In *D. boliviensis* and *D. dactylinus,* head and body length is about 300 mm, tail length is 400 to 430 mm, the upper parts are olivaceous gray or grayish, often with a rusty suffusion, and the underparts are white. In *D. peruanus,* head and body length is about 250 mm, tail length is about 320 mm, the back is yellowish brown, the sides are lighter yellowish brown, most of the head and the basal part of the tail are gray, and the underparts are white.

Makalata armatus, painting by R. van Assen, Rijksmuseum van Natuurlijke Historie, Leiden.

The soft fur of *Dactylomys* lacks spines and bristles. In *D. boliviensis* and *D. dactylinus* the tail is naked, except for the well-haired basal part. In *D. peruanus* the pelage of the body is thicker and the tail is hairy, being bushy on the basal half and more or less tufted at the end. In each species, the third and fourth digits of all four feet are elongated and broadened,

the palate is constricted anteriorly, and the main lobes of the upper cheek teeth are not united by enamel bridges.

These rodents have been found in dense vegetation near water. The type specimen of *L. peruanus* was taken at an elevation of 1,800 meters. Coro-coros are apparently arboreal and nocturnal. They grasp branches between the third

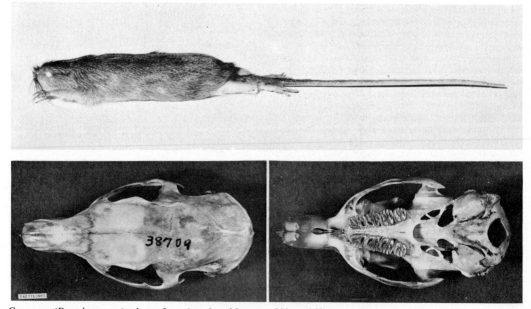

Coro-coro (*Dactylomys* sp.), photos from American Museum of Natural History.

and fourth digits and can run swiftly. LaVal (1976:403) described the call of *D. dactylinus* as follows: "The rather low-pitched notes, all having approximately the same frequency range, might be described as a long succession of 'boop . . . boop . . . boop.'" He reported that the call was restricted to the night, mainly from 1800 to 0100 hrs, and could be heard all year. He suggested that the call is territorial in function. The call was heard mostly in disturbed forest, where bamboo was common, but seemed to originate from broad-leaved evergreens in the canopy. One specimen was collected and its stomach contents "appeared to consist entirely of finely ground plant parts."

RODENTIA; ECHIMYIDAE; *Genus KANNABATEOMYS*
Jentink, 1891

Rato de Taquara

The single species, *K. amblyonyx*, occurs in southeastern Brazil, Paraguay, and northeastern Argentina (Cabrera 1961). There are specimens in the British Museum (Natural History) and the Departamento de Biologia, San Paulo, Brazil.

Head and body length is about 250 mm, and tail length is about 320 mm. The upper parts are dull buffy yellowish, sometimes with an orangish cast, and the underparts are paler, being buffy or whitish. The thick, soft fur lacks spines

and bristles, and the tail is haired. The third and fourth digits of all four limbs are elongated and broadened. *Kannabateomys* differs structurally from *Dactylomys* in that the palate is not appreciably constricted and the main lobes of the upper cheek teeth are united by narrow enamel ridges.

In Brazil this rodent usually lives in bamboo thickets along stream banks. By grasping the bamboo shoots and stalks between its third and fourth digits, it climbs the plants and eats the succulent parts of the bamboo during the night. A pregnant female was noted in November, the usual number of young per birth appears to be one. A captive specimen lived for a year and seven months (Marvin L. Jones, Zoological Society of San Diego, pers. comm.).

RODENTIA; ECHIMYIDAE; *Genus THRINACODUS*
Günther, 1879

There are two species (Cabrera 1961):

T. albicauda, northwestern and central Colombia;
T. edax, northwestern Venezuela and probably adjacent parts of northern Colombia.

Head and body length is 180 to 240 mm, and tail length is 250 to 350 mm. The upper parts are reddish brown or yellowish brown, and the underparts are yellowish white or white. A yellow line may occur along the sides. The fur is thick and soft, there are no spines or bristles, and the tail is haired. The

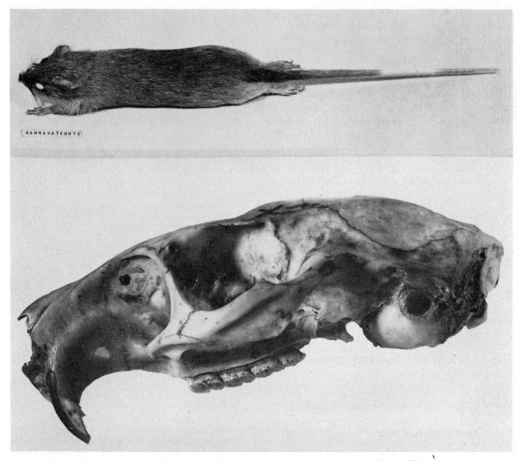

Rato de Taquara (*Kannabateomys amblyonyx pallidior*), photos from British Museum (Natural History).

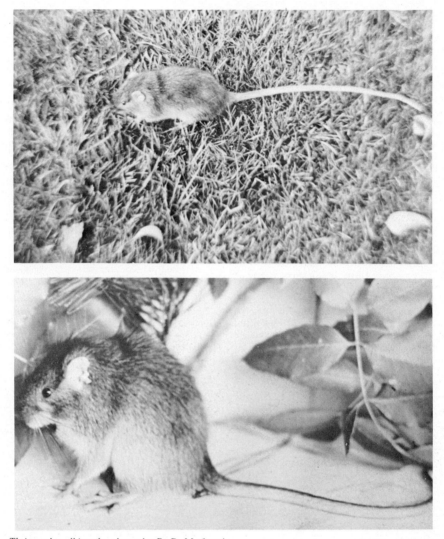

Thrinacodus albicauda, photos by R. B. Mackenzie.

third and fourth digits of the limbs are elongated and broadened, but not to a marked degree. In *T. edax* the tail is completely white below and on the terminal half of the dorsal surface.

Specimens are in the American Museum of Natural History and the British Museum (Natural History). Specimens have been collected at 2,000 and 2,800 meters elevation. In the mountains of Colombia these rodents live in thickets of the bamboo *Chusquea.* They emit a whistling cry.

RODENTIA; ECHIMYIDAE; **Genus HETEROPSOMYS**
Anthony, 1916

There are three subgenera and six species, all probably extinct (Hall 1981):

subgenus *Heteropsomys*

H. insulans, known only by skeletal remains from the type locality near Utuado in Puerto Rico;

H. antillensis, known only by skeletal remains from Puerto Rico;

subgenus *Brotomys*

H. voratus, known only by skeletal remains from the Dominican Republic;

H. contractus, known only by skeletal remains from the type locality in north-central Haiti;

subgenus *Boromys*

H. offella, known only by skeletal remains from Cuba and the Isle of Pines.

H. torrei, known only by skeletal remains from Cuba and the Isle of Pines.

These rodents can be described only on the basis of their skulls and teeth. In the subgenus *Heteropsomys* the skull is somewhat smaller than that of an agouti (*Dasyprocta*), having a total length of about 70 mm. The upper incisors are long but weak. The upper cheek teeth have a single conspicuous internal fold with three separate, transverse, enamel-sur-

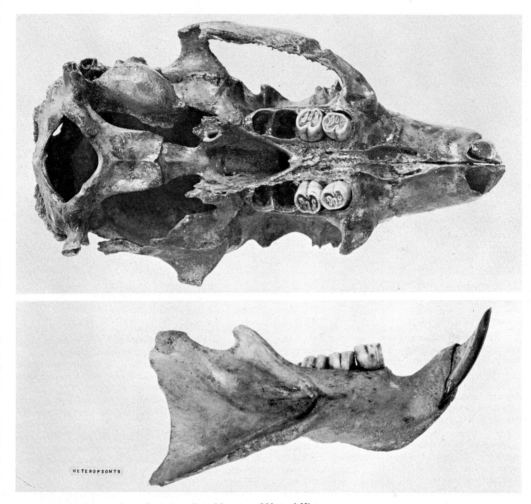

Heteropsomys insulans, photos from American Museum of Natural History.

rounded lakes. In the subgenus *Brotomys* the skull is about the same size as that of *Proechimys,* but has a shorter and broader rostrum. The infraorbital canal is very large and lacks an accessory canal for the passage of a nerve. The teeth are weakly developed. Each of the upper cheek teeth has three short roots and a deep enamel infolding from about the middle of each side. The subgenus *Boromys* closely resembles *Brotomys,* but differs in having a swelling on the bone over the end of the root of the upper incisors, a channel for the passage of a nerve on the floor of the antorbital foramen, and two inner and two outer reentrant folds on each of the cheek teeth. The incisors of *Boromys* are orange yellow in color (Allen 1942; Hall 1981).

The subgenus *Heteropsomys* of Puerto Rico is thought to have survived until relatively recent times, but has not been found in direct association with human artifacts. The subgenus *Brotomys* was widely distributed on Hispaniola, and its remains have been found both in cave deposits, probably made by owls, and in abundance in aboriginal kitchen middens. Early Spanish writings suggest that it was a prized food source of the natives. Its extinction, at some point after the arrival of Europeans, may have been associated with introduction of *Rattus.* The subgenus *Boromys* evidently occurred over much of Cuba and the Isle of Pines. The species *H. offella* was probably used as food by people, as its bones

were first discovered during excavation of an old Indian village. The remains of both this species and *H. torrei* also have been found in abundance in cave deposits. Some of the bones are remarkably fresh, and *Boromys* probably survived until the latter half of the 19th century (Allen 1942; Goodwin and Goodwin 1973).

RODENTIA; **Family THRYONOMYIDAE; Genus THRYONOMYS** *Fitzinger, 1867*

Cane Rats

The single Recent genus, *Thryonomys,* contains two species (Meester and Setzer 1971; Yalden, Largen, and Kock 1976):

T. swinderianus, Gambia to southern Sudan, and south to northern Namibia and eastern South Africa;

T. gregorianus, Cameroon and western Ethiopia to Rhodesia.

Head and body length is about 350 to 610 mm, and tail length is 65 to 260 mm. Adults generally weight 4 to 7 kg, but a few individuals approach 9 kg. The pelage is coarse. The

Cane rat (*Thryonomys swinderianus*), photo by Ernest P. Walker.

bristlelike hairs are flattened, grooved longitudinally along their upper surface, and usually grow in groups of five or six. Underfur is lacking. The general coloration of the upper parts is speckled yellowish brown or grayish brown, and that of the underparts is grayish or whitish. The tail is scantily covered by short, bristly hairs, with scales between the hairs; its color is brownish above and buffy white below.

The body is large and heavyset. The ears are short and rounded, and barely extend above the spiny pelage. The forefoot has three well-developed central digits, but the first digit (thumb) is reduced and the fifth digit is small and nearly functionless. The digits of the hind foot are somewhat larger, though the first digit is absent. The palms and soles are naked, and the claws are thick and heavy. The massive skull is strongly ridged. The dental formula is: (i 1/1, c 0/0, pm 1/1, m 3/3) x 2 = 20. The orange-colored incisors are broad and powerful. The upper incisors have three deep grooves. The cheek teeth are rooted and moderately high crowned. Females have two or three pairs of lateral mammae (Weir 1974b).

While both species of *Thryonomys* may occur in the same general area, Kingdon (1974b) explained that there are distinct primary ecological niches. *T. swinderianus* is a semiaquatic inhabitant of marshes and reedbeds, and *T. gregorianus* lives mainly on dry ground in moist savannahs. Both species depend on grass for cover and food. They may shelter only in tall grass, but sometimes use rock crevices, termite mounds, or the abandoned holes of the aardvark (*Orycteropus*) and porcupine (*Hystrix*). If none of these is available, *Thryonomys* sometimes excavates a shallow burrow. Well-defined paths lead from the shelters to the feeding sites, which may be 50 meters away. Cane rats are generally nocturnal, but are sometimes active by day. They apparently cannot see well, but have good senses of smell and hearing. Despite their appearance, they are fast and agile. If startled while above ground, they bolt through the matted vegetation with sudden and amazing speed, often toward water. They swim and dive with ease. The diet consists mainly of grass and cane, and also includes bark, nuts, fallen fruit, and cultivated crops. Rocks, bones, and ivory are sometimes gnawed, apparently to hone the incisors rather than as a source of nutrients.

Kingdon (1974b) reported that during a brush-clearing operation in Uganda a count of cane rats indicated an overall population density of nearly one per ha. During part of the year *T. swinderianus* seems to live in groups of mixed ages and sexes, but in the dry season there is reportedly a division into solitary males and groups of females. *T. gregorianus* lives in small family groups, each of which may have a territory of 3,000–4,000 sq meters. Captive males live peacefully with females and young but fight other adult males. Ewer (1971) described the highly specialized fighting technique of *Thryonomys*. The two contestants have a nose-to-nose pushing duel. If one relaxes pressure for a moment, the other may whip the rump around to try to knock him off balance.

In East Africa both species of *Thryonomys* apparently have two breeding seasons, which occur during the wetter months from March to November (Kingdon 1974b). Under favorable conditions a female can produce two litters in this period. In South Africa the young are born from June to August. In Ghana, Asibey (1974b) determined that *T. swinderianus* breeds throughout the year but has peaks in January–March and July–September. In this country, females probably have two litters annually. The gestation period was calculated to vary from 137 to 172 days and to average 155 days. Embryo counts ranged from one to eight, and mean litter size was four young. *Thryonomys* makes a special natal nest, usually scooping out a hollow depression in a sheltered area and lining it with grass and leaves. The newborn weigh about 129 grams each, are covered with hair, have their eyes open, and are soon able to run about. The young begin to breed when about 1 year old (Kingdon 1974b). A captive *T. swinderianus* lived for 4 years and 4 months (Marvin L. Jones, Zoological Society of San Diego, pers. comm.).

Cane rats may do considerable damage in sugar cane fields, and, in an effort to reduce their numbers on plantations, pythons are often protected. Cane rats also become pests by eating maize, millet, cassava, sweet potatoes, and pumpkins (Kingdon 1974b). The flesh of *Thryonomys* is an important source of protein for many native people. There are intensive organized hunts with dogs and spears, and investigations into domestication have been made. In Ghana, where *Thryonomys* is known as the "grasscutter," its meat sells for nearly twice the cost of beef, mutton, and pork (Asibey 1974a). From July 1970 to June 1971 the markets of Accra are estimated to have sold nearly 200,000 kg of cane rat meat, with a value of about U.S. $220,000.

The geological range of the family Thryonomyidae is early Miocene to Recent in Africa, and Pliocene in Europe and Asia (Packard 1967).

RODENTIA; *Family PETROMYIDAE; Genus PETROMUS* A. *Smith, 1831*

Dassie Rat

The single known genus and species, *Petromus typicus,* is found only in southern Angola, Namibia, and northwestern South Africa (Meester and Setzer 1971). No fossils have been discovered (Packard 1967).

Head and body length is 140 to 200 mm, and tail length is 130 to 180 mm. The pelage is soft and silky, but the underfur is absent, so the hairs stand out separately and have a wiry appearance on the living animal. The hairs grow in clusters of three to five, and the facial whiskers are long and black. When this rodent lies flattened out on a rock, its only distinguishing mark is the yellowish nose, as the coloration of *Petromus* blends with that of the rocks on which it lives. Above, the foreparts are dark gray, light gray, yellowish, or pale buffy. The lower back is tawny, orangish tawny, or yellowish. Occasionally the upper parts are grayish throughout. The underparts are grayish or yellowish. The skin is very soft, and tears easily when a specimen is being prepared. The tail is covered with scattered long hairs; it has soft joints and breaks readily, generally at the base.

This rodent is somewhat squirrellike in external appearance, but the tail is not bushy. The limp body can be greatly compressed, so that with its flat skull this rodent can squeeze into narrow crevices. The ribs are so flexible that *Petromus* can be pressed almost flat without injury to the body. The ears are short. The feet are narrow. The forefoot has four digits and the hind foot has five. The claws are short, and some stiff, bristlelike hairs are associated with the claws of the hind foot. The dental formula is: (i 1/1, c 0/0, pm 1/1, m 3/3) x 2 = 20. The incisors are fairly narrow, and the cheek

Dassie rat (*Petromus typicus*), photo by C. T. Stuart.

teeth are high crowned and rooted. Females usually have three pairs of mammae, but sometimes there are only two pairs. The mammae are lateral, placed high up on a level with the shoulder blades, enabling the young to suckle from the side when hiding in narrow rock crevices.

The dassie rat is associated with rocky areas on hills and mountains. It generally emerges from its rock shelter during the day, especially in the early morning and late afternoon, but occasionally forages after sunset. While resting or sunning itself, it generally selects a spot under a projecting rock that provides protection from attack by birds of prey. *Petromus* usually moves on the rocks by running, rather than jumping, but does occasionally spring from one rock to another. When jumping, it spreads its flattened body, somewhat in the manner of flying squirrels during a glide. The dassie rat seeks food on the ground or in bushes. It is a vegetarian, eating a variety of green plant material, seeds, and berries.

Petromus usually travels alone or in pairs. When undisturbed, it is playful, often whisking about and playing with plant stems, but when alarmed it darts quickly to shelter, often giving a warning note (a whistling call) upon reaching safety. Mating occurs during the early summer months (November–December), and births take place near the end of the year. Litters usually contain one or two young, which are born in an advanced stage, being quite large and covered with hair.

RODENTIA; *Family BATHYERGIDAE*

African Mole-rats, Blesmols

This family of five Recent genera and eight species occurs in Africa to the south of the Sahara Desert. The sequence of genera presented here follows that of Simpson (1945).

Head and body length is 80 to 330 mm, and tail length is 10 to 70 mm. The genus *Heterocephalus* is practically hairless, but the other genera have a thick, soft, and woolly or velvety pelage. The external form resembles that of other fossorial mammals. The body is stocky, the tail and limbs are short, and the eyes and ears are minute. Blesmols apparently see indistinctly for a very short distance, and generally close the eyelids when the head is touched. External ears are represented by a small circle of bare skin around the ear opening. The hands and feet are large, and the palms and soles are naked. The five fingers and five toes have either long or short claws, depending on the genus. The hind claws are usually hollow on the underside and shorter than the foreclaws.

The skull is stoutly built and modified to support the large incisor teeth. In all genera except *Bathyergus,* the lower jaws are not tightly joined in front. The upper incisors are grooved in some forms, and the cheek teeth are high crowned and rooted. The dental formula is: (i 1/1, c 0/0, pm 2–3/2–3, m 0–3/0–3) x 2 = 12 to 28. The cheek teeth (premolars plus molars) number 4/4 in *Georychus, Cryptomys,* and *Bathyergus;* 2/2 or 3/3 in *Heterocephalus;* and 6/6 in *Heliophobius,* though all are not usually in place simultaneously (Meester and Setzer 1971). The increase over the usual rodent number of cheek teeth in *Heliophobius* may result from retention of the milk dentition (Packard 1967).

The usual habitat of African mole-rats is an area of loose, sandy soil. All species feed primarily on subterranean bulbs and roots that are found by tunneling from 10 to 300 mm below the surface. The depth of the tunnel is apparently directly related to the looseness of the soil. The burrow systems include large chambers for sleeping and storing food.

The excavated soil is thrown on the surface at intervals in the form of mounds. In some areas these rodents may so honeycomb the ground that a person sinks deep into the sand at almost every step. It is sometimes difficult to distinguish between the mounds of the different genera. *Cryptomys*, *Georychus*, and *Bathyergus* are reported to occur together in parts of the Cape Province of South Africa. This close association is unusual, as only one genus of a particular family of strictly burrowing mammals generally occupies a single habitat. Presumably, the different sizes of the tunnels eliminate actual contact between the genera. Bathyergids can orientate well; when their burrows are destroyed, they dig new tunnels directly to their exits or special chambers. They seldom emerge above ground, except when flooded out or when seeking new quarters.

In *Bathyergus* the foreclaws are the principal tools for digging. In the other four genera the incisor teeth are used to a greater extent than the claws. Blesmols use their hind limbs and the flattened rows of bristles on each side of the tail to remove the soil excavated by the foreclaws or incisors. Soil is prevented from entering the mouth during digging by fusion and apposition of lateral folds of the lips.

These rodents are reported to make interesting pets, but some species bite unexpectedly. Even the newborn can inflict severe bites. Blesmols often destroy the tuber crops of native people, but like all burrowing animals, they help to furrow and aerate the soil.

The geological range of this family is Miocene, and Pleistocene to Recent, in Africa, and Oligocene in Mongolia (Kingdon 1974*b*; Packard 1967). *Gypsorhychus*, an extinct giant mole-rat allied with the Recent genus *Georychus*, is represented by skull and dental remains from Taungs, in Mongolia, and from Namibia.

Head and body length is 150 to 205 mm, and tail length is 15 to 40 mm. The soft pelage is so thick, fluffy, and long that it practically conceals the tail in many individuals. The tail has a flattened appearance, because it is haired mainly on the sides. The general coloration of the upper parts is buffy to buffy orange, frequently with a brownish tinge. The underparts are somewhat paler than the upper parts. The head has various black or dark brown markings, with white spots. The hands, feet, and tail are white. Partial albinism is not uncommon.

The general form of *Georychus* is similar to that of North American pocket gophers (Geomyidae), but the Cape mole-rat lacks cheek pouches and is smaller. The external ears of the Cape mole-rat are represented by round openings, surrounded with thickened skin around the edges. The claws are only moderately developed and are relatively weak. The incisors are white, ungrooved, and project forward to an unusual degree. Normally, there are three pairs of mammae, but it is not unusual for a female to have four pairs.

Georychus is strictly a burrower and spends almost its entire life underground. Apparently the incisors are used more than the claws for digging, especially in hard soil. The earth from the tunnels is thrown out at intervals, thereby plainly marking the line of excavation. The burrows are occasionally near the surface and branch into blind tunnels. The main burrow eventually leads into a smooth-walled, somewhat globular chamber, in which the mole-rat stores tubers, roots, and bulbs. It is said that *Georychus* bites off the buds ("eyes") of bulbs and tubers to prevent them from sprouting. *Georychus* is destructive to tuber crops. It makes an interesting pet, but has the habit of biting unexpectedly. One captive lived for three years (Marvin L. Jones, Zoological Society of San Diego, pers. comm.).

RODENTIA; BATHYERGIDAE; *Genus GEORYCHUS*
Illiger, 1811

Cape Mole-rat

The single species, *G. capensis*, occurs in the southwestern and southern parts of Cape of Good Hope Province in South Africa. There are also relictual populations in Natal and possible the Transvaal (Meester and Setzer 1971).

RODENTIA; BATHYERGIDAE; *Genus CRYPTOMYS*
Gray, 1864

Common Mole-rats

There are three species (Meester and Setzer 1971; Ansell 1978; Kingdon 1974*b*):

C. ochraceocinereus, Ghana to Sudan and northern Uganda;

Cape mole-rat (*Georychus capensis*), photo by G. B. Rabb, Chicago Zoological Park, Brookfield, Illinois.

Common mole-rat (*Cryptomys* sp.), photo by Ernest P. Walker.

C. mechowi, Angola, southern Zaire, Zambia, Malawi, southern Tanzania;

C. hottentotus, southern Zaire and Tanzania to Namibia and South Africa.

Head and body length is 100 to 215 mm, and tail length is 10 to 30 mm. Reported weights are 46 to 131 grams for *C. hottentotus* (Smithers 1971) and 200 grams for *C. ochraceocinereus* (Kingdon 1974b). The coloration is variable—whitish, yellowish, clay, fawn, grayish, brown, reddish brown, cinnamon buff, and blackish. There may or may not be a white spot on the head. The fur is thick and often velevety. The eyes and external ears are very small, but the cornea of the eye is sensitive to air currents and the animals are remarkably receptive of sounds and vibrations (Kingdon 1974b). The large lower incisors of *Cryptomys* are separately movable. Female *C. ochraceocinereus* were reported to have two pairs of mammae by Weir (1974b) and three or four pairs by Kingdon (1974b).

According to Kingdon (1974b), *Cryptomys* occupies a wide variety of soils in woodlands, savannahs, and secondary forests. *C. hottentotus* occurs at elevations of up to 2,200 meters. Tunnel depth is adapted to soil consistency; the looser the soil, the deeper the burrow. In seasonally flooded areas, *Cryptomys* constructs an extensive mound, in which living chambers and food stores are above the high-water mark. Even in other areas, the main living and storage rooms are located in higher ground, and from there tunnels extend in several directions. Fresh tunnels are characterized by mounds of excavated earth at intervals of about 1.25 to 6.25 meters. In one chamber is a large communal nest, made of vegetation and used for sleeping. Hickman (1979) excavated four burrow systems of *C. hottentotus*. Each contained a single functional nest at a depth of 17 cm. The total length of the systems varied from 58 to 340 meters, and indicates that *Cryptomys* may have the longest constructed and maintained burrow of any animal.

Cryptomys digs with the protuberant lower incisors and the forefeet, unlike *Bathyergus*, which uses mainly the forefeet. The ground excavated from the main tunnel is pushed with the hind feet and tail up a side tunnel, until the earth comes up in a cone and topples over to form a mound. Most burrowing activity occurs during the wet season (Kingdon 1974b; Smithers 1971). At this time the mole-rats extend new tunnels to feeding areas and carry back food for stoarge. In the dry season new mounds are not made, but soil is redistributed in old, unused tunnels. Surface activity is rare, but the animals evidently sometimes go above ground to gather nesting materials and dig up seeds. The diet consists mainly of roots, bulbs, tubers, and aloe leaves. Invertebrates, such as earthworms, cockchafer larvae, and white ants, are occasionally eaten.

Cryptomys lives in small colonies of 2 to 12 individuals, which share a sleeping nest and burrow system. Captives from the same system live peacefully together, but are extremely aggressive to individuals originating in other colonies. *Cryptomys* can bite savagely and has a variety of squeaks, grunts, and growls. There is probably a single annual litter per female, and its timing probably varies by location. In Botswana, pregnant females were collected in February and July, and very young individuals were found in January, February, July, August, and November; in eastern Uganda a pregnant female was taken in December; in the Kalahari Desert two pregnant females were taken in April; and in Angola young were produced in January and February. Litter size is 1 to 5 young (Kingdon 1974b; De Graaf 1972; Smithers 1971; Hickman 1979). A captive specimen lived for two years and five months (Marvin L. Jones, Zoological Society of San Diego, pers. comm.).

Common mole-rats often raid the tuber crops of native people, and become pests in vegetable and flower gardens by destroying bulbs. These animals are sometimes hunted for use as human food (Kingdon 1974b).

Sand rat (*Heliophobius argenteocinereus*), photo from Nyasaland Museum through P. Hanney. Insets: photos from *Naturw. Reise nach Mossambique*, W. Peters.

RODENTIA; BATHYERGIDAE; *Genus HELIOPHOBIUS Peters, 1846*

Silvery Mole-rat, Sand Rat

The single species, *H. argenteocinereus*, occurs in southern Kenya, Tanzania, southeastern Zaire, eastern Zambia, Malawi, and Mozambique (Meester and Setzer 1971; Kingdon 1974b; Ansell 1978).

Head and body length is 100 to 200 mm, and tail length is about 15 to 40 mm. Average weight is 160 grams (Kingdon 1974b). The pelage is short and dense. The upper parts are pale sandy, reddish, or grayish, and the underparts are usually paler. Some individuals have a white frontal spot and white spots on the belly. Modifications for fossorial life include extreme reduction of the eyes and external ears; a short, stubby tail bearing a stiff fringe of hairs; a cylindrical body; and short limbs with large, broad hands and feet, the latter also fringed with stiff hairs (Jarvis and Sale 1971). The incisors are prominent and folds of the lips extend downward and prevent the entrance of soil into the mouth during digging.

The following ecological information was taken largely from Jarvis (1973a) and Jarvis and Sale (1971). *Heliophobius* is usually found in sandy areas with low rainfall (250 to 630 mm per year). It constructs burrows consisting of a single main tunnel, about 47 meters long, 15 to 25 cm deep, and 5 cm wide, and numerous side branches. Mounds of

excavated soil mark the course of the main tunnel. The nest is located in a small circular chamber, 8 to 10 cm in diameter and 30 cm below the ground. The nest chamber is not used for food storage or defecation. Bolt holes for emergency refuge may extend from the main tunnel to depths of about 50 cm.

In burrowing, *Heliophobius* uses its powerful incisors to loosen the soil, which is then pushed under the abdomen with the forefeet and kicked behind the animal with the hind feet. When enough soil has accumulated to the rear of the mole-rat, the animal backs up, thrusts its nose and upper incisors into the top of the burrow for support, and uses its hind feet to force the plug of earth back through the tunnel and out onto the surface.

Heliophobius appears unreceptive to light stimuli, but is said to have acute senses of smell and hearing. It spends about half of the day outside of its nest chamber. It may be active at any time, the peak period being 0900–2200 hrs. It rarely forages aboveground, but apparently sometimes moves about on the surface, as its skulls have been found in owl pellets and one individual was captured when seemingly attracted to a campfire at night. Food storage might occur, but was not evident during the studies of Jarvis and Sale. Instead, *Heliophobius* fed on underground portions of plants along the foraging tunnels, and allowed them to continue growing in place. The diet consists mainly of tubers and bulbs.

This mole-rat is solitary and extremely aggressive to others of its kind. The burrows excavated by Jarvis and Sale

(1971) contained a single individual each. A female, captured on 24 May 1967 and subsequently held in isolation, gave birth to two blind and hairless young on 19 August 1967 (Jarvis 1969). According to Kingdon (1974*b*), the one annual litter is reportedly produced at the beginning of the long rains, pregnant females containing two or three embryos were collected in Kenya in May, and overall litter size is one to four. An adult female held captive for two years by Jarvis (1973*a*) was released for a month and then recaptured, and was still alive and healthy a year later.

Heliophobius reportedly sometimes damages cultivated crops, such as potatoes. In many areas this animal is eaten by the native people.

RODENTIA; BATHYERGIDAE; **Genus BATHYERGUS**
Illiger, 1811

Dune Mole-rats

There are two species (Meester and Setzer 1971):

B. janetta, extreme southwestern Namibia, coastal part of northwestern South Africa;
B. suillus, coastal parts of southwestern and southern South Africa.

Head and body length is 175 to 330 mm, and tail length is 40 to 70 mm. Adult males weigh about 1,500 grams and adult females weigh about 850 grams (Jarvis 1969). The pelage is thick and rather woolly. The upper parts are cinnamon, drab gray, or silvery buff. In *B. janetta* there is a broad, dark dorsal band. In most forms the underparts are colored like the upper parts, but in one form the ventral surface is black. White and piebald varieties of *Bathyergus* are common.

These mole-rats have short legs and five well-developed foreclaws, that of the second digit being the longest. The

Dune mole-rat (*Bathyergus suillus*), photo by John Visser.

third toe of the hind foot also has an especially long claw. The pale brown tail has a flat, featherlike appearance resulting from the fringe of long, pale slate hairs growing out from either side. The eyes are very small and there are no external ears. The incisor teeth are white; the upper ones are heavily grooved and the lower ones are not grooved. Both sets of incisors project forward to a remarkable degree, but unlike those of all other bathyergid genera, the two lower ones cannot be moved independently, as the lower jaws are ossified at the mandibular symphysis. Females have six mammae.

The habitat of *Bathyergus* appears restricted to sand dunes and sandy flats, especially near the coast, but up to elevations of about 1,500 meters. Dune mole-rats spend their lives in extensive burrow systems that they themselves have excavated. They are the only bathyergids that dig mainly with the forefeet rather than the incisor teeth (Jarvis 1969). They push soil in front of them through the burrows and then throw it out through short side tunnels leading to the surface. A captive individual made a chattering noise while burrowing and would turn and bite if interrupted. The diet consists principally of bulbs and fleshy roots. In agricultural areas, *Bathyergus* may collect a great surplus of food and store it within the burrow.

Small groups of dune mole-rats probably occupy the same burrow system and form a close-knit colony. Females are probably monestrous and mating occurs toward the end of the winter rains (winter extends from May to August in South Africa). Pregnant females have been found from midwinter to early spring (July–October), with a peak in August. The gestation period apparently lasts about 2 to 2½ months. Embryo counts in 34 pregnant females averaged 2.4 and ranged from 1 to 4. Births seem to be timed so that the young are weaned, and establish their own burrows, mainly during the early spring when the rainfall is slight, but the soil is still easily worked and the vegetation is lush (Jarvis 1969; Van der Horst 1972).

Dune mole-rats are sometimes considered pests, because they raid cultivated crops, especially potatoes. Their burrows are a serious menace to persons riding horseback, as the ground is weakened to such an extent that it collapses under the weight. Railroad ties may also be undermined, causing the rails to drop under the weight of a passing train.

RODENTIA; BATHYERGIDAE; **Genus HETEROCEPHALUS** *Rüppell, 1842*

Naked Mole-rat, Sand Puppy

The single species, *H. glaber*, occurs in central and eastern Ethiopia, central Somalia, and Kenya (Meester and Setzer 1971).

Head and body length is 80 to 92 mm, tail length is 28 to 44 mm, and weight is 30 to 80 grams (Kingdon 1974*b*). This is the smallest bathyergid, the average weight being only about 35 grams (Jarvis 1978). The animal appears to be naked at first glance, but a few pale-colored hairs are scattered about the body and tail, there are prominent vibrissae on the lips, and a fringe of fine hairs is present along the edges of the feet. This fringe helps to collect and sweep back loose soil during digging. The skin is wrinkled, and is pinkish or yellowish in color.

Heterocephalus has a short and chunky head, minute eyes and external ears, and very prominent incisor teeth. The front feet are large and have five broad, flattened digits, each of which has a small, conical claw. The hind feet also have five

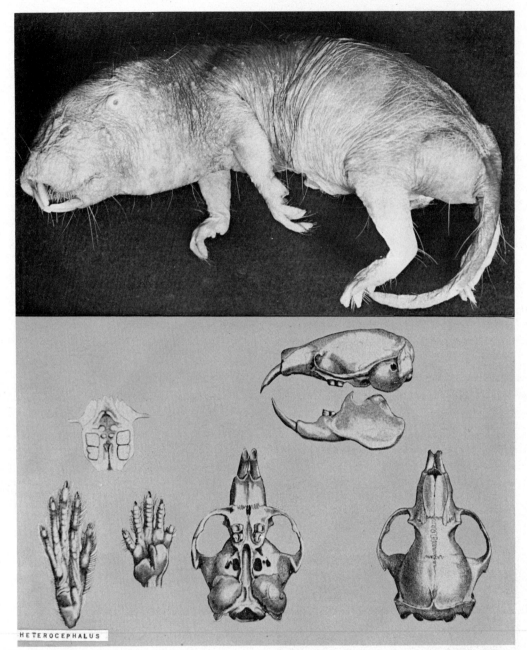

Naked mole-rat (*Heterocephalus glaber*), photo by Dietrich Starck. Anatomy (*H. g. phillipsi*), photo from *Proc. Zool. Soc. London*.

clawed digits. The body lacks sweat glands and the normal mammalian layer of subcutaneous fat (Jarvis 1978). One female was found to have seven functional pairs of mammae (Jarvis 1969).

The naked mole-rat is completely fossorial, is adapted exclusively to arid conditions, and is found in a variety of soil types (Kingdon 1974*b*). The elevational range in Ethiopia is 400 to 1,500 meters (Yalden, Largen, and Kock 1976). The investigations of Jarvis (1969, 1978) and Jarvis and Sale (1971) have added substantially to our knowledge of this remarkable genus, and the remainder of this account is based largely on those studies.

Heterocephalus constructs an extensive and complex burrow system. There is a large communal nest chamber, which is either unlined or partly floored with dry vegetation. Radiating outward are a series of foraging tunnels, usually about 3 cm wide and 15 to 40 cm below the surface. One system was estimated to have more than 300 meters of such passages. There is much branching of the tunnels in the vicinity of food supplies. Bolt holes, reaching depths of about 70 cm, are sometimes present and may serve as emergency refuges. Above the burrow are the mounds of excavated earth, which are uniquely volcanolike in shape and in the presence of a central, unplugged hole.

Digging is done primarily with the incisor teeth, and the feet are used to kick and push the loosened earth. An individual near the surface may work alone in much the same manner as *Heliophobius,* but deeper burrowing is done by relay. One animal remains for a while at the earth face, loosening the soil and kicking it backward. Behind this animal is a chain of several others. When enough earth has accumulated, the first animal of the chain pushes the soil backward with its hind feet through the tunnel to a point near the surface. When this is accomplished the animal takes a place at the end of the chain. Meanwhile, the other animals of the chain have advanced in position by straddling the one moving to the rear. Near the surface another individual remains for a while to collect the soil brought up by the others and kick it out through the central hole of the forming mound. At irregular intervals, this individual and the one doing the drilling are relieved by other members of the team.

Heterocephalus has the poorest capacity for thermoregulation of any known mammal. Its body temperature is relatively low, only 32° C. Nonetheless, the temperature within its burrow is maintained at 30° to 32° C, and the humidity at 90 percent, regardless of external conditions. The animals avoid extreme temperatures near the surface by restricting burrowing activity to early morning and late afternoon. Most burrowing seems to be done following the rains, when the soil is more easily worked. Therefore, the important task of locating new food supplies is accomplished mainly during energetically favorable periods. The tubers, roots, and corms, which compose most of the diet, are generally left growing in place after discovery, and are visited by individuals as the need arises.

Heterocephalus lives in well-organized colonies, which usually comprise about 20 to 30 individuals, though there have been reports of up to 100 members. As colonies may be located less than 100 meters apart in suitable habitat, it is possible that the reports of large groups have resulted from simultaneous observation of several colonies. In any event, detailed studies of captive colonies indicate that they are led by a single large female, and that there are both working and nonworking classes. In one group the dominant female weighed 53 grams, the 3 nonworkers (2 males and 1 female) had an average weight of 38 grams, and the 10 workers (mixed sexes) averaged 32 grams. Although the working class comprised mostly younger animals, the weight difference did not appear to be associated mainly with age variation, as all animals were fully adult and over 3½ years old. The dominant female and the nonworkers spent most of their time in the nest chamber, except when coming out to urinate or defecate. The workers cooperated in burrowing, gathered nest materials, and brought food to the nest for the dominant female and nonworkers. All animals huddled together to sleep. Before full development of the social hierarchy, females coming into estrus fought and frequently killed one another. Once the dominant female established her position, she suppressed breeding by the other females. Only she mated, and she initiated courtship, generally with a nonworking male.

There is evidently a well-defined breeding season, with births occurring mainly from February to April during the long rains. If the first litter survives, the female apparently breeds only once a year, but if not, she mates again. One female gave birth to three litters in a period of only six months. Recorded litter size is 3 to 11 young. The newborn are blind and, of course, naked. Their average weight is 1.9 grams. The young mature slowly, their eyes do not open for several weeks, and they do not reach adult size for at least a year. Nonetheless, they enter the working class at an age of two to three months. Many captives have lived for at least five years.

RODENTIA; *Family CTENODACTYLIDAE*

Gundis

This family of four Recent genera and five species occurs in northern Africa. The sequence of genera presented here follows that of George (1979).

Head and body length is 160 to 240 mm, and tail length is 10 to 50 mm. The fur is soft and the tail is fully haired. Gundis resemble guinea pigs (*Cavia*) in external appearance, having thickset and compact bodies. There are four digits on each foot; the two inner digits of the hind foot have comblike bristles. The claws are not enlarged, but are very sharp. The skull is flattened and broad posteriorly. The bullae are inflated. The dental formula is: (i 1/1, c 0/0, pm 1–2/1–2, m 3/3) x 2 = 20 or 24. The upper incisors are slightly grooved (*Felovia*) or are ungrooved, and the cheek teeth are rootless (ever growing). This family is unusual in that females have a cervical pair of mammae in addition to a single pair placed laterally on the anterior thorax (Barbara J. Weir, Wellcome Institute, Zoological Society of London, pers. comm.).

According to George (1974), all four genera are found in desert or semidesert habitats at elevations of sea level to 2,400 meters. They never excavate or occupy burrows, but dwell in caves or rocky crevices. Ideal living sites have permanent shelters, temporary shelters, and ledges for sunbathing. Nests are not constructed. All genera are diurnal and are not known to hibernate or estivate. When the weather is cold, windy, or wet, activity is restricted, and the animals may not emerge at all. When the temperature is high, however, they come out at first light and remain active for five hours. Activity declines during the hot, middle part of the day, but increases in the two to four hours before dusk. There are three main kinds of activity: foraging, sunbathing, and general (play, chasing, and exploring). The diet includes leaves, stalks, seeds, and flowers. Animal food is not known to be eaten. Food is not stored and fat reserves are not accumulated by the body. The young are born fully furred and open eyed, after a gestation period of about 55 days.

The systematic affinities of the Ctenodactylidae are uncertain, and the family perhaps should be placed within the suborder Sciuromorpha rather than the Hystricomorpha. The earliest fossils assigned to the family date from the middle Eocene of Asia. There are also Oligocene, Miocene, and Pliocene remains from Asia; Miocene and Pleistocene remains from Sardinia and Sicily; and Miocene and Pleistocene remains from North Africa. The living genera evidently originated in Africa and are known only from the Recent of that continent (Wood 1977; George 1979).

RODENTIA; CTENODACTYLIDAE; *Genus* CTENODACTYLUS *Gray, 1828*

Gundis

There are two species (Meester and Setzer 1971):

C. gundi, Morocco to northwestern Libya;
C. vali, southern Morocco to northwestern Libya.

Head and body length is about 160 to 200 mm, and tail length is 10 to 20 mm. The average weight of captive adult *C. vali* is 174.5 grams (George 1978). The upper parts are buffy in color, occasionally pinkish buff, and the underparts are paler, usually whitish or slaty. The generic name, like that of the tuco-tucos (*Ctenomys*), is derived from the comblike bristles on the toes of the hind foot. *Ctenodactylus* resembles

Gundi (*Ctenodactylus gundi*): A. Photo by Schomber-Kock; B & C. Right hind and front foot, photos by P. F. Wright of specimen in U.S. National Museum; D. Photo from the Archives of the Zoological Garden Berlin-West.

Pectinator externally, but differs in having a tail that is shorter than the hind foot. Females have two pairs of mammae.

The basic ecology is as given above in the family account. Gundis live on rocky slopes within warm temperate and subtropical deserts. The normal movement on a level surface is a quick run, with the belly almost touching the ground. On sloping surfaces, however, *C. gundi* presses its body against the wall and uses the slightest irregularities in the obstruction to ascend almost perpendicularly. Flat stones or logs are preferred for sunning and resting. It has been reported that *Ctenodactylus* will "play possum" when threatened and may show "fear paralysis" for up to 12 hours. Such behavior, however, has never been observed by Wilma George, who has studied gundis in the wild and in captivity for many years. Gundis rely mainly on speed and ability to squeeze into very small cracks to escape from predators. These rodents seem to feed solely on plant material. They do not drink, but obtain sufficient water from the plants that they eat. The main call is a birdlike whistle, and they also thump with their hind feet when excited or alarmed.

In Algeria, according to George (1978), semicaptive *C. vali* produced litters mainly in March and April, but local nomads reported the overall breeding season to extend from February to June. In Tunisia a female *C. gundi* was found with a juvenile that probably had been born in January. Females evidently produce only one litter per year and do not

experience a postpartum estrus. The estrous cycle averages 28.7 days in *C. gundi* and 23.4 days in *C. vali*. Anestrus extends from September to May in *C. gundi* and May to December in *C. vali*. A single gestation period in *C. vali* was determined to be 56 days. Litter size is one to three young. The newborn weigh about 20 grams each, are fully furred, have their eyes open, and are able to run. They are capable of feeding by themselves after a few days, but usually nurse for several weeks. They may require 9 to 12 months to reach full adult size and sexual maturity. One of the females kept by George lived for 5 years.

RODENTIA; CTENODACTYLIDAE; **Genus PECTINATOR**
Blyth, 1856

Speke's Pectinator

The single species, *P. spekei*, occurs in coastal and eastern Ethiopia, Afars and Issas, and Somalia (Meester and Setzer 1971).

Head and body length is 140 to 190 mm, and tail length is 40 to 60 mm (Funaioli 1971). Adult females weigh about 178 grams (George 1978). The pelage is quite soft. The upper

Speke's pectinator (*Pectinator spekei*), photo by Wilma George.

parts are ashy gray, suffused with black or brown, and the sides are grayish. The hands, feet, and underparts are grayish white. Individuals taken from near the coast and from an elevation of 1,800 meters seem to differ only slightly in color and thickness of fur. The whiskers are fairly long and the tail is bushy. The skin is said to be extremely thin and easily torn. The digits of the hind feet have brushes of comblike bristles. Females have two pairs of mammae.

The basic ecology is as given above in the family account. In Ethiopia, according to Yalden, Largen, and Kock (1976), *Pectinator* is a denizen of rocky cliffs in desert or semidesert areas, and is often found together with the hyrax (*Procavia*). All reliable records in that country lie between sea level and approximately 1,200 meters elevation. *Pectinator* shelters in rock crevices, emerges shortly after sunset, often basks in the sun, and feeds only on vegetation. In a study in Ethiopia, George (1974) found three colonies to live within well-defined boundaries encompassing areas of 1,500 to 2,000 sq meters. A whistling call was heard during the approach of a goshawk. The call has been described as a long, drawn-out "whee, whee."

George (1978) observed many young in the Danakil Desert of Ethiopia, and these probably had been born from late August to mid-September. Captive females were in anestrus only in July, and perhaps *Pectinator* is a more opportunistic breeder than *Ctenodactylus* and *Massoutiera*. The estrous cycle averages 22.7 days. Five captive-born litters contained a single young each, and one litter contained two young. A female was held in captivity for 4 years.

RODENTIA; CTENODACTYLIDAE; *Genus* *MASSOUTIERA* Lataste, 1885

The single species, *M. mzabi*, occupies the central Sahara Desert in southeastern Algeria, northern Niger, northwestern Chad, and probably southwestern Libya (Meester and Setzer 1971).

Head and body length is 170 to 240 mm, and tail length is about 35 mm. The weight of adults held in captivity by George (1978) averaged about 172 grams for males and 194 grams for females. The coloration ranges through different yellow and brown shades. The ears are extraordinary in being flat against the head and immovable. Like the eyes, they are rounder in *Massoutiera* than in *Ctenodactylus*. There is a fringe of hairs around the inner margin of the ear, like that in jerboas and other desert rodents, which protects the meatus from wind-blown sand. Females have two pairs of mammae.

The basic ecology is as given above in the family account. *Massoutiera* is found mainly in mountainous areas that rise above the Sahara Desert. It lives in rock crevices, is diurnal, does not emerge in cold or stormy weather, often sunbathes, and feeds on plant material. In studies in Algeria, George (1974) found that colonies used many temporary shelters. *Massoutiera* thumps with its hind feet, when excited or alarmed.

George (1978) found young in the wild from March to June. In April and May two females that were collected were pregnant, but four were not pregnant. Captive females had an estrous cycle averaging 24.9 days and were in anestrus from October to March. Four litters of two young, and one of three young, were produced. Two newborn weighed 20 and 21 grams. They were fully furred and had their eyes open. Of seven captive females held by George, three remained with a male for approximately 2 years, two were paired each with a single male for varying periods, and two lived together.

RODENTIA; CTENODACTYLIDAE; *Genus FELOVIA* Lataste, 1886

The single species, *F. vae*, is known from Mauritania, Senegal, and Mali (Meester and Setzer 1971).

Head and body length is about 170 to 230 mm, and tail length is about 20 to 30 mm. The mean weight of 10 captive adult males was about 186 grams (George 1978). The upper parts are dark yellowish red and the underparts are paler red. *Felovia* resembles *Massoutiera* externally, but can be distinguished by its slightly grooved upper incisors and less prominent hair fringe on the inner edge of the ear. Females have two pairs of mammae.

The basic ecology is as given above in the family account. George (1974) stated that *Felovia* lives on the edge of very dry tropical forest, but was never seen to climb a tree. She studied a colony in western Mali in 1972 that was present on the same site where it had been discovered in 1885. In this area, George (1978) found young in mid-March that probably had been born from mid-December to January, 2 to 3 months after the rains. A captive female had an estrous cycle of 23 days. Four litters contained a single young each. According to Barbara J. Weir (Wellcome Institute, Zoological Society of London, pers. comm.), George's captives appeared very nervous and would vocalize with a birdlike trill when disturbed.

Massoutiera mzabi, photo by Jean-Marie Baufle through F. Petter.

Felovia vae, photo by Wilma George.

ORDER CETACEA

Whales, Dolphins, Porpoises

This order of wholly aquatic mammals occurs in all the oceans and adjoining seas of the world, and in certain lakes and river systems. Living cetaceans are traditionally divided into two suborders: the Odontoceti, individuals of which have teeth (generally all of one kind) and an asymmetrical skull, and which comprise the families Platanistidae, Delphinidae, Monodontidae, Physeteridae, and Ziphiidae; and the Mysticeti, individuals of which have plates of baleen instead of teeth, and a symmetrical skull, and which comprise the families Eschrichtidae, Balaenopteridae, and Balaenidae. Rice (1967, 1977) considered the Odontoceti and Mysticeti to be distinct orders, because it is questionable whether the two groups had a common origin and because the differences between them are as great as those between some of the universally recognized orders of mammals. Certain authorities, such as Jones, Carter, and Genoways (1979) and Ellis (1980), have followed Rice's procedure. Gaskin (1976), however, summarized data supporting a monophyletic origin for cetaceans, and treated the Odontoceti and Mysticeti as suborders. Some other authorities, such as Hall (1981), also have continued to place all cetaceans in a single order. There are 8 Recent cetacean families, 39 genera, and 79 species. Rice's (1977) sequence of these taxa is generally followed here, except that the odontocete families are here placed before the mysticete families.

The length of the cetacean head and body is taken in a straight line from the tip of the snout to the notch between the tail flukes. Head and body length varies from about 1.2 to 31 meters, and weight ranges from 23 to 160,000 kg. The tail flukes are set in a horizontal plane, thereby immediately distinguishing cetaceans from fish, the tail fins of which are in a vertical position. Other external features conspicuous in cetaceans are the torpedo-shaped body, front limbs that are modified into flippers (pectoral fins) and ensheathed in a covering, the absence of hind limbs, and the usual presence of a dorsal fin. There is no covering of fur, but all whales have some hairs in the embryonic stage and a few bristles persist around the mouth in adult mysticetes. Cetaceans lack sweat glands and sebaceous glands. They have a fibrous layer (blubber) filled with fat and oil just beneath the skin, which assists in heat regulation. There are no external ears or ear muscles, and no scales or gills.

The nostrils open externally, usually at the highest point of the head. There is a single blowhole in the odontocetes and a double blowhole in the mysticetes. The asymmetry of the skull of the toothed whales is in correlation with the reduction of one of the nasal passages. There is a direct connection between the blowhole and the lungs, so that a suckling calf cannot get milk into its lungs. The blowhole is closed by valves when the animals are submerged. Cetaceans do not blow liquid water out of the lungs. When the animals exhale, the visible spout is from the condensation of water vapor entering the air from the lungs, and possibly from the discharge of the mucous oil foam that fills the air sinuses.

The bones are spongy in texture and the cavities are filled with oil. The vertebrae are fused in the neck of some genera, but all lack the complex articulation of vertebrae in land mammals, and present a graded series from head to tail. There are no bony supports for the dorsal fin and the tail

flukes. The pelvic girdle, represented by two small bones embedded in the body wall and free from the backbone, serves only as the attachment for the muscles of the external reproductive organs.

Propulsion is obtained by means of up and down movements of the tail, in such a way that the flukes present an inclined surface to the water at all times. The force generated at right angles to the surface of the flukes is resolvable into two components, one raising and lowering the body and the other driving the animal forward. The fins serve as balancing and steering organs. Cetaceans are the swiftest animals in the sea; some dolphins can maintain a sustained speed of 26 to 33 km per hr (Rice 1967).

Certain questions remain unanswered regarding the physiological adaptations of cetaceans in diving and temperature tolerance, but some of the general mechanisms are known. Before diving, a cetacean expels the air from its lungs. Some of the adaptations that make long dives possible are: (1) the oxygen combined with the hemoglobin of the blood and with the myoglobin of the muscles accounts for 80 to 90 percent of the oxygen supply utilized during prolonged diving; (2) arterial networks seem to act as shunts, maintaining the normal blood supply to the brain but effecting a reduced supply to the muscles and an oxygen debt that the animal can repay when it again surfaces; (3) a decreased heartbeat further economizes the available oxygen; and (4) the respiratory center in the brain is relatively insensitive to an accumulation of carbon dioxide in the blood and tissues. The hydrostatic pressures encountered at great depths are alleviated by not breathing air under pressure and by the permeation of the body tissues with noncompressible fluids. The only substances in the body of a cetacean that can be compressed appreciably by the pressure of great depths are the free gases, found mainly in the lungs.

The back of a humpback whale (*Megaptera novaeangliae*) which has surfaced to breathe. The two openings on top of the head are the ''blow holes,'' which the whale closes before submerging. These are comparable to the nostrils of other mammals. Photo by Vincent Serventy.

The collapse of these gases drives them into the more rigid, thick-walled parts of the respiratory system. The body temperature is regulated by the insulation of the blubber, which retains body heat when the animals are in cold water, and by the thin-walled veins associated with arteries in the fins and flukes.

Some cetaceans are the only animals, other than elephants, that have brains larger than those of people; brain weight varies from about 0.2 kg in *Pontoporia blainvillei* to 9.2 kg in *Physeter catodon* (Rice 1967; Kamiya and Yamasaki 1974). In adult *Homo sapiens* the weight is about 1.3 kg. Like those of humans, the brains of cetaceans are highly convoluted. Moreover, cetaceans evolved brains the size of people's 30,000,000 years ago, whereas human brains have been their present size for only about 100,000 years (Lilly 1977). These factors, along with the ability of captive odontocetes to learn rapidly and to form social bonds with humans, suggest the existence of a high level of intelligence in the Cetacea.

Most cetaceans have eyes well adapted for underwater vision, and can also see well above water. A greasy secretion of the tear glands protects the eyes against the irritation of salt water. The sense of smell is vestigial in mysticetes and absent in odontocetes. Cetaceans have good directional hearing underwater (Rice 1967).

Hydrophonic studies have shown that cetaceans produce numerous underwater sounds and that some genera depend to a large extent on echolocation for orientation and securing food (Rice 1967; McNally 1977; Caldwell and Caldwell 1977, 1979; Thompson, Winn, and Perkins 1979). Odontocetes commonly produce two kinds of sounds: clicklike pulses, which last 0.001 to 0.01 second, have a frequency band of 100 to over 200,000 cycles per second, and are uttered in series of five to several hundred per second; and whistles, which last about half a second and have a frequency band of 4,000 to 20,000 cycles per second. Clicks are used in echolocation, and sometimes in communication, while whistles may be highly individualized and are used primarily for communication. There is still some question as to where sounds are produced, but there is increasing evidence that both clicks and whistles originate in a series of air sacs that lie above the bony cranium in the soft tissues around the blowhole on top of the head. Sounds are reflected off the concave dorsal surface of the skull, and are then focused and directed by the melon, a large pocket of fat on the forehead of most odontocetes. This pocket is especially well developed in genera that regularly feed in the lightless bathypelagic zone (such as *Hyperoodon*) and those that live in turbid rivers (such as *Platanista*). The enormous square head of the sperm whale (*Physeter*) is filled in large part by the spermaceti organ, which probably has about the same function as the melon in smaller odontocetes. Mysticetes seem to employ only a crude form of echolocation, but do produce a variety of moans and squeals, which are sometimes combined into elaborate ''songs'' and are primarily for communication. It is probable that the large size of the cetacean brain is associated with the development of a precise sense of hearing and the analysis of complex echolocative and communicative signals.

Odontocetes generally feed on fish, cephalopods (such as squid and octopus), and crustaceans. The conical teeth seize slippery prey, but are not adapted for chewing. The killer whale (*Orcinus*) regularly takes warm-blooded prey, such as penguins, pinnipeds, and other cetaceans. Mysticetes feed on many different kinds of small animals, mostly crustaceans, which are collectively referred to as zooplankton. Mysticetes are often called filter feeders or grazers. Their baleen is composed of modified mucous membrane and arranged in thin plates, one behind the other and suspended from the palate. There are two rows of plates, one on each side of the mouth. The plates hang at right angles to the longitudinal axis of the head. The outer borders of the plates are smooth, whereas the inner parts are frayed into brushlike fibers. These plates act as sieves or strainers. Most mysticetes feed by swimming with their mouth open through swarms of plankton. They then use their huge tongue to force water out through the lowered baleen, within which food organisms are trapped.

Most cetaceans are gregarious to some extent, and most have a relatively long period of parental care and maturation. The gestation period ranges from 9½ to 17 months. There is almost always a single offspring, which at birth is usually one-fourth to one-third the length of the mother. Immediately after being born in the water, baby cetaceans must reach the surface for a supply of air. Most, if not all, cetacean mothers probably push their offspring to the surface. When nursing, the mother floats on her side so that the calf can breathe. Later the calf can suckle underwater. The teats of the mammary glands lie within paired slits on either side of the reproductive opening. The mammary glands have large reservoirs in which the milk collects, and the contraction of the body muscles forces the milk, by way of the teats, into the mouth of the young. The rapid growth rate of most cetaceans is at least partly related to the high calcium and phosphorus content of the milk. Because cetaceans live in an aquatic environment, and need not support their own weight, they can attain great size. Maximum size is sometimes not reached until many years after sexual maturity. Most species have a potentially long life span, and some individuals are thought to have lived over 100 years.

Cetaceans were spared from the wave of extinction of large mammals that swept the earth toward the close of the Pleistocene (about 10,000 years ago) and that was probably associated with the spread of advanced human hunters (Martin and Wright 1967). Some whales and dolphins were killed by people in ancient times, but generally in small numbers and near the shore. By about the year 1000 A.D., the Basques of coastal France and Spain had developed a fishery involving the pursuit and harpooning of the right whale (*Balaena glacialis*) from small boats. As the whales became fewer, the Basques went farther and farther out to sea, perhaps reaching waters off Newfoundland just before the arrival of Columbus in the West Indies. Over the next 500 years, ships became bigger and faster, hunting and processing techniques steadily improved, human population and demand for cetacean products increased, and whale populations declined (Allen 1942; D. O. Hill 1975; Ellis 1980; McHugh 1974; Grzimek 1975; Small 1971; U.S. National Marine Fisheries Service 1981; Committee for Whaling Statistics 1980).

For centuries, one of the most valuable cetacean products was baleen (sometimes erroneously called ''bone''), thin strips of which were used to stiffen various articles of clothing and for other purposes requiring a combination of strength and flexibility. Changes in fashion and the development of spring steel and plastics greatly reduced the demand for baleen. The most consistently important objective of whaling was the oil derived from blubber, which was a major fuel for lamps. Some approximate average yields of barrels of oil (1 barrel=about 105 liters) for individuals of certain species of whales are: sperm, 35; fin, 40; blue, 75; humpback, 30; and bowhead, 90. The invention of kerosene and eventually of the electric light helped to alleviate the pressure on whale populations in the latter half of the 19th century. In 1905, however, a new process was developed for hardening fat, and led to the use of baleen whale oil in the production of margarine. Whale oil also came to be used in the manufacture of soaps, lubricants, waxes, explosives, and numerous other products. The true bones of whales were ground up and used to make glue, gelatin, and fertilizer.

Bottle-nosed dolphin (*Tursiops truncatus*), a baby being born tail first, photo from Miami Seaquarium.

Whale meat was eaten by the people of some nations, and was used as dog food and (when ground up) as cattle feed.

For many years the right whale, rich in oil and baleen, remained the prime target of the industry. It was hunted from bases on both sides of the Atlantic and in Iceland. As it became rare, the fishery slackened, finally terminating around 1700 in Europe and 1800 in America. Meanwhile, however, whalers had been turning increasingly toward arctic waters in search of the bowhead (*Balaena mysticetus*). From the early 17th to the late 19th centuries, first at Spitsbergen and then in the waters off northern North America and Russia, one population after another was devastated. Both the right whale and the bowhead, as well as the gray (*Eschrichtius robustus*) and the humpback (*Megaptera novaeangliae*), which were also sometimes taken in the early days of whaling, tended to remain near the coast and so could be easily killed and processed. With the decline of these species in the Northern Hemisphere, deep-sea whaling intensified, lengthy voyages became the rule, and the sperm whale (*Physeter catodon*) became the main target. The pursuit of this species was led by the United States from the early 18th to the mid-19th centuries. Populations of the right whale in the North Pacific and the Southern Hemisphere also came under intensive exploitation, starting about 1800. When the industry reached its peak in 1847, nearly 700 American whaling ships were operating throughout the world. Over the next 20 years the industry declined, with some suggested causes being a reduction in whale numbers, increasing expenses, destruction of much of the U.S. whaling fleet in the Civil War, and the development of kerosene, which could be used in place of whale oil for illumination.

American whaling never again attained the importance of the early 1800s, but developments elsewhere allowed the industry to recover and even expand. During the 1860s in Norway, a gun was perfected that fired a harpoon with an explosive head. Subsequently, people learned how to inflate the carcass of a dead whale with air so that it would not sink. These techniques encouraged large-scale exploitation of the blue (*Balaenoptera musculus*), fin (*B. physalus*), and sei (*B. borealis*) whales, which had previously been too big or too fast to approach, harpoon by hand, and keep afloat. In the early 1900s, enormous stocks of these species, as well as the humpback, were discovered in the waters around Antarctica. Starting in 1904, whaling stations were established on islands in this region. The humpback, often found in the vicinity of land, was quickly and drastically reduced. In the 1920s came the first floating factories, huge ships with slip sternways for drawing up dead whales, no matter how large, and with all the facilities for processing them. Accompanied by a fleet of swift, diesel-powered catcher boats, the factories could remain for lengthy periods in the midst of whale populations in remote waters. Eventually the whalers began to use such sophisticated methods as sonar and aircraft observation to locate their objective.

The value of cetacean products, and improvements in human technology, resulted in the 20th century's becoming the most destructive period in whaling history. From 1904 to 1939, 580,931 blue, fin, and humpback whales were recorded killed in the Southern Hemisphere. The peak annual tonnage taken was in the 1930/31 season (a season refers to the austral summer, December–March), when 29,410 blue, 10,017 fin, and 576 humpback whales were killed in antarctic waters. The rate of kill fell sharply during World War II, but rapidly built up by 1947. With the decline of the blue whale, pressure increased on the fin and sei whales, and, once again, on the sperm whale. There also was renewed interest in whaling in the Northern Hemisphere, especially in the Pacific Ocean. The peak annual catch of individual whales was reached during the 1961/62 season, when the world-wide kill was 65,966 blue, fin, sei, humpback, and sperm whales. During that season, 21 factory ships and 269 catcher boats operated in the Antarctic. The major whaling nations were Japan, Norway, the Soviet Union, Great Britain, and the Netherlands. As the larger species became rare, and as international regulations became more effective, the harvest fell. The blue and humpback whales received complete protection in 1966. The recorded kill of fin, sei, and sperm whales was 41,640 in the 1968/69 season and 9,429 in 1978/79. There was, however, a corresponding rise in the take of the smaller minke (*Balaenoptera acutorostrata*) and Bryde's (*B. edeni*) whales, the total catch of which was 4,238 in 1968/69 and 10,777 in 1978/79.

International control of whaling was first seriously discussed in 1927 at a meeting of the League of Nations. In 1935 the United States, Norway, Great Britain, and several other nations entered into an agreement under which the right and bowhead whales were protected (effective in 1937) and certain other whaling regulations were set forth. In 1946 the International Whaling Commission, a regulatory body now consisting of 35 member nations, was formed. Although the commission established various limits and refuge areas, its objectives were long thwarted by lack of cooperation by those members most involved in the whaling industry. Finally, as all the large species of whale approached or passed commercial extinction, regulations became more meaningful. The total kill quota fell from 45,673 whales in the 1973/74 season to 14,523 in 1980/81, only 2,121 of which were from the larger species (fin, sei, sperm). Only Japan and the Soviet Union still conduct major whaling operations, each employing one or two floating factories. Several other countries have small, land-based whale fisheries, and illegal "pirate" whaling ships are sometimes active. The U.S. government, reacting to immense public concern, has long advocated a total ban on commercial whaling. In 1972 the U.S. Congress passed the Marine Mammal Protection Act, which essentially ended the taking and importation of cetaceans and their products by persons subject to U.S. jurisdiction. All cetaceans are on appendix 2 of the CITES, except for those species placed on appendix 1.

One of the immediate goals of the U.S. Marine Mammal Protection Act was to reduce the number of small cetaceans (notably *Stenella* and *Delphinus*) being killed or injured by commercial fishing operations. The most critical problem was the incidental catch of these mammals in seines intended for tunafish, especially in the eastern Pacific. Although many kinds of small cetaceans had been intentionally hunted by people, and while some populations had declined, there had been no world-wide pursuit and devastation comparable to what befell the larger whales (Mitchell 1975a, 1975b). In the 1960s, however, improved types of tuna seines began trapping large numbers of dolphins that commonly are found in association with the fish. The number of dolphins killed or seriously injured in 1972 was estimated at 368,600 for U.S. fishing vessels and 55,078 for vessels of other countries. Subsequent regulations, modifications of fishing gear, and cooperation by fishermen reportedly reduced the respective figures for 1979 to 17,938 and 6,837.

The known geological range of the Cetacea is middle Eocene to Recent. Possibly, however, the order diverged from an extinct order of terrestrial carnivores, the Creodonta, at the end of the Cretaceous, and entered the sea in the Paleocene (Gaskin 1976). The earliest known fossil cetaceans are assigned to an extinct suborder, the Archaeoceti. In this group the skull is symmetrical and the teeth are clearly differentiated into incisors, canines, premolars, and molars. The Odontoceti first appear in upper Eocene strata, and the Mysticeti first appear in the lower Oligocene.

CETACEA; *Family PLATANISTIDAE*

Fresh Water or River Dolphins

This family of four Recent genera and six species is found in certain river systems of south-central Asia, China, and South America, and in coastal waters and estuaries off eastern South America. Rice (1967, 1977) recognized three subfamilies: Iniinae, with the genera *Inia* and *Lipotes;* Pontoporiinae, with *Pontoporia;* and Platanistinae, with *Platanista.* He noted that some earlier workers had given familial rank to the Iniinae and Platanistinae, and had transferred the Pontoporiinae to the family Delphinidae. Kasuya (1973) gave familial rank to all three groups, but placed them together in a superfamily, the Platanistoidea. Zhou, Qian, and Li (1979) thought that *Lipotes* should also be in a monotypic family, the Lipotidae.

These rather small cetaceans are approximately 150 to 300 cm in length and weigh about 20 to 160 kg. The beak is long, narrow, and clearly distinguishable from the rest of the head. For most of their length, the lower jaws are fused or at least closely appressed. The forehead is usually blunt, and a distinct external neck occurs in *Inia* and *Platanista.* The pectoral fin is short and broad; the upper arm bone is longer than the lower arm bones. The dorsal fin is small in *Inia* and *Platanista* but well developed in *Lipotes* and *Pontoporia.* The eyes of all genera are very small, and those of *Platanista* lack a lens. The skull is nearly symmetrical. There are 41 to 45 vertebrae, those of the neck being free. The number of teeth ranges from 210 to 242 in *Pontoporia* and from about 100 to 140 in the other genera.

These curious and inoffensive dolphins often swim around fishing boats. They generally travel singly or in groups of 2 to 12 individuals. They are not as active as the members of the family Delphinidae. Their dives seldom last more than a few minutes. All genera appear to make local migrations. The diet consists mainly of aquatic, bottom-dwelling creatures, such as mud-frequenting fishes and fresh-water crustaceans. Much of the food is obtained by probing in the mud with the sensitive snout.

Rice (1967) listed the following geological ranges: subfamily Platanistinae, middle Miocene in North America, late Miocene in Europe, and Recent in south-central Asia; subfamily Iniinae, early Miocene to Recent in South America, early Pliocene in North America, and Recent in eastern Asia; and subfamily Pontoporiinae, middle Pliocene and Pleistocene in North America, and Pliocene and Recent in South America.

CETACEA; PLATANISTIDAE; *Genus INIA* D'Orbigny, 1834

Boutos, Amazon Dolphins

There are two species (Pilleri and Gihr 1977a; Rice 1977):

I. boliviensis, upper Madeira River system of Bolivia;

I. geoffrensis, Amazon and Orinoco basins of South America.

Amazon dolphins (*Inia geoffrensis*): A. Photo from Fort Worth Zoological Park through Lawrence Curtis; B. Photo by James N. Layne.

Head and body length is about 170 to 300 cm, and expanse of the tail flukes is about 51 cm. Two adult males weighed 59 and 122 kg, and an adult female weighed 71 kg (Harrison and Brownell 1971). The variable coloration of this genus is apparently associated with age. All of the younger animals kept by Trebbau (1975) had dark bluish metallic gray upper parts, and paled to silvery gray on the lateral and ventral parts. Older and larger individuals were much lighter and generally pink in color. The lateral and ventral parts were clear pinkish gray, the flukes and most of the snout were pinkish, and the melon (forehead) was bluish and pink. The darkest parts of the older animals were the regions around the blowhole, eyes, neck, and middorsum. There were no distinct borders to the colors.

A long, slender, slightly downcurved beak is characteristic. The teeth number about 33 or 34 on each side of each jaw, making a total of 132 to 136; the back 8 or 9 teeth have a distinct keel. The rounded head bears the blowhole on its summit. The dorsal fin is long and ridgelike. The wide snout is covered with short, stiff bristles.

These dolphins are restricted to fresh water. According to Trebbau and Van Bree (1974), they generally are found in brownish, turbid, and slow-moving or temporarily stagnant streams. They may migrate into flooded forests, and small streams and lakes, during periods of high water. They then sometimes become trapped in lakes during the dry season,

but are able to survive by preying on fish that are also trapped. If fish are abundant in an area, the dolphins can be seen there for weeks at a time. They surface to breathe every 30 to 60 seconds. Sometimes only the blowhole and top of the head emerge, but frequently the dorsal fin and ridge of the back are also exposed. When swimming rapidly, and apparently when feeding, *Inia* rolls to breathe. This genus seems to be less active than the Delphinidae, but will occasionally leap out of the water to heights of 125 cm.

The senses of touch and hearing are probably acute; *Inia* probes for food on the bottom with its sensitive snout, and may use echolocation to detect underwater obstacles and prey. The eyes are small, but Caldwell, Caldwell, and Evans (1966) found vision to be acute and apparently to be the preferred method of environmental investigation. They also recorded 12 types of sounds, which generally were less varied, lower in intensity, and of slightly lower frequencies than those heard in most other odontocetes. The diet consists of fish, usually less than 30 cm long. Captive adults ate four to five kg of fish a day (Trebbau and Van Bree 1974).

Trebbau (1975) found *I. geoffrensis* mostly in small groups, which had a tendency to occupy a defined territory. When an individual was taken captive, others would come to its assistance. Up to 8 captives were held together, evidently without strife. Caldwell and Caldwell (1969) observed wild *I. geoffrensis* in schools of 12 to 15, occasionally up to 20,

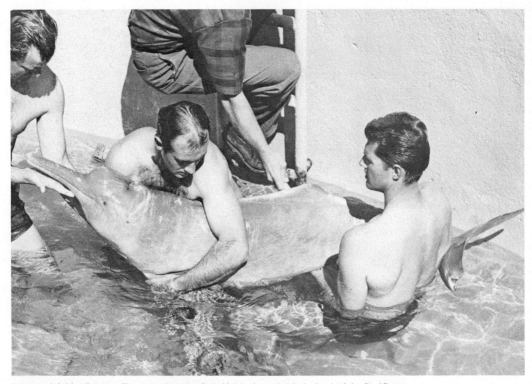

Amazon dolphin (*Inia geoffrensis*), photo by Bob Noble through Marineland of the Pacific.

individuals, and noted that captives tended to cluster together. Pilleri and Gihr (1977*a*) reported *I. boliviensis* usually to occur alone or in pairs. Harrison and Brownell (1971) collected a pregnant female *I. geoffrensis* in February, and pregnant and lactating females, and a calf, in April. They suggested that implantation can occur in October and November, and probably earlier, and that births take place from July to September in the upper Amazon. Length at birth is 76 to 80 cm. According to Stephen Spotte (Mystic Marinelife Aquarium, pers. comm.), one specimen was still alive after 16½ years in captivity.

Pilleri (1979) stated that *I. geoffrensis* seems common in the Orinoco Basin and is not hunted there, but is threatened by motorboats, pollution, and dam construction. Pilleri and Gihr (1977*a*) reported that *I. boliviensis*, which is isolated from *I. geoffrensis* by a 400-km stretch of rapids, has been severely reduced in numbers, mainly through hunting by people for its hide and fat.

CETACEA; PLATANISTIDAE; **Genus LIPOTES** Miller, 1918

Baiji, Whitefin Dolphin

The single species, *L. vexillifer,* has been recorded from the mouth of the Chang Jiang (Yangtze) to a point about 1,900 km up that river, from Dongting and Poyang lakes, and from the Quiantang River just south of the mouth of the Chang Jiang (Zhou, Qian, and Li 1977). Early reports, indicating that the species was restricted to Dongting Lake and some nearby waters, were incorrect.

Head and body length is 200 to 250 cm, and one specimen weighed 160 kg. Coloration is pale blue gray above and whitish below. The long, beaklike snout is curved upward, and there are 32 to 36 teeth in each side of each jaw. The eye is greatly reduced, but is functional. *Lipotes* resembles *Inia*, but differs in having an upcurved rostrum, a relatively smaller and more rounded pectoral fin, a longitudinal and somewhat rectangular blowhole, and a larger and more triangular dorsal fin (Brownell and Herald 1972).

Zhou, Qian, and Li (1977) considered *Lipotes* to be a fluvatile and estuarine odontocete, contrary to early reports that the genus was primarily an inhabitant of shallow lakes. During periods of high water, as in the late spring and summer, *Lipotes* also goes upstream into lakes and small rivers. The long beak is used to probe muddy bottoms for food. Dives are short, usually lasting only 10 to 20 seconds (Zhou, Pilleri, and Li 1979). The diet consists of fish, and the stomach of the type specimen contained 1.9 liters of an eellike catfish (Brownell and Herald 1972).

Zhou, Pilleri, and Li (1979) found an overall average of one *Lipotes* every four km along part of the Chang Jiang. The genus usually occurs in pairs, which in turn make up a larger social unit of about 10 individuals. According to the IUCN (1976), births occur in January and February in the deeper waters of the Chang Jiang. Brownell and Herald (1972), however, stated that a lactating female was captured on 21 December.

Lipotes was long protected by custom, though if an individual was killed accidentally its meat would be eaten and its fat used for medicinal purposes. In recent decades, numbers and distribution have declined seriously through hunting, accidental catching by fishermen, collision with motorboats, and development of irrigation facilities (Zhou, Qian, and Li 1977; Zhou, Pilleri, and Li 1979; Chen et al. 1979). The

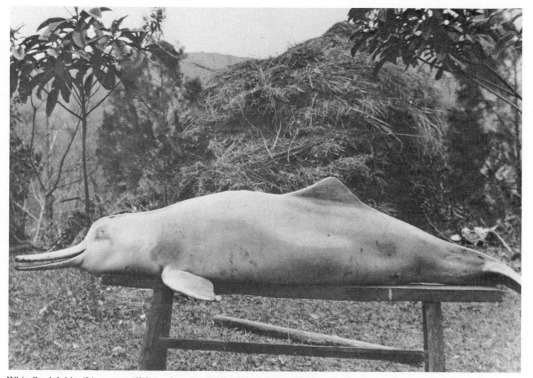

Whitefin dolphin (*Lipotes vexillifer*), photo by Clifford Pope through Robert L. Brownell, Jr.

genus no longer occurs in Dongting Lake, because of sedimentation. Legal protection has been provided in China since 1975. The whitefin dolphin is classified as indeterminate by the IUCN (1976) and is on appendix 1 of the CITES.

CETACEA; PLATANISTIDAE; **Genus PONTOPORIA**
Gray, 1846

Franciscana, La Plata Dolphin

The single species, *P. blainvillei,* occurs in the coastal waters and estuaries off southeastern South America, from the Tropic of Capricorn near Ubatuba, Brazil, south to the Valdez Peninsula, Argentina (Brownell 1975a).

Head and body length is 130 to 175 cm, and weight is 20 to 61 kg. Females are generally larger than males. The color is grayish above and paler below. The young are usually brownish. The beak is long, slender, and slightly down-curved. There are 48 to 61 teeth on each side of each jaw (Kasuya and Brownell 1979), and the total number of teeth per individual has been counted at 210 to 242. The dorsal fin is prominent and triangular in shape.

This is the only member of the family Platanistidae that occurs in salt water. It is found in the estuary of the Rio de la Plata, but has not been recorded from the adjoining Parana and Uruguay rivers (Brownell 1975a). According to Kellogg (1940), *Pontoporia* is rarely seen in the Rio de la Plata during winter, perhaps because most schools then migrate out to sea or northward along the Brazilian coast. *Pontoporia* presumably locates its prey by echolocation and by probing the bottom with its long snout. Examination of stomach contents suggests that most food is taken at or near the bottom. The diet includes fish, squid, and shrimp (Fitch and Brownell 1971).

Examination of specimens taken off the coast of Uruguay (Kasuya and Brownell 1979; Harrison and Brownell 1971; Brownell 1975a) indicates that females have a 2-year reproductive cycle, with mating from December to February, births from September to December, lactation until the following August or September, and then a rest of several months. The gestation period is about 10½ months. The single young is generally 70 to 75 cm long at birth, starts to take solid food when 3 months old, and is weaned by the age of 8 or 9 months. Both males and females become sexually mature when 2 to 3 years old, and physically mature 1 or 2 years later. One male specimen was 16 years old at the time of death.

The Franciscana is taken regularly off the coast of Uruguay in the nets of fishermen who are primarily seeking sharks (Mitchell 1975a; Kasuya and Brownell 1979; Pilleri 1971). About 1,500 to 2,000 individual *Pontoporia* are killed annually in this manner. They are used for pig feed and as a source of oil.

CETACEA; PLATANISTIDAE; **Genus PLATANISTA**
Wagler, 1830

Ganges and Indus Dolphins, Susus

There are two species (Pilleri and Gihr 1971; Rice 1977):

P. gangetica, Ganges-Brahmaputra-Meghna river system and the Karnaphuli River of India, Bangladesh, and Nepal;

La Plata dolphin (*Pontoporia blainvillei*), photos by Cory T. de Carvalho.

P. indi, Indus River system of Pakistan and India.

Van Bree (1976) argued that the name *P. minor* has priority over *P. indi,* but Pilleri and Gihr (1977*b*) rejected this view.

Head and body length is variable, but adults are usually 200 to 300 cm long. The female is the larger sex, there being a questionable report of one that measured about 400 cm. The expanse of the tail flukes is about 46 cm. The back is dark lead gray to lead black, and the belly is somewhat lighter. The snout is slender, slightly upcurved, well differentiated from the steeply rising forehead, and about 18 to 21 cm long. The blowhole is longitudinal and slitlike, the neck is distinctly constricted, the dorsal fin is low and ridgelike, and the pectoral fin is cut off squarely at the end (Kellogg 1940). There are 28 or 29 teeth on each side of each jaw. A unique feature is the presence of two plates of bone, one on each side of the skull, that project outward and nearly meet in front of the blowhole.

Susus occur only in fresh water, but are found from the tidal limits of rivers to the foothills of the Himalayas. They may undergo local migrations, moving into small tributaries during the monsoons but returning to large rivers in the dry winter (Kasuya and Haque 1972). *Platanista* rises to breathe about every 30 to 120 seconds, usually plunging out of the

water in an upward or forward direction but occasionally exposing only the blowhole. Captives normally swim on their sides, apparently because the tiny, deeply set eye can receive light only at such an angle (Purves and Pilleri 1974–1975; Herald et al. 1969). Although the eye of *Platanista* lacks a lens and is usually not visible externally, and the genus is sometimes referred to as being blind, the eye does seem to function as a direction-finding device. Captives swim about continuously over a 24-hour period, perhaps because the streams they naturally inhabit flow swiftly during the monsoons and constant activity is necessary to avoid injury (Pilleri et al. 1976). There is also a continuous transmission of sound over a 24-hour period. Mizue, Nishiwaki, and Takemura (1971) found 87 percent of the sounds to be clicks for echolocation and 5 percent to be communicative. To find food, *Platanista* probably uses echolocation, and also probes with its sensitive snout for fish, shrimp, and other organisms in the bottom mud.

There are reports that *Platanista* travels and feeds in schools of 3 to 10 or more individuals. In 90 percent of the sightings of *P. gangetica* made by Kasuya and Haque (1972), however, only a single animal was seen. Births of *P. gangetica* may occur at any time of the year but are mostly from October to March, with a peak in December and January at the beginning of the dry season (Kasuya 1972*b*). The

Susus (*Platanista* sp.), photos by G.Pilleri.

single young is 70 cm long at birth and is weaned within a year. Sexual maturity is attained at about 10 years. Two males were reported to be still growing at 16 and 28 years of age.

Both species of *Platanista* are on appendix 1 of the CITES. According to the U.S. National Marine Fisheries Service (1978), however, *P. gangetica* is still fairly numerous and is not endangered. A few are captured incidentally in the nets of fishermen, who attempt to release the dolphins alive. The species has apparently disappeared in the Karnaphuli River, which empties into the Bay of Bengal to the east of the Ganges. *P. indi* has declined drastically in range and numbers, and is classified as endangered by the IUCN (1976). The main threat seems to be construction of numerous barrages—barriers to impound water for subsequent use in irrigation. Therefore, dolphin populations are split up, seasonal movements are restricted, and water quality deteriorates. The animals have also been hunted for their meat and oil. In 1974 the total population was estimated to contain only 450 to 600 individuals, most of which were in a 130-km stretch of the Indus between Sukkur and Guddu barrages (Kasuya and Nishiwaki 1975). Although protected by law, *P. indi* is still regularly taken by fishermen (Pilleri and Pilleri 1979*a*).

CETACEA; *Family DELPHINIDAE*

Dolphins and Porpoises

This family of 20 Recent genera and 40 species inhabits all the oceans and adjoining seas of the world, as well as the estuaries of many large rivers. Some species occasionally ascend rivers. Rice (1967) separated the families Stenidae (with the genera *Steno, Sousa,* and *Sotalia*) and Phocoenidae (with the genera *Phocoena, Neophocaena,* and *Phocoenoides*) from the Delphinidae, and also divided what remained of the Delphinidae into four subfamilies. Later, Rice (1977) returned the Stenidae and Phocoenidae to the Delphinidae, and did not refer to subfamilies. Hall (1981) also did not recognize any familial or subfamilial divisions within the group here called the Delphinidae.

Head and body length ranges from as little as 120 cm in *Phocoena* to as much as 980 cm in *Orcinus,* and weight ranges from 23 to about 9,000 kg. The common name "dolphin" is generally applied to small cetaceans having a beaklike snout and a slender, streamlined body form, whereas the term "porpoise" refers to those small cetaceans with a blunt snout and a rather stocky body form. The blowhole is located well back from the tip of the beak or the front of the head. The pectoral and dorsal fins are sickle shaped, triangular, or broadly rounded, and the dorsal fin is located near the middle of the back. The genera *Lissodelphis* and *Neophocaena* lack a dorsal fin.

In the Delphinidae the vertebrae number 50 to 100 and the first 2 neck vertebrae are fused, whereas in the Platanistidae all the neck vertebrae are free. Delphinids also have a lesser number of double-headed ribs than do the platanistids. In the

Pacific white-sided dolphins (*Lagenorhynchus obliquidens*), which have been trained to leap unusually high. Many of the smaller cetaceans regularly leap out of the water but rarely go more than a foot or two above the surface and re-enter the water at a distance of only a few feet. Photo from Marineland of the Pacific.

Delphinidae, unlike the Platanistidae, the skull lacks a crest. The fusion of the lower jaws in delphinids does not exceed one-third of their length, but this fusion is more than half the length in platanistids. There are usually many functional teeth in both the upper and lower jaws of delphinids, the maximum number being about 260. In the genus *Grampus*, however, there are only 4 to 14 teeth, and these are confined to the lower jaw. In the genera *Phocoena*, *Neophocaena*, and *Phocoenoides*, the teeth are spade shaped.

The spout of dolphins and porpoises is usually not well defined, but that of *Globicephala* extends about 150 cm. Breathing is often accompanied by a low hissing or puffing noise. The respiration rate of bottlenosed dolphins (*Tursiops*), taken in an aquarium where they could swim freely, was 1.5 to 4 respirations per minute. This genus has a heart rate of 81 to 137 beats per minute (the average is 100 per minute). Bottlenosed dolphins have been observed sleeping in calm water about 30 cm below the surface; slight movements of the tail brought the head above water so that the animals could breathe.

The Delphinidae include the most agile and some of the most speedy cetaceans. They commonly surface several times a minute, frequently leap clear of the water, and are capable of sustained speeds of 26 to 33 km per hr (Rice 1967). Many species follow ships and seem to frolic about the bow. They have remarkable group precision and regularity of movement. Migration is known to occur in some species.

The ability to perceive objects by means of reflected sound (echolocation) was demonstrated in *Tursiops* in 1958, the first time for a cetacean. The sense of vision is not as well developed as that of hearing, but *Tursiops* can see moving objects in the air at least 15 meters away. Observations of captive dolphins indicate that these animals are very intelligent. Adult delphinids will engage in complex play with various objects and can be trained to perform tricks.

Dolphins and porpoises usually associate in schools of five to several hundred individuals, though they are sometimes seen alone or in pairs. Groups may assemble when frightened and will attack intruders. They sometimes kill large sharks by

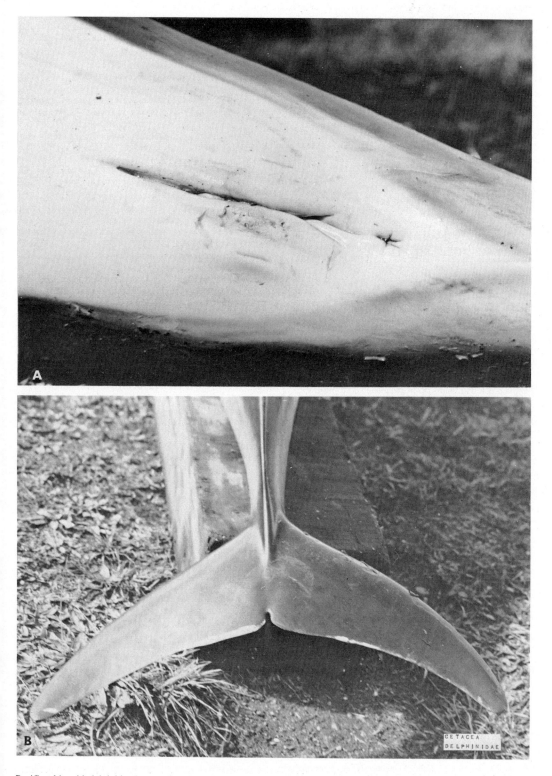

Pacific white-sided dolphin (*Lagenorhynchus obliquidens*): A. Urogenital region of a young female, note the mammary slits on each side of the urogenital opening, the anus is to the right; B. Dorsal view of caudal flukes showing the deep tail notch typical of the Delphinidae and the prominent ridge on the top of the tail; Photos by Warren J. Houck.

ramming them. They utter a wide variety of underwater sounds that appear to function in communication. Cooperative behavior has often been observed; one or more individuals will come to the aid of another that is injured, sick, or giving birth, pushing it to the surface so that it can breathe.

The known geological range of the Delphinidae is lower Miocene to Recent in North America and Europe, upper Pliocene to Recent in Japan, Pleistocene to Recent in New Zealand, and Recent in all oceans.

CETACEA; DELPHINIDAE; *Genus STENO Gray, 1846*

Rough-toothed Dolphin

The single species, *S. bredanensis,* occurs in tropical and warm temperate waters of all oceans and adjoining seas (Rice 1977). On the basis of reported hybridization between *Steno* and *Tursiops* and between *Tursiops* and *Grampus,* Van Gelder (1977*b*) recommended making *Steno* a synonym of *Grampus* (the earliest available name of the three).

Head and body length is 175 to 275 cm, pectoral fin length is about 30 cm, and dorsal fin height is about 15 cm. A male specimen, 216 cm long, weighed 114 kg (Perrin and Walker 1975). The upper parts are slate colored or purplish black, with scattered spots and markings. The belly is pinkish white or rose colored, with slaty spots. The pectoral fin, dorsal fin, and tail flukes are dark. The beak is white, slender, and compressed from side to side. There are 20 to 27 teeth on each side of each jaw. The surfaces of the teeth are roughened and furrowed by vertical ridges and wrinkles.

Steno is found mainly in tropical and subtropical waters, but a few individuals have stranded in colder areas outside the normal range. Stomach contents have included fish and octopus. Group size is usually 50 individuals or less, but one school of over 100 animals reportedly stranded. *Steno* has been found in schools together with *Tursiops* and *Stenella.* Captives are easily trained and are reportedly even more intelligent than *Tursiops.* Four females, taken off West Africa in May, contained fetuses 60 to 87 cm long (Rice 1967; Leatherwood, Caldwell, and Winn 1976; Mitchell 1975*a;* Perrin and Walker 1975). One specimen was still living after 15 years in captivity (Marvin L. Jones, Zoological Society of San Diego, pers. comm.). There is little human effort to hunt this genus, but some individuals are caught and used for food in Japan, Africa, and the Caribbean, and some are captured accidentally in the tropical Pacific tuna fishery (Mitchell 1975*b*).

CETACEA; DELPHINIDAE; *Genus SOUSA Gray, 1866*

Humpback Dolphins

Rice (1977) and Mitchell (1975*a*) recognized two species:

S. chinensis, coastal waters of eastern and southern Africa, southern Asia, the East Indies, and Australia;
S. teuszii, coastal waters from Mauritania to Angola.

Rice suggested that *S. teuszii* might be only a subspecies of *S. chinensis.* Some authorities, including Medway (1977) and Lekagul and McNeely (1977), consider *Sousa* to be part of *Sotalia,* and have divided *S. chinensis* into a number of separate species.

Head and body length is 120 to 250 cm, pectoral fin length is 30 cm or less, dorsal fin height is about 15 cm, and width of the tail flukes is about 45 cm. A specimen of *S. teuszii,* 230 cm long, weighed 139 kg. Coloration is variable; most forms are brown, gray, or black above and paler beneath, but some populations are whitish, speckled with gray, or freckled with brown spots.

The skull of *Sousa* differs from that of other dolphins in the rather long symphysis of the jaws and in the widely separated pterygoid bones that do not close together behind the palate. There are 23 to 50 teeth on each side of each jaw. *Sousa* resembles *Tursiops* and *Steno,* but is distinguished by having more teeth and 10 to 15 fewer vertebrae. The rostrum of *Sousa* is always narrower than that of *Tursiops.*

These dolphins are found in both salt and fresh water, inhabiting seas, estuaries, and the mouths of rivers. They sometimes ascend rivers, and have been reported 1,200 km up the Chang Jiang (Yangtze) in China (Lekagul and McNeely 1977). One population of about 500 individuals seems to remain all year in muddy mangrove creeks in the delta of the Indus River (Pilleri and Pilleri 1979*b*). A population off South Africa also does not appear to migrate, but its habitat consists of deep waters over sand and reefs (Saayman and Tayler 1979).

Sousa sometimes leaps out of the water to a height of 120 cm, but seems to be slower than most dolphins. It rolls to breathe, and appears to roll faster when feeding. Saayman and Tayler (1979) found *Sousa* to ride waves rarely, to remain submerged for up to three minutes, and to hunt individually, generally in the vicinity of reefs. The diet evidently consists of fish.

Sousa often is seen alone or in pairs, but Saayman and Tayler (1979) reported group size off South Africa to average about 7, and range up to 20, individuals. In that area births occurred all year, with a peak in summer (December–February). The number of offspring per birth is 1.

Small numbers of humpback dolphins are caught in the Arabian Sea, Red Sea, and Persian Gulf, and are used for human consumption (Mitchell 1975*a*). The population in the Indus Delta is not deliberately fished, but is threatened by accidental catching and industrial pollution (Pilleri and Pilleri 1979*b*). *Sousa* is on appendix 1 of the CITES.

CETACEA; DELPHINIDAE; *Genus SOTALIA Gray, 1866*

Tucuxi, River Dolphin

Rice (1977) recognized a single species, *S. fluviatilis,* with two subspecies: *S. f. fluviatilis,* in the Amazon River and its tributaries; and *S. f. guianensis,* in coastal waters and the lower reaches of rivers from northwestern Venezuela to southern Brazil. Husson (1978) treated *guianensis* as a full species. *Sotalia* may also occur along the Atlantic coast of Central America (Mitchell 1975*a*).

Head and body length is 91 to 165 cm, dorsal fin height is 11 to 13 cm, and greatest girth is 70 to 98 cm. An adult male, 160 cm long, weighed about 47 kg. Another male, 148 cm long and sexually active, weighed about 33 kg (Harrison and Brownell 1971). In *S. f. guianensis* the upper parts range from pale bluish gray to brown or blackish. In at least some populations the color of the back extends to a circle around the eye, onto the pectoral fin, and to the sides of the tail. In some populations the sides are yellowish orange and there is a bright yellow patch on each side of the dorsal fin near the top. The underparts of *S. f. guianensis* are white, pinkish, or grayish. In *S. f. fluviatilis* the upper parts are bluish or pearl gray, the color being darker anteriorly. Larger animals are noticeably paler above. The pectoral fin, both above and below, is the same color as the back. The underparts are pinkish white to white. A prominent band of the ventral

Rough-toothed dolphin (*Steno bredanensis*): Top, photo by Gary L. Friedrichsen through National Marine Fisheries Service; Bottom, photos by Michael S. Sinclair.

coloration extends upward on the sides of the body to slightly above the level of the eye. The dorsal color, however, extends down to a distinct line from the corner of the mouth to the base of the pectoral fin and includes the eye. The tip of the beak and the apex of the dorsal fin are conspicuously white.

Sotalia is often found in the same area as *Inia*, but the latter has a more prominent dorsal fin, a longer beak, and a more prominently bulging forehead. From *Steno* and *Tursiops*, *Sotalia* is distinguished by the separation of its pterygoid bones and by having fewer caudal vertebrae and more teeth. There are 26 to 35 teeth on each side of each jaw.

Sotalia is found in both salt and fresh water. *S. f. guianen-*

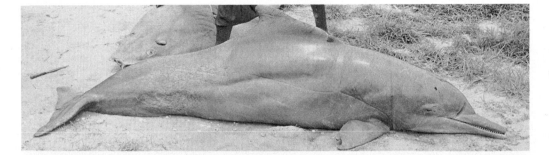

Humpback dolphin: Top, *Sousa teuszii*, photo by J. Cadenat; Middle and bottom, *S. chinensis*, photos by Michael M. Bryden.

River dolphins (*Sotalia fluviatilis guianensis*), photos by W. Gewalt.

sis occurs along the coast, but often enters large rivers. *S. f. fluviatilis* occurs in much of the Amazon Basin, as far west as the Andes (Mitchell 1975a). The latter subspecies has been reported to be more active but less inquisitive than *Inia*. *Sotalia* swims rather slowly and rarely leaps clear of the water, but has been seen to jump 120 cm above the surface. When breathing, it always rolls out of the water, the head and trunk usually appearing in smooth sequence and a short puff being the only sound heard. The reported interval between breaths is 5 to 85 seconds, with a mean of 33 seconds. *Sotalia* is most active in the early morning and late afternoon. Captives have been found to emit echolocation-type clicks (Caldwell and Caldwell 1970) and have been trained to perform various acrobatics (Gewalt 1979a). The diet consists mostly of fish, but shrimp are also eaten.

The tucuxi has regular movements within a home range, and is gregarious. Groups often swim and roll in tight formation and in nearly perfect synchrony, the individuals almost touching sides when they appear at the surface. No social interaction between *Sotalia* and *Inia* has been observed, though the two genera are often in close proximity. Husson (1978) reported *Sotalia* to be rather common in the mouths of large rivers in Surinam, where it occurs in groups of about 10 individuals. A female taken in this area between 15 February and 13 April 1971 had a fetus about 60 cm long. A lactating female was collected in the upper Amazon in October (Harrison and Brownell 1971).

Some South American natives regard *Sotalia* as a sacred animal, consider it a good friend and protector, and believe that it will bring the bodies of drowned persons to shore. *Sotalia*, however, is often taken accidentally in the nets of fishermen (Mitchell 1975a). The genus is on appendix 1 of the CITES.

CETACEA; DELPHINIDAE; **Genus TURSIOPS** *Gervais, 1855*

Bottlenosed Dolphins

There are apparently three species (Rice 1977; Ross 1977; Hall 1981; Lekagul and McNeely 1977; Ellerman and Morrison-Scott 1966; Gaskin 1968):

T. truncatus, primarily in temperate and tropical waters of the Atlantic Ocean and adjoining seas;

T. aduncus, primarily in temperate and tropical waters of the Indian, South Pacific, and western and southern North Pacific oceans, and adjoining seas;

T. gillii, primarily in temperate waters of the eastern North Pacific Ocean and Gulf of California.

Many authorities, including Rice (1977), have referred all *Tursiops* to the single species *T. truncatus*. Hall (1981) used the name *T. nesarnack* in place of *T. tursiops*, and treated *T. aduncus* as a subspecies thereof. Van Gelder (1977b) recommended making *Tursiops* a synonym of *Grampus*, on the basis of reported hybridization between the two genera.

Head and body length is 175 to 400 cm, pectoral fin length is 30 to 50 cm, dorsal fin height is about 23 cm, and width of the tail flukes is about 60 cm. Adults usually weigh 150 to 200 kg, but weights in excess of 650 kg have been reported (Leatherwood, Caldwell, and Winn 1976). The upper parts are usually dark gray or slaty blue, the flippers and flukes are darker, and the underparts are paler. The genus is distinguished by the short, well-defined snout or beak, about 8 cm long and supposedly resembling the top of an old-fash-

Bottlenosed dolphin (*Tursiops truncatus*), photo from Marineland of the Pacific. Inset: *T. gillii*, photo by Warren J. Houck.

ioned gin bottle. There are 20 to 28 teeth on each side of each jaw. Each tooth is about 1 cm in diameter.

Tursiops is found mainly in coastal waters, often in bays and lagoons, and sometimes ascends large rivers. In certain areas, however, it ranges as far offshore as the edge of the continental shelf. Many researchers recognize two forms of *Tursiops,* the smaller staying in shallow waters near the mainland and the larger occurring farther offshore. The genus appears to be migratory in some areas, and movements may relate to seasonal changes in temperature or food distribution. *Tursiops* sometimes rides the surf or the bow wave of vessels, and may jump to heights of six meters above the surface (Leatherwood, Caldwell, and Winn 1976; Leatherwood and Reeves 1978; Mitchell 1975*b*). In 26 separate observations along a tidal creek on the Georgia coast, Hoese (1971) saw *Tursiops* rush its entire body, except the tail, up onto a mudbank to capture small fish that had been driven in front of it.

Tursiops is commonly seen along the coasts of the United States and is also familiar to people through its displays of agility and sagacity under captive conditions. Compared to most wild animals it is easily trained to perform acrobatics, locate hidden objects, and play with balls. It is also used widely in research work involving cetacean physiology, psychology, and sociology. Such studies have demonstrated that *Tursiops* has a high degree of what humans commonly refer to as intelligence. Some of these dolphins were even able to learn complex procedures quickly merely by watching other individuals perform (Adler and Adler 1977).

Tursiops is also noted for having a brain that is larger than that of *Homo* and a variety of vocalizations. Certain authorities have argued that these dolphins, and probably other cetaceans as well, have a complex language and may eventually be able to communicate meaningfully with people. The preponderance of current evidence, however, suggests that the large cetacean brain is associated mainly with development of the sense of hearing. Each individual *Tur-*

siops appears to have its own "signature" whistle, by which it can communicate a limited amount of information on its identity, location, and condition to others of its kind (Caldwell and Caldwell 1977, 1979; Hickman and Grigsby 1978). Other sounds and visual signals are also used in communication. Both the whistles and the high-frequency, clicklike pulses used for echolocation are thought to be produced not in the larynx but in the nasal sacs within the forehead (Hollien et al. 1976). Echolocation—the detection of objects by bouncing sound waves off of them—is utilized by *Tursiops* to find obstacles and food underwater. The diet consists mainly of bottom-dwelling fish. Captive dolphins eat six or seven kg of food per day (Mitchell 1975*a*).

Tursiops may form large aggregations, which usually comprise smaller groups of not over 12 individuals (Leatherwood, Caldwell, and Winn 1976). In the eastern Pacific, group size is usually 15 or less near the shore but 25 to 50 far offshore (Leatherwood and Reeves 1978). Schools of up to 600 were reported off North Carolina in the late nineteenth century, and sexes were about equally represented there (Mead 1975). In 93 sightings of *T. aduncus* in the Indian Ocean, Saayman and Tayler (1973) found the number of individuals per group to range from 3 to 1,000 and to average about 140. At least some groups in the eastern Pacific appear to have limited home ranges that are associated with particular islands (Leatherwood and Reeves 1978).

In European waters the young of *T. truncatus* are born in midsummer, and most births off Florida occur from February to May (Mitchell 1975*a*). *T. aduncus* has a prolonged reproductive season off South Africa, with most births occurring in late spring and summer (Ross 1977). The normal interval between calves is two years, but if the young dies at birth, another offspring may be produced a year later. The gestation period is 12 months. The single newborn weighs about 9 to 12 kg and is 100 to 125 cm long. Lactation usually lasts 12 to 18 months, but the young may begin to take some solid food when less than 6 months old. Sexual maturity is

Bottlenosed dolphins (*Tursiops truncatus*), photo by Bob Noble through Marineland of the Pacific.

reportedly attained at some point from 5 to 12 years of age in females and 9 to 13 years of age in males. Specimens taken from Florida waters have lived at least 25 years, while some taken in Japanese waters have lived 30 to 35 years (Mitchell 1975a; Odell 1975; Sergeant, Caldwell, and Caldwell 1973; Ross 1977).

Bottlenosed dolphins have been deliberately hunted by people in many parts of the world (Mitchell 1975a, 1975b; Mead 1975). Products include meat for human consumption, fertilizer, body oil for cooking and illumination, and jaw oil for use as a lubricant in watches and precision instruments. An intensive fishery in the Black and Azov seas led to a sharp reduction in the number of *Tursiops,* and finally to a ban on dolphin catching by the Soviet Union in 1966. Several commercial fisheries existed in the eastern United States until the early twentieth century. The largest was that on Cape Hatteras, North Carolina, where 2,000 or more dolphins were taken annually during the peak years 1885 to 1890. There are still small directed fisheries in West Africa, Sri Lanka, Indonesia, and Japan. *Tursiops* also is occasionally taken accidentally in nets set for tunafish, and is sometimes shot by fishermen, who consider it a competitor. Many bottlenosed dolphins have been taken alive in waters off the United States, Japan, and Italy for purposes of display and human entertainment. Under the U.S. Marine Mammal Protection Act of 1972, such collecting requires a permit, and all other taking and importation of cetaceans and their products is prohibited. The U.S. National Marine Fisheries Service (1981) estimated that there were 3,000 to 10,000 bottlenosed dolphins off the east coast of the United States.

CETACEA; DELPHINIDAE; **Genus STENELLA** *Gray,* *1866*

Spinner, Spotted, and Striped Dolphins

There seem to be five species (Rice 1977; Perrin 1975a, 1975b; Perrin et al. 1981; Hall 1981; Brownell and Praderi

1976; Hubbs, Perrin, and Balcomb 1973; Mullen 1977; Jones, Carter, and Genoways 1979):

S. longirostris (pantropical spinner dolphin), tropical and warm temperate waters of the Atlantic, Indian, and Pacific oceans, and adjoining seas;

S. clymene (Atlantic spinner dolphin), tropical and subtropical waters of the Atlantic Ocean and adjoining seas;

S. coeruleoalba (striped dolphin), tropical and temperate waters of the Atlantic, Indian, and Pacific oceans, and adjoining seas;

S. attenuata (bridled or pantropical spotted dolphin), tropical and warm temperate waters of the Atlantic, Indian, and Pacific oceans, and adjoining seas;

S. plagiodon (Atlantic spotted dolphin), tropical and warm temperate waters of the Atlantic Ocean and adjoining seas.

Although Hall (1981) suspected that there are only two species of spotted dolphins, he, along with Schmidly, Beleau, and Hildebran (1972), and Leatherwood, Caldwell, and Winn (1976), recognized a third species, *S. frontalis,* to take the place of *S. attenuata* in the Atlantic and adjoining seas. Hall also listed a fourth specific name that might apply to certain spotted dolphin populations, *S. dubia,* and he used the name *S. pernettensis* in place of *S. plagiodon.*

Head and body length is 150 to 350 cm, pectoral fin length is 15 to 20 cm, dorsal fin height is 15 to 30 cm, width of the tail flukes is about 35 to 50 cm, and weight is about 60 to 165 kg. Males average larger than females (Perrin 1975b). There is much variation in color pattern. In general, spinner dolphins are dark gray to black above, light gray or tan on the sides, and paler below; mature spotted dolphins have a sprinkling of pale spots on a dark background above and a sprinkling of dark spots on a pale background below; and the striped dolphin is grayish above, has mostly pale underparts and sides, and has one dark stripe extending from each eye to the anus and another stripe extending from the eye to the pectoral fin.

There is a distinct beak, the rostrum is long and narrow, the palate is not grooved on the inner side of the tooth row, and the union of the two branches of the lower jaw is rela-

Spotted dolphin (*Stenella attenuata*), photo by R. L. Pitman through National Marine Fisheries Service.

tively short. There are 29 to 65 teeth on each side of each jaw, and 68 to 81 vertebrae (Hall 1981; Leatherwood, Caldwell, and Winn 1976).

These dolphins seem to prefer the deeper, clearer offshore waters, mainly in tropical and subtropical parts of the world. *S. plagiodon* is generally found over 8 km from the shore, but, in the Gulf of Mexico, may move closer during spring and summer (Leatherwood, Caldwell, and Winn 1976). *S. longirostris* is often found far out to sea, where its distribution fluctuates in accordance with changing oceanographic conditions (Mitchell 1975a). *S. coeruleoalba* apparently is migratory in the western Pacific, with some populations approaching Japan in the autumn, swimming along the coast, and then moving back out to sea the following spring (Nishiwaki 1975).

These dolphins can swim at speeds of 22 to 28 km per hr. They often seem to frolic about ships and ride the bow waves. They sometimes jump clear of the water, and the spinner dolphins derive their name from their habit of leaping above the surface and turning two or more times on their longitudinal axis (Leatherwood, Caldwell, and Winn 1976). *S. plagiodon* has been found to produce two general categories of sounds: whistles, probably highly individualized, for communication; and regular trains of clicks for exploring the environment and finding food through echolocation (Caldwell and Caldwell 1971a; Caldwell, Caldwell, and Miller 1973). *S. attenuata* feeds near the surface, but *S. longirostris* may sometimes feed at depths of 250 meters or more (Fitch and Brownell 1968; Mitchell 1975a). The diet of *Stenella* consists mainly of squid and small fish.

Group size is 5 to 30 individuals in *S. attenuata*, up to several hundred in *S. longirostris*, and up to 100, but more commonly 6 to 10, in *S. plagiodon* (Leatherwood, Caldwell, and Winn 1976). Schools of up to 3,000 *S. coeruleoalba* are found off Japan. In this area the young of *S. coeruleoalba* evidently leave their mothers at about the age of 2 or 3 years and join in their own schools until reaching sexual maturity, when they return to the main group (Nishiwaki 1975). Estrous females also sometimes form a separate school (Kasuya 1972a). Much the same pattern has been reported in Japanese populations of *S. attenuata* (Kasuya, Miyazaki, and Dawbin 1974), but no evidence of age or sex segregation has been found in schools of this species in the eastern tropical Pacific (Perrin, Coe, and Zweifel 1976).

Strandings of *S. longirostris* in Florida indicate a calving season from May to July; otherwise little is known about the biology of *Stenella* in Atlantic waters (Mead et al. 1980). Within the last decade, however, studies of many individuals killed through direct exploitation off Japan and incidental catching by the tuna fishery of the eastern Pacific have greatly increased our knowledge of the reproduction and life history of *Stenella* (Perrin, Coe, and Zweifel 1976; Perrin, Holts, and Miller 1977; Perrin, Miller, and Sloan 1977; Kasuya 1972a, 1976; Kasuya, Miyazaki, and Dawbin 1974; Nishiwaki 1975). Births may occur at any time , but there are pronounced spring and fall peaks on both sides of the Pacific. The interval between giving birth is about 26 months in *S. longirostris* in the eastern Pacific and about 28 months in *S. coeruleoalba* off Japan. The birth interval of *S. attenuata* is about 48 months off Japan, where there is little disturbance, but only 26 months in the eastern Pacific, perhaps reflecting a greater compensatory reproductive effort in the latter area, where there have been heavy losses to the tuna fishery.

Reported gestation periods are about 10.6 months in *S. longirostris*, 11.2–11.5 months in *S. attenuata*, and 11–12 months in *S. coeruleoalba*. There is normally a single young, but twins have been reported to occur on rare occasion. The mean length of the newborn is about 77 cm in *S. longirostris*, 83 cm in *S. attenuata*, and 100 cm in *S. coeruleoalba*. Lactation continues for about 10.1 months in *S. longirostris*, 11.2–12.4 months in *S. attenuata*, and 18 months in *S. coeruleoalba*. A small percentage of females have been found to be simultaneously lactating and pregnant. In *S. longirostris*, males reach sexual maturity when 6–11½ years old and females probably at 5 years. In *S. attenuata*, males reach puberty at 6–11 years of age and females at 4½–8 years. In *S. coeruleoalba*, both sexes become sexually mature when 5–9 years old and physically mature when 14–15 years old. Natural annual mortality is less than 14 percent for both adults and juveniles, and life span is relatively long. Maximum age has been estimated at 46 years for *S. attenuata* and 50 years for *S. coeruleoalba*.

There are fisheries directed against *Stenella* in various parts of the world, the most important being that of Japan (Mitchell 1975a, 1975b; Nishiwaki 1975; Kasuya 1976). Fishermen of that country have annually taken about 20,000 *S. coeruleoalba* and 500–2,000 *S. attenuata* by driving and harpooning the animals. The dolphins are used locally for

Spinner dolphins (*Stenella longirostris*), photo by Gary L. Friedrichsen through National Marine Fisheries Service.

human consumption. The population of *S. coeruleoalba* off Japan is estimated to have contained 400,000 to 600,000 individuals originally, but excessive exploitation has reduced this number by about half.

Stenella has been the genus most seriously affected through the incidental killing of dolphins by tuna fisheries (Mitchell 1975*a*, 1975*b*; U.S. National Marine Fisheries Service 1978, 1981). Fishermen know that *Stenella* often associates with schools of tuna, and therefore set their nets around groups of dolphins. Deployment of a new type of purse seine in the 1960s led to the accidental deaths of great numbers of dolphins, which became entangled as the nets were closed. The most serious losses were caused by American vessels, to the species *S. attenuata* and *S. longirostris* in the eastern tropical Pacific. In 1972, the worst year on record, the estimated number of dolphins taken by U.S. fishermen was 368,600, of which 178,000 were *S. attenuata* and 42,000 were *S. longirostris*. That same year, however, the U.S. Marine Mammal Protection Act was passed, and it prohibited the killing and importation of cetaceans by persons subject to U.S. jurisdiction, except under permit. To avoid economic hardship, the U.S. government issued permits allowing a certain number of dolphins to be taken acci-

dentally, but the set quota was reduced each year, dropping from 78,000 in 1976 to 31,150 in 1980. Regular inspection, development of improved fishing gear, and cooperation by fishermen has led to a marked reduction in mortality. In 1980 the U.S. fishing fleet was estimated to have accidentally taken only 15,000 dolphins of all kinds. Current population estimates for the eastern tropical Pacific are: *S. longirostris*, 2,000,000 individuals; *S. attenuata*, 3,900,000; and *S. coeruleoalba*, 248,000.

CETACEA; DELPHINIDAE; *Genus **DELPHINUS***
Linnaeus, 1758

Common or Saddleback Dolphin

Rice (1977) recognized only a single species, *D. delphis*, occurring throughout the tropical and warm temperate oceans and adjoining seas of the world. Banks and Brownell (1969) recognized an additional species, *D. bairdii*, based on specimens from waters off California and Baja California. Neither Van Bree and Purves (1972) nor Hall (1981) accept-

Common dolphin (*Delphinus delphis*), photo by M. Scott Sinclair through National Marine Fisheries Service.

ed *D. bairdii* as a distinct species. Van Bree and Gallagher (1978), however, suggested that *D. tropicalis* of the northern Indian Ocean and adjoining waters is a species distinct from *D. delphis,* which occurs in the same region.

Head and body length is 150 to 250 (rarely to 260) cm, pectoral fin length is about 30 cm, dorsal fin height is about 40 cm, and width of the tail flukes is about 50 cm. Lekagul and McNeely (1977) listed weight as 60 to 75 kg. *Delphinus* is among the most colorful of cetaceans. Over most of its range, the back is brown or black, the belly is whitish, and the sides have bands and stripes of gray, yellow, and white. Some populations, however, lack the side markings. The well-defined beak is narrow and sharply set off from the forehead by a deep V-shaped groove. There are 40 to 50 teeth on each side of each jaw.

Delphinus is a pelagic genus, generally occurring well out to sea, but it has been reported to enter fresh water occasionally. Populations off southern California evidently shift northward and farther offshore during the spring and summer (U.S. National Marine Fisheries Service 1978). Intensive radio-tracking studies have shown that the common dolphin tends to occur in places with significant bottom relief—canyons, escarpments, sea mounts—where currents are interrupted and the resultant upswell supports high levels of oceanic productivity. Schools tend to divide into small parties to feed in the afternoon and night, and then regroup at dawn. Dives generally last two to three minutes and reach depths of up to 250 meters (Leatherwood and Reeves 1978). *Delphinus* is among the swiftest of cetaceans; although it usually travels at about 10 km per hr, it is capable of reaching over four times that speed. It often rides the bow waves of ships and seems to frolic about them. It sometimes feeds in the company of *Lagenorhynchus.* The diet includes cephalopods and small fish, including flying fish.

Delphinus is among the most gregarious of mammals, often being seen in groups of 1,000 individuals or more (Leatherwood and Reeves 1978). Schools of up to 300,000 individuals sometimes formed over concentrations of fish in the Black Sea (Mitchell 1975*b*). During the spring and summer the large schools may break up into groups of 50 to 200 animals, and the sexes may segregate between mating seasons (U.S. National Marine Fisheries Service 1978). Individuals have been seen aiding wounded members of their group, supporting them in the water and pushing them to the surface to breathe (Lekagul and McNeely 1977).

In the northeastern Pacific there are apparently two periods of mating, January–April and August–November, and two periods of calving, March–May and August–October. Each adult female gives birth every 2 or 3 years. Gestation lasts 10–11 months. The offspring is 75 to 85 cm long at birth and is weaned after 5–6 months, but may remain with the mother until it is 1 year old. Sexual maturity is attained at 3 to 5 years of age, and potential longevity is over 20 years (Leatherwood and Reeves 1978; U.S. National Marine Fisheries Service 1978; Rice 1967).

Delphinus has long been hunted by people in various parts of the world. Perhaps the largest fishery was in the Black Sea, where the population was estimated to have been 1,000,000 individuals and where up to 200,000 were taken annually in the 1930s. This population subsequently underwent a precipitous decline and its hunting was banned by the Soviet government in 1966 (Mitchell 1975*b*). Turkish fishermen, however, continued to catch large numbers, perhaps 88,000 in 1971 alone (Berkes 1977). In the eastern Pacific, *Delphinus* has been accidentally taken in the tuna fishery, in the same manner described in the account of *Stenella.* The kill was around 22,000 individuals in 1973, but has subsequently declined in response to regulation. The total current population of the common dolphin in the eastern tropical Pacific is estimated at 1,500,000 animals (U.S. National Marine Fisheries Service 1978, 1981).

CETACEA; DELPHINIDAE; *Genus LAGENODELPHIS*
Fraser, 1956

Shortsnouted Whitebelly Dolphin, Fraser's Dolphin

The single species, *L. hosei,* occurs in the tropical and warm temperate waters of the Atlantic, Indian, and Pacific oceans, and adjoining seas (Rice 1977; Caldwell, Caldwell, and Walker 1976). Until about a decade ago, *Lagenodelphis* was known to science only by a single skeleton collected before 1895 at the mouth of the Lutong River in Borneo.

An adult male and an adult female, taken off South Africa, had the following measurements: head and body length, 264 and 236 cm; pectoral fin length, 29 and 27 cm; dorsal fin height, 22 and 17 cm; width of the tail flukes, 59 and 57 cm; and weight, 209 and 164 kg (Perrin et al. 1973). The back is gray and the belly is white. Along each side, extending from the region of the eye to the anus, are three stripes, a creamy white one above, a blackish one in the middle, and another white one below. The body is robust, the pectoral and dorsal fins are relatively small, and the beak is short and indistinct. There are 39 to 44 teeth on each side of each jaw (Leatherwood, Caldwell, and Winn 1976). The skull is similar to that of *Lagenorhynchus,* but the facial part has a pair of deep palatal grooves and the premaxillary bones are fused dorsally in the midline. This fusion resembles that seen in *Delphinus,* but is not so extensive.

Most records are from well offshore. *Lagenodelphis* appears to be a deep diver, and when it surfaces to breathe it often charges up, creating a spray from its head. It has been reported to leap clear of the water. *Lagenodelphis* occurs in groups of up to 500 individuals and is occasionally seen in the company of *Stenella attenuata* (Leatherwood, Caldwell, and Winn 1976). Recorded stomach contents include deep-sea fish, squid, and shrimp (Caldwell, Caldwell, and Walker 1976; Mitchell 1975*a*). A pregnant female was collected off South Africa on 17 February 1971 (Perrin et al. 1973). Length of the single newborn is approximately 100 cm. Many of the known specimens have been taken accidentally in the tuna fishery, in the same manner described in the account of *Stenella* (U.S. National Marine Fisheries Service 1978).

CETACEA; DELPHINIDAE; *Genus LAGENORHYNCHUS*
Gray, 1846

White-sided and White-beaked Dolphins

There are six species (Rice 1977):

L. albirostris (white-beaked dolphin), North Atlantic from Davis Strait and Newfoundland to Barents Sea and North Sea;

L. acutus (Atlantic white-sided dolphin), North Atlantic from southern Greenland and Massachusetts to western Norway and the British Isles;

L. obliquidens (Pacific white-sided dolphin), waters from southeastern Alaska to Baja California, and off Japan and the Kuril Islands;

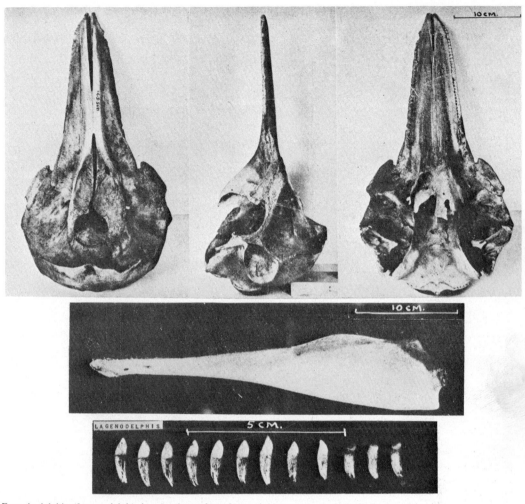

Fraser's dolphin (*Lagenodelphis hosei*), photos from *Sarawak Museum Journal*, F. C. Fraser.

L. obscurus (dusky dolphin), temperate waters near the coasts of South America, South Africa, Kerguelen Island, southern Australia, and New Zealand;

L. australis (Peale's or blackchin dolphin), temperate waters off southern South America and the Falkland Islands;

L. cruciger (hourglass dolphin), temperate waters of the Southern Hemisphere.

Head and body length is 150 to 310 cm, pectoral fin length is about 30 cm, dorsal fin height is up to 50 cm, and the width of the tail flukes is 30 to 60 cm. Sergeant, St. Aubin, and Geraci (1980) stated that adult weight of *L. acutus* is up to 234 kg in males and 182 kg in females. Banfield (1974) gave the weight of *L. obliquidens* as 82 to 124 kg. Most species have a gray or black back, a paler belly, and various bands and stripes on the sides. In *L. albirostris* the beak is white, and in *L. obliquidens* there are conspicuous yellowish brown and grayish streaks on the sides. The dorsal and pectoral fins are pointed. The beak is poorly defined and only about 5 cm long. There are 73 to 92 vertebrae, and 22 to 45 teeth on each side of each jaw (Hall 1981).

These dolphins generally occur in cool waters. *L. acutus*

and *L. cruciger* are primarily pelagic species, while *L. obscurus* and *L. australis* are usually found near coasts (Rice 1977; Mitchell 1975a). In the western part of its range, *L. albirostris* moves north into Davis Strait during the spring and summer, then turns back in the autumn and spends the winter as far south as Cape Cod (Leatherwood, Caldwell, and Winn 1976). *L. obliquidens* may move close to the southern California shore in the winter and spring, but then moves north and farther out to sea in the summer and fall (Leatherwood and Reeves 1978). Such movements are probably related to the availability of food.

Neither *L. acutus* nor *L. albirostris* commonly rides the bow waves of ships (Leatherwood, Caldwell, and Winn 1976). *L. obliquidens* often leaps clear of the water, and is the only dolphin of the eastern Pacific known to turn complete somersaults under natural conditions (Leatherwood and Reeves 1978). *L. obscurus* also is highly acrobatic, its displays apparently serving to communicate social information (Würsig and Würsig 1980). The diet of *Lagenorhynchus* includes herring, mackerel, capelin, anchovies, hake, squid, crustaceans, and whelks (gastropods).

These dolphins are gregarious. Group size may be up to 1,000 individuals in *L. acutus* and 1,500 in *L. albirostris*, but

Pacific white-sided dolphin (*Lagenorhynchus obliquidens*), photo from New York Zoological Society.

Lagenorhynchus sp., photo by Robert Pitman.

Dusky dolphin (*Lagenorhynchus obscurus*), photo by Stephen Leatherwood.

years old), but that immature animals predominated in individual strandings and occasionally formed schools of their own. *L. obliquidens* is found in groups of up to several thousand individuals (Leatherwood and Reeves 1978), but segregates by age and sex during the breeding season (Banfield 1974). In a study of *L. obscurus* off Argentina, Würsig and Würsig (1980) observed group size to range from 6 to 300, but noted that the larger schools were feeding aggregations, that nonfeeding groups usually contained only 6 to 15 animals, and that the latter individuals included both sexes and sometimes young.

All species that have been investigated seem to give birth in the late spring and summer. Gestation periods of 10 to 12 months have been estimated. There is normally a single young per birth, but there is at least one record of a pregnant female with two large embryos. Birth size is about 76 cm and 14 kg in *L. obliquidens*, 100 cm in *L. acutus*, and 122 cm in *L. albirostris*. In *L. acutus*, nursing lasts about 18 months, the young leave the mother at an age of 2 years, and maximum longevity is estimated to be 22 years for males and 27 years for females. Captive *L. obliquidens* have been maintained for approximately 20 years (Sergeant, St. Aubin, and Geraci 1980; Mitchell 1975a; Würsig and Würsig 1980; Banfield 1974; U.S. National Marine Fisheries Service 1978).

Fishermen have deliberately taken small numbers of *L. acutus* and *L. albirostris* off Newfoundland, Norway, and the British Isles. *L. obliquidens* has been taken both intentionally and incidental to other fisheries off Japan (Mitchell 1975a, 1975b). The population of the latter species in Japanese waters has been estimated at 30,000 to 50,000 individuals, but this figure may not be reliable (U.S. National Marine Fisheries Service 1981).

CETACEA; DELPHINIDAE; *Genus*
CEPHALORHYNCHUS Gray, 1846

Southern or Piebald Dolphins

There are four species (Rice 1977):

C. commersonii, coastal waters of southeastern South America, the Falkland Islands, South Georgia, and Kerguelen Island;

is usually much smaller, especially in the western Atlantic (Leatherwood, Caldwell, and Winn 1976; Mitchell 1975a). Off Canada, *L. acutus* is generally seen in groups of only 6 to 8 individuals (Banfield 1974). Investigations based on strandings (Sergeant, St. Aubin, and Geraci 1980) determined that large groups contained a great excess of adult females over adult males, and no immature animals (2½ to 6

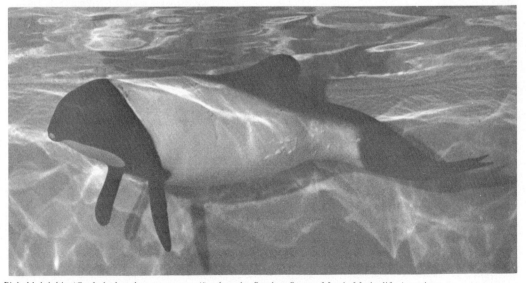

Piebald dolphin (*Cephalorhynchus commersonii*), photo by Stephen Spotte, Mystic Marinelife Aquarium.

Piebald dolphin (*Cephalorhynchus commersonii*), photo by Stephen Spotte, Mystic Marinelife Aquarium.

C. eutropia (black dolphin), coastal waters of southern Chile;
C. heavisidii, coastal waters of southwestern Africa;
C. hectori, coastal waters of New Zealand.

Head and body length is 120 to 180 cm, pectoral fin length is 15 to 40 cm, dorsal fin height is about 10 cm, and width of the tail flukes is about 30 cm. Gewalt (1979b) gave the weight of *C. commersonii* as 30 to 40 kg. The color pattern consists of various amounts of sharply contrasting black and white. Generally the head, pectoral and dorsal fins, tail region, and posterior part of the back are black, and the chin, sides, belly, and anterior part of the back are white. Sometimes the upper parts are entirely black. There is no beak, and there are 25 to 32 teeth on each side of each jaw.

These dolphins are found in coastal waters (Rice 1977). According to Baker (1978), *C. hectori* usually occurs within 8 km of the coast of New Zealand, often in shallow, muddy waters. It is often attracted to boats, and briefly rides the bow waves and follows in the wake. It rarely jumps clear of the water. It feeds on the sea floor, taking various kinds of fish and perhaps such invertebrates as squid and crustaceans. Groups usually contain 2 to 8 individuals, but occasionally over 20 are seen together and there is one record of an aggregation of 200 to 300. There is some evidence that the young are born early in the austral summer.

Less seems to be known about the other species, though *C. commersonii* has recently been brought into captivity (Gewalt 1979b; Spotte, Radcliffe, and Dunn 1979). Prior to capture, schools of four to eight individuals were commonly observed, and a captive group of six was placed together without strife. Each dolphin consumed three to four kg of food per day. Watkins, Schevill, and Best (1977) recorded three kinds of sounds made by all four species of *Cephalorhynchus*: clicks at slow rates, short bursts of clicks, and a "cry."

Small numbers of *Cephalorhynchus* are taken accidentally by fishermen, and the genus is considered especially vulnerable to coastal purse seining (Mitchell 1975a, 1975b). There have been reports of a decline in the New Zealand population, perhaps because of trawl fishing and pollution, but Baker (1978) did not consider these reports substantiated.

CETACEA; DELPHINIDAE; **Genus LISSODELPHIS**
Gloger, 1841

Right Whale Dolphins

There are two species (Rice 1977):

L. borealis, temperate waters of the North Pacific, from Japan and the Kuril Islands to British Columbia and California;
L. peronii, temperate waters of the oceans of the Southern Hemisphere.

In *L. borealis*, head and body length is up to 307 cm in males and 230 cm in females, pectoral fin length is about 28 cm, and width of the tail flukes is about 35 cm. A mature male, 282 cm long, weighed 113 kg. A mature female, 226 cm long, weighed 81 kg (Leatherwood and Walker 1979). *L. peronii* is smaller, reaching a length of about 210 to 240 cm at maturity (Gaskin 1968). *L. borealis* is mostly black or dark brown, but there is a broad white patch on the chest and a narrow midventral white band from the chest to the tail. In *L. peronii*, the top of the head and back are black or bluish black, and the beak, flippers, tail, sides, and underparts are white. The beak is short but distinct. There are 36 to 47 sharp, pointed teeth on each side of each jaw. The vernacular name refers to the lack of a dorsal fin in both *Lissodelphis* and the right whales (*Balaena*).

These dolphins are primarily oceanic, usually remaining well offshore. Their numbers apparently increase over the continental shelf of California during the fall and winter, a period corresponding with the maximum abundance of squid in the region. In the late spring and summer, the dolphins move farther north and out to sea. *Lissodelphis* is quick and active, and sometimes travels by smooth, low leaps above the surface, each of which may cover up to 7 meters. Speeds as great as 45 km per hr may be briefly attained; a boat traveling at 33 km per hr was easily outdistanced during a 10-minute chase. Bow waves of vessels are usually avoided. Entire schools have been observed to dive for as long as 6 minutes and 15 seconds. Otoliths found in the stomach of an adult *L. borealis* indicated that the animal had been feeding

Right whale dolphins (*Lissodelphis borealis*): Top and middle, photos by M. Scott Sinclair through National Marine Fisheries Service; Bottom, photo by K. C. Balcomb.

on fish that dwell at depths of at least 200 meters. The diet consists of various kinds of fish, especially lanternfish, and cephalopods, such as squid. *L. borealis* produces series of clicks, much like those used by other genera for echolocation. Whistles, apparently for communication, are also emitted, but are not common. Groups contain up to 2,000 individuals, but usually number 10–50. A pregnant female *L. borealis,* with a 26-cm fetus, was collected on 9 March. Newborn are about 60 cm long. Fishermen take small numbers of *L. peronii* off Chile and of *L. borealis* off Japan. The eastern Pacific population of *L. borealis* has been estimated to contain 17,800 animals (Leatherwood and Walker 1979; U.S. National Marine Fisheries Service 1978; Mitchell 1975a; Fitch and Brownell 1968).

CETACEA; DELPHINIDAE; *Genus GRAMPUS Gray, 1828*

Risso's Dolphin, Gray Grampus

The single species, *G. griseus,* occurs in all temperate and tropical oceans and adjoining seas (Rice 1977).

Head and body length is 360 to 400 cm, pectoral fin length is about 60 cm, dorsal fin height is about 40 cm, and width of the tail flukes is about 76 cm. An adult male stranded in Florida was estimated to weigh between 400 and 450 kg (Paul 1968). Coloration varies with age, but adults are usually slaty or black above, tinged with blue or purple, and are paler beneath. The fins and tail are black. There are usually whitish streaks on the body, probably healed scars from attacks by other *Grampus* and perhaps by squid.

Grampus is distinguished from other dolphins in having only three to seven pairs of teeth. These pairs are in the front end of the lower jaw, but occasionally one or two vestigial teeth are present in the upper jaw. A beak is lacking, and the front of the head rises almost vertically from the tip of the upper jaw. Field marks include the blunt snout and the pale gleam of the back in front of the high, pointed, recurved dorsal fin.

Grampus generally occurs well out to sea in waters deeper than 180 meters (Mitchell 1975a). It probably migrates northward during the warmer months (U.S. National Marine Fisheries Service 1978). It is frolicsome, sometimes riding the bow waves of vessels and leaping clear of the water. It may fall back into the water head up, and occasionally waves the back third of its body in the air. The diet includes cephalopods and fish. *Grampus* usually is seen in groups of

Gray grampus (*Grampus griseus*), photo by Warren J. Houck.

fewer than 12 individuals, but several such schools may associate in one area. Aggregations of several hundred individuals have been reported (Leatherwood, Caldwell, and Winn 1976). A pregnant female, with a near-term embryo, was taken in December. The newborn is approximately 150 cm long (Mitchell 1975a).

The gray grampus has been taken sporadically by small whale fisheries in Newfoundland, the Lesser Antilles, Japan, and Indonesia (Mitchell 1975a, 1975b). The famous dolphin "Pelorus Jack," which received lifelong protection from the government of New Zealand, is believed to have been a *Grampus*. This individual had the habit of playing about ships and seemed to guide them into Pelorus Sound. Observations of Pelorus Jack extended from about 1896 to 1916, thus giving some idea of the longevity of *Grampus*.

CETACEA; DELPHINIDAE; *Genus* **PEPONOCEPHALA**
Nishiwaki and Norris, 1966

Many-toothed Blackfish, Melon-headed Whale

The single species, *P. electra,* occurs in tropical waters of the Atlantic, Indian, and Pacific oceans, and also is known from Japan and southeastern Australia (Rice 1977; Perrin 1976; Caldwell, Caldwell, and Walker 1976). This species was long placed in the genus *Lagenorhynchus,* but Nishiwaki

and Norris (1966) considered it to resemble *Feresa* and *Pseudorca* more closely, and placed it in a distinct genus.

Head and body length at physical maturity is about 270 to 280 cm, pectoral fin length is about 50 cm, dorsal fin height is about 25 cm, and width of the tail flukes is about 60 cm. The upper parts are black and the belly is slightly paler. The lips and an oval patch on the abdomen are often white or unpigmented. The body is rather long and slim, and the tail stock is long. Unlike *Lagenorhynchus,* there is no beak, the forehead being rounded, curving smoothly from the anterior tip of the rostrum to the blowhole, and overhanging the lower jaw to some extent. *Peponocephala* resembles *Pseudorca,* but is smaller, has a sharper appearance to the snout, and has more teeth. There are 20 to 26 teeth on each side of each jaw. The anterior three cervical vertebrae are fused (Bryden, Harrison, and Lear 1977; Nishiwaki and Norris 1966; Caldwell, Caldwell, and Walker 1976).

Most of what is known about the natural history of this genus was summarized by Bryden, Harrison, and Lear (1977). *Peponocephala* seems normally to stay well out to sea, but may be seen near land at any time of the year. One group was observed to swim at 20 km per hr, and many individuals leaped almost clear of the water. The diet consists of fish, squid, and possibly pelagic mollusks. *Peponocephala* is probably a social animal. Schools containing 15 to 500 individuals have been observed. One stranding of 53 specimens in Australia included 16 males, 35 females, and 2 individuals of undetermined sex. Limited data suggest that in

Melon-headed whale (*Peponocephala electra*), photo by Michael M. Bryden.

the Southern Hemisphere births occur from August to December, and that gestation lasts 12 months. A newborn individual was found in Hawaii in June. Length at birth is about 94 cm.

CETACEA; DELPHINIDAE; *Genus FERESA Gray, 1871*

Pygmy Killer Whale

The single species, *F. attenuata,* occurs in tropical and warm temperate waters of the Atlantic, Indian, and Pacific oceans, and adjoining seas (Rice 1977; Leatherwood, Caldwell, and Winn 1976).

Head and body length is usually 240 to 270 cm, and dorsal fin height is usually about 20 to 30 cm (Leatherwood, Caldwell, and Winn 1976). A female collected in Japan had a head and body length of 235 cm, a pectoral fin length of 44 cm, a dorsal fin height of 24 cm, and a tail fluke width of 36 cm. The back is black or dark gray, the sides are often paler, and the chin is often white. There also are commonly white areas around the lips and on the belly. The forehead is rounded. There are 10 to 13 teeth on each side of each jaw. There is some resemblance to *Orcinus,* but that genus is far larger.

Until 1952, *Feresa* was known only by two skulls, but subsequently the genus was found to be widely distributed, though perhaps rare. Most records are from places with little or no continental shelf. *Feresa* apparently is a year-round resident off Japan, Hawaii, and South Africa, but short,

Pygmy killer whales (*Feresa attenuata*), photos by Kenneth S. Norris (top) and by Gary L. Friedrichsen through National Marine Fisheries Service (bottom).

seasonal migrations may occur (Caldwell and Caldwell 1971b). The diet is not well known, but a captive in Japan ate sardines, horse mackerel, sauries, and squid (Nishiwaki 1966). According to the U.S. National Marine Fisheries Service (1978, 1981), *Feresa* feeds on other kinds of dolphins, especially the young. Captives are aggressive, and other kinds of dolphins kept with them show fright reactions. *Feresa* usually travels in groups of 5 to 10 individuals, but aggregations of several hundred have been seen.

CETACEA; DELPHINIDAE; *Genus PSEUDORCA* Reinhardt, 1862

False Killer Whale

The single species, *P. crassidens*, occurs in all the temperate and tropical oceans and seas of the world (Rice 1977). The species was described in 1846 on the basis of a fossilized skull found in the Lincolnshire Fens in England. Stranded animals, collected in 1862 in the Bay of Kiel, allowed assessment of the external morphology (Allen 1942).

Head and body length is up to 610 cm in males and 490 cm in females. Dorsal fin height is about 40 cm and weight reaches 1,360 kg (Scheffer 1978b; U.S. National Marine Fisheries Service 1978). The coloration is black throughout. *Pseudorca* bears some resemblance to *Orcinus,* but can be distinguished by its uniformly dark color, more slender build, more tapering head, smaller and more backwardly curving dorsal fin, and tapering flippers, which average about one-tenth of the head and body length. There are usually 8 to 11 teeth on each side of each jaw.

Pseudorca is usually found well out to sea. It often rides the bow waves of ships and jumps completely out of the water (Leatherwood, Caldwell, and Winn 1976). The diet is known to include squid, medium to large fish, and young dolphins (U.S. National Marine Fisheries Service 1981). A captive, however, was not aggressive toward other kinds of dolphins in the same pool, and formed a close social relationship with them. This captive also was friendly to people and learned tricks rapidly. The wild school from which it was taken contained about 300 individuals that comprised small subgroups of 2 to 6 animals each. They emitted a variety of vocalizations, including a piercing whistle (Brown, Caldwell, and Caldwell 1966). Schools of up to 835 individuals have stranded, and these have included animals of all ages and sexes (U.S. National Marine Fisheries Service 1978). There does not appear to be a restricted breeding season (Scheffer 1978b). The young are usually 170 to 200 cm long at birth. Gestation lasts 15½ months, and sexual maturity is

False killer whales (*Pseudorca crassidens*): A. Photo by K. C. Balcomb; B. Photo from Marineland of the Pacific.

attained at an age of 8 to 12 years (Purves and Pilleri 1978). One specimen was still living after 14 years in captivity (Marvin L. Jones, Zoological Society of San Diego, pers. comm.).

Pseudorca is notorious to fishermen in various parts of the world, because of its habit of stealing fish from lines (Leatherwood, Caldwell, and Winn 1976; Scheffer 1978*b*). *Pseudorca* is taken in small numbers for human consumption and incidental to the tuna fishery of the eastern tropical Pacific (Mitchell 1975*a*, 1975*b*).

CETACEA; DELPHINIDAE; *Genus GLOBICEPHALA*
Lesson, 1828

Pilot Whales, Blackfish

There appear to be two species (Rice 1977; Hall 1981; Van Bree 1971; Reilly 1978; Mitchell 1975*a*; Kasuya 1975):

G. melaena, cool temperate waters of the Southern Hemisphere and the North Atlantic, also occurred in the North Pacific off Japan until at least the 10th century A.D.;
G. sieboldii, tropical and temperate waters of the Atlantic, Indian, and Pacific oceans, and adjoining seas.

The latter species has often been referred to as *G. macrorhynchus*.

Head and body length is 360 to 850 cm, pectoral fin length is about one-fifth head and body length, dorsal fin height is about 30 cm, and width of the tail flukes is about 130 cm. Males are usually larger than females. Banfield (1974) stated that the average weight of *G. melaena* is 800 kg, and that the largest recorded male weighed 2,750 kg. The coloration is black throughout, except for a white area often present below the chin. The head is swollen so the forehead bulges above the upper jaw. The pectoral fin is narrow and tapering, and the dorsal fin is located just in front of the middle of the back. There are 7 to 11 teeth on each side of each jaw.

According to Banfield (1974), *G. melaena* may occur either near the shore or well out to sea. During the colder months a population of this species remains in the Gulf Stream to the south of the Grand Banks of Newfoundland. In the summer this population shifts shoreward and northward to the Gulf of St. Lawrence, the Labrador Sea, and Greenland. Such movements are evidently in response to the migrations of the squid *Illex*, the major source of food. There is an apparently separate population of *G. melaena* off northwestern Europe (Mitchell 1975*b*). In the eastern Pacific, *G. sieboldii* seems to become more abundant near the shore during the winter (Reilly 1978).

Pilot whales sleep by day and feed by night. They normally swim at about 8 km per hr, but can attain speeds of 40 km per hr (U.S. National Marine Fisheries Service 1978; Banfield 1974). A radio-tagged individual was tracked to a depth of 610 meters, and a captive was used in a U.S. Navy experiment to recover objects about 500 meters below the surface. Pilot whales are among the most intelligent and affable of cetaceans, adapting well to captivity and being easily trained. They have been demonstrated to use echolocation efficiently and to have a variety of sounds for communication, including an individualized "signature" whistle (Reilly 1978; Taruski 1979; Banfield 1974). The preferred diet is squid, up to 27 kg of which may be consumed daily. Other cephalopods and small fish are also eaten.

Group size varies. The U.S. National Marine Fisheries Service (1978) stated that *G. melaena* occurs in schools of hundreds and thousands, but Banfield (1974) gave average group size as only 20 individuals. In the eastern Pacific, three kinds of herds of *G. sieboldii* have been identified (Reilly

Pacific pilot whale (*Globicephala sieboldii*), photos from Marineland of the Pacific.

1978): the traveling-hunting group, a large and well-organized unit that may be divided into harems of females and young led by one or a few large males, and other subgroups based on age and sex; the feeding group, a loose aggregation of animals that come together only to exploit a source of food; and the loafing group, a cluster of 12 to 30 animals that float in one area for such purposes as mating and nursing young. Polygamy seems to be the rule in both species of *Globicephala*. Adult males are often scarred, presumably from battles over control of a harem. Fighting involves biting, butting with the large head, and slapping with the tail.

Social organization seems to be highly developed in *Globicephala*, and this factor sometimes works to the disadvantage of the animals. Individuals harpooned by people usually rush forward in panic and can be driven toward the shore. The rest of the group seems to respond to the movements and cries of the wounded animals, and to follow them into shallow water, where all may be easily killed. The same tendency may be associated with mass stranding, an especially common occurrence in *Globicephala*. It is possible that if a leader's echolocation mechanism fails to function properly, because of pathology or unusual environmental conditions, it will guide the entire group onto the shore. The cries of the first animals to become stranded attract others to the same fate (Reilly 1978; Banfield 1974).

In the western Atlantic, mating takes place mainly in April and May, before departure from the wintering area. The gestation period is 15–16 months, and the birth peak is in August. The single newborn is 150 to 210 cm long. Lactation lasts 21–22 months. Sexual maturity is attained at an age of 6 or 7 years by females and 12 years by males. Longevity is up to 40 years in females and 50 years in males (Reilly 1978; Banfield 1974; Leatherwood, Caldwell, and Winn 1976).

Pilot whales have been among the more heavily exploited of the small cetaceans (Mitchell 1975a, 1975b; Banfield 1974; Reilly 1978; U.S. National Marine Fisheries Service 1978). There have been active fisheries in Japan, the Caribbean, the northeastern United States, Newfoundland, Ireland, Norway, and several islands north of Britain. The main products are meat for human and domestic animal consumption, blubber oil, and oil from the bulbous head. Records of the Faeroe Islands fishery extend back nearly 400 years. The cumulative kill there was 117,546 animals from 1584 to 1883. Despite this large take, the annual average has not

declined and the population of *Globicephala* in the northeastern Atlantic appears to be in balance with the harvest. In contrast, the western Atlantic population, estimated to have once contained 60,000 animals, evidently declined sharply following commercial organization of the Newfoundland fishery in 1947. About 40,000 pilot whales were killed there from 1951 to 1959, but only 6,902 were taken from 1962 to 1973.

CETACEA; DELPHINIDAE; **Genus ORCINUS** *Fitzinger,*
1860

Killer Whale

The single species, *O. orca,* occurs in all the oceans and adjoining seas of the world, but chiefly in coastal waters and cooler regions (Rice 1977).

Orcinus is the largest member of the dolphin family. Sexual maturity is attained at a length of about 670 cm in males and 490 cm in females (Mitchell 1975a). Subsequently, as in all cetaceans, growth continues for some years. Maximum head and body length is about 980 cm in males and 850 cm in females. Maximum weight is about 9,000 kg in males and 5,500 kg in females (Scheffer 1978a). In males, the body becomes stocky with age and there is a disproportionate increase in the size of certain organs: the broad, round pectoral fin reaches a length of 200 cm; the erect, triangular dorsal fin reaches a height of 180 cm; and the tail flukes reach an expanse of 275 cm. In females, the body remains less stocky and the dorsal fin is smaller and backwardly curved. There is a sharply contrasting color pattern. The upper parts are black, except usually for a light gray area behind the dorsal fin; the underparts are white; and there are white patches on the sides of the head and on the posterior flanks. The snout is bluntly rounded. There are 10 to 14 large teeth on each side of each jaw.

The killer whale seems to prefer coastal waters, and often enters shallow bays, estuaries, and the mouths of rivers. Some populations appear to reside permanently in relatively small areas, but others migrate, probably in association with food supplies. In the North Atlantic, during the spring and summer, there appear to be movements both closer to the

Killer whale (*Orcinus orca*), photo by Scanlan. Skull: photo by Warren J. Houck.

Killer whale (*Orcinus orca*), photo by Dr. Bubenik through Bernhard Grzimek.

southeastern Canadian coast and northward into arctic waters. *Orcinus* usually swims at about 10 to 13 km per hr, but can exceed 45 km per hr. Its dives usually last 1 to 4 minutes, but one individual is known to have remained underwater for 21 minutes. Another was found entangled in a submarine cable at a depth of about 1,000 meters. *Orcinus* produces clicklike sounds, such as those known to be used by other delphinids for echolocation, and also emits "screams" that are apparently used in communication. *Orcinus* adjusts to captivity and learns quickly. Within two months of its first training session, one individual was performing before the public, and after five months its trainer could safely place his head into its mouth (Scheffer 1978a; Leatherwood, Caldwell, and Winn 1976; Banfield 1974).

Orcinus received its common name because it is the world's largest predator of warm-blooded animals, though in relative terms it is no more of a "killer" than an insectivorous frog or a beef-eating human. And, while marine mammals are evidently the preferred prey in certain areas at certain times, an examination of the stomach contents of 364 *Orcinus* taken off Japan showed that fish and cephalopods predominated over cetaceans and pinnipeds (Scheffer 1978a). *Orcinus* does have a throat large enough to swallow seals, young walrus, and the smaller cetaceans. A group of killer whales sometimes joins in an attack on a large baleen whale, literally tearing it to pieces. Penguins and other aquat-

ic birds are also seized. When *Orcinus* detects a seal or bird near the edge of the ice in arctic waters, it may dive deeply and then rush to the surface, breaking ice up to a meter thick and dislodging its prey into the water. The approach of a group of killer whales usually seems to panic other marine vertebrates. The U.S. National Marine Fisheries Service (1981) stated that in Argentina *Orcinus* has been seen flopping well up onto the land to drive pinnipeds into the water.

Schools of up to 250 individuals have been reported (Hoyt 1977). According to Scheffer (1978a), however, groups usually consist of 2 to 40 animals. In the enclosed waters from northeast of Vancouver Island to Puget Sound, a survey found a permanent population of 210 killer whales, divided into family groups of about 10 members each. Groups are generally well organized and led by a large male, though females and young may sometimes separate from the older males. Individuals have been seen to go to the assistance of others that are trapped or injured. A group observed by Cousteau and Diole (1972) contained 1 large alpha male, 1 large female, 7 or 8 medium-sized females, and 6 to 8 calves.

Breeding can occur at any time of the year, though in the Northern Hemisphere mating may peak from May to July and births take place mainly in the autumn. The gestation period is not known, but probably is about 13 to 16 months. The single newborn weighs about 180 kg and is about 240 cm long (U.S. National Marine Fisheries Service 1978, 1981). Maximum longevity is at least 50 years (Haley 1978).

There has long been controversy about whether the killer whale is a threat to human swimmers and boaters. Documented records of attacks are rare and usually involve provocation by the persons involved (Leatherwood, Caldwell, and Winn 1976; Scheffer 1978a). In contrast, people often kill or harass *Orcinus*, because it is considered a competitor to fisheries. In the 1950s the U.S. Navy reportedly machine-gunned hundreds of killer whales at the request of the government of Iceland (Mitchell 1975b). *Orcinus* is also killed by people in such places as Norway, Greenland, and Japan to obtain meat and oil. During the 1979/80 season, the whaling fleet of the Soviet Union took 916 killer whales (Committee for Whaling Statistics 1980). *Orcinus* has been regularly live-captured off the coasts of British Columbia and Washington, mainly for eventual use in human entertainment (Bigg and Wolman 1975).

CETACEA; DELPHINIDAE; **Genus ORCAELLA** Gray, *1866*

Irrawaddy River Dolphin

The single species, *O. brevirostris*, occurs in coastal waters from the Bay of Bengal to New Guinea and northern Australia, and ascends far up the Ganges, Irrawaddy, Mekong, and other rivers (Rice 1977).

Head and body length is 180 to 275 cm (Lekagul and McNeely 1977). The coloration is slaty blue or slaty gray throughout, or the underparts may be slightly paler. There is a bulging forehead, a short and shelflike beak, and 12 to 19 teeth in each side of each jaw. The pectoral fin is broadly triangular. The dorsal fin is small, sickle shaped, and located on the posterior half of the back. There is great flexibility of the neck and tail (Mitchell 1975a).

Orcaella lives in warm, tropical, and often silty waters. It regularly enters rivers, has been recorded from 1,440 km up the Irrawaddy, and can live permanently in fresh water. It often accompanied river steamboats. It breathes at intervals of 70 to 150 seconds; the head appears first and then disappears, and then the back emerges, but the tail is rarely seen.

Irrawaddy River dolphin (*Orcaella brevirostris*), photo from *Anatomical and Zoological Researches Comprising an Account of the Two Expeditions to Western Yunnan in 1868 and 1875* . . . , John Anderson.

The diet consists of fish and crustaceans, often found on the bottom (Lekagul and McNeely 1977; Mitchell 1975a). A pregnant female, about 210 cm in length, contained an embryo 90 cm long.

This dolphin is occasionally taken accidentally, and its oil has been reportedly used as a remedy for rheumatism in parts of India, but it is generally unexploited. According to Thein (1977), the fishermen of Burma attract *Orcaella* by tapping the sides of their boats with oars. The dolphin swims around the boat in ever-diminishing circles, thereby forcing the fish into nets. The fishermen share their catch with *Orcaella* and consider it a friend that is not to be harmed.

CETACEA; DELPHINIDAE; **Genus PHOCOENA**

G. Cuvier, 1817

Common Porpoises, Harbor Porpoises

There are four species (Rice 1977; Villa 1976; Baker 1977; Brownell and Praderi in press):

P. phocoena, coastal waters from Davis Strait to Delaware, from Iceland and the White Sea to Senegal, from northern Alaska to Japan and Baja California, and in the Black Sea;

P. sinus, Gulf of California;

P. dioptrica, coastal waters of Uruguay, Argentina, the Falkland Islands, South Georgia, and the Auckland Islands;

P. spinipinnis, coastal waters from northern Peru and Uruguay to Tierra del Fuego.

Head and body length is about 120 to 200 cm, pectoral fin length is 15 to 30 cm, dorsal fin height is 15 to 20 cm, and width of the tail flukes is 30 to 65 cm. In *P. phocoena,* the mean head and body length is about 150 or 160 cm and the usual weight is 45 to 60 kg. The respective maximums are 186 cm and 90 kg (Gaskin, Arnold, and Blair 1974). The sexes are approximately equal in size. Coloration is variable, but usually is dark gray or black above and white below, or is entirely dark. The conical head is not beaked, and the dorsal fin is usually triangular and located just behind the middle of the back. There are 16 to 28 teeth on each side of each jaw.

The spade-shaped teeth are entirely crowned or bear crowns with two or three lobes.

Common porpoises frequent coastal waters, bays, estuaries, and the mouths of large rivers. They sometimes ascend the rivers. In the western Atlantic, at least, *P. phocoena* apparently moves well out to sea at the end of summer and reappears in the spring. This species generally swims quietly near the surface, rising about four times per minute to breathe. It seems less frolicsome than most delphinids, seldom jumps out of the water, ignores boats, and rarely rides bow waves. When pursued, it can attain a speed of about 22 km per hr. When diving for food, it stays underwater for an average of about four minutes. One individual was caught in a fish trap at a depth of 79 meters. *P. phocoena* produces clicklike sounds such as those known to be used by other delphinids for echolocation. Its diet consists mainly of smooth, nonspiny fish about 10 to 25 cm long, such as herring, pollack, mackerel, sardines, and cod (Gaskin, Arnold, and Blair 1974). Captives do not seem to differ much from *Tursiops* in trainability (Andersen 1976).

Although *P. phocoena* may sometimes travel in schools of nearly 100 individuals, it is usually seen in pairs or in groups of 5 to 10 (Leatherwood, Caldwell, and Winn 1976). The reproductive season of this species could be extensive, but mating seems to occur mainly during the summer, and most births take place by late in the following spring. Gestation probably lasts 10 to 11 months. The single newborn weighs 6 to 8 kg and is 70 to 100 cm long. The young is brought to a sheltered cove by the mother, and nurses for perhaps 8 months. Sexual maturity is attained at an age of 3 or 4 years. The life span is short for a cetacean, usually about 6 to 10 years and rarely exceeding 13 years (Gaskin 1977; Gaskin, Arnold, and Blair 1974; Gaskin and Blair 1977).

Little is known about the life history of the other species of *Phocoena*. Pregnant female *P. dioptrica* were collected in July and August (Brownell 1975b). A pregnant female *P. spinipinnis,* with a near-term fetus, was collected off Uruguay in late February or early March (Mitchell 1975a).

Phocoena has been heavily exploited in various areas, its meat used for human and domestic animal consumption, and its oil used for lamps and lubricants. Indian tribes in eastern Canada formerly caught several thousand *P. phocoena* yearly and sold the oil. This fishery ended by the late 19th cen-

Common porpoise (*Phocoena phocoena*), photo by Stephen Spotte, Mystic Marinelife Aquarium.

tury, but the population may still be jeopardized by accidental catches in nets set for fish. Another population, in the Baltic Sea, seems to have declined drastically following centuries of exploitation, during which catches reached up to 3,000 per year off Denmark alone. The isolated population of *P. phocoena* in the Black Sea, once estimated to contain 25,000 to 30,000 individuals, sustained an annual harvest of up to 2,500, but this fishery was terminated by order of the Soviet government in 1966. *P. spinipinnis* is taken commercially off the coasts of Chile and Peru. In Peru, about 113,500 kg of meat are marketed annually (Mitchell 1975a, 1975b; Gaskin 1977; U.S. National Marine Fisheries Service 1978).

The species *P. sinus* is classified as vulnerable by the IUCN (1978) and is on appendix 1 of the CITES. Villa (1976) considered it nearly extinct. It formerly may have occurred throughout the Gulf of California, but the few recent records are from the upper part of the gulf. *P. sinus* was considered abundant in the early 20th century, but declined in conjunction with the intensification and modernization of commercial fisheries, starting in the 1940s. A main problem may be accidental catching in nets set for fish and shrimp. In addition, fishing might be reducing food supplies, and dams in the southwestern United States and western Mexico may be cutting off the flow of nutrients to the Gulf of California.

CETACEA; DELPHINIDAE; *Genus NEOPHOCAENA*
Palmer, 1899

Finless Porpoise

Rice (1977) recognized a single species, *N. phocaenoides*, occurring in warm coastal waters and certain rivers from Pakistan to Korea, Japan, Borneo, and Java. Pilleri and Gihr (1975) considered that there were actually three species: *N. phocaenoides* of southern Asia (as far west as the Persian Gulf) and the East Indies, *N. asiaeorientalis* of the Chang Jiang (Yangtze) River Valley in China, and *N. sunameri* of Japan and Korea.

This is the smallest cetacean. Head and body length is 120 to 160 cm, pectoral fin length is about 28 cm, and expanse of

the tail flukes is about 55 cm. Weight is 25 to 40 kg (Lekagul and McNeely 1977). Although there are numerous literary references indicating that the color is black or dark plumbeous gray, observations of live animals by Pilleri, Zbinden, and Gihr (1976) show that the upper parts are actually pale gray, with a bluish tinge on the back and sides, and that the underparts are whitish. Apparently, however, the skin quickly darkens after death. Distinctive features are the small size, the abruptly rising forehead, and the absence of a dorsal fin. The teeth are spade shaped and number from 15 to 21 on each side of each jaw.

In southern Asia and the East Indies, *Neophocaena* seems closely associated with mangroves and salt marshes (Pilleri and Gihr 1975). In China the genus is found in shallow coastal waters, in the middle and lower Chang Jiang (Yangtze), and in Dongting Lake, often in association with *Lipotes* (Chen et al. 1979; Zhou, Pilleri, and Li 1979). Part of the Japanese population is migratory, concentrating in the Inland Sea in the spring and moving out to the Pacific coast from late summer to midwinter (Kasuya and Kureha 1979). *Neophocaena* moves rather slowly, rolls to the surface to breathe, and does not jump out of the water. The diet includes fish, shrimp, and small squid.

About half of all sightings made in Japan by Kasuya and Kureha (1979) involved solitary animals, and most of the others were of two animals, generally a cow and calf or a mated pair. There is usually a 2-year breeding cycle in Japan, with gestation lasting approximately 11 months, births occurring from April to August, and weaning taking place from September to June. Sexual maturity may possibly be attained by the age of 2 years. Some individuals are estimated to have lived up to 23 years.

In China, *Neophocaena* is usually seen in groups of 3 to 5 individuals, and occasionally in schools of up to 20. Births occur in April and May. The calf clings with its flippers and is carried on the back of the female in a spot where her skin is roughened. The extent of this roughened area, however, is considerably less in the Chinese population than in that of southern Asia. Sometimes a group of as many as 6 females is seen, each carrying its calf in this manner (Chen et al. 1979; Pilleri and Chen 1979).

Neophocaena has been taken regularly in nets set for shrimp and fish, and is also deliberately hunted for its meat,

Finless porpoise (*Neophocaena phocaenoides*), photos by T. Kasuya.

largest in Japanese waters, contains about 4,900 individuals (Kasuya and Kureha 1979). *Neophocaena* is on appendix 1 of the CITES.

CETACEA; DELPHINIDAE; **Genus PHOCOENOIDES**
Andrews, 1911

Dall Porpoise

The single species, *P. dalli*, occurs from Japan, around the rim of the North Pacific, to Baja California (Rice 1977; Mitchell 1975a; Kasuya 1978; Morejohn 1979). The name *truei* has sometimes been used in a specific, subspecific, or other categorical sense for most of the animals found off the Pacific coast of Japan. This name, however, seems to be based on a color variant of *P. dalli*.

Head and body length is about 170 to 236 cm, pectoral fin length is 18 to 25 cm, dorsal fin height is 13 to 20 cm, and expanse of the tail flukes is 40 to 56 cm. Weight is usually 80 to 125 kg, but the U.S. National Marine Fisheries Service (1978) reported a maximum of 218 kg. Both sexes are about the same size (Morejohn 1979). The upper parts of *Phocoenoides* are black. In most populations there is a large white area on each side of the body, but it does not extend far in front of the anterior margin of the dorsal fin. In the population off the Pacific coast of Japan, to which the name *truei* is often applied, the white area usually extends laterally to above the flipper, and there also may be a white throat patch. Some individuals have intermediate color patterns, a few are entirely black, and a few are mostly brownish or gray.

The head of *Phocoenoides* is sloping, and the lower jaw projects slightly beyond the upper. The dorsal fin is low and triangular. There are 97 to 98 vertebrae, compared to only 64 or 65 in *Phocoena* (Hall 1981). The teeth are the smallest of any delphinid (Morejohn 1979). They are spade shaped and number 19 to 27 on each side of each jaw. An unusual feature is the presence of horny protuberances of the gums between the teeth. These "gum teeth" function as gripping organs and probably wear down with use to expose the teeth. The true teeth thus seem to function mainly in older animals.

Phocoenoides occurs mainly in cool waters, sometimes near land but generally well offshore in seas more than 180 meters deep. In the western Pacific there is a well-defined annual migration, in which most Japanese animals shift northward for the summer to the Sea of Okhotsk and the Kuril Islands. In the eastern Pacific there are apparently no large-

skin, and oil (Mitchell 1975b). For these reasons, and also perhaps because of pollution and collision with motorboats, its numbers have declined in some areas. Modification of fishing methods in Japan has reduced the accidental catch there. The population using the Inland Sea, perhaps the

Dall porpoise (*Phocoenoides dalli*): A. Color pattern of most populations; B. Color pattern commonly found off Pacific coast of Japan; Photos by Warren J. Houck.

scale migrations, and large numbers of *Phocoenoides* remain throughout the year from Alaska to California. There is, however, a tendency for concentration near the shore and to the south during the fall and winter, and offshore and to the north in the spring and summer. Such seasonal movements are probably related to distributional changes in prey organisms (Morejohn 1979; Kasuya 1978; Mitchell 1975a; Leatherwood and Reeves 1978; U.S. National Marine Fisheries Service 1978).

Phocoenoides, unlike *Phocoena*, often plays about ships and leaps out of the water, and also emerges farther from the water when rolling. It is reputedly the fastest cetacean, sometimes briefly attaining speeds of 50 km per hr. It does not bother to ride bow waves unless the vessel is moving at least 27 km per hr. It commonly preys on animals that live at depths in excess of 180 meters. The diet includes fish, mainly small unarmed kinds like herring and anchovies, and squid (Morejohn 1979; Leatherwood and Reeves 1978).

Groups of 2–15 individuals are usual in the southern part of the range, but aggregations of up to 200 have been seen off Alaska (Leatherwood and Reeves 1978). Limited evidence indicates that births occur throughout the year in the eastern Pacific (Morejohn 1979). More detailed studies in Japan (Kasuya 1978) have shown that births in the western Pacific take place in August and September, and that females apparently give birth at intervals of 3 years. The gestation period averages 11.4 months and lactation lasts 2 years. The single newborn is about 100 cm long. The mean age of sexual maturity is 7.9 years in males and 6.8 years in females. A few individuals over 16 years old have been collected in the western Pacific.

The meat of *Phocoenoides* has been used for human con-

sumption, its blubber is a source of oil, and its bones have been ground up and used for fertilizer (Banfield 1974). Direct exploitation has been especially heavy in Japan, and the population there appears to have declined as a result. An even greater problem in this area, however, has been the accidental kill by fishermen using gillnets to catch salmon. This kill has approached 20,000 porpoises annually, and seems to involve mainly immature animals (Mitchell 1975a, 1975b; Kasuya 1978; U.S. National Marine Fisheries Service 1978). The current estimate of the total number of *Phocoenoides* in the Pacific is 920,000 (U.S. National Marine Fisheries Service 1981).

CETACEA; Family MONODONTIDAE

Beluga and Narwhal

This family of two known genera, each with one species, occurs in the Arctic Ocean and nearby seas. The family has sometimes been considered part of the Delphinidae.

Head and body length (not including the tusk of the narwhal) is usually 280 to 490 cm. The body form resembles that of the Delphinidae. The forehead is high and globose, the snout is blunt, and there is no beak. The blowhole is located well back from the tip of the snout. There are no external grooves on the throat. The pectoral fin is short and rounded, and there is no dorsal fin. The skull lacks crests. There are 50 to 51 vertebrae. The cervical vertebrae are not fused, allowing the neck to be more flexible than in most cetaceans.

These animals live in cold waters. They migrate in response to shifting pack ice and hard winters. Individuals that become trapped in ice fields can sometimes break through the ice by ramming it from the underside. The cushion on top of the head lessens the shock to the animal. The small spout is not well defined. Both genera sometimes ascend rivers. They emit various sounds, and are sensitive to sounds in the water, but apparently disregard noises originating on land. They seem to feed mainly on the bottom. The diet includes fish, cephalopods, and crustaceans. Groups sometimes consist of more than 100 individuals. Both genera are extensively hunted by natives of the Arctic.

The geological range of the Monodontidae is Pleistocene to Recent.

CETACEA; MONODONTIDAE; *Genus*
DELPHINAPTERUS Lacépède, 1804

Beluga, White Whale

The single species, *D. leucas,* occurs primarily in the Arctic Ocean and adjoining seas, the Sea of Okhotsk, the Bering Sea, the Gulf of Alaska, Hudson Bay, and the Gulf of St. Lawrence (Rice 1977). It also ascends large rivers, such as the Amur, Anadyr, Ob, Yenesei, Yukon, Churchill, and St. Lawrence. Individuals have appeared as far south as Japan, Washington, Connecticut, New Jersey, Ireland, Scotland, the Rhine River, and the Baltic Sea (Ellerman and Morrison-Scott 1966; Hall 1981; Banfield 1974; Ellis 1980). The use of the name "beluga" has sometimes caused confusion, since the term is also applied to the great white sturgeon, a fish that is one of the principal sources of caviar.

According to Fay (1978), head and body length is usually about 340 to 460 cm in males and 300 to 400 cm in females, and average weight is about 1,500 kg in males and 1,360 kg in females. Kleinenberg et al. (1969) listed maximum size as 700 cm and 2,000 kg. Pectoral fin length is about 20 to 45 cm. The general adult color is creamy white. The young are dark gray, black, or bluish in the first year of life; then become yellowish, mottled brown, or pale gray; and finally attain the white coloration at around five years of age. The lightening in color is caused by a reduction in the melanin of the skin.

The body is fusiform, the tail is strongly forked, there is a constriction at the neck, and the snout is blunt. The upper jaw protrudes slightly ahead of the bulbous melon. There is no dorsal fin, but a low dorsal ridge is present on the back. There are 8 to 10 teeth on each side of each jaw.

The beluga occurs both along coasts and in deep offshore waters. Some populations, such as that of the Gulf of Alaska, seem to reside permanently in relatively small areas, while others undergo extensive migrations. Wintering sites may be either north or south of summering sites, and may be either nearer to or farther from shore. The environmental conditions that prevail in the range of a particular population, such as prey distribution and extent of the pack ice, apparently affect the choice of seasonal movements. One population concentrates in the deep waters of the Beaufort Sea during the summer, but is distributed mainly in shallow coastal areas, bays, and river mouths around the Bering Sea in winter. Another population concentrates in river estuaries adjoining southern Hudson Bay in the summer but moves northward and into the open waters of the bay in winter. In the waters off northern Eurasia there seems a general tendency to move away from shore in the autumn and toward shore in the spring, but there are exceptions, and times of migration vary from year to year. An advantage of being close to shore in the summer is that rivers are unfrozen and may be entered for purposes of finding food and rearing young. Some individuals have ascended rivers in Siberia for distances of up to 2,000 km (Kleinenberg et al. 1969; Sergeant 1973; Harrison and Hall 1978; Banfield 1974; Fay 1978).

Delphinapterus, a very supple animal, can scull with its tail and thus swim backward. When swimming normally, it moves at about 3 to 9 km per hr, and when pursued it can attain a speed of 22 km per hr. It usually surfaces to breathe every 30 to 40 seconds, but feeding dives last 3 to 5 minutes, and one individual is known to have remained underwater for 15 minutes. The beluga is capable of covering a distance of 2 or 3 km while underwater, and therefore the presence of surface ice is not necessarily a problem (Kleinenberg et al. 1969). The beluga cannot remain in waters extensively covered by thick ice, however, because it is unable to break

Beluga whale (*Delphinapterus leucas*), photos by Warren J. Houck.

Beluga whales (*Delphinapterus leucas*), photo by Warren J. Houck.

through more than a very thin layer to make breathing holes (Harrison and Hall 1978).

Delphinapterus is known to produce a variety of sounds. One, "a low liquid trill, like the cries of curlews in the spring," has given rise to the vernacular name "sea canary." Many other animals produce underwater noises, but few such sounds can be heard so readily above water as those of the beluga. Also emitted are whistles and clicks, comparable to those known to be used by certain other cetaceans, respectively, for communication and echolocation (Kleinenberg et al. 1969; Fay 1978). That *Delphinapterus* employs echolocation extensively is also suggested by its huge frontal melon, an organ thought to be used in the directing of sonar signals, and by the fact that this genus lives in regions that lack sunlight for months at a time, and in waters that may be partly covered by ice. Echolocation presumably is used to avoid obstacles and to search for prey on the bottom. The diet includes a great variety of fish, such as cod and herring, as well as octopus, squid, crabs, and snails (Fay 1978; Kleinenberg et al. 1969).

Schools of as many as 10,000 individuals have been reported, but such aggregations are formed only temporarily for migration or in the presence of abundant food sources. Permanent social units consist of only about 10 animals and seem to be led by a large male (Kleinenberg et al. 1969). There is conflicting information on reproduction, but studies in arctic Canada (Sergeant 1973; Brodie 1971) indicate that the calving season peaks around the end of June and extends to mid-August, that each adult female gives birth every 2 or 3 years, and that the gestation period is about 14 months. The newborn is about 160 cm long and weighs about 80 kg. Lactation lasts for 20 months. Sexual maturity is attained at an age of 4 to 7 years in females and 8 or 9 years in males. Potential longevity is 25–30 years.

The beluga has been exploited by the native people of the Arctic since ancient times, and has been subject to commercial fishing operations in the twentieth century. It is often caught in nets set across migratory routes. The meat is used for human and domestic animal consumption. The 10- to 25-cm-thick layer of blubber yields 100 to 300 liters of oil,

which is used in the production of soap, lubricants, and margarine. The fat of the head is rendered into a high-quality lubricant. The bones are ground up for fertilizer, and the skin is used to make boots and laces. The tanned hide of the beluga is sometimes called "porpoise leather." In recent decades the demand for these products has declined. From the 1950s to the 1970s, annual catches fell from about 1,000 to 500 in Canada and from about 4,000 to 700 in the Soviet Union. The current world population is estimated at 62,000 to 80,000 individuals, of which at least 30,000 are in North American waters (Sergeant and Brodie 1975; Mitchell 1975*a*, 1975*b*; Kleinenberg et al. 1969; U.S. National Marine Fisheries Service 1981).

CETACEA; MONODONTIDAE; **Genus MONODON** *Linnaeus, 1758*

Narwhal

The single species, *M. monoceros*, occurs in the Arctic Ocean and nearby seas, from 65° to 85° N, mainly between 70° and 80° N. There have been occasional records as far south as the Alaska Peninsula, Newfoundland, Britain, and Germany (Reeves and Tracey 1980).

Head and body length, exclusive of the tusk, is 360 to 620 cm, pectoral fin length is 30 to 40 cm, and expanse of the tail flukes is 100 to 120 cm. According to Reeves and Tracey (1980), average head and body length is about 470 cm in males and 400 cm in females, and average weight is 1,600 kg in males and 900 kg in females. About one-third of the weight is blubber. Coloration becomes paler with age. Adults have brownish or dark grayish upper parts and whitish underparts, with a mottled pattern of spots throughout. The head is relatively small, the snout is blunt, and the flipper is short and rounded. There is no dorsal fin, but there is an irregular ridge, about 5 cm high and 60 to 90 cm long, on the posterior half of the back. The posterior margins of the tail

Narwhals (*Monodon monoceros*), painting by Richard Ellis.

flukes are strongly convex, rather than concave or straight as in most cetaceans.

There are only two teeth, both in the upper jaw. In females the teeth usually are not functional and remain embedded in the bone. In males, the right tooth remains embedded but the left tooth erupts, protrudes through the upper lip, and grows forward in a spiral pattern to form a straight tusk. The tusk is about a third to a half as long as the head and body, and sometimes reaches a length of 300 cm and a weight of 10 kg. In rare cases the right tooth also forms a tusk, but both tusks are always twisted in the same direction. Occasionally, one or even two tusks develop in a female. The distal end of the tusk has a polished appearance, and the remainder is usually covered by a reddish or greenish growth of algae. There is an outer layer of cement, an inner layer of dentine, and a pulp cavity that is rich in blood. Broken tusks are common, but the damaged end is filled by a growth of reparative dentine. The form of this plug sometimes gives the erroneous impression that another narwhal has jammed its own tusk directly into the damaged one (Reeves and Tracey 1978).

It has traditionally been suggested that the tusk of *Monodon* is used to break through surface ice, to probe the bottom for food, or to skewer prey (Ellis 1980). More recent hypotheses are that the tusk serves to radiate heat or to enhance echolocation (Reeves and Mitchell 1981). New evidence, however, supports an old view that the tusk is exactly what it looks like—a weapon. According to Silverman and Dunbar (1980), the presence of many scars on the heads of adult males, the large proportion of broken tusks seen, the discovery of a tusk tip embedded in the jaw of a male, and actual observations of narwhals crossing and striking their tusks strongly indicate that the tusk is used in intraspecific aggression, most probably during the mating season. It also may serve to communicate dominance rank, or in ritualized displays, much like the horns and antlers of some ungulates.

Monodon has the most northerly distribution of any mammal; its normal range is almost entirely above the Arctic Circle. It is generally found in deeper waters than those used by *Delphinapterus*, and tends to avoid shallow seas like those around Alaska. It seems to be most common in the eastern Canadian Arctic and in waters off Greenland. One or more major populations winter in the open waters of Baffin Bay

and Davis Strait. When the ice breaks up in the summer, there are large-scale migrations to the north, east, and west, into the fjords and inlets of Greenland and the islands of the eastern Canadian Arctic. Another group winters in the Greenland Sea and summers along the east coast of Greenland and around Svalbard (Reeves and Tracey 1980; Newman 1978).

The narwhal is a rapid swimmer. It has been reported to dive to depths as great as 370 meters, and has been timed to stay underwater for 15 minutes. It maintains breathing holes in pack ice by upward thrusts of the large melon on the forehead (not the tusk), and can break through a layer of ice 18 cm thick. Sometimes several individuals cooperate to break the ice. *Monodon* emits a wide variety of sounds. The deep groans, roars, and gurgles that have been reported are probably associated with respiration. There are also screeches and whistles, probably for communication, and series of clicks, much like those known to be used by some other cetaceans for echolocation (Reeves and Tracey 1980; Ford and Fisher 1978). There has been no conclusive demonstration that *Monodon* actually does echolocate, but such an ability would seem useful in finding obstacles and prey in ice-covered waters or on long arctic nights. The diet consists of fish, cephalopods, and crustaceans.

Up to 2,000 individuals may join in a migratory school, but permanent groups contain 3 to 20. Most of the smaller groups are apparently families, and each seems to have only 1 large male. There are also groups consisting entirely of males or of females and young. Births occur from June to August, the period when the break up of pack ice allows entrance to protected bays and fjords. Each female apparently produces a single calf every 3 years. Gestation is thought to last about 14½ months. The newborn is around 160 cm long and weighs just over 80 kg. The period of lactation and the time of sexual maturity are unknown, but are presumed to be about the same as in *Delphinapterus*. Maximum longevity evidently approaches 40 years (Reeves and Tracey 1980; Best and Fisher 1974; Mansfield, Smith, and Beck 1975; Ellis 1980).

The narwhal has long been hunted for subsistence by the native people of the Arctic. The animals are traditionally harpooned from kayaks when they enter inlets during the summer. Occasionally a large number can be easily killed when they become trapped by the rapid formation of pack ice at the end of summer. The sinews are used for thread, the meat is sometimes eaten by humans and sled dogs, and the skin, high in vitamin C, is probably what prevented scurvy in the arctic peoples (Reeves 1977).

Commercial hunting of *Monodon* seems to have begun in the 10th century A.D., soon after the establishment of Viking colonies in Greenland. The tusk was long sold at high prices, as it was said to be the horn of the mythical unicorn and to have magical properties, such as rendering poison harmless. By the 17th century the true origin and general availability of the tusk were recognized, and prices fell accordingly. Some demand continued, especially in the Orient, where the tusk was used in decorative work and medicinal preparations. Skins were imported to Europe to make boots, laces, and gloves. In the 1960s, as in the case of other kinds of ivory, the value of the narwhal's tusk again began to increase. The price went from about U.S. $2.75 per kg in 1965 to nearly $100 per kg in 1979. Most tusks, however, go whole to collectors and decorators, who have paid up to $4,500 for a choice specimen. Such a market caused an intensification of hunting, which is now often done with motorboats, explosive harpoons, and high-powered rifles. In 1976 the Canadian government established regulations that set quotas for various areas and required full utilization of the carcass, as well as the tusk, but these rules are nearly impossible to enforce. The annual kill of *Monodon* in Canada and Greenland is now thought to be well over 1,000. There are probably at least 20,000 narwhals to the west of Greenland and several thousand more to the east. It is not known if overall populations in Canada and Greenland are being reduced by human hunting, but there apparently have been severe declines in Soviet waters (Reeves 1977; Reeves and Mitchell 1981; Reeves and Tracey 1980; Ellis 1980; Davis et al. 1978; Mansfield, Smith, and Beck 1975; Newman 1978).

CETACEA; *Family PHYSETERIDAE*

Sperm Whales

This family of two Recent genera and three species occurs in all the oceans and adjoining seas of the world. The genus *Kogia* is sometimes placed in a subfamily or family, the Kogiidae, distinct from that containing the genus *Physeter*. The sequence of genera presented here follows that of Hall (1981) and not that of Rice (1977).

The characters common to both genera include: a broad, flat rostrum; a great depression in the facial part of the skull to accommodate a highly developed spermaceti organ, which is bounded posteriorly by a high occipital crest; an S-shaped blowhole, located on the left side of the snout; no preorbital or postorbital processes on the skull; a simple air sinus system; a left nasal passage for respiration and a right one apparently modified for sound production; numerous short grooves on the throat; a small, narrow lower jaw that ends well short of the anterior end of the snout; functional teeth only in the lower jaw; and relatively small pectoral appendages (Rice 1967; Hall 1981). The two genera differ markedly in size and certain other morphological characters. There are also major differences in known aspects of natural history.

The geological range of the Physeteridae is middle Miocene to Pleistocene in North America, early Miocene in South America, early Miocene to middle Pleistocene in Europe, late Pliocene in Japan, early Pliocene in Australia, and Recent in all oceans (Rice 1967).

CETACEA; PHYSETERIDAE; *Genus KOGIA* Gray, 1846

Pygmy and Dwarf Sperm Whales

There are two species (Rice 1977; Hall 1981):

K. breviceps (pygmy sperm whale), world-wide in tropical and warm temperate waters;

K. simus (dwarf sperm whale), known to occur in the seas adjacent to South Africa, India, Sri Lanka, Japan, Hawaii, California, Baja California, the eastern and Gulf coasts of the United States, Cuba, and St. Vincent Island in the Lesser Antilles.

Head and body length is 210 to 270 cm in *K. simus* and 270 to 340 cm in the larger *K. breviceps*. Pectoral fin length is about 40 cm and expanse of the flukes is about 61 cm. The usual weight has been estimated at 136 to 272 kg for *K. simus* and 318 to 408 kg for *K. breviceps*, but the U.S. National Marine Fisheries Service (1981) indicated that the latter species sometimes exceeds 772 kg in weight. Males are generally larger and heavier than females. Both species are dark steel gray on the back, shade to a lighter gray on the sides, and gradually fade to a dull white on the belly (Leatherwood, Caldwell, and Winn 1976).

The head is only about one-sixth the total length, and resembles that of some delphinid genera in external outline, but the blunt snout projects beyond the lower jaw. The facial part of the skull is among the shortest of any cetacean. The blowhole of *Kogia*, like that of *Physeter*, is on the left side of the head, but is located on the forehead rather than toward the tip of the snout. *Kogia* resembles *Physeter* in having a spermaceti organ in the head and in having the functional teeth confined to the lower jaw. The teeth are sharp and curved; the number on each side of the lower jaw is 8 to 11 in *K. simus* and 12 to 16 in *K. breviceps* (Hall 1981). A few smaller, nonfunctional teeth may be present in the upper jaw of *K. simus*. The dorsal fin is hooked and curved like a sickle; it is relatively high and located near the middle of the back in *K. simus*, and is much lower and located posterior to the middle of the back in *K. breviceps*.

A few observations of live animals indicate that *Kogia* is rather sluggish and sometimes floats motionless on the surface. Otherwise, the genus is known largely on the basis of strandings. Some of these records suggest that *K. breviceps* moves away from the shores of the eastern United States and South Australia during the summer, and that *K. simus* migrates toward Japan in the summer. South African records, however, indicate that both species are present throughout the year and that neither makes major seasonal movements. Examination of stomach contents suggests that *K. breviceps* normally inhabits oceanic waters beyond the edge of the continental shelf, and that *K. simus* tends to occur closer to shore, but sometimes feeds on organisms that dwell at depths of 250 meters or more (G.J.B. Ross 1979; Fitch and Brownell 1968). The diet includes mostly cephalopods, and also a variety of fish and some crustaceans.

According to Rice (1978c), *K. breviceps* is usually seen in pairs—perhaps a female and a calf—or in groups of 3 to 5 individuals. Data summarized by G.J.B. Ross (1979) suggest that the school size of *K. simus* is 10 animals or less, that females with calves may group together, that immature animals may form their own groups, and that sexually mature males and females can be found in the same school. Intraspecific fighting has been witnessed. Ross also analyzed records of small and near-term fetuses in *K. breviceps*, which indicate that in both the Northern and Southern hemispheres the mating and calving seasons extend over a period of approximately seven months, from autumn to spring. There were too few records of *K. simus* to determine if there is a

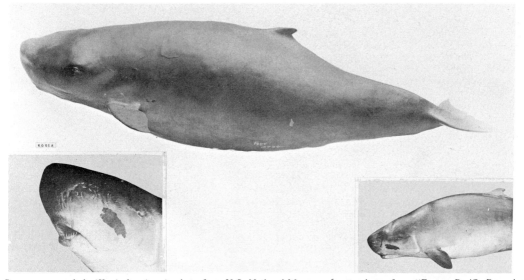

Pygmy sperm whale (*Kogia breviceps*), photo from U.S. National Museum. Insets: photos from "Eastern Pacific Records and General Distribution of the Pygmy Sperm Whale," Carl Hubbs, *J. Mamm.*

distinct breeding season in that species, though there does appear to be an extended calving season of at least five or six months. In *K. breviceps* the gestation period may last over 11 months. A number of females have been found to be pregnant and lactating at the same time, and it seems likely that many females give birth every year. The single newborn is about 100 cm long in *K. simus* and 120 cm long in *K. breviceps*.

Pygmy and dwarf sperm whales seem to be rare, but are occasionally taken by fishermen and utilized. The rarity may reflect an early, unrecorded exploitation. *Kogia* is lethargic and easy to approach, and a few records indicate that this genus was sometimes harpooned by whalers (Mitchell 1975*b*; Ellis 1980).

CETACEA; PHYSETERIDAE; *Genus PHYSETER*
Linnaeus, 1758

Sperm Whale

The single species, *P. catodon*, occurs in all oceans and adjoining seas of the world, except in polar ice fields (Rice 1977). Most recent authorities have followed Husson and Holthuis (1974) in using the name *P. macrocephalus* for this species, but Hall (1981) considered *P. catodon* the proper designation.

This is the largest toothed mammal in the world and is the most sexually dimorphic of all cetaceans. Sexual maturity is attained at a length of about 12.2 meters in males and 8.5 meters in females. Growth subsequently continues for a number of years. Maximum head and body length is nearly 20 meters in males, though individuals over 15.2 meters are rare. Head and body length in females seldom exceeds 12 meters. Physically mature males usually weigh 35,000 to 50,000 kg, while females weigh only about one-third as much. The pectoral fin is about 200 cm long, and the expanse of the tail flukes is usually 400 to 450 cm. The color is gray to dark bluish gray or black. With increasing age, males may become paler and sometimes piebald. Most specimens have some white in the genital and anal regions, and on the lower jaw (Best 1974, 1979; Ellis 1980; Leatherwood, Caldwell, and Winn 1976).

The head is enormous and squarish, especially in males. The skull is the most asymmetrical of any mammal. The blowhole is located toward the tip of the snout and on the left side. The lower jaw is extraordinarily slender and has 16 to 30 strong, conical teeth on each side. These teeth may be more than 20 cm long. They fit into sockets in the palate when the mouth is closed. Smaller, nonfunctional teeth are present in the upper jaw. *Physeter* is the only cetacean with a gullet large enough to swallow a man. There is no dorsal fin, but there is a longitudinal row of bumps on the posterior half of the back. The vertebrae number 50 to 51, and most of the neck vertebrae are fused. The blubber is up to 35 cm thick. One specimen of *Physeter*, 13 meters long and weighing 20,000 kg, had a heart that weighed 116 kg. The brain of *Physeter*, the largest of any mammal, weighs about 9.2 kg (Rice 1967).

The most striking morphological feature, and the one that gives *Physeter* its vernacular name, is the huge spermaceti organ in the head, filled with up to 1,900 liters of waxy oil. At the anterior and posterior ends of the organ are air sacs. The left nasal passage goes along the left side of the spermaceti organ, leads directly to the blowhole, and also connects to the anterior air sac. The much narrower right nasal passage goes through the lower part of the spermaceti organ, opens into the anterior air sac through large internal lips, and also connects to the posterior air sac. This general structure has been known for centuries, but its exact functions are still being debated (Best 1979). It may either assist in evacuating the lungs prior to a dive and in absorbing nitrogen at extreme pressures, regulate buoyancy during deep diving, or reverberate and focus sounds. Clarke (1979), noting that *Physeter* apparently lies still at great depths to await and snap up squid, suggested that buoyancy is then controlled by varying the temperature and thus the density of the spermaceti oil. Cooling might be accomplished by drawing water into the nasal passages and around the oil. Norris and Harvey (1972) opposed this theory, and suggested that the spermaceti organ is involved mainly in the production of burst pulses for long-range echolocation. According to this view, air goes through the right nasal passage, produces sounds at the lips leading into the anterior air sac, goes back up the left nasal passage, and is then pumped around again. The two air sacs may serve as sound mirrors.

Sperm whale (*Physeter catodon*), photos from *British Mammals,* Archibald Thorburn (top), and R. L. Pitman (bottom).

Physeter has a variety of sounds, which have been described in such terms as groans, whistles, chirps, pings, squeaks, yelps, and wheezes (Ellis 1980). The voice is very loud and can be heard many kilometers away by persons with proper underwater listening equipment. The most common sounds are series of clicklike pulses, which are generally much lower in frequency than those of the Delphinidae. The spermaceti organ may be used in the directional beaming, as well as the production of these pulses, which are apparently involved in both echolocation and communication. In addition to other sounds, each whale has its own individualized "coda," a stereotyped, repetitive sequence of 3 to 40 or more clicks, which is heard only when one animal meets another (Rice 1978c; Norris and Harvey 1972).

The sperm whale is generally found in waters conducive to the production of squid—at least 1,000 meters deep and with cold-water upwellings (Ellis 1980). The best areas are off the coasts of South America and Africa, in the North Atlantic and Arabian Sea, between Australia and New Zealand, in the western North Pacific, and all along the Equator. Most animals stay between 40° N and 40° S, but during the summer the bachelor males of medium size move to between 40° and 50°, and at least some of the older males venture beyond 50° into or near arctic and antarctic waters. Although the other animals do not wander as far, there are northward shifts of concentration in the boreal summer and southward shifts in the austral summer as the movements of squid are followed (Rice 1978c; Berzin 1972; Best 1974, 1979). Certain major

Sperm whale (*Physeter catodon*), photo from American Museum of Natural History.

populations or stocks seem to move together on a seasonal basis, but migrations are not as predictable as those of the baleen whales. *Physeter* normally swims at a maximum of around 10 km per hr, but when pursued it can attain speeds of up to 30 km per hr (Berzin 1972). It sometimes lifts its head vertically out of the water, apparently to look or listen. Before a dive it gives a spout, characteristic in being directed obliquely forward to the left. It then lifts the tail flukes high in the air and descends almost vertically. There is commonly a prolonged initial submergence at around 360 meters for 20 to 75 minutes, and then a series of shallow dives. Females do not generally go as deep as the larger males (Ellis 1980). There are at least 14 instances on record of sperm whales becoming entangled in underwater communication cables, one at a depth of 1,135 meters. *Physeter* has been tracked by sonar to a depth of 2,500 meters. One individual, killed after an 82-minute dive in waters 3,200 meters deep, was found to have just consumed a kind of shark known to dwell on the bottom (Rice 1978c). Several species of shark and other fish are included in the diet, but the predominant food is squid. Most of the squid taken are less than 1 meter long, but the stomach of one sperm whale contained a squid about 10½ meters long. It has been estimated that each whale eats about 3 percent of its weight in squid per day (Ellis 1980). *Physeter* often bears the scars of combat with large squid.

Hundreds of individuals sometimes join in migratory schools, and there is one record of 3,000–4,000 being seen together off Patagonia. The basic social unit, however, is the mixed school, which has a year-round membership of adult females, calves, and some juveniles, usually 20 to 40 individuals in all. These units are stable, and there is evidence that the bonds between females last many years. If nursing calves are present, the group may be referred to as a nursery school. Shortly after weaning, young males and some young females begin to leave and to join in juvenile schools, usually with 6 to 10 individuals. The females eventually return to a mixed school before reaching puberty. Males from the juvenile schools, and some coming directly from mixed schools, form bachelor schools. When the bachelors are small, these schools contain 12 to 15 whales, though several groups may aggregate. As the bachelors grow larger, they divide into smaller groups, and then often into pairs and lone animals. Even if an association is maintained by older males, the individuals may stay some distance from each other. During the breeding season, each mixed school is commonly joined by 1 to 5 large males. The groups then become known as breeding or harem schools, and there usually is about 1 adult male for every 10 adult females. The exact relationship between these adult males is not well understood. There is apparently some fighting for the right to join a breeding school, and the scars from such battles frequently cover the head of males. Nonetheless, it may be that several males establish a dominance hierarchy and that they share the females of one or more mixed schools. Such males may even have composed an organized bachelor school, prior to the breeding season. In any event, only 10 to 25 percent of the fully adult males in a population are able to get into a breeding school. The others move toward the Arctic and Antarctic. It is not known if some or all of the adult males in a breeding school also eventually spend some time in polar waters (Best 1979; Ohsumi 1971; Gaskin 1970; Ellis 1980).

The breeding season varies, but mating generally peaks in the spring and calving in the fall in both the Northern and Southern hemispheres. Gestation has been variously calculated at 14 to 17 months. The single newborn is about 400 cm long and weighs nearly 1,000 kg. Nursing usually lasts just over 2 years, but may possibly continue in some instances for up to 13 years. Sexual maturity is attained by females at an age of 7 to 9 years. Males may have the physiological ability to produce offspring when less than 19 years old, but do not reach social maturity—the ability to enter a breeding school—until 25 to 27 years old. Full physical maturity comes at age 35 years in males and 28 years in females. Some individuals are estimated to have lived up to 77 years (Best 1968, 1970a, 1974, 1979; Rice 1978c; Berzin 1972; Caldwell, Caldwell, and Rice 1966; Ohsumi 1965, 1966; Haley 1978; Frazer 1973).

The sperm whale has been regularly hunted by people since 1712. The meat is not generally used for human consumption, and of course there is no yield of baleen. The large teeth of *Physeter* were valued, however, as a medium for the artistic form of engraving and carving known as scrimshaw. A product unique to the sperm whale is ambergris, a waxy substance probably formed in the intestines from solid wastes coalescing around a matrix of indigestible material. It is used as a fixative and has the property of retaining the fragrance of

Sperm whale (*Physeter catodon*), photo by Warren J. Houck. Teeth of *P. catodon*, photo by P. F. Wright of specimen in U.S. National Museum.

perfumes. Apparently the heaviest mass of ambergris from a single animal weighed about 450 grams. Such an amount of the substance would have sold for about $10 to $50 in the early 1960s, depending on color, but prices have fallen in recent years because of the development of synthetic fixatives. The most important product of the sperm whale is oil, though unlike that of baleen whales it cannot be used in the manufacture of margarine. Sperm whale oil was once the major source of fuel for lamps, and has recently served as a lubricant and as the base for skin creams and cosmetics. The oil of the spermaceti organ solidifies into a white wax upon exposure to air, and is used in making ointments and fine, smokeless candles. It also has served as a high-quality lubri-

cant for precision instruments and machinery, and as a component of automatic transmission fluid.

An intensification of sperm whale hunting came around 1750, with the invention of the spermaceti candle and the development of onboard processing facilities that allowed whaling vessels to remain at sea until their holds were filled with oil (Ellis 1980). For the next 100 years, Americans dominated the industry, probably taking up to 5,000 sperm whales annually. This kill may not have been enough to reduce overall populations, but in the second half of the 19th century both sperm whale hunting and American whaling declined. Some of the suggested reasons are increasing costs, replacement of whale oil by kerosene, destruction of much of the U.S. fleet in the Civil War, and opening of the western Arctic and North Pacific to hunting of the bowhead whale (*Balaena mysticetus*). The development of the harpoon gun in the 1860s brought more hunting of the large baleen whales and less emphasis on *Physeter*.

By the 1930s, a decline in some of the baleen species had resulted in renewed interest in sperm whale hunting. Floating factory ships could remain for lengthy periods within the prime habitat of *Physeter*, especially the North Pacific. In the 1936/37 season, for the first time in many years, the annual kill rose above 5,000. Subsequent efforts at international regulation were largely unsuccessful. In the 1950/51 season the take was 18,264 sperm whales, in 1963/64 it peaked at 29,255, and it remained above 20,000 in all but one season until 1975/76. About one-fourth of the total catch was made by shore-based operations, and the remainder by pelagic expeditions. Finally, in response to immense scientific and public concern, the International Whaling Commission began to reduce quotas substantially. The kill in the 1978/79 season was 8,536, the quota set for 1980/81 was only 1,849, and no kill was authorized for 1981/82. Moreover, the use of floating factories has now been banned in the hunting of *Physeter*, and most shore-based sperm whale fisheries have closed down. The species is listed as endangered by the USDI (1980), but is still the most numerous of the great whales. In 1946 the estimated number of males over 9.2 meters long (the minimum legal limit for catching) and sexually mature females was about 1,100,000. The current estimate for these categories is just over 700,000, and the total population is about twice as great (U.S. National Marine Fisheries Service 1978, 1981; Ellis 1980; McHugh 1974; Gulland 1974; Rice 1978c; Committee for Whaling Statistics 1980).

CETACEA; *Family ZIPHIIDAE*

Beaked Whales

This family of 6 Recent genera and 18 species occurs in all the oceans and adjoining seas of the world.

Head and body length is about 4 to 13 meters, and weight ranges from 1,000 to over 11,000 kg. The vernacular name is derived from the long, narrow snout, which is sharply demarked from the high, bulging forehead in *Berardius* and *Hyperoodon*, and which forms a continuous smooth profile with the head in *Ziphius*, *Tasmacetus*, and *Mesoplodon* (the external appearance of the sixth genus, *Indopacetus*, is not known). The pectoral fin is rather small and ovate. The dorsal fin is small, usually sickle shaped, and located on the posterior half of the back. The genus *Ziphius* has a low median keel from the dorsal fin to the tail. The tail flukes of beaked whales are not notched in the center as in other cetaceans. There is a pair of grooves on the throat, that converge anteriorly to form a V pattern at the chin. The stomach has 4 to 14 chambers, but no esophageal chamber (Rice 1967).

The genus *Tasmacetus* has one pair of large functional teeth in the lower jaw, and also 17 to 28 small functional teeth on each side of both the upper and lower jaws. In the other genera there are only one or two pairs of large functional teeth, and these are in the lower jaw. These teeth push through the gums sooner in males than in females, and often never erupt in the latter sex. There also are frequently series of small nonfunctional teeth in the upper and lower jaws. The bones of the skull are asymmetrical, except in *Berardius*. Certain cranial bones are crested or elevated, forming large ridges in *Hyperoodon* and lesser ridges in some of the other genera. There are 43 to 49 vertebrae, and those of the neck tend to fuse.

Beaked whales are the least known of cetacean families. They usually remain well out to sea, avoid ships, and dive to great depths to secure cephalopods and fish. The presence of a well-developed melon on the forehead suggests that ziphiids are echolocators (U.S. National Marine Fisheries Service 1981). Some appear to be solitary, some travel in groups of 2 to 12 individuals, and some, particularly *Ziphius*, associate in schools of 40 or more. They generally swim and dive in unison. The hides of many individuals are scratched from intraspecific fighting. The newborn are about one-third as long as the mother.

The geological range of this family is early Miocene to Pliocene in South America, early Miocene to middle Pliocene in Europe, late Miocene to Pliocene in North America, and Recent in all oceans (Rice 1967).

CETACEA; ZIPHIIDAE; *Genus BERARDIUS* Duvernoy, *1851*

Giant Bottlenosed Whales

There are two species (Rice 1977; Goodall 1978):

B. arnuxii, known from waters off South Africa, Australia, New Zealand, Argentina, Tierra del Fuego, the Falkland Islands, South Georgia, the South Shetlands, and the Antarctic Peninsula;

B. bairdii, the North Pacific from the Bering Sea to Japan and California.

These are the largest ziphiids. *B. bairdii* attains sexual maturity at a head and body length of about 10 meters, and growth continues to a maximum of 12.8 meters in females and 12 meters in the slightly smaller males. A female *B. bairdii*, 11.1 meters long, weighed about 11,380 kg (Rice 1967). Pectoral fin length is about 100 cm, height of the dorsal fin is about 30 cm, and expanse of the tail flukes is 250 to 300 cm. *B. arnuxii* is smaller than *B. bairdii*, not being known to exceed 9.9 meters in head and body length (McCann 1975). Coloration is uniformly blackish brown or dark gray, sometimes with white blotches on the underparts. The skin is always extensively covered with pairs of parallel scratches, probably made by the teeth of conspecifics.

The skull is more nearly symmetrical than in the other beaked whales. The snout is tapered and the forehead is well defined. Both sexes have two pairs of large teeth in the lower jaw. The anterior pair is completely visible, even when the mouth is closed, because the lower jaw protrudes well beyond the upper.

Giant bottlenosed whales usually stay well offshore in waters over 1,000 meters deep. Substantive natural history data are available only for *B. bairdii*. This species is alleged to have dived to depths as great as 2,400 meters, after being harpooned. It raises its flukes in the air before diving. It is said to be alert and hard to capture. It normally stays underwater for 15 to 20 minutes at a time, but has remained submerged for up to an hour. Its diet consists mostly of squid, and also includes octopus, crustaceans, and deep-water fish (Rice 1967, 1978d; Ellis 1980).

Groups of *B. bairdii* are tightly organized and contain 3 to 30 individuals of all ages and sexes (Rice 1978d; Ellis 1980). Studies of this species in Japan (Kasuya 1977) indicate that mating peaks in October and November, and births occur mainly in March and April. The gestation period is about 17 months, apparently the longest of any cetacean. The newborn is around 460 cm long. Sexual maturity is attained at an age of 8 to 10 years and physical maturity at over 20 years. Some individuals are thought to have lived as long as 70 years.

Giant bottlenosed whale (*Berardius bairdii*): A & B. Photos from the Fisheries Research Board of Canada through I. B. MacAskie; C. Head showing teeth in lower jaw, photo from Tokyo Whales Research Institute through Hideo Omura; D & E. Photos by Warren J. Houck.

Hunting of *B. bairdii* has been undertaken in Japan since at least 1612. The dried meat is considered a delicacy in some parts of the country. For many years the annual kill amounted to only a few individuals, but following World War II modern fishing methods resulted in a sharply increased harvest. The peak kill was 382 whales in 1952. A subsequent decline in the take, which averaged 69 per year from 1969 to 1977, may be partly associated with a reduction in the whale population (Mitchell 1975*a*, 1975*b*; Rice 1978*d*; Nishiwaki and Oguro 1971; Ellis 1980).

CETACEA; ZIPHIIDAE; *Genus* **ZIPHIUS** *G. Cuvier, 1823*

Goosebeaked Whale

The single species, *Z. cavirostris*, occurs in the temperate and tropical waters of all oceans and adjoining seas (Rice 1977).

Females reach sexual maturity at a head and body length of 6.1 meters and then continue to grow to a maximum of 7 meters. Males reach sexual maturity at 5.4 meters and then

continue to grow to a maximum of 6.7 meters. A frequently cited report of an individual 8.5 meters long is erroneous (Mitchell 1975*a*). Pectoral fin length is about 50 cm, dorsal fin height is about 40 cm, and expanse of the tail flukes is about 150 cm. An adult female, 6.6 meters long, weighed 2,952.5 kg. Coloration is variable, but two frequently observed patterns are: face and upper back creamy white, remainder of body black; and entire body grayish fawn, with some small blotches of slightly darker gray below.

The beak is short and blends into the sloping forehead. The opening of the mouth is relatively small. Males have two functional teeth, which protrude from the tip of the lower jaw. These teeth are usually not visible in females. Rows of small rudimentary teeth are usually present in both jaws. As in other ziphiids, the tail flukes generally lack a median notch, but some specimens of *Ziphius* do have such a notch.

According to Leatherwood, Caldwell, and Winn (1976), *Ziphius* seems primarily tropical in distribution, but moves northward into temperate waters during the summer. There are also records from as far north as the Aleutian Islands and the Gulf of Alaska (C. S. Harrison 1979; Rice 1978*d*). *Ziphius* is generally found well offshore, and often dives to great depths, remaining underwater for 30 minutes or more.

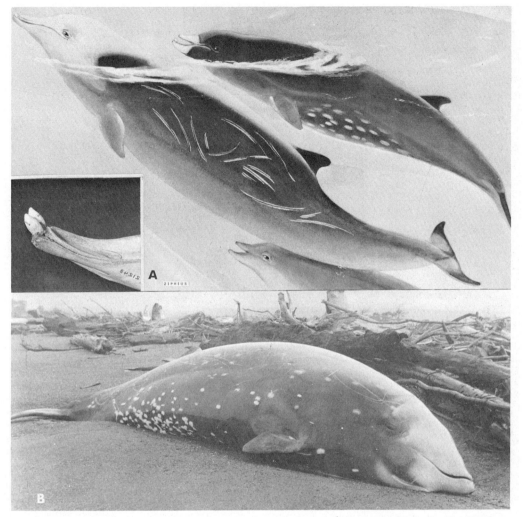

Goosebeaked whales (*Ziphius cavirostris*): A. Photo from National Geographic Society of painting by Else Bostelmann. Inset: lower jaw showing the two teeth, photo by P. F. Wright of specimen in U.S. National Museum; B. Photo by Warren J. Houck.

Before diving, it raises its tail flukes straight above the surface. It has been observed to leap clear of the water. Its diet consists primarily of squid and also includes deep-water fish.

Groups of up to 40 individuals have been reported, but schools usually contain less than half that number. Some white-headed adults, possibly old males, are solitary (Rice 1978*d*). The members of a group often travel, dive, and feed together in fairly close association. Banfield (1974) stated that births occur in late summer or early autumn. A gestation period of about 1 year has been reported. The newborn is about 200 to 300 cm long (Mitchell 1975*a*). One female specimen is thought to have lived for over 20 years.

CETACEA; ZIPHIIDAE; **Genus TASMACETUS** Oliver, *1937*

Shepherd's Beaked Whale

The single species, *T. shepherdi*, is known by 10 specimens stranded in New Zealand, Stewart and Chatham islands near New Zealand, the Juan Fernandez Islands off central Chile, the Valdez Peninsula of east-central Argentina, and Tierra del Fuego; and by a probable sighting of a live individual off New Zealand (Goodall 1978; Mead and Payne 1975; Brownell, Aguayo, and Torres 1976; Watkins 1976).

On the basis of four specimens, Mead and Payne (1975) provided the following data. Head and body length is 6.1 to 7 meters, pectoral fin length is 69 cm, dorsal fin height is 34 cm, and expanse of the tail flukes is 135 to 152 cm. The general coloration is thought to be dark, but the ventral surface is pale, there is a light area anterior to the pectoral fin, and there are two light stripes along part of the side. In males there are two large teeth at the tip of the lower jaw, but in the one known female specimen these teeth did not erupt. In contrast to all other ziphiids, there are also numerous smaller functional teeth. The total tooth count is 17 to 21 on each side of the upper jaw and 18 to 28 on each side of the lower jaw. The stomach of one specimen contained a number of fish, and also a small crab and a small squid beak that may have been eaten by the fish. The stomach contents suggest that *Tasmacetus* had been feeding on the bottom, in fairly deep water.

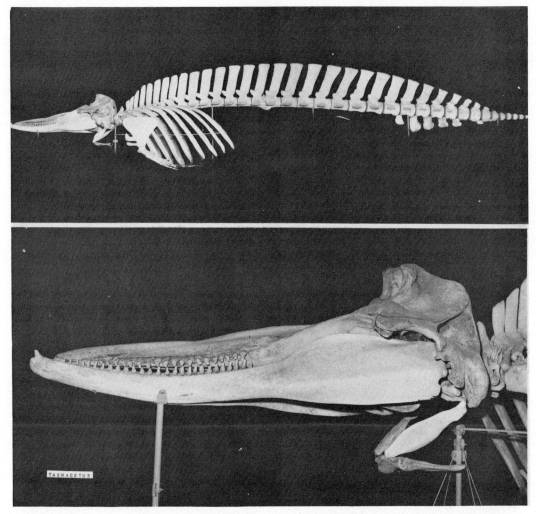

Shepherd's beaked whale (*Tasmacetus shepherdi*), photos by Eldon V. Burkett through Wanganui Public Museum.

CETACEA; ZIPHIIDAE; *Genus INDOPACETUS Moore, 1968*

Indo-Pacific Beaked Whale

The single species, *I. pacificus,* is known only by two skulls, one found at Mackay on the east coast of Queensland and the other at Danane on the east coast of Somalia (Moore 1968).

The first skull to be found was nearly 122 cm long and was thought to have come from a fully mature animal, about 7.6 meters in length (Ellis 1980). Moore (1968, 1972) listed the following cranial characters that distinguish the genus: (1) the alveoli of the developed teeth are a single pair, apical on the mandible, and in an old adult male become progressively at least as shallow as 30 mm; (2) the frontal bones occupy an area of the synvertex of the skull approximating or exceeding that occupied by the nasal bones; (3) there is almost no posterior process of the premaxillary crest extending posteriorly on the synvertex between the nasal and maxillary bones or between the frontal and maxillary bones; (4) in the lateral extension of the maxillary bone over the orbit there is a deep groove about half as long as the orbit; (5) at about the mid-length of the beak there is a swelling caused by the lateral margins proceeding forward a short distance without convergence, or even with a little divergence and then convergence again; (6) fusion of a considerable length of the mesethmoid bone to both premaxillary rims of the mesorostral canal; and (7) proliferation of bone from the vomer (and distally the premaxillae) into the mesorostral canal at the onset of adulthood is absent or minimal in the one adult specimen.

CETACEA; ZIPHIIDAE; *Genus HYPEROODON Lacépède, 1804*

Bottlenosed Whales

There are two species (Rice 1977; Ellis 1980):

H. ampullatus, the North Atlantic from Davis Strait and Novaya Zemlya to Rhode Island and the English Channel;

H. planifrons, known from waters off Australia, New

Bottlenosed whale (*Hyperoodon ampullatus*), photo from *Endeavour*.

Zealand, Brazil, Argentina, Tierra del Fuego, the Falkland Islands, South Georgia, the South Orkney Islands, South Africa, and the Pacific and Indian ocean sectors of Antarctica.

Moore (1968) placed *H. planifrons* in its own subgenus, *Frasercetus*.

Maximum head and body length is 9.1 meters in males and 7.5 meters in females. The dorsal fin is at least 30 cm high and is distinctly hooked (Leatherwood, Caldwell, and Winn 1976). One female, 6 meters long, weighed 2,500 kg. Coloration becomes lighter with age. Calves are grayish brown to black, immature animals are often spotted yellowish brown and white, and old individuals may be completely yellowish white. There are usually two large teeth in the lower jaw of young males, but in older males one or both may be lost. In females the functional teeth are smaller or do not emerge at all. Rows of vestigial teeth are often present in the lower and upper jaws. In females and young males the forehead slopes rather smoothly into the beak (Ellis 1980). In older males the forehead rises abruptly from the beak and may become so bulbous that, when viewed in profile, it protrudes forward at the top. This development results in part from the enlargement of crests on the maxillary bones. The forehead encloses a huge melon, filled with oil that is much the same in appearance, and probably in function, as that of the spermaceti of *Physeter*.

Bottlenosed whales generally stay well out to sea. They seem to occur mainly in cooler waters and may approach the polar ice packs during the summer. Substantive natural history data are available only for *H. ampullatus*. This species is usually found in waters over 1,000 meters deep. During the spring and early summer it migrates to the northern parts of its geographic range, and in the late summer and autumn it returns south. It dives suddenly and with great speed. Although dives usually last about 30 to 45 minutes, there are reports that this whale has remained underwater for 2 hours,

which, if accurate, would be a record for the Cetacea. One individual, after being harpooned, is said to have dived vertically and pulled out 1,000 meters of line. Feeding is probably done at great depths. The diet consists mainly of squid, and also includes fish and bottom-dwelling echinoderms (Benjaminsen and Christensen 1979; Leatherwood, Caldwell, and Winn 1976; Rice 1967).

A variety of sounds has been reported for *H. ampullatus*, most notably series of clicks and whistles, such as are known to be used by certain other odontocetes for, respectively, echolocation and communication (Winn, Perkins, and Winn 1970). Groups usually consist of two to four individuals, generally of the same age and sex. Larger groups are sometimes seen and usually contain mature animals of both sexes. Solitary whales are common and are often young individuals. Groups are said not to abandon an injured member (Benjaminsen and Christensen 1979).

Both the mating and calving seasons of *H. ampullatus* peak in April. Females probably give birth every 2 years. The gestation period is thought to last 12 months. The single newborn is about 300 to 330 cm long, and lactation may last about a year. Sexual maturity is attained at the age of 8–12 years in females and 7–11 years in males. Maximum longevity is at least 37 years (Benjaminsen and Christensen 1979).

A large bottlenosed whale can yield up to 200 kg of spermaceti oil, the uses of which are given in the account of *Physeter*, as well as about 2,000 kg of blubber oil (Ellis 1980). Following the decline of the bowhead whale (*Balaena mysticetus*) in the late nineteenth century, there was increasing commercial interest in *H. ampullatus*, especially in Norway. From 1882 to 1920, approximately 50,000 individuals were taken off northwestern Europe. The annual kill fell from 2,000–3,000 in the 1890s to 20–100 in the 1920s. After some years of reduced hunting, and perhaps of recovery by the whale population, the fishery again intensified. The annual kill peaked at around 700 individuals in 1965, but then declined to near zero. The population in the eastern Atlantic is estimated to have originally contained 40,000 to 100,000 whales (Benjaminsen and Christensen 1979; U.S. National Marine Fisheries Service 1978; Ellis 1980). The current number is much lower, and the western Atlantic population is also small, but exact status is unknown. *H. ampullatus* is classified as vulnerable by the IUCN (1976).

CETACEA; ZIPHIIDAE; *Genus MESOPLODON Gervais, 1850*

beaked Whales

There are 3 subgenera and 11 species (Rice 1977, 1978*d*; Moore 1968; Ellis 1980; Goodall 1978):

subgenus *Mesoplodon*

M. hectori, known by 11 specimens from southern California, Tierra del Fuego, the Falkland Islands, South Africa, Tasmania, and New Zealand;

M. mirus, temperate waters from Nova Scotia and Florida to the British Isles, and off South Africa;

M. europaeus, western North Atlantic from New York to the West Indies, one record from the English Channel;

M. ginkgodens, known by 10 specimens from Sri Lanka, Taiwan, Japan, and southern California;

M. grayi, known from the Netherlands, South Africa, southern Australia, New Zealand, the Chatham Islands, Chile, Tierra del Fuego, Argentina, and the Falkland Islands;

M. carlhubbsi, temperate waters of the North Pacific from Japan to British Columbia and California;

M. bowdoini, known from Western Australia, Victoria, Tasmania, New Zealand, Campbell Island, and Kerguelen Island;

M. stejnegeri, subarctic waters from the Bering Sea to Japan and Oregon;

M. bidens, cool temperate waters of the North Atlantic from Newfoundland and Massachusetts to southern Norway and the Bay of Biscay;

subgenus *Dolichodon*

M. layardii, known from South Africa, Australia, Tasmania, New Zealand, Tierra del Fuego, Uruguay, and the Falkland Islands;

subgenus *Dioplodon*

M. densirostris, tropical and warm temperate waters of all oceans.

Hall (1981) used the generic name *Micropteron* Eschricht, 1849 in place of *Mesoplodon.*

Head and body length is 3 to 7 meters, pectoral fin length is 20 to 70 cm, dorsal fin height is about 15 to 20 cm, and expanse of the tail flukes is about 100 cm. Color is variable, but is usually slaty black to bluish black above and paler below. Only two teeth become well developed, one on each side of the lower jaw, and these may be lost in old age. In females the functional teeth are much smaller than those of males and often do not erupt above the gums. There may be small nonfunctional teeth in both jaws. A row of such vestigial teeth, on each side of the upper jaw, seems to be characteristic of *M. grayi.*

The structure of the pair of functional teeth in males varies remarkably (Ellis 1980; Rice 1978*d*). They can be briefly described as follows: *M. hectori,* shaped like laterally flattened triangles, located at tip of jaw; *M. mirus,* small and angled slightly forward, located at tip of jaw; *M. europaeus,* shaped like laterally flattened triangles, located one-third of the way from tip to apex of mouth; *M. ginkgodens,* shaped in profile like the leaf of the ginkgo tree, located one-third of the

True's beaked whales (*Mesoplodon mirus*), photo from National Geographic Society of painting by Else Bostelmann. Inset: *M.* sp., photo from American Museum of Natural History.

way from tip to apex of mouth; *M. grayi,* shaped in profile like a flattened onion, located in middle of mouth; *M. carlhubbsi,* large and straight sided, located near front of mouth; *M. bowdoini,* large and flattened, located in raised sockets near front of mouth; *M. stejnegeri,* large and tusklike, located in back of mouth; *M. bidens,* small and pointed slightly to rear, located near middle of mouth; *M. layardii,* shaped somewhat like the tusks of a boar (Suidae), come out of mouth and curve over upper jaw; and *M. densirostris,* set at a point where the lower jaw becomes extremely high and broad, so the teeth are exposed and extend above the upper jaw.

Mesoplodon is apparently a pelagic genus, staying mainly in deep waters far from shore. The diet includes squid, other cephalopods, and fish. Rice (1978*d*) observed a group of 10 to 12 *M. densirostris* dive, and then waited 45 minutes but did not see them come back up. Most other sightings of the genus that he has made have been of schools of 2 to 6 individuals. There is one record of a stranding of 28 *M. grayi* (Ellis 1980). Usually, however, *Mesoplodon* is encountered alone or in small groups (Leatherwood, Caldwell, and Winn 1976; Rice 1967). Reproductive data are very limited (Rice 1978*d*; Ellis 1980). In *M. bidens,* mating usually takes place in late winter and spring, gestation lasts about a year, the newborn is about 213 cm long, and weaning occurs after about a year of life. A female *M. hectori* and a young calf stranded near San Diego in May. A female *M. europaeus* and a young calf stranded in Florida in October. Very small numbers of *Mesoplodon* are taken by commercial fisheries (Mitchell 1975*b*).

CETACEA; *Family ESCHRICHTIDAE; Genus ESCHRICHTIUS Gray, 1864*

Gray Whale

The single known genus and species, *Eschrichtius robustus,* occurs in coastal waters from the Sea of Okhotsk to southern Korea and Japan, and from the Chukchi and Beaufort seas to the Gulf of California (Rice 1977). Other populations apparently once existed in the North Atlantic, there being records from the east coast of the United States, western Europe, the Baltic Sea, and possibly Iceland (Mitchell and Mead 1977; Fraser 1970). Some of the Atlantic records are based on material that is subfossil, but from historical times. Geologically, the family Eschrichtidae is known only from the Recent. Hall (1981) used the name *E. gibbosus* in place of *E. robustus.*

At sexual maturity head and body length is about 11.1 meters, at physical maturity the length averages 13 meters in males and 14.1 meters in females, and the maximum reliably recorded length is 14.3 meters in males and 15 meters in females (Rice and Wolman 1971). Pectoral fin length is approximately 200 cm and expanse of the tail flukes is about 300 cm. There is no dorsal fin, but there are 8 to 14 low humps along the midline of the lower back. Adults weigh 20,000 to 37,000 kg. Coloration is black or slaty gray, with many white spots and skin blotches, some being discolored patches of skin and others being areas of white barnacles.

The snout is high and rigid, and the throat has two or three (rarely four) short, shallow, curved furrows. *Eschrichtius* is a baleen whale and has no teeth. The baleen is yellowish white in color, and is arranged in series of 130 to 180 plates on each side of the mouth. The largest plates are 34 to 45 cm long. The two rows of plates do not meet in front, as they usually do in the family Balaenopteridae. There are 56 vertebrae in *Eschrichtius,* and the neck vertebrae are separate.

The gray whale is found primarily in shallow water, and generally remains closer to shore than any other large cetacean. The eastern Pacific population makes an annual migration of over 18,000 km. From late May to early October this population concentrates in the shallow waters of the northern and western Bering Sea, the Chukchi Sea (between northern Alaska and Siberia), and the Beaufort Sea (off northeastern Alaska). Some individuals, however, spend the summer farther south, scattered along the coast of Oregon and northern California. From October to January the main part of the population moves down the east side of the Bering Sea, goes through Unimak Pass in the Aleutians, and proceeds down the west coast of North America. In January and February most whales are along the west coast of Baja California and on the eastern side of the Gulf of California off Sonora and Sinaloa. In these areas are five bays and lagoons, within which the young are born. From late February to June the population migrates northward to its summering sites in the Arctic (Rice and Wolman 1971; Wolman and Rice 1979).

During the southward phase of the migration of the eastern Pacific population, the animals are more concentrated and may move somewhat nearer to the shore than during the northward phase. For example, most individuals pass San Diego during a six-week period from late December to early February, and in certain past years 95 percent of the population traveled from 3 to 5 km offshore. At other points the whales sometimes move within 1 km of land. The migration is therefore witnessed by large numbers of people. Unfortunately, perhaps, increasing boat traffic seems to have forced the whales to stay farther from shore (Wolman and Rice 1979). Observations by Leatherwood (1974) indicated that half the population passed more than 64 km off San Diego.

Considerably less is known about the western Pacific population, which has been reduced to very small numbers. Apparently it spent the summer in the Sea of Okhotsk and then migrated through the Sea of Japan to wintering sites off the southern coast of Korea and in the Inland Sea of Japan. The extinct population of the eastern Atlantic probably summered in the Baltic Sea and wintered along the Atlantic and Mediterranean coasts of southern Europe and North Africa (Rice and Wolman 1971; Omura 1974).

The normal swimming speed of the gray whale is 7 to 9 km per hr (Rice and Wolman 1971), but when pursued it may reach about 13 km per hr. Migrating individuals usually submerge for four or five minutes and then surface to blow three to five times. The spout rises to a height of about 300 cm. The tail flukes usually appear above the water just before a deep dive but not before a shallow dive. The back is not arched before diving, as in the humpback whale (*Megaptera*). The gray whale often seems to play in heavy surf and in shallow water along the shore, sometimes throwing itself clear of the water and falling back on its back or side. It occasionally becomes stranded in less than 1 meter of water, but refloats on the next tide, without injury. Visual stimuli appear to be involved in orientation during migration. The habit of spyhopping—lifting the head vertically out of the water and appearing to look around—is commonly observed in *Eschrichtius* and may be involved in social interaction (Samaras 1974). A variety of sounds has been reported, including grunts, groans, rumbles, whistles, and clicks. The clicks seem to be of a frequency too low for use in precise echolocation, but may possibly assist in sonar navigation or the detection of dense concentrations of food organisms (Fish, Sumich, and Lingle 1974).

Feeding takes place primarily on the bottom, with individuals apparently sometimes plowing their heads sideways through the mud or sand to stir up prey (Ellis 1980). Water and organisms are sucked into the mouth, and then the water is forced out, leaving the food trapped within the baleen. The

Gray whales (*Eschrichtius robustus*), painting by Richard Ellis.

diet consists mainly of gammaridean amphipods (small crustaceans), especially *Ampelisca macrocephala,* a creature about 25 mm long and found on sandy bottoms at depths of 5 to 300 meters (Rice and Wolman 1971). Other crustaceans, certain mollusks and worms, and small fish are also eaten. A captive yearling female consumed and apparently thrived on 900 kg of squid per day (Ray and Schevill 1974). Available evidence suggests that there is little or no feeding, except when the animals are in the northern part of their annual migratory route, and that fasting lasts up to six months (Rice 1978b). Individuals moving south past San Francisco lost 11 to 39 percent of their body weight by the time they moved north past the city (Rice and Wolman 1971).

Aggregations of up to 150 individuals have been seen in the arctic feeding waters (Rice 1967), but *Eschrichtius* is not a particularly social cetacean. It usually migrates alone or in groups of 2 or 3, though sometimes up to 16 travel together. Moreover, there is segregation by age and sex during migrations. Generally females precede males and adults precede immature animals. Newly pregnant females are the first to leave on the northward migration (Rice and Wolman 1971; Rice 1978b). Segregation also has been reported at the wintering sites, with calving and nursing females remaining well within protected lagoons and males positioning themselves at the entrances (Ellis 1980). Individuals have been observed to aid others that are injured or giving birth by pushing them to the surface so they can breathe. Females are reportedly very protective of their young, even to the point of attacking whaling boats.

Females have a 2-year reproductive cycle. Most enter estrus and mate during a 3-week period in late November and early December, while still migrating south, but some do not mate until in the wintering lagoons or even on the northward migration. Births occur mainly from late December to early February. Gestation lasts about 13 months, lactation lasts about 7 months, and there are then 3 to 4 months of anestrus. The single newborn averages about 490 cm in length and 500 kg in weight. It is weaned around August in the summer feeding area, after it has grown to a length of 850 cm. The average age of puberty is 8 (5–11) years in both sexes, and full physical maturity comes at 40 years (Rice and Wolman 1971; Rice 1978b). Maximum estimated longevity is 70 years (Haley 1978).

Like various other large cetaceans, the gray whale has been killed by people for its oil, meat, hide, and baleen. It was hunted in ancient times by the native people of northwestern North America and eastern Siberia, and probably also by Europeans and Japanese. Because of its migrations close to shore, it was comparatively simple to locate and secure. The European population may have disappeared around 500 A.D., though there is a possible record for Iceland in the early seventeenth century. The western Atlantic population apparently survived until the early 1700s (Rice and Wolman 1971; Mitchell and Mead 1977; Fraser 1970).

The eastern Pacific population seems to have always been rather small. The branch that bred in the Inland Sea of Japan probably contained fewer than 1,000 individuals. It was gone by 1910, the last survivors perhaps being driven away by increasing boat traffic and industrialization (Omura 1974). The branch that bred off southern Korea may have numbered

Gray whale (*Eschrichtius robustus*): Top, photo by David Withrow; Bottom, photo by David Rugh.

1,000 to 1,500, when modern whaling began in that area around the turn of the century. By 1933 the Korean population appeared to be extinct, but there were 67 known kills from 1948 to 1966, and sightings as late as 1974 (Rice and Wolman 1971; Brownell and Chun 1977; Wolman and Rice 1979).

The wintering lagoons of the eastern Pacific population were discovered by American whalers in 1846. Shore-whaling stations were established in the area, and from 1846 to 1874 the known kill was 10,800 animals. By about the turn of the century, regular shore whaling stopped, and the population seemed extinct, but there may still have been several thousand individuals left. Shortly thereafter, factory ships came into use for the hunting of the gray whale. From 1921 to

1947, 1,153 individuals are known to have been killed, mainly by Norwegian, Japanese, and Russian vessels. Since 1946 the species has received protection under the International Whaling Convention, except that about 160 individuals have been taken legally each year by Siberian Eskimos. The eastern Pacific population evidently increased steadily in response to protection, and currently is estimated to contain 16,000 whales, about the same number as before exploitation began. The main current problem is disturbance of the animals and their habitat, especially in the calving lagoons, by industrialization, shipping, and even well-meaning tourists, who follow the whales in motorboats (Rice and Wolman 1971; Wolman and Rice 1979; Rice 1978b; Storro-Patterson 1977; U.S. National Marine Fisheries Service

1981). The gray whale is listed as endangered by the USDI (1980) and is on appendix 1 of the CITES.

CETACEA; *Family BALAENOPTERIDAE*

Rorquals

This family of two Recent genera and six species occurs in all the oceans and adjoining seas of the world.

Head and body length of sexually mature adults is 6.7 to 31 meters, and weight is estimated to range up to 160,000 kg (Rice 1967). In each species the females are larger than the males. The pectoral fin is tapering and the dorsal fin is located on the posterior part of the back. The rostrum is broad and flat, and the sides of the lower jaw bow outward. The Balaenopteridae are baleen whales and have no teeth past the embryonic stage. There is one row of baleen plates on each side of the top of the mouth, the two rows usually joining anteriorly. Each row may consist of more than 300 plates, which range in vertical length from 20 to 101 cm. The vertebrae number 42 to 65; the neck vertebrae are generally separate.

The term "rorqual" is Norwegian and means "tube whale" or "furrow whale" (Ellis 1980). It refers to the longitudinal folds or pleats, 10 to 100 in number and 25 to 50 mm deep, that are present on the throat, chest, and, in some species, belly. These furrows allow the throat to expand enormously, and so greatly increase the amount of material that the whale can take in when feeding.

The Balaenopteridae are distinguished from the Balaenidae by the presence of the throat and chest furrows, the more elongate and streamlined body form, the relatively smaller head, the more tapering pectoral fin, the softer and less massive tongue, and the usually shorter and less flexible baleen. Unlike the genera *Balaena* and *Eschrichtius,* the Balaenopteridae have a dorsal fin.

The feeding habits and dietary preferences of the Balaenopteridae are associated with the characteristics of the head, mouth, tongue, and baleen plates. The sei whale (*Balaenoptera borealis*) has finely meshed baleen fringes, and sometimes uses the feeding method known as skimming, as do the Balaenidae (Ellis 1980). In skimming the animal swims slowly through swarms of tiny food organisms, with its mouth open and its head above water to just behind the

Baleen from a whale (*Balaenoptera* sp.), photo by P. F. Wright of specimen in U.S. National Museum.

nostrils. When a mouthful of organisms has been filtered from the water by the baleen plates, the whale dives, closes its mouth, and swallows the food. The other rorquals use the feeding method known as swallowing. The animal may turn on its side and often has part of its head above the water. After taking in a great amount of food organisms and water, the whale forces the water out with its tongue, leaving the food trapped in the baleen.

All species are migratory, most giving birth in warm areas and feeding predominantly in colder waters. The diet depends partly on location. The most important foods are shrimplike creatures of the family Euphausiidae (mainly the genera *Euphausia, Thysanoessa,* and *Meganyctiphanes*) and copepods (*Calanus* and *Metrida*). These crustaceans form part of the zooplankton, the mass of minute floating and weak-swimming animals found near the surface of the ocean. In colder areas there are fewer kinds of plankton, but the numbers of each kind are greater. During the summer, shoals of these organisms concentrate in the upper layers of polar waters, and the whales are attracted to them. In the Antarctic the diet of the minke (*B. acutorostrata*), fin (*B. physalus*), blue (*B. musculus*), and humpback (*Megaptera*) whales consists mainly of the shrimp *Euphausia superba* (commonly called krill). Outside of the Antarctic the blue whale feeds predominantly on other euphausiids, but the fin, minke, and humpback take substantial amounts of copepods and fish. The fine, fleecy baleen of the sei whale is suited for trapping smaller organisms than those eaten by other rorquals. The sei feeds on *Euphausia superba* in the Antarctic, but also on the amphipod *Parathemisto gandichaudi,* and in other areas it seems to prefer copepods. In contrast, the Bryde's whale (*B. edeni*) has coarse baleen, does not enter the Antarctic, and feeds mainly on fish (Ellis 1980).

All rorquals have been hunted by people for oil, meat, baleen, and other products. The most intensive exploitation has taken place in antarctic waters, where large numbers of most species gather for part of the year to feed on plankton. The whaling activity in this region followed a systematic pattern (Gulland 1974, 1976; D. G. Chapman 1974; McHugh 1974). The first species to be depleted was the humpback, which has a localized, coastal distribution and could be easily taken from the island whaling stations established in the first years of the 20th century. The introduction in the 1920s of factory ships with slip sternways, which allowed whalers to operate independently of land facilities, led to increased killing of other species.

The blue, fin, and sei whales, in that order, become smaller in size and concentrate farther from the antarctic ice pack. Exploitation proceeded in this same order. The blue whale, the most valuable species in relation to individual hunting effort, received initial maximum attention and was evidently in decline by the late 1930s. Whalers then centered their activity somewhat farther north, and the fin whale became the mainstay of the industry for the next 30 years. Meanwhile, efforts at protection were largely ineffective. The International Whaling Commission established quotas on the basis of "blue whale units." Each unit equaled 1 blue whale, 2 fin whales, 2½ humpbacks, or 6 sei whales. Since quotas were assigned by units rather than by species, it remained more profitable to hunt the larger species. By the 1960s, however, the blue whale seemed nearly extinct and the fin whale greatly reduced in numbers, and attention was then directed still farther north to the smaller sei whale. Another factor in the switch to the sei was a lower demand for whale oil, along with an increase in the value of the meat. The sei has less blubber and relatively more and better meat than the larger species. Concern for the future of the sei, and the setting of lower quotas specifically for it and the larger whales, contributed to an increased kill of the minke whale,

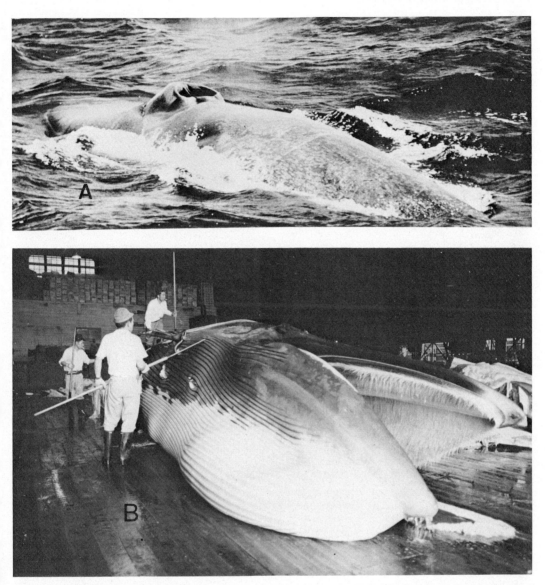

A. Blue whale (*Balaenoptera musculus*), photo by K. C. Balcomb. B. Sei whale (*B. borealis*), photo by Warren J. Houck.

smallest of the rorquals, in the 1970s. The minke currently remains the most heavily hunted of whales, and there also has been a recent rise in the take of Bryde's whale, a species considerably larger than the minke but one that does not migrate to the Antarctic.

The geological range of the Balaenopteridae is late Miocene to Pleistocene in North America, early Pliocene to Pleistocene in Europe, Miocene in Asia, and Recent in all oceans (Rice 1967).

CETACEA; BALAENOPTERIDAE; **Genus**
BALAENOPTERA *Lacépède, 1804*

Minke, Bryde's, Sei, Fin, and Blue Whales

There are five species (Rice 1977):

B. acutorostrata (minke whale), all oceans and adjoining seas;

B. edeni (Bryde's whale), tropical and warm temperate waters of the Atlantic, Indian, and Pacific oceans, and adjoining seas;

B. borealis (sei whale), all oceans and adjoining seas except in tropical and polar regions;

B. physalus (fin whale), all oceans and adjoining seas, but rare in tropical waters and among pack ice;

B. musculus (blue whale), all oceans and adjoining seas.

In addition to the characters set forth in the familial account, *Balaenoptera* is distinguished by a relatively slender body, a compressed tail stock that abruptly joins the flukes, a pointed snout, a relatively short pectoral fin (usually 8 to 11 percent of head and body length), and a usually sickle-shaped dorsal fin. Additional information is provided separately for each species.

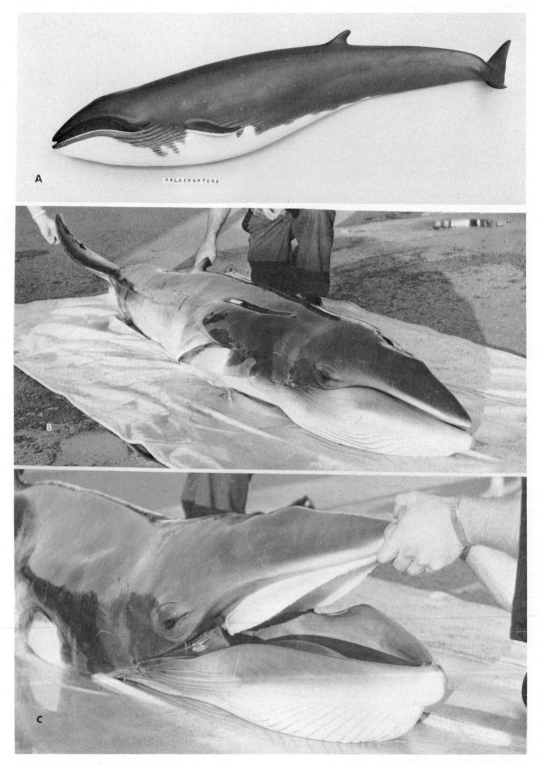

A. Fin whale (*Balaenoptera physalus*), photo from Field Museum of Natural History. B & C. Minke whale (*B. acutorostrata*), photos from Marineland of the Pacific.

Balaenoptera acutorostrata.   Head and body length at sexual maturity averages about 7.3 meters in males and 7.9 meters in females. At physical maturity the respective averages are about 8.3 and 8.8 meters (Ohsumi and Masaki 1975; Ohsumi, Masaki, and Kawamura 1970; Mitchell 1975*b*). Maximum size is around 10.2 meters in length and 10,000 kg in weight. As in all rorquals, females average slightly larger than males. The upper parts are dark gray to black and the underparts are white. There is also a white patch on the pectoral fin. The rostrum is markedly triangular in shape and is shorter than that of any other rorqual. The dorsal fin is high and strongly curved back. There are 50 to 70 ventral grooves, and these do not extend as far as the navel. On each side of the upper part of the mouth is a row of about 300 baleen plates, mostly yellowish white in color (Ellis 1980; Leatherwood, Caldwell, and Winn 1976).

The minke whale is not as coastal an animal as *Eschrichtius,* but seldom goes farther than 160 km from land and often enters bays and estuaries. It also moves farther into polar ice fields than any other rorqual. There are a number of discrete populations, at least some of which make extensive migrations. The populations of the Northern Hemisphere winter in tropical waters, but during the spring and summer they spread northward, often beyond the Arctic Circle. While these populations are in the north, those of the Southern Hemisphere are in the tropics for the austral winter. When the northern populations return to the tropics in the boreal autumn, the southern populations set out in the direction of Antarctica (Ellis 1980; Mitchell 1978; U.S. National Marine Fisheries Service 1978).

The minke whale is a fast swimmer and is the most acrobatic member of its genus. It has a tendency to approach ships and often leaps completely out of the water (Leatherwood, Caldwell, and Winn 1976; Mitchell 1978). Its spout, however, is invisible. It has a wide variety of vocalizations, including a series of "thumps" that may be distinctive for each individual (Thompson, Winn, and Perkins 1979). In the Antarctic the diet consists almost exclusively of plankton (see familial account), but in the Northern Hemisphere *B. acutorostrata* also eats squid, herring, cod, sardines, and various other kinds of small fish (Ellis 1980).

This rorqual is generally seen alone or in groups of two to four individuals, but several hundred sometimes gather in the vicinity of abundant food (Leatherwood, Caldwell, and Winn 1976; Mitchell 1975*a*). There is some segregation by age and sex off Newfoundland, with adult males evidently remaining farther from shore than females and young (Mitchell and Kozicki 1975).

There is some question about the reproductive cycle. Mitchell (1978) gave the calving interval as only 1 year, and Mitchell and Kozicki (1975) found over 85 percent of mature females off Newfoundland to be pregnant. The U.S. National Marine Fisheries Service (1978), however, stated that females are now known to give birth once every 2 years. The breeding period is extensive (Mitchell 1975*a*). In the North Atlantic mating occurs from December to May and calving from October to March. In the North Pacific reproduction occurs throughout the year, with mating peaks in January and June and birth peaks in December and June. The gestation period is 10 to 10½ months. There is normally a single offspring, about 280 cm long. Lactation is thought to last less than 6 months. Sexual maturity is attained at an age of 7 to 8 years, and physical maturity when males are 18–20 and females 20–22 years old (Ohsumi and Masaki 1975; Ohsumi, Masaki, and Kawamura 1970). Maximum longevity is estimated at 47 years (Haley 1978).

The minke whale has been hunted in Norway since the Middle Ages, and was regularly taken by local fisheries there and in some other countries from the 1920s to 1970s, but was

long not considered of major commercial importance (Mitchell 1975*b*). With the decline of the larger rorquals in the 1960s, the minke became subject to more intensive hunting, especially in the Antarctic. In the 1976/77 season, the world-wide kill peaked at 12,398 individuals, surpassing the take of any other species (Committee for Whaling Statistics 1980). The quota set by the International Whaling Commission for the 1981/82 season is 12,017 minkes out of a total of 14,070 whales of all kinds, not including those to be taken by aboriginal peoples (*Marine Mammal News* 7, no. 7 [1981]:1–3). The U.S. National Marine Fisheries Service (1981) reported the current population estimate of minke whales to be 200,000 in the Southern Hemisphere and 120,000 in the North Atlantic. Ohsumi (1979), however, gave an estimate of 416,700 for the Southern Hemisphere, and suggested that this figure represented an eightfold increase since the 1930s, because the minke whale had multiplied in the absence of competition from the larger rorquals.

Balaenoptera edeni.   Head and body length at sexual maturity averages 12.2 meters in males and 12.5 meters in females (U.S. National Marine Fisheries Service 1978). Maximum head and body length is 14 meters, and dorsal fin height is up to 46 cm (Leatherwood, Caldwell, and Winn 1976). The coloration is dark blue gray above and paler below. *B. edeni* resembles *B. borealis,* but can be distinguished by the presence of two dorsal ridges running from the tip of the snout to beside the blowhole, in addition to another ridge down the middle of the snout. The dorsal fin is strongly sickle shaped. There are approximately 45 ventral grooves and these extend to the navel. On each side of the mouth is a row of about 300 baleen plates, and often the rows do not join anteriorly. The plates are relatively short; the front ones are whitish and the rear ones are dark (Ellis 1980).

Certain populations inhabit waters of high productivity near the shore, while others are mainly pelagic. Unlike other rorquals, there are no large-scale movements toward the poles in the spring. Tropical populations may be sedentary, but those in temperate waters seem to make limited migrations in response to shifting food sources. *B. edeni* has a tendency to approach ships and is a relatively deep diver. Its diet consists mainly of small schooling fish, such as anchovies and sardines, but in some areas it also feeds extensively on shrimplike crustaceans of the family Euphausiidae (Leatherwood, Caldwell, and Winn 1976; Ellis 1980; U.S. National Marine Fisheries Service 1978).

Bryde's whale is often seen in groups, usually of 2 to 10 individuals but sometimes of over 100. The reproductive season is usually winter, but in some regions, as in the waters off South Africa, it extends throughout the year. Females generally give birth every other year to a single offspring about 335 cm long. The age of sexual maturity is 7–10 years and that of physical maturity is 15–18 years. Some individuals are estimated to have lived up to 72 years (Mitchell 1978; U.S. National Marine Fisheries Service 1978, 1981; Ellis 1980; Haley 1978).

Bryde's whale has long been hunted along the coasts of Japan, South Africa, and Baja California, but did not receive major commercial attention until the 1970s, subsequent to the decline of most other rorquals. The maximum recorded kill was 1,882 individuals in the 1973/74 season (Committee for Whaling Statistics 1980). The 1981/82 quota set by the International Whaling Commission is 1,392 (*Marine Mammal News* 7, no. 7 [1981]:1–3). The population of the North Pacific is estimated to have contained 20,000 individuals originally and now to contain 16,000 (U.S. National Marine Fisheries Service 1978, 1981). Estimates for other regions are not available.

Sei whale (*Balaenoptera borealis*), painting by Richard Ellis.

Balaenoptera borealis. Head and body length at sexual maturity averages 13.1 meters in males and 13.7 meters in females (U.S. National Marine Fisheries Service 1978). The biggest specimen on record, a female, was 20 meters long, but most individuals now measure 12.2 to 15.2 meters. The general coloration is dark steel gray, with irregular white markings ventrally. The dorsal fin is set farther back than it is on *B. acutorostrata,* and is 25 to 61 cm in height. There are 38 to 56 ventral grooves, and these end well before the navel. On each side of the upper part of the mouth is a row of 300 to 380 baleen plates. The color of the plates is ashy black, but the fine inner bristles are whitish (Ellis 1980; Leatherwood, Caldwell, and Winn 1976).

The sei whale is a pelagic species, normally occurring far from shore. Like most rorquals, it feeds in temperate and subpolar regions in summer and migrates to subtropical waters for the winter. It does not, however, go as far into the Antarctic as do the blue and fin whales. It is one of the fastest cetaceans, reportedly being able to reach speeds of up to 50 km per hr. It is not usually a deep diver. Periods of submergence generally last 5 to 10 minutes. When feeding, the sei whale remains near the surface, often twisting on its side as it swims through swarms of prey. Unlike other rorquals, it obtains food by the skimming method (see familial account). It generally eats smaller organisms than do the other rorquals, mostly copepods and amphipods, but also takes euphausiids and small fish. The daily consumption per whale is approximately 900 kg (Mitchell 1978; Ellis 1980; Banfield 1974; Kawamura 1973, 1974). *B. borealis* is not known to use echolocation to search for prey, but has been heard to emit a "sonic burst of 7–10 metallic pulses" (Thompson, Winn, and Perkins 1979).

Groups usually consist of two to five individuals, but sometimes thousands may gather in the vicinity of abundant food (Leatherwood, Caldwell, and Winn 1976; Ellis 1980). According to Banfield (1974), there is a protracted mating season that extends from November to February in the Northern Hemisphere and from May to July in the Southern Hemi-

sphere. Females generally give birth every other year, but Masaki (1978) reported that the percentage of pregnant females being taken has recently increased, perhaps as a natural compensatory response to human exploitation. The gestation period was reported as 10½ months by Masaki (1976), 11½ months by Frazer (1973), and 12 months by Ellis (1980). There are rare cases of multiple fetuses, but normally a single offspring, about 450 cm long, is produced. Lactation lasts from 6 to 7 months. Sexual maturity is attained at around 10 years of age and physical maturity at 25 years (Mitchell 1978). Some individuals are estimated to have lived up to 74 years (Haley 1978).

The sei whale has been regularly taken by people since the 1860s, but did not achieve major commercial importance for another 100 years. The annual kill in antarctic waters did not exceed 1,000 individuals until 1950. There was a pronounced increase in the kill during the 1950s and 1960s, in conjunction with a declining harvest of blue and fin whales (see familial account). The world-wide take of the sei whale peaked at 25,454 in the 1964/65 season (McHugh 1974). The take then steadily dropped to only 150 in 1978/79 (Committee for Whaling Statistics 1980). The quota set by the International Whaling Commission for the 1981/82 season is 100 (*Marine Mammal News* 7, no. 7 [1981]:1–3). The original number of sexually mature sei whales, exclusive of those in the North Atlantic, is estimated at 200,000, and the current number is estimated at 57,000. The total population, including immature animals, would be about 50 percent larger (U.S. National Marine Fisheries Service 1978). *B. borealis* is listed as endangered by the USDI (1980). The North Pacific stock, and the stock from 0° to 70° E and from the Equator to Antarctica, are on appendix 1 of the CITES. All other stocks, like most cetaceans, are on appendix 2 of the CITES.

Balaenoptera physalus. Head and body length at sexual maturity averages 17.7 meters in males and 18.3 meters in females. At physical maturity the respective averages are about 19 and 20 meters. Maximum known size is 25 meters in length and over 45,000 kg in weight. The strongly curved dorsal fin is up to 61 cm high, and the expanse of the tail flukes is about 25 percent of the head and body length. The general coloration is brownish gray above and white below. The pattern, however, is asymmetrical, especially in that the lower jaw is white on the right and dark on the left. *B. physalus* is slimmer than *B. musculus;* if two individuals, one of each species, are the same length, the *B. musculus* will be much heavier. Indeed, even though *B. physalus* is the second longest cetacean, it often weighs less than the thickset *Balaena* and *Physeter*. The rostrum of *B. physalus* is sharply pointed. There is an average of 85 ventral grooves, and these terminate at the level of the navel. On each side of the upper part of the mouth is a row of 350 to 400 baleen plates. Those plates in the forward third of the right row are whitish, but those in the rear two-thirds of the right row, and all of those in the left row, are dark grayish blue (Ellis 1980; Leatherwood, Caldwell, and Winn 1976; Mitchell 1978; U.S. National Marine Fisheries Service 1978).

The fin whale is a pelagic species, seldom found in water less than 200 meters deep. Numerous discrete populations have been identified, most of which are known to be highly migratory. In the spring and early summer, populations generally move into cold temperate and polar waters to feed, and in the autumn they return to warm temperate and tropical regions. Since the seasons in the northern half of the world are opposite those of the southern half, the fin whale populations of one hemisphere do not meet those of the other in equatorial waters (Ellis 1980; D. G. Chapman 1974; Banfield 1974).

The fin whale is among the fastest of cetaceans, being able to sustain a speed of around 37 km per hr. It is probably a deeper diver than the blue and sei whales, sometimes reaching depths of at least 230 meters and remaining underwater for up to 15 minutes. It occasionally leaps completely out of the water. A swallowing method of feeding is employed, during which the throat becomes distended to nearly double normal diameter (see familial account). There is very little, if any, feeding during the fall and winter, when the whales are in lower latitudes. In the Antarctic the diet consists almost entirely of small, shrimplike crustaceans of the family Euphausiidae. In northern waters *B. physalus* also eats other crustaceans and various kinds of small fish (Ellis 1980; Leatherwood, Caldwell, and Winn 1976). Banfield (1974) indicated that the fin whale uses echolocation to find its food, but this activity has not actually been demonstrated. The species is known to produce a great variety of low-frequency sounds, and probably also some high-frequency pulses (Thompson, Winn, and Perkins 1979).

The fin whale may be monogamous and is regularly seen in pairs. Usual group size is 6 or 7 individuals, there are often up to 50, and occasionally as many as 300 travel together on migrations. Both mating and calving occur during the winter, when the whales are in warm waters. Each mature female gives birth every 2 or 3 years. The gestation period is about 11½ months. There are rare records of as many as six fetuses, but normally a single offspring is produced. The newborn is about 650 cm long and weighs 1,800 kg. It nurses for about 6 months, until it is around 12.2 meters long, and then travels with the female to the polar feeding areas. Sexual maturity is attained at 6 to 10 years of age (Ellis 1980; Banfield 1974; Mitchell 1978; Frazer 1973). Some individuals are estimated to have lived up to 114 years (Haley 1978).

Following development of the harpoon gun in the 1860s, *B. physalus* was regularly hunted by Norwegian whalers in the North Atlantic. Exploitation of the vast antarctic populations began after establishment of the first island whaling station in that region in 1904, and intensified after introduction of floating factories there in the 1920s. As the blue whale

declined, the fin received increased hunting emphasis. In the 1937/38 season, the take of *B. physalus* in the Antarctic was 28,009 individuals, nearly twice as great as that of *B. musculus*. After a lull during World War II, large-scale pelagic whaling resumed in the Antarctic and increased in other areas, especially the North Pacific. The world-wide kill of the fin whale exceeded 10,000 animals in every season from 1946/47 to 1964/65, and averaged around 30,000 annually from 1952 to 1962. Such a harvest was far in excess of sustainable yield, and by 1960 there was evidence of seriously reduced populations. Initial conservation efforts were largely unsuccessful, but by the mid-1970s the International Whaling Commission had drastically lowered annual quotas and had completely banned hunting in some regions. The kill dropped to 5,320 in 1968/69 and 743 in 1978/79 (McHugh 1974; Gulland 1974; Ellis 1980; Committee for Whaling Statistics 1980). The quota for the 1981/82 season is 561 (*Marine Mammal News* 7, no. 7 [1981]:1–3).

The number of mature *B. physalus* in the world prior to exploitation is estimated at 470,000, of which 400,000 were in the Southern Hemisphere. The current number is estimated at 107,000, including 80,000 in the Southern Hemisphere. Inclusion of immature individuals would increase these figures by roughly 50 percent (U.S. National Marine Fisheries Service 1978). *B. physalus* is classified as vulnerable by the IUCN (1976) and as endangered by the USDI (1980). All stocks are also on appendix 1 of the CITES, except those off Iceland and Newfoundland and those from 60° to 120° W and from 40° S to Antarctica, which are on appendix 2.

Balaenoptera musculus. The blue whale is the largest animal ever known to have existed. There is, however, a pygmy subspecies (*B. m. brevicauda*), described by Ichihara (1966) and reported to inhabit a restricted zone of the Southern Hemisphere to the north of 54° S and between 0° and 80° E. Most authorities now seem to accept this subspecies as valid (Hall 1981; Rice 1977; Ellis 1980), but others, such as Small (1971), thought it to represent only young individuals of another subspecies. That part of the body of the pygmy blue whale posterior to the dorsal fin is described as being relatively shorter than that of other populations, and hence overall head and body length is reduced. Sexual maturity in this subspecies is attained at an average length of 19.2 meters, and maximum length is about 24.4 meters (Ellis 1980).

Disregarding the pygmy subspecies, head and body length of the blue whale at sexual maturity averages 22.5 meters in males and 24 meters in females (U.S. National Marine Fisheries Service 1978). Average length at physical maturity for antarctic specimens is 25 meters in males and 27 meters in females (Banfield 1974). Animals from the Northern Hemisphere are somewhat smaller. There is some question about maximum size, and it is possible that individuals grew larger prior to the period of intensive exploitation that began in the 1920s. Hall (1981) gave maximum length as 32 meters, and Rice (1978a) indicated that the largest specimen ever reliably measured, a female, was about 30 meters long. Rice (1967) stated that a 27.1-meter female weighed 136,400 kg, and that the weight of a 30-meter individual would be about 160,000 kg.

The general coloration is slate or grayish blue, mottled with light spots, especially on the back and shoulders. Although the underparts have about the same basic color as the upper parts, they sometimes acquire a yellowish coating of microorganisms. Thus, the vernacular name "sulphurbottom" is often applied to *B. musculus*. The dorsal fin is relatively small, usually reaching a height of only 33 cm. The rostrum is less sharply pointed than that of any other member of the genus. There are approximately 90 ventral grooves,

Blue whale (*Balaenoptera musculus*), painting by Richard Ellis.

and these extend to the navel. On each side of the upper part of the mouth is a row of 300 to 400 baleen plates. They are black in color and range in length from 50 cm in front to 100 cm in back (Ellis 1980; Leatherwood, Caldwell, and Winn 1976).

At least 90 percent of all blue whales, prior to exploitation, lived in the Southern Hemisphere, but there are populations in the North Atlantic and North Pacific. Banfield (1974) described the species as a pelagic denizen of polar and temperate seas. Mitchell (1978) noted that whereas the blue whale goes farther into the Antarctic than do the other large rorquals, it does not move as close to the northern polar ice fields as does the fin whale. Populations generally spend the winter in temperate and subtropical zones, migrate toward the poles in the spring, feed in high latitudes during the summer, and move back toward the equator in the fall. Because of the difference in seasons between the northern and southern parts of the earth, all migrating populations move in roughly the same direction at the same time, and thus those of the Northern Hemisphere do not meet those of the Southern Hemisphere. There seem to be a number of distinct stocks, and though these may overlap to some extent in the summer feeding areas, they separate and return to their own discrete breeding sites each year (Gulland 1974; Mackintosh 1966). The stock of the pygmy subspecies around Kerguelen, Crozet, and Heard islands does not appear to migrate to the same extent as other populations (Ellis 1980).

The blue whale normally swims at a speed of around 22 km per hr, but may hit 48 km per hr if alarmed. It usually feeds at depths of less than 100 meters, but harpooned individuals have gone deeper than 500 meters below the surface. Dives normally last 10 to 20 minutes, and are followed by a series of 8 to 15 blows. The spout reaches a height of 9.1 meters. The blue whale is the only member of its genus that commonly lifts its tail flukes out of the water before a dive. To feed, *B. musculus* takes in large amounts of water and organisms, greatly distending its throat, and then forces the water out, leaving the food trapped in the inner fibers of the baleen (see familial account). The highly restricted diet consists almost entirely of shrimplike crustaceans of the family Euphausiidae. These organisms, some of which are known as krill, are generally less than 5 cm long. When in the summer feeding areas, the daily consumption of each whale is probably about 40,000,000 individual euphausiids, with a total weight of 3,600 kg. During the rest of the year—a period of up to eight months—the blue whale apparently does not eat at all and lives off of stored fat (Rice 1978a; Leatherwood, Caldwell, and Winn 1976; Mitchell 1978; Ellis 1980). *B.*

musculus emits deep, low-frequency sounds, and series of clicks that may possibly be used in the echolocation of swarms of krill (Thompson, Winn, and Perkins 1979).

Off the coast of California, aggregations of up to 60 blue whales are common (Rice 1978a). Usually, however, the species is seen alone or in groups of two or three individuals (Banfield 1974). Mating and calving take place in the late spring and summer. Females give birth every 2 or 3 years (U.S. National Marine Fisheries Service 1978). The gestation period, usually reported at between 10 and 12 months, is surprisingly short for so large an animal. Frazer (1973) listed gestation as only about 9.6 months, and explained that if the period lasted much longer, the young would be born at a disadvantageous time—just before or during the season spent in cold waters, and before it had accumulated much protective blubber. Twins have been reported on rare occasion, but there is normally a single offspring. At birth it is about 7 meters long and weighs 2,000 kg. It gains 90 kg per day and is weaned after 8 months, when it is about 15 meters long. Sexual maturity may not be attained until an age of 23–30 years (Ellis 1980; Banfield 1974). Some individuals are estimated to have lived for up to 110 years (Haley 1978).

Because of its size, speed, strength, and remote habitat, the blue whale was generally considered too difficult a target in the early days of whaling. This situation was changed by a series of developments from the 1860s to 1920s, as described in the account of the order Cetacea. Regular hunting of *B. musculus* began off Norway, steadily spread across the Northern Hemisphere as one population after another was depleted, and finally came to center in the vast feeding waters of the Antarctic. The annual kill went up dramatically following the introduction of factory ships with slip sternways. The total recorded kill in the twentieth century is approximately 350,000 individuals, of which over 90 percent were taken in the Antarctic. The peak season was 1930/31, when the kill was 29,410 in the Antarctic and 239 in other parts of the world. There subsequently was a general decline in the world-wide seasonal harvest, to 12,559 in 1939/40, 6,313 in 1949/50, and 1,465 in 1959/60. Various international efforts to set size limits, sanctuary areas, and quotas had been attempted since the 1930s, but were inadequate and did not receive full compliance. By the early 1960s, it was clear to almost all concerned persons that the blue whale was nearing extinction, and groups of scientists were recommending that the International Whaling Commission establish total protection for the species. Because of a continued lack of cooperation from whaling interests, such protection did not come until the year after the 1965/66 season, when 613 blue whales were killed, only 20 of them in the Antarctic (McHugh 1974; Allen 1942; Small 1971; Gulland 1974; Ellis 1980).

The estimated numbers of blue whales prior to exploitation are: Southern Hemisphere, 200,000; North Pacific, 5,000; and western North Atlantic, 1,100 (there was also a small but undetermined number in the eastern North Atlantic). Current estimated numbers are: Southern Hemisphere, 9,000 (about half of which are the pygmy subspecies); North Pacific, 1,700; and North Atlantic, a few hundred (U.S. National Marine Fisheries Service 1978, 1981). There have been indications of slowly increasing numbers during the last few years, and populations off the east and west coasts of North America appear to be larger than once feared (Berzin 1978; Leatherwood, Caldwell, and Winn 1976). Although international protection against direct killing seems to be working, there are other concerns, especially regarding proposals to harvest antarctic krill for use as human food (Gulland 1974; McWhinnie and Denys 1980). The blue whale is classified as endangered by the IUCN (1972) and the USDI (1980), and is on appendix 1 of the CITES.

CETACEA; BALAENOPTERIDAE; *Genus MEGAPTERA* Gray, 1846

Humpback Whale

The single species, *M. novaeangliae*, occurs in all oceans and adjoining seas of the world (Rice 1977).

Head and body length at sexual maturity averages 11.6 meters in males and 11.9 meters in females (U.S. National Marine Fisheries Service 1978). Average length at physical maturity for North Pacific specimens is 12.5 meters in males and 13 meters in females (Banfield 1974). Individuals over 15 meters long are very rare. Average weight is around 30,000 kg. The pectoral fin is about one-third as long as the head and body, and is the largest of any cetacean, both absolutely and relatively. The dorsal fin varies from bumplike to strongly curved in shape, and from 15 to 60 cm in height. The expanse of the tail flukes is also about one-third the length of the head and body. The general coloration is usually black above and white below, but there is much variation. Clusters of white barnacles are often present. The body is stocky compared to that of *Balaenoptera*. There are 10 to 36 grooves extending from beneath the tip of the snout to the area of the navel. Irregular knobs and protuberances occur along the snout and lower lip. On each side of the upper part of the mouth is a row of about 340 baleen plates, gray to black in color.

The humpback whale is basically oceanic, but enters shallow, tropical waters for the winter breeding season. At this time, each of the discrete populations utilizes a certain archipelago or stretch of continental coastline. During the spring there are movements, along well-defined migratory routes, toward the high-latitude feeding areas. In the Northern Hemisphere some populations go as far as the Chukchi Sea and Spitsbergen. In the Southern Hemisphere there are summer concentrations off the southern ends of continents and around such subantarctic islands as South Georgia. In the autumn there are return migrations toward the equator. Because of the reversal of seasons, the populations of the Northern Hemisphere are not in equatorial waters at the same time as those of the Southern Hemisphere (Dawbin 1966; Wolman 1978).

Despite its common name, *Megaptera* is a graceful swimmer and is among the most acrobatic of large cetaceans. It often makes a somersault by leaping completely out of the water with its belly up, plunging back headfirst, and then circling underwater to the original position. It normally travels at about 7 km per hr, but maximum speed has been estimated at 27 km per hr. Just before a dive it emits a series of three to six short, broad spouts, and raises its tail flukes high above the water. Feeding is usually accomplished by taking in a large amount of water and organisms, and then forcing out the water to strain the food in the baleen (see familial account). There have been observations of one or two whales diving beneath a school of fish and then spiraling upward while emitting bubbles. The bubbles rose in a cylindrical pattern around the fish and seemed to form a barrier, through which they would not pass. The whales then rushed into the school of fish to feed. Small fish evidently form a major part of the diet, but shrimplike crustaceans, especially of the family Euphausiidae, are the most important foods in the Antarctic. As is the general case with rorquals, most, if

Humpback whales (*Megaptera novaeangliae*), painting by Richard Ellis.

not all, feeding is done during the summer in high latitudes (Wolman 1978; Ellis 1980).

Experiments with a captive humpback suggested to Beamish (1978) that the genus does not use echolocation to find food. *Megaptera* is, however, among the most loquacious of cetaceans. It emits a great variety of sounds, certain of which are sometimes combined into an elaborate song. Winn and Winn (1978) reported the song to evolve through "moans and cries," to "yups or ups and snores," to "whos or wos and yups," to "ees and oos," to "cries and groans," and finally to various "snores and cries." These authorities considered the song to be one of the longest, and possibly the most patterned, in the animal kingdom. It lasts 6 to 35 minutes, but the highly rhythmic pattern is repeated over and over for hours and possibly months at a time. It is heard only in the tropics during the winter breeding season. It is produced only by single individuals, which are thought to be young but sexually mature males. It may serve to space the males, to attract females, or to locate group members. There apparently are dialects, in that the song of the whales in one area differs to some extent from that heard in other areas. Perhaps the most remarkable feature of the song is that in each area it seems to change from year to year, and that the new versions are effectively spread throughout the singing population.

There is segregation by age and sex during migrations (Dawbin 1966). In the autumn the order of progression is: females with their recently weaned calves, then independent juveniles, then mature males and females that are not reproductively active, and finally females in late pregnancy. During the spring the order is: females in early pregnancy, then independent juveniles, then mature males and females that are not reproductively active, and finally females in the early stage of lactation. This arrangement insures that pregnant females spend a maximum of time in the feeding waters, and that young calves spend a maximum of time in warm regions. In feeding areas *Megaptera* is sometimes seen in aggregations of up to 150 individuals. In the breeding season, however, animals are usually found alone or in groups of 2 to 9. A female and young calf are commonly accompanied by another adult, apparently a male (Ellis 1980; Herman and Antinoja 1977).

The mating and calving season is October to March in the Northern Hemisphere and April to September in the Southern Hemisphere. Females are seasonally polyestrous and usually give birth every 2 years to a single young, about 400 cm in length and 1,350 kg in weight. The gestation period is 12 to 13 months, and lactation lasts about 11 months. Sexual maturity is attained at an age of 6 to 12 years (U.S. National Marine Fisheries Service 1978; Banfield 1974). Some individuals are estimated to have lived up to 77 years (Haley 1978).

The humpback has been regularly hunted for a much longer period than other large rorquals. It is relatively easy to secure, because it tends to stay near the shore in the breeding season, is a slow swimmer, often approaches vessels, and is highly visible. It has been considered a valuable source of oil, meat, and baleen. It was taken by aboriginal peoples in northwestern North America in ancient times, and has recently been hunted for subsistence on certain West Indian and Pacific islands, and in Greenland. Coastal fisheries began in Japan in the 1600s and in eastern North America by the 1700s. Major commercial exploitation of the genus spread over the North Atlantic and North Pacific in the 19th and early 20th centuries. The most intensive period of hunting, however, began with the establishment of whaling stations on the islands around Antarctica in the early 1900s. The total kill in the Southern Hemisphere from 1904 to 1939 was 102,298 individuals. Populations were quickly and dras-tically reduced. Humpbacks composed 96.8 percent of the catch of whales at South Georgia in the 1910/11 season, but only 9.3 percent in 1916/17. Nonetheless, large-scale exploitation continued for many years. The annual world-wide kill was above 2,400 in every season from 1948/49 to 1963/64. By this time there was general realization that the genus was nearing extinction. From 1964 to 1966 the International Whaling Commission extended protection to all populations, except that a small aboriginal quota continued (Ellis 1980; Allen 1942; McHugh 1974; Gulland 1974). The quota for the 1981/82 season is 10 whales (*Marine Mammal News* 7(7):1–3; 1981).

Original populations are estimated at 100,000 animals in the Southern Hemisphere and 15,000 in the North Pacific. The current world-wide estimate is 5,700 to 6,800. The largest group in the North Pacific, several hundred individuals, winters in the waters around Hawaii, but, interestingly, this population seems to have become established only in the last 200 years. It now may be jeopardized by increasing boat traffic and disturbance from tourists. There are similar problems in the summer waters of Glacier Bay, Alaska. There is evidence that the North Atlantic population has increased since 1915, but individuals sometimes become entangled in fishing nets off eastern Canada (U.S. National Marine Fisheries Service 1978, 1981; Herman 1980; Wolman 1978; Winn, Edel, and Turuski 1975). The humpback is classified as endangered by the IUCN (1972) and the USDI (1980), and is on appendix 1 of the CITES.

CETACEA; *Family BALAENIDAE*

Right Whales

This family of two Recent genera and three species is found in all oceans and adjoining seas of the world, except in tropical and south polar regions (Rice 1967).

The two genera differ greatly in size and certain other characters. *Caperea* is the smallest baleen whale and *Balaena* includes two of the largest. Females average larger than males. Both genera have a relatively large head, accounting for one-fourth to one-third of the total length. The rostrum is narrow and highly arched, resulting in the cleft of the mouth being curved. The lower jaw is massive and has a large, fleshy lip. There are no throat and chest grooves, as in the Balaenopteridae. The pectoral fin is short, broad, and rounded. Only *Caperea* has a dorsal fin. The tongue is heavy and muscular. This family, like the Balaenopteridae and Eschrichtidae, has baleen instead of teeth. In the Balaenidae the baleen plates are usually longer and narrower than in the other two families. There is a row of plates on each side of the upper part of the mouth, and the two rows do not join anteriorly. The plates fold in the floor of the closed mouth and straighten when the mouth opens. The vertebrae number 54 to 57 in *Balaena* and 43 in *Caperea;* the 7 neck vertebrae are fused into a single unit.

The diet consists mainly of the small forms of animal life known collectively as zooplankton (and sometimes called krill). These organisms include crustaceans, such as the shrimplike euphausiids and copepods, and the free-swimming mollusks known as pteropods. Unlike most species of the Balaenopteridae, the Balaenidae use the feeding method called skimming. The whales swim through swarms of prey, with the mouth open and the head above water to just behind the nostrils. When a sufficient mouthful of organisms has been filtered from the water by the inner bristles of the baleen plates, the whales force out the water, dive, and swallow the food.

Dr. Roy Chapman Andrews, who was 6 ft. 1 in. in height, standing beside the skull of a right whale (*Balaena glacialis*). The plates of baleen are smooth on the outside, but on the inside they are fringed to form an effective strainer. Photo from American Museum of Natural History.

The common name "right" whale refers to the consideration of *Balaena*, in the early days of commercial whaling, as the proper or best kind of whale to hunt. It was slower and less active than the sperm whale and most other baleen whales, and it often came closer to land. Unlike the others, it had greater buoyancy and was less likely to sink after being killed. Moreover, it yielded the most valuable products. The amount of oil usually derived from an adult was 80–100 barrels (1 barrel=about 105 liters), more even than from a blue whale (*Balaenoptera musculus*). Some large individual *B. glacialis* and *B. mysticetus* were said to have produced several hundred barrels each. The oil was used mainly as a fuel for lamps and in cooking, but also served as a lubricant, in leather tanning, and in the manufacture of soap and paint.

The baleen of *B. glacialis* and *B. mysticetus* was the finest available, and the plates were far longer than those of other species. The usual yield from an adult *B. mysticetus* was 680 to 760 kg. The hard outer portion of the plates was split into strips that could be used in the manufacture of umbrella ribs, fishing rods, carriage springs, whips, and numerous other items that required a combination of strength and elasticity. Baleen strips were especially useful as a stiffening in fashionable garments, such as the farthingale (a framework for supporting the elaborate dresses of the Elizabethan era) and the hoop skirt of the 19th century. The fine inner fibers of the baleen plates were also important in fashion. According to Gilmore (1978), they were woven into fabrics, giving a stiffness and rustle to taffeta and crinoline, and could be washed without softening or loss of elasticity.

The value of whale products fluctuated over the years. Typical 19th-century prices in the United States, however, were around $30 a barrel for oil and $10 a kg for baleen. At such rates the average *B. mysticetus* could bring over $10,000, and this was at a time when the cost of living was about 5 percent of what it is today. By the early 20th century the last great concentrations of *Balaena* had been eliminated, whale oil was no longer being used for illumination, and spring steel was replacing baleen. Although whaling was about to enter its most intensive era, *Balaena* was not to be of major commercial importance.

The geological range of the family Balaenidae is early Miocene to Pleistocene in South America, early Pliocene to Pleistocene in Europe, and Recent in all oceans (Rice 1967).

CETACEA; BALAENIDAE; *Genus BALAENA* Linnaeus, *1758*

Right and Bowhead Whales

There are two species (Rice 1977; Banfield 1974):

B. glacialis (right whale), mainly temperate parts of the Atlantic, Indian, and Pacific oceans, and adjoining seas;
B. mysticetus (bowhead or Greenland right whale), Arctic Ocean and adjoining seas, Sea of Okhotsk, Hudson Bay, Gulf of St. Lawrence.

The species *B. glacialis* has sometimes been placed in a separate genus, *Eubalaena* Gray, 1864. In addition to the characters set forth in the familial account, *Balaena* is distinguished by a massive body, a constricted tail stock that tapers into the flukes, an enormous lower jaw, the longest baleen plates in the Cetacea, and no dorsal fin. Additional information is provided separately for each species.

Balaena glacialis.   Head and body length at physical maturity is usually 13.6 to 16.6 meters. Banfield (1974) listed average size as 13.7 meters and 22,000 kg for males, and 14 meters and 23,000 kg for females. Rice (1967) stated that maximum length is 18.3 meters and that one individual, 17.1 meters long, weighed 67,197 kg. Pectoral fin length is 180 to 210 cm. The coloration is usually black throughout, but there are sometimes large white patches, especially on the belly. Entirely white calves have been observed (Best 1970b). Around the head are series of horny protuberances, representing accumulations of cornified layers of skin and commonly infested with barnacles and parasitic crustaceans. The most conspicuous of these callosities is located on the tip of the upper jaw and is known as the bonnet. The callosities are present from the time of birth, but their function is unknown.

The head of *B. glacialis* is smaller than that of *B. mysticetus,* about one-fourth of the total length, and the upper jaw is not so strongly arched. The upper jaw is narrow and is practically concealed by the high, massive lower jaw, when the mouth is closed. On each side of the upper part of the mouth is a row of 225 to 250 baleen plates. These plates are 180 to 220 cm in vertical length, and from dark gray or brown

Right whales (*Balaena glacialis*), photo by Randall R. Reeves.

to black in color. The two blowholes are set well apart, and thus there are two spouts that form a V pattern (Leatherwood, Caldwell, and Winn 1976; Gilmore 1978).

The right whale is primarily a species of temperate waters, though some individuals apparently move just north of the Arctic Circle and just south of the Tropic of Cancer. It is usually found closer to land than are most large whales, especially in the breeding season. Although it was largely eliminated by whalers before much scientific information could be gathered, available evidence indicates that there were well-defined migrations to higher latitudes for summer feeding, and back to lower latitudes for winter breeding. Populations of the Southern Hemisphere wintered mainly between 30° and 50° S, off the coasts of South America, southern Africa, Australia, and New Zealand, and then moved toward the Antarctic for the summer. The population of the eastern North Atlantic wintered in the Bay of Biscay and summered between Britain and Iceland. In the western North Atlantic, from about November to March, the whales remained from Cape Cod to Bermuda and the Gulf of Mexico. In late winter they began to move north, and by spring they were concentrated off the northeastern United States. Some remained in the latter region, but the major summer feeding waters were around Newfoundland and in the Labrador Sea. Pacific populations seem to have passed Japan in April and fed off southern Alaska from May to September; wintering waters are unknown, but may possibly have included those around Hawaii (Allen 1942; Reeves, Mead, and Katona 1978; Gilmore 1978; Banfield 1974; Herman et al. 1980; Gaskin 1968).

The right whale is a relatively slow swimmer, averaging about eight km per hr, but frequently leaps clear of the water and engages in other acrobatics. It commonly makes a series of five or six shallow dives, and then lifts its tail flukes above the water and submerges for about 20 minutes. It is usually not wary of boats and can be easily approached (Leatherwood, Caldwell, and Winn 1976). It appears to frolic in stormy seas and even to use its tail to sail in the wind (Payne 1976). To feed it employs the skimming method, generally swimming near the surface with its mouth open to strain out organisms from the water with its baleen (see familial account). Its diet consists mainly of copepods—crustaceans only a few millimeters in diameter—and also includes euphausiids, pteropods, and small fish (Gilmore 1978).

According to Banfield (1974), *B. glacialis* was sometimes found in aggregations of 100 individuals or more. Social activity, however, is generally limited to mating and the relationship between mother and young. A male may attempt to mate with several females, and a female may sometimes be courted by several males at once (Payne 1976; Ellis 1980). *B. glacialis* emits a number of low-frequency sounds, mostly during courtship (Gilmore 1978). Females give birth to a single offspring every 2 or 3 years. Both mating and calving occur in winter, and gestation probably lasts 12 months. The young is about 350 to 550 cm long at birth, and nurses for about 7 months.

Because of its coastal habitat and economic value (see familial account), *B. glacialis* was among the first of the large whales to be extensively exploited. It may have been hunted as early as the 9th century A.D. off Norway, and was

Right whale (*Balaena glacialis*): Top photo shows the baleen in place and the very large tongue; Inset: a smaller piece of baleen of lighter color; Bottom photo shows the tongue and the rough surface of the top of the front portion of the head (the baleen has been removed); photos by G. C. Pike through I. B. MacAskie.

regularly being taken in the Bay of Biscay in the 10th century. By the late 15th century it had become rare in the latter area, and the Basque whalers shifted their emphasis to other waters, eventually reaching Newfoundland. The right whale became the basis of a major industry in the American colonies during the 17th century, with hunting initially being done from small boats in such places as Delaware Bay and Cape Cod Bay, and later involving extended voyages in large ships. By 1700 in Europe and 1800 in the United States, the right whale had become too rare to be of commercial interest. Toward the end of the 18th century, however, large stocks of the species were discovered in the wintering waters of the Southern Hemisphere and in the summer feeding areas of the North Pacific. By this time the industry was dominated by American companies. On the basis of oil imports into the United States, it has been estimated that 193,522 right whales were killed from 1804 to 1877 (see Best 1970b for an explanation of these figures). By the latter year the right whale was rare throughout its range. Some hunting continued in the early 20th century, but takes were usually considerably

fewer than 100 per year. In 1937, in accordance with the International Agreement for the Regulation of Whaling, *B. glacialis* received complete protection (Allen 1942; Banfield 1974; Reeves, Mead, and Katona 1978).

There are no estimates for the original number of right whales in the world, but the species has become one of the rarest of large mammals. According to the U.S. National Marine Fisheries Service (1978, 1981), current estimated numbers are 3,200 in the Southern Hemisphere, 220 in the North Pacific, and 200 in the North Atlantic. There is evidence that the populations of the Southern Hemisphere have increased since the granting of protection. There have been suggestions that an increase also has occurred in the North Atlantic, but Reeves, Mead, and Katona (1978) cautioned that recovery has been modest at best, and that the species may be jeopardized by such factors as competition for food with the sei whale, accidental trapping in fish nets, pollution, and boat collisions. *B. glacialis* is classified as endangered by the IUCN (1972) and the USDI (1980), and is on appendix 1 of the CITES.

Bowhead whale (*Balaena mysticetus*), accompanied by belugas, painting by Richard Ellis.

Balaena mysticetus.   Head and body length at sexual maturity averages 11.6 meters in males and 12.2 meters in females (U.S. National Marine Fisheries Service 1978). At physical maturity the usual length is 15 to 18 meters. The largest specimens reach about 19.8 meters in length and 69,000 kg in weight (Ellis 1980). Pectoral fin length is about 200 cm and expanse of the tail flukes is 550 to 790 cm. The adult coloration is black, except that the anterior part of the lower jaw is cream colored, the belly occasionally has white patches, and the junction of the body and tail is sometimes gray.

The skull is about 40 percent as long as the entire animal (Hall 1981). When seen from the front, the enormous lower jaw forms a U around the relatively narrow upper jaw. When seen from the side, both jaws are highly arched and, if the mouth is closed, the lower jaw may partly conceal the upper. On each side of the mouth is a row of about 300 baleen plates. These plates, the largest of any whale, are 300 to 450 cm in vertical length and are black in color. As in *B. glacialis,* there are two separated blowholes that produce a V-shaped spout. According to Ellis (1980), the blubber of *B. mysticetus* is 25 to 50 cm thick.

The bowhead is primarily an arctic species, seldom occurring south of about 45° N, but in 1969 a specimen was discovered off Japan at 33° 28' N (Nishiwaki and Kasuya 1970). Normally the bowhead is found in association with ice floes, appearing to move seasonally in response to the melting and freezing of the ice. During the summer it frequents bays, straits, and estuaries. Prior to exploitation there seem to have been four major populations. One wintered off the Kuril Islands and summered in the Sea of Okhotsk. Another win-

tered in the western Bering Sea, migrated north through the Bering Strait in the spring and early summer, remained in the Chukchi and Beaufort seas from late summer to early autumn, and then returned south. A third group apparently wintered along the east coast of Canada as far south as the Gulf of St. Lawrence. In the spring and summer this group moved northward and westward into Davis Strait, Hudson Bay, Baffin Bay, and adjoining waters. The fourth population apparently wintered off southeastern Greenland and migrated to and beyond the Spitsbergen region during the spring and summer (Rice 1977; Banfield 1974; W. G. Ross 1979; Marquette 1978).

During migration the bowhead swims at about 6 km per hr; its maximum speed is around 15 km per hr. It normally surfaces for up to 2 minutes, blows four to nine times, and then submerges for 5 to 10 minutes. Harpooned individuals are said to have stayed underwater for over an hour. Before a dive the bowhead lifts its tail flukes above the surface. It sometimes leaps almost entirely out of the water. To feed, it employs the skimming method (see familial account). The diet consists mainly of zooplankton—copepods, amphipods, euphausiids, and pteropods. The estimated daily intake during the summer feeding season is 1,800 kg. At other times of the year, like most baleen whales, the bowhead lives off of stored fat reserves and eats little or nothing (Marquette 1978; Banfield 1974).

The bowhead is usually found alone or in groups of two or three individuals, but larger groups may form during migration, and in the past such schools often included several hundred whales. There is sometimes segregation by age and sex. During the spring migration of the Bering Sea population, young animals move north first, and then come large males and females with calves. The voice of *B. mysticetus* has been described as a drawn-out hooting or humming sound. Both mating and calving occur mainly in the spring and early summer. The female gives birth every 2 or 3 years, normally to a single calf. The gestation period is thought to last 12 or 13 months. The young is 300 to 450 cm long at birth, and is apparently weaned after about 6 months. Sexual maturity may come at 4 years of age. There are questionable reports, based on the finding of old harpoons in newly taken whales, of individuals having lived about 40 years, and also of having made a transpolar passage from the Atlantic to the Pacific side of the Arctic (Marquette 1978; Banfield 1974; Ellis 1980).

Like *B. glacialis,* the bowhead was one of the "right" whales of commerce. Indeed, because of the length of its baleen and the thickness of its blubber, it was the most economically valuable of all cetaceans. Regular hunting by Europeans began in the 16th century off Greenland. In the early 17th century Spitsbergen became the center of the industry. Whaling bases were established there by expeditions from the Netherlands, England, France, Spain, Denmark, and Germany. By the early 1700s the bowhead had become so rare in this region that hunting was no longer profitable. The focus of exploitation then shifted to the west of Greenland. Intensive whaling began in Davis Strait in 1719, Baffin Bay and adjacent sounds in 1818, and Hudson Bay in 1860. In these regions whaling was dominated by the Dutch for most of the 18th century, and by the British and Americans in the 19th century. In some years several hundred ships were involved. By the middle of the 19th century the bowhead was becoming rare in most areas from Greenland to eastern Canada, and after 1887 not more than 10 ships per year hunted in these waters. The last 2 vessels to try, in 1912 and 1913, did not take a single whale. Meanwhile, however, starting in 1848, American whalers had sailed north through the Bering Strait to exploit the last major summer concentration of bowheads, that of the Beaufort and Chukchi seas. This population was probably approaching extinction by the turn of the century, but the price of baleen per pound fell from $5 in 1907 to 7½¢ in 1912, and hunting ceased. The total kill of bowheads had been about 20,000 in the western Arctic since 1848, and about 37,000 in the waters from northeastern Canada to Greenland since 1719 (Allen 1942; Ellis 1980; W. G. Ross 1974, 1979; Bockstoce 1980; Banfield 1974).

The bowhead whale received protection under the International Agreement for the Regulation of Whaling in 1937, the International Whaling Convention of 1946, the United States Marine Mammal Protection Act of 1972, and the U.S. Endangered Species Act of 1973. In each case, however, an exception was made for subsistence hunting by aboriginal peoples. Certain groups of Eskimos in northwestern Alaska had been whaling since ancient times. Their cultural and economic status is still associated with the pursuit and kill of the bowhead during its annual migrations. They once utilized the entire animal, including the blubber for fuel and the baleen for making tools, but now the meat (around 10,000 kg per whale) is the main economic objective.

Until about 1970 the usual take by the Eskimos was 10 to 15 whales per year. Subsequently the kill rose, apparently because the petroleum industry had made money available for native Alaskans to purchase the boats, guns, and other equipment necessary for the culturally prestigious venture of whaling. In 1976, 48 bowheads were killed and another 46 were known to have been wounded. In 1977, the respective figures were 29 and 82. Because of growing concern that such exploitation would jeopardize the survival of the remaining bowhead population, the International Whaling Commission decided in 1977 to rescind the exemption for Eskimo whaling. The United States government, in a rever-

sal of its general stand against whaling, then urged that some hunting be allowed. The commission thus began to set quotas for the bowhead, just as for commercially harvested species. The quota for 1980 was 18 whales landed or 26 struck, whichever came first ("struck" means having been hit by a harpoon but then lost). Some Eskimos argued for a higher kill and announced that they would not abide by the commission's quota. Indeed, the quota was exceeded during the 1980 hunts, 16 whales having been landed and 18 others struck and lost (U.S. National Marine Fisheries Service 1978, 1981; Bockstoce 1980; Ellis 1980; Marquette 1978, 1979; Evans and Underwood 1978; R. Rau 1978).

There is disagreement regarding how many bowheads there are and were. Reported estimates of the western arctic population in the mid-19th century range from 4,000 to 40,000 individuals. The most commonly mentioned figures, however, are 18,000 to 30,000 (Ellis 1980; Evans and Underwood 1978; Marquette 1978). The U.S. National Marine Fisheries Service (1981) gave the best estimate of the current population in the western Arctic as 2,264 whales, and noted that there is evidence of a decline. There have been only a few recent reports of the bowhead from the Spitsbergen area and the Sea of Okhotsk. The species is classified as endangered by the IUCN (1972) and the USDI (1980), and is on appendix 1 of the CITES.

CETACEA; BALAENIDAE; *Genus CAPEREA* Gray, 1864

Pygmy Right Whale

The single species, *C. marginata,* is known from strandings or sightings in western and southern Australia, Tasmania, New Zealand, the Falkland Islands, the South Atlantic Ocean, South Africa, the Crozet Islands, and possibly Argentina and Chile (Ross, Best, and Donnelly 1975).

Head and body length is 610 to 640 cm. An adult female specimen measured as follows: pectoral fin length, 66 cm; dorsal fin height, 25 cm; and expanse of the tail flukes, 181 cm. The coloration is black or dark gray above and paler below. Young animals may be lighter than adults, and one juvenile appears to have been albinistic (Ross, Best, and Donnelly 1975). The tongue and interior part of the mouth are pure white.

The head is about one-fourth of the total length. The jaws are arched. In profile, the ventral outline of the anterior part of the throat is concave (Ross, Best, and Donnelly 1975). On each side of the mouth is a row of about 230 baleen plates (Ellis 1980). The plates are up to 70 cm long, and are ivory white in color, but have a dark outer margin. The dorsal fin is sickle shaped and located on the posterior part of the back.

Caperea has 34 ribs, more than any other cetacean genus. These ribs also extend farther posteriorly than those of any other cetacean, so only two vertebrae without ribs intervene between those with ribs and the tail. The ribs become increasingly flattened and widened toward the tail, and thereby probably provide additional protection to the internal organs.

The pygmy right whale is known only by 71 specimens and a small number of sightings. Available information on natural history was summarized by Ross, Best, and Donnelly (1975) and Mitchell (1975a). The genus appears restricted to temperate waters of 5° to 20° C in the Southern Hemisphere. Although it is primarily pelagic, stranding records suggest a movement, mainly by juveniles, toward the shore in spring and summer. At such times *Caperea* is frequently found in sheltered, shallow bays. Records indicate that this genus occurs throughout the year around Tasmania, but only from August to December off Australia, and from December to February off South Africa.

Pygmy right whale (*Caperea marginata*), photos by Mr. T. Dicks through Peter B. Best.

Old reports suggested that *Caperea* dives deeply and may remain for long periods near the bottom. More recent observations indicate that it stays underwater only for three or four minutes at a time, at a depth of two or three meters. It is a relatively slow swimmer, averaging around seven km per hr, and proceeds with a flexing of the entire body. The diet evidently consists mainly of the minute crustaceans known as copepods.

Schools of up to eight individuals have been observed. Records of pregnant females suggest a mating and calving season of several months. At least some young are thought to be born during the autumn or winter, apparently well away from land. Theoretical length at birth is 190 cm. Nursing is presumed to last about five or six months and to be followed by a movement of the young toward shore in the spring and summer.

ORDER CARNIVORA

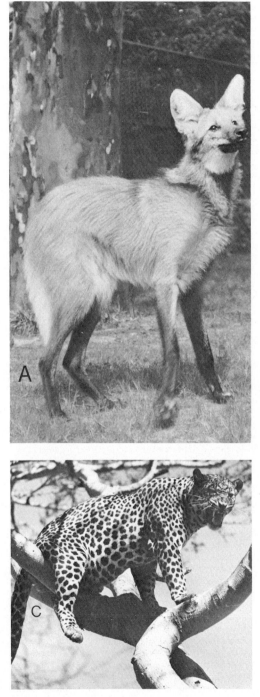

Dogs, Bears, Raccoons, Weasels, Mongooses, Hyenas, Cats

This order of 7 families, 92 genera, and 238 species occurs naturally throughout the world, except for Australia, New Guinea, New Zealand, Antarctica, and many oceanic islands. One species, *Canis familiaris,* apparently was introduced into Australia by human agency in prehistoric time, and subsequently established wild populations on that continent. Two living superfamilies of carnivores are usually recognized (Hall 1981; Stains 1967): the Arctoidea (or Canoidea), with the families Canidae, Ursidae, Procyonidae, and Mustelidae; and the Aeluroidea (or Feloidea), with the families Viverridae, Hyaenidae, and Felidae. Some authorities, such as Simpson (1945), have considered that these

A. Maned wolf (*Chrysocyon brachyurus*). B. Linsang (*Prionodon linsang*). C. Leopard (*Panthera pardus*). D. Giant panda (*Ailuropoda melanoleuca*). Photos by Bernhard Grzimek.

7 families are comprised by 1 suborder, the Fissipedia, and that there is a second living carnivore suborder, the Pinnipedia. Other authorities, such as Tedford (1976) and Rice (1977), place the pinnipeds in the arctoid group of the Carnivora.

The smallest living carnivore is the least weasel (*Mustela nivalis*), which has a head and body length of 135 to 185 mm, a tail length of 30 to 40 mm, and a weight of 35 to 70 grams. The largest is the grizzly or brown bear, some individuals of which, particularly along the coast of southern Alaska, attain a head and body length of 2,800 mm and a weight of 780 kg.

Carnivores have four or five clawed digits on each limb. The first digit (pollex and hallux) is not opposable, and is sometimes reduced or absent. Some carnivores, including canids and felids, are digitigrade, walking only on their toes. Others, such as ursids, are plantigrade, walking on their soles with the heels touching the ground. The brain has well-developed cerebral hemispheres, and the skull is heavy, with strong facial musculature. The articulation of the lower jaw is such as to permit only open and shut (not side to side) movements. The stomach is simple. Males have a baculum. The number of mammae in females is variable; they are located on the abdomen, except that in the Ursidae some are pectoral.

The teeth are rooted. The small, pointed incisors number 3/3 in all species except *Ursus ursinus*, which has 2/3, and *Enhydra lutris*, which has 3/2. The first incisor is the smallest and the third is the largest, the difference in size being most marked in the upper jaw. The canine teeth are strong, recurved, pointed, elongate, and round to oval in section. The premolars are usually adapted for cutting, and the molars usually have four or more sharp, pointed cusps. The last upper premolar and the first lower molar are called the carnassials, and often work together as a specialized shearing mechanism. The carnassials are most highly developed in the Felidae, which have a diet consisting almost entirely of meat, and are least developed in the omnivorous Ursidae and Procyonidae.

Most carnivores are terrestrial or climbing animals. Two genera, *Potos* and *Arctictis*, have prehensile tails. Apparently, all carnivores can swim if necessary, but the polar bear (*Ursus maritimus*) and the river otters (*Lutra, Aonyx, Pteronura*) are semiaquatic, and the sea otter (*Enhydra*) spends practically its entire life in the water. Land-dwelling carnivores shelter in caves, crevices, burrows, and trees. They may be either diurnal or nocturnal.

Most species of the Canidae, Mustelidae, Viverridae, and Felidae live solely or mainly on freshly killed prey. Their whole body organization and manner of living are adapted for predation. The diet may vary by season and locality. Hunting is done by scent and sight, and the prey is captured by a surprise pounce from concealment (*Panthera pardus*), a stalk followed by a swift rush (*Mustela frenata*), or a lengthy chase (*Canis lupus*). Some species regularly eat carrion. *Arctictis* is largely frugivorous, and the diet of certain other viverrids consists partly of fruit. The Hyaenidae include two genera (*Crocuta* and *Hyaena*) that both hunt large animals and feed on carrion, and one genus (*Proteles*) that is largely insectivorous. The Procyonidae and Ursidae, except for the carnivorous polar bear, are omnivorous, eating a wide variety of plant and animal life.

Carnivores are solitary, or associate in pairs or small groups. Females commonly produce a single litter each year, but those of a few species may give birth two or three times annually, and those of some large species usually mate at intervals of several years. Most species have gestation periods of about 49 to 113 days. Delayed implantation of the fertilized egg occurs in ursids and some mustelids, so the period from mating to birth is considerably longer than average. Litter size commonly ranges from 1 to 13. The offspring

are usually born blind and helpless, but with a covering of hair. They are cared for solicitously by the mother and, in some species, by the father. There is often a lengthy period of parental care and instruction.

The Carnivora were once considered to include the Creodonta, an extinct group dating back to the late Cretaceous, as a suborder. It now appears, however, that the Carnivora arose independently from ancestral insectivore stock and were initially represented by the family Miacidae of the late Paleocene and Eocene (Hall 1981; Romer 1968).

CARNIVORA; *Family CANIDAE*

Dogs, Wolves, Coyotes, Jackals, Foxes

This family of 16 Recent genera and 36 species has a natural distribution that includes all land areas of the world, except the West Indies, Madagascar, Taiwan, the Philippines, Borneo and islands to the east, New Guinea, Australia, New Zealand, Antarctica, and most oceanic islands. There are wild populations of the species *Canis familiaris* in Australia and New Guinea, but these apparently originated through introduction by human agency. The living Canidae have traditionally been divided, based mainly on dentition, into three subfamilies: Caninae, with the genera *Canis, Alopex, Vulpes, Fennecus, Urocyon, Nyctereutes, Dusicyon, Cerdocyon, Atelocynus,* and *Chrysocyon;* Simocyoninae, with *Speothos, Cuon,* and *Lycaon;* and Otocyoninae, with *Otocyon.*

Recent studies have indicated that subfamilial distinction for the Simocyoninae and Otocyoninae is not warranted, and also have revealed considerable controversy regarding the systematics of the Caninae. Langguth (1975b) referred most species of the South American *Dusicyon* to a subgenus (*Pseudalopex*) of *Canis,* referred one species (*D. vetulus*) to the genus *Lycalopex,* and retained only one species (*D. australis*) in *Dusicyon.* Clutton-Brock, Corbet, and Hills (1976) considered the genus *Vulpes* to include *Fennecus* and *Urocyon;* and considered the genus *Dusicyon* to include *D. australis,* the species referred by Langguth to *Pseudalopex* and *Lycalopex,* and also *Cerdocyon* and *Atelocynus.* Van Gelder (1978) expanded the genus *Canis* to contain the following as subgenera: *Dusicyon,* with the single species *D. australis; Pseudalopex,* with most species traditionally assigned to *Dusicyon; Lycalopex,* with the species formerly called *Dusicyon vetulus; Cerdocyon; Atelocynus; Vulpes,* including *Fennecus* and *Urocyon;* and *Alopex.*

Because of the controversy, a conservative position has been taken here, and all traditionally recognized genera have been maintained. In addition, *Pseudalopex* and *Lycalopex* have been given generic rank. The sequence of genera presented here is based partly on information given by Langguth (1975b) and Nowak (1978, 1979), indicating that *Vulpes* and the other foxes are more primitive than *Canis.* Further systematic comments are made in the generic accounts.

In wild species, head and body length is 357 to 1,600 mm, tail length is 125 to 560 mm, and weight is 1 to 80 kg. *Fennecus zerda* is the smallest species and *Canis lupus* is the largest. In a given population, males generally average larger than females. Most species are uniformly colored or speckled, but one species of jackal (*Canis adustus*) has stripes on the sides of its body, and *Lycaon* is covered with blotches.

Canids have a lithe, muscular, deep-chested body; usually long, slender limbs; a bushy tail; a long, slender muzzle; and

A. Maned wolf (*Chrysocyon brachyurus*), photo from Los Angeles Zoological Society. B. Red fox pup (*Vulpes vulpes*), photo by Leonard Lee Rue III.

Bush dog (*Speothos venaticus*), photo by Bernhard Grzimek.

large, erect ears. There are four digits on the hind foot and five on the forefoot, except in *Lycaon*, which has four on both feet. The claws are blunt. Males have a well-developed baculum, and females generally have three to seven pairs of mammae.

The skull is elongate. The bullae are prominent, but usually not highly inflated. The dental formula in all but three species is: (i 3/3, c 1/1, pm 4/4, m 2/3) × 2 = 42. The molars are 1/2 in *Speothos*, 2/2 in *Cuon*, and 3/4 or 4/4 in *Otocyon*.

Canids occur from hot deserts (*Fennecus*) to arctic ice fields (*Alopex*). For dens they may use burrows, caves, crevices, or hollow trees. These alert, cunning animals may be either diurnal, nocturnal, or crepuscular. They are generally active throughout the year. They walk, trot tirelessly, amble, or canter, either entirely on their digits or partly on more of the foot. At full speed, they gallop. The gray foxes (*Urocyon*) often climb trees, an unusual habit for canids. The senses of smell, hearing, and sight are acute. Prey is captured by an open chase or by stalking and pouncing. The diet may vary by season, and vegetable matter is important to some species at certain times.

Some canids, especially the larger species, occur in packs of up to 30 members, and seek prey animals that are larger than themselves. Most smaller canids hunt alone or in pairs, preying on rodents and birds. There is usually a regular home range, part or all of which may be an exclusive territory. Females generally give birth once a year. Litters usually contain 2 to 13 young. Gestation averages around 63 days. The offspring are blind and helpless at birth, but are covered with hair. They are cared for solicitously by the mother, and often by the father and other group members as well. Sexual maturity comes after 1 or 2 years. Potential longevity is probably at least 10 years in all species.

The geological range of this family is late Eocene to Recent in North America and Europe, early Oligocene to Recent in Asia, early Pleistocene to Recent in South America, Pliocene to Recent in Africa, and late Pleistocene to Recent in Australia (Stains 1967; Macintosh 1975; Langguth 1975*b*).

CARNIVORA; CANIDAE; **Genus VULPES** Brisson, 1762

Foxes

There are 10 species (Clutton-Brock, Corbet, and Hills 1976; Ellerman and Morrison-Scott 1966; Corbet 1978; Meester and Setzer 1971; Hall 1981; Roberts 1977):

V. vulpes (red fox), Eurasia except southeastern tropical zone, northern Africa, most of Canada and the United States;

V. corsac (corsac fox), dry steppe and subdesert zone from the lower Volga River to Manchuria and Tibet;

V. ferrilata (Tibetan sand fox), high plateau country of Tibet, Nepal, and north-central China;

V. cana (Blanford's fox), mountain steppe zone of southern Turkmen, Iran, Pakistan, and Afghanistan;

V. velox (swift fox), southern Alberta and North Dakota to northwestern Texas;

V. macrotis (kit fox), southern Oregon to Baja California and north-central Mexico;

V. bengalensis (Bengal fox), Pakistan, India, Nepal;

V. rueppelli (sand fox), desert zone from Morocco and Niger to Afghanistan and Somalia;

V. pallida (pale fox), savanna zone from Senegal to northern Sudan and Somalia;

V. chama (Cape fox), dry areas of southern Angola, Namibia, Botswana, western Rhodesia, and South Africa.

Treatment of *Vulpes* as a distinct genus that does not include *Fennecus, Urocyon,* or *Alopex* is in keeping with the arrangements of such authorities as Hall (1981), Meester and Setzer (1971), Rosevear (1974), and Jones, Carter, and Genoways (1979). Other views have been to consider *Vulpes* a full genus that includes *Alopex* (Youngman 1975); a full genus that includes *Fennecus* and *Urocyon,* but not *Alopex* (Clutton-Brock, Corbet, and Hills 1976); and a subgenus of *Canis* that includes *Fennecus* and *Urocyon,* but not *Alopex* (Van Gelder 1978). The North American red fox has sometimes been designated a separate species, *V. fulva,* but most authorities now consider it conspecific with the Palearctic *V. vulpes.* Hall (1981) treated *V. macrotis* as being conspecific with *V. velox,* because Rohwer and Kilgore (1973) had reported interbreeding between these two kinds of fox where their ranges meet in eastern New Mexico and western Texas. Thornton and Creel (1975), however, argued for continued recognition of the two as separate species.

Vulpes is characterized by a rather long, low body; relatively short legs; a long, narrow muzzle; large, pointed ears; and a bushy, rounded tail that is at least half as long, and often fully as long, as the head and body. The pupils of the eyes generally appear elliptical in strong light. Some species have a pungent "foxy" odor, arising mainly from a gland located on the dorsal surface of the tail, not far from its base. Females usually have six or eight mammae. Additional information is provided separately for each species.

Vulpes vulpes.   Head and body length is 455 to 900 mm, tail length is 300 to 555 mm, and weight is 3 to 14 kg. Average weights in North America are 4.1 to 4.5 kg for females and 4.5 to 5.4 kg for males (Ables 1975). The usual weight in central Europe is 8 to 10 kg (Haltenorth and Roth 1968). The typical coloration ranges from pale yellowish red to deep reddish brown on the upper parts, and is white, ashy, or slaty on the underparts. The lower part of the legs is usually black, and the tail is generally tipped with white or black. Color variants, known as the cross fox and silver fox, represent, respectively, about 25 and 10 percent of the species. The cross fox is reddish brown in color, and gets its name from the cross formed by one black line down the middle of the back and another across the shoulders. The color of the silver fox, which has the most prized fur of any fox, ranges from strong silver to nearly black. The general color effect depends on the proportion of white or white-tipped to black hairs. An individual with only a few white hairs is sometimes called a black fox.

The red fox rivals the gray wolf (*Canis lupus*) for having the greatest natural distribution of any living terrestrial mammal besides *Homo sapiens.* Habitats range from deep forest to arctic tundra, open prairie, and farmland, but the red fox prefers areas of highly diverse vegetation and avoids large homogeneous tracts (Ables 1975). Elevational range is sea level to 4,500 meters (Haltenorth and Roth 1968). Daily rest may be taken in a thicket or any other protected spot, but each individual or family group usually has a main earthen den and one or more emergency burrows within the home range. An especially large den may be constructed during the late winter and subsequently used to give birth and rear the young. Some dens are used for many years by one generation of foxes after another. The preferred site is a sheltered, well-drained slope with loose soil. Often a marmot burrow is taken over and modified. Tunnels are up to 10 meters long and lead to a chamber 1 to 3 meters below the surface. There is sometimes only a single entrance, but there may be as many as 19. A system of pathways connects the dens, other resting sites, favored hunting areas, and food storage holes (Ables 1975; Banfield 1974; Stanley 1963; Haltenorth and Roth 1968).

The red fox is terrestrial, normally moving by a walk or trot. It has great endurance and can gallop for many kilome-

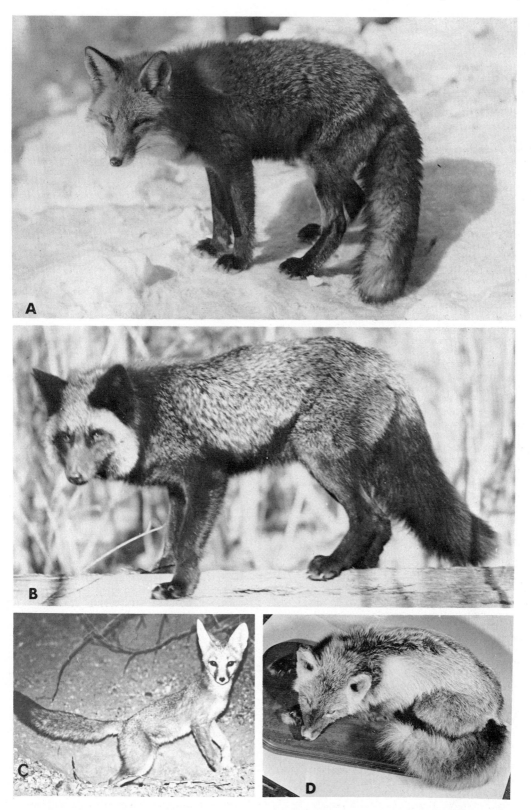

A. Red Fox (*Vulpes vulpes*), photo from New York Zoological Society. B. Silver fox (*V. vulpes*), photo from Fromm Brothers, Inc. C. Kit fox (*V. macrotis*), photo by O. J. Reichman. D. Cross fox (*V. vulpes*), photo by Howard E. Uible of mounted specimen in U.S. National Museum.

ters if pursued. It can run at speeds of up to 48 km per hr, can leap fences 2 meters high, and can swim well (Haltenorth and Roth 1968). It has keen senses of sight, smell, and hearing. Its ability to survive in the close proximity of people, and often to elude human hunters and their dogs, has given it a reputation for cunning and intelligence. Most activity is nocturnal and crepuscular. Individuals cover up to 8 km per night as they move on circuitous routes through the home range (Banfield 1974). During the autumn, the young born the previous spring disperse from the parental home range. The usual distance traveled at this time is about 40 km for males and 10 km for females; maximum known is 394 km (Ables 1975; Storm et al. 1976). Once the young animals establish themselves in a new area, they generally remain there for life.

The diet is omnivorous, consisting mostly of rodents, lagomorphs, insects, and fruit. To hunt mice, the red fox stands motionless, listens and watches intently, and then leaps suddenly, bringing its forelegs straight down to pin the prey. Rabbits are stalked and then captured with a rapid dash (Ables 1975). Daily consumption is around 0.5 to 1.0 kg. Sometimes a hole is dug and excess prey placed therein and covered over, to be eaten at a later time (Haltenorth and Roth 1968).

The most favorable areas usually support an average of one or two adults per sq km (Ables 1975; Insley 1977; Haltenorth and Roth 1968). Home range size varies with habitat conditions and food availability, and becomes larger in winter and smallest around the time of the arrival of newborn (Ables 1975). According to Grzimek (1975), the home range is 5 to 12 sq km in good habitat and 20 to 50 sq km in poor habitat. V. vulpes is apparently territorial. There is little overlap of home ranges, and individuals on different ranges avoid one another (Storm and Montgomery 1975). Captive males were found continually to harass and chase foxes newly introduced to the enclosure, while females seldom became involved in such interaction (Preston 1975). In the breeding season, however, females do exhibit territorial behavior (Haltenorth and Roth 1968).

A home range is typically occupied by an adult male, one or two adult females, and their young (Storm and Montgomery 1975). Occasionally, two females have litters in the same den (Pils and Martin 1978). Males may fight one another during the breeding season. A vixen sometimes mates with several males, but later establishes a partnership with just one of them (Haltenorth and Roth 1968). For a period extending from shortly before birth to several weeks thereafter, the female remains in or very near the den. The male then brings it food, but does not actually enter the maternal den.

The mating season varies with latitude. In Europe, it is December–January in the south, January–February in central regions, and February–April in the north (Haltenorth and Roth 1968). In North America, mating occurs over about the same period (Ables 1975; Storm et al. 1976). Females are monestrous, have an estrus of 1 to 6 days, and have a gestation period of 49 to 56, usually 51 to 53, days. Litter size is 1 to 13 young, but averages about 5 throughout the range of the fox (Ables 1975; H. G. Lloyd 1975). The young weigh 50–150 grams each at birth, open their eyes after 9 to 14 days, emerge from the den at 4 to 5 weeks of age, and are weaned when 8 to 10 weeks old. They may be moved to a new den at least once. The family remains together until the autumn. Sexual maturity comes at about the age of 10 months. Potential natural longevity is around 12 years, though few individuals live more than 3 or 4 years, at least where the species is heavily hunted and trapped (Ables 1975).

The red fox is killed by people for sport, to protect domes-

tic animals and game, to prevent the spread of rabies, and to obtain the valuable pelt. Sport hunting may involve an elaborate daytime chase by large numbers of riders and dogs, or a nocturnal effort by one person to lure the fox with a call imitating that of a wounded rabbit. In Britain, V. vulpes is traditionally valued as a game animal, but H. G. Lloyd (1975) noted that it is also the only mammal in the country subject to a government-approved bounty. The red fox is often considered a threat to poultry, but depredations are generally localized, and many of the birds eaten are taken in the form of carrion. Studies have indicated that the red fox has little effect on wild pheasant populations (Ables 1975). Rabid foxes are said to be a serious menace in some areas, especially Europe, and intensive persecution there may be threatening the species in certain parts of its range; 180,000 individuals are taken every year in Germany alone (Grzimek 1975). Problems caused by the fox are perhaps more than balanced by its control of rodent populations, which might otherwise multiply and damage human interests.

From 1900 to 1920 in North America, and to some extent in other parts of the world, catching wild foxes and raising them in captivity developed into an important industry. In the early stages of the breeding effort, choice animals often sold for more than $1,000 each. Through selective breeding, strains were developed that nearly always produce silver-colored offspring. The number of foxes being raised for their fur now exceeds that of all other normally wild animals, except possibly the mink (Mustela vison). One fox farm permanently employed about 400 people and sold pelts worth more than $18,000,000. The fur is used in coats, stoles, scarves, and trimming. The value of fox pelts has varied widely, depending on fashions, availability, and economic conditions. According to Banfield (1974), the average price of a silver fox skin was $246.46 in 1919/20, but only $17.94 in 1971/72. The average price of a wild-caught United States red fox skin rose from $12 in the 1970/71 season to about $48 in 1976/77. The reported number of red foxes trapped for their fur during the latter season in the United States and Canada was 421,705 (Deems and Pursley 1978).

Despite human persecution, V. vulpes has maintained or even increased its numbers in many parts of its range. There are now probably more in Britain than there were in medieval times, because of improved habitat conditions resulting from the establishment of hedgerows and crop rotation (H. G. Lloyd 1975). This species is able to carry on its mode of life in intensively farmed areas, and sometimes even in large cities (Grzimek 1975; Ables 1975). It has also been successfully introduced in some areas, especially by persons of English background, who desired to continue traditional fox hunting. The species was brought to Australia in 1868 and subsequently spread over much of that continent, to the lasting detriment of the native fauna (Clutton-Brock, Corbet, and Hills 1976; Ride 1970).

Introductions from England also were made in eastern North America in colonial times. The species was naturally present in this region, but apparently was not abundant. It subsequently increased in numbers and became established in areas not previously occupied, mainly because of the breaking up of the homogeneous forests by people and continuous introduction by hunting clubs. In the 20th century, the red fox has greatly extended its range in the southeastern United States, and has occupied Baffin Island and moved as far north as the southern coast of Ellesmere Island. It has spread westward across the Great Plains, possibly in response to a human-caused reduction of coyote (Canis latrans) numbers (Hall 1981; Banfield 1974; Lowery 1974). The only major North American population that may be in trouble from a conservation viewpoint is that of the Sierra

Corsac foxes (*Vulpes corsac*), photo from Amsterdam Zoo.

Nevada of California, where surveys indicate that the native subspecies (*V. v. necator*) is very rare and evidently declining (Schempf and White 1977).

Vulpes corsac.   Head and body length is 500 to 600 mm, and tail length is 250 to 350 mm. The fur is thick and soft. The general coloration of the upper parts is pale reddish gray, or reddish brown with silvery overtones. The underparts are white or yellow. *V. corsac* is externally similar to *V. vulpes*, but has relatively longer legs. Its ears are large, pointed, and very broad at the base (Novikov 1962).

The corsac fox is a typical inhabitant of steppes and semi-desert. It avoids forests, thickets, plowed fields, and settled areas. It lives in a burrow, often taken over from another mammal, such as a marmot or badger. Self-excavated burrows are simple, usually very shallow, and sometimes found in groups (Novikov 1962). Although usually reported to be nocturnal in the wild, *V. corsac* is active by day in captivity; it is said to be an excellent climber (Grzimek 1975). It runs with only moderate speed and can be caught by a slow dog, but has excellent senses of vision, hearing, and smell (Stroganov 1969). Most reports indicate that it is nomadic and does not keep to a fixed home range (Ognev 1962). It may migrate southward when deep snow and ice make hunting difficult (Stroganov 1969). The diet consists mostly of small rodents, but also includes pikas, birds, insects, and plant material.

This species is more social than other foxes, with several individuals sometimes living together in the same burrow (Ognev 1962). Small hunting packs are said to form in the winter (Stroganov 1969), though perhaps these represent mated pairs and their grown young of the previous spring. Males fight one another during the breeding season, but then remain with the family (Novikov 1962; Grzimek 1975). Mating occurs from January to March, gestation lasts 50–60 days, and litters usually contain 2–11 young (Stroganov 1969). Females in the Berlin Zoo did not reach sexual maturity until their third year of life (Grzimek 1975).

The corsac fox lacks the penetrating odor of most *Vulpes*, and was frequently kept as a pet in 18th-century Russia (Grzimek 1975). Its warm and beautiful fur led to large-scale commercial trapping; up to 10,000 pelts were sold annually in the west Siberian city of Irbit in the late 19th century. For this reason, and also because of the settlement and plowing of the steppes, the corsac fox has disappeared in much of its range (Stroganov 1969; Ognev 1962).

Vulpes ferrilata.   Head and body length is 575 to 700 mm, tail length is 400 to 475 mm, and males weigh up to 7 kg (Mitchell 1977). The fur is soft and thick, and the tail is bushy. The upper parts are pale gray agouti or sandy, with a tawny band along the dorsal region. The underparts are pale, the fronts of the legs are tawny, and the tip of the tail is white. The skull is peculiarly elongated and has a very narrow maxillary region (Clutton-Brock, Corbet, and Hills 1976).

Mitchell (1977) found this fox on barren slopes and in stream beds at 3,000 to 4,000 meters in the Mustang District of Nepal. In this area, dens are made in boulder piles or in burrows under large rocks. The diet consists of rodents, lagomorphs, and ground birds. Mitchell observed pairs hunting along stream beds, on boulder heaps, and in wheat fields. Mating occurs in late February, and two to five young are born in April or May.

Vulpes cana.   Novikov (1962) reported head and body length to be less than 500 mm and tail length to be 330 to 410 mm. Roberts (1977) stated that one specimen had a head and body length of 422 mm and a tail length of 298 mm. Clutton-Brock, Corbet, and Hills (1976:155) described *V. cana* as "a small fox with extremely soft fur and a long very bushy tail. The colouring is blotchy black, grey and white with a dark tip to the tail and a dark patch over the tail gland. There is an almost black mid-dorsal line and the hind legs may be dark. . . . The underparts are white, the ears are grey, and there is a small dark patch between the eyes and nose."

According to Roberts (1977), the habitat of *V. cana* is mountain steppe. It is reportedly more frugivorous than the other foxes of Pakistan, being fond of ripe melons and seedless grapes, and sometimes damaging crops. Its skin is valued in commerce and is heavily hunted. Novikov (1962) called *V. cana* one of the rarest predators of the Soviet Union. It is on appendix 2 of the CITES.

Vulpes velox.   Head and body length is 375 to 525 mm, tail length is 225 to 350 mm, and weight is 1.8 to 3 kg. Males average larger than females. The winter coat is long and dense; the upper parts are dark buffy gray; the sides, legs, and lower surface of the tail are orange tan; and the underparts are buff to pure white. In summer the coat is shorter, harsher, and more reddish. *V. velox* differs from the closely related *V. macrotis* in having smaller ears, a broader snout, and a shorter tail (Egoscue 1979).

The swift fox inhabits prairies, especially those with grasses of short and medium height. For shelter it depends on burrows, which are either self-excavated or taken over from another mammal. The burrows are usually simple and located on high, well-drained ground. The tunnels may be 350 cm long, and lead to a chamber as much as 150 cm below the surface. There are one to seven or more entrances. The swift fox is primarily nocturnal, but sometimes suns itself near the den. Its diet consists mostly of lagomorphs, and also includes rodents, birds, lizards, and insects (Egoscue 1979; Kilgore 1969).

The usual social unit is a mated pair and their young, but occasionally a male will live with two adult females. The mating season in Oklahoma is late December to early January, and most young are born in March or early April.

Females are monestrous. Litters consist of three to six young. Their eyes open after 10 to 15 days, weaning occurs after 6 or 7 weeks, and they probably remain with the parents until August or early September. A captive lived for 12 years and 9 months (Egoscue 1979; Kilgore 1969).

The swift fox is not as cautious as *V. vulpes* and seems to take poison baits readily. In the middle and late 19th century, intensive poisoning was carried out on the Great Plains, mainly to eliminate wolves, coyotes, and other predators, and many swift foxes were accidentally killed. Subsequently, much habitat was lost as the prairies were converted to agriculture. The swift fox was also taken for its fur, though it has never been of major commercial importance. By the 1920s, the northern subspecies, *V. velox hebes,* had apparently disappeared, though occasional reports continued in Canada, and the southern subspecies, *V. velox velox,* survived only in Colorado, New Mexico, and western Texas. For reasons not fully understood, a moderate comeback seems to have occurred in the last few decades. The species reappeared in Oklahoma, Kansas, Nebraska, and Wyoming in the 1950s, and in South Dakota, North Dakota, and Montana in the 1960s and 1970s (Egoscue 1979; Kilgore 1969; Moore and Martin 1980; Floyd and Stromberg 1981). *V. velox hebes* is classified as endangered by the IUCN (1976) and is on appendix 1 of the CITES. The USDI (1980) lists the subspecies as endangered, but this designation officially applies only in Canada. The swift fox populations now on the northern plains of the United States may be descended from animals that moved north from the range of *V. velox velox.*

Vulpes macrotis.   The account of this species is based in large part on McGrew (1979). Head and body length is 375 to 500 mm and tail length is 225 to 323 mm. Average weight is 2.2 kg for males and 1.9 kg for females. The back is generally light grizzled or yellowish gray, the shoulders and sides are buffy to orange, and the underparts are white. The ears are relatively the largest of any North American canid, and are set closely together. The eyes are usually more slitlike than those of *V. velox.*

The kit fox is closely associated with steppe and desert habitat, generally with a covering of shrubs or grasses. Dens

Kit fox (*Vulpes macrotis*), photo from San Diego Zoological Society.

usually have multiple entrances, the number varying from 2 to 24. There are groups of dens in favorable areas, and a fox family may move from one to another during the year, leaving most vacant at any given time. *V. macrotis* is nocturnal and may travel several kilometers per night during hunts. Its diet consists largely of rodents, such as kangaroo rats, and lagomorphs.

Optimal habitat in Utah was found to support two adults per 259 ha. Other areas had densities of one fox per 471 to 1,036 ha. In the San Joaquin Valley, Morrell (1972) found that each fox apparently spent its entire life in an area of 260 to 520 ha. Home ranges overlap extensively and apparently there is no definite territory. Usually an adult male and female live together, though not necessarily permanently, and a second female is sometimes present. When a female is nursing young, it rarely leaves the den, and the male supplies it with food. Several vocalizations are known, including a bark by mothers to recall the young. Females are monestrous. Mating occurs from December to February and the young are born in February and March. There are usually four or five offspring, each of which weighs about 40 grams. The young emerge from the den when 1 month old, and begin to accompany the parents at 3 or 4 months of age. The family splits up in the fall, with the young dispersing beyond the parental home range. According to Marvin L. Jones (Zoological Society of San Diego, pers. comm.), one specimen was still living after 20 years in captivity.

The kit fox is not a particularly cautious animal, and its numbers have been greatly reduced in some areas by poisoning, trapping, and shooting. Habitat disruption has also led to declines, especially in California. The subspecies *V. m. devia* of the southwestern corner of the state disappeared by 1910. The San Joaquin Valley subspecies, *V. m. mutica,* is classified as endangered by the USDI (1980) and as rare by the California Department of Fish and Game (1978). It has declined because of conversion of areas of natural vegetation to irrigated agriculture.

Vulpes bengalensis. The account of this species is based largely on Roberts (1977). Head and body length is 450 to 600 mm and tail length is 250 to 350 mm. Males weigh 2.7 to 3.2 kg and females weigh less than 1.8 kg. The upper parts are yellowish gray or silvery gray. The underparts are paler, and the backs of the ears and the tip of the tail are dark.

The Bengal fox is generally found in open country with a scattering of trees. It avoids deserts and mountains. It digs its own burrow and hunts mainly by night. A captive could climb low trees and a vertical wire net. The diet is omnivorous, and includes small vertebrates, insects, and fruit. *V. bengalensis* is adept at catching frogs and digging lizards out of their burrows. The young, usually four to a litter, are born from February to April.

Vulpes rueppelli. Head and body length is 400 to 520 mm and tail length is 250 to 350 mm (Roberts 1977). Weight is about 2.7 kg. The coat is very soft and dense. The upper parts are silvery gray, the sides are grayish buff, and the underparts are whitish. *V. rueppelli* is much more lightly built than *V. vulpes;* it has rather short legs and broad ears (Dorst and Dandelot 1969).

The usual habitat is stony or sandy desert. Activity is mainly nocturnal. The diet consists largely of insects. The sand fox seems rather gregarious; parties of three to five individuals have been noted. A captive female had a litter of three young, and a litter of two young was evidently born in March (Dorst and Dandelot 1969; Roberts 1977).

Vulpes pallida. Head and body length is 406 to 455 mm, tail length is 270 to 286 mm, and weight is 1.5 to 3.6 kg. The upper parts are pale sandy fawn, variably suffused with blackish. The flanks are paler and the underparts are buffy white. The tail is long and bushy (Dorst and Dandelot 1969; Rosevear 1974).

The habitat is savannah. Burrows are large, with tunnels extending 10 to 15 meters and opening into chambers lined

Sand fox (*Vulpes rueppelli*), photo from Antwerp Zoo.

Cape fox (*Vulpes chama*), photo from San Diego Zoological Society.

with dry vegetation. Activity is mainly nocturnal. The diet includes rodents, small reptiles, birds, eggs, and vegetable matter. The pale fox is gregarious (Dorst and Dandelot 1969; Meester and Setzer 1971). Three captive adults, a female and two males, seemed to get along amicably. The female gave birth to a litter of four young in June 1965 (Bueler 1973).

Vulpes chama. Head and body length is about 560 mm, tail length is about 330 mm, and weight is about 4 kg. The upper parts are silvery gray and the underparts are pale buff. The tail is very bushy and has a black tip. The ears are pointed. The muzzle is short, but pointed (Dorst and Dandelot 1969).

The Cape fox inhabits dry country, mainly open plains and karroo. It is nocturnal, hiding by day under rocks or in burrows in sandy soil. The diet consists mainly of small vertebrates and insects. This species lives alone or in pairs. Its call is a yell followed by several yaps. The breeding season is September–October, gestation lasts 51–52 days, and litters contain three to five young. Human persecution has been responsible for numerical and distributional declines, though studies have shown that *V. chama* is not a harmful predator (Dorst and Dandelot 1969; Bekoff 1975; Bothma 1966).

CARNIVORA; CANIDAE; **Genus FENNECUS** Desmarest, *1804*

Fennec Fox

The single species, *F. zerda,* occurs in the desert zone from southern Morocco and Niger to Egypt and Sudan (Meester and Setzer 1971). Only two occurrences in Asia have been confirmed—one in Sinai and one in Kuwait—but the species probably has a much wider distribution in the Arabian region (Harrison 1968). *Fennecus* was included in the genus *Vulpes* by Clutton-Brock, Corbet, and Hills (1976), and Corbet (1978), and was included in the subgenus *Vulpes* of the genus *Canis* by Van Gelder (1978).

This is the smallest canid, but it has the relatively largest ears in the family. Head and body length is 357 to 407 mm, tail length is 178 to 305 mm, and weight is about 1 to 1.5 kg. The ears are 100 to 150 mm long. The coloration of the upper parts, the palest of all foxes, is reddish cream, light fawn, or almost white. The underparts are white and the tip of the tail is black. The coat is thick, soft, and long. The tail is heavily furred. The bullae are exceedingly large and the dentition is weak (Clutton-Brock, Corbet, and Hills 1976). The feet have hairy soles, enabling the animal to run in loose sand (Bekoff 1975).

Fennecus occurs in arid regions and usually lives in burrows several meters long in the sand. It digs so rapidly that it has gained the reputation of being able to sink into the ground. It is nocturnal and quite agile; a captive could spring 60 to 70 cm upward from a standing position, and could jump about 120 cm horizontally. Some food is apparently obtained by digging, as evidenced by the pronounced scratching or raking habit of captives. The diet includes plant material, small rodents, birds and their eggs, lizards, and insects, such as the noxious migratory locusts. Although an abundance of fennec tracks around some waterholes indicates that this fox drinks freely when the opportunity arises, travelers also report tracks in the desert far from oases. Laboratory studies suggest that *Fennecus* can survive without free water for an indefinite period (Banholzer 1976).

The fennec lives in groups of up to 10 individuals. Males mark territories with urine and become aggressive during the breeding season. One captive male became dominant over a group at the age of 4 years, and then killed its 8-year-old father. Females are aggressive and defend the nest site when they have newborn offspring. Males remain with their mates after the young are born, and defend them, but do not enter the maternal den. Mating occurs in January and February in captivity, and the young are born in late winter and early spring. Females normally give birth once a year, but if the first litter is lost, another may be produced 2½ to 3 months later. The gestation period is about 50 to 52 days. There are 2 to 5 young per litter. The young are weaned at 61 to 70 days of age, and become sexually mature at 11 months (Koenig 1970; Bekoff 1975). One captive lived for 14 years and 3 months (Marvin L. Jones, Zoological Society of San Diego, pers. comm.).

Although it does no harm to human interests, the fennec is intensively hunted by the native people of the Sahara. It has become rare in some parts of northwestern Africa (Grzimek 1975).

Fennec foxes (*Fennecus zerda*), photo from New York Zoological Society.

CARNIVORA; CANIDAE; *Genus UROCYON* Baird, 1858

Gray Foxes

There are two species (Hall 1981; Cabrera 1957):

U. cinereoargenteus, Oregon and southeastern Canada to western Venezuela:

U. littoralis, San Miguel, Santa Rosa, Santa Cruz, Santa Catalina, San Nicolas, and San Clemente islands off southwestern California.

Urocyon was included in the genus *Vulpes* by Clutton-Brock, Corbet, and Hills (1976), and in the subgenus *Vulpes* of the genus *Canis* by Van Gelder (1978). All other authorities cited in this account treated *Urocyon* as a full genus.

In *U. cinereoargenteus,* head and body length is 483 to 685 mm and tail length is 275 to 445 mm. In *U. littoralis,* head and body length is 480 to 500 mm and tail length is 110 to 290 mm. Usual weight in the genus is 2.5 to 7 kg. The face, upper part of the head, back, sides, and most of the tail are gray. The throat, insides of the legs, and underparts are white. The sides of the neck, lower flanks, and ventral part of the tail are rusty. The hairs along the middle of the back and top of the tail are heavily tipped with black, which gives the effect of a black mane. Black lines also occur on the legs and face of most individuals. A concealed mane of stiff hairs occurs on top of the tail. The pelage is coarse. The skull of *Urocyon* is distinguished from that of *Vulpes* in having a deeper depression above the postorbital process, much more pronounced and more widely separated ridges extending from the postorbital processes to the posterior edges of the parietals, and a conspicuous notch toward the rear of the lower edge of the mandible (Lowery 1974).

Gray foxes frequent wooded and brushy country, often in rocky or broken terrain, and are possibly most common in the arid regions of the southwestern United States and Mexico. The preferred habitat in Louisiana is mixed pine-oak woodland bordering pastures and fields with patches of weeds (Lowery 1974). Several sheltered resting sites may be used on different days. The main den is in a pile of brush or rocks, a crevice, a hollow tree, or a burrow that is either self-excavated or taken over from another animal. Dens in hollow trees have been found up to 9.1 meters above the ground. Dens used for giving birth may be lined with vegetation (Trapp and Hallberg 1975).

Urocyon is sometimes called a tree fox, because it frequently climbs trees, a rather unusual habit for a canid. It lacks the endurance of the red fox and, when pursued, often will seek refuge in a tree. It also climbs without provocation, shinnying up the trunk and then leaping from branch to branch. Most activity is nocturnal and crepuscular. Nightly movements cover about 200 to 700 meters (Trapp and Hallberg 1975). The diet includes many kinds of small vertebrates as well as insects and vegetable matter. *Urocyon* seems to take plant food more than do other foxes, and its diet may consist mostly of fruits and grains at certain seasons and places.

Reported population densities are about 0.4 to 10 foxes per sq km (Trapp and Hallberg 1975). Reported home range size varies from 0.13 to 7.7 sq km. Four females in central California, followed by radio-tracking for various periods from January to July, used ranges of 0.3 to 1.85 sq km (Fuller 1978). Apparently, each family group uses a separate area. The normal social unit is an adult pair and their young. There is one litter per year. Mating occurs from late December to March in the southeastern United States, and from mid-January to late May in New York. The gestation period has never been precisely determined, but has been variously estimated at 51 to 63 days. Litters usually contain about 4 young, but may have from 1 to 10. The young weigh around 100 grams each at birth, are blackish in color, and open their eyes after 9 to 12 days. They can climb vertical tree trunks after 1 month of life, and begin to take solid food in 6 weeks. They forage independently by late summer or early fall, but apparently remain in the parental home range until January or February. Most females breed in their first year of life (Trapp and Hallberg 1975). A captive lived for 13 years and 8 months (Marvin L. Jones, Zoological Society of San Diego, pers. comm.).

If captured when small, gray foxes tame readily, are as affectionate and playful as domestic dogs, and make more satisfactory pets than do red foxes. The attractive skins of

Gray fox (*Urocyon cinereoargenteus*), photo from San Diego Zoological Society.

Urocyon are used commercially, but are not classed as fine furs. For much of the 20th century the price per pelt averaged around 50¢, but it began to rise in the 1960s (Lowery 1974). For the 1970/71 season the reported harvest in the United States was 26,109 skins with an average value of $3.50. By the 1976/77 season the reported take had grown to 225,277 pelts, and the average price to $34.00 (Deems and Pursley 1978). The island fox, *U. littoralis,* is classified as rare by the California Department of Fish and Game (1978), and is fully protected by state law.

CARNIVORA; CANIDAE; **Genus ALOPEX** Kaup, 1829

Arctic Fox

The single species, *A. lagopus,* occurs on the tundra and adjacent lands and ice-covered waters of northern Eurasia, North America, Greenland, and Iceland (Hall 1981; Corbet 1978; Chesemore 1975; Banfield 1974). *Alopex* was included within the genus *Vulpes* by Youngman (1975) and was considered a subgenus of *Canis* by Van Gelder (1978), but was treated as a full genus by Clutton-Brock, Corbet, and Hills (1976), as well as by all other authorities cited in this account.

Head and body length is 458 to 675 mm, tail length is 255 to 425 mm, shoulder height is about 280 mm, and weight is 1.4 to 9 kg. The dense, woolly coat gives *Alopex* a heavy appearance. There are two color phases. Individuals of the "white" phase are generally white in winter and brown in summer, but may remain fairly dark throughout the year in areas of less severe climate. Individuals of the "blue" phase are pale bluish gray in winter and dark bluish gray in summer. The blue phase constitutes less than 1 percent of the arctic fox population of most of mainland Canada, and less than 5 percent of that of Baffin Island, but makes up 50 percent of the population of Greenland. The winter pelage develops in October and is shed in April. In addition to its coloration, *Alopex* differs from *Vulpes* in having short, rounded ears, and long hairs on the soles of its feet.

Alopex is found primarily in arctic and alpine tundra, usually in coastal areas. It generally makes its den in a low mound, 1 to 4 meters high, on the open tundra. Dens usually have 4 to 12 entrances, and a network of tunnels covering about 30 sq meters. Some dens may be used for centuries, by many generations of foxes, and eventually become very large, with up to 100 entrances. The organic matter that accumulates in and around a den stimulates a much more extensive growth of vegetation than is found on most of the tundra. *Alopex* sometimes dens in a pile of rocks at the base of a cliff. It may use its den throughout the year, but often seems not to have a fixed home site, except when rearing young. It is active at any time of the day and throughout the year. It moves easily over snow and ice, and swims readily (Chesemore 1975; Macpherson 1969; Stroganov 1969; Banfield 1974). During blizzards it may shelter in burrows dug in the snow. Several individuals were noted in winter on the Greenland icecap, more than 450 km from the nearest ice-free land, and with the temperature below −50° C. Captives have survived experimental temperatures as low as −80° C.

The arctic fox makes the most extensive movements of any terrestrial mammal other than *Homo sapiens*. Over much of its range, including Alaska, there is a basic seasonal pattern, probably associated with food availability (Chesemore 1975). In the fall and early winter the animals shift toward the shore and out onto the pack ice. They are capable of remarkably long travels across the sea ice, having been sighted 640 km north of the coast of Alaska. One individual reportedly reached a latitude of 88° N, at a point 800 km from the nearest land in the Soviet Union. Another fox, tagged 8 August 1974 on Banks Island, was trapped on 15 April 1975 on the northeastern mainland of the Northwest Territories, after having covered a straight-line distance of 1,530 km (Wrigley and

Arctic foxes (*Alopex lagopus*): A. White and blue phases; B. Blue phase; C. Summer coat; photos by Alwin Pedersen; D. Winter coat, photo by Ernest P. Walker.

Hatch 1976). Still another animal, tagged in northeastern Alaska, reached a point on Banks Island, 945 km away (Eberhardt and Hanson 1978). Some foxes have been carried by ice floes as far as Cape Breton and Anticosti islands in the Gulf of St. Lawrence (Banfield 1974).

In some areas, though apparently not in Alaska, there are inland migrations to the forest zone during the winter, in addition to or instead of a shift onto the sea ice. These movements sometimes involve large numbers of foxes and seem especially extensive following a crash in lemming populations, a major food source (Chesemore 1975; Banfield 1974; Pulliainen 1965). *Alopex* has occasionally traveled overland to south-central Ontario and the St. Lawrence River. The deepest known penetration in North America was made by an individual taken in December 1974 in southern Manitoba, nearly 1,000 km south of the tundra. In the Soviet Union,

however, inland movements have extended as far as 2,000 km (Wrigley and Hatch 1976).

Alopex takes any available animal food, alive or dead, and frequently stores food for later use. It may follow polar bears on the pack ice, and wolves on land, in order to obtain carrion (Banfield 1974). Its winter diet includes the remains of marine mammals, invertebrates, sea birds, and fish (Chesemore 1975). It also is known to be an important predator of the ringed seal (*Phoca hispida*), digging through the snow to reach the pups in their subnivean lairs (T. G. Smith 1976).

In winter, for those populations that move inland, and in summer, throughout most of the range of *Alopex*, the diet consists mainly of lemmings. Other small mammals, ground-nesting birds, and stranded marine mammals are also utilized, but arctic fox populations seem to be associated with those of lemmings in a three- to five-year cycle (Chesemore

1975; Wrigley and Hatch 1976). Following a lemming crash, population density has been estimated at only 0.086 foxes per sq km (Banfield 1974).

Although a number of dens may be found in a favorable area, most are not utilized at any one time. Occupied dens are at least 1.6 km apart and are usually distributed such that there is only one family in every 32 to 70 sq km (Macpherson 1969). *Alopex* is monogamous and may mate for life (Chesemore 1975). A number of adults sometimes gather temporarily around a food source, such as a stranded whale, but they then may fight one another. Vocalizations include barks, screams, and hisses (Banfield 1974).

Mating occurs from February to May and births take place from April to July. Females are monestrous, and have an estrus of 12–14 days and an average gestation period of 52 (49–57) days. The number of young per litter varies, depending on environmental conditions, but ranges from 2 to as many as 25. The usual number seems to be about 6 to 12. The young weigh an average of 57 grams each at birth. They emerge from the den, and are weaned, when around 2 to 4 weeks old. Subsequently, for a brief period, they are brought food by both parents, but by autumn they disperse. They are capable of breeding at an age of 10 months. Most young do not survive their first 6 months of life, and few animals live more than several years in the wild (Macpherson 1969; Banfield 1974; Chesemore 1975; Stroganov 1969). A captive, however, lived for 15 years (Marvin L. Jones, Zoological Society of San Diego, pers. comm.).

The raising of blue-phase foxes has been an important industry, as the undressed skins have sold for up to $300 each. However, prices have sometimes fallen below $10, thereby forcing many fox farmers out of business. Some operations have pens, where selective breeding is practiced, but others are on small islands, where the animals run at liberty. Fox farming began in Alaska in 1865, and blue-phase animals were extensively introduced on the Aleutian Islands (Chesemore 1975). These animals increased in numbers and jeopardized or exterminated some populations of native birds. *Alopex* apparently did not originally occur in the Aleutians, except for the islands at the extreme western end of the chain (Murie 1959).

White-phase foxes also have desirable furs, and have been subject to intensive trapping. The skins may be left the natural white or dyed one of many colors, especially "platinum" or "blue" imitation. The arctic fox has long been important to the economy of the native people living within its range. Fluctuations in fox numbers and fur prices have frequently combined to cause hardship for native trappers. Trade in white fox skins developed in northern Alaska during the 19th century, in conjunction with the whaling industry. In the 1920s, fox trapping was the most important source of income in the area, and the price per pelt averaged $50. The Depression destroyed the market, and by 1931 skins sold for $5 or less (Chesemore 1972). Recently, prices began to rise again. During the 1976/77 season, the number of skins marketed and the average price per pelt were: Alaska, 4,261 and $36.00; and Canada, 36,482 and $54.20 (Deems and Pursley 1978).

CARNIVORA; CANIDAE; *Genus **LYCALOPEX***
Burmeister, 1854

Hoary Fox

Langguth (1975*b*) considered *Lycalopex* a full genus with a single species, *L. vetulus,* occurring in the states of Mato Grosso, Goias, Minas Gerais, and Sao Paulo in south-central Brazil. *Lycalopex* was treated as a subgenus of *Dusicyon* by Cabrera (1957) and as a subgenus of *Canis* by Van Gelder (1978). *L. vetulus* was included within *Dusicyon* by Clutton-Brock, Corbet, and Hills (1976), who noted, however, that it was the most foxlike member of that genus.

Head and body length is about 585 to 640 mm, tail length is 280 to 320 mm, and weight is about 4 kg. The coat is short. The upper parts are a mixture of yellow and black, giving an overall gray tone. The ears and the outsides of the legs are reddish or tawny. The tail has a black tip and a marked dark stripe along the dorsal line. The underparts are cream to fawn. Compared to *Dusicyon,* the muzzle is short, the skull and teeth are small, the carnassials are reduced, and the molars are broad (Bueler 1973; Clutton-Brock, Corbet, and Hills 1976; Grzimek 1975).

The habitat is grassy savannah on smooth uplands, or savannahs with scattered trees. A deserted armadillo burrow may be used for shelter and for the natal nest. The diet consists of small rodents, birds, and insects, especially grasshoppers. The dentition suggests substantial dependence on insects, but *Lycalopex* is persecuted by people because of

Hoary fox (*Lycalopex vetulus*), photo by E. K. P. da Silveira.

South American fox (*Pseudalopex culpaeus*), photo by Ernest P. Walker.

presumed predation on domestic fowl (Langguth 1975b). It is usually timid, but courageously defends itself and its young. Births occur in the spring, there usually being two to four young per litter (Bueler 1973; Grzimek 1975).

CARNIVORA; CANIDAE; **Genus PSEUDALOPEX**
Burmeister, 1856

South American Foxes

There are four species (Langguth 1975b; Van Gelder 1978; Clutton-Brock, Corbet, and Hills 1976):

P. sechurae, arid coastal zone of southwestern Ecuador and northwestern Peru;

P. griseus, plains and low mountains of Chile, western Argentina, and Patagonia;

P. gymnocercus, humid grasslands of southern Brazil, Paraguay, northern Argentina, and Uruguay;

P. culpaeus, the Andes region from Ecuador to Patagonia.

Pseudalopex was considered a subgenus of *Canis* by Langguth (1975b) and Van Gelder (1978). The four species listed above were included in the genus *Dusicyon* by Clutton-Brock, Corbet, and Hills (1976), along with the type species of that genus, *D. australis.* Information provided by these various authorities suggests that *D. australis* is at least as different from the species of *Pseudalopex* as it is from *Canis,* and also that *Pseudalopex* is at least as different from *Canis* as it is from *Vulpes.* Therefore, since *Dusicyon* (for the species *D. australis*) and *Vulpes* are here being maintained as separate genera, it is also advisable to give generic rank to *Pseudalopex.*

Head and body length is 600 to 1,200 mm, tail length is 300 to 500 mm, and weight is 4 to 13 kg. *P. culpaeus* is the largest species, and *P. sechurae* the smallest. The coat is usually heavy, with a dense underfur and long guard hairs. The upper parts are generally gray agouti with some ochraceous or tawny coloring (Clutton-Brock, Corbet, and Hills 1976). The head, ears, and neck are often reddish. The underparts are usually pale. The tail is long, bushy, and black tipped. There is some resemblance to a small coyote (*Canis latrans*). The dentition, however, is more foxlike than doglike; the molars are well developed, and the carnassials are relatively short (Clutton-Brock, Corbet, and Hills 1976).

Habitats include sandy deserts for *P. sechurae;* low, open grasslands and forest edge for *P. griseus;* pampas, hills, deserts, and open forests for *P. gymnocercus;* and dry rough country and mountainous areas, up to 4,500 meters in elevation, for *P. culpaeus* (Langguth 1975b; Crespo 1975; Grzimek 1975). Dens are usually among rocks, under bases of trees and low shrubs, or in burrows made by other animals, such as viscachas and armadillos. Most activity is nocturnal, but some individuals are occasionally active during the day. *P. gymnocercus* sometimes collects and stores objects, such as strips of leather and cloth. This species may freeze and remain motionless upon the appearance of a human being; in one case it reportedly did not move even when approached and struck with a whip handle. The omnivorous diet includes rodents, lagomorphs, birds, lizards, frogs, insects, fruit, and sugar cane.

Studies of stomach contents indicate that *P. gymnocercus* takes an equal amount of plant and animal food. *P. culpaeus* seems more carnivorous than other species, and reportedly sometimes preys heavily on introduced sheep and European hares. In western Argentina, during the spring, part of the population of *P. culpaeus* shifts 15 to 20 km into the higher mountains, in response to the seasonal movements of the sheep and hares. Its normal home range in that area is 4 km in diameter (Crespo 1975).

The voice of *Pseudalopex* has been described as a howl or a series of barks and yaps. It is heard mainly at night, especially during the breeding season. According to Crespo (1975), *P. culpaeus* and *P. gymnocercus* mate from August to October and give birth from October to December (the austral spring). Females are monestrous. The gestation period is 55 to 60 days. Embryo counts range from one to eight, with averages of about four in *P. gymnocercus* and five in *P. culpaeus*. The male helps to provide food to the family. After 2 to 3 months of life, the young begin to hunt with the parents. Few individuals live more than several years in the wild, but a captive *P. gymnocercus* survived 13 years and 8 months (Marvin L. Jones, Zoological Society of San Diego, pers. comm.).

These canids are killed by people, because they are alleged to prey on domestic fowl and sheep, and because their fur is desirable. Populations of *Pseudalopex* have thus declined in some areas, such as in Buenos Aires Province, Argentina. Crespo (1975) noted, however, that *P. culpaeus* occurred in relatively low numbers in Neuquen Province, western Argentina until the early 20th century, and then greatly increased in response to the introduction of sheep and European hares. *P. culpaeus* and *P. griseus* are on appendix 2 of the CITES.

CARNIVORA; CANIDAE; *Genus DUSICYON*
Hamilton Smith, 1839

Falkland Island Wolf

This genus is here considered to include a single species, *D. australis*, which formerly occurred on West and East Falkland Islands off the southeastern coast of Argentina. Additional species have usually been assigned to *Dusicyon*. Cabrera (1957) recognized two subgenera: *Dusicyon*, with the species here placed in the genus *Pseudalopex;* and *Lycalopex*, which is here treated as a full genus. Clutton-Brock, Corbet, and Hills (1976) included within *Dusicyon* all of those species that are here placed in the genera *Lycalopex, Pseudalopex, Dusicyon, Cerdocyon*, and *Atelocynus*. These authorities indicated, however, that *D. australis* is at least as close, systematically, to some species of *Canis* as it is to the species here assigned to *Pseudalopex*. They even discussed the possibility that *D. australis* is a form of *Canis familiaris*. Both Langguth (1975b) and Van Gelder (1978) considered *Dusicyon* to be a subgenus of *Canis* and to include only *D. australis*.

Only 11 specimens are known, and not all include skins. In one specimen, head and body length was 970 mm and tail length was 285 mm. The upper parts are brown with some rufous and a speckling of white, and the underparts are pale brown. The coat is soft and thick. The tail is short, bushy, and tipped with white. The face and ears are short, and the muzzle is broad. The skull is large and has inflated frontal sinuses, more like the situation of *Canis* than of *Pseudalopex* (Clutton-Brock, Corbet, and Hills 1976; Allen 1942).

Dusicyon was the only mammal found on the Falkland Islands by the early explorers. Its natural diet consisted mainly of birds, especially geese and penguins, and also included pinnipeds. Its presence on the islands, about 400 km from the mainland, is something of a mystery. A land connection may have existed during part of the Pleistocene, but Clutton-Brock, Corbet, and Hills (1976) thought it much more likely that *Dusicyon* had been taken to the Falklands as a domestic

animal by prehistoric Indians. Possible descent from either *Pseudalopex* or *Canis* was suggested.

Dusicyon demonstrated remarkable tameness toward people. Individuals waded out to meet landing parties. Later they came into the camps in groups, carried away articles, pulled meat from under the heads of sleeping men, and stood about while their fellow animals were being killed. The dogs were often killed by a man holding a piece of meat as bait in one hand and stabbing the dog with a knife in the other hand.

Although *Dusicyon* was discovered in 1690, it was still common, and still behaving in a very tame manner, when Darwin visited the Falklands in 1833. In 1839, however, large numbers were killed by fur traders from the United States. In the 1860s, Scottish settlers began raising sheep on the islands. *Dusicyon* preyed on the sheep and was therefore intensively poisoned. The genus was very rare by 1870, and the last individual is said to have been killed in 1876 (Allen 1942).

CARNIVORA; CANIDAE; *Genus CERDOCYON*
Hamilton Smith, 1839

Crab-eating Fox

The single species, *C. thous*, has been recorded from Colombia, Venezuela, Guyana, Surinam, eastern Peru, Bolivia, Paraguay, Uruguay, northern Argentina, and most of Brazil outside of the lowlands of the Amazon Basin (Langguth 1975b; Husson 1978; Grimwood 1969). *Cerdocyon* was recognized as a distinct genus by Cabrera (1957) and Langguth (1975b), as a part of *Dusicyon* by Clutton-Brock, Corbet, and Hills (1976), and as a subgenus of *Canis* by Van Gelder (1978).

Head and body length is 600 to 700 mm, tail length is about 300 mm, and weight is 6 to 7 kg. The coloration is variable, but the upper parts are usually grizzled brown to gray, often with a yellowish tint, and the underparts are brownish white. The ears are short, and ochraceous or rufous. The tail is fairly long, bushy, and either totally dark or black tipped (Clutton-Brock, Corbet, and Hills 1976). The legs may be tawny and are relatively short and robust (Brady 1979).

The information for the remainder of this account was taken from Brady (1978, 1979). *Cerdocyon* inhabits woodlands and savannahs, and is mainly nocturnal. On the llanos of Venezuela, during the wet season, it uses high ground and shelters by day under brush. During the dry season it occupies lowlands and spends the day in clumps of matted grass. These grass shelters have several entrances, are used repeatedly, and may serve as natal dens. *Cerdocyon* forages from about 1800 to 2400 hrs. It stalks and pounces on small vertebrates, and apparently listens for crabs in tussocks of grass. In the dry season the percentage composition of its diet is: vertebrates, 48; crabs, 31; insects, 16; carrion, 3; and fruit, 2. In the wet season the percentage breakdown is: insects, 54; vertebrates, 20; fruit, 18; and carrion, 7.

Three pairs occupied home ranges of approximately 54, 60, and 96 ha. The ranges overlapped to some extent, but were regularly marked with urine. Tolerance of neighbors was shown in the wet season, but aggression increased in the dry season. Mated pairs form a lasting bond and commonly travel together, but do not usually hunt cooperatively. Breeding may take place throughout the year, but births peak in January and February on the llanos of Venezuela. Captive females produced two litters annually at intervals of about 8 months. The gestation period is 52–59 days and litter size is three to six young. The young weigh 120–160 grams at birth,

Crab-eating fox (*Cerdocyon thous*), photo from San Diego Zoological Garden.

open their eyes at 14 days, begin to take some solid food at 30 days, and are completely weaned by about 90 days. Both parents guard and bring food to the young. Independence comes at 5 or 6 months of age, and sexual maturity at about 9 months.

CARNIVORA; CANIDAE; **Genus ATELOCYNUS** *Cabrera, 1940*

Small-eared Dog

The single species, *A. microtis*, inhabits the Amazon, upper Orinoco, and upper Parana basins in Brazil, Peru, Ecuador, Colombia, and probably Venezuela. *Atelocynus* was recognized as a distinct genus by Cabrera (1957) and Langguth (1975*b*), as a part of *Dusicyon* by Clutton-Brock, Corbet, and Hills (1976), and as a subgenus of *Canis* by Van Gelder (1978).

Head and body length is 720 to 1,000 mm, tail length is 250 to 350 mm, shoulder height is about 356 mm, and weight is around 9 or 10 kg. The ears, rounded and relatively shorter than those of any other canid, are only 34 to 52 mm in length. The upper parts are dark gray to black, and the underparts are rufous mixed with gray and black. The thickly haired tail is black, except for the paler basal part on the underside. It has been reported to sweep the ground, if hanging perpendicularly. However, when captives at the Brookfield Zoo were standing, they curved the tail forward and upward

against the outer side of a hind leg, so that the terminal hairs did not drag on the ground. The temporal ridges of the skull are strongly developed, and the frontal sinuses and cheek teeth are relatively large (Clutton-Brock, Corbet, and Hills 1976).

Atelocynus inhabits tropical forests from sea level to about 1,000 meters. It moves with a catlike grace and lightness not observed in any other canid. A male, when in captivity in Bogota, ate raw meat, shoots of grass, and the common foods that people eat. This male and a female were brought to the Brookfield Zoo in Chicago. They proved to be completely different in temperament. The male was exceedingly friendly and docile, but the female exhibited constant hostility. Although the male was shy in captivity before being sent to Brookfield, and growled and snarled when angry or frightened, it did not show any unfriendly actions at the Chicago zoo and, in fact, became very tame. It permitted itself to be hand fed and petted by persons it recognized, responding to petting by rolling over on its back and squealing. This male came to react to attention from familiar people by a weak but noticeable wagging of the back part of its tail. The female, on the other hand, when under direct observation, emitted a continuous growling sound, without opening its mouth or baring its teeth.

The odor from the anal glands was strong and musky in the male, but was scarcely detectable in the female. In both sexes the eyes glowed remarkably in dim light. The male, though smaller, was dominant in most activities. Although some snapping was observed between the two animals, no biting or fighting was noted, and they occupied a common sleeping

Small-eared dog (*Atelocynus microtis*), photo from Field Museum of Natural History.

box when not active. According to Marvin L. Jones (Zoological Society of San Diego, pers. comm.), one captive lived for 11 years.

The IUCN (1976) classifies *Atelocynus* as rare. It may be at risk from indirect influence of human intrusion throughout its range. It is protected by law in Brazil and Peru.

CARNIVORA; CANIDAE; **Genus CANIS** Linnaeus, *1758*

Dogs, Wolves, Coyotes, Jackals

There are eight species (Clutton-Brock, Corbet, and Hills 1976; Nowak 1979; Corbet 1978; Meester and Setzer 1971; Kingdon 1977; Hall 1981):

C. simensis (Simien jackal), mountains of central Ethiopia;
C. adustus (side-striped jackal), open country from Senegal to Somalia, and south to northern Namibia and eastern South Africa;
C. mesomelas (black-backed jackal), open country from Sudan to South Africa;
C. aureus (golden jackal), Balkan Peninsula to Thailand and Sri Lanka, Morocco to Egypt and northern Tanzania;
C. latrans (coyote), Alaska to New Brunswick and Costa Rica;
C. rufus (red wolf), central Texas to southern Pennsylvania and Florida;
C. lupus (gray wolf), Eurasia except tropical forests of southeastern corner, Alaska, Canada, Greenland, conterminous United States except southeastern quarter and most of California, highlands of Mexico;
C. familiaris (domestic dog), world-wide in association with people, extensive feral populations in Australia and New Guinea.

Van Gelder (1977*b*, 1978) considered *Canis* to comprise *Vulpes* (including *Fennecus* and *Urocyon*), *Alopex*, *Lycalopex*, *Pseudalopex*, *Dusicyon*, *Cerdocyon*, and *Atelocynus* as subgenera. *C. simensis* has sometimes been placed in a separate genus or subgenus, *Simenia* Gray, 1868. *C. familiaris* is sometimes included in *C. lupus*. The feral populations of *C. familiaris* in Australia are sometimes considered to represent a distinct species, *C. dingo* or *C. antarcticus*. Lawrence and Bossert (1967, 1975) suggested that *C. rufus* is not more than subspecifically distinct from *C. lupus*.

Canis is characterized by a relatively high body, long legs, and a bushy, cylindrical tail. The pupils of the eyes generally appear round in strong light. Although most species have a scent gland near the base of the tail, it does not produce as strong an odor as that of *Vulpes*. The skull has large frontal sinuses, and temporal ridges that are close together, often uniting to form a sagittal crest. The facial region of the skull, except in *C. simensis*, is relatively shorter than in *Vulpes* and *Pseudalopex* (Clutton-Brock, Corbet, and Hills 1976). Females have 8 or 10 mammae. Additional information is provided separately for each species.

Canis simensis. Head and body length averages about 990 mm, tail length about 250 mm, and shoulder height about 600 mm. The general coloration of the upper parts is tawny rufous with pale ginger underfur. The chin, insides of the ears, chest, and underparts are white. There is a distinctive white band around the ventral part of the neck. The tail is rather short, the facial region of the skull is elongated, and the teeth, especially the upper carnassials, are relatively small (Clutton-Brock, Corbet, and Hills 1976).

The present habitat is montane grassland and moorland at elevations of approximately 3,000 to 4,000 meters. *C. simensis* may have been found at lower elevations before becoming subject to severe human persecution (Yalden, Largen, and Kock 1980). The remainder of the account of this species is based largely on studies by Morris and Malcolm (1977) in the Bale Mountains of south-central Ethiopia. In this area the Simien jackal was found to be relatively

Top, golden jackal (*Canis aureus*); Bottom, dingos (*Canis familiaris*); photos by Lothar Schlawe.

numerous in grasslands supporting large rodent populations. It was uncommon in scrub and was not seen in forests at low altitudes. It was primarily diurnal, but there was some evidence of activity at night, especially in moonlight. Farther north, in the Simien Mountains, the few remaining jackals were reported to be almost entirely nocturnal, probably because of human disturbance. The animals made little effort to find shelter, and individuals slept in the open or in places with slightly longer grass than usual, even when temperatures were as low as −7° C.

The diet consisted mainly of diurnal rodents, especially *Tachyoryctes* and *Otomys*. Hunting was done mainly by walking slowly through areas of high rodent density, investigating holes and listening carefully, and then stealthily creeping toward the prey and capturing it with a final dash. Individuals generally hunted alone and did not come together during the day, though there was apparently considerable overlap in hunting ranges. Highest population density of *C. simensis* was perhaps two individuals per sq km. No direct evidence of territoriality was observed, but urine marking was frequent, and some animals seemed to avoid others. Around dawn and dusk, most jackals came together in groups of up to seven members. There was then much friendly social interaction and noise. Vocalizations varied from a howl to a staccato scream interspersed with barks, and also included yips and whines. Pairs appeared to be forming in January, but births apparently did not occur until May or June.

The Simien jackal was reported from most Ethiopian provinces in the 19th century. It subsequently declined because of agricultural development in its range and environmental disruption that reduced prey populations. Also, it was incorrectly believed to be a predator of sheep and so was frequently shot. At present only about 40 individuals of the subspecies *C. s. simensis* remain in the Simien Mountains and nearby parts of north-central Ethiopia. There are estimated still to be 350 to 475 individuals of the subspecies *C. s. citerni* in the Bale Mountains, and a few others in neighboring highlands. There is a national park in the Simien Mountains and another planned for the Bale Mountains, but these areas are still being used for grazing. *C. simensis* is protected by law in Ethiopia, and is classified as endangered by the IUCN (1978) and the USDI (1980).

Canis adustus.   Head and body length is 650 to 810 mm, tail length is 300 to 410 mm, shoulder height is 410 to 500 mm, and weight is 6.5 to 14 kg (Kingdon 1977). Males are larger than females. The coat is long and soft. The upper parts are generally mottled gray. On each side of the body is a line of white hairs, followed below by a line of dark hairs. The underparts and tip of the tail are white (Clutton-Brock, Corbet, and Hills 1976).

According to Kingdon (1977), the side-striped jackal is widespread in the moister parts of savannahs, thickets, forest edge, cultivated areas, and rough country up to 2,700 meters in elevation. Favored denning sites include old termite mounds, abandoned aardvark holes, and burrows dug into hillsides. The animal is strictly nocturnal in areas well settled by people. It is a more omnivorous scavenger than other kinds of jackal, taking a variety of invertebrates, small vertebrates, carrion, and plant material. Social groups are well spaced and usually consist of a mated pair and their young. In East Africa births occur mainly in June–July and September–October, gestation lasts 57–70 days, and litters consist of three to six young. Rosevear (1974) wrote that captives have lived up to 10 years.

Canis mesomelas.   Head and body length is 680 to 745 mm, tail length is 300 to 380 mm, shoulder height is 300 to 480 mm, and weight is 7 to 13.5 kg. Males average about 1 kg heavier than females. There is a dark saddle extending the length of the back to the black tip of the tail. The sides, head, limbs, and ears are rufous. The underparts are pale ginger. The build is slender and the ears are very large (Kingdon 1977; Clutton-Brock, Corbet, and Hills 1976; Bekoff 1975).

The black-backed jackal is found mainly in dry grassland, brushland, and open woodland. It generally dens in an old termite mound or aardvark hole. It is partly diurnal and crepuscular in undisturbed areas, but becomes nocturnal in places intensively settled by people. The diet includes a substantial proportion of plant material, insects, and carrion, but *C. mesomelas* frequently hunts rodents and is capable of killing antelopes up to the size of *Gazella thomsoni*. Some food may be cached (Kingdon 1977; Smithers 1971; Bothma 1971b).

Home ranges may be several kilometers across, and a portion is a territory that is marked with urine and defended by both sexes. Although 20 or 30 individuals may gather around a lion kill, the basic social unit is a mated pair and their young (Kingdon 1977; Bekoff 1975). Observations by Moehlman (1978) indicate that some young remain with the parents and assist in feeding and guarding the next litter.

Births occur mainly from July to October. The gestation period is about 60 days. Average litter size is four young and the range is one to eight. The young emerge from the den after 3 weeks and may then be moved several times to new sites. They are weaned when 8 to 9 weeks old, and the adults then regurgitate food for them. After 3½ months they no longer use a den, and by the age of 6 months they are hunting on their own. Captives have lived up to 14 years (Kingdon 1977; Moehlman 1978; Bekoff 1975; Bothma 1971a).

Unlike *C. adustus*, the black-backed jackal is considered a serious predator of sheep and is therefore intensively hunted and poisoned (Clutton-Brock, Corbet, and Hills 1976). According to Bothma (1971b), *C. mesomelas* is the most impor-

Black-backed jackal (*Canis mesomelas*), photo by Ernest P. Walker.

Golden jackal (*Canis aureus*), photo by Cyrille Barrette.

tant problem animal in the sheep-farming areas of the Transvaal.

Canis aureus.   Head and body length is 600 to 1,060 mm, tail length is 200 to 300 mm, shoulder height is 380 to 500 mm, and weight is 7 to 15 kg (Kingdon 1977; Lekagul and McNeely 1977). The fur is generally rather coarse and not very long. The dorsal area is mottled black and gray. The head, ears, sides, and limbs are tawny or rufous. The underparts are pale ginger or nearly white. The tip of the tail is black (Clutton-Brock, Corbet, and Hills 1976).

The golden jackal is found mainly in dry, open country. It is strictly nocturnal in areas inhabited by people, but may be partly diurnal elsewhere. Its opportunisitc diet includes rodents, hares, ground birds and their eggs, reptiles, frogs, fish, insects, and fruit. It takes carrion on occasion, but is a capable hunter (Kingdon 1977; Van Lawick and Van Lawick-Goodall 1971).

The basic social unit is composed of a mated pair and young. Sometimes the young of the previous year remain in the vicinity of the parents, and even mate and bear their own litters. The resulting associations are probably responsible for the reports of large packs hunting together. The usual hunting range of a family is about 2 or 3 sq km. A portion of this area is a territory that is marked with urine and defended against intruders. The young are born in a den within the territory. Births occur mainly in January and February in East Africa, and in April and May in the Soviet Union, but take place throughout the year in tropical Asia. The gestation period is 63 days. Litters contain one to nine, usually two to four, young. Their eyes open after about 10 days and they begin to take some solid food when about 3 months old. Both parents provide food and protection. Sexual maturity comes at 11 months of age. Captives have lived up to 16 years (Kingdon 1977; Van Lawick and Van Lawick-Goodall 1971; Lekagul and McNeely 1977; Rosevear 1974; Novikov 1962).

Canis latrans.   Head and tail length is 750 to 1,000 mm and tail length is 300 to 400 mm. In a given population, males

are generally larger than females. Bekoff (1977) listed weights of 8 to 20 kg for males and 7 to 18 kg for females. Northern animals are usually larger than those to the south. Gier (1975) gave average weight as 11.5 kg in the deserts of Mexico and 18 kg in Alaska. The largest coyotes of all are found in the northeastern United States, where the population apparently has been modified through hybridization with *C. lupus* (Nowak 1979). Pelage characters are variable, but usually the coat is long, the upper parts are buffy gray, and the underparts are paler. The legs and sides may be fulvous and the tip of the tail is usually black (Clutton-Brock, Corbet, and Hills 1976). From *C. lupus* the coyote is distinguished by smaller size, narrower build, proportionally longer ears, and a much narrower snout.

The coyote is found in a variety of habitats, mainly open grasslands, brush country, and broken forests. The natal den is located in such places as brush-covered slopes, thickets, hollow logs, rocky ledges, and burrows made by either the parents themselves or other animals. Tunnels are about 1.5 to 7.5 meters long, and lead to a chamber about 0.3 meter wide and 1 meter below the surface. Activity may take place at any time of day, but is mainly nocturnal and crepuscular. The average distance covered in a night's hunting is 4 km. The coyote is one of the fastest terrestrial mammals in North America, sometimes running at speeds of up to 64 km/hr. There may be a migration to the high country in the summer and a return to the valleys in the fall. Some of the young disperse from the parental range in the fall and winter, generally covering a distance of 80 to 160 km (Bekoff 1977; Banfield 1974; Gier 1975). In a tagging study in Iowa, Andrews and Boggess (1978) found individuals to travel an average of about 31 km, but up to 323 km, from the point of capture.

About 90 percent of the diet is mammalian flesh, mostly jackrabbits, other lagomorphs, and rodents. Such small animals are usually taken by stalking and pouncing (Bekoff 1977; Gier 1975). Larger mammals, especially deer, are also eaten, often in the form of carrion, but sometimes after a chase, in which several coyotes work together. In northwestern Wisconsin, deer represent 21.3 percent of the diet (Niebauer and Rongstad 1977), and in northern Minnesota,

Coyote (*Canis latrans*), photo from U.S. Fish & Wildlife Service.

deer, mainly as carrion, is the main food of the coyote (Berg and Chesness 1978). In Alberta, half of the food is carrion (Nellis and Keith 1976). Analyses of coyote predation on large mammals generally indicate that most of the individuals taken are immature, aged, or sick (Bekoff 1977). Livestock is killed by some coyotes, but the diet of the species seems largely neutral or beneficial with respect to human interests. *C. latrans* sometimes forms a "hunting partnership" with the badger (*Taxidea*). The two move together, the coyote apparently using its keen sense of smell to locate burrowing rodents and the badger digging them up with its powerful claws. Both predators then share in the proceeds.

Population density is generally 0.2 to 0.4 individuals per sq km, but may be up to 2 per sq km under extremely favorable conditions (Knowlton 1972; Bekoff 1977). Reported home range size is 8 to 80 sq km. The ranges of males tend to be large and to overlap one another considerably. The ranges of females are smaller and do not overlap (Bekoff 1977). In northern Minnesota, Berg and Chesness (1978) found home range size to average 68 sq km in males and 16 sq km in females.

Social structure and territoriality vary depending on habitat conditions and food supplies. *C. latrans* is found alone, in pairs, or in larger groups, but generally is less social than *C. lupus* (Bekoff 1977). Several males may court a female during the mating season, but she eventually seems to select one for a lasting relationship. The pair then live and hunt together, sometimes for years. Gier (1975) suggested that pair territories tend to be bordered by natural features, cover about 1 sq km or less, and are defended only during the denning season. Some of the offspring of a pair disperse, but others, especially females, may remain with the parents and thus form the basis of a pack. Studies in northwestern Wyoming indicate that packs are larger and more stable if a major,

clumped food resource, such as ungulate carrion, is readily available (Bekoff and Wells 1980). In this area, on the National Elk Refuge, Camenzind (1978) found that 61 percent of the coyotes composed resident packs of 3 to 7 members, 24 percent were resident mated pairs, and 15 percent were nomadic individuals. The latter ranged through the territories of the resident animals and were constantly chased away. The packs had well-defined social hierarchies, and marked and defended territories. The sizes of two pack territories were 5 and 7.2 sq km. Sometimes, more than one pair of a pack mated and bore young. Occasionally, up to 22 individuals would come together around carrion, but such aggregations usually lasted less than an hour.

The coyote has a great variety of visual, auditory, olfactory, and tactile forms of communication (Lehner 1978). At least 11 different vocalizations have been identified. These are associated with alarm, threats, submission, greeting, and contact maintenance. The best-known sounds are the howls given by one or more individuals, apparently to announce location. Group howls may have a territorial and spacing function.

Mating usually occurs from January to March, and births take place in the spring. Females are monestrous. Estrus averages 10 (4–15) days, gestation averages 63 (58–65) days, and litter size averages about 6 (2–12) young. Larger numbers of offspring are sometimes found in one den, but apparently represent the litters of two different females. The young weigh about 250 grams each at birth, open their eyes after 14 days, emerge from the den when 2–3 weeks old, and are fully weaned at 5–7 weeks. The mother, and perhaps the father and older siblings, begins to regurgitate some solid food for the young when they are about 3 weeks old. Adult weight is attained at about the age of 9 months. Some individuals mate in the breeding season following that in which they were born, but others wait another year. Maximum known longevity in the wild is 14½ years, but few animals survive that long. In one unexploited population, 70 percent of the animals were under 3 years old (Bekoff 1977; Kennelly 1978; Gier 1975). A captive lived for 21 years and 10 months (Marvin L. Jones, Zoological Society of San Diego, pers. comm.).

The coyote has long been considered a serious predator of domestic animals, especially sheep. It is also sometimes alleged to spread rabies and to damage populations of game birds and mammals. Bounties against the coyote began in the United States in 1825, and are still paid in some areas, but have never been effective in lasting prevention of depredations. In 1915 the U.S. government initiated a large-scale program of predator control, mainly in the western part of the country. The resulting known kill of *C. latrans* often exceeded 100,000 individuals per year. From 1971 to 1976, the annual take in 13 western states averaged 71,574 (Evans and Pearson 1980). Some state governments and private organizations also have control programs. Killing methods include trapping, poisoning, shooting from aircraft, and destroying the young in dens. In 1972 the federal government banned the general use of poison on its lands and in its programs.

Few wildlife issues have been as controversial as the question of coyote control (Bekoff 1977; Bekoff and Wells 1980; Sterner and Shumake 1978; Wade 1978; Hall 1946). No authorities deny that predation is a problem to some farmers and ranchers, but almost since the beginning of the federal program there have been serious doubts as to the extent of claimed livestock losses and degree of response. There was particular resentment against the widespread application of poison on western ranges, which was thought to kill not only far more coyotes than necessary but also many other carnivorous mammals. It has been argued that sheep raising is declining in the United States, regardless of the coyote and its control, and that damage by predators represents a relatively small part of overall losses to the industry. Surveys by trained biologists have generally indicated that losses are substantial in some areas, but are lower than often claimed. The stated intention of most current control operations is elimination of the specific individual coyotes that may be causing problems, and not the destruction of entire populations.

Information compiled by the U.S. Department of Agriculture (Gee et al. 1977) indicates that in 1974 losses attributed to the coyote in the western states amounted to 728,000 lambs (more than 8 percent of those born and one-third of the total lamb deaths) and 229,000 adult sheep (more than 2 percent of inventory and one-fourth of all adult deaths), valued at $27,000,000. In that same year the amount expended for coyote control in the west was $7,000,000 (Gum, Arthur, and Magleby 1977). In 1974, 71,522 coyotes were killed by the federal control program. Data compiled by Pearson (1978) suggest that the total kill in the 17 western states that year (including the take by fur trappers and sport hunters) was around 300,000 individuals, approximately the same as in 1946.

Coyote fur has varied in price. In 1974, skins taken in the United States sold for an average of $17.00 each. For the 1976/77 season, 30 states of the United States reported a total harvest of 320,323 pelts, with an average value of $45.00; in 6 Canadian provinces, the take was 65,819 pelts, with an average value of $59.76 (Deems and Pursley 1978).

It is sometimes said that *C. latrans,* as a species, can easily sustain human-inflicted losses, because of its adaptability, cunning, and high reproductive potential. Such may not necessarily be the case. In the past, intensive control programs eliminated the species from central Texas, much of North Dakota, and certain large sheep-raising sections of Colorado,

Wyoming, Montana, Utah, and Nevada (Gier 1975; Nowak 1979).

It does appear that *C. latrans* has substantially extended its range since the arrival of European colonists in North America (Nowak 1978, 1979; Gier 1975; Hilton 1978). The newly occupied regions may include Alaska, the Yukon, and Central America, though some evidence suggests that the species was already there, at least intermittently, in prehistoric times. The main expansion was eastward, the two main factors apparently being the creation of favorable habitat through the breaking up of the forests by settlement, and human elimination of the wolves (*C. rufus* and *C. lupus*) that might have competed with the coyote. In the late 19th and early 20th centuries, the coyote moved from its prairie homeland, through the Great Lakes region, and into southeastern Canada. From the 1930s to the 1960s, the species established itself in New England and New York, and it has now spread down the Appalachians as far as West Virginia. Other coyote populations pushed into southern Missouri and Arkansas in the 1920s, overran Louisiana in the 1950s, crossed into Mississippi by the 1960s, and subsequently became established as far as Georgia. The expanding coyote populations were modified by hybridization with the remnant pockets of wolves that they encountered, *C. lupus* in southeastern Canada and *C. rufus* in the south-central United States. Therefore, the coyotes of eastern North America are larger and generally more adapted for forest life than are their western relatives.

Canis rufus. Head and body length is 1,000 to 1,300 mm, tail length is 300 to 420 mm, shoulder height is 660 to 790 mm, and weight is 20 to 40 kg. While the red element of the fur is sometimes pronounced, the upper parts are usually a mixture of cinnamon buff, cinnamon, or tawny, with gray or black; the dorsal area is generally heavily overlaid with black. The muzzle, ears, and outer surfaces of the limbs are usually tawny. The underparts are whitish to pinkish buff, and the tip of the tail is black. There was reported to be a locally common dark or fully black color variant in the forests of the Southeast. From *C. lupus,* the red wolf is distinguished by its narrower proportions of body and skull, shorter fur, and relatively longer legs and ears.

Habitats include upland and bottomland forests, swamps, and coastal prairies. Natal dens are located in the trunks of hollow trees, stream banks, and sand knolls. Dens are either self-excavated or taken over from some other animal. They average about 2.4 meters in length and are usually no deeper than 1 meter below the surface. The red wolf is primarily nocturnal, but may increase its daytime activity during the winter. It hunts over a relatively small part of its home range for about 7 to 10 days, and then shifts to another area. Reported foods include nutria, muskrats, other rodents, rabbits, deer, hogs, and carrion (Nowak 1972; Carley 1979; Riley and McBride 1975).

Home range in southeastern Texas has been reported: (1) to average 44 sq km for seven individuals (Shaw and Jordan 1980); (2) to cover 65 to 130 sq km over 1 to 2 years (Riley and McBride 1975); and (3) to average 116.5 sq km for males and 77.7 sq km for females (Carley 1979). The basic social unit is apparently a mated, territorial pair. Groups of 2 or 3 individuals are most common, though larger packs have frequently been reported. Vocalizations are intermediate to those of *C. latrans* and *C. lupus* (Paradiso and Nowak 1972). Mating occurs from January to March, and offspring are produced in the spring. The gestation period is 60 to 63 days. Litters contain up to 12 young, but usually have about 4 to 7. Few individuals survive more than 4 years in the wild, but potential longevity in captivity is at least 14 years (Carley 1979).

Gray wolf (*Canis lupus*), photo from Zoological Society of London.

Like most large carnivores, the red wolf was considered a threat to domestic livestock, if not to people themselves. It was therefore intensively hunted, trapped, and poisoned after the arrival of European settlers in North America. In addition, disruption of its habitat and reduction of its numbers allowed *C. latrans* to invade its range from the west and north, and evidently stimulated interbreeding between the two species. This process led to the genetic swamping of the small pockets of red wolves that survived human persecution. By the 1960s, the only pure populations of *C. rufus* were in the coastal prairies and swamps of southeastern Texas and southern Louisiana. Conservation efforts, especially by the U.S. Fish and Wildlife Service, were not successful. The hybridization process continued to spread, and by 1975 the prevailing view was that the species could be saved only by securing and breeding some of the remaining animals that appeared to represent unmodified *C. rufus*. About 31 individuals were placed in captivity, mainly at the Point Defiance Zoo in Tacoma, and offspring have been produced since 1977. Experimental reintroduction of a pair was made on Bulls Island, South Carolina, in December 1976 and January 1978. Both pairs were recaptured, but plans for permanent reintroductions are being made (Carley 1979; McCarley and Carley 1979; Nowak 1972, 1974, 1979). The red wolf is classified as endangered by the IUCN (1978) and the USDI (1980).

Canis lupus.   This species includes the largest wild individuals in the family Canidae. Head and body length is generally about 1,000 to 1,600 mm and tail length is 350 to 560 mm. However, there are several small subspecies along the southern edge of the range of the gray wolf, especially *C.*

lupus arabs of the Arabian Peninsula. An adult male of that subspecies had a head and body length of only 820 mm and a tail length of 320 mm (Harrison 1968). In North America the largest animals are found in Alaska and western Canada, and the smallest in Mexico. In a given population, males are larger on the average than females. Overall mean (and extreme) weights are about 40 (20–80) kg for males and 37 (18–55) kg for females (Mech 1970, 1974*b*). The pelage is long; the upper parts are usually light brown or gray, sprinkled with black, and the underparts and legs are yellow white. Entirely white individuals occur frequently in tundra regions and occasionally elsewhere. Black individuals are also common in some populations.

The gray wolf has the greatest natural range of any living terrestrial mammal other than *Homo sapiens*. It is found in all habitats of the Northern Hemisphere, except tropical forests and arid deserts. Dens are used only for the rearing of young. They may be in rock crevices, hollow logs, or overturned stumps, but are usually in a burrow, either dug by the parents themselves or initially made by another animal and enlarged by the wolves. Several such burrows are sometimes excavated in the same season, perhaps as far apart as 16 km. A den may be used year after year. The animals prefer an elevated site near water. Tunnels are about 2 to 4 meters long, and lead to an enlarged underground chamber with no bedding material. There may be several entrances, each marked by a large mound of excavated soil. After the young are about 8 weeks old, they are moved to a rendezvous site, an area of about 0.5 ha., usually near water and marked by trails, holes, and matted vegetation. Here the young romp and play, and the other members of the pack gather for daily rest during the summer. Such sites are frequently changed, but some may be used for one or two months (Mech 1970, 1974b; Peterson 1977).

Movements are extensive, and usually at night, but diurnal activity may increase in cold weather. During the summer, the pack usually sets out in the early evening and returns to the den or rendezvous site by morning. In winter the animals wander farther and do not necessarily return to a particular location. They tend to move in single file along regular pathways, roads, streams, and ice-covered lakes. The daily distance covered ranges from a few to 200 km (Mech 1970). In Finland, the mean daily movement was determined to be 23 km (Pulliainen 1975). On Isle Royale, Peterson (1977) found that packs averaged 11 km per day or 33 km per kill. When individuals permanently disperse from a pack, they move much farther than normal; one traveled 206 km in 2 months (Mech 1974b), and another covered a straight line distance of 670 km in 81 days (Van Camp and Gluckie 1979). Five dispersing wolves, tracked by Fritts and Mech (1981), emigrated from 20 to 390 km, though others remained in the vicinity of the parental groups. Packs that depend on barren ground caribou make seasonal migrations with their prey and move as far as 360 km (Mech 1970, 1974b; Kuyt 1972).

The gray wolf usually moves at about 8 km/hr, but has a running gait of 55 to 70 km/hr. It can cover up to 5 meters in a single bound, and can maintain a rapid pursuit for at least 20 minutes. Prey is located by chance encounter, direct scenting, or following a fresh scent trail. Odors can be detected up to 2.4 km away. A careful stalk may be used to get as close as possible to the prey. If the objective is large and healthy and stands its ground, the pack usually does not risk an attack. Otherwise, a chase begins, which usually covers 100 to 5,000 meters. If the wolves can not quickly close with the intended victim, they generally give up. There seems to be a continuous process of testing the individual members of the prey population to find those that are easily captured. Most hunts are unsuccessful. On Isle Royale, for example, only about 8 percent of the moose tested are actually killed (Mech 1970, 1974b). In one exceptional case, the chase of a deer went for 20.8 km (Mech and Korb 1978). Once the wolves overtake an ungulate, they strike mainly at its rump, flanks, and shoulders.

The gray wolf is primarily a predator of mammals larger than itself, such as deer, wapiti, moose, caribou, bison, muskox, and mountain sheep. The smallest consistent prey is beaver. Following a drastic decline of deer in central Ontario, beaver remains were found in most wolf droppings (Voight, Kolenosky, and Pimlott 1976). Kill rates vary from about one deer per individual wolf every 18 days to one moose per wolf every 45 days (Mech 1974b). The usual condition is that a pack of several wolves will make a kill, consume a large amount of food, and then make another kill some days later. An adult can eat about 9 kg of meat in one feeding. Studies in Minnesota indicate that an average daily consumption is about 2.5 kg of deer per wolf, and that pack members generally remain in the vicinity of a kill for several days, eventually utilizing nearly the entire carcass, including much of the hair and bones (Mech et al. 1971). Most analyses of predation on wild species show that immature, aged, and otherwise inferior individuals constitute most of the prey taken by wolves (Mech 1974b).

There has long been controversy regarding the effects of the wolf on overall prey populations. It was once generally thought, even by some experienced zoologists, that the wolf could and did eliminate its prey. For example, Bailey (1930) wrote that "wolves and game animals can not be successfully maintained on the same range." Subsequently, a popular view developed that the wolf and the animals on which it depends are in precise balance, with the wolf taking just enough to prevent substantial population increases. Mech

(1970) tentatively concluded that wolf predation is the major controlling factor where prey-predator ratios are about 11,000 kg of prey per wolf or less, as is the case on Isle Royale in Lake Superior. More recently it became apparent that the wolf is not regulating the moose herd on Isle Royale, but merely cropping part of the annual surplus production (Mech 1974a; Peterson 1977). On the other hand, studies by Mech and Karns (1977) indicate that wolf predation was a major contributing factor in a serious decline of deer in the Superior National Forest of Minnesota from 1968 to 1974. Primary causes in the decline here, and in other parts of the Great Lakes region, were maturation of the forests that had been cut around the turn of the century, and a series of severe winters.

Wolf population density varies considerably, ranging as low as one individual per 520 sq km in parts of Canada. In Alberta, Fuller and Keith (1980) found densities of 1 per 73 to 1 per 273 sq km. Several studies in the Great Lakes region found apparently stable densities of around 1 per 26 sq km, and suggested that this level was the maximum allowed by social tolerance (Mech 1970; Pimlott, Shannon, and Kolenosky 1969). Subsequent investigations determined that the wolf could attain densities nearly twice as great in areas where prey concentrate for the winter (Kuyt 1972; Parker 1973; Van Ballenberghe, Erickson, and Byman 1975). On Isle Royale, an area of 544 sq km, the number of wolves remained relatively stable at a midwinter average of 22 from 1959 to 1973. There was an increase to 44 in 1976, evidently in response to rising moose numbers, thus indicating that food supply is the main factor in density level (Peterson 1977).

Home range size depends on food availability, season, and number of wolves. The largest pack range, found in Alaska during winter, was 13,000 sq km, and the smallest, in southeastern Ontario during summer, was 18 sq km (Mech 1970; Pimlott, Shannon, and Kolenosky 1969). In Minnesota, ranges of 52 to 555 sq km have been reported (Harrington and Mech 1979; Van Ballenberghe, Erickson, and Byman 1975; Fritts and Mech 1981). One radio-collared pack in Minnesota used summer ranges of 117 to 132 sq km and winter ranges of 123 to 183 sq km (Mech 1977d). Alberta packs used ranges of 195 to 629 sq km in summer and 357 to 1,779 sq km in winter (Fuller and Keith 1980).

The home range of a wolf pack usually corresponds to a defended territory. There is generally little or no overlap between the ranges of neighboring packs. Lone wolves that split off from a pack may disperse a considerable distance, or are continuously pursued by the resident packs and forced to shift about (Mech 1974b; Rothman and Mech 1979; Peterson 1977). Territories are relatively stable, some known to have been used at least 10 years (Mech 1979). Buffer zones, areas of little use, tend to develop between territories. Lone wolves may sometimes center their activities there. In Minnesota, deer have been found to have higher densities in such border areas, and to survive longer there than elsewhere during times of general population declines (Rogers et al. 1980; Mech 1977b, 1979). Under natural conditions, packs avoid areas in which they might encounter other packs. When food shortages induce stress, however, wolves move into the buffer zones to hunt, and eventually trespass into the territories of other packs. Meetings between packs are agonistic and may result in chases, savage fighting, and mortality (Mech 1970, 1977c; Harrington and Mech 1979; Van Ballenberghe and Erickson 1973).

Although packs are hostile to one another, the gray wolf is among the most social of carnivores. Groups usually contain 5 to 8 individuals, but have been reported to have as many as 36 (Mech 1974b). Such associations are probably essential for consistent success in the pursuit and overpowering of large prey. The number of wolves in a pack tends to increase with the size of the usual prey, being 7 or less where deer is the only important food, 6–14 where both deer and wapiti are eaten, and 15–20 on Isle Royale, where moose is the primary prey (Peterson 1977). In certain southern parts of its range, such as Mexico, Italy, and Arabia, *C. lupus* seems less gregarious than in the north (McBride 1980; Zimen 1981; Harrison 1968). This situation may be partly unnatural, resulting from intense human persecution and forced dependence on easily captured domestic animals and garbage.

A pack is essentially a family group, consisting of an adult pair, which may mate for life, and their offspring of one or more years (Mech 1970). The leader is usually a male, often referred to as the alpha male. He initiates activity, guides movements, and takes control at critical times, such as during a hunt. The males and females of a pack may have separate dominance hierarchies, reinforced by aggressive behavior and elaborate displays of greeting and submission by subordinate members. Generally, only the most dominant pair mate, and they inhibit sexual activity in the others. Social status is rather consistent, and a leader may retain its position for years, but roles can be reversed. Intragroup strife, perhaps resulting from increasing membership or declining food supplies, can result in a division into two packs or the splitting off of individuals. The latter may maintain a loose association with the parental group, sometimes following at a distance and feeding on scraps left behind, or disperse to a new area to seek a mate and begin a new pack (Mech 1970, 1974b; Peterson 1977; Zimen 1975; Wolfe and Allen 1973).

Studies by Fritts and Mech (1981) have contributed to our understanding of new pack formation in an area of an expanding wolf population. A young individual evidently does not remain with its parental pack past breeding age (about 22 months), unless it becomes a breeder itself upon the death of an alpha animal of the same sex. As it approaches maturity, it may actively explore the fringes of the parental territory. After dispersal, it either joins another lone wolf to search for a new territory or establishes itself in an area and awaits the arrival of an animal of the opposite sex. A new pack territory is relatively small, but may incorporate a portion of the parental territory. Therefore, as population size increases, average territory size decreases.

The gray wolf has a variety of visual, olfactory, and auditory means of communication. Vocalizations include growls, barks, and howls—continuous sounds usually lasting 3 to 11 seconds (Mech 1974b). Different individuals have distinctive howls (Peterson 1977). Humans can hear howls 16 km away on the open tundra, and wolves probably respond to howls at distances of 9.6 to 11.2 km. Howling functions to bring packs together and as an immediate, long-distance form of territorial expression (Harrington and Mech 1979).

Territories are also maintained by scent marking, via scratching, defecation, and, especially, urination. Scent marking differs from ordinary elimination in that there is a regular pattern of deposition at certain repeatedly used points. Peters and Mech (1975) determined that as a pack moves through its territory—visiting most parts at least every three weeks—sign is left at average intervals of 240 meters. Rothman and Mech (1979) found that scent marking also is important in bringing new pairs together for breeding and in helping established pairs achieve reproductive synchrony.

Threats and attacks by the dominant members of a pack probably prevent sexual synchronization in subordinates, and thus only the highest-ranking female normally bears a litter during the reproductive season (Zimen 1975). Mating may occur anytime from January, in low latitudes, to April, in high latitudes. Births take place in the spring. Courtship

may extend for days or months, estrus lasts 5–15 days, and the gestation period is usually 62–63 days. Mean litter size is 6 young and the range is 1 to 11. The young weigh about 450 grams each at birth, and are blind and deaf. Their eyes open after 11–15 days, they emerge from the den at 3 weeks of age, and they are weaned when around 5 weeks old. The mother usually stays near the den for a period, while the father and other pack members hunt and bring food for both her and the pups. The young are commonly fed by regurgitation. When 8 to 10 weeks old, the young are shifted to the first in a series of rendezvous sites, each up to 8 km from the other. If in good condition, the young begin to travel with the pack in early autumn. Sexual maturity generally comes at around 22 months of age, but social restrictions often prevent mating at that time. A captive pair successfully bred when only 10 months old. Mortality is highest among the young. In times of sharply declining food supplies, all pups may be severely underweight or die from malnutrition, and reproduction may even cease. For adults in such a situation, the primary mortality factor has been found to be intraspecific strife. Annual survival of adults in a population not under nutritional stress or human exploitation has been calculated at 80 percent. In the wild, 10 years is considered an old age for a wolf, but potential longevity is at least 16 years (Mech 1970, 1974b, 1977c, 1977d; Medjo and Mech 1976; Van Ballenberghe, Erickson, and Byman 1975; Van Ballenberghe and Mech 1975).

The wolf is often believed to be a direct threat to people. In Eurasia, attacks are unusual, but evidently have occurred, sometimes resulting in deaths (Pulliainen 1980; Ricciuti 1978). In North America there appear to be only two well-documented attacks by wild wolves, the first involving a probably rabid individual and the second being simply an aggressive leap that made contact with the person but caused no injury (Mech 1970; Munthe and Hutchison 1978).

A far more substantive basis for the age-old warfare between people and the gray wolf is depredation by the latter on domestic animals, notably cattle, sheep, and reindeer. The wolf also has been persecuted, especially in the 20th century, because of its alleged threat to populations of the wild ungulates that are desired by some persons for sport and subsistence hunting. The wolf was long taken by various kinds of traps and snares, and by pursuit with packs of specially trained dogs. In the 19th century, poison came into widespread use, and in the mid-20th century, hunting from aircraft became popular, especially on the open tundra.

The last wolves in the British Isles were exterminated in the 1700s. By the early 20th century the species, except for occasional wandering individuals, had disappeared in most of western Europe and in Japan. Modest comebacks occurred in Europe during World Wars I and II, but currently the only substantial populations on the continent, west of Russia, are in the Balkans. There are also very small remnant groups in Portugal, Spain, Italy, Czechoslovakia, Poland, and Scandinavia. *C. lupus* survives over much of its former range in southwestern and south-central Asia, but is generally rare. There were estimated to be 150,000 to 200,000 wolves in the Soviet Union after World War II, but these became subject to an intensive government control program. The annual kill was 40,000 to 50,000 individuals from 1947 to 1962, and subsequently dropped to about 15,000. In the mid-1970s the estimated number of wolves in the Soviet Union was 50,000, about two-thirds of them in the Central Asian republics (Smit and Van Wijngaarden 1976; Pimlott 1975; Grzimek 1975; Roberts 1977; Pulliainen 1980; Zimen 1981). Recently, wolf numbers again increased in the Soviet Union, and there is now a bounty of up to 100 rubles (Bibikov 1980).

The decline of the gray wolf was even more sweeping in the New World than in the Old. The species was largely

Gray wolf (*Canis lupus*), photo from New York Zoological Society.

eliminated along the east coast and in the Ohio Valley by the mid-19th century. By 1914 the last gray wolves had been killed in Canada south of the St. Lawrence River and in the eastern half of the United States, except for northern parts of Minnesota and Wisconsin and the upper peninsula of Michigan. In the following year, the federal government began a large-scale program to destroy predators. Partly as a result of this campaign, resident populations of the gray wolf apparently disappeared from the western half of the conterminous United States by the 1940s. The rate of decline subsequently slowed, mainly because the wolf had been eliminated from nearly all parts of the continent that were conducive to the raising of livestock. Conflict with agricultural interests does continue all along the lower edge of the current major range of the gray wolf, which extends across southern Canada, from coast to coast, and dips down into northern Minnesota (Nowak 1975a).

The situation in Minnesota has led to serious controversy (see *Endangered Species Tech. Bull.* 2, no. 3 [1977]:1–4; 3, no. 8 [1978]:1–4; and Mech 1977a). The wolf still occurs over much of the northern part of the state, primarily in boreal forest and bog country that is not suitable for agriculture, but also in adjoining lands that are well settled. The species is protected by federal law, except that individuals preying on domestic animals may be taken by government agents. Many persons in Minnesota have argued that current protective regulations are overly restrictive and do not allow adequate control of depredations, and that the wolf should be subject to sport hunting and commercial trapping. Recent studies in northwestern Minnesota, however, indicate that wolves infrequently attack domestic livestock, even when living nearby (Fritts and Mech 1981).

In Alaska the wolf is legally open to sport hunting and fur trapping, but there is also controversy (see *National Wildlife*

15, no. 5 [1977]:4–13). Shooting from aircraft was a major hunting method for many years. In 1972 the practice was banned by federal law, except for government predator control programs. Starting in the mid-1970s, the Alaska Department of Fish and Game authorized aerial hunting of wolves in an effort to increase the numbers of moose and caribou in certain parts of the state. A series of legal battles followed, involving opposing conservationist organizations, the state government, and the federal agencies that owned much of the land where the activity was to take place. Some of the hunts were halted, but others were allowed to proceed. The taking of the wolf in Alaska has been stimulated in part by rising fur prices. For the 1976/77 season, the reported state harvest was 1,076 pelts with an average value of $200 (Deems and Pursley 1978).

The total number of gray wolves in North America can only be guessed at. A figure of around 10,000 has sometimes been given for Alaska in recent years, and there may be several times that many in Canada. Overall populations do not appear to have declined since the 1950s, but there is concern that excessive hunting, and oil and mineral exploitation, could adversely affect prey species, especially the caribou of tundra regions. In Minnesota there are about 1,200 wolves, probably as many as at any time in decades (Mech 1977a). Resident populations may have been eliminated in Wisconsin and Michigan (except for Isle Royale) in the 1950s, but a few individuals have subsequently returned (Robinson and Smith 1977; Mech and Nowak 1981). In a 1974 experiment, 4 wolves were captured in northern Minnesota and released in the upper peninsula of Michigan, but within eight months all 4 had been killed by human agency (Weise et al. 1975). Since the 1940s there have been regular reports of wolves from the northwestern conterminous United States, especially along the Rocky Mountains between

Dingo (*Canis familiaris*), photo from New York Zoological Society.

Glacier and Yellowstone national parks (Ream 1980; Weaver 1978). Such records probably represent individuals that have wandered from Canada. Likewise, the occasional wolves that have appeared during the same period in the southwestern United States probably originated in Mexico. Mainly because of persecution by cattle ranchers, however, there may now be no more than 50 wolves left in Mexico itself (McBride 1980).

The IUCN (1972, 1974) classifies *C. lupus* as vulnerable, except for the northern Rocky Mountain subspecies (*irremotus*), which is considered endangered. The USDI (1980) lists all populations of *C. lupus* in the conterminous United States and Mexico as endangered, except for that in Minnesota, which is designated as threatened. *C. lupus* is on appendix 2 of the CITES, except for the populations of Pakistan, India, Nepal, and Bhutan, which are on appendix 1.

Canis familiaris. There are approximately 400 breeds of domestic dog, the chihuahua being the smallest and the Irish wolfhound the largest (Grzimek 1975). According to the National Geographic Society (1981), head and body length is 360 to 1,450 mm, tail length is 130 to 510 mm, shoulder height is 150 to 840 mm, and weight is 1 to 79 kg. In the wild subspecies of Australia (*C. f. dingo*), head and body length is 1,170 to 1,240 mm, tail length is 300 to 330 mm, shoulder height is about 500 mm, and weight is 10 to 20 kg. The dingo is usually tawny yellow in color, but some individuals are white, black, brown, rust, or other shades. The feet and tail tip are often white (Clutton-Brock, Corbet, and Hills 1976). From other forms of *C. familiaris* of comparable size and shape, the dingo can be distinguished by its longer muzzle, larger bullae, more massive molariform teeth, and longer, more slender canine teeth (Newsome, Corbett, and Carpenter 1980).

There have been various ideas regarding the origin of the domestic dog. The current consensus is that it was derived from one of the small south Eurasian subspecies of *C. lupus*, and subsequently spread throughout the world in association with people (Nowak 1979). Some authorities, including Clutton-Brock, Corbet, and Hills (1976), think that the direct ancestor is probably the Indian wolf, *C. lupus pallipes*, but

Olsen and Olsen (1977) argued in favor of the Chinese wolf, *C. l. chanco*. The wolf and dog still hybridize readily, but only when brought together in captivity or under very unusual natural conditions. The oldest well-documented remains of *C. familiaris*, dating from about 11,000 and 12,000 years ago, were found, respectively, in Idaho and Iraq. However, Beebe (1978) reported a specimen from the northern Yukon, with a minimum age of 20,000 years.

At present, from the Balkans and North Africa to Southeast Asia, there are dogs known as pariahs which lead a semidomestic or even feral existence around villages (Grzimek 1975; Bueler 1973). These dogs are generally primitive in physical appearance, and are probably closely related to the earliest dogs, as well as to the Australian dingo. The oldest definitely known fossils of the dingo date from about 3,000 years ago, but other remains may be as old as 8,600 years. People arrived in Australia at least 30,000 years ago. The dingo evidently was brought in long afterward, but before true domestication had been achieved, and it was able to establish wild populations (Macintosh 1975). There are also wild dog populations in New Guinea, which are related to the primitive pariah-dingo group (Troughton 1971).

The dingo was once found throughout Australia, in forests, mountainous areas, and plains (Bueler 1973). Studies by Corbett and Newsome (1975) were carried out in an arid region of deserts and grasslands in central Australia, where the dingo depends in part on waterholes made for cattle. Natal dens were found in caves, hollow logs, and modified rabbit warrens, usually within 2 or 3 km of water. Most activity in this region is nocturnal. In good seasons, the diet consists mainly of small mammals, especially the introduced European rabbit (*Oryctolagus*). In times of drought, the dingo takes kangaroos and cattle, mostly calves. According to Grzimek (1975), there may be regular seasonal migrations.

Corbett and Newsome (1975) learned that the dingo is basically solitary, but that individuals in a given area form a loose, amicable association and sometimes come together. Fighting may develop between strangers. The dingo is not a particularly vocal canid, but has a variety of sounds. Howls, which probably function to locate friends and repel strangers, are heard frequently in the single annual breeding season. Pups are born in late winter and spring, following a gestation period of 63 days. Litter size is usually four or five young, but ranges from one to eight (Bueler 1973). Yearlings may assist an older pair to raise their pups. Independence is generally achieved by the age of 3 or 4 months, but the young animals often then associate with a mature male (Corbett and Newsome 1975). Maximum known longevity is 14 years and 9 months (Grzimek 1975).

Although the dingo is said to be regularly captured and tamed by the natives and other people of Australia, Macintosh (1975) argued that it has never been successfully domesticated. It does have a close relationship with some groups of aborigines, but is evidently not used in hunting and is not intentionally fed. Its main function may be to sleep in a huddle with persons and thereby provide protection from the cold.

Agriculturalists in Australia generally have a low opinion of the dingo, mainly because of its predation on sheep. Intensive persecution began in the 19th century and has continued to the present. During the 1960s, about 30,000 dingos were killed for bounty annually in Queensland alone. Nearly 10,000 km of fencing has been constructed in eastern Australia in an effort to keep the wild dogs off the sheep ranges. Nonetheless, Macintosh (1975) suggested that depredations have been greatly exaggerated, and Grzimek (1975) pointed out that the dingo has actually aided agriculture by destroying introduced rabbits. Examination of stomach contents in Western Australia indicated that domestic stock was not a significant part of the dingo's diet, even though sheep and cattle were common in the study area (Whitehouse 1977).

Aside from the dingo, *C. familiaris* is one of the least known canids, with respect to its behavior and ecology under noncaptive conditions. Beck (1973, 1975) estimated that up to half of the 80,000 to 100,000 dogs in Baltimore are free ranging, at least at times, giving an average density of about 230 per sq km. They shelter in vacant buildings and garages, and under parked cars and stairways. They are active mainly from 0500 to 0800 and 1900 to 2200 hrs in the summer, remaining out of sight during the midday heat. Their diet, consisting mostly of garbage but also including rats and ground-nesting birds, seems adequate to maintain weight and good health.

Fully feral dogs sometimes occur in the countryside. Scott and Causey (1973) found Alabama packs to use moist flood plains in warm weather and dry uplands in cool weather, and to cover distances of 0.5 to 8.2 km per day. Nesbitt (1975) determined that the females on a wildlife refuge in Illinois did not dig a den, but gave birth in heavy cover. Activity in that area occurred anytime, but was mainly nocturnal and crepuscular. The dogs traveled single file along roads, trails, and crop rows, and, if frightened, took cover among trees and bushes. They fed on crippled waterfowl and deer, road kills and other carrion, small animals, some vegetation, and garbage.

Home range was found to be about 2.6 ha. each for 2 full-time free-ranging individuals in Baltimore (Beck 1975), 444 to 1,050 ha. each for three feral packs of 2 to 5 dogs (Scott and Causey 1973), and 28,500 ha. for a feral pack of 5 to 6 individuals (Nesbitt 1975). The latter group was observed to have a dominance hierarchy and to be led by a female. In the Baltimore study, half of the animals seen were solitary, 26 percent were in pairs, and the rest were in groups of up to 17 members. Spring and fall breeding peaks there were suggested by fluctuations in reports of unwanted dogs.

Female dogs enter estrus twice a year, usually in late winter or early spring and in the fall. Heat lasts about 12 days. At such times, males tend to leave the homes of their owners, mark territories, and fight rivals. The gestation period averages 63 days, litter size is usually 3–10 young, and nursing lasts about 6 weeks. Males often remain with the females and

young. Sexual maturity comes after 10 to 24 months of life, and old age generally after 12 years, but a few individuals live for 20 years (Grzimek 1975; Bueler 1973; Asdell 1964).

There are estimated to be 33,000,000 owned dogs in the United States, and there are many more that lack owners. Although these animals have abundant uses and values, they may cause problems for some people. In Baltimore, for example, dogs have been implicated in the spread of several diseases (in addition to rabies), they may benefit rats by overturning garbage cans, and there are about 7,000 reported attacks on people each year (Beck 1973, 1975). Fatal attacks are rare, but are reported in the United States with some regularity. *C. familiaris* also is often considered a serious predator of livestock and game animals, especially deer. Field studies, however, have indicated that feral dogs do not significantly affect deer populations, and may even have a sanitary function in eliminating carrion and crippled animals (Nesbitt 1975; Scott and Causey 1973; Gipson and Sealander 1977).

CARNIVORA; CANIDAE; **Genus CHRYSOCYON**
Hamilton Smith, 1839

Maned Wolf

The single species, *C. brachyurus*, occurs in central and eastern Brazil, eastern Bolivia, Paraguay, northern Argentina, and Uruguay (Cabrera 1957; Langguth 1975*b*).

Head and body length is 1,245 to 1,320 mm, tail length is 280 to 405 mm, and shoulder height is about 740 mm (Bueler 1973). Weight is 20 to 23 kg (Grzimek 1975). The overall appearance gives the impression of a red fox (*Vulpes vulpes*) on stilts (Clutton-Brock, Corbet, and Hills 1976). The general coloration is yellow red. The hair along the nape of the neck and middle of the back is especially long and may be dark in color. The muzzle and lower parts of the legs are also dark, almost black. The throat and tail tuft may be white. The coat is fairly long, is somewhat softer than that of *Canis*, and has an erectile mane on the back of the neck and top of the shoulders. The ears are large and erect, the skull is elongate, and the pupils of the eyes are round.

The maned wolf inhabits grasslands, savannahs, and swampy areas (Langguth 1975*b*). The natal nest is located in thick, secluded vegetation (Bueler 1973). Activity is thought to be mainly nocturnal and crepuscular. There have been suggestions that the remarkably long legs are adaptations for fast running or for movement through swamps, but the actual function is probably to allow seeing above tall grass. *Chrysocyon* is not an especially swift canid, does not pursue prey for long distances, and generally stalks and pounces like a fox (Kleiman 1972). Its omnivorous diet includes rodents, other small mammals, birds, reptiles, insects, fruit, and other vegetable matter.

Observations in the wild indicate that the maned wolf is solitary, outside of the reproductive season (Meritt 1973). Captives can sometimes be kept together without apparent strife, though there is usually an initial period of fighting and then establishment of a dominance hierarchy (Brady and Ditton 1979). The three main vocalizations are: a deep-throated single bark, heard mainly after dusk; a high-pitched whine; and a growl during agonistic behavior (Kleiman 1972).

Births in captivity have occurred in July and August in South America, and in January and February in the Northern Hemisphere. Females are monestrous, heat lasts about 5 days, the gestation period is 62 to 66 days, and litter size is two to four young. The young weigh about 350 grams each at

Maned wolf (*Chrysocyon brachyurus*), photo by Bernhard Grzimek.

birth, open their eyes after 8 or 9 days, begin to take some regurgitated food at 4 weeks, and are weaned by 15 weeks of age (Faust and Scherpner 1967; Da Silveira 1968; Brady and Ditton 1979). One captive lived for 13 years (Marvin L. Jones, Zoological Society of San Diego, pers. comm.).

The maned wolf is not extensively hunted for its fur, but is sometimes persecuted because of an unjustified belief that it kills domestic livestock (Meritt 1973; Grzimek 1975). It disappeared from Uruguay in the 19th century, and is now threatened in other regions by the annual burning of its grassland habitat, and by hunting and live capture. It is classified as vulnerable by the IUCN (1976) and as endangered by the USDI (1980), and is on appendix 2 of the CITES.

CARNIVORA; CANIDAE; **Genus NYCTEREUTES**
Temminck, 1839

Raccoon Dog

The single species, *N. procyonoides,* originally occurred in the woodland zone from southeastern Siberia to northern Viet Nam, and on all the main islands of Japan (Corbet 1978).

Head and body length is 500 to 680 mm and tail length is 130 to 250 mm. Weight is 4 to 6 kg in the summer, but is 6 to 10 kg prior to winter hibernation (Novikov 1962). The pelage is long, especially in winter. The general color is yellowish brown. The hairs of the shoulders, back, and tail are tipped with black. The limbs are blackish brown. The facial markings resemble those of raccoons (*Procyon*). There is a large dark spot on each side of the face, beneath and behind the eye. *Nyctereutes* is somewhat like a fox in external appearance, but has proportionately shorter legs and tail.

According to Novikov (1962), the raccoon dog occurs mainly in forests and in thick vegetation bordering lakes and streams. It usually dens in a hole initially made by a fox or badger, or in a rocky crevice, but sometimes digs its own burrow. It is primarily nocturnal, but may wander about by day if pressed by hunger. *Nyctereutes* is the only canid that hibernates, though the process is neither profound for individuals nor universal for the species. In northern parts of the range, animals that are well nourished hibernate from as early as November to as late as March, but may awaken occasionally to forage on warm days. In the southern parts of the range there is no winter sleep. Poorly nourished individuals do not hibernate, even in the north. Successful hibernation may be preceded by a period of intensive eating, which increases weight by nearly 50 percent. The summer diet may consist in large part of frogs, and also includes rodents, reptiles, fish, insects, mollusks, and fruit. In the fall, vegetable matter, such as berries, seeds, and rhizomes, becomes important. Northern individuals that do not hibernate have a difficult time in the winter, but may subsist on small mammals, carrion, and human refuse.

In the Ussuri region of southeastern Siberia, Kucherenko and Yudin (1973) found population density to average 1 to 3 individuals per 1,000 ha., and to reach 20 per 1,000 ha. in the best habitat. Home ranges of 100 to 200 ha. have been reported for the introduced European populations, but were found to average only 2.8 ha. in a study in western Kyushu (Ikeda, Eguchi, and Ono 1979). *Nyctereutes* is evidently a social canid, living in small groups that consist basically of a mated pair and their offspring. Vocalizations include growls and whines, but not barks.

Raccoon dog (*Nyctereutes procyonoides*), photo by Ernest P. Walker.

Mating occurs from January to March, estrus lasts about 4 days, and the gestation period is usually 59 to 64 days. There are commonly 5 to 8 young, but as many as 19 have been reported. They weigh 60 to 90 grams each at birth, open their eyes after 9 or 10 days, and nurse for up to 2 months. Both parents, however, begin to bring them solid food after only 25 or 30 days. They are capable of an independent existence by the age of 4 or 5 months, and attain sexual maturity at 9 to 11 months (Novikov 1962; Valtonen, Rajakoski, and Mäkelä 1977). A captive specimen lived for 10 years and 8 months (Marvin L. Jones, Zoological Society of San Diego, pers. comm.).

In Japan the flesh of the raccoon dog has been used for human consumption, and the bones for medicinal preparations. The skin, known commercially as "Ussuri raccoon," is used widely in the manufacture of such items as parkas, bellows, and decorations on drums. Because of excessive killing by people, *Nyctereutes* is now rare in Japan. Populations also have declined in southeastern Siberia, through overhunting and habitat disturbances (Kucherenko and Yudin 1973).

From 1927 to 1957, over 9,000 raccoon dogs were released by people in regions to the west of the natural range, especially in European Russia. The hope was to create new and valuable fur-producing populations. *Nyctereutes* did become established, eventually spreading almost throughout European Russia, and did increase in importance in the Soviet fur trade. It also, however, extended its range westward, reaching Finland in 1935, Sweden in 1945, Romania in 1951, Poland in 1955, Czechoslovakia in 1959, and Germany and Hungary in 1962 (Mikkola 1974; Kubiak 1965; Novikov 1962). It now occurs throughout West Germany (Roben 1975), and an individual was recently caught in England (*Oryx* 13 [1977]:434). *Nyctereutes* is generally consid-

ered a nuisance to the west of the Soviet Union. It destroys small game animals and fish, and its fur, which does not become as long as in the native habitat, is almost worthless.

CARNIVORA; CANIDAE; **Genus OTOCYON** *Müller, 1836*

Bat-eared Fox

The single species, *O. megalotis*, is found from Ethiopia and southern Sudan to Tanzania, and from southern Angola and Rhodesia to South Africa (Meester and Setzer 1971). According to Ansell (1978), it does not occur in Zambia. *Otocyon* has sometimes been placed in a separate subfamily, the Otocyoninae, on the basis of its unusual dentition. Clutton-Brock, Corbet, and Hills (1976), however, considered *Otocyon* simply an aberrant fox with systematic affinities to *Urocyon* (which they included in *Vulpes*) and some behavioral similarities to *Nyctereutes*.

Head and body length is 460 to 660 mm, tail length is 230 to 340 mm, shoulder height is 300 to 400 mm, and weight is 3 to 5.3 kg. The upper parts are generally yellow brown and have gray agouti guard hairs. The throat, underparts, and insides of the ears are pale. The outsides of the ears, mask, lower legs, feet, and tail tip are black. In addition to coloration, distinguishing characters include the relatively short legs and the enormous ears (114 to 135 mm long).

Otocyon has more teeth than any other placental mammal that has a heterodont condition (the teeth being differentiated into several kinds). Whereas in all other canids there are no more than two upper and three lower molars, *Otocyon* has at least three upper and four lower molars. This condition is sometimes held to be primitive, but it more likely represents

Bat-eared fox (*Otocyon megalotis*), photo by Bernhard Grzimek.

the results of a mutation that caused the appearance of the extra molar teeth in what had been a fox population with normal canid dentition (Clutton-Brock, Corbet, and Hills 1976).

The bat-eared fox is found in arid grasslands, savannahs, and brush country. It seems to prefer places with much bare ground or where the grass has been kept short by burning or grazing. When the grass again grows high, *Otocyon* may depart and wander about in search of a new place of residence. It is a capable digger, and either excavates its own den or enlarges the burrow of another animal. A family may have more than one den in its home range, each with multiple entrances and chambers and several meters of tunnels. In the Serengeti 85 percent of activity occurs at night, but in South Africa *Otocyon* is mainly diurnal in winter and nocturnal in summer. One female was observed to forage over about 12 km per night. The diet consists predominantly of insects, most notably termites, and also includes other arthropods, small rodents, the eggs and young of ground-nesting birds, and vegetable matter (Lamprecht 1979; Nel 1978; Smithers 1971; Kingdon 1977).

In South Africa, Nel (1978) found home ranges to overlap extensively, and observed no territorial defense or marking. Up to 15 individuals, representing four groups, were seen foraging within less than 0.5 sq km. In the Serengeti, Lamprecht (1979) found resident families to occupy largely exclusive home ranges of 0.25 to 1.5 sq km, and to mark them with urine. Groups usually consisted of a mated adult pair, which were accompanied by their young of the year for a lengthy period. A few observations suggested that there may sometimes be two adult females with a male. Strangers of the same sex generally were hostile. Contact between members of a group was maintained by soft whistles.

In both the Serengeti and Botswana, births occur mainly from September to November (Lamprecht 1979; Smithers 1971), but pups have been recorded in Uganda in March (Kingdon 1977). Gestation is usually reported as 60–70

days, but Rosenberg (1971) calculated the period at 75 days for a birth in captivity. Litters contain two to six young. They are suckled for 15 weeks and then begin to forage with the parents. Regurgitation is evidently rare. The young are probably full grown by the age of 5 or 6 months, and separate from the parents prior to the breeding season. According to Marvin L. Jones (Zoological Society of San Diego, pers. comm.), a captive lived for 13 years and 9 months.

Otocyon has declined in settled parts of South Africa (Meester and Setzer 1971). Nonetheless, it is apparently extending its range eastward into Mozambique and into previously unoccupied parts of Rhodesia and Botswana (Pienaar 1970).

CARNIVORA; CANIDAE; **Genus SPEOTHOS** Lund, 1839

Bush Dog

The single species, *S. venaticus*, occurs in Panama, Colombia, Venezuela, the Guianas, eastern Peru, Brazil, eastern Bolivia, Paraguay, and extreme northeastern Argentina (Cabrera 1957; Hall 1981; IUCN 1976). *Speothos* was first described from fossils collected in caves in Brazil.

Head and body length is 575 to 750 mm, tail length is 125 to 150 mm, and shoulder height is about 300 mm. Two males weighed 5 and 7 kg. The head and neck are ochraceous fawn or tawny, and this color merges into dark brown or black along the back and tail. The underparts are as dark as the back, though there may be a light patch behind the chin on the throat (Clutton-Brock, Corbet, and Hills 1976). The body is stocky, the muzzle is short and broad, and the legs are short. The tail is short and well haired, but not bushy.

Speothos inhabits forests and wet savannahs, often near water. It seems to be mainly diurnal, and to retire to a den at night, either in a burrow or in a hollow tree trunk. It is reportedly semiaquatic, and one captive could dive and swim

Bush dog (*Speothos venaticus*), photo from San Diego Zoological Garden.

underwater with great facility. It preys mainly on relatively large rodents, such as *Agouti* and *Dasyprocta* (Langguth 1975*b*; Kleiman 1972; Husson 1978).

Speothos is a highly social canid, living and hunting cooperatively in packs of up to 10 individuals. Several captives of the same or opposite sex can stay confined together without fighting, though a dominance hierarchy may be established. There are a number of vocalizations, the most common of which is a high-pitched squeak that appears to help maintain contact as the group moves about in dense forest (Kleiman 1972; Clutton-Brock, Corbet, and Hills 1976).

Husson (1978) wrote that litters of two to three young are produced during the rainy season. Kleiman (1972) stated that there is evidence of two estrous cycles per year, and that births in the Northern Hemisphere have occurred in December and February. Kitchener (1971) reported that a captive female gave birth to three young, 83 days from its first and 76 days from its last observed mating. Collier and Emerson (1973) noted that a litter of six young was born in the Los Angeles Zoo on 4 November 1971. Jantschke (1973) reported that a female in the Frankfurt Zoo produced litters of six young each on 11 December 1969, 5 December 1970, and 4 August 1971. According to Marvin L. Jones (Zoological Society of San Diego, pers. comm.), a captive lived for 10 years and 4 months.

Speothos is classified as rare by the IUCN (1976) and is on appendix 1 of the CITES. It is still widespread, but is scarce, and it seems to disappear as settlement progresses.

CARNIVORA; CANIDAE; **Genus CUON** Hodgson, 1838

Dhole

The single species, *C. alpinus*, is found from southern Siberia and Soviet Central Asia to India and the Malay Penin-

sula, and on the islands of Sumatra and Java, but not Sri Lanka. Except as otherwise noted, the information for this account was taken from the review papers by Cohen (1977, 1978) and Davidar (1975).

Head and body length is 880 to 1,130 mm, tail length is 400 to 500 mm, and shoulder height is 420 to 550 mm. Males weigh 15 to 21 kg and females weigh 10 to 17 kg. The coloration is variable, but generally the upper parts are rusty red, the underparts are pale, and the tail is tipped with black. In the northern parts of the range the winter pelage is long, soft, dense, and bright red, and the summer coat is shorter, coarser, sparser, and less vivid in color. *Cuon* resembles *Canis* externally, but the skull has a relatively shorter and broader rostrum. Females have 12 to 16 mammae.

The dhole occupies many types of habitat, but avoids deserts. In the Soviet Union it occurs mainly in alpine areas, and in India it is found in dense forest and thick scrub jungle. The preferred habitat in Thailand is dense montane forest at elevations of up to 3,000 meters (Lekagul and McNeely 1977). *Cuon* may excavate its own den, enlarge a burrow made by another animal, or use a rocky crevice. It may be active at any time, but mainly in early morning and early evening. Cohen et al. (1978) reported a major peak of activity at 0700–0800 hrs and a lesser peak at 1700–1800 hrs.

The dhole hunts in packs and is primarily a predator of mammals larger than itself. Prey is tracked by scent and then pursued, sometimes for a considerable distance. When the objective is overtaken, it is surrounded and attacked from different sides. Prey animals include deer, wild pigs, mountain sheep, gaur, and antelope. The chital (*Axis axis*) is probably the major prey in India, though Cohen et al. (1978) found remains of this deer to occur less frequently than those of *Lepus* in the droppings of *Cuon*. The diet also includes rodents, insects, and carrion. Reports of predation on tigers, leopards, and bears are generally not well documented, but those carnivores are sometimes driven from their kills by

Dhole (*Cuon alpinus*), photo from New York Zoological Society.

packs of dholes. There are numerous records of leopards being treed.

There are usually 5 to 12 individuals in a pack, but up to 40 have been reported. A pack apparently consists of a mated pair and their offspring. Although social structure has not been closely studied, there seems to be a leader, a dominance hierarchy, and submissive behavior by lower-ranking animals. Intragroup fighting is rarely observed. More than one female sometimes den and rear litters together. Vocalizations include nearly all of those made by the domestic dog, except loud and repeated barking. The most distinctive sound is a peculiar whistle that probably serves to keep the pack together during pursuit of prey.

In India, mating occurs from September to November, and births from November to March. In the Moscow Zoo, mating occurred in February and gestation lasted 60–62 days (Sosnovskii 1967). Litters usually contain four to six young, but up to nine embryos have been recorded. In the wild, both mother and young are provided with regurgitated food by other pack members. Longevity in the Moscow Zoo is 15–16 years.

Although the dhole only rarely takes domestic livestock, and was only once reported to attack a person, it has been intensively poisoned and hunted throughout its range. This situation seems based mainly on dislike by hunters, who see *Cuon* as a competitor for game and who are repulsed by its method of predation. Some of the village people of India actually welcome the dhole, and follow it to expropriate its kills. Because of direct persecution, elimination of natural prey, and destruction of its forest habitat, the dhole has declined seriously in range and numbers. It is classified as vulnerable by the IUCN (1976), as endangered by the USDI (1980), and as rare by the Union of Soviet Socialist Republics Ministry of Agriculture (1978), and is on appendix 2 of the CITES.

CARNIVORA; CANIDAE; **Genus LYCAON** Brookes, 1827

African Hunting Dog

The single species, *L. pictus,* occurs in most of Africa south of the Sahara Desert. It formerly occurred in suitable parts of the Sahara and in Egypt (Meester and Setzer 1971; Kingdon 1977).

Head and body length is 760 to 1,120 mm, tail length is 300 to 410 mm, shoulder height is 610 to 780 mm, and weight is 17 to 36 kg (Kingdon 1977). Males and females are about the same size (Frame et al. 1979). There is great variation in pelage, the mottled black, yellow, and white occurring in almost every conceivable arrangement and proportion

African hunting dog (*Lycaon pictus*), photo by Bernhard Grzimek.

of color. In most individuals, however, the head is dark and the tail has a white tip or brush. The fur is short and scant, sometimes so sparse that the blackish skin is plainly visible. The ears are long, rounded, and covered with short hairs. The legs are long and slender, and there are only four toes on each foot. The jaws are broad and powerful. *Lycaon* has a strong, musky odor. Females have 12 to 14 mammae (Van Lawick and Van Lawick-Goodall 1971).

Lycaon inhabits grassland, savannah, and open woodland (Meester and Setzer 1971). Its den, usually an abandoned aardvark hole, is occupied only to bear young (Kingdon 1977). The pack does not wander very far from the den, when pups are present (Frame et al. 1979). Otherwise, movements are generally correlated with hunting success; if prey is scarce the entire home range may be traversed in two or three days. Hunts take place in the morning and early evening. Schaller (1972) recorded peaks of activity from 0700 to 0800 and 1800 to 1900 hrs. Prey is apparently located by sight, approached silently, and then pursued at speeds of up to 66 km/hr for 10 to 60 minutes (Kingdon 1977). Van Lawick and Van Lawick-Goodall (1971) observed a pack to maintain a speed of about 50 km/hr for 5.6 km. Their investigation indicated this distance to be about the maximum that *Lycaon* would usually follow before giving up. They observed 91 chases, 39 of which were successful. In all but 1 of the latter cases, the quarry was killed within five minutes of being caught. In his study, Schaller (1972) observed 70 percent of chases to be successful. Groups of *Lycaon* generally cooperate in hunting large mammals, but individuals sometimes pursue hares, rodents, or other small animals. The main prey seems to vary by area, being bushduiker and reedbuck in the Kafue Valley of Zambia, impala in Kruger National Park, and Thomson's gazelle and wildebeest in the Serengeti (Kingdon 1977). Certain packs in the Serengeti, however, specialize in the capture of zebra (Malcolm and Van Lawick 1975).

From 1970 to 1977, population density in the Serengeti declined from one adult *Lycaon* per 35 sq km to one per 200 sq km. Pack home range in this area is generally 1,500 to 2,000 sq km (Frame et al. 1979). A pack in South Africa reportedly used a home range of about 3,900 sq km (Van Lawick and Van Lawick-Goodall 1971). Range contracts when there are small pups at a den; at such time one Serengeti pack used an area of only 160 sq km for 2½ months (Schaller 1972). The home range of a pack overlaps by about 10 to 50 percent with those of several neighboring packs (Frame and Frame 1976). Territoriality does not seem well developed, and hundreds of individuals may once have gathered temporarily in response to migrations of the formerly vast herds of springbok in southern Africa (Kingdon 1977).

Studies on the Serengeti Plains of Tanzania have revealed that *Lycaon* has an intricate and unusual social structure (Frame et al. 1979; Malcolm 1980; Frame and Frame 1976; Schaller 1972; Van Lawick and Van Lawick-Goodall 1971). Groups were found to contain averages of 9.8 (1–26) individuals, 4.1 (0–10) adult males, and 2.1 (0–7) adult females. This sexual proportion is unlike the usual condition in social mammals. Some packs have as many as 8 adult males with only a single adult female. Moreover, in a reversal of the usual mammalian process, females emigrate from their natal group far more than do males. Commonly, several sibling females, 18 to 24 months old, leave their pack and join another that lacks sexually mature females. Following the transfer, one of the females achieves dominance, and its sisters may then depart. Whereas no female seems to stay in its natal pack past the age of 2½ years, about half of young males do remain. The other males emigrate, usually in sibling groups. The typical pack thus consists of several related males, often representing more than one generation, and one or more females that are genetically related to each other, but not to the males. Some male lineages within a pack are known to have lasted at least 10 years.

There are separate dominance hierarchies for each sex. Only the highest-ranking male and female normally breed, and they inhibit reproduction by subordinates. There is intensive rivalry among the females for the breeding position. If a subordinate female does bear pups, the dominant one may steal them. Females sometime fight savagely, and the loser may leave the group and perish. Aside from this aspect of the social life of *Lycaon*, packs are remarkably amicable, with little overt strife. Food is shared, even by individuals that do not participate in the kill. An animal with a broken leg was allowed to feed, when it hobbled up after the others, throughout the time required for its leg to mend. Pups old enough to take solid food are given first priority at kills, eating even before the dominant pair. There are several vocalizations, the most striking of which is a series of wailing hoots that probably serves to keep the pack together during pursuit of prey.

Births may occur at any time of year, but peak from March to June during the second half of the rainy season. The interval between births is normally 12 to 14 months, but may be as short as 6 months if all the young perish. The gestation period is 60 to 80 days. Litter size averages about 10 and ranges from 6 to 16 young. When they are about 3 weeks old, the young emerge from the den and begin to take some solid food. Weaning is normally completed by the age of 11 weeks. All adult pack members regurgitate food to the young. Once, when a mother died, the males of the pack were able successfully to raise her 5-week-old pups. When the pack is hunting, 1 or 2 adults remain at the den to guard the pups. After 3 months, the young begin to follow the pack, and at 9–11 months of age they can kill easy prey, but they are not proficient until they are about 12–14 months old. Social restrictions blur the actual time of sexual maturity. Five males were observed to first mate at 1¾, 2¾, 3, 3, and 5 years of age. The youngest female to give birth was 22 months old at the time. Maximum observed longevity is 11 years (Frame et al. 1979; Kingdon 1977; Malcolm 1980; Schaller 1972; Van Lawick and Van Lawick-Goodall 1971).

Lycaon has only rarely been reported to attack people, but is widely persecuted as a predator of domestic livestock and game. It has been wiped out in South Africa, except in the vicinity of Kruger National Park, and has declined greatly in distribution in Namibia, Rhodesia, Tanzania, and Kenya (Kingdon 1977; Skinner, Fairall, and Bothma 1977; Lensing and Joubert 1977). Although it still occurs over much of its original range, there are probably fewer than 7,000 individuals in existence (Malcolm 1980). *Lycaon* is classified as vulnerable by the IUCN (1976).

CARNIVORA; *Family URSIDAE*

Bears

This family of three Recent genera and eight species occurred historically almost throughout Eurasia and North America, in the Atlas Mountains of North Africa, and in the Andes of South America. Hall (1981) recognized three living subfamilies: Tremarctinae, with the genus *Tremarctos;* Ursinae, with *Ursus;* and Ailuropodinae, with *Ailuropoda*. The sequence of genera presented here basically follows that of Simpson (1945), though he, like many other authorities, recognized additional genera.

Head and body length is 1,000 to approximately 2,800 mm, tail length is 65 to 210 mm, and weight is 27 to 780 kg. Males average about 20 percent larger than females. The coat is long and shaggy, and the fur is generally unicolored, usually brown, black, or white. Some genera have white or buffy crescents or semicircles on the chest. *Tremarctos*, the spectacled bear of South America, typically has a patch of white hairs encircling the eyes. *Ailuropoda*, the giant panda, has a striking black and white color pattern.

Bears have a big head; a large, heavily built body; short,

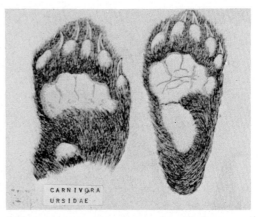

American black bear (*Ursus americanus*), right forepaw and right hind foot, photo from *Proc. Zool. Soc. London*.

powerful limbs; a short tail; and small eyes. The ears are small, rounded, and erect. The soles are hairy in species that are mainly terrestrial, but naked in bears that climb considerably, such as *Ursus malayanus*. All limbs have five digits. The claws are strong, recurved, and used for tearing and digging. The lips are free from the gums.

The skull is massive and the tympanic bullae are not inflated. In most genera the dental formula is: (i 3/3, c 1/1, pm 4/4, m 2/3) × 2 = 42. The species *Ursus ursinus*, however, has only 2 upper incisors and a total of 40 teeth. Ursid incisors are not specialized, the canines are elongate, the first 3 premolars are reduced or lost, and the molars have broad, flat, and tubercular crowns. The carnassials are not developed as such.

Habitats range from arctic ice floes to tropical forests. Those populations that occur in open areas often dig dens in hillsides. Others shelter in caves, hollow logs, or dense vegetation. Bears have a characteristic shuffling gait. They walk plantigrade, with the heel of the foot touching the ground. They are capable of walking on their hind legs for short distances. When need be, they are surprisingly agile and careful in their movements. Their eyesight and hearing are not particularly good, but their sense of smell is excellent. Bears are omnivorous, except that the polar bear (*Ursus maritimus*) feeds mainly on fish and seals.

During the autumn, in most parts of the range of the family, bears become fat. With the approach of cold weather they cease eating and go into a den that they have prepared in a protected location. Here they sleep through the winter, living mainly off stored fat reserves. With certain exceptions, especially pregnant females, the polar bear does not undergo winter sleep. Some authorities prefer not to call this process hibernation, as body temperature is not substantially reduced, body functions continue, and bears can usually be easily aroused. Sometimes they awaken on their own during periods of mild weather. Folk, Larson, and Folk (1976), however, found that heart rate of a hibernating bear drops to less than half of normal and that other physiological changes occur. They concluded that bears do experience true mammalian hibernation.

Bears live alone, except for courting pairs and females with young. Litters are produced at intervals of 1 to 4 years. In most regions, births occur from November to February, while the mother is hibernating. The period of pregnancy is commonly extended 6 to 9 months by delayed implantation of the fertilized egg. Litter size is one to four young. The young are relatively tiny at birth, ranging from 225 to 680

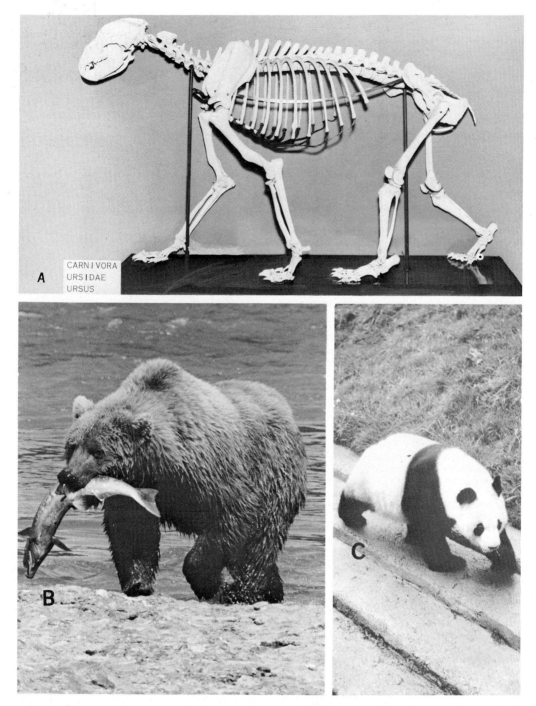

A. Brown bear (*Ursus arctos*), skeleton from U.S. National Museum. B. Photo of *U. arctos* by Leonard Lee Rue III. C. Giant panda (*Ailuropoda melanoleuca*), photo by Elaine Anderson.

grams each. They remain with the mother at least through their first autumn. They become sexually mature at 2½ to 6 years of age, and normally live 15 to 30 years in the wild.

Bears are usually peaceful animals that try to avoid conflict. However, if they consider themselves, their young, or their food supply threatened, they can become formidable adversaries. Only a small proportion of the stories of un-provoked attacks by bears on people are true. When such cases are carefully investigated, it is usually found that there was provocation. Nonetheless, bears have been persecuted almost throughout their range because of alleged danger to humans, and because they are sometimes considered serious predators of domestic livestock.

The geological range of the Ursidae is late Pliocene to

Recent in North America, Pleistocene to Recent in South America, middle Miocene to Recent in Europe, middle Pliocene to Recent in Asia, Recent in North Africa, and Pliocene in South Africa (Stains 1967; Hendey 1977).

CARNIVORA; URSIDAE; *Genus TREMARCTOS Gervais, 1855*

Spectacled Bear

The single species, *T. ornatus,* inhabits the mountainous regions of western Venezuela, Colombia, Ecuador, Peru, and western Bolivia (Cabrera 1957).

Head and body length is usually 1,200 to 1,800 mm, tail length is about 70 mm, and shoulder height is 700 to 800 mm. One male, 1,740 mm in length, weighed 140 kg. Peyton (1980) reported that a male, 2,060 mm in total length, weighed 175 kg. Grzimek (1975) gave the weight of females as 60 to 62 kg. The entire body is uniformly black or dark brown, except for large circles or semicircles of white around the eyes and a white semicircle on the lower side of the neck, from which lines of white extend onto the chest. The common name is derived from the white around the eyes. The head and chest markings are variable, however, and may be completely lacking in some individuals.

In the Andes of Peru, Peyton (1980) found *Tremarctos* to occupy a wide variety of habitats from 457 to 3,658 meters in elevation. The preferred habitats are humid forests between 1,900 and 2,350 meters, and coastal thorn forests, when water is available. High-altitude grasslands are also utilized. *Tremarctos* is apparently mainly nocturnal and crepuscular. During the day it beds down between or under large tree roots, on a tree trunk, or in a cave. It frequently climbs large trees to obtain fruit. While in a tree it may assemble a large platform of broken branches, on which it positions itself to eat and to reach additional fruit. Peyton found one such platform at a height of 15 meters.

Tremarctos feeds extensively on fruit, moving about in response to seasonal ripening. It also depends on plants of the family Bromeliaceae, especially when ripe fruit is not available. It tears off the leaves of large bromeliads to feed on the white bases, and obtains the edible hearts of small bromeliads by ripping the entire plant off the substrate. In addition, this bear climbs large cacti to get the fruits at top, tears into the green stalks of young palms to eat the unopened inner leaves, and strips bark off trees to feed on the cortex. The diet also includes bamboo hearts, corn, rodents, and insects. Only about 4 percent of the food was found to be animal matter (Peyton 1980).

In Peru, Peyton (1980) received reports that a male often enters a cornfield with one or more females, and sometimes with yearling animals, during the months of March to July. According to Grzimek (1975), *Tremarctos* has a "striking shrill voice." In the Buenos Aires Zoo young were produced in July, while in European zoos births have occurred from late December to March. The length of pregnancy is 6½ to 8½ months, and apparently involves delayed implantation. Litters contain one to three young, each weighing about 320 grams (Grzimek 1975; Bloxam 1977; Gensch 1965). One captive lived for 36 years and 5 months (Marvin L. Jones, Zoological Society of San Diego, pers. comm.).

Mittermeier et al. (1977) reported the meat of *Tremarctos* to be highly esteemed in northern Peru, and that this bear also is killed by people to obtain its skin and fat. Grimwood (1969) warned that *Tremarctos* had become rare and endangered in Peru, through intensive hunting by sportsmen and landowners, who consider it a predator of domestic live-

Spectacled bear (*Tremarctos ornatus*), photo from San Diego Zoological Garden.

stock. Peyton (1980) did not think *Tremarctos* to be in immediate danger of extinction, because it is adapted to a diversity of habitats, some of which are largely inaccessible to people. He did note, however, that some bears become habituated to raiding cornfields, and that these animals are frequently shot. *Tremarctos* is classified as vulnerable by the IUCN (1974) and is on appendix 1 of the CITES.

CARNIVORA; URSIDAE; *Genus URSUS Linnaeus, 1758*

Black, Brown, Polar, Sun, and Sloth Bears

There are six species (Hall 1981; Ellerman and Morrison-Scott 1966; Corbet 1978; Lekagul and McNeely 1977; Laurie and Seidensticker 1977; Simpson 1945; Kurten 1973; Lay 1967):

U. thibetanus (Asiatic black bear), Afghanistan, southeastern Iran, Pakistan, Himalayan region, Burma, Thailand, Indochina, China, Manchuria, Korea, extreme southeastern Siberia, Japan, Taiwan, Hainan;

U. americanus (American black bear), Alaska, Canada, conterminous United States, northern Mexico;

U. arctos (brown or grizzly bear), western Europe and Palestine to eastern Siberia and Himalayan region, Atlas

Mountains of northwestern Africa, Hokkaido, Alaska to Hudson Bay and northern Mexico;

U. maritimus (polar bear), primarily on arctic coasts, islands, and adjacent sea ice of Eurasia and North America;

U. malayanus (Malayan sun bear), Burma, Thailand, Indochina, Malay Peninsula, Sumatra, Borneo, possibly parts of southern China;

U. ursinus (sloth bear), India, Nepal, Bangladesh, Sri Lanka.

Each of these species has often been placed in its own genus or subgenus: *Selenarctos* Heude, 1901, for *U. thibetanus; Euarctos* Gray, 1864, for *U. americanus; Ursus* Linnaeus, 1758, for *U. arctos; Thalarctos* Gray, 1825, for *U. maritimus; Helarctos* Horsfield, 1825, for *U. malayanus;* and *Melursus* Meyer, 1793, for *U. ursinus*. Hall (1981), however, placed all of these names in the synonymy of *Ursus*. This arrangement is supported in part by the captive production of viable offspring through hybridization between several of the above species (Van Gelder 1977*b*).

The systematics of the brown or grizzly bear have caused considerable confusion. Old World populations have long been recognized to compose a single species, with the scientific name *U. arctos* and the general common name brown bear. In North America the name ''grizzly'' is applied over most of the range, while the term ''big brown bear'' is often used on the coast of southern Alaska and nearby islands, where the animals average much larger than those inland. Hall (1981) listed 77 Latin names that have been used in the specific sense for different populations of the brown or grizzly bear in North America. No one now thinks that there are actually so many species, but some authorities, such as Burt and Grossenheider (1976), have recognized the North American grizzly (*U. horribilis*) and the Alaskan big brown bear (*U. middendorfi*) as species distinct from *U. arctos* of the Old World. Other authorities (Erdbrink 1953; Rausch

1953, 1963; Kurten 1973), based on limited systematic work, have referred the North American brown and grizzly to *U. arctos*. This procedure is being used by most persons now studying or writing about bears, and is followed here. Kurten (1973) distinguished three North American subspecies: *U. a. middendorfi*, on Kodiak and Afognak islands; *U. a. dalli*, on the south coast of Alaska and the west coast of British Columbia; and *U. a. horribilis*, in all other parts of the range of the species.

From *Tremarctos, Ursus* is distinguished in having the masseteric fossa on the lower jaw not divided by a bony septum into two fossae. From *Ailuropoda*, it is distinguished in having an alisphenoid canal (Hall 1981). Additional information is provided separately for each species.

Ursus thibetanus.   Head and body length is 1,200 to 1,800 mm, and tail length is 65 to 106 mm. Stroganov (1969) listed weight as 110 to 150 kg for males and 65 to 90 kg for females. Roberts (1977) stated that an exceptionally large male weighed 173 kg, but that an adult female weighed only 47 kg. The coloration is usually black, but is sometimes reddish brown or rich brown. There is some white on the chin, and a white crescent or V on the chest.

The Asiatic black bear frequents moist deciduous forests and brushy areas, especially in the hills and mountains. It ascends to elevations as high as 3,600 meters in the summer and descends in the winter. It swims well. According to Lekagul and McNeely (1977), this bear is generally nocturnal, sleeping during the day in hollow trees, caves, or rock crevices. It is also seen abroad by day when favored fruits are ripening. It climbs expertly to reach fruit and beehives. It usually walks on all fours, but often stands on its hind legs so that its forepaws can be used in fighting. The diet includes fruit, buds, invertebrates, small vertebrates, and carrion. Domestic livestock is sometimes taken, and animals as large as adult buffalo are killed by breaking the neck. Individuals become fat in late summer and early fall before hibernation,

Asiatic black bear (*Ursus thibetanus*), photo from New York Zoological Society.

but some populations do not undergo winter sleep, or do so only for brief periods of severe weather. Roberts (1977) stated that in the Himalayas *U. thibetanus* hibernates, sometimes in a burrow of its own making, but that in southern Pakistan there is no evidence of hibernation. According to Stroganov (1969), hibernation in Siberia begins in November and lasts four or five months. Dens in that area are usually in tree holes. The bears are easily aroused during the first month, but sleep more deeply from December to February.

In Siberia, individual home range is 500 to 600 ha., only one-third or one-fourth the size of that of *U. arctos*. Mating in Siberia occurs in June or July, and births take place from late December to late March, mostly in February (Stroganov 1969). In Pakistan, mating is thought to occur in October and the young are born in February (Roberts 1977). According to Lekagul and McNeely (1977), pregnancy lasts 7 or 8 months, and usually two cubs are born in a cave or hollow tree in early winter. The eyes open when the cubs are about 1 week old, and shortly thereafter the young begin to follow the female as she forages. They are weaned when about 3½ months old, but remain with the mother until they are 2 or 3 years old. Females have been seen with two sets of cubs. Sexual maturity comes at about 3 years of age, and longevity in captivity may be as much as 33 years.

The Asiatic black bear sometimes raids cornfields and attacks domestic livestock. It also has been occasionally reported to kill humans. For these reasons *U. thibetanus* is hunted by people, and it also has declined because of the destruction of its forest habitat (Cowan 1972). It is on appendix 1 of the CITES, and the subspecies *U. t. gedrosianus* of southern Pakistan is classified as endangered by the IUCN (1972).

Ursus americanus.   Head and body length is 1,500 to 1,800 mm, tail length is about 120 mm, and shoulder height is up to 910 mm. Banfield (1974) listed weights of 92 to 140 kg for females and 115 to 270 kg for males. The most common color phases are black, chocolate brown, and cinnamon brown. Different colors may occur in the same litter. A white phase is generally rare, and never in the majority, but seems to be most common on the Pacific coast of central British Columbia. A blue black phase is also generally rare, but occurs frequently in the St. Elias Range of southeastern Alaska. Compared to *U. arctos*, *U. americanus* has a shorter and more uniform pelage, shorter claws, and shorter hind feet. Females have three pairs of mammae (Banfield 1974).

The American black bear occurs mainly in forested areas. It may originally have avoided open country, because of the lack of trees, in which to escape *U. arctos*. The latter species is known to be a competitor with, and sometimes a predator upon, *U. americanus* (Jonkel 1978). The black bear now appears to have extended its range northward onto the tundra, possibly in response to the decline of the barren ground grizzly (Jonkel and Miller 1970). Following the extermination of the grizzly in the mountains of southern California, *U. americanus* moved into the area (Hall 1981).

The usual locomotion is a lumbering walk, but *U. americanus* can be quick when the need arises. It swims and climbs well. It may move about at any hour, but is most active at night (Banfield 1974). Like other bears that sleep through the winter, it becomes fat with the approach of cold weather, finally ceases eating, and goes into a den in a protected location. The shelter may be under a fallen tree, in a hollow tree or log, or in a burrow. In the Hudson Bay area, individuals may burrow into the snow. During hibernation, body temperature drops from 38° to 31–34° C, the respiration slows, and the metabolic rate is depressed (Banfield 1974). The winter sleep is interrupted by excursions outside during periods of relatively warm weather. Such emergences are more numerous at southern latitudes. Hibernation begins as

A

American black bear (*Ursus americanus*), photo from San Diego Zoological Garden.

early as October and may last until May. In Washington the average period is 126 days, while three Louisiana bears slept for 74–124 days each (Lindzey and Meslow 1976; Lowery 1974). At least 75 percent of the diet consists of vegetable matter, especially fruits, berries, nuts, acorns, grass, and roots. In some areas sapwood is important, and to reach it the bear peels bark from trees, thereby causing forest damage (Poelker and Hartwell 1973). The diet also includes insects, fish, rodents, carrion, and, occasionally, large mammals.

Banfield (1974) suggested an overall population density of about one bear to every 14.5 sq km. Field studies in Alberta, Washington, and Montana, however, indicate a usual density of one per 2.6 sq km (Kemp 1976; Poelker and Hartwell 1973; Jonkel and Cowan 1971). Still higher densities have been reported: one per 1.3 sq km in southern California (Piekielek and Burton 1975) and one per 0.67 sq km on Long Island off southwestern Washington (Lindzey and Meslow 1977a). In the latter area, home range was found to average 505 ha. for adult males and 235 ha. for adult females (Lindzey and Meslow 1977b). Farther north in Washington, however, Poelker and Hartwell (1973) determined home range to average about 5,200 ha. for males and 520 ha. for females. The ranges of males did not overlap one another, but the ranges of females overlapped with those of males and, occasionally, with each other. In Idaho, Amstrup and Beecham (1976) found home ranges to vary from 1,660 to 13,030 ha., to remain stable from year to year, and to overlap extensively. Despite such overlap, individuals tend to avoid one another and to defend the space being used at a given

time. A number of bears sometimes congregate at a large food source, such as a garbage dump, but they try to keep out of each other's way. More tolerance is shown to familiar individuals than to strangers (Jonkel 1978; Jonkel and Cowan 1971; Banfield 1974). There is a variety of vocalizations. When startled, the ordinary sound is a "woof." When cubs are lonely or frightened, they utter shrill howls.

The sexes come together briefly during the mating season, which generally peaks from June to mid-July. Females remain in estrus throughout the season until they mate. They usually give birth every other year, but sometimes wait 3 or 4 years. Pregnancy generally lasts about 220 days, but there is delayed implantation. The fertilized eggs are not implanted in the uterus until the autumn, and embryonic development occurs only in the last 10 weeks of pregnancy. Births occur mainly in January and February, commonly while the female is hibernating. The number of young per litter ranges from one to five and is usually two or three. At birth the young weigh 225 to 330 grams each, and are naked and blind. They are usually weaned at around 6 to 8 months of age, but remain with the mother and den with her during their second winter of life. Upon emergence in the spring they usually depart in order to avoid aggression of the adult males in the breeding season. Females reach sexual maturity at 4 to 5 years of age, and males about a year later. One female is known to have lived 26 years and to have been in estrus at that age (Banfield 1974; Poelker and Hartwell 1973; Jonkel 1978).

The black bear is generally harmless to people, except when wounded or attempting to protect its young. In areas of

American black bear (*Ursus americanus*), photo by J. Perley Fitzgerald.

total protection, such as national parks, the species has become accustomed to humans. It thus can be easily seen and is a popular attraction, but is sometimes a nuisance, raiding campsites or begging for food along roads. Physical attacks are rare, but occur with some regularity, often because the involved persons disregard safety regulations (Cole 1976; Jonkel 1978; Pelton, Scott, and Burghardt 1976). Black bears have killed people on occasion, most recently in Alberta in 1980 (John R. Gunson, Alberta Fish and Wildlife Division, pers. comm.).

People have intensively killed *U. americanus,* because of fear, to prevent depredations on domestic animals and crops, for sport, and to obtain fur and meat. According to Lowery (1974), attacks on livestock are negligible, but the bear does serious damage to cornfields and honey production. The economic loss caused to beekeepers in the Peace River Valley of Alberta was estimated at $200,000 in 1973, and a government control program is directed against the bear in that area (Gilbert and Roy 1977). In most of the states and provinces occupied by the black bear, it is treated as a game animal, subject to regulated hunting. An estimated 30,000 individuals are killed annually in North America (Jonkel 1978). Relatively few skins go to market now, as regulations sometimes forbid commerce and there is no great demand. The average price per pelt in the 1976/77 season was about $44 (Deems and Pursley 1978).

The distribution of the black bear has declined substantially, but the species is still common in Alaska, Canada, the western conterminous United States, the upper Great Lakes region, northern New England and New York, and parts of the Appalachians. Populations also survive in the Ozark region and in coastal lowlands from the Dismal Swamp of Virginia to the Okefenokee of Georgia. Data compiled by Cowan (1972) indicate the presence of about 170,000 black bears in the conterminous United States.

The subspecies *U. a. floridanus* is widespread in Florida, but is considered threatened through habitat loss and persecution by beekeepers (Layne 1978). The subspecies *U. a. luteolus,* formerly found from eastern Texas to Mississippi, was by the mid-20th century reduced to a few individuals along the Mississippi and Atchafalaya rivers in eastern Louisiana. It is now probably in imminent danger of extinction, if it survives at all. During the 1960s the wildlife agencies of both Louisiana and Arkansas imported a number of bears from Minnesota (within the range of the subspecies *U. a. americanus*) to their respective states (Lowery 1974; Sealander 1979).

Ursus arctos.   Head and body length is 1,700 to 2,800 mm, tail length is 60 to 210 mm, and shoulder height is 900 to 1,500 mm. In any given population, adult males are larger, on the average, than adult females. The largest individuals—indeed, the largest of living carnivores—are found along the coast of southern Alaska and on nearby islands, such as Kodiak and Admiralty. In this area weight is as great as 780 kg. Size rapidly declines to the north and east. In southwestern Yukon, for example, Pearson (1975) found average weights of 139 kg for males and 95 kg for females. In the Yellowstone region, Knight, Blanchard, and Kendall (1981) found weights of full-grown animals to average 181 kg and to range from 102 to 324 kg. In Siberia and northern Europe, weight is usually 150 to 250 kg. In parts of southern Europe, average weight is only 70 kg (Grzimek 1975). Coloration is usually dark brown, but varies from cream to almost black. In the Rocky Mountains, the long hairs of the shoulders and back are often frosted with white, thus giving a grizzled appearance and the common name grizzly or silvertip. From *U. americanus, U. arctos* is distinguished in having a prominent hump on the shoulders, a snout that rises more abruptly into the forehead, longer pelage, and longer claws.

The brown bear has one of the greatest natural distributions of any mammal. It occupies a variety of habitats, but in the New World seems to prefer open areas, such as tundra, alpine meadows, and coast lines. It was apparently common

Alaskan brown bear (*Ursus arctos*), photo by Ernest P. Walker.

on the Great Plains, prior to the arrival of European settlers. In Siberia the brown bear occurs primarily in forests (Stroganov 1969). Surviving European populations are restricted mainly to mountain woodlands (Van Den Brink 1968). Even when living in generally open regions, *U. arctos* needs some areas with dense cover (Jonkel 1978). It shelters in such places by day, sometimes in a shallow excavation, and moves and feeds mainly during the cool of the evening and early morning. Egbert and Stokes (1976) noted that activity in coastal Alaska occurs throughout the day, but peaks from 1800 to 1900 hrs. Seasonal movements are primarily toward major food sources, such as salmon streams and areas of high berry production (Jonkel 1978). In Siberia, individuals may travel hundreds of kilometers during the autumn to reach areas of favorable food supplies (Stroganov 1969).

According to Banfield (1974), the usual gait is a slow walk. *U. arctos* is capable of moving very quickly, however, and can easily catch a black bear. Its long foreclaws are not adapted for climbing trees. It has excellent senses of hearing and smell, but relatively poor eyesight. The brown bear has great strength. Banfield saw one drag a carcass of a horse about 90 meters. In another case, a 360-kg grizzly killed and dragged a 450-kg bison.

Hibernation begins from October to December and ends from March to May. The exact period depends on the location, weather, and condition of the animal. In certain southerly areas, hibernation is very brief or does not take place at all. In most cases the brown bear digs its own den and makes a bed of dry vegetation. The burrow is often located on a sheltered slope, either under a large stone or among the roots of a mature tree. The bed chamber has an average volume of around two cubic meters. A den is sometimes used repeatedly, year after year. During winter sleep there is a marked depression in heart rate and respiration, but only a slight drop in body temperature. The animal can be aroused rather easily and can make a quick escape, if necessary (Craighead and Craighead 1972; Stroganov 1969; Slobodyan 1976; Ustinov 1976; Grzimek 1975).

The diet consists mainly of vegetation (Jonkel 1978). Early spring foods include grasses, sedges, roots, moss, and bulbs. In late spring, succulent, perennial forbs become important. During the summer and early autumn, berries are essential, and bulbs and tubers are also taken. Banfield (1974) wrote that *U. arctos* consumes insects, fungi, and roots at all times of the year, and also digs mice, ground squirrels, and marmots out of their burrows. In the Canadian Rockies, the grizzly is quite carnivorous, hunting moose, elk, mountain sheep and goats, and even black bears. In Mount McKinley National Park, Alaska, Murie (1981) found *U. arctos* to feed mostly on vegetation, but to also eat carrion whenever available, and occasionally to capture young calves of caribou and moose. During the summer, when salmon are moving upstream along the Pacific coasts of Canada, southern Alaska, and northeastern Siberia, brown bears gather to feed on the vulnerable fish (Banfield 1974; Egbert and Stokes 1976; Kistchinski 1972). Perhaps because of this abundant food supply, the bears of these areas are larger and are found at greater densities than anywhere else.

Some approximate reported population densities are: Carpathian Mountains, one bear per 20 sq km; Lake Baikal area, one per 60 sq km; coast of Sea of Okhotsk, one per 10 sq km; Kodiak Island, one per 1.5 sq km; Mount McKinley National Park, one per 30 sq km; northern parts of Alaska and Northwest Territories, one per 150 sq km; and Glacier National Park, Montana, one per 21 sq km (Harding 1976; Martinka 1974, 1976; Slobodyan 1976; Dean 1976; Kistchinski 1972; Ustinov 1976). In the Yellowstone region of the western United States, overall average density is about one bear per 88 sq km. In summer, however, individuals have concentrated by night at feeding sites, so densities have reached about one per 0.05 sq km. Daytime dispersal has reduced density to about one per 0.36 sq km. In the Yellowstone, individual home range averages about 80 sq km, and varies from about 20 to 600 sq km, with respect to the area used in the course of a year. Lifetime individual range has been as great as 2,600 sq km (Craighead 1976; Knight, Blanchard, and Kendall 1981). The home ranges of males are generally substantially larger than those of females. In the northern Yukon, Pearson (1976) found averages of 414 sq km for males and 73 sq km for females.

Home ranges overlap extensively and there is no evidence of territorial defense (Craighead 1976; Murie 1981). Although generally solitary, the grizzly is the most social of North American bears, occasionally gathering in large numbers at major food sources, and often forming family foraging groups with more than one age class of young (Jonkel 1978). At a salmon stream in southern Alaska, Egbert and Stokes (1976) sometimes observed more than 30 bears at one time. Considerable intraspecific tolerance was demonstrated in such aggregations, but dominance hierarchies were enforced by aggression. The highest-ranking animals were the large adult males, which most other bears attempted to avoid. The most aggressive animals were females with young, and the least aggressive were adolescents. Overt fighting was usually brief, and no infliction of serious wounds was observed, but the researchers suspected that killing of young individuals by adult males was a factor in population regulation.

There are no lasting social bonds, except those between females and young. During the breeding season males may fight over females. Successful males attend one or two females for 1 to 3 weeks. Mating takes place from May to July, implantation of the fertilized eggs in the uterus is usually delayed until October or November, and births generally occur from January to March, while the mother is in hibernation. The total period of pregnancy may last from 180 to 266 days. Females remain in estrus throughout the breeding season until mating, and do not again enter estrus for at least 2, usually 3 or 4, years. The number of young in a litter averages about two and ranges from one to four. Each weighs 340 to 680 grams at birth, and is naked and blind. They are weaned at about 5 months of age. They remain with the mother at least until their second spring of life, and usually until their third or fourth. Litter mates sometimes maintain an association for 2 or 3 years after leaving the mother. Puberty comes around 4 to 6 years of age, but growth then continues. Males in southern Alaska may not reach full size until they are 10 to 11 years old. Females in the Yellowstone region are known to have lived 25 years and still be capable of reproduction (Craighead, Craighead, and Sumner 1976; Egbert and Stokes 1976; Murie 1981; Jonkel 1978; Glenn et al. 1976; Pearson 1976; Slobodyan 1976). Potential longevity in captivity may be as great as 50 years (Stroganov 1969).

The brown bear has the reputation of being the most dangerous animal in North America. If we disregard venomous insects, disease-spreading rodents, domestic animals, and people themselves, this may be true. Three persons were killed by grizzlies in Glacier National Park, Montana, in 1980 (*Washington Post*, 25 July 1980, p. A-14; 9 October 1980, p. A-54; 20 October 1980, p. A-5). Another was killed in Canada (John R. Gunson, Alberta Fish and Wildlife Division, pers. comm.). That was an unusually tragic year, but injuries and an occasional death have been reported from the western national parks since around 1900 (Herrero 1970, 1976; Cole 1976). During the 19th century, attacks on people apparently occurred with some regularity in California (Storer and Tevis 1955). According to Ustinov (1976), over 70 attacks and 17 deaths have been attributed to *U. arctos* in the Lake Baikal area of Siberia. Many such incidents have probably been provoked by an effort to shoot or harass the animal, as the brown bear normally tries to avoid humans. It is unpredictable, however, if startled at close quarters, especially when accompanied by young or engrossed in a search for food. Jonkel (1978) cautioned that there may be more difficulties as recreational and commercial activity increases in areas occupied by the grizzly. He suggested that problems could be reduced by improved education and planning, and by not locating campsites, trails, and residential facilities in places that are regularly used by bears.

The brown bear has long been persecuted as a predator of domestic livestock, especially cattle and sheep. Those parts of North America from which it has been eliminated correspond closely to areas of intensive ranching and grazing. In the 19th and early 20th centuries there were apparently some remarkably destructive bears (Storer and Tevis 1955; Hubbard and Harris 1960), and their activities earned the entire species the lasting enmity of cattle ranchers and sheepherders. The brown bear also has been widely sought as a big game trophy, and is currently subject to regulated sport hunting in most of its range. There is now little commercial demand for its meat and hide (Banfield 1974).

The original eastern limits of *U. arctos* in North America are not certain. A skull found in a Labrador Eskimo midden dating from the late 18th century supports earlier stories that the grizzly used to occur to the east of Hudson Bay (Spiess 1976). The decline of the species on the Great Plains may have begun when the Indians of that region obtained the horse and hence an improved hunting capability. A precipitous drop in grizzly numbers came in the 19th century as settlers and livestock filled the West, thereby setting up confrontations that usually ended to the detriment of the bear. This process was intensified by logging, mining, and road construction, which increased human presence in remote areas. In the early 19th century there may have been 100,000 grizzlies in the western conterminous United States. There are now probably fewer than 1,000. Of these, about 200 are in Glacier National Park (Martinka 1974) and several hundred more are in nearby parts of northwestern Montana, northern Idaho, and extreme northeastern Washington. The population in Yellowstone National Park and vicinity has been estimated at 136 (Craighead, Varney, and Craighead 1974) and 247 (Knight, Blanchard, and Kendall 1981). There may be very small remnant groups in south-central Colorado and northwestern Mexico. The grizzly has been extirpated from the Great Plains of Canada, except for an isolated group in west-central Alberta (Banfield 1974). The species also has declined on the barrens of the Northwest Territories (Macpherson 1965). In the mountainous regions of western Canada, and in Alaska, *U. arctos* is still relatively common, perhaps numbering about 30,000 individuals (Cowan 1972). In Eurasia there are an estimated 100,000 brown bears, about 70,000 of them in the Soviet Union (Vereschagin 1976). To the west of Russia, the only substantial surviving populations are in the Balkans and northern Scandinavia. There are small, isolated groups in Poland, Czechoslovakia, Austria, Italy, southern France, northern Spain, and southern Norway (Smit and Van Wijngaarden 1976; Elgmork 1978). *U. arctos* apparently disappeared from northwestern Africa around the middle of the 19th century (Harper 1945).

Appendix 2 of the CITES includes all North American populations of *U. arctos* except *U. a. nelsoni*, the Mexican grizzly bear. Appendix 1 includes *U. a. nelsoni*, *U. a. pruinosus* of Tibet, *U. a. isabellinus* of the mountains of Central Asia, and those populations occurring in Italy. The IUCN (1972) classifies *U. a. nelsoni* as endangered and *U. a. richardsoni*, the barren ground grizzly of the Northwest Territo-

ries, as rare. The USDI (1980) lists *U. a. nelsoni*, *U. a. pruinosus*, and the Italian populations of *U. arctos* as endangered. Those populations of *U. arctos* in the conterminous United States are listed as threatened by the USDI.

Ursus maritimus.   Head and body length is 2,000 to 2,500 mm, tail length is 76 to 127 mm, and shoulder height is up to 1,600 mm. DeMaster and Stirling (1981) gave the weight as 150 to 300 kg for females and 300 to 800 kg for males. Banfield (1974), however, wrote that males usually weigh 420 to 500 kg. The color is often pure white following the molt, but may become yellowish in the summer, probably because of oxidation by the sun. The pelage also sometimes appears gray or almost brown, depending on season and light conditions. The neck is longer than that of other bears, and the head is relatively small and flat. The forefeet are well adapted for swimming, being large and oarlike. The soles are haired, probably for insulation from the cold and traction on the ice. Females have four functional mammae (DeMaster and Stirling 1981).

The polar bear is often considered a marine mammal. It is distributed mainly in arctic regions around the north pole. The southern limits of its range are determined by distribution of the pack ice. It has been recorded from as far north as 88° N and from as far south as the Pribilof Islands in the Bering Sea, the island of Newfoundland, the southern tip of Greenland, and Iceland. There also are permanent populations in James Bay and the southern part of Hudson Bay. Although found generally in coastal areas or on ice hundreds of kilometers from shore, individuals have wandered up to 200 km inland (Stroganov 1969).

According to DeMaster and Stirling (1981), the preferred habitat is pack ice that is subject to periodic fracturing by wind and sea currents. The refreezing of such fractures provides places where hunting by the bear is most successful. Some animals spend both winter and summer along the lower edge of the pack ice, perhaps undergoing extensive north-south migrations as this edge shifts. Others move onto land for the summer, and disperse across the ice as it forms along the coast and between islands during winter. The bears of the Labrador coast sometimes move north to Baffin Island, and some individuals have traveled as far as 1,050 km to the islands of northern Hudson Bay (Stirling and Kiliaan 1980). The population that summers along the southern shore of Hudson Bay spreads all across the partly ice-covered bay in November and returns to shore in July or August (Stirling et al. 1977). Despite such movements, the polar bear is not a true nomad. There are a number of discrete populations, each with its own consistently used areas for feeding and breeding (Stirling, Calvert, and Andriashek 1980).

The polar bear can outrun a reindeer for short distances on land, and can attain a swimming speed of about 6.5 km/hr. It swims rather high, with head and shoulders above the water. If killed in the water, it will not immediately sink. According to DeMaster and Stirling (1981), it has been reported to swim for at least 65 km across open water. It is capable of diving under the ice and surfacing in holes utilized by seals. It seems to be most active during the first third of the day and least active in the final third. From July to December in the James Bay region, when a lack of ice prevents seal hunting, *U. maritimus* spends about 87 percent of its time resting, apparently living off of stored fat (Knudsen 1978). Depressions or complete earthen burrows are sometimes excavated on land during the summer, in order to avoid the sun and keep cool (Jonkel et al. 1976).

Any individual bear may make a winter den for temporary shelter during severe weather, but only females, especially those that are pregnant, generally hibernate for lengthy periods. As with other bears, winter sleep involves a depressed

Polar bear (*Ursus maritimus*), photo from New York Zoological Society. Insets: A. Forefoot; B. Hind foot; photos from *Proc. Zool. Soc. London;* C. Young, 24 hours old, photo by Ernest P. Walker.

Polar bears (*Ursus maritimus*), photo by Sue Ford, Washington Park Zoo.

respiratory rate and a slightly lowered body temperature, but not deep torpor. Most pregnant females evidently do not spend the winter along the pack ice, but hibernate on land from October or November to March or April. Maternal dens are usually found within 8 km of the coast, but in the southern Hudson Bay region they are concentrated 30 to 60 km inland. They are excavated in the snow to depths of 1 to 3 meters, often on a steep slope. They usually consist of a tunnel, several meters long, which leads to an oval chamber of about 3 cubic meters. Some dens have several rooms and corridors (Harington 1968; Uspensky and Belikov 1976; DeMaster and Stirling 1981; Stirling, Calvert, and Andriashek 1980; Stirling et al. 1977; Larsen 1975).

The polar bear feeds primarily on the ringed seal (*Phoca hispida*) (DeMaster and Stirling 1981). The bear either remains still until a seal emerges from the water or stealthily stalks its prey on the ice (Stirling 1974). It may also dig out the subnivean dens of seals to obtain the young (Stirling, Calvert, and Andriashek 1980). During summer and autumn in the southern Hudson Bay region, *U. maritimus* often swims among sea birds and catches them as they sit on the water (Russell 1975). The diet also includes the carcasses of stranded marine mammals, small land mammals, reindeer, fish, and vegetation. Berries become important for some individuals during summer and autumn (Jonkel 1978).

Reported population densities range from 1 bear per 37 sq km to 1 per 139 sq km (DeMaster and Stirling 1980). Although *U. maritimus* is generally solitary, large aggregations may form around a major source of food (Jonkel 1978). Up to 40 individuals have been seen at one time in the vicinity of the Churchill garbage dump, on the southern shore of Hudson Bay (Stirling et al. 1977). Wintering females evidently tolerate one another well, as dens on Wrangel Island are sometimes found at densities of one per 50 sq meters (Uspensky and Belikov 1976). High concentrations of summer dens also have been reported (Jonkel et al. 1976). Adult females with young are not subordinate to any other age or sex class, but tend to avoid interaction with adult males, presumably because the latter are potential predators of the cubs (DeMaster and Stirling 1981).

The sexes usually come together only briefly during the mating season, which lasts from March to June. Delayed implantation apparently extends the period of pregnancy to 195–265 days. The young are born from November to January, while the mother is in her winter den. Females give birth every 2 to 4 years. The number of young per litter averages about two and ranges from one to four. They weigh about 600 grams each at birth, and have some fur, but are blind. Upon emergence from the den in March or April, the cubs weigh 10 to 15 kg each. They usually leave the mother when 24 to 28 months old. The age of sexual maturity averages about 5 or 6 years. Adult weight is attained at about 5 years of age by females, but not until 10 or 11 years by males. Females are known to have been still capable of reproduction when 21 years old. Annual adult mortality in a population is about 8 to 16 percent. Potential longevity in the wild is estimated at 25 to 30 years (DeMaster and Stirling 1981; Stirling, Calvert, and Andriashek 1980; Uspensky and Belikov 1976). According to Marvin L. Jones (Zoological Society of San Diego, pers. comm.), a captive polar bear lived for 34 years and 7 months, and a hybrid *U. maritimus* × *U. arctos* lived for 38 years and 1 month.

The polar bear is often considered dangerous to people, though usually the two species are not found in close proximity. An exception developed during the 1960s in the vicinity of the town of Churchill, on the southern shore of Hudson Bay (Stirling et al. 1977). Bears apparently increased in this area, because of a decline in hunting. At the same time, more people moved in and several large garbage dumps were established. A number of persons were attacked and one was killed. Many bears were shot or translocated by government personnel.

The native peoples of the Arctic have long hunted the polar bear for its fat and fur. Sport and commercial hunting increased in the 20th century. The pelt of *U. maritimus* is the most valuable of any North American mammal that is now regularly marketed. During the 1976/77 season, 530 skins from Canada were sold at an average price of $585.22 (Deems and Pursley 1978). Some individual prime pelts have brought over $3,000 each (Smith and Jonkel 1975).

The use of aircraft to locate polar bears and to land trophy hunters in their vicinity developed in Alaska in the late 1940s. The annual kill by such means increased to about 260 bears by 1972. In that year, however, the killing of *U. maritimus*, except for native subsistence, was prohibited by the United States Marine Mammal Protection Act. Canada and Denmark (for Greenland) also limit hunting to resident natives, and the Soviet Union and Norway (for Spitsbergen) provide complete protection. In 1973, the above five nations drafted an agreement calling for the restriction of hunting, the protection of habitat, and the carrying out of cooperative research on polar bears. The agreement was ratified by the United States in 1976. The yearly world-wide kill is now estimated at around 1,000 animals. The total number of polar bears in the wild is perhaps 20,000, and populations are generally thought to be stable or increasing. *U. maritimus*, however, may be threatened by the exploitation of oil and gas reserves in the Arctic, especially with respect to development in the limited areas suitable for denning by pregnant females (U.S. Fish and Wildlife Service 1980; DeMaster and Stirling 1981; Stirling and Kiliaan 1980). The species is classified as vulnerable by the IUCN (1974) and is on appendix 2 of the CITES.

Ursus malayanus. This is the smallest bear. Head and body length is 1,000 to 1,400 mm, tail length is 30 to 70 mm, shoulder height is about 700 mm, and weight is 27 to 65 kg. The general coloration is black. There is a whitish or orange breast mark, a grayish or orange muzzle, and, occasionally, pale-colored feet. The breast mark is often U shaped, but is variable and is sometimes wholly lacking. The body is stocky, the muzzle is short, the paws are large, and the claws are strongly curved and pointed. The soles are naked.

The sun bear inhabits dense forests at all elevations (Lekagul and McNeely 1977). It is active at night, usually sleeping and sunbathing by day in a tree, two to seven meters above the ground. Tree branches are broken or bent to form a nest and lookout post. *U. malayanus* has a curious gait in that all the legs are turned inward while walking. The species is usually shy and retiring, and does not hibernate. It is an expert tree climber, and is cautious, wary, and intelligent. A young captive observed the way in which a cupboard, containing a sugar pot, was locked with a key. It then later opened the cupboard by inserting a claw into the eye of the key and turning it. Another captive scattered rice from its feeding bowl in the vicinity of its cage, thus attracting chickens, which it then captured and ate.

The diet is omnivorous, and the front paws are used for most of the feeding activity. Trees are torn open in search of nests of wild bees and for insects and their larvae. The soft growing point of the coconut palm, known as palmite, is ripped apart and consumed. After digging up termite colonies, the animal places its forepaws alternately in the nest and licks the termites off. Jungle fowl, small rodents, and fruit juices also are included in the diet.

Births may occur at any time of the year. In the East Berlin Zoo, a female produced one litter on 4 April 1961 and another on 30 August 1961. The gestation period for six births

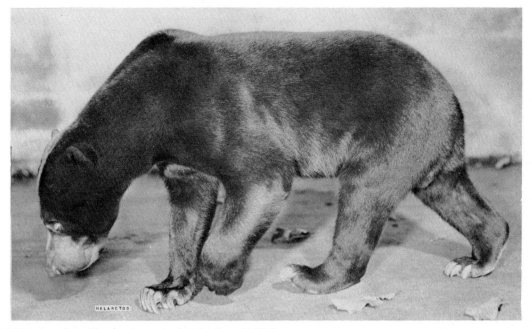

Malayan sun bear (*Ursus malayanus*), photo by Ernest P. Walker.

at that zoo was 95–96 days (Dathe 1970). At the Fort Worth Zoo, however, three pregnancies lasted 174, 228, and 240 days, evidently because of delayed fertilization or implantation (McCusker 1974). Litters usually contain one or two young, each weighing about 325 grams. They remain with the mother until nearly full grown (Lekagul and McNeely 1977). A captive lived for 24 years and 9 months (Marvin L. Jones, Zoological Society of San Diego, pers. comm.).

Young individuals make interesting pets, but become unruly within a few years. In the wild, *U. malayanus* is said to be one of the most dangerous animals within its range (Lekagul and McNeely 1977). It sometimes does great damage to coconut plantations. It is thought to be declining in some areas, because of forest destruction (Cowan 1972). It is on appendix 1 of the CITES.

Ursus ursinus.   Head and body length is 1,400 to 1,800 mm, tail length is 100 to 125 mm, shoulder height is 610 to 915 mm, and weight is 55 to 145 kg. The shaggy black hairs are longest between the shoulders. The overall black colora-

Sloth bear (*Ursus ursinus*): A. Photo by Hans Jurg Kuhn; B. Photo from New York Zoological Society.

tion is often mixed with brown and gray, but cinnamon and red individuals also have been noted. The chest mark, typically shaped like a V or Y, varies from white or yellow to chestnut brown.

The sloth bear has a number of structural modifications associated with an unusual method of feeding. The lips are protrusible, mobile, and naked; the snout is mobile; the nostrils can be closed at will; the inner pair of upper incisors is absent, thus forming a gap in the front teeth; and the palate is hollowed. These features enable the bear to feed on termites (white ants) in the following manner: the nest is dug up, the dust and dirt blown off, and the occupants sucked up in a "vacuum cleaner" action. The resulting noises can be heard for over 185 meters, and often lead to the bear's detection by hunters.

The sloth bear inhabits moist and dry forests, especially in areas of rocky outcrops. It may be active at any hour, but is mainly nocturnal. During cool weather it spends the day in dense vegetation or shallow caves. The sense of smell is well developed, but sight and hearing are relatively poor. Hibernation is not known to occur. Termites are the most important food for most of the year, but the diet also includes other insects, grubs, honey, eggs, carrion, grass, flowers, and fruit.

In a study in the Royal Chitawan National Park of Nepal, Laurie and Seidensticker (1977) found a minimum density of about 0.1 individual per sq km. Most observations were of lone bears or of females with cubs. Vocalizations, heard mainly in association with intraspecific agonistic encounters, included roars, howls, screams, and squeals. Births apparently occurred mostly from September to January.

Previous observations indicate that breeding takes place mainly in June in India and over most of the year in Sri Lanka. Pregnancy lasts about 6 or 7 months. The young, usually one or two and rarely three, are born in a ground shelter. They leave the den at an age of 2 or 3 months and often ride on the mother's back. They remain with the mother until they are almost full grown, possibly 2 or 3 years. Captives have lived for 40 years.

The sloth bear is normally not aggressive, but is held in great respect by some of the people that inhabit its range. Apparently because of its poor eyesight and hearing, it is sometimes closely approached by humans. It may then attack in what it considers to be self-defense, and inflict severe wounds. Since it is thought to be dangerous, and since it sometimes damages crops, it has been extensively hunted. It also seems not to tolerate regular human disturbance, and is losing habitat to agriculture, logging, settlement, and hydroelectric projects. Fewer than 10,000 individuals are estimated to survive in India and Sri Lanka. The status of *U. ursinus* is considered indeterminate by the IUCN (1978).

CARNIVORA; URSIDAE; Genus AILUROPODA
Milne-Edwards, 1870

Giant Panda

The single species, *A. melanoleuca*, is known from the central Chinese provinces of Kansu, Shensi, and Szechwan, and may also occur in eastern Tsinghai and northern Yunnan (Chorn and Hoffmann 1978). In prehistoric times the species occupied much of eastern China (Schaller 1981). Although it is sometimes considered a relative of *Ailurus* and placed with that genus in the family Procyonidae, most authorities now treat *Ailuropoda* as a bear. Hall (1981) put it in the ursid subfamily Ailuropodinae, but Chorn and Hoffmann (1978) referred it to the otherwise extinct ursid subfamily Agri-

otheriinae. While recognizing it as an offshoot of the Ursidae, Thenius (1979) suggested that the giant panda represents a distinct family, the Ailuropodidae.

Head and body length is 1,200 to 1,500 mm, tail length is about 127 mm, and weight is 75 to 160 kg. The coat is thick and woolly. The eye patches, ears, legs, and band across the shoulders are black, sometimes with a brownish tinge. The remainder of the body is white, but may become soiled with age. There are scent glands under the tail.

Ailuropoda resembles other bears in general appearance, but is distinguished by its striking coloration and certain characters associated with its diet. The head is relatively massive, because of the expanded zygomatic arches of the skull and the well-developed muscles of mastication. The second and third premolar teeth, and the molars, are relatively larger and broader than those of other bears. The forefoot has an unusual modification, thought to aid in the grasping of bamboo stems. The pad on the sole of each forepaw has an accessory lobe, and the pad of the first digit—and to a lesser extent the pad of the second digit—can be flexed onto the summit of this accessory lobe and its supporting bone.

The giant panda is found in montane forests with dense stands of bamboo. Its usual elevational range is 2,700 to 3,900 meters, but it may descend to as low as 800 meters in the winter. It does not make a permanent den, but takes shelter in hollow trees, rock crevices, and caves. It lives mainly on the ground, but evidently can climb trees well. Activity is largely crepuscular and nocturnal. *Ailuropoda* does not hibernate, but descends to lower elevations in the winter and spring (Chorn and Hoffmann 1978).

The diet consists mainly of bamboo shoots, up to 13 mm in diameter, and bamboo roots. *Ailuropoda* spends 10 to 12 hours a day feeding, usually in a sitting position with the forepaws free to manipulate the bamboo. Other plants, such as gentians, irises, crocuses, and tufted grasses, are also taken. *Ailuropoda* occasionally hunts for fish, pikas, and small rodents.

The following data on reproduction and life history were taken largely from Chorn and Hoffmann (1978). Individuals usually stay in a single ravine, within an area of about 2.5 sq km, but may wander farther in the mating season. Captives are known to scent mark with secretions from the anal region. *Ailuropoda* is generally solitary, but two or three individuals may come together in the breeding season, which extends from March to May. Estrus usually lasts about 10 days. Births usually occur in September, following 122 to 163 days of pregnancy. It has been suggested that fertilization or implantation may be delayed. The number of young per litter is usually one or two, and occasionally three, but only a single cub is normally raised. At birth the offspring averages about 104 grams in weight, is covered with sparse white fur, and has a tail that is about one-third as long as the head and body. Adult coloration is attained by the end of the first month, and the eyes open after 40 to 60 days. Independent feeding begins at 3 to 4 months of age, separation from the mother comes at 6 months, and sexual maturity is usually reached after 6 or 7 years. According to Marvin L. Jones (Zoological Society of San Diego, pers. comm.), one specimen was still living after 26 years in captivity.

The range of the giant panda began to decline in the late Pleistocene, because of both climatic changes and the spread of people (Wang 1974). At present, about 1,000 individuals are thought to survive, and they are apparently divided into three isolated groups. In the mid-1970s, about 100 pandas starved when an important food plant died over a large area. The species receives complete legal protection, and cooperative field investigations were recently begun by the Chinese government and the World Wildlife Fund (Schaller 1981). The giant panda is classified as rare by the IUCN (1976). It is

Giant panda (*Ailuropoda melanoleuca*), photo from New York Zoological Society. Insets: A. Right hind foot; B. Right forefoot; photos from *Proc. Zool. Soc. London;* C. Skull showing dentition, photos by P. F. Wright of specimen in U.S. National Museum.

among the most popular of zoo animals, but has been extremely difficult to breed. According to the "Census of Rare Animals in Captivity" in the 1980 *International Zoo Yearbook,* the number of giant pandas in captivity is about 40 in China and 13 in other countries.

CARNIVORA; *Family PROCYONIDAE*

Raccoons and Relatives

This family contains 7 Recent genera and 19 species. There are 2 subfamilies: Ailurinae, with the single genus *Ailurus* (lesser panda), which is found in the Himalayas and adjacent parts of eastern Asia; and Procyoninae, with the other 6 genera, which occur in temperate and tropical areas of the Western Hemisphere. *Ailurus* has sometimes been placed together with *Ailuropoda* (giant panda) in a separate family, the Ailuropodidae or Ailuridae.

Head and body length is 305 to 670 mm, tail length is 200 to 690 mm, and weight is about 0.8 to 12 kg. Males are about one-fifth larger and heavier than females. The pelage varies from gray to rich reddish brown. Facial markings are often present, and the tail is usually ringed with light and dark bands. The face is short and broad. The ears are short, furred, erect, and rounded or pointed. The tail is prehensile in the aboreal kinkajou (*Potos*) and is used as a balancing and semi-prehensile organ in the coatis (*Nasua*). Each limb bears five digits, the third being the longest. The claws are short, compressed, recurved, and, in some genera, semiretractile. The

soles are haired in several genera. Males have a baculum.

The dental formula is usually: (i 3/3, c 1/1, pm 4/4, m 2/2) $\times 2 = 40$. The premolars, however, number 3/3 in *Potos* and 3/4 in *Ailurus* (Stains 1967). The incisors are not specialized, the canines are elongate, the premolars are small and sharp, and the molars are broad and low crowned. The carnassials are developed only in *Bassariscus*.

Procyonids walk on the sole of the foot, with the heel touching the ground, or partly on the sole and partly on the digits. The gait is usually bearlike. They are good climbers, and one genus (*Potos*) spends nearly its entire life in trees. Most procyonids shelter in hollow trees, on large branches, or in rock crevices. Most become active in the evening, but *Nasua* may be primarily diurnal. The diet is omnivorous, though *Potos* and *Bassaricyon* seem to depend largely on fruit and *Ailurus* feeds mainly on bamboo. Most genera travel in pairs or family groups, and give birth in the spring.

The geological range of the Procyonidae is late Oligocene to Recent in North America, Pliocene to Recent in South America, late Miocene to early Pliocene in Europe, and early Pliocene to Recent in Asia (Stains 1967).

CARNIVORA; PROCYONIDAE; *Genus AILURUS*
F. Cuvier, 1825

Lesser Panda

The single species, *A. fulgens,* occurs in Nepal, Sikkim, northern Burma, and the provinces of Yunnan and Szechwan

Lesser panda (*Ailurus fulgens*), photo by Arthur Ellis, *Washington Post*.

in south-central China (Ellerman and Morrison-Scott 1966).

Head and body length is 510 to 635 mm, tail length is 280 to 485 mm, and weight is usually 3 to 4.5 kg. The coat is long and soft, and the tail is bushy. The upper parts are rusty to deep chestnut, being darkest along the middle of the back. The tail is inconspicuously ringed. Small, dark-colored eye patches are present, and the muzzle, lips, cheeks, and edges of the ears are white. The back of the ears, the limbs, and the underparts are dark reddish brown to black. The head is rather round, the ears are large and pointed, the feet have hairy soles, and the claws are semiretractile. The tail is non-prehensile and is about two-thirds as long as the head and body. There are glandular sacs in the anal region. Females have four mammae.

The lesser panda inhabits mountain forests and bamboo thickets at elevations of 1,800 to 4,000 meters. It seems to prefer colder temperatures than does the giant panda (*Ailuropoda*). It is nocturnal and crepuscular, sleeping by day in a tree. When sleeping, it generally curls up like a cat or dog, with the tail over the head, but it may also sleep while sitting on top of a limb, with the head tucked under the chest and between the forelegs, as the American raccoon (*Procyon*) does at times. Although *Ailurus* is a capable climber, it seems to do most feeding on the ground. The diet consists mostly of bamboo sprouts, grasses, roots, fruits, and acorns. It also occasionally takes insects, eggs, young birds, and small rodents.

In the wild the lesser panda sometimes travels in pairs or small family groups. Its disposition is mild; when captured it does not fight, tames readily, and is gentle, curious, and generally quiet. The usual call is a series of short whistles or squeaking notes; when provoked, it utters a sharp, spitting hiss or a series of snorts while standing on its hind legs. A musky odor is emitted from the anus when the animal is excited. According to Grzimek (1975), there is territorial scent marking by rubbing the anal region against objects.

Births take place in the spring in a hollow tree or rock crevice. The length of pregnancy can vary from 90 to 150 days, probably because there is sometimes delayed implantation. The number of young per litter is usually one or two and rarely three or four. In the National Zoo in Washington,

D.C., cubs were born in late June, weighed about 200 grams when 1 week old, opened their eyes after 17–18 days, attained full adult coloration by 90 days, and took their first solid food at 125–35 days (Roberts 1975). The young seem to stay with the mother, or both parents, for about 1 year, or until the next litter is about to be born. According to Marvin L. Jones (Zoological Society of San Diego, pers. comm.), one individual lived for 13 years and 5 months in captivity.

The lesser panda is a very popular zoo animal, and is frequently involved in the animal trade. It is on appendix 2 of the CITES.

CARNIVORA; PROCYONIDAE; **Genus BASSARISCUS**
Coues, 1887

Ringtails, Cacomistles

There are two species (Hall 1981):

B. astutus, southwestern Oregon and eastern Kansas to Baja California and southern Mexico;

B. sumichrasti, southern Mexico to western Panama.

The latter species was formerly often placed in a separate genus, *Jentinkia* Trouessart, 1904.

In *B. astutus,* head and body length is 305 to 420 mm, tail length is 310 to 441 mm, and shoulder height is about 160 mm. Armstrong, Jones, and Birney (1972) listed weights of 824 to 1,338 grams. The upper parts are buffy, with a black or dark brown wash, and the underparts are white or white washed with buff. The eye is ringed by black or dark brown, and the head has white to pinkish buff patches. The tail is bushy, longer than the head and body, and banded with black and white for its entire length. Females have four mammae.

In *B. sumichrasti,* head and body length is 380 to 470 mm, and tail length is 390 to 530 mm. One individual weighed 900 grams. The color is usually buffy gray to brownish, and the tail is ringed with buff and black. From *B. astutus, B. sumichrasti* is distinguished in having pointed (rather than rounded) ears, a longer tail, naked (rather than hairy) soles,

Ringtail (*Bassariscus astutus*), photo by Ernest P. Walker.

nonretractile (rather than semiretractile) claws, and low (rather than high) ridges connecting the cusps of the molariform teeth.

A variety of habitat is utilized by *B. astutus*, but this animal seems to prefer rocky, broken areas, often near water. It dens in rock crevices, hollow trees, the ruins of old Indian dwellings, and the upper parts of cabins. Activity occurs mainly at night, *B. sumichrasti* is found in tropical forests and appears to be more arboreal than *B. astutus*. The latter, however, is a good climber, and travels quickly and agilely among cliffs and along ledges. In a study of captive *B. astutus*, Trapp (1972) determined that the hind foot can rotate at least 180°, permitting a rapid, head-first descent and great dexterity. One individual traveled upside down along a cord 5 mm in diameter. Ringtails sometimes climb in a crevice by pressing all four feet on one wall and the back against the other. They also maneuver by ricocheting off of smooth surfaces to gain momentum to continue to an objective. The diet includes insects, rodents, birds, fruit, and other vegetable matter.

According to Grzimek (1975), recorded population densities are 1 individual for every 8 sq km in California and 10 per 1.5 sq km on the Edwards Plateau of Texas. Home range is about 3.2 sq km at most, but is usually substantially smaller, and individuals tend to stay in one area. *Bassariscus* is generally solitary, except during the mating season. *B. astutus* scent marks its territory by regularly urinating at certain sites. *B. sumichrasti*, however, eliminates at random (Poglayen-Neuwall 1973). Both species have a variety of vocalizations. Adult *B. astutus* may emit an explosive bark, a piercing scream, and a long, plaintive, high-pitched call.

Female *B. sumichrasti* enter estrus in winter, spring, or summer, but late winter appears to be the main breeding season. *B. astutus* mates from February to May and gives birth from April to July. Heat lasts only 24 hours and the gestation period is about 51 to 54 days. The number of young per litter is one to five and is usually two to four. The young

Central American cacomistle (*Bassariscus sumichrasti*), photo by I. Poglayen-Neuwall.

Central American cacomistle (*Bassariscus sumichrasti*), immature, photo from Jorge A. Ibarra through the Museo Nacional de Historia Natural, Guatemala City.

are born in a nest or den, and may be given care by both parents. They weigh about 28 grams each at birth, open their eyes when 31 to 34 days old, begin taking some solid food at 4 weeks of age, begin to forage with the adults at 2 months, are completely weaned at 4 months, and disperse in early winter. Both sexes attain sexual maturity when approximately 10 months old (Poglayen-Neuwall and Poglayen-Neuwall 1980; Grzimek 1975; Leopold 1959). A captive *B. astutus* lived for 14 years and 3 months (Marvin L. Jones, Zoological Society of San Diego, pers. comm.).

Ringtails, especially females obtained when young, make charming pets. They were sometimes kept about the homes of early settlers as companions and to catch mice. The coat is not of particularly high quality. It is known commercially as "California mink" or "civet cat," but there is no scientific basis for either name. In the 1976/77 trapping season, 88,329 pelts of *B. astutus* were reported taken in the United States, and sold for an average price of $5.50 (Deems and Pursley 1978). *B. astutus* apparently extended its range into Kansas, Arkansas, and Louisiana in the 20th century, and has even been reported from Alabama and Ohio (Hall 1981). These occurrences might result partly from the habit of boarding railroad cars (Sealander 1979).

CARNIVORA; PROCYONIDAE; **Genus PROCYON** Storr, *1780*

Raccoons

Two subgenera and seven species are currently recognized (Hall 1981; Cabrera 1957; Gardner 1976):

subgenus *Procyon*

P. lotor, southern Canada to Panama;
P. insularis, Tres Marias Islands off western Mexico;
P. maynardi, New Providence Island (Bahamas);
P. pygmaeus, Cozumel Island off northeastern Yucatan;
P. minor, Guadeloupe Island (Lesser Antilles);
P. gloveralleni, Barbados (Lesser Antilles);

subgenus *Euprocyon*

P. cancrivorus (crab-eating raccoon), eastern Costa Rica to eastern Peru and Uruguay.

Lotze and Anderson (1979) suggested that several of the designated species of the subgenus *Procyon* might be conspecific with *P. lotor*.

Head and body length is 415 to 600 mm, tail length is 200 to 405 mm, shoulder height is 228 to 304 mm, and weight is usually 2 to 12 kg. Generally, males are larger than females, and northern animals are larger than southern ones. Five adult males in the Florida Keys averaged 2.4 kg. Mean weights in Alabama were 4.31 kg for males and 3.67 kg for females. Means in Missouri were 6.76 kg for males and 5.94 kg for females (Lotze and Anderson 1979; Johnson 1970). In Wisconsin the normal weight range is about 6 to 11 kg, but there is one record of a male weighing 28.3 kg (Jackson 1961).

The general coloration is gray to almost black, sometimes with a brown or red tinge. There are 5 to 10 black rings on the rather well-furred tail, and a black "bandit" mask across the face. The head is broad posteriorly and has a pointed muzzle. The toes are not webbed and the claws are not retractile. The front toes are rather long and can be widely spread. The footprints resemble those of people. Females have four pairs of mammae (Banfield 1974).

Raccoons frequent timbered and brushy areas, usually near water. They are more nocturnal than diurnal, and are good climbers and swimmers. The den is usually in a hollow tree, with an entrance more than 3 meters above the ground (Banfield 1974). The den may also be in a rock crevice, overturned stump, burrow made by another animal, or human building. Urban (1970) found most raccoons in a marsh to den in muskrat houses. Except when sequestered during severe winter weather, or in cases of females with newborn, each den is usually occupied for only one or two days. The average distance between dens has been reported as 436 meters, and the general movements of *Procyon* are not extensive, but one individual was found to have traveled 266 km (Lotze and Anderson 1979).

Raccoons do not hibernate. In the southern parts of their range they are active throughout the year. In northern areas they may remain in a den for much of the winter, but will emerge during intervals of relatively warm weather. While in winter sleep, their heart beat does not decline, their body temperature stays above 35° C, and their metabolic rate remains high. They do, however, live mostly off of fat reserves accumulated the previous summer and fall, and may lose up to 50 percent of their weight (Lotze and Anderson 1979).

Procyon has a well-developed sense of touch, especially in

Raccoon (*Procyon lotor*), photo from Zoological Society of Philadelphia.

the nose and forepaws. The hands are regularly used almost as skillfully as monkeys use theirs. Food is generally picked up with the hands and then placed in the mouth. Although raccoons have sometimes been observed to dip food in water, especially under captive conditions, the legend that they actually wash their food is without foundation (Lowery 1974). The omnivorous diet consists mainly of crayfish, crabs, other arthropods, frogs, fish, nuts, seeds, acorns, and berries.

As many as 167 raccoons have been found in an area of 41 ha., but more typical population densities are 1 individual per 5 to 43 ha. (Lotze and Anderson 1979). Reported home range size varies from 0.2 to 4,946 ha., but seems to be typified by the situation found by Lotze (1979) on St. Catherine's Island, Georgia. He reported an annual average of 65 ha. for males and 39 ha. for females, but indicated that there was much variation and that different study methods might give different results. About the smallest population density (0.5 to 1.0 residents per 100 ha.) and largest home range (means of 1,139 ha. for males and 806 ha. for females) were reported by Fritzell (1978) for the prairies of North Dakota. His study indicated that the ranges of adult males are largely exclusive of one another, but do commonly overlap the ranges of 1 to 3 adult females and up to 4 yearlings. This and other studies (Lotze and Anderson 1979; Schneider, Mech, and Tester 1971) suggest that female ranges are often not exclusive, and that territorial defense is not well developed in *Procyon*, but that unrelated animals tend to avoid one another. Nonetheless, as many as 23 individuals have been found in the same winter den, and about the same number have congregated around artificial feeding sites (Lotze and Anderson 1979; Lowery 1974). Raccoons have a variety of vocalizations, most with little carrying power.

In the United States the reproductive season extends from December to August. Mating peaks in February and March, and births from April to June (Lotze and Anderson 1979; Johnson 1970). The breeding season of *P. cancrivorus* is July to September (Grzimek 1975). If a female *P. lotor* loses a newborn litter, she may ovulate a second time during the season (Sanderson and Nalbandov 1973). The gestation period averages 63 days and ranges from 60 to 73. The number of young per litter is one to seven, usually three or four. Captives in New York weighed 71 grams each at birth. The eyes open after about 3 weeks (Banfield 1974), and weaning takes place from 7 weeks to 4 months of age (Lotze and Anderson 1979). In Minnesota, Schneider, Mech, and Tester (1971) found that the young were kept in a den in a hollow tree until they were 7 to 9 weeks old, and then were moved to one or a series of ground beds. At the age of 10–11 weeks, they were taken on short trips by the mother, and after another week the family began to move together. In November the members denned either together in one hollow tree or individually in nearby trees. The young usually separate from the mother at the end of winter. They may attain sexual maturity when about 1 year old, but most do not mate until the following year. Few wild raccoons live more than 5 years, but some are estimated to have survived for 13 to 16 years (Lotze and Anderson 1979). One captive was still living after 20 years and 7 months (Marvin L. Jones, Zoological Society of San Diego, pers. comm.).

Raccoons sometimes damage cornfields and other crops, but usually not to a serious extent (Jackson 1961). They make good pets and are interesting to observe in the wild, but carry pathogens known to cause such human diseases as leptospirosis, tularemia, and rabies. *P. lotor* is currently the most valuable wild fur bearer in the United States. The harvest in 44 states during the 1976/77 season was 3,832,802 skins, which sold at an average price of $26.00 (Deems and Pursley 1978). Because of its commercial value, *P. lotor* was introduced in France, the Netherlands, Germany, and various parts of the Soviet Union, but now is sometimes considered a nuisance in those areas (Corbet 1978; Grzimek 1975). *P. lotor* seems to have extended its range and increased in numbers in certain parts of North America since the 19th century. The various insular species are rare, however, and some may be extinct (Lotze and Anderson 1979).

CARNIVORA; PROCYONIDAE; **Genus NASUA** *Storr, 1780*

Coatis, Coatimundis

There are two species (Hall 1981; Cabrera 1957):

N. nasua, Arizona to Argentina;
N. nelsoni, Cozumel Island off northeastern Yucatan.

Head and body length is 410 to 670 mm, tail length is 320 to 690 mm, and shoulder height is up to 305 mm. Grzimek (1975) gave the weight as 3 to 6 kg. Males are usually larger than females. *N. nelsoni* has short, fairly soft, silky hair, but in *N. nasua* the fur is longer and somewhat harsh. In both species the general color is reddish brown to black above, and yellowish to dark brown below. The muzzle, chin, and throat are usually whitish, and the feet blackish. Black and

gray markings are present on the face, and the tail is banded. The muzzle is long and pointed, and the tip is very mobile. The forelegs are short, the hind legs are long, and the tapering tail is longer than the head and body.

Coatis are found mainly in wooded areas. They forage in trees, as well as on the ground, using the tail as a balancing and semiprehensile organ. While moving along the ground, the animals usually carry the tail erect, except for the curled tip. The long, highly mobile snout is well adapted for investigating crevices and holes. Adult males are often active at night, but coatis are primarily diurnal. They move about 1,500 to 2,000 meters a day in the search for food, and usually retire to a roost tree at night (Kaufmann 1962). The diet includes both plant and animal matter. When fruit is abundant, coatis are almost exclusively frugivorous. At other times, females and young forage for invertebrates on the forest floor, and adult males tend to prey on large rodents (Smythe 1970).

Reported population densities are 26 to 42 individuals per 100 ha. on Barro Colorado Island in the Panama Canal Zone, and 1.2 to 2 per 100 ha. in Arizona (Lanning 1976). Home ranges of four solitary males in Arizona were 70 to 270 ha. each (Kaufmann, Lanning, and Poole 1976). On Barro Colorado Island, groups had overlapping, undefended home ranges of 35 to 45 ha. Each group, however, spent about 80 percent of its time in an exclusive core area within its home range (Kaufmann 1962).

The social system of *N. nasua,* as studied on Barro Colorado Island, is an interesting example of the interrelationship of ecology and behavior (Kaufmann 1962; Smythe 1970; Rus-

Coatimundi (*Nasua nasua*), photo from New York Zoological Society.

sell 1981). All females, and males up to two years old, are found in loosely organized bands, usually of 4 to 20 individuals each. Males over two years old become solitary, except during the breeding season. They are usually excluded from membership in the bands by the collective aggression of the adult females, sometimes supported by the juveniles. In the breeding season, an adult male is accepted into each group, but he is completely subordinate to the females. The breeding season corresponds with the period of maximum abundance of fruit. At this time there is thus a minimum of competition for food between the large males and the other animals. Moreover, at other times of the year, when the males become carnivorous, they may attempt to prey on young coatis, and so would then be dangerous to the survival of the group.

There is a single reproductive season, with births in Panama occurring from April to June at the beginning of the rains. The gestation period lasts 10 to 11 weeks. Pregnant females separate from the group and construct a tree nest, where they give birth to a litter of two to seven young. When the young are 5 weeks old, they leave the nest, and they and the mother join the group. The young weigh 100 to 180 grams each at birth, open their eyes after 11 days, are weaned at 4 months, reach adult size at 15 months, and attain sexual maturity when 2 years old (Kaufmann 1962; Grzimek 1975). A captive specimen was still living after 17 years and 8 months (Marvin L. Jones, Zoological Society of San Diego, pers. comm.).

Coatis seem to have extended their range northward in the 20th century. Numbers reached a peak in Arizona in the late 1950s, crashed in the early 1960s, and then began a slow recovery (Kaufmann, Lanning, and Poole 1976). Coatis rarely damage crops and only infrequently take chickens. They are hunted for their meat by natives, who sometimes have dogs trained for this purpose. If cornered by a dog on the ground, a coati can inflict serious wounds with its large and sharp canine teeth. Coatis can be tamed, and make interesting and inquisitive pets.

CARNIVORA; PROCYONIDAE; Genus NASUELLA
Hollister, 1915

Mountain Coati

The single species, *N. olivacea*, is known from the Andes of western Venezuela, Colombia, and Ecuador (Cabrera 1957). The species was originally placed in *Nasua*, and some authorities, such as Grzimek (1975), still assign it to that genus.

Nasuella resembles *Nasua*, but is usually smaller and has a shorter tail. A specimen of a male from Colombia in the U.S. National Museum of Natural History has a head and body length of 383 mm and a tail length of 242 mm. Respective measurements for a female are 394 and 201 mm (John Miles, U.S. National Museum of Natural History, pers. comm.). The general color is grayish sooty brown. The tail is ringed with alternating yellowish gray and dark brown bands. The skull of *Nasuella* is smaller and more slender than that of *Nasua*, the middle part of the facial portion is greatly constricted laterally, and the palate extends farther posteriorly.

Handley (1976) collected seven specimens in Venezuela at elevations of 2,000 to 3,020 meters. Of these, all were taken on the ground, four were taken at dry and three at moist sites, and four were taken in cloud forest and three in páramo. Like *Nasua*, *Nasuella* probably feeds on insects, small vertebrates, and fruit.

CARNIVORA; PROCYONIDAE; Genus POTOS
E. Geoffroy St.-Hilaire and G. Cuvier, 1795

Kinkajou

The single species, *P. flavus*, is found from southern Tamaulipas in eastern Mexico to the Mato Grosso of central Brazil (Hall 1981; Cabrera 1957). Confusion sometimes results from the application of the vernacular name "potto" to this species, because the same term is used for the African primate *Perodicticus*.

Head and body length is 405 to about 760 mm, tail length is 392 to 570 mm, shoulder height is up to 254 mm, and weight is 1.4 to 4.6 kg. Males are generally larger than females (Kortlucke 1973). The upper parts and upper surface of the tail are tawny olive, yellow tawny, or brownish. Some individuals have a black middorsal line. The underparts and undersurface of the tail are tawny yellow, buff, or brownish yellow. The muzzle is dark brown to blackish. The hair is soft and woolly.

The kinkajou has a rounded head, a short face, a long, prehensile tail, and short, sharp claws. The hind feet are longer than the forefeet. The tongue is narrow and greatly extensible. *Potos* is similar to *Bassaricyon*, but differs in its round, tapering, short-haired, prehensile tail; in its stockier body form; and in its face not being grayish. Females have two mammae (Grzimek 1975).

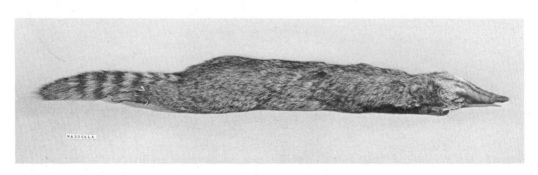

Mountain coati (*Nasuella olivacea*), photo by Howard E. Uible of skin in the U.S. National Museum.

Kinkajou (*Potos flavus*), photo from New York Zoological Society.

The kinkajou inhabits forests and is almost entirely arboreal. It spends the day in a hollow tree, sometimes emerging on hot, humid days to lie out on a limb or in a tangle of vines. By night it forages among the branches. Although it moves rapidly through a single tree, progress from one tree to another is made cautiously and relatively slowly. An individual probably returns to the same trees night after night. The long tongue of *Potos* is an adaptation for a frugivorous diet. It eats mainly fruit, but also takes honey, insects, and small vertebrates (Husson 1978).

A population density of 0.74 individuals per ha. was found in Tikal National Park, Guatemala (Walker and Cant 1977). *Potos* travels alone or in pairs. Small groups may form, but usually only temporarily in a fruit-bearing tree. Territories are not defended, though individuals do scent mark, probably as a sexual signal (Grzimek 1975). *Potos* barks when disturbed, and emits a variety of other vocalizations, but the usual call, given while feeding during the night, seems to be "a rather shrill, quavering scream that may be heard for nearly a mile" (Dalquest 1953:182).

According to Grzimek (1975), there is no particular breeding season, but Husson (1978) wrote that in Surinam births reportedly take place in April and May. The gestation period is 112–18 days. The number of young is usually one and is rarely two. They are born in a hollow tree. They weigh 150–200 grams at birth, open their eyes at 7–19 days, and begin to take solid food, and can hang by their tails, at 7 weeks. Sexual maturity comes at 1½ years of age for males and 2¼ years for females. One pair lived together for 9 years and had their first litter at 12½ years of age. One individual lived for 23 years and 7 months in the Amsterdam Zoological Gardens.

If captured when young and treated kindly, the kinkajou becomes a good pet. It is often sold under the name "honey bear." Its meat is said to be excellent (Husson 1978), and its pelt is used in making wallets and belts. It does little damage to cultivated fruit.

CARNIVORA; PROCYONIDAE; **Genus BASSARICYON**
J. A. Allen, 1876

Olingos

Five species are currently recognized (Hall 1981; Cabrera 1957; Handley 1976):

B. gabbii, Nicaragua to Ecuador and Venezuela;
B. pauli, known only from the type locality in western Panama;
B. lasius, known only from the type locality in central Costa Rica;
B. beddardi, Guyana, and possibly adjacent parts of Venezuela and Brazil;
B. alleni, Ecuador, Peru, possibly Venezuela.

All of the above species may be no more than subspecies of *B. gabbii* (Hall 1981; Cabrera 1957; Grimwood 1969).

Head and body length is 350 to 475 mm, and tail length is 400 to 480 mm. Grzimek (1975) gave the weight as 970 to 1,500 grams. The fur is thick and soft. The upper parts are pinkish buff to golden, mixed with black or grayish above. The underparts are pale yellowish. The tail is somewhat flattened and more or less distinctly annulated along the median portion. The general body form is elongate, the head is flattened, the snout is pointed, and the ears are small and

Olingo (*Bassaricyon gabbii*), photo from New York Zoological Society. Inset: olingo (*B. pauli*), photo by Hans Jurg Kuhn.

rounded. The limbs are short, the soles are partly furred, and the claws are sharply curved. The tail of *Bassaricyon,* unlike that of *Potos,* is long haired and not prehensile. Females have a single pair of inguinal mammae.

The olingo is found in tropical forests from sea level to 2,000 meters. It is primarily arboreal and nocturnal. It is thought to spend the day in a nest of dry leaves in a hollow tree. According to Grzimek (1975), it is an excellent climber and jumper, and can leap three meters, from limb to limb, without difficulty. It feeds mainly on fruit, but also hunts for insects and warm-blooded animals. Poglayen-Neuwall (1966) found that captives required considerably more meat than did *Potos.*

In the wild, *Bassaricyon* lives alone or in pairs. It is sometimes found in association with *Potos,* as well as with opossums and douroucoulis (*Aotus*). It scent marks with urine, but the particular function is not known (Grzimek 1975). In studies at the Louisville Zoo, Poglayen-Neuwall (1966) found that *Bassaricyon* is less social than *Potos,* and that two males could not be kept together. There was no definite breeding season, gestation lasted 73–74 days, and each of four births yielded a single offspring. The young weighed about 55 grams at birth, opened their eyes after 27 days, began taking solid food after 2 months, and attained sexual maturity at 21 months of age. According to Marvin L. Jones (Zoological Society of San Diego, pers. comm.), one captive was still living after 17 years and 1 month.

CARNIVORA; *Family MUSTELIDAE*

Weasels, Badgers, Skunks, Otters

This family of 23 Recent genera and 64 species occurs in all land areas of the world, except the West Indies, Madagascar, Celebes and islands to the east, most of the Philippines, New Guinea, Australia, New Zealand, Antarctica, and most oceanic islands. One genus, *Enhydra,* inhabits coastal waters of the North Pacific Ocean. The sequence of genera presented here follows basically that of Simpson (1945), who recognized five subfamilies: Mustelinae (weasels), with *Mustela, Vormela, Martes, Eira, Galictis, Lyncodon, Ictonyx, Poecilictis, Poecilogale,* and *Gulo;* Mellivorinae (honey badger), with *Mellivora;* Melinae (badgers), with *Meles, Arctonyx, Mydaus, Taxidea,* and *Melogale;* Mephitinae (skunks), with *Mephitis, Spilogale,* and *Conepatus;* and Lutrinae (otters), with *Lutra, Pteronura, Aonyx,* and *Enhydra.*

The smallest member of this family is the least weasel (*Mustela nivalis*), which has a head and body length of 114 to 260 mm and a weight as low as 25 grams. The largest members are the otters *Pteronura* and *Enhydra,* which have a head and body length of around 1 meter or more and a weight of 22 to 45 kg. Male mustelids are about one-fourth larger than females. The pelage is either uniformly colored, spotted, or striped. Some species of *Mustela* turn white in winter in the

Giant otter (*Pteronura brasiliensis*), photo by Bernhard Grzimek. Skeleton of mink (*Mustela vison*), photo by P. F. Wright of specimen in U.S. National Museum.

northern parts of their range. The body is long and slender in most genera, but stocky in the wolverine (*Gulo*) and in badgers. The ears are short, and either rounded or pointed. The limbs are short and bear five digits each. The claws are compressed, curved, and nonretractile. The claws of badgers are large and heavy for burrowing. The digits of otters are usually webbed for swimming. Well-developed anal scent glands are present in most genera. Males have a baculum.

The skull is usually sturdy and has a short facial region. The dental formula is: (i 3/2–3, c 1/1, pm 2–4/2–4, m 1/1–2) × 2 = 28 to 38. *Enhydra* is the only genus with 2 lower incisors. The incisors of mustelids are not specialized, the canines are elongate, the premolars are small and sometimes reduced in number, and a constriction is usually present between the lateral and medial halves of the upper molar. The carnassials are developed. The second lower molar, if present, is reduced to a simple peg.

Mustelids are either nocturnal or diurnal, and often shelter in crevices, burrows, and trees. Badgers usually dig elaborate burrows. Mustelids move about on their digits, or partly on their digits and partly on their soles. The smaller, slender forms usually travel by means of a scampering gait, interspersed with a series of bounds. The larger, stocky forms proceed in a slow, rolling, bearlike shuffle. Mustelids often sit on their haunches to look around. Many genera are agile climbers, and otters and minks are skillful swimmers. Mustelids are mainly flesh eaters. They hunt by scent, though the senses of hearing and sight are also well developed. Some

species occasionally feed on plant material, a few are omnivorous, and otters subsist mainly on aquatic animals. The storage of food has been reported for *Mustela*, *Poecilogale*, *Gulo*, and certain badgers.

Many mustelids use the secretions of their anal glands as a defensive measure. Some genera have a contrasting pattern of body colors; for example, skunks have black and white stripes. This pattern is thought to be a form of warning coloration associated with the fetid anal gland secretion, and to be a reminder that the animal is better left alone. Some of the forms with contrasting body colors, such as the marbled polecat (*Vormela*) and certain skunks, expose and emphasize this contrasting pattern by means of bodily movements when they are alarmed.

Delayed implantation of the fertilized eggs in the uterus occurs in many genera. The actual period of developmental gestation is about 30 to 65 days, but with delayed implantation the total period of pregnancy is as long as 12½ months in *Lutra canadensis*. There is usually a single litter per year. The offspring are usually tiny and blind at birth. The young of *Enhydra*, however, are born with their eyes open and in a more advanced stage than the young of other mustelids. The young of most genera can care for themselves at about 2 months of age and are sexually mature after a year or 2. The potential longevity in the wild is generally 5 to 20 years.

Mustelids sometimes kill poultry, but they also help to keep rodents in check. Many species are widely hunted for their fur.

The geological range of this family is early Oligocene to Recent in North America, Europe, and Asia; middle Pliocene to Recent in Africa; and Pliocene to Recent in South America (Stains 1967).

CARNIVORA; MUSTELIDAE; *Genus MUSTELA* Linnaeus, 1758

Weasels, Ermines, Stoats, Minks, Ferrets, Polecats

There are 4 subgenera and 16 species (Hall 1951, 1981; Ellerman and Morrison-Scott 1966; Corbet 1978; Lekagul and McNeely 1977; Izor and de la Torre 1978; Meester and Setzer 1971):

subgenus *Grammogale*

M. felipei (Colombian weasel), known by two specimens from southwestern Colombia;

M. africana (tropical weasel), Amazon Basin of Brazil, eastern Ecuador, and northeastern Peru;

subgenus *Mustela*

M. erminea (ermine or stoat), Scandinavia and Ireland to northeastern Siberia and the western Himalayan region, Japan, Alaska and northern Greenland to northern New Mexico and Maryland;

M. nivalis (least weasel), western Europe and Asia Minor to northeastern Siberia and Korea, parts of China and possibly Indochina, Britain, several Mediterranean islands, Japan, northwestern Africa, Egypt, Alaska, Canada, north-central conterminous United States, Appalachian region;

M. frenata (long-tailed weasel), southern Canada to Guyana and Bolivia;

M. altaica (mountain weasel), southern Siberia to the Himalayan region and Korea;

M. kathiah (yellow-bellied weasel), Himalayan region to southern China;

M. sibirica (Siberian weasel), eastern European Russia to eastern Siberia and Thailand, Japan, Taiwan, Java;

M. strigidorsa (back-striped weasel), Nepal to Thailand;

M. nudipes (Malaysian weasel), Malay Peninsula, Sumatra, Borneo;

subgenus *Lutreola*

M. lutreola (European mink), France to western Siberia and the Caucasus;

M. vison (American mink), Alaska, Canada, conterminous United States except parts of southwest;

M. macrodon (sea mink), formerly on Atlantic coast from New Brunswick to Massachusetts;

subgenus *Putorius*

M. putorius (European polecat), western Europe to Ural Mountains;

M. eversmanni (steppe polecat), steppe zone from Austria to Manchuria and Tibet;

M. nigripes (black-footed ferret), plains region from Alberta and Saskatchewan to northeastern Arizona and Texas.

Grammogale Cabrera, 1940 was considered a full genus by Cabrera (1957) and Stains (1967), but was given only subgeneric rank by Hall (1951) and Izor and de la Torre (1978). Although Hall (1981) also recognized *Lutreola* and *Putorius* as subgenera, he noted that some members of these two taxa closely resemble some members of the subgenus *Mustela*, and that distinction from the latter might not be warranted. Corbet (1978) did not use any subgeneric designations. *M. macrodon* was considered a subspecies of *M. vison* by Manville (1966), but was given specific rank by Hall (1981). Several authorities have suggested that *M. nigripes* of North America is conspecific with *M. eversmanni* of the Old World (Anderson 1973; Corbet 1978). Both *M. erminea* and *M. putorius* have been reported, with much doubt, from northwestern Africa (Meester and Setzer 1971).

There is considerable variation in size, but *Mustela* includes the smallest species in the Mustelidae. The body is usually long, lithe, and slender. The tail is shorter than the head and body, often less than half as long. The legs are short, and the ears are small and rounded. The skull has a relatively shorter facial region than does that of *Martes*. The bullae are long and inflated. Additional information is given separately for each species.

Mustela felipei.   In the type specimen, an adult male, head and body length is 217 mm and tail length is 111 mm. The fur is relatively long, soft, and dense. The upper parts and the entire tail are uniformly blackish brown. The lower parts are light orange buff. All four plantar surfaces are naked, and there is extensive interdigital webbing. These features are thought to be adaptations for semiaquatic life. The two

Weasel (*Mustela* sp.), photo by Ernest P. Walker.

Ermine or stoat (*Mustela erminea*) in winter coat, photo by Bernhard Grzimek.

known specimens were collected in riparian areas at elevations of 1,750 and 2,700 meters (Izor and de la Torre 1978).

Mustela africana.   Head and body length is 240 to 380 mm, and tail length is about 190 to 210 mm. The upper parts are reddish to chocolate. The underparts are pale and have a longitudinal median stripe of the same color as the upper parts. All four plantar surfaces are nearly naked, and there is extensive interdigital webbing. The tropical weasel seems restricted to humid riparian forests (Izor and de la Torre 1978). It is reported to be a good swimmer and climber.

Mustela erminea.   Except as noted, the information for the account of this species was taken from Jackson (1961), Banfield (1974), Novikov (1962), and Stroganov (1969). Head and body length is 170 to 325 mm, tail length is 42 to 120 mm, and weight is 42 to 258 grams. Old World animals average larger than those of North America, and males average larger than females. Except in certain southern parts of its range, the ermine changes color during three- to five-week molts in April–May and October–November. In summer, the back, flanks, and outer sides of the limbs are rich chocolate brown, the tip of the tail is black, and the underparts are white. In winter, the entire coat is white, except for the tip of the tail, which remains black. The winter pelage is longer and denser than that of the summer. Females have eight mammae.

The ermine is found in many habitats, from open tundra to deep forest, but seems to prefer areas with vegetative or rocky cover. It makes its den in a crevice, among tree roots, in a hollow log, or in a burrow taken over from a rodent. Its nest is lined with dry vegetation, or the fur and feathers of its prey. It is primarily terrestrial, but climbs and swims well. It generally hunts in a zigzag pattern, progressing by a series of leaps of up to 50 cm each. It can easily run over the snow, and, if pursued, may move under the snow. It may travel 10 or 15 km in a night. Activity takes place at any hour, but is primarily nocturnal. The ermine is swift, agile, and strong,

and has keen senses of smell and hearing. Its slender body allows it to enter and move quickly through the burrows of its prey. It generally kills by biting at the base of the skull. It sometimes attacks animals considerably larger than itself, such as adult hares. The diet consists mainly of small rodents, and also includes birds, eggs, frogs, and insects. Food may be stored underground for the winter.

Population density fluctuates with prey abundance. Under good conditions there may be an ermine for every 10 ha. Individual home range is up to about 20 ha. and is generally larger for males than for females. There are several vocalizations, including a loud and shrill squeaking. In a radio-tracking study in southern Sweden, Erlinge (1977) found home range sizes of 2 to 3 ha. in females and 8 to 13 ha. in males. The ranges of the males included portions of those of the females. Resident animals of both sexes maintained exclusive territories. Boundaries were regularly patrolled and scent marked, and neighbors usually avoided one another. Adult males were dominant over females and young. Females usually spent their lives in the vicinity of their birthplace, but juvenile males wandered extensively in the spring to find a territory.

Females are polyestrous, but produce only one litter per year. Mating occurs in late spring or early summer, but implantation of the fertilized eggs in the uterus is delayed until around the following March, and birth takes place in April or May. Pregnancy thus lasts about 10 months, but embryonic development only a little over 1 month. Litter size is 3 to 18 young. The average is about 6 in the New World and 8 or 9 in the Old. The young are blind and helpless at birth, but are covered with fine white hair. They grow rapidly and are able to hunt with the mother by the time they are 8 weeks old. Females attain sexual maturity at an age of 2 or 3 months, and sometimes can mate in their first summer of life.

The ermine rarely molests poultry, and is valuable to human interests by its destruction of mice and rats. Its white winter fur has long been used in trimming coats and making stoles. The reported number of ermine pelts taken in eight

A. Least weasel (*Mustela nivalis*). B. Least weasel changing into winter coat. Photos by Ernest P. Walker.

Canadian provinces during the 1976/77 trapping season was 55,216, and the average price was $1.03 (Deems and Pursley 1978).

Mustela nivalis.   Except as noted, the information for the account of this species was taken from Banfield (1974), Jackson (1961), and Stroganov (1969). The least weasel is the smallest carnivore. Head and body length is 114 to 260 mm, tail length is 17 to 78 mm, and weight is 25 to 250 grams. Old World animals average larger than those of North America, and males average larger than females. Except in certain southern parts of its range, the least weasel changes color during the spring and fall. In summer, the upper parts are brown and the underparts are white. In winter, the entire coat is white, though there may be a few black hairs at the tip of the tail.

The least weasel is much like *M. erminea* in details of habitat, nest construction, and movements. It is, however, even more agile, and does not travel over such large areas. It feeds almost entirely on small rodents and may store food for the winter. In an area of 27 ha. in England, King (1975) determined that not more than four adult males were resident at one time. Home range size was 7 to 15 ha. for males and 1 to 4 ha. for females.

Births apparently may occur at any time of the year. Females are polyestrous and are capable of bearing more than one litter annually. Delayed implantation does not occur, and the gestation period is 35 to 37 days. The number of young per litter averages 5 and ranges from 3 to 10. The offspring are weaned when 24 days old and attain sexual maturity at 4 months.

The least weasel is rare and of little or no commercial value. It is not known to prey on domestic animals, and is beneficial to people through its destruction of mice and rats. It has evidently been introduced by human agency on certain Mediterranean islands, on the Azores, and on Sao Tomes off west-central Africa (Corbet 1978).

Mustela frenata.   In Canada and the United States, females have a head and body length of 203 to 228 mm, a tail length of 76 to 127 mm, and a weight of 85 to 198 grams; and males have a head and body length of 228 to 260 mm, a tail length of 102 to 152 mm, and a weight of 198 to 340 grams (Burt and Grossenheider 1976). In Mexico, the head and body length is 250 to 300 mm, tail length is 140 to 205 mm, and weight of one male was 365 grams (Leopold 1959). Except as noted, the information for the remainder of this account was taken from Jackson (1961), Banfield (1974), and Lowery (1974). In Canada and the northern United States, a color change occurs in the course of 25- to 30-day molts from early October to early December and from late February to late April. During the summer in these regions, and throughout

Immature long-tailed weasel (*Mustela frenata*), photo by Ernest P. Walker.

the year farther south, the upper parts are brown, the under-parts are ochraceous or buff, and the tip of the tail is black. During the winter in these regions, the entire coat is pure white, except that the terminal quarter of the tail is black. Females have eight mammae.

The long-tailed weasel occurs in a variety of habitats, but shows preference for open, brushy or grassy areas near water. It dens in a hollow log or stump, among rocks, or in a burrow taken over from a rodent. Its den is lined with the fur of its victims. It is primarily nocturnal, but is frequently active by day. It can climb and swim, but apparently not as well as *M. erminea*. Its long and slender shape allows it to follow a mouse to the end of a burrow or to enter a chicken coop through a knothole. This shape also prevents curling into a spherical resting posture to conserve heat, and thus the metabolic rate of *M. frenata* (and presumably of other weasels) is 50 to 100 percent higher than that of "normally" shaped mammals of the same weight (Brown and Lasiewski 1972). There is thus a voracious appetite, but also speed, agility, and determination. *M. frenata* seizes its prey with its claws and teeth, and usually kills by a bite to the back of the neck. It may kill animals larger than itself, and has even been known to attack humans who get between it and its prey. The diet, however, consists mainly of rodents and other small mammals. Although weasels are sometimes said to suck blood, this behavior has not been scientifically documented.

Population density varies from about one individual per 2.6 ha. to one per 260 ha., and home range from 4 to 120 ha. Home ranges may overlap, but individuals seldom meet, except during the reproductive season. The voice of *M. frenata* has been reported to consist of a "trill, screech, and squeal" (Svendsen 1976).

Females are monestrous. Mating occurs in July and August, but implantation of the fertilized eggs in the uterus is delayed to around the following March. Embryonic development then proceeds for approximately 27 days until birth in April or May. The total period of pregnancy is 205 to 337 days. The mean number of young per litter is six and the range is three to nine. The young weigh about 3.1 grams each at birth, open their eyes after 35 to 37 days, and are weaned at

3½ weeks. The father has been reported to assist in the care of the offspring. Females attain sexual maturity at 3 to 4 months of age, but males do not mate until the year following the one in which they are born.

The long-tailed weasel is more prone to raid hen houses than are other species of *Mustela,* but is generally beneficial in the vicinity of poultry farms, because of its destruction of the rats that prey on young chickens. The white winter pelage, known collectively with that of *M. erminea* as "ermine," has varied in value, with prime pelts sometimes bringing up to $3.50 each. The reported number of pelts taken in the United States and Canada during the 1976/77 trapping season was 61,175, and the average price paid was about $1.00 (Deems and Pursley 1978).

Mustela altaica. The information for the account of this species was taken from Stroganov (1969). In males, head and body length is 224 to 287 mm, tail length is 108 to 145 mm, and weight is 217 to 350 grams. In females, head and body length is 217 to 249 mm, tail length is 90 to 117 mm, and weight is 122 to 220 grams. *M. altaica* resembles *M. sibirica,* but is smaller and has shorter fur and a less luxuriant tail. There are spring and fall molts. The winter pelage is yellowish brown above and pale yellow below. In summer the coat is gray to grayish brown.

This weasel occurs in highland steppes and forests up to elevations of 3,500 meters. It nests in a rock crevice, among tree roots, or in an expropriated rodent burrow. It is quick, agile, and chiefly nocturnal or crepuscular. It feeds mainly on rodents, pikas, and small birds. In Kazakh, mating occurs in February and March. The gestation period is 40 days, litters contain two to eight young, and lactation lasts 2 months. Following independence, litter mates remain together until autumn. *M. altaica* is of little importance in the fur trade, and is considered to be beneficial to agricultural interests.

Mustela kathiah. The information for the account of this species was taken from Mitchell (1977). Head and body length is 250 to 270 mm and tail length is 125 to 150 mm. The tail is more than half and sometimes nearly two-thirds as long as the head and body. The upper parts are dark brownish and the underparts are deep yellow. This weasel inhabits pine forests and also occurs above the timber line. The elevational range is 1,800 to 4,000 meters. The diet includes birds, rodents, and other small mammals.

Mustela sibirica. Except as noted, the information for the account of this species was taken from Stroganov (1969). In males, head and body length is 280 to 390 mm, tail length is 155 to 210 mm, and weight is 650 to 820 grams. In females, head and body length is 250 to 305 mm, tail length is 133 to 164 mm, and weight is 360 to 430 grams. In winter the upper parts are bright ochre to straw yellow, and the flanks and underparts are somewhat paler. The summer pelage is darker, shorter, coarser, and sparser. Females have four pairs of mammae.

The Siberian weasel dwells mainly in forests, especially along streams, but sometimes enters towns and cities. It dens in tree hollows, under roots or logs, between stones, in a modified rodent burrow, or in a building. It lines its nest with fur, feathers, and dried vegetation. It is swift and agile, has good senses of smell and hearing, and can climb and swim well. It is mainly nocturnal and crepuscular, and has been observed to cover a distance of 8 km in one night. In the fall it may move from upland areas to valleys. There are reports of mass migrations associated with food shortages. The diet consists mainly of small rodents, but also includes pikas, birds, eggs, frogs, and fish. Food may be stored for winter use.

Several males may pursue and fight over a single female. Mating occurs in late winter and early spring, and births take place from April to June. The gestation period is 28–30 days, litter size is 2–12 young, the offspring open their eyes after 1 month of life, and lactation lasts 2 months. The young leave their mother by the end of August, but litter mates may travel together through the autumn. A captive lived for 8 years and 10 months (Marvin L. Jones, Zoological Society of San Diego, pers. comm.).

The Siberian weasel is important in the fur trade. It occasionally attacks domestic fowl, but is generally considered beneficial through its destruction of noxious rodents. It has been introduced on Sakhalin and Iriomote islands (Corbet 1978).

Mustela strigidorsa. The information for the account of this species was taken from Lekagul and McNeely (1977) and Mitchell (1977). Head and body length is 250 to 325 mm and tail length is 130 to 205 mm. The general coloration is dark brown, but the upper lips, cheeks, chin, and throat are pale yellow. There is a narrow whitish stripe down the middle of the back and another along the venter. As in *M. nudipes,* the area around the foot pads is entirely naked. Females have two pairs of mammae.

The back-striped weasel inhabits evergreen forest at elevations of 1,200 to 2,200 meters. Only eight specimens are known. One was taken in a tree hole, 3 or 4 meters above the ground. An individual was observed to attack a bandicoot rat three times its own size.

Mustela nudipes. The information for the account of this species was taken from Lekagul and McNeely (1977). Head and body length is 300 to 360 mm and tail length is 240 to 260 mm. The body coloration varies from pale grayish white to reddish brown, and the head is much paler than the body. The soles of the feet are naked around the pads. Females have two pairs of mammae. This species is not well known, but its habits are thought to be like those of other weasels. A litter of four young has been recorded.

Mustela lutreola. Except as noted, the information for the account of this species was taken from Novikov (1962) and Stroganov (1969). In males, head and body length is 280 to 430 mm, tail length is 124 to 190 mm, and weight is up to 739 grams. In females, head and body length is 320 to 400 mm, tail length is 120 to 180 mm, and weight is up to 440 grams. The general coloration is reddish brown to dark cinnamon. The underparts are somewhat paler than the back, and there may be some white on the chin, chest, and throat. The pelage is dense, but is short, even in winter.

The European mink inhabits the densely vegetated banks of creeks, rivers, and lakes. It is rarely found more than 100 meters from fresh water. It may excavate its own burrow, take one from a water vole (*Arvicola*), or den in a crevice, among tree roots, or in some other sheltered spot. It swims and dives well. Activity is mainly nocturnal and crepuscular. The summer is generally spent in an area of 15 to 20 ha., but there may be extensive fall and winter movements to locate swift, nonfrozen streams. The chief prey is the water vole. The diet also includes other small rodents, amphibians, mollusks, crabs, and insects. Food is often stored.

Mating occurs from February to March, and births in April and May. Pregnancy lasts 35 to 72 days, the variation probably resulting from delayed implantation in some females. The number of young per litter is two to seven, usually four or five. The young open their eyes after 4 weeks, are weaned at 10 weeks of age, disperse in the autumn, and attain sexual maturity the following year. Longevity is 7 to 10 years.

Although its fur is not as valuable as that of *M. vison,* the European mink has been widely trapped for commercial purposes. It has also been killed as a predator, has lost much habitat through hydroelectric developments and water pollution, and has suffered from competition with the introduced *M. vison.* To the west of the Soviet Union, it now survives only in Finland, eastern Poland, parts of the Balkans, and western France (Smit and Van Wijngaarden 1976).

Mustela vison. Except as noted, the information for the account of this species was taken from Banfield (1974), Lowery (1974), Jackson (1961), and Burt and Grossenheider (1976). In males, head and body length is 330 to 430 mm, tail length is 158 to 230 mm, and weight is 681 to 2,310 grams. In females, head and body length is 300 to 400 mm, tail length is 128 to 200 mm, and weight is 790 to 1,089 grams. The pelage is soft and luxurious. Its general color varies from rich brown to almost black, but the ventral surface is paler and may have some white spotting. Captive breeding has produced a number of color variants. Anal scent glands emit a strong, musky odor, considered by some persons to be more obnoxious than that of skunks. Females have three pairs of mammae.

The mink is found along streams and lakes, and in swamps and marshes. It prefers densely vegetated areas. It dens under stones or the roots of trees, in an expropriated beaver or muskrat house, or in a self-excavated burrow. Such a burrow may be about 3 meters long and 1 meter beneath the surface, and have one or more entrances just above water level. The mink is an excellent swimmer, can dive to depths of 5 or 6 meters, and can swim underwater for about 30 meters. It is primarily nocturnal and crepuscular, but is sometimes active by day. It is normally not wide ranging, but may travel up to about 25 km in a night during times of food shortage. The most important dietary components are small mammals, fish, frogs, and crayfish. Other foods include insects, worms, and birds.

Population densities of about one to eight individuals per sq km have been recorded. Females have home ranges of about 8 to 20 ha. The ranges of males are larger, sometimes up to 800 ha. Individuals are generally solitary and hostile to one another, except when opposite sexes come together for

American mink (*Mustela vison*), photo by Ernest P. Walker.

breeding. Females are polyestrous, but have only one litter per year. Mating occurs from February to April, and births in late April and early May. Because of a varying period of delay in implantation of the fertilized eggs, pregnancy may last from 39 to 78 days. Actual embryonic development takes 30 to 32 days. The number of young per litter averages 5 and ranges from 2 to 10. The young are born in a nest lined with fur, feathers, and dry vegetation. They are blind and naked at birth. They open their eyes after 5 weeks, are weaned when 5 to 6 weeks old, leave the nest and begin to hunt at 7 to 8 weeks, and separate from the mother in the autumn. Females reach adult weight at 4 months of age and sexual maturity at 12 months. Males reach adult weight at 9–11 months and sexual maturity at 18 months. Potential longevity is 10 years.

Most of the mink fur used in commerce is produced on farms. The preferred breeding stock is a result of crossing the large Alaskan and dark Labrador forms. Selective breeding has led to development of strains that regularly yield such colors as black, white, platinum, and blue (Grzimek 1975). In Canada, during the 1971/72 season, 72,674 wild and 1,155,020 farm-raised mink pelts were sold (Banfield 1974). During the 1976/77 season, the reported number of wild mink taken and the average price per pelt were 116,537 and $19.67 in Canada, and 320,823 and $14.00 in the United States (Deems and Pursley 1978). *M. vison* has been deliberately introduced in many parts of the Soviet Union, and escaped animals have established populations in Ireland, Britain, Scandinavia, and Germany (Corbet 1978).

Mustela macrodon. The information for the account of this extinct species was taken from Manville (1966) and Allen (1942). The sea mink was said to resemble *M. vison,* but to be much larger, to have a coarser and more reddish fur, and to have an entirely different odor. Head and body length has been estimated at 660 mm and tail length at 254 mm. No complete specimen is known to exist, and descriptions are based only on recorded observations and numerous bone fragments and teeth found at Indian middens along the New England coast.

The sea mink reportedly made its home among the rocks along the ocean. Its den had two entrances. The diet consisted mainly of fish and probably also included mollusks. An adult and four young, estimated to be three or four weeks old, were seen along a beach in August. Because of the large size of *M. macrodon,* its pelt brought a higher price than that of *M. vison* and was persistently sought. Some persons pursued the species from island to island, using dogs to locate individuals on ledges and in rock crevices, and then digging or smoking them out. The sea mink apparently had been exterminated by about 1880, and its range seems subsequently to have been occupied by *M. vison.*

Mustela putorius. Except as noted, the information for the

account of this species was taken from Grzimek (1975), Novikov (1962), and Stroganov (1969). Males have a head and body length of 350 to 460 mm, a tail length of 115 to 190 mm, and a weight of 500 to 1,500 grams. Females have a head and body length of 290 to 340 mm and a tail length of 85 to 123 mm. The general coloration is dark brown to black. The underfur is pale yellow and is clearly seen through the guard hairs. The area between the eye and the ear is silvery white.

The European polecat is most common in open forests and meadows. It dens in such places as crevices, hollow logs, and burrows made by other animals. It sometimes enters settled areas and buildings occupied by people. It is nocturnal and terrestrial, but is capable of climbing. The diet consists of small mammals, frogs, and fish.

Mating occurs from March to June, heat lasts 3 to 5 days, and the gestation period is about 40 days. The number of young per litter is 2 to 12, usually 3 to 7. The young open their eyes and are weaned after about a month. Sexual maturity may not be attained until the second year of life. Longevity in the wild is usually up to 5 or 6 years.

The domestic ferret, sometimes given the subspecific name *M. putorius furo,* is thought to be a descendant of the European polecat. It was bred in captivity as early as the fourth century B.C. It is usually tame and playful, and is used to control rodents and to drive rabbits from their burrows. It is now found in captivity in much of the world. Unlike the wild polecat, it is generally white or pale yellow in color. According to Asdell (1964), females may have two or three litters annually. Both the domestic ferret and the polecat were apparently introduced in New Zealand, and large feral populations are now established there. These animals are trapped for their fur in New Zealand, as well as in their original range. The pelt is sometimes referred to as ''fitch.''

Mustela eversmanni. The information for the account of this species was taken from Stroganov (1969). Males have a head and body length of 370 to 562 mm, a tail length of 80 to 183 mm, and a weight of up to 2,050 grams. Females have a head and body length of 290 to 520 mm, a tail length of 70 to 180 mm, and a weight of up to 1,350 grams. There is much variation in color pattern, but generally the body is straw yellow or pale brown, being somewhat darker above than below. There is a dark mask across the face. The chest, limbs, groin area, and terminal third of the tail are dark brown to black. Some individuals have a striking resemblance to the North American black-footed ferret (*M. nigripes*).

The steppe polecat is found in open grassland and semidesert. It usually expropriates the burrow of a ground squirrel or some other animal, and modifies the home for its own use. Some burrow systems, especially those of females, may be occupied for several years and become rather complex. *M.*

European polecat (*Mustela putorius*), photo from Zoological Society of London.

eversmanni is quick and agile, and has keen senses, especially of smell and hearing. It moves by leaps of up to a meter, and constantly changes direction during hunts. It is nocturnal and has been known to cover up to 18 km in the course of a winter night. Local migrations may occur in response to extreme snow depth or food shortage. The diet consists mainly of pikas, voles, marmots, hamsters, and other rodents. Food is sometimes stored for later use.

Mating usually occurs from February to March, and births from April to May. If a litter is lost, however, the female may produce a second later in the year. The gestation period is 38–41 days. The average number of young per litter is about 8–10, and the range is 4–18. The young weigh 4–6 grams each at birth, open their eyes after 1 month, are weaned and start hunting with the mother at 1½ months of age, disperse at 3 months, and attain sexual maturity at 9 months.

The steppe polecat is considered to be beneficial to agriculture because of its destruction of rodents. Its fur has commercial importance, but is not as valuable as that of *M. putorius*.

Mustela nigripes.   Head and body length is 380 to 500 mm and tail length is 114 to 150 mm. The linear measurements of males are about 10 percent greater than those of females. Two males weighed 964 and 1,078 grams, and two females weighed 764 and 854 grams. The body is generally colored yellow buff, and becomes palest on the underparts. The forehead, muzzle, and throat are nearly white. The top of the head and middle of the back are brown. The face mask, feet, and terminal fourth of the tail are black. Females have three pairs of mammae (Henderson, Springer, and Adrian 1969; Hillman and Clark 1980; Burt and Grossenheider 1976).

The black-footed ferret is found mainly on short and midgrass prairies. It is closely associated with prairie dogs (*Cynomys*), and utilizes their burrows for shelter and travel. It is primarily nocturnal, and is thought to have keen senses of hearing, smell, and sight. It depends largely on prairie dogs for food, but captives have readily accepted other small mammals. *M. nigripes* is solitary, except during the breeding season, and males apparently do not assist in the rearing of young (Hillman and Clark 1980; Henderson, Springer, and Adrian 1969).

Studies in South Dakota (Hillman, Linder, and Dahlgren 1979) indicate that *M. nigripes* kills only enough to eat. A prairie dog town of 14 ha. was occupied by a single ferret for six months, but the prey population was not severely reduced. Females raising litters require relatively large prairie dog towns. Whereas overall average town size was found to be 8 ha., the average town size occupied by females with young was 36 (10–120) ha. The mean distance between a town and the nearest neighboring town was 2.4 km. The mean distance between two towns occupied by ferrets was 5.4 km.

Captives have mated in March and early April. The gestation periods of one female in two seasons were 42 and 45

Steppe polecat (*Mustela eversmanni*), photo by Ernest P. Walker.

Black-footed ferret (*Mustela nigripes*), photo by Luther C. Goldman.

days. The number of young per litter in the wild averages 3.5 and ranges from 1 to 5. The young emerge from the burrow in early July, and separate from the mother in September or early October (Hillman and Clark 1980). Captives are estimated to have lived up to about 12 years (Hillman and Carpenter 1980).

The black-footed ferret may once have been common on the Great Plains, but its subterranean and nocturnal habits made it difficult to locate and observe. According to Clark (1976), there have been about 1,000 reports of the species since 1851, and there are about 100 specimens in museums. During the 20th century, *M. nigripes* apparently declined in association with the extermination of prairie dogs by human agency. In Kansas, for example, the area occupied by prairie dog towns has been reduced by 98.6 percent since 1905. The ferret became so rare that some persons considered it extinct. Reports continued in several states, however, and in 1964 a population was discovered in South Dakota. This group was regularly observed and studied until 1974, when confirmed records ceased. Several captives were taken, but eventually all died without leaving surviving offspring. There again was fear that the species was extinct, though Clark (1978) gathered a number of reliable reports in Wyoming, and Boggess, Henderson, and Choate (1980) found the skull of a ferret, which may have died as recently as 1977, in Kansas. Then, on 25 September 1981, a ferret was killed by dogs on a ranch in northwestern Wyoming. Agents of the U.S. Fish and Wildlife Service quickly live-captured, radio-collared, and released another individual in the same area, and found evidence of several additional animals (*Endangered Species Tech. Bull.* 6, no. 12 [1981]:6). *M. nigripes* is classified as endangered by the IUCN (1978) and the USDI (1980), and is on appendix 1 of the CITES.

CARNIVORA; MUSTELIDAE; *Genus VORMELA Blasius, 1884*

Marbled Polecat

The single species, *V. peregusna*, occurs in the steppe and subdesert zones from the Balkans and Palestine to Inner Mongolia and Pakistan (Corbet 1978).

Head and body length is 290 to 380 mm, tail length is 150 to 218 mm, and weight is 370 to 715 grams (Stroganov 1969; Grzimek 1975). *Vormela* resembles *Mustela putorius*, but differs in its broken and mottled color pattern on the upper parts and in its long claws. The mottling on the back is reddish brown, and white or yellowish, and the tail is usually whitish with a dark tip. The underparts are dark brown or blackish, and the facial mask is dark brown. Females have five pairs of mammae (Roberts 1977).

Like most mustelids, *Vormela* possesses anal scent glands, from which a noxious-smelling substance is emitted. When this animal is threatened, it throws the head back, bares the teeth, erects the body hairs, and bristles and curls the tail over the back. This behavior results in the fullest display of the contrasting body colors, and the pattern thus exposed is thought to be a warning associated with the fetid anal gland secretion.

The marbled polecat seems to prefer steppes and foothills. With its strong paws and long claws, it excavates deep, roomy burrows. It may also shelter in the burrows of other animals. It is chiefly nocturnal and crepuscular, but is sometimes active by day. It is a good climber, but feeds mainly on the ground. It preys on rodents, birds, reptiles, and other animals.

Vormela is solitary, except during the breeding season. Births occur from February to March, after a gestation period of about 9 weeks. Litter size is four to eight young (Stroganov 1969). Only the mother cares for the young, which are reared in a grass and leaf nest within a burrow. A captive specimen was still living after 8 years and 11 months (Marvin L. Jones, Zoological Society of San Diego, pers. comm.).

The fur of the marbled polecat has been sought at certain times, but is not of major commercial importance (Grzimek 1975; Stroganov 1969). *Vormela* sometimes preys on poultry and has been eliminated in parts of its range. It is classified as rare by the Union of Soviet Socialist Republics Ministry of Agriculture (1978).

CARNIVORA; MUSTELIDAE; *Genus MARTES Pinel, 1792*

Martens, Fisher, Sable

There are three subgenera and eight species (Anderson 1970; Ellerman and Morrison-Scott 1966; Corbet 1978; Hall

Marbled polecat (*Vormela peregusna*), photo by Bernhard Grzimek.

1981; Lekagul and McNeely 1977; Chasen 1940; Pilgrim 1980):

subgenus *Martes*

M. foina (beech or stone marten), Denmark and Spain to Mongolia and the Himalayas, Crete, Rhodes, Corfu;

M. martes (European pine marten), western Europe to western Siberia and the Caucasus, Ireland, Britain, Corsica, Sardinia, Sicily;

M. zibellina (sable), originally the entire taiga zone from Scandinavia to eastern Siberia and North Korea, Sakhalin, Hokkaido;

M. melampus (Japanese marten), South Korea, Honshu, Kyushu, Shikoku, Tsushima;

M. americana (American pine marten), Alaska to Newfoundland, south in mountainous areas to central California and northern New Mexico, Great Lakes region, New England;

subgenus *Pekania*

M. pennanti (fisher or pekan), southern Yukon to Labrador, south in mountainous areas to central California and Utah, Great Lakes and Appalachian regions, New England;

subgenus *Charronia*

M. flavigula (yellow-throated marten), southeastern Siberia to Malay Peninsula, Himalayan region, Taiwan, Sumatra, Bangka Island, Java, Borneo;

M. gwatkinsi (Nilgiri marten), Nilgiri Hills of extreme southern India.

Anderson (1970) stated that additional study might show that *M. martes, M. zibellina, M. melampus,* and *M. americana* are conspecific. Corbet (1978) suggested that *M. gwatkinsi* is conspecific with *M. flavigula.*

From *Mustela, Martes* is generally distinguished by a larger and heavier body, a longer and more pointed nose, larger ears, longer limbs, and a more bushy tail. Additional information is provided separately for each species.

Martes foina.   Except as noted, the information for the account of this species was taken from Anderson (1970), Stroganov (1969), Grzimek (1975), and Waechter (1975). Head and body length is about 400 to 540 mm, tail length is 220 to 300 mm, and weight is 1.1 to 2.3 kg. The general coloration is pale grayish brown to dark brown. Most specimens have a prominent white or pale yellow neck patch. The fur is coarser than that of *M. martes,* and the tail is relatively longer. The soles are covered with sparse hairs, through which the pads stand out markedly.

The stone marten is less dependent on forests than is *M. martes,* and prefers rocky and open areas. It is found in mountains at elevations of up to 4,000 meters. It often enters towns and may occupy buildings. Natural nest sites include rocky crevices, stone heaps, abandoned burrows of other animals, and hollow trees. *M. foina* is a good climber, but rarely goes high in trees. It is nocturnal and crepuscular. The diet consists of rodents, birds, eggs, and berries. Vegetable matter forms a major part of the summer food in some areas. The average home range in Alsace was found to be 80 ha.

Mating occurs in midsummer, but, because of delayed implantation of the fertilized eggs in the uterus, births do not occur until the following spring. The total period of pregnancy is 230–75 days. Litters usually contain three to four young, but may have up to eight. A captive was still living after 18 years and 1 month (Marvin L. Jones, Zoological Society of San Diego, pers. comm.).

Martes martes.   The information for the account of this species was taken from Stroganov (1969) and Grzimek

(1975). Head and body length is 450 to 580 mm, tail length is 160 to 280 mm, and weight is 800 to 1,800 grams. The general coloration is chestnut to gray brown. There is a light yellow patch on the chest and lower neck. The winter pelage is luxuriant and silky, the summer pelage shorter and coarser. The paws are covered by dense hair. Females have four mammae.

The pine marten dwells in forests, both coniferous and deciduous. It is better adapted than *M. zibellina* for an aboreal life. An individual has several nests, located preferably in hollow trees. Activity is mainly nocturnal, and 20 to 30 km may be covered in a night's hunting. The diet consists of murids, sciurids, other small mammals, honey, fruit, and berries. There are several food storage sites within the home range, which may be 5 km in diameter.

Mating occurs in midsummer, but, because of delayed implantation, births do not take place until March or April. Pregnancy lasts 230 to 275 days. The number of young per litter is two to eight, usually three to five. The young weigh about 30 grams each at birth, open their eyes after 32–38 days, are weaned at 6 or 7 weeks of age, separate from the mother in the autumn, and usually attain sexual maturity when 2 years old. Maximum known longevity is 17 years.

The fur of the pine marten is more valuable than that of *M. foina*. Wild populations were excessively trapped and greatly reduced during the 20th century, but the species is not in immediate danger of extinction. Efforts at captive breeding have had only limited success.

Martes zibellina.   The information for the account of this species was taken from Novikov (1962), Stroganov (1969), and Grzimek (1975). Males have a head and body length of 380 to 560 mm, a tail length of 120 to 190 mm, and a weight of 880 to 1,800 grams. Females have a head and body length of 350 to 510 mm, a tail length of 115 to 172 mm, and a weight of 700 to 1,560 grams. The winter pelage is long, silky, and luxurious. Coloration varies, but is generally pale gray brown to dark black brown. The summer pelage is shorter, coarser, duller, and darker. The soles are covered with extremely dense, stiff hairs. The body is very slender, long, and supple.

The sable dwells in both coniferous and deciduous forests, sometimes high in the mountains, and preferably near streams. It is mainly terrestrial, but can climb. An individual may have several permanent and temporary dens, located in holes among or under rocks, logs, or roots. A burrow several meters long may lead to the enlarged nest chamber, which is lined with dry vegetation and fur. The sable hunts either by day or by night. It tends to remain in one part of its home range for several days and then move on. It sometimes stays in its nest for several days during severe winter weather. There may be migrations to higher country in summer, and also large-scale movements associated with food shortage. The diet consists mostly of rodents, but also includes pikas, birds, fish, honey, nuts, and berries.

Reported population densities vary from 1 sable per 1.5 sq km in some pine forests to 1 per 25 sq km in larch forests. Individual home range is usually several hundred hectares, but may be as great as 3,000 ha. in more desolate parts of Siberia. At least part of the home range is defended against intruders, but a male may sometimes share its territory with a female. Mating occurs from June to August, and births usually in April or May. Actual embryonic development takes perhaps 25–40 days, but, because of delayed implantation, the total period of pregnancy is 250 to 300 days. The number of young per litter ranges from one to five and is usually three or four. The young weigh 30–35 grams each at birth, open their eyes after 30–36 days, emerge from the den at 38 days, are weaned at about 7 weeks, and attain sexual maturity when 15–16 months old. Maximum known longevity is 15 years.

The sable is one of the most valuable of fur bearers. Its pelt has been avidly sought since ancient times. Several hundred thousand skins were traded annually during the late 18th century in the west Siberian city of Irbit. Because of excessive trapping, this figure dropped to 20,000 to 25,000 by 1910–13, and the sable had then disappeared from much of its range. Subsequently, through programs of protection and reintroduction, the species increased in numbers and distribution.

Martes melampus.   According to Anderson (1970), head and body length is 470 to 545 mm and tail length is 170 to 223 mm. The general coloration is yellowish brown to dark brown, and there is a whitish neck patch. Little is known about the habits of the Japanese marten. It is decreasing in numbers through excessive hunting for its fur, and because of the harmful effects of agricultural insecticides.

Martes americana.   Males have a head and body length of 400 to 440 mm, a tail length of 150 to 200 mm, and a weight of 700 to 1,300 grams. Females have a head and body length of 350 to 400 mm, a tail length of 135 to 182 mm, and a weight of 600 to 775 grams. The pelage is long and lustrous. The upper parts vary in color from dark brown to pale buff, the legs and tail are almost black, the head is pale gray, and the underparts are pale brown with irregular cream or orange spots. The body is slender, the ears and eyes are relatively large, and the claws are sharp and recurved. Females have eight mammae (Banfield 1974; Burt and Grossenheider 1976).

The American pine marten is found mainly in coniferous forest. It dens in hollow trees or logs, in rock crevices, or in burrows. The natal nest is lined with dry vegetation. The marten is primarily nocturnal, and is partly arboreal but spends considerable time on the ground. It can swim and dive well. It is active all winter, but may descend to lower elevations if living in a mountainous area. The diet consists mostly of rodents and other small mammals, and also includes birds, insects, fruit, and carrion.

Population density varies from about 0.5 to 1.7 individuals per sq km of good habitat (Banfield 1974). In Maine, Soutiere (1979) found densities of 1.2 resident marten per sq km in undisturbed and partly harvested forests, but only 0.4 per sq km in commercially clear-cut forests. In a radio-tracking study in northeastern Minnesota, Mech and Rogers (1977) determined home ranges to be 10.5, 16.6, and 19.9 sq km for 3 males, and 4.3 sq km for 1 female. There was considerable overlap between the ranges of 2 of the males. The marten is primarily a solitary species, but Herman and Fuller (1974) regularly observed what seemed to be an adult male and female together, sometimes apparently with two of their offspring, even though it was not the breeding season.

Mating occurs in July and August. Implantation of the fertilized eggs in the uterus is delayed until February, and embryonic development then proceeds for about 28 days. The total period of pregnancy is 220–75 days. The average number of young per litter is 2.6 and the range is 1–4. The young weigh 28 grams each at birth, open their eyes after 39 days, are weaned at 6 weeks of age, reach adult size at 3½ months, and attain sexual maturity when 15–24 months old (Banfield 1974). Captives have lived up to 17 years.

The fur of the marten is valuable and is sometimes referred to as "American sable." In the 1940s, marten pelts were worth as much as $100 each. In the mid-19th century, the Hudson's Bay Company traded as many as 180,000 Canadian skins each year (Banfield 1974). By the early 20th century, excessive trapping had severely depleted *M. americana* in Alaska, Canada, and the western conterminous United States. Protective regulations subsequently allowed the species to make a comeback in some areas, but in the eastern

American pine marten (*Martes americana*), photo by Howard E. Uible.

United States the marten survives only in small parts of Minnesota, New York, and Maine (Mech 1961; Mech and Rogers 1977; Blanchard 1974; Yocom 1974). Reintroductions have been attempted in northern Michigan and Wisconsin (Knap 1975). During the 1976/77 trapping season, 27,898 marten skins were taken in the United States, mostly in Alaska, and sold for an average price of $14.00. The respective figures for Canada were 102,632 skins and $19.92 (Deems and Pursley 1978).

Martes pennanti. Head and body length is 490 to 630 mm and tail length is 253 to 425 mm. Weight is 2.6 to 5.5 kg in males and 1.3 to 3.2 kg in females. The head, neck, shoulders, and upper back are dark brown to black. The underparts

are brown, sometimes with small white spots. The thick pelage is coarser than that of *M. americana*. The body is slender, but rather stocky for a weasel. Females have eight mammae.

Except as noted, the information for the remainder of this account was taken from Powell (1981). The fisher inhabits dense forests with an extensive overhead canopy, and avoids open areas. It generally lacks a permanent den, but seeks temporary shelter in hollow trees and logs, brush piles, abandoned beaver lodges, holes in the ground, and snow dens. Activity may take place at any hour. The fisher is adapted for climbing, but is primarily terrestrial. It is capable of traveling long distances; one individual covered 90 km in three days. Usual daily movement in New Hampshire, however, was

Fisher (*Martes pennanti*), photo from San Diego Zoological Garden.

Yellow-throated marten (*Martes flavigula*), photo from Zoological Society of London.

found to be 1.5 to 3.0 km. The fisher generally forages in a zigzag pattern, constantly investigating places where prey might be concealed.

The diet consists mainly of small to medium-sized birds and mammals, and carrion. It has been calculated that a fisher requires either 1 porcupine every 10 to 35 days, 1 snowshoe hare every 2.5 to 8 days, 1 kg of deer carrion every 2.5 to 8 days, 1 to 2 squirrels per day, or 7 to 22 mice per day. Hares are killed with a quick rush and a bite to the back of the neck. Porcupines are taken only on the ground, and are killed by repeatedly circling and biting at the face.

Population density in preferred habitat is 1 fisher per 2.6 to 7.5 sq km, but in other areas may be as low as 1 per 200 sq km. Individual home range size is 15 to 35 sq km, and is generally larger for males. There is little overlap between the ranges of animals of the same sex, but extensive overlap between the ranges of opposite sexes. *M. pennanti* is solitary, except during the breeding season.

Mating occurs from March to May, but implantation of the fertilized eggs in the uterus is delayed until the following January to early April, and births occur from February to May. Postimplantation embryonic development lasts about 30 days, but the total period of pregnancy is 10 or 11 months. Litters contain an average of about 3 young, but there may be as many as 6. Weaning begins at the age of 8 to 10 weeks, and separation from the mother occurs in the fifth month of life. Females reach adult weight after 6 months, and males after 1 year. Sexual maturity comes at the age of 1 to 2 years. Longevities of about 10 years have been recorded for both wild and captive animals.

In the 19th and early 20th centuries, the fisher declined over most of its range, because of excessive fur trapping and habitat destruction through logging. The species was almost totally eliminated in the United States. Widespread closed seasons and other protective regulations were initiated in the 1930s, and reintroductions were made in various areas in the 1950s and 1960s. To the south of Canada, populations of *M. pennanti* are now present in Washington, Oregon, northern California, Idaho, Montana, Minnesota, Wisconsin, Michigan, New York, Vermont, New Hampshire, Maine, Massachusetts, and West Virginia (Cottrell 1978; Coulter 1974; Handley 1980*b*; Mohler 1974; Mussehl and Howell 1971;

Olterman and Verts 1972; Penrod 1976; Petersen, Martin, and Pils 1977; Schempf and White 1977; Yocom and McCollum 1973). In the last six of these states, during the past decade, the fisher has been taken legally for the fur market. In the late 1970s, however, populations again declined, thus forcing reclosing or limiting of seasons in New York, New Hampshire, and Maine. The average price of a fisher pelt was $100 in the 1920s. The price subsequently declined, but then again increased. During the 1976/77 trapping season, 12,557 skins were taken in the United States and Canada, and sold for an average of about $95 each (Deems and Pursley 1978).

Martes flavigula. The information for the account of this species was taken from Stroganov (1969) and Lekagul and McNeely (1977). Head and body length is 450 to 650 mm, tail length is 370 to 450 mm, and weight is 2 to 3 kg. The fur is short, sparse, and coarse. There is much variation in color, but generally the top of the head and neck, the tail, the lower limbs, and parts of the back are dark brown to black. The rest of the body is pale brown, except for a bright yellow patch from the chin to the chest. Females have four mammae.

The yellow-throated marten is generally found in forests. It climbs and maneuvers in trees with great agility, but often comes to the ground to hunt. Activity is primarily diurnal. The diet includes rodents, pikas, eggs, frogs, insects, honey, and fruit. In the northern parts of its range, *M. flavigula* evidently preys heavily on the musk deer (*Moschus*) and the young of other ungulates. Individuals often hunt in pairs or family groups, and lifelong pair bonding has been suggested. The number of young per litter is usually two or three, and may be up to five. Maximum known longevity is about 14 years. The pelt of this marten has little commercial value. The subspecies on Taiwan, *M. f. chrysospila*, has become very rare and is classified as endangered by the USDI (1980).

Martes gwatkinsi. According to Anderson (1970), no external measurements of this species are available. As in *M. flavigula*, the general coloration is dark brown and there is a yellowish patch from the chin to the chest. The dorsal profile of the skull of *M. gwatkinsi* is flat, not convex as in *M. flavigula*. The Nilgiri marten is found only in hill forests, seldom occurring below 900 meters.

Tayra (*Eira barbara*), photo by Ernest P. Walker.

CARNIVORA; MUSTELIDAE; *Genus EIRA* Hamilton Smith, 1842

Tayra

The single species, *E. barbara,* is found from southern Sinaloa (west-central Mexico) and southern Tamaulipas (east-central Mexico) to northern Argentina, and on the island of Trinidad (Hall 1981; Cabrera 1957; Goodwin and Greenhall 1961).

Head and body length is 560 to 680 mm, tail length is 375 to 470 mm, and weight is 4 to 5 kg (Grzimek 1975; Leopold 1959). The short, coarse pelage is gray, brown, or black on the head and neck, has a yellow or white spot on the chest, and is black or dark brown on the body. There also is a rare, light-colored form, which is pale buffy and has a darker head. The tayra has a long and slender body, short limbs, and a long tail. The head is broad, the ears are short and rounded, and the neck is long. The soles are naked and the strong claws are nonretractile.

The tayra is a forest dweller. It nests in a hollow tree or log, a burrow made by another animal, or tall grass. It can climb, run, and swim well. When pursued by dogs, it may run on the ground for some distance, then climb a tree, and then leap through the trees for about 100 meters before descending to the ground again. It thus gains time, while the dogs are trying to pick up the trail. Leopold (1959) observed a group of four animals springing swiftly through the trees with incredible agility. *Eira* is active both at night and, especially when there is cloud cover, in the morning. The diet seems to consist mostly of rodents, but also includes rabbits, birds, small deer (*Mazama*), honey, and fruit.

Eira is often seen alone, in pairs, or in small family groups. Hall and Dalquest (1963) wrote that the genus is not social to any extent, but Leopold (1959) referred to an old record of hunting troops with 15 to 20 individuals. He also cited a report that 2 young are born in February, that their eyes open after 2 weeks, and that they forage with their mother by the time they are 2 months old. Other reports suggest that the young may be born in any season. Poglayen-Neuwall (1975) determined that six gestation periods lasted 63–65 days. According to Vaughn (1974), litters of 2 young

each were produced in captivity on 4 March and 22 July, following respective gestation periods of about 67 to 70 days.

The tayra can live 18 years in captivity, loves to play, and can be tamed. It reportedly was used long ago by the Indians to control rodents (Grzimek 1975). It is of no particular importance as a fur bearer or predator of game. It may occasionally eat poultry, but does no substantial damage (Leopold 1959). It also has been accused of raiding corn and sugarcane fields (Hall and Dalquest 1963).

CARNIVORA; MUSTELIDAE; *Genus GALICTIS* Bell, 1826

Grisóns

There are two species (Hall 1981; Cabrera 1957):

G. vittata (greater grisón), southern Mexico to central Peru and southeastern Brazil;

G. cuja (little grisón), central and southern South America.

Galictis has often been referred to as *Grison* Oken, 1816. Cabrera placed *G. cuja* in the subgenus *Grisonella* Thomas, 1912.

In *G. vittata,* the head and body length is 475 to 550 mm, tail length is about 160 mm, and weight is 1.4 to 3.3 kg. In *G. cuja,* the head and body length is 400 to 450 mm, tail length is 150 to 190 mm, and weight is about 1 kg. The color pattern is striking. In both species the black face, sides, and underparts, including the legs and feet, are sharply set off from the back. The back is smoky gray in *G. vittata* and yellow gray or brownish in *G. cuja.* A white stripe extends across the forehead and down the sides of the neck, separating the black of the face from the gray or brown of the back. Short legs and slender bodies give the animals of this genus somewhat the appearance of *Mustela,* but the color pattern immediately distinguishes them.

Grisóns are found in forests and open country, from sea level to 1,200 meters. They live under tree roots or rocks, in hollow logs, or in burrows made by other animals, such as the viscachas of South America. They are probably capable burrowers themselves. They are quick and agile, are good

Greater grisón (*Galictis vittata*), photo by Ernest P. Walker.

climbers and swimmers, and are active both by day and by night. The diet includes small mammals, birds and their eggs, cold-blooded vertebrates, invertebrates, and fruit.

Grisóns are sometimes seen in pairs or groups, and playing together. They have a number of vocalizations, including sharp, growling barks when threatened. Various reports indicate that offspring have been produced in March, August, September, and October (Leopold 1959; Grzimek 1975). Litter size is two to four young. A captive *G. vittata* was still living after 10 years and 6 months (Marvin L. Jones, Zoological Society of San Diego, pers. comm.).

Young grisóns tame readily and make affectionate pets. In early-19th-century Chile, grisóns reportedly were domesticated by natives, and used in the same manner as ferrets to enter the crevices and holes of chinchillas to drive the latter out (Osgood 1943).

CARNIVORA; MUSTELIDAE; **Genus LYNCODON**
Gervais, 1844

Patagonian Weasel

The single species, *L. patagonicus*, is found in Argentina and southern Chile (Cabrera 1957).

Head and body length is usually 300 to 350 mm and tail length is 60 to 90 mm. The coloration on the back is grayish brown with a whitish tinge. The top of the head is creamy or white, and this color extends as a broad stripe on either side to each shoulder. The nape, throat, chest, and limbs are dark brown, and the rest of the lower surface is lighter brown varied with gray. The color pattern is characteristic and quite attractive. *Lyncodon* is somewhat similar externally to *Galictis*, but differs in such features as coloration and the shorter

Patagonian weasel (*Lyncodon patagonicus*), photo by Tom Scott of mounted specimen in Royal Scottish Museum.

tail. Internally, *Lyncodon* has fewer teeth than *Galictis*, there being only two upper and two lower premolars on each side, and one upper and one lower molar. Like most mustelids, *Lyncodon* has a slender body and short legs.

The Patagonian weasel inhabits the pampas. Its habits are little known, but Ewer (1973:177) wrote: "The reduced molars and cutting carnassials of *Lyncodon* strongly suggest that it is a highly carnivorous species." This weasel reportedly was sometimes kept in the houses of ranchers for the purpose of destroying rats.

CARNIVORA; MUSTELIDAE; *Genus ICTONYX Kaup, 1835*

Zorilla, Striped Polecat

The single species, *I. striatus*, occurs from Mauritania to Sudan, and south to South Africa (Meester and Setzer 1971).

Head and body length is 280 to 385 mm, tail length is 200 to 305 mm, and weight is 420 to 1,400 grams. Males are generally larger than females. The body is black with white dorsal stripes, the tail is more or less white, and the face has white markings. The appearance is somewhat like that of the spotted skunks (*Spilogale*) of North America, and early writ-

ers sometimes confused the two genera. *Ictonyx* also has some resemblance to the other African genera *Poecilictis* and *Poecilogale*, but may be recognized by its color pattern, long hair, and bushy tail. Females have two pairs of mammae (Rowe-Rowe 1978a).

The zorilla is found in a variety of habitats, but avoids dense forest. It is mainly nocturnal, resting during the day in rock crevices, or in burrows excavated by itself or some other animal. It occasionally shelters under buildings and in outhouses in farming areas. It is terrestrial, but can climb and swim well. The usual pace is an easy trot, slower than that of a mongoose, and with the back slightly hunched. The diet consists mainly of small rodents and large insects, but also includes eggs, snakes, and other kinds of animals. The zorilla may take a chicken on occasion, but is often useful in eliminating rodents from houses and stables.

At the sight of an enemy, such as a dog, a zorilla may erect its hair and tail, and perhaps emit its anal gland secretion. Such behavior would seem to make the zorilla more formidable than it really is. When actually attacked, it usually emits fluid into the face of the enemy and then feigns death. The ejected fluid may vary in potency with the individual animal, and perhaps with age and the time of year. Some writers have remarked that the fluid is much less pungent than that of the American skunks, while others have stated that it is most repulsive, acrid, and persistent.

Zorilla (*Ictonyx striatus*), photo from Zoological Society of London.

Ictonyx is solitary. Captive males are totally intolerant of one another, and even adults of opposite sexes are amicable only at the time of mating (Rowe-Rowe 1978b). There are several adult vocalizations associated with threat, defense, and greeting (Channing and Rowe-Rowe 1977). In a study of captives, Rowe-Rowe (1978a) determined the reproductive season to extend from early spring to late summer. All births occurred from September to December. There was usually a single annual litter, but if all young died at an early stage, the female sometimes mated and gave birth a second time. Gestation periods of 36 days were recorded. Litter size was one to three young. They weighed about 15 grams each at birth, began to take solid food after about 32 days, opened their eyes at 40 days, were completely weaned at 18 weeks, and were almost full grown at 20 weeks. A male first mated at 22 months of age, and a female bore its first litter when 10 months old. According to Marvin L. Jones (Zoological Society of San Diego, pers. comm.), a captive zorilla lived for 13 years and 4 months.

CARNIVORA; MUSTELIDAE; *Genus POECILICTIS*
Thomas and Hinton, 1920

North African Striped Weasel

The single species, *P. libyca,* occurs from Morocco and Senegal to the Red Sea (Rosevear 1974). There is some question as to whether *Poecilictis* warrants generic distinction from *Ictonyx*. The two taxa are sometimes confused in northern Nigeria and Sudan, where their ranges overlap.

Head and body length is 200 to 285 mm and tail length is 100 to 180 mm. Three males weighed 200 to 250 grams each. The snout is black, the forehead is white, and the top of the head is black. The back is white with a variable pattern of black bands. The tail is white, but becomes darker toward the tip. The underparts and limbs are black. From *Ictonyx, Poecilictis* differs in color pattern, in smaller size, in larger bullae, and in having hairy soles except for the pads. There are well-developed anal glands, capable of ejecting a malodorous fluid (Rosevear 1974).

According to Rosevear (1974), this weasel is restricted to the edges of the Sahara and the contiguous arid zones. It is nocturnal, sheltering throughout the daylight hours in single subterranean burrows. These it digs for itself, either down into level surfaces or in the sides of dunes. Maternal dens consist of a single gallery ending in an unlined chamber. The diet apparently consists of rodents, young ground birds, eggs, lizards, and insects.

Births reportedly occur from January to March. Rosevear (1974) wrote that gestation may be as short as 37 or as long as 77 days. Litters usually contain two or three young. At birth they are blind and covered with short hair. Some captives have lived for 5 years or more. *Poecilictis* probably does not make a good pet, as it constantly has a disagreeable smell and is aggressive toward people.

CARNIVORA; MUSTELIDAE; *Genus POECILOGALE*
Thomas, 1883

African Striped Weasel

The single species, *P. albinucha,* is found from Zaire and Uganda to South Africa (Meester and Setzer 1971).

Head and body length is 250 to 360 mm and tail length is 130 to 230 mm. Rowe-Rowe (1978b) listed weights of 283 to 380 grams for males and 230 to 290 grams for females. From the white on the head and nape, four whitish to orange yellow stripes and three black stripes extend on the black back toward the tail, which is white. The legs and underparts are black. There is little variation in pattern.

Poecilogale is smaller and more slender than *Ictonyx,* and has narrower back stripes. It resembles *Mustela* in its remarkably slender and elongate body and short legs. Like other weasels, it is able to enter any burrow that it can get its head into. Females have two pairs of mammae (Rowe-Rowe 1978a).

This weasel is found in a variety of habitats, including forest edge, grassland, and marsh. It is almost entirely nocturnal, usually spending the day in a burrow that is either self-excavated or taken over from another animal (Kingdon

North African striped weasel (*Poecilictis libyca*), photo by Robert E. Kuntz.

African striped weasel (*Poecilogale albinucha*): Top, photo of mounted specimen from Field Museum of Natural History; Bottom, photo by D. T. Rowe-Rowe.

1977). It can climb well, but spends most of its time on the ground. It cannot run rapidly, and patiently trails prey by scent. The diet consists mainly of small mammals and birds, but also includes snakes and insects.

Like most weasels, *Poecilogale* attacks by grabbing the throat or neck and hanging on and chewing until the victim is dead. Fair-sized prey, such as springhare, may run some distance before dropping from exhaustion, with the weasel retaining its hold. *Poecilogale* kills venomous snakes in much the same manner as the mongoose *Herpestes*. It repeatedly provokes the snake to strike, until the reptile is tired and slower in recovery. Then, after awaiting an opportunity, it seizes the snake in back of the head.

Poecilogale is generally solitary, but groups of two to four individuals—apparently family parties—have been observed on occasion. Adults of opposite sexes lived together quite amicably in captivity, but adult males fought each other at every encounter (Rowe-Rowe 1978*b*). The area around the den is marked by defecation (Kingdon 1977).

When attacked or under stress, *Poecilogale* emits a noxious odor from its anal glands. Although nauseating, this odor is not as strong or persistent as that of the American skunks or the African *Ictonyx*. Normally, *Poecilogale* is silent, but when alarmed it utters a loud sound, described as being between a growl and a shriek. Several adult vocalizations, associated with threat, defense, and greeting, were recorded by Channing and Rowe-Rowe (1977).

In studies of captives and wild-caught animals in South Africa, Rowe-Rowe (1978*a*) determined that births occurred from September to April. Females were polyestrous and would mate a second time in the season if their first litter was lost. Gestation periods of 31 to 33 days were recorded. Litters contained one to three young. They weighed four grams each at birth, started taking solid food after 35 days, opened

Wolverine (*Gulo gulo*), photo by James K. Drake.

their eyes at 51–54 days of age, were completely weaned at about 11 weeks, and were nearly full grown at 20 weeks. A male first mated at 33 months of age, and a female had her first litter when 19 months old. According to Smithers (1971), one individual lived for 5 years and 2 months after capture.

This weasel is considered rare, but not endangered, in South Africa (Meester 1976). Although accused of killing poultry on occasion, *Poecilogale* is generally beneficial to human interests, because of its destruction of rats, mice, and springhares. It also eats quantities of locusts, when available, and digs their larvae out of the ground. Some African tribes use the skins of *Poecilogale* in ceremonial costumes or as ornaments, and it is said that some medicine men use parts of this animal in their rituals.

CARNIVORA; MUSTELIDAE; Genus **GULO** Pallas, 1780

Wolverine

The single species, *G. gulo*, originally occurred from Scandinavia and Germany to northeastern Siberia, throughout Alaska and Canada, and as far south as central California, southern Colorado, Indiana, and Pennsylvania (Corbet 1978; Hall 1981). Although Hall treated the North American populations as a separate species, *G. luscus*, he indicated that there was evidence supporting their recognition as a subspecies of *G. gulo*.

Head and body length is 650 to 1,050 mm, tail length is 170 to 260 mm, and weight is 7 to 32 kg. Females average 10 percent less than males in linear measurements and 30 percent less in weight (Hall 1981). The fur is long and dense.

The general coloration is blackish brown. A light brown band extends along each side of the body, from shoulder to rump, and joins its opposite over and across the base of the tail. *Gulo* gives the impression of a giant marten, with a heavy build, a large head, relatively small and rounded ears, a short tail, and massive limbs (Stroganov 1969). Females have two pairs of mammae (Grzimek 1975).

The wolverine occurs in the tundra and taiga zones. It may be found in forests, mountains, or open plains. For shelter, it may construct a rough bed of grass or leaves in a cave or rock crevice, in a burrow made by another animal, or under a fallen tree (Banfield 1974; Stroganov 1969). Most maternal dens in Finland were found in holes under the snow (Pulliainen 1968). *Gulo* is mainly terrestrial, the usual gait being a sort of loping gallop, but it can climb trees with considerable speed and is an excellent swimmer. It has a keen sense of smell, but apparently poor eyesight and indifferent hearing. It seems to be unexcelled in strength among mammals of its size, and has been reported to drive bears and cougars from their kills. It is largely nocturnal, but is occasionally active in daylight. In the far north, where there are extended times of light or darkness, the wolverine reportedly alternates three- to four-hour periods of activity and sleep. In winter it has been known to cover up to 45 km in a day. It can maintain a gallop for a lengthy period, sometimes moving 10–15 km without rest (Stroganov 1969).

The diet includes carrion, the eggs of ground-nesting birds, lemmings, and berries. Large mammals, such as reindeer, roe deer, and wild sheep, are taken mainly in winter, when the snow cover allows *Gulo* to travel faster than its prey. Most large mammals, however, are obtained in the form of carrion. Caches of prey or carrion are covered with earth or snow, or sometimes wedged in the forks of trees.

Gulo occurs at relatively low population densities (Van

Zyll de Jong 1975). In Scandinavia the estimates vary from one individual per 200 sq km to one per 500 sq km. Each animal has a home range that may cover as much as 2,000 sq km in winter. At least part of the range is a territory, within which individuals of the same sex are not tolerated. Territories are regularly marked, mainly by secretions from anal scent glands but also with urine (Pulliainen and Ovaskainen 1975; Grzimek 1975).

The wolverine is solitary, except during the breeding season. Females are monestrous. Mating usually occurs from late April to July, but implantation of the fertilized eggs in the uterus is delayed until the following December to March, and births occur from January to April. The usual number of young per litter is two to four, and the range is one to five. The young weigh 90 to 100 grams at birth, nurse for 8 to 10 weeks, separate from the mother in the autumn, and attain adult size after 1 year. Sexual maturity comes in the second or third year of life (Rausch and Pearson 1972; Stroganov 1969; Banfield 1974; Pulliainen 1968; Grzimek 1975). A captive female, approximately 10 years of age, gave birth after a pregnancy of 272 days (Mehrer 1976). Another captive wolverine lived for 17 years and 4 months (Marvin L. Jones, Zoological Society of San Diego, pers. comm.).

The pelt of the wolverine is not used widely in commerce, but is valued for parkas by persons living in the Arctic because it accumulates less frost than other kinds of fur. During the 1976/77 trapping season, 1,922 skins were reported taken in Canada, Alaska, and Montana. The average selling price was about $182 (Deems and Pursley 1978). Fur trapping has contributed to a decline in the numbers and distribution of the wolverine, but a more important factor may be human consideration of the genus as a nuisance. In Scandinavia it was intensively hunted, often for bounty, because of alleged predation on domestic reindeer. Throughout its range the wolverine came into conflict with people by following traplines and devouring the fur bearers it found, and by breaking into cabins and food caches and spraying the contents with its strong scent.

In Europe, *Gulo* is now found only in parts of Scandinavia and the northern Soviet Union (Smit and Van Wijngaarden 1976). The genus has disappeared over most of southeastern and south-central Canada (Van Zyll de Jong 1975). By the early 20th century, the wolverine also had been nearly eliminated in the conterminous United States. One population, however, held out and subsequently increased in the mountains of northern and eastern California, and another population reestablished itself in western Montana. Since 1960

there have been numerous reliable reports from Washington, Oregon, Idaho, Wyoming, and Colorado, and a few questionable records in Minnesota, Iowa, and South Dakota (Nowak 1973; Birney 1974; Johnson 1977; Schempf and White 1977; Field and Feltner 1974).

CARNIVORA; MUSTELIDAE; *Genus MELLIVORA* Storr, *1780*

Honey Badger, Ratel

The single species, *M. capensis,* originally occurred from Palestine and the Arabian Peninsula to Turkmen and eastern India, and from Morocco and lower Egypt to South Africa (Corbet 1978; Meester and Setzer 1971).

Head and body length is 600 to 770 mm, tail length is usually 200 to 300 mm, and shoulder height is usually 250 to 300 mm. Kingdon (1977) listed weight as 7 to 13 kg. The upper parts, from the top of the head to the base of the tail, vary from gray to pale yellow or whitish, and contrast sharply with the dark brown or black of the underparts. Completely black individuals, however, have been found in Africa, particularly in the Ituri Forest of northern Zaire. The color pattern of the honey badger has been interpreted as being a warning coloration, because it makes the animal easily recognizable. Females have two pairs of mammae.

The body is heavily built, the legs and tail are relatively short, the ears are small, and the muzzle is blunt. The forefeet are large, and armed with very large and strong claws. The hair is coarse and quite scant on the underparts. The skin is exceedingly loose on the body and very tough. The skull is massive and the teeth are robust. There are anal glands that secrete a vile-smelling liquid. This combination of characters provides an effective system of deterrence and defense.

Mellivora is very difficult to kill. The skin is so tough that a dog can make little impression, except on the belly. The ratel can twist about in its skin, so it can even bite an adversary that has seized it by the back of the neck. Porcupine quills and bee stings have little effect, and snake fangs are rarely able to penetrate. *Mellivora* seems devoid of fear, and it is doubtful that any animal of equivalent size can regularly kill it. It may rush out from its burrow and charge an intruder, especially in the breeding season. Horses, antelope, cattle, and even buffalo have been attacked and severely wounded in this manner.

Honey badger or ratel (*Mellivora capensis*), photo by Bernhard Grzimek.

The honey badger occupies a variety of habitats, mainly in dry areas, but also in forests and wet grasslands. It lives among rocks, in hollow logs or trees, and in burrows. Its powerful limbs and large claws make it a capable and rapid digger. It is primarily terrestrial, but can climb, especially when attracted by honey. It travels by a jog-trot, but is tireless and trails its prey until the prey is run to the ground. The diet includes small mammals, the young of large mammals, birds, reptiles, arthropods, carrion, and vegetation. Honey and bees are important foods at certain times of the year.

A remarkable association has developed, at least in tropical Africa, between a bird—the honey guide (*Indicator indicator*)—and *Mellivora*. The association is mutually beneficial in the common exploitation of the nests of wild bees. In the presence of any mammal, even people, the honey guide has the unusual habit of uttering a series of characteristic calls. If a honey badger hears these calls, it follows the bird, which invariably leads it to the vicinity of a beehive. The badger breaks open the nest, and the bird obtains enough of the honey and insects to pay for its work.

Mellivora may travel alone or in pairs. It is generally silent, but may utter a very harsh, grating growl, when annoyed. The sparse data on reproduction were summarized by Kingdon (1977), Smithers (1971), and Rosevear (1974). Mating has been noted in South Africa in February, June, and December. A lactating female was found in Botswana in November, and newborn were recorded in Zambia in December. Seasonal breeding has been reported in Turkmen, with mating occurring in autumn and births in the spring. The gestation period is thought to last about 6 months. The number of young in a litter is commonly two, and the range is one to four. The young are born in a grass-lined chamber, and evidently remain close to the burrow for a long time. *Mellivora* appears to thrive in captivity. One specimen lived for 26 years and 5 months (Marvin L. Jones, Zoological Society of San Diego, pers. comm.).

If captured before half grown, the honey badger can become a satisfactory pet, as it is docile, affectionate, and active. It is, however, incredibly strong and energetic, and can wreck cages and damage property in its explorations. Wild individuals sometimes prey on poultry, tearing through wire netting to effect entry (Smithers 1971). Destruction of commercial beehives in East Africa seems to be a significant problem, and *Mellivora* has thus been intensively poisoned, trapped, and hunted (Kingdon 1977). The overall range of the genus has declined, probably through human persecution. The ratel is considered rare in South Africa (Meester 1976) and Soviet Central Asia (Union of Soviet Socialist Republics Ministry of Agriculture 1978).

CARNIVORA; MUSTELIDAE; **Genus MELES** Brisson, *1762*

Old World Badger

The single species, *Meles meles,* originally occurred throughout Europe, including the British Isles and several Mediterranean islands, and in Asia as far south as Palestine, Iran, Tibet, and southern China (Corbet 1978).

Head and body length is 560 to 900 mm, tail length is 115 to 202 mm, and weight is usually 10 to 16 kg. Novikov (1962), however, wrote that old males attain weights of 30 to 34 kg in the late fall. The upper parts are grayish, and the underparts and limbs are black. On each side of the face is a dark stripe that extends from the tip of the snout to the ear and encloses the eye; white stripes border the dark stripe. Like other badgers, *Meles* has a stocky body, short limbs, and a short tail. From *Arctonyx,* it is distinguished by its black throat and shorter tail that is the same color as the back. Females have three pairs of mammae (Grzimek 1975).

The Old World badger is found mainly in forests and densely vegetated areas. It usually lives in a large communal burrow system that covers about one-quarter hectare. There are numerous entrances, passages, and chambers. Nests may be located 10 meters from an entrance and 2 to 3 meters below the surface of the ground, and have a diameter of 1.5 meters. A burrow system may be used for decades or centuries, by one generation of badgers after another; it continually increases in complexity and may eventually cover several hectares (Novikov 1962; Grzimek 1975). The animals occupying a system may utilize one nest for several

Old World badger (*Meles meles*), photo from Paignton Zoo.

months and then suddenly move to another part of the burrow. The living quarters are kept quite clean. Bedding material, in the form of dry grass, brackens, moss, or leaves, is dragged backward into the den. Occasionally, this bedding is brought up and strewn around the entrance to air for an hour or so in the early morning. Around the burrows are dung pits, sunning grounds, and areas for play (badgers play all sorts of games, including leapfrog). Well-defined foraging paths may extend outward for 2 or 3 km (Novikov 1962). Kruuk (1978a) distinguished two kinds of burrows: "main setts," with an average of 10.5 entrances; and small "outliers," usually with only a single entrance.

Meles usually does not emerge from its burrow until after sundown. During periods of very cold weather and high snow, it may spend days or weeks in the den. Such intervals of winter sleep extend to several months in northern Europe and up to seven months in Siberia. There is no substantial drop in body temperature, and the badger can be easily aroused, but the animal lives off of fat reserves accumulated in the summer and fall. The omnivorous diet includes almost any available food—small mammals, birds, reptiles, frogs, mollusks, insects, larvae of bees and wasps, carrion, nuts, acorns, berries, fruits, seeds, tubers, rhizomes, and mushrooms. Earthworms have been found to be of major dietary importance in some areas (Kruuk 1978b; Skoog 1970).

In a study in England, Kruuk (1978a) found badgers to be organized into clans of up to 12 individuals. The minimum distance between the main burrows of clans was 300 meters. Most clans used ranges of 50 to 150 ha., and there was little overlap. The ranges were marked by defecation and secretions from subcaudal glands. Several fights were observed along territorial boundaries. Most clans had more females than males, but one, which used a range of only 21 ha., consisted solely of males (it was suggested that in other parts of the range of *Meles* the bachelors may be nomadic). Individuals moved around alone within the clan range. Adult males always slept in the main setts, but females sometimes slept in the outliers, especially in the summer.

Mating occurs from late winter to midsummer. Development of the fertilized eggs stops at the blastocyst stage, and implantation in the uterus is delayed for about 10 months. The time of implantation seems to be controlled by conditions of light and temperature. Following implantation, embryonic development proceeds for 6 to 8 weeks, and births occur mainly from February to March. The total period of pregnancy may thus extend for about 1 year. Females may experience a postpartum estrus. The number of young in a litter is two to six, usually three or four. The young weigh 75 grams each at birth, open their eyes after about 1 month, nurse for 2½ months, and usually separate from the female in the autumn. Both males and females apparently attain sexual maturity at the age of 1 year (Novikov 1962; Grzimek 1975; Ahnlund 1980; Canivenc and Bonnin 1979). One specimen lived in captivity for 16 years and 2 months (Marvin L. Jones, Zoological Society of San Diego, pers. comm.).

Meles sometimes damages ripening grapes, corn, and oats. Its hair is used to make various kinds of brushes, and its skin, at least formerly, has been used in northern China to make rugs.

CARNIVORA; MUSTELIDAE; Genus *ARCTONYX*
F. Cuvier, 1825

Hog Badger

The single species, *A. collaris*, occurs from Sikkim and northeastern China to peninsular Thailand, and on the island of Sumatra (Corbet 1978; Lekagul and McNeely 1977).

Head and body length is 550 to 700 mm, tail length is 120 to 170 mm, and weight is usually 7 to 14 kg. The back is yellowish, grayish, or blackish, and there is a pattern of white and black stripes on the head. The dark stripes run through the eyes, and are bordered by white stripes that merge with the nape and with the white of the throat. The ears and tail also are white, and the feet and belly are black. The body form is stocky. This badger is distinguished from *Meles* by its white, rather than black, throat, and by its long and mostly white tail, as distinct from the short tail, colored the same as the back, in *Meles*. Another external difference is that in *Arctonyx* the claws are pale colored, while in *Meles* they are dark. The common name of *Arctonyx* refers to the

Hog badger (*Arctonyx collaris*), photo by Ernest P. Walker.

Stink badger (*Mydaus javanensis*), photo of mounted specimen from National Museum of Ireland.

long, truncate, mobile, and naked snout, which is often compared to that of a pig (*Sus*).

The hog badger is usually found in forested areas up to elevations of 3,500 meters (Lekagul and McNeely 1977). It is nocturnal, spending the day in natural shelters, such as rock crevices, or in self-excavated burrows. The snout is thought to be used in rooting for the various plants and small animals that compose the diet.

The general habits probably resemble those of *Meles*. As with *Meles* and *Mellivora*, the color pattern has been interpreted as a means of warning potential enemies that the animal so marked is best left alone. Like these other genera, *Arctonyx* is a savage and formidable antagonist. It has thick and loose skin, powerful jaws, fairly strong teeth, well-developed claws, and potent anal gland secretions.

A female from northern China had four newborn young in April. Parker (1979) reported that a captive pair from China was received at the Toronto Zoo in July 1976. The female gave birth to two cubs in February 1977. One cub survived and reached approximate adult size at an age of 7½ months. In February 1978 the same female gave birth to four young. Matings had been observed from April to September 1977, but delayed implantation was suspected, and true gestation was postulated at less than 6 weeks. According to Marvin L. Jones (Zoological Society of San Diego, pers. comm.), a captive lived for 13 years and 11 months.

CARNIVORA; MUSTELIDAE; **Genus MYDAUS** F. *Cuvier, 1821*

Stink Badgers

Long (1978) recognized two subgenera and two species:

subgenus *Mydaus*

M. javanensis, Sumatra, Java, Borneo, Natuna Islands;

subgenus *Suillotaxus*

M. marchei, Palawan and Calamian Islands (Philippines).

Suillotaxus Lawrence, 1939 has often been given generic rank.

In *M. javanensis*, the head and body length is 375 to 510 mm, tail length is usually 50 to 75 mm, and weight is usually 1.4 to 3.6 kg. The coloration is blackish, except for a white crown and a complete or partial narrow white stripe down the back onto the tail. In *M. marchei*, the head and body length is 320 to 460 mm, tail length is 15 to 45 mm, and weight is about 2.5 kg. The upper parts are brown to black, with a scattering of white or silvery hairs on the back and sometimes on the head, and the underparts are brown.

Both species have a pointed face, a somewhat elongate and mobile snout, short and stout limbs, and well-developed anal scent glands. Compared to *M. javanensis*, *M. marchei* has smaller ears, a shorter tail, and larger teeth.

M. javanensis is nocturnal, residing by day in holes in the ground, dug either by itself or by the porcupines with which it sometimes lives. The burrows are usually not more than 60 cm deep. Captives have consumed worms, insects, and the entrails of chickens.

Grimwood (1976) wrote that *M. marchei* is active both by day and by night. It is common, leaving its tracks and scent along roads and paths. It moves with a rather ponderous, fussy walk. One individual shammed death when first touched, and allowed itself to be carried, but finally squirted a jet of yellowish fluid from its anal glands into the lens of a camera about a meter away.

M. javanensis may growl and attempt to bite when handled. If molested or threatened, it raises the tail and ejects a pale greenish fluid. This vile-smelling secretion is reported by natives sometimes to asphyxiate dogs, or even blind them if they are struck in the eye. The old Javanese sultans used this fluid, in suitable dilution, in the manufacture of perfumes.

Some natives eat the flesh of *Mydaus*, removing the scent glands immediately after the animals are killed. Others mix shavings of the skin with water and drink the mixture as a cure for fever or rheumatism.

CARNIVORA; MUSTELIDAE; **Genus TAXIDEA** *Waterhouse, 1839*

American Badger

The single species, *T. taxus*, is found from northern Alberta and southern British Columbia to Ohio, central Mexico, and Baja California (Hall 1981).

American badger (*Taxidea taxus*), photo by E. P. Haddon from U.S. Fish & Wildlife Service.

Head and body length is 420 to 720 mm, tail length is 100 to 155 mm, and weight is 4 to 12 kg. The upper parts are grayish to reddish, and a dorsal white stripe extends rearward from the nose. In the north this stripe usually reaches only to the neck or shoulders, but in the south the stripe usually extends to the rump (Long 1972c). Black patches are present on the face and cheeks. The chin, throat, and midventral region are whitish. The underparts are buffy, and the feet are dark brown to black. The hairs are longest on the sides. *Taxidea* can be recognized by its flattened and stocky form, large foreclaws, distinctive black and white head pattern, long fur, and short, bushy tail. There are anal scent glands. Females have eight mammae (Jackson 1961).

The American badger is usually found in relatively dry, open country. It is a remarkable burrower, and can quickly dig itself out of sight. The usual signs of its presence are the large holes that it digs when after rodents. For shelter, it either excavates a burrow or modifies one initially made by another animal. The burrow can be as long as 10 meters and as deep as 3 meters below the surface. A bulky nest of grass is located in an enlarged chamber, and the entrances are marked by mounds of earth (Banfield 1974).

Taxidea may be active at any hour, but is mainly nocturnal. Its movements are restricted, especially in winter, and it shows a strong attachment to a home area. In the summer, however, the young animals disperse over a considerable distance; one traveled 110 km. The badger is active all year, but may sleep in its den for several days or weeks during severe winter weather. One female was found to have emerged only once during a 72-day period (Messick and Hornocker 1981).

Most food is obtained by excavating the burrows of fossorial rodents, such as ground squirrels. Also eaten are other small mammals, birds, reptiles, and arthropods (Banfield 1974; Messick and Hornocker 1981). Food is sometimes buried and used later. If a sizable meal, such as a rabbit, is obtained, the badger may dig a hole, carry in the prey, and remain below ground with it for several days. There are reports that a badger sometimes forms a "hunting partnership" with a coyote (see account of *Canis latrans*).

On the basis of a radio-tracking study in southwestern Idaho, Messick and Hornocker (1981) estimated a population density of up to five residents per sq km, and found average home ranges of 2.4 sq km for males and 1.6 sq km for females. The ranges overlapped, but individuals were solitary, except during the reproductive season. A female, radio-tracked in Minnesota, used an area of 752 ha. during the summer. She had 50 dens within this area and was never found in the same den on two consecutive days. In the fall she shifted to an adjacent area of 52 ha. and often reused dens. In the winter she used a single den, and traveled only infrequently within an area of 2 ha. (Sargeant and Warner 1972).

Mating occurs in summer and early autumn, but implantation of the fertilized eggs in the uterus is delayed until December–February, and births take place in March and early April (Long 1973b). The total period of pregnancy is thus about 7 months, but actual embryonic development lasts only about 6 weeks (Ewer 1973). The litter of one to five, usually two, young is born and raised in a nest of dry grass within a burrow. The young are weaned at about 6 weeks of age and disperse shortly thereafter. The study by Messick and Hornocker (1981) indicated that 30 percent of the young females mate in the first breeding season following birth, when they are about 4 months old, but that males wait until the following year. The oldest wild badger caught in this study had attained an age of 14 years. A captive badger lived for 26 years (Marvin L. Jones, Zoological Society of San Diego, pers. comm.).

Taxidea is generally beneficial to human interests, as it destroys many rodents and its burrows provide shelter for many kinds of wildlife, including cottontail rabbits. Its burrows and holes, however, may constitute a hazard to cattle, horses, and riders, and thus ranchers have often killed badgers. Many badgers also have been killed by poison put out for coyotes (Hall 1955). Although *Taxidea* has declined in numbers in some areas, it has extended its range eastward in the 20th century. It now occupies most of Ohio and is invading southeastern Ontario (Long 1978; Hall 1981).

For many years badger fur was used to make shaving brushes, but it has now been largely replaced by synthetic materials. The fur is still used for trimming garments. In the 1976/77 trapping season, 49,807 pelts were reported taken in

Ferret badger (*Melogale moschata*), photo by Gwilym S. Jones.

the United States and Canada, and were sold at an average price of about $38 (Deems and Pursley 1978).

CARNIVORA; MUSTELIDAE; *Genus MELOGALE*
I. Geoffroy St.-Hilaire, 1831

Ferret Badgers

There are three species (Long 1978):

M. moschata, Assam to southern China and northern Indochina, Taiwan, Hainan;
M. personata, Nepal to Indochina, Java;
M. everetti, Borneo.

Long (1978) noted that *M. moschata sorella* of southeastern China and *M. personata orientalis* of Java might possibly be full species.

Head and body length is 330 to 430 mm, tail length is 145 to 230 mm, and weight is 1 to 3 kg. The general coloration of the upper parts is gray brown to brown black. The underparts are somewhat paler. A white or reddish dorsal stripe is usually present. *Melogale* is distinguished by the striking coloration of the head, which combines black with patches of white or yellow. The tail is bushy. The limbs are short, and the feet are broad and have long, strong claws for digging (Lekagul and McNeely 1977).

Ferret badgers are found in wooded country and grassland. They reside in burrows and natural shelters during the day, and are active at dusk and during the night. They climb on occasion. *M. moschata*, in Taiwan, is reported to be a good climber and often to sleep on the branches of trees. Ferret badgers are savage and fearless when provoked or pressed, and have an offensive odor. The conspicuous markings on the head have been interpreted as being a warning signal. The diet is omnivorous, and is known to include small vertebrates, insects, earthworms, and fruit. A ferret badger is sometimes welcome to enter a native hut, because of its destruction of insect pests.

The young, usually one to three per litter, are born in a burrow in May and June. They are apparently dependent on the milk of the mother for some time, as two nearly full grown suckling animals and their mother were once found in a burrow.

CARNIVORA; MUSTELIDAE; *Genus SPILOGALE* Gray, *1865*

Spotted Skunks

There are two species (Hall 1981):

S. pygmaea, Pacific coast of Mexico from southern Sinaloa to Oaxaca;
S. putorius, southern British Columbia and Pennsylvania to Costa Rica and Florida.

Head and body length is 115 to 345 mm, tail length is 70 to 220 mm, and weight is usually 200 to 1,000 grams. Of the three genera of skunk, *Spilogale* has the finest fur. The hairs are longest on the tail and shortest on the face. The basic color pattern consists of six white stripes, which extend along the back and sides, and these are broken into smaller stripes and spots on the rump. There is a triangular patch in the middle of the forehead, and the tail is usually tipped with white. The variations are infinite, as no two individuals have ever been found with exactly the same pattern. This genus may be distinguished from the other two genera of skunk by its small size, forehead patch, and pattern of stripes and spots, the white never being massed. There is a pair of scent glands under the base of the tail, from which a jet of strong-smelling fluid can be emitted through the anus. Females have 10 mammae (Lowery 1974).

Spotted skunks occur in a variety of brushy, rocky, and wooded habitats, but avoid dense forests and wetlands. They generally remain under cover more than do striped skunks (*Mephitis*). They usually den underground, but can climb well and sometimes shelter in trees. The dens are lined with dry vegetation. Spotted skunks are largely nocturnal and are active all year. The omnivorous diet consists mainly of vegetation and insects in the summer, and rodents and other small animals in the winter.

The white-plumed tail is used to warn other animals that spotted skunks should not be molested. If, however, the sudden erection of the tail is not sufficient deterrence, *Spilogale* may stand on its forefeet and sometimes even advance toward its adversary. Finally, the fluid from the anal glands is discharged at the enemy, usually after the skunk has returned its forefeet to the ground and assumed a horseshoe position (Lowery 1974).

Spotted skunk (*Spilogale putorius*), photo by Ernest P. Walker.

According to Banfield (1974), population densities reach 5 individuals per sq km in good agricultural land, and winter home range is approximately 64 ha. In spring the males wander over an area of about 5 to 10 sq km, but females have smaller ranges. Spotted skunks are very playful with one another. As many as eight individuals sometimes share a den.

The reproductive pattern is not the same in all parts of the range (Mead 1968*a*, 1968*b*; Foresman and Mead 1973). Populations in South Dakota and Florida apparently mate mainly in March and April. Implantation of the fertilized eggs in the uterus occurs only 14 to 16 days later, and births take place in late May and June. Pregnancy is estimated to last 50 to 65 days. Populations farther to the west mate in September and October, but implantation is delayed until the following March or April and births occur from April to June. The total period of pregnancy thus lasts 230 to 250 days, but actual embryonic development takes only 28 to 31 days. The number of young per litter is two to nine, usually three to six. The young weigh about 22.5 grams each at birth, have adult coloration after 21 days, open their eyes at 32 days, can spray musk at 46 days, are weaned at about 54 days, and attain adult size at about 15 weeks of age. A captive specimen lived for 9 years and 10 months (Egoscue, Bittmenn, and Petrovich 1970).

Spotted skunks have been reported to carry rabies and occasionally to take poultry and eggs, but are generally beneficial to people through their destruction of rodents and insects. The pelts are very attractive and durable, but generally sold for well under $1.00 each until about 1970 (Lowery 1974; Jackson 1961). In the 1976/77 trapping season, the reported harvest in the United States was 41,952 skins and the average selling price was $4.00 (Deems and Pursley 1978).

CARNIVORA; MUSTELIDAE; **Genus MEPHITIS**
E. Geoffroy St.-Hilaire and G. Cuvier, 1795

Striped and Hooded Skunks

There are two species (Hall 1981):

M. mephitis (striped skunk), southern Canada to northern Mexico;

M. macroura (hooded skunk), Arizona and southwestern Texas to Nicaragua.

Head and body length is 280 to 380 mm, tail length is 185 to 435 mm, and weight is 700 to 2,500 grams. Both species have black and white color patterns, but with considerable variation. *M. mephitis* usually has white on top of the head and on the nape extended posteriorly and separated into stripes. In some individuals of this species the top and sides of the tail are white, while in others the white is limited to a small spot on the forehead. The white areas are composed entirely of white hairs, with no black hairs intermixed. *M. macroura* has a white-backed color phase and a black-headed color phase. In the former, there are some black hairs mixed with the white hairs of the back. In the latter color phase, the two white stripes are widely separated and are situated on the sides of the animal, instead of being narrowly separated and situated on the back, as in *M. mephitis*. Female *Mephitis* have 10 to 14 mammae (Leopold 1959; Jackson 1961).

According to Lowery (1974), the well-known scent of

Striped skunk (*Mephitis mephitis*): A. Facing danger that does not appear to be imminent; B. Aimed toward the enemy ready to spray its scent; photos by Ernest P. Walker.

Mephitis is expelled from two tiny nipples, located just inside the anus, which mark the outlets of the two ducts leading from glands lying adjacent to the anus. This musk is discharged either as an atomized spray or as a short stream of rain-sized drops. The skunk usually employs this weapon only after much provocation. When confronted by an antagonist, it arches its back, elevates the tail, erects the hairs thereon, and sometimes stamps its feet on the ground. Finally, it turns the body in a U-shaped position with the head and tail facing the intruder. The musk usually travels 2 or 3 meters, but the smell can be detected up to 2½ km downwind. Lowery observed that one squirt is sufficient to send the most ferocious dog yelping in agony from burning eyes and nostrils, and retching with nausea.

These skunks are found in a variety of habitats, including woods, grasslands, and deserts. They generally are active at dusk and through the night, and spend the day in a burrow, under a building, or in any dry, sheltered spot. In Minnesota, Houseknecht and Tester (1978) found a general shift from underground, upland dens in winter to aboveground, lowland dens in summer. Although skunks tended to remain at a single den for a long time in winter, females with young able to travel changed dens every one or two days. Bjorge, Gunson, and Samuel (1981) reported that females moved a minimum daily distance of 1.5 km between dens. In summer, juveniles were found to disperse up to 22 km.

In northern parts of its range, *M. mephitis* stays in one den and sleeps through much of the winter. Males tend to sleep for shorter periods than females, and to become active more

readily during intervals of mild weather (Banfield 1974). The degree of lethargy achieved during the winter is not well understood. It does not appear to be deep torpor, but some skunks are known to have remained underground for over 100 consecutive days (Sunquist 1974). In Alberta, the overall period of female hibernation is 120–50 days, while in Illinois it is 62–87 days (Gunson and Bjorge 1979). A striped skunk may become very fat in the autumn before hibernation. The diet is omnivorous and includes rodents, other small vertebrates, insects, fruit, grains, and green vegetation.

In Alberta, Bjorge, Gunson, and Samuel (1981) found population densities of 0.7 to 1.2 striped skunks per sq km. They noted, however, that densities reported for other areas, 0.4 to 27 per sq km, were generally higher. The home ranges of 6 females, radio-tracked for 45 to 105 days each in Alberta, averaged 208 ha. and varied from 110 to 370 ha. Males wandered over larger areas, most notably in the fall. *Mephitis* is generally solitary, but there is a tendency for individuals to den together, especially in the north, as a means of optimizing winter survival and reproductive success. In Alberta, Gunson and Bjorge (1979) found that only males, both adults and juveniles, denned alone during the winter. Communal winter dens contained an average of 6.7 (2–19) individuals. There was usually only a single adult male per den and an average of 5.8 females. A male apparently wanders in search of a group of females during the fall, and then keeps other males away. *Mephitis* is usually silent, but makes several sounds, such as low churrings, shrill screeches, and birdlike twitters (Lowery 1974).

Mating takes place from mid-February to mid-April, and births occur in May and early June. The period of pregnancy is 59 to 77 days, and delayed implantation may be involved. Females are usually monestrous, but sometimes have a second estrus and parturition subsequent to the normal period if their first pregnancy is not successful. Litters contain 1 to 10 young, usually about 4 or 5. The young weigh about 30 grams at birth, open their eyes after 3 weeks, are weaned when 8 to 10 weeks old, and separate from the mother by the fall. Females may bear their first litters at the age of 1 year (Wade-Smith and Richmond 1975, 1978; Lowery 1974). The average longevity in captivity is about 6 years, but one individual was still living after 12 years and 11 months (Marvin L. Jones, Zoological Society of San Diego, pers. comm.).

Striped skunks are generally beneficial to human interests because of their destruction of rodents and insects. *Mephitis,* however, sometimes attacks poultry and is reportedly the principal carrier of rabies among North American wildlife (Wade-Smith and Richmond 1975). The fur is durable and of good texture, but demand and value have varied widely (Lowery 1974; Jackson 1961). During the 1976/77 season, 175,884 skins were reported taken in the United States and Canada, and sold at an average price of about $2.25 (Deems and Pursley 1978).

CARNIVORA; MUSTELIDAE; *Genus CONEPATUS Gray, 1837*

Hog-nosed Skunks

Five species are now recognized (Hall 1981; Ewer 1973; Kipp 1965; Cabrera 1957; Pine, Miller, and Schamberger 1979):

C. mesoleucus, southern Colorado and southeastern Texas to Nicaragua;

C. leuconotus, southern Texas, eastern Mexico;

C. semistriatus, southern Mexico to northern Peru and eastern Brazil;

C. chinga, central and southern Peru, Bolivia, Chile, northwestern Argentina;

C. humboldti, Paraguay, southeastern Brazil, Uruguay, Argentina.

Statements by the various authorities cited above suggest that some, perhaps all, of the listed species are conspecific.

Head and body length is 300 to 490 mm, tail length is 160 to 410 mm, and weight is usually 2.3 to 4.5 kg. *Conepatus* has the coarsest fur of all skunks. There are two main color patterns, with variations. In one, the top of the head, the back, and the tail are white, and the remainder of the animal is black. This coloration occurs most commonly in areas where the ranges of *Conepatus* and *Mephitis* overlap. In the other pattern, the pelage is black except for two white stripes, beginning at the nape and extending on the hips, and a mostly white tail. This coloration resembles that of *Mephitis mephitis,* and seems to be most common in areas where *Conepatus* is the only kind of skunk present. In all cases, hog-nosed skunks lack the thin white stripe down the center of the face that is present in *Mephitis. Conepatus* may be distinguished from the other two genera of skunks by its nose, which is bare, broad, and projecting. Females have three pairs of mammae (Leopold 1959).

Hog-nosed skunks are found in both open and wooded areas, but avoid dense forests. They occur at all elevations up to at least 4,100 meters (Grimwood 1969). Dens are located in rocky places, hollow logs, or burrows made by other animals. Like other skunks, *Conepatus* is mainly nocturnal, is generally slow moving, does not ordinarily climb, and defends itself by expelling musk from anal scent glands.

The diet may consist principally of insects and other invertebrates, though fruit and small vertebrates, including snakes, probably are also eaten. Hog-nosed skunks may turn over the soil in a considerable area with their bare snouts and their claws when in search of food. Like *Mephitis,* they also

Hog-nosed skunk (*Conepatus mesoleucus*), photo by Lloyd G. Ingles.

Canadian river otters (*Lutra canadensis*), photos by Ernest P. Walker.

pounce on insects. At least in the Andes, hog-nosed skunks are resistant to the venom of pit vipers. There is some evidence that the spotted skunks (*Spilogale*) also are resistant to rattlesnake venom. Since the musk of skunks produces an alarm reaction in rattlers (the same reaction that they exhibit in the presence of king snakes, which prey on them), it may be that skunks feed on rattlesnakes quite extensively.

According to Davis (1966*b*), *C. mesoleucus* is not as social as *Mephitis*, and usually only one individual lives in a den. The breeding season in Texas begins in February, most mature females are pregnant by March, and births occur in late April or early May. Gestation lasts approximately 2 months. Of six pregnant females on record, three contained three embryos each, and three had two each. By August most of the young are weaned and foraging for themselves. Available evidence indicates that in Mexico the young are also born in the spring (Leopold 1959; Hall and Dalquest 1963). The gestation period of one South American species is 42 days, and litter size is usually two to five young. A captive *C. chinga* lived for 6 years and 7 months (Marvin L. Jones, Zoological Society of San Diego, pers. comm.).

The pelt of *Conepatus* is inferior in quality to that of *Mephitis*, but large numbers have been marketed from Texas (Davis 1966*b*). Some natives use the skins for capes or blankets, and others consider the meat to have curative properties. *C. humboldti* is on appendix 2 of the CITES.

CARNIVORA; MUSTELIDAE; **Genus LUTRA** Brünnich, *1772*

River Otters

There are four subgenera and eight species (Van Zyll de Jong 1972; Ellerman and Morrison-Scott 1966; Hall 1981; Rosevear 1974; Meester and Setzer 1971; Corbet 1978; Lekagul and McNeely 1977; Chasen 1940):

subgenus *Lutra*

L. lutra, western Europe to northeastern Siberia and Korea, Asia Minor and certain other parts of southwestern Asia, Himalayan region, extreme southern India, China, Burma, Thailand, Indochina, northwestern Africa, British Isles, Sri Lanka, Sakhalin, Japan, Taiwan, Hainan, Sumatra, Java;

L. sumatrana, Indochina, Thailand, Malay Peninsula, Sumatra, Bangka, Java, Borneo;

subgenus *Hydrictis*

L. maculicollis, Sierra Leone to Ethiopia, and south to South Africa;

subgenus *Lutrogale*

L. perspicillata, southern Iraq, Pakistan to Indochina and Malay Peninsula, Sumatra;

subgenus *Lontra*

L. canadensis, Alaska, Canada, conterminous United States;
L. longicaudis, northwestern Mexico to Uruguay;
L. provocax, Chile, southern Argentina;
L. felina, Pacific coast from northern Peru to Tierra del Fuego.

Van Zyll de Jong (1972) considered *Lontra* Gray, 1843 and *Lutrogale* Gray, 1865 to be full genera, but this procedure was not followed by Hall (1981), Corbet (1978), or Jones, Carter, and Genoways (1979). Van Zyll de Jong also suggested a close relationship between *L. sumatrana,* which was referred to the subgenus *Lutra* by Ellerman and Morrison-Scott (1966), and *L. maculicollis* of the subgenus *Hydrictis.* Rosevear (1974), however, stated that there is a strong case for regarding *Hydrictis* as a full genus.

Head and body length is 460 to 820 mm, tail length is 300 to 500 mm, and weight is 3 to 14 kg. Males average larger than females (Van Zyll de Jong 1972). The upper parts are brownish and the underparts are paler; the lower jaw and throat may be whitish. The fur is short and dense. The head is flattened and rounded; the neck is short and about as wide as the skull; the trunk is cylindrical; the tail is thick at the base, muscular, flexible, and tapering; the legs are short; and the digits are webbed. The small ears and the nostrils can be closed when the animal is in the water. Female *L. canadensis* have four mammae (Jackson 1961).

These aquatic mammals inhabit all types of inland waterways, as well as estuaries and marine coves. They are excellent swimmers and divers, and usually are found no more than a few hundred meters from water. They may shelter temporarily in shallow burrows, or in piles of rocks or driftwood, but also have at least one permanent burrow beside the water (Stroganov 1969; Jackson 1961). The main entrance may open underwater and then slope upward into the bank to a nest chamber that is above the high-water level. Erlinge (1967) found *L. lutra* to utilize the following types of facilities in southern Sweden: "dens," generally with several passages and a chamber lined with dry leaves and grass; "rolling places," bare spots near water, where the otters roll and groom themselves; "slides," either on slopes or in level places, but most common on winter snow; "feeding places," including holes kept open through winter ice; "runways," well-defined paths on land that connect waterways and other facilities; and "sprainting spots and sign heaps," prominent points of land where the animals mark by scratching and elimination.

Otters swim by movements of the hind legs and tail, and can remain underwater for six to eight minutes (Grzimek 1975). When traveling on ground, snow, or ice, they may use a combination of running and sliding. Although normally closely associated with water, river otters sometimes move many kilometers overland to reach different river basins and to find ice-free water in winter (Stroganov 1969). When running on land, they can attain speeds of up to 29 km per hr (Banfield 1974). They may be either diurnal or nocturnal, but are generally more active at night.

With the possible exception of the Old World badger (*Meles*), river otters are the most playful of the Mustelidae. Some species engage in the year-round activity of sliding down mud and snow banks, and individuals of all ages participate. Sometimes they tunnel under snow to emerge some distance beyond.

The diet consists largely of fish, frogs, crayfish, crabs, and other aquatic invertebrates. Birds and land mammals, such as rodents and rabbits, are also taken. Studies indicate that the fish consumed are mainly nongame species. River otters capture their prey with the mouth, not the hands (Rowe-Rowe 1977).

Investigations by Erlinge (1967, 1968) in southern Sweden indicate that *L. lutra* occurs at population densities of 0.7 to 1 individual per sq km of water area, or 1 individual for every 2 or 3 km of lakeshore or 5 km of stream. The straight-line length of a home range, including land area, was found to average about 15 km for adult males and 7 km for females with young. The ranges of males constitute territories, which may overlap the ranges of 1 or more adult females and from which other males are excluded. Females also defend their ranges against individuals of the same sex. Territories are marked with scent, and fights occasionally take place. Apparently, males form a dominance hierarchy, with the highest-ranking animals occupying the most favorable ranges. Erlinge noted that the males generally are solitary and ignore the females and young.

Observations in other areas suggest that the male is excluded from the vicinity of the female, when the female has small young, but joins the family when the cubs are about six months old (Banfield 1974; Jackson 1961). On the Malay Peninsula, *L. perspicillata* typically occurs in groups consisting of a mated adult pair and up to four young, which have a territory of 7 to 12 km of river (Lekagul and McNeely 1977). On Lake Victoria, *L. maculicollis* may undergo a regular cycle of aggregation and dispersal, with males and females each forming their own groups. The males' groups grow larger after the mating season, when males are not tolerated by the females with young. These groups may contain 8 to 20 individuals from January to May, but then become smaller from June to August, when the older males leave to pair with the females (Kingdon 1977).

Within the wide geographic range of *Lutra,* there is considerable variation in reproductive pattern (Duplaix-Hall 1975; Ewer 1973; Banfield 1974; Kingdon 1977; Stroganov 1969; Liers 1966). Female *L. lutra* are polyestrous. In some areas (England, for example), mating and birth may occur at any time of the year. In areas with a more severe climate (such as Sweden and Siberia), mating is in late winter or early spring and births take place in April or May. The gestation period of this species is 60 to 63 days. The length of pregnancy is about the same in *L. perspicillata* and *L. maculicollis.* In *L. canadensis* of North America, however, there is delayed implantation of the fertilized eggs in the uterus. Mating occurs in the winter or spring, and births take place the following year. The total period of pregnancy has been reported to vary from 245 to 380 days, though actual embryonic development is about 2 months, the same as in other kinds of river otters. Litter size in the genus is one to five young, usually two or three. The young weigh about 130 grams at birth, open their eyes after 1 month, emerge from the den and begin to swim at 2 months of age, nurse for 3 or 4 months, separate from the mother when about 1 year old, and attain sexual maturity in the second or third year of life. A female *L. canadensis* lived in captivity for 23 years.

As a group, otters have suffered severely through habitat destruction, water pollution, misuse of pesticides, excessive fur trapping, and persecution as supposed predators of game and commercial fish. *L. canadensis* has disappeared or become rare throughout the conterminous United States, except in the northwest, the upper Great Lakes region, New York, New England, and the states along the Atlantic and Gulf coasts. *L. lutra* has declined drastically in such diverse places as Britain (Chanin and Jefferies 1978), Germany (Roben 1974), southeastern Siberia (Kucherenko 1976), and Japan (Mikuriya 1976). The IUCN (1976, 1978) classifies *L. lutra lutra,* the subspecies found in Europe and most of Asia, as vulnerable, and the following as endangered: the populations

European river otter (*Lutra lutra*), photo by Annelise Jensen.

of *L. longicaudis* in southern Brazil, Paraguay, northern Argentina, and Uruguay; *L. provocax;* and *L. felina.* The USDI (1980) lists *L. longicaudis, L. provocax,* and *L. felina* as endangered. *L. lutra, L. longicaudis, L. provocax,* and *L. felina* are on appendix 1 of the CITES, and the remaining species of *Lutra* are on appendix 2.

The beautiful and durable fur of river otters is used for coat collars and trimming. During the 1976/77 trapping season, 32,846 pelts were reported taken in the United States, and the average selling price was $53.00. Respective figures for Canada that season were 19,932 pelts and $69.04 (Deems and Pursley 1978).

CARNIVORA; MUSTELIDAE; **Genus PTERONURA** *Gray, 1867*

Giant Otter

The single species, *P. brasiliensis,* is found in Colombia, Venezuela, the Guianas, eastern Ecuador and Peru, Brazil, Bolivia, Paraguay, Uruguay, and northeastern Argentina (Cabrera 1957; IUCN 1976).

The remainder of this account is based largely on Duplaix's (1980) study of *Pteronura* in Surinam. Head and body length is 864 to 1,400 mm and tail length is 330 to 1,000 mm. Males weigh 26 to 34 kg and females weigh 22 to 26 kg. The fur is short, and generally appears brown and velvetlike when dry, and shiny black chocolate when wet. On the lips, chin, throat, and chest there are often creamy white to buff splotches, which may unite to form a large white "bib." The feet are large, and thick webbing extends to the ends of the five clawed digits. The tail is thick and muscular at the base, but becomes dorsoventrally flattened with a noticeable bilateral flange. There are subcaudal anal glands for secretion of musk.

The giant otter is found mainly in slow-moving rivers and creeks within forests, swamps, and marshes. It prefers waterways with gently sloping banks that have good cover. At certain points along a stream, areas of about 50 sq meters are cleared and used for rest and grooming. Some of these sites have dens, which consist of one or more short tunnels leading to a chamber about 1.2 to 1.8 meters wide. *Pteronura* seems clumsy on land, but may move a considerable distance between waterways. When swimming slowly or remaining stationary in the water, it paddles with all four feet. When swimming at top speed, it depends largely on undulations of the tail and uses the feet for steering. It is entirely diurnal. During the dry season, when cubs are being reared, activity is generally restricted to one portion of a waterway. In the wet season, movements are far more extensive, and spawning fish are followed into the flooded forest. Prey is caught with the mouth and then may be held in the forepaws while being consumed. Small fish may be eaten in the water, but larger prey is taken to shore. The diet consists mainly of fish and crabs.

Group home range, including land area, measures about 12 km in both length and width. During the dry season, at least, several kilometers of stream form a defended territory. Both sexes regularly patrol and mark the area, but groups tend to avoid one another and fighting is evidently rare. *Pteronura* is more social than *Lutra.* A population includes both resident groups and solitary transients. Up to 20 individuals have reportedly been seen together, but groups of 4 to 8 are usually observed. A group consists of a mated adult pair, 1 or more subadults, and 1 or more young of the year. There is a high degree of pair bonding and group cohesiveness. A male and female stay close together and share the same den, even when cubs are present. *Pteronura* is much noisier than *Lutra.* Nine vocalizations have been distinguished, including "screams" of excitement, often given while swimming with the forepart of the body steeply out of the water, and "coos" upon close intraspecific contact.

Although data are scanty, the young are apparently born from late August to early October, at the start of the dry season. If the first litter is lost, a second is sometimes pro-

Giant otter (*Pteronura brasiliensis brasiliensis*), photo from New York Zoological Society.

duced from December to April. The gestation period is 65 to 70 days. The number of young per litter is one to five, usually one to three. The cubs weigh about 200 grams each at birth and are eating solid food by 3 or 4 months of age. The young remain with the parents at least until the birth of the next litter, and probably for some time afterward. A captive lived for 12 years and 10 months (Marvin L. Jones, Zoological Society of San Diego, pers. comm.).

The giant otter is classified as endangered by the IUCN (1976) and the USDI (1980), and is on appendix 1 of the CITES. It has become very rare or has entirely disappeared over vast parts of its range. The main factor in its decline is excessive hunting by people for its large and valuable pelt. Because of its noise, diurnal habits, and tendency to approach intruders, it is relatively easy to locate and kill.

CARNIVORA; MUSTELIDAE; **Genus AONYX** Lesson, *1827*

Clawless Otters

There are three subgenera and three species (Meester and Setzer 1971; Ellerman and Morrison-Scott 1966; Lekagul and McNeely 1977; Rosevear 1974; Chasen 1940):

subgenus *Aonyx*

A. capensis, Senegal to Ethiopia, and south to South Africa;

subgenus *Paraonyx*

A. congica, southeastern Nigeria and Gabon to Uganda and Burundi;

subgenus *Amblonyx*

A. cinerea, northwestern India to southeastern China and Malay Peninsula, southern India, Hainan, Sumatra, Java, Borneo, Rhio Archipelago, Palawan.

Paraonyx Hinton, 1921 and *Amblonyx* Rafinesque, 1832 have sometimes been treated as full genera.

In the African species, *A. capensis* and *A. congica,* head and body length is 600 to 1,000 mm, tail length is 400 to 710 mm, and weight is 13 to 34 kg. In the smaller *A. cinerea* of Asia, head and body length is 450 to 610 mm, tail length is 250 to 350 mm, and weight is 1 to 5 kg. The general coloration is brown, with paler underparts and sometimes white markings on the face, throat, and chest.

Aonyx differs from *Lutra* and *Pteronura* in having webbing that either does not extend to the ends of the digits or is entirely lacking, and in having much smaller claws. In *A. congica,* all the toes bear small, blunt claws. In *A. cinerea,* the claws of adults are only minute spikes that do not project beyond the ends of the digital pads. In *A. capensis,* there are

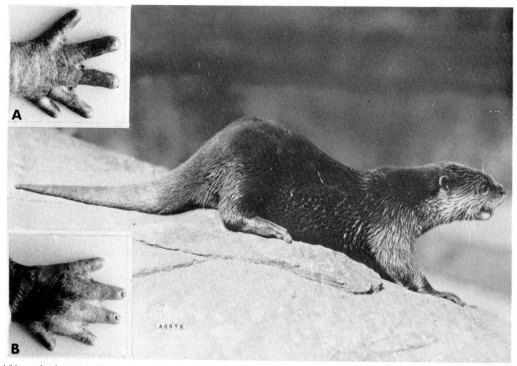

African clawless otter (*Aonyx capensis*), photo from the Zoological Society of London. Insets: A. Forefoot; B. Hind foot; photos by U. Rahm.

no claws, except for tiny ones on the third and fourth toes of the hind feet. In association with these adaptations, *Aonyx* has developed very sensitive forepaws and considerable digital movement. *A. capensis* and *A. cinerea* have relatively large, broad cheek teeth, apparently for purposes of crushing the shells of crabs and mollusks. *A. congica* has lighter and sharper dentition, more adapted to cutting flesh.

The general habitat of *A. capensis* varies from dense rain forest to open coastal plain and semiarid country. The species is usually found near water, preferring quiet ponds and sluggish streams, but may sometimes wander a considerable distance overland. It is mainly nocturnal, but may be active by day in areas remote from human disturbance. It dens under boulders or driftwood, in crannies under ledges, or in tangles of vegetation. It apparently does not dig its own burrow. *A. cinerea* occurs in rivers, creeks, estuaries, and coastal waters (Lekagul and McNeely 1977). *A. congica* is seemingly found only in small, torrential mountain streams, within heavy rain forest.

These otters use their sensitive and dexterous forepaws to locate prey in mud or under stones. Captive *A. capensis* usually take food with the forepaws and do not eat directly off the ground. In the wild this species also catches most of its food with the forefeet, not with the mouth, as do *Lutra* and *Pteronura* (Rowe-Rowe 1978b). The diet of both *A. capensis* and *A. cinerea* seems to consist mainly of crabs, other crustaceans, mollusks, and frogs. Fish are relatively unimportant. There is thus apparently little competition for food between *Aonyx* and the fish-eating *Lutra*, where both genera occur together. Piles of cracked crab and mollusk shells are signs of the presence of *A. capensis*. Donnelly and Grobler (1976) observed this species to use hard objects as anvils, on which to break open mussel shells.

Little has been recorded about the habits of *A. congica*.

Because of its scanty hair, weakly developed facial vibrissae, digital structure, and dental features, there has been speculation that it is more terrestrial than other otters. It is thought to feed mainly on relatively soft matter, such as small land vertebrates, eggs, and frogs. If this supposition is correct, there would be little competition for food between *A. congica* and the other kinds of otter that may occur in the same region.

According to Timmis (1971), *A. cinerea* lives in loose family groups of about 12 individuals and has a vocabulary of 12 or more calls, not including basic instinctive cries. Rowe-Rowe (1978b) stated that *A. capensis* occurs in groups of up to 5 individuals, each of which includes 1 or 2 adults. *A. capensis* emits powerful, high-pitched shrieks, when disturbed or trying to attract attention.

Births of *A. capensis* have been recorded in July and August in Zambia, and young have been found in March and April in Uganda (Kingdon 1977). There is probably no set breeding season in West Africa (Rosevear 1974). This species has a gestation period of 63 days and a litter size of two young. The young remain with the parents for at least 1 year. Female *A. cinerea* have an estrous cycle of 24 to 30 days, with an estrus of 3 days. They may produce two litters annually. The gestation period is 60–64 days, and litters contain one to six young, usually one or two. Their eyes open after 40 days, they first swim at 9 weeks of age, and they take solid food after 80 days (Lekagul and McNeely 1977; Leslie 1971; Timmis 1971; Duplaix-Hall 1975). Little is known of the reproduction of *A. congica*, but it probably has a gestation period of about 2 months, gives birth to two or three young, and attains sexual maturity at about 1 year of age. One specimen of *A. capensis* was still living after 11 years in captivity (Marvin L. Jones, Zoological Society of San Diego, pers. comm.).

If captured when young, these otters make intelligent and

Oriental small-clawed otter (*Aonyx cinerea*), photo by Lim Boo Liat.

charming pets. *A. cinerea* has been trained to catch fish by Malay fishermen. The fur of *Aonyx* is not as good as that of *Lutra*. Nonetheless, the subspecies *A. congica microdon* of Nigeria and Cameroon has declined seriously through uncontrolled commercial hunting. It is classified as endangered by the IUCN (1976) and the USDI (1980), and is on appendix 1 of the CITES. The other species of *Aonyx* are on appendix 2 of the CITES.

CARNIVORA; MUSTELIDAE; **Genus ENHYDRA** *Fleming, 1822*

Sea Otter

The single species, *E. lutris,* was originally found in coastal waters off Hokkaido, Sakhalin, Kamchatka, the Commander Islands, the Pribilof Islands, the Aleutians, southern Alaska, British Columbia, Washington, Oregon, California, and western Baja California (Estes 1980).

Head and body length is usually 1,000 to 1,200 mm and tail length is 250 to 370 mm. Males weigh 22 to 45 kg and females weigh 15 to 32 kg (Estes 1980). The color varies from reddish brown to dark brown, almost black, except for the gray or creamy head, throat, and chest. Albinistic individuals are rare. The head is large and blunt, the neck is short and thick, and the legs and tail are short. The ears are short, thickened, pointed, and valvelike. The hind feet are webbed and are flattened into broad flippers. The forefeet are small and have retractile claws. *Enhydra* is the only carnivore with only four incisor teeth in the lower jaw. The molars are broad, flat, and well adapted to crushing the shells of such prey as crustaceans, snails, mussels, and sea urchins. Unlike most mustelids, the sea otter lacks anal scent glands. Females have two abdominal mammae (Estes 1980).

The sea otter differs from most marine mammals in that it lacks an insulating subcutaneous layer of fat. For protection against the cold water, it depends entirely on a layer of air trapped among its long, soft fibers of hair. If the hair becomes soiled, as if by oil, the insulating qualities are lost and the otter may perish. The underfur is very dense and about 25 mm long. It is protected by a scant coat of guard hairs.

Although the sea otter is a marine mammal, it rarely ventures more than 1 km from shore. According to Estes (1980), it forages in both rocky and soft-sediment communities, on or near the ocean floor. Off California, *Enhydra* seldom enters water of greater depth than 20 meters, but in the Aleutians it commonly forages at depths of 40 meters or more. The maximum confirmed depth of a dive was 97 meters. The usual period of submergence is 52 to 90 seconds, and the longest on record is 4 minutes and 25 seconds. The sea otter is capable of spending its entire life at sea, but sometimes rests on rocks near the water. It walks awkwardly on land. When supine on the surface of the water, it moves by paddling with the hind limbs and sculling with the tail. For rapid swimming and diving, it uses dorsoventral undulations of the body. It can attain velocities of up to 1½ km/hr on the surface and 9 km/hr underwater.

The sea otter is generally diurnal. It often spends the night in a kelp bed, lying under strands of kelp to avoid drifting while sleeping. It sometimes sleeps with a forepaw over the eyes. There may be local seasonal movements, but no extensive migrations.

The diet consists mainly of slow-moving fish and marine invertebrates, such as sea urchins, abalones, crabs, and mollusks (Estes 1980). Prey is usually captured with the forepaws, not the jaws. *Enhydra* floats on its back while eating, and uses its chest as a "lunch counter." It is one of the few mammals known to use a tool. While floating on its back, it places a rock on its chest and then employs the rock as an anvil for breaking the shells of mussels, clams, and large sea snails, in order to obtain the soft internal parts. The sea otter requires a great deal of food, and must eat 20 to 25 percent of its body weight every day. It obtains about 23 percent of its water needs from drinking sea water, and most of the rest from its food.

According to Estes (1980), *Enhydra* is basically solitary, but sometimes rests in concentrations of up to 2,000 individuals. Animals of opposite sex usually come together only

Sea otters (*Enhydra lutris*): A. Juvenile walking; B. *E. lutris* floating on its back eating the head of a large codfish; C. Mother floating on her back with newborn pup; photos by Karl W. Kenyon.

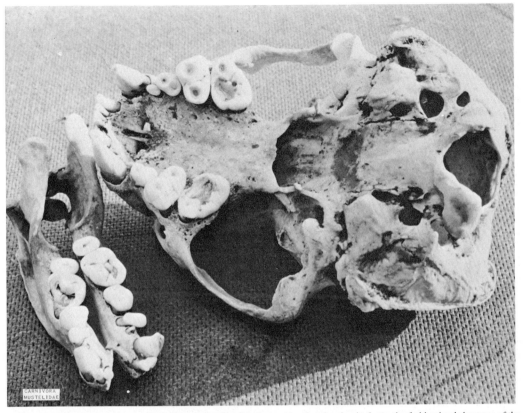

Skull and lower jaw of a sea otter (*Enhydra lutris*), showing the cavities that develop in the teeth of old animals because of the hard, rough materials that they eat, photo by H. Robert Krear.

briefly for courtship and mating. At most times there is sexual segregation, with males and females occupying separate sections of coastline. Males usually occur at higher densities. Males have been observed to establish territories during the breeding season in California waters. These territories average 30 ha. in size and the boundaries are patrolled, but fights are rare. Loughlin (1980) found individual home range to average 35 ha. for males and 80 ha. for females. Kenyon (1969) described nine vocalizations of *Enhydra*, including "screams" of distress and "coos" of contentment.

Reproductive data have been summarized by Estes (1980). Breeding occurs throughout the year, but births peak in late May and June in the Aleutians, and from December to February off California. Males may mate with more than one female during the season. Females are capable of giving birth every year, but usually do so at greater intervals. If a litter does not survive, the female may experience a postpartum estrus. Females are known to adopt and nurse orphaned pups. Reports of the period of pregnancy range from 6½ to 9 months, and delayed implantation is probably involved. Births probably occur most often in the water. There is normally a single offspring. About 2 percent of births are multiple, but only one young can be successfully reared. The pup weighs 1.4 to 2.3 kg at birth. While still small, it is carried, nursed, and groomed on the mother's chest, as the mother swims on her back. The pup begins to dive in the second month of life. It may take some solid food shortly after being born, but may nurse almost until it attains adult size. The period of dependency on the mother is thought to be about 6 to 8 months. Females become sexually mature at about 4 years of age. Males are capable of mating when 5 to 6 years

old, but usually do not become active breeders until several more years have passed. According to Marvin L. Jones (Zoological Society of San Diego, pers. comm.), a captive sea otter was estimated to have been 19 years old at time of death.

The fur of the sea otter may be the most valuable of any mammal. During the 1880s, prices on the London market ranged from $105 to $165 each. By 1903, when the species had become scarce, large, high-quality skins sold for up to $1,125. Pelts taken in Alaska in the late 1960s, during a brief reopening of commercial activity, sold for an average of $280 each (Kenyon 1969).

Intensive exploitation of the sea otter was begun by the Russians in 1741. Hunting was uncontrolled until 1799, when some conservation measures were established. Unregulated killing resumed in 1867, when Alaska was purchased by the United States. By 1911, when the sea otter was protected by a treaty among the United States, Russia, Japan, and Great Britain, probably only 1,000 to 2,000 of the animals survived world-wide (Kenyon 1969). Under protection of the treaty, state and national laws, and finally the United States Marine Mammal Protection Act of 1972, the sea otter has steadily increased in numbers and distribution. There are now probably at least 100,000 individuals in the major populations from the Kuril Islands to south-central Alaska. Reintroduced populations have apparently been established off southeastern Alaska (now numbering 500 to 1,000 animals) and Vancouver Island (about 100 animals). Reintroduced groups off Washington and Oregon, and in the Pribilof Islands, do not seem to have done well (Estes 1980).

The southern sea otter (subspecies *E. l. nereis*), which originally ranged from Baja California to at least Washington, and perhaps to south-central Alaska, was generally considered extinct by 1920. Apparently, however, a group of 50 to 100 individuals survived off central California in the vicinity of Monterey. In 1938 this population became generally known to be present, and subsequently it grew to include about 1,800 animals. The sea otter now regularly occurs along about 300 km of the central California coast, and there have been scattered reports of individuals from southern California and northern Baja California. As this population increased, there was concern that stocks of abalone and other shellfish were being depleted. There is now controversy regarding whether the sea otter population should be completely protected or subject to control. There is also fear that an oil spill, associated with either the extensive tanker traffic in the area, or offshore drilling, could devastate the population (U.S. Fish and Wildlife Service 1980; Armstrong 1979; Leatherwood, Harrington-Coulombe, and Hubbs 1978). The southern sea otter is listed as threatened by the USDI (1980) and is on appendix 1 of the CITES.

CARNIVORA; *Family VIVERRIDAE*

Civets, Genets, Linsangs, Mongooses, Fossas

This family of 36 Recent genera and 70 species is found in southwestern Europe, southern Asia, the East Indies, Africa, and Madagascar. Certain genera have been introduced to areas in which the family does not naturally occur. The following sequence of genera, grouped in seven subfamilies, is based on the classifications of Ewer (1973), Meester and Setzer (1971), and Ellerman and Morrison-Scott (1966):

Subfamily Viverrinae
(civets, genets, linsangs)

Viverra	Genetta	Poiana
Civettictis	Osbornictis	Prionodon
Viverricula		

Subfamily Paradoxurinae
(palm civets)

Nandinia	Paradoxurus	Macrogalidia
Arctogalidia	Paguma	Arctictis

Subfamily Hemigalinae
(palm and otter civets)

Hemigalus	Chrotogale	Cynogale

Subfamily Fossinae
(Malagasy civets)

Fossa	Eupleres

Subfamily Galidiinae
(Malagasy mongooses)

Galidia	Mungotictis	Salanoia
Galidictis		

Subfamily Herpestinae
(mongooses)

Herpestes	Dologale	Atilax
Mungos	Bdeogale	Cynictis
Crossarchus	Rhynchogale	Paracynictis
Liberiictis	Ichneumia	Suricata
Helogale		

Subfamily Cryptoproctinae
(fossa)

Cryptoprocta.

These are the characteristic small and medium-sized carnivores of the Old World. Head and body length is 170 to about 1,000 mm, tail length is 120 to 900 mm, and adult weight is 0.45 to approximately 14 kg. *Helogale* is the smallest genus and *Arctictis* is the largest. Viverrids have a variety of striped, spotted, and uniform color patterns. In some genera the tail is banded or ringed. The body is long and sinewy, with short legs and generally a long, bushy tail. One genus (*Arctictis*) has a truly prehensile tail. The head is elongate and the muzzle is pointed. Most genera have five toes on each foot, *Cynictis* has only four digits on the hind foot, and *Suricata* has only four digits on all feet (Stains 1967). The claws are semiretractile in some genera. Female viverrids usually have two or three pairs of abdominal mammae. Males have a baculum.

Most viverrids have scent glands in the anal region, which secrete a nauseous-smelling fluid as a defensive measure. In mongooses, and some other members of the family, these glands open into a pouch or saclike depression, outside the anus proper, in which the secretion is stored. The conspicuous pattern of pelage in some genera has been interpreted as being a warning that the fetid anal gland secretion is present. Such a color pattern is also found in skunks and certain other members of the family Mustelidae. The secretion of the scent glands, when rubbed on various objects, is recognized by other individuals of the same species, and is probably used to communicate various information.

The skull is usually long and flattened. The dental formula is: (i 3/3, c 1/1, pm 3–4/3–4, m 1–2/1–2) × 2 = 32 to 40. The second lower incisor is raised above the level of the first and third, the canines are elongate, and the carnassials are developed.

Viverrids are essentially forest inhabitants, but they also live in dense brush and thick grass. They are either diurnal or nocturnal, and shelter in any convenient retreat, usually a hole in a tree, a tangle of vines, ground cover, a cave, a crevice, or a burrow. A few species dig their own burrows. Those species living near people sometimes seek refuge under rafters or in the drains of houses.

Those viverrids that walk on their digits (such as *Genetta*) have a gait described as "a waltzing trot," whereas the members of the family that walk on the sole, with the heel touching the ground (such as *Arctictis*), have a bearlike shuffle. Many genera are agile and extremely graceful in their movements. A number of species are skillful climbers; some apparently spend most of their lives in trees. Some genera take to water readily and swim well; three genera, *Osbornictis, Atilax,* and *Cynogale,* are semiaquatic. Sight, hearing, and smell are acute.

Viverrids may fight when cornered. They seek their prey in trees and on the ground, either by stalking it or by pouncing upon it from a hiding place. They eat small vertebrates and various invertebrates, and occasionally consume vegetable matter, such as fruit, bulbs, and nuts. Carrion is taken by some species.

Viverrids are solitary, or live in pairs or groups. Several genera, including *Cynictis* and *Suricata,* live in colonies in ground burrows. Some genera of mongoose associate in bands, and take refuge as a group in any convenient shelter. Breeding may occur seasonally or throughout the year. A number of genera have two litters annually. The one to six offspring are born blind, but haired. Most species probably have a potential longevity of 5–15 years.

The secretion of the scent glands, known as civet, is obtained from several genera (*Civettictis, Viverra,* and *Viverricula*) for both perfumery and medicinal purposes. Some viverrids are tamed and kept to extract the musk. They may also be kept as pets. Viverrids occasionally kill poultry, but also prey on rodents. Mongooses, particularly of the genus *Herpestes,* have been introduced into several areas to check

Oriental civet (*Viverra tangalunga*), photo by Ernest P. Walker.

the numbers of rodents and venomous snakes. Such introductions, however, generally have not proven beneficial, as the mongooses quickly multiply and destroy many desirable forms of mammals and birds.

The geological range of this family is late Eocene to Recent in Europe, early Pliocene to Recent in Asia, Pleistocene to Recent in Africa, and Recent in Madagascar (Stains 1967).

CARNIVORA; VIVERRIDAE; **Genus VIVERRA** Linnaeus, *1758*

Oriental Civets

There are two subgenera and three species (Ellerman and Morrison-Scott 1966; Chasen 1940; Medway 1978; Taylor 1934):

subgenus *Viverra*

V. zibetha, Nepal and eastern India to southeastern China and Malay Peninsula;

V. tangalunga, Malay Peninsula, Sumatra, Bangka, Borneo, Rhio Archipelago, Philippines;

subgenus *Moschothera*

V. megaspila, peninsular India, Burma to Indochina and Malay Peninsula.

Civettictis is sometimes regarded as a third subgenus of *Viverra* (see Meester and Setzer 1971).

Head and body length is 585 to 950 mm, tail length is 300 to 482 mm, and weight is 5 to 11 kg. The fur, especially in winter, is long and loose. It is usually elongated in the median line of the body, forming a low crest or mane. The color pattern of the body is composed of black spots on a grayish or tawny ground color. The sides of the neck and throat are marked with black and white stripes—usually three black and two white collars. The crest is marked by a black spinal stripe, which runs from the shoulders to the tail, and the tail is banded or ringed with black and white. The feet are black. In *V. zibetha* and *V. tangalunga*, the third and fourth digits of the forefeet are provided with lobes of skin which act as protective sheaths for the retractile claws. *Viverra* is distinguished from *Viverricula* by larger size, by the presence of a dorsal crest of erectile hairs, and by the insertion of the ears, the inner edges of which are set farther apart on the forehead.

Oriental civets occur in a wide variety of habitats in forest, brush, and grassland. They stay in dense cover by day and come out into the open at night. They are mainly terrestrial and often live in holes in the ground dug by other animals. They apparently can climb readily, but seldom do so. Like *Viverricula*, they are often found near villages and are common over most of their range. Like most civets, they are easily trapped. They are vigorous hunters, killing small mammals, birds, snakes, frogs, and insects, and taking eggs, fruit, and some roots. The species *V. zibetha* has been observed fishing in India, and the remains of crabs have been found in the stomachs of two individuals from China.

Viverra is generally solitary. *V. zibetha* is said to breed all year and to bear two litters annually (Lekagul and McNeely 1977). The number of young per litter is one to four, usually two or three. The young are born in holes in the ground or in dense vegetation. The young of *V. zibetha* open their eyes after 10 days, and weaning begins at the age of 1 month (Medway 1978). A captive specimen of *V. zibetha* lived for 15 years and 4 months (Marvin L. Jones, Zoological Society of San Diego, pers. comm.).

Viverra is one of the sources of civet, a substance used commercially in producing perfume. Because of this function, *V. tangalunga* has been introduced to several islands of the East Indies, apparently including Celebes (Groves 1976; Laurie and Hill 1954). Civet is also obtained from the genera *Viverricula* and *Civettictis*. The subspecies *V. megaspila civettina* of peninsular India is classified as endangered by the IUCN (1972) and the USDI (1980). It has evidently become very rare through persecution by people and loss of habitat to agriculture.

African civet (*Civettictis civetta*), photo from New York Zoological Society.

CARNIVORA; VIVERRIDAE; **Genus CIVETTICTIS**
Pocock, 1915

African Civet

The single species, *C. civetta*, is found from Senegal to Somalia, and south to Namibia and eastern South Africa. Recognition of *Civettictis* as a distinct genus is in keeping with Ewer (1973), Rosevear (1974), and Kingdon (1977). Some other authorities, such as Meester and Setzer (1971) and Rowe-Rowe (1978b) have included *C. civetta* in the genus *Viverra*.

Head and body length is 680 to 890 mm, tail length is 445 to 463 mm, and weight is 7 to 20 kg (Kingdon 1977). The color is black with white or yellowish spots, stripes, and bands. The hair is long and coarse, and is thick on the tail. The perineal glands under the tail contain the oily scented matter used commercially in making perfume. All the feet have five claws, and the soles are hairy. From *Viverra*, *Civettictis* is distinguished by much larger molar teeth and a far broader lower carnassial (Rosevear 1974).

The African civet is widely distributed in both forests and savannahs, wherever long grass or thickets are sufficient to provide daytime cover (Ewer and Wemmer 1974). It seems to use a permanent burrow or nest only to bear young. It is nocturnal and almost completley terrestrial, but takes to water readily and swims well. The omnivorous diet includes carrion, rodents, birds, eggs, reptiles, frogs, crabs, insects, fruit, and other vegetation. Poultry and young lambs are sometimes taken (Rosevear 1974).

Civettictis is generally solitary, but has a variety of visual, olfactory, and auditory means of communication. Individuals may have defined and well-marked territories. The scent glands have a major social role, leaving scent along a path to convey information, such as whether a female is in estrus (Kingdon 1977). There are three agonistic vocalizations—the "growl," "cough-spit," and "scream"—but the most commonly heard sound is the "ha-ha-ha" used in making contact (Ewer and Wemmer 1974).

Available data on reproduction (Ewer and Wemmer 1974; Rosevear 1974; Kingdon 1977; Mallinson 1974) suggest that breeding occurs throughout the year. There may be two or even three litters annually. The gestation period is usually 60 to 72 days, but is occasionally extended to as much as 81 days, perhaps because of delayed implantation. The number of young per litter is one to four, usually two or three. The young are born fully furred, open their eyes within a few days, cease suckling when 14 to 20 weeks old, and attain sexual maturity at about 1 year of age. According to Marvin L. Jones (Zoological Society of San Diego, pers. comm.), a captive lived for 28 years.

In Ethiopia, and to a lesser extent in other parts of Africa, the natives keep civets in captivity and remove the musk from them several times a week. An average animal yields 3 or 4 grams weekly. The natives do not raise the civets, however, but merely capture wild ones. In 1934, Africa produced about 2,475 kg of musk with a value of U.S. $200,000. In that same year the United States imported 200 kg of musk. The production of civet musk is an old industry; King Solomon's supply came from East Africa. Rosevear (1974) reported that the trade in civet musk has now diminished considerably.

CARNIVORA; VIVERRIDAE; **Genus VIVERRICULA**
Hodgson, 1838

Lesser Oriental Civet, Rasse

The single species, *V. indica*, occurs from Pakistan to southeastern China and the Malay Peninsula, and on Sri Lanka, Taiwan, Hainan, Sumatra, Java, and Bali (Ellerman and Morrison-Scott 1966; Roberts 1977). The name *V. malaccensis* was used for this species by Medway (1978) and Lekagul and McNeely (1977).

Head and body length is 450 to 630 mm, tail length is 300 to 430 mm, and weight is usually 2 to 4 kg. The fur is harsh, rather coarse, and loose. The body color is buffy, brownish, or grayish, and the feet are black. Small spots are present on the forequarters, and larger spots, tending to run into longitudinal lines, are present on the flanks. There are six to eight dark stripes on the back, and the tail is ringed black and white by six to nine rings of each color.

Lesser oriental civet (*Viverricula indica*), photo by Ernest P. Walker.

Viverricula is distinguished from *Viverra* by smaller size, by the absence of a dorsal crest of erectile hairs, and by the insertion of the ears, the inner edges of which are set closer together on the forehead than those of *Viverra*. The muzzle is also shorter and more pointed. Internally, the two genera differ in a number of cranial and dental features.

The rasse inhabits grasslands or forests. It probably excavates its own burrow (Roberts 1977), but may also shelter in thick clumps of vegetation, buildings, or drains (Lekagul and McNeely 1977). It is generally nocturnal, but may be seen hunting by day in areas not populated by humans. It is mainly terrestrial, but is said to climb well. It usually tries to escape from dogs by dodging and twisting through the underbrush. The diet consists of small vertebrates, carrion, insects and their grubs, fruit, and roots.

Viverricula is usually solitary, but occasionally associates in pairs. It breeds throughout the year in Sri Lanka. The two to five young are born in a shelter on the ground. A captive lived for 10 years and 6 months (Marvin L. Jones, Zoological Society of San Diego, pers. comm.).

The rasse is kept in captivity by natives for the purpose of extracting the civet that is secreted and retained in sacs close to the genitals in both sexes. The removal of this secretion is accomplished by scraping the inside of the sac with a spoon-like implement. In India this secretion is used as a perfume to flavor the tobacco that is smoked by some natives. *Viver-*

ricula was introduced by people to Socotra, the Comoro Islands, the Philippines, and Madagascar, probably for the production of civet. Its presence on the island of Sumbawa, to the east of Bali, is thought to have resulted from introduction (Laurie and Hill 1954).

CARNIVORA; VIVERRIDAE; **Genus GENETTA** Oken, *1816*

Genets

There are 3 subgenera and 10 species (Meester and Setzer 1971; Schlawe 1980; Crawford-Cabral 1981; Ansell 1978):

subgenus *Pseudogenetta*

G. thierryi, savannah zone from Senegal to area south of Lake Chad;

G. abyssinica, highlands of Ethiopia;

subgenus *Paragenetta*

G. johnstoni, Liberia, Ghana;

subgenus *Genetta*

G. servalina, southern Nigeria to western Kenya;

Genets (*Genetta tigrina*): Top, photo by John Markham; Bottom, photo by John Visser.

G. *victoriae*, northern and eastern Zaire, Uganda;

G. *genetta*, France, Spain, Portugal, Balearic Islands, Palestine, southwestern Arabian Peninsula, northwestern Africa, savannah zones of Africa south of the Sahara;

G. *angolensis*, southern Zaire, central and northeastern Angola, western Zambia, northern Mozambique, probably southern Tanzania;

G. *pardina*, Gambia to Cameroon;

G. *tigrina*, South Africa, Lesotho;

G. *rubiginosa*, Senegal to Somalia, and south to Namibia and eastern South Africa.

Corbet (1978) suggested that the European populations of G. *genetta* are the result of introduction by human agency. Schlawe (1980) restricted the name G. *genetta* to Europe and northwestern Africa. He referred the populations to the south of the Sahara and in the southwestern Arabian Peninsula to a separate species, G. *felina*, and he questioned the occurrence of *Genetta* in Palestine.

Head and body length is usually 420 to 580 mm, tail length is 390 to 530 mm, and weight is 1 to 3 kg. Coloration is variable, but the body is generally grayish or yellowish, with brown or black spots and blotches on the sides, tending to be arranged in rows. A row of black erectile hairs is usually present along the middle of the back. The tail has black and white rings. Melanistic individuals seem to be fairly common. Genets have a long body, short legs, a pointed snout, prominent and rounded ears, short and curved retractile claws, and soft and dense hair. They have the ability to emit a musky-smelling fluid from their anal glands. Females have two pairs of abdominal mammae.

Genets inhabit forests, savannahs, and grasslands. They are active at night, usually spending the day in a rock crevice, in a burrow excavated by some other animal, in a hollow tree, or on a large branch. They seem to return daily to the same shelter. They climb trees to prey on nesting and roosting birds, but much of their food is taken on the ground. They are silent and stealthy hunters; when stalking prey, they crouch until the body and tail seem to glide along the ground. At the same time, the body seems to lengthen. Genets can go through any opening the head can enter, because of the slender and loosely jointed body. The diet consists of any small animals that can be captured, including rodents, birds, reptiles, and insects. Genets sometimes take game birds and poultry.

Genets travel alone or in pairs. They communicate with one another by a variety of vocal, olfactory, and visual signals. Breeding seems to correspond with the wet seasons in both West and East Africa. In Kenya, for example, pregnant and lactating females have been taken in May and from September to December. A pair of G. *genetta* in the National Zoo in Washington, D.C., regularly produced two litters per

Aquatic genet (*Osbornictis piscivora*), photo from *Bull. Amer. Mus. Nat. Hist.*

year, one in April–May and another in July–August. Gestation periods ranging from 56 to 77 days have been reported. The number of young per litter is one to four, usually two or three. The young weigh 61 to 82 grams each at birth, begin to take solid food at 2 months of age, and attain adult weight when 2 years old. One female *G. genetta* became sexually mature at about 4 years, and produced young regularly until she died at the age of 13 years (Rosevear 1974; Wemmer 1977; Kingdon 1977).

CARNIVORA; VIVERRIDAE; **Genus OSBORNICTIS**
J. A. Allen, 1919

Aquatic Genet

The single species, *O. piscivora,* is known only by 16 specimens taken in northeastern Zaire (Hart and Timm 1978; Meester and Setzer 1971).

An adult male had a head and body length of 445 mm, a tail length of 340 mm, and a weight of 1,430 grams; an adult female weighed 1,500 grams (Hart and Timm 1978). The body is chestnut red to dull red, and the tail is black. There is a pair of elongated white spots between the eyes. The front and sides of the muzzle and the sides of the head below the eyes are whitish. Black spots and bands are absent, and the tail is not ringed. The pelage is long and dense, especially on the tail. The palms and soles are bare, not furred as in *Genetta* and other related genera. The skull is long and lightly built, and the teeth are relatively small and weak.

Osbornictis is among the rarest genera of carnivore. All specimens probably originated in areas of dense forest at elevations of 500 to 1,500 meters (Hart and Timm 1978). The genus is generally thought to be semiaquatic, as several specimens have been taken in or near streams, and available evidence suggests that fish constitute a major part of the diet. Hart and Timm, for example, noted the following: the stomach of one specimen contained the remains of fish; natives of the area indicated that fish is the favored prey; the dentition of *Osbornictis* seems adapted to deal with slippery vertebrate prey, such as fish and frogs; and the bare palms may be an adaptation allowing the genet to feel for fish in muddy holes and then handle the prey. *Osbornictis* is apparently solitary. A pregnant female with a single embryo, 15 mm in length, was taken on 31 December.

CARNIVORA; VIVERRIDAE; **Genus POIANA** *Gray, 1864*

African Linsang, Oyan

The single species, *P. richardsoni,* occurs from Sierra Leone to northern Zaire, and on the island of Fernando Poo (Meester and Setzer 1971). The generic name reflects the occurrence on this island. Rosevear (1974) recognized the populations in the western part of the range of *Poiana* as a distinct species, *P. leightoni.*

The average head and body length is 384 mm, and the average tail length is 365 mm (Rosevear 1974). The general color effect is light brownish gray to rusty yellow; dark brown to black spots and rings are present. Some individuals have alternating broad and narrow black bands on the tail, whereas others have only the broad bands. This genus differs from the Asiatic linsangs (*Prionodon*) in that the spots are smaller and show no tendency to run into bands or stripes, except in the region of the head and shoulder. It also differs from them, and resembles *Genetta,* in having a narrow bare line on the sole of each hind foot.

African linsang (*Poiana richardsoni*), photo from *Proc. Zool. Soc. London.*

The oyan is a forest animal and is nocturnal. According to Dr. Hans-Jürg Kuhn (Anatomisches Institut der Universität Frankfurt am Mein, pers. comm.), *Poiana* builds a round nest of green material, in which several individuals sleep for a few days, and then moves on and builds a new nest. The nests are at least two meters above the ground, usually higher. Although it has been reported that the oyan sleeps in the abandoned nests of squirrels, reliable hunters say that the reverse is true—the squirrels sleep in abandoned nests of *Poiana*. The diet includes cola nuts, other plant material, insects, and young birds. In the Liberian hinterland, natives make medicine bags from the skins of *Poiana*.

Observations by Charles-Dominique (1978) in northeastern Gabon indicate a population density of one oyan per sq km. A lactating female has been noted in October. As in some other genera of viverrid, there may be two litters per year. The number of young per birth is two or three. A captive oyan lived for five years and four months (Marvin L. Jones, Zoological Society of San Diego, pers. comm.).

CARNIVORA; VIVERRIDAE; **Genus PRIONODON**
Horsfield, 1822

Oriental Linsangs

There are two subgenera and two species (Ellerman and Morrison-Scott 1966; Lekagul and McNeely 1977; Chasen 1940):

subgenus *Prionodon*

P. linsang (banded linsang), western and southern Thailand, Malay Peninsula, Sumatra, Bangka, Java, Borneo;

subgenus *Pardictis*

P. pardicolor (spotted linsang), Nepal to Indochina.

Head and body length is 350 to 450 mm, and tail length is 304 to 420 mm. Medway (1978) listed the weight of *P. linsang* as 598 to 798 grams. On the average, *P. pardicolor* is slightly smaller. In *P. linsang* the ground color varies from whitish gray to brownish gray and becomes creamy on the underparts. The dark pattern consists of four or five broad, transverse black or dark brown bands across the back; there is one large stripe on each side of the neck. The sides of the body and legs are marked with dark spots, and the tail is banded. Some individuals of *P. pardicolor* have a ground color of orange buff, whereas others are pale brown. Black spots on the upper parts are more or less arranged in longitudinal rows, and the tail has 8 to 10 dark rings.

These animals are extremely slender, graceful, and beautiful. The fur is short, dense, and soft; it has the appearance of and feels like velvet. The claws are retractile; claw sheaths are present on the forepaws, and protective lobes of skin are present on the hind paws. The skull is long, low, and narrow, and the muzzle is narrow and elongate. Unlike many viverrids, *Prionodon* seems to be free from odor.

Oriental linsangs dwell mainly in forests. They are nocturnal and generally arboreal, but frequently come to the ground in search of food (Lekagul and McNeely 1977). *P. linsang*

Banded linsang (*Prionodon linsang*), photo by Ernest P. Walker.

constructs a nest of sticks and leaves; in one case a nest was located in a burrow at the base of a palm. This species is also said to live in tree hollows. The diet includes small mammals, birds, eggs, and insects.

The limited data on reproduction suggest that *P. linsang* has no clear breeding season (Lekagul and McNeely 1977). Two pregnant females, one with two embryos and the other with three, were collected in May, and two lactating females were found in April and October. *P. pardicolor* is said to breed in February and August, and to have litters of two young. A captive *P. linsang* lived for 10 years and 8 months (Marvin L. Jones, Zoological Society of San Diego, pers. comm.).

The species *P. pardicolor* is listed as endangered by the USDI (1980) and is on appendix 1 of the CITES. *P. linsang* is on appendix 2 of the CITES.

CARNIVORA; VIVERRIDAE; Genus NANDINIA Gray, 1843

African Palm Civet

The single species, *N. binotata*, occurs from Guinea-Bissau to southern Sudan, and south to northern Angola and eastern Rhodesia (Meester and Setzer 1971).

Head and body length is 440 to 580 mm, and tail length is 460 to 620 mm. Kingdon (1977) listed weight as 1.7 to 2.1 kg, but Charles-Dominique (1978) reported that males weighed as much as 5 kg. Coloration is quite variable, but is usually grayish or brownish, tinged with buffy or chestnut. Two creamy spots are often present between the shoulders, and obscure dark brown spots are present on the lower back and top of the tail. The tail is somewhat darker than the body; it is the same color above and below, and has a variable pattern of black rings. The throat tends to be grayish, and the underparts are grayish, tinged with yellow. The pelage is short and woolly, but coarse tipped. The ears are short and rounded, the tail is fairly thick, the legs are short, and the claws are sharp and curved. There are scent glands on the palms, between the toes, on the lower abdomen, and possibly on the chin (Kingdon 1977).

In a radio-tracking study in Gabon, Charles-Dominique (1978) found *Nandinia* to be largely arboreal and to occur mainly 10 to 30 meters above the ground in various types of forest. It was nocturnal, and slept by day in a fork, on a large branch, or in a bundle of lianas. Stomach contents consisted of 80 percent fruit, on the average, but also included remains of rodents, bird eggs, large beetles, and caterpillars.

Charles-Dominique found a population density of five palm civets per sq km in his study area. Adult females established territories averaging 45 ha. They allowed immature females on these areas, but did not tolerate trespassing by other adult females. Large, dominant adult males had territories averaging about 100 ha., which overlapped a number of female territories. The large males drove away other animals of the same size and sex, but allowed smaller adult males to remain; however, the small adult males were not permitted access to the females. Territories were marked with scent. Fighting was severe, sometimes resulting in death. Loud calls were exchanged during courtship.

In West Africa, breeding apparently can occur during the wet or dry season (Rosevear 1974). Records from East Africa suggest that there are two birth peaks or seasons, May and October. The gestation period is 64 days. The number of young is usually two, but up to four (Kingdon 1977). As soon as they are weaned, young males leave the territories of their mothers. Sexual maturity comes in the third year of life (Charles-Dominique 1978). One individual was still alive after 15 years and 10 months in captivity (Marvin L. Jones, Zoological Society of San Diego, pers. comm.).

The African palm civet is easily tamed and will drink milk in captivity. It is said to be quite clean and to keep houses free of rats, mice, and cockroaches.

CARNIVORA; VIVERRIDAE; Genus ARCTOGALIDIA Merriam, 1897

Small-toothed or Three-striped Palm Civet

The single species, *A. trivirgata,* is found from Assam to Indochina and the Malay Peninsula, and on Sumatra, Bangka, Java, Borneo, and numerous small nearby islands of

African palm civet (*Nandinia binotata*), photo by Ernest P. Walker.

the East Indies (Ellerman and Morrison-Scott 1966; Chasen 1940).

Head and body length is 432 to 532 mm, tail length is 510 to 660 mm, and weight is usually 2 to 2.5 kg. The upper parts, proximal part of the tail, and outside of the limbs are usually tawny, varying from dusky grayish tawny to bright orangish tawny. The head is usually darker and grayer, and the paws and distal part of the tail are brownish. There is a median white stripe on the muzzle, and three brown or black longitudinal stripes on the back. The median stripe is usually complete and distinct, whereas the laterals may be broken up into spots or may be almost absent. The undersides are grayish white or creamy buff, with a whitish patch on the chest.

Only the females of this genus possess the civet gland, which is located near the opening of the urinogenital tract. *Arctogalidia* closely resembles *Paradoxurus* in external form, and in the length of the legs and tail, but differs externally in characters of the feet. Internally, the skull differs from that of *Paradoxurus,* and the back teeth are smaller.

Arctogalidia inhabits dense forests. In some areas it frequents coconut plantations, though Lekagul and McNeely (1977) reported that it avoids human settlements. It is nocturnal, resting by day in the upper branches of tall trees (Medway 1978). It is arboreal, climbing actively and leaping from branch to branch with considerable agility. The omnivorous diet includes squirrels, birds, frogs, insects, and fruit.

Three animals, representing both sexes, occupied an emp-

Small-toothed palm civet (*Arctogalidia trivirgata*), photo by Lim Boo Liat.

Palm civet (*Paradoxurus hermaphroditus*), photo by Ernest P. Walker.

CARNIVORA; VIVERRIDAE; **Genus PARADOXURUS**
F. Cuvier, 1821

Palm Civets, Musangs, Toddy Cats

There are three species (Ellerman and Morrison-Scott 1966; Chasen 1940; Laurie and Hill 1954):

P. hermaphroditus, Kashmir and peninsular India to southeastern China and Malay Peninsula, Sri Lanka, Hainan, Sumatra, Java, Borneo, Philippines, Celebes, many small islands of the East Indies;
P. zeylonensis, Sri Lanka;
P. jerdoni, southern India.

Taylor (1934) referred the populations of *Paradoxurus* in the Philippines to three species—*P. philippinensis, P. torvus,* and *P. minax*—but Lekagul and McNeely (1977) indicated that the only species in the Philippines is *P. hermaphroditus.*

Head and body length is 432 to 710 mm, tail length is 406 to 660 mm, and weight is 1.5 to 4.5 kg. The ground color is grayish to brownish, but is often almost entirely masked by the black tips of the guard hairs. There is a definite pattern of dorsal stripes and lateral spots, at least in the new coat, but this is sometimes concealed by the long black hairs. The pattern is most plainly shown in the species *P. hermaphroditus,* where it consists of longitudinal stripes on the back, and spots on the shoulders, sides, and thighs, and sometimes on the base of the tail. A pattern may also be present on the head of this species; it consists of white patches and a white band across the forehead.

The species *P. hermaphroditus* can always be distinguished from *P. zeylonensis* and *P. jerdoni* by the backward direction of the hairs on the neck. In the other species the hairs on the neck grow forward from the shoulders to the head.

Paradoxurus differs from *Arctogalidia* and *Paguma* in color pattern, and in characters of the skull and teeth. Ac-

ty nest of *Ratufa bicolor,* about 20 meters above the ground in a tree. Mewing calls and light snarls, accompanied by playful leaps and chases, have been noted in a male and female at night in the wild. The young are reared in hollow trees. According to Lekagul and McNeely (1977), breeding probably continues throughout the year, there may be two litters annually, and litter size is two or three young. Batten and Batten (1966) reported that a female was captured in Borneo in August 1961 at an age of about 2 weeks. It entered estrus for the first time in December 1962 and then again at intervals of 6 months. In August 1964 it gave birth to its first litter (three young) after a gestation period of approximately 45 days. The young opened their eyes at 11 days of age and were suckled for over 2 months. The father was reintroduced to the family when the young were 2½ months old, and was soon accepted by the others. According to Marvin L. Jones (Zoological Society of San Diego, pers. comm.), one individual was still living after 15 years and 10 months in captivity.

cording to Lekagul and McNeely (1977), the teeth of *Paradoxurus* are less specialized for meat eating than those of most viverrids, having low, rounded cusps on rather square molars. There are well-developed anal scent glands in both sexes. Females have three pairs of mammae.

Musangs are nocturnal forest dwellers. They are expert climbers and spend most of their time in trees, where they utilize cavities or secluded nooks. They are often found about human habitations, probably because of the presence of rats and mice. Under such conditions, they shelter in thatched roofs and in dry drain tiles and pipes. They eat small vertebrates, insects, fruits, and seeds. They are fond of the palm juice or "toddy" collected by the natives—thus one of the vernacular names.

Reproduction occurs throughout the year, though Lekagul and McNeely (1977) stated that the young of *P. hermaphroditus* seem to be seen most often from October to December. Litter size is two to four young. Sexual maturity is attained at the age of 11 to 12 months in *P. hermaphroditus*. A captive of this species lived for 22 years and 5 months (Medway 1978).

Groves (1976) noted that *P. hermaphroditus* has been carried about from island to island by people, for use as a rat catcher. Such activity is probably responsible for the presence of this species on Celebes, Timor, and various other islands.

CARNIVORA; VIVERRIDAE; *Genus* **PAGUMA** *Gray, 1831*

Masked Palm Civet

The single species, *P. larvata,* occurs from Kashmir to Indochina and the Malay Peninsula, in much of eastern and southern China, and on the Andaman Islands, Taiwan, Hainan, Sumatra, and Borneo (Ellerman and Morrison-Scott 1966; Chasen 1940).

Head and body length is 508 to 762 mm, tail length is usually 508 to 636 mm, and weight is usually 3.6 to 5 kg. In the facial region there is generally a mask, which consists of a median white stripe from the top of the head to the nose, a white mark below each eye, and a white mark above each eye extending to the base of the ear and below. The general color is gray, gray tinged with buff, orange, or yellowish red. There are no stripes or spots on the body, and no spots or bands on the tail. The distal part of the tail may be darker than the basal part, and the feet are blackish. This genus differs externally from *Paradoxurus* and *Arctogalidia* in the absence of the striping and spotting. Like *Paradoxurus*, *Paguma* has a potent anal gland secretion, which it uses to ward off predators. The conspicuously marked head has been interpreted as a warning signal of the presence of the secretion. Females have two pairs of mammae (Lekagul and McNeely 1977).

The masked palm civet frequents forests and brush country. It is reported to raise its young in tree holes, at least in Nepal. It is arboreal and nocturnal (Roberts 1977). The omnivorous diet includes small vertebrates, insects, and fruit.

Paguma is solitary, and apparently most young in the western parts of its range are born in spring and early summer (Roberts 1977). Births in Borneo have taken place in October (Banks 1978). Data on captives indicate that there may be two breeding seasons, in early spring and late autumn. Litters contain one to four young. They open their eyes at 9 days of age and are almost the size of adults by 3 months. Maximum known longevity is 15 years and 5 months (Medway 1978; Lekagul and McNeely 1977).

In Tenasserim, *Paguma* is reported to be a great ratter and not to destroy poultry. Medway (1978), however, wrote that it has been known to raid hen runs. The genus has been introduced on central Honshu, Japan (Corbet 1978).

Masked palm civet (*Paguma larvata*), photo from New York Zoological Society.

Celebes palm civet (*Macrogalidia musschenbroeki*), photo by Christen Wemmer.

CARNIVORA; VIVERRIDAE; **Genus *MACROGALIDIA***
Schwarz, 1910

Celebes Palm Civet

The single species, *M. musschenbroeki*, occurs only on Celebes (Laurie and Hill 1954).

Head and body length is about 1,000 mm and tail length is about 600 mm. The upper parts are light brownish chestnut to dark brown. The underparts range from fulvous to whitish, with a reddish breast. The cheeks and a patch above the eye are usually buffy or grayish. Faint brown spots and bands are usually present on the sides and lower back, and the tail is ringed with dark and pale brown. The tail has more bands than that of *Arctogalidia* or *Paradoxurus*. Other distinguishing characters are the short, close fur, and a whorl in the neck with the hairs directed forward.

Macrogalidia has been reported from both montane and lowland forests, and in scrubby grassland. It is known to climb trees and to feed on rodents and fruit. Several individuals have been killed recently while raiding chicken roosts. Although once thought to be extinct or restricted to the northern peninsula of Celebes, and currently classified as rare by the IUCN (1978), recent observations indicate that *Macrogalidia* occurs in most parts of the island and is relatively common in some places (World Wildlife Fund, *Conservation Indonesia* 2, no. 3 [1978]:2; 4, nos. 3–4 [1980]:12).

CARNIVORA; VIVERRIDAE; **Genus *ARCTICTIS***
Temminck, 1824

Binturong

The single species, *A. binturong*, occurs from Burma, and possibly from Nepal, to Indochina and the Malay Peninsula, and on Sumatra, Bangka, Java, the Rhio Archipelago, Borneo, and Palawan (Ellerman and Morrison-Scott 1966; Chasen 1940).

Head and body length is 610 to 965 mm, tail length is 560 to 890 mm, and weight is usually 9 to 14 kg. The fur is long and coarse, that on the tail being longer than that on the body. The hairs are black and lustrous, often with gray, fulvous, or buff tips. The head is finely speckled with gray and buff, and the edges of the ears and the whiskers are white. The ears have long hairs on the back that project beyond the tips and produce a fringed or tufted effect. The tail is particularly muscular at the base and is prehensile at the tip. The only other carnivore with a truly prehensile tail is the kinkajou (*Potos*), which the binturong resembles in habits to some extent. Females have two pairs of mammae.

The binturong lives in dense forests and is nowhere abundant. It is mainly arboreal and nocturnal. When resting, it usually lies curled up with the head tucked under the tail. It has never been observed to leap; rather, it progresses slowly but skillfully, using the tail as an extra hand. Its movements, at least during daylight hours, are rather slow and cautious,

Binturong (*Arctictis binturong*), photo from New York Zoological Society.

the tail slowly uncoiling from the last support as the animal moves carefully forward. According to Medway (1978), the binturong is reported to dive, swim, and catch fish. The diet also includes birds, carrion, fruit, leaves, and shoots.

Medway (1978) stated that *Arctictis* occurs either alone or in small groups of adults with immature offspring. Captives are very vocal, uttering high-pitched whines and howls, rasping growls, and, when excited, a variety of grunts and hisses. Lekagul and McNeely (1977) wrote that breeding seems to occur throughout the year. Numerous observations in captivity (Bulir 1972; Gensch 1963b; Grzimek 1975; Kuschinski 1974; Xanten, Kafka, and Olds 1976) indicate that females are nonseasonally polyestrous and may give birth to two litters annually. Wemmer and Murtaugh (1981) reported the following data for reproduction in captivity: breeding occurs year-round, but with a pronounced birth peak from January to March; the estrous cycle averages 81.8 days; the mean gestation period is 91.1 days, and the range is 84 to 99 days; the number of young per litter averages 1.98 and ranges from 1 to 6; the young weigh an average of 319 grams each at birth and begin to take solid food when 6 to 8 weeks old; the mean age of first mating is 30.4 months in females and 27.7 months in males; and both sexes can remain fertile until at least 15 years old. According to Marvin L. Jones (Zoological Society of San Diego, pers. comm.), one binturong was still living after 22 years and 8 months in captivity.

Arctictis is sometimes kept as a pet. It is said to be easily domesticated, to become quite affectionate, and to follow its master like a dog.

CARNIVORA; VIVERRIDAE; **Genus HEMIGALUS**
Jourdan, 1837

Banded Palm Civets

There are two species (Ellerman and Morrison-Scott 1966; Medway 1977; Lekagul and McNeely 1977; Chasen 1940):

H. derbyanus, Tenasserim, Malay Peninsula, Sumatra and certain small islands to the west, Borneo;

H. hosei, Borneo.

The latter species has sometimes been placed in a separate genus, *Diplogale* Thomas, 1912.

In *H. derbyanus,* head and body length is 410 to 510 mm, tail length is 255 to 383 mm, and weight is usually 1.75 to 3 kg. On the head there is a narrow, median dark streak extending from the nose to the nape, and on each side of this there is a broader dark stripe that encircles the eye and passes backward over the base of the ear. Two broad stripes, sometimes more or less broken into shorter stripes or spots, run backward from the neck and curve downward to the elbow. Behind these are two shorter stripes. The back behind the shoulders is marked with four or five broad transverse stripes separated by pale, usually narrower spaces, and there are two imperfect stripes at the base of the tail. The ground color is whitish to orange buff, usually lighter and more buffy below, and the tail is usually black. The hair on the back of the neck is reversed, in that the tips of the hairs point forward. The five-toed feet have strongly curved claws that are retractile like those of cats. Small scent glands are present.

In *H. hosei,* head and body length is about 600 mm and tail length is about 300 mm. Coloration is dark brown or black above, and grayish, yellowish white, or slightly rufescent below. The ears are thinly haired and white inside. A buffy gray patch extends from above the eye to the cheek, and terminates where it meets the white of the lips and throat. The inner sides of the limbs near the body are grayish, while the remainder of each limb is black. The tail is not banded, but is dark throughout.

On the Malay Peninsula, *H. derbyanus* is restricted to tall forest and apparently is largely terrestrial (Medway 1978). This species, however, is at least partly arboreal and climbs well (Lekagul and McNeely 1977). *H. hosei* of Borneo is found mainly in montane forest and is largely terrestrial (Medway 1977). All specimens of *Hemigalus* taken by Davis (1962) in Borneo were collected on the ground. He observed that the animals were exclusively nocturnal and apparently foraged on the forest floor, picking up food from the surface. Orthopterans and worms made up 80 percent of the contents of 12 stomachs, and the rest of the food was mostly other invertebrates.

Banded palm civet (*Hemigalus derbyanus*), photo from Duisburg Zoo.

Gangloff (1975) reported that captive *H. derbyanus* were fond of fruit, did not construct nests, and marked with scent. A pregnant female *H. derbyanus,* with one embryo, was taken in Borneo in February. A captive female in the Wassenaar Zoo, Netherlands, had two young. They weighed 125 grams each at birth, opened their eyes after 8 to 12 days, and first took solid food when about 70 days old (Ewer 1973). One specimen was still living after 12 years in captivity (Marvin L. Jones, Zoological Society of San Diego, pers. comm.). *H. derbyanus* is on appendix 2 of the CITES.

CARNIVORA; VIVERRIDAE; **Genus CHROTOGALE**
Thomas, 1912

Owston's Palm Civet

The single species, *C. owstoni,* occurs in Laos and northern Viet Nam (Ellerman and Morrison-Scott 1966). The 15 known specimens are in the French National Museum of Natural History in Paris, the British Museum (Natural History) in London, and the Field Museum of Natural History in Chicago.

Head and body length is 508 to 635 mm and tail length is 381 to 482 mm. The body and base of the tail have alternating and sharply contrasting dark and light transverse bands, and longitudinal stripes are present on the neck. The pattern of stripes and bands resembles that of *Hemigalus derbyanus,* but it is supplemented by black spots on the sides of the neck,

the forelimbs and thighs, and the flanks. Four seems to be the maximum number of dorsal bands. It has been suggested that this striking pattern serves as a warning signal, as *Chrotogale* is thought to possess a particularly foul-smelling anal gland secretion. The underparts are pale buffy, and a narrow orange midventral line runs from the chest to the inguinal region. The terminal two-thirds of the tail are completely black.

Form and markings are strikingly like those of *Hemigalus derbyanus,* but the hairs on the back of the neck of *Chrotogale* are not reversed in direction. *Chrotogale* is also distinguished by cranial and dental characters. The incisor teeth are remarkable in that they are broad, close set, and arranged in practically a semicircle, a type unique among carnivores and only approached in certain of the marsupials. The other teeth and the skull are also peculiar, and indicate habits and a mode of life different from that of other genera in the family Viverridae. As yet, however, there has been no substantive investigation of the natural history of the genus. The stomachs of two specimens were found to contain earthworms.

CARNIVORA; VIVERRIDAE; **Genus CYNOGALE** *Gray,*
1837

Otter Civet

The single species, *C. bennettii,* occurs in northern Viet Nam, the Malay Peninsula, Sumatra, and Borneo (Ellerman

Owston's palm civet (*Chrotogale owstoni*), photo from *Field Mus. Nat. Hist. Publ.*, "Mammals of the Kelley-Roosevelt and Delacour Asiatic Expeditions," W. H. Osgood.

and Morrison-Scott 1966; Lekagul and McNeely 1977). The one known specimen from Viet Nam was originally referred to a separate species, *C. lowei*.

Head and body length is 575 to 675 mm, tail length is 130 to 205 mm, and weight is usually 3 to 5 kg. The form is somewhat like that of the otters (*Lutra*). The underfur is close, soft, and short. It is pale buff near the skin and shades to dark brown or almost black at the tip. The longer, coarser guard hairs are usually partially gray, which gives a frosted

or speckled effect on the head and body. The lower side of the body is lighter brown and not speckled with gray. The whiskers are remarkably long and plentiful; those on the snout are fairly long, but those on a patch under the ear are the longest. The newly born young lack dorsal speckling; they have some gray on the forehead and ears, and two longitudinal stripes down the sides of the neck extending under the throat.

Because of the deepening and expansion of the upper lip, the rhinarium occupies a horizontal position with the nostrils

Otter civet (*Cynogale bennettii*), photo from San Diego Zoological Society.

Otter civet (*Cynogale bennettii*), photo from San Diego Zoological Society.

opening upward on top of the muzzle. The nostrils can be closed by flaps, an adaptation for aquatic life. The ears can also be closed. Although the webbing on the feet does not extend farther toward the tips of the digits than in such genera as *Paradoxurus,* it is quite broad and the fingers are capable of considerable flexion. A glandular area, merely three pores in the skin, is located near the genitals and secretes a mild scent material. The premolar teeth are elongate and sharp, adapted for capturing and holding prey, while the molars are broad and flat, for crushing. Females have four mammae.

The otter civet is usually found near streams and swampy areas. It can climb well, and, when chased by dogs, often takes refuge in a tree. While walking, it usually carries its head and tail low and arches its back. Although it is partly adapted for an aquatic life, its tail is short and lacks special muscular power, and the webbing between the digits is only slightly developed. *Cynogale* is thus probably a slow swimmer and cannot turn quickly in the water. It probably captures aquatic animals only after they have taken shelter from the chase, and catches some birds and mammals as they come to drink. It cannot be seen by its prey, because it is submerged with only the tip of the nose exposed above the surface of the water. The diet includes crustaceans, perhaps mollusks, fish, birds, small mammals, and fruit.

There are records of pregnant females with two and three embryos. Young, still with the mother, have been noted in May in Borneo. A captive specimen lived for five years (Marvin L. Jones, Zoological Society of San Diego, pers. comm.).

Cynogale is on appendix 2 of the CITES.

CARNIVORA; VIVERRIDAE; **Genus FOSSA** Gray, 1864

Malagasy Civet

The single species, *F. fossa,* was originally found throughout the forested parts of Madagascar (IUCN 1972). Although the generic name is the same as the vernacular term for another Malagasy viverrid (*Cryptoprocta*), the two animals are distinctly different.

Head and body length is 400 to 450 mm and tail length is 210 to 230 mm (Albignac 1972). Males weigh up to 2 kg, and females up to 1.5 kg (Meester and Setzer 1971). The ground color is grayish, washed with reddish. There are four rows of black spots on each side of the back, and a few black spots on the backs of the thighs. These spots may merge to form

Malagasy civets (*Fossa fossa*), photo from U.S. National Zoological Park.

stripes, and the gray tail is banded with brown. The underparts are grayish or whitish and more or less obscurely spotted. The limbs are slender, perhaps being adapted for running. There are no anal scent glands, but there probably are marking glands on the cheeks and neck (Albignac 1972).

The Malagasy civet inhabits evergreen forests, and shelters in hollow trees or crevices. It is nocturnal, and may occur in trees or on the ground. The preferred foods are crustaceans, worms, small eels, and frogs. Other kinds of animal matter and fruit are also taken (Meester and Setzer 1971).

Fossa lives in pairs, which share a territory. Vocalizations include cries, groans, and a characteristic "coq coq," heard only in the presence of more than one individual. Births have been recorded from October to January. The gestation period is 3 months and a single offspring is born. It weighs 65 to 70 grams at birth, is weaned after 2 months, and probably attains adult weight when 1 year old (Albignac 1970*a*, 1972; Meester and Setzer 1971). A captive specimen lived for 11 years (Marvin L. Jones, Zoological Society of San Diego, pers. comm.).

The Malagasy civet is classified as vulnerable by the IUCN (1972) and is on appendix 2 of the CITES. Loss of habitat and excessive hunting have restricted its range to the eastern and northwestern rain forests of Madagascar.

CARNIVORA; VIVERRIDAE; **Genus EUPLERES** Doyère, *1835*

Falanouc

The single species, *E. goudotii*, is found in the coastal forests of Madagascar (Meester and Setzer 1971).

Head and body length is 450 to 650 mm, tail length is 220 to 250 mm, and weight is 2 to 4 kg (Albignac 1974). The subspecies *E. g. goudotii* is fawn colored above and lighter below. In the subspecies *E. g. major* the males are brownish and the females are grayish. The pelage is woolly and soft, being made up of a dense underfur and longer guard hairs. The tail is covered by rather long hairs that give a bushy appearance.

Eupleres has a pointed muzzle, a narrow and elongate head, and short, conical teeth. It resembles the civets in some structural features and the mongooses in others. The small teeth are similar to each other and resemble those of insectivores, rather than carnivores. Indeed, *Eupleres* was classified as an insectivore before its somewhat obscure relationship with the mongooses was detected. The claws are relatively long and not retractile, or are only imperfectly so. The feet are peculiar in the comparatively large size and low position of the great toe and thumb.

The falanouc inhabits humid, lowland forests. It is crepuscular and nocturnal, resting by day in crevices and burrows. It is terrestrial and, if threatened, may either run or remain motionless. In autumn, up to 800 grams of fat can be accumulated in the tail, and the suggestion has been made that *Eupleres* hibernates during winter. Observations of active individuals, however, have been made in winter. The diet consists mainly of earthworms and also includes other invertebrates and frogs, but apparently not reptiles, birds, rodents, or fruit (Albignac 1974; Meester and Setzer 1971).

Eupleres may live alone or in small family groups. It has several vocalizations and other means of communication. Mating probably takes place in July or August (winter), and a birth was observed in November. Litters contain one or two young. Weight of the newborn is 150 grams, and weaning occurs at nine weeks of age (Albignac 1974; Meester and Setzer 1971).

The falanouc is classified as vulnerable by the IUCN (1976) and is on appendix 2 of the CITES. It has declined in numbers and distribution, because of excessive hunting, hab-

Falanoucs (*Eupleres goudotii*), photos by R. Albignac.

Malagasy ring-tailed mongoose (*Galidia elegans*), photo from National Zoological Park.

itat destruction, and possibly competition from the introduced *Viverricula indica*.

CARNIVORA; VIVERRIDAE; *Genus GALIDIA*
I. Geoffroy St.-Hilaire, 1837

Malagasy Ring-tailed Mongoose

The single species, *G. elegans,* is found in eastern, west-central, and extreme northern Madagascar (Albignac 1969).

Head and body length is about 380 mm and tail length is about 305 mm. Weight is 700 to 900 grams (Meester and Setzer 1971). The general coloration is dark chestnut brown, and the tail is ringed with dark brown and black. *Galidia* has some of the structural features of civets, and some of mongooses. The feet differ from those of *Galidictis* in having shorter digits, fuller webbing, shorter claws, and more hairy soles. The lower canine teeth are smaller. From *Salanoia,* *Galidia* is distinguished by its ringed tail and very small second upper premolar. Dissection of two individuals indicates the presence of a scent gland, closely associated with the external genitalia, in males but not in females.

Galidia occurs in humid forests. It shelters in burrows, which it digs very rapidly, and probably also in hollow trees. It is mainly diurnal, but may also be active at night. It is more arboreal than most mongooses, and is able to climb and descend on vertical trunks only 4 cm in diameter. It can also swim. The diet consists mainly of small mammals, birds and their eggs, and frogs, and also includes fruit, fish, reptiles, and invertebrates (Meester and Setzer 1971; Albignac 1972).

Galidia is less social than most mongooses, being found alone or in pairs. Mating in Madagascar occurs from April to November, and births from July to February. The gestation period is 79 to 92 days. A single young is produced, which weighs 50 grams at birth. Physical maturity is reportedly attained at 1 year of age, and sexual maturity at 2 years (Meester and Setzer 1971; Larkin and Roberts 1979). A captive animal was still living after 13 years and 2 months (Marvin L. Jones, Zoological Society of San Diego, pers. comm.).

CARNIVORA; VIVERRIDAE; *Genus GALIDICTIS*
I. Geoffroy St.-Hilaire, 1839

Malagasy Broad-striped Mongoose

The single species, *G. fasciata,* is found in eastern Madagascar (Meester and Setzer 1971).

Head and body length is 320 to 340 mm, and tail length is 280 to 300 mm (Albignac 1972). The general body color is pale brown or grayish. In the subspecies *G. fasciata striata* there are usually 5 longitudinal black bands or stripes on the back and sides, and the tail is whitish. In *G. fasciata fasciata* there are usually 8 to 10 stripes, and the tail is bay colored and somewhat bushy.

Galidictis differs from other viverrids in the color pattern, and in cranial and dental characters. The feet differ from those of *Galidia* in having longer digits, less extensive webbing, and longer claws. A scent pouch is present in females.

This mongoose occurs in forests, and is nocturnal and crepuscular. The diet consists mainly of small vertebrates, especially rodents, and also includes invertebrates. *Galidictis* is found in pairs or small social groups. It apparently produces one young per year, during the summer (Albignac

Malagasy broad-striped mongoose (*Galidictis fasciata*), photo from *Zoologie de Madagascar*, G. Grandidier and G. Petit.

1972; Meester and Setzer 1971). Young individuals are said to tame readily, to follow their master, and even to sleep in his lap.

CARNIVORA; VIVERRIDAE; Genus MUNGOTICTIS Pocock, 1915

Malagasy Narrow-striped Mongoose

The single species, *M. decemlineata,* is found in western and southwestern Madagascar (Meester and Setzer 1971).

Head and body length is 250 to 350 mm, tail length is 230 to 270 mm, and weight is 600 to 700 grams. The fur is rather dense and generally gray beige in color. There are usually 8 or 10 dark stripes on the back and flanks. The underparts are pale beige. The soles of the feet are naked, and the digits are partly webbed. There are glands on the side of the head and on the neck, which appear to be for marking with scent. Females have a single pair of inguinal mammae (Albignac 1971, 1972, 1976).

Mungotictis is found on sandy, open savannahs. It is diurnal, and both arboreal and terrestrial. It spends the night in tree holes during the wet summer and in ground burrows during the dry winter. It can swim well. The diet consists mostly of insects, but also includes small vertebrates, birds' eggs, and invertebrates. To break an egg or the shell of a snail, *Mungotictis* lies on one side, grasps the object with all four feet, and throws it abruptly until it breaks and the contents can be lapped up (Albignac 1971, 1972, 1976).

In a radio-tracking study, Albignac (1976) determined that 22 individuals inhabited 300 ha. The animals were divided into two stable social units, which sometimes engaged in agonistic encounters, where their ranges met. Each group contained several adults of both sexes, as well as juveniles and young of the year. Group cohesiveness and intrarelationships varied. Generally, adult males and females came together in the summer. In the winter there was division into small units, such as temporary pairs, maternal family parties, all-male groups, and solitary males.

The breeding season extends from December to April, peaking in February and March (summer). The gestation period is 90 to 105 days. There is usually a single offspring, which weighs 50 grams at birth. Weaning occurs at 2 months of age, and separation from the mother comes after 2 years (Albignac 1972, 1976).

CARNIVORA; VIVERRIDAE; Genus SALANOIA Gray, 1864

Malagasy Brown-tailed Mongoose, Salano

The single species, *S. concolor,* is found in northeastern Madagascar (Meester and Setzer 1971).

Head and body length is 250 to 300 mm and tail length is 200 to 250 mm (Albignac 1972). The general coloration is brown, and there are either dark or pale spots. The tail is the same color as the body and is not ringed. The claws are not strongly curved, the ears are broad and short, and the muzzle is pointed.

The salano occurs only in the evergreen forests on the northeastern part of the central plateau of Madagascar. It is typically diurnal, sheltering at night in tree trunks or burrows. It feeds mainly on insects and fruit, but also on frogs, small reptiles, and rodents. It occurs individually or in pairs, depending on the season. The young are born mainly during the summer (Meester and Setzer 1971; Albignac 1972).

Malagasy narrow-striped mongoose (*Mungotictis decemlineata*); Inset: head (*M. decemlineata*); photos by Don Davis.

Malagasy brown-tailed mongoose (*Salanoia concolor*), photo of mounted specimen in Field Museum of Natural History.

A. African mongoose (*Herpestes ichneumon*), photo by Ernest P. Walker. B. Crab-eating mongoose (*H. urva*), photo by Robert E. Kuntz.

CARNIVORA; VIVERRIDAE; *Genus HERPESTES Illiger, 1811*

Mongooses

There are 3 subgenera and 14 species (Ellerman and Morrison-Scott 1966; Meester and Setzer 1971; Corbet 1978; Chasen 1940; Medway 1977; Sanborn 1952; Lynch 1981):

subgenus *Herpestes*

H. ichneumon, southern Spain and Portugal, Asia Minor, Palestine, Morocco to Tunisia, Egypt, possibly eastern Libya, most of Africa south of the Sahara;

H. javanicus, Thailand, Indochina, Malay Peninsula, Java;

H. auropunctatus, Iraq to Malay Peninsula, Hainan;

H. edwardsi, eastern Arabian Peninsula to Assam, Sri Lanka;

H. smithi, India, Sri Lanka;

H. fuscus, southern India, Sri Lanka;

H. vitticollis, southern India, Sri Lanka;

H. urva, Nepal to southeastern China and peninsular Thailand, Taiwan, Hainan;

H. brachyurus, Malay Peninsula, Sumatra, Borneo, Palawan;

H. hosei, known only by the type specimen from Sarawak in northern Borneo;

H. semitorquatus, Sumatra, Borneo;

subgenus *Xenogale*

H. naso, southeastern Nigeria to eastern Zaire;

subgenus *Galerella*

H. pulverulentus, southern Namibia, South Africa, Lesotho;

H. sanguineus, savannah and semiarid regions of Africa south of the Sahara.

An additional species, *H. palustris,* has been described from the vicinity of Calcutta in eastern India (Ghose 1965; Ghose and Chaturvedi 1972). Although said to be most closely related to *H. auropunctatus,* Ewer (1973) suggested that it may be a subspecies of *H. urva.* Lekagul and McNeely (1977) considered *H. auropunctatus* to be conspecific with *H. javanicus.* Some authorities, such as Rosevear (1974), treat *Xenogale* J. A. Allen, 1919 and *Galerella* Gray, 1865 as distinct genera.

Head and body length is 230 to 650 mm, tail length is 230 to 510 mm, and weight is 0.4 to 4.0 kg. Coloration varies considerably. Some forms are greenish gray, yellowish brown, or grayish brown. Others are finely speckled with white or buff. The underparts are generally lighter than the back and sides, and are white in some species. The fur is short and soft in some species, and rather long and coarse in others. The body is slender, the tail is long, there are five digits on each limb, the hind foot is naked to the heel, and the foreclaws are sharp and curved. Small scent glands are situated near the anus, and some species can eject a vile-smelling secretion. Females have four or six mammae.

Mongooses occupy a wide variety of habitats, ranging from densely forested hills to open, arid plains. They shelter in hollow logs or trees, holes in the ground, or rock crevices. They may be either diurnal or nocturnal. They are basically terrestrial, but are very agile, and some species can climb skillfully (Grzimek 1975). During the morning, they frequently stretch out in an exposed area to sun themselves. The diet includes insects, crabs, fish, frogs, snakes, birds, small mammals, fruit, and other vegetable matter. Some species kill cobras and other venomous snakes. Contrary to popular belief, mongooses are not immune to the bites of these reptiles. Rather, they are so skillful and quick in their movements that they avoid being struck by the snake and almost invariably succeed in seizing it behind the head. The battle usually ends with the mammal's eating the snake.

Data cited by Ewer (1973) indicate that in Hawaii the individual home range diameter of *H. auropunctatus* is about 1.6 km for males and 0.8 km for females. Mongooses are found alone, in pairs, or in groups of up to 14 individuals. Ewer noted that family parties of *H. pulverulentus* den together but forage individually. Taylor (1975) wrote that *H. sanguineus* travels alone or in pairs, that its home range may be as small as 1 sq km but is much larger in desert areas, that it may be territorial, and that it is generally silent.

Some species breed throughout the year, with females giving birth two or three times annually. One female *H. edwardsi* produced five litters in 18 months. Gestation periods range from about 42 days in *H. auropunctatus* to 84 days in *H. ichneumon.* Litters contain one to four young. The young of *H. auropunctatus* are weaned after 4 or 5 weeks. Captive *H. ichneumon* have lived over 20 years (Ewer 1973; Grzimek 1975; Roberts 1977; Kingdon 1977; Lekagul and McNeely 1977).

Mongooses have been widely introduced by people to kill rats and snakes. The presence of *H. ichneumon* in Spain and Portugal probably results from introduction in ancient times. The species also has been brought to central Italy, Yugoslavia, and Madagascar. Populations of *H. auropunctatus* have been established on most Caribbean islands, the northeastern coast of South America, Mafia Island off East Africa, and the Hawaiian and Fiji islands. Several individuals of this species have been taken on the mainland of North America. *H. edwardsi* has been introduced on the Malay Peninsula, Mauritius, and the Ryukyu Islands (Corbet 1978; Nellis et al. 1978; Van Gelder 1979; Gorman 1976). In addition to killing rats and snakes, mongooses have destroyed harmless birds and mammals, and have contributed to the extinction or endangerment of many desirable species of wildlife. They also have become pests by preying on poultry. The importation or possession of mongooses is therefore now forbidden by law in some countries.

CARNIVORA; VIVERRIDAE; *Genus MUNGOS*
E. Geoffroy St.-Hilaire and G. Cuvier, 1795

Banded and Gambian Mongooses

There are two species (Meester and Setzer 1971):

M. gambianus (Gambian mongoose), savannah zone from Gambia to Nigeria;

M. mungo (banded mongoose), Gambia to northeastern Ethiopia, and south to South Africa.

The name *Mungos* was formerly sometimes used for many other species of mongoose, including those now assigned to *Herpestes.*

Head and body length is 300 to 450 mm, tail length is 230 to 290 mm, and weight is 1 to 2.2 kg. *M. mungos* is brownish gray with dark brown and well-defined yellowish or whitish bands across the back. The banded pattern is produced by hair markings of the same type that produce the ground color of the remainder of the body. The hairs are alternately ringed with dark and light bands; the color rings of the individual hairs coincide with like colors on adjacent hairs. *M. gambianus* lacks the transverse bands, but has a dark streak on the side of the neck (Meester and Setzer 1971).

The pelage is coarse; compared with other mongooses, there is little underfur. Although the tail is not bushy, it is covered with coarse hair and is tapered toward the tip. The foreclaws are elongate, the soles are naked to the wrist and heel, and there are five digits on each limb. There is no naked grooved line from the tip of the nose to the upper lip. Females have six mammae.

The information in the remainder of this account applies to *M. mungos,* and was taken from Rood (1974, 1975), Simpson (1964), Neal (1970), Rosevear (1974), and Kingdon (1977). The banded mongoose is found in grassland, brushland, woodland, and rocky, broken country. It dens mainly in old termite mounds, but also in such places as erosion gullies, abandoned aardvark holes, and hollow logs. Dens are communal and consist of one to nine entrance holes, a central sleeping chamber of about 1 or 2 cubic meters, and perhaps several smaller chambers. Most dens are used only for a few days, but some favorite sites may be occupied for as long as 2 months. *M. mungos* is terrestrial and diurnal, and has excellent senses of vision, hearing, and smell. A group generally emerges from the den around 0700–0800 hrs, forages for several hours, rests in a shady spot during the hottest part of the day, forages again, and returns to a den before sunset. More time is spent in the vicinity of the den, if young

Banded mongoose (*Mungos mungo*), photo from Zoological Garden Berlin-West through Ernst von Roy.

are present therein. A group generally covers 2 or 3 km per day, moving in a zigzag pattern and searching among rocks and vegetation for food. The diet consists largely of invertebrates, especially beetles and millipedes, and also includes small vertebrates. To break an egg and obtain the contents, *M. mungos* grasps the item with its forefeet and propels it backward between its hind feet and against a hard object.

In the Ruwenzori National Park of Uganda, population density was found to be about 18 individuals per sq km, and group home range varied from about 38 to 130 ha. In the Serengeti, home range may be over 400 ha. Ranges overlap, but intergroup encounters are generally noisy and hostile, and sometimes involve chasing and fighting. Groups contain up to 40 individuals, usually about 10 to 20, including sever-

al adults of both sexes. Captive females have been seen to dominate males. The animals sleep in contact and forage in a fairly close, but not bunched formation. Groups are cohesive, but some splitting off has been observed and mating between members of different groups sometimes occurs. Individuals may mark one another with scent from anal glands. The most common vocalization is a continuous birdlike twittering that probably serves to keep the group together during foraging. There are also various agonistic growls and screams, and an alarm chitter.

In East Africa, at least, reproduction continues throughout the year. Breeding is synchronized within a given group, with several females bearing litters at approximately the same time. Groups breed up to four times per year, though it

Banded mongooses (*Mungos mungo*), photo by J. Rood.

Cusimanse (*Crossarchus obscurus*), photo by Ernest P. Walker.

cannot be said that any one female has that many litters annually. A period of mating often begins within 1 or 2 weeks after the birth of young. The gestation period is about 2 months. The number of young per litter seems usually to be about two or three, but may be as many as six. The young weigh about 20 grams each at birth, but grow rapidly. They are apparently kept together and raised commonly by the group. They suckle indiscriminantly from any lactating female, and are usually guarded by one or two adult males, while the rest of the group forages. They begin to travel with the others when about 1 month old. Females attain sexual maturity at the age of 9 or 10 months. Maximum known longevity in captivity is 11 years.

CARNIVORA; VIVERRIDAE; *Genus CROSSARCHUS*
F. Cuvier, 1825

Cusimanses

There are three species (Meester and Setzer 1971; Kingdon 1977):

C. *ansorgei*, northern Angola, possibly east-central and southern Zaire;
C. *obscurus*, Sierra Leone to Cameroon;
C. *alexandri*, northern Zaire, Uganda.

The latter two species may be conspecific.

Head and body length is 305 to 450 mm, tail length is 150 to 255 mm, and weight is 450 to 1,350 grams. The body is covered by relatively long, coarse hair, which is a mixture of browns, grays, and yellows. The head is usually lighter colored than the remainder of the body, while the feet and legs are usually the darkest. The legs are short, the tail is tapering, the ears are small, and the face is sharp.

Cusimanses live in forests and swampy areas. Various reports suggest that they may be active either by day or by night (Kingdon 1977). They travel about in groups, seldom remaining in any one locality longer than two days and taking temporary shelter in any convenient place. While seeking food, they scratch and dig in dead vegetation and in the soil. The diet consists principally of insects, larvae, small reptiles, crabs, tender fruits, and berries. It is said that the shells of snails and eggs are cracked by being hurled with the forepaws back between the hind feet and against some hard object.

Groups contain 10 to 24 individuals. These probably represent 1 to 3 family units, each with a mated pair and the surviving members of 2 or 3 litters. There is no evidence of seasonal breeding in either East or West Africa. Observations in captivity indicate that there may be several litters annually, that the gestation period is about 10 weeks, and that litter size is 4 young (Rosevear 1974; Kingdon 1977). One captive lived for 8 years and 2 months (Marvin L. Jones, Zoological Society of San Diego, pers. comm.).

Cusimanses tame easily and make good pets. They are affectionate, playful, clean, and readily housebroken, but sometimes mark objects with their anal scent glands (Rosevear 1974).

CARNIVORA; VIVERRIDAE; *Genus LIBERIICTIS*
Hayman, 1958

Liberian Mongoose

The single species, *L. kuhni*, is known by 2 whole specimens and 10 skulls from northeastern Liberia (Schlitter 1974).

Head and body length of an adult male is 423 mm, tail length is 197 mm, and weight is 2.3 kg. The predominant color of the pelage is dark brown. A dark stripe, bordered above and below by a pale stripe, is present on the neck. The throat is pale, the tail is slightly bicolored, and the legs are dark. From *Crossarchus*, *Liberiictis* is distinguished externally by the presence of neck stripes, a more robust body, and apparently longer ears (Schlitter 1974). The skull of *Liberiictis* is larger than that of *Crossarchus*, the rostrum and nasals are more elongate, the teeth are proportionally smaller and weaker, and there is an additional premolar in both the upper and lower jaws.

Schlitter (1974) noted that the long claws of the front feet, the long mobile snout, and the weak dentition of *Liberiictis* indicate that it is a terrestrial animal with a primarily insectivorous diet. His two specimens were taken in a densely forested area transversed by numerous streams. People native to the area said that *Liberiictis* is diurnal and found only on the ground. An adult male was taken in a snare on the ground. A juvenile female was excavated from a burrow, associated with a termite mound, on 29 July. An adult of unknown sex was also in the burrow, but was not preserved.

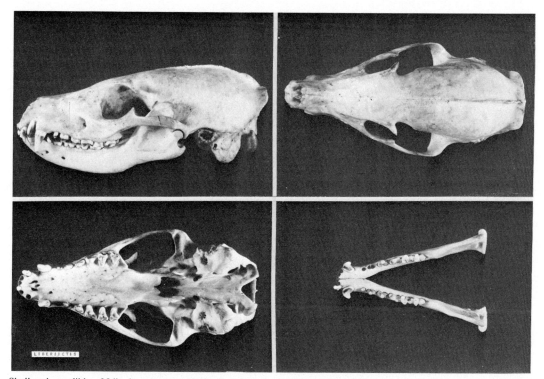

Skull and mandible of Liberian mongoose (*Liberiictis kuhni*), photos by Hans-Jurg Kuhn.

Schlitter stated that *Liberiictis* is eaten by human hunters. In a letter of 26 June 1963 to Ernest P. Walker, Dr. Hans-Jürg Kuhn, who had traveled in Liberia, wrote that native people say that *Liberiictis* is usually found in tree holes and lives in groups of three to five individuals.

CARNIVORA; VIVERRIDAE; *Genus HELOGALE* Gray, 1861

Dwarf Mongooses

There are two species (Meester and Setzer 1971; Yalden, Largen, and Kock 1980):

H. parvula, Ethiopia to Angola and eastern South Africa;
H. hirtula, southern Ethiopia, southern Somalia, northern Kenya.

Head and body length is 180 to 260 mm, tail length is 120 to 200 mm, and weight is 230 to 680 grams (Kingdon 1977). Coloration is variable, but generally the upper parts are speckled brown to grayish. The lower parts are only slightly paler, and the tail and lower parts of the legs are dark. In some individuals there is a rufous patch on the throat and breast, and the basal portion of the lower side of the tail is reddish brown. Other individuals are entirely black.

Dwarf mongooses are found in savannahs, woodlands, brush country, and mountain scrub, from sea level to about 1,800 meters elevation. They are mainly terrestrial and diurnal. They seek shelter at dusk in deserted or active termite mounds, among gnarled roots of trees, and in crevices. Their slender bodies enable them to squeeze into small openings. On occasion, they may excavate their own burrows. Dens are changed frequently, sometimes on a daily basis (Rood 1978).

Most of the day is spent in an active and noisy search for food among brush, leaves, and rocks. The diet consists mainly of insects (Rasa 1977; Rood 1980) and also includes small vertebrates, eggs, and fruit.

Helogale is found in organized groups that are thought to use a definite home range. A portion of the range is occupied for two or three months, and then there is a shift to another part, presumably because food supplies are depleted (Rasa 1977). One group, observed for seven years, stayed mainly within an area of 2 ha.; another group used 2 ha. during the dry season but moved away with the coming of the rains (Kingdon 1977). In the Serengeti, Rood (1978) found group home ranges to average 30 ha., to overlap by 5 to 40 percent with the ranges of one to four neighboring packs, and to contain 10 to 20 dens each. Captives have been observed to mark the vicinity of dens with secretions from cheek and anal glands, and to show considerable intergroup aggression (Rasa 1973, 1977).

The social organization of *Helogale* is unique among mammals (Rasa 1972, 1973, 1975, 1976, 1977; Rood 1978, 1980; Kingdon 1977). There are as many as 40 individuals in a group, but usually about 10 or 12. The groups are matriarchal families, founded and led by an old female. She initiates movements and has priority to food. The second highest ranking member of the group is her mate, an old male. These two dominant animals are monogamous, only they usually produce offspring, and they suppress sexual activity in other group members. The latter form a hierarchy, with the *youngest* individuals ranking highest. This arrangement probably serves to allow the young animals to obtain sufficient food, without competition from older and stronger mongooses, during the unusually long period of growth in this genus. Within any age class in the group, females are dominant over males.

Despite the rigid class structure, or perhaps because of it,

Dwarf mongoose (*Helogale parvula*), photo by Bernhard Grzimek.

intragroup relations are generally harmonious and severe fights are rare. Subordinate adults clean, carry, warm, and bring food to helpless young, and take turns "baby sitting," while the rest of the group forages. Females, in addition to the mother, sometimes nurse the young. The youngest mobile animals seem to have the role of watching for danger and alerting the others by means of visual signals or a shrill alarm call. Often, a single animal occupies an exposed position, where it serves as a group guard. One series of observations showed that when a low-ranking male became sick, it was allowed a higher than normal feeding priority, and was also warmed by other group members. Individuals communicate by depositing scent from the cheek and anal glands, and also sometimes mark one another as an apparent sign of acceptance. As surviving subordinate animals grow older, they seem not to leave the group, even though they are not allowed to mate. If the dominant female dies, however, the group may split up.

In the Serengeti National Park of Tanzania, births occur mainly in the rainy season from November to May, and the alpha female usually has three litters per year (Rood 1978, 1980). In a captive colony in Europe, the young are born regularly in spring and autumn (Rasa 1972). According to Rasa (1977), females there normally give birth twice a year, entering estrus 4 to 7 days after lactation ceases. If the newborn die, however, females may quickly remate, and they thus have the potential of producing five litters annually. The gestation period is 49 to 56 days. The number of young per litter averages about four and ranges from one to seven.

Nursing lasts for at least 45 days, but group members begin to bring solid food to the young before weaning is complete. The young start to forage with the group by the time they are 6 months old. Females may reach physiological sexual maturity as early as the age of 107 days (Zannier 1965), but social restrictions normally delay breeding for several years. Apparently, full physical maturity is not attained until the age of 3 years (Rasa 1972). One dominant pair was observed to mate first when about 3 years old, and then to continue to breed for 7 years (Kingdon 1977).

CARNIVORA; VIVERRIDAE; *Genus DOLOGALE Thomas, 1926*

African Tropical Savannah Mongoose

The single species, *D. dybowskii,* is found in the Central African Republic, northeastern Zaire, southern Sudan, and western Uganda (Meester and Setzer 1971). This species was originally assigned to *Crossarchus* and has sometimes been referred to *Helogale.*

Head and body length is about 250 to 330 mm and tail length is 160 to 230 mm. Kingdon (1977) listed the weight as approximately 300 to 400 grams. Stripes are lacking. The head and neck are black, grizzled with grayish white. The back, tail, and limbs are lighter in color and have brownish spots. The underparts are reddish gray. The fur is short,

African tropical savannah mongoose (*Dologale dybowskii*), photo by F. Petter of mounted specimen in Museum National d'Histoire Naturelle.

Black-legged mongoose (*Bdeogale* sp.), photo by Don Davis.

even, and fine, in contrast to the loose, coarse pelage of *Crossarchus*. The snout is not lengthened like that of *Crossarchus*. *Dologale* closely resembles *Helogale*, but does not have a groove on the upper lip, and has weaker teeth (Meester and Setzer 1971).

Kingdon (1977) stated that the few known specimens of *Dologale* suggest adaptation to a variety of habitats—thick forest, savannah-forest, and montane forest grassland. A little evidence indicates that the genus is at least partly diurnal. It has robust claws, suggesting digging habits as in *Mungos*. Asdell (1964) wrote that a litter of four young had been noted in Zaire.

CARNIVORA; VIVERRIDAE; *Genus BDEOGALE* Peters, 1850

Black-legged Mongooses

There are two subgenera and three species (Meester and Setzer 1971):

subgenus *Bdeogale*

B. crassicauda, southern Kenya to central Mozambique;

subgenus *Galeriscus*

B. nigripes, southeastern Nigeria to northern Zaire and northern Angola;
B. jacksoni, southeastern Uganda, central Kenya.

Rosevear (1974) considered *Galeriscus* Thomas, 1894 a separate genus. Kingdon (1977) suggested that *B. jacksoni* is only a subspecies of *B. nigripes*.

Head and body length is 375 to at least 600 mm and tail length is 175 to 375 mm. Kingdon (1977) listed weight as 0.9 to 3.0 kg. There is considerable variation in color, both within and between species. The predominant general coloration is some shade of gray or brown, and the legs are usually black. The fur of adults is rather close, dense, and short, while that of the young is nearly twice as long and is lighter in color.

Bdeogale resembles *Ichneumia* in having black feet, soft underfur, and long, coarse hair over the upper parts of the body. It differs in lacking the first or inner toe on each foot, and in having larger premolar teeth. *Bdeogale* differs from *Rhynchogale* in having a naked groove from the nose to the upper lip. The foreparts of the feet of *Bdeogale* are naked, but the hind parts are well haired.

According to Kingdon (1977), *B. crassicauda* inhabits woodland and moist savannah, while the other species live in tropical forest. *B. crassicauda* feeds almost entirely on insects, especially ants and termites, but may also take crabs and rodents. *B. nigripes* seems to prefer ants, but also eats small vertebrates and carrion. These mongooses are not infrequently seen in pairs. A female and a quarter-grown young were taken in December on the coast of Kenya, a pregnant female with a large fetus was also taken in December, and a female with a newborn infant was found in southeastern Tanzania in late November. Information compiled by Rosevear (1974) suggests that adults are basically solitary in the wild, but not quarrelsome when kept together in captivity; that births in West Africa occur from November to January; and that litters normally contain a single young. According to Marvin L. Jones (Zoological Society of San Diego, pers. comm.), a captive *B. nigripes* lived for 15 years and 10 months.

CARNIVORA; VIVERRIDAE; *Genus RHYNCHOGALE* Thomas, 1894

Meller's Mongoose

The single species, *R. melleri*, occurs from southern Zaire and Tanzania to eastern South Africa and possibly northeastern Angola (Meester and Setzer 1971).

Head and body length is 440 to 485 mm and tail length is usually 300 to 400 mm. Kingdon (1977) listed weight as 1.7 to 3.0 kg. The general coloration is grayish or pale brown, the head and undersides are paler, and the feet are usually darker. *Rhynchogale* resembles *Ichneumia* in having coarse guard hairs protruding from the close underfur, and the same dental formula. *Rhynchogale* differs from *Ichneumia* in the frequent reduction of the hallux and the lack of a naked crease from the nose to the upper lip. *Rhynchogale* has hind soles that are hairy to the roots of the toes. Females have two abdominal pairs of mammae.

According to Kingdon (1977), the habitat appears restricted to the woodland belt, and possibly to moister and more heavily grassed or wooded areas, such as drainage lines and rock outcrops. Available information suggests that *Rhynchogale* is terrestrial, nocturnal, and solitary. The diet includes wild fruit, termites, and probably small vertebrates. A litter of two newborn, with eyes still unopened, was found in a small cave on a rocky hill in Zambia in December. A

Meller's mongoose (*Rhynchogale melleri*), photo from *Proc. Zool. Soc. London.*

pregnant female, containing two embryos, was found in the same area, also in December. In Rhodesia, births occur around November and litters contain up to three young.

CARNIVORA; VIVERRIDAE; **Genus ICHNEUMIA**
I. Geoffroy St.-Hilaire, 1837

White-tailed Mongoose

The single species, *I. albicauda,* occurs in the southern part of the Arabian Peninsula, and in most of Africa south of the Sahara (Corbet 1978; Meester and Setzer 1971; Nader 1979).

Head and body length is 470 to 710 mm, tail length is 355 to 470 mm, and weight is 1.8 to 5.2 kg (Kingdon 1977; Taylor 1972). Long, coarse, black guard hairs protrude from a yellowish or whitish close, woolly underfur, producing a grayish general body color. The four extremities, from the elbows and knees, are black. The basal half of the tail is of the general body color. The terminal portion is usually white, but is occasionally black. *Ichneumia* is characterized by its large size; bushy, tapering tail; having the soles of the forelimbs naked to the wrist; and the division of the upper lip by a naked slit from the nose to the mouth. Females have four mammae.

The white-tailed mongoose is found mainly in savannahs and grassland. It prefers areas of thick cover, such as forest edge and bush-fringed streams. It is basically terrestrial and nocturnal, sheltering by day in porcupine or aardvark burrows or in cavities under roots or rocks. The diet consists mainly of insects and also includes snakes, other small vertebrates, and fruit. *Ichneumia* breaks eggs by hurling them back between its hind legs and against some hard object. It may defend itself by ejecting a particularly noxious secretion from its anal scent glands (Kingdon 1977).

Individual home range is about 8 sq km (Taylor 1972). *Ichneumia* is normally seen in pairs or family groups, and sometimes forms larger aggregations. It is highly vocal, the most unusual sound being a doglike yap that may be associ-

White-tailed mongoose (*Ichneumia albicauda*), photo by Ernest P. Walker.

Marsh or water mongoose (*Atilax paludinosus*), photo by Ernest P. Walker.

ated with sexual behavior. The two to four young are born in a burrow (Kingdon 1977). A captive was still living after 10 years (Marvin L. Jones, Zoological Society of San Diego, pers. comm.).

When the white-tailed mongoose dwells near a poultry raiser, it may prove to be a pest. In captivity it is the shyest of mongooses, but, if captured young, it is said to become a pleasing pet.

CARNIVORA; VIVERRIDAE; *Genus ATILAX F. Cuvier, 1826*

Marsh or Water Mongoose

The single species, *A. paludinosus,* is found from Guinea-Bissau to Ethiopia, and south to South Africa (Rosevear 1974).

Head and body length is 460 to 620 mm, tail length is 320 to 530 mm, and weight is 2.5 to 4.1 kg (Kingdon 1977). The pelage is long, coarse, and generally brown in color. A sprinkling of black guard hairs often gives a dark effect. In some individuals, light rings on the hairs are prevalent and impart a grayish tinge. The head is usually lighter than the back, and the underparts are still paler.

Atilax is fairly heavily built. Although it is more aquatic than any other mongoose, it is the only one with toes that completely lack webbing. This feature may be associated with the habit of feeling for aquatic prey in mud or under stones. There are five digits on each limb, the soles are naked, and the claws are short and blunt. The anal area is large and naked, and there is a narrow, naked slit between the nose and the upper lip. Females usually have two pairs of mammae.

Atilax is found in a variety of general habitat types, but its basic requirement seems to be permanent water bordered by dense vegetation (Rosevear 1974). Favored haunts are marshes, reed-grown stream beds, and tidal estuaries.

Grassy patches and floating masses of vegetation often serve as feeding places and as dry resting spots. Like other mongooses, *Atilax* does little or no climbing, but does run up leaning tree trunks or other inclines easy of access. It is an excellent swimmer and diver. When hard pressed, it submerges, leaving only the tip of the nose exposed for breathing. Normally, when swimming, the head and part of the back are exposed. Food is sought in the water and in travels on regular pathways along the borders of streams and marshes. Activity is usually said to be nocturnal and crepuscular, but Rowe-Rowe (1978b) referred to *Atilax* as diurnal, and stated that it does much of its hunting while walking in shallow water.

The diet consists of almost any form of animal life that can be caught and killed. Regular foods include insects, mussels, crabs, fish, frogs, snakes, eggs, small rodents, and fruit (Rosevear 1974; Kingdon 1977). *Atilax* may throw such creatures as snails and crabs against hard surfaces to break the shell. A captive was seen to take a piece of beef rib between its forefeet, rear onto its hind feet with its forefeet held high, and then forcefully throw the bone to the floor of the cage, in an effort to break it.

Kingdon (1977) wrote that *Atilax* is usually seen alone, and that individuals are widely spaced and undoubtedly highly territorial. The young are said to be born in burrows in stream banks or on masses of vegetation gathered into heaps among reed beds. There is no evidence of a particular breeding season in West Africa (Rosevear 1974). Young have been found in South Africa in June, August, and October (Asdell 1964; Rowe–Rowe 1978b). The number of young per litter has been reported as one to three, and usually seems to be two. One water mongoose was still living after 17 years and 5 months in captivity (Marvin L. Jones, Zoological Society of San Diego, pers. comm.).

Rosevear (1974) observed that within the last 50 years *Atilax* has probably declined substantially in the drier parts of its range, because of human destruction of the available riverine habitat. *Atilax* is also widely hunted by people, as it is reputed to be a poultry thief.

Yellow mongoose (*Cynictis penicillata*), photo by Hans-Jurg Kuhn.

digger, constructing extensive underground tunnels and chambers with a number of entrance holes. Certain places within the burrow system are used for deposit of body wastes.

Cynictis is mainly diurnal, but may be active at night when living near people. It basks in the sun and sits up on its haunches to obtain a better view of the surroundings. It is agile and is capable of traveling at considerable speed. It seldom wanders more than a kilometer from the burrow. Apparently, pairs and perhaps entire colonies seek new homes when food becomes scarce in the vicinity of a burrow. The diet consists mainly of insects and other invertebrates (Herzig-Straschil 1977; Smithers 1971). Other reported foods of *Cynictis* include lizards, snakes, birds, the eggs of birds and turtles, small rodents, and even mammals as large as itself.

The number of individuals in a group has been reported to be as many as 50 or more, but Ewer (1973) wrote that such large colonies would be exceptional. Smithers (1971) indicated that maximum group size is 8 individuals. Although *Cynictis* dens communally, it usually forages alone or in pairs. According to Ewer, mating in South Africa occurs from September to December, births take place from October to January, estimates of gestation range from 45 to 57 days, litters contain 1 to 4 (usually 2) young, and weaning occurs at 6 weeks of age. Smithers (1971) stated that there may be sporadic breeding throughout the year in Botswana, with a peak at some interval from October to April, and that the number of embryos per pregnant female averages 3.2 and ranges from 2 to 5. One yellow mongoose lived in captivity for 15 years and 2 months (Marvin L. Jones, Zoological Society of San Diego, pers. comm.).

CARNIVORA; VIVERRIDAE; *Genus CYNICTIS* Ogilby, 1833

Yellow Mongoose

The single species, *C. penicillata,* is found in southern Angola, Namibia, Botswana, extreme southwestern Rhodesia, and South Africa (Meester and Setzer 1971).

Head and body length is 270 to 380 mm, and tail length is 180 to 280 mm. Smithers (1971) listed weights of 440 to 797 grams. The hair is fairly long, especially on the tail, which is somewhat bushy. The general color is dark orange yellow to light yellow gray. The underfur is rich yellow, the chin is white, the underparts and limbs are lighter than the back, and the tail is tipped with white. The individual guard hairs are usually yellowish in the basal half, followed by a black band and a white tip. There is a seasonal color change in the coat. The summer pelage, typical in January and February, is reddish, short, and thin. The winter coat, typical from June to August, is yellowish, long, and thick. The transitional coat, typical in November and December, is pale yellow. The hands have five fingers and the feet have four toes. The first or inner digit on the hand is small and is above the level of the other four, so it does not touch the ground. Females have three pairs of mammae (Ewer 1973).

The yellow mongoose frequents open country, preferably with loose soil, but, when disturbed, may take refuge among rocks or in brush along the banks of streams. It lives in communal burrows that may cover 50 sq meters or more. It sometimes uses holes made by other animals, such as *Pedetes,* but usually excavates its own burrows. It is an energetic

CARNIVORA; VIVERRIDAE; *Genus PARACYNICTIS* Pocock, 1916

Gray Meerkat, Selous' Mongoose

The single species, *P. selousi,* occurs in Angola, Zambia, Malawi, northern Namibia, Botswana, Rhodesia, Mozambique, and eastern South Africa (Meester and Setzer 1971). This species was originally referred to *Cynictis.*

Head and body length is 390 to 470 mm, and tail length is 280 to 400 mm. Smithers (1971) listed weights of 1.4 to 2.2 kg. The upper parts are dull buff gray, the belly is buffy, the feet are black, and the tail is white tipped. There is no rufous in the coloring, nor any spots or stripes. Unlike *Cynictis,* there are only four digits on all four limbs. The claws are long and slightly curved. These features are associated with a strong digging ability. *Paracynictis* can defend itself by expelling a strong-smelling secretion from its anal glands; its white-tipped tail, which makes the animal visible at night, may serve as a warning of this capability.

Paracynictis seems to prefer open scrub and woodland (Smithers 1971). It resides in labyrinthine burrows of its own construction. It is terrestrial and nocturnal, but has been seen aboveground by day. It has been described as shy and retiring. The diet consists of insects, other arthropods, frogs, lizards, and small rodents (Smithers 1971).

Apparently, each individual constructs its own burrow system and there is less social activity than in *Cynictis.* On the basis of limited data, Smithers (1971) suggested that the young are born in the warm, wet months, probably from September to March. Two pregnant females each contained two embryos.

Selous' mongoose (*Paracynictis selousi*), photo by Don Carter.

CARNIVORA; VIVERRIDAE; *Genus SURICATA*
Desmarest, 1804

Suricate, Slender-tailed Meerkat

The single species, *S. suricatta,* occurs in southwestern Angola, Namibia, Botswana, and South Africa (Meester and Setzer 1971).

Head and body length is 250 to 350 mm and tail length is 175 to 250 mm. Smithers (1971) listed weights of 626 to 797 grams for males and 620 to 969 grams for females. The coloration is a light grizzled gray. The rear portion of the back is marked with black transverse bars. The bars result from the alternate light and black bands of individual hairs coinciding with similar markings of adjacent hairs. The head is almost white, the ears are black, and the tail is yellowish with a black tip. The coat is long and soft, and the underfur is

Suricates (*Suricata suricatta*), photo by Bernhard Grzimek.

dark rufous in color. The body is quite slender, though this feature is difficult to see because of the long fur. There are scent glands, peripheral to the anus, which open into a pouch that presumably stores the secretion (Ewer 1973). The forefeet have very long and powerful claws. Females have six mammae (Grzimek 1975).

The suricate inhabits dry, open country, commonly with hard or stony ground (Smithers 1971). It is an efficient digger. Colonies on the plains may excavate their own burrows or share the holes of African ground squirrels (*Xerus*). Colonies in stony areas live in crevices among the rocks. Outside activity is almost entirely diurnal. The suricate seems to enjoy basking in the sun, lying in various positions or sitting up on its haunches like a prairie dog (*Cynomys*). If food supplies run low, a colony may establish a new den, one or two kilometers from the original site (Grzimek 1975). Individuals generally forage near the burrow, turning over stones and rooting in crevices. The diet is primarily insectivorous (Ewer 1973), but also includes small vertebrates, eggs, and vegetable matter.

According to Ewer (1973), *Suricata* is highly social. Groups usually have 2 or 3 family units and a total of 10 to 15 individuals. Each family contains a pair of adults and their young. The female may be larger than and dominate the male. At least 10 vocalizations have been identified, including a threatening growl and an alarm bark.

There is normally a single annual litter. Mating generally occurs in September and October, and births in November and December. The gestation period is 77 days, possibly less. The number of young per litter is two to five, usually four. The young weigh 25 to 36 grams each at birth, open their eyes after 10 to 14 days, and are weaned when they are 7 to 9 weeks old (Ewer 1973). Sexual maturity is attained by 1 year of age (Grzimek 1975). One suricate was still living after 12 years and 6 months in captivity (Marvin L. Jones, Zoological Society of San Diego, pers. comm.).

Suricata tames readily, is affectionate, and enjoys the warmth of snuggling close to its master. It is sensitive to cold. It is often kept about homes in South Africa to kill mice and rats. It should not be confused with the yellow or thick-tailed meerkat (*Cynictis*), with which it often associates. *Cynictis* is not so winning in its ways nor so pleasing as a pet.

CARNIVORA; VIVERRIDAE; *Genus* **CRYPTOPROCTA** *Bennett, 1833*

Fossa

The single species, *C. ferox*, is found in Madagascar (Meester and Setzer 1971). *Cryptoprocta* has sometimes been placed in the cat family (Felidae).

This is the largest carnivore of Madagascar. Head and body length is 610 to 800 mm, tail length is approximately the same, and shoulder height is about 370 mm. Weight is 7 to 12 kg (Albignac 1975). The fur is short, smooth, thick, soft, and usually reddish brown in color. Some black individuals have been captured. The mustache hairs are as long as the head. Like some other viverrids, *Cryptoprocta* has scent glands in the anal region which discharge a strong, disagreeable odor when the animal is irritated.

The general appearance is much like that of a large jaguarundi (*Felis yagouaroundi*) or small cougar (*Felis concolor*). The curved claws are short, sharp, and retractile like those of a cat, but the head is relatively longer than in the Felidae. *Cryptoprocta* walks in a flat-footed manner on its soles, as do bears, rather than on its toes, like cats.

The fossa dwells in forests and woodland savannahs, from coastal lowlands to mountainous areas at elevations of 2,000 meters (Meester and Setzer 1971). It is mainly nocturnal and crepuscular, is occasionally active by day, and often shelters in caves (Albignac 1972). A maternal den was located in an old termite mound and contained an unlined chamber about 70 cm deep, 100 cm wide, and 30 cm high (Albignac 1970b). The fossa is an excellent climber and pursues lemurs through the trees. Its diet consists mainly of small mammals and

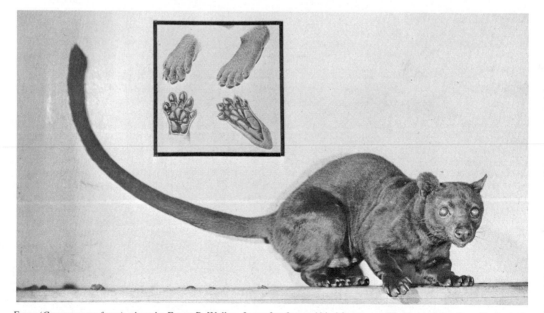

Fossa (*Cryptoprocta ferox*), photo by Ernest P. Walker. Inset: forefeet and hind feet, top and bottom, photos from *Zoologie de Madagascar*, G. Grandidier and G. Petit.

birds, but also includes reptiles, frogs, and insects (Meester and Setzer 1971; Albignac 1972).

Cryptoprocta is solitary, except during the reproductive season. Mating occurs in September and October, and births take place in the austral summer. The gestation period lasts about 3 months. There are usually two young, but sometimes three or four, each of which weighs about 100 grams at birth, Observations in captivity indicate that sexual maturity is not attained until the age of 4 years (Albignac 1972, 1975; Meester and Setzer 1971). A captive lived for 17 years (Marvin L. Jones, Zoological Society of San Diego, pers. comm.).

The fossa is a powerful predator, but has an exaggerated reputation for savagery and destructiveness. It normally flees at the sight of a human, though it may be dangerous if wounded. It sometimes preys on poultry, and some accounts claim that it attacks wild hogs and even oxen. It is widely hunted by people and is now depleted in numbers. It is classified as vulnerable by the IUCN (1972) and is on appendix 2 of the CITES.

CARNIVORA; *Family HYAENIDAE*

Aardwolf and Hyenas

This family of three Recent genera and four species is found in Africa, and southwestern and south-central Asia. The sequence of genera presented here follows that of Simpson (1945), who recognized two subfamilies: Protelinae, with the single genus *Proteles* (aardwolf); and Hyaeninae, with *Crocuta* and *Hyaena*. Some authorities, such as Meester and Setzer (1971), and Yalden, Largen, and Kock (1980), treat the Protelinae as a separate family.

Head and body length is 550 to 1,658 mm, tail length is 187 to 470 mm, and weight is 9 to 86 kg. The color pattern is striped in *Proteles* and *Hyaena hyaena,* spotted in *Crocuta,* and unmarked in *Hyaena brunnea,* except for its barred feet and pale head. A well-developed mane is present in *Proteles* and *Hyaena.* In all genera the guard hairs are coarse and the tail is bushy.

The head and forequarters are large, but the hind quarters are rather weak. The forelimbs, which are slender in *Proteles* but powerfully built in *Crocuta* and *Hyaena,* are longer than the hind limbs. *Proteles* has five digits on the forefoot and four on the hind foot, while *Crocuta* and *Hyaena* have four digits on each foot. The blunt claws are not retractile. Scent glands are present in the anal region, males do not have a baculum, and females have one to three pairs of mammae.

In *Proteles* the skull and jaws are weak, the cheek teeth are reduced and widely spaced, and the carnassials are not developed, but the canines are sharp and fairly powerful. This genus may have as few as 24 teeth (Ewer 1973). In *Crocuta* and *Hyaena,* on the other hand, the skull and jaws are strong and the teeth, including the carnassials, are powerfully developed for crushing bones. The dental formula of *Crocuta* and *Hyaena* is: (i 3/3, c 1/1, p 4/3, m 1/1) × 2 = 34. In all genera of the family the incisors are not specialized and the canines are elongate.

Hyenas and the aardwolf generally inhabit grassland or bush country, but may also occur in open forest. They live in caves, dense vegetation, or the abandoned burrows of other animals. They are mainly nocturnal. They move about on their digits and seem to trot tirelessly. The aardwolf feeds primarily on insects, while hyenas are efficient scavengers and predators of large mammals. Individuals may occur alone, in pairs, or in groups. Females normally bear a single litter per year.

The family is thought to have evolved from a branch of the Viverridae. The geological range of the Hyaenidae is Miocene to Recent in Asia, Pliocene to Pleistocene in Europe and North America, and Pliocene to Recent in Africa (Stains 1967; Kurten 1968).

CARNIVORA; HYAENIDAE; *Genus PROTELES*
I. Geoffroy St.-Hilaire, 1824

Aardwolf

The single species, *P. cristatus,* is found from the southern border of Egypt to central Tanzania, and from southern Angola and southern Zambia to the Cape of Good Hope (Meester and Setzer 1971; Corbet 1978; Kingdon 1977).

Head and body length is 550 to 800 mm, tail length is usually 200 to 300 mm, and shoulder height is usually 450 to 500 mm. Kingdon (1977) listed weight as 9 to 14 kg. The underfur is long, loose, soft, and wavy, and is interspersed with larger, coarser guard hairs. The body is yellow gray with black stripes. The legs are banded with black, and the part below the knee and hock is entirely black. The tail is bushy and tipped with black, and the hair along the back is long and crestlike, as evidenced in the Boer vernacular of "manhaarjakkal," or maned jackal. The jaws are weak and the cheek teeth are vestigial and widely spaced, but the canine teeth are sharp and reasonably powerful.

The aardwolf is most common on open, sandy plains or in bush country. It dens in holes in the ground, usually abandoned burrows of the aardvark (*Orycteropus*). It is primarily nocturnal and has been observed to cover just over one kilometer per hour while foraging (Bothma and Nel 1980). The diet consists almost entirely of termites and insect larvae that are dug out of the ground. Although a captive juvenile killed a number of birds (Kingdon 1977), there are no substantive data on such behavior in the wild. *Proteles* has been accused of preying on chickens and lambs, but evidence to the contrary is overwhelming, and it is known that the aardwolf can scarcely be induced to eat meat unless it is finely ground or cooked. When seen near carrion, *Proteles* is usually there to pick up carrion beetles, maggots, and other insects.

Proteles is generally solitary, with individuals well spaced. Bothma and Nel (1980) usually found a single animal per den, but noted that dens could be less than 500 meters apart. Pairs and family groups of five or six individuals are sometimes observed. Several females have been found, together with their young, in a single den (Kingdon 1977). Both sexes mark their ranges with secretions from anal scent glands. The hairs of the mane can be erected to make the aardwolf look twice its normal size, for purposes of intraspecific display or defense against predators. Surprisingly loud growls and roars can be produced under stress (Smithers 1971). If attacked by dogs, the aardwolf ejects musky fluid from its anal glands, and may fight effectively with its formidable canine teeth.

The following data on reproduction are available: gestation is thought to last 90 to 110 days, an animal several months old was taken in Uganda in July, a birth in Kenya occurred in late May, the number of young per litter is usually two or three and ranges from one to five, and reports of larger litters may represent the utilization of one den by two mothers (Kingdon 1977); litters of two to four young are born in November and December in southern Africa (Ewer 1973); and in Botswana, lactating females have been taken in January and April and pregnant females in July and October (Smithers 1971). A captive aardwolf was still living after 14

Aardwolf (*Proteles cristatus*), photo by R. Pucholt.

years and 3 months (Marvin L. Jones, Zoological Society of San Diego, pers. comm.).

Proteles is threatened in at least some areas by human hunting and habitat destruction. It was classified as rare in South Africa by Skinner, Fairall, and Bothma (1977).

CARNIVORA; HYAENIDAE; *Genus CROCUTA Kaup, 1828*

Spotted Hyena

In historical time, the single species, *C. crocuta*, was found throughout Africa south of the Sahara, except in equatorial rain forests (Meester and Setzer 1971). Through the end of the Pleistocene, the same species occurred in much of Europe and Asia (Kurten 1968).

Head and body length is 950 to 1,658 mm, tail length is 255 to 360 mm, shoulder height is 700 to 915 mm, and weight is 40 to 86 kg. On the average, females are 120 mm longer and 6.6 kg heavier than males (Kingdon 1977). The hair is coarse and woolly. The ground color is yellowish gray, and the round markings on the body are dark brown to black. There is no mane, or only a slight one. The jaws are probably the most powerful, in proportion to size, of any living mammal. From *Hyaena*, *Crocuta* differs in its larger size, shorter ears, paler and spotted coat, larger and more swollen braincase, and greatly reduced upper molar tooth.

The external genitalia of the female so closely resemble those of the male that the two sexes are practically impossible to distinguish in the field (Kruuk 1972; Kingdon 1977). The clitoris looks like the penis, occupies the same position, and is capable of elongation and erection. In addition, the female has a pair of sacs, filled with nonfunctional fibrous tissue,

that looks very much like the scrotum and is located in the same place. Both sexes have two anal scent glands that empty into the rectum. Females usually have a single pair of mammae.

The spotted hyena originally occupied nearly all the more open habitats of Africa south of the Sahara (Kingdon 1977). It is especially common in dry acacia bush, open plains, and rocky country. It is found at elevations of up to 4,000 meters. *Crocuta* is probably the most numerous of the large African predators, because of its ability quickly to eat and digest entire carcasses, including skin and bones, and because the plasticity of its behavior allows it to function effectively either as a solitary scavenger and predator of small animals or as a group-living hunter of ungulates. Dens vary greatly in size, sometimes accommodating entire communities of hyenas. They are usually in abandoned aardvark holes, but may also be in natural caves. Most activity is nocturnal or crepuscular. Up to 80 km may be covered in a night's foraging. *Crocuta* has keen senses of sight, hearing, and smell.

Except as noted, the information for the remainder of this account was taken from Kruuk's (1972) report of his studies in the Serengeti National Park and the Ngorongoro Crater of northern Tanzania. In these areas the spotted hyena shelters in holes on the plains or in shady places with bushy vegetation on hillsides. Dens may have 12 or more entrances. Activity occurs mainly in the first half of the night, then declines, and then increases again toward dawn. The Ngorongoro population remains in one general area throughout the year, but the Serengeti hyenas follow the seasonal migrations of their prey.

The diet consists mainly of medium-sized ungulates, especially wildebeest (*Connochaetes*), and mostly very young, very old, or otherwise inferior animals. One hyena commonly forces a herd of wildebeest to run, watches for a weak individual, and then begins a chase, which is soon joined by

Spotted hyena (*Crocuta crocuta*), photo by E. L. Button.

other hyenas. Zebras are hunted in a more organized manner by packs of 10 to 25 hyenas. Chases usually go for less than 2 km, with *Crocuta* averaging 40–50 km/hr. Maximum speed is about 60 km/hr. Roughly one-third of the hunts are successful. Far more animals are killed than are consumed as carrion. Although the spotted hyena is sometimes said to be a scavenger of the lion, most dead prey, on which both hyenas and lions were seen feeding, had been killed by the hyenas. The specialized teeth and digestive system of *Crocuta* allow it to crush and utilize much more bone than do other predators. *Crocuta* can consume 14.5 kg of food in one meal.

Population densities were estimated at 0.12 hyenas per sq km in the Serengeti and 0.16–0.24 per sq km in Ngorongoro. In the latter area, *Crocuta* is clearly organized into large communities or "clans," each with up to 80 individuals and each occupying a territory of about 30 sq km. Clans are usually divided into smaller hunting packs and individuals, but all members are evidently recognizable to one another. The members mark the borders of their territory with secretions from their anal scent glands, and defend the area from other clans. Groups that move about together commonly contain 2 to 10 individuals, but larger groups form to hunt zebra. In the Serengeti, the clan system is not so well developed and average group size is smaller. Within any group of

Crocuta, females are dominant over males and adults are dominant over young. The animals eat together and may compete by shoving one another, but there is generally no fighting. The spotted hyena is extremely vocal. The well-known laughing sound is emitted by an animal that is being attacked or chased. A whoop or howl is usually given spontaneously by a lone individual with the head held close to the ground. It begins low and deep, and increases in volume as it runs to higher pitches.

There is apparently no permanent pair bonding. Breeding may occur at any time of the year, perhaps with a peak in the wet season. Females are polyestrous and have an estrous cycle of 14 days. Gestation lasts about 110 days. There are usually two young per birth, occasionally one or three. At birth they weigh about 1.5 kg and have their eyes open. Each clan has a single central denning site, where all females bear their cubs, but each mother suckles its own young. Food is not carried to the den or regurgitated for the young. Weaning comes after 12 to 16 months of life, when the young are nearly full grown. Sexual maturity is attained at about 2 years of age by males and 3 years by females. A captive *Crocuta* lived for 41 years and 1 month (Marvin L. Jones, Zoological Society of San Diego, pers. comm.).

Human relationships with the spotted hyena have varied

Spotted hyena (*Crocuta crocuta*), photo by E. L. Button.

(Kruuk 1972; Kingdon 1977). Some native peoples protected it as a valuable scavenger, while others regarded it with superstitious dread. Certain tribes put their dead out for hyenas to consume. Fatal attacks on living humans have occurred. In the 20th century, *Crocuta* has been considered a predator of domestic livestock and game, and has been widely hunted, trapped, and poisoned. It has declined in numbers over much of its range, and has been eliminated in parts of East and South Africa. Kurten (1968) suggested that the disappearance of *Crocuta* in Eurasia in the early postglacial period was associated with the development of agriculture.

CARNIVORA; HYAENIDAE; **Genus HYAENA** Brünnich, *1772*

Striped and Brown Hyenas

There are two species (Meester and Setzer 1971; Corbet 1978):

H. hyaena (striped hyena), open country from Morocco and Senegal to Egypt and Tanzania, and from Asia Minor and Arabian Peninsula to southern Soviet Central Asia and eastern India;

H. brunnea (brown hyena), Namibia, Botswana, western and southern Rhodesia, southern Mozambique, South Africa.

Both species have pointed ears and erectile manes of long hairs, up to 305 mm in length, on the neck and back (in *Crocuta* the ears are rounded and there is little or no mane). In *Hyaena* the diameter of the upper molar tooth is at least twice that of the first upper premolar, and there is a small metaconid on the first lower molar (in *Crocuta* the upper molar is much smaller than the first premolar, and there is no metaconid on the first lower molar). The sexual organs of female *Hyaena*, unlike those of female *Crocuta*, do not closely resemble those of males. Like *Crocuta*, *Hyaena* has scent glands leading to an anal pouch that can be extruded for depositing secretions. *Hyaena* also has powerful jaws and teeth that can crush the largest bones of cattle, and a digestive system that can effectively utilize all parts of a carcass. Each front and hind foot has four digits. Additional information on each species is provided separately.

Hyaena hyaena. Except as noted, the information for the account of this species was taken from the review papers by Rieger (1979, 1981). Head and body length is 1,036 to 1,190 mm, tail length is 265 to 470 mm, shoulder height is 600 to 942 mm, and weight is 25 to 55 kg. Males and females are

Brown hyena (*Hyaena brunnea*), top, this specimen has a tumor near the left eye, photo from U.S. National Zoological Park. Striped hyena (*H. hyaena*), bottom, photo by Klaus Kussmann.

about the same size. The general coloration is gray to pale brown, with dark brown to black stripes on the body and legs. The mane along the back is more distinct than in *H. brunnea*. The hairs of the mane are up to 200 mm long, while those of the rest of the body are about 70 mm long. Females have two or three pairs of mammae.

The striped hyena prefers open or rocky country, and has an elevational range of up to 3,300 meters. It avoids true deserts and requires the presence of fresh water within 10 km. It is mainly crepuscular or nocturnal, resting by day in a temporary lair, usually under overhanging rocks (Kruuk 1976). Cubs are reared in natural caves, rocky crevices, or holes that are dug or enlarged by the parents. When searching for food, the striped hyena moves in a zigzag pattern at 2 to 4 km/hr (Kruuk 1976). The diet varies by season and area, but seems to consist mainly of mammalian carrion. The larger, more northerly subspecies prey on sheep, goats, donkeys, and horses. Other important foods are small vertebrates, insects, and fruit. Some food is stored in dense vegetation for later use. Food is often brought to the den, and eventually a large number of bones may accumulate (Skinner, Davis, and Ilani 1980).

The striped hyena may have a small defended territory around its breeding den, surrounded by a larger home range. In a radio-tracking study in the Serengeti, Kruuk (1976) determined home range to be about 44 sq km for a female and 72 sq km for a male. This species seems to be mainly solitary in East Africa, but there are indications of greater social activity farther north, where the animals are more prone to have a predatory life. In over half of his sightings of *H. hyaena* in Israel, Macdonald (1978) saw more than one animal. An adult pair and their offspring may sometimes forage together, and a family unit may persist for several years. Adult females are evidently intolerant of one another and are dominant over males. Scent marking with the anal glands is an important means of communication. Aggressive displays involve erection of the long hairs of the mane and tail. *H. hyaena* is much less vocal than *Crocuta*, but growls, whines, and makes several other sounds.

Females are polyestrous and breed throughout the year. The estrous cycle may be 40 to 50 days long. Estrus lasts 1

Young striped hyena (*Hyaena hyaena*), photo by Reginald Bloom.

day and may follow birth by 20–21 days. The gestation period is 88–92 days. The number of young per litter is 1–5, averaging 2.4. The young weigh about 700 grams each at birth, open their eyes after 5–9 days, first take solid food when 30 days old, and nurse for at least 4–5 months. Food is brought to the den for the young (Skinner and Ilani 1979). Sexual maturity usually comes at 2–3 years of age, and maximum longevity in captivity is 23–24 years.

The striped hyena has been known to attack and kill people, especially children. It can be tamed, however, and is said to become loyal and affectionate. Many of its parts are believed by some persons to have medicinal value. It is notorious in Israel for its destruction of melons, dates, grapes, apricots, peaches, and cucumbers (Kruuk 1976). The North African subspecies, *H. h. barbara,* is classified as endangered by the IUCN (1972) and the USDI (1980). It has declined through habitat loss and persecution as a predator. Its range is now restricted to the highlands of Morocco, Algeria, and Tunisia.

Brown hyenas (*Hyaena brunnea*), photo by Bernhard Grzimek.

Hyaena brunnea. Head and body length is 1,100 to 1,356 mm, tail length is 187 to 265 mm, shoulder height is 640 to 880 mm, and weight is 37 to 47.5 kg (Skinner 1976; Smithers 1971). Males average slightly larger than females. The coat is long, coarse, and shaggy. The general coloration is dark brown. The head is gray, the neck and shoulders are tawny, and the lower legs and feet are gray with dark brown bars.

The brown hyena is found largely within arid habitat— open scrub, woodland savannah, grassland, and semidesert. It is mainly nocturnal and crepuscular, sheltering by day in a rocky lair, dense vegetation, or an underground burrow (Eaton 1976). It commonly uses an aardvark hole as a den, but is capable of digging its own burrow. A favored maternal den site may be utilized for years (Owens and Owens 1979*a*). Foraging is done in a zigzag pattern along regular pathways. Nightly movements cover 10 to 20 km in the wet season and 20 to 30 km in the dry season. Food is located mainly by scent, but hearing and night vision are also keen (Owens and Owens 1978: Mills 1978).

The brown hyena is primarily a scavenger of the remains of large mammals that are killed by other predators (Owens and Owens 1978; Mills and Mills 1978; Skinner 1976). Other important foods include rodents, insects, eggs, and fruit. At certain times of the year, vegetation may constitute up to half of the diet of some individuals. *H. brunnea* is also known to frequent shorelines and feed on dead crabs, fish, and seals. It often stores excess food in shrubs or holes, and usually recovers it within 24 hours. Individuals have been seen to raid ostrich nests and to cache the eggs at scattered locations.

Eaton (1976) wrote that maximum population density is about one individual per 130 sq km, but Skinner (1976) found at least six adults and two cubs within 20 sq km. Social behavior evidently varies to some extent according to season (Owens and Owens 1978, 1979*a*, 1979*b*; Mills 1978). The animals in a given area are organized into a "clan" and are recognized by one another. Foraging is usually done alone. In the dry season, individual home range averages about 40 sq km. Individuals do not maintain territories, but use common hunting paths and frequently meet and exchange greetings. At times, especially during the wet season, up to six clan members join to exploit carrion. Each clan has a central breeding den and defends a surrounding territory of approximately 170 sq km.

A clan typically includes a dominant male, three or four subordinate males, four to six adult females, and associated young animals. A clan is not a strictly closed system, and emigration has been observed, especially among the young. There is a stable rank order, maintained by ritualized displays of aggression which commonly involve biting at the back of the neck, where the skin is tough. Individuals regularly deposit scent from their anal glands as they move about. This activity seems mainly for communication of information to other clan members, rather than demarcation of territory. Identified vocalizations include a squeal, growl, yell, scream, and squeak, all of which seem associated with conflict or submissive behavior.

Females are apparently seasonally polyestrous. Mating is thought to occur mainly from May to August, and births from August to November. The gestation period is about 3 months. The number of young per litter is two to five, usually three. The young open their eyes after 8 days and emerge from the den after 3 months (Eaton 1976; Skinner 1976). The young of several litters are raised together in a communal den, and each suckles from any lactating female. All clan females participate in bringing food to the den; there is no regurgitation. The cubs do not leave the vicinity of the den until they are about 14 months old (Owens and Owens 1979*a*, 1979*b*).

The brown hyena is classified as vulnerable by the IUCN (1976) and as endangered by the USDI (1980), and is on appendix 1 of the CITES. Although protected and still widely distributed in Botswana, it has declined drastically in range and numbers in Namibia and South Africa (Eaton 1976). The main problem is killing by people, who consider it a predator of domestic animals. The damage that it does seems to have been greatly exaggerated; for example, on one large cattle ranch in the Transvaal, where *H. brunnea* was studied, no depredations were known to have occurred in 15 years (Skinner 1976).

CARNIVORA; **Family FELIDAE**

Cats

This family of 4 Recent genera and 37 species has a natural distribution that includes all land areas of the world, except the West Indies, Madagascar, Japan, most of the Philippines, Celebes and islands to the east, New Guinea, Australia, New Zealand, Antarctica, and most Arctic and oceanic islands. Recognition here of 4 genera—*Felis, Neofelis, Panthera,* and *Acinonyx*—is based on Stains (1967), Ellerman and Morrison-Scott (1966), Corbet (1978), Meester and Setzer (1971), Lekagul and McNeely (1977), Chasen (1940), Medway (1977, 1978), Taylor (1934), Hall (1981), and Cabrera (1957). There is, however, a great diversity of opinion as to how the cats should be classified. Some authorities, such as Romer (1968), divide the family into only 2 genera, *Felis* and *Acinonyx*. Others, including some who have recently done extensive research on the subject, recognize many more genera. The following table shows the systematic arrangements of three of these authorities, plus a composite listing based on the sources cited above. The domestic cat (*Felis catus*) is not included in the table, but is thought to be most closely related to *F. silvestris.*

Leopard cat (*Felis bengalensis*) and young, photo from U.S. National Zoological Park.

Leyhausen (1979)	Hemmer (1978)	Ewer (1973)	Sources Cited Above
Genus *Acinonyx*	Genus *Acinonyx*	Genus *Felis*	Genus *Felis*
A. *jubatus*	A. *jubatus*	F. *silvestris*	Subgenus *Felis*
Genus *Prionailurus*	Genus *Puma*	F. *libyca*	F. *silvestris*
P. *bengalensis*	P. *concolor*	(including *ornata*)	(including *libyca* and *ornata*)
P. *rubiginosus*	Genus *Herpailurus*	F. *chaus*	F. *bieti*
P. *viverrinus*	H. *yagouaroundi*	F. *bieti*	F. *chaus*
P. *planiceps*	Genus *Prionailurus*	F. *margarita*	F. *margarita*
P. *iriomotensis*	Subgenus *Prionailurus*	(including *thinobia*)	(including *thinobia*)
Genus *Leopardus*	P. *bengalensis*	F. *nigripes*	F. *nigripes*
L. *pardalis*	P. *rubiginosus*	Genus *Leptailurus*	Subgenus *Otocolobus*
L. *wiedii*	P. *viverrinus*	L. *serval*	F. *manul*
L. *tigrinus*	Subgenus *Mayailurus*	Genus *Prionailurus*	Subgenus *Lynx*
L. *guigna*	P. *iriomotensis*	P. *bengalensis*	F. *lynx*
L. *geoffroyi*	Subgenus *Ictailurus*	P. *rubiginosus*	(including *canadensis*)
Genus *Lynchailurus*	P. *planiceps*	P. *viverrinus*	F. *pardina*
L. *pajeros*	Genus *Profelis*	Genus *Mayailurus*	F. *rufus*
(=*Felis colocolo*)	P. *aurata*	M. *iriomotensis*	Subgenus *Caracal*
Genus *Oreailurus*	Genus *Catopuma*	Genus *Ictailurus*	F. *caracal*
O. *jacobita*	C. *temmincki*	I. *planiceps*	Subgenus *Leptailurus*
Genus *Herpailurus*	(including *tristis*)	Genus *Otocolobus*	F. *serval*
H. *yagouaroundi*	C. *badia*	O. *manul*	Subgenus *Pardofelis*
Genus *Felis*	Genus *Leopardus*	Genus *Pardofelis*	F. *marmorata*
F. *silvestris*	L. *pardalis*	P. *marmorata*	F. *badia*
F. *libyca*	L. *wiedii*	P. *badia*	Subgenus *Profelis*
F. *ornata*	Genus *Oncifelis*	Genus *Profelis*	F. *temmincki*
F. *thinobia*	Subgenus *Oncifelis*	P. *temmincki*	(including *tristis*)
F. *margarita*	O. *tigrinus*	(including *tristis*)	F. *aurata*
F. *bieti*	O. *geoffroyi*	P. *aurata*	Subgenus *Prionailurus*
F. *nigripes*	O. *guigna*	Genus *Caracal*	F. *bengalensis*
F. *manul*	Subgenus *Lynchailurus*	C. *caracal*	F. *rubiginosus*
F. *chaus*	O. *pajeros*	Genus *Puma*	F. *viverrinus*
Genus *Leptailurus*	(=*Felis colocolo*)	P. *concolor*	F. *planiceps*
L. *serval*	Subgenus *Oreailurus*	Genus *Leopardus*	Subgenus *Mayailurus*
Genus *Lynx*	O. *jacobita*	L. *pardalis*	F. *iriomotensis*
L. *lynx*	Genus *Pardofelis*	L. *tigrinus*	Subgenus *Lynchailurus*
L. *canadensis*	P. *marmorata*	L. *wiedii*	F. *colocolo*
L. *pardina*	Genus *Neofelis*	L. *geoffroyi*	Subgenus *Leopardus*
L. *rufus*	N. *nebulosa*	Genus *Oncifelis*	F. *pardalis*
Genus *Pardofelis*	Genus *Uncia*	O. *guigna*	F. *wiedii*
P. *marmorata*	U. *uncia*	Genus *Lynchailurus*	F. *tigrinus*
P. *badia*	Genus *Panthera*	L. *colocolo*	F. *geoffroyi*
Genus *Profelis*	Subgenus *Tigris*	Genus *Oreailurus*	F. *guigna*
P. *temmincki*	P. *tigris*	O. *jacobita*	Subgenus *Oreailurus*
P. *tristis*	Subgenus *Panthera*	Genus *Herpailurus*	F. *jacobita*
P. *aurata*	P. *onca*	H. *yagouaroundi*	Subgenus *Herpailurus*
P. *caracal*	P. *pardus*	Genus *Lynx*	F. *yagouaroundi*
P. *concolor*	P. *leo*	L. *lynx*	Subgenus *Puma*
Genus *Uncia*	Genus *Leptailurus*	(including *canadensis*)	F. *concolor*
U. *uncia*	L. *serval*	L. *pardina*	Genus *Neofelis*
Genus *Neofelis*	Genus *Caracal*	L. *rufus*	N. *nebulosa*
N. *nebulosa*	C. *caracal*	Genus *Panthera*	Genus *Panthera*
N. *tigris*	Genus *Felis*	P. *leo*	Subgenus *Uncia*
Genus *Panthera*	Subgenus *Lynx*	P. *tigris*	P. *uncia*
P. *pardus*	F. *rufus*	P. *pardus*	Subgenus *Tigris*
P. *onca*	F. *canadensis*	P. *onca*	P. *tigris*
P. *leo*	F. *lynx*	P. *uncia*	Subgenus *Panthera*
	F. *pardina*	Genus *Neofelis*	P. *pardus*
	Subgenus *Otocolobus*	N. *nebulosa*	Subgenus *Jaguarius*
	F. *manul*	Genus *Acinonyx*	P. *onca*
	Subgenus *Felis*	A. *jubatus*	Subgenus *Leo*
	F. *nigripes*		P. *leo*
	F. *margarita*		Genus *Acinonyx*
	(including *thinobia*)		A. *jubatus*
	F. *chaus*		
	F. *bieti*		
	F. *silvestris*		
	(including *libyca* and *ornata*)		

The preceding table indicates that there is little disagreement at the specific level. The main controversy involves the division of what is here considered the genus *Felis* into genera and subgenera. Several authorities, including Meester and Setzer (1971), Hall (1981), and Chasen (1940), do not employ subgenera. All of the authorities cited here accept *Acinonyx* as a distinct genus, nearly all accept *Neofelis* (though with some disagreement as to content), and most do not accept *Uncia* as a full genus. *Panthera* is recognized as a separate genus by most, but not by Hall (1981) or Van Gelder (1977*b*). Hall, like most American authorities, treated *Lynx* as a distinct genus, and he included *Caracal* within *Lynx*. The preponderance of available information, however, suggests that if the genus *Felis* is accepted as comprising most species of the Felidae, *Lynx* should be included within *Felis* as no more than a subgenus, and *Caracal* should be a separate subgenus.

The living Felidae are sometimes divided into subfamilies. Stains (1967) recognized the Acinonychinae, for *Acinonyx* (cheetah); the Felinae, for *Felis;* and the Pantherinae, for *Neofelis* and *Panthera.* Leyhausen (1979) accepted the Acinonychinae, for *Acinonyx,* but put all other cats in the Felinae. The cheetah is usually placed last on lists of cats to express its aberrant characters. Actually, however, it seems in some respects to be little changed from the primitive stock that gave rise to all other cats. Location at the beginning of a systematic account, as was done by Leyhausen (1979) and Hemmer (1978), is thus fully appropriate. Moreover, Hemmer's placement of *Felis (Puma) concolor* next to *Acinonyx* may be fitting, as Adams (1979) has presented evidence that these two cats evolved from a common ancestor in North America.

In the family Felidae, head and body length is 337 to 2,800 mm, tail length is 51 to 1,100 mm, and weight is 1.5 to 306 kg. The color varies from gray to reddish and yellowish brown, and there are often stripes, spots, or rosettes. The pelage is soft and woolly. Its beautiful, glossy appearance is maintained by frequent cleaning with the tongue and paws. The tail is well haired, but not bushy, and the whiskers are well developed.

Cats have a lithe, muscular, compact, and deep-chested body. The limbs range from short to long and sinewy. The forefoot has five digits and the hind foot has four. In most species the claws are retractile (to prevent them from becoming blunted), large, compressed, sharp, and strongly curved (to aid in holding living prey). In the cheetah (*Acinonyx*), however, they are only semiretractile and are relatively poorly developed. Except for the naked pads, the feet are well haired to assist in the silent stalking of prey. The baculum is vestigial or absent. Females have two to four pairs of mammae.

The head is rounded and shortened, the ears range from rounded to pointed, and the eyes have pupils that contract vertically. The tongue is suited for laceration and retaining food within the mouth, its surface being covered with sharp-pointed, recurved, horny papillae. The dental formula is: (i 3/3, c 1/1, p 2–3/2, m 1/1) × 2 = 28 or 30. The incisors are small, unspecialized, and placed in a horizontal line. The canines are elongate, sharp, and slightly recurved. The carnassials, which cut the food, are large and well developed. The upper molar is small.

The dentition of the Felidae, with emphasis on the teeth that are used for seizing and cutting rather than on those used for grinding, reflects the highly predatory life style of the

Fishing cat (*Felis viverrinus*), photo from U.S. National Zoological Park.

A. Lion (*Panthera leo*), photo by Leonard Lee Rue III. B. Clouded leopard (*Neofelis nebulosa*), photo by Bernhard Grzimek. C. Leopard (*Panthera pardus*), photo by Leonard Lee Rue III.

family. Cats prey on almost any mammal or bird they can overpower, and occasionally on fish and reptiles. They stalk their prey, or lie in wait, and then seize the quarry with a short rush. The cheetah (*Acinonyx*), however, is adapted for a more lengthy pursuit at high speeds. Cats walk or trot on their digits, often placing the hind foot in the track of the forefoot. They are agile climbers and good swimmers, and have acute hearing and sight. Many are nocturnal, but some are active mainly during daylight. They shelter in trees, hollow logs,

caves, crevices, abandoned burrows of other animals, or dense vegetation. They defend themselves with fang and claw, or flee, sometimes seeking refuge in trees.

Cats are usually solitary, but are sometimes found in pairs or larger groups. The females of most species are polyestrous and give birth once a year, some can have two litters annually, and those of the larger species sometimes breed only every 2 or 3 years. Gestation periods of 55 to 119 days have been reported, and litter size is usually 1 to 6 young. At birth the kittens are usually blind and helpless, but are haired and often spotted. They remain with their mother until they can hunt for themselves. Potential longevity is probably at least 15 years for most species, and some individuals have lived over 30 years.

Some of the larger species of cat have occasionally become a serious menace to human life in localized areas. Various felids are considered by some people to be a threat to domestic animals. Certain species, especially those with spotted or striped skins, are of considerable value in the fur trade. The big cats, and some of the smaller ones, are sought as trophies by hunters. For all of these reasons the Felidae have been extensively hunted and killed by people. As a result, many species and subspecies have become rare or endangered in at least parts of their ranges. The entire family has been placed on appendix 2 of the CITES, except for those species on appendix 1 and the domestic cat (*Felis catus*).

The geological range of the Felidae is late Eocene or early Oligocene to Recent in North America and Eurasia, Miocene to Recent in Africa, and Pleistocene to Recent in South America. There are several extinct groups, including the saber-toothed cats (subfamily Machairodontinae), which persisted through the end of the Pleistocene.

CARNIVORA; FELIDAE; **Genus FELIS** Linnaeus, 1758

Small Cats, Lynxes, and Cougar

There are 14 subgenera and 30 species (Guggisberg 1975; Ewer 1973; Ellerman and Morrison-Scott 1966; Corbet 1978; Meester and Setzer 1971; Lekagul and McNeely 1977; Chasen 1940; Medway 1977, 1978; Taylor 1934; Hall 1981; Cabrera 1957; Roberts 1977):

subgenus *Felis*

F. *silvestris* (wild cat), France and Spain to north-central China and central India, Britain, Balearic Islands, Sardinia, Corsica, Crete, woodland and savannah zones throughout Africa;
F. *catus* (domestic cat), world-wide in association with people;
F. *bieti* (Chinese desert cat), southern Mongolia, central China;
F. *chaus* (jungle cat), Volga River Delta and Egypt to Sinkiang and Indochina, Sri Lanka;
F. *margarita* (sand cat), desert zone from Morocco and northern Niger to Soviet Central Asia and Pakistan;
F. *nigripes* (black-footed cat), Namibia, Botswana, South Africa;

subgenus *Otocolobus*

F. *manul* (Pallas's cat), Caspian Sea and Iran to southeastern Siberia and Tibet;

subgenus *Lynx*

F. *lynx* (lynx), western mainland Europe to eastern Siberia and Tibet, possibly Sardinia, Sakhalin, Alaska, Canada, northern conterminous United States;
F. *pardina* (Spanish lynx), Spain, Portugal;
F. *rufus* (bobcat), southern Canada to Baja California and central Mexico;

subgenus *Caracal*

F. *caracal* (caracal), Arabian Peninsula to Aral Sea and northwestern India, most of Africa;

subgenus *Leptailurus*

F. *serval* (serval), Morocco, Algeria, most of Africa south of the Sahara;

subgenus *Pardofelis*

F. *marmorata* (marbled cat), Nepal to Indochina and Malay Peninsula, Sumatra, Borneo;
F. *badia* (bay cat), Borneo;

subgenus *Profelis*

F. *temmincki* (Asian golden cat), Tibet and Nepal to southeastern China and Malay Peninsula, Sumatra;
F. *aurata* (African golden cat), Senegal to Kenya and northern Angola;

subgenus *Prionailurus*

F. *bengalensis* (leopard cat), Ussuri region of southeastern Siberia, Manchuria, Korea, Quelpart and Tsushima islands (between Korea and Japan), eastern China, Taiwan, Hainan, Pakistan to Indochina and Malay Peninsula, Sumatra, Java, Bali, Borneo, several islands in the western and central Philippines;
F. *rubiginosus* (rusty-spotted cat), southern India, Sri Lanka;
F. *viverrinus* (fishing cat), Pakistan to Indochina, Sri Lanka, Sumatra, Java;
F. *planiceps* (flat-headed cat), Malay Peninsula, Sumatra, Borneo;

subgenus *Mayailurus*

F. *iriomotensis* (Iriomote cat), Iriomote Island (southern Ryukyu Islands);

subgenus *Lynchailurus*

F. *colocolo* (pampas cat), Ecuador and Mato Grosso region of Brazil to central Chile and Patagonia;

subgenus *Leopardus*

F. *pardalis* (ocelot), Arizona and Texas to northern Argentina;
F. *wiedii* (margay), northern Mexico and possibly southern Texas to northern Argentina and Uruguay;
F. *tigrinus* (little spotted cat), Costa Rica to northern Argentina;
F. *geoffroyi* (Geoffroy's cat), Bolivia and extreme southern Brazil to Patagonia;
F. *guigna* (kodkod), central and southern Chile, southwestern Argentina;

subgenus *Oreailurus*

F. *jacobita* (mountain cat), the Andes of southern Peru, southwestern Bolivia, northeastern Chile, and northwestern Argentina;

subgenus *Herpailurus*

F. *yagouaroundi* (jaguarundi), southern Arizona and southern Texas to northern Argentina;

subgenus *Puma*

F. *concolor* (cougar, puma, mountain lion), southern Yukon and Nova Scotia to southern Chile and Patagonia.

All of the above subgenera are sometimes treated as full genera, and still other genera have been used for certain species here assigned to *Felis*. In contrast, some authorities include all living cats, except *Acinonyx*, in the genus *Felis*,

A. Marbled cat (*Felis marmorata*), photo from *Jour. Bombay Nat. Hist. Soc*. B. Serval (*F. serval*), photo by Bernhard Grzimek. C. Golden cat (*F. temmincki*), photo by Ernest P. Walker. D. Pallas's cat (*F. manul*), photo by Howard E. Uible. E. Pumas (*F. concolor*), photo by Ernest P. Walker.

Top, black-footed cat (*Felis nigripes*), photo by John Visser. B. Little spotted cat (*F. tigrinus*). C. Jungle cat (*F. chaus*), photos by Ernest P. Walker. D. European wild cat (*F. silvestris*), photo by Bernhard Grzimek. E. Jaguarundi (*F. yagouaroundi*), photo by Ernest P. Walker.

and some do not employ subgenera (see account of the family Felidae).

North American authorities, such as Hall (1981), commonly recognize *Lynx* Kerr, 1792 as a distinct genus and the New World populations of *Lynx lynx* as a separate species, *Lynx canadensis*. Jones, Carter, and Genoways (1979), however, used the name *Felis lynx* instead of *Lynx canadensis*. There is also controversy regarding the Spanish lynx, *F. pardina*. It is recognized as a full species by most authorities, including Leyhausen (1979), Hemmer (1978), and Ewer (1973), but was considered only a subspecies of *F. lynx* by Ellerman and Morrison-Scott (1966) and Corbet (1978). In a recent evolutionary study, Werdelin (1981) treated *Lynx* as a full genus, and *F. canadensis* and *F. pardina* as species distinct from *F. lynx*.

The African and most Asian populations of *F. silvestris* have been assigned to a separate species, *F. libyca*, by numerous authorities, including Meester and Setzer (1971). There is substantial evidence, however, that *F. silvestris* and *F. libyca* intergrade in the Middle East, and there is also much doubt as to whether *F. bieti* of East Asia is distinct from *F. silvestris* (Corbet 1978). Leyhausen (1979) listed *F. silvestris* and *F. libyca* as separate species, and considered the populations found from Iran to India to represent still another species, *F. ornata*. He united all three, however, in a single superspecies. Leyhausen (1979) also recognized *F. thinobia* of Soviet Central Asia and *F. tristis* of Tibet as species distinct from, respectively, *F. margarita* and *F. temmincki*.

Except for *F. concolor* and some individuals of *F. lynx*, the members of the genus *Felis* are smaller than those of the three other genera of cat. Otherwise, the characters of *Felis* are the same as those set forth for the family Felidae. From *Neofelis*, *Felis* is distinguished by having relatively shorter canine teeth and a smaller gap between the canines and the cheek teeth. From *Panthera*, it is distinguished by having a completely ossified hyoid apparatus without an elastic ligament. From *Acinonyx*, it is distinguished by having a greater gap between the canines and the cheek teeth, and usually by having shorter limbs and fully retractile claws. Additional information is provided separately for each species. Except as noted, the information for the following accounts was taken from Guggisberg (1975).

Felis silvestris.   Head and body length is usually 500 to 750 mm, tail length is 210 to 350 mm, and weight is usually 3 to 8 kg. Males are larger, on the average, than females. The fur is long and dense. In European populations the ground color is yellowish gray, the underparts are paler, and the throat is white. There are four or five longitudinal stripes from the forehead to the nape, merging into a dorsal line that ends near the base of the tail. The tail has several dark encircling marks and a blackish tip. The legs are transversely striped. In African and Asian populations the general coloration varies from pale sandy to gray brown and dark gray; there may be a pattern of distinct spots or stripes. European animals are generally about one-third larger than domestic cats, and have longer legs, a broader head, and a relatively shorter, more bluntly ending tail. Females have eight mammae.

The wild cat occupies a variety of forested, open, and rocky country. It is mainly nocturnal and crepuscular, spending the day in a hollow tree, thicket, or rock crevice. It climbs with great agility and seems to enjoy sunning itself on a branch. It normally stays in one area, within which it has several dens and a system of hunting paths. It usually stalks its prey, attempting to approach to within a few bounds. The diet consists mainly of rodents and other small mammals, and also includes birds, reptiles, and insects.

The wild cat is usually solitary, each individual having a well-defined home range of about 50 ha. Males defend these areas, but may wander outside of their usual ranges during times of food shortage or to locate estrous females. Mating occurs in Europe and Central Asia from about January to March. Females are polyestrous, with heat lasting 2 to 8 days. Several males collect around a female in heat; there is considerable screeching and other vocalization, and sometimes violent fighting. There is usually only a single litter per year, though a second is occasionally produced in the summer. Births in East Africa may occur at any time of year, but tend to peak during and in southern Africa during the wet season (Kingdon 1977; Smithers 1971). The gestation period averages 66 days in Europe and about 1 week less in Africa. Litters usually contain two or three young in the wild. At the Berne Zoo, Meyer-Holzapfel (1968) observed that births occurred from March to August, and that litter size averaged four and ranged from one to eight. The young weigh about 40 grams each at birth, open their eyes after about 10 days, nurse for about 30 days, emerge from the den at 4 or 5 weeks of age, begin to hunt with the mother at 12 weeks, probably separate from her at 5 months, and attain sexual maturity when around 1 year old. According to Kingdon (1977), captives have lived up to 15 years.

The wild cat once occupied most of Europe, but withdrew from Scandinavia and most of Russia by the Middle Ages, because of climatic deterioration. In modern times, especially during the 19th century, the species was intensively hunted by persons, who considered it a threat to game and domestic animals, and so was eliminated from much of western and central Europe. Diversion of human activity during World Wars I and II apparently stimulated recovery in such places as Scotland and West Germany. *F. silvestris* is now protected and encouraged in several nations (Smit and Van Wijngaarden 1976).

Felis catus.   According to the National Geographic Society (1981), there are more than 30 different breeds of domestic cat, and the average measurements of several popular breeds are: head and body length, 460 mm, and tail length, 300 mm. E. Jones (1977) found that feral males on Macquarie Island, south of Australia, averaged 522 mm in head and body length, 269 mm in tail length, and 4.5 kg in weight, and that females there averaged 478 mm in head and body length, 252 mm in tail length, and 3.3 kg in weight. On Macquarie, 90 percent of the cats were orange or tabby, and the remainder were black or tortoiseshell. Female *F. catus* have four pairs of mammae.

The domestic cat is evidently descended primarily from the wild cat of Africa and extreme southwestern Asia, *F. silvestris libyca*. The latter may have been present in towns in Palestine as long ago as 7,000 years, and actual domestication occurred in Egypt about 4,000 years ago. Introduction to Europe began around 2,000 years ago, and some interbreeding occurred there with the wild subspecies *F. silvestris silvestris*. Domestication seems originally to have had a religious basis (Grzimek 1975; Kingdon 1977). The cat was the object of a passionate cult in ancient Egypt, where a city, Bubastis, was dedicated to its worship. The followers of Bast, the goddess of pleasure, put bronze statues of cats in sanctuaries and carefully mummified the bodies of hundreds of thousands of the animals.

There have been relatively few detailed field studies of *F. catus*, but there is no reason to think that its behavior and ecology under noncaptive conditions differ greatly from what has been found for *F. silvestris*. On Macquarie Island, where the cat population has been feral since 1820, E. Jones (1977) obtained specimens in a variety of habitats by both day and night. The cats sheltered in rabbit burrows, thick vegetation, or piles of rocks. The diet consisted largely of rabbits (also introduced on the island), and also included rats,

A. Leopard cat (*Felis bengalensis*), photo by Lim Boo Liat. B. Margay (*F. wiedii*), photo by Ernest P. Walker. C. Pampas cat (*F. colocolo*), photo from San Diego Zoological Garden. D. Fishing cat (*F. viverrinus*), photo from New York Zoological Society. E. Geoffroy's cat (*F. geoffroyi*), photo by Ernest P. Walker. F. African wild cat (*F. silvestris libyca*), photo by Bernhard Grzimek.

Jungle cat (*Felis chaus*), photo from San Diego Zoological Society.

mice, birds, and carrion. Population density was estimated at two to seven cats per sq km.

In a rural area of southern Sweden, Liberg (1980) found a population density of 2.5 to 3.3 per sq km. About 10 percent of the cats were feral, and the rest, including all of the females, were associated with human households. Adult females lived alone or in groups of up to 8 usually closely related individuals. Each member of a group had a home range of 30 to 40 ha., which overlapped extensively with the ranges of other members of the same group but not with the ranges of the cats in other groups. Most females spent their lives in the area in which they were born, seldom wandering more than 600 meters away. Nonferal males remained in their area of birth, along with females, until they were 1½ to 3½ years old, but then left and tried to settle somewhere else. Males living in the same group had separate home ranges. There were 6 to 8 feral males in the study area; their home ranges were 2 to 4 km across, partly overlapped one another, and sometimes included the areas used by several groups of females.

According to Ewer (1973), the house cat is basically solitary, but individuals in a given area seem to have a social organization and hierarchy. A male, newly introduced to an area, normally must undergo a series of fights before its position is stabilized in relation to other males. Both males and females sometimes gather within a few meters of each other, without evident hostility. A male and female may form a bond that extends beyond the mating process. Females are polyestrous and normally produce two litters annually. They may mate with more than one male per season, and, if a litter is lost, soon enter estrus again. The gestation period averages 65 days. The number of young per litter averages four and ranges from one to eight. Kittens weigh 85 to 110 grams each at birth, open their eyes after 9 to 20 days, are weaned at 8 weeks, and attain independence at about 6 months of age. Hemmer (1976) listed age of sexual maturity in females as 7 to 12 months.

Although the cat has sometimes been venerated, it has also been associated with evil. Certain superstitions concerning it have persisted to modern times (Grzimek 1975). The species is now generally looked upon with favor by most cultures, but free-ranging cats are often considered one of the greatest decimators of native wildlife, especially song birds (Lowery 1974).

Felis bieti.   Head and body length is 685 to 840 mm, and tail length is 290 to 350 mm. The general coloration is yellowish gray, the back is somewhat darker, and there are practically no markings on the flanks. The tail is tipped with black and has three or four subterminal blackish rings. Despite the common name "desert cat," this species has been found only in steppe country, and in mountains covered with brush and forest.

Felis chaus.   Head and body length is 500 to 750 mm, tail length is 250 to 290 mm, and weight is 4 to 16 kg (Novikov 1962; Lekagul and McNeely 1977). The general coloration varies from sandy or yellowish gray to grayish brown and tawny red, usually with no distinct markings on the body. The tail has several dark rings and a black tip. Lekagul and McNeely (1977) noted that the legs are proportionally the longest of any felid in Thailand, and are thus helpful in running down prey.

The jungle cat is found in either woodland or open country, from sea level to elevations of 2,400 meters. It dens in thick vegetation or in the abandoned burrow of a badger, fox, or porcupine. It is active either by day or by night. The diet consists mainly of hares and other small mammals, and also includes birds, frogs, and snakes. Mating occurs in Soviet Central Asia in February and March, but 3-week-old kittens have been found in Assam in January and February. The gestation period is 66 days and the usual litter size is three to five young. Sexual maturity comes at the age of 18 months.

Felis margarita.   Head and body length is 450 to 572 mm, and tail length is 280 to 348 mm. The general coloration is

Sand cat (*Felis margarita*), photo by Lothar Schlawe.

pale sandy to gray straw ochre, the back is slightly darker, and the belly is white. A fulvous reddish streak runs across each cheek from the corner of the eye. The tail has two or three subterminal rings and a black tip. The pelage is soft and dense. The soles of the feet are covered with dense hair. The limbs are short and the ears are set low on the head.

The sand cat is adapted to extremely arid terrain, such as shifting dunes of sand. The padding on its soles facilitates progression over loose, sandy soil. Activity is mainly nocturnal and crepuscular. The day is spent in a shallow burrow. Prey consists of jerboas and other rodents, and occasionally hares, birds, and reptiles. The sand cat is apparently able to subsist without drinking free water.

In Pakistan there may be two litters annually, as kittens have been found in both March–April and October (Roberts 1977). The gestation period is 59–63 days, litters contain two to four young, average weight at birth is 39 grams, and the eyes open after 12 to 16 days (Hemmer 1976).

The sand cat seems to be generally rare. The subspecies in Pakistan, *F. m. scheffeli,* was not discovered until 1966 and is now classified as endangered by the IUCN (1978). It apparently declined drastically through uncontrolled exploitation by commercial animal dealers from 1967 to 1972.

Felis nigripes. This is the smallest cat. Head and body length is 337 to 500 mm, tail length is 150 to 200 mm, and weight is 1.5 to 2.75 kg. The general coloration is dark ochre to pale ochre or sandy, being somewhat darker on the back and paler on the belly. There is a bold pattern of dark brown to black spots arranged in rows on the flanks, throat, chest, and belly. There are two streaks across each cheek, two transverse bars on the forelegs, and up to five transverse bars

on the haunches. The tail has a black tip and two or three subterminal bands. The bottoms of the feet are black (since the animal walks on its toes, much of the black is usually visible).

The black-footed cat inhabits dry, open country. It shelters in old termite mounds and the abandoned burrows of other mammals. It is mainly nocturnal, but in captivity has been found to be more active by day than are most other small cats. The diet probably includes rodents, birds, and reptiles.

This felid seems to be highly unsocial. Even opposite sexes evidently come together only for 5 to 10 hours. Gestation lasts 59 to 68 days and litters contain one to three young. The kittens leave the nest after 28 or 29 days and take their first solid food a few days later. Captive females have not initially entered heat until 15 and 21 months old (Grzimek 1975).

Felis manul. Head and body length is 500 to 650 mm, tail length is 210 to 310 mm, and weight is 2.5 to 3.5 kg. The general coloration varies from light gray to yellowish buff and russet; the white tips of the hairs produce a frosted silvery appearance. There are two dark streaks across each side of the head and four rings on the dark-tipped tail. The coat is relatively longer and more dense than that of any other wild species of *Felis*. The fur is especially long near the end of the tail, and on the underparts of the body is almost twice as long as on the back and sides. Such an arrangement provides good insulation for an animal that spends much time lying on frozen ground and snow. The body is massive, the legs are short and stout, the head is short and broad, and the ears are very short, bluntly rounded, and set low and wide apart.

Pallas's cat inhabits steppes, deserts, and rocky country,

Pallas's cat (*Felis manul*), photo by Bernhard Grzimek.

up to elevations of over 4,000 meters. It dens in a cave, crevice, or burrow dug by another animal. It is usually nocturnal, but is occasionally seen by day. It feeds on pikas and other small mammals. According to Stroganov (1969), the young are born in Siberia in late April and May, and litter size is five or six.

Felis lynx. In Eurasian populations, head and body length is 800 to 1,300 mm, tail length is 110 to 245 mm, shoulder height is 600 to 750 mm, and weight is 8 to 38 kg (Van Den Brink 1968; Novikov 1962). In North American populations, head and body length is about 800 to 1,000 mm, tail length is 51 to 138 mm, and weight is 5.1 to 17.2 kg (Banfield 1974; Burt and Grossenheider 1976). In any given area, males are generally larger than females. The coloration varies, but is commonly yellowish brown, the upper parts may have a gray frosted appearance, the underparts are more buffy, and there is often a pattern of dark spots. The markedly short tail may have several dark rings and is tipped with black. The fur is long, lax, and thick. It is especially long on the lower cheeks in winter, and gives the impression of a ruff around the neck. The triangular ears are tipped by tufts of black hairs, about 40 mm long. The legs are relatively long. The paws are large and densely furred, an adaptation for moving over winter snow. Females have four mammae (Banfield 1974).

The lynx is generally found in tall forests with dense undergrowth, but may also enter open forest, rocky areas, or tundra. For shelter it constructs a rough bed under a rock ledge, fallen tree, or shrub (Banfield 1974). It is mainly nocturnal; reports of average nightly movement range from 5 to 19 km (Ewer 1973). The lynx usually keeps to one area, but may migrate under adverse conditions. The maximum known movement was by a female that was marked on 5 November 1974 in northern Minnesota and trapped on 20 January 1977 in Ontario, 483 km from the point of release (Mech 1977e). The lynx climbs well and is a good swimmer, sometimes crossing wide rivers. It hunts mainly by eye and also has well-developed hearing. It usually stalks its prey to within a few bounds, or it may wait in ambush for hours

Canada lynx (*Felis lynx*), photo from San Diego Zoological Garden.

(Stroganov 1969). Reports from much of the range of the species indicate that leporids form a major part of the diet. The snowshoe rabbit (*Lepus americanus*) is of particular importance in North America. Deer and other ungulates are utilized heavily in certain areas, especially during the winter. Other foods include rodents, birds, and fish.

In Canada the numbers of lynx seem to fluctuate, together with those of the snowshoe rabbit, in a regular cycle. Numerical peaks occur at an average interval of 9.6 years, though not at the same time all over Canada. Several authorities have suggested that this phenomenon is actually an artifact, perhaps associated with the intensity of human trapping, but Finerty (1979) upheld the traditional view. Studies in central Alberta (Brand and Keith 1979; Brand, Keith, and Fischer 1976; Nellis, Wetmore, and Keith 1972) indicate that the main direct cause of the cyclic drop in lynx numbers is postpartum mortality of kittens, through lack of food. There is also a reduced rate of pregnancy in females. Population density is this area varied from a low of 2.3 individuals per 100 sq km in the winter of 1966/67 to a peak of 10 per 100 sq km in the winter of 1971/72. Maximum population density in eastern Europe and the Soviet Union is 5 per 100 sq km (Ewer 1973).

The reported size of individual home range varies from 11 to 300 sq km (Ewer 1973; Guggisberg 1975; Brand, Keith, and Fischer 1976; Mech 1980). The lynx is probably territorial, but the ranges of females may overlap one another to some extent, and the range of a male may include that of a female and her young. Adults tend to avoid one another, except during the breeding season. Females are monestrous and bear a single litter per year (Ewer 1973; Banfield 1974). Mating occurs mainly in February and March, gestation lasts 9 or 10 weeks, and litters contain one to five, usually two or three, young. Banfield (1974) gave birth weight as 197 to 211 grams. Lactation lasts for 5 months, but some meat is eaten by 1-month-old kittens. The young usually remain with their mother until the winter mating season, and siblings may stay together for a while afterward. The age of sexual maturity is 21 months in females and 33 months in males. Captives have lived up to 21 years (Grzimek 1975).

The lynx has been widely hunted by people, because it is considered a predator of game and domestic animals and because its pelt has sometimes been of commercial value. It disappeared from much of Europe during the 19th century. Major populations there now occur only in the Soviet Union, Scandinavia, and the Carpathian region. Small, isolated groups are present in northeastern Poland, the southern Balkans, and possibly southern France. Reintroductions have been carried out recently in parts of Germany, Austria, Switzerland, Italy, and Yugoslavia (Smit and Van Wijngaarden 1976). The lynx may have occurred in Palestine before that area was largely deforested; in the Arabian region the species is now found only in northern Iraq, where it is rare (Harrison 1968).

The range of *F. lynx* once extended at least as far south as Oregon, southern Colorado, Nebraska, southern Indiana, and Pennsylvania (Hall 1981). The only substantial population now remaining in the conterminous United States is that of western Montana and nearby parts of northern Idaho and northeastern Washington. On occasion, especially during periods of cyclically high populations in Canada, the species is found in the states of the northern plains and upper Great Lakes regions. There are small numbers in northern New England and Utah, and possibly in Oregon, Wyoming, and Colorado (Deems and Pursley 1978; Cowan 1971; Gunderson 1978; Olterman and Verts 1972; Armstrong 1972; Hunt 1974).

The lynx has been exterminated in parts of southeastern Canada, but still occurs regularly over most of that country.

Its numbers there seem to be affected more by the cyclical availability of food than by human hunting pressure. Populations evidently were relatively high until the early 1900s, declined until the mid-20th century, and then again increased (Cowan 1971). Subsequent highs and lows in the reported number of pelts taken in Canada are: 1949/50 trapping season, 3,734; 1954/55, 14,427; 1956/57, 8,748; 1962/63, 51,376; 1966/67, 13,038; 1971/72, 53,589; 1975/76, 13,162; and 1979/80, 34,366. The average price per pelt rose dramatically from $3.62 in 1953/54 to $30.52 in 1968/69 and to a peak of $336.36 in 1978–79 (Statistics Canada 1981).

Felis pardina. Head and body length is 850 to 1,100 mm, tail length is 125 to 130 mm, and shoulder height is 600 to 700 mm. The upper parts are yellowish red and the underparts are white. There are round black spots on the body, tail, and limbs. The ears have tufts and the face has a prominent fringe of whiskers (Smit and Van Wijngaarden 1976; Van Den Brink 1968).

The Spanish lynx inhabits open forests and thickets. Its general habits are much like those of *Felis lynx*. It feeds mainly on rabbits. Home range diameter is 4 to 10 km in summer and somewhat larger in winter. Mating is thought to occur mainly in January. According to the IUCN (1978), the gestation period is 63–73 days and litter size is two to three young.

The Spanish lynx is classified as endangered by the IUCN (1978) and the USDI (1980). It formerly occurred throughout the Iberian Peninsula, but now is restricted to scattered mountainous areas and the Guadalquivir Delta. The total number of animals is estimated at 1,000 to 1,500. There was a major decline during the 1950s and 1960s, when the disease myxomatosis hit the rabbit populations. The decline is continuing as suitable rabbit and lynx habitat is replaced by cereal cultivation and forest plantations.

Felis rufus. Head and body length is 650 to 1,050 mm, tail length is 110 to 190 mm, and shoulder height is 450 to 580 mm. Banfield (1974) gave the weight as 4.1 to 15.3 kg. There are various shades of buff and brown, spotted and lined with dark brown and black. The crown is streaked with black and the backs of the ears are heavily marked with black. The short tail has a black tip, but only on the upper side. *F. rufus* resembles *F. lynx*, but is usually smaller and has more slender legs, smaller feet, shorter fur, and ears that are tufted less conspicuously or not at all. As in *F. lynx*, there is a ruff of fur extending from the ears to the jowls, giving the impression of sideburns. Females have four mammae (Lowery 1974).

The bobcat is more ubiquitous than the lynx, occurring in forests, mountainous areas, semideserts, and brushland. Its den is usually concealed in a thicket, hollow tree, or rocky crevice. It is mainly nocturnal and terrestrial, but climbs with ease. Nightly movements of about 3 to 11 km have been reported. Prey is usually stalked with great stealth and patience, and then seized after a swift leap. The diet consists mainly of small animals, such as rabbits, rodents, and birds. Larger prey, such as deer, is sometimes taken, especially in the winter. A study in Massachusetts found deer to be the most common winter food (McCord 1974).

Reported maximum population densities are one bobcat per 18.4 sq km in the western United States and one per 2.6 sq km in the southeast (Jachowski 1981). Average minimum home range in Louisiana was found to be about 5 sq km for males and 1 sq km for females (Hall and Newsom 1981). In a study in southeastern Idaho, Bailey (1974) found average (and extreme) home range size to be 42.1 (6.5–107.9) sq km for males and 19.3 (9.1–45.3) sq km for females. Female ranges were almost exclusive of one another, but the ranges

Spanish lynx (*Felis pardina*), photo by Lothar Schlawe.

of males overlapped each other, as well as those of females. There was a land tenure system, seemingly based on prior right—no neighboring resident or transient permanently settled in an area already occupied by a resident. All observed changes in resident home ranges were attributed to the death of a resident. Territoriality was pronounced, especially in females, and scent marking was accomplished by use of feces, urine, scrapes, and anal gland secretions. Individuals were solitary, avoiding each other even in areas of range overlap, except during the mating season. The bobcat is usually silent, but may emit loud screams, hisses, and other sounds during courtship.

Females are apparently seasonally polyestrous; they usually produce a single annual litter, during the spring, but there is evidence of a second birth peak in late summer and early autumn, perhaps involving younger females or ones that lost their first litters (Banfield 1974). According to Fritts and Sealander (1978), mating may occur as early as November or as late as August. The gestation period is 60–70 days (Hemmer 1976). The number of young per litter is one to six, commonly three. Weight at birth is 283 to 368 grams (Banfield 1974). The young open their eyes after 9 or 10 days, nurse for about 2 months, and begin to travel with the mother when 3 to 5 months old. They separate from the mother in the winter (Jackson 1961), probably in association with the mating season. Females may reach sexual maturity by 1 year of age, but males do not mate until their second year of life. The oldest individuals captured in a Wyoming study were 12 years of age and were still sexually active (Crowe 1975). A

captive bobcat lived for 32 years and 4 months (Marvin L. Jones, Zoological Society of San Diego, pers. comm.).

The bobcat occasionally preys on small domestic mammals and poultry, and has thus been hunted and trapped by people. It has been exterminated in much of the Ohio Valley, upper Mississippi Valley, and southern Great Lakes region (Deems and Pursley 1978). The bobcat is uncommon in central Mexico (Leopold 1959); the subspecies there, *F. rufus escuinapae*, is listed as endangered by the USDI (1980) and is on appendix 1 of the CITES.

The value of bobcat fur has varied widely, depending on fashion and economic conditions. The average price per pelt rose from about $10 in the 1970/71 season (Deems and Pursley 1978) to $145 in 1978/79 (Jachowski 1981). There was a corresponding increase in the number of bobcats being harvested. The total known annual kill in the United States was approximately 92,000 individuals in the late 1970s. About two-thirds of these animals were being taken primarily for their fur, and most of the pelts were being exported. There was concern that bobcat populations were being seriously reduced in some areas, and that little was being done to manage the resource. Jachowski (1981), however, reported substantial improvement in management since *F. rufus* (except *F. r. escuinapae*) was placed on appendix 2 of the CITES in 1977. Prior to that time, few states had closed seasons and many had bounties. Presently, there are no bounties, 11 states provide complete protection, and the rest allow a regulated harvest during a limited season. Jachowski estimated that there are between 725,000 and 1,020,000 bobcats

Caracals (*Felis caracal*), photo by Bernhard Grzimek.

in the United States, and suggested that current known harvest levels are not jeopardizing overall populations in the country.

Felis caracal.   Head and body length is 600 to 915 mm, tail length is 230 to 310 mm, shoulder height is 380 to 500 mm, and weight is 13 to 19 kg (Kingdon 1977). The pelage is dense but relatively short, and there are no side whiskers as in *F. lynx*. The general coloration is reddish brown. There is white on the chin, throat, and belly, and a narrow black line from the eye to the nose. The ears are narrow, pointed, black on the outside, and adorned with black tufts up to 45 mm long. *F. caracal* is smaller than *F. lynx,* has a long and slender body, and a tapering tail that is approximately one-third the length of the head and body.

The caracal is found mainly in dry country—woodland, savannah, and scrub—but avoids sandy deserts. Maternal dens are located in porcupine burrows, rocky crevices, or dense vegetation. This cat is largely nocturnal, but sometimes is seen by day. It climbs and jumps well. It is mainly terrestrial and is apparently the fastest feline of its size. Prey is stalked and then captured after a quick dash or leap. The diet includes birds, rodents, and small antelope.

The caracal has been reported to be territorial and to mark with urine; vocalizations include miaows, growls, hisses, and coughing calls (Kingdon 1977). The species is usually seen alone, but Rowe-Rowe (1978*b*) reported a group of two adults and five young. Various observations suggest that the young may be born at any time of year. According to Kingdon (1977), the gestation period is 69–78 days, and the number of young per litter is one to six, usually three. The kittens open their eyes after 10 days, are weaned when 10–25 weeks old, and attain sexual maturity between the ages of 6 and 24 months. Captives have lived up to 17 years.

The caracal is easily tamed and has been used to assist human hunters in Iran and India. It sometimes raids poultry, however, and has thus been killed by people. It has apparently become scarce in North Africa, South Africa, and parts of Asia. The subspecies in Soviet Central Asia, *F. c. michaelis,* is classified as rare by the IUCN (1978). All Asian populations are on appendix 1 of the CITES.

Felis serval.   Head and body length is 670 to 1,000 mm, tail length is 240 to 450 mm, shoulder height is 540 to 620 mm, and weight is 8.7 to 18 kg. Males are generally larger than females (Kingdon 1977). The general coloration of the upper parts ranges from off white to dark gold, and the underparts are paler, often white (Smithers 1978). The entire pelage is marked either with small, dark spots or with large spots that tend to merge into longitudinal stripes on the head and back. The tail has several rings and a black tip. The build is light, the legs and neck are long, and the ears are large and rounded.

According to Smithers (1978), the serval is generally a species of the savannah zone, and is found in the vicinity of streams with densely vegetated banks. It is primarily nocturnal and may move 3 or 4 km per night. It is mainly terrestrial and can run or bound swiftly for short distances. Prey is apparently located both by sight and hearing. Birds up to 3 meters above the ground may be captured by remarkable leaps, but the diet seems to consist mostly of murid rodents.

The serval is basically solitary. It has a shrill cry, and also growls and purrs. There is no definite mating season, but Smithers (1978) reported that births in Rhodesia occurred mainly in the warm months from September to April. Kingdon (1977) suggested that there are two birth peaks in East Africa, in March–April and September–November. Observations at the Basle Zoo (Wackernagel 1968) show that females can give birth twice a year, with a minimum normal interval of 184 days. Estrus usually lasted only one day, and the average gestation period was determined to be 74 days. The number of young in 20 litters averaged 2.35 and ranged from 1 to 4. Five newborn weighed 230 to 260 grams each, and one opened its eyes when 9 days old. One female at the

Asian golden cat (*Felis temmincki*), photo from San Diego Zoological Society.

Basle Zoo gave birth to her last litter when 14 years old and died at an age of about 19 years and 9 months.

The serval is hunted for its skin in East Africa and now no longer occurs in areas heavily populated by people (Kingdon 1977). The species has been mercilessly hunted in farming areas of South Africa, and is now considered rare in that country (Skinner, Fairall, and Bothma 1977). The subspecies *F. s. constantina* of Algeria is listed as endangered by the USDI (1980).

Felis marmorata. Head and body length is 450 to 530 mm, tail length is 475 to 550 mm, and weight is 2 to 5 kg (Lekagul and McNeely 1977). The ground color is brownish gray to bright yellow or rufous brown. The sides of the body are marked with large, irregular, dark blotches, each margined with black. There are solid black dots on the limbs and underparts. The tail is spotted, tipped with black, long, and bushy. The pelage is thick and soft, and the ears are short and rounded.

The marbled cat is a forest dweller, apparently nocturnal and partly arboreal in habit. It is thought to prey mostly on birds, to some extent on squirrels and rats, and possibly on lizards and frogs. Because of human disturbance and habitat

destruction, it has evidently declined and become very rare in much of its range. It is classified as indeterminate by the IUCN (1978) and as endangered by the USDI (1980), and is on appendix 1 of the CITES.

Felis badia. Head and body length is 500 to 600 mm and tail length is 350 to 400 mm. The pelt is bright chestnut above and paler on the belly, with some obscure spots on the underparts and limbs. The tail is long and tapering, has a whitish median streak down the middle of its lower surface, and becomes pure white at the tip. The bay cat has been reported to inhabit dense forests or areas of rocky limestone on the edge of the jungle. It is apparently known only by a few specimens and is classified as rare by the IUCN (1978).

Felis temmincki. Head and body length is 730 to 1,050 mm and tail length is 430 to 560 mm. Lekagul and McNeely (1977) gave the weight as 12 to 15 kg. The pelage is of moderate length, dense, and rather harsh. The general coloration varies from golden red to dark brown and gray. Some specimens, especially from the northern parts of the range, have a pattern of spots on the body. The face is marked with white and black streaks, and the underside of the terminal

Leopard cat (*Felis bengalensis*), photo from New York Zoological Society.

third of the tail is white. The ears are short and rounded.

According to Lekagul and McNeely (1977), the Asian golden cat occurs in deciduous forests, tropical rain forests, and occasionally more open habitats. It is usually terrestrial, but is capable of climbing. The diet includes hares, small deer, birds, lizards, and domestic livestock. This cat often hunts in pairs, and the male is said to play an active role in rearing the young. There is no confirmed breeding season. If a litter is lost, the female may produce another within 4 months. The gestation period is 95 days. Litters usually contain one or two young, which weigh about 250 grams each at birth. Individuals have lived nearly 18 years in captivity.

Because of habitat destruction and inability to adjust to the presence of human activity, *F. temmincki* has declined in much of its range. It is classified as indeterminate by the IUCN (1978) and as endangered by the USDI (1980), and is on appendix 1 of the CITES.

Felis aurata.   Head and body length is 616 to 1,016 mm, tail length is 160 to 460 mm, shoulder height is 380 to 510 mm, and weight is 5.3 to 16 kg; males are generally larger than females (Kingdon 1977). The overall coloration varies from chestnut through fox red, fawn, gray brown, silver gray, and blue gray to dark slaty. The cheeks, chin, and underparts are white. Some specimens are marked all over with dark brown or dark gray dots, while in others the spots are restricted to the belly and insides of the limbs. Black specimens have been recorded. The legs are long, the head is small, and the paws are large.

According to Kingdon (1977), the African golden cat is found mainly in forests, sometimes in mountainous areas. It is said to be active both by day and by night. It is mainly terrestrial, but climbs well. Prey is taken by stalking and rushing. The diet includes birds, small ungulates, and do-

mestic animals. *F. aurata* is normally solitary. A pregnant female was taken in Uganda in September.

Felis bengalensis.   Head and body length is 445 to 1,070 mm, tail length is 230 to 440 mm, and weight is 3 to 7 kg (Lekagul and McNeely 1977; Stroganov 1969). There is much variation in color, but the upper parts are usually pale tawny and the underparts are white. The body and tail are covered with dark spots. There are usually four longitudinal black bands running from the forehead to behind the neck and breaking into short bands and rows of elongate spots on the shoulders. The tail is indistinctly ringed toward the tip. The head is small, the muzzle is short, and the ears are moderately long and rounded.

The leopard cat is found in many kinds of forested habitat at both high and low elevations. It dens in hollow trees or small caves, or under overhangs or large roots. It is mainly nocturnal, but is often seen by day. It is an excellent swimmer, and has populated many offshore islands (Lekagul and McNeely 1977). It apparently hunts on the ground, as well as in trees, and feeds on hares, rodents, young deer, birds, reptiles, and fish.

Births have been reported to occur in May in both Siberia and India. According to Lekagul and McNeely (1977), however, breeding continues throughout the year in Southeast Asia. If one litter is lost, the female may mate and produce another within 4 or 5 months. The gestation period is 65–72 days. The number of young per litter is one to four, usually two or three. The father may participate in rearing the young. The latter open their eyes when about 10 days old, and reach sexual maturity at an age of 18 months.

The subspecies *F. b. bengalensis* of peninsular India and Southeast Asia is listed as endangered by the USDI (1980) and is on appendix 1 of the CITES. It is, however, not particularly intolerant of human activity and is often found near villages, where it may raid poultry (Lekagul and McNeely 1977).

Felis rubiginosus. Head and body length is 350 to 480 mm and tail length is 150 to 250 mm (Grzimek 1975). The upper parts are grizzled gray, with a rufous tinge of varying intensity, and marked with lines of brown, elongate blotches. The belly and insides of the limbs are white with large, dark spots. There are two dark streaks on the face and four dark streaks run from the top of the head to the nape.

On the mainland of India, the rusty-spotted cat seems to prefer scrub, dry grassland, and open country. In Sri Lanka, however, it occurs in humid mountain forests. It is nocturnal, and preys on birds and small mammals. Births occur in the spring in India (Grzimek 1975).

Felis viverrinus. Head and body length is 750 to 860 mm, tail length is 255 to 330 mm, shoulder height is 380 to 406 mm, and weight is 7.7 to 14 kg. The ground color is grizzled gray, sometimes tinged with brown, and there are elongate dark brown spots arranged in longitudinal rows. Six to eight dark lines run from the forehead, over the crown, and along the neck. The fur is short and rather coarse, the head is big and broad, and the tail is short and thick. The forefeet have moderately developed webbing between the digits. The claw sheaths are too small to allow the claws to retract completely.

The fishing cat is found in marshy thickets, mangrove swamps, and densely vegetated areas along creeks. It often wades in shallows and does not hesitate to swim in deep water. It catches prey fish by crouching on a rock or sand bank and using its paw as a scoop. The diet also includes crustaceans, mollusks, frogs, snakes, birds, and small mammals.

Kittens have been seen in the wild in April and June. At the Philadelphia Zoo, births occurred in March and August (Ulmer 1968). Observations there indicate that the gestation period is 63 days, birth weight is about 170 grams, the eyes of the young open by 16 days of age, the first meat is eaten at 53 days, and adult size is attained at 264 days. According to Grzimek (1975), the number of young per litter is one to four.

Felis planiceps. Head and body length is 410 to 500 mm, and tail length is 130 to 150 mm (Grzimek 1975). Two specimens from Malaysia weighed 1.6 and 2.1 kg (Muul and Lim 1970). The body is dark brown with a silvery tinge. The underparts are white, generally spotted and splashed with brown. The top of the head is reddish brown. The face below the eyes is light reddish, there are two narrow dark lines running across each cheek, and a yellow line runs from each eye to near the ear. The coat is thick, long, and soft. The pads of the feet are long and narrow, and the claws cannot be completely retracted. The skull is long, narrow, and flat; the nasals are short and narrow, and the orbits are placed well forward and close together (Lekagul and McNeely 1977).

The flat-headed cat is thought to be nocturnal and to hunt for frogs and fish along river banks. Observations of a captive kitten (Muul and Lim 1970) suggest that *F. planiceps* is a fishing cat. The kitten seemed to enjoy playing in water, took pieces of fish from the water, and captured live frogs, but ignored live birds. Moreover, the long, narrow rostrum and the well-developed first upper premolar of the species would seem to be efficient for seizing slippery prey. *F. planiceps* is classified as indeterminate by the IUCN (1978) and as endangered by the USDI (1980), and is on appendix 1 of the CITES. It appears to be very rare, and some authorities consider it endangered in Thailand and Indonesia.

Felis iriomotensis. Head and body length is 600 mm and tail length is 200 mm. The ground color is dark dusky brown,

and there are spots that are arranged in longitudinal rows and tend to merge into bands. Five to seven lines run from the back of the neck to the shoulders. The body is relatively elongate, the legs and tail are short, and the ears are rounded.

This cat dwells only on Iriomote, a Japanese-owned island of 292 sq km to the east of Taiwan. Its presence was not known to science until 1967. According to the IUCN (1978), *F. iriomotensis* appears to be restricted to lowland, subtropical rain forest. It is nocturnal and preys on small rodents, water birds, crabs, and mud skippers. It is territorial and has a home range not exceeding 2 sq km. Although totally protected by Japanese law, the Iriomote cat is rapidly losing habitat to agricultural development. Its numbers have declined drastically since 1974, and only 40 to 80 individuals probably remain. The species is classified as endangered by the IUCN (1978) and the USDI (1980).

Felis colocolo. Head and body length is 567 to 700 mm, tail length is 295 to 322 mm, and shoulder height is 300 to 350 mm. The coloration ranges from yellowish white and grayish yellow to brown, gray brown, silvery gray, and light gray. There are transverse bands of yellow or brown running obliquely from the back to the flanks. Two bars run from the eyes across the cheeks and meet beneath the throat. The coat is long, the tail is bushy, the face is broad, and the ears are pointed.

The pampas cat inhabits open grassland in some areas, but also enters humid forests and mountainous regions. Grimwood (1969) reported it to occur throughout the Andes of Peru. It is mainly terrestrial, but may climb a tree if pursued. It hunts at night, killing small mammals, especially guinea pigs (*Cavia*), and ground birds. Litters are said to contain one to three young.

Felis pardalis. Head and body length is 550 to 1,000 mm and tail length is 300 to 450 mm (Grzimek 1975; Leopold 1959). Weight is 11.3 to 15.8 kg. The ground color ranges from whitish or tawny yellow to reddish gray and gray. There are dark streaks and spots, which are arranged in small groups around areas that are darker than the ground color. There are two black stripes on each cheek, and one or two transverse bars on the insides of the legs. The tail is either ringed or marked with dark bars on the upper surface.

The ocelot occurs in a great variety of habitats, from humid tropical forests to fairly dry scrub country. It is generally nocturnal, sleeping by day in a hollow tree, in thick vegetation, or on a branch. It is mainly terrestrial, but climbs, jumps, and swims well. The diet includes rodents, rabbits, young deer and peccaries, birds, snakes, and fish.

A pair is thought to share a territory, but to hunt separately. Communication is by mewing and, during courtship, by yowls not unlike those of *F. catus*. There is probably no seasonal breeding in the tropics (Grzimek 1975). In Mexico and Texas, births are reported to occur in fall and winter (Leopold 1959). The gestation period is 70 days. There are usually two young, but may be as many as four. A captive lived for 20 years and 3 months (Marvin L. Jones, Zoological Society of San Diego, pers. comm.).

The ocelot is classified as vulnerable by the IUCN (1978) and as endangered by the USDI (1980). The subspecies *F. p. mearnsi* of Central America, and *F. p. mitis* of central and eastern Brazil are on appendix 1 of the CITES. The subspecies *F. p. albescens* probably once ranged over most of Texas and at least as far east as Arkansas and Louisiana (Lowery 1974), but is currently restricted to the border region of extreme southern Texas and northeastern Mexico. Fewer than 1,000 individuals are thought to survive, and fewer than 100 of these are in Texas. The clearing of brush country for agricultural purposes is the main problem in this area. The ocelot still occurs over most of the remainder of its

Flat-headed cat (*Felis planiceps*), photo from San Diego Zoological Society.

original range, but is declining because of habitat loss and hunting for its fur. Prior to 1972, when importation into the United States was prohibited, enormous numbers of skins were brought into the country—133,069 in 1969 alone. As late as 1975, Britain imported 76,838 skins. The ocelot is also reported to be in much demand for use as a pet, a live animal selling for $800. The species is now protected by the laws of most countries in which it lives, and its trade is partly regulated by the CITES, but enforcement is difficult (IUCN 1978).

Felis wiedii.   Head and body length is 463 to 790 mm, and tail length is 331 to 510 mm. The ground color is yellowish brown above and white below. There are longitudinal rows of dark brown spots, the centers of which are paler than the borders. The margay closely resembles the ocelot, but is smaller, has a more slender build, and has a relatively longer tail.

The margay is mainly, if not exclusively, a forest dweller. It is much more arboreal than the ocelot and is thought to forage in trees. The only specimen known from Texas was taken prior to 1852 and is thought to represent an individual that strayed far from the normal habitat (Leopold 1959). The arboreal acrobatics and effortless climbing of the margay are partly the result of limb structure (Grzimek 1975). The feet

are broad and soft, and have mobile metatarsals. The hind foot is much more flexible than that of other felids, being able to rotate 180°. There is probably no particular breeding season in the tropics, litters contain one or two young, and captives have lived over 13 years (Grzimek 1975).

The margay is listed as endangered by the USDI (1980). The subspecies *F. w. nicaraguae* and *F. w. salvinia* of Central America are on appendix 1 of the CITES. Leopold (1959) referred to the margay as "an exceedingly rare animal." Grimwood (1969) stated that exports of pelts from Peru were increasing. Paradiso (1972) indicated that at least 6,701 margays, including both live animals and skins, were imported into the United States in 1970.

Felis tigrinus.   Head and body length is 400 to 550 mm, and tail length is 250 to 400 mm (Grzimek 1975). A captive male weighed 2.75 kg and a captive female weighed 1.75 kg. The upper parts vary in color from light to rich ochre and have rows of large, dark spots. The underparts are paler and less spotted. The tail has 10 or 11 rings and a black tip. One-fifth of all specimens are melanistic.

The little spotted cat lives in forests, and its habits in the wild are not known. Captive females have an estrus of several days, a gestation period of 74–76 days, and litters of one or two young. The kittens develop slowly, opening their eyes at

Geoffroy's cat (*Felis geoffroyi*), photo by K. Rudloff through Berlin Zoo.

17 days of age and starting to take solid food at 55 days.

Like the ocelot and margay, the little spotted cat was, until recently, subject to uncontrolled commerce. At least 3,170 individuals, including both live animals and skins, were imported into the United States in 1970 (Paradiso 1972). The species is now listed as endangered by the USDI (1980) and importation is banned. The subspecies *F. t. oncilla* of Central America is on appendix 1 of the CITES.

Felis geoffroyi.   Head and body length is 450 to 700 mm and tail length is 260 to 350 mm. The ground color varies widely, from brilliant ochre in the northern parts of the range to silvery gray in the south. The body and limbs are covered with small black spots. There may be several black streaks on the crown, two on each cheek, and one between the shoulders. The tail is ringed.

Geoffroy's cat inhabits scrubby woodland and open bush country. Ximinez (1975) reported the elevational range to be sea level to 3,300 meters, and that the species is nocturnal. According to Grzimek (1975), it is a good climber and swimmer; it sleeps in trees and readily enters water. It preys on small mammals, birds, and possibly fish. Hemmer (1976) listed a gestation period of 74–76 days. The single annual litter contains two or three young, and in Uruguay is produced from December to May (Ximinez 1975).

Felis guigna.   Head and body length is 390 to 493 mm, and tail length is 195 to 230 mm. The coat is buff or gray brown, and is heavily marked with rounded, blackish spots on both the upper and lower parts. The tail has blackish rings. Melanistic individuals are not uncommon.

The kodkod is a nocturnal forest dweller. Some authorities have stated that it is an expert climber, living mainly in trees, while others have called it terrestrial. It probably preys mainly on small mammals, but has been reported to raid hen houses.

Felis jacobita.   Head and body length is 600 mm and tail length is 350 mm. The coat is soft and fine. The upper parts are silvery gray and marked by irregular brown or orange yellow spots and transverse stripes. The underparts are whitish and have blackish spots. The tail is bushy, ringed with black or brown, and light tipped.

The mountain cat is found in the arid and semiarid zone of the Andes. Elevational range is up to about 5,000 meters (Grzimek 1975). It preys on small mammals, such as chinchillas and viscachas. It is classified as rare by the IUCN (1978) and as endangered by the USDI (1980), and is on appendix 1 of the CITES.

Felis yagouaroundi.   Head and body length is 550 to 770 mm, tail length is 330 to 600 mm, and weight is 4.5 to 9 kg. There are two color phases: blackish to brownish gray, and fox red to chestnut. The body is slender and elongate, the head is small and flattened, the ears are short and rounded, the legs are short, and the tail is very long. This cat is sometimes said to resemble a weasel or otter in external appearance.

The jaguarundi inhabits lowland forests and thickets. It hunts in the morning and evening, and is much less nocturnal than most cats. It forages mainly on the ground, but is an agile climber. The diet includes birds and small mammals. *F.*

yagouaroundi has been reported to live in pairs in Paraguay, but to be solitary in Mexico. According to Grzimek (1975), there is no definite reproductive season in the tropics, but in Mexico young are produced around March and August. It is not known whether one female gives birth in both seasons. The gestation period is 63–70 days and litters contain two to four young.

The pelt is of poor quality and of little value (Leopold 1959). The jaguarundi is widespread and not subject to commercial exploitation (Paradiso 1972). Nonetheless, four subspecies—*cacomitli, tolteca, fossata,* and *panamensis*—which range from southern Texas and Arizona to Panama, are listed as endangered by the USDI (1980) and are on appendix 1 of the CITES. The jaguarundi is not now known to be resident in Arizona, and only a few individuals survive in Texas. During the late Pleistocene the species occurred as far as Florida, and a small population may have become established there recently through introduction by human agency.

Felis concolor. The names cougar, puma, panther, and mountain lion are used interchangeably for this species, and various other vernacular terms are applied in certain areas. This is by far the largest species in the genus *Felis,* averaging about the same size as the leopard (*Panthera pardus*). In males, head and body length is 1,050 to 1,959 mm, tail length is 660 to 784 mm, and weight is 67 to 103 kg. In females, head and body length is 966 to 1,517 mm, tail length is 534 to 815 mm, and weight is usually 36 to 60 kg. Shoulder height is 600 to 700 mm. Generally, the smallest animals are in the tropics and the largest are in the far northern and southern parts of the range. There are two variable color phases. One ranges from buff, cinnamon, and tawny to cinnamon rufous and ferrugineous. The other ranges from silvery gray to bluish and slaty gray. The body is elongate, the head is small, the face is short, and the neck and tail are long. The limbs are powerfully built; the hind legs are larger than the forelegs. The ears are small, short, and rounded. Females have three pairs of mammae (Banfield 1974).

The cougar has the greatest natural distribution of any mammal in the Western Hemisphere, except *Homo sapiens.* It can thrive in montane coniferous forests, lowland tropical forests, swamps, grassland, dry brush country, or any other area with adequate cover and prey. The elevational range extends from sea level to at least 3,350 meters in California and 4,500 meters in Ecuador. There is usually no fixed den, except as used by females to rear young. Temporary shelter is taken in such places as dense vegetation, rocky crevices, and caves. The cougar is agile and has great jumping power. It may leap from the ground to a height of up to 5.5 meters in a tree. It swims well, but commonly prefers not to enter water. Sight is the most acute sense, and hearing is also good, but smell is thought to be poorly developed. Activity may be either nocturnal or diurnal. The cougar hunts over a large area, sometimes taking a week to complete a circuit of its home range (Leopold 1959). In Idaho, Seidensticker et al. (1973) found residents to occupy fairly distinct, but usually contiguous, winter-spring and summer-fall home areas. The latter area was generally larger and at a higher elevation, reflecting the summer movements of ungulate herds.

The cougar carefully stalks its prey and may leap upon the victim's back or seize it after a swift dash. Throughout the range of the cougar, the most consistently important food is deer—*Odocoileus* in North America, and *Blastoceros, Hippocamelus,* and *Mazama* in South America. Estimates of kill frequency vary from one deer per week to one every three or four weeks, with intermittent feeding on smaller animals (Russell 1978). The diet also includes other ungulates,

beaver, porcupines, and hares. The kill is usually dragged to a sheltered spot and then partly consumed. The remains are covered with leaves and debris, and then visited for additional meals over the next several days.

Detailed studies in central Idaho (Hornocker 1969, 1970; Seidensticker et al. 1973) showed that the cougar population depends almost equally on mule deer (*Odocoileus hemionus*) and elk (*Cervus canadensis*). The study area contained 1 cougar for every 114 deer and 87 elk. About half of the animals killed were in poor condition. Deer and elk populations increased during a four-year study period, evidently being affected more by food availability than by cougar predation. Nonetheless, predation was thought to moderate prey oscillations and to remove less fit individuals. During the same period, the cougar population remained stable, its numbers being regulated mainly by social factors rather than food supply. Density was about 1 adult cougar per 35 sq km. The total area used by individuals varied from 31 to 243 sq km in winter-spring and from 106 to 293 sq km in summer-fall. There was little overlap in the areas occupied by resident adult males, but the areas of resident females often overlapped one another completely and were overlapped by resident male areas. Young, transient individuals of both sexes moved through the areas used by residents. There was a land tenure system based on prior right; transients could not permanently settle in an occupied area unless the resident died. Dispersal and mortality of young individuals, unable to establish an area for themselves, seemed to limit the size of the cougar population.

According to Russell (1978), population densities in other areas have been found to range from one cougar per 26 sq km to one per 261 sq km. The usual area for activity for individual residents is 65 to 90 sq km for males and 40 to 80 sq km for females, but one animal in British Columbia used 650 sq km. The cougar is generally solitary, with individuals deliberately avoiding one another except during the brief period of courtship. Several males may fight over a female. There are a number of vocalizations, which resemble those of the domestic cat but are louder. A very loud scream is apparently emitted on rare occasion, but its function is not known.

There is no specific breeding season, but most births in North America occur in late winter and early spring. Females are seasonally polyestrous and usually give birth every other year (Banfield 1974). Estrus lasts about 9 days and the gestation period is 90–96 days. The number of young per litter is one to six, commonly three or four. The kittens weigh 226 to 453 grams at birth and are spotted until they are about 6 months old. They nurse for 3 months or more, but begin to take some meat at 6 weeks of age. If born in the spring, they are able to accompany the mother by autumn and to make their own kills by the end of winter. Nonetheless, they usually remain with their mother for several more months or even another year. Litter mates stay together for 2 or 3 months after leaving the mother (Russell 1978). Sexual maturity is attained by females at about 2½ years of age, but males may not mate until at least 3 years old (Banfield 1974). Regular reproductive activity does not begin until a young animal establishes itself on a permanent home area (Seidensticker et al. 1973). Captive cougars have lived over 19 years (Grzimek 1975).

The remainder of the account of this species is based largely on Nowak (1976). The cougar is generally considered harmless to people, but attacks have occurred, and seven persons are known to have been killed in the United States and Canada since 1900. The cougar has long been viewed as a threat to domestic animals, such as horses and sheep, and is also sometimes thought to reduce populations of game. The species has thus been intensively hunted since the arrival of European colonists in the Western Hemisphere. Most suc-

Cougar (*Felis concolor*): Top, photo by Bernhard Grzimek; Bottom, photo by Don Deminick, Colorado Division of Wildlife.

cessful hunting is done by using dogs to pursue the cat until it seeks refuge in a tree, where it can easily be shot.

By the early 20th century, the cougar appeared to have been eliminated everywhere to the north of Mexico, except in the mountainous parts of the West, in southern Texas, and in Florida. The species still occupies the same regions, and about 16,000 individuals may be present. California pro-

vides almost complete protection to the cougar, Texas still allows it to be killed at any time, and all other western states and provinces permit regulated hunting for sport and predator control. Public antipathy seems to have moderated in the last two decades, and the general pattern of decline may have been halted, but loss of habitat and conflict with agricultural interests are still problems.

Since the late 1940s, evidence has accumulated that indicates the presence of small surviving cougar populations in south-central Canada, New Brunswick, the southern Appalachians, and the Ozark region and adjoining forests of Arkansas, southern Missouri, eastern Oklahoma, and northern Louisiana. In southern Florida, several individuals have been killed or live-captured in the last decade (Belden and Forrester 1980; Shapiro 1981). The subspecies *F. c. coryi*, which formerly occurred from eastern Texas to Florida, and the subspecies *F. c. couguar* of the northeastern United States and southeastern Canada, are classified as endangered by the IUCN (1978). These two subspecies, along with *F. c. costaricensis* of Central America, are also listed as endangered by the USDI (1980) and are on appendix 1 of the CITES.

CARNIVORA; FELIDAE; *Genus NEOFELIS Gray, 1867*

Clouded Leopard

The single species, *N. nebulosa*, occurs from Nepal to southeastern China and the Malay Peninsula, and on Taiwan, Hainan, Sumatra, and Borneo. Most authorities, including Ellerman and Morrison-Scott (1966), Hemmer (1978), Ewer (1973), and Guggisberg (1975), treat *Neofelis* as a distinct genus with one species. Simpson (1945) considered *Neofelis* a subgenus of *Panthera*. Leyhausen (1979) listed *Neofelis* as a full genus, but included within it not only the clouded leopard but also the tiger (here referred to as *Panthera tigris*).

Head and body length is about 616 to 1,066 mm, tail length is 550 to 912 mm, and weight is usually 16 to 23 kg. The coat is grayish or yellowish, with dark markings ("clouds") in such forms as circles, ovals, and rosettes. The markings on the shoulders and back are darker on their posterior margins than on their front margins, thus suggesting that stripes can be evolved from blotches or spots. The forehead, legs, and base of the tail are spotted, and the remainder of the tail is banded. Melanistic specimens have been reported (Medway 1977).

The tail is long, the legs are stout, the paws are broad, and the pads are hard. The skull is long, low, and narrow (Guggisberg 1975). The upper canine teeth are relatively longer than those of any other living cat, having a length about three times greater than the basal width at the socket. The first upper premolar is greatly reduced or absent, leaving a wide

gap between the canine and the cheek teeth. Unlike that of *Panthera,* the hyoid of *Neofelis* is ossified (Guggisberg 1975).

The clouded leopard inhabits various kinds of forest, perhaps to elevations of up to 2,500 meters. It is usually said to be highly arboreal, to hunt in trees, and to spring on ground prey from overhanging branches. Information compiled by Guggisberg (1975), however, suggests that *Neofelis* is more terrestrial and diurnal than is generally assumed. It has been reported to feed on birds, monkeys, pigs, cattle, young buffalo, goats, deer, and even porcupines.

Reproduction is known only through observations in captivity (Fellner 1965; Fontaine 1965; Guggisberg 1975; Murphy 1976). Births in Europe and Texas have occurred from March to August. The gestation period seems normally to be 86 to 93 days. The number of young per litter ranges from one to five, but is commonly two. The young weigh about 140 to 170 grams each at birth, open their eyes after 12 days, take some solid food at 10½ weeks of age, and nurse for 5 months. The coat is initially black in the patterned areas, and full adult coloration is attained at about 6 months of age. Captives have lived over 17 years (Grzimek 1975). Many captives are gentle and playful, and like to be petted by their custodians.

The clouded leopard is classified as vulnerable by the IUCN (1978) and as endangered by the USDI (1980), and is on appendix 1 of the CITES. The main problem is loss of forest habitat to agriculture. The species has also been excessively hunted in some areas for its beautiful pelt. It may already have disappeared from Hainan, and in Taiwan is restricted to the wildest and most inaccessible parts of the central mountain range.

CARNIVORA; FELIDAE; *Genus PANTHERA Oken, 1816*

Big Cats

There are five subgenera and five species (Ellerman and Morrison-Scott 1966; Guggisberg 1975; Corbet 1978; Meester and Setzer 1971; Hall 1981; Cabrera 1957):

subgenus *Uncia*

P. uncia (snow leopard), mountainous areas from Afghanistan to Lake Baikal and eastern Tibet;

Clouded leopard (*Neofelis nebulosa*), photo from San Diego Zoological Garden.

A. Lion (*Panthera leo*), photo from P. D. Swanepoel. B. Lioness and cub (*P. leo*), photo by Ernest P. Walker. C. Lion cubs (*P. leo*), photo from San Diego Zoological garden. D. Leopard (*P. pardus*), photo from the U.S. National Zoological Park.

Chinese tiger (*Panthera tigris amoyensis*), top. Melanistic jaguar (*Panthera onca*), bottom. Photos from East Berlin Zoo.

subgenus *Tigris*

P. tigris (tiger), eastern Turkey to southeastern Siberia and Malay Peninsula, Sumatra, Java, Bali;

subgenus *Panthera*

P. pardus (leopard), western Turkey and Arabian Penin-sula to southeastern Siberia and Malay Peninsula, Sri Lanka, Java, Kangean Islands;

subgenus *Jaguarius*

P. onca (jaguar), southern United States to northern Argentina;

Snow leopard (*Panthera uncia*), photo by Ernest P. Walker.

subgenus *Leo*

P. leo (lion), found in historical time from the Balkan and
Arabian peninsulas to central India, and in almost all
of Africa.

The use of the name *Panthera* for this genus is in keeping
with Ellerman and Morrison-Scott (1966), Corbet (1978),
Mazak (1981), Hemmer (1978), Leyhausen (1979), and
most of the other authorities cited herein. A few authorities,
such as Cabrera (1957), consider *Panthera* to be invalid for
technical reasons of nomenclature, and prefer to use the generic term *Leo* Brehm, 1829. Some authorities, including
Hall (1981), do not consider *Panthera* to be generically distinct from *Felis*. *Uncia* Gray, 1854 was treated as a separate
genus by Hemmer (1972, 1978), Leyhausen (1979), and
Guggisberg (1975), but not by Ewer (1973), Ellerman and
Morrison-Scott (1966), and Corbet (1978). Leyhausen
placed *P. tigris* in the genus *Neofelis*. Both Hemmer (1978)
and Mazak (1981) divided *Panthera* into only two subgenera: *Tigris*, with *P. tigris;* and *Panthera*, with *P. pardus*,
P. onca, and *P. leo*.

The members of the genus *Panthera* have an incompletely
ossified hyoid apparatus, with an elastic cartilaginous band
replacing the bony structure found in other cats (Grzimek
1975). This elastic ligament is usually considered the anatomical feature that allows roaring, but limits purring to
times of exhaling. Other cats can purr both when exhaling
and inhaling. The snow leopard (*P. uncia*) possesses an elastic ligament, but does not roar. According to Paul Leyhausen
(Max-Planck Institut für Verhaltensphysiologie, pers.
comm.), certain small cats, such as *Felis nigripes,* produce a
full roar, but do not sound as formidable as the lion or tiger,
simply because of the size difference. *Panthera* is also distinguished from *Felis* by having hair that extends to the front
edge of the nose (Grzimek 1975). Additional information is
provided separately for each species.

Panthera uncia.   Except as noted, the information for the
account of this species was taken from Hemmer (1972) and
Guggisberg (1975). Head and body length is 1,000 to 1,300
mm, tail length is 800 to 1,000 mm, shoulder height is about
600 mm, and weight is 25 to 75 kg. The ground color varies

from pale gray to creamy smoke gray, and the underparts are
whitish. On the head, neck, and lower limbs are solid spots.
On the back, sides, and tail are large rings or rosettes that
often enclose some small spots. The coat is long and thick,
and the head is relatively small.

The snow leopard is found in the high mountains of Central Asia. In summer it occurs commonly in alpine meadows
and rocky areas at elevations of 2,700 to 6,000 meters. In the
winter it may follow its prey down into the forests below
1,800 meters. It sometimes dens in a rocky cavern or crevice.
It is often active by day, especially in the early morning and
late afternoon. It is graceful and agile, and has been reported
to leap as far as 15 meters. Prey is either stalked or ambushed. The diet includes mountain goats and sheep, deer,
boar, marmots, pikas, and domestic livestock.

A large home range is utilized and transversed in the
course of about a week. It is possible that a pair shares a
range. Socially, the snow leopard is thought to be like the
tiger, essentially solitary but not unsociable. It does not roar,
but has several vocalizations, including a loud moaning that
is associated with attraction of a mate. Births usually occur
from April to June, both in the wild and in captivity. The
gestation period is 90 to 103 days. The young are born in a
rocky shelter, lined with the mother's fur. The number of
young per litter is one to five, usually two or three. The cubs
weigh about 450 grams each at birth, open their eyes after 7
days, eat their first solid food at 2 months of age, and follow
their mother at 3 months. They hunt with the mother at least
through their first winter of life, and attain sexual maturity
when about 2 years old. A captive lived for 15 years and 8
months (Marvin L. Jones, Zoological Society of San Diego,
pers. comm.).

The snow leopard is classified as endangered by the IUCN
(1978), the USDI (1980), and the Union of Soviet Socialist
Republics Ministry of Agriculture (1978), and is on appendix
1 of the CITES. It has declined in numbers through hunting
by people, because it is considered a predator of domestic
stock, it is valued as a trophy, and its fur is in demand by
commerce. Contributing to the problem in Nepal, where
perhaps 150 to 300 individuals remain, are a reduction in
natural prey and increased use of alpine pastures by people
and stock (Jackson 1979).

Tiger (*Panthera tigris*), photo from East Berlin Zoo.

Panthera tigris. Head and body length is 1,400 to 2,800 mm and tail length is 600 to 950 mm (Grzimek 1975). The subspecies found in southeastern Siberia and Manchuria, *P. t. altaica,* is the largest living cat. The other mainland subspecies are also large, but those of the East Indies are much smaller. In *P. t. altaica,* males weigh 180 to 306 kg and females weigh 100 to 167 kg; in *P. t. tigris* of India and adjoining countries, males weigh 180 to 258 kg and females weigh 100 to 160 kg; in *P. t. sumatrae* of Sumatra, males weigh 100 to 140 kg and females weigh 75 to 110 kg; and in *P. t. balica* of Bali, males weigh 90 to 100 kg and females weigh 65 to 80 kg (Mazak 1981). The ground color of the upper parts ranges from reddish orange to reddish ochre, and the underparts are creamy or white. The head, body, tail, and limbs have a series of narrow black, gray, or brown stripes. On the flanks, the stripes generally run in a vertical direction. In some specimens, the stripes are much reduced on the shoulders, forelegs, and anterior flanks (Guggisberg 1975).

Except as noted, the information for the remainder of this account was taken from Mazak (1981). The tiger is tolerant of a wide range of environmental conditions, its only requirements being adequate cover, water, and prey. It is found in such habitats as tropical rain forests, evergreen forests, mangrove swamps, grasslands, savannahs, and rocky country. An individual may have one or more favored dens within its territory, located in such places as caves, hollow trees, and dense vegetation. The tiger usually does not climb trees, but is capable of doing so. It has been reported to cover up to 10 meters in a horizontal leap. It seems to like water and can swim well, easily crossing rivers 6 to 8 km wide and sometimes swimming up to 29 km. It is mainly nocturnal, but may be active in daylight, especially in winter in the northern part of its range. Siberian animals have moved up to 60 km per day. In Nepal, Sunquist (1981) determined the usual daily movement to be 10 to 20 km.

To hunt, the tiger depends more on sight and hearing than

on smell. It usually carefully stalks its prey, approaching from the side or rear and attempting to get as close as possible. It then leaps upon the quarry, and tries simultaneously to throw it down and grab its throat. Killing is by strangulation or a bite to the back of the neck. The carcass is often dragged to an area within cover or near water. One individual dragged an adult gaur (*Bos gaurus*) 12 meters, and later 13 men tried to pull the carcass, but could not move it (Grzimek 1975). After eating its fill, the tiger may cover the remains with grass or debris, and return for additional meals over the next several days. The diet consists mainly of large mammals, such as pigs, deer, antelope, buffalo, and gaur. Smaller mammals and birds are occasionally taken. A tiger can consume up to 40 kg of meat at one time, but individuals in zoos are given 5 or 6 kg per day. Although the tiger is an excellent hunter, it fails in at least 90 percent of its attempts to capture animals. The tiger thus cannot eliminate entire prey populations. Indeed, Sunquist (1981) found that prey numbers, with the exception of one species, were not even being limited by tiger predation.

In Kanha National Park, central India, Schaller (1967) determined that 10 to 15 adult tigers were regularly resident in an area of about 320 sq km. In Royal Chitawan National Park, Nepal, Sunquist (1981) found an overall population density of 1 adult per 36 sq km. Observations in these and other areas indicate much variation in home range size and social behavior, evidently depending on habitat conditions and prey availability. In India, individual home range seems usually to be 50 to 1,000 sq km. In Manchuria and southeastern Siberia the usual size is 500 to 4,000 sq km, and the maximum reported is 10,500 sq km. In Nepal, Sunquist found home range size to be 60 to 72 sq km for males and 16 to 20 sq km for females. These ranges essentially corresponded to defended territories, in that there was no overlap between those of adults of the same sex. A male range, however, overlapped the ranges of several females. Schaller's studies indicate that the same kind of situation may exist in central India, but also that females are not territorial, that adults of the same sex sometimes share a home range, and that there are transient animals that lack an established range. Both Schaller and Sunquist suggested the presence of a land tenure system based on prior right, by which a resident animal is never replaced on its range until its death. Territorial boundaries are not patrolled, but individuals do visit all parts of their ranges over a period of days or weeks. Avoidance, rather than fighting, seems to be the rule for tigers. Nonetheless, one individual, transplanted from its normal range to a different area, was evidently soon killed by another tiger (Seidensticker et al. 1976).

The tiger is essentially solitary, except for courting pairs and females with young. Even individuals that share a range usually keep 2 to 5 km apart (Sunquist 1981). The tiger is not unsociable, however, and the animals in a given area (probably close relatives) may know one another and have a generally amicable relationship (Schaller 1967). Several adults may come together briefly, especially to share a kill. Limited evidence suggests that a tiger roars to announce to its associates that it has made a kill. An additional function of roaring seems to be the attraction of the opposite sex. There are a number of other vocalizations, such as purrs and grunts, and the tiger also communicates by marking with urine, feces, and scratches.

Mating may occur at any time, but is most frequent from November to April. Females usually give birth every 2 to 2½ years, and occasionally wait 3 or 4 years. If all the newborn are lost, however, another litter can be produced within 5 months (Schaller 1967). Females enter estrus at intervals of 3 to 9 weeks, and receptivity lasts 3 to 6 days. The gestation period is usually 104 to 106 days, but ranges from 93 to 111 days. Births occur in a cave, a rocky crevice, or dense vegetation. The number of young per litter is usually two or three and ranges from one to six. The cubs weigh 780 to 1,600 grams at birth, open their eyes after 6 to 14 days, nurse for 3 to 6 months, and begin to travel with the mother when 5 or 6 months old. They are taught how to hunt prey, and apparently are capable killers at the age of 11 months (Schaller 1967). They usually separate from the mother when 2 years old, but may wait another year. Sexual maturity is attained at 3 or 4 years of age by females, and at 4 or 5 years by males. About half of all cubs do not survive more than 2 years, but maximum known longevity is about 26 years in both the wild and captivity.

The tiger has probably been responsible for more human deaths, through direct attack, than has any other wild mammal. About 1,000 people were reportedly killed each year in India during the early 1900s. Guggisberg (1975) questioned the accuracy of these statistics, but there seems little doubt that some tigers have preyed extensively, or almost exclusively, on people. One individual is said to have killed 430 persons in India. Although such man-eaters have declined with the general reduction in tiger numbers in the 20th century, the problem does persist. In 1972, for example, India's production of honey and beeswax dropped by 50 percent when at least 29 persons who gathered these materials were devoured (Mainstone 1974). Tigers seem to be especially dangerous in the Sundarbans mangrove forest at the mouth of the Ganges River. Hendrichs (1975c) reported that 129 persons were killed in this area from 1969 to 1971, but noted that only 1 percent of the tigers there actually seem to seek out human prey.

Because it is considered a threat to human life and domestic livestock, and also because it is valued as a big game trophy, the tiger has been relentlessly hunted, trapped, and poisoned. Some European hunters and Indian maharajahs killed hundreds of tigers each. After World War II, hunting became even more widespread than previously (Guggisberg 1975). The commercial trade in tiger skins intensified in the 1960s, and by 1977 a pelt brought as much as U.S. $4,250 in Britain (IUCN 1978). Perhaps the greatest threat to the survival of the tiger is destruction of its habitat. With the expansion of human populations, the logging of forests, the elimination of natural prey, and the spread of agriculture, there is continuous conflict between people and the tiger, and the latter species is almost always the loser. In 1920 there are estimated to have been about 100,000 tigers in the world; current estimates range as low as 4,000 (Fisher 1978; Jackson 1978). The tiger is classified as endangered by the IUCN (1978) and the USDI (1980), and is on appendix 1 of the CITES.

Except as noted, the information for the following summaries of the status of the eight subspecies of *P. tigris* was taken from the IUCN (1978) and Mazak (1981):

P. t. virgata (Caspian tiger), formerly occurred from eastern Turkey and the Caucasus to the mountains of Soviet Central Asia and Afghanistan, a few individuals still present in Turkey in the 1970s, now possibly extinct;

P. t. tigris (Bengal tiger), originally found from Pakistan to western Burma, exterminated in Pakistan by 1906 (Roberts 1977), about 2,500 individuals now thought to survive;

P. t. corbetti (Indochinese tiger), still found from eastern Burma to Viet Nam and the Malay Peninsula;

P. t. amoyensis (Chinese tiger), formerly occurred throughout eastern China, now confined largely to the Yangtze Valley and apparently near extinction;

P. t. altaica (Siberian tiger), formerly found from Lake Baikal to the Pacific coast and Korea, apparently now very rare in Manchuria and Korea, protection in the Ussuri region of the Soviet Union seems to have resulted in an increase in

Siberian tiger (*Panthera tigris altaica*), top. Sumatran tiger (*P. t. sumatrae*), bottom. Photos from East Berlin Zoo.

numbers and distribution there—about 200 individuals are now present (Prynn 1980);

P. t. sumatrae (Sumatran tiger), about 1,000 animals estimated to exist on the island, but population declining rapidly;

P. t. sondaica (Javan tiger), almost all suitable habitat destroyed, only four or five individuals thought to survive;

P. t. balica (Bali tiger), probably extinct, last known specimen taken in 1937.

Panthera pardus. Head and body length is 910 to 1,910 mm, tail length is 580 to 1,100 mm, and shoulder height is 450 to 780 mm. Males weigh 37 to 90 kg and females weigh 28 to 60 kg. There is much variation in color and pattern. The ground color ranges from pale straw and gray buff to bright fulvous, deep ochre, and chestnut. The underparts are white. The shoulders, upper arms, back, flanks, and haunches have dark spots arranged in rosettes, which usually enclose an area darker than the ground color. The head, throat, and chest are marked with small black spots, and the belly with large black blotches. Melanistic leopards (black panthers) are common,

especially in moist, dense forests (Guggisberg 1975; Kingdon 1977).

The leopard can adapt to almost any habitat that provides it with sufficient food and cover. It occupies lowland forests, mountains, grasslands, brush country, and deserts. A specimen was found at an elevation of 5,638 meters on Kilimanjaro. The leopard is usually nocturnal, resting by day on the branch of a tree, in dense vegetation, or among rocks. It may move 25 km in a night, or up to 75 km if disturbed. It generally progresses by a slow, silent walk, but can briefly run at speeds of over 60 km/hr. It has been reported to leap over 6 meters horizontally and over 3 meters vertically. It climbs with great agility and can descend head first. It is a strong swimmer, but is not as fond of water as is the tiger. Vision and hearing are acute, and the sense of smell seems to be better developed than in the tiger (Guggisberg 1975).

Hunting is accomplished mainly by stalking and stealthily approaching as close as possible to the quarry. Larger animals are seized by the throat and killed by strangulation. Smaller prey may be dispatched by a bite to the back of the

Leopard (*Panthera pardus*), photo from Zoological Society of London.

neck. The diet is varied, but seems to consist mainly of whatever small or medium-sized ungulates are available, such as gazelles, impala, wildebeest, deer, wild goats and pigs, and domestic livestock. Monkeys and baboons are also commonly taken. If necessary, the leopard can switch to such prey as rodents, rabbits, birds, and even arthropods. Food is frequently stored in trees for later use. Such is the strength of the leopard that it can ascend a tree carrying a carcass larger than itself (Guggisberg 1975; Kingdon 1977).

Population density is usually about one leopard per 20 to 30 sq km, but under exceptionally favorable conditions may be as high as one per sq km. Reported home range size is 8 to 63 sq km. Individuals apparently keep to a restricted area, which they usually defend against others of the same sex. Territories are marked with urine, and severe intraspecific fighting has been observed. The range of a male may include the range of one or two females. Several males sometimes follow and fight over a female. Apparently, the leopard is normally a solitary species, but there have been reports of males remaining with females after mating, and even helping to rear the young. There are a variety of vocalizations, the most common of which are a coughing grunt and a rasping sound, which seem to function in communication (Kingdon 1977; Guggisberg 1975; Schaller 1972; Muckenhirn and Eisenberg 1973; R. M. Smith 1977; Grzimek 1975; Myers 1976).

Breeding occurs throughout the year in Africa and India, though there may be peaks in some areas. In Manchuria and southeastern Siberia the mating season is apparently in January and February. A female may give birth every 1 or 2 years. The estrous cycle averages about 46 days and heat lasts 6 or 7 days. The gestation period is 90 to 105 days. Births occur in a cave, crevice, hollow tree, or thicket. The number of young per litter is one to six, usually two or three. The cubs weigh 500 to 600 grams each at birth, open their eyes after 10 days, are weaned at 3 months of age, and usually separate from the mother when 18 to 24 months old. Full size and sexual maturity are attained at around 3 years of age. Maximum longevity in captivity is over 23 years (Guggisberg 1975; Grzimek 1975; Kingdon 1977).

The leopard seems more adaptable to the presence and activities of people than is the tiger, and still occurs over a greater portion of its original range. Nonetheless, it is confronted by the same problems—persecution as a predator, value as a trophy, commercial demand for its beautiful fur, and loss of habitat. Man-eating leopards, though representing only a tiny percentage of the species, have undeniably been a menace in some areas. One in India, for example, is said to have killed over 200 people (Guggisberg 1975). Intensification of agriculture, along with elimination of natural prey, sets up conflicts between herders and the leopard, to the usual detriment of the latter. The pelt of the leopard has been sought since ancient times, but in the 1960s there was a substantial increase in the world-wide market for the furs of spotted cats. Illegal hunting became rampant and some leopard populations were decimated. As many as 50,000 leopard skins were marketed annually, of which nearly 10,000 were imported into the United States in some years (Myers 1976; Paradiso 1972). National laws and international agreements subsequently seem to have reduced this traffic. The leopard is still relatively common in parts of East and Central Africa, but has become rare in most of West and North Africa and in most of Asia.

The IUCN (1972, 1976) classifies *P. pardus* generally as

Jaguar (*Panthera onca*), photo from Zoological Society of London.

vulnerable, and the following subspecies as endangered: *P. p. orientalis* of southeastern Siberia, Manchuria, and Korea; *P. p. tulliana* of Asia Minor and adjacent areas; *P. p. nimr* of the Arabian Peninsula, Jordan, and Israel; *P. p. jarvisi* of the Sinai Peninsula; and *P. p. panthera* of northwestern Africa. The USDI (1982) lists *P. pardus* as endangered, except in that part of Africa south of a line corresponding with the southern borders of Equatorial Guinea, Cameroon, Central African Republic, Sudan, Ethiopia, and Somalia. In the region so demarcated, the leopard is listed as threatened, and regulations allow the importation into the United States, under the provisions of the CITES, of trophy leopards taken in this region for purposes of sport hunting. Importation for commercial purposes in prohibited. *P. pardus* is on appendix 1 of the CITES.

Panthera onca.   Head and body length is 1,120 to 1,850 mm and tail length is 450 to 750 mm. Reported weights range from 36 to 158 kg. The ground color varies from pale yellow through reddish yellow to reddish brown, and pales to white or light buff on the underparts. There are black spots on the head, neck, and limbs, and large black blotches on the underparts. The shoulders, back, and flanks have spots, forming large rosettes that enclose one or more dots in a field darker than the ground color. Along the midline of the back is a row of elongate black spots that may merge into a solid line. Melanistic individuals are common, but the spots on such animals can still be seen in oblique light. *P. onca* averages larger than *P. pardus,* and has a relatively shorter tail, a more compact and powerfully built body, and a larger and broader head (Guggisberg 1975; Grzimek 1975).

The jaguar is commonly found in forests and savannahs, but, at the northern extremity of its range, may enter scrub country and even deserts. It seems usually to require the presence of much fresh water, and is an excellent swimmer. It climbs well and is almost as arboreal as the leopard. Most hunting, however, is done on the ground and at night. Prey is stalked or ambushed, and carcasses may be dragged some distance to a sheltered spot. The most important foods are peccaries and capybaras. Tapirs, crocodilians, and fish are also taken (Guggisberg 1975; Grzimek 1975).

In a radio-tracking study in southwestern Brazil, Schaller (1980a, 1980b) found a population density of about 1 jaguar per 25 sq km. There was a land tenure system much like that of the cougar and tiger. Females had home ranges of 25 to 38 sq km, which overlapped one another. Resident males used areas twice as large, which overlapped the ranges of several females. The jaguar seems to be basically solitary and territorial. It marks with urine and has a variety of vocalizations, including roars, grunts, and mews (Guggisberg 1975).

There is no specific breeding season in most of the range of *P. onca,* but mating usually occurs in the spring in the extreme north. The gestation period is 93 to 105 days, and litters contain one to four young. The offspring weigh 700 to 900 grams each at birth, open their eyes after 13 days, stay with the mother about 2 years, and attain full size and sexual maturity at 3 to 4 years of age. Captives have lived up to 22 years (Grzimek 1975; Guggisberg 1975).

Until the end of the Pleistocene, *P. onca* occurred throughout the southern United States, and seems to have been especially common in Florida. There is some evidence that the species still inhabited the southeastern United States in historical time (Nowak 1975b; Daggett and Henning 1974). By the early 20th century, resident populations had disappeared from the United States, though for many years wanderers continued to enter the country from Mexico, and were usually quickly shot. The jaguar has now been exterminated in most of Mexico, much of Central America, and, at the other end of its range, in Uruguay and all but the northernmost parts of Argentina.

The jaguar declined because of the same factors that affected other large cats—persecution as a predator, habitat loss, and commercial fur hunting. The jaguar is thought to have killed people on rare occasion, but no individuals are known to have become systematic man-eaters. The species does have a reputation as a serious menace to domestic cattle; indeed, it may actually have increased in colonial times, when livestock was introduced to the savannahs of South America, a continent that lacks vast natural herds of ungulates (Guggisberg 1975). As land was cleared and opened to ranching and other human activity, inevitable conflicts arose between the jaguar and people. As in the case of the leopard and other spotted cats, there was a great increase in the commercial demand for jaguar skins in the 1960s. An estimated 15,000 jaguars were then being killed annually in the Amazonian region of Brazil alone. The recorded number of pelts entering the United States reached 13,516 in 1968. Subsequent national and international conservation measures seem to have reduced the kill, but the problem continues. *P. onca* is classified as vulnerable by the IUCN (1978) and as endangered by the USDI (1980), and is on appendix 1 of the CITES.

Panthera leo.   Except as noted, the information for the account of this species was taken from Schaller (1972), Guggisberg (1975), Kingdon (1977), and Grzimek (1975). In males, head and body length is 1,700 to 2,500 mm, tail length is 900 to 1,050 mm, shoulder height is about 1,230 mm, and weight is 150 to 250 kg. In females, head and body length is 1,400 to 1,750 mm, tail length is 700 to 1,000 mm, shoulder height is about 1,070 mm, and weight is 120 to 182 kg. The coloration varies widely, from light buff and silvery gray to yellowish red and dark ochraceous brown. The underparts and insides of the limbs are paler, and the tuft at the end of the tail is black. The male's mane, which apparently serves to protect the neck in intraspecific fighting, is usually yellow, brown, or reddish brown in younger animals, but tends to darken with age, and may be entirely black.

The preferred habitats of the lion are grassy plains, savannahs, open woodlands, and scrub country. It sometimes enters semideserts and forests, and has been recorded in mountains at elevations of up to 5,000 meters. It normally walks at about 4 km/hr, and can run for a short distance at 50 to 60 km/hr. Leaps of up to 12 meters have been reported. The lion readily enters trees by jumping, but is not an adept climber. Senses of sight, hearing, and smell are all thought to be excellent. Activity may occur at any hour, but is mainly nocturnal and crepuscular. In places where the lion is protected from human harassment, it is commonly seen by day. The average period of inactivity is about 20 or 21 hours a day. Nightly movements in Nairobi National Park were found to cover 0.5 to 11.2 km (Rudnai 1973). In the Serengeti, most lions remain throughout the year in one area, but about one-fifth of the animals are nomads, which follow the migrations of ungulate herds.

The lion usually hunts by a slow stalk, alternately creeping and freezing, and utilizing every available bit of cover. It then makes a final rush and leaps upon the objective. If the intended victim cannot be caught in a chase of 50 to 100 meters, the lion usually tires and gives up, but pursuits of up to 500 meters have been observed. Small prey may be dispatched by a swipe of the paw. Large animals are seized by the throat and strangled, or are suffocated by clamping the jaws over the mouth and nostrils. Two lions sometimes approach prey from opposite directions; if one misses, the other tries to capture the victim as it flees by. An entire pride may fan out and then close in on the quarry from all sides. Groups have about twice the chance of lone individuals in capturing prey. Most hunts fail; of 61 stalks observed by Rudnai (1973), only 10 were successful. The lion eats anything it can catch and kill, but depends mostly on animals that weigh 50 to 300 kg. Important prey species are wildebeest, impala, other antelopes, giraffe, buffalo, wild hogs, and zebra. Carrion is readily taken. Up to 40 kg of meat can be consumed by an adult male at one meal. After making a kill, a lion may rest in the vicinity of the carcass for several days. About 10 to 20 large animals are killed per lion each year.

Population density in the whole Serengeti ecosystem of East Africa was determined to be one lion for every 10 to 12.7 sq km. Reported densities in other areas vary from one per 2.6 to one per 17 sq km. Nonmigratory lions in the Serengeti live in prides, each with a home range of 20 to 400 sq km. All or part of the range is a territory that is vigorously defended against other lions. Nomadic lions have ranges of up to 4,000 sq km; there is much overlap in these areas and individuals behave amicably. Nomads are commonly found in groups of two to four animals, and membership changes freely.

The basis of a resident pride is a group of related females and their young. These associations may persist for many

African lions (*Panthera leo*), photo by Bernhard Grzimek.

years, being generally closed to strange females. Daughters of group members are recruited into the pride, but young males depart as they approach maturity. Several adult males often come together. Such a group, or a single male, joins a pride of females and young for an indefinite period. The males cooperatively defend the pride against the approach of outside males. Some males associate with and defend several prides. Eventually, usually within three years, the pride males are driven off by another group of males (Bertram 1975).

In the Serengeti, the average number of lions in a pride was found to be 15, the range was 4 to 37, and the number of adult males present was 2 to 4. Prides were often divided into widely scattered smaller groups with about 4 individuals each. In Nairobi National Park, Rudnai (1973) observed a single adult male to be associated with four prides of females. There was a rank order among the females, and a female led each group, even when the male was present, but the male was dominant with respect to access to food. When males live within a pride, they allow the females to do almost all of the hunting, but arrive subsequent to a kill and sometimes drive the others away. Lions appear to behave asocially at kills, there being much quarreling and snapping, and little tolerance shown to subordinates and cubs.

The lion has at least nine distinct vocalizations, including a series of grunts that apparently serve to maintain contact as a pride moves about. The roar, which can be heard by people up to 9 km away, is usually given shortly after sundown for about an hour, and then again following a kill and after eating. It apparently has a territorial function. The lion also proclaims its territory by scent marking through urination, defecation, and rubbing its head in a bush.

Breeding occurs throughout the year in India and in Africa south of the Sahara. In any one pride, however, females tend to give birth at about the same time (Bertram 1975). Females are polyestrous and heat lasts about 4 days. A female normally gives birth every 18 to 26 months, but, if an entire litter is lost, may mate again within a few days. The gestation period is 100 to 119 days. Litters contain one to six young, usually three or four. The newborn weigh about 1,300 grams each. The eyes may be open at birth or take up to 2 weeks to open. Cubs follow their mother after 3 months, suckle from any lactating female in the pride, and are usually weaned by the age of 6 or 7 months. They do not participate in kills until they are about 11 months old, are fully dependent on the adults for food until 16 months old, and probably are not capable of surviving on their own until at least 2½ years of age. Sexual maturity is attained at around 3 or 4 years, and growth continues to about the age of 6 years. The average longevity in zoos is about 13 years, but some captives have lived nearly 30 years.

With the exception of people, their domestic animals, and their commensals, the lion attained the greatest geographical distribution of any terrestrial mammal. Various populations, known from fossils by such names as *Panthera atrox* and *P. spelaea*, are now regarded as conspecific with *P. leo* (Hemmer 1974). About 10,000 years ago, the lion apparently occurred in most of Africa, in all of Eurasia except probably the southeastern forests, throughout North America, and at least in northern South America. The lion is thought to have disappeared from most of Europe, because of the development there of dense forests (Guggisberg 1975). It probably vanished from the Western Hemisphere, when many of the large mammals on which it preyed were exterminated through the spread of advanced human hunters at the close of the Pleistocene. It was eliminated in the Balkan Peninsula, its last major stronghold in Europe, about 2,000 years ago, and in Palestine at the time of the Crusades.

The continued decline of *P. leo* in modern times has re-

sulted primarily from the expansion of human activity and domestic livestock, and the consequent persecution of the lion as a predator. A few lions became regular man-eaters—for example, a pair killed 124 people in Uganda in 1925—and thus gave a sinister reputation to the entire species. Hunting for sport was also a major factor in some areas—for example, one person killed over 300 lions in India in the mid-19th century. At that time *P. leo* was still common from Asia Minor to central India, and in northern Africa. By about 1940 the species had been eliminated throughout these regions, except in the Gir Forest, Gujarat State, western India. The animals there have been under continuous pressure from livestock interests, but vigorous conservation efforts have been made, and the number of lions seems to have stabilized at around 180. The subspecies involved, *P. leo persica*, is classified as endangered by the IUCN (1978) and the USDI (1980), and is on appendix 1 of the CITES.

To the south of the Sahara, the lion has become rare in West Africa (Rosevear 1974) and has been exterminated in most of South Africa and much of East Africa. It is still common over a large region, but is widely hunted and poisoned by persons owning livestock. Myers (1975a) wrote that since 1950 the number of lions in Africa may have been reduced by a half, to as few as 200,000 or less. He cautioned that the species was rapidly losing ground to agriculture, and that by the end of the century it might number only a few thousand individuals and survive only in major parks and reserves.

CARNIVORA; FELIDAE; *Genus ACINONYX Brookes, 1828*

Cheetah

The single species, *A. jubatus*, originally occurred from Palestine and the Arabian Peninsula to Tadzhik and central India, and throughout Africa, except in the tropical forest zone and the central Sahara (Meester and Setzer 1971; Kingdon 1977; Guggisberg 1975; Ellerman and Morrison-Scott 1966).

Except as noted, the information for the remainder of this account was taken from Eaton (1974), Kingdon (1977), Guggisberg (1975), and Grzimek (1975). Head and body length is 1,120 to 1,500 mm, tail length is 600 to 800 mm, shoulder height is 700 to 900 mm, and weight is 35 to 72 kg. On the average, males are larger than females. The ground color of the upper parts is tawny or pale buff or grayish white. The underparts are paler, often white. The pelage is generally marked by round, black spots, set closely together and not arranged in rosettes. A black stripe extends from the anterior corner of the eye to the mouth. The last third of the tail has a series of black rings. The coat is coarse. The hair is somewhat longer on the nape than elsewhere, forming a short mane. In young cubs, the mane is much more pronounced and extends over the head, neck, and back. *Acinonyx* has a slim body, very long legs, a rounded head, and short ears. The pupil of the eye is round. The paws are very narrow, compared to those of other cats, and look something like those of dogs. The claws are blunt, only slightly curved, and only partly retractile.

An additional species, *A. rex* (king cheetah), was described in 1927. It was based on specimens that differed from other cheetahs in having longer and softer hair, and partial replacement of the normal spots by dark bars. Only 13 skins have been recorded, all from Rhodesia and adjacent areas. It is now generally accepted that these specimens represent merely a variety of *A. jubatus* (Hills and Smithers 1980).

Cheetah (*Acinonyx jubatus*), photo by Bernhard Grzimek.

The habitat of the cheetah varies widely, from semidesert through open grassland to thick bush. Activity is mostly diurnal and shelter is sought in dense vegetation. Recorded daily movements are about 3.7 km for a female with cubs and 7.1 km for adult males. The cheetah is capable of climbing and often plays about in trees. It is the fastest terrestrial mammal. Reported estimates of maximum speed range from 80 to 112 km/hr. Such velocities, however, can not be maintained for more than a few hundred meters. Unlike most cats, the cheetah does not usually ambush its prey or approach to within springing distance. It stalks an animal and then charges from about 70 to 100 meters away. It is seldom successful, if it attacks from a point over 200 meters distant, and it can only continue a chase for about 500 meters. Most hunts fail. If an animal is overtaken, it is usually knocked down by the force of the cheetah's charge and then seized by the throat and strangled. The diet consists mainly of gazelles, impalas, other small and medium-sized ungulates, and the calves of large ungulates. A female with cubs may kill such an animal every day, while lone adults hunt every two to five days. Hares, other small mammals, and birds are sometimes taken. The cheetah seems to work harder for its living than do the other big cats of Africa, and thus may be more vulnerable to environmental changes brought about by human disturbance.

Population density in good habitat varies from about one cheetah per 5 sq km to one per 100 sq km. In marginal habitat, density may be only one per 250 sq km or less (Myers 1975*b*). Reported home range size is about 50 to 130 sq km. The cheetah occurs alone or in small groups. The groups seem usually to be a female with cubs, or two to four related adult males. There is no substantive evidence of territorial defense, but groups avoid one another and mark the area they are using at a given time. Marking is accomplished by regular urination on prominent objects. Such activity also serves to communicate sexual information. The cheetah is normally amicable toward others of its kind, but several males sometimes gather near and fight over an estrous female. There are a number of antagonistic vocalizations, purrs of contentment, a chirping sound made by a female to its cubs, and an explosive yelp that can be heard by people 2 km away.

Births are reported to occur from January to August in East Africa, November to January in Namibia, and November to March in Zambia. Wild females normally give birth at intervals of 17 to 20 months. If all young are lost, however, the mother may soon mate and bear another litter. Estrus lasts about 2 weeks and the gestation period is 90 to 95 days. The number of young per litter is one to eight, usually three to five. The cubs weigh 150 to 300 grams each at birth, open their eyes after 4 to 11 days, and are weaned when 3 to 6 months old. They begin to follow the mother after about 6 weeks, and the family may then shift its place of shelter on an almost daily basis. The cubs are taught by the mother to hunt, separate from her at an age of 15 to 17 months, and attain sexual maturity at 21 to 22 months. Captives have lived up to 19 years.

People have tamed the cheetah and used it to run down game for at least 4,300 years. It was employed in ancient Egypt, Sumeria, and Assyria, and more recently by the royalty of Europe and India. It is usually hooded, like a falcon, when taken out for the chase, and freed when the game is in sight. If the hunt is successful, the cheetah is rewarded with a portion of the kill. If the cheetah should attempt to escape, it soon tires and can be easily caught by persons on horseback. Tame individuals are usually playful and affectionate.

The removal of live cheetahs from the wild has contributed to a decline in the species. Other factors are excessive hunting of both the cheetah and its prey, the spread of people and their livestock, and the fur market. The cheetah seems much less adaptable to the presence of people than is the leopard. It has evidently disappeared in Asia, except for Iran and possibly adjacent parts of Pakistan, Afghanistan, and Turkmen. In the mid-1970s, the population in Iran was estimated to include over 250 individuals and was considered well protected (IUCN 1976). In Africa the species is still widely distributed, but has become very rare in the northern and western parts of the continent. It has been extirpated in most of South Africa and much of East Africa. The total number of individuals remaining in Africa has been estimated at roughly 15,000 (Myers 1975*b*). The IUCN (1976) classifies the cheetah generally as vulnerable and the Asiatic subspecies (*A. j. venaticus*) as endangered. The entire species is listed as endangered by the USDI (1980) and is on appendix 1 of the CITES.

ORDER PINNIPEDIA

Seals, Sea Lions, Walrus

This order of aquatic mammals occurs along ice fronts and coastlines, mainly in polar and temperate parts of the oceans and adjoining seas of the world, but also in some tropical areas and in certain inland bodies of water. The Pinnipedia are traditionally regarded as a full order, and some authorities, including Corbet (1978) and Hall (1981), continue to treat them as such. Other authorities, such as Simpson (1945), have considered the Pinnipedia only a suborder of the order Carnivora. There is now substantial evidence that the pinnipeds belong within the arctoid division of the Carnivora, and are biphyletic in origin, the families Otariidae (eared seals, sea lions) and Odobenidae (walrus) having arisen from bearlike ancestors, and the family Phocidae (earless seals) being an early offshoot of the line leading to the otters (Tedford 1976; McLaren 1960; Rice 1977; Jones, Carter, and Genoways 1979). There are also, however, immunological and chromosomal data that support a monophyletic origin for the Pinnipedia (Aranson 1974; Sarich 1969). The following sequence of 3 Recent pinniped families, 18 genera, and 34 species generally follows that of Rice (1977).

Pinnipeds are measured in a straight line from the tip of the nose to the tip of the tail. Total length varies from 120 to 600 cm; a short or vestigial tail between the hind limbs grows very little after birth. Adults weigh from about 35 kg to 3,700 kg, with *Phoca* containing some of the smallest species and *Mirounga* the largest. Pinnipeds have a streamlined, torpedo-shaped body, with all four limbs modified into flippers. The arm and leg bones are similar to those in the Carnivora, but the bases of the limbs to or beyond the elbows and the knees are deeply enclosed within the body. The hands and feet are long and flattened, hence the name Pinnipedia, which means feather footed. Each limb has five broadly webbed, oarlike digits, which form the flipper. In most species, the head is flattened and the face shortened to aid in rapid propulsion through the water. External ears are small or entirely lacking, and the nostrils are slitlike; the ears and the nose can be tightly closed when the animal is under water. The eyes, which are set in deep protective cushions of fat, are also adapted for underwater use. The cornea is flattened, and the pupil is capable of great enlargement to enable better sight in dark water. The neck in pinnipeds is generally thick and muscular, yet quite flexible. There is a reduction in the interlocking processes of the vertebrae which enables these animals to bend backward to a greater degree than most other mammals. The overall design of the body is fluid, with great power and grace evident in the movements. This allows for absorbing the shock of the impact of ocean waves, hauling out on ice or rocky coasts, or executing agile maneuvers to capture prey at sea.

Pinnipeds are less modified for aquatic life than are the wholly aquatic cetaceans. Like the cetaceans, pinnipeds have a thick coat of subcutaneous blubber to provide energy, buoyancy, and insulation, but most also have hairy coats to protect them from sand and rocks when ashore. All pinnipeds have whiskers and a hairy covering (although this is almost lacking in *Odobenus*), which is kept lubricated by secretions from sebaceous glands. Although all have a coarse coat of guard hairs, the fur seals (*Callorhinus* and *Arctocephalus*) also possess a dense layer of underfur. This underfur traps

small bubbles and keeps the skin dry, while the stiffer guard hairs protect the body from abrasion. Most pinnipeds are born with a woolly coat called "lanugo" which is white in some species and jet black in others. Molting in pinnipeds usually occurs after the breeding season, and is most spectacular in the elephant seals and monk seals, whose outer layer of skin is shed in patches along with the fur.

Pinnipeds are clumsy on land, but in the water they are skillful divers and swimmers. They swim by means of the flippers and by sinuous movements of the trunk. In the Otariidae and the Odobenidae, locomotion is accomplished mainly by use of the forelimbs; in the Phocidae, the hind limbs provide most of the thrust.

Expert diving by pinnipeds is dependent upon the animals' efficient use of oxygen, by means of which they remain submerged longer than terrestrial mammals without sustaining brain damage or the "bends." According to King (1964), just before diving, the seal exhales; when breathing stops, the heartbeat slows, thus conserving oxygen. Adult seals can slow the heart rate from a normal speed of 55–120 beats per minute to 4–15 beats per minute. This phenomenon is known as bradycardia, and it develops more rapidly and lasts longer as the seal grows older. During the dive, the seal's peripheral blood vessels are constricted and circulation is reserved for the heart and brain, thus reducing oxygen consumption by one-third. In addition, pinnipeds have a high tolerance for carbon dioxide and lactic acid build-up in the blood. On surfacing, the seal's heart regains its normal beat within 5–10 minutes, and the blood is reoxygenated by the increased beat and a few deep breaths of air. This efficient use of oxygen enables pinnipeds to make long, deep dives. Diving ability is known to be at least 600 meters in some species, and submergence time can be in excess of 43 minutes.

Pinnipeds, unlike cetaceans, must keep some sort of link to the land, since they can mate and give birth to the young only on shore or on ice. The habitats in which they gather for mating and pupping vary from floe ice in the Arctic and Antarctic to ragged cliffs, sandy beaches, and lava caves elsewhere. The major necessity appears to be isolation from humans and other predators (Haley 1978).

Pinnipeds are carnivorous and consume a wide variety of animal matter, ranging from krill and other crustaceans to mollusks and fish. They usually eat the common seafood of an area, swallowing moderate-sized species whole and head-first. Larger catches are shaken into bite-sized pieces (Haley 1978). Some pinnipeds seek food at night, and certain polar seals feed in total darkness for four months out of the year. A 100-kg seal eats approximately 5–7 kg of food daily when not fasting.

Some species, such as the Ross seal (*Ommatophoca*), live alone during the winter, but most pinnipeds are much more gregarious than land carnivores. A breeding colony of pinnipeds ranges from a few individuals to more than 1 million animals within a radius of 50 km. These mammals tend to frequent small, isolated breeding grounds and are polygamous (Otariidae and Odobenidae) or mostly monogamous (Phocidae). All give birth ashore, on land or ice, and mate once a year. The period of pregnancy is 8 to 15 months, with delayed implantation occurring in many species. Delayed implantation may represent an adaptation that allows the births to take place at approximately the same time of the year, an important feature for colonial and, in some cases, migratory species. Single births are the rule; twins are the exception. Newborn pinnipeds can swim, but the pups of some species do not have enough blubber to provide buoyancy and insulation until they are several weeks old. Growth during the nursing period is rapid, for the mother's milk is particularly rich, about 50 percent fat. The adult pelage is usually acquired near the end of the first summer. Pinnipeds are sexually mature at 2 to 5 years and may live to 40 years in the wild. Predators include large sharks, killer whales (*Orcinus*), leopard seals (*Hydrurga*), and polar bears (*Ursus*).

Seals have been greatly valued by humans for centuries because of their fur, oil, and ivory, or for food or fertilizer. They have generally been easy targets, because of their tendency to congregate in large numbers in specific areas for breeding and giving birth (Haley 1978). Seals have been hunted for hundreds of years, and sealing expeditions have been responsible for the slaughter of millions of animals. For example, the IUCN (1978) reported that the Juan Fernandez fur seal (*Arctocephalus philippii*) was abundant off the coast

A. Guadalupe fur seal (*Arctocephalus philippii*), photo by Warren J. Houck. B. Leopard seal (*Hydrurga leptonyx*), photo by Michael C. T. Smith. C. A young northern elephant seal (*Mirounga angustirostris*) on a sandy beach. The remarkable modification of the hind limbs and body which adapts it for an aquatic existence are well illustrated here. The front flippers are partially hidden by the loose sand; photo by Julio Berdegué.

of Chile during the 16th and 17th centuries, but by 1824, after millions of seals had been killed by sealers, the species was commercially extinct. By the early 1960s, scientists thought this seal was totally extinct, but remnant populations, discovered on Alejandro Selkirk and Robinson Crusoe islands in 1965, enabled the species to survive to the present. The Caribbean monk seal (*Monachus tropicalis*), on the other hand, was particularly vulnerable to human disturbance and is now thought to be extinct. It was seriously exploited in the 18th century and was already rare by 1850, large numbers being killed without difficulty due to their sluggish and unsuspicious nature when on land. No Caribbean monk seals have been observed in the wild in decades (IUCN 1978). Allen (1942) reported that in 1824 ten vessels secured between 70,000 and 80,000 fur seals on New Zealand and adjacent islands alone. In 1806, a single vessel bound for London from Sydney, Australia, had 30,000 seal skins aboard.

Today, the Soviet Union permits seals and walruses to be taken in the Bering and Chukchi seas. There is a regulated take of fur seals along the Pacific coast of Canada and Alaska, and young harp seals and hooded seals are commercially harvested annually off Newfoundland and in the Gulf of St. Lawrence for their soft pelage. This harvest of young seals has been highly controversial and strongly protested by conservation groups in recent years. The pelts, known as ''white coats'' in the harp seals and ''blue coats'' in the hooded seals, are valuable items in commercial trade; Barzdo (1980) reported that 208,759 seals of both species were killed during the 1977 season. Elsewhere, seals are little exploited commercially today; the United States has prohibited the take of seals in its waters (except for certain specific purposes) since the passage of the Marine Mammal Protection Act of 1972.

The known geological range of the Pinnipedia is early Miocene to Recent in North America, Pliocene to Recent in South America and Europe, late Pliocene in Egypt, and Pleistocene to Recent in New Zealand, Australia, and Japan.

PINNIPEDIA; *Family OTARIIDAE*

Eared Seals, Fur Seals, Sea Lions

This family of 7 Recent genera and 14 species occurs in coastal waters of northeastern Asia, western North America, South America, South Africa, southern Australia, New Zealand, and many, predominantly southern oceanic islands. King (1964) recognized two subfamilies: the Otariinae (sea lions), with the genera *Phocarctos, Otaria, Zalophus, Neophoca*, and *Eumetopias;* and the Arctocephalinae (fur seals), with *Callorhinus* and *Arctocephalus.* Hall (1981) did not consider the Arctocephalinae a separate subfamily, but did regard the Odobenidae as only a subfamily of the Otariidae, and used the name Rosmarinae for this subfamily. Various other authorities, including Rice (1977), Cabrera (1957), and Ellerman and Morrison-Scott (1966), do not employ subfamilial divisions within the Otariidae.

A northern fur seal (*Callorhinus ursinus*), member of the family Otariidae, which, like the Odobenidae, is able to move its foreflippers with great freedom and turn its hind limbs forward, thus enabling it to assume an erect posture. Photo by Victor B. Scheffer through U.S. Fish and Wildlife Service.

Total length of these seals ranges from 150 to 350 cm, the males always being much larger then the females; weight ranges from about 35 to 1,100 kg. The tail in members of this family is small, but always distinct. Body form in all genera is slender and elongated; there are long, oarlike flippers that bear rudimentary nails. These flippers are thick and cartilaginous, thickest at the leading edge, with the surfaces being smooth and leathery. External ears are present, but they are small and entirely cartilaginous. Members of this family, along with the Odobenidae, differ from the Phocidae in that the hind limbs can be turned forward to help support the body so that all four limbs can be used for traveling on land. In the Phocidae, the hind limbs cannot be moved ahead, and the animals must wiggle and hunch to travel on land. Members of both the Otariidae and the Odobenidae walk or run on land in a somewhat doglike fashion.

Sea lions have blunt snouts; coats of short, coarse guard hairs covering only a small amount of underfur; the first digit of the foreflipper longer than the second; and the outer digit of the hind flipper longer than the inner. Fur seals have more pointed snouts, long guard hairs concealed by very thick underfur of considerable commercial value, the first digit of the foreflipper of approximately the same length as or slightly shorter than the second, and the digits of the hind flipper more nearly of the same length (King 1964). The hair of the newborn of all eared seals is silky, never woolly. Adult coloration varies from yellowish or red brown to black; there are no stripes or sharp markings.

The normal dental formula is: (i 3/2, c 1/1, pm 4/4, m 1–2/1) × 2 = 34 to 36. The first and second upper incisors are small and divided by a deep groove into two cusps, and the third (outer) upper incisor is caninelike. The canine teeth are large, conical, pointed, and recurved, and the premolars and molars are similar, with one main cusp. The number of upper molars varies within and among the genera. The skull is somewhat elongated, though quite bearlike.

Eared seals inhabit arctic, temperate, and subtropical waters. Their breeding habitat is exclusively marine, never fresh water. They shelter in quiet bays, on rocky isolated islands, and along sea coasts. In this family (and the Odobenidae) the swimming mechanism is centered near the forepart of the body and locomotion in the water is accomplished mainly by use of the forelimbs. The animals are both diurnal and nocturnal, and have acute sight but poorly developed smell and hearing. They defend themselves by tearing an adversary with their canine teeth, by hurling their weight against the adversary, or by diving and swimming away. Eared seals feed mainly on fish, but also eat cephalopods and crustaceans. Harem bulls generally fast during the breeding season.

Otariids are highly gregarious, especially during the breeding season. The males arrive first on the breeding grounds, where they establish definite territories. The females come somewhat later and give birth shortly after arrival. The members of this family are polygamous, each bull having from 3 to over 40 females in its harem; the size of the harem depends on the species, and on the strength and ferocity of the bull. The bulls mate with the females on land soon after the females have given birth to their young. Pregnancy lasts about 250–365 days, and, in a number of species, is known to include a period of delayed implantation. Only the mother cares for the pups. The young usually do not swim for at least 2 weeks; they are generally weaned in about 6 to 12 months.

The known geological range of this family is early Miocene to Recent in North America; Pliocene to Recent in South America; Pleistocene to Recent in Australia, New Zealand, and Japan; and Recent in other parts of the current range (Stains 1967).

New Zealand sea lion (*Phocarctos hookeri*), photo by D. J. Griffiths.

PINNIPEDIA; OTARIIDAE; *Genus PHOCARCTOS Peters, 1866*

Hooker's, New Zealand, or Auckland Sea Lion

The single species, *P. hookeri,* is known from the coastal waters of New Zealand and the subantarctic islands to the south. It regularly breeds at Carnley Harbor and on Enderby Island in the Aucklands, and rarely at Campbell Island. It hauls out on Snares Island, Macquarie Island, and the South Island of New Zealand. It occurred on the North Island of New Zealand less than 1,000 years ago (Rice 1977). *P. hookeri* has sometimes been placed in the genus *Neophoca,* and has even been considered conspecific with *N. cinerea,* but, based on morphological and behavioral distinctions, most authorities now treat *Phocarctos* as a full genus (Crawley and Cameron 1972; Marlow 1975; Marlow and King 1974).

The remainder of this account is based largely on King (1964). Large adult males are about 300 cm in total length, while females are much smaller, only about 190 cm. Males are blackish brown in color and have a well-developed mane of long, dark hairs. Females are lighter, being tawny or buff in color.

Phocarctos prefers sandy beaches, but has been seen as much as one kilometer inland, among brush and tussock grass. It feeds on flounders, other small fish, crabs, and mussels. It has been observed eating red crabs (*Nectocarcinus antarcticus*) while at sea, and coming ashore to digest the meal; a ball of undigested crab legs and shell was then regurgitated. It has been seen to chase penguins along the beach and to take them out to sea to eat, first tearing them apart and later discarding bones and skin.

In early October, adult bulls move onto the beaches, take up territories, and defend them against other bulls. Cows arrive in early November, and each bull takes about a dozen cows into a harem. After the pups are born in December and

New Zealand sea lions (*Phocarctos hookeri*): Top, adult male; Bottom, adult female; photos by Basil Marlow.

January, mating takes place. The bulls then allow the cows greater freedom to go into the sea to feed; the cows return at intervals to nurse the young. By the end of January or the beginning of February, most cows have mated and the pups have taken to the sea; harems break up, and the bulls, which have been fasting since gathering the harems, go to sea to feed.

The pups at birth have a coat of fine-textured, chestnut brown hair, which is molted within the next few months. The pups suckle for seven months, and may remain with their mothers for about one year. When they are nearly one month old, the pups are coaxed into the water by the mothers; after that they spend much of their time playing in the water and chasing each other through muddy pools.

In the 19th century, when sealing activities in the Southern Hemisphere were at their height, *Phocarctos* was included in the general slaughter, the blubber being used to obtain oil and the skin used as leather. This seal is not killed commercially today, and the Auckland Islands have been made a reserve for it. Ling (1979) reported that this animal was rare, but the population appeared to be stable at about 3,000 to 4,000 animals.

PINNIPEDIA; OTARIIDAE; *Genus OTARIA* Peron, *1816*

Southern or South American Sea Lion

The single species, *O. flavescens,* occurs in the coastal waters of South America, from Brazil and Peru, south to the Strait of Magellan and the Falkland Islands (Rice 1977). Wellington and DeVries (1976) reported that in October of 1973 a dead individual was found in the Galapagos, indicating that the range of the species may extend to that area. Some authorities, such as King (1978), have used the name *O. byronia* for this species.

The total length of adult males is about 245 cm and the usual weight is 230 to 320 kg. Females are about 200 cm in total length and weigh about 140 kg. The general coloration is dark brown, but there is considerable variation in shade. The mane of the male is usually paler than the body, and the belly lightens to a dark yellow. Females have the back of the head and neck dull yellow, or sometimes the whole head is yellow (King 1964; Pilleri and Gihr 1977*c*).

The South American sea lion is usually found in and near salt water, but has been noted 300 km up the Rio de la Plata. On the Verde Islands, off the coast of Uruguay, this sea lion lived on the rocky parts and cared for its young in the small pools among the granite blocks that protected it from the surf (Pilleri and Gihr 1977*c*). Although some populations of *Otaria* make limited movements, there is apparently no true migration.

Except as noted, the information for the remainder of this account was taken from King (1964). The main food of *Otaria* is squid and the crustacean *Munida*, both of which are very abundant around the Falkland Islands and along the Argentine coast. Small fish are also eaten, and occasionally penguins. Stones are swallowed sometimes, and as much as nine kilograms of pebbles and sharp-angled stones have been removed from the stomach of a single animal. Whether these

South American sea lions (*Otaria flavescens*), male and female, photo by Lothar Schlawe.

stones are harmful to the seal or serve some sort of digestive function is not known.

There is usually no segregation of any part of a herd, but as the breeding season approaches (November–December), the breeding animals congregate and are surrounded by a fringe of immature animals and idle bulls. All suitable stretches of the coast are occupied, and it is only because certain stretches of the beach are unsuitable that the herd is divided into distinct rookeries, separated from each other by the unoccupied areas; each rookery may contain several harems. The average number of cows per harem is nine, but the harems are often so close to each other that the limits of each harem are known only to the harem master himself; harem masters are very quick to repel any trespassers onto their territory. Harem bulls do not leave their territories, even if a high tide should almost cover them.

Females mate a few days after giving birth to the young, and are then allowed by the harem bulls to enter the sea, from which they return at intervals to nurse the young. After all the cows have mated, the harem disintegrates. The harem bulls have spent at least two months defending their territories, during which they do not feed. Consequently, at the close of the mating season, they are very lean and emaciated, and spend much of their time sleeping. In the next six months they feed heavily, and recover their strength so that they are in good condition for the next breeding season. Already by August, they are beginning to take an interest in the females again.

The pups are born from about 25 December to 15 February, most being born in early January. At birth, the pups are about 80 cm in total length, and have jet black fur that soon fades to chocolate brown and finally to brownish yellow by the end of the first year (King 1964; Pilleri and Gihr 1977c). A few pups remain with their mothers, possibly still suckling, until the next pup is born, but most have been weaned and are independent by the time they are six months of age.

As the females return to the sea after the mating season, the pups wander about on shore, and gather in groups called "pods" at the edge of the water. They do not attempt to go into deeper water until coaxed by the mother, and even then seem to be hesitant about swimming. Males become sexually mature in their sixth year of life, when the mane begins to appear. Females become sexually mature in their fourth year, probably producing their first young in the fifth year. Potential longevity in the wild is thought to be about 20 years. According to Marvin L. Jones (Zoological Society of San Diego, pers. comm.), a captive was estimated to have been 24 years and 10 months old at time of death.

O. flavescens has been exploited commercially since the 16th century for meat, hides, and oil; an adult male can yield about 50 liters of oil in September, before the mating season. The U.S. National Marine Fisheries Service (1981) estimated that 273,000 of these seals now occur throughout their entire range, of which 228,000 are distributed on the Pacific side of the South American continent and 45,000 on the Atlantic side.

PINNIPEDIA; OTARIIDAE; *Genus ZALOPHUS Gill, 1866*

California Sea Lion

The single species, *Z. californianus*, occurs in three isolated populations, one in the Sea of Japan, one off the western coast of North America from Vancouver Island to Sinaloa (northwestern Mexico), and one breeding on the Galapagos Islands (Rice 1977).

In males, total length is 200 to 250 cm and weight is 200 to 300 kg. In females, total length is 150 to 200 cm and weight is generally 50 to 100 kg (Peterson and Bartholomew 1967). Both sexes are chocolate brown in color, although shades may vary. The adult male has a very noticeable raised forehead because of the extremely high sagittal crest on the underlying skull; the top of the animal's head often gets lighter with age.

According to King (1964), *Zalophus* is a shore-living and coastal seal, not being found farther than about 160 km out to sea. Along the California and Mexican coasts it occurs chiefly on small islands, such as those in the Santa Barbara group, Los Coronados, Guadalupe, or San Benito Island. Bigg (1973) found this sea lion to prefer the inner reefs as hauling out areas, rather than the outer rocks preferred by *Eumetopias*. Although present year-round in this area, there is a certain amount of seasonal movement, with many bulls moving north in the winter to Washington and British Columbia (U.S. National Marine Fisheries Service 1981).

Peterson and Bartholomew (1967) stated that *Zalophus* has little visual acuity in the air, but may possess the ability to echolocate under water. When swimming, it frequently leaps from the water and reenters headfirst in a shallow arc (porpoising). Groups of 5 to 20 young sometimes swim in a single file or leap out of the water one after the other. Terrestrial locomotion is variable and includes walking, galloping, and striding with the front flippers only. *Zalophus* grooms itself extensively with its flippers. It feeds day and night, mainly on squid, octopus, and a variety of fishes such as herring, sardines, rockfish, hake, and ratfish (U.S. National Marine Fisheries Service 1978).

The population off the western coast of North America breeds from San Miguel Island, California, to Punta Estrada, Baja California. At other times, according to Peterson and Bartholomew (1967), there is a conspicuous, but incomplete sexual segregation. Adult and subadult males move northward as soon as the breeding season ends, and return to the rookeries in the spring. Females and young either remain in the vicinity of the rookeries all year or move south in the winter. In the breeding season, males establish a territory and gather a harem of 5 to 20 females thereon. The females are not tightly controlled, and move freely from one territory to the next. Peterson and Bartholomew (1967) stated that the typical rookery has the territorial bulls spaced fairly uniformly in a single line along the beach, with females in dense aggregations at irregular intervals among the males. Almost all bull territories extend to the water's edge, many being partly aquatic or even wholly aquatic offshore. The territories are poorly defined and vary in size, but many are spaced at 10- to 15-meter intervals. Bulls establish the territories by forcing others out. Those on the territories bark incessantly, while the intruding males are silent. Once established on a territory, the bulls do very little fighting; the territory is maintained primarily by loud and frequent vocalizations and displays with ritualized postures and movements.

Births occur from mid-May to June along the California coast, and from October to December in the Galapagos Islands. Pregnancy lasts 11½ to 12 months and the period of delayed implantation is probably about 3½ months. Females usually mate 15 to 30 days after giving birth. The single pup weighs 5 or 6 kg at birth and may nurse for more than 1 year; the exact age of weaning is unknown. Longevity records of 20 to 25 years in captivity are common (U.S. National Marine Fisheries Service 1978, 1981).

Zalophus is the pinniped that is most commonly trained for use in circuses. According to King (1964), training is accomplished by constantly rewarding the sea lion with fish. Ball balancing may be achieved by constantly throwing a ball at the animal until it is accidentally balanced, or by holding a ball on its nose until the animal realizes what is wanted.

California sea lion (*Zalophus californianus*), photo by Victor B. Scheffer. Inset: photo by Daniel K. Odell.

Zalophus is intelligent and has a good memory. Although training may take a year before the sea lion is fully ready for performances, the animal will be able to perform perfectly later, even after a complete rest of 3 months. The performing life of a sea lion may last 8 to 12 years.

Fishermen often consider *Zalophus* a serious competitor. In California it has been accused of damaging the salmon fishing industry, but Briggs and Davis (1972) reported that available evidence is inconclusive. They found that from mid-April to mid-September 1969, in Monterey Bay, *Zalophus* accounted for a loss of 4.1 percent of the sport and commercial salmon catch, but they noted that their study achieved only a 0.21 percent sampling level.

The U.S. National Marine Fisheries Service (1981) estimated the Mexican and Californian populations of *Zalophus* to number about 45,000 animals each. The Galapagos population seems to have stabilized at around 20,000, after recovering from sealing operations at the turn of the century. The population in the Sea of Japan may have been completely extirpated by the 1950s, mainly because of persecution by fishermen, but there have been reports of its continued existence on Dakto Island. This population (subspecies *japonicus*) is classified as endangered by the IUCN (1976).

PINNIPEDIA; OTARIIDAE; **Genus NEOPHOCA Gray, 1866**

Australian Sea Lion

The single species, *N. cinerea*, occurs in the coastal waters from Houtman Rocks, off west-central Western Australia, south and west to Kangaroo Island, South Australia (Rice 1977). The genus *Phocarctos* has sometimes been included in *Neophoca*.

Neophoca is a big sea lion; adult males reportedly measure 300 to 350 cm in total length, and may weigh up to 300 kg (King 1964; Ling 1977). Females are much smaller, however, having a recorded length of 132 to 181 cm, and a weight of 63 to 104 kg (Walker and Ling 1980). Adult males are dark brown in color, with a mane of coarse yellowish hairs. Younger males have a paler area on the neck, but no mane. Females are rich brown dorsally and yellowish fawn underneath (King 1964).

Stirling (1972a) reported that at South Neptune Island and at Kangaroo Island, *Neophoca* was present at breeding and nonbreeding colonies throughout the year. Where there was a choice, beaches and smooth rock areas were preferred to

Australian sea lions (*Neophoca cinerea*): A. Photo by Vincent Serventy; B. Photo by Eric Lindgren.

boulder beaches, jagged promontories, or cliffs. King (1964) stated that on land, *Neophoca* is noted for its ability to climb cliffs, and has been found as far as 10 km inland. According to Ling (1977), fish, crayfish, and squid compose the main diet for *Neophoca*. King (1964) noted that penguins are sometimes eaten. Males apparently do not fast during the breeding season, but catch penguins. Stones, ranging in size from a walnut to a tennis ball, have been found in their stomachs (King 1964).

The following reproductive data were taken from Stirling (1972a), Ling and Walker (1976, 1977), and Ling (1977). The breeding biology is something of a puzzle because of its irregularity. Observations suggest that this seal may breed every one and a half years, which is unlike the breeding cycle of most other pinnipeds. Only one pup is born to each cow, and family groups do not form. A breeding bull may, however, assemble a harem of from three to five cows, which he protects against intruders. Each bull controls a territory, which assures a fairly even spacing of breeding groups in the colony, thereby reducing crowding. Territorial bulls do not fixate on one area, but relocate at intervals depending on environmental factors. Guttural threats, growls, and high-pitched rapid barking are recorded in agonistic encounters between males, but the incessant vocalizing of *Zalophus* is absent in *Neophoca*. Breeding sites tend to be in rather inaccessible areas on distant islands and reefs, or isolated coves that are easily approached from the sea.

Harrison (1969) suggested a gestation period of 11 months in *N. cinerea;* delayed implantation has not yet been demonstrated in this species. Cows nurse their young right up to the birth of the next pup. The pups weigh about 7 kg at birth, but gain weight rapidly, and weigh up to 30 kg at one year of age. Newborn pups are closely attended by their mothers at the breeding sites for several weeks before moving out onto more open beaches, from which they can reach the water. At about 2 months of age the young begin to play in the shallow water and pools, and actually may go into the sea for short distances. However, they are almost helpless in the waves, and it is only after an active training period, instituted by the cows, that they are able to fend for themselves in the sea.

Neophoca practices the nanny system in breeding rookeries. This involves one or two cows looking after several other pups besides their own while the real mother is away feeding. When the feeding mother returns, it looks after the young of the cow that was protecting its calf. These cows will aggressively defend not only their own calf but also the calf of the mother that is temporarily away feeding.

Ling (1977, 1979) estimated the population of *Neophoca* at 3,000 to 5,000 individuals. If this estimate is valid, *Neophoca* would be one of the rarest pinnipeds in the world today. However, Ling pointed out that early reports indicate that the genus may never have been very numerous, even before the days of Australian colonization. Moreover, the fact that, in spite of recent human disturbance, relatively small numbers of sea lions are able to sustain viable populations suggests that the genus may never have been very abundant during its recent history.

PINNIPEDIA; OTARIIDAE; *Genus EUMETOPIAS Gill, 1866*

Northern or Steller Sea Lion

The single species, *E. jubatus,* breeds on islands in the Sea of Okhotsk, on the Kuril Islands, along the eastern coast of Kamchatka, on the Aleutian and Pribilof islands, and along the western coast of North America from the Alaska Penin-

sula to San Miguel Island, California; the postbreeding range extends farther north in the Bering Sea (Rice 1977).

Eumetopias is the largest sea lion, males being about 330 to 350 cm in total length and weighing over 910 kg. Females are much smaller, averaging about 240 cm in total length and about 275 kg in weight (King 1964). Both sexes are a variable yellowish buff color, and adult males develop a heavy, muscular neck bearing a mane of long, coarse hair (King 1964).

According to Kenyon and Rice (1961), *Eumetopias* tends to remain in the water during stormy weather, when a heavy surf is running. Most individuals keep in a compact group a few meters beyond the beaches. In mild weather, especially when the wind is light and the sky clear, most animals haul out onto the beaches. During daylight, more sea lions are on land in the afternoon than in the morning hours. Observations of *E. jubatus* in Alaska and California indicate that this sea lion undertakes seasonal movements, both latitudinal and local, or longitudinal. The Pribilof Islands are the northernmost breeding grounds, but a number of animals, probably adult and subadult males, migrate north in late summer and early fall. As many as a thousand regularly reach St. Lawrence Island, and a few reach the Bering Strait. The animals move south again with the advance of ice in the fall, but ice does not cause them to desert the Pribilof Islands. Certain hauling grounds within the breeding range are used seasonally.

The northern sea lion may dive 110 to 146 meters below the surface of the water, and sometimes reach a depth of 183 meters. A study of 382 stomachs indicated the following diet: squid, clams, octopus, sand lance, rockfish, crabs, flounders, halibut, greenling, and lumpfish (U.S. National Marine Fisheries Service 1981). Adult northern sea lions utter a prolonged, deep-throated roar, and often make coughing and grunting sounds.

Eumetopias favors isolated localities for breeding, with some shelter, free access to the sea, and freedom from human harassment. Breeding colonies become established on rocky outcrops, and boulder, cobblestone, and coarse sand beaches (U.S. National Marine Fisheries Service 1981). According to King (1964), the breeding season really begins early in May when the adult bulls arrive at the breeding sites to take up their territories. Harems of 10 to 20 cows are formed as the females arrive a few weeks later, and the pups are born a few days thereafter. Sandegren (1976) found that the female plays an active and elaborate role in courtship. She often climbs over the male, and bites him to attract attention, and then rushes away as if to enter the sea, or another male's territory.

Pitcher and Calkins (1981) reported that in the Gulf of Alaska, births occur from mid-May to mid-July, mating is from late May to middle or late July, and implantation of the fertilized egg in the uterus is delayed until September or October. The total period of pregnancy thus lasts about 1 year, including a delayed implantation period of about 4 months. The female-pup bond also usually lasts about 1 year, but young up to 3 years old have been seen to suckle. Females mature sexually between 3 and 4 years of age, and reach a breeding peak of one pup per year between 7 and 14 years of age. Males mature sexually at 4 to 5 years, but do not become effective breeders until their 9th to 15th year. This large sea lion is seldom seen in zoos, and because of its belligerent nature it is not amenable to circus training (King 1964). One specimen lived in captivity for 17 years and 3 months (Marvin L. Jones, Zoological Society of San Diego, pers. comm.).

The U.S. National Marine Fisheries Service (1981) stated that northern sea lions are probably at or near the carrying capacity of their ecosystem. The Alaska populations are esti-

Steller sea lions (*Eumetopias jubatus*): A. Rookery, photo by Victor B. Scheffer; B & C. Male, female, and pup, photos by Karl W. Kenyon.

mated at over 200,000 animals. Russian populations are thought to be between 20,000 and 50,000. The British Columbia population numbers about 5,000 animals; Washington, about 600; Oregon, about 2,000; and California, about 5,000 to 7,000. Braham, Everitt, and Rugh (1980), however, reported that a significant decline in northern sea lions occurred in the eastern Aleutians between 1957 and 1977. They estimated that the population was below 25,000 in 1977, whereas in the late 1950s it exceeded 50,000. No specific cause for the decline was apparent, but several possibilities exist, including a shift in distribution of the population to the west, exposure of the population to the pathogen *Leptospira*, and decline in prey fishes as commercial fisheries activity in the area increased.

Northern fur seals (*Callorhinus ursinus*): A. Harem bull with females and young; B. Courtship between fur seals; photos by Victor B. Scheffer.

PINNIPEDIA; OTARIIDAE; *Genus CALLORHINUS Gray, 1859*

Northern Fur Seal

The single species, *C. ursinus*, occurs in a great arc across the North Pacific from the Sea of Japan through the Sea of Okhotsk and the Bering Sea to the Channel Islands of southern California (Rice 1977; U.S. National Marine Fisheries Service 1981).

According to Baker, et al. (1970), large males may measure up to 213 cm in total length and weigh 181.4 to 272.2 kg. Females are much smaller, a large animal measuring 142 cm in length and weighing 43.1 to 49.9 kg. The size dif-

ference between males and females is evident even at birth, when males average 66 cm in length and females are 2.5 cm smaller. The color of adult males is dark, rich brown, the intermingling of white hairs with the dark ones on the neck and shoulders giving a grayish tinge. The females are dark gray dorsally, and lighter gray, with a tinge of chestnut, ventrally. In both sexes, there is a characteristic light patch across the chest, and the rich underfur is chestnut colored. When born, the pups have coarse black hair, but this is shed at about eight weeks of age for a coat that is steel gray dorsally and creamy white underneath. Juveniles assume adult pelage in their second year.

The northern fur seal spends most of its life in the water. It travels relatively well on land, but must rest frequently. At sea, it feeds primarily at night, because most of the important prey species rise to the surface at that time. It will also feed during the day, however, if prey is available. It is a rapid and skillful swimmer and an expert diver when searching for food. The deepest dive recorded by Kooyman, Gentry, and Urquhart (1976) was 190 meters, and the longest submersion was 5.6 minutes. Baker et al. (1970) reported *Callorhinus* to feed mainly on small schooling fish, such as anchovy, capelin, and herring, but to take whatever species are available. Squid is a mainstay almost everywhere. Salmon occurred in only 239 of 9,580 stomachs examined from the Pacific Northwest, suggesting that *Callorhinus* is not detrimental to the salmon fishery, as has often been claimed.

The northern fur seal is highly migratory (King 1964; Rice 1977; U.S. National Marine Fisheries Service 1978, 1981). Its breeding range apparently was once much more extensive than now, extending far down the western coast of North America, but at present the only southerly breeding colony is at San Miguel Island off southern California. Currently, there are four main breeding populations, the largest on the Pribilof Islands in the eastern Bering Sea and smaller ones on the Commander Islands in the western Bering Sea, Robben Island in the Sea of Okhotsk, and the Kuril Islands. Adult females and young animals range the farthest in the non-breeding season. Migration from the Pribilof Islands begins in October, and by December most animals have left. They head mainly to the southeast, through the Aleutian passes, and in December are found as far south as San Francisco. From January to April they may be found anywhere along the migration route from Alaska to California, but in April they begin to move north again. By May, large numbers are in the Gulf of Alaska, and from June to October most are back in the vicinity of the Pribilofs. *Callorhinus* travels not in large herds but alone or in groups of up to 10 individuals.

In the western Pacific, *Callorhinus* is first seen off Hokkaido in October, and numbers there reach a maximum in December. The animals then move south to Honshu, where they congregate until returning to the breeding grounds in July. Recoveries of banded and tagged individuals have shown that thousands of seals from Alaskan waters migrate into Japanese waters and mingle there with Asian seals, and that seals born in Alaskan rookeries occasionally return north to Asian breeding grounds as subadults.

Bulls have a strong homing instinct, however, and usually return to the rookery of their birth to establish a territory and to mate. In the eastern Pacific, bulls begin arriving on the breeding grounds in mid-June, and cows follow them in late June and early July. The bulls gather harems of 1 to over 100 cows, the average being about 40. Location influences the size of a bull's harem and the rapidity with which it attracts females; generally, a territory nearest the water will attract the most females. Females seem to seek a locality rather than a specific male, and the most vigorous efforts of a bull are not sufficient to keep a cow if she is determined to leave. Bulls do not eat during the breeding season, but subsist on their own blubber at this time.

Within three days of the cow's arrival, she gives birth to a single pup; she mates with the harem bull a few days later. Pregnancy lasts about 1 year, including a delayed implantation period of about 4 months. The pup is born headfirst; delivery takes usually less than 10 minutes. It weighs about 5 kg. For a few days, the mother is very protective of the pup, driving all other females away. Then begins a cycle of nursing-feeding of 2 days on land and 8 days at sea within a 330-km radius of the breeding rookery. Nursing of the pups is ended when the female begins its migration southward in the fall. In this species at least, the female apparently plays no part in teaching the young to swim. The pup can swim from the moment of its birth and never receives lessons from its mother. Males take no part in raising or protecting the young,

and generally pay no attention to them (King 1964; Baker et al. 1970; U.S. National Marine Fisheries Service 1978, 1981).

Pups shed their birth coat in late summer for the adult pelage. The females become sexually mature in 3 to 5 years. The males become sexually mature as young as 4 to 5 years, but do not become members of the organized breeding structure until age 10. Females have been known to reach 22 years of age, but few males live beyond 15 years (U.S. National Marine Fisheries Service 1978).

The northern fur seal has been hunted by sealers for centuries for its very fine pelt. According to Baker et al. (1970), early pelagic sealing had a devastating effect on the northern fur seal. Almost 1 million skins were taken on the high seas from 1879 to 1909, and many of the seals shot or speared were not recovered. The effect on the Alaska herd was disastrous, because females made up 60 to 80 percent of the pelagic catch. In addition to the enormous harvest at sea, thousands of fur seals were taken on their rookeries. After the purchase of Alaska by the United States, some 300,000 were taken during the first sealing season alone. As a result of this excessive kill over the years, a census in 1912 showed only about 216,000 seals remaining in the Alaska herd. An international treaty abolishing sealing on the high seas for 15 years was signed in 1911 by the United States, Great Britain (representing Canada), Russia, and Japan to help prevent the continuing decline of this species. Under this treaty, the United States and Russia, as guardians of the seal herds, agreed to pay to both Great Britain and Japan, for their relinquishment of pelagic sealing, 15 percent of the proceeds from the land sealing conducted by these two nations. Japan, in turn, agreed to pay to the United States, Great Britain, and Russia, 10 percent of the land catch of the herd under its jurisdiction. The treaty endured until 1941. Following 16 years of temporary arrangements, a new four-nation treaty was signed in 1957. The Pribilofs are a special government reservation—landings can be made only with government permission or because of bad weather conditions. Such measures allowed a remarkable numerical recovery.

At present, the populations of *Callorhinus* that breed on Soviet islands are estimated to contain 585,000 individuals, the population breeding on the Pribilofs is estimated at 1,250,000, and that using San Miguel Island off southern California contains more than 3,000. The U.S. National Marine Fisheries Service (1978, 1981) considers the northern fur seal to be at or near the carrying capacity of the ecosystem, and this agency strictly regulates the annual commercial harvest on the two Pribilof Islands, St. Paul and St. George. The mean yearly kill was 65,000 from 1939 to 1955 and 48,000 from 1956 to 1972. Only bachelor males, mainly three and four years old, have been taken since 1968, and since 1973 there has been no commercial take on St. George Island. In 1980, a total of 24,278 males was harvested.

PINNIPEDIA; OTARIIDAE; **Genus ARCTOCEPHALUS**
E. Geoffroy St.-Hilaire and F. Cuvier, 1826

Southern Fur Seals

There are eight species (Rice 1977; King 1964; Ling 1979):

A. pusillus, coastal waters of Angola, Namibia, South Africa, southern Australia, and Tasmania;

A. gazella, to the south of the Antarctic Convergence on South Shetland, South Orkney, South Sandwich, South Georgia, Bouvet, Kerguelen, Heard, and McDonald islands;

A. tropicalis, to the north of the Antarctic Convergence on

Southern fur seal (*Arctocephalus forsteri*), photo by John Warham.

Tristan da Cunha, Gough, Marion, Prince Edward, Crozet, Amsterdam, and St. Paul islands;

A. *forsteri*, coastal waters of southern Australia, New Zealand, and subantarctic islands south of New Zealand;

A. *australis*, coasts of South America from central Peru and southeastern Brazil to Tierra del Fuego, Falkland Islands;

A. *galapagoensis*, Galapagos Islands;

A. *philippii*, Juan Fernandez Islands off central Chile, probably also formerly bred on San Felix and San Ambrosio islands off northern Chile;

A. *townsendi*, Guadalupe Island and occasionally other islands off southern California and Baja California.

Both King (1964) and Hall (1981) considered A. *townsendi* a subspecies of A. *philippii*, and King also treated A. *gazella* as a subspecies of A. *tropicalis*.

Total length for males of this genus ranges from about 150 to 270 cm, and weights range from 140 to 700 kg; females are much smaller in all species, measuring from 130 to 180 cm in length and weighing generally between 32 and 120 kg. A. *tropicalis* is one of the smallest species, while A. *pusillus* is the largest. The coloration of all species of *Arctocephalus* is very similar. The underfur and the short hairs at the bases of the flippers are a rich, chestnut brown color. The males have a thick mane of long guard hairs that are frequently black with white tips, giving the mane a frosted, grayish appearance (King 1964; Stirling and Warneke 1971; Miller 1975a).

Peterson et al. (1968) found that A. *townsendi* generally occurs on shores characterized by solid rock and large lava blocks, usually at the base of towering cliffs. These authors stated that perhaps the most unusual feature of this species is its tendency to frequent caves and recesses while on land. By looking into vents and windows of these caves, it was possible to detect the seals at distances of at least 25 meters from the entrances. Crawley and Wilson (1976) reported that fur

seals of the genus *Arctocephalus* are nearly always found on exposed, rocky coastlines. Specifically, for A. *forsteri*, they recognized three major types of habitat: tumbledown (talus) beaches, terraced rocky ledges, and small islets. Bonner (1968) reported that A. *gazella* on South Georgia Island is found on beaches composed of pebbles and cobbles, with many large rocks. Hubbs and Norris (1971) stated that A. *philippii* generally hauls out on solid rock at the base of cliffs, on ledges, or in various recesses and caves.

There does not appear to be true migration in any of the species of *Arctocephalus*, although Crawley and Wilson (1976) reported that bulls of A. *forsteri* in New Zealand move north in the winter from the breeding grounds. In most species, there may be movement between different islands, and some individuals may wander widely in search of food. In fact, recoveries of tagged pups show that southern African A. *pusillus* may move throughout the whole range of the southern African population, as much as 1,300 km in one direction; individuals have been seen as much 160 km from the nearest land.

Bonner (1968) found the crustacean *Euphausia superba* (krill) to be the staple diet of A. *gazella* on South Georgia Island; fish and cephalopods were also taken. Crawley and Wilson (1976) reported that A. *forsteri* in New Zealand waters fed on squid, octopus, barracuda, red cod, whiptail, kahawai, horse mackerel, butterfish, lamprey, blind eel, ling, blue cod, flounder, crayfish, crab, and penguin. Squid and octopus were the most important food items, followed by penguin and surface-feeding fish. This species apparently feeds mainly at night in surface waters. King (1964) said that the diet of southern African A. *pusillus* included small shoaling fish such as maasbankers and pilchards, cephalopods, rock lobsters (in small amounts), and various other small crustaceans.

Peterson et al. (1968) described some nonbreeding behavior of A. *townsendi* which may be common to most other species of the genus. In all seasons, play is conspicuous among the young animals. They play in shallow tide pools, lunging at one another's heads and foreflippers and chasing one another into the water. They have been observed to position themselves in the water vertically, head down, while the exposed hindflippers wave slowly in the air. On shore, these seals, especially when concealed, commonly allow themselves to be approached by a human observer. After diving into the sea, they almost invariably follow the shoreline closely, often "porpoising." They seldom rear their head out of the water to scrutinize the observer as does *Zalophus*. Adult fur seals of this species behaved aggressively toward a human diver during one of the expeditions.

For the Guadalupe fur seal (*A. townsendi*), Peterson et al. (1968) described the following vocalizations: a bark, usually sounded when the animals move or play; a high-pitched roar, which was often directed at humans when they approached closely; and a guttural cough, used when one animal threatened another. Adult males on territories seemed to utilize the bark more often than any other sort of vocalization. When startled by a human, these fur seals typically began to utter roars. Females with pups made a prolonged bawl, apparently during interactions with the pups. Many other types of vocalization have been identified in other species of this genus. Crawley and Wilson (1976) categorized calls of *A. forsteri* into the following types: male's full threat (readiness to attack), male's low-intensity threat (does not precede attack), male's guttural challenge, male's and female's submissive call, female's threat, male's bark (indicates sexual interest), female's pup-attraction call (mother to straying pup), pup's female-attraction call (pup to attract mother's attention), male's moan (habitual use, function unknown). There is some evidence that the threat calls of bulls are learned, and differ somewhat from one population to another of the same species; thus, they may serve in assisting individual females in recognizing their own intraspecific group at a particular geographic location. Stirling and Warneke (1971), in a study of vocalizations of *A. pusillus* and *A. forsteri,* postulated that if such a situation exists for populations within a species,

where differences in vocalizations are slight, then it may serve as an effective reproductive isolating mechanism between species where the vocalizations differ more markedly.

The following account of the annual cycle of *A. forsteri* in New Zealand (Crawley and Wilson 1976:13–14) may be considered as generally typical for the cycle of other members of the genus *Arctocephalus* (except for dates, which may differ in other species):

"Seals are present in the rookeries and on the hauling grounds throughout the year, although the proportions of the various sex and age classes vary seasonally. Few bulls frequent the rookeries in the March–September period, the resident population comprising small subadults, yearlings and cows. Yearling numbers decrease steadily between June and November and few remain when the first pups are born in November. From mid-October to mid-November the number of bulls ashore increases steadily, while cows arrive in large numbers in late November and throughout December. From mid-November to late December the number of bulls increases slowly. The relative absence of cows ashore in late November may be because pregnant females feed heavily at sea in the weeks before giving birth. Most cows arrive about two weeks after the main influx of bulls, by which time territory establishment is already well advanced. The numbers of both bulls and cows in the rookeries decrease sharply in early January.

"Pupping is from late November to mid-January with a peak in mid-December. Cows stay with their newborn pups for about ten days and during that period they mate with the nearest male, usually a territorial bull. Their first feeding trip is of three to five days and on their return they suckle the pups for two to four days. As the pups grow older the cows leave them alone for longer, and the pups congregate into unstable small groups called pods. Each pup leaves the pod to suckle its mother on her return from each feeding trip.

"The decline in the numbers of bulls and cows ashore during January is a consequence of the gradual breakdown of the territorial system following the main birth and mating period. The bulls may have spent up to ten weeks ashore

Southern fur seals (*Arctocephalus forsteri*), photo by Graham J. Wilson.

without food or water, and they depart for the feeding areas to regain their strength. The cows may have been ashore for two to three weeks, during which time they have given birth, mated, and suckled their young; they move away for their first feeding trip. In February there may be no bulls ashore, and possibly only half as many cows as in January. The feeding trips of the cows are longer than in December and January. As the adult population declines there is an influx of subadults and yearlings on to the main breeding rocks. Many of the subadults occupy offshore rocks during the breeding period, while some of the larger subadult males even infiltrate the edges of the rookery before being chased by the territorial bulls.

''During the March–September non-breeding period the rookeries are occupied by pups, yearlings, small subadults of both sexes and females still suckling young. Many of the bulls and large numbers of subadult males move north for the winter. Cows probably spend long periods at sea feeding when they are not at the rookeries suckling their young.''

Stirling (1971a) noted that the ratio of territorial males to adult females in pupping colonies of A. forsteri is difficult to calculate. Individual females have no bonds to specific males, and may pass freely from one bull territory to the next. He noted that none of the species of Arctocephalus approaches the extreme polygamy of Callorhinus, for whom a ratio of harem bulls to breeding females averages 1:40 and a harem of 153 females has been recorded for a single bull. King (1964) recorded Arctocephalus harem sizes as 3 to 5 for A. australis, 4 or 5 for A. tropicalis, 6 or 7 for a Tasmanian population of A. pusillus, and up to 10 for A. forsteri. The U.S. National Marine Fisheries Service (1978) reported harems of as many as 10 cows for A. townsendi. King (1964) reported delayed implantation for all species of this genus for which data are available. In A. australis, which mates in December, the blastocyst was still free in a female killed in April, although an embryo of 15 cm was found in a female in July. The pregnancy period averages about 259 days for A. pusillus and about 330 days for A. australis (including a delayed implantation period in both cases). Crawley and Wilson (1976) found that at birth, pups of A. forsteri were about 55 cm long and weighed 3.5 kg. At 6 weeks of age, male pups averaged 5.1 kg and were significantly larger than females, which weighed on the average about 4.5 kg; both male and female pups averaged about 65 cm in total length at 6 weeks of age. A specimen of A. pusillus lived in captivity for 21 years (Marvin L. Jones, Zoological Society of San Diego, pers. comm.).

Most species of southern fur seal were subjected to catastrophic exploitation in the late 18th and 19th centuries. Payne (1978) stated that at the peak of sealing activities, 17 ships took 112,000 skins of A. gazella at South Georgia Island in a single season, and the species was almost exterminated there. Hubbs and Norris (1971) reported that A. philippii occurred in teeming numbers during the 16th and 17th centuries on the Juan Fernandez and San Felix islands off the mainland coast of Chile; the numbers may have been as high as 3 million to 3½ million seals (U.S. National Marine Fisheries Service 1978). Intensive sealing for the China trade began shortly before the end of the 18th century and expanded ruthlessly during the first quarter of the 19th century. Early accounts indicate that several million seals were slaughtered on these islands, and that by 1824 the species was approaching commercial extinction. A few animals were taken in the late 1800s and some were seen in 1917, but subsequently the species was thought to be extinct. Commercial harvesting reduced the population of A. townsendi to the point that it was also believed extinct for a long period of time (U.S. National Marine Fisheries Service 1978). Stirling

Southern fur seal (*Arctocephalus pusillus*), photo by John Visser.

(1972c) noted that A. forsteri was also decimated by the early 1800s by ruthless commercial sealers. Every species of this genus was adversely affected by the great commercial sealing fleets of the last two centuries.

The situation today is much more encouraging. Commercial take of these seals has virtually ended, and protection has brought about some encouraging recoveries. Payne (1978) reported that the population of A. gazella at South Georgia Island increased rapidly during the past 40 years, and that the picture now is one of a population growing at a rate unprecedented in pinnipeds. Hubbs and Norris (1971) stated that as early as 1958, reports suggested some survival of A. philippii on the Juan Fernandez Islands. In 1965, the survival of the species was confirmed, and the population was said to be between 700 and 800 animals in 1973 (U.S. National Marine Fisheries Service 1978). Recovery of A. townsendi has also been noted; the U.S. National Marine Fisheries Service (1981) reported that observers have counted 1,100 of these seals on Guadalupe Island and that there have been occasional sightings in the offshore waters of Baja California and southern California. Stirling (1972c) said that small numbers of A. forsteri remained in South Australia after the slaughter of the 18th century, and these remnants multiplied and eventually repopulated the offshore islands, although recolonization has been very slow. Ling (1979) stated that this species is still rare in Australia, but the populations are now stable; in New Zealand, A. forsteri is abundant and increasing. Budd (1970, 1972) found a rapid population growth (as much as a sixfold increase) of A. gazella at Heard Island in the early 1970s. There now may be over 10,000 A. galapagoensis, though the species was nearly exterminated in the 19th century. All species of Arctocephalus are on appendix 2 of CITES, except for A. townsendi, which is on appendix 1. The IUCN (1978) classifies A. philippii and A. galapagoensis as vulnerable.

PINNIPEDIA; *Family ODOBENIDAE; Genus ODOBENUS Brisson, 1762*

Walrus

The single living genus and species, *Odobenus rosmarus*, occurs in the Arctic Ocean and adjoining seas. There are late Pleistocene and prehistoric Recent records from as far south as California, Michigan, North Carolina, and France (Rice 1977). Hall (1981) considered the Odobenidae only a subfamily of the Otariidae, used the name Rosmarinae for this subfamily and the name *Rosmarus* in place of *Odobenus*, and indicated that the historical range of the walrus extended to Massachusetts.

The walrus has a thick, swollen body form; a rounded head and muzzle; a short neck; a tough, wrinkled, thinly haired skin; and large tusks. The eyes are small and piglike, and the external ears are represented by only a low wrinkle of skin. Pedersen (1962) observed that the nostrils of the Pacific subspecies (*O. r. divergens*) are not visible when viewed from the front. In the Atlantic subspecies (*O. r. rosmarus*), however, the nostrils are visible when viewed from the front. The foreflippers are long and oarlike, and are about one-quarter the length of the body. All the flippers are thick and cartilaginous, being thickest on the forward (leading) edge. The flippers have five digits, and the palms and soles are bare, rough, and warty for traction on ice. In the family Odobenidae, as in the Otariidae, the hind limbs can be turned forward to aid the body in maneuvering on all four limbs on land; the members of these two families walk or run with a somewhat upright carriage. In the Phocidae, on the other hand, the hind limbs cannot be so manipulated, and these animals must twist and move forward awkwardly on land.

Bulls have a total length of 300 to 370 cm and a height of as much as 150 cm; the weight is usually more than 900 kg, and can be as much as 1,600 kg. Cows are only about one-third of the size of the bulls. There is no external tail in the walrus. The skin is 1.5 to 5 cm thick, with an underlying layer of blubber that can measure about 15 cm in thickness. The wrinkled, yellowish gray hide, which appears to be bay colored when dry, is naked except for a few scattered stiff hairs. There is also a conspicuous mustache, consisting of about 400 thick bristles richly supplied with blood vessels and nerves. Males have a baculum that is about 63 cm in length. Females have four mammae.

The skull is rounded, and the halves of the lower jaw are solidly fused in the adult. The permanent dental formula is variable, but usually is: (i 1/0, c 1/1, pm 3/3, m 0/0) × 2 = 18. The upper canines are an outstanding feature of adult walruses. These teeth grow throughout life in both sexes, developing into tusks that may attain a length of 100 cm in the male and 60 cm in the female; about 80 percent of the tusk is exposed above the gumline. A single tusk in an old male may weigh 5.35 kg. The tusks of the female are more recurved in the middle and more slender than those of the male. Walrus skulls with three upper tusks are fairly common. The tusk is peculiar in that it bears enamel only at the tip for a short time after it is erupted; the entire crown in the adult consists of dentin (ivory). The tusks of both sexes are used for defense, breaking through ice, hooking over ice for stability while sleeping in water, and as aids to hauling out and locomotion on ice; contrary to popular belief, the tusks are not used for digging food from the ocean floor (Miller 1975*b*).

The walrus is usually found in open waters near the edge of polar ice. It shelters on isolated rocky coasts, islands, and ice floes. As in the Otariidae, the swimming mechanism of *Odobenus* is centered near the forepart of the body, and progress in the water is made by using the foreflippers. Swimming speed is up to 24 km/hr. The walrus has fair vision, but poor senses of smell and hearing. The heavy bristles on the snout are sensory organs, and may act as a filter for food on muddy ocean bottoms. Activity is both diurnal and nocturnal. Most individuals migrate south in the winter with the advance of the arctic ice, and move north in the spring as the ice retreats. They ride ice floes in migration, but their movements are sufficiently directional that they can abandon the ice when it strays from their intended course. Stormy weather may force them to travel on snow as far as 33 km overland.

Odobenus lives mainly in coastal waters, as it usually forages at depths of less than 90 meters. The U.S. Fish and Wildlife Service (1980) reported that clams are the most important food item for the Pacific walrus. Also eaten are echinoderms, annelids, coelenterates, sipunculids, echiurids, priapulids, arthropods, and tunicates. The shells of its prey are seldom swallowed. The food is worked into the mouth by the muzzle pads. Remains of young ringed seals and bearded seals, and even of young walrus, have been recorded from stomachs of *Odobenus*, but it is believed that these are eaten only when other foods are scarce. Habitual seal-eating walrus, usually males, are less frequently encountered; walrus stomachs have been found to contain narwhal remains, but it is not known whether the walrus actually killed them (King 1964).

Odobenus generally associates in mixed herds of cows, calves, and bulls. Groups contain over 100, and sometimes over 1,000, individuals. Bulls compete and fight with one another for females, and are polygamous. During the nonbreeding season, males form large aggregations but continue to be antagonistic toward each other. Large males, with long tusks, are generally dominant and aggressive toward others. The tusks are used in threatening displays and in fighting, such as for a space to haul out on land, but serious injuries are rare (Miller 1975*b*). The walrus has a bellow that sounds like the voice of a large dog, and also utters a loud, elephantlike trumpeting sound.

Mating is exclusively marine and occurs mainly in February or March. Pregnancy lasts about 15 months and includes an approximately 3-month-long period of delayed implantation (U.S. Fish and Wildlife Service 1980; Reeves 1978). Females do not normally mate more than once every other year. The young are born from April to June. The single calf at birth is approximately 125 cm long, 45 to 68 kg in weight, gray in color, and capable of swimming. There is no external evidence of teeth. The mother defends and guards her young against danger; a small calf travels on the mother's neck even when she swims and dives. The offspring nurses for 18 to possibly 24 months. Females are capable of bearing young at 4 to 5 years of age and the bulls are sexually mature at about 7 years. A walrus probably can live 40 years in its native habitat.

Practically every part of the walrus is used by the Eskimos for either food, material for boats or shelters, oil, or charms. The Eskimos may actually fish for walrus, using strong lines baited with chunks of blubber, in addition to hunting them. According to the U.S. Fish and Wildlife Service (1980), the Pacific walrus population increased during the past several decades, following a decline in abundance caused by overexploitation. The population may have numbered as few as 40,000 to 50,000 by about 1950. However, over 96,000 walrus were counted in 1975–76 at coastal hauling areas along the Soviet coastline, and another 30,000 to 40,000 were found along the ice edge west of the international dateline. Very few walrus remain in the eastern North Atlantic, where the total population numbered in at least the high

Walruses (*Odobenus rosmarus*): A. A herd of walruses, photo from U.S. Navy; B. Walrus bulls, photo by Leonard Lee Rue III; C. Young walrus, photo from Zoologisk Have, Copenhagen.

tens of thousands in historic times. Fewer than 500 were counted at Novaya Zemlya and around Franz Josef Land in 1969–70. The species was virtually exterminated in Svalbard. A total population of about 200 walrus in Greenland may be stable. Exploitation of walrus diminished in Canada in recent years owing to changes in technology and culture. The northern Hudson Bay herds, estimated at 3,000 in 1961, are probably stable. The population in Foxe Basin appears to have grown, although no estimates of its numbers are available. The Laptev walrus (*O. r. laptevi*), found to the north of Siberia, is classified as indeterminate by the IUCN (1972). This subspecies seems to have been reduced by excessive human hunting, and contained about 4,000 to 5,000 individuals in 1975 (Reeves 1978).

The known geological range of the family Odobenidae is late Miocene to Pleistocene in Atlantic North America, mid-

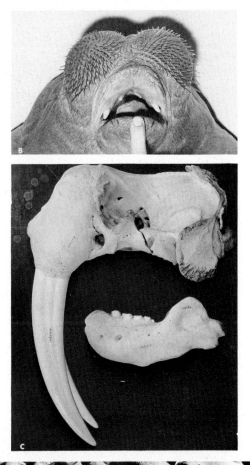

A. Mounted head of an adult male walrus from eastern Siberia (*Odobenus rosmarus divergens*), photo by Grancel Fitz through New York Zoological Society. B. Yearling male walrus (*O. rosmarus rosmarus*), photo by Victor B. Scheffer. C. Skull of Pacific walrus (*O. rosmarus divergens*), photo from U.S. National Museum. D. Walrus bulls (*O. rosmarus*), photo by Leonard Lee Rue III.

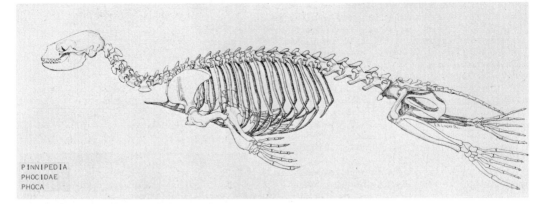

PINNIPEDIA
PHOCIDAE
PHOCA

Skeleton of harbor seal (*Phoca vitulina*), photo from the American Museum of Natural History.

dle Pliocene to Pleistocene in Europe, and Recent in arctic regions.

PINNIPEDIA; *Family PHOCIDAE*

True, Earless, or Hair Seals

This family of 10 Recent genera and 19 species occurs along ice fronts and coastlines, mainly in polar and temperate parts of the oceans and adjoining seas of the world but also in some tropical areas and in certain inland bodies of water. Some authorities, including Stains (1967) and Hall (1981), recognize three subfamilies: Phocinae, with the genera *Phoca*, *Halichoerus*, and *Erignathus;* Monachinae, with *Monachus*, *Lobodon*, *Ommatophoca*, *Hydrurga*, and *Leptonychotes;* and Cystophorinae, with *Cystophora* and *Mirounga*. Other authorities, such as Burns and Fay (1970), do not recognize the Cystophorinae, and relegate *Cystophora* to the Phocinae and *Mirounga* to the Monachinae. Still other authorities, such as Corbet (1978) and Rice (1977), do not use subfamilial designations.

These seals generally measure between 120 and 600 cm in total length and weigh 90 to 3,700 kg, but some individuals of the genus *Phoca* are considerably smaller and lighter. The fur in adults is stiff and lacks appreciable amounts of underfur (adult *Mirounga* have almost no hair). The eyebrow vibrissae are well developed, and the vibrissae in the mustache are often beaded. A number of species have spotted color patterns, and some of the species of *Phoca* are the only pinnipeds with a banded pattern and sexual differences in the pattern. In some genera the newborn are covered with dense, soft, woolly, often white coats, and some members of the family have three distinct coats: newborn, subadult, and adult. As do other pinnipeds, the phocids have extensive amounts of blubber; approximately 113 kg of blubber were removed from a *Leptonychotes*, more than one-fourth of the total weight of the animal. The baculum is well developed in the males of this family, and females have four mammae.

The foreflippers, which are placed far forward, are smaller than the hind flippers and are much less than one-fourth of the body length. The flippers are flexible, of nearly uniform thickness, and equipped with five digits. Phocids cannot turn their hind limbs forward; therefore, they wriggle and hunch along in order to travel on land. Such locomotion is rather laborious, so whenever possible the animals roll or slide. In the Phocidae the external ear is represented by only a faint wrinkling of the skin, and, as in the Odobenidae, there is no

supporting cartilage. The dental formula is: (i 2–3/1–2, c 1/1, pm 4/4, m 0–2/0–2) × 2 = 26 to 36. The upper incisors have simple pointed crowns, the canines are elongate, and the postcanines usually have three or more distinct cusps.

Phocids shelter on rocky or sandy coasts, islands, and ice floes, with some members of the family staying on land or ice much of the time. They generally congregate in groups, but a few forms tend to be solitary except during the breeding and molting seasons. Earless seals do not congregate in large rookeries as the eared seals do, and there are no large harems. They go onto rocky or sandy beaches singly or in small groups, where ice floes or ledges provide easy access onto the ice.

Members of this family propel themselves in water by moving the hind flippers in a vertical plane, as the swimming mechanism is centered near the hind part of the body, not close to the forebody as in the Otariidae and Odobenidae. Phocids frequently swim on their backs, at least in captivity, and regularly stand upright in the water, maintaining their position by "treading" with their foreflippers. Some forms migrate, others make local movements corresponding to fluctuations of the ice, and long dispersals from the home colony are common in some species. The Weddell seal (*Leptonychotes*) is known to have stayed submerged for up to an hour, and to have descended to a depth of at least 600 meters; it frequently dives to 300 to 400 meters, and usually remains submerged for 6 to 15 minutes (Stirling 1971*b*). Phocids defend themselves by opening their mouths, uttering menacing cries, and advancing against the enemy, or else they may flee to the water and dive. Despite their hunching and wiggling form of progression on land, phocids can move fast when necessary.

The members of this family are diurnal and nocturnal, and have good sight and a fair sense of smell, but poor hearing. Most forms eat fish, shellfish, and cephalopods; some species feed mainly on invertebrates and the weaned young of other seal species. The leopard seal (*Hydrurga*) is the only seal that preys regularly on penguins and other seals. Some members of this family can fast for long periods of time. The southern elephant seal, for example, fasts in the wild on land during the rearing of the young for more than two months; a monk seal in captivity has been known to fast for more than four months.

The breeding habitat of the phocids is usually marine, but often at a river mouth. Males establish territories on land during the breeding season. Most species live in pairs during this period, but males of some, such as *Halichoerus* and *Mirounga*, may mate with a number of females. Delayed implantation is known to occur in most species. The pregnan-

A. Southern elephant seals (*Mirounga leonina*), a non-breeding group assembled on Macquarie Island. In such places they are helpless against human predators, and this group is composed of the descendants of a few that remained after the species was almost exterminated in the Antarctic up to 1918. Photo from Australian News and Information Bureau. B & C. Weddell seals (*Leptonychotes weddelli*), about 14 days after birth of the pup, the mother tries to get it in the water by going in herself and calling it. While the pup gains courage, she saws a ramp in the ice with her teeth so that the pup can slide into the water. Photos from U.S. Navy.

cy period for the group ranges from 270 to 350 days (including delayed implantation). The pups at birth may weigh from 9 to almost 40 kg; in *Phoca* the pups can swim from birth, but in some other genera the young often shun the water for several weeks. Young phocids nurse for at least 10 days, usually for several weeks; they reach sexual maturity at an age of 2 to 8 years. Some are known to have lived over 40 years in captivity.

The commercial sealing industry was carried on for many years from both the European and northeastern American coasts. Seals were taken in the North Atlantic, where they were usually killed by being hit on the head with a club as they were lying on the ice floes, although some were shot. Both the newly born young in the white woolly coats and adults were taken. The dwindling population and the increased cost of operations finally caused the industry to decline, but there is still some commerical sealing.

The skins are obtained for leather and for decorative purposes (when the hair if left on). The white woolly skins of the young are used for ornamental purposes. Seal fat was formerly an important source of oil.

The known geological range of this family is early Miocene to Recent in North America, late Pliocene in Egypt, middle Miocene to Recent in Europe, and Recent in nearly all seas and oceans.

PINNIPEDIA; PHOCIDAE; *Genus PHOCA* Linnaeus, 1758

Harbor, Ringed, Harp, and Ribbon Seals

There are four subgenera and seven species (Burns and Fay 1970; King 1964; Rice 1977; Shaughnessy and Fay 1977):

subgenus *Phoca*

P. vitulina (harbor or common seal), coastal waters from Hokkaido around the North Pacific to Baja California, from northwestern Greenland and Hudson Bay to South Carolina, and from Iceland and northern Norway to the Baltic Sea and Portugal;

P. largha (spotted or larga seal), coastal waters and ice floes from northeastern Siberia and northern Yukon to northeastern China;

subgenus *Pusa*

P. hispida (ringed seal), Arctic Ocean and adjoining seas, Sea of Okhotsk, Sea of Japan, Baltic Sea, several inland lakes in southern Finland and adjacent parts of the Soviet Union;

P. sibirica (Baikal seal), Lake Baikal, a body of fresh water in south-central Siberia;

P. caspica (Caspian seal), Caspian Sea;

subgenus *Pagophilus*

P. groenlandica (harp seal), primarily in coastal waters and pack ice from Hudson Bay and the Gulf of St. Lawrence to northwestern Siberia;

subgenus *Histriophoca*

P. fasciata (ribbon seal), mainly in pack ice from northern Hokkaido and the Sea of Okhotsk to northern Alaska.

Pusa Scopoli, 1777, *Pagophilus* Gray, 1844, and *Histriophoca* Gill, 1873 have sometimes been considered distinct genera. Hall (1981) continued to treat them as such, whereas Corbet (1978) recognized only *Histriophoca* and *Pagophilus* as genera separate from *Phoca*.

Seals of this genus usually measure 120 to 190 cm in total length and weigh about 65 to 180 kg. Males average a little larger than females. *P. groenlandica* is the largest species, while *P. sibirica* is the smallest. Coloration in these seals varies enormously. *P. vitulina* and *P. largha* are basically gray colored, blotched with black spots to varying degrees. *P. hispida* is also spotted with black, but many of the spots are surrounded by ring-shaped, lighter marks, which give the animal its vernacular name of ringed seal. *P. caspica* is grayish yellow, irregularly spotted with black. *P. sibirica* is dark silvery gray, shading to lighter yellowish gray ventrally. *P. groenlandica* males are light gray, with the head to just behind the eyes being black; females are similar, but the black facial mask is fainter, or absent. *P. fasciata* is uniquely colored. Males are chocolate brown, with wide, white or yellowish, ribbonlike bands around the neck, hind end, and each flipper; females are paler, with the bands less distinct.

Most species in the genus *Phoca* are ice associated. Bigg (1969) found that *P. vitulina* in British Columbia occurs mainly along the coast, in mud flats, estuaries, and reefs. Mansfield (1967) reported that in the Canadian Arctic this species resides in swift-flowing water, both tidal and fresh, at the edge of fast ice. Boulva and McLaren (1979) found that the harbor seal in eastern Canada generally occurs on islets, on reefs, and in inlets. This species is found even in freshwater lakes and rivers, but Beck, Smith, and Mansfield (1970) reported that these fresh-water habitats are not isolated from the marine environment. *P. hispida* is found on the seasonally shifting ice pack, and on fast ice around the arctic region (T. G. Smith 1973). The primary adaptation of this species to an ice-covered environment throughout the winter is its ability to maintain breathing holes by abrading the sea ice with the claws of its foreflippers. These breathing holes are kept open and are used the year around. The holes are cone shaped, with the apex of the cone at the surface of the ice. *P. largha* also has a strong association with the ice (Naito and Nishiwaki 1975). These seals haul out in winter months on ice floes, and do not haul out on the rocky shores that they are forced to utilize in other seasons when there is little or no ice.

The habitat of *P. sibirica* is in the deepest fresh-water lake in the world, Lake Baikal. This lake averages over 900 meters in depth, and has a maximum depth of over 1,600 meters. King (1964) stated that in winter and the beginning of spring, the Baikal seal is more numerous in the northern parts of the lake. In early summer it is more evenly distributed throughout the lake, but in June it comes onto the shore in herds for breeding. As winter approaches, it leaves the land and goes back into the water, where it spends the winter.

Seals of the genus *Phoca* feed on all sorts of oceanic animal life. Specifically, Sergeant (1976) reported that *P. groenlandica* feeds on larger zooplankton, pelagic fish, benthic crustacea, and benthic fishes; in some areas, capelin is an important food item, while at high latitudes polar cod is taken. Boulva and McLaren (1979) stated that herring was found in 24.2 percent of 279 stomachs of *P. vitulina*. Other important food items for this species were squid (20.6 percent), flounder (14.1 percent), and alewife, hake, smelt, mackerel, sand lance, capelin, and cod. Boulva and McLaren reported that *P. vitulina* can swallow fish whole, or bite them into mouthfuls before swallowing them; the heads of fish are usually bitten off and seldom eaten. *P. sibirica* feeds on gobies and deep-water fishes; *P. caspica* eats gobies, small crustaceans, sprats, and occasionally herring (King 1964).

Members of this genus are rapid swimmers, and can dive deeply in search of food. Sergeant (1976) reported that the maximum depth recorded for *P. groenlandica* is 250 meters. The U.S. National Marine Fisheries Service (1981) stated

Hair or harbor seals (*Phoca vitulina*): A. Adult male, photo by Erna Mohr; B. Two females, photo from Miami Seaquarium; C. Immature seals, photo by Alwin Pedersen; D. Head, showing nostrils closed, photo by Erwin Kulzer.

that *P. vitulina* often dives to depths of over 90 meters, and has been recorded to stay submerged for 23 minutes.

Stirling (1973) found that ringed seals emit vocalizations underwater that are not readily audible to humans standing on the ice above them. He distinguished four such vocalizations for this species: a high-pitched bark, a low-pitched bark, a yelp, and a chirp. He thought that these sounds probably facilitate communication and social organization.

None of the seals of this genus gathers a harem. Sergeant (1976) reported the following information on reproduction for *P. groenlandica* in northern Canada. From December to February, intensive feeding takes place and the pregnant females become fat in preparation for the metabolic demands of lactation. Whelping takes place on the pack ice, and the pup is born with a yellow white lanugo coat that lasts about 2½ weeks. Born weighing 9 kg, the pup increases in weight through intensive suckling to about 35 kg at 16 to 18 days of age. It is then abandoned by the female, who mates with one or more males, usually in the water (adult males have been present at the whelping sites for some days). The females

Ribbon seal (*Phoca fasciata*), photo from San Diego Zoological Society.

then leave the area and feed heavily for some weeks. The young lose their fetal hair gradually, and are fully molted at 4 weeks of age. Young animals actively seek ice, and slowly move north in May and June; at this stage they are solitary. The U.S. National Marine Fisheries Service (1981) reported that the period of pregnancy for *P. groenlandica* is 11½ months, including a 4½ month period of delayed implantation.

Bigg (1969) reported that in British Columbia the main event in the reproductive cycle of the adult female *P. vitulina* is a pupping season of 2 to 2½ months a year, with a peak in late July. The lactation period lasts 5 to 6 weeks; ovulation occurs at the end of lactation. Adult males are reproductively active from March to November and inactive (having no sperm in the epididymis) from December to February. Females mature at 2 to 5 years of age, with most maturing at 3 to 4 years. Males mature between 3 and 6 years of age, with most maturing at 5 years. On the average, 88 percent of the females produce a pup each season. Prenatal males and females grow at similar rates from implantation to birth. At birth, both sexes are similar in size, averaging about 81.6 cm in total length and weighing 10.2 kg. By weaning time, the

Harp seal (*Phoca groenlandica*): A. Adult female; B. Young, 2 years old; photos by Alwin Pedersen.

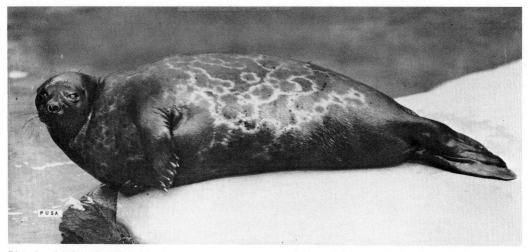

Ringed seal (*Phoca hispida*), photo from the Zoological Garden Berlin-West through Ernst von Roy.

weight is 24.0 kg. From birth to 5 years of age, both sexes grow at about the same rate, but males continue to grow until they are 9 or 10 years of age, long after females have ceased to grow. On the Channel Islands of California, *P. vitulina* gives birth to young in March and April (Odell 1971). The pregnancy period for the species is 10½ to 11 months, including a delayed implantation period of 2 months (U.S. National Marine Fisheries Service 1981).

For *P. sibirica,* King (1964) reported that pregnant females do not spend the winter in water. They come onto the ice and make a lair under snow, where the pups are born in February and March; the mothers feed the young for about three months. Both mating and molting take place in May and June, and the seals are sexually mature in their fourth year.

Female *P. hispida* also construct subnivean lairs for giving birth. These dens are established above the winter breathing holes in the ice and below 20 to 150 cm of snow. A single pup is born there in March or April off northwestern Canada (T. G. Smith and Stirling 1975). *P. hispida* has a pregnancy period of 10½ to 11 months, including a delayed implantation period of 3½ months (U.S. National Marine Fisheries Service 1981).

According to Marvin L. Jones (Zoological Society of San Diego, pers. comm.), a specimen of *P. vitulina* lived in captivity for over 32 years, and a *P. hispida* for 15 years. *P. caspica* has survived in the wild for up to 50 years (Eibatov 1976).

Eskimos hunt these seals and use practically every portion of the carcass for clothing, food, fuel, harnesses, sewing thread, etc. Killer whales (*Orcinus*) and polar bears (*Ursus*) are the greatest natural enemies of these seals.

Seals of the genus *Phoca* have been hunted for commercial purposes for centuries. They have been highly prized for their meat, oil, and pelts, particularly the pelts of young animals in lanugo. In European waters, hunting and pollution have brought about declines in *P. vitulina, P. hispida,* and *P. groenlandica* in many places. Smit and Van Wijngaarden (1976) said that the hunting of *P. groenlandica* was particularly easy and rewarding since the animals gather at only a few places during the breeding season. In 1925, hunting of this seal in the White Sea reached a maximum when 500,000 animals were killed by Russian and Norwegian sealers. In 1951 and 1952 alone over 550,000 were killed, and in the 1960s up to 180,000 pups were killed annually for their

lanugo coats. Pups of the harp seal (along with those of the hooded seal) are commercially harvested in eastern Canada. The harvest is regulated, but each year it has been strongly opposed by conservation groups on humanitarian grounds.

According to the U.S. National Marine Fisheries Service (1981), *P. largha* appears to be stable and near the carrying capacity of its ecosystem; there are regulated hunts for these seals in Soviet waters. *P. vitulina* is considered abundant throughout most of its range, and may be close to the carrying capacity of its ecosystem. *P. hispida* is hunted for subsistence needs throughout most of its range, but is the most abundant pinniped in the arctic basin. Norwegian and White Sea populations of *P. groenlandica* appear to be increasing. From 1961 to 1967, large numbers of *P. fasciata* were taken in waters off the Soviet Union; in 1968, the take was reduced because of a noticeable drop in the population, but by the mid-1970s, the numbers were increasing again. Total world populations of the species of the genus *Phoca* are as follows: *P. largha,* 335,000 to 450,000; *P. vitulina,* 380,000 to 399,000; *P. hispida,* 6 to 7 million; *P. sibirica,* 40,000 to 50,000; *P. caspica,* 500,000 to 600,000; *P. groenlandica,* 1,300,000 to 2,300,000; *P. fasciata,* 200,000 to 250,000. The IUCN (1972) classifies the subspecies *P. hispida saimensis* of the Saimaa Lake System in Finland as rare; only about 200–250 individuals survive.

PiNNIPEDIA; PHOCIDAE; **Genus HALICHOERUS** *Nilsson, 1820*

Gray Seal

The single species, *H. grypus,* occurs on the temperate coasts of the North Atlantic. There are three breeding populations: one in the western Atlantic from Newfoundland to Massachusetts, another in the eastern Atlantic from the British Isles (rarely France) and Iceland to the White Sea, and a third in the Baltic Sea (Rice 1977).

The adult male gray seal ranges up to 290 cm in total length and weighs about 280 kg; females are smaller, reaching about 220 cm in total length and weighing about 250 kg. The adult coat coloration varies enormously, and all shades of gray, brown, and silver may be found. Both sexes have a darker back and lighter belly, and may be spotted to a degree.

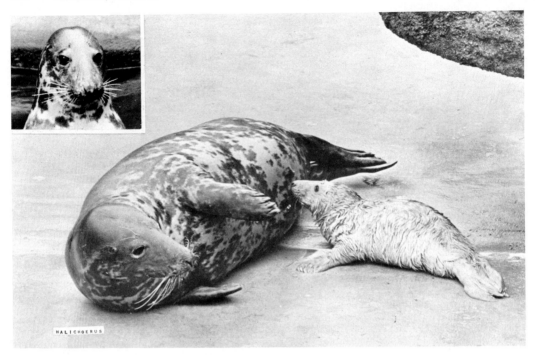

Gray seals (*Halichoerus grypus*), mother suckling young, photo from North American Newspaper Alliance, Inc. Inset: young adult (*H. grypus*), photo by Erna Mohr.

The sexes differ as to the distribution of the darker and lighter color tones. In bulls, the darker tone, whether brown, black, or gray, forms a continuous background upon which may be greater or lesser amounts of irregular spotting of the lighter tone. In females, the lighter tone forms the background upon which there are scattered spots of the darker tone (King 1964).

The adult male gray seal has a high, arched "Roman nose" that is prolonged above the mouth to a much greater extent than that of most other seals; this character is not found in females. Miller and Boness (1979) suggested that this large snout in the males has a substantial visual signal effect. The snout is displayed prominently in numerous short-range agonistic encounters during the breeding season. Gray seals of both sexes have ear pinnae that are more pronounced than is usual for other phocids (King 1964).

In general, the gray seal prefers rocky coasts, cliffs, and caves when ashore. Bonner (1972) stated that North Rona Island, off the coast of northern Scotland, offers typical breeding habitat for the gray seal in the eastern Atlantic. It has no beaches sheltered from the oceanic swells, so the seal resorts to vegetation-covered slopes above the rocky shores for breeding. On some other offshore islands in the eastern Atlantic, there is a certain amount of protected beach on which the seal can haul out and deliver pups, often backed by high cliffs where access to the interior may be impossible. In the western Atlantic and the Baltic Sea, breeding takes place on the ice or on rocky islets. Outside of the breeding season, this seal spends its life in the sea.

The gray seal eats a great variety of fish. Bonner (1972) cited a study in which fish were the most important prey in the northeastern Atlantic. In 21 percent of the stomachs examined, cod was found; salmonids, particularly *Salmo salar*, were found in 26.1 percent of the stomachs. Cephalopods also comprised a large percentage of the food remains. Bonner reported that the gray seal appears to be a generalized

coastal feeder, eating most of the fish that are abundant and easy to catch. The U.S. National Marine Fisheries Service (1981) stated that this species, in the western Atlantic, eats halibut, pollack, lamprey, salmon, and herring.

The following information on the reproductive biology of the gray seal was taken from King (1964), Mansfield (1966), Bonner (1972), Anderson, Burton, and Summers (1975), and the U.S. National Marine Fisheries Service (1981). There are two different breeding seasons, the Baltic and St. Lawrence populations pupping in February and March, while the British and associated populations pup in the autumn between September and November or December. The breeding season in the British population is very long: two or three months elapse between the births of the first and last pups in any one colony. About one month before the breeding season begins, large numbers of pregnant cows and bulls of all ages assemble on the breeding beaches. The birth of the first pup seems to be the signal for the beginning of territorial activities. *Halichoerus* forms polygamous breeding aggregations in British waters, and this imposes a definite social organization throughout the breeding season in this area (it is not definitely known if western Atlantic populations actively maintain territories and defend harems). The older bulls take up their territories, usually starting inland, with the best territories for attracting cows being occupied first. In the beginning, the territories are very large, and it is only later, when the dominant bulls are distracted by mating activities, that fringe areas of the large territories are occupied by bulls of lesser status. The number of cows covered by a single bull averages 7.5.

The birth of the pup is rapid, and the cow becomes receptive to the bull within a fortnight of pupping. Mating may take place either in the water or on land, depending on the bull's territory. Implantation of the blastocyst is delayed for about 3 months. The period of pregnancy is 11½ months, including the delayed implantation period.

At birth, the pup is between 76 and 77 cm in total length and weighs about 16 kg. It is covered with long, creamy white, woolly fur, known as the lanugo coat. At about 3 weeks of age, the pup acquires a coat of short hair, blue gray dorsally and paler ventrally. The pup is able to swim at birth, but usually does not enter the water until the first molt is completed. Suckling lasts for about 2 or 3 weeks, and the pup is deserted by its mother to fend for itself thereafter. During the time it is suckling, the pup grows quickly, and the mother, which fasts during lactation, loses weight. Cows produce their first pup when 3 years old, and bulls are sexually mature at 6 to 7 years of age. According to Marvin L. Jones (Zoological Society of San Diego, pers. comm.), a captive gray seal lived for 41 years.

Smit and Van Wijngaarden (1976) wrote that the gray seal has been persecuted in the past in European waters, especially by fishermen along the Norwegian coast and in the Baltic. According to these authors, in Sweden and Finland there has been a bounty system in effect on this species for many years, and Finland still has one in force (1976). King (1964) noted that the gray seal is not often killed commercially, as its skin is of little value. The U.S. National Marine Fisheries Service (1981) estimated that the total world population of *Halichoerus* is 88,000 to 94,000 animals.

PINNIPEDIA; PHOCIDAE; *Genus ERIGNATHUS Gill, 1866*

Bearded Seal

The single species, *E. barbatus,* occurs along coasts and ice floes in the Arctic Ocean and adjoining seas, and ranges as far south as the Sea of Okhotsk, Hokkaido, Hudson Bay, and the Gulf of St. Lawrence (King 1964; Rice 1977).

Adult males and females are about the same size. Total length is approximately 250 cm, and weight ranges from 250 kg in summer to 325 kg in winter. Coloration is also the same in the two sexes: gray with a brownish or reddish tinge on the head. A darker brown area extends from the top of the head down the middle of the back. The most noticeable feature of *Erignathus* is the prominent, bushy mustache, composed of long, flattened bristles, on each side of the muzzle, and which is responsible for the common name of the species. The bearded seal resembles *Monachus* in having four mammae, whereas all other phocid seals have two. The Norwegian sealers' name of ''square flipper'' for this seal reflects the square tips of the foreflippers, caused by the third digit of these flippers being slightly longer than the others. The nails on the foreflippers are strong, while those on the hind flippers are slender and pointed (King 1964).

The bearded seal prefers shallow waters near the coasts that are free of fast ice in winter; gravel beaches and ice floes that are not too far out to sea are also favored (King 1964). It generally associates with the moving ice pack as it advances and retreats each winter and summer. The ice habitat of this seal brings occasional stragglers as far south in the Atlantic Ocean as the British Isles and Normandy (King 1964). It does not make long seasonal migrations, though it may drift a little on the floating ice in the spring. This seal is essentially solitary, except during the mating season, when small groups may gather (U.S. National Marine Fisheries Service 1981).

Food for the bearded seal consists almost entirely of bottom-living animals such as shrimp, crabs, holothurians, clams, welks, snails, octopus, and bottom fishes such as sculpin, flounder, and polar cod. It is possible that both *Erignathus* and the walrus (*Odobenus rosmarus*) use their

Bearded seal (*Erignathus barbatus*), photo from the Zoological Garden Berlin-West.

mustachial whiskers to help sort out the small food animals on which they prey (King 1964).

Adult bearded seals have been heard whistling to their pups while under water, but they apparently do not make this noise when on land. Ray, Watkins, and Burns (1969) described the underwater ''song'' of the bearded seal as being a long, oscillating warble that may last for more than one minute, followed by a short, low-frequency moan. Apparently this sound is made only by mature males during the spring courtship season.

The following reproductive data are taken from King (1964) and the U.S. National Marine Fisheries Service (1981). The pups are born in the open on ice floes in April or May. An average pupping date is 1 May for the eastern Canadian Arctic; it may be a little later in the high arctic areas and as early as March in the more southern latitudes. At birth, the pup is about 130 cm in total length and weighs about 37 kg. It is covered with a short, woolly coat of grayish brown hairs, with a lighter region often found down the middle of the back. After a short time, probably at the end of lactation, the pup molts and exchanges this woolly coat for one of stiff hairs like that of the adult. A pup follows its mother for a long time, but the nursing period is only 12 to 18 days. One pup is produced every second year, whereas for most phocids the rule is one pup per year. Mating takes place in mid-May in the Canadian Arctic, and the blastocyst becomes implanted in August, after a delay of some 2½ months; pregnancy lasts 10½ to 11 months, including the delayed implantation period. Males may breed when they are 7 years old, females at about 6 years of age. The bearded seal has rarely been kept in captivity.

Polar bears prey heavily on the bearded seal; killer whales also feed on this species. *Erignathus* is not of great commercial importance because of its scattered distribution, but a small number of animals are taken by Norwegian sealers, and they are important in the life of the Eskimo. The hide is strong, durable, and elastic, and it is used to make boot soles, heavy ropes, dog harnesses, etc. The flesh is eaten by some Eskimos, but others dislike it and feed it only to their dogs. The liver frequently contains sufficient amounts of vitamin A to be poisonous to humans (King 1964). The oil from the blubber is often used by the Eskimos in lamps. The U.S. National Marine Fisheries Service (1981) reported that the total world population of the bearded seal exceeds 500,000 animals. About 10,000 to 13,000 are taken annually by subsistence hunters in Norwegian, Russian, and U.S. waters.

PINNIPEDIA; PHOCIDAE; *Genus CYSTOPHORA Nilsson, 1820*

Hooded Seal

The single species, *C. cristata,* occurs mainly along ice floes from Baffin Island and Newfoundland to Novaya Zemlya. There are also records from as far west as northern Alaska, as far north as Ellesmere Island, and as far south as Florida and the Bay of Biscay (King 1964; Rice 1977; Hall 1981; Burns and Gavin 1980).

Males have a total length of 210 to 350 cm and weigh about 410 kg. Females measure up to nearly 310 cm and weigh

Hooded seal (*Cystophora cristata*): A. Adult male, photo by Bernhard Grzimek; B. Bladder nose fully inflated while swimming; C. Adult male with hood inflated; D. Young in stage called ''blueback''; photos by Erna Mohr.

about 270 kg. The color of the adults is gray, covered with black patches of irregular shape and size (two or three inches square) on the back; much smaller spots appear on the neck and abdomen. The face and muzzle of the males are black. There are strong nails on both the hind flippers and foreflippers—particularly well developed on the foreflippers.

The most striking feature of this seal is the enlargement of the nasal cavity to form an inflatable hood on the top of the head. When not inflated, the hood hangs down in front of the snout like a kind of proboscis. This hood is found only in adult males. It appears as a fleshy "Roman nose" in young males, becoming progressively larger with increasing body size and age. Males larger than 220 cm in body length and more than 13 years of age always have large hoods (Berland 1966). The hood consists of mucous membrane overlaid by muscle fibers in a support of connective tissue and fat that is covered by skin. It is about 25 cm when fully inflated and holds about 6.3 liters of water. In addition to the hood, male hooded seals have the ability to blow out "balloons" through one side of the nose. This balloon, or bladder, is an inflation of the highly elastic mucous nasal septum, bright red in color and 15 to 18 cm in length. The hood and the balloon apparently are secondary sexual characters, and may be used to impress or frighten other males during the breeding season (Berland 1966).

The hooded seal prefers deep water and thick, drifting ice floes; it is not often found on firm ice. Little is known of its food preferences, but King (1964) said that it has been reported to feed on octopus, squid, rose fish, herring, capelin, and cod. Shrimps, mussels, and starfish, for which the seal must dive to the very bottom of the sea, are also eaten.

The following information on the social and reproductive habits of the hooded seal was taken from King (1964), Sergeant (1974), and Øritsland (1975). Cystophora is generally solitary, but during the reproductive season in March, it concentrates in four general areas: (1) in the White Sea, (2) around Jan Mayen Island in the Norwegian Sea, (3) near Newfoundland and in the Gulf of St. Lawrence, and (4) in Davis Strait between Canada and Greenland. In the breeding season, family groups of a bull, cow, and calf are formed, and both adults defend the young fiercely. Pups are born between the end of March and the beginning of April. At birth they are about 100 cm in total length and weigh about 20 kg. When born, they have a beautiful coat of silvery, blue gray fur, darker on the face and separated by a sharp line of demarcation from a creamy white ventral surface. The hairs do not lie quite as flat as those of other seals, and this gives a soft, furry appearance to the coat. Animals in this juvenile coat are known as "blue backs." Pups are suckled for about two weeks, during which time they grow rapidly. At the end of lactation, the adults mate, and then return to the sea, while the young remain on the drifting ice floes before they eventually take to the sea.

The next aggregations form between the middle of June and the middle of July, when the animals assemble for molting in the Denmark Strait area. After molting, they leave the area and apparently roam around the northern seas until they converge in the spring again on the mating grounds.

Polar bears are known to prey on the hooded seal, but mankind is by far the greatest enemy (King 1964). Cystophora and the harp seal (Phoca groenlandica) sometimes travel together, and also breed in the same general area. Both have been subject to heavy commercial exploitation; mainly the newborn pups are killed as they drift south on the ice floes. The pelt of the baby hooded seal is extremely beautiful and valuable. Each year a highly controversial harvest of young Cystophora (as well as juvenile Phoca groenlandica) is conducted in Canada, Greenland, and Norway. The harvest in Canada has been vigorously protested in recent years by conservation groups for humanitarian reasons. In 1977, over 6,000 skins of baby hooded seals entered international trade through Canada, 21,000 through Norway, and 3,000 through Greenland (Barzdo 1980). Although skins of the juveniles are by far the most important product of the hooded seal, it is also hunted for its oil and for meat. Native hunters on the coast of Greenland take the hooded seal for subsistence purposes, and Russian sealers also exploit the species; seals taken by the Russians are not thought to enter international trade. The U.S. National Marine Fisheries Service (1981) estimated the world population of Cystophora to be around 500,000 to 600,000 animals.

PINNIPEDIA; PHOCIDAE; **Genus MONACHUS** Fleming, 1822

Monk Seals

There are three species (Rice 1977; King 1964; Hall 1981):

M. monachus (Mediterranean monk seal), Mediterranean and Black seas, Atlantic coast of Morocco and Western Sahara, Madeira and Canary islands;
M. tropicalis (Caribbean monk seal), originally found throughout the West Indies, and along the coasts of Florida, Yucatan, and eastern Central America;
M. schauinslandi (Hawaiian monk seal), Hawaiian Islands.

Each species is at least 5,000 km from its nearest neighbor. Repenning and Ray (1977) suggested that the Hawaiian monk seal may have been separated from its ancestral form (probably a Caribbean or North American population) more than 15 million years ago when a Central American seaway separating North and South America closed. The unspecialized features of M. schauinslandi are more primitive than those of the oldest known monachine seal, approximately 14.5 million years old. M. schauinslandi appears to be the modern representative of the most ancient of living phocid lineages, and as such might be characterized as a "living fossil."

According to King (1964), M. monachus is probably the largest of the three species, with animals of up to 300 cm in total length being known. Adults of this species are chocolate brown dorsally, the brown hairs being tipped with yellow and the ventral surface being gray, sometimes with a centrally placed white patch. M. tropicalis averages about 200 cm in total length, and is grayish brown on the back, shading ventrally to yellowish white. M. schauinslandi averages a little over 200 cm in total length, and weighs up to 200 kg; it is light silver gray ventrally and slate gray dorsally. In all three species, females average somewhat larger and heavier than males.

This genus contains the only truly tropical species of seal, although it shows little apparent adaptation to warm temperature (King 1964). It is usually found along sandy beaches and in the shoreline vegetation. Sergeant et al. (1978) reported that the Mediterranean monk seal is found on two types of coastline: archipelagoes, especially those with small islands, often uninhabitable by humans; and cliff-bound mainland coastlines. These are the only types of coast that are relatively inaccessible or unattractive to humans. Whether the species was originally present on sandy coastlines and islands is uncertain, but seems probable judging from reports and records from early Greek and Roman times. Kenyon (1972) reported that Hawaiian monk seals avoid any beaches that are disturbed often by humans. King (1964) wrote that this species spends most of its time in relatively shallow

Hawaiian monk seals (*Monachus schauinslandi*): A. Photo from San Diego Zoological Garden; B. Photo by Karl W. Kenyon.

water around the reefs, or on the sandy beaches in the sun or in the shade of shrubs.

Seals of this genus are apparently not migratory, but do move about from island to island. Sergeant et al. (1978) reported that *M. monachus* seems to be diurnal, or at least indifferent to daylight. This species was observed to feed by day, and apparently does so most intensively in the early morning. The U.S. National Marine Fisheries Service (1981) reported that *M. schauinslandi* feeds among the coral reefs on a wide variety of fish and invertebrates, including eels, octopus, and lobsters. King (1964) noted that *M. tropicalis* preyed on fish. Evidence suggests that the Mediterranean monk seal feeds mainly at depths above 20 to 30 meters. Seal damage to nets in the Aegean Sea occurred mostly within a depth of 20 meters, and the seals took few fish from nets set deeper than 30 meters (Sergeant et al. 1978).

The following information on reproduction is taken from King (1964), Sergeant et al. (1978), and the U.S. National Marine Fisheries Service (1981). Mediterranean monk seal pups are born between September and October; they are about 91 or 92 cm in total length at birth. The pups have a black, woolly coat, which is molted at weaning when the pup is about 6 weeks old. Mating is believed to take place about 2 months after the birth of the pups. The gestation period is 11 months; females reach sexual maturity when they are about 210 cm in total length. Pups of the Caribbean monk seal are born about the beginning of December with a soft, woolly, black coat; there is no information on mating, lactation, or molting in this species. With the Hawaiian monk seal, mating probably takes place in the water; pups are born from December through July, with a peak birth time in April and May. New born are about 98 cm in total length and weigh

about 18 kg. Some females apparently give birth annually, but many give birth every other year.

According to Marvin L. Jones (Zoological Society of San Diego, pers. comm.), one *M. monachus* lived in captivity for almost 24 years. Kenyon and Fiscus (1963) stated that *M. schauinslandi* has the potential for living to at least 20 years of age in the wild.

All three species of monk seal have been greatly reduced in numbers and range, apparently because of human interference in their lives. These seals are very sluggish and unsuspicious by nature, and are easy for humans to approach and kill. This, combined with their restricted habitat and gregarious habits, made them highly vulnerable. Sergeant et al. (1978) reported that during the 20th century *M. monachus* was totally exterminated along the mainland coasts of Spain, southern France, and the Crimea, areas that included rocky coastlines and islands. Extirpation has also occurred along the sandy coasts of Palestine and Egypt. In all of these areas, the causes were clearly the direct pressure of people.

The U.S. National Marine Fisheries Service (1981) stated that *M. schauinslandi* is highly intolerant of human disturbance, and readily abandons breeding and haul out areas if molested. In order to protect this species, the U.S. military installations in Hawaii have restricted the movements of their personnel to minimize disturbance of remaining populations. Kenyon (1972) pointed out that this species avoids beaches where it has been disturbed often by people, and has a reduced rate of reproduction where such disturbance is frequent. In addition, survival of young is apparently reduced where the seals are forced to leave preferred pupping grounds. *M. schauinslandi* may once have bred throughout the Hawaiian Islands, but now breeds only on the Leeward

Chain, from French Frigate Shoals to Kure Atoll, and only rarely wanders as far as Hawaii.

M. tropicalis is probably extinct (IUCN 1978). This seal was very seriously overexploited in the 18th century, and was already rare by 1850, large numbers being killed without difficulty because of their trustfulness and sluggish habits. In addition, the large and rapidly growing human population of the Caribbean Islands has disturbed or destroyed many areas of good habitat for the species. Fishermen considered the species a competitor and probably killed any individuals encountered. Although *M. tropicalis* may have disappeared by the 1950s, sightings were reported in the southeastern Bahamas in the 1960s and 1970s, and there is a slight chance that the species still survives (*Oryx* 15 [1980]:426).

The U.S. National Marine Fisheries Service (1981) reported a world population of about 500 animals for *M. monachus* and another 500 for *M. schauinslandi*. All three species of *Monachus* are classified as endangered by the IUCN (1978) and the USDI (1980), and are on appendix 1 of the CITES.

PINNIPEDIA; PHOCIDAE; **Genus *LOBODON* Gray, 1844**

Crabeater Seal

The single species, *L. carcinophagus*, is found primarily along the coasts and pack ice of Antarctica, and occasionally wanders as far north as the southern tips of Australia, Tasmania, New Zealand, South America, and South Africa (Rice 1977; King 1964).

Adult animals have a total length of about 250 to 260 cm and weigh 200 to 225 kg; the largest females slightly exceed the largest males in size. They are slim, lithe animals, and can move swiftly over the ice. In color, the adults are silvery, brownish gray dorsally, with very variable, chocolate brown ring markings on the sides and shoulders, shading to a pale gray ventral surface. The coat color fades throughout the year, particularly in the summer, to an overall creamy white (King 1964). As with other genera of antarctic phocids, *Lobodon* has cusps on each of the premolar and molar teeth, in contrast to the cone-shaped teeth of most other pinnipeds.

When the upper and lower teeth are brought together, the rows of teeth form a closely knit filter structure for straining krill from the sea (U.S. National Marine Fisheries Service 1981).

This seal is an animal of the antarctic drifting pack ice, and is found at its greatest abundance at the edge of this pack ice. In the late antarctic summer, the crabeater seal also comes relatively close to the shoreline of the continent (U.S. National Marine Fisheries Service 1981).

Stirling and Kooyman (1971) reported that, unlike *Leptonychotes*, the crabeater does not roll onto its back when approached, and thus can be sexed only when found sleeping on its sides on the ice. When approached on the ice, a crabeater seal bares its teeth and snorts loudly by expelling air through its nostrils. When caught, the seal rolls over many times; this could be a natural reaction of this species to escape the killer whale (*Orcinus orca*), inasmuch as scars (presumably from killer whale attacks) on the back of crabeater seals usually curve around the body from anterior to posterior. Stirling and Kooyman found that a group of crabeater seals was difficult to disperse on the ice; possibly this is a flocking reaction that might be used in the water to decrease vulnerability to attack by killer whales.

The use of breathing holes of *Leptonychotes* by *Lobodon*, apparently without strife between the two genera, indicated to Stirling and Kooyman that there may be a marked tolerance between them. When both genera were seen together during summer on fast or pack ice, they generally ignored each other.

Lobodon is perhaps the fastest seal on land, obtaining speed from powerful blows against the ice with the hind flippers, and also striking the ice with backward blows from the front flippers. Land speeds of at least 24 km per hour can be achieved in this manner. The common name of *Lobodon* is hardly appropriate, as this seal feeds almost exclusively on the small shrimplike animals known as krill. It catches these animals by swimming into a shoal of them with its mouth open, probably sucking them in, and then sieving the krill from the water through the spaces between the cusps of their complicated cheek teeth.

Lobodon apparently is not migratory, and any movement of groups is probably subject to only the advance and retreat of the ice pack. Siniff et al. (1979) distinguished three groups of crabeater seal association: (1) family groups, (2) mated

Teeth of crabeater seal, postcanines 3, 4, and 5, right mandible, lingual aspect, photo by Victor B. Scheffer.

Crabeater seal (*Lobodon carcinophagus*), photos by University of Minnesota Antarctic Seal Research Program.

pairs, and (3) fast ice concentrations. Family groups consisted of an adult female, its pup, and an adult male. After weaning of the pup, the adult male and female formed a mated pair, which remained together until copulation. Fast ice concentrations, primarily composed of immature animals, ranged from 50 to 1,000 seals, and inhabited bays where seasonal ice persisted.

Siniff et al. (1979) speculated that pregnant females find a suitable ice floe for giving birth and then are joined by a male either just before or just after parturition. As early as 7 October, they observed lone females that were unusually fat and suspected to be pregnant. Males most likely scent and find estrous females, but may also use pup odor to locate proestrous females. Observations place the pupping season from September to October, with a peak in mid-October. Female crabeaters use small, separate ice floes for birth of pups, and they come into estrus after weaning. Mating, contrary to that of most ice seals, probably occurs on the ice surface. Mating takes place 2 to 3 weeks after the female gives birth; the period of pregnancy is 11 months, including a delayed implantation period of 2 to 3 months. Newborn pups are about 120 cm in total length and weigh about 20 kg. Bengtson and Siniff (1981) found that females attain sexual maturity at an average age of 3.8 years.

The chief enemy of the crabeater appears to be the killer whale, and most adults bear the scars of encounters with this whale. Leopard seals also prey on crabeaters. *Lobodon* may live as long as 33 years in the wild (King 1964; U.S. National Marine Fisheries Service 1981).

The crabeater seal is the most abundant pinniped and one of the most populous large mammals on earth. The U.S. National Marine Fisheries Service (1981) estimated its total numbers at around 15 million animals.

PINNIPEDIA; PHOCIDAE; **Genus OMMATOPHOCA**
Gray, 1844

Ross Seal

The single species, *O. rossii,* is circumpolar in pack ice of the Antarctic (Rice 1977).

This is a large, plump seal, with a short, wide head that can be pulled back into rolls of blubber around the neck. Adult males average around 200 cm in length and weigh about 170 kg; females are larger, measuring about 230 cm in length and weighing around 200 kg. The Ross seal is dark gray, slightly darker along the middle of the back and shading to a whitish color ventrally, with paler gray streaks and lines running from the side of the neck and shoulders. The flippers in this genus are large; the mouth is rather small for the size of the animal, and the protruding eyes indicate the large size of the underlying eyeballs (King 1964; U.S. National Marine Fisheries Service 1981).

Ommatophoca prefers the heavy consolidated ice pack around the edge of the antarctic continent. It is among the least known of any pinniped. It is apparently solitary in habits, and nonmigratory. King (1964) stated that cephalopods are its main food, though krill and fish have also been found in its stomach. It feeds under the great thickness of the heavy pack ice where there is little light (U.S. National Marine Fisheries Service 1981). For catching the swift-moving and slippery prey, this seal is equipped with sharp, but delicate, recurved teeth, and can swim very fast, with rapid maneuverability. The large eyes are adapted for perception of movement in the dimly lit water under heavy ice.

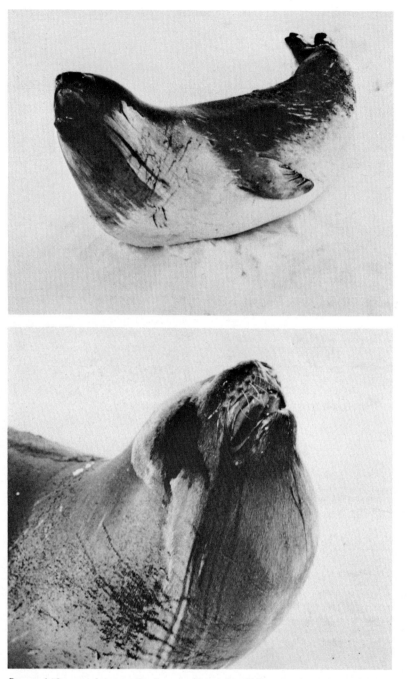

Ross seal (*Ommatophoca rossii*), photos by University of Minnesota Antarctic Seal Research Program.

Perkins (1945:280) commented that: "Although these seals are sluggish in their movements on the ice, and although they exhibited the usual defenselessness of the other seals, they seemed to resent the presence of man even more than the crabeaters. They gave a little birdlike chirping call similar to that of Weddell's seal; and when disturbed, they raised their heads, filled their lungs with air, and exhaled suddenly." King (1964) noted that the curious trilling and cooing noises made by the Ross seal are said to be achieved by blowing out the expandible posterior end of the long, soft palate and using it rather like a bagpipe. When the mouth is shut, a clucking noise is produced, and the remains of the air are blown out with a snort from the nostrils.

Almost nothing is known of the reproductive habits, but it seems unlikely that *Ommatophoca* collects in large breeding rookeries. Scars often found on individuals may have been caused by fights during the breeding season. Newborn are about 104 cm in length and weigh about 27 or 28 kg. Females

mature sexually at three to four years of age, and males between two and seven years (King 1964; U.S. National Marine Fisheries Service 1981).

The U.S. National Marine Fisheries Service (1981) stated that there is no reason to think that this seal has experienced a decline in numbers. Antarctic sealing activities did not involve species from the pack ice to any great extent. The present population estimate is around 220,000 animals, and this is probably a low figure.

PINNIPEDIA; PHOCIDAE; *Genus HYDRURGA Gistel, 1848*

Leopard Seal

The single species, *H. leptonyx*, occurs in the waters around Antarctica, and has been recorded from most subantarctic islands, as well as southern Australia, New Zealand, the Cook Islands, southern South America, and South Africa (Rice 1977).

Adult males are about 300 to 310 cm in length, and weigh about 270 kg; females are larger, and may reach 360 or 370 cm and weigh up to 380 kg. The adults are dark gray dorsally and very light gray ventrally, spotted on the throat, shoulders, and sides with black, dark gray, and light gray. The long, slim body, large head (proportionally larger than in any other species of pinniped), wide gape, and curiously reptilian appearance of the head make the leopard seal easy to recognize. At close quarters, the large cheek teeth with their three long and distinct cusps (the middle one being high, pointed, and curved backward) make identification certain (King 1964).

The leopard seal normally occurs in the outer fringes of the ice pack. It is migratory and ranges widely to most of the subantarctic islands in the winter, although some individuals are believed to stay in the pack ice all the year around. The younger animals seem to range more widely than the adults.

This seal is a major predator of penguins and seals throughout the antarctic region (U.S. National Marine Fisheries Service 1981). King (1964) said that *Hydrurga* has a wide variety of food preferences: it will eat fish, squids, penguins, and bits of carcasses of whales and seals that have been killed, and it has been known to attack pups of other species. Penguins are frequently chased and caught underwater, the body then being brought to the surface and so vig-

Leopard seals (*Hydrurga leptonyx*): A. Photo by U.S. Office of Territories through the National Archives; B. Photo by John Warham.

orously shaken that sometimes complete penguin skins are found, turned inside out, on the ice. Siniff and Bengtson (1977) found that *Hydrurga* also preys extensively on the crabeater seal (*Lobodon*). The U.S. National Marine Fisheries Service (1981) reported that krill is an important food item for this species, especially for young animals. Leopard seals feed in the water rather than on the ice.

When at sea, *Hydrurga* is usually solitary, but groups of up to 250 individuals regularly haul out in some areas. Mating may not occur until January or March, the gestation period is 240 days, and the pups are born on the pack ice in November and December. The newborn are about 150 cm long and weigh about 30 kg. The soft and thick natal coat is dark gray above, there is a darker central stripe down the middle of the back, and the sides and ventral surface are almost white, irregularly spotted with black. Lactation lasts about 2 months. Females reach sexual maturity at 2 to 6 years of age and males at 3 to 7 years. Potential longevity is estimated at more than 26 years (U.S. National Marine Fisheries Service 1978, 1981; King 1964).

The leopard seal is probably at the carrying capacity of its ecosystem. It is not hunted and is under the regulatory regimes of international treaties. The total world population is estimated to be about 500,000 animals (U.S. National Marine Fisheries Service 1981).

Weddell seals (*Leptonychotes weddelli*), photo from U.S. Navy.

PINNIPEDIA; PHOCIDAE; **Genus LEPTONYCHOTES**
Gill, 1872

Weddell Seal

The single species, *L. weddelli*, is found in areas of fast ice attached to Antarctica and subantarctic islands, and occasionally wanders as far north as Uruguay (Rice 1977; Stirling 1971b).

Adult males have a total length of about 280 cm (maximum recorded length 297 cm); females are larger, measuring about 295 cm in total length (maximum recorded length 329 cm). Adults weigh 325 to 550 kg (King 1964; Stirling 1971b; U.S. National Marine Fisheries Service 1981). King (1964) wrote that a typical Weddell seal is black dorsally, with white streaks and splashes increasing in number and extent from the dorsolateral area, where white predominates. Ventrally, the coat is gray with white streaks. During the summer, the coat fades to rusty grayish brown, but never bleaches to the whiteness often found in the crabeater seal (*Lobodon*). The upper incisor teeth are markedly unequal, the outer being about four times longer than the inner; the postcanines are strong and functional.

Bertram (1940) reported that the Weddell seal normally inhabits the inshore waters of the antarctic continent and adjacent islands. It spends much of the time in water, but emerges at intervals to lie out on the beaches or on the fast ice. During the winter, when individuals spend most of their time in water, they can be heard calling to each other under the ice. *Leptonychotes* is not a seal of the antarctic pack ice,

and it is rarely found on isolated floes or far from land. It does not migrate and is to be found in the most southerly parts of its range throughout the year, though in winter it is less evident as it spends more time in the water where the temperature is more constant (King 1964). Local movements may be stimulated by ice conditions. During winter, *Leptonychotes* uses its canine teeth to abrade the sea ice and maintain breathing holes in areas where physical factors produce natural cracks. Breathing holes away from natural cracks are rare. On land or ice, the Weddell seal moves slowly, in a humping fashion. In water, it swims using both its foreflippers and hind flippers, and estimates of its speed range from 9 to 13 km/hr. It can dive to depths of 600 meters, and can stay underwater for up to an hour (Kooyman 1968; Stirling 1971b). It is, however, a lethargic animal, not easily aroused, and spends much of its time sleeping on its side on the ice.

The diet of *Leptonychotes* consists chiefly of nototheniid fishes and squids; crustaceans and holothurians are also taken. This seal apparently devours its food under ice, and is probably able to meet its water requirements from its diet and by metabolizing sea water, but individuals have been observed eating snow (Stirling 1971b).

The following information on social life and reproduction was taken from King (1964), Stirling (1971b), and the U.S. National Marine Fisheries Service (1981). Concentrations of pregnant cows begin toward the end of August, and the pups are born during September and October. During the pupping season, seals on the ice are more widely spaced than they are later in the summer, and nonbreeding seals and subadults are excluded from these areas. Males spend much of their time underwater, and establish territories there. Threat calls are used, and vigorous fighting may take place. Newborn are about 150 cm in total length and weigh about 27 or 28 kg; the natal coat is woolly, and grayish in coloration. It is molted by the time the pup is four weeks old, and the new coat is similar to that of an adult. The pups enter water early, even while in the natal coat. Lactation lasts six to seven weeks, during which time the mother does not feed. Cows are concerned about their young, and will aggressively attack intruders. After the birth and weaning of the pups, the cows are ready to breed again, and mating takes place, in the water, in December. Pregnancy lasts about 11 months, including a delayed implantation period of 2 months. Both sexes mature between 3 and 6 years of age, though males probably do not mature socially until 7 or 8 years. *Leptonychotes* is known to live in the wild for at least 20 years.

Mainly because of its inaccessibility to humans, this seal has not been greatly reduced in numbers; some have been killed by the personnel of antarctic stations for dog food. The total population of the Weddell seal is estimated to be about 750,000 animals, and it is considered stable (U.S. National Marine Fisheries Service 1981).

PINNIPEDIA; PHOCIDAE; **Genus MIROUNGA** *Gray, 1827*

Elephant Seals

There are two species (King 1964; Rice 1977):

M. leonina, most subantarctic islands, coast of southern Argentina, waters south to edge of antarctic ice at 78° S, occasionally as far north as Saint Helena (16° S) and Mauritius (20° S);

M. angustirostris, coastal waters from southeastern Alaska to Baja California.

These are the largest pinnipeds. Adult males measure from 550 to over 600 cm in total length and weigh up to about 3,700 kg. Females are much smaller, ranging from 310 to 370 cm in length and weighing about 900 kg. The sex discrepancy in size in *Mirounga* is probably the greatest of its kind in mammals (Bryden 1969). Just after the molt, the males are dark gray, and a little brighter ventrally, but this coloration fades to grayish brown through the year. There is usually intensive scarring in the neck region of the bulls from fighting, and this makes the skin in that region extremely tough and thick. Females are generally darker and browner than the males, with a light-colored yoke around the neck resulting from the many scars they receive from bites during mating. The hair is short, stiff, and harsh with no underfur (King 1964).

The common name of these seals is derived from the large, trunklike proboscis. This proboscis is seen in its full development only in adult males; full development is obtained at about eight years of age. Its tip overhangs the mouth in front, so the nostrils open downward. It is an enlargement of the nasal cavity, and is internally divided into two parts by the nasal septum. The proboscis is flattened, and less obvious outside of the mating season. When breeding, *Mirounga* can erect this organ partly by muscular action, partly by blood pressure, assisted by inflation, to form a high, bolster-shaped cushion on top of the snout, the tip hanging down in front of the open mouth. At its maximum inflation, the proboscis is about 28 cm in length. It may act as a resonating chamber, and the roar of a big bull may carry for several miles (King 1964).

Both species are similar in external appearance. *M. leonina* is longer and heavier, but the proboscis is better developed in *M. angustirostris*. *M. leonina* can bend its body into a U shape at a much greater angle than can *M. angustirostris*. In addition, the two species differ in cranial and skeletal details, but few behavioral differences have been reported.

The species *M. angustirostris* currently breeds on islands off the coast of California and Baja California and at one point on the mainland south of San Francisco. *M. leonina* breeds on many subantarctic islands and on the coast of southern Argentina (King 1964; Le Boeuf, Ainley, and Lewis 1974; Le Boeuf and Panken 1977; U.S. National Marine Fisheries Service 1978). Adults are present in these areas during the spring breeding season and the summer molting season. At other times, elephant seals are pelagic, going out to sea to feed and not coming to shore again in any numbers until the next breeding season; occasionally adult cows may come ashore to rest, but adult bulls usually remain at sea. These seals are deep divers, feeding on fish below 100 meters as well as on fish that occur in shallower depths (U.S. National Marine Fisheries Service 1981). DeLong (1978) wrote that available information indicates that the northern elephant seal feeds on deep-water, bottom-dwelling marine life, such as ratfish, swell sharks, spiny dogfish, cusk eels, various species of rockfish, and squid.

The following information on the life history and reproduction of *Mirounga* was taken from Carrick et al. (1962), King (1964), Le Boeuf and Peterson (1969), Le Boeuf and Petrinovich (1974), Condy (1978), Christenson and Le Boeuf (1978), DeLong (1978), Van Aarde (1980), and the U.S. National Marine Fisheries Service (1981). Female *M. angustirostris* arrive on the breeding rookeries in late December and form compact aggregations on the beach. For *M. leonina* on Macquarie and Heard Islands, this occurs in August. Until the female aggregation exceeds 40 to 50 animals, the most dominant adult male associates with the female group; but at over this number of females, a single male can no longer exclude other males, and a second and sometimes a third male gains access to the female group. The growth of the rookeries proceeds in this way until early January. The term "harem" is not accurate to describe the social organization of elephant seals. One male does not keep a fixed number of females which it herds and protects, but rather the social unit comprises a variable number of females and a few mature bulls, one of which is dominant. A social hierarchy among bulls, established through threats and fighting, determines the access a bull will have to a group of cows. Thus, *Mirounga* differs from those pinnipeds in which a few males defend specific sites and females gather in harems. For *Mirounga*, males establish social hierarchies in which bulls of highest rank remain near breeding females but do not defend specific sites. During the breeding season, the bulls frequently make a deep, trumpeting sound to warn off rivals.

About a week after arriving on land, each female bears a single pup weighing 32 to 36 kg and measuring about 120 cm in length. The pup is nursed for about 28 days and frequently triples its weight during this time. When the pup is weaned, it molts its black natal pelage and grows a sleek gray coat.

At about the time of weaning, the female comes into estrus, and mates with one or more of the dominant bulls. She then leaves the rookery to feed and recover some of the weight lost in the fast during the nursing period. An adult male *M. angustirostris* will depart in February, but may remain on land during March after the breeding season, presumably to rest; it returns to the sea in late March. *M. leonina* returns to the sea in late October and November. Pregnancy lasts about 11.3 months, including a delayed implantation period of 3 months. Females mature sexually between 3 and

Northern elephant seals (*Mirounga angustirostris*): A. Male, photo from Allan Hancock Foundation; B. Mother and baby, photo by Julio Berdegué; C. Seal shedding its skin, photo from Colorado Museum of Natural History through Alfred M. Bailey and R. J. Niedrach; D. Adult male with proboscis turned downward into mouth, photo by Julio Berdegué; E. Group of females, animal on left is shedding, photo by Rupert R. Bonner; F. South Atlantic elephant seal (*M. leonina*), photo from the Zoological Garden, Berlin-West.

5 years of age; males mature between 4 and 5 years, and become territorial breeding bulls at age 9 or 10. *Mirounga* comes ashore again in numbers only to molt in the summer. This genus and *Monachus* are the only pinnipeds that have an epidermal molt. In this type of molt, the pelage and large pieces of the outer epidermis are shed and replaced by new epidermis and sleek fur. When molting, these seals generally sit quietly on the beach and rest (Delong 1978).

Elephant seals are relatively short lived. According to Marvin L. Jones (Zoological Society of San Diego, pers. comm.), a specimen lived in captivity for 15 years. Natural enemies of elephant seals are large sharks, particularly the white shark, and killer whales (DeLong 1978).

Elephant seals have been exploited by humans since the 19th century. About 350 liters of fine oil can be obtained from a large bull. Some 75 to 80 years ago, *M. angustirostris* was thought to be virtually extinct off the California and Mexican coasts because of heavy hunting pressure. It had been hunted there from as early as 1818. *M. leonina* met a similar fate on South Georgia Island and elsewhere. By 1892, *M. angustirostris* survived only on an isolated beach at Guadalupe Island off Baja California, where about 100 individuals were present. These few animals were protected, and the species has now made a remarkable recovery. With *M. leonina*, the seals were hunted until their numbers were so reduced as to provide no incentive for commercial exploitation. After the stopping of commercial exploitation, numbers gradually recovered until they reached the point where harvesting was again possible, though it is now strictly regulated and controlled. The world population of *M. angustirostris* is estimated at about 60,000 animals. The world estimate for *M. leonina* is about 600,000 animals, equally divided between the Atlantic coast of South America and the subantarctic region of the Southern Ocean (King 1964; Carlisle 1973; DeLong 1978; U.S. National Marine Fisheries Service 1981).

ORDER TUBULIDENTATA

TUBULIDENTATA; *Family ORYCTEROPODIDAE;*
Genus ORYCTEROPUS E. Geoffroy St.-Hilaire, 1795

Aardvark, Ant Bear

This order contains one family, Orycteropodidae, with the single living genus and species *Orycteropus afer,* which occurs throughout Africa, south of the Sahara, wherever suitable habitat is available (Meester and Setzer 1971).

The aardvark has a massive body, with a long head and snout that terminates in a round, blunt, piglike muzzle that is pierced by circular nostrils, from which grow many curved whitish hairs 25 to 50 mm long. The ears are tubular, and 150 to 210 mm in length; they fold back to exclude dirt when the animal is burrowing and can be moved independently of one another. They are waxy and smooth like scalded pig's skin. The tapering tongue often hangs out of the mouth with the end coiled like a clock spring. The neck is short, the forequarters are low, and the back is arched. The strong and muscular tail is thick at the base and tapers to a point. The legs are short and stocky; the forefoot has four digits and the hind foot has five, the digits being webbed at the base. The long, straight, strong, blunt claws are suited to burrowing. This animal resembles a medium-sized to large pig. Head and body length is 1,000 to 1,580 mm, tail length is 445 to 610 mm, and shoulder height is approximately 600 to 650 mm. Adults weigh as much as 82 kg, although most individuals weigh from about 50 to 70 kg. The thick skin is scantily covered with bristly hair that varies in color from dull brownish gray to dull yellowish gray. The hair on the legs is often darker than that on the body. Numerous vibrissae occur on the face around the muzzle and about the eyes. The dull pinkish gray skin is so tough that it sometimes saves the aardvark from the attacks of other animals.

The females have two pairs of inguinal mammae. In the males the penis has a fold of skin which covers scent glands at its base.

The teeth in the embryo are numerous and traversed by a number of parallel vertical pulp canals. The milk teeth do not break through the gums. In the adult the teeth are only in the posterior part of the jaw. The usual dental formula is: (i 0/0, c 0/0, pm 2/2 [additional vestigial premolars are sometimes present], m 3/3) ×2=20. Teeth do not grow simultaneously. Those nearest the front of the jaw develop first and fall out about the time the animal reaches maturity; they are succeeded by others farther back. Each cheek tooth, which is covered externally by a layer of cement, resembles a flat-crowned column and is composed of numerous hexagonal prisms of dentine surrounding tubular pulp cavities, hence the ordinal name Tubulidentata ("tubule toothed"). The teeth of aardvarks grow continuously and lack enamel. The skull is elongate, and the lower jaw is straight, bladelike anteriorly, and swollen at the molars.

According to Smithers (1971), the aardvark occurs in a wide variety of habitats, including grassy plains, bush country, woodland, and savannah. In such areas it appears to prefer sandy soils. The presence of sufficient quantities of termites and ants is apparently the main factor in distribution. The word "aardvark" means "earth pig" in the Afrikaans language, an appropriate name, as *Orycteropus* looks somewhat like a pig and is an extraordinarily active burrower. If overtaken away from its den, it digs into the ground with amazing speed. It can dig faster in soft earth than can several persons with shovels, and even the hardest sun-baked ground is no obstacle to its powerful forefeet. An aardvark, when

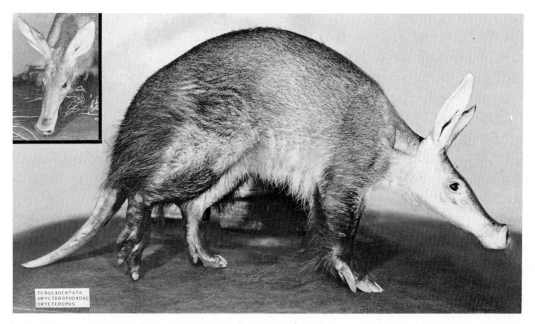

Aardvark (*Orycteropus afer*), photos by Ernest P. Walker.

digging, pushes the ground backward under its body while resting on its hind legs and tail. When a sufficient amount of soil has accumulated, it is shoved back or to one side with the hind feet, sometimes with the aid of the tail. *Orycteropus* excavates extensive burrows about 3 meters long. The tunnel ends in a chamber large enough for the aardvark to turn in, as the animal generally enters and leaves its burrow headfirst. A number of aardvarks may burrow in the same vicinity; there is one record of 60 entrances in an area approximately 300 by 100 meters. The burrows, when abandoned, are used by dozens of other kinds of animals. *Orycteropus* sometimes occupies a termite nest as a temporary shelter; its thick skin seems impervious to insect bites. It occasionally digs its holes in areas that seasonally flood, and thus may have to evacuate at certain times of the year.

The aardvark tends to walk on its claws, and its tail often leaves a track on soft ground. It is an extremely powerful animal. In one case, a man with a firm grip on the tail of an aardvark in its den was slowly drawn into the burrow up to his waist and finally had to relinquish his hold, despite the additional leverage afforded by two other persons holding onto his legs. The aardvark's movements are awkward and slow, but the animal is said to escape with surprising rapidity if alarmed. Its hearing is acute, and at the least alarm it seeks a burrow. Eyesight does not appear good, since the aardvark frequently crashes into bushes, tree trunks, and other obstructions when running (Smithers 1971). It avoids enemies by digging or running, but if cornered it will fight by striking with the tail or shoulders, by rearing on the hind legs and slashing with the forefeet, or by rolling on the back and slashing with all feet.

The aardvark is mainly nocturnal, but sometimes goes abroad by day and occasionally suns itself in the early morning at the burrow entrance. It usually sleeps during the day, curled up in a tight circle, with the snout protected by the hind limbs and tail. By night, it may move up to 16 km in the search for food, moving from one termite nest to another but rarely visiting the same mound on successive nights (Dorst and Dandelot 1969). The diet consists principally of ants and termites, which are obtained by digging, tearing into nests, or seeking out the insects as they are on the march. A moving column of termites may contain tens of thousands of individuals and be about 40 meters long, and may be detected by the aardvark by sound and smell. *Orycteropus* gathers ants and termites with its long, sticky tongue, which can be extended up to 300 mm. Other insects are occasionally taken, and Smithers (1971) reported apparent predation on the fat mouse (*Steatomys pratensis*). In captivity the aardvark accepts meal worms, boiled rice, meat, eggs, and milk, and seems to thrive when some carbohydrate is included in the diet.

According to Dorst and Dandelot (1969), females are attached to a particular place, to which they come back regularly, whereas males are more vagabond. *Orycteropus* is generally solitary, but the young accompanies its mother for a long time. Births in Central Africa reportedly occur at the beginning of the second rainy season in October or November. In southern Africa the young are produced during the cooler, drier months of the year from about May to August (Smithers 1971). The gestation period is about 7 months (Dorst and Dandelot 1969). There is usually a single offspring, occasionally two. It is naked and flesh colored, remains in the burrow for about 2 weeks, and then begins to accompany its mother on nightly excursions. For the next several months the mother and young occupy a series of burrows, moving from one to another. The young can dig for itself after about 6 months. A male aardvark lived in captivity for about 23 years and was still reproductively active toward the end of his life (*Animal Kingdom* 83, no. 2 [1980]:2).

The flesh of *Orycteropus* has the appearance of coarse beef and is prized by some persons, but others say that it is strong smelling and as tough as leather. In some areas the hide is made into straps and bracelets, and the claws are worn as good luck charms. *Orycteropus* has been greatly reduced in numbers and distribution, and is now on appendix 2 of the CITES.

The geological range of the Tubulidentata is Eocene in North America, Eocene to Oligocene and Pliocene in Europe, Pleistocene in Madagascar, and Miocene and Recent in Africa.

ORDER PROBOSCIDEA

PROBOSCIDEA; *Family ELEPHANTIDAE*

Elephants

This order contains one living family, Elephantidae, with two living genera and species, *Elephas maximus* of southern Asia and *Loxodonta africana* of Africa.

The most conspicuous external feature of living elephants is the trunk, which is flexible and muscular. The trunk is actually a great elongation of the nose, the nostrils being located at the tip. The fingerlike extremity of the trunk is used to pick up small objects, such as peanuts.

The head is huge, the ears are large (especially in *Loxodonta*) and fan shaped, the neck is short, the body is long and massive, and the tail is of moderate length. Living members of the Proboscidea have a maximum height of nearly 400 cm and a weight of up to 7,500 kg (males). The limbs are long, massive and columnar. All the limb bones are well developed and separate; lacking marrow cavities, they are filled with spongy bone. The feet are extremely short and broad, columnar in shape. The weight of the animal rests on a pad of elastic tissue. There are five toes on each foot, but the outer pair may be vestigial, so some digits do not have hooves (nails). The Asiatic elephant has four hooves (occasionally five) on the hind foot and five on the forefoot, and the African elephant has three on the hind foot and five on the forefoot.

The skin of adult elephants is sparsely haired. The glands associated with the hair follicles in most mammals (sebaceous glands), which soften and lubricate the hair and skin, are not present in the Elephantidae. Females have two nipples just behind the front legs; the young nurse with the mouth, as do young of other mammals. Males retain the testes permanently within the abdomen.

The dental formula for Recent species of Elephantidae is: (i 1/0, c 0/0, pm 3/3, m 3/3) × 2 = 26. The single upper incisor grows throughout life into a large tusk (up to 330 cm long in the African elephant). The tusk, which is usually absent in females of the Asiatic elephant, has enamel only on the tip, where it is soon worn away (some extinct proboscideans had lower tusks and others had longitudinal bands of enamel on their tusks). The grinding teeth are generally large, high crowned, and with a complex structure. Each tooth is composed of a large number of transverse plates of dentine covered with enamel; the spaces between the ridges of enamel are filled with cement. Ridges do not show on an unworn tooth since it is covered with cement. The grinding teeth increase in size and in number of ridges from front to back. These teeth do not succeed one another vertically in the usual mammalian pattern, but come in successively from behind, the series thus moving obliquely forward. When the foremost tooth is so worn down as to be of no further use, it is pushed out, mostly in pieces. As these teeth are very large and the jaws are fairly short, only one tooth on each side, above and below, is in use at the same time (part of a second tooth may also be in use).

The skull is huge and shortened. The premaxillary bones have been converted into sheaths for the tusks, and the nasal bones are extremely shortened. All the bones forming the braincase are greatly thickened and, at the same time, lightened by the development of an extensive system of communicating air cells and cavities. The brain chamber is hidden

Indian and African elephants (*Elephas maximus* and *Loxodonta africana*), the photo by Elaine Anderson clearly shows the difference between the two species.

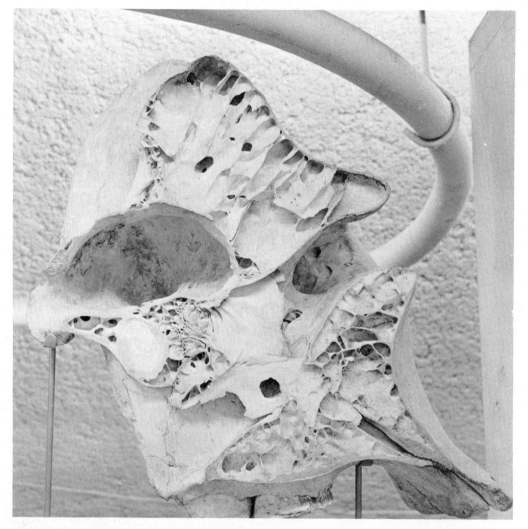

View of vertical median section through skull of Asiatic elephant (*Elephas maximus*), which shows the great thickness of bone with large cells or sinuses which almost completely surround the brain cavity. Photo by P. F. Wright of specimen in U.S. National Museum.

within the middle of the huge mass of the skull. The skeleton of a proboscidean is massive, comprising 12 to 15 percent of the body weight in Recent forms.

The living proboscideans occupy a variety of habitats, but generally live in forests, savannahs, and river valleys. They are vegetarians that may consume more than 225 kg of forage a day. Modern elephants are gregarious herd animals, with a potential life span of about 80 years. The Asiatic elephant is commonly used as a beast of burden, and the African is sometimes used as a draft animal. Ivory is obtained from the tusks of both species.

According to Sikes (1971), there are four extinct families of elephants: Moeritheriidae, Deinotheriidae, Gomphotheriidae, and Mammutidae. At one time the Proboscidea ranged over most of the land areas of the world. The geological range of the family Elephantidae is late Pliocene to Recent in Asia, Pliocene and Pleistocene in Europe, Pleistocene to Recent in Africa, and Pleistocene to early Recent in the New World.

Some extinct proboscideans were contemporary with humans. The mammoths (*Mammuthus*) are known from the northern hemisphere of both the Old and the New worlds (several species have been described). In *M. primigenius*, the woolly mammoth, the body was covered with a dense coat of woolly hair, and long, coarse outer hair, as a protection against the cold. This is the species found in the frozen ground in Siberia, complete with hide and hair. Paleolithic humans left drawings of woolly mammoths on the walls of caves. The tusks in *Mammuthus*, although quite variable in form, had a tendency to spiral, first downward and outward, then upward and inward, and the grinding teeth had numerous ridges of enamel. Mammoths stood about 300 cm high at the shoulders. Mastodons (*Mammut*) became extinct in the Old World before the end of the Pliocene, but the American species, *M. americanum*, is thought to have outlived the now extinct species of elephants in North America, probably persisting until well after the coming of humans to this hemisphere. Mastodons were about the same size as mammoths, the distinguishing feature of the genus *Mammut* being the low-crowned and fairly small grinding teeth, which had three or four prominent transverse ridges of enamel. The tusks in the American species were directed nearly straight forward and were almost parallel with each other; another group of mastodons had two fully developed lower tusks.

The following genera of proboscideans are known from the Pleistocene of the Old World and are now extinct. *Anancus* is known from England, France, Germany, Austria, Hungary, and Russia. The generic name refers to the straight upper tusk that occurred in some forms. Fossils of *Stegolophodon* have been found in India, Burma, and Borneo. The tusks were straight or slightly curved, and the grinding teeth were large. *Stegodon,* from Asia, had a larger skull and lower crowned teeth than present-day elephants. The unusual *Deinotherium,* known from the lower Miocene to the middle Pliocene of Europe and Asia, and the lower Miocene to the Pleistocene of Africa, is characterized by its large size, a pair of lower tusks directed downward, relatively small grinding teeth, a long, flattened skull, and probably a proboscis of some sort. This animal may have been aquatic in habits.

PROBOSCIDEA; ELEPHANTIDAE; *Genus ELEPHAS*
Linnaeus, 1758

Asiatic or Indian Elephant

According to Olivier (1978), the single species, *E. maximus,* occurred in historical time from Syria and Iraq east across Asia south of the Himalayas to Indochina and the Malay Peninsula, north in China to at least the Yangtze River, and on Sri Lanka, Sumatra, and possibly Java. In addition, the species apparently occurred on Borneo in the Pleistocene, and living populations have inhabited that island for at least several hundred years, but the latter may have resulted from introduction by human agency. There is also a small introduced population on the Andaman Islands. There have been suggestions that the elephants of ancient Syria actually represented *Loxodonta* or even *Mammuthus* (Sikes 1971; Corbet 1978).

Head and body length is 550 to 640 cm, tail length is 120 to 150 cm, shoulder height is 250 to 300 cm, and weight is up to about 5,000 kg. The hair covering is scant; the hairs are long, stiff, and bristly. There is a tuft of hair at the tip of the tail. The coloration of the skin is dark gray to brown, often mottled about the forehead, ears, base of the trunk, and chest with flesh-colored blotches that are perhaps caused by some skin disease. True coloration of the skin is often masked by the color of the soil on which the animal lives, as it constantly throws dirt over its back and wallows in the mud. *Elephas* has a heart with a double apex which is often confused with a ''double heart.''

Elephas is distinguished from *Loxodonta* in that the former genus has considerably smaller ears, usually 4 hooves (nails) on the hind foot, 19 pairs of ribs, and 33 caudal vertebrae. In the Asiatic genus, the forehead is flat and the top of the head is the highest point of the animal. In the African genus, the ears are large, there are generally 3 hooves on the hind foot, 21 pairs of ribs, and a maximum of 26 caudal vertebrae. The forehead is more convex, and the back more sloping, so the shoulders are the highest point. In addition, the trunk of the African elephant has 2 fingerlike processes at its tip in contrast to the single tip of the trunk of *Elephas*.

This elephant occurs from thick jungles to grassy plains. In Sri Lanka, McKay (1973) learned that a wide variety of grasses is consumed. In addition, large amounts of bark, roots, leaves, and small stems of a number of species of trees, vines, and shrubs are eaten. Cultivated crops, such as bananas, paddy, and sugar cane are favored foods, with the

Asiatic elephants (*Elephas maximus*), photo by Lothar Schlawe.

Asiatic elephants (*Elephas maximus*), photo by Cyrille Barrette.

result that the elephant often becomes a pest in agricultural regions. An adult Asiatic elephant appears to have a daily food intake of approximately 150 kg net weight per day. The elephant feeds during the morning, evening, and night; it rests during the middle of the day. When feeding on long grasses, the Asiatic elephant uses its trunk as a "hand" to grasp a number of stems, which are pulled up and inserted directly into the mouth. When grasses are too short to be picked up in this way, the elephant scrapes the ground with its forefeet until a loose pile of grasses is formed, and then sweeps the pile into its mouth with the "hand" of the trunk. When feeding on shrubs, the end of the trunk is used to break off small twigs and branches. Bark is removed from larger branches by inserting the branch into the mouth and then, with a turning motion of the trunk tip, rotating the branch against either the molars or a tusk, thus stripping the bark.

Elephas is gregarious, and, though bulls sometimes live alone, cows are always found in herds. In Sri Lanka, bulls may live in temporary all-male groups of up to seven animals, but cows and young are found in herds of sometimes great size (Kurt 1974). In the 19th century, these herds usually consisted of 30 to 50 animals, but much larger groups, with up to even 100 individuals, were not uncommon (McKay 1973). A herd is a family group consisting of mothers, daughters, and sisters, and not a group drawn together by accident or attachment. McKay (1973), working with elephants in Sri Lanka, found that within the herds there are two classes of groups: females with their nursing infants, and females with juveniles. He termed these "nursing units" and "juvenile care units." Within the two groups, there is a certain flexibility in that females may associate at any one time, or divide into smaller subunits. The totality of these unit members that occupy a common range is considered a herd. These herds, now numbering in Sri Lanka between 15

and 40 animals, tend to remain distinct from all other herds that may occur in the same general area. The units and subunits of each herd remain in close proximity, but range out from each other when foraging. Each herd has a rainy season range and a dry season range, the total of which adds up to the herd range.

In 40 percent of the herds observed by McKay (1973) in Sri Lanka, at least one adult male was present; his observations indicate that an adult male is associated with any one herd from 25 to 30 percent of the time. Males stay with the herd when one or more females is in estrus. When adult males are not associated with females in the herds, they usually remain solitary and dispersed over relatively small, widely overlapping home ranges; sometimes they gather together in small but temporary bull herds. Adult males do not seem to be territorial, and there is a great amount of toleration between them, except possibly when the cows are in estrus. Young males appear to leave the herds and become solitary at about the time of puberty.

Eisenberg and Lockhart (1972) and McKay (1973) concluded that, at least in Wilpattu in Sri Lanka, there is no evidence of reproductive seasonality for the Asiatic elephant. Eisenberg, McKay, and Jainudeen (1971) reported that the estrous cycle for the elephant in Sri Lanka averaged 22 days, with the duration of estrus being 4 days. Dittrich (1966) said that 32 births in European zoos and circuses showed a gestation period for male calves of 644 days, with extremes of 615 to 668 days; 17 births of female calves showed gestations of 628 to 668 days, with an average of 648 days. A single calf is born, which weighs from 50 to 150 kg (averaging about 107 kg). At birth, baby elephants have a coat of widely spaced brown hairs that produces a halo effect as it stands out from the body. As the animals grow older, the hairy coat becomes less noticeable, until adult elephants appear to be without hair, but throughout life they have a scattered coat of hair. Females attain sexual maturity when 9 to 12 years old (Grzimek 1975). Males are capable of reproduction at 10 to 17 years of age, but are still too young to dominate older cows and are severely hampered in reproduction (Eisenberg, McKay, and Jainudeen 1971). One captive was estimated to have been 69 years old at time of death (Marvin L. Jones, Zoological Society of San Diego, pers. comm.).

The Asiatic elephant has been domesticated for centuries; it is intelligent and docile when well treated. It is valued in Asia as a draft animal and also is used for transportation and hunting purposes. To capture it from the wild, a strong fence, made of high posts set close together, is usually built in a circular form, with wings leading outward from the front to make a V-shaped entrance to the enclosure. When a group of elephants is in the proper position, it is driven into the V and the corral. Then the entrance is closed, and the process of selecting and taming is started. Tame elephants that have been trained are of great assistance in this work. Wealthy Indians and royalty used to own elephants and ride on them, and elaborate howdahs instead of saddles provided comfortable seats for the riders. Tame elephants were extensively used in handling heavy items such as teak logs.

The Asiatic elephant is classified as endangered by the IUCN (1978) and the USDI (1980), and is on appendix 1 of the CITES. Like *Loxodonta* in Africa, it has been hunted ruthlessly for centuries because of its valuable ivory. In addition, habitat disruption has had a very detrimental effect on populations. *Elephas* disappeared centuries ago from southwestern Asia and most of China. Olivier (1978) estimated that only 28,000 to 42,000 wild Asiatic elephants remain throughout the entire range of the genus. Of these, 9,950 to 15,050 occur on the Indian subcontinent, 11,000 to 14,600 are in continental southeast Asia, and 7,330 to 12,330 occur on the Malay Peninsula, Sri Lanka, Sumatra, Borneo, and the Andaman Islands.

PROBOSCIDEA; ELEPHANTIDAE; *Genus LOXODONTA*
F. Cuvier, 1827

African Elephant

In historical time, the single species, *L. africana*, occurred throughout Africa, from the Mediterranean Sea to the Cape of Good Hope, except in parts of the Sahara and some other desert regions (Laursen and Bekoff 1978; Meester and Setzer 1971). Sikes (1971) suggested that *Loxodonta* also occurred in southwestern Asia within the last 2,000 years. Over the years, many researchers have reported that another species of elephant lives in the dense jungles from Sierra Leone to Zaire. This animal is said to be very small, and to be more solitary than *L. africana*. Most authorities, however, believe that the so-called pygmy elephants are merely small individuals of *L. africana*.

This is the largest living terrestrial mammal. Head and body length (including the trunk, which is really an elongated nose capable of varied uses) is 600 to 750 cm, tail length is 100 to 130 cm, shoulder height is 300 to 400 cm, and weight is 2,200 to 7,500 kg.

Sparsely scattered with black bristly hairs, the skin is dull brownish gray in color. The flattened end of the tail has a tuft of coarse, crooked hairs 38 to 76 cm long. The African elephant wallows in streams and pools and tosses dirt or mud onto its back; thus its coloration is usually similar to that of the soil it frequents. In both sexes, one incisor tooth on each side of the upper jaw is greatly developed to form a tusk. The largest known tusk measures about 350 cm and weighs about 107 kg. The largest female tusk weighs about 18 kg (the average is about 7 kg). The tusks are used for fighting, digging, feeding, and marking (Laursen and Bekoff 1978). The end of the long, tubular, muscular, and very sensitive trunk has a fingerlike projection both above and below. These projections are skillfully used to pick up food or other articles and to examine objects. *Loxodonta* has much larger ears than does *Elephas*, and often has a hole through the lower portion of the ear lobe caused by injury. The ears are sometimes 150 cm in length from top to bottom.

Loxodonta lives in many types of habitat, including deep forests, open savannahs, wet marshes, thornbush, and semidesert scrub. It occurs from sea level to elevations of over 5000 meters.

Rodgers and Elder (1977) stated that nearly all seasonal movements of African elephants have the same general pattern, that is, a mass migration from permanent water sources at the start of the rainy season, followed by a movement back to permanent water when the rain ends and water holes dry up. Shorter, nonseasonal movements of elephants may involve travel between water and feeding areas.

Most of the following information on food habits of *Loxodonta* was taken from Hanks (1979). Various studies show that males eat more than females; one such study estimated that an adult male elephant consumed a mean of about 170 kg of food each day, while a female consumed about 150 kg. African elephants feed on grasses, trees, bark, and shrubs. There seems to be a pronounced seasonal variation in the elephants' diet, with grass intake increasing during the rainy season but falling down to low levels during the dry season. Feeding takes place at all hours, but it was found that in Uganda there was a reduction in feeding activity between 0400 and 0700 hrs. Because of its enormous size, the African elephant can be destructive to vegetation, pushing over trees to obtain edible twigs and leaves and sometimes modifying the habitat over large areas. In the past, when the elephants could roam more freely over the African continent, this presented few problems, and in fact may have been necessary for the maintenance of ecosystems that could support large and diversified populations of animals. As human populations increased in Africa, however, and large areas of the continent were cleared for agricultural purposes, elephants became more and more confined to restricted areas such as parks and reserves. Within these restricted areas, elephant numbers sometimes became very high, and their feeding habits were detrimental to the restricted environments. In such cases, elephant control programs have sometimes been necessary to prevent undue damage to the habitat.

According to Hanks (1979), elephant society is essentially matriarchal; it is organized around a family unit of cows and their calves. The simplest group is a single adult female with one or more of its offspring, but two or three such groups may be permanently associated. At puberty, male elephants are usually driven out of the family groups by the older females, and join or form bachelor groups. Nearly all females are part of a family unit; solitary females are extremely rare, usually very old and at the end of their reproductive life. Solitary males, however, are common. Douglas-Hamilton (1973) found that sometimes two to four family units, numbering up to 50 animals, may come together into slightly less stable associations, which he termed "kinship groups." Laursen and Bekoff (1978) referred to such groups as "clans," gave their membership as 6 to 70 individuals, and noted that they were led by a large female, the dominance of which was undisputed.

Sometimes, very large associations of elephants, numbering several hundred animals or more, are observed, mainly in East Africa. These associations are composed of family units

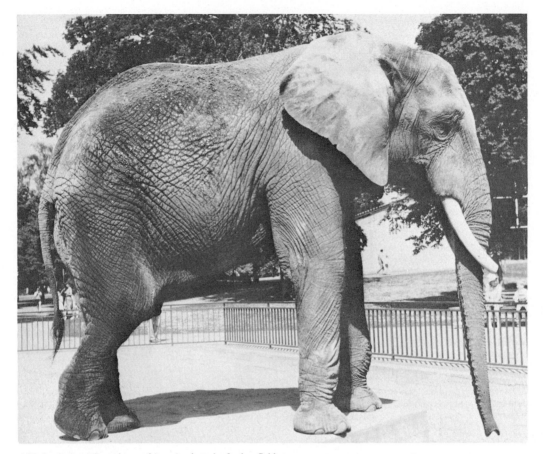

African elephant (*Loxodonta africana*), photo by Lothar Schlawe.

and bachelor herds, and are generally thought to occur as a result of disruption of the normal pattern of social life, in particular, when a herd matriarch has been shot. Douglas-Hamilton (1973) suggested that this bunching could provide a collective defense against predators in time of social stress or disturbance.

Bull elephant groups are very temporary associations between similar animals that find themselves in the same place at the same time. They apparently have little stability and are constantly changing in composition as bulls leave to become solitary or form new alliances.

There is no convincing evidence from anywhere in Africa of territorial behavior in *Loxodonta*, though both bulls and family units do have a fairly distinct home range (Hanks 1979). Little information is available on the size of these home ranges, but published data indicate that the range varies tremendously in different types of habitat. Douglas-Hamilton (1973) found that in the Lake Manyara area, the home ranges of family units and of individual bulls were widely overlapping and varied in size between 15 and 52 sq km. Leuthold and Sale (1973) recorded mean home ranges of 530 sq km in Tsavo West and 1,580 sq km in Tsavo East. Food quantity and quality and availability of water are the most important factors in determining when and how far elephants move. If food, water, and shade all remain available, elephants or elephant units will not venture far; when these items are scarce, large-scale movements may occur. Hanks (1979) found that, on the whole, elephants in southern Africa had small home ranges, but some moved considerable distances for short periods of time.

Hanks (1979), studying elephants in Zambia, concluded that females first ovulate at about 14 years of age. The average gestation period is 22 months, with a recorded range of 17 to 25 months. In Zambia, the breeding season seems to correspond to the wet season, with nearly 88 percent of all conceptions occurring between November and April (when almost all the rain falls). Smith and Buss (1973) reported that in Uganda, breeding also occurs primarily during the rainy season, but it may occur year-round. It appears that a change in the elephants' diet from browse and dry grass with a low crude protein content at the end of the dry season to fresh green grass with a high protein content at the height of the rains may stimulate ovulation and fertile mating (Hanks 1979). According to Laursen and Bekoff (1978), there is normally a single young, but twins occur in 1 or 2 percent of births. The newborn weighs 90 to 120 kg, can stand after ½ hour, and suckles for 2 or 3 years. The herd waits until the young have the strength to roam with them (usually about 2 days). A female may bear four to five young in its lifetime, and the life expectancy is similar to that of humans, about 50 to 70 years.

The African elephant is intelligent, and not difficult to tame, but has not been used as extensively for motive power and as a beast of burden as has the Asiatic elephant. Populations in North Africa were the source of war elephants for the ancient Carthaginians, Romans, and Ethiopians. *Loxodonta* disappeared from north of the Sahara by about the sixth century A.D. It was subsequently eliminated in much of western and southern Africa (Laursen and Bekoff 1978; Meester and Setzer 1971).

African elephants (*Loxodonta africana*), photos by Bernhard Grzimek.

Elephant ivory has been a valuable commercial commodity for centuries. Ricciuti (1980) stated that over the past decade the price of ivory has soared on the world market. Elephant populations, already on the decline because of extensive habitat disruption, have been subjected to unprecedented exploitation as well. Biologists estimate that Kenya lost more than half of its 120,000 elephants between 1970 and 1977, and that the rate continued at perhaps 25 percent a year thereafter. In Uganda, herds in some of the national parks have been reduced to mere remnants of what they once were. As many as 30,000 elephants may have been destroyed in recent years in the Central African Republic alone. Probably fewer than 1,500,000 elephants survive in all of Africa, and in many, if not most, areas the population is down drastically. Illegal poaching and habitat disruption are threatening this elephant nearly everywhere it occurs outside of parks and reserves, and even within many of these it is difficult to control the illegal killing. *L. africana* is classified as vulnerable by the IUCN (1978) and as threatened by the USDI (1980), and is on appendix 2 of the CITES.

ORDER HYRACOIDEA

HYRACOIDEA; *Family PROCAVIIDAE*

Hyraxes, Dassies

This order contains the single Recent family Procaviidae, which comprises three Recent genera and seven species and which occurs in Lebanon, Israel, Jordan, Sinai, the Arabian Peninsula, and most of Africa.

These animals are comparable in size and external appearance to rodents and lagomorphs. The length of the head and body is 300 to 600 mm; the tail is 10 to 30 mm in length or lacking. Adults sometimes weigh as much as 4.5 kg. The pelage consists of fine underhairs and coarser guard hairs. Scattered bristles, presumably tactile, are located mainly on the snout. A gland on the back is covered with hair of a different color from that on the rest of the body. The eye is unique in that a portion of the iris above the pupil bulges slightly into the aqueous humor, thus cutting off light from almost directly above the animal. Hyraxes have a short snout, a cleft upper lip, short ears, and short, sturdy legs. The vertebral column is convex from the neck to the tip of the tail. The forefoot has four digits, with flattened nails resembling hooves. The hind foot has three digits; the inner toe, the second digit, has a long curved claw and the other digits short, flattened, hooflike nails. The soles have special naked pads for traction; these pads are kept continually moist by a glandular secretion and have a muscle arrangement that retracts the middle of the sole. This forms a hollow, which is a suction cup of considerable clinging power.

The dental formula of the deciduous teeth is (i 2/2, c 1/1, pm 4/4) ×2 = 28; but the formula for the permanent teeth is (i 1/2, c 1/1, pm 4/4, m 3/3) ×2 = 38. The single pair of upper incisors grows continuously and is long and curved. The upper incisors are triangular in cross section and semicircular in form; the flattened back surfaces are without enamel and so produce pointed cutting edges. The lower incisors are chisel shaped; the first pair has three cusps and the second pair only one. There is a wide space between the incisors and the cheek teeth. The milk canines are rarely persistent. The premolars resemble the molars, for they are arranged in a continuous series with the molars that are low crowned to high crowned. Each has four roots and the lower ones bear two crescents similar to the corresponding teeth of rhinos and horses. The skull is stout and the roof is flattened.

The terrestrial genera (*Procavia* and *Heterohyrax*) inhabit rocky areas, arid scrub, and open grassland, whereas tree hyraxes (*Dendrohyrax*) are usually arboreal, and are found in

Tree hyrax (*Dendrohyrax dorsalis*), the light-colored hairs indicate the location of glands that exist on the back of hyraxes, photo by Jean-Luc Perret.

forested areas, though in eastern Africa they inhabit lava flows. The elevational range of the family is sea level to 4,500 meters.

Although only *Dendrohyrax* is arboreal, all the genera apparently can climb well; *Heterohyrax*, for example, sometimes suns itself in a tree. Hyraxes move quickly and are extremely agile on rugged and steep surfaces, running and jumping with skill and gaining traction by means of specialized foot pads and probably the inner claw on the foot. This claw is apparently also used to groom the hair. Hyraxes travel on the sole of the foot, with the heel touching the ground, or partly on the digits. The terrestrial forms are similar in habits to the pikas, *Ochotona*, sheltering in colonies of 5 to about 50 individuals, usually among rocks. They are active mainly during the daylight hours and are fond of basking in the sun and rolling in the dust. Tree hyraxes shelter singly or in family groups, using tree hollows and dense foliage; they are active during the night. They are less gregarious than the terrestrial forms and are usually more tractable. The terrestrial hyraxes whistle, scream, and chatter; the tree hyraxes utter a series of loud cries.

Hyraxes have acute sight and hearing. They feed mainly on vegetation, and are both grazers and browsers. Tree hyraxes feed on the ground as well as in trees. All the species habitually work their jaws in a manner reminiscent of cud chewing. Generally these animals are not limited in their distribution by a lack of water to drink. They sometimes travel more than 1.3 km for food. The hyraxes are preyed on mainly by rock pythons, eagles, and leopards.

Hyraxes, referred to in the Bible as conies, are in an order that originated in Africa and evolved for some time before the Oligocene. They probably never spread beyond Africa and the Mediterranean regions. The geological range of this order and of the single Recent family is early Oligocene to Recent in Africa, early Pliocene in Europe, and Recent in southwestern Asia. There are two extinct families. No fossils have been found of the living genera.

HYRACOIDEA; PROCAVIIDAE; *Genus PROCAVIA* Storr, *1780*

Rock Dassie or Hyrax

Ellerman and Morrison-Scott (1966), Roche (1972a), and Corbet (1979) each recognized only a single species, *P. capensis*, occurring in Lebanon, Israel, Jordan, Sinai, the Arabian Peninsula, and almost all of Africa. Some other authorities, including Meester and Setzer (1971), consider *P. capensis* restricted to southern Africa, and treat the following as distinct species: *P. welwitschii*, southwestern Angola, Namibia; *P. ruficeps*, southern Algeria and Senegal to Central African Republic; *P. johnstoni*, northeastern Zaire and central Kenya to Malawi; and *P. syriacus*, Egypt to Kenya, and the southwest Asian portion of the range of the genus.

The length of the head and body is 305 to 550 mm; an external tail is lacking; the height of the shoulder is 202 to 305 mm; and the weight of males is about 4 kg, and of females about 3.6 kg. The hair is short and rather coarse and the general coloration of the upper parts is brownish gray. The flanks are somewhat lighter, and the underparts are

A. Gray hyraxes (*Heterohyrax* sp.), photo by C. A. Spinage. B. Rock dassie (*Procavia* sp.), photo by Roy Pinney. C. Skull of rock dassie (*P. capensis*), photo by P. F. Wright of specimen in U.S. National Museum. D. Palmer surface of hand (*P. capensis*); E. Plantar surface of foot (*P. capensis*); photos by Ernest P. Walker.

Rock dassie (*Procavia capensis*), photo from San Diego Zoological Garden.

creamy. There is, however, considerable variation in color and intensity. The black whiskers may be as long as 180 mm. In general appearance *Procavia* is similar to a large pika (*Ochotona*) or a tailless woodchuck (*Marmota*). However, the similarities go no further than the superficial appearance. The soles are moist and rubberlike, which gives the animals traction on smooth surfaces and steep slopes. This genus is distinguished from *Heterohyrax* and *Dendrohyrax*, by a black dorsal patch. This patch covers a gland which is exposed if the surrounding hairs are erected. Erection of these hairs occurs when the animal is angry or frightened. Females have six mammae.

Procavia frequents rocky, scrub-covered habitat, wherever there are suitable shelters in, between, or under rocks, or where it can dig burrows of its own. The information for the remainder of this account was taken in large part from Hoeck (1975), Kingdon (1971), Mendelssohn (1965), Millar (1971), and Smithers (1971). Individuals may be found feeding at any time of the day, provided it is warm and sunny; they are reluctant to leave their rocky shelters if the weather is overcast or cold, and do not emerge at all during rainy periods. In spite of its rather heavy build, *Procavia* is active and agile, running up steep, smooth rock surfaces with ease. Its senses are keen, and when persecuted it becomes quite shy. When alarmed, it quickly dashes into a rocky cranny or its own burrow. Although not a large animal, it can put up a vigorous fight in self-defense, biting savagely. It is predominantly a browser, though in some parts of Africa, especially in the wet season, it does a considerable amount of grazing. It feeds on practically any plant, including some that are poi-

sonous to most other animals, such as plants of the families Solanaceae and Euphorbiaceae.

Procavia lives in colonies that contain from 4 to as many as 60 individuals, depending on the amount of available habitat. The basic family unit consists of 1 adult territorial male and several adult females and young. The adult male is the leader and most watchful member of the colony. He often stays on a high rock or branch while the others feed, and, if danger threatens, he gives an alarm call at which the others take to cover. Fourie (1977) distinguished 21 adult vocalizations that apparently are important for information transfer. Males other than the group leader may live in bachelor groups of their own.

There is a definite breeding season for *Procavia,* which seems to vary with the geographic locality. Mating has been reported in Israel from August to September, in South Africa in the second half of the summer (February and March), and in Tanzania at the end of the wet season from April to June. The gestation period is 205 to 245 days. The average number of young per litter is 3.2 and the range is 1 to 6. The young are born fully covered with hair, and with the eyes open; birth usually occurs in a crevice. Sexual maturity is reached at about 16 to 17 months of age. A captive rock dassie lived for 11 years (Marvin L. Jones, Zoological Society of San Diego, pers. comm.).

The flesh is prized by some natives. The young make interesting pets and will eat practically any vegetable matter. The chief natural enemies of *Procavia* are leopards and eagles, but foxes, weasels, and mongooses prey on them also.

Gray hyrax (*Heterohyrax* sp.), photo by W. Leuthold.

HYRACOIDEA; PROCAVIIDAE; *Genus **HETEROHYRAX***
Gray, 1868

Gray Hyraxes, Yellowspotted Hyraxes

There are three species (Meester and Setzer 1971):

H. antineae, Ahaggar Mountains of southern Algeria;
H. brucei, southeastern Egypt to central Angola and
    northeastern South Africa;
H. chapini, east-central Zaire.

Both Ellerman and Morrison-Scott (1966) and Roche (1972a) regarded *Heterohyrax* as only a subgenus of *Dendrohyrax*, but Hoeck (1978) showed that generic distinction is warranted.

The length of the head and body is 305 to 380 mm; an external tail is absent; the height of the shoulder is 305 mm; and the weight is 475 grams to 4.5 kg. The body is covered with thick, short, rather coarse hair and the general coloration is brown and whitish, suffused with black; the underparts are white. The general appearance is much like a large guinea pig (*Cavia*). Members of this genus are distinguished by a patch of yellowish or whitish hair in the middle of the back. This patch covers a gland which is exposed when the surrounding hairs are erected in anger. Females usually have one pair of pectoral mammae and two pairs of inguinal mammae.

According to Sclater (1900): "The soles, which are naked, are covered by a very thick epithelium which is kept constantly moist by the secretion of the sudorific glands there present in extraordinary abundance; furthermore, a special arrangement of muscles enables the sole to be contracted so as to form a hollow air-tight cup which, when in contact with the rock, gives the animal great clinging power, so much so that even when shot dead it remains attached to almost perpendicular surfaces as if fixed there."

Smithers (1971) reported that in Botswana, *H. brucei* is confined to areas of rocky kopjes, rocky hillsides, krantzes, and piles of loose boulders, particularly where there is a cover of trees and bushes on which it can feed. This species is more strictly confined to the larger areas of this type of habitat, and is less prone to colonize smaller outlying kopjes and piles of boulder than is *Procavia*. The altitudinal distribution of *Heterohyrax* is from sea level to at least 3,800 meters in mountains.

Yellowspotted hyraxes are diurnal. Their habits are very much like those of *Procavia*, with which genus they occur in close association, sometimes inhabiting the same rocky crevices. *Heterohyrax* is more of a browser than *Procavia*, some 80 percent of the area to prepare poisoned arrows. According to this animal is known to go up in trees sometimes in order to feed. Food consists of many types of bushes and trees, even those that are poisonous to most other mammals. Dobroruka (1973) reported that at the Numba Caves in Zambia, *H. brucei* fed almost exclusively on the leaves of wild bitter yam (*Dioscorea dumetorum*), a plant which is often used by natives of the area to prepare poisoned arrows. According to Hoeck (1975), *H. brucei* in Serengeti National Park had feeding peaks in the morning and evening, and, though primarily a browser, it grazed more grass in the wet season than in the dry season.

Yellowspotted hyraxes are sharp sighted, keen of hearing, and quite aggressive, prepared to bite anything that attacks them. Their presence is often detected by their calls, which resemble shrill screams. These animals are noted for their wariness; they are ever alert, and seek shelter upon the slightest alarm, but are extremely curious and will soon show themselves again. They enjoy laying on rocky ledges during the afternoon, and like to play and chase one another among the rocks.

Heterohyrax lives in colonies of sometimes hundreds of animals. The basic family unit consists of an adult territorial male, several females, and young (Hoeck 1975). The male defends the territory from intruders, mostly which are peripheral males that harrass the territorial male by rushing into his territory and mating with one of the females.

Hoeck (1975) reported that there is a definite breeding season for *Heterohyrax* in Serengeti National Park, which is generally at the end of the wet season from April to June. Up

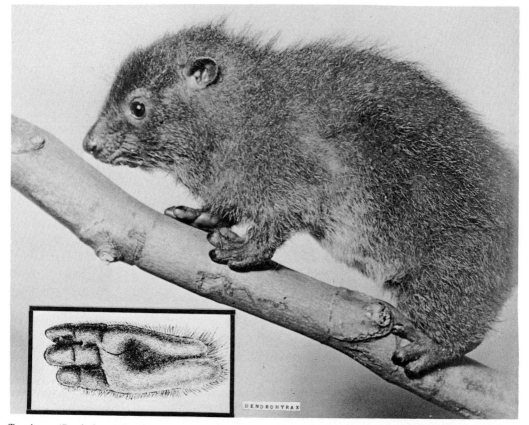

Tree hyrax (*Dendrohyrax dorsalis*), photo by John Markham. Inset: underside of right hind foot, photo from *Proc. Zool. Soc. London.*

to four young, usually two, are born after a gestation period of 7½ months. The young are born in a fur-lined nest, and within a few hours after birth they are able to scamper around with the adults. Potential longevity is probably about 7 years.

Native peoples eat hyrax flesh when other types of meat are scarce, but it is said to be tough, dry, and not very desirable. The chief foe of hyraxes is the rock python. They are also preyed upon by leopards, birds of prey, and probably mongooses and other small carnivores.

HYRACOIDEA; PROCAVIIDAE; *Genus DENDROHYRAX* Gray, 1868

Tree Hyraxes, Bush Hyraxes

There are three species (Meester and Setzer 1971):

D. dorsalis, forest zone from Gambia to Uganda and extreme northwestern Angola, Fernando Poo;

D. arboreus, eastern Zaire and central Kenya to eastern South Africa;

D. validus, eastern mainland of Tanzania, and the islands of Zanzibar, Pemba, and Tumbatu.

Head and body length is 400 to 600 mm, tail length is 10 to 30 mm, and weight is usually 1,500 to 4,500 grams. The hair on the upper parts is brown, tipped with gray or yellow; black hairs are also present on the back. Some forms have two color

phases, the darker coat and a yellow pelage. A white patch of hair, marking the location of a gland, is present on the back, and the ears are edged with white hairs. The underparts are usually brownish. The fur is longer and slightly more silky than that of other hyraxes.

Tree hyraxes are found in forests at elevations of up to 4,500 meters. According to Jones (1978), *D. dorsalis* occurs most frequently from sea level to about 1,500 meters elevation, but has been recorded at up to 3,000 meters. It inhabits upland and riverine forests within a vegetation type described as either tropical rainforest, or tropical closed forest. Forests with lianas contain more *D. dorsalis* than do other habitats. It is arboreal, but frequently descends and moves about on the ground. This species is strictly herbivorous, feeding on leaves, fruits, twigs, and bark, usually from the upper canopy of the forest. It is nocturnal, with most activity taking place soon after dark, and just before dawn. According to Kingdon (1971), *D. arboreus* can be very noisy, and the calls often seem to follow a frequency pattern; most vocal activity seems to occur following intensive feeding periods, between 2000 and 2300 hrs and between 0300 and 0500 hrs. The cry is made by both sexes, but it is much more powerful in the male, and may have both sexual and territorial functions.

Jones (1978) stated that *D. dorsalis* is usually solitary, but that groups of two and occasionally three animals are to be found. Each animal uses a small area in the forest, usually centered around a single tree. Both mating and birth of this species are reported to occur during the dry season. Birth of the young is primarily during March and April in Gabon and Cameroon, and from May to August in the western and southern parts of Zaire. In the eastern parts of the range of the species, young are born throughout the year. A single baby, occasionally two, is produced after a gestation period of about eight months; the young are precocious. At about 120 days of age, the young reach adult length, and little growth occurs thereafter. Sexual maturity is attained at more than 1 year of age. One *D. dorsalis* lived in captivity for 5½ years. According to Marvin L. Jones (Zoological Society of San Diego), a specimen of *D. arboreus* was still living after 12 years and 3 months of captivity.

Predators of *D. dorsalis* include hawk eagles, leopards, golden cats, genets, servals, and pythons (Jones 1978). In addition, tree hyraxes are hunted by people for food and pelts throughout their range. These animals often assume a characteristic defensive position when threatened by a predator; the back and the rump are turned toward the enemy, and the hairs around the dorsal gland are spread out and separated from each other so that the naked glandular area is exposed. They are said to be less irritable than the other dassies and to tame readily.

ORDER SIRENIA

Manatees, Dugong, Sea Cow

This order of aquatic mammals contains two Recent families: Dugongidae for the genera *Dugong* (dugong, one species) and *Hydrodamalis* (Steller's sea cow, one species, extinct); and Trichechidae for the single genus *Trichechus* (manatees, three species). The dugong inhabits coastal regions in the tropical parts of the Old World, but some individuals go into the fresh water of estuaries and up rivers. Steller's sea cow occurred in the Bering Sea, the only Recent member of this order adapted to cold waters; and manatees live along the coast and in coastal rivers in the southeastern United States, the West Indies, northern South America, and western Africa.

These massive, fusiform (spindle-shaped) animals have paddlelike forelimbs, no hind limbs or dorsal fin, and a tail in the form of a horizontally flattened fin. The adults of the living forms generally are from 250 to 400 cm in length and weigh up to 908 kg. The skin is thick, tough, wrinkled, and nearly hairless; stiff, thickened vibrissae (tactile hairs) are present around the lips. The head is rounded, the mouth is small, and the muzzle is abruptly cut off. The nostrils, which are valvular and separate, are located on the upper surface of the muzzle. The eyelids, although small, are capable of contraction, and a well-developed nictitating membrane is present. An external ear flap is lacking. The neck is short.

The females have two mammae, located in the chest region. The testes in males are abdominal (borne permanently within the abdomen).

The skeleton is dense and heavy, and is characterized by an absence of pneumatic cavities (pachyostosis). The increased specific gravity is probably an adaptation to remaining submerged in shallow waters (Kaiser 1974). The skull is large in proportion to the size of the body; the nasal opening is located far back on the skull and directed posteriorly. The lower jaws are heavy and united for a considerable distance. The forelimb has a well-developed skeletal support, but there is no trace of the skeletal elements of the hind limb. In *T. senegalensis*, at least, there is no movement in the wrist joint (Kaiser 1974). The pelvis consists of one or two pairs of bones suspended in muscle. The vertebrae are separate and distinct throughout the spinal column; one genus (*Trichechus*) has six neck vertebrae (nearly all other mammals have seven).

The dentition is highly modified and often reduced (functional teeth are lacking in Steller's sea cow). The tusklike incisors are reduced or absent, and canines are present only in certain fossil species. When the incisors are present, there is a space between them. The cheek teeth, which are arranged in a continuous series, number from 3 to 10 in each half of each jaw. The anterior part of the palate and the corresponding surface of the lower jaw are covered with rough, horny plates, presumably used as an aid in chewing food; the small, fixed tongue is also supplied with rough plates.

The genera differ in a number of ways. In *Dugong* (and in *Hydrodamalis*) the tail fin is deeply notched, but in *Trichechus* the tail fin is more or less evenly rounded. The upper lip is more deeply cleft in manatees than in *Dugong* (and in *Hydrodamalis*). *Dugong* has one pair of tusklike incisors (in the males but concealed in bone in the females) and 3/3 functional cheek teeth; in *Trichechus* functional incisors are not present, but the cheek teeth are numerous and indefinite in number (up to 10 in each half of each jaw). The cheek teeth in *Trichechus* are replaced consecutively from the rear, as in many proboscideans; an individual cheek tooth is worn down as it moves forward.

Sirenians are solitary, travel in pairs, or associate in groups of three to about six individuals. Generally slow and inoffensive, they spend all their life in the water. They are vegetarians and feed on various water plants. This is the only mammalian order that has evolved to exploit plant life in the sea margin (Anderson 1979). The ordinal name Sirenia is related to the supposed mermaidlike nursing of dugongs (the Sirens of mythology) and manatees. The only reliable observations of nursing in manatees, however, have revealed that the young suckle while the mother is underwater in a horizontal position, belly downward; the dugong probably nurses in the same position.

These aquatic mammals were apparently more abundant in previous times, that is, during the Miocene and early Pliocene. Their comparative scarcity at the present time is probably due to persecution by humans for food, hides, and

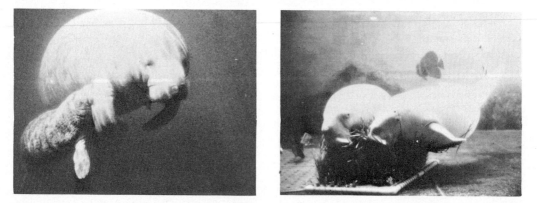

Left, Manatees (*Trichechus manatus*), female and calf, photo by James A. Powell, Jr., U.S. Fish & Wildlife Service. Right, dugongs (*Dugong dugon*), female on right and male on left, photo by T. Kataoka, Toba Aquarium, Japan.

oil, and perhaps most importantly, to climatic changes in the recent past. The geological range of the order Sirenia is Eocene to Recent.

SIRENIA; *Family DUGONGIDAE*

Dugong, Sea Cow

This family contains two Recent subfamiles: Dugonginae, with the single Recent genus *Dugong* (dugong), found in coastal regions of the tropics of the Old World; and Hydrodamalinae, with the single Recent genus *Hydrodamalis* (Steller's sea cow), formerly found in the Bering Sea (Jones and Johnson 1967).

Dugong is generally 250 to 350 cm in length, whereas Steller's sea cow measured almost 750 cm. The flippers in Steller's sea cow were curiously bent and were said to have been used to pull an individual along the bottom of the ocean as it foraged. Members of the family Dugongidae lack nails on their flippers. The deeply notched tail fin has two pointed, lateral lobes (the tail fin in manatees is more or less evenly rounded). The upper lip, which is more deeply cleft in the adult dugong than in the young, is not so deeply cleft as that of the manatees. The nostrils are located more dorsally than in other sirenians. In the dugong (and apparently all sirenians) the eyelids contain a number of glands which produce an oily secretion to protect the eye against water.

The ribs of Steller's sea cow consisted entirely of dense bone; those of *Dugong* contain some porous bone. The skull of a dugong has a downwardly bent rostrum (for holding the incisors), whereas the skull of a Steller's sea cow had a rostrum that was only slightly inclined. Functional teeth were lacking in *Hydrodamalis* but occur in *Dugong*. *Dugong* has one pair of tusklike incisors (in the males but concealed in bone in the females), which are directed downward and forward and are partially covered with enamel, and 3/3 functional cheek teeth (up to 6/6 may be present in an individual's lifetime). The cheek teeth are columnar, are covered with cement, lack enamel, and have simple, open roots. *Hydrodamalis* had no teeth, but did possess functional, rough oral plates.

The geological range of this family is middle Eocene to Recent in Africa and Europe, Eocene to Recent in North America, and Oligocene to Recent in Asia (Husar 1978a).

SIRENIA; DUGONGIDAE; *Genus DUGONG Lacépède, 1799*

Dugong

The single species, *D. dugon*, originally occurred in the Red Sea and Gulf of Aqaba, along the eastern coast of Africa as far south as Mozambique, along the southern coast of Asia, around Madagascar and many of the islands in the Indian Ocean, throughout the East Indies, as far north as Taiwan and the Ryukyu Islands, along the coast of Australia except in the south, and in the Pacific Ocean at least as far east as the Caroline Islands and the New Hebrides (Husar 1978a; Nishiwaki et al. 1979).

The total length of adults is 240 to 406 cm, with weights ranging from 230 to 908 kg. The usual weight is 230 to 360 kg (Husar 1978a). The coloration is variable, but usually dull, brownish gray above, and somewhat lighter below. The skin is thick, tough, wrinkled, and covered with widely scattered hairs. The front limbs are modified into flippers about 35 to 45 cm in length; they are used for propulsion by the young, but adults propel themselves by use of the flukelike tail, and use the flippers only for steering. When the dugong grazes on the ocean floor, the flippers are used for "walking," and not to carry food into the mouth. As with other grazing mammals, the dugong has a large, complex stomach, but also has unique adaptations for feeding on marine grasses. The upper lip protrudes considerably beyond the lower; it is deeply cleft, forming a large U-shaped, muscular pad, which overhangs the small, downwardly opening mouth. On the sides of this facial pad are two ridges that bear short, sturdy, blunt bristles. The lower lip and the distal portion of the palate have rough, horny pads that are used to grasp sea grasses during feeding; the whole upper lip is extended and curved around the base of a plant, which is then grasped with the mouth pads and pulled up by the roots. The dental formula is: (i 1/0, c 0/0, pm 0/0, m 2–3/2–3) × 2 = 10–14. The upper incisors are rootless and straight, forming short, thick tusks in males. In females, the incisors usually do not pierce the gum, which implies that the incisors have some sex-related function, rather than a food-gathering function. Adult males invariably carry conspicuous scars that seem to have been made by competing males. The molars are circular in section, thick, rootless, and lacking in enamel. As in proboscideans, the molars are replaced by continuous growth from behind as the front ones wear away. The last molar tooth is a double, rather than a single, cylinder (Lekagul and McNeely 1977).

The dugong occurs in the shallow waters of coastal regions of tropical seas, where there is an abundance of vegetation; it is more strictly marine than manatees, and is seldom found in fresh-water localities (Jones and Johnson 1967). It generally rests in deep water during the day, and moves toward the shore to feed at night. Long-distance migration is unknown for the dugong, but seasonal changes in abundance in local coastal waters are apparent in East African, Indian, and Philippine waters (Husar 1978a). Average swimming speed is 10 km per hour, but animals can double this rate if pressed.

Dugong is mostly herbivorous, and historic distribution broadly coincided with the tropical Indo-Pacific distribution of food plants, the phanerogamous sea grasses of the families Potomogetonaceae and Hydrocharitaceae. Although sea grasses are the primary food of the dugong, it has been noted feeding on brown algae following a cyclone that damaged the sea grass bed; green algae, marine algae, and some crabs have also been taken from dugong stomachs (Husar 1978a).

The following information on social life and reproduction was taken from Husar (1978a) and Lekagul and McNeely (1977). The dugong is now usually found only in small groups or singly, but in the past, huge herds, numbering sometimes thousands of animals, were observed. Neither fights nor aggressive behavior has been recorded. *Dugong* has been reported to make whistling sounds when frightened; calves make a bleating, lamblike cry. These vocalizations are thought to be used only for short-range communication. Breeding seems to occur throughout the year, with no well-defined season. Pair bonds between mated dugongs have been reported, and it is possible that family groups are formed within the larger population. The gestation period is between 11 and 12 months, and the single young is born underwater, swimming immediately to the surface for its first breath of air. The baby clings to its mother's back as she browses through the shoals of sea grass, submerging when the mother submerges and rising when she rises. The breast is under the flipper, and the young probably reach around to suckle from the rear, usually underwater. Calves begin grazing within the first 3 months after birth. Young may remain with the mother for more than 1 year, and calves as long as 183 cm in length have been recorded accompanying females.

Dugongs (*Dugong dugon*): Top, picture from *Extinct and Vanishing Mammals of the Western Hemisphere*, Glover M. Allen;
Bottom, photo by Paul K. Anderson.

A captive pair were maintained for over 10 years, and longevity may be as much as 40 years.

The dugong has been hunted for food throughout its range; the meat of this animal has been likened to tender veal. The hide has been used to make a good grade of leather. The species has also been taken for its oil (24 to 56 liters for an average adult), and its bones and tusks, which have been used to make ivory artifacts and a good grade of charcoal for sugar refining. Commercial dugong fisheries once operated from Sri Lanka, and several Asian cultures have prized the dugong for its supposed medicinal and aphrodisiac properties. Large gill nets and harpooning are used to capture the

dugong (Husar 1978a). Other than humans, sharks are probably the only enemy of the dugong, but individuals have been seen to "gang up" on sharks in shallow water, and drive them off by butting them with the head (Lekagul and McNeely 1977).

In many areas, the dugong has declined greatly in numbers, and fears have been expressed that it might be exterminated by continued hunting pressure. It is classified as vulnerable by the IUCN (1976) and as endangered by the USDI (1980), and is on appendix 1 of the CITES, except for the Australian population, which is on appendix 2. No specific population estimates are available for dugongs, except for

Dugongs (*Dugong dugon*), male above and female below, photo by T. Kataoka, Toba Aquarium, Japan.

the northeastern Australian region, where 1,000 to 2,000 animals are thought to occur (U.S. Fish and Wildlife Service 1980). Nishiwaki et al. (1979) estimated that roughly 30,000 dugongs may inhabit the whole Indo-Pacific region.

SIRENIA; DUGONGIDAE; **Genus HYDRODAMALIS**
Retzius, 1794

Steller's or Great Northern Sea Cow

The single Recent species, *H. gigas,* occurred in historical time around Bering and Copper islands in the Commander Islands of the western Bering Sea (Rice 1977). Pleistocene remains of the same species have been found on Amchitka Island in the Aleutians and in Monterey Bay off California. Pliocene remains of a related species, *H. cuestae,* have been found in Japan, California, and Baja California (Domning 1978).

This animal was the largest of the sirenians. From notes and data taken from observations of them by Steller (from Stejneger 1936), the famous German naturalist of Bering's trip, we can get an idea of the size and appearance of the animal in life. It measured 752 cm from the tip of the nose to the point of the tail flipper. Its greatest circumference measured 620 cm. Steller estimated that an adult female weighed about 4,000 kg. The head was small in proportion to the body. The tail had two pointed lobes forming a caudal flipper. The forelimbs were very small, flipperlike, and somewhat truncated. There were no hind limbs externally, no

Steller's sea cow (*Hydrodamalis gigas*): A. Photo from *Extinct and Vanishing Mammals of the Western Hemisphere,* Glover M. Allen; Inset: Steller's sea cow's palate, from *Symbolae Sirenologicae,* drawing by J. F. Brandt; B. Outer surface of skin; C. Flesh side of skin; photos by Erna Mohr.

trace even remaining of the pelvic elements. The skin was naked and was covered with a very thick, barklike, extremely uneven appearing epidermis, from which the German name for this animal, *Borkèntier,* was derived. The skin was a dark brown to gray brown color, occasionally spotted or streaked with white. The forelimbs were covered with short, brushlike hairs. Some dried pieces of this skin are still preserved, one in Leningrad and one in Hamburg. These animals were parasitized by two or three species of small crustacean that burrowed into the rough skin. Steller said that he found some of the holes made by the small crablike crustacean *Cyamus* to ooze a thin serum.

Hydrodamalis was the only Recent sirenian adapted to cold waters. All that we know about its habits comes from Steller's account. It was quite numerous in the shallow bays and inlets around the coast of Bering Island when first discovered. It fed upon the kelp beds and the extensive growths of various marine algae which grew in the shallower waters. It was slow moving, was utterly fearless of humans, and was said to be affectionate toward others of its kind, as individuals apparently tried to aid others that were either wounded or in distress.

The following are excerpts from Steller's own account (from Stejneger 1936):

"Usually entire families keep together, the male with the female, one grown offspring and a little, tender one. To me they appear to be monogamous. They bring forth their young at all seasons, generally however in autumn, judging from the many new-born seen at that time; from the fact that I observed them to mate preferably in the early spring, I conclude that the fetus remains in the uterus more than a year. That they bear not more than one calf I conclude from the shortness of the uterine cornua and the dual number of mammae, nor have I ever seen more than one calf about each cow.

"These gluttonous animals eat incessantly, and because of their enormous voracity keep their heads always under water with but slight concern for their life and security, so that one may pass in the very midst of them in a boat even unarmed and safely single out from the herd the one he wishes to hook. All they do while feeding is to lift the nostrils every four or five minutes out of the water, blowing out air and a little water with a noise like that of a horse snorting. While browsing they move slowly forward, one foot after the other, and in this manner half swim, half walk like cattle or sheep grazing. Half the body is always out of water . . . . In winter these animals become so emaciated that not only the ridge of the backbone but every rib shows.

"Their capture was effected by a large iron hook, . . . the other end being fastened by means of an iron ring to a very long, stout rope, held by thirty men on shore . . . . The harpooner stood in the bow of the boat with the hook in his hand and struck as soon as he was near enough to do so, whereupon the men on shore grasping the other end of the rope pulled the desperately resisting animal laboriously towards them. Those in the boat, however, made the animal fast by means of another rope and wore it out with continual blows, until, tired and completely motionless, it was attacked with bayonets, knives and other weapons and pulled up on land. Immense slices were cut from the still living animal, but all it did was shake its tail furiously."

In the late Pleistocene, *Hydrodamalis* evidently occurred all around the rim of the North Pacific from Japan to California. Subsequent hunting by primitive peoples probably eventually restricted it to the vicinity of the Commander Islands (Domning 1978). This last population was discovered in 1741 when a Russian expedition led by Captain Vitus Bering was stranded on Bering Island. At that time the total number of sea cows probably did not exceed 1,000 to 2,000. Bering's crew, mostly sick with scurvy, and subsequent visitors to the area, slaughtered the animals relentlessly. The meat of *Hydrodamalis* was used for food, and the hide for making skin boat covers and shoe leather. The genus probably became extinct by 1768, though there were later reports of its presence in various areas.

SIRENIA; *Family TRICHECHIDAE; Genus*
TRICHECHUS Linnaeus, 1758

Manatees

The single Recent genus, *Trichechus*, contains three species (Husar 1977, 1978*b*; 1978*c*; Rice 1977):

T. manatus (West Indian manatee), coastal waters and some connecting rivers from North Carolina around the Gulf of Mexico and Caribbean Sea to northeastern Brazil, Greater Antilles, Bahamas;

T. senegalensis (West African manatee), coastal waters and connecting rivers from Senegal to Angola, occurs as far as 2,000 km up the Niger River and in tributaries of Lake Chad;

T. inunguis (Amazonian manatee), throughout the Amazon Basin of northern South America, possibly also in the Orinoco Basin and the coastal waters of eastern Brazil.

These animals have a rounded body, a small head, and a squarish snout. The upper lip is deeply split and each half is capable of moving independently of the other. The nostrils are borne at the tip of the muzzle, the eyes are small, and there are no external ears. Vestigial nails occur on the flippers, and the tail fin is more or less evenly rounded (not notched as in the dugong and sea cow). Stout bristles appear on the upper lip, and bristlelike short hairs are scattered singly over the body at intervals of about 1.25 cm. These sirenians attain a maximum length of nearly 450 cm and a weight of 680 kg; most individuals are 250 to 400 cm long and 140 to 360 kg in weight. Coloration is a uniform dull gray to black. The skin is 5.1 cm thick.

The stout ribs consist wholly of dense bone. Manatees have only six neck vertebrae, whereas nearly all other mammals have seven. The single pair of mammae are pectoral.

Nasal bones are present in the skull (these bones are absent or vestigial in the dugong and sea cow). Manatees have 2/2 incisors, which are concealed beneath horny plates and are lost before maturity. The cheek teeth number up to 10 in each half of each jaw (although more than 6/6 are rarely present at any one time). They are low crowned, enameled, divided into cuspidate cross-crests, lack cement, and have closed roots. These teeth are replaced horizontally; they form at the back of the jaw and wear down as they move forward. This may be an adaptation to eating food mixed with sand, as a similar replacement of teeth occurs in many proboscideans and other mammals.

The species *T. manatus* inhabits shallow coastal waters, bays, estuaries, lagoons, and rivers, occupying both fresh- and salt-water habitats; water temperature of 21°C is generally the minimum favored, but individuals of this species have been observed in waters as cold as 13.5°C (Husar 1978*c*; Hartman 1979). *T. senegalensis* occurs in both shallow coastal waters and fresh-water rivers, but seems to prefer large, shallow estuaries and weedy swamps; its range is apparently limited to water with a temperature above 18°C (Husar 1978*b*). *T. inunguis* favors backwater lakes, oxbows, and lagoons; it has been maintained successfully in waters with temperatures from 22° to 30°C (Husar 1977).

Except as noted, the information for the remainder of this account was taken from Hartman (1979), and refers to *T. manatus* in Florida waters. Manatees use their tails not only to propel themselves through the water but also as rudders by means of which they control roll, pitch, and yaw. The flippers are used in precise maneuvering, and in minor corrective movements to stabilize, position, and orient the animal. Speed normally ranges from 3 to 7 km per hour, but can reach 25 km per hour when the animal is pressed.

Populations of manatees appear to be concentrated in select estuarine and riverine habitats, and there is evidence of long-range, offshore migrations between population centers. Migration seems to be both seasonal, in response to changes in air temperature, and nonseasonal. In their movements, manatees follow established travel routes, preferring channels that are two meters or more in depth and shunning those that are less than one meter in depth. Animals generally swim from one to three meters below the surface of the water. Salinity, tides, and currents affect manatee activities. Ac-

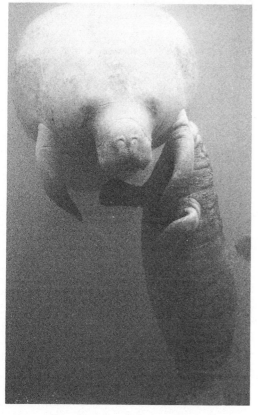

West Indian manatees (*Trichechus manatus*), mother and calf, photos by James A. Powell, Jr.

emergent and floating vegetation is sometimes taken. Staple foods of *T. manatus* in the rivers and estuaries of Citrus County, Florida, are *Hydrilla, Vallisneria, Ceratophyllum, Myriophyllum, Ruppia,* and *Diplanthera.* In salt water, *T. manatus* prefers *Syringodium* and *Thalassia.* According to Husar (1977, 1978*a*, 1978*b*), captive *T. manatus* reportedly eat 30 to 50 kg of food daily, captive adults of the smaller *T. inunguis* consume 9 to 15 kg of leafy vegetables a day, and a free-ranging adult *T. senegalensis* might be expected to consume 8,000 kg of aquatic vegetation in one year. Wild *T. manatus* in Florida usually feed in discrete sessions, during which an animal focuses its attention on a single kind of food plant. There is evidence that manatees time their movements to coincide with the availability of food. They also seem to require fresh water for osmoregulation.

Manatees have exceptional acoustic sensitivity; sound is doubtless the major directional determinant in social interactions. They also make extensive use of their eyes; in clear water, their preferred method of environmental exploration is visual. The prevalence of mouthing in social interaction suggests that manatees possess a chemoreceptive sense by which they can recognize odors in the water. Manatees rest by hanging suspended near the surface of the water, or by lying prone on the bottom. In both positions, the animals lapse into a somnolent state, with eyes closed and body motionless. Comfort activities include stretching, scratching, rubbing, mouth cleaning, and rooting.

Manatees are normally silent, but can emit high-pitched squeals, chirp-squeals, and screams under conditions of fear, aggravation, protest, internal conflict, male sexual arousal, and play. Their vocalizations are probably nonnavigational, and lack ultrasonic signals, pulsed emissions, or directional sound fields; they seem to be more impulsive than communicative. The only predictable vocal exchange between manatees is the alarm duet between a cow and its calf.

Manatees are weakly social, essentially solitary animals. The family unit consists of a cow with its calf. All other

cording to Husar (1978*c*), average submergence time is 259 seconds, but a dive of 980 seconds has been recorded.

Manatees are herbivorous, though some invertebrates and small fish may be incidentally ingested (Powell 1978). The diet consists mainly of submerged, vascular plants, but

ns, with the exception of an estrous female accom-
by courting bulls, are temporary and loosely orga-
d. Groups are randomly made up of juveniles and adults
both sexes. There is no evidence of communal defense or
mutual aid, and little or no indication of a social hierarchy.
Manatees indulge in what appears to be play. Animals exchange gentle nibbles, kisses, and embraces that are age and
sex independent.

Manatees in Florida appear to have no definite breeding
season. Barring the death of an infant, cows probably breed
every 2 to 2½ years. A cow in heat is accompanied by courting bulls for periods of less than 1 week to more than 1
month. During this time, bulls court the cows persistently,
but the cows appear to be receptive for only brief intervals;
while receptive, a cow is promiscuous. Copulation is effected when the bull rolls on his back and grasps the cow to
his stomach from underneath. Male competition for a position next to the estrous cow seems to be the sole source of
aggression among manatees. The gestation period for *T.
manatus* is about 13 months, and the cow seeks the shelter of
a backwater to give birth. The calves at birth are about 100
cm in length and weigh 18 to 27 kg; they are pinkish in
coloration. Cows are not protective of young, uttering only
alarm calls and fleeing when danger threatens. A cow suckles
its young from birth to the dissolution of the parent-offspring
bond, a period of about 1 to 2 years. Males are sexually
mature at 9 to 10 years of age, females at 8 to 9 years (Odell,
Forrester, and Asper 1978). A specimen of *T. manatus* was
still living after 30 years in captivity (Marvin L. Jones, Zoological Society of San Diego, pers. comm.).

All three species of manatee have been ruthlessly hunted
for their meat and hides. According to the IUCN (1976), *T.
inunguis* was originally taken commercially, but when such
hunting was no longer profitable local subsistence hunters
continued to reduce populations. As recently as 1950, over
38,000 manatees were killed for commercial purposes in the
state of Amazonas, Brazil, alone. The West African manatee
has never been subject to commercial hunting, but local subsistence hunting has been persistent, and some animals are
known to be accidentally taken in fixed shark nets. Commercial hunting of the Caribbean manatee began in the 17th
century, and, combined with extensive subsistence hunting,
has resulted in delcines in most areas. In the United States,
where the species is strictly protected by the Marine Mammal
Protection Act and the Endangered Species Act, the major
loss of manatees is now from injuries caused by vandals and
by propellers of power boats.

The U.S. Fish and Wildlife Service (1980) estimated that
about 1,000 manatees occur along Florida coasts and rivers,
and that the population is stable; everywhere else, the populations of this species are reduced and declining. For *T.
inunguis*, the service reported that all evidence indicates declining populations, and the extinction of the species may
occur within the next few decades if hunting pressures continue. For *T. senegalensis*, the service reported that populations seem stable in the lower Niger, the Benue River, and
the Anambra system of creeks; the lower Congo River reportedly still has numerous animals. Elsewhere, populations
have been extirpated or are declining.

The IUCN (1976) classifies *T. senegalensis* and *T. manatus* as vulnerable, and *T. inunguis* as endangered; the USDI
(1980) lists *T. manatus* and *T. inunguis* as endangered, and
T. senegalensis as threatened. *T. inunguis* and *T. manatus*
are on appendix 1 of the CITES and *T. senegalensis* is on
appendix 2.

The geological range of the family Trichechidae is
Pleistocene to Recent in North America, Miocene to Recent
in South America, and Recent in Africa.

West Indian manatees (*Trichechus manatus*), mother and calf, photos by James A. Powell, Jr.

emergent and floating vegetation is sometimes taken. Staple foods of *T. manatus* in the rivers and estuaries of Citrus County, Florida, are *Hydrilla, Vallisneria, Ceratophyllum, Myriophyllum, Ruppia,* and *Diplanthera*. In salt water, *T. manatus* prefers *Syringodium* and *Thalassia*. According to Husar (1977, 1978a, 1978b), captive *T. manatus* reportedly eat 30 to 50 kg of food daily, captive adults of the smaller *T. inunguis* consume 9 to 15 kg of leafy vegetables a day, and a free-ranging adult *T. senegalensis* might be expected to consume 8,000 kg of aquatic vegetation in one year. Wild *T. manatus* in Florida usually feed in discrete sessions, during which an animal focuses its attention on a single kind of food plant. There is evidence that manatees time their movements to coincide with the availability of food. They also seem to require fresh water for osmoregulation.

Manatees have exceptional acoustic sensitivity; sound is doubtless the major directional determinant in social interactions. They also make extensive use of their eyes; in clear water, their preferred method of environmental exploration is visual. The prevalence of mouthing in social interaction suggests that manatees possess a chemoreceptive sense by which they can recognize odors in the water. Manatees rest by hanging suspended near the surface of the water, or by lying prone on the bottom. In both positions, the animals lapse into a somnolent state, with eyes closed and body motionless. Comfort activities include stretching, scratching, rubbing, mouth cleaning, and rooting.

Manatees are normally silent, but can emit high-pitched squeals, chirp-squeals, and screams under conditions of fear, aggravation, protest, internal conflict, male sexual arousal, and play. Their vocalizations are probably nonnavigational, and lack ultrasonic signals, pulsed emissions, or directional sound fields; they seem to be more impulsive than communicative. The only predictable vocal exchange between manatees is the alarm duet between a cow and its calf.

Manatees are weakly social, essentially solitary animals. The family unit consists of a cow with its calf. All other

cording to Husar (1978c), average submergence time is 259 seconds, but a dive of 980 seconds has been recorded.

Manatees are herbivorous, though some invertebrates and small fish may be incidentally ingested (Powell 1978). The diet consists mainly of submerged, vascular plants, but associations, with the exception of an estrous female accompanied by courting bulls, are temporary and loosely organized. Groups are randomly made up of juveniles and adults of both sexes. There is no evidence of communal defense or mutual aid, and little or no indication of a social hierarchy. Manatees indulge in what appears to be play. Animals exchange gentle nibbles, kisses, and embraces that are age and sex independent.

Manatees in Florida appear to have no definite breeding season. Barring the death of an infant, cows probably breed every 2 to 2½ years. A cow in heat is accompanied by courting bulls for periods of less than 1 week to more than 1 month. During this time, bulls court the cows persistently, but the cows appear to be receptive for only brief intervals; while receptive, a cow is promiscuous. Copulation is effected when the bull rolls on his back and grasps the cow to his stomach from underneath. Male competition for a position next to the estrous cow seems to be the sole source of aggression among manatees. The gestation period for *T. manatus* is about 13 months, and the cow seeks the shelter of a backwater to give birth. The calves at birth are about 100 cm in length and weigh 18 to 27 kg; they are pinkish in coloration. Cows are not protective of young, uttering only alarm calls and fleeing when danger threatens. A cow suckles its young from birth to the dissolution of the parent-offspring bond, a period of about 1 to 2 years. Males are sexually mature at 9 to 10 years of age, females at 8 to 9 years (Odell, Forrester, and Asper 1978). A specimen of *T. manatus* was still living after 30 years in captivity (Marvin L. Jones, Zoological Society of San Diego, pers. comm.).

All three species of manatee have been ruthlessly hunted for their meat and hides. According to the IUCN (1976), *T. inunguis* was originally taken commercially, but when such hunting was no longer profitable local subsistence hunters continued to reduce populations. As recently as 1950, over 38,000 manatees were killed for commercial purposes in the state of Amazonas, Brazil, alone. The West African manatee has never been subject to commercial hunting, but local subsistence hunting has been persistent, and some animals are known to be accidentally taken in fixed shark nets. Commercial hunting of the Caribbean manatee began in the 17th century, and, combined with extensive subsistence hunting, has resulted in delcines in most areas. In the United States, where the species is strictly protected by the Marine Mammal Protection Act and the Endangered Species Act, the major loss of manatees is now from injuries caused by vandals and by propellers of power boats.

The U.S. Fish and Wildlife Service (1980) estimated that about 1,000 manatees occur along Florida coasts and rivers, and that the population is stable; everywhere else, the populations of this species are reduced and declining. For *T. inunguis,* the service reported that all evidence indicates declining populations, and the extinction of the species may occur within the next few decades if hunting pressures continue. For *T. senegalensis,* the service reported that populations seem stable in the lower Niger, the Benue River, and the Anambra system of creeks; the lower Congo River reportedly still has numerous animals. Elsewhere, populations have been extirpated or are declining.

The IUCN (1976) classifies *T. senegalensis* and *T. manatus* as vulnerable, and *T. inunguis* as endangered; the USDI (1980) lists *T. manatus* and *T. inunguis* as endangered, and *T. senegalensis* as threatened. *T. inunguis* and *T. manatus* are on appendix 1 of the CITES and *T. senegalensis* is on appendix 2.

The geological range of the family Trichechidae is Pleistocene to Recent in North America, Miocene to Recent in South America, and Recent in Africa.

ORDER PERISSODACTYLA

Odd-toed Ungulates (Hoofed Mammals)

This order of 3 Recent families, 6 genera, and 17 species is native to southeastern Europe, central and southern Asia, parts of the East Indies, Africa, and the region from southern Mexico to Argentina. Introduction by human agency has led to establishment of wild-living populations of two species (*Equus caballus* and *E. asinus*) in certain areas where the order does not naturally occur. Simpson (1945) divided the living Perissodactyla into two suborders: Hippomorpha, for the family Equidae (horses, asses, and zebras); and Ceratomorpha, for the familes Tapiridae (tapirs) and Rhinocerotidae (rhinoceroses).

These are medium-sized to large animals adpated to running (especially members of the family Equidae). All the Recent families are quite distinct, with tapirs and rhinos resembling one another more than either family resembles the horses. The main feature common to all is that the weight of the body is borne on the central digits, with the main axis of the foot passing through the third digit, which is the longest on all four feet. In the horses, only the third digit of each foot is functional; in tapirs, four digits are developed on the forefoot and three on the hind foot; in rhinos, three digits are present on all four feet. The first digit is not present in any Recent forms; it was vestigial in certain fossil species. The terminal digit bones are flattened and triangular, with evenly rounded free edges, and are encased by hooves (some members of the extinct family Chalicotheriidae had clawed digits). Perissodactyls progress on their hooves or on their digits, never on the sole of the foot with the heel touching the ground. The ulna and the fibula are reduced, so the movement of these bones is reduced or lacking. The ankle bone, or astragalus, has only a single, deeply grooved, pulleylike surface for the tibia, and its lower end is nearly flat; the calcaneum, or heel bone, which has a widened lower end, does not articulate with the fibula.

The skin is usually thickened, and sparsely to densely haired. The mammae are located in the region of the groin, and the males do not possess a baculum.

The dental formula for the order is as follows: (i 0–3/0–3, c 0–1/0–1, pm 2–4/2–4, m 3/3) × 2 = 20–44; for the Recent species it is: (i 0–3/0–3, c 0–1/0–1, pm 3–4/3–4, m 3/3) × 2 = 22–44. The canines, when present, are never tusklike in the Recent species. The cheek teeth are arranged in a continuous series; the premolars (at least the rear members of the series) are molarlike in the Recent species; the first cheek tooth is a persistent milk premolar. The grinding teeth are usually complex in structure, massive, and low crowned to high crowned; prominent transverse ridges are present in the cheek teeth of tapirs and rhinos, whereas the cheek teeth of horses, which are grazers rather than browsers, develop high crowns with four main columns and various infoldings. Some fossil species in this order had tubercles on the crowns of the grinding teeth. The skull is usually elongate with an abrupt slope in the back. The nasal bones are expanded posteriorly. Characteristic of the order is the arrangement of openings in the skull by which nerves and blood vessels enter and leave the braincase. The Recent species lack horns with

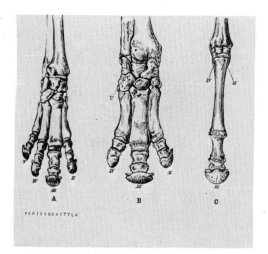

Bones of the forefeet: A. Tapir (*Tapirus indicus*); B. Rhinoceros (*Dicerorhinus sumatrensis*); C. Horse (*Equus caballus*); photos from *Mammalia*, Beddard.

true bony cores, although roughened cushions on the nasal bones of the skull bear horns in rhinos.

The development of the foot is a specialization that, in its highest form in the horses, enables the animals to be swift and strong runners. The foot is not developed to such an extent in the rhinos and tapirs. However, rhinos can run rapidly for short distances, and tapirs can also run well, although they usually inhabit a type of terrain that permits them to plunge into dense cover or water to escape their enemies.

Rhinos and horses usually live on grassy plains or in open scrub country, while tapirs are found in humid tropical forests. Members of this order are strict herbivores, and can be classified as either browsers or grazers depending on how they feed. The structure of their lips and teeth facilitates the obtaining and chewing of coarse vegetable food. Adulthood in these animals is attained in four to six years, and individuals may live five to seven times that long.

Except for the horses, which are the only members of this group that have been domesticated, perissodactyls are not very numerous, with regard to either species or individual animals, in the wild today. The living species represent mere remnants of a group that at one time was extremely diversified and widely distributed over the world. Various members of the Equidae, however, have been second probably only to the cattle (Bovidae) in the economic life of humans.

Perissodactyls flourished in early and middle Tertiary times. The genera still living represent the end of a major evolutionary line, and each of these genera has one or more species that may be considered endangered. Twelve families of living and extinct forms have generally been recognized; nine of these have become extinct. The geological range of the order is early Eocene to Recent.

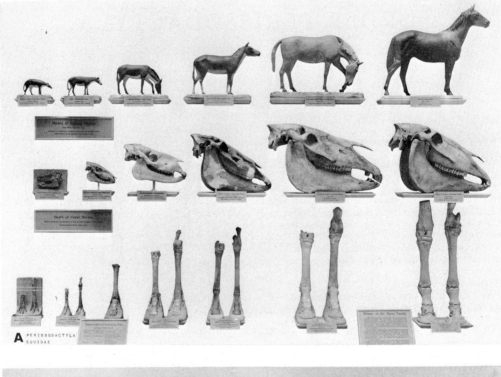

A. Exhibition in the Chicago Natural History Museum showing the evolution of horses. B. A restoration of the primitive three-toed horse (*Mesohippus* sp.) that inhabited western America and other regions. Photos from Field Museum of Natural History.

PERISSODACTYLA; *Family EQUIDAE; Genus EQUUS*
Linnaeus, 1758

Horses, Zebras, Asses

The single Recent genus, *Equus,* contains four subgenera and eight species (Corbet 1978; Ellerman and Morrison-Scott 1966; Meester and Setzer 1971):

subgenus *Equus*

E. caballus (horse), probably once found in the wild throughout the steppe zone from Poland and Hungary to Mongolia, now occurs in domestication throughout the world and feral populations established in many areas;

subgenus *Asinus*

E. hemionus (kulan, onager), desert and dry steppe zone from Syria and Iraq to Manchuria and western India;
E. kiang (kiang), Tibet and adjacent highland regions;
E. asinus (African wild ass, donkey, burro), probably once found in the wild from Morocco to Somalia and possibly the Arabian Peninsula, now occurs in domestication throughout the world and feral populations established in some areas;

subgenus *Dolichohippus*

E. grevyi (Grevy's zebra), southern and eastern Ethiopia, Somalia, northern Kenya;

subgenus *Hippotigris*

E. zebra (mountain zebra), southwestern Angola, Namibia, western and southern South Africa;
E. quagga (quagga), formerly found in South Africa;
E. burchelli (Burchell's zebra), open country from southern Ethiopia to central Angola and eastern South Africa.

Bennett (1980) concluded that the horse, quagga, and zebras should be united in a single subgenus, *Equus,* and also treated *E. onager* of Turkmen, Iran, Pakistan, and Afghanistan as a species distinct from *E. hemionus.* In contrast, Groves and Willoughby (1981) divided *Equus* into six subgenera; *Equus,* for the species *E. caballus; Asinus,* for *E. asinus; Hemionus,* for *E. hemionus* (with *onager* treated as a subspecies); *Dolichohippus,* for *E. grevyi; Hippotigris,* for *E. zebra;* and *Quagga,* for *E. quagga* and *E. burchelli.* The names *E. caballus* and *E. asinus* are based on domestic animals, and some authorities, such as Corbet (1978), prefer to use the names *E. ferus* for the wild horse and *E. africanus* for the wild African ass. Both R. E. Rau (1978) and Grubb (1981) suggested that *E. quagga* and *E. burchelli* may be conspecific.

The general body form ranges from thick headed, short legged, and stocky to slender headed and graceful. The size is variable, especially among the domesticated forms. The wild species measure from 1,000 to 1,500 mm high at the shoulders. The tail is moderately long, with the hairs reaching at least to the middle of the leg when the tail is hanging down. Equids are heavily haired, but the length of the hair is variable. Most species have a mane on the neck, and a lock of hair on the forepart of the head, known as the forelock. The two mammae of the female are located in the groin region.

The Recent species have only one functional digit, the third. The terminal digit bone on each foot is widened and evenly rounded or spade shaped; equids walk on the tips of their toes. The radius and ulna are united, although the ulna is greatly reduced in size, so all the weight is born on the radius. In the hind leg, the enlarged tibia supports the weight and the fibula is reduced and fused to the tibia.

The dental formula in the Recent forms is: (i 3/3, c 1/1, pm 3–4/3, m 3/3) $\times$ 2 = 40 to 42. The incisors are shaped like chisels, the enamel on the tips folding inward to form a pit, or "mark," that is worn off in early life. The first permanent incisors appear at about 2½ to 3 years of age in domestic horses. The canine teeth are vestigial or absent in female equids. The premolars are similar to the molars, and they are permanent, except for the first, which, when present, represents a persistent milk tooth. This first upper premolar is often reduced or entirely absent. The cheek teeth have a complex structure. They are high crowned, with four main columns and various infoldings with much cement. The grinding teeth of certain fossil species had four tubercles and little cement. Age in horses is often estimated by the degree of wear of the surface pattern on the cheek teeth, but the rapidity of wear depends to a great extent on the abrasive character of the food. The skull is long, and the nasal bones are long and narrow, freely projecting anteriorly to points; they are always hornless. The eye socket is behind the teeth, thus avoiding pressure on the eye by the long molars.

All members of the Equidae are relatively swift runners, although asses may be particularly slow in their movements, especially domesticated forms. In the wild, they are alert for enemies, and since they are not very effective fighters, they usually run away. In fighting among themselves, or in attempting a defense, they kick with the hind feet, occasionally strike with the forefeet, and sometimes bite. Their teeth, although not adapted for lacerating or tearing (except in male zebras, which have pointed canines), can give quite hard pinches. Unfortunately, the curiosity of some of the species has led to their extermination or depletion.

Equids, which have been termed "admirable running machines," are active day and night, but mainly during the evening. The wild species are known to associate and interbreed with domesticated forms. The members of this family are entirely vegetarian in habits, feeding mainly on grass, although some browsing is done. Most forms drink water daily, but they can go without water for long periods of time.

In the African zebras (*E. grevyi, E. zebra, E. quagga,* and *E. burchelli*), head and body length is 2,000 to 2,400 mm, tail length is 470 to 570 mm, and shoulder height is 1,200 to 1,400 mm. The weight is about 350 kg. All of these zebras are characterized by their coloration, which consists of variously arranged dark bands on portions of, or all over, their bodies. All living species of zebra are well distinguished from each other by external characters such as ear size and shape, body size, and patterns of striping. *E. quagga* had much less pronounced striping than other zebras, occasionally being almost stripeless; *E. grevyi* has very narrow, close stripes that extend down the legs to the hooves; *E. zebra* has very broad stripes that extend onto the legs but not onto the belly; and *E. burchelli* has a widely varying striping pattern depending on the subspecies involved.

The species *E. burchelli* and *E. zebra* live in coherent family groups consisting of one stallion with one to several mares and their young, and in stallion groups. These social units are not territorial, and the animals move freely in rather large home ranges that they share with conspecifics and whose size depends on the ecological conditions of the area. In both species, the family groups were found to be permanent units as far as the adult members are concerned. Mares normally stay in the same family group for their lifetime, even when they are very old and sick. Old stallions may be replaced by younger, healthier ones; the old stallion then joins a stallion group. The stallion groups are more variable in their composition than the family groups. Some individuals, however, stay together for many years (Klingel 1974; Penzhorn 1979).

Klingel (1974) found the social organization of *E. grevyi*

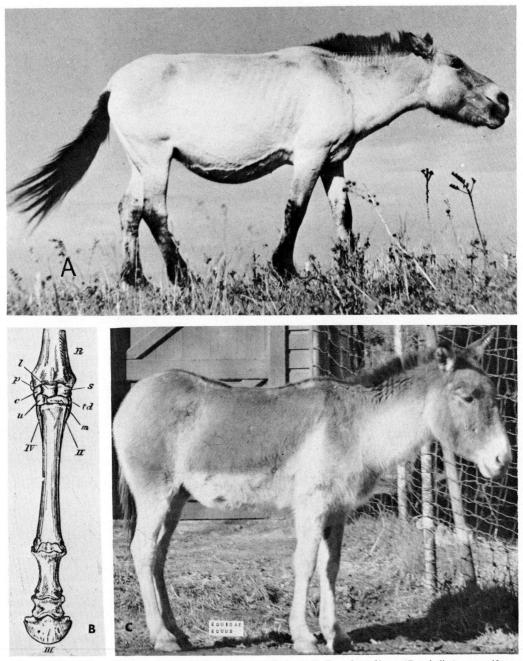

A. Przewalski's wild horse (*Equus caballus*), photo by Bernhard Grzimek. B. Front foot of horse (*E. caballus*): c. cuneiform; l. lunar; m. magnum; p. pisiform; r. radius; s. scaphoid; td. trapezoid; u. unciform; II & IV. rudimentary metacarpals; photo from *Mammalia,* Beddard. C. Kiang (*E. kiang*), photo by Ernest P. Walker.

to be completely different from that of *E. burchelli* and *E. zebra.* In this species, there are no permanent bonds between any two or more adult animals. Instead, these zebras are solitary, or live in a variety of different associations, for example, stallion groups, groups of mares and foals, and mixed herds. The only strong bond in these associations is that between a mare and her foal or foals. Some of the solitary stallions are territorial. Klingel found the territories of these stallions to be among the largest ever recorded for any her-

bivore (2.7 to 10.5 sq km in northern Kenya). Another unusual feature in *E. grevyi* is that the territorial stallion is highly tolerant of other stallions, and permits them to enter his territory. These intruding stallions, however, respect the territorial stallion, and do not interfere with him when he is engaged in mating activity.

The asses (*E. asinus* and *E. hemionus*) are found mostly in desert plains, sparsely covered with low shrub. Klingel (1977) found that their social organization is virtually identi-

Grevy's zebra (*Equus grevyi*), photo by Leonard Lee Rue III.

cal to that of *E. grevyi*. Territories for both species are very large, measuring as much as 20 sq km for *E. asinus* and probably even larger for *E. hemionus*. Members of both species live in unstable groups of variable composition, and there is no indication of permanent bonds between any adult individuals.

Asses can go for longer periods without water than can other species of equid, and they are remarkably capable of surviving on a minimum of food and of working under hot and difficult conditions. Because they are sure-footed, persistent, and endurable, they are useful as pack animals. They are capable of carrying well over 100 kg for days with little food. *E. asinus* has been domesticated for centuries. Its movements are much slower than those of other members of *Equus*, but in portions of the world where horses do not thrive, or where poverty prevents local people from obtaining horses, asses have been extensively used for riding and as beasts of burden. Asses are frequently mentioned in the Bible. They have been used in the western United States by prospectors to pack supplies into regions that were difficult of access. Mules are the hybrid offspring of a male ass and a female horse; the cross between a female ass and a male horse is known as a hinny. Both mules and hinnies are usually sterile.

The horse (*E. caballus*) appears not to be territorial, and to have a social organization similar to that of *E. zebra* and *E.*

burchelli. In the horse, there are bachelor stallion groups, and groups consisting of a stallion and one or more mares. Feist and McCullough (1976) reported that 270 wild horses in the Pryor Mountain Wild Horse Range in Wyoming and Montana were organized into 44 harems, with a dominant stallion, 1 to 3 mares, and offspring, in each. There were also 23 bachelor groups averaging 1.8 but up to 8 stallions. Harem groups were long lasting, most changes involving immature females. All groups used home ranges, but were not territorial, and there was much overlap between groups in the home ranges. In harem groups, the stallion fought to get mares from other groups, but almost always stallion masters managed to retain their mares. The bachelor groups also have a lead stallion who is dominant over the rest and treats his group of males as if they were a harem. He will fight other males who come too close, and within the stallion group itself there is a clear dominance hierarchy.

The tarpan, the wild horse that formerly occurred on the steppes of southeastern Europe, is believed to be the ancestral form of the domesticated horse; it disappeared in the western part of its range in the early Middle Ages, and finally was exterminated from the wild in southern Russia by the 19th century. It was an animal of moderate size, gray colored with a black tail and mane. In the 1930s a group of dedicated horse lovers began a program to try to ''breed back'' to the tarpan from existing domestic horses. They selected breeds

African wild asses (*Equus asinus*), top, photo by Lothar Schlawe. Domestic asses (*E. asinus*), bottom, photo from San Diego Zoological Garden.

Kulan (*Equus hemionus*), photo by Lothar Schlawe.

that most closely resembled descriptions of the original tarpan and by selective breeding produced an animal that probably closely approximates the original wild European horse. Today, these tarpanlike horses are established in some European zoos and in the United States, and give an idea of what the ancestral horse looked like.

Przewalski's horse of Mongolia and Sinkiang is the only true wild horse still in existence in its natural habitat. It is a heavily built animal, light yellow in coloration, with a pronounced, stiff black mane. In the 1930s, this animal declined almost to the point of extinction because of poor pasture, ice-encrusted ground, and excessive hunting by local tribes.

Female equids come into heat several times a year, or until they become pregnant. Each period of estrus in *E. burchelli* lasts 2 to 9 days, and females of that species attain sexual maturity in their second or third year of life (Grubb 1981). Some species give birth every two years; the number of young is usually one (twin births are generally abortive), and the gestation period is 11 to 13 months. The potential life span is 25 to 35 years.

The geological range of the family Equidae is early Eocene to Pleistocene in North America, early Eocene to Recent in Europe, Miocene to Recent in Asia, Pliocene to Recent in Africa, and Pleistocene in South America. The geological range of the genus *Equus* is late Pliocene to Recent. Although equids passed through most of their evolutionary history in North America, none of the Pleistocene species left any descendants. Horses did not appear again in North America until after the Spanish expeditions of De Soto and Coronado in 1541. American Indians probably first acquired horses from the early 17th-century stock-raising settlements of the Southwest. Vast populations of feral horses became established in western North America by the 19th century, but subsequently declined drastically through hunting and habitat loss.

Beginning with such forms as *Eohippus* (from the Eocene) and proceeding through *Mesohippus, Protohippus*, and others, culminating in *Equus* at present, the following modifications may be noted in the evolution of this line. The early species were browsers, and had teeth well suited for that purpose; the grinding teeth bore conical cusps and were low crowned, rooted, and free of cement. When grazing forms replaced the browsers, the teeth became progressively more high crowned, prismatic, and cement covered. *Mesohippus*, from the Oligocene of North America, was the first equid in which the "mark" or enamel pit became established (only in the upper incisors of this genus). During the course of equid evolution, the size of the forms became larger, the neck was extended, the arched back became straightened, the limbs became elongated, and the feet were lengthened, the third (middle) digit growing and changing shape until it bore all the weight. The first digit was lost initially, then the fifth; the second and fourth digits were reduced to dewclaws, and finally to splints. An animal with long legs must have a long enough neck to enable the mouth to reach the ground; in the evolution of the equids, the elongation of the head and neck paralleled the lengthening of the legs.

Equids are among those groups of mammals most seriously affected by people. As mentioned above, the wild horse was exterminated in Europe by the 19th century, but the form known as Przewalski's horse still survives in small numbers along the border between Mongolia and Sinkiang. It has also been successfully bred in captivity, and, according to the "Census of Rare Animals in Captivity" in the 1980 *International Zoo Yearbook*, there are now 338 individuals in zoos. Przewalski's horse is classified as endangered by the IUCN (1976) and the USDI (1980), and is on appendix 1 of the CITES.

The wild asses of Asia and Africa have declined drastically through such factors as excessive hunting, habitat deterioration, transmission of disease from livestock, and interbreeding with the domestic donkey. *E. asinus* now apparently survives only in small parts of Ethiopia and Somalia. The IUCN (1972, 1976) classifies *E. asinus* as endangered and *E.*

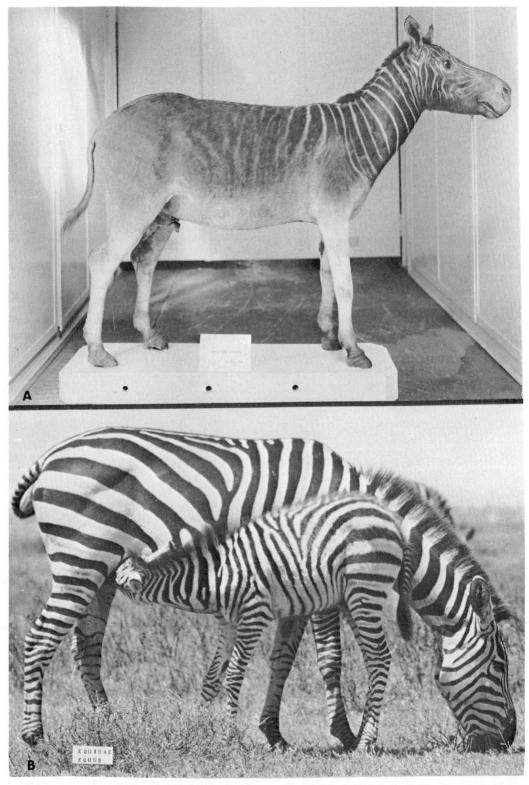

A. Quagga (*Equus quagga*), photo from Rijksmuseum van Natuurlijke Historie, Leiden. B. Burchell's zebras (*E. burchelli*), photo by C. A. Spinage.

hemionus as vulnerable, though it specifies the subspecies *E. h. khur* (Pakistan and western India) and *E. h. hemihippus* (Syria) as endangered; the latter subspecies may actually be extinct. The USDI (1980) lists *E. asinus* and *E. hemionus* as endangered. *E. hemionus* is on appendix 2 of the CITES, except for the subspecies *E. h. khur* and *E. h. hemionus* (Mongolia and Manchuria), which are on appendix 1.

The zebras of Africa also have declined, mainly because of hunting by people for their skins and competition for habitat with domestic livestock. Both *E. quagga* and the subspecies *E. burchelli burchelli* of South Africa are extinct, the last survivors having died in zoos in 1883 and 1909, respectively (Goodwin and Goodwin 1973). *E. zebra* is classified as vulnerable by the IUCN (1978). It has two subspecies: *E. z. zebra* (South Africa) is listed as endangered by the USDI (1980) and is on appendix 1 of the CITES, and *E. z hartmannae* (Angola and Namibia) is listed as threatened by the USDI and is on appendix 2 of the CITES. *E. grevyi* apparently underwent a severe decline within the last decade, because of demand for its particularly attractive skin, and has been entirely exterminated in Somalia. It is classified as endangered by the IUCN (1978) and as threatened by the USDI (1980), and is on appendix 1 of the CITES.

PERISSODACTYLA; *Family TAPIRIDAE; Genus TAPIRUS Brisson, 1762*

Tapirs

The single Recent genus, *Tapirus*, contains three subgenera and four species (Hall 1981; Cabrera 1961; Ellerman and Morrison-Scott 1966; Grzimek 1975; IUCN 1972, 1973):

subgenus *Tapirus*

T. terrestris, Colombia and Venezuela to northern Argentina and southern Brazil;
T. pinchaque, the Andes from northwestern Venezuela to northwestern Peru;

subgenus *Tapirella*

T. bairdii, southern Mexico to Ecuador;

subgenus *Acrocodia*

T. indicus, southern Burma and Thailand, Malay Peninsula, Sumatra.

According to Medway (1977), *T. indicus* was also present on Borneo until within the last 8,000 years, and possibly survived there into historical time.

The general form of all tapirs is rounded in back and tapering in front, well suited for rapid movement through thick underbrush. These animals are about the size of a donkey. Head and body length is 1,800 to almost 2,500 mm, tail length is 50 to 100 mm, shoulder height is 735 to 1,030 mm, and weight is usually 225 to about 300 kg. Short, bristly hairs are scattered on the body; they are thickest on the mountain tapir (*T. pinchaque*). A low, narrow mane, which is not always conspicuous, is present in *T. bairdii* and *T. terrestris*. The skin is quite thin in the mountain tapir but thick in the other species. The Asiatic tapir (*T. indicus*) is readily distinguished by its color pattern; in this species, the front half of the body and the hind legs are black, and the rear half, above the legs, is white. This black and white pattern renders the animal practically invisible at night in the jungle, when moonlight on the vegetation assumes the same black and white pattern. *T. terrestris*, *T. pinchaque*, and *T. bairdii* are dark brown to reddish brown above, and often paler below.

The snout and upper lips of all tapirs are projected into a short, fleshy proboscis; the transverse nostrils are located at its tip. The proboscis is more elongated in the New World species than in the Asiatic tapir. The eyes are small and flush with the side of the head; the ears are oval, erect, and not very mobile. The legs are rather short and slender. The radius and ulna are separate and about equally developed, and the fibula is complete. The forefoot has three main digits and a smaller one (the fifth), for a total of four; the small digit is functional only on soft ground. The hind foot has three digits. The tail is short and thick. Female tapirs have a single pair of mammae, located in the region of the groin.

The dental formula in Recent species is: (i 3/3, c 1/1, pm 4/3–4, m 3/3) $\times$ 2 = 42–44. The incisors are shaped like chisels; the third upper incisor is shaped like a canine, but is larger than the true canine, and the third lower incisor is reduced in size. The canines are conical, and separated from the premolars by a space. In Recent species the posterior three premolars resemble the true molars in size and shape; in extinct species they were usually simpler than the molars. All of the cheek teeth lack cement and are low crowned, with a series of transverse ridges and cusps. Horns are absent in all species. The skull is relatively short and laterally compressed, with a high braincase and a convex profile. The nasal bones are short, arched, and freely projecting.

The altitudinal range of tapirs is from sea level to about 4,500 meters. *T. pinchaque* has never been taken at altitudes of under 2,000 meters (Schauenberg 1969). Tapirs may live in nearly any wooded or grassy habitat where there is a permanent supply of water. They usually shelter in forests and thickets by day, and emerge at night to feed in bordering grassy or shrubby areas. They are agile in closed or open habitat, and in or under water. They are good hillclimbers, runners, sliders, waders, divers, and swimmers. They generally walk with the snout close to the ground, and are fond of splashing in water or wallowing in mud. Tapirs are generally shy and docile, and seek refuge in water or crash off into the brush when threatened, but they can and will defend themselves by biting. They possess keen senses of hearing and scent.

These hoofed mammals wear paths to permanent bodies of water in areas where their populations are dense, and human engineers sometimes follow their trails up the sides of mountains in the construction of roads. Tapirs consume aquatic vegetation and the leaves, buds, twigs, and fruits of low-growing terrestrial plants, but in any particular habitat they eat mainly the green shoots of the most common browsing plants. They also graze for food, and at times have been known to damage young corn and other grains, especially in Mexico and Central America.

Except for females with young, all tapirs are solitary animals. Breeding in the Asiatic species occurs in April or May, and a single young, weighing about 6 to 7 kg, is born after a gestation period of 390–95 days. The young stays with its mother for at least 6 to 8 months, by which time it is nearly adult size. A female in her prime probably produces one calf every second year (Lekagul and McNeely 1977). In the South American species, breeding apparently takes place at any time of year. The gestation period is 390–400 days, and the number of offspring is one, rarely two. Young tapirs of all species are dark reddish brown, with yellow and white stripes and spots; this juvenile pattern is usually lost by the sixth month of age. A captive *T. terrestris* lived for 35 years (Marvin L. Jones, Zoological Society of San Diego, pers. comm.).

In some areas, tapirs are hunted extensively for food and

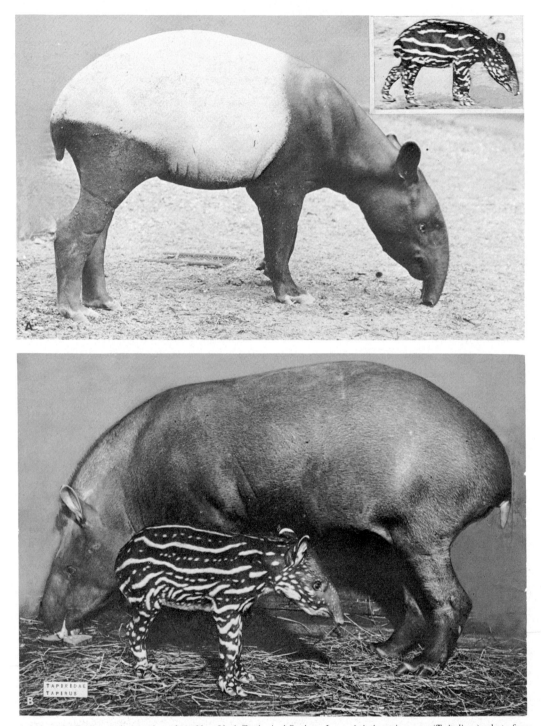

Asiatic tapir (*Tapirus indicus*), photo from New York Zoological Society. Inset: Asiatic tapir young (*T. indicus*), photo from Zoological Society of London. B. Brazilian tapir (*T. terrestris*) and 8-day-old young, photo by Ernest P. Walker.

sport (some Indian tribes, however, do not kill tapirs for religious reasons). In addition, the populations of all species have declined in recent years, mainly because of the clearing of the forests by humans for agricultural purposes. The IUCN (1972, 1973) classifies *T. pinchaque*, *T. bairdii*, and *T. terrestris* as endangered. The USDI (1980) lists all four species as endangered. *T. pinchaque*, *T. bairdii*, and *T. indicus* are on appendix 1 of the CITES, and *T. terrestris* is on appendix 2.

The discontinuous distribution of modern tapirs suggests that they represent the remnants of a once widespread family. The fossil record shows that tapirs originated in the Northern

Hemisphere and at various times occupied the land masses between where the present-day Asiatic and South American forms exist. The fact that tapirs are now living in both the American and Asiatic tropics supports the many pieces of evidence that the two continents were connected rather recently as measured by geological time, and that during the period when the two were joined the climate was mild to warm in the northern portion of these continents, making conditions favorable for animals to move from one continent to the other. Subsequently, the continents were separated at the Bering Strait, or any other land bridge that might have existed, and the climate changed so the animals were prevented from moving between the two continents by the strait and by colder climatic conditions.

The geological range of the Tapiridae is early Eocene to Recent in North America, early Oligocene to Pleistocene in Europe, Pleistocene to Recent in South America, and Miocene to Recent in Asia. The geological range of the genus *Tapirus* is late Miocene to Recent. *Megatapirus* is the only extinct tapir that has been found in Pleistocene deposits of the Old World; it is known from Szechwan Province of China. It was much larger than any Recent tapir and had a shorter and deeper skull.

Young Asiatic tapir (*Tapirus indicus*) in a position unusual among perissodactyls, photo by Ernest P. Walker.

A. Mountain tapir (*Tapirus pinchaque*); B. Brazilian tapir (*T. terrestris*); photos by Bernhard Grzimek.

PERISSODACTYLA; *Family RHINOCEROTIDAE*

Rhinoceroses

This family of four Recent genera and five species occurred in historical time in most of Africa south of the Sahara, perhaps in parts of North Africa, and in south-central and southeastern Asia. The sequence of genera presented here follows that suggested by Groves (1967*b*, 1975*b*), who recognized two subfamilies: Rhinocerotinae, for *Dicerorhinus* and *Rhinoceros;* and Dicerotinae, for *Diceros* and *Ceratotherium.*

Rhinoceroses have a massive body, a large head, one or two horns, a short neck, a broad chest, and short, stumpy legs. The radius and ulna, and the tibia and fibula, are only slightly movable, but are well developed and separate. The forefoot has three digits (four in some fossil forms), and the hind foot also has three; the hooves are distinct and separate

Sumatran rhinoceros (*Dicerorhinus sumatrensis*), top; Indian rhinoceros (*Rhinoceros unicornis*), bottom; photos by Lothar Schlawe.

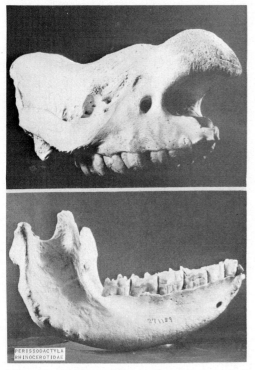

African black rhinoceros (*Diceros bicornis*), photos by P. F. Wright of skull in U.S. National Museum.

for each digit. The upper lip is prehensile in two genera (*Rhinoceros* and *Diceros*). The small eyes are located on the side of the head, midway between the nostrils and the ears; the ears are faily short, but prominent and erect. The thick skin, which is scantily haired and wrinkled, is furrowed or pleated, having the appearance of riveted armor plate in some species. The tail bears stiff bristles.

The length of the head and body is 200 to 420 cm, the length of the tail is 60 to 75 cm, and the height at the shoulders is 100 to 200 cm. Females are smaller than males. Adults weigh from 1,000 to 3,500 kg. The coloration is grayish to brownish, but the true color is often concealed by a coating of mud.

The dental formula in the family is: (i 0–2/0–1, c 0/0–1, pm 3–4/3–4, m 3/3) × 2 = 24–34. The incisors and canines are vestigial. The premolars resemble the molars (except the small first premolar). The cheek teeth, which are high crowned in *Ceratotherium* (the only species of Recent rhino that grazes rather than browses) and fairly low crowned in the other Recent genera, are marked with transverse ridges of enamel. The skull, which is elongate and elevated posteriorly, has a small braincase. The nasal bones project freely beyond the skull. One or two median conical horns are present in rhinos, although they may be short or obscure in some forms (they were not present in some extinct species). If there is only one horn, it is a borne on the nasal bones; if there are two horns, the posterior one is over the frontal bones of the skull. These horns are dermal in origin; although solid, they are composed of compressed keratin of a fibrous nature.

Rhinos generally inhabit savannahs, shrubby regions, and dense forests in tropical and subtropical regions. The African species usually live in more open areas than do the Asiatic forms. Members of this family are active mainly during the evening, through the night, and in the early morning, resting during the day in heavy cover that may be several km from the waterholes. They penetrate dense thorn thickets by sheer force. Rhinos sleep in both standing and recumbent positions, and are fond of wallowing in muddy pools and sandy river beds. They run with a cumbersome motion, reaching their top speed at a canter, that is, at a gait resembling a gallop but with moderate and easy bounds or leaps. *Diceros* can attain speeds of up to 45 km per hour for short distances. Rhinos are usually timid, but can be ferocious at bay. They sometimes charge an enemy, although their attack is often poorly directed. They may grunt or squeal when excited. Vision is poor, but smell and hearing are apparently acute.

These mammals are often accompanied by tick birds and egrets, which act as sentinels and feed on external parasites that often infest rhinos. The large cats prey on young rhinos, but the adults apparently have no enemies other than humans.

Rhinos are generally restricted to areas where a daily trip to water is possible. Their paths between the watering and feeding places often pass through tunnels in the brush. These animals eat a variety of vegetation, but succulent plants comprise the bulk. Rhinos drop their dung in well-defined piles and often furrow the area around the piles with their horns; these piles may be scattered afterward. They are believed to act as "sign posts" or territory markers (urination spots and rubbing sticks also seem to serve this purpose).

During the breeding season, a pair of rhinos may be together for 4 months; females may give birth every 2 years. The gestation period is about 420 to 570 days. The single offspring is active soon after birth, and remains with the mother until the next youngster is born. The mother sometimes guides the baby with her horn. Rhinos have a potential lifespan of almost 50 years.

Humans have hunted rhinoceroses extensively because nearly all parts of the animal are used in folk medicine. Nearly all species of the family are threatened at the present time; some are verging on extinction.

The geological range of this family is middle Eocene to Pleistocene in Europe, late Eocene to Recent in Asia, Miocene to Recent in Africa, and late Eocene to Pliocene in North America. The living genera are known from at least the Pleistocene; *Dicerorhinus* is recorded from the lower Oligocene of Europe and Asia. The only genus of extinct rhinoceros known from the Pleistocene of the Old World is *Elasmotherium,* a huge animal from Siberia. This family was more dominant in earlier geological epochs than it is at present. At least 30 genera referable to this family are known from past eopchs, 1 of them (*Baluchitherium*) being the largest land mammal yet known.

PERISSODACTYLA; RHINOCEROTIDAE; **Genus**
DICERORHINUS *Gloger, 1841*

Sumatran Rhinoceros, Hairy Rhinoceros

The single species, *D. sumatrensis,* originally occurred from Assam and southeastern Bangladesh to the Malay Peninsula and possibly Viet Nam, and on Sumatra and Borneo (Groves and Kurt 1972; Lekagul and McNeely 1977; Rookmaaker 1977, 1980; Van Strien 1975).

This is a short-bodied, two-horned rhinoceros, with the frontal horn often so inconspicuous that it appears to be single horned. The nasal (posterior) horn is generally short, the record length being 381 mm. Head and body length is 235 to 320 cm, shoulder height is 110 to 150 cm, and estimated weight is 1,000 to 2,000 kg. The facial skin is characteristically wrinkled around the eye, but the muzzle is rounded and unwrinkled due to heavy keratinization. The skin on

Sumatran rhinoceros (*Dicerorhinus sumatrensis*), photo by Erna Mohr.

the body of this animal is folded, as in *Rhinoceros,* so it has the appearance of being armor plated. The ears are fringed with hair, and the body hair is plentiful, coarse, and bristlelike. Coloration is brown or dark gray.

According to Groves and Kurt (1972), the habitat of the Sumatran rhinoceros is mainly hilly country, near water; it needs high humidity for survival. This animal seems to be attracted to secondary growth, where it feeds sometimes on cultivated plants. It inhabits both tropical rain forest and mountain moss forest where it can climb steep slopes with agility. *Dicerorhinus* swims well, and has even been known to swim in the sea. Van Strien (1975) concluded that *D. sumatrensis* is an animal that can live in a wide variety of habitats, from swamps at sea level to high in the mountains. Older literature, however, indicated that the hill tracts were probably the preferred habitat of the species.

According to Groves and Kurt (1972), the Sumatran rhino engages in seasonal movements, keeping to hilly country when the lowlands are flooded during the rains and descending when the weather has become cool near the end of the rains. In March, the rhino returns to the high ground, possibly to escape the hoards of horse flies that abound at lower levels during the dry season.

Dicerorhinus feeds on fruit, leaves, twigs, and bark; wild mangoes, bamboos, and figs are especially favored. All species of plants found in second growth are eaten; the rhino uses its horn to break down small trees. *Dicerorhinus* is fond of salt, and the genus is known to visit salt licks. Feeding is generally before dawn and after sunset, mostly at night. Much of the day is spent in wallows.

Male Sumatran rhinos seem to be more nomadic than females. Females apparently live in territories, the diameters of which may be some 500 to 700 meters. Each territory is surrounded by feeding grounds, which are visited by several different animals. Within the territory is a dense system of tracks leading to and from a wallow, which is usually located on a mountaintop or a catchment area of a small stream.

Home range for the female is 2 to 3.5 km in diameter. Both sexes usually wander about singly (Van Strien 1975).

Van Strien (1975) wrote that reproductive data are incomplete, but that all authorities agree that *Dicerorhinus* is a slow breeder and has only one calf at a time. The gestation period has often been said to be 8 months, but Groves and Kurt (1972) thought this unlikely, considering the 15- to 18-month gestation period of other rhinos. A newborn young weighed about 23 kg and measured 914 mm in length. The horn of this baby was 20 mm in height, the coloration of the body was nearly black, and the hair over the whole body was short, crisp, and black. A captive specimen lived for 32 years and 8 months (Marvin L. Jones, Zoological Society of San Diego, pers. comm.).

The Sumatran rhino is considered endangered by the IUCN (1976) and the USDI (1980), and is on appendix 1 of the CITES. It has been exterminated in much of its original range, and the world population is probably in the low hundreds. The principal cause of the decline has been overhunting for supposed aphrodisiac and medicinal products that are made from its horn and other parts of the carcass by some peoples of the Orient.

PERISSODACTYLA; RHINOCEROTIDAE; **Genus**
***RHINOCEROS** Linnaeus, 1758*

Asian One-horned Rhinoceroses

There are two species (Ellerman and Morrison-Scott 1966; Rookmaaker 1980; IUCN 1976, 1978):

R. unicornis (Greater Indian rhinoceros), originally found in northern Pakistan, much of northern India, Nepal, and Assam;

R. sondaicus (Javan rhinoceros), originally found from

Greater Indian rhinoceroses (*Rhinoceros unicornis*), photo by Dorothy Y. Mackenzie.

Sikkim and eastern India to Viet Nam and possibly southern China, and on the Malay Peninsula, Sumatra, and Java.

Head and body length is 210 to 420 cm, tail length is 60 to 75 cm, shoulder height is 110 to 200 cm, and weight is 2,000 to 4,000 kg. *R. unicornis* is very much larger than *R. sondaicus*. The skin is practically naked except for a fringe of stiff hairs around the ears and the tip of the tail. The skin of *R. unicornis* has large convex tubercles, whereas that of *R. sondaicus* is covered with small, polygonal, scalelike disks. Coloration is gray to black, with a pinkish cast on the undersurfaces and on the margins of the skin folds. These rhinos are large, awkward-looking creatures, with large heads, short, tubular legs, small eyes, and wide nostrils. Members of this genus have a single horn on the nose, which is composed of agglutinated hairs and has no firm attachment to the bones of the skull.

These rhinos may be distinguished from their African relatives by their skin, which has a number of loose folds, giving the animal the appearance of wearing armor. The African rhinos lack such folds. *R. unicornis* has a fold of skin that does not continue across the back of the neck; *R. sondaicus*, on the other hand, has a fold that continues across the midline of the back.

These rhinos live in tall grass and reed beds in swampy jungles. They remain more or less solitary throughout their lives. They usually seek to escape rather than attack an enemy. When wounded or when a calf is threatened, they may charge. In such defensive charges, contrary to popular belief, they use their sharp-pointed lower tusks, not the horn. They remain near water, in which they bathe daily; they also enjoy wallowing in mud. Mornings and evenings are the chief feeding periods, and the remainder of the day is spent in slumber. Their diet consists of grass, reeds, and twigs.

The birth of a single young takes place between the end of February and the end of April, about 16 months after mating occurs. The young have a head and body length of 100 to 120 cm, shoulder height of 60 cm, and weight of 34 to 75 kg. The young nurse for 2 years. The life span may be 50 years or more.

Both species have been greatly reduced in numbers and distribution, mainly through habitat loss and ruthless hunting for their horns, which many persons believe to have medicinal properties. *R. unicornis* is now restricted to a few places in Nepal and eastern India, and numbers about 1,000 individuals. *R. sondaicus*, among the world's most critically endangered species, may now survive only in the Udjung Kulon Reserve in western Java, where about 50 individuals are present. There also, however, have been reports of the species in southern Laos and Tenasserim (Lekagul and McNeely 1977; Rookmaaker 1980). Both species are classified as endangered by the IUCN (1976, 1978) and the USDI (1980), and are on appendix 1 of the CITES.

PERISSODACTYLA; RHINOCEROTIDAE; **Genus DICEROS**
Gray, 1821

Black Rhinoceros

The single species, *D. bicornis*, originally occurred throughout eastern and southern Africa, and in the north ranged as far as northeastern Sudan and at least as far west as northeastern Nigeria (Meester and Setzer 1971).

Head and body length is 300 to 375 cm, tail length is about 70 cm, shoulder height is 140 to 150 cm, and weight is 1,000 to 1,800 kg. The anterior horn is larger than the posterior one, averaging about 50 cm in length; sometimes the beginning of a third posterior horn is present. Both this rhino and the white rhino (*Ceratotherium*) are dark in color, but the black rhino is somewhat darker. Its coloration is dark yellow brown to dark brown. An external feature more clearly distinguishing these genera is the upper lip: in *Diceros* it protrudes slightly in the middle and its tip is prehensile, whereas in *Ceratotherium* it is squared and nonprehensile.

Diceros inhabits the transitional zone between grassland and forest, generally in thick thorn bush or acacia scrub but also in more open country (Schenkel and Schenkel-Hulliger 1969). Joubert and Eloff (1971) reported that, in Etosha National Park in Namibia, the most important factor influencing black rhino distribution is the presence of many natural, permanent water holes. *Diceros* is a browser and lives on a variety of bushes and shrubs. Joubert and Eloff (1971) stated that in Etosha National Park, the rhino browses on many different kinds of plants, yet seems to prefer only a few genera, particularly *Acacia*, which forms the bulk of the diet.

Black rhinoceroses (*Diceros bicornis*): Top, photo by Bernhard Grzimek; Bottom, photo from Zoological Society of London.

The black rhino is less gregarious than the white rhino. Schenkel and Schenkel-Hulliger (1969) found that in Tsavo National Park East, in Kenya, *Diceros* is not territorial. Joubert and Eloff (1971) also found that this animal is not territorial in Etosha National Park. The only stable bond among black rhinos seems to be that between mother and calf, which lasts even past the next birth (but probably breaks soon afterward). Intolerance between bulls does occur, but it is not a predominant feature of the social system. Bulls in Tsavo were observed to form temporary groups, moving and feeding together, but these associations were shortlived.

Breeding apparently occurs throughout the year. A premating bond develops between the bull and the cow, and the pair remain together during resting and feeding; they even sleep in contact with each other. Jarvis (1967) tabulated data from six zoos, which indicate that the first successful mating for a female *Diceros* is in its sixth year. The estrous cycle is 17 to 60 days, and the gestation period is from 419 to 476 days. According to Grzimek (1975), there is a single calf, which weighs about 20 kg at birth, nurses for about 2 years, and usually remains with the mother for 3½ years. One black rhino was still living after 45 years in captivity (Marvin L. Jones, Zoological Society of San Diego, pers. comm.).

The black rhino is unpredictable and can be a dangerous animal, sometimes charging a disturbing sound or smell. It has tossed people in the air with the front horn, and regularly charges vehicles and campfires. Catching the scent of humans, it usually crashes off through the brush and runs upwind at speeds of up to 45 km per hour, sometimes for several kilometers, before stopping. Apparently the sense of smell is the primary method of detecting danger. Schenkel and Schenkel-Hulliger (1969) found that human scent alone causes great alarm among black rhinos. On the other hand, if they detect no scent, rhinos will show no interest in a motionless person or car unless it is closer than 20 to 30 meters.

The black rhino has been greatly reduced in numbers and distribution in recent years, because of excessive hunting and habitat disruption. Probably fewer than 30,000 animals survive in all of Africa. The price of rhino horn (used as an aphrodisiac in the Orient and made into dagger handles in the Middle East) has escalated dramatically and poaching of the rhino has been widespread and very destructive. *Diceros* is classified as vulnerable by the IUCN (1978) and is listed as endangered by the USDI (1980). It is on appendix 1 of the CITES.

PERISSODACTYLA; RHINOCEROTIDAE; *Genus*
CERATOTHERIUM Gray, 1868

White or Square-lipped Rhinoceros

In the 19th century the single species, *C. simum*, inhabited two widely separated regions of Africa (Groves 1972*b*; Meester and Setzer 1971). The subspecies *C. s. cottoni* occurred in southern Chad, the Central African Republic, southwestern Sudan, northeastern Zaire, and northwestern Uganda. The subspecies *C. s. simum* occurred in southeastern Angola, possibly southwestern Zambia, central and southern Mozambique, Rhodesia, Botswana, eastern Namibia, and northern and eastern South Africa. About 2,000 years ago, the range of *Ceratotherium* extended up the Nile Valley into southern Egypt, and probably covered much of northwestern Africa.

Except for *Elephas, Loxodonta,* and perhaps *Hippopotamus, Ceratotherium* is the largest living genus of land mammal. Head and body length is 360 to 500 cm, shoulder height is 160 to 200 cm, and weight is usually 2,300 to 3,600 kg. Coloration is yellowish brown to slaty gray. This mammal is naked except for the ear fringes and tail bristles. Hairs are present in the skin, but do not protrude. The front horn averages about 60 cm in length but can reach more than 150 cm.

External features distinguishing *Ceratotherium* from *Diceros* are as follows: usually a lighter coloration; a squared upper lip with no trace of a proboscis; elongated and pointed ear conchae with a few bristly hairs at the tips, compared to rounded conchae with hairy edges in the black rhino; a more sloping, less sharply defined forehead; a shoulder hump; and less conspicuous skin folds on the body.

In South Africa, *Ceratotherium* inhabits primarily the bushveldt zones; in the Nile region, it lives in open *Combretum* forest and the nearby plains. Steeply undulating country is traversed but not permanently inhabited. In Uganda, the white rhino enters swampy country in the dry season; when the rains come, it moves 10 or more km inland from the Nile, especially to an area known as the Bibba Ridge. It is active in the morning and in the evening; other hours are spent wallowing or resting. On warm, windless days, however, animals may be actively feeding at all hours. During rest periods, they lie in the shade of a tree or a termite hill. *Ceratotherium* differs from other rhinos in that it is entirely a grazer. It feeds largely on such grasses as *Pennisetum, Panicum, Urochloa,* and *Digitaria* (Groves 1972*b*; Owen-Smith 1975).

In Zululand, South Africa, Owen-Smith (1974) found that cohesive social groups of *C. simum* included cow-calf pairs, adolescent groups, cow-cow groups, and adult male singletons; the largest group numbered six animals. Adult males occupy territories of about 2 sq km in size for several years. Olfactory marking is carried out by dung scattering and urine spraying. Cows have overlapping basic home ranges covering some 10 to 12 sq km, but at times they may wander farther afield. Groves (1972*b*) wrote that the bull is more vocal than the cow, making noises that include snorts, bellows, and loud cries like an elephant's trumpet.

The following data on reproduction were taken from Groves (1972*b*) and Owen-Smith (1974). Mating is year-round, but is most common from July to September in Zululand and from February to May in Uganda. Males fight fiercely and bloodily for the females, charging each other head on; death often may occur as a result of these encounters. Several males follow a female in estrus, and one finally succeeds in driving away the others, and mates with the female. In Zululand, estrous females form consort pairs with territorial bulls which may last for 2 to 3 weeks. The gestation period is estimated to be between 480 and 550 days. The newborn calf weighs about 50 kg and remains shaky for 2 to 3 days. When alarmed, it runs ahead of the cow. Weaning commences at 2 months of age, but nursing may continue until well over 1 year. The older calf is driven off at the birth of a new calf. Females have their first calf at age 6½ to 7 years, but a male is probably over 12 years of age before he can claim a territory and mate. A wild, 36-year-old female was still reproductively active.

The northern subspecies of the white rhino (*C. s. cottoni*) has been greatly reduced in numbers through indiscriminate hunting for the horns, which are considered by some peoples of the Orient to have aphrodisiac properties. The IUCN (1978) reported that the total number of animals for this subspecies may be less than 500, and certainly less than 1,000. The southern population (*C. s. simum*), however, has made a remarkable comeback from near extinction in recent years. It was greatly overexploited in the 19th century, but since the turn of the century has been carefully protected on reserves. The animals on these reserves multiplied to the point where management problems were evident in some areas. Surplus animals were removed from these overpopulated areas and used in reintroduction programs elsewhere. Today, this southern population is considered safe, and represents a major conservation success story. *C. s. cottoni* is classified as endangered by the IUCN (1978) and the USDI (1980), and is on appendix 1 of the CITES. *C. s. simum* is on appendix 2 of the CITES.

White or square-lipped rhinoceroses (*Ceratotherium simum*): Top, photo from Société Royale de Zoologie d'Anvers through Walter Van den Bergh; Bottom, photo by K. Rudloff through East Berlin Zoo.

ORDER ARTIODACTYLA

Even-toed Ungulates (Hoofed Mammals)

This order of 9 Recent families, 79 genera, and 192 species is native to all land areas of the world, except the West Indies, New Guinea and associated islands, Australia, New Zealand, Antarctica, and most oceanic islands. Introduction by human agency has led to the establishment of wild populations of some species in certain areas where the order does not naturally occur. Simpson (1945) divided the living Artiodactyla into the following groups:

Suborder Suiformes
 Infraorder Suina
  Family Suidae (pigs)
  Family Tayassuidae (peccaries)
 Infraorder Ancodonta
  Family Hippopotamidae (hippopotamuses)
Suborder Tylopoda
  Family Camelidae (llamas, camels)
Suborder Ruminantia
 Infraorder Tragulina
  Family Tragulidae (mouse deer)
 Infraorder Pecora
  Superfamily Cervoidea
   Family Cervidae (deer)
  Superfamily Giraffoidea
   Family Giraffidae (giraffes)

Superfamily Bovoidea
 Family Antilocapridae (pronghorn)
 Family Bovidae (antelopes, cattle, goats, sheep).

The smallest genus in the order is *Tragulus,* which has a head and body length of 400 to 750 mm, and a weight of 0.7 to 8.0 kg. Maximum size is represented by *Hippopotamus,* which weighs up to 4,500 kg, and *Giraffa,* which attains a height of up to 5.8 meters. The principal distinguishing feature of the order is the foot, which has an even number of well-developed digits, except in the genus *Tayassu* (collared and white-lipped peccaries), in which the hind foot has three digits. A first digit (pollex of forelimb, hallux of hind limb) occurs only in certain fossil artiodactyls. The second and fifth (lateral) digits are more slender than the third and fourth (median) digits, or are vestigial or absent. The main axis of the foot passes between the third and fourth digits, and the body weight is borne thereon. In the two-toed or "cloven-hoofed" genera of artiodactyl, the central wrist or ankle bones are fused to form a "cannon bone" and the lateral wrist or ankle bones are absent. The humerus is usually shorter than the forearm, there being some few exceptions, and the radius and ulna are separated or fused. The ankle bone, or astragalus, has a rolling surface above the joint and a pulley surface below, giving free movement to the ankle. The fibula articulates with the heel bone, is usually slender or incomplete, and in some cases is fused with the tibia.

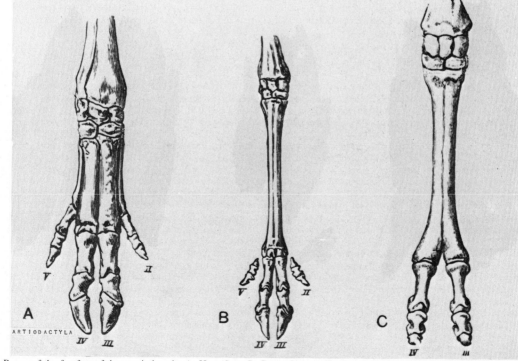

Bones of the forefeet of three artiodactyls: A. Hog (*Sus*); B. Deer (*Cervus*); C. Camel (*Camelus*); photos from *Mammalia,* Beddard.

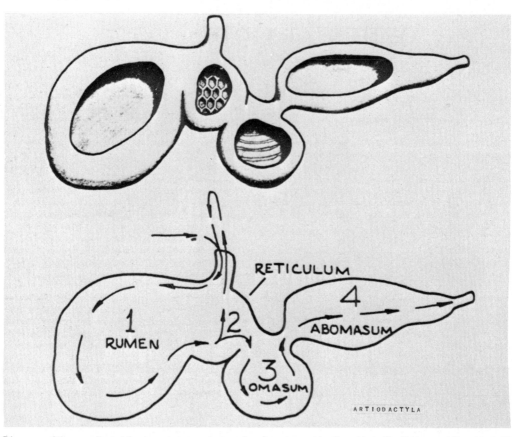

Diagrams of the complicated four-compartmented stomachs of ruminants (the Cervoidea, Giraffoidea, and Bovoidea). The lower diagram shows the route traversed by the food from the time it is first swallowed until it is returned to the mouth for further mastication and then swallowed again to go into the second chamber and the remainder of the digestive tract. The Camelidae and Tragulidae also ruminate or chew their cud, but their stomachs are slightly different and have only three compartments. Photo from *Detroit Zoo Guide Book.*

Many species have frontal appendages, commonly called horns or antlers. The nasal bones are not expanded posteriorly as they are in the order Perissodactyla. There is no alisphenoid canal. The upper incisor teeth are reduced or absent. The canines are usually reduced or lost, though in some species they are enlarged and tusklike. The molars, which are more complex than the premolars, are low crowned with cusps in the suborder Suiformes, and high crowned with crescents in the suborders Tylopoda and Ruminantia.

In the infraorder Suina the stomach is two chambered and nonruminating, in the infraorder Ancodonta it is three chambered and nonruminating, in the suborder Tylopoda and the infraorder Tragulina it is three chambered and ruminating, and in the infraorder Pecora it is four chambered and ruminating. All the ruminants, or "cud chewers," crop or graze vegetable food, such as grasses and woody material, in which there is a relatively low amount of nutriment. They swallow the food rapidly, with little chewing, and then retire to some secluded spot to digest it more thoroughly. In the ruminants, when the food is first swallowed, it enters the rumen or paunch, and, after undergoing a softening process there, it is regurgitated into the mouth, where it is chewed again and further mixed with salivary juices. The food is then

swallowed a second time, entering the second compartment of the stomach (the reticulum or honeycomb bag). It then progresses to the third stomach (the manyplies, omasum, or psalterium), and then to the fourth stomach or digesting chamber (the reed or abomasum), where the greatest digestive activity takes place. In this manner, cud-chewing animals can quickly consume a large quantity of low-grade food, and, when no danger threatens, impart to it the thorough grinding and chemical treatment necessary to convert it to their use. Bacterial action is also involved in the breakdown of food by a ruminant.

Because most artiodactyls are massive and have large bones that resist decay and other destructive forces, there exists a good record of fossil forms going back to the Eocene. The order Artiodactyla includes 16 extinct families. The Old World seems to have been the center of evolution of the order, whereas the perissodactyls flourished mainly in North America. The earliest artiodactyls apparently had a full complement of teeth (a total of 44), four distinct digits on each foot, separate foot bones, no frontal appendages, and a simple, nonruminating stomach. Division into bunodont and selenodont dentition lines seems to have taken place at an early period, and there is evidence that intermediate forms existed.

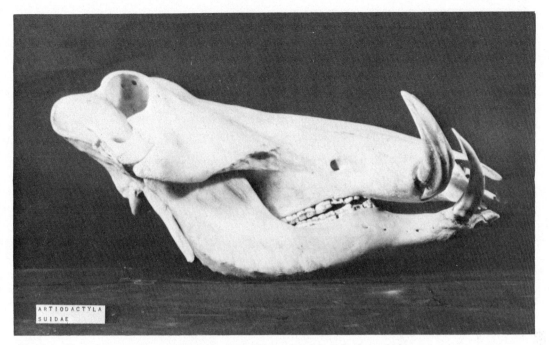

Wart hog (*Phacochoerus aethiopicus*), photo by P. F. Wright of skull in U.S. National Museum.

ARTIODACTYLA; *Family SUIDAE*

Pigs or Hogs

This family of five Recent genera and eight species originally occurred in Eurasia to the south of 48°N, on most associated continental islands as far to the southeast as the Philippines and Celebes, throughout Africa, and in Madagascar. Introduction by human agency has led to the establishment of wild populations of one species, *Sus scrofa*, in parts of North America, in New Guinea and on many nearby islands, and in New Zealand. This species also has given rise to the domestic pig, which occurs throughout the world in association with people. The sequence of genera presented here follows that of Simpson (1945).

Head and body length is approximately 500 to 1,900 mm, tail length is about 35 to 450 mm, and adult weight is as much as 350 kg. The thick skin is usually sparsely covered with coarse bristles or bristly hairs, and some suids are almost naked. A mane occurs on some forms, and the tail has bristly hairs at its tip. The young are striped, except in *Babyrousa* and domestic *Sus scrofa*. The females of most genera have three or six pairs of mammae, but there is only one pair in *Babyrousa*. Suids have a two-chambered, simple, non-ruminating stomach.

Suids are medium-sized mammals with a long and pointed head, a short neck, and a stocky, barrellike body. The mobile snout is truncated terminally and has a disklike cartilage in the tip. This cartilaginous snout, used for turning up surface soil, is strengthened by an unusual bone, the prenasal, situated below the tip of the nasal bones of the skull. Grooving of the snout occurs in only one genus, *Babyrousa*. The nostrils are terminal, the eyes are small, and the ears are fairly long, often with a tassel of hairs near the tip. Some members of this family have warts or ridges on the face, which are skin growths without a bony support or core. The forelegs are half

as long as the height at the shoulders. The foot bones are separate (not fused) and the feet are narrow. The first digit is absent, so each foot has four toes; the middle two (the third and fourth) are flattened and have hooves, whereas the other two toes (the second and fifth) are located higher up on the limb, do not reach to the ground in the ordinary walking position, and have smaller hooves than do the third and fourth digits.

The most striking feature of the skull in the family Suidae is the elevation and backward slope of the occipital crest, formed by the union of the supraoccipital and parietal bones. In *Potamochoerus, Sus,* and *Hylochoerus* the dental formula is: (i 3/3, c 1/1, pm 4/4, m 3/3) × 2 = 44. In *Babyrousa* it is: (i 2/3, c 1/1, pm 2/2, m 3/3) × 2 = 34. In *Phacochoerus* it is: (i 1/3, c 1/1, pm 3/2, m 3/3) × 2 = 34. The upper incisors decrease in size from the first to the third, and the lower incisors are long, narrow, set closely together, and almost horizontal in position. The incisors and the canines have sharp lateral edges. The large upper canines grow outward and backward, and the lower canines grow upward and backward, tending to form a complete circle. Usually, however, the canine teeth wear against each other, thus producing the sharp edges. These tusks, which are most prominent in males, reach their greatest development in *Babyrousa;* in this genus the upper canines are directed upward through the skin, never entering the mouth, and curve backward and downward, often touching the forehead. The cheek teeth are cuspidate, and the upper premolars are simpler in structure than the molars. With age, the enamel wears away and all the teeth disappear except for the canines and back molars. The third molars of *Phacochoerus*, which are often the only teeth in that genus, are unlike those of any other mammal; they are composed of a number of closely set cylinders of dentine embedded in cement.

Pigs generally live in forests and woodlands. They shelter in tall grass or reedbeds, and in burrows that are either self-

African bush pig (*Potamochoerus porcus*): A. Photo by Bernhard Grzimek; B. Photo from New York Zoological Society of immature animal.

excavated or abandoned by other animals. Pigs are sure-footed and rapid runners, are good swimmers, and are fond of mud baths. When cornered or wounded, they frequently fight courageously. In such battles their tusks are deadly weapons. They are active mainly at night, especially in areas where they are molested by people. Some forms use their snout, and a few use their tusks, to dig for food. The omnivorous diet includes fungi, leaves, roots, bulbs, tubers, fruit, snails, earthworms, reptiles, young birds, eggs, small rodents, and carrion. Contrary to popular belief, a wild pig will rarely overeat.

The geological range of this family is early Oligocene to Recent in Europe, early Miocene to Recent in Africa, and late Miocene to Recent in Asia.

ARTIODACTYLA; SUIDAE; **Genus POTAMOCHOERUS**
Gray, 1854

African Bush Pig, Red River Hog

The single species, *P. porcus*, occurs nearly throughout Africa south of the Sahara. It is also found in Madagascar and on Mayotte Island in the Comoros, probably through introduction by human agency many years ago (Meester and Setzer 1971).

Head and body length is 1,000 to 1,500 mm, tail length in one specimen was 316 mm, shoulder height is 585 to 965 mm, and weight is about 75 to 130 kg. Coloration varies from reddish brown to black, often with a heavy mixture of white or yellowish hairs. Some individuals are almost white or mottled white and black or brown. The young are longitudinally striped with pale yellow or buff on a dark brown ground color. The long, pointed ears often have long tufts or streamers of hairs at the tips, and there is a pronounced light-colored mane along the top of the neck and back.

The average upper tusk length is 76 mm, and the lower tusk measures 165 to 190 mm. The upper tusks point downward and wear against the lower ones. The male has warts in front of the eye, and, although they protrude 40 mm, they are frequently not conspicuous, as they are often concealed by facial hair. This genus resembles *Phacochoerus* and *Sus*, but differs from the former in having more hair and a greater number of teeth, and from the latter in having ears that are more tufted. Females have three pairs of mammae.

The African bush pig is found in a variety of habitats. It is most active at night, and rests by day in a self-excavated burrow within an area of dense vegetation. It is a fast runner and good swimmer. When cornered or wounded, it exhibits considerable courage and frequently attacks. It is wary of traps; when a trap is discovered, *Potamochoerus* will avoid the area for weeks. The omnivorous diet consists mainly of roots, berries, and fruit. Reptiles, eggs, and, occasionally, young birds are also eaten. The snout is used as a plough to "root" up subsoil vegetation, and in a short period of time a herd can do much damage to crops.

In Uganda, Laws, Parker, and Johnstone (1975) estimated the average population density to be 1.29 individuals per sq km. *Potamochoerus* is gregarious and travels in groups of up to 20 individuals. In the Transvaal, Skinner, Breytenbach, and Maberly (1976) found most groups, or "sounders," to contain 4 to 6 pigs, and to be led by a dominant male that defended the area being utilized at a given time. The tusks were used to mark trees along regularly traveled paths.

In the southern part of Africa, births seem to occur mainly from September to April, peaking during the warm, wet summer. The gestation period is about 4 months. The number of young per litter is one to eight, usually about three or four. Females evidently attain sexual maturity at about 3 years of age (Skinner, Breytenbach, and Maberly 1976; Smithers 1971). A captive specimen lived for 20 years (Marvin L. Jones, Zoological Society of San Diego, pers. comm.).

The African bush pig has greatly increased in numbers, because of the destruction of leopards by people and the creation of more favorable habitat by the spread of agriculture. *Potamochoerus* now damages many kinds of cultivated plants, and has been known to wipe out entire peanut crops. It is widely hunted for this reason, and also for its highly palatable flesh (Smithers 1971; Grzimek 1975).

European wild hog (*Sus scrofa*), photo by Ernest P. Walker and William J. Schaldach, Jr.

ARTIODACTYLA; SUIDAE; **Genus SUS** *Linnaeus, 1758*

Pigs, Hogs, Boars

There are two subgenera and four species (Grzimek 1975; Ellerman and Morrison-Scott 1966; Corbet 1978; Lekagul and McNeely 1977; Medway 1977, 1978; Chasen 1940; Laurie and Hill 1954; Taylor 1934; Sanborn 1952; Meester and Setzer 1971):

subgenus *Sus*

S. scrofa (wild boar, pig), originally found from southern Scandinavia and Portugal to southeastern Siberia and Malay Peninsula, from Western Sahara to Egypt, and on Britain, Ireland, Corsica, Sardinia, Sri Lanka, Andaman Islands, Japan, Taiwan, Hainan, Sumatra, Java, and many small islands of the East Indies;
S. barbatus (bearded pig), Malay Peninsula, Sumatra, Bangka, Borneo, Rhio Archipelago, Palawan and Balabac islands (Philippines);
S. verrucosus (Javan pig), Java, Celebes, Molucca Islands, Philippines;

subgenus *Porcula*

S. salvanius (pygmy hog), Nepal, Sikkim, Bhutan, and adjacent parts of northeastern India.

In *S. salvanius,* head and body length is 500 to 650 mm, tail length is about 30 mm, and shoulder height is 250 to 300 mm. In the other three species, head and body length is 900 to 1,800 mm, tail length is about 300 mm, shoulder height is 550 to 1,100 mm, and weight is 50 to 350 kg. Some domestic breeds of *S. scrofa* may attain a weight of 450 kg (Grzimek

1975). Males are larger, on the average, than females. Coloration of wild forms is dark gray to black or brown. The body is covered with stiff bristles and usually some finer fur, but the pelage is often quite scant and the tail is only lightly covered with short hairs. Many individuals have side whiskers and a mane on the nape. The young are striped. Hogs have four continually growing tusks, two in each jaw. There are three pairs of mammae in female *S. salvanius,* and six pairs in females of the other species.

Pigs live in many kinds of habitat, but generally where there is some dense vegetation for cover. Individuals construct crude shelters by cutting grass and spreading it over a given area. They crawl under the grass and then raise themselves to lift the grass mat, which then attaches to uncut grass and forms a canopy. They wallow in mud and will do so for hours if the opportunity affords. They are swift runners and good, strong swimmers. Activity is mainly nocturnal and crepuscular. A great distance may be traveled during the night in the search for food. The omnivorous diet includes fungi, roots, tubers, bulbs, green vegetation, grains, nuts, cultivated crops, invertebrates, small vertebrates, and carrion.

In a study of feral *S. scrofa* in coastal South Carolina, Wood and Brenneman (1980) found a population density of 10 to 20 individuals per sq km, and an average annual home range of 226 ha. for males and 181 ha. for females. In a nearby area, Kurz and Marchinton (1972) found home range to average about 400 ha., and seldom observed groups of more than 3 individuals. In the Old World, *S. scrofa* has been seen in herds, or "sounders," of over 100, though average size seems to be about 20 (Lekagul and McNeely 1977). According to Fradrich (1974), in both wild and feral *S. scrofa* the basic social unit is the female and her litter. After the

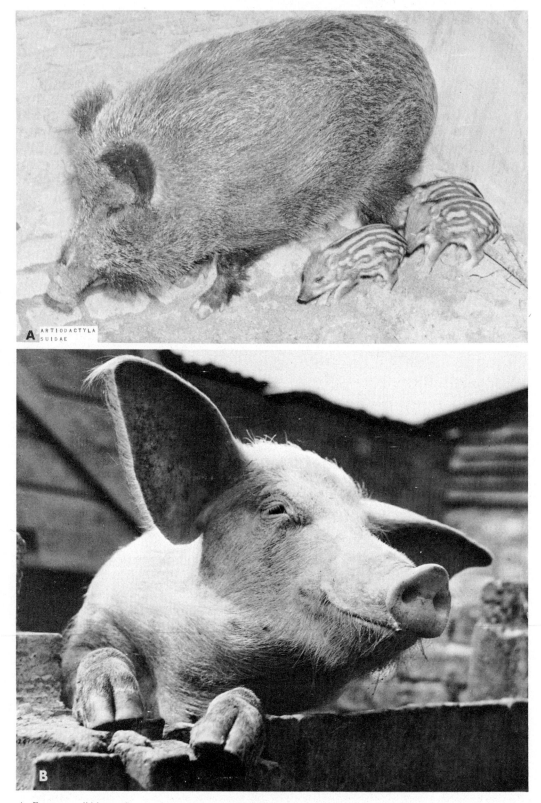

A. European wild hogs (*Sus scrofa*), mother and young, photo by Eric Parbst through Zoologisk Have, Copenhagen. B. Domestic hog (*S. scrofa*), photo by Roy Pinney.

young are weaned, 2 or more families may come together. This association remains stable until the beginning of the next mating season, when the previously solitary adult males join in fighting over the females. A male usually wins control of 1 to 3 females, but sometimes obtains as many as 8. After mating, the males depart.

Breeding occurs throughout the year in the tropics, but births peak shortly before or just after the rains. In temperate regions, the young are born in the spring. Females have an estrous cycle of about 21 days, are receptive for 2 or 3 days, and generally produce one litter annually. The gestation period is 100 to 140 days. The number of young per litter is 1 to 12, usually 4 to 8. Unlike the young of most ungulates, piglets are born in a nest and remain there following birth. They are weaned after about 3 or 4 months, and may leave the mother prior to birth of the next litter, but young females often remain longer. Sexual maturity may be attained as early as 8 to 10 months of age, but females usually do not mate until around 1½ years old. Males generally are not able to compete successfully for mating privileges until they reach full size, at around 5 years of age (Fradrich 1974; Grzimek 1975; Lekagul and McNeely 1977; Henry 1968). Average longevity is about 10 years, but some pigs have lived up to 27 years.

Wild pigs have been extensively hunted by people for use as food, for sport, and because they are sometimes destructive to crops. Their size and sharp tusks make them potentially dangerous antagonists, but they usually do not attack unless molested. *S. scrofa* was exterminated long ago in many parts of its original range, including the British Isles, Scandinavia, and Egypt. It has been reintroduced, in the wild form, to Britain, and established by human agency in New Guinea and many adjacent islands, New Zealand, and parts of North America. There is sometimes difficulty, however, in determining whether the free-ranging pigs in these areas represent originally wild stock or are the feral descendants of domestic animals (Grzimek 1975; Corbet 1978; Laurie and Hill 1954; Wood and Barrett 1979).

According to Lekagul and McNeely (1977), domestication of *S. scrofa* took place in China around 4900 B.C., and may have occurred as early as 10,000 B.C. in Thailand. Many breeds have since been developed, especially in Europe. Pigs are valuable to agricultural economies, because they mature sooner than do other domestic ungulates, have larger litters, and can feed on human refuse. They have thus contributed to the spread of human populations to New Guinea and the Pacific islands.

The first pigs in the United States were those brought by the Polynesians to Hawaii around 1000 A.D., and those introduced by the Spanish to the Southeast in the early 16th century. Several valuable farm breeds of *S. scrofa* were developed in the United States, but large feral populations also became established. European wild boar were introduced in several places for purposes of sport hunting, and these interbred with the feral animals already present. Free-ranging pigs now occur from Texas to Florida and the Carolinas, throughout California, on eight of the major Hawaiian Islands, and on Puerto Rico and the Virgin Islands. They are valued as game animals in some areas, and over 100,000 are taken annually by sport hunters. Many persons, however, consider feral pigs to be detrimental to agriculture, forestry, and native wildlife (Wood and Barrett 1979; Grzimek 1975).

The pygmy hog (*S. salvanius*) is classified as endangered by the IUCN (1978) and the USDI (1980), and is on appendix 1 of the CITES. Its range now appears restricted to northwestern Assam, where an estimated 100 to 150 individuals survive. It has declined mainly because of the modification and usurpation of its limited habitat by people. It is protected by law in India, but is illegally hunted.

ARTIODACTYLA; SUIDAE; Genus PHACOCHOERUS
F. Cuvier, 1817

Wart Hog

The single species, *P. aethiopicus,* occurs in most of Africa south of the Sahara (Meester and Setzer 1971).

Head and body length is about 900 to 1,500 mm, tail length is 250 to 450 mm, shoulder height is 635 to 850 mm, and weight is 50 to 150 kg. The length of the upper tusks is 255 to 635 mm in males and 152 to 255 mm in females. A long, thin mane of coarse hair extends from the nape to the middle of the back, where it is broken by a bare space, and then continues on the rump. The remainder of the body is covered with bristles. The color of both the skin and the hair is dark brown to blackish. The immature are reddish brown. A long, ridgelike fold on the cheek bears white hairs. The warts, which are prominent only on the males, are skin growths and have no bony support or core; they are located on the side of the head and in front of the eye. When an animal moves slowly, the tail hangs limply, but when it runs, the tail is carried in an upright position with the tufted tip hanging over.

The wart hog is usually found in savannah and lightly forested country. In contrast to most suids, it is usually diurnal. In places where it is molested by people, however, it may become almost nocturnal. To sleep, to rear young, and to find refuge from predators, *Phacochoerus* depends on holes, either natural ones or those made by the aardvark (*Orycteropus*). It usually backs into such holes, perhaps to confront a pursuer with its formidable tusks, but sometimes enters headfirst (Cumming 1975). It is usually inoffensive, but will defend itself, when cornered, and can inflict severe wounds with its tusks. Fradrich (1974) wrote that the main weapons of defense are not the large upper tusks but their lower counterparts, which are smaller but sharper. Like other hogs, *Phacochoerus* enjoys mud baths. Its maximum running speed is 55 km per hr (Schaller 1972). Although its eyesight seems to be poor, its senses of hearing and smell are acute.

The wart hog feeds on grass, roots, berries, the bark of young trees, and, occasionally, carrion. While feeding, it drops on its padded wrists and frequently shuffles along in this position. In Rhodesia, Cumming (1975) found *Phacochoerus* to be almost entirely graminivorous, being specialized both for grazing on short, seasonally succulent grasses and for digging grass rhizomes with its powerful rhinarium in hard, dry soils.

According to Cumming (1975), reported population densities vary from about 0.2 to 20.0 wart hogs per sq km. In Rhodesia, the maximum density of 10.6 per sq km was found in grassland. Home range there varied from 64 to 374 ha. The population was divided into "clans," each consisting of several bands, or "sounders," and associated lone animals. Most bands contained 4 to 6 individuals, though groups of up to 40 have been reported. Most groups represent a temporary aggregation of young males or an association of adult females that may have lasting bonds and occupy an area for years. The different groups within a clan have overlapping ranges and share holes and other resources. Adult males are usually solitary, but join the female groups briefly for mating. At such times, males engage in highly ritualized battles, in which they push and strike with the head and the blunt upper tusks. The warts on the sides of the head serve to cushion the blows, and injuries are rare. There seems to be no territorial defense, but there is sometimes competition for resources, such as waterholes, and the area being used at a given time is marked with saliva and secretions from glands

Wart hog (*Phacochoerus aethiopicus*): Top, photo from New York Zoological Society; Bottom, photo by Karl H. Maslowski.

around the eyes. Vocalizations include various grunts, growls, snorts, and squeals, which are used for greeting, contact maintenance, threats, warning, and submission.

There are clearly defined breeding seasons. In Rhodesia, mating occurs in May and June, and births in October and November. The gestation period is 171 to 175 days. The number of young per litter is one to eight, usually two or three. The piglets begin to accompany their mother regularly at about 50 days of age, and are completely weaned by 21 weeks. The young are temporarily driven away when the female is about to bear a new litter, but may subsequently rejoin the family. Males separate from their mother by the

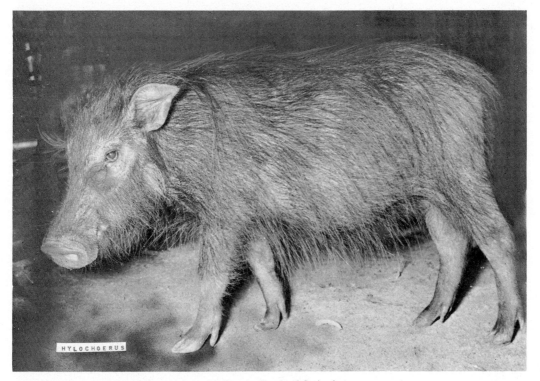

Giant forest hog (*Hylochoerus meinertzhageni*), photo by Bernhard Grzimek.

age of 15 months. Females stay longer, perhaps in permanent association. Sexual maturity comes at the age of 18 to 20 months, but males usually do not mate until they are around 4 years old (Cumming 1975). A captive wart hog lived for 18 years and 9 months (Marvin L. Jones, Zoological Society of San Diego, pers. comm.).

Phacochoerus has been eliminated from most of South Africa, but still occurs throughout the remainder of its original range (Meester and Setzer 1971). It is widely hunted by people for use as food. It is much less destructive to native crops than are the other wild pigs of Africa.

ARTIODACTYLA; SUIDAE; *Genus HYLOCHOERUS*
Thomas, 1904

Giant Forest Hog

The single species, *H. meinertzhageni*, is found in the forest zone from Liberia to southwestern Ethiopia and northern Tanzania (Meester and Setzer 1971).

Head and body length is 1,500 to 1,900 mm, tail length is about 300 mm, shoulder height is 762 to 1,100 mm, and weight is 160 to 275 kg. The pelage is long, coarse, and black, and becomes sparse with age. The skin is blackish gray. The skin in front of each eye and on the upper part of the cheek below the eye is almost naked. Below and behind each eye there are two movable cutaneous thickenings or facial excrescences. One of these forms a wide ridge on each naked area of skin. A preorbital gland is present, marked externally by a slit on the naked area of the face in front of each eye. There are no facial protuberances, as in *Phacochoerus*, and the upper canines are set horizontally, not at an angle, as in *Phacochoerus*. The large skull of *Hylochoerus* contains a

depression in the roof capable of holding nearly a cup of water.

This hog occurs in tropical forest and savannah at both low and high elevations. According to D'Huart (1976), it is diurnal in areas where it is protected from human molestation, and uses holes in the ground for nightly rest. *Hylochoerus* also makes tunnels and pathways, and wallows in water and swamps. It feeds mainly on shrubs and tall, lush grass, and does not do much digging or grubbing for food. It travels in groups, or "sounders," of up to 20 individuals. D'Huart (1976) reported that groups maintain a territory and are led by an old male. Litters of 2 to 6 young are born in the bedding-down place, following an average gestation period of 125 days.

The giant forest hog was one of the last mammals of its size to become known to science. Its range may still not be completely known, but it apparently has disappeared from western Liberia (Meester and Setzer 1971). People seem to fear its more than *Potamochoerus*. Males often charge without warning or provocation, presumably to protect the sounder. *Hylochoerus* sometimes raids native shambas at the forest edge and causes much destruction. Certain tribes make war shields from the hide of this hog.

ARTIODACTYLA; SUIDAE; *Genus BABYROUSA Perry,*
1811

Babirusa

The single species, *B. babyrussa*, occurs on Celebes, the nearby Togian and Sula Islands, and Buru Island in the Moluccas (Laurie and Hill 1954). Its presence on Buru and

Babirusa (*Babyrousa babyrussa*), photo from New York Zoological Society.

possibly Sula is evidently the result of introduction by human agency (Groves 1980*a*).

Head and body length is usually 875 to 1,065 mm, tail length is 275 to 320 mm, shoulder height is 650 to 800 mm, and weight is up to 100 kg. The skin is either rough and brownish gray, or smooth and sparsely covered with short whitish gray to yellowish hairs. The underside of the body and inner sides of the legs are sometimes lighter than the rest of the body. Frequently, this whitish color extends along the sides of the upper lip. The skin usually hangs in loose folds. Females have two pairs of mammae.

In most wild swine the tusks grow from the sides of the jaw. In *Babyrousa,* however, the upper tusks grow through the top of the muzzle and then curve backward toward the forehead. Although the tusks would thus appear to have little use as weapons, Mackinnon (1981) suggested that they are important to males for fighting. The upper tusks apparently have a general defensive function, and the daggerlike lower tusks are used offensively. Since the lowers are not honed by wearing against the uppers, however, the male actively sharpens them on trees. The wear pattern of upper tusks from Celebes indicates that these teeth are used to interlock and hold the opponent's lower tusks. On Buru Island, though, the upper tusks have evidently lost this function and are probably used for butting. According to a native legend, this hog hangs itself by its tusks from a tree limb at night. The natives also say that the tusks are like the antlers of a deer; hence the name "babirusa," which means "pig deer."

The preferred habitats of the babirusa are moist forests, canebrakes, and the shores of rivers and lakes. This pig is a swift runner and often swims in the sea to reach small islands. Its senses of hearing and smell are acute. It has been reported to be nocturnal, but according to the IUCN (1978) it may be active by day. It seems not to root with its snout, as does *Sus,* and probably feeds on foliage and fallen fruit.

The babirusa travels about in small parties and reveals its presence by low grunting moans. Offspring are produced in the early months of the year and are not striped like the young of most pigs. The gestation period is 125 to 150 days and there are usually two young per litter. Captives have lived up to 24 years (Grzimek 1975).

The babirusa is frequently captured young and tamed by native people. The natives hunt it regularly for food, erecting an enclosure of poles and nets into which the animal is driven. Even though the babirusa has long been protected by law, it has declined substantially in numbers because of excessive hunting and habitat loss. It is classified as vulnerable by the IUCN (1978) and as endangered by the USDI (1980), and is on appendix 1 of the CITES.

ARTIODACTYLA; *Family TAYASSUIDAE*

Peccaries

This family of two Recent genera and three species occurs from the southwestern United States to central Argentina. The sequence of genera presented here is based on the suggestions by Wetzel (1977*a*, 1977*b*, 1981) that *Catagonus* is more primitive than *Tayassu.*

Head and body length is 750 to 1,112 mm, and tail length is 15 to 102 mm. Peccaries have only 6 to 9 tail vertebrae, whereas suids have 20 to 23. The pelage is bristly, and there is a mane of long, stiff hairs on the middorsal line from the crown to the rump. The body form is piglike, but the legs are long and slim, and the hooves are small. There are four digits on the forefoot, the two lateral ones being reduced and not touching the ground. There are two functional digits on the hind foot. There is a vestigial, median digit on the back of the

Chacoan peccary (*Catagonus wagneri*), photo from West Berlin Zoo.

hind foot in *Tayassu,* but not in *Catagonus* (Wetzel 1977*b*). The third and fourth foot bones, which are completely separate in the Suidae, are united at their proximal ends in the Tayassuidae, as in the ruminants. The snout is the same as in the Suidae: elongate, mobile, and cartilaginous, with a nearly naked terminal surface, in which the nostrils are located. The ears are ovate and erect. Both genera have a scent gland on the rump in front of the tail. The gland is about 75 mm in diameter and 125 mm thick. When a peccary is excited, the hairs on the neck and back bristle, and the dorsal gland emits a musky secretion, the odor of which can be detected many meters away. The stomach is two chambered and nonruminating, but is more complex than that of the Suidae. Females have four mammae.

The dental formula is: (i 2/3, c 1/1, pm 3/3, m 3/3) × 2 = 38. The upper canines form tusks, but these are directed downward, not outward or upward as in the Suidae, and are smaller than those in pigs, the average length being about 40 mm. There is a space between the canines and the premolars. The premolars and molars form a continuous series of teeth, gradually increasing in size from the first to the last. The last premolar is nearly as complex as the molars. The molars have square crowns with four cusps.

The geological range of this family is early Oligocene to late Miocene in Europe, early Pliocene in Asia, Pleistocene to Recent in South America, and early Oligocene to Recent in North America.

ARTIODACTYLA; TAYASSUIDAE; **Genus CATAGONUS**
Ameghino, 1904

Chacoan Peccary

The single species, *C. wagneri,* occurs in the Gran Chaco region of southeastern Bolivia, Paraguay, and northern Argentina (Wetzel 1977*b*). *Catagonus* was long known by fossil material from the Pleistocene, but the living population was not scientifically discovered until the early 1970s (Wetzel et al. 1975).

Except as noted, the information for the remainder of this account was taken from Wetzel (1977*b*, 1981). Head and body length in five specimens was 923 to 1,112 mm, tail length in three specimens was 70 to 102 mm, and weight of one female was 37 kg. The general coloration is brownish gray, and there is a faint collar of lighter hairs across the shoulders. From *Tayassu, Catagonus* differs externally in having larger overall size, longer and paler-colored hair on the ears and legs, a larger head, a longer snout, and longer ears, legs, and tail. There is no median dewclaw on the posterior side of the hind foot. The skull of *Catagonus* differs from that of *Tayassu* in having extreme development of the rostrum, nasal chamber, and sinuses, and more posteriorly placed orbits. The canine teeth of *Catagonus* are more slender than those of *Tayassu,* and the molars are high crowned (hypsodont), rather than low crowned (bunodont).

The Chacoan peccary inhabits semiarid thorn forest and steppe. It is cursorial and diurnal, with peak activity in the late morning. Morphological characters indicate that it is superior to *Tayassu* in speed, distance vision, and olfaction, but that it has less mental capacity and depends more on browsing for food. Legume seeds, roots, and cacti are thought to be important components of the diet.

Catagonus is usually found in groups of about five individuals. Most young appear in September (early spring). Mayer and Brandt (1978) found the number of young per litter to be one to four, usually two or three, and estimated minimum breeding age of females to be three years.

The Chacoan peccary is classified as vulnerable by the IUCN (1978). The species has long been distinguished from other peccaries by the native Indians of the Chaco, who hunt it for its meat and hide. It is also coming under increasing hunting pressure from ranchers, trappers, sportsmen, and military personnel. The main problem, however, is the clearing of natural vegetation to provide pastures for cattle. There is concern that this most recently discovered of large mammals could disappear by the end of the century.

ARTIODACTYLA; TAYASSUIDAE; **Genus TAYASSU**
Fischer, 1814

Collared and White-lipped Peccaries, Javelinas

There are two species (Hall 1981; Cabrera 1961):

T. tajacu (collared peccary), Arizona and Texas to northern Argentina;

T. pecari (white-lipped peccary), southern Mexico to northeastern Argentina.

Hall (1981) placed *T. tajacu* in a separate genus, *Dicotyles*, but Wetzel (1977*b*, 1981) considered *Dicotyles* congeneric with *Tayassu*.

Head and body length is 750 to about 1,000 mm, tail length is 15 to 55 mm, shoulder height is 440 to 575 mm, and weight is 14 to 30 kg. *T. pecari* is considerably larger, on the average, than *T. tajacu*. Males and females are about the same size. *T. pecari* is dark reddish brown to black, and has white on the sides of the jaws. *T. tajacu* is generally dark gray and has a whitish collar on the neck; its young are reddish and have a blackish stripe along the back. The name "javelina" is derived from the Spanish word for javelin or spear, and refers to the sharp tusks of these animals.

Peccaries live in a great variety of habitats, including desert scrub, arid woodland, and rain forest. They usually shelter in a thicket or under a large boulder. Limestone caves often serve as winter quarters in some areas. Most activity is at night or in the cooler hours of the day. Peccaries are fairly sedentary and do not seem to travel far from their place of birth. Their speed, agility, and group defense render them more than a match for dogs, coyotes, and even bobcats. They often clash their canine teeth together as a warning. When fleeing danger, they move with a fast running gait. They have poor vision and fair hearing. Their sense of smell is keen enough to locate a small covena bulb, 5 to 8 cm underground, before the new shoots are visible.

Collared peccary and young (*Tayassu tajacu*), photo from New York Zoological Society.

Peccaries, like pigs, are not "dirty" animals; on the contrary, they are quite clean. Their habit of pawing sand against the belly with the front feet is thought to be a cleansing action. They grub for food with their snout. They are mainly vegetarian, feeding on cactus fruit, berries, tubers, bulbs, and rhizomes. They also consume grubs, and, occasionally, snakes and other small vertebrates. Like some pigs, they do not seem to be harmed by rattlesnake bites. They frequent waterholes, or, in the tropics, stay near running streams.

Peccaries are gregarious. *T. pecari* is sometimes found in herds of several hundred individuals (Leopold 1959). In *T. tajacu,* group size ranges from 2 to 50, but is usually 5 to 15. Both sexes and all ages are found in a group. There is a rank order, with females usually dominating males. Each group of *T. tajacu* has a home range of about 0.5 to 8.0 sq km. The central part of the range is an exclusive territory, while the peripheral area is shared with other herds. The dorsal scent gland is rubbed against tree trunks and other objects for territorial marking, and also seems to identify group members and coordinate movements (Sowls 1974, 1978; Schweinsburg 1971; Bigler 1974).

Breeding apparently can occur throughout the year. In Arizona, however, *T. tajacu* usually mates in February and March, and gives birth during the summer. The gestation period in this species is around 115 days, while in *T. pecari* it is 156 to 162 days (Roots 1966). If a litter is lost, the female may quickly mate again and bear a second annual litter (Sowls 1978). The number of young per litter is one to four, usually two. The young are born in a thicket, hollow log, cave, or burrow dug by another animal. They can run in a few hours, and accompany their mother 1 day or so after being born, when she rejoins the herd. In *T. tajacu,* lactation lasts 6 to 8 weeks; the milk of this species is lower in fat and total solids than that of domestic *Sus scrofa.* Young peccaries, at least of *T. tajacu,* reach the teats from the rear of the sow, instead of standing parallel to the side. They remain with the mother for 2 or 3 months. Captives may breed in their first year of life (Sowls 1978). One captive *T. tajacu* lived for 24 years and 7 months (Marvin L. Jones, Zoological Society of San Diego, pers. comm.).

If unmolested, peccaries ordinarily do not bother humans, but if a member of a band is wounded or pursued, the entire herd may counterattack. Captive adults are sometimes unpredictable, but a pet kept at the U.S. National Zoo in Washington, D.C., would come to the fence when it recognized a friend. It knew its name, would come promptly when called, and enjoyed being scratched. Wild peccaries are hunted for their meat and skin. There has been an extensive and wasteful trade in hides in parts of Latin America (Leopold 1959; Grimwood 1969). In the United States, *T. tajacu* is classified as a game animal and is subject to regulated sport hunting (Sowls 1978). Both species of *Tayassu* were introduced to Cuba in 1930 (Hall 1981).

ARTIODACTYLA; *Family HIPPOPOTAMIDAE*

Hippopotamuses

This family of two Recent genera, each with a single species, originally occurred in suitable habitat throughout Africa south of the Sahara, and all along the Nile River.

The two genera, *Hippopotamus* and *Choeropsis,* differ greatly in size, but both are characterized by a broad snout, a large mouth, a short barrellike body, and short, stocky legs. The belly is carried only a short distance above the ground. The nostrils are located on top of the snout and can be closed. The sparsely haired body contains special pores, which secrete a pinkish substance known as "blood sweat." This material is thick, oily, and protective in nature, allowing the animal to remain in water or in a dry atmosphere on land for extended periods. The skin contains a layer of fat, which in *Hippopotamus* is 50 mm thick. The short tail is rather bristled. The foot bones are separate. All four toes on each foot support the body weight; the two lateral digits are nearly as

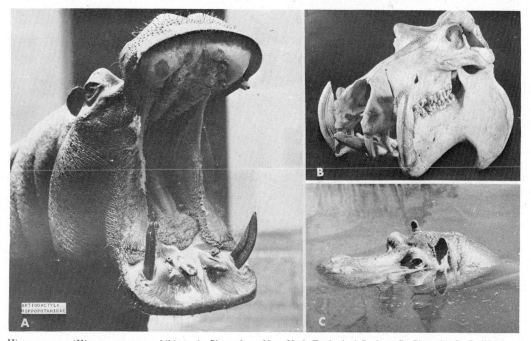

Hippopotamus (*Hippopotamus amphibius*): A. Photo from New York Zoological Society; B. Photo by P. F. Wright of specimen in U.S. National Museum; C. Photo from Castle Films through U.S. National Zoological Park.

well developed as the median digits. The terminal digital bones have naillike hooves. The stomach is complex and three chambered, but nonruminating. Females have two mammae.

The large skull has an elongate facial region, but the braincase is relatively small. The dental formula is: (i 2–3/1–3, c 1/1, pm 4/4, m 3/3) × 2 = 38–42. There is usually only one pair of lower incisors in *Choeropsis,* but two or three pairs in *Hippopotamus.* The incisors and canines are tusklike and grow continuously. The incisors are rounded, smooth, and widely separated. The upper incisors are quite short and project downward. The lower ones are longer, especially the inner pair (in *Hippopotamus*), and project forward and only slightly upward. The lower canines, which are the largest of the tusklike teeth, project upward and outward. The premolars are usually single cusped, though this condition may vary even on opposite sides in the same animal. Two pairs of cusps are developed in the molars, except in the third molar, which has three pairs. These cusps wear down to various trefoil, figure eight, or dumbell shaped enamel patterns.

The geological range of this family is middle Pliocene to Pleistocene in Asia and Sri Lanka, Pleistocene in Europe and Madagascar, and middle Pliocene to Recent in Africa.

ARTIODACTYLA; HIPPOPOTAMIDAE; *Genus*
HIPPOPOTAMUS Linnaeus, 1758

Hippopotamus

The single species, *H. amphibius,* originally occurred throughout Africa south of the Sahara, in areas with suitable waterways, and down the Nile River to its delta (Meester and Setzer 1971).

Head and body length is 3,750 to 4,600 mm, tail length is about 560 mm, shoulder height is about 1,500 mm, and weight is 3,000 to 4,500 kg. The body is so scantily covered with short, fine hairs that it appears naked. The usual color of the skin is a slaty copper brown, which shades to dark brown above and purplish below. The skin is glandular and exudes droplets of moisture that contain red pigment; light reflected from the skin through these droplets appears red, thus giving rise to the belief that *Hippopotamus* ''sweats blood.'' The eyes are protruding, and the ears (up to 100 mm long) are set high up and far back on the head. The upper canine teeth are as much as 230 mm or more in circumference. The lower canines are about 600 mm long and weigh as much as 3 kg.

The preferred habitat of the hippopotamus is an area of deep water with adjacent reed beds and grassland. This is an amphibious mammal, and is a good swimmer and diver. When submerging, it closes the slitlike nostrils and ears. If it wants to see and breathe without exposing itself, it can keep the protruding eyes and nostrils out of the water and remain submerged. It normally stays underwater for 3 to 5 minutes, but can remain under longer, perhaps for 30 minutes. Its specific gravity is such that it can walk about on the bottom. It sometimes enters salt water adjacent to the mouth of a river. Apparently, hearing, sight, and smell are well developed (Grzimek 1975).

The hippopotamus spends practically the entire day sleeping and resting in or near water. If molested, it will lie in deep water in reed beds. At night it may cover 33 km of water in search of food. It usually emerges to feed. According to Lock (1972), the grazing range extends about 3.2 km from water, though some individuals may move farther. Vegetation is cropped with the heavy front teeth and the lips. The diet consists mostly of grass.

Hippopotamus sometimes attains very dense populations. In an 88-km stretch of the Nile River in Uganda, including 3 km of land on both sides of the river, the average density was 19.2 individuals per sq km. The recommended density for the habitat was 7.7 per sq km (Laws, Parker, and Johnstone 1975). *Hippopotamus* may occur alone or in groups of up to 30 individuals. Apparently, the larger groups consist mostly of adult females and their young. Adult males compete with one another for control of a herd and the territory it occupies. The losers may become solitary and live in areas of marginal habitat. Fights between bulls are vicious and may last two hours. The chief weapons are the large lower canines. Hides of mature males invariably have numerous scars. Deaths are not infrequent. Bulls have a loud roar that can be heard over a

Hippopotamus (*Hippopotamus amphibius*), photo from New York Zoological Society.

Hippopotamuses (*Hippopotamus amphibius*), photo from Amsterdam Zoo.

great distance. They mark the land portion of their territory by defecation. Several females may separate from a herd and form a temporary nursery unit for the bearing of young (Grzimek 1975).

Breeding evidently can occur at any time of year, but there may be seasonal peaks in some areas. In Uganda, most mating is in February and August, and most births in October and April, months of maximum rainfall (Grzimek 1975). Females have an estrus of 3 days, and mate again within 12 to 16 days of weaning the previous offspring. The gestation period is 227 to 240 days. Normally a single calf is born; twins are rare. The weight at birth is 27 to 50 kg. One individual weighed 250 kg at the end of its first year. The young can swim before they can walk, and they nurse underwater. The cow is a devoted mother; the calf may scramble onto her back and sun itself while she is floating at the surface. Such behavior may afford some protection against crocodiles. Data compiled by Dittrich (1976) indicate that sexual maturity comes at the age of 3 to 4 years in captivity, but that mating in the wild does not occur until males are 6 to 13 years old and females are 7 to 15 years old. Average longevity in a protected wild population is 41 years (Grzimek 1975). A captive hippopotamus lived for 54 years and 4 months (Marvin L. Jones, Zoological Society of San Diego, pers. comm.).

The hippopotamus has been extensively hunted by people for its highly prized flesh, its abundance of fat (about 90 kg per individual), the superior ivory of its teeth, and its hide. It has also been killed for sport, and because it sometimes enters cultivated fields and does extensive damage by both eating and trampling crops. It may be aggressive toward people, and is said to be the most dangerous of wild artiodactyls. It became very rare in the Nile Delta during the 18th century, and the last in Egypt was killed about 1816 (Corbet 1978). In the Nile Valley, the hippopotamus now occurs only as far north as Khartoum. It has also disappeared in most of

West and South Africa, and has become rare in much of the remainder of its range. There are still large populations in the upper Nile Valley of East Africa, and numbers have even been considered excessive in some areas (Grzimek 1975).

ARTIODACTYLA; HIPPOPOTAMIDAE; *Genus CHOEROPSIS* Leidy, 1853

Pygmy Hippopotamus

The single species, *C. liberiensis*, has a discontinuous range in the lowland forest zone from Sierra Leone to Nigeria (Meester and Setzer 1971).

Head and body length is 1,500 to 1,750 mm, tail length is about 156 mm, shoulder height is 750 to 1,000 mm, and weight is 160 to 270 kg. The body is without hair except for a few bristles on the lips and tail. The color above is a slaty greenish black, the sides are more gray, and the underparts are grayish white to yellowish green. Superficially, this animal appears to be a miniature *Hippopotamus*, but there are notable structural differences. The head is rounder and not so broad or flat, the nostrils are large and almost circular, the eyes are set on the side of the head and do not protrude, and the toes are well separated and have sharp nails. In addition, *Choeropsis* usually has only a single pair of lower incisors, as compared to two or three pairs in *Hippopotamus*. The exudation from the pores of the skin of *Choeropsis* is a clear, viscous material that makes the animal sleek to the touch. In certain lights the reddish brown tone of the skin is reflected in the beads of secretion, which then look almost blood red. Erroneously, many people have believed the animal to be "sweating blood."

The pygmy hippopotamus is found along streams, and in

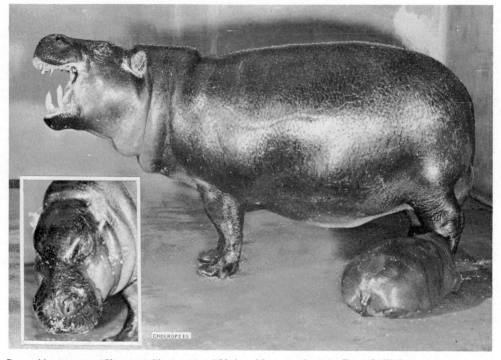

Pygmy hippopotamus (*Choeropsis liberiensis*) and 23-day-old young, photos by Ernest P. Walker.

wet forests and swamps. It is less aquatic than *Hippopotamus*, but, contrary to early reports, it does seek refuge in the water, when there is potential danger (Grzimek 1975). It sleeps by day, and wanders in the forest at night, seeking tender shoots, leaves, and fallen fruit. *Choeropsis* is found alone or in pairs. One young is born at a time. The gestation period ranges from 192 to 210 days; the average is 199 days (Grzimek 1975). Weight at birth is 3.4 to 6.4 kg (Stroman and Slaughter 1972). A captive was still living after 38 years and 10 months (Marvin L. Jones, Zoological Society of San Diego, pers. comm.).

The pygmy hippopotamus seems always to have been rare. A person may travel for days within its range and see no sign of it. *Choeropsis* is not an unduly vicious animal, but, when disturbed, it can be dangerous. It adjusts easily to captivity and some individuals breed well. It is hunted for its flesh, which is said to taste like that of *Sus*. Its survival is jeopardized by uncontrolled hunting and destruction of habitat by logging. It is classified as vulnerable by the IUCN (1978) and is on appendix 2 of the CITES.

ARTIODACTYLA; *Family CAMELIDAE*

Camels, Guanaco, Llama, Alpaca, Vicuña

This family of three Recent genera and six species apparently was found during historical time, in the wild state, from the Arabian Peninsula to Mongolia, and in western and southern South America. Through human agency, there has been a drastic reduction in the range of wild camelids, but domesticated members of the family have spread over much of the world. The sequence of genera presented here follows that of Simpson (1945), though he did not give *Vicugna* generic rank.

There is a striking difference in size between the Old World genus *Camelus* and the New World *Lama* and *Vicugna*. All genera, however, are characterized by a long and thin neck, a small head, and a slender snout with a cleft upper lip. The hind part of the body is contracted. The stomach is three chambered and ruminating. The Camelidae differ from all other mammals in the shape of their red blood corpuscles, which are oval instead of circular.

Each foot of Recent camelids has only two digits (the third and fourth). The proximal digital bones are expanded distally. The middle digital bones are wide, flattened, and embedded in a broad, cutaneous pad that forms the sole of the foot. The distal digital bones are small, not flattened on the inner surface, and not encased in hooves, bearing nails on the upper surface only. The digits are spread nearly flat on the ground. The feet are broad in *Camelus*, and slender in *Lama* and *Vicugna*. The foot bones are united to form a cannon bone. In *Vicugna*, glands are associated with the cannon bone of the hind limb. In the forearm the ulna is reduced distally, and in the shank the fibula is reduced. The knee joint is low in position, because of the long femur and its vertical placement. All the limbs are long. The forelimbs have naked callosities in the guanaco, and prominent knee pads are present in camels.

The skull is low and elongate, and there are no horns or antlers. The usual dental formula in *Camelus* is: (i 1/3, c 1/1, pm 3/2, m 3/3) × 2 = 34. In *Lama* and *Vicugna*, it is: (i 1/3, c 1/1, pm 2/1, m 3/3) × 2 = 30. However, there is some variation. The premaxillary bones of the skull bear the full number of upper incisors in the young, but only the outer incisor persists in the adult. The incisors, which are spatulate, are located in a forward, somewhat upward position. The lower incisors of *Vicugna* are unique among the Recent Artiodactyla—ever growing, with the enamel on only one side. The canines are nearly erect and pointed, and are sometimes absent in the lower jaw. The front premolars are simple

The head of a Bactrian camel (*Camelus bactrianus*), showing the closable, slitlike nostrils and the heavy eyebrows and eyelashes, which provide valuable protection in severe sandstorms. The slightly divided upper lip is also shown. Photo by Ernest P. Walker.

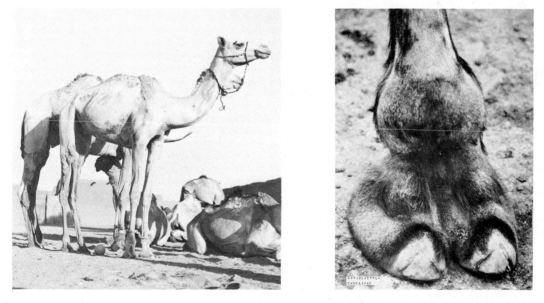

Camel (*Camelus dromedarius*), left, photo by Bernhard Grzimek. Right, front foot of *C. bactrianus* showing the hard ends of the toes and the broad foot with soft pads, which is well adapted to traversing soft sands; photo by Constance P. Warner.

A. Alpaca (*Lama pacos*), photo by Ernest P. Walker. B. Llamas (*L. glama*), photo from the U.S. National Zoological Park. C. Leg of llama. D. Guanaco (*L. guanicoe*).

and usually separated from the other cheek teeth. The molars have crescentic ridges of enamel on their crowns.

Wild camelids inhabit semiarid to arid plains, grasslands, and deserts. They are diurnal. *Camelus* and *Lama* run with a swinging stride, as the front and hind legs move in unison on each side of the body. Camelids lie down to rest and sleep. They are grazers, feeding on many kinds of grass, though camels, when hungry, will eat a wide variety of food. They have the habit of spitting the contents of their stomach at annoying objects, including incautious zoo visitors.

The geological range of this family is late Eocene to Pleistocene in North America, Pleistocene in Europe and North Africa, late Pliocene to Recent in Asia, and Pleistocene to Recent in South America (Koopman 1967).

ARTIODACTYLA; CAMELIDAE; *Genus LAMA* G. *Cuvier, 1800*

Guanaco, Llama, and Alpaca

There are three species (Cabrera 1961):

L. guanicoe (guanaco), southern Peru to eastern Argentina and Tierra del Fuego;
L. glama (llama), found in domestication from southern Peru to northwestern Argentina;
L. pacos (alpaca), found in domestication in southern Peru and western Bolivia.

The llama and alpaca are evidently domesticated descendants of *L. guanicoe*. Some authorities do not consider them more than subspecifically distinct from *L. guanicoe* (Koopman 1967; Grzimek 1975).

Head and body length is 1,200 to 2,250 mm, tail length is 150 to 250 mm, shoulder height is 900 to 1,300 mm, and weight is 48 to 140 kg. In *L. guanicoe* the upper parts are dark fawn brown, the underparts are white, and the face is blackish. The woolly coat is longest on the flanks, chest, and thighs. The limbs and neck are slender and the body is trim. Females have four mammae. *L. glama* is brown to black, or is white, usually irregularly blotched with a darker color. The body has fairly long, dense, fine wool, and the hair on the head, neck, and limbs is shorter than it is elsewhere. *L. pacos* is somewhat smaller than *L. glama;* its hairs are up to 500 mm long on the body.

The guanaco inhabits dry, open country, either in mountains or on plains. The other two species are found mainly in the Andes. The guanaco can run at a speed of 56 km per hr. It enjoys standing and even lying in mountain streams, and is said to be a good swimmer. When about to lie down, it first gets on the front knees (wrists), collapses the hindquarters, and then drops onto the chest, with the legs tucked under the body. Both the guanaco and the llama are graceful in their movements. The llama is capable of carrying a load of 96 kg at a rate of 26 km per day over rugged mountain terrain at an elevation of 5,000 meters. The hemoglobin of *Lama* has a much greater affinity for oxygen than does that of other mammals, and the blood contains more red corpuscles. These factors are partly responsible for the ability to function well at high altitudes. These animals are grazers, the diet consisting almost wholly of grass.

Alpacas at livestock fair, Arequipa, Peru, 1952. This is the best wool breed of alpacas. Photo by Hilda Heller.

The guanaco is usually found in herds of 4 to 10 females led by a single male. It is said that if the male is shot, the females will not run, but will nudge him in an effort to get him to his feet. When fighting a rival, the male emits a high-pitched scream, which changes to a low growl. Certain excretion sites are used by the guanaco and llama, as is indicated by dung heaps up to 2.4 meters in diameter and about 31 cm deep, composed of small, dry pellets.

Females give birth every other year. Mating occurs in August and September, and 10 to 11 months later a single young is born. Immediately following its birth, the young can run with surprising endurance. Lactation lasts for 6 to 12 weeks. As the young males develop, they are driven from the herd. A captive guanaco lived for 28 years and 4 months (Marvin L. Jones, Zoological Society of San Diego, pers. comm.).

The guanaco is usually gentle and can become a good pet, though an adult can be aggressive. It has been exterminated by people in most of the eastern, lowland parts of its range. It is now on appendix 2 of the CITES. The llama was evidently first domesticated in Peru about 4,500 years ago (Grzimek 1975). Both it and the alpaca have been bred for certain characters for such a long period that it may never be possible to determine if they are species distinct from *L. guanicoe*.

The llama is used primarily as a beast of burden, and is apparently the only animal with such a function to be domesticated by the native peoples of the New World. At the time of the Spanish conquest, 300,000 llamas were being used by the Incas at their silver mines. In addition, the meat of the llama is used for food, its fleece is woven into clothing, its hide is made into sandals, its fat is used in making candles, its hairs are braided into rope, and its dried excrement is used as a fuel. The alpaca is selectively bred for its wool, which is the finest of any animal. It was formerly woven into the robes worn by the Inca royalty.

ARTIODACTYLA; CAMELIDAE; *Genus VICUGNA Lesson, 1842*

Vicuña

The single species, *V. vicugna,* is found in the Andes of southern Peru, western Bolivia, northwestern Argentina, and northern Chile (Cabrera 1961). There are reports that it once may have occurred as far north as Ecuador.

Head and body length is 1,250 to 1,900 mm, tail length is 150 to 250 mm, and shoulder height is 700 to 1,100 mm (Grzimek 1975). Weight is 35 to 65 kg. The upper parts are tawny brown, the underparts are paler, and there is a white or yellowish red bib on the lower neck and chest. *Vicugna* resembles *Lama guanicoe* in general form, but is about one-fourth smaller, is paler in color, lacks the dark face, and has no callosities on the inner sides of the forelimbs. The lower

Vicuña (*Vicugna vicugna*), photo by C. B. Koford.

incisor teeth are unique among living artiodactyls, being rodentlike—ever growing, with enamel on only one side.

The vicuña inhabits semiarid rolling grasslands and plains at elevations of 3,500 to 5,750 meters. It is capable of running 47 km per hr at an elevation of 4,500 meters. To dust its fleece and scratch its body, the vicuña paws and then vigorously rolls on the ground. Vision is the best-developed sense, hearing is only moderately acute, and olfaction is poor. *Vicugna* is extremely graceful in its movements, perhaps surpassing any other hoofed mammal in this respect. Juveniles often graze while lying prone, their legs tucked under the body, and both young and adults chew cud while resting. The vicuña is a grazer, the diet consisting almost entirely of low perennial grasses.

In a study in southern Peru, Franklin (1974) found the vicuña to be one of the few ungulates to defend a year-round feeding territory and a separate sleeping territory. The basic group consisted of a single dominant adult male, several adult females, and any associated young. The usual size of such a family was 5 to 10 individuals. Territories ranged from 7 to 30 ha., the two main parts being connected by an undefended corridor. The male led the group, determined the extent of the territory, regulated group membership, and kept other males away. In addition to these groups, there were families that lacked permanent territories, and also lone males and all-male groups, usually consisting of 15 to 25 animals. In times of danger, a male with a family will warn the females with an alarm trill, and interpose itself between the source of alarm and the other animals as they retreat. Members of a group urinate and defecate in a communal dung heap.

Mating in the wild occurs in March and April, and births take place in February and March. The gestation period is 11 months, and litters contain a single young. The young can stand and walk about 15 minutes after being born. It sleeps by the side of its mother until at least 8 months old, and nurses several times a day until at least 10 months old. The dominant male of the family drives young males away when they are about 1½ years old. Young females are forced out somewhat later, but they are usually accepted into another group. Most females mate when about 2 years old, and some were still reproductively active at the age of 19 years. Maximum known longevity in captivity is 24 years and 9 months (Franklin 1974; Schmidt 1975).

During the time of the Incas, vicuñas were periodically rounded up, sheared of their wool, and then released. There were then an estimated 1,000,000 to 1,500,000 of the animals. Subsequent to the destruction of the Inca Empire, the vicuña began to be simply slaughtered in large numbers to obtain its wool and meat. In the 1950s there were still thought to be 400,000 left, but intensification of hunting pressure, commercial demand, and the spread of domestic livestock resulted in a drastic decline. By 1967 there were only about 10,000 vicuñas. Conservation efforts have since allowed numbers to increase to 80,000. The vicuña is classified as vulnerable by the IUCN (1978) and as endangered by the USDI (1980), and is on appendix 1 of the CITES.

ARTIODACTYLA; CAMELIDAE; **Genus CAMELUS**
Linnaeus, 1758

Camels

There are two species (Corbet 1978; Ellerman and Morrison-Scott 1966):

C. bactrianus (Bactrian, or two-humped camel), formerly found throughout the dry steppe and semidesert zone from Soviet Central Asia to Mongolia;
C. dromedarius (dromedary, or one-humped camel), probably once found as a wild animal throughout the Arabian region, but known with certainty only in the domestic state.

The name *C. ferus* is often used in place of *C. bactrianus*.

Head and body length is 2,250 to 3,450 mm, tail length is about 550 mm, shoulder height is 1,800 to 2,100 mm, and weight is 450 to 690 kg. The color varies from deep brown to dusty gray. In *C. dromedarius,* which has only one hump on the back, the pelage is relatively short, soft, fine, and woolly. In *C. bactrianus,* which has two humps, the long hairs (255 mm) are thickest on the head, neck, humps, forelegs, and tip of the tail. Camels shed their winter coat so rapidly that it comes off in large masses, giving the animal the ragged appearance shown in the photograph of *C. bactrianus* that accompanies this account. The skin has almost no sweat glands.

Both species have a long head and neck, and a relatively short tail. The eyes have heavy lashes, the ears are small and haired, and the upper lip is deeply divided. The slitlike nostrils can be closed to keep out dust and sand. There is a groove from each nostril to the cleft upper lip, so that any moisture from the nostrils can be caught in the mouth. The feet have two toes and undivided soles. The long, slender legs have prominent knee pads. *C. bactrianus* has shorter legs than *C. dromedarius,* and is therefore not as tall; it also has shorter and harder feet, and is more docile, slower, and easier to ride.

Camels live in arid areas. *C. bactrianus* is found on the Gobi Steppe along rivers, but moves to the desert as soon as the snow melts. Camels can withstand extreme heat and cold, and are said to be good swimmers. Their characteristic rolling gait is accomplished by simultaneously bringing up both legs on the same side. Speeds of over 65 km per hr have been recorded in a pursuit of *C. bactrianus* (Dash et al. 1978). Over a four-day period, a camel is able to carry 170 to 270 kg at a rate of 47 km per day and 4 km per hr. Sight is keen and the sense of smell is extremely good.

Contrary to popular legend, there is no evidence that camels store water in the stomach. Although they are adapted for conservation of water, they will lose weight and strength if they go for long periods without drinking. According to Gauthier-Pilters (1974), domestic *C. dromedarius* in the Sahara can obtain sufficient water for months from desert vegetation, and can survive a water loss exceeding 40 percent of its body weight. Camels can and will drink brackish and even salt water. In drinking as much as 57 liters of water, camels restore the normal amount of body fluid. They eat practically any vegetation that grows in the desert or semiarid regions. If forced by hunger, they will eat fish, flesh, bones, and skin. They thrive on salty plants that are wholly rejected by other grazing mammals. Camels are said to need halophytes in their diet and to lose weight if they are lacking. When camels are well fed, the hump is erect and plump, but when they do not have adequate food, the hump shrinks and often leans to one side.

Wild *C. bactrianus* is found alone or in groups, sometimes with over 30 individuals (Dash et al. 1978). In the Sahara, domestic *C. dromedarius* is left on its own for four or five months of the year, including the mating season (Gauthier-Pilters 1974). At that time, three kinds of herds form: groups of bachelor males; groups of adult females with their newborn; and groups of up to 30 adult females, along with their one-and two-year-old offspring, all led by a single adult male.

A. Camel (*Camelus bactrianus*), the unusually ragged appearance is due to spring shedding of its winter coat, photo by Fritz Grogl. B. Camel (*C. dromedarius*), photo from U.S. National Zoological Park. Inset: *C. bactrianus*, photo by Ernest P. Walker.

The peak birth season is March to April in *C. bactrianus* and February to May in *C. dromedarius* (Schmidt 1973). A female camel usually gives birth every other year. The gestation period is 370 to 440 days. Litters contain a single offspring, rarely two. The calf has soft fleece, utters a gentle "bah," and lacks the knee pads of the adult. By the end of the first day it can move about freely. At 4 years the young camel becomes wholly independent. Full growth is attained at 5 years. Potential longevity is up to 50 years.

Camels are well known as beasts of burden. In addition, they supply milk, meat, wool, hides, sinews, and bone. Chemicals can be extracted from their dung and urine. The milk is fermented to produce kumiss, a common intoxicating liquor.

Camelus dromedarius may have persisted in Arabia as a wild animal until about 2,000 years ago (Corbet 1978). It was domesticated at least by 1800 B.C., and perhaps as early as 4000 B.C., and subsequently spread throughout southwestern Asia and North Africa. At various times, populations also became established in Spain, Namibia, and Australia (Grzimek 1975). In the middle of the 19th century, camels were introduced into the southwestern United States as a potential source of efficient transportation. Enthusiastic advocates even predicted that because of their versatility camels would supplant beef cattle in the Southwest. Due to various factors, such as the building of railroads, the experiment was a failure.

In Central Asia, *C. bactrianus* was domesticated as early as the third and fourth centuries B.C. Through human agency, its range was extended from Asia Minor to northern China (Grzimek 1975). Wild populations also remained common until the 1920s, but subsequently became restricted to relatively small areas of southwestern Mongolia and northwestern China. About 300 to 500 individuals now survive. The wild camel is classified as vulnerable by the IUCN (1976) and endangered by the USDI (1980).

ARTIODACTYLA; *Family TRAGULIDAE*

Chevrotains, Mouse Deer

This family of two Recent genera and four species occurs in West and Central Africa, and south-central and southeastern Asia.

These small, graceful animals bear some resemblance to deer (Cervidae), but also look externally like agoutis (Rodentia: Dasyproctidae). The head is small and the snout is pointed; the narrow, slitlike nostrils are located in the naked nasal surfaces. There are no facial or foot glands. The legs are long, thin, and delicate, being about the size of a lead pencil. In the genus *Tragulus* the pairs of metacarpal and metatarsal bones in the limbs are united into a single unit, known as a cannon bone. In *Hyemoschus*, a cannon bone is not formed until old age. Each foot has four well-developed digits, though the lateral digits of *Hyemoschus* do not touch the ground when the animal stands. The stomach is three chambered and ruminating. Females have four mammae.

There are no horns or antlers. The dental formula is: (i 0/3, c 1/1, pm 3/3, m 3/3) × 2 = 34. The upper canines are well developed (especially in the males), protrude below the lips, and are narrow, curved, and pointed. The lower canines look like incisors. All the premolars, except the last, function in cutting the food; there are no premolars that resemble the canines. The molars bear crescentic ridges of enamel on their crowns. All the cheek teeth are arranged in a continuous series.

The geological range of this family is late Miocene to early Pliocene in Europe, Pleistocene to Recent in Africa, and late Miocene to Recent in Asia.

ARTIODACTYLA; TRAGULIDAE; *Genus HYEMOSCHUS* Gray, 1845

Water Chevrotain

The single species, *H. aquaticus*, occurs in the lowland forest zone from Sierra Leone to western Uganda (Meester and Setzer 1971).

Head and body length is 750 to 850 mm, tail length is 100 to 150 mm, shoulder height is 305 to 355 mm, and weight is 10 to 15 kg. Coloration is a rich brown, marked along the body with several longitudinal rows of white spots that become fused to form broken lines on the flanks. A white stripe runs along the edge of the jaw and the sides of the neck. The underside of the tail is white. The color and the arrangement of the spots are not constant. *Hyemoschus* has a small head, a pointed snout, a short tail, and a high body set on long,

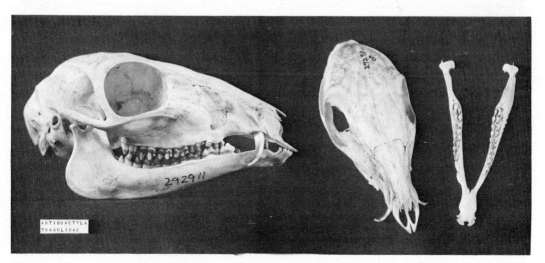

Asiatic mouse deer (*Tragulus* sp.), photos by P. F. Wright of specimen in the U.S. National Museum.

Water chevrotains (*Hyemoschus aquaticus*), photo by Ernest P. Walker.

slender legs. Neither sex has antlers, but the upper canines of the male are developed into tusks similar to those of the musk deer (*Moschus*).

The information for the remainder of this account was taken from Dubost's (1978) report of his study in Gabon. *Hyemoschus* lives in evergreen forest, usually within 250 meters of water. However, it enters the water only when there is danger. It is nocturnal, resting by day in a hidden place, such as a pile of vegetation. The diet consists almost exclusively of fruit that has fallen to the ground.

Population density is 7.7 to 28 individuals per sq km. Groups consist of an adult female and her young of one or two years. Such a group occupies a home range averaging about 13 to 14 ha. The young are usually found separately from the female, except for brief periods of suckling. Females are sedentary and remain in an area all their lives, once a home range is established. Males are solitary and have home ranges of 20 to 30 ha., which overlap the ranges of two females. There seems to be a continuous replacement of males in any given area, none having been found in the same range for more than a year. There is no evident territoriality, but individuals are well spaced and seldom in contact.

Females bear a single young annually. Gestation lasts 6 to 9 months, and lactation 3 to 6 months. The young reach maturity when they are 9 to 26 months old, and then leave the home range of their mother. Some individuals live to an age of 11 to 13 years, which is considered old age. Hunting by people is evidently a serious cause of mortality, and overall numbers are decreasing as a result.

ARTIODACTYLA; TRAGULIDAE; **Genus TRAGULUS**
Brisson, 1762

Asiatic Chevrotains or Mouse Deer

There are two subgenera and three species (Ellerman and Morrison-Scott 1966; Lekagul and McNeely 1977; Chasen 1940; Medway 1977):

subgenus *Moschiola*

T. meminna, peninsular India, Sri Lanka;

subgenus *Tragulus*

T. napu, southern Thailand and Indochina, Malay Peninsula, Sumatra, Borneo, many small islands of the East Indies, Balabac Island (extreme southwestern Philippines);

T. javanicus, Thailand, Indochina, Malay Peninsula, Sumatra, Java, Borneo, many small islands of the East Indies.

These are the smallest artiodactyls. Head and body length is 400 to 750 mm, tail length is 25 to 125 mm, shoulder height is 200 to 350 mm, and weight is 0.7 to 8.0 kg. The hair is short, even, and close, and is shortest on the head and underparts. All species are colored some shade of brown and have whitish underparts. The white expanse of the chin and throat is variously, though symmetrically, cut by patches of brown. Certain forms have longitudinal rows of spots or stripes of white or buff on the back and sides. There are no horns, but there are canine teeth in both jaws. The upper canines are especially well developed, and are enlarged into tusks in males. Females have four mammae.

These mouse deer generally live among undergrowth on the edges of heavy lowland forests. They are seldom found far from water. They are present in large numbers, but are seldom seen, as they are shy, retiring, and nocturnal. They customarily traverse tiny tunnellike jungle trails. Lekagul and McNeely (1977) noted that *T. javanicus* may shelter in a rock crevice, a hollow tree, or dense vegetation. The diet consists chiefly of grass, the leaves of low bushes, and fallen jungle fruit and berries.

Tragulus is solitary, except during the breeding season. Mating occurs throughout the year in some areas. Females may mate again within 48 hours of giving birth. Gestation periods of 140 to 177 days have been reported. There is usually a single offspring, occasionally two. The young is fully formed and active at birth, and is able to stand in 30 minutes. Sexual and physical maturity comes by the age of 5 months (Medway 1978; Lekagul and McNeely 1977). A specimen of *T. napu* was still living after 14 years in captivity (Marvin L. Jones, Zoological Society of San Diego, pers. comm.).

These animals are preyed upon by a large number of carnivorous mammals and snakes, and also are widely sought by native people for use as food. They tame readily and make

Asiatic chevrotain: Top, *Tragulus meminna*, photo by Cyrille Barrette; Bottom, *T. napu*, photo from New York Zoological Society.

good pets, but are very delicate creatures. The mouse deer plays the same role in Malay folklore as the red fox in European folklore.

ARTIODACTYLA; *Family CERVIDAE*

Deer

This family of 17 Recent genera and 38 species occurs nearly throughout North America, South America, and Eurasia, on most associated continental islands, and in northern Africa. Wild populations of deer have been established through introduction by people in Cuba, New Guinea, Australia, New Zealand, and certain other places where the family does not naturally occur. The sequence of genera presented here is based on the classifications of Simpson (1945) and Ellerman and Morrison-Scott (1966), in which five subfamilies are recognized: Moschinae, for the genus *Moschus;* Hydropotinae, for *Hydropotes;* Muntiacinae, for *Muntiacus* and *Elaphodus;* Cervinae, for *Dama, Axis, Cervus,* and *Elaphurus;* and Odocoileinae, for all other genera. Some authorities, such as Corbet (1978), place *Moschus* in an entirely separate family, the Moschidae.

Head and body length is 720 to 2,900 mm, the tail is relatively short, shoulder height is 320 to 1,900 mm, and weight is 7 to 825 kg. *Pudu* is the smallest genus in the family and *Alces* is the largest. Females are generally slightly smaller and more delicately built than males; their neck in particular is not so full as that of males during the rutting season, and the hair on the neck is not so heavy. The general coloration of deer is brownish; the young of most species and the adults of some species are spotted. The winter coats are darker than the summer coats, at least in those species found in temperate latitudes.

These slim, long-legged artiodactyls are best characterized by the presence of antlers, which are lacking in only two genera, *Moschus* and *Hydropotes*. Antlers are borne only by the males, except for *Rangifer,* in which both sexes have antlers. Male deer are illustrated more than females in the following pages, inasmuch as the antlers are often the principal external character by which the genera can be distinguished. Antlers are appendages of the skull, composed of a solid bony core and supported on permanent skin-covered pedicels. In the temperate zones the antlers begin to grow early in the summer, during which time they are well supplied with blood. They are soft and tender and are covered with a thin skin, which bears short fine hairs and has the appearance of velvet. By late summer the antlers have attained their maximum size. The blood then gradually recedes, and the thin skin with the velvety hair dries, loosens, and is rubbed off. Before the velvet is shed, all circulation of blood has ceased and, when shedding takes place, there is no bleeding and probably not even discomfort for the animal. After the velvet is rubbed off, the antlers serve as sexual ornaments and weapons. In the Northern Hemisphere the antlers are shed each year from January to April, following the mating season. The species in temperate climates usually complete the shedding process in two to three weeks. In a single deer, both antlers are usually shed within several hours or days of each other. At least some of the tropical deer do not have a fixed breeding season, and these forms do not shed their antlers at any certain time. Deer usually grow their first set of antlers when one or two years old, and these first-year horns are generally short, almost straight spikes. The antlers become larger and acquire more points in succeeding years until the animal is mature, when the antlers attain the shape typical of the species. Normal antler growth is dependent upon an adequate diet, for if certain minerals or vitamins are lacking, the antlers may be stunted or dwarfed.

Nearly all deer have facial glands, located in a depression in front of the eye. The pit is lined with a continuation of the skin of the face. Glands also occur on the limbs. There is a cannon bone that is formed by the fusion of the two main foot bones. The ulna and fibula are reduced. As in other artiodactyls, the first digit is absent from each foot. The third and fourth digits are well developed and bear the weight of the animal, and the second and fifth digits are small. The stomach is four chambered and ruminating. A gall bladder is lacking, except in *Moschus*. Females usually have two pairs of mammae, but *Moschus* has only one pair.

The dental formula is: (i 0/3, c 0–1/1, pm 3/3, m 3/3) $\times$ 2 = 32–34. The upper canines are saber shaped and enlarged in the antlerless species, and are absent in most others. The lower canines resemble the incisors. The low cheek teeth bear crescentic ridges of enamel on their crowns.

The extremely varied habitats of deer include forests, swamps, brush country, deserts, and arctic tundra. Deer are usually good swimmers, some species being semiaquatic. Some undergo seasonal migrations. The diet is herbivorous and may include grass, bark, twigs, and young shoots. Most deer associate in groups. The stags often fight among themselves for possession of a harem. Females in tropical regions may enter estrus several times a year, but in temperate climates there is a definite mating season in the late fall or winter. The gestation period ranges from about 160 days in *Moschus* to 10 months in *Capreolus,* the latter being the only cervid genus known to have delayed implantation. The number of young is usually one or two, but three or four are sometimes born.

The geological range of this family is early Oligocene to Recent in Asia, late Oligocene to Recent in Europe, early Miocene to Recent in North America, and Pleistocene to Recent in South America. The extinct Pleistocene giant deer (*Megaloceros*), many complete skeletons of which have been found in peat bogs in Ireland, had antlers that sometimes measured over 3½ meters from tip to tip.

Père David's deer (*Elaphurus davidianus*), photo by D. W. Yalden.

A. Central American white-tailed deer (*Odocoileus virginianus*); B. Père David's deer (*Elaphurus davidianus*); photos by Ernest P. Walker. C. Red deer (*Cervus elaphus*), photo by Erna Mohr. D. Père David's deer (*Elaphurus davidianus*) showing deep preorbital lachrymal pits, photo by Ernest P. Walker.

Musk deer (*Moschus moschiferus*): Top, photo by Bernhard Grzimek; Inset: photo from *Guide to the Great Game Animals*, British Museum (Natural History); Right, photo from East Berlin Zoo.

ARTIODACTYLA; CERVIDAE; **Genus MOSCHUS**
Linnaeus, 1758

Musk Deer

Groves (1975*a*, 1980*b*) recognized three species:

M. moschiferus, Siberia, northern Mongolia, Manchuria, Sakhalin;

M. chrysogaster, Himalayan region, central and southern China, northern Viet Nam;

M. sifanicus, Himalayan region, central China, Burma.

Head and body length is 800 to 1,000 mm, tail length is 38 to 60 mm, shoulder height is 510 to 610 mm, the rump is about 50 mm higher than the shoulder, and weight is 7 to 17 kg. The body is covered with long, thick, bristly hairs, which are usually white at the base; they are pithy and protect the animal from the severe weather encountered at high elevations. Coloration appears to be quite variable, possibly as a result of age and season, but generally it is a rich dark brown, mottled and speckled with light gray above, and paler beneath. The chin, inner borders of the ears, and insides of the thighs are whitish, and occasionally there is a spot of white on each side of the throat.

Although antlers are lacking, the upper canine teeth are developed as tusks in the male, about 75 mm in length. Those of the female are smaller. A musk gland in the abdomen of

the male (three years of age or older) secretes a brownish waxlike substance. About 28 grams of this secretion can be obtained from a single animal. Unlike other Cervidae, *Moschus* has a gall bladder. Females have two mammae.

The preferred habitat of musk deer is forest and brushland

at elevations of 2,600 to 3,600 meters. These animals are most active in the morning and evening, and sleep in "forms," like those of hares (*Lepus*), during the day. The diet consists of a variety of vegetation, such as grass, moss, and tender shoots. In winter, twigs, buds, and lichens are taken.

Musk deer are shy, timid, and generally solitary. Almost never are more than two or three individuals found together. Mating usually takes place in January, the gestation period is 160 days, and the number of offspring is one, rarely two. The young are spotted, and sexual maturity is attained at about 1 year of age. A captive lived for 13 years and 11 months (Marvin L. Jones, Zoological Society of San Diego, pers. comm.).

The substance obtained from the gland of the male is one of the musks used extensively in the manufacture of perfume and soap. Because of the small quantity of musk obtained per animal, a high price is paid for this product, thus making *Moschus* much sought after by native and European hunters. Also, the pouch in which the musk is stored is highly valued. Many females and young that do not produce musk are captured in the traps set for the adult males. Because of excessive hunting, musk deer have declined seriously in numbers and have been extirpated in some areas. The populations of the Himalayan region are classified as vulnerable by the IUCN (1974) and as endangered by the USDI (1980), and are on appendix 1 of the CITES.

ARTIODACTYLA; CERVIDAE; **Genus HYDROPOTES**
Swinhoe, 1870

Chinese Water Deer

The single species, *H. inermis,* is found in the lower Yangtze Basin of east-central China, and in Korea (Corbet 1978). The scientific name means "unarmed water drinker."

Head and body length is 775 to 1,000 mm, tail length is 60 to 75 mm, shoulder height is 450 to 550 mm, and weight is 9 to 15 kg. The hair is generally thick and coarse, and is longest on the flanks and rump. The top of the face is grayish to reddish brown, the chin and upper throat are whitish, and the back and sides are usually a uniform yellowish brown, finely stippled with black. The underparts are white. Both sexes lack antlers, but the upper canine teeth, especially in the males, are enlarged, forming fairly long, slightly curved tusks. There is a small inguinal gland present on each side in both sexes; this is the only known case of such glands in the Cervidae. Females have four mammae.

Hydropotes lives among tall reeds and rushes along rivers, and also frequents tall grass on mountains and cultivated fields. When disturbed, it humps its back and travels by a series of leaps, much as do *Lepus* and *Sylvilagus*. The diet includes reeds, coarse grasses, and vegetables.

Hydropotes rarely congregates in herds. The young are born in the spring, and are marked with white spots and stripes. The gestation period is 6 months (Grzimek 1975). Females are sometimes said to give birth to up to eight young at a time, more than are produced by any other kind of deer. In a survey of zoos, however, Dobroruka (1970a) found that there were usually only two offspring per birth, occasionally three. One litter contained five young, but two of these were born dead. One specimen of *Hydropotes* was still living after 11 years and 5 months in captivity (Marvin L. Jones, Zoological Society of San Diego, pers. comm.).

The Chinese water deer has been extensively bred in captivity. Many individuals escaped from the Duke of Bedford's Woburn Park in England, and established a wild-living population. *Hydropotes* also has been introduced in France (Corbet 1978).

Chinese water deer (*Hydropotes inermis*), photo by Ernest P. Walker. Insets: head, photo by Erna Mohr; Fawn, photo from New York Zoological Society; Skull, photo from *Guide to the Great Game Animals*, British Museum (Natural History).

ARTIODACTYLA; CERVIDAE; *Genus MUNTIACUS*
Rafinesque, 1815

Muntjacs, Barking Deer

There are five species (Ellerman and Morrison-Scott 1966; Lekagul and McNeely 1977; Medway 1977; Corbet 1978; Chasen 1940; Sheng and Lu 1980):

M. muntjak, India and Nepal to southern China and Malay Peninsula, Sri Lanka, Hainan, Sumatra, Bangka, Java, Bali, Kangean Islands, Rhio Archipelago;

M. rooseveltorum, known only from the type locality in Laos;

M. reevesi, southern China, Taiwan;

M. crinifrons, east-central China;

M. feae, Tenasserim and adjacent parts of Thailand.

Head and body length is 800 to 1,130 mm, tail length is 110 to 240 mm, shoulder height is 450 to 580 mm, and weight is 14 to 28 kg. The body is covered with short, soft hairs, except for the ears, which are sparsely haired. Coloration varies from deep brown to yellowish or grayish brown with creamy or whitish markings. In one species the head is decidedly lighter than the body. Antlers are carried only by males and are shed annually. They rarely exceed 125 to 152 mm in length, and are carried on long, bony, hair-covered pedicels. The upper canine teeth of the males are elongated into tusks, which curve strongly outward from the lips; they are thus capable of inflicting serious injuries to dogs and other animals. The females have small bony knobs and tufts of hair where the antlers occur in the males.

Muntjacs are usually found in forests and areas of dense vegetation, from sea level to medium elevations in hilly country. They generally do not occur far from water. Both diurnal activity and nocturnal activity have been reported in the wild, and observations in captivity indicate that *M. reevesi* is mainly crepuscular (Yahner 1980). Barking deer are dainty little creatures; they are always on the alert and lift their feet high when walking. Their diet includes grasses, low-growing leaves, and tender shoots.

In a study of *M. muntjak* in Sri Lanka, Barrette (1977) found fallen fruit to be a major food item. Population density in the study area was 2.5 individuals per sq km when the animals were concentrated during a drought, but may have been 1.5 per sq km when the animals were evenly distributed. Most sightings were of lone individuals, and no group of over 4 was seen. Reproduction occurred throughout the year and the gestation period was about 210 days. Muntjacs emit a deep, barklike sound when they are alarmed or during the breeding season. If they sense a predator in the area, they will "bark" for 1 hour or more.

Medway (1978) stated that there was no evidence of a specific breeding season in the Malay Peninsula, but Lekagul and McNeely (1977) wrote that mating in Thailand occurs in December and January, and births in June or July. Sheng and Lu (1980) reported that newborn *M. crinifrons* are seen in April. Chaplin and Dangerfield (1973) reported that births of captive *M. reevesi* in England occurred mostly from March to August, and that gestation was approximately 209 to 220 days. Young muntjacs are usually born in dense jungle growth, where they remain hidden until they can move about with the mother. At birth the young weigh about 550 to 650 grams. A specimen of *M. muntjak* was still living after 17 years and 7 months in captivity (Marvin L. Jones, Zoological Society of San Diego, pers. comm.).

Barking deer are hunted for their meat and skins. They thrive in captivity and are found in many zoos. They are considered a nuisance in some areas, because they destroy trees by ripping off the bark. *M. feae,* restricted to a small area and subject to uncontrolled hunting, is classified as endangered by the IUCN (1972) and the USDI (1980). *M. crinifrons,* long known by only three specimens, is classified as indeterminate by the IUCN (1972) but was found living in four provinces of east-central China in the 1970s (Sheng and Lu 1980). Introduction by human agency has resulted in the establishment of wild populations of *M. reevesi* in England and France (Corbet 1978).

ARTIODACTYLA; CERVIDAE; *Genus ELAPHODUS*
Milne-Edwards, 1871

Tufted Deer

The single species, *E. cephalophus,* occurs in eastern and southern China, and in northern Burma (Ellerman and Morrison-Scott 1966).

This dainty deer is somewhat larger than *Muntiacus,* having a head and body length of 1,100 to 1,600 mm, a tail length of 70 to 150 mm, a shoulder height of 500 to 700 mm, and a weight of 17 to 50 kg. The body is covered with coarse, almost spinelike hairs that give a shaggy appearance. The general color of the upper parts is deep chocolate brown, the underparts are white, and the head and neck are gray. In some cases, a pale streak extends forward from the pedicel and above the eye, and the tuft on the forehead is blackish brown. There are white markings on the back tips of the ears and on the underside of the tail.

Muntjac (*Muntiacus muntjak*), photo from East Berlin Zoo.

Muntjacs (*Muntiacus muntjak*): Top, photo from East Berlin Zoo; Bottom, photo by Cyrille Barrette.

Elaphodus is similar to *Muntiacus*. Although the canines are comparable in size, the antlers of *Elaphodus* are much smaller, frequently completely hidden by the tuft on the forehead. Other distinguishing characteristics are that the bony pedicels, on which the antlers grow, are shorter than those of *Muntiacus*, do not form heavy ridges on the forehead, and converge at the tips.

The tufted deer inhabits areas of dense vegetation at elevations of 900 to 2,600 meters, and is said always to be found near water. When feeding, it carries its tail high; the tail flops

Tufted deer (*Elaphodus cephalophus*), photo from *Deer of All Lands*, Lydekker.

with every bounce, displaying the white underside, much as does that of *Odocoileus virginianus*. The diet includes grasses and other vegetation.

Elaphodus is usually solitary, but occasionally travels in pairs. Both sexes "bark" when suddenly alarmed and during the mating season in April and May. One or two young are born after a gestation period of about six months. The young are colored like their parents, but have a row of spots along each side of the midline of the back. One specimen lived for seven years at the London Zoo.

ARTIODACTYLA; CERVIDAE; **Genus DAMA** *Frisch, 1775*

Fallow Deer

The single species, *D. dama*, originally occurred in the Mediterranean region of southern Europe, from Asia Minor and Palestine to Iran, and probably in northern Africa and Ethiopia (Ellerman and Morrison-Scott 1966; Corbet 1978; Meester and Setzer 1971; Chapman and Chapman 1975; Harrison 1968). The populations in Asia, to the east and south of Turkey, and in Africa, have sometimes been referred to a separate species, *D. mesopotamica*. Opinion seems about evenly divided as to whether *Dama* should be considered a distinct genus or included within *Cervus*.

Head and body length is 1,300 to 1.750 mm, tail length is 150 to 230 mm, shoulder height is 800 to 1,050 mm, and weight is 40 to 100 kg. Chapman and Chapman (1975) identified four main color varieties, more than in any other kind of deer: (1) common—upper parts rich brown with white spots and underparts whitish in summer, dark grayish brown with spots barely detectable in winter; (2) menil—pale fawn, heavily spotted with white all year; (3) white—not truly albino and probably restricted to parks; and (4) black—not jet black, but dark, especially in summer, with barely detectable gray brown spots. In the common phase, the summer coat is acquired in May and the winter coat in October.

Only the males have antlers; their peculiar shape is the basis for placing this deer in a genus distinct from *Cervus*. The antlers are flattened and palmate with numerous points. The males develop a single unbranched horn the second year, and each succeeding year the horns are larger and have more points, until the fifth or sixth year. The antlers are generally shed in April, and the new ones are full grown and free of velvet in August. The front outer curve of the antlers is 635 to 940 mm, and the tip-to-tip measurement is 305 to 762 mm. There are no upper canine teeth.

There is a great variety of habitats, but generally some

Fallow deer (*Dama dama*): A. Male in spring pelage, photo from Zoological Society of London; B. Female in winter pelage, photo from New York Zoological Society.

forest is required. Feeding occurs mainly in the early morning and from late afternoon to evening. The fallow deer is predominantly a grazer, but also commonly browses trees and shrubs (Chapman and Chapman 1975).

Reported population densities range from 8 to 43 individuals per 100 ha. Social behavior varies; in some areas *Dama* does not appear gregarious, but in others herds of up to 30 individuals are commonly seen throughout the year. There does not seem to be a dominance hierarchy, but a doe is usually the group leader. During the mating season the adult males establish small territories, centered at least 100 meters apart (Chapman and Chapman 1975). Throughout this season, the males do a dancelike ritual and bellow in a deep voice, which presumably attract the females that gather around them. There is much fighting between rival males, but it mostly involves ritualized shoving with the antlers.

Mating takes place in September or October in the Northern Hemisphere, and births occur in the spring. The estrous cycle lasts about 24 to 26 days, and the mean gestation period has been reported as 229 days in Germany and 237 days in New Zealand (Chapman and Chapman 1975). There is usually a single fawn, which is slightly darker than commonly colored adults and is spotted with white. Females attain sexual maturity at about 16 months of age and subsequently breed each year. The life span is about 15 years.

There is uncertainty regarding the natural distribution of this genus in Recent times. *Dama* occurred over much of Europe in the Pleistocene, but by the end of that epoch was restricted to the Mediterranean region, or possibly only to Asia. It probably also occurred naturally from Morocco to Egypt and in Ethiopia. Within historical time, it was widely reintroduced in Europe by the Phoenicians and Romans, and apparently was brought to Britain by the Normans. Wild populations now occur as far north as Sweden and western Russia. Free-living herds have also been established by people in the United States, Canada, the West Indies, Argentina, Chile, Peru, Uruguay, South Africa, Madagascar, Japan,

Australia, Tasmania, New Zealand, and Fiji (Chapman and Chapman 1975, 1980; Grzimek 1975).

While the fallow deer was spreading into new areas, however, it was disappearing from its original range, because of both excessive hunting and climatic changes. The last evidence of its presence in Ethiopia dates from about 1,000 years ago. The genus apparently disappeared from Africa in the 19th century, from the mainland of Greece in the early 1900s, and from Sardinia in the 1950s. At the same time, it became very rare in the Asian parts of its range (Chapman and Chapman 1975, 1980; Meester and Setzer 1971).

The subspecies that formerly occurred from Palestine to Iran, *D. d. mesopotamica*, is classified as endangered by the IUCN (1972) and the USDI (1980), and is on appendix 1 of the CITES. This subspecies was once considered extinct, but in the 1950s a small population, probably containing fewer than 50 individuals, was found along several rivers in western Iran, near the border with Iraq. In the late 1970s, prior to recent disturbances in the region, this population was reportedly well protected and starting to increase in numbers; reintroduction and captive breeding projects were also under way (Chapman and Chapman 1980).

ARTIODACTYLA; CERVIDAE; **Genus AXIS**
Hamilton Smith, 1827

Axis Deer

There are two subgenera and two species (Ellerman and Morrison-Scott 1966; Grzimek 1975; IUCN 1972, 1976):

subgenus *Axis*

A. axis (chital, spotted deer), India, Nepal, Sikkim, Sri Lanka;

Axis deer (*Axis axis*), photo from Zoological Society of London.

subgenus *Hyelaphus*

A. porcinus (hog deer), northern India to Indochina, Bawean Island (Java Sea), Calamian Islands (Philippines).

These species have been placed in the genus *Cervus* by such authorities as Koopman (1967), Lekagul and McNeely (1977), and Corbet (1978). The isolated populations of *A. porcinus* on Bawean Island and the Calamian Islands, respectively, have often been treated as distinct species, *A. kuhli* and *A. calamianensis*.

Head and body length is 1,000 to 1,750 mm, tail length is 128 to 380 mm, shoulder height is 600 to 1,000 mm, and weight is 27 to 110 kg. The hair is coarse and longest on the flanks. There is no mane on the throat or neck. Coloration varies not only between the species but also from season to season. The general color is a bright rufous fawn or is yellowish brown to brownish. During part of the year the upper parts of *A. axis* are beautifully marked with small white spots. Both species have a dark dorsal stripe, white underparts, and a whitish tail undersurface. Albino strains have been developed in captivity.

The species *A. axis* is slender, graceful, and of medium body build, whereas *A. porcinus* is stocky and has shorter legs. The antlers are carried on pedicels and are three tined, the brow tine nearly forming a right angle with the beam. Upper canine teeth are usually absent, and the tail is usually relatively long and slender.

Axis deer frequent grasslands and open forest, seldom penetrating into heavy jungles. Normally they rest during the hotter part of the day and move about in the early morning and late afternoon. They may become nocturnal in the summer, or when molested by people. They take readily to water and are said to be good swimmers. Lekagul and McNeely (1977) noted that the term "hog deer" refers to the habit of *A. porcinus* of running through the underbrush with its head held low, in the manner of a hog, rather than leaping over obstacles, as do other deer. Both species are predominantly grazers, though they occasionally browse and are fond of various fallen flowers and fruits of forest trees.

Axis deer (*Axis axis*), photo from San Diego Zoological Garden.

In a study of *A. axis* in central India, Schaller (1967) found a population density of 23 deer per sq km. Observations suggested that individuals spend their entire lives in a relatively small area. Home ranges during about half of the year were approximately 500 ha. for a male and 180 ha. for a female. The animals are gregarious, usually being found in herds of 5 to 10 individuals and sometimes forming aggregations of 100 to 200. Some groups seen by Schaller contained all ages and sexes, while others had only adult males or only adult females and young. Group composition changed constantly, the only lasting association being that of a mother and its young. Herd movements were usually led by an adult female, but in social interaction the male was dominant. During the mating season, individual males roamed about in search of estrous females; they defended such females while courting but did not prevent other males from entering the herd, and no territorial behavior was evident. When a few males came together, a rank order was established, based on displays and sparring with the antlers. Larger males, with the longest antlers, attained dominance and had a disproportionately large share of the females.

Schaller (1967) called *A. porcinus* essentially solitary, but stated that up to 40 individuals may assemble in certain feeding areas or during the mating season. At such congregation sites, each male seems to try to acquire an estrous female, and then leaves with her. The mating season of *A. porcinus* peaks in September and October.

In *A. axis,* reproduction continues throughout the year, but mating peaks from March to June and most births occur in the cool season from January to May. If a fawn dies, the mother may mate again and give birth a second time in the course of a year. The gestation period is 8 to 8½ months. A single offspring is the rule; twins are rare. The young remain hidden in dense cover, with the mother feeding nearby, until it has sufficient strength to roam with the herd. It may become independent after 12 months, but young females sometimes stay with the mother for nearly 2 years. Females often attain

sexual maturity at 14 to 17 months of age (Schaller 1967). One specimen of *A. axis* lived for 20 years and 9 months in captivity (Marvin L. Jones, Zoological Society of San Diego, pers. comm.).

People have established wild-living populations of *A. axis* in Yugoslavia and New Guinea, and of *A. porcinus* in Sri Lanka. Human hunting, however, has caused axis deer to become rare in many parts of their natural range. The subspecies *A. porcinus annamiticus* of Thailand and Indochina, *A. p. kuhli* of Bawean Island, and *A. p. calamianensis* of the Calamian Islands are listed as endangered by the USDI (1980) and are on appendix 1 of the CITES. The IUCN (1972, 1976) classifies *A. p. kuhli* as rare and *A. p. calamianensis* as vulnerable.

ARTIODACTYLA; CERVIDAE; **Genus CERVUS** Linnaeus, *1758*

Red Deer, Wapiti

There are seven subgenera and nine species (Ellerman and Morrison-Scott 1966; Grzimek 1975; Chasen 1940; Laurie and Hill 1954; Lekagul and McNeely 1977; Medway 1977; Corbet 1978; Hall 1981):

subgenus *Rusa*

C. unicolor (sambar), India to southeastern China and Malay Peninsula, Sri Lanka, Taiwan, Hainan, Sumatra, Borneo, and many small nearby islands;

C. timorensis (Sunda sambar), Java, Celebes, Timor, Moluccas, and many small nearby islands;

C. mariannus (Philippine sambar), Philippines;

subgenus *Rucervus*

C. duvauceli (barasingha), India;

A. North American wapiti (*Cervus elaphus*), photo from U.S. Fish & Wildlife Service. B. European red deer (*C. elaphus*), photo by Hans-Jurg Kuhn. C. Swamp or barasingha deer (*C. duvauceli*), photo from Zoological Society of London. D. Sika deer (spotted form) (*C. nippon*), photo by Alwin Pedersen. E. Sika deer (unspotted form) (*C. nippon*), photo by Leonard Lee Rue III.

subgenus *Thaocervus*

C. schomburgki (Schomburgk's deer), south-central Thailand;

subgenus *Panolia*

C. eldi (thamin, brow-antlered deer), Assam to Indochina, Hainan;

subgenus *Sika*

C. nippon (sika deer), Ussuri district of southeastern Siberia, Manchuria, Korea, eastern China, northern Viet Nam, Japan, Ryukyu Islands, Taiwan;

subgenus *Przewalskium*

C. albirostris (Thorold's deer), Tibet, central China;

subgenus *Cervus*

C. elaphus (red deer, wapiti, elk), Europe, Asia Minor, Caucasus, Central Asia, southern Siberia, Mongolia, Manchuria, Korea, northern and western China, Himalayan region, northwestern Africa, southern Canada, most of the conterminous United States.

Some authorities, such as Corbet (1978) and Lekagul and McNeely (1977), include *Dama* and *Axis* in *Cervus*. Taylor (1934) listed eight separate species of the subgenus *Rusa* in the Philippines. In contrast, Ellerman and Morrison-Scott

(1966) suggested that the Philippine populations of *Rusa* are conspecific with *C. unicolor*. The North American populations of *C. elaphus* were formerly often placed in a separate species, *C. canadensis*.

In the smallest species, *C. mariannus*, head and body length is 1,000 to 1,510 mm, tail length is 80 to 120 mm, shoulder height is 550 to 700 mm, and weight is 40 to 60 kg. In the largest species, *C. elaphus*, head and body length is 1,650 to 2,650 mm, tail length is 100 to 270 mm, shoulder height is 750 to 1,500 mm, and weight is 75 to 340 kg (Grzimek 1975). The upper parts of most species are some shade of brown, and the underparts are paler. In some species, including *C. elaphus*, there may be a prominent pale-colored patch on the rump and buttocks. Some forms of *C. nippon* and *C. duvauceli* have many small white spots on the back and sides. The pelage is coarse; males may have a long, dense mane. The antlers are well developed; in *C. elaphus* they measure up to about 1,750 mm along the beam.

These deer are found in a wide variety of habitats, from open grassland to dense jungle. *C. elaphus* is usually active during the early morning and late afternoon. In some areas it moves into higher country during the spring and returns to the lowlands in the fall. Up to 64 km may be covered in such migrations (Boyd 1978). *C. elaphus* feeds largely on grass and herbs, but also browses foliage from deciduous trees. *C. unicolor* is more of a browser than a grazer (Lekagul and McNeely 1977).

In a study of *C. unicolor* in central India, Schaller (1967) found a population density of about 0.8 individuals per sq km. Herds usually consisted of fewer than 6 animals, mainly females and young. Adult males roamed about alone during the mating season, and then established a small territory, where they were joined by females for various periods. Population density of *C. duvauceli* in this same region was about 0.2 per sq km. Herds of that species usually comprised 13 to 19 animals, and one had at least 500. Some groups consisted of only males or of only females and young, but in the mating season loose associations of both sexes formed. Males established a dominance hierarchy, with the highest-ranking individuals having priority access to estrous females.

The most widely distributed species, *C. elaphus*, is highly gregarious. Discrete herds are formed, each usually occupying a definite area. For most of the year the sexes stay in separate herds. In South Dakota, Varland, Lovaas, and Dahlgren (1978) found three herds of cows and calves to be distributed as follows: 170 animals on 20.7 sq km, 90 animals on 25.9 sq km, and 40 animals on 11.7 sq km. Small groups of adult males also were present in the study area. Boyd (1978) described an annual progression in the social life of *C. elaphus*. Following parturition in the late spring, the mother and newborn live alone for several weeks. Subse-

Top, thamin (*Cervus eldi*), photo by Lothar Schlawe. Bottom: B. Philippine sambar (*C. mariannus*), photo from U.S. National Zoological Park; C. Schomburgk's deer (*C. schomburgki*), photo from *Proc. Zool. Soc. London*; D. Thorold's deer (*C. albirostris*), photo by Joseph F. Rock.

quently, the cows and their young, along with immature animals of both sexes, begin to congregate. By mid-July, herds of up to 400 individuals have formed and are led by older cows. Meanwhile, full-grown bulls live alone or in groups of up to 6 individuals. By September, the bulls have shed their velvet and begun to emit a deep, powerful bellow to attract the females. The older, more powerful bulls drive off the young males, and gather harems of 15 to 20 cows each. No particular territories are defended. After mating is completed, by about mid-October, the males separate from the female groups. Later, aggregations of up to 1,000 animals may concentrate on the limited winter range.

Feldhamer (1980) summarized social and reproductive data on *C. nippon*. Individuals are usually found alone or in small groups. During the summer, adult males begin to establish territories, the maximum size of which has been estimated at 2 or 12 ha. The boundaries of the territories are marked by urination and thrashing of the ground. In the mating season, each successful male gathers as many as 12 females on his territory. Other males are driven away and serious fighting may occur. *C. nippon* is highly vocal: 10 different sounds have been recorded, ranging from soft whistles between females to loud screams by the males. Mating occurs in September and October and births usually in May and June. The gestation period is about 30 weeks. There is

usually a single young, which weighs 4.5 to 7.0 kg. Sexual maturity is attained at 16 to 18 months of age. Weight increases to the age of 4 to 6 years in females and 7 to 10 years in males.

In tropical areas, there may be a long mating season. *C. eldi,* for example, mates from February to April (Lekagul and McNeely 1977). In India, the mating season of *C. duvauceli* and *C. unicolor* may extend from September to April (Schaller 1967). *C. elaphus* generally mates in September and October and gives birth in the late spring. The gestation period of this species is 8 to 9 months. There is usually a single offspring, which weighs 13 to 18 kg at birth. The calf stands quickly, follows the cow after 3 days, grazes at 4 weeks of age, and is weaned and loses its spots after 3 months. Females attain sexual maturity when about 28 months old. Males are capable of mating in their second year of life, but usually must wait considerably longer, because of competition from older bulls. A captive *C. elaphus* lived for 26 years and 8 months (Marvin L. Jones, Zoological Society of San Diego, pers. comm.).

Few mammalian genera have been so extensively affected by people. On one hand, various species of *Cervus* have been introduced to many areas beyond their natural range; for example, *C. nippon* in Europe and America, *C. elaphus* in New Zealand, and *C. timorensis* in Borneo. Even within

North American wapiti (*Cervus elaphus*), photo by Leonard Lee Rue III.

European red deer (*Cervus elaphus*), photo from Zoological Society of London.

Sambar (*Cervus unicolor*), photo by Cyrille Barrette.

natural ranges of species, there has been much transplantation of populations and manipulation of herds in an effort to improve big game hunting. On the other hand, excessive hunting and habitat modification have resulted in drastic declines in the natural distribution and numbers of most, if not all, species. *C. elaphus* has disappeared from much of Europe; existing populations there are generally well protected, but some do not represent the original stock of the area in which they are found. The subspecies of eastern, central, and southwestern North America, *C. elaphus canadensis* and *C. e. merriami*, are extinct. The tule elk of California, *C. e. nannodes*, was nearly exterminated in the 19th century, but a few protected herds have been maintained, and the IUCN (1972) considers the subspecies out of danger. The remaining subspecies of western North America also declined drastically in the 19th century. By 1900 only 41,000 individuals remained in the United States. Subsequent conservation measures allowed numbers to increase to nearly 1,000,000 (Vogt 1978), and the wapiti is now generally subject to limited, legal sport hunting.

At present, the most seriously jeopardized populations of *Cervus* are found mainly in the less-developed parts of the world. Problems vary, but generally involve uncontrolled hunting for food and commerce, and usurpation of habitat by the growing number of people. One species, *C. schomburgki* of Thailand, has apparently been extinct since about 1938. Both the IUCN (1972, 1974, 1976, 1978) and the USDI (1980) classify the following as endangered: *C. duvauceli*, *C. eldi*, *C. nippon taioanus* (Taiwan), *C. n. keramae* (Ryukyu Islands), *C. n. mandarinus* (northern China), *C. n. grassianus* (northern China), *C. n. kopschi* (east-central China), *C. elaphus corsicanus* (Corsica, Sardinia), *C. e. wallichi* (Tibet, Bhutan), *C. e. barbarus* (northwestern Africa), *C. e. hanglu* (Kashmir), *C. e. yarkandensis* (Sinkiang), and *C. e. bactrianus* (Central Asia). In addition, the IUCN designates *C. albirostris* and *C. elaphus macneilli* (Sinkiang, Tibet) as indeterminate, and the USDI lists the latter subspecies as endangered. *C. duvauceli*, *C. eldi*, and *C. elaphus hanglu* are on appendix 1 of the CITES, and *C. elaphus bactrianus* is on appendix 2.

Père David's Deer

The single species, *E. davidianus*, originally occurred in the lowlands of northeastern China. Fully wild animals disappeared from this region about 1,800 years ago, but a herd was maintained in the large Imperial Hunting Park, south of Peking, until about 1900. Captives, descended from that herd, are now present in many parts of the world (Grzimek 1975; Corbet 1978; Cao 1978). Also, Dobroruka (1970*b*) reported that two skins, collected on Hainan in 1869, appear to represent *Elaphurus*.

Head and body length is about 1,500 mm, tail length is about 500 mm, shoulder height is about 1,150 mm, and weight is 150 to 200 kg (Grzimek 1975). The summer pelage, reddish tawny and mixed with gray, is much shorter than the grayish buff winter pelage. A mane is present on the neck and throat. The antlers have a single, long, straight tine pointing backward, while the main beam extends almost directly upward and usually forks only once. The antlers are shed from October through December, and a new set begins to grow immediately, even though it may take six months to attain full size. The hooves are large and spreading, like those of *Rangifer*.

Père David's deer may have originally inhabited swampy, reed-covered marshlands. Although it supplements its grass diet with water plants in the summer, it is essentially a grazing animal. At Woburn Park in England, where a large herd is now maintained, the sexes are together for half of the year. Adult males keep to themselves for about 2 months before and 2 months after the mating season, which begins in June. In that month, hinds begin to bunch together in several groups in definite areas. A stag joins each group of females and then fasts for several weeks, while he engages in mock combat and serious fights with rival males. In fighting, *Elaphurus* not only uses its antlers and teeth but also rises on its hind legs and boxes, in much the same manner as *Cervus*

Père David's deer (*Elaphurus davidianus*), photo from Cleveland Zoological Society.

elaphus. In a process that continues to the end of the rut in August, males are successively ousted and replaced by other stags. After leaving a harem, a male begins feeding again and quickly regains weight. The one or two spotted fawns are born in April or May, following a gestation period of 250 to 270 days. Sexual maturity is usually attained at an age of 2¼ years (Grzimek 1975). Potential longevity is at least 20 years.

This deer is named for Abbé Armand David, who procured two skins in 1865 by bribing guards at the Imperial Hunting Park. Subsequently, a number of live animals were sent from China to various zoos in Europe. The imperial herd was largely destroyed by a flood in 1894, and by subsequent hunting during a famine and the Boxer Rebellion. The last member of this herd survived at the Peking Zoo until 1922. Meanwhile, however, the captives from several European zoos had been pooled at the Duke of Bedford's estate, Woburn Abbey, in England. The herd thus established increased in numbers, and individual deer were later distributed to other areas. In 1956, two pairs were sent to Peking, and in 1957 the first calf was born there (Bower 1979; Grzimek 1975). According to the "Census of Rare Animals in Captivity" in the 1980 *International Zoo Yearbook,* there are now 801 living Père David's deer in 94 collections.

ARTIODACTYLA; CERVIDAE; **Genus ODOCOILEUS**
Rafinesque, 1832

White-tailed and Mule Deer

There are two species (Hall 1981; Cabrera 1961):

O. hemionus (mule deer), southern Yukon and Manitoba to Baja California and northern Mexico;

O. virginianus (white-tailed deer), southern Canada, conterminous United States except parts of the Southwest, Mexico to Peru and northeastern Brazil.

Hall (1981) used the name *Dama* Zimmermann, 1780 for this genus.

Head and body length is 850 to 2,100 mm, tail length is 100 to 350 mm, shoulder height is 550 to 1,100 mm, and weight is 22 to 215 kg. In winter the upper parts are brownish gray and the underparts are lighter. In summer the coat is reddish brown above and is spoken of as being "in the red." In *O. virginianus* the tail is brown above, and white laterally and below. In *O. hemionus* the tail is usually smaller, white or black above, and tipped with black. The hairs, especially those of the winter coat, are tubular and somewhat stiff and brittle. For this reason, the winter skins float in water and have at times been used as life preservers. In North America the antlers are shed from January to March and the new ones being to grow about April or May, losing their velvet in August or September. The antlers attain full size by the fourth or fifth year of life. The antler of the mule deer branches into two nearly equal parts, whereas the antler of the white-tailed deer has one main beam with minor branches.

These deer occur in a great variety of habitats. They prefer areas with enough vegetation for concealment, but usually avoid dense forests. They walk about cautiously, flee from danger with a series of bounds, can run at speeds of up to 64 km per hr, and are excellent swimmers (Banfield 1974). They are generally most active at dawn and dusk. In some areas there is a fall migration to lower elevations or to the vicinity of favorable winter food supplies. The diet includes grass, weeds, shrubs, twigs, mushrooms, nuts, and lichens. Both species browse for part of the year, but *O. hemionus* is predominantly a grazer in the summer.

Hirth (1977) reported population densities of one *O. virginianus* per 4 ha. in a study area in Michigan, and one per 2 ha. in another area in Texas. Individual home range varies widely in this species, but most reported figures fall within the extremes of those found in Texas: 24.3 to 137.6 ha. for females, and 97.1 to 356.1 ha. for males (Halls 1978). Unlike *Cervus* and some other kinds of deer, *Odocoileus* does

Mule deer (*Odocoileus hemionus*): A. Photo by Leonard Lee Rue IV; B. Photo by Leonard Lee Rue III.

A. White-tailed deer (*Odocoileus virginianus*), male in velvet, photo by Earl W. Craven of U.S. Fish & Wildlife Service. B. White-tailed deer (*O. virginianus*), two-year-old doe with fawn, photo by Rex Gary Schmidt of U.S. Fish & Wildlife Service.

not usually gather in large herds. A basic social unit is an adult doe, her yearling daughter, and the two fawns of the season. Adult males occur either alone or in small groups for most of the year. In the mating season, the males compete with one another, through ritualized combat, for the right to associate temporarily with the females. A male does not attempt to gather a group of females or to defend a territory, but associates with one doe until he mates with her or is displaced by another male. During the winter, animals of both sexes and all ages may aggregate in a favorable area. Such groups appear to be generally larger and more organized in *O. hemionus* (Hawkins and Klimstra 1970; Banfield 1974; Halls 1978; Kucera 1978).

In Canada and the United States, mating occurs from October to January, usually peaking in November, and births take place from April to September. Females are seasonally polyestrous, with an estrous cycle of about 28 days and an estrus of 24 hours. The gestation period is 195 to 212 days (Banfield 1974; Halls 1978). Females usually give birth to a single fawn in their first litter, and subsequently to two, or occasionally three or four. The offspring are beautifully spotted and weigh 1.5 to 3.5 kg at birth, and are able to stand after a few hours. They are left hidden in dense vegetation for 1 month or so and are nursed about every 4 hours by the mother. They nibble on vegetation after a few days, can run when 3 weeks old, and are completely weaned at an age of about 4 months. Young females may follow their mother for 2 years, but the males usually leave after 1 year. In a study of *O. virginianus* in Iowa, Haugen (1975) determined that 65 to 74 percent of young females successfully bred within their first year of life. Most other reports indicate that both sexes of *Odocoileus* do not usually mate until the second year. Some captives have lived up to 20 years; wild individuals seldom survive more than 10 years, or 4½ years if in a regularly hunted herd (Halls 1978).

These deer have long been hunted by people. The venison is said to be tender, juicy, and of excellent flavor. Buckskin, a leather originally tanned by the Indians, is made from the skin of *Odocoileus*. Following the arrival of European settlers in North America, hunting became so intense that legal protection was established in some areas in colonial times. Such measures were largely unsuccessful, and by 1900 both species had been exterminated in most of their range. At that time, the number of individuals remaining in each species is estimated to have been about 500,000. Subsequent regulations, management efforts, and environmental changes stimulated a great increase in numbers and distribution. In some places, such as the Great Lakes region, where logging created favorable habitat, deer apparently became far more numerous than they had been prior to European colonization. Currently in the United States there are about 12,450,000 *O. virginianus* and 2,250,000 *O. hemionus*. Respective annual kills by sport hunters are 2,000,000 and 500,000 (Halls 1978; Wallmo 1978; Vogt 1978; Banfield 1974).

Despite the general recovery of *Odocoileus*, three subspecies are listed as endangered by the USDI (1980). The tiny key deer (*O. virginianus clavium*), which is considered out of danger by the IUCN (1972), inhabits the western Florida Keys. Because of excessive hunting and loss of habitat, fewer than 30 individuals were estimated to survive by 1949. Full legal protection, and establishment of a refuge, allowed numbers to increase to a current estimate of 350–400. The Columbian white-tailed deer (*O. v. leucurus*), which is classified as endangered by the IUCN (1976), formerly occurred throughout western Washington and Oregon. It now exists only on a national wildlife refuge and some nearby islands and lowlands along the lower Columbia River. The Cedros Island mule deer (*O. hemionus cerrosensis*), also considered endangered by the IUCN (1973), is confined to a small island off Baja California. Only a few animals are thought to survive, and they are subject to illegal hunting and repeated burning of their habitat.

ARTIODACTYLA; CERVIDAE; *Genus BLASTOCERUS*
Wagner, 1844

Marsh Deer

The single species, *B. dichotomus,* originally occurred in the southern half of Brazil, Bolivia, southeastern Peru, Paraguay, northeastern Argentina, and Uruguay (Cabrera 1961; IUCN 1976). Some authorities, such as Grzimek (1975), place this species in the genus *Odocoileus*.

This is the largest deer of South America. Head and body length is 1,800 to 1,950 mm, tail length is 100 to 150 mm, shoulder height is 1,100 to 1,200 mm, and weight is 100 to 150 kg (Grzimek 1975). The coat is long and coarse. In summer the coloration is bright rufous chestnut. In winter it is brownish red, becoming lighter on the flanks, neck, and chest. The lower legs are black. The tail is yellowish rusty red above and black below. The antlers of adult males are usually doubly forked, that is, each of the two branches has a single fork, for a total of four points. The hooves can be spread widely and are bound by a strong membrane in the inner part of the divergence point. This represents an adaptation against sinking into soft ground.

Blastocerus prefers marshes and wet savannahs with high grass and wooded islands, and damp forest edges. It enters water frequently. It remains in seclusion most of the day and comes out into the clearings to feed in the evening and through the night. The diet consists of grass, reeds, and numerous aquatic plants.

Schaller and Vasconcelos (1978) found population densities ranging from one deer per 3.8 sq km to one per 42 sq km in the Mato Grosso. According to Grzimek (1975), *Blastocerus* lives alone or in groups of up to six individuals, which usually include an older male, and two females and their young. It has been reported that the males do not fight for possession of the females, and that there is not a definite season in which the antlers are dropped. Newborn fawns have been noted at various times of the year. In the Mato Grosso the birth season extends at least from May to September (Schaller and Vasconcelos 1978). The gestation period is about nine months, and a single young is usually born.

The marsh deer is classified as vulnerable by the IUCN (1976) and endangered by the USDI (1980), and is on appendix 1 of the CITES. Numbers and distribution have declined substantially through loss of habitat to agriculture, marsh drainage, and uncontrolled hunting. During the wet season, when the grasslands are flooded, the deer are forced onto a limited amount of high ground with livestock. There is resultant competition, and thus persecution by stockmen. A survey by Schaller and Vasconcelos (1978) indicated that deer populations in the Mato Grosso are declining, perhaps in part because of disease contracted from cattle.

ARTIODACTYLA; CERVIDAE; *Genus OZOTOCEROS*
Ameghino, 1891

Pampas Deer

The single species, *O. bezoarticus,* is found in Brazil to the south of the Amazon, eastern Bolivia, Paraguay, Uru-

Marsh deer (*Blastocerus dichotomus*): Top, photo of mounted group in Denver Museum of Natural History; Bottom, photo from West Berlin Zoo.

place in the evening. When startled, *Ozotoceros* may bound off at considerable speed.

Ozotoceros travels alone, in pairs, in parties of three or four individuals, or in small herds. It tends to live alone or in pairs during the winter, and in larger groups in the spring. Males are often solitary for most of the year. The young are said to be born in April on the plains of Argentina. In other areas there does not seem to be a definite breeding season. In Paraguay, for example, newly born young have been noted in May, June, October, and possibly January. The number of young is usually one. When a mother with a fawn is surprised by hunters, she may remain motionless until the young has managed to conceal itself. The mother may then move off slowly, occasionally with a limping gait, to focus attention on herself and away from the young.

The pampas deer is classified as indeterminate by the IUCN (1972) and as endangered by the USDI (1980), and is on appendix 1 of the CITES. Numbers and distribution have declined drastically through uncontrolled hunting and conversion of natural habitat to agricultural fields. The subspecies *O. b. celer*, formerly found in much of Argentina, is designated as endangered by the IUCN (1976). Probably no more than 200 individuals of this subspecies survive.

guay, and northern and central Argentina (Cabrera 1961). Some authorities, such as Grzimek (1975), place this species in the genus *Odocoileus*. The generic name *Blastoceros* Fitzinger, 1860, which is disliked because it may be confused with *Blastocerus* Wagner, 1844, has sometimes been used in place of *Ozotoceros*.

Head and body length is 1,100 to 1,300 mm, tail length is 100 to 150 mm, shoulder height is 700 to 750 mm, and weight is 30 to 40 kg. The prevailing color of the upper parts and limbs is reddish brown or yellowish gray, the face is somewhat darker, the underparts are white, and the tail is dark brown above and white below. Fawns are spotted. The lower or front prong of the main fork of the antler is not divided, but the upper or posterior prong is. The usual number of tines is three. Males have glands on the posterior hooves, which emit a strong odor capable of being detected at a distance of 1.5 km.

The pampas deer originally occurred mainly in open country and sought cover in tall grass. Much of its habitat has been modified by agriculture, however, and some surviving populations inhabit forest or rough country. Most grazing takes

ARTIODACTYLA; CERVIDAE; **Genus HIPPOCAMELUS** *Leuckart, 1816*

Guemals, Huemuls

There are two species (Cabrera 1961; IUCN 1976):

H. antisensis, the Andes of Ecuador, Peru, western Bolivia, northeastern Chile, and northwestern Argentina;
H. bisulcus, the Andes of central and southern Chile and Argentina.

Head and body length is 1,400 to 1,650 mm, tail length is 115 to 130 mm, shoulder height is 775 to 795 mm, and weight is 45 to 65 kg. The pelage is coarse and brittle, and is longest on the forehead and tail. Each individual hair has a whitish base. Generally, the speckled yellowish gray brown coloration of *H. antisensis* is uniform during all seasons and in both sexes. This species has a dark brown tail with a white undersurface, and there is a black Y-shaped streak on the

Pampas deer (*Ozotoceros bezoarticus*): Top, photo of mounted group in Denver Museum of Natural History; Bottom, photo from West Berlin Zoo.

Huemul (*Hippocamelus antisensis*): A. With antlers in velvet; B. With fully developed antlers; photos by Heinz-Georg Klos through the Zoological Garden Berlin-West.

face. *H. bisulcus* is colored much the same, but is paler beneath. This species has a brown spot on the rump, and its tail has a brown underside. The males of both species possess small, simple antlers, which usually branch only once, the front prong being the smaller. Both sexes of both species have canine teeth similar to those of *Moschus* and *Hydropotes*, but the tusks of *Hippocamelus* do not project beyond the lips. Metatarsal glands are absent.

Huemuls inhabit grassy hills and dense forests at elevations of 3,300 to 5,000 meters. They tend to live at higher altitudes in the summer, move down the mountains in the fall, and spend the winter in forested valleys. Although *Hippocamelus* has been reported to be nocturnal, Roe and Rees (1976) observed all activity during the day in southern Peru. The diet in this area consisted mainly of grasses and sedges. Groups contained up to 8 individuals, included adults and yearlings of both sexes, and appeared to be led by a female. Mating in southern Peru takes place from June to August, and births are thought to occur around February or March. A captive specimen of *H. antisensis* lived for 10 years and 7 months (Marvin L. Jones, Zoological Society of San Diego, pers. comm.).

Both species of *Hippocamelus* are listed as endangered by the USDI (1980) and are on appendix 1 of the CITES. The IUCN (1976) classifies *H. antisensis* as vulnerable and *H. bisulcus* as indeterminate. Numbers and distribution have declined through excessive hunting by people for use as food, and competition from livestock and introduced *Cervus elaphus*.

ARTIODACTYLA; CERVIDAE; *Genus MAZAMA*
Rafinesque, 1817

Brocket Deer

There are four species (Cabrera 1961; Hall 1981):

M. americana, eastern Mexico to northern Argentina;
M. gouazoubira, Isla San Jose off southern Panama, Colombia and Venezuela to northern Argentina and Uruguay;

M. rufina, the Andes from western Venezuela to Ecuador, southeastern Brazil, Paraguay, northeastern Argentina;
M. chunyi, the Andes of southern Peru and Bolivia.

Head and body length is 720 to 1,350 mm, tail length is 50 to 200 mm, shoulder height is 350 to 750 mm, and weight is 8 to 25 kg. The hair on the face radiates in all directions from two whorls. The pelage of the adult is generally uniformly colored, with the species varying in color from light to dark brown. Usually, the body is bright reddish brown above and lighter below, and the underside of the tail is white. Some species are especially light colored. The body is stout, the limbs are slender, and the back is arched. Although the antlers ·.ary in size, they lack a brow tine and are usually simple spikes in all species. The upper canines may be present or absent. The metatarsal gland is absent. Females have four mammae.

Brocket deer are usually found in woodlands and forests from sea level to elevations of 5,000 meters. They are relatively sedentary, sometimes staying in an area only a few hundred meters in circumference. Diurnal, nocturnal, and crepuscular activity has been reported. These small deer are seldom seen, however, because of their shyness, protective coloration, and habit of "freezing" when danger is sensed. If they are flushed, they frequently scamper a short distance and stop to look back at the pursuer. They do not seem to have the endurance of some other kinds of deer, and can be overtaken and killed by a common dog. They enjoy water and are good swimmers. The diet includes many kinds of plants, preferred items being grasses, vines, and tender green shoots.

Brocket deer are usually solitary, the sexes coming together only briefly during courtship (Grzimek 1975). According to Gardner (1971*b*), breeding of *M. americana* continues throughout the year in Peru, but mating peaks from July to September and most young fawns are found from February to April. A female, taken on 30 July in Peru, was both lactating and pregnant, which suggests that a postpartum estrus had occurred. Continuous reproduction has also been reported in other areas. The gestation period is about 225 days, a single young is usually produced, and weight at birth is 510 to 567 grams (Grzimek 1975; Thomas 1975). A

Brocket deer (*Mazama* sp.): Top, photo of mounted group in Denver Museum of Natural History; Bottom, photo from Gladys Porter Zoo.

captive *M. americana* lived for 13 years and 10 months (Marvin L. Jones, Zoological Society of San Diego, pers. comm.).

Brocket deer are intensively hunted by people for use as food and because they frequently damage bean and corn crops. Leopold (1959) noted, however, that *Mazama* is far superior to *Odocoileus* in being able to avoid human hunters.

ARTIODACTYLA; CERVIDAE; **Genus PUDU** Gray, 1852

Pudus

There are two subgenera and two species (Cabrera 1961; IUCN 1978):

subgenus *Pudu*

P. pudu, southern Chile, southwestern Argentina;

subgenus *Pudella*

P. mephistophiles, the Andes of Colombia, Ecuador, and northern Peru.

These are the smallest deer. Head and body length is 775 to 930 mm, tail length is 25 to 35 mm, shoulder height is 320 to 415 mm, and weight is 7 to 10 kg. The hairs of the coat are long, coarse, and brittle. The general coloration of *P. pudu* ranges from rufous to dark brown or gray. The legs and feet are generally tawny. The fawns of this species have three rows of spots running from the shoulder to the base of the tail, plus additional spots on the shoulders and flanks. *P. mephistophiles* has a generally rich brown body coloration. The face and legs are nearly black, long white hairs line the short

Pudu: Top, *Pudu pudu*, photo by Lothar Schlawe; Bottom, *P. mephistophiles*, photo from West Berlin Zoo.

ears, and the inner sides of the legs and the abdomen are covered with long yellow hairs.

Both species possess short, spikelike antlers, tusks do not occur in the upper jaw, and an external tail is practically lacking. *P. mephistophiles* differs from *P. pudu* in having narrow and pointed hooves, a smaller rhinarium, no preorbital glands, and premaxillary bones of the skull that do not reach the nasals.

The species *P. pudu* seems to favor forests from sea level to moderate elevations in the Andes. *P. mephistophiles* inhabits forests and areas of dense vegetation at 2,000 to 4,000 meters. When *P. pudu* is pursued, it generally heads for water, but it is not a strong swimmer. Pudus live in small groups. The females normally produce a single fawn, occasionally twins, from November to January (IUCN 1978). Adult size is attained after only three months of life, and sexual maturity after a year (Grzimek 1975).

The species *P. pudu* is listed as endangered by the USDI (1980) and is on appendix 1 of the CITES. *P. mephistophiles* is classified as indeterminate by the IUCN (1978) and is on appendix 2 of the CITES. It has been relentlessly hunted throughout its range and its habitat is being destroyed through burning.

ARTIODACTYLA; CERVIDAE; Genus ALCES Gray, 1821

Moose (North American term), Elk (European term)

The single species, *A. alces,* originally occurred from northern Europe and the Caucasus to eastern Siberia, and from Alaska to northern Colorado and the northeastern United States (Corbet 1978; Pulliainen 1974; Hall 1981).

This is the largest member of the deer family. Head and body length is 2,400 to 3,100 mm, tail length is 50 to 120 mm, shoulder height is 1,400 to 2,350 mm, and weight is 200 to 825 kg. The summer pelage is dark above, varying from black to dark brown, reddish brown, or grayish brown, and the underparts and lower legs are lighter. The winter pelage is grayer. The young are reddish brown and, unlike most immature members of this family, are not spotted. *Alces* is easily identified by its broad overhanging muzzle, massive antlers, heavy mane, and characteristic pendulant flap of skin beneath the throat, known as the "bell." Franzmann (1981) noted that the maximum recorded antler spread is 2,048 mm.

The moose generally occurs in wooded areas that have a seasonal snow cover (Franzmann 1981). A favorite habitat is

Alaskan moose (*Alces alces*), photo by Bernhard Grzimek.

North American moose (*Alces alces*), juvenile, photo by Eugene Maliniak.

a moist area with abundant willows and poplars. Although vision is poorly developed, the senses of smell and hearing are acute. According to Banfield (1974), the moose normally walks carefully and quietly through the underbrush, can reach a speed of 56 km per hr, and is a strong swimmer. Activity occurs throughout the day, especially in the winter, but there are definite peaks at dawn and dusk. Some populations make regular seasonal movements between areas of favorable food supplies. Several distinct home ranges, separated by a considerable distance, are sometimes used in the course of a year. Migrations reportedly cover up to 179 km in North America and 300 km in northeastern Europe (LeResche 1974; Pulliainen 1974).

The moose frequents marshy and timbered regions in search of the shrubs and trees on which it browses. It also feeds on water vegetation by wading into lakes and streams, often submerging entirely to obtain the roots and stems of plants on the bottom. By necessity, it eats twigs and the bark of trees, and paws through the snow for small plants in the winter. Franzmann (1981) stated that *Alces* depends mainly on woody vegetation in an early developmental stage, and requires 19.5 kg of food per day.

Normal population densities reportedly vary from about 0.1 to 1.1 moose per sq km, but local concentrations sometimes reach 100 to 200 per sq km (Bouchard and Moisan 1974; Filonov and Zykov 1974; Krefting 1974; Peek, LeResche, and Stevens 1974; Syroechkovskiy and Rogacheva 1974). An individual seasonal home range gener-

ally covers 2.2 to 16.9 sq km, but much larger areas may be used in the course of a year or a lifetime (Franzmann 1981; LeResche 1974). Many moose may aggregate in a small, favorable area during late fall or winter (Peek, Urich, and Mackie 1976), but the genus is essentially solitary. In the breeding season, males compete for one female at a time, engaging in elaborate displays, shoving matches with the antlers, and, occasionally, serious fighting. The female also takes an active role at this time, moving about independently, attempting to attract males through sound and scent, and behaving aggressively to others of her sex. The mating call of the female moose is a long moan, while that of the male is described as a "croak" (Lent 1974; Franzmann 1981).

Mating occurs in September and October, and births take place the following spring. Females are seasonally polyestrous, having an estrous cycle of 20–22 days and a true estrus of less than 24 hours. A litter is usually produced every year. The gestation period is 226 to 264 days. The number of young per litter is most often one, twins are common, and triplets are rare. The calf weighs 11 to 16 kg at birth, is able to browse and follow the mother after 2 or 3 weeks, and is completely weaned by the fifth month of life. It remains with the female at least 1 year, is driven away when the mother gives birth again, and may subsequently rejoin the family for some additional months. Both sexes are physiologically capable of mating in their second year of life. Females often do so at this time, and continue to breed until they are 18 years old, but their maximum reproductive potential is reached between the ages of 4 and 12 years (Franzmann 1978, 1981; Verme 1970; Stringham 1974). Maximum known longevity is 27 years (Peterson 1974b).

Because of uncontrolled hunting by people for the meat, leather, and bone of *Alces,* the distribution of the genus has been reduced substantially. By the 13th century A.D. it had disappeared from western Europe except for the Scandinavian Peninsula, by the early 19th century it was gone from the Caucasus, and by 1900 it had been all but wiped out in the conterminous United States. Subsequent measures for protection and management allowed the moose to rebuild its numbers in some regions in which it had been depleted. Small populations have also returned to New England, the upper Great Lakes region, and the central Rocky Mountains. Human habitat modifications, such as burning and logging, have sometimes benefited the moose by providing more abundant food, and seem to have stimulated a northward range expansion in Alaska and parts of Canada. *Alces* was introduced by people on the island of Newfoundland, and a large population is now present there. Available numerical estimates suggest that in the 1970s there were about one million moose in North America and another million in Eurasia (Banfield 1974; Bouchard and Moisan 1974; Cumming 1974; Filonov and Zykov 1974; Hall 1973; Kelsall 1972; Krefting 1974; Markgren 1974; Mercer and Manuel 1974; Pulliainen 1974; Rausch, Somerville, and Bishop 1974; Ritcey 1974; Vogt 1978).

The moose is considered a game animal in most parts of its range, and is subject to limited sport and subsistence hunting. The most productive populations in this regard are those of Sweden, where 94,000 individuals were killed in 1978. This harvest represented 2 to 3 percent of the total meat production of Sweden (Franzmann 1981). In the Soviet Union, *Alces* has been domesticated for the production of meat and milk, and also for use as a farm draft animal (Knorre 1974; Grzimek 1975).

ARTIODACTYLA; CERVIDAE; Genus *RANGIFER*
Hamilton Smith, 1827

Caribou (North American term), Reindeer (European term)

In historical time, the single species, *R. tarandus,* occurred in Ireland, Scotland, Germany, Poland, Svalbard, Russia north of 50°N, Mongolia, northeastern China, Sakhalin, Alaska, Canada, the extreme northern conterminous United States, and Greenland (Hall 1981; Corbet 1978; Smit and Van Wijngaarden 1976; Banfield 1961).

Head and body length is 1,200 to 2,200 mm, tail length is 70 to 210 mm, shoulder height is 870 to 1,400 mm, and weight is 60 to 318 kg. Average weight in Canada is 110 kg for males and 81 kg for females (Banfield 1974). As protection against the rigorous climate in which it lives, the caribou has a heavy coat of woolly underfur, and straight, stiff, tubular guard hairs. Coloration varies widely, but most individuals are predominantly brownish or grayish above, and have white or pale underparts, inner legs, and buttocks. The winter pelage is somewhat paler. Some animals are dark brown or almost black, and some are quite pale. The subspecies *R. t. pearyi,* of northern Greenland, Ellesmere Island, and nearby islands, is almost pure white.

Rangifer is the only genus of the deer family in which both sexes are antlered. Although there is great diversity in the shape of the antlers, they are usually long, sweeping beams with forwardly projecting brow tines. Length is 520 to 1,300 mm in males and 230 to 500 mm in females (Banfield 1974). The broad, flat, and deeply cleft hooves aid in walking on soft ground and snow. When the caribou walks, a clicking noise is produced by a tendon slipping over a bone in the foot.

Most caribou inhabit arctic tundra and surrounding boreal coniferous forest. Some populations are found in moun-

Caribou (*Rangifer tarandus*), photo of mounted group in Field Museum of Natural History.

Reindeer (*Rangifer tarandus*): Top, photo by Bernhard Grzimek; Bottom, photo from Antwerp Zoo.

tainous areas farther south. *Rangifer* is active primarily during the day, and is then almost constantly on the move. Its maximum running speed is between 60 and 80 km per hr. Its eyesight is apparently poor, its hearing is adequate, and it depends in large part on the sense of smell to detect food and danger. Most populations have seasonal shifts in range. Those to the south may simply move to lower elevations for

the winter or to areas of favorable food supplies and shelter. Northern populations, however, make extensive spring and fall migrations, sometimes traveling over 1,000 km between the summer range on the tundra and the wintering grounds in timbered areas. The rate of movement during migration is 19 to 55 km per day (Bergerud 1978; Banfield 1974). The populations on the arctic islands of Canada may make seasonal interisland movements of up to 450 km (Miller, Russell, and Gunn 1977). The diet comprises a wide variety of plants, including new growth in the spring, green leaves of deciduous shrubs in the summer, lichens and evergreen leaves in the fall, and fine twigs and greens dug through the snow in the winter (Bergerud 1978).

The original overall population density in North America is estimated to have been 0.4 to 0.6 individuals per sq km. The maximum known density within the regular range of a particular herd is 1.5 to 1.9 individuals per sq km. During migration, however, when a herd concentrates, there may be up to 19,000 animals per sq km. *Rangifer* is highly gregarious. The animals in a given region of the world are divided into populations or herds, which move about as a unit, keep to the same general ranges and migration routes over the years, and do not mix extensively with other herds. About 30 discrete herds have been identified in North America. The smallest, that of northern Idaho and adjacent areas, contains only 15 to 20 individuals. The seven largest herds occur in a sequence all across the northern parts of Alaska and Canada. Although size fluctuates, each of these herds has generally comprised 50,000 to 200,000 individuals in recent years. At most times the members of a large herd are divided into groups of 10 to 1,000 animals each, often with only one sex represented in the smaller groups. Typically, a large herd comes together for the spring migration, the females separate to give birth in late May and June, there is another concentration immediately after fawning, the animals disperse into small groups for the summer, the herd again masses for the fall migration, and there is dispersal for the winter (Banfield 1974; Bergerud 1978). Vocalizations include the snort of an alarmed adult, the bawl of a fawn, and the grunting roar of a rutting male.

In the mating season, males compete and fight for possession of females. Each male may either attempt to control a harem of 5 to 15 females or simply move about and try to mate with as many females as possible. Females are seasonally polyestrous, enter heat at intervals of 10 to 12 days, and usually give birth once a year. Most mating occurs in October and most young are born in late May and early June. The gestation period averages about 227–229 days. There is almost always a single offspring, but twins have been reported. The young weighs 5 to 9 kg at birth, is not spotted, and is remarkably precocious. It is able to follow its mother after 1 hour of life and can outrun a human when 1 day old. It nurses for at least 1 month and sometimes until the winter. Sexual maturity is usually attained between the ages of 29 and 41 months. In the wild, average longevity is 4½ years and the maximum known is 13 years (Banfield 1974; Bergerud 1978). A captive lived for 20 years and 2 months (Marvin L. Jones, Zoological Society of San Diego, pers. comm.).

Rangifer is extensively hunted by people to obtain the meat, skin, antlers, and other parts. The genus survived in Germany until Roman times, in the British Isles until the Middle Ages, and in Poland until the 16th century (Banfield 1961). Small populations still exist in Scandinavia, and there are about 800,000 wild reindeer in the Soviet Union, but the range is shrinking (Andreev 1978; Smit and Van Wijngaarden 1976). In North America, uncontrolled hunting led to the disappearance of the caribou in much of the southern portion of its range by the early 20th century. The only population that still regularly occurs in the conterminous United States is a group of 15 to 20 animals that wanders through the mountains of northern Idaho and adjacent parts of Washington and British Columbia. The great barren ground herds of northern Canada declined drastically after 1900 as human activity increased in their ranges and the native people obtained modern firearms. The total number of caribou in North America is estimated to have been 3,500,000 originally and 1,100,000 in 1977 (Bergerud 1974, 1978). There has been much concern about how the remaining herds may be affected by oil and gas exploitation, and other environmental

modifications in the Arctic. The most immediate problem, however, continues to be excessive hunting by people. Mostly because of that reason, the largest herd in North America, that of northwestern Alaska, declined from 242,000 individuals in 1970 to a minimum of 75,000 in 1976. Stricter limits on hunting contributed to an increase to about 113,000 by 1979 (Davis, Valkenburg, and Reynolds 1980). The unique white, island-hopping subspecies, *R. t. pearyi*, has evidently disappeared in Greenland, has declined to only 10–15,000 individuals in the Canadian Arctic Archipelago, and is considered threatened with extinction (Gunn, Miller, and Thomas 1981).

The domestication of reindeer may have begun in the Old World about 3,000 years ago, and subsequently spread all across northern Eurasia. Domestic animals are generally allowed to live in herds and to move about freely, but selective breeding is practiced to some extent. Domestic reindeer have been introduced in Iceland, the Orkney Islands, Scotland, South Georgia, the Kerguelen Islands, Alaska, Canada, and Greenland (Banfield 1961; Smit and Van Wijngaarden 1976). The Alaskan animals, originally brought over in 1891, numbered around 500,000 in 1930, but subsequently declined to about 30,000. Currently, there are about 3,000,000 domestic reindeer in the world, most of them in the Soviet Union. The annual production in that country is 32,000 tons of meat and 650,000 hides (Andreev 1978).

ARTIODACTYLA; CERVIDAE; **Genus CAPREOLUS** *Gray, 1821*

Roe Deer

The single species, *C. capreolus,* occurs from Britain and Spain to southeastern Siberia and central China (Corbet 1978).

Head and body length is 950 to 1,350 mm, tail length is about 20 to 40 mm, shoulder height is 650 to 775 mm, and weight is 15 to 50 kg. Coloration varies with the season. In summer the coat is reddish above, the ears are black, and the underparts are white. In winter the general color is buff or dark brown, the throat and rump patch are white, and the face, chest, and legs are tawny. *Capreolus* is small and

Roe deer (*Capreolus capreolus*), female with fawn born in the spring, and male with winter coat and antlers, photos from the Zoological Garden Berlin-West.

graceful. The antlers are slightly roughened at the base, only about 230 mm long, and seldom more than three tined. The tail is inconspicuous.

The roe deer generally avoids thick forests, preferring sparsely wooded valleys and the lower slopes of mountains, usually not exceeding 2,400 meters in elevation. It is found in almost any area that furnishes a reasonable amount of cover, and is able to live in parklike places among dense human populations. In the early morning and late evening, it emerges from cover and seeks open grassland, where it grazes. It is shy, but curious, and has well-developed senses. It is an excellent swimmer. The diet consists of grass, herbs, and cultivated crops.

Capreolus lives alone or in small groups. Females establish territories during the mating season and when they have young fawns. If two females meet at such times, they may fight. Males have individual territories in the late spring and early summer, but subsequently wander in search of females (Grzimek 1975). Unlike most deer, male *Capreolus* usually escort only one female. They will savagely fight any intruding male. During the mating season, males often chase females in circles, leaving a track referred to as a "witch circle." Such chases also occur in other seasons, sometimes between individuals of the same sex. The roe deer utters a barking sound when disturbed. Females emit a screeching sound during the breeding season and to call fawns.

Mating usually occurs in July and August, and implantation of the fertilized egg in the uterus is subsequently delayed about 4 months. A few females, however, do not mate until November or December, do not experience delayed implantation, and have a total gestation of about 5½ months (Grzimek 1975). Births occur in the spring. When about to bear its young, the doe chases away the offspring of the previous season and then retires to the forest, returning in about 10 days with the new family. The number of young per birth is usually two, occasionally one or three. Twins are born and suckled not at the same spot but usually 10 to 20 meters from each other. The fawns have three longitudinal rows of white spots. Sexual maturity is attained at about 1 year and 4 months of age. *Capreolus* is delicate in captivity. The average life span of 11 specimens in the London Zoo was 40 months, and the maximum was 7 years. Normal longevity in the wild is 10 to 12 years, and the maximum may be 17 years.

The roe deer has adapted well to civilization, and its distribution and numbers seem to have increased in Europe. About 500,000 are killed annually in West Germany, but the population there remains stable (Grzimek 1975).

ARTIODACTYLA; **Family GIRAFFIDAE**

Okapi and Giraffe

This family of two Recent genera, each with a single species, was found over most of Africa in historical time. The sequence of genera presented here follows that of Simpson (1945), who recognized two subfamilies: Paleotraginae, with the genus *Okapia;* and Giraffinae, with *Giraffa.*

Giraffids have large eyes and ears, long and thin lips, and an extensible tongue. The neck, legs, and terminally tufted tail are long. The back inclines upward from the loins to the withers. The feet are large and heavy. There are two hoofed digits, the third and the fourth; the lateral digits are not developed. The stomach is four chambered and ruminating, and a gall bladder is not present. Females have two or four mammae.

Front foot of a giraffe (*Giraffa camelopardalis*), from photo by Constance P. Warner.

The horns of the Giraffidae are unlike those of any other mammal. They are present at birth as cartilaginous knobs, which rapidly ossify and grow slowly throughout life. They consist of a bony core, which is at first separate from, but later fuses with, the skull. In *Giraffa* they arise over the anterior part of the parietal bones behind the eyes, the forward base growing over the frontoparietal suture. In *Okapia* they arise on the frontals above the orbit. They are covered with skin and hair throughout life; the hair is worn away from the apex, but not the skin as is commonly supposed. Horns occur in both sexes, but growth is less vigorous in the female. *Giraffa* also bears a median horn on the forepart of the frontal bones and the back of the nasal bones, which develops to a greater or lesser extent and is identical in its mode of growth with the posterior horns.

The skull of *Giraffa* is characterized by large pneumatic sinuses above the nasal bones and cranium. These are developed to a much lesser extent in *Okapia* and do not extend back over the cranium. The dental formula is: (i 0/3, c 0/1, pm 3/3, m 3/3) × 2 = 32. The low-crowned molars are characteristically rugose in giraffids, being unlike those of all other mammals, in which the enamel is always smooth. The upper molars lack inner accessory columns.

The geological range of this family is early Miocene to Pleistocene in Asia, early Pliocene in eastern Europe, and Pliocene to Recent in Africa. *Sivatherium* was a large, heavily built member of this family known from the Pleistocene of southern Eurasia and Africa. The male often developed a variety of horns, frequently being four in number and greatly branched. This creature was more bovine in appearance than the Recent giraffe, both the neck and legs being shorter than those of *Giraffa.*

ARTIODACTYLA; GIRAFFIDAE; **Genus OKAPIA**
Lankester, 1901

Okapi

The single species, *O. johnstoni,* is known to occur in the equatorial forests of northern, central, and eastern Zaire (Meester and Setzer 1971).

Okapis (*Okapia johnstoni*), photo by Bernhard Grzimek.

Head and body length is about 2,000 to 2,100 mm, tail length is about 300 to 420 mm, shoulder height is 1,500 to 1,700 mm, and weight is about 200 to 250 kg. The general coloration is purplish, deep reddish, or almost black. The sides of the buttocks and upper portions of the limbs are transversely barred with black and white stripes of varying width. The shanks are white to maroon, and the facial markings are light. The hair is short and sleek, and the tail has a terminal tuft. The body is short and compact, the neck and legs are long, the ears are relatively large, and the eyes are large and dark. The tongue is so long that it can be used to clean the eyes. Males have small, hair-covered horns.

The preferred habitat of the okapi is dense, damp forest. It is diurnal and normally moves along well-trodden paths through the jungle. It is extremely wary and secretive, dashing through the forest at the least suspicion of danger. Hearing is the best developed of its senses. It is a browser, feeding on the leaves, fruit, and seeds of many plants.

Okapia is found alone, in pairs, or in small family parties, but never in herds. The young are born from August to October, which is the period of maximum rainfall. The single offspring is about 790 mm tall and 16 kg in weight at birth, and starts to nurse within 6 to 12 hours. According to Gijzen and Smet (1974), estimated gestation is 425 to 491 days. Females attain sexual maturity at the age of 1 year and 7 months. One captive female gave birth to 12 calves, the last when she was 26 years old. One captive okapi is estimated to have been 33 years old at time of death (Marvin L. Jones, Zoological Society of San Diego, pers. comm.).

Although the okapi was long hunted by the pygmies of Zaire, it did not become known to science until 1900. The first live specimen reached Europe in 1918. The okapi is generally considered rare, and has been protected in Zaire since 1933, but its densely forested habitat and elusive nature may conceal its true status (Grzimek 1975).

ARTIODACTYLA; GIRAFFIDAE; *Genus GIRAFFA*
Brünnich, 1772

Giraffe

In historical time, the single species, *G. camelopardalis*, occurred in most of the open country of Africa. For about the last 1,400 years, the species has been restricted to areas south of the Sahara (Dagg 1971; Meester and Setzer 1971).

This is the tallest living terrestrial animal. Average height is 5,300 mm for males and 4,300 mm for females. The record is 5,880 mm. In seven males, head and body length was 3,810 to 4,724 mm, and tail length was 787 to 1,041 mm (Dagg and Foster 1976). Shoulder height is 2,500 to 3,700 mm, and weight is 550 to 1,800 kg. Average adult weight is 800 kg (Dagg 1971). The color pattern varies, but consists essentially of dark reddish to chestnut brown blotches of various shapes and sizes on a buff ground color. The underparts are generally light and unspotted. The coloration darkens with age. The long neck is maned with short hair, and the long tail is terminally tufted.

Like most other mammals, the giraffe has seven neck vertebrae, but they are greatly elongated. Both sexes possess two to four blunt, short, hornlike structures on top of the head. In certain forms there is another protuberance, sometimes only a knob, situated more or less between the eyes. Such a giraffe is referred to as five horned by some people. The lower canine teeth are peculiarly flattened, and are deeply grooved at right angles to the plane of flattening. The tongue is long, capable of being extended up to 456 mm, flexible, and used in plucking leaves from trees. The lips are prehensile and hairy. The nostrils can be closed at will. The eyes are large, dark brown, and shaded by long black lashes. The feet are large and heavy. The great height of *Giraffa*

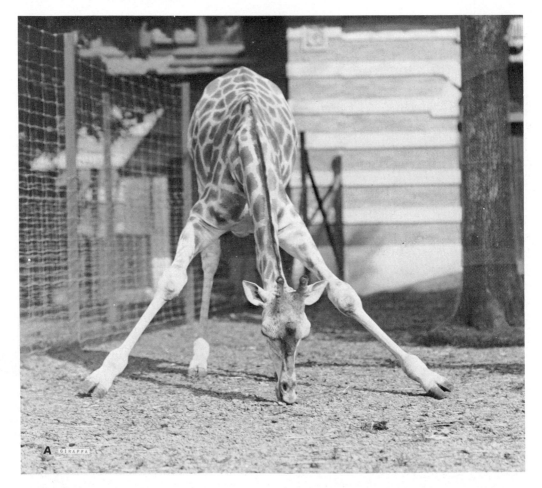

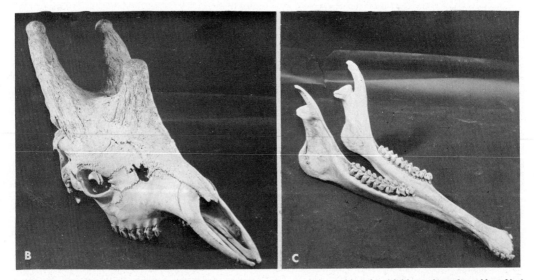

A. Giraffe (*Giraffa camelopardalis*) with legs spread in a characteristic position for drinking, photo from New York Zoological Society. B & C. Skull of *G. camelopardalis*, photos by P. F. Wright of specimen in U.S. National Museum.

Nubian giraffes (*Giraffa camelopardalis*), photo from San Diego Zoological Garden.

requires a series of valves to regulate the flow of blood to the head. Females have four mammae.

The giraffe dwells mainly on dry savannahs and in open woodland. It is usually associated with scattered acacia growth. In a study in Tsavo National Park, Kenya, Leuthold and Leuthold (1978) found animals to concentrate near rivers in the dry season and to disperse into deciduous woodland during the rains. Such seasonal movements generally covered 20 to 30 km. The giraffe is active mainly in the evening and early morning, and rests during the heat of the day. It usually sleeps standing up, but occasionally lies down. In real sleep, a giraffe rests its head on the lower part of one hind leg, its neck forming an impressive arch. While dozing, which is more common, it rests on its withdrawn legs, but the neck remains outstretched, the eyes are half closed, and the ears continue twitching. In order to drink or to pick food from the ground, the giraffe spreads its forelegs widely and well to the front, or bends at the knee until its head can reach the ground or water level.

The record running speed of *Giraffa* is 56 km per hr (Dagg and Foster 1976). Over moderate distances, it can scarcely be overtaken by a good horse. If not crowded, a giraffe can lope for long distances without tiring. While the giraffe is running, the hind feet are swung forward of the forefeet, the head and neck swing widely, almost in a figure eight, and the tail is raised over the back. While it is walking, however, both feet on a side are carried forward simultaneously. Although the giraffe has large feet, it is able to walk only on firm earth, since the long legs supporting such a heavy body would soon become bogged down in swampy terrain. As a result, large rivers are usually barriers.

The giraffe is shy, timid, and alert. It is especially vulnerable to lion predation when lying down, ground feeding, or drinking. A panicky movement indicates danger. When forced to defend itself, the giraffe kicks with its forefeet. In addition, it frequently uses the head to give blows, particularly to another giraffe. The senses of smell, hearing, and vision are acute. Probably this animal has the keenest sight of any African big game species, and its height gives it the greatest range of vision of any terrestrial creature. The giraffe is a browser, feeding almost entirely on leaves from acacia, mimosa, and wild apricot trees. It takes branches in its mouth and tears off the leaves by pulling its head away. It may chew its cud at any time of day. If water is available, it will take an occasional drink (about 7½ liters a week). However, it is able to go without water for many weeks, if not months, at a time.

Reported normal population density varies from about 0.1 to 3.4 individuals per sq km. Mean individual home range in different areas varies from about 23 to 163 sq km. The average range size is about the same for each sex; there is considerable overlap and no evidence of territoriality. *Giraffa* is usually observed alone or in small, loosely organized groups. When the genus was more common, herds of over 100 individuals were recorded, but most groups now contain 2 to 10 animals. There seems to be little evidence of a dominance hierarchy, except when several males are in the presence of an estrous female. At such times the males may fight one another by swinging their necks and striking with their heads from the side. Females with young sometimes aggregate, and one or two of the adults may remain with all of the calves, while the rest of the adults go off to browse or drink. *Giraffa* has a variety of sounds, but is rarely heard. It may grunt or snort when alarmed, a female may whistle to call its young, and calves bleat (Dagg and Foster 1976; Langman 1973, 1977; Leuthold 1979; Leuthold and Leuthold 1978; Berry 1978).

In at least some parts of Africa, such as Zambia, breeding may occur throughout the year (Berry 1973). Females give birth at a mean interval of about 20–23 months (Leuthold and

Leuthold 1978; Hall-Martin and Skinner 1978). The average gestation period is 457 days (Skinner and Hall-Martin 1975). The number of offspring is almost always one, though twins are known. At birth the young weigh 47 to 70 kg and are 1,700 to 2,000 mm tall. They are able to stand on their wobbly legs about 20 minutes after being born, and begin to suckle within an hour. Contrary to early reports, there appears to be a strong bond between mother and calf. The calf may nurse for up to 13 months and then remain with the mother for another 2 to 5 months (Langman 1977; Leuthold 1979). The age of sexual maturity is 3½ years in females and 4½ years in males. Full size is attained at 5 years by females and 7 years by males. Females are capable of reproduction until they are at least 20 years old, and maximum known longevity in the wild is 26 years (Dagg and Foster 1976). A captive giraffe lived for 36 years and 2 months (Marvin L. Jones, Zoological Society of San Diego, pers. comm.).

The native peoples of Africa sometimes take giraffes in snares and pitfalls. They use the strong sinews for bowstrings and musical instruments, and the thick hide as a covering for shields. The meat, although tough, has a good flavor. European settlers killed the animals in great numbers for their hides, which were used to make traces, long reins, whips, and other items. A combination of excessive hunting and climatic change has caused a great reduction in the distribution and numbers of *Giraffa* in historical time. The genus disappeared in Egypt about 2600 B.C., but may have survived in Morocco to 600 A.D. In the 20th century, it was wiped out in most of western and southern Africa. The only remaining large populations are in Tanzania and some adjacent areas (Dagg 1971; Dagg and Foster 1976).

ARTIODACTYLA; *Family ANTILOCAPRIDAE; Genus ANTILOCAPRA Ord, 1818*

Pronghorn

The single living genus and species, *Antilocapra americana*, originally occurred in open country from eastern Washington and southern Manitoba to Baja California and northeastern Mexico (Hall 1981). Most authorities continue to regard the Antilocapridae as a distinct family, but O'Gara and Matson (1975) suggested that the group be ranked only as a subfamily of the Bovidae.

Head and body length is 1,000 to 1,500 mm, tail length is 75 to 178 mm, shoulder height is 810 to 1,040 mm, and weight is 36 to 70 kg. Males average approximately 10 percent larger than females (Hall 1981). The woolly undercoat is overlaid with fairly long, straight, coarse, pithy, and brittle guard hairs. By flexing certain skin muscles, the pronghorn can maintain its pelage at different angles. Cold air is excluded when the hairs lie smooth and flat, but the hairs may be erected in the desert sun to allow air movement to cool the skin. The upper parts are reddish brown to tan, the neck has a black mane, and the underparts, the rump, and two bands across the neck are white. In the male, the face and a patch on the side of the neck are black, the horns are longer than the ears, and the nose is pointed downward slightly when running. In the female, the mask and patch are lacking, or nearly so; the horns, if present, seldom exceed the ears in length; and the nose is held more nearly horizontal when the animal is running. Females have four mammae.

Males, and most females, carry horns that consist of a permanent, laterally flattened bone core covered with a keratinous sheath that is shed annually after each breeding season. The new sheath grows upward under the old sheath. The shedding process is usually considered a major distinguish-

Pronghorn (*Antilocapra americana*), photo by Bernhard Grzimek.

ing feature between the Antilocapridae and the Bovidae, but O'Gara and Matson (1975) observed that a similar process occurs in at least one species of each of the five bovid subfamilies. The horns of *Antilocapra* are erect, and backwardly curved at the tips. Those of males are about 250 mm long and have a forward-directed prong arising from the upper half. Those of females, if present, are seldom over 120 mm long and seldom have prongs.

Antilocapra has large eyes (approximately 50 mm in diameter), and long, pointed ears. Only two digits are developed, the third and the fourth; the lateral toes are lacking. The hooves, especially those of the forefeet, are supplied with cartilaginous padding. The dental formula is: (i 0/3, c 0/1, pm 3/3, m 3/3) × 2 = 32. The high-crowned cheek teeth have crescentic ridges of enamel.

Habitats include grasslands and deserts. The elevational range is sea level to 3,353 meters (Yoakum 1978). The pronghorn is the swiftest terrestrial mammal in the New World. Kitchen (1974) clocked herds moving at 64 to 72 km per hr, and observed a maximum speed of 86.5 km per hr. Such velocities can be attained only on hard ground, and involve leaps of 3.5 to 6 meters. The cruising speed is approximately 48 km per hr. Fast runs of 5 to 6 km are common, but then exhaustion occurs rapidly. The pronghorn shakes the body after a fast run. The front feet carry most of the weight while running. The pronghorn is a good swimmer.

It can see objects several kilometers away, but apparently lacks visual acuity. A motionless person, only 10 to 15 meters away, may be ignored (Kitchen 1974). The pronghorn is a curious animal, and may approach a strange object if it does not cause alarm by scent or sudden movement.

Activity is both diurnal and nocturnal, with slight peaks just after sunset and before sunrise (Kitchen 1974). Daily movement is usually 0.1 to 0.8 km in spring and summer, and 3.2 to 9.7 km in fall and winter (Yoakum 1978). The range may shift several times in a year, for purposes of obtaining food and water. The distance between the summer and winter range may be as much as 160 km (O'Gara 1978). The pronghorn both browses and grazes on a wide variety of shrubs, forbs, grasses, cacti, and other plants. It uses its front feet to dig food buried under the snow, and also to scratch depressions for deposit of droppings. If water is available, it will drink freely, but, if necessary, it can derive sufficient moisture from plants.

According to O'Gara (1978), studies in Wyoming indicated that 48.6 ha. of desert could support 8 pronghorns year-round. Home range in Wyoming during the summer and early autumn was found to be 2.6 to 5.2 sq km. Yoakum (1978) stated that most herds have an overall range 8 to 16 km wide. During the fall and winter, *Antilocapra* forms large, loose aggregations of all age and sex classes. In the northern part of the range, such groups contain up to 1,000 individuals (Yoakum 1978).

From late March to early October, groups are smaller and are segregated by sex (Kitchen 1974; O'Gara 1978). At this time, males over three years old compete with one another for possession of territories 0.23 to 4.34 sq km in size. A territory usually contains a permanent source of water, often has prominent physical borders, and may be separated by a no man's land of up to 0.8 km from other territories. An old male may return to the same territory each year after the herd breaks up and moves out of its winter range. He scent marks the area with urine, feces, and secretions from subauricular glands. He constantly attempts to keep groups of females within his territory and to keep other males out. When a territorial male spots a rival, he will first stare for a period. A stare is considered aggressive, and avoidance of a stare is considered submissive. If the other male holds his ground, there may be loud vocalization, an aggressive approach, a chase, and, occasionally, a fight. The sharp horns evidently often cause serious injury.

Old males that are not able to hold a territory may wander about alone. Younger males form bachelor herds with up to 36 members. These groups have a loose hierarchy, and move about on the edges of the areas controlled by the territorial males. Female groups contain up to 23 members, have a permanent linear hierarchy, and move about freely on the territories of the dominant males. A female group seldom remains on a single territory through the entire period from March to October. Females are pursued by bachelor herds, but try to avoid them. Individual does may separate from the group to give birth. Following the mating season, the horn sheaths are cast and social distinctions become obscured, though the female associations may persist as subgroupings of the main herd.

The pronghorn has a variety of vocalizations. Mothers call calves with individually recognizable grunts, fawns bleat, and males sometimes roar during agonistic encounters. Mature animals of both sexes reveal anger or anxiety by forcefully expelling air through their nostrils. The hairs of the white rump patch are raised when danger is sensed, a conspicuous warning signal to other pronghorns. The white flash can be seen by a human for a distance of at least 4 km.

Mating occurs during a period of about 3 weeks between July and early October, and births take place in the spring. The gestation period averages 252 days (O'Gara 1978). A female generally produces a single young in her first pregnancy, and thereafter usually gives birth to twins, rarely triplets. At birth a fawn usually weighs 2 to 4 kg and has a beautiful, wavy, grayish pelage. The mother's milk is extremely rich in solids. At the age of 4 days the fawn can outrun a human, at 3 weeks it is nibbling at vegetation, and before 3 months it has acquired its first adultlike pelage. Apparently, the young may separate at least temporarily from the mothers during the mating season and form small groups of their own, but there may subsequently be a reassociation with the mothers until the following spring or summer (Autenrieth and Fichter 1975; O'Gara 1978). Females usually reach sexual maturity when 15 or 16 months old. Males are capable of mating at this same age, but generally are not able to do so until they are 3 years old. A captive specimen lived for 11 years and 2 months (Marvin L. Jones, Zoological Society of San Diego, pers. comm.).

There are estimated to have been about 35,000,000 pronghorns in North America prior to the arrival of European explorers. Subsequent uncontrolled hunting for meat and sport, and human usurpation of habitat, resulted in a decline to fewer than 20,000 in the 1920s. Conservation efforts have since allowed numbers to expand to about 500,000 in the United States and Canada. The pronghorn is now subject to limited sport hunting in most parts of its range, and the annual harvest is about 40,000. In Mexico, however, only about 1,200 individuals survive and the number is declining through illegal hunting and habitat modification (O'Gara 1978; Yoakum 1978). The subspecies *A. a. sonoriensis* of extreme southern Arizona and northwestern Mexico and *A. a. peninsularis* of Baja California are classified as endangered by the IUCN (1972, 1976) and the USDI (1980), and are on appendix 1 of the CITES. *A. a. mexicana* of Arizona, New Mexico, Texas, and northern Mexico is on appendix 2 of the CITES.

The geological range of the family Antilocapridae is middle Miocene to Recent in North America. Thirteen now-extinct genera were present during the Pliocene and Pleistocene (O'Gara 1978).

ARTIODACTYLA; *Family BOVIDAE*

Antelopes, Cattle, Bison, Buffalo, Goats, Sheep

This family of 45 Recent genera and 128 species has a natural distribution covering all of Africa, most of Eurasia and North America, and some islands of the Arctic and East Indies. The great majority of genera are native to Africa, and southern and Central Asia. Wild-living populations of some species have been introduced by human agency in New Guinea, New Zealand, Australia, and surrounding islands. Certain species have been domesticated and are now found throughout the world in association with people. The sequence of genera presented here follows basically that of Simpson (1945), who recognized five subfamilies:

Subfamily Bovinae

Tragelaphus	*Tetracerus*	*Bos*
Taurotragus	*Bubalus*	*Bison*
Boselaphus	*Syncerus*	

Subfamily Cephalophinae

Cephalophus	*Sylvicapra*

Subfamily Hippotraginae

Kobus	*Hippotragus*	*Damaliscus*
Redunca	*Oryx*	*Alcelaphus*
Pelea	*Addax*	*Connochaetes*

Subfamily Antilopinae

Oreotragus	*Dorcatragus*	*Litocranius*
Ourebia	*Antilope*	*Gazella*
Raphicerus	*Aepyceros*	*Antidorcas*
Neotragus	*Ammodorcas*	*Procapra*
Madoqua		

Subfamily Caprinae

Pantholops	*Rupicapra*	*Capra*
Saiga	*Budorcas*	*Pseudois*
Nemorhaedus	*Ovibos*	*Ammotragus*
Capricornis	*Hemitragus*	*Ovis.*
Oreamnos		

Simpson's list of genera has been modified extensively here, in accordance with various sources cited in the generic accounts that follow. There remains much controversy regarding the classification of the Bovidae. Ansell (in Meester and Setzer 1971), for example, recognized four additional subfamilies: Reduncinae, with the genera *Redunca* and *Kobus;* Alcelaphinae, with *Connochaetes, Alcelaphus,* and

Damaliscus; Aepycerotinae, with *Aepyceros;* and Peleinae, with *Pelea*.

Shoulder height ranges from as low as 255 mm in the pygmy antelope (*Neotragus*) to as great as 2,200 mm in some species of *Bos*. The pelage varies from smooth and sleek to rough and shaggy. The ulna and fibula are reduced, and the main foot bones are fused into a cannon bone. The front and hind feet are not the same length. On each foot, only two digits, the third and fourth, are well developed. The second and fifth digits are either absent or small and form the so-called dew hooves or dew claws. The stomach is four chambered and ruminating, and a gall bladder is usually present. Females have one or two pairs of functional mammae.

Horns are carried by all adult male bovids, and by the females of most genera. The horns are composed of a bony core, attached to the frontal bones of the skull, and a hard sheath of horny material. *Tetracerus* is unique among bovids in having four horns; all other genera have only two. The dental formula is: (i 0/3, c 0/1, pm 3/2–3, m 3/3) × 2 = 30–32. The lower incisors project more or less forward, and the cheek teeth are low crowned or high crowned, with crescentic ridges of enamel on the crowns.

Most bovids inhabit grassland, scrubby country, or desert, but some live in forests, swamps, or arctic tundra. Goats and sheep generally occur in rocky or mountainous areas. Bovids graze or browse, and are ruminants, that is, they chew the cud. The food is brought up from the first compartment of the stomach and chewed while the animal is at leisure, before being swallowed a second time for thorough digestion. Bovids feed by twisting grass, stems, or leaves around the tongue and cutting the vegetation off with the lower incisors.

According to Estes (1974), most species of Bovidae are gregarious and territorial. Exceptions include most species of the subfamily Bovinae, which are either gregarious and nonterritorial or, in the case of *Tragelaphus scriptus* and *T. spekei*, solitary and nonterritorial. The genera *Cephalophus, Redunca, Oreotragus, Ourebia, Raphicerus, Neotragus, Madoqua*, and *Dorcatragus* are generally solitary and territorial. Three social classes are found in all gregarious bovids: nursery herds of females and young, bachelor herds, and solitary adult males. If the species is also territorial, each adult male usually defends an exclusive area for all or part of the year. Glands on the hooves of the gregarious species release a substance onto the ground having a characteristic scent that an isolated animal can follow back to the herd.

Wild bovids have been extensively hunted by people for meat, hides, and sport. Many species and subspecies have become rare or threatened with extinction. Domestic cattle, sheep, and goats have all been derived from Eurasian species. The domestication of sheep and goats probably began in southwestern Asia 8,000 to 9,000 years ago.

The geological range of this family is early Miocene to Recent in Europe, early Pliocene to Recent in Asia and Africa, and Pleistocene to Recent in North America.

ARTIODACTYLA; BOVIDAE; **Genus TRAGELAPHUS**
De Blainville, 1816

Bongo, Sitatunga, Bushbuck, Nyalas, and Kudus

There are two subgenera and seven species (Meester and Setzer 1971; Harrison 1972):

subgenus *Tragelaphus*

T. buxtoni (mountain nyala), highlands of southern Ethiopia;

T. spekei (sitatunga), Gambia to southern Sudan, and south to northern Botswana;

T. angasi (nyala), southern Malawi, Mozambique, Rhodesia, eastern South Africa;

T. scriptus (bushbuck), most of Africa south of the Sahara;

T. strepsiceros (greater kudu), southern Chad to Somalia, and south to South Africa;

T. imberbis (lesser kudu), South Yemen, southeastern Sudan, Ethiopia, Somalia, northeastern Uganda, Kenya, eastern Tanzania;

subgenus *Boocercus*

T. euryceros (bongo), forest zone from Sierra Leone to Kenya.

As explained by Meester and Setzer (1971) and Van Gelder (1977a, 1977b), there has been considerable disagreement regarding the classification of this genus. On the one hand, *Boocercus* Thomas, 1902 has often been treated as a full genus, and some of the species listed above in the subgenus *Tragelaphus* have sometimes been separated into the genera or subgenera *Strepsiceros* Hamilton Smith, 1827 and *Limnotragus* Pocock, 1900. On the other hand, *Taurotragus* has occasionally been considered only a subgenus of *Tragelaphus*.

According to Meester and Setzer (1971), both *Tragelaphus* and *Taurotragus* comprise medium-sized to large antelopes with a variously developed face and body pattern of white spots and stripes. Males are distinctly larger than females. The tail is relatively short in the subgenus *Tragelaphus*, but is long and tufted in the subgenus *Boocercus* and the genus *Taurotragus*. The horns are twisted and are present in only the males of the subgenus *Tragelaphus*, but in both sexes of *Boocercus* and *Taurotragus*. The skull has an ethmoid fissure and lacks a preorbital fossa. Females have two pairs of mammae. From *Taurotragus*, the genus *Tragelaphus* differs in having the horns in an open spiral, no

Sitatunga (*Tragelaphus spekei*), photo by N. W. Labanoff through Bernhard Grzimek.

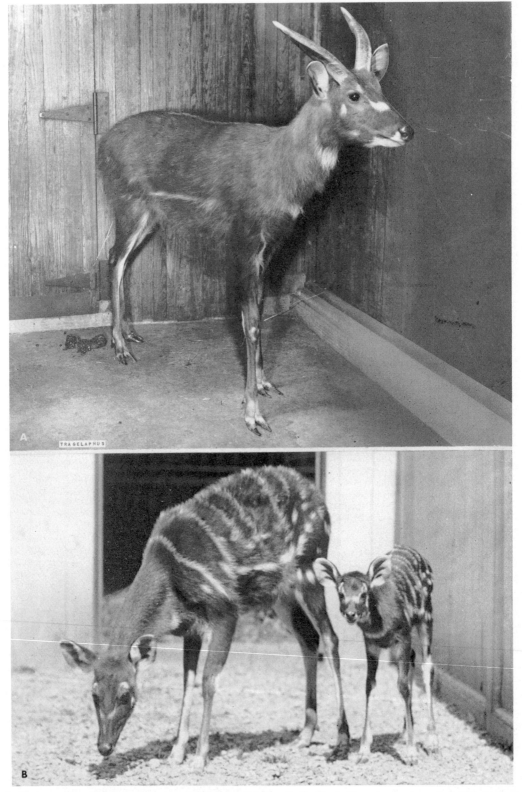

A. Sitatunga (*Tragelaphus spekei*), photo by Ernest P. Walker. B. Bushbuck (*T. scriptus*), photo from U.S. National Zoological Park.

dewlap, and feet that are nonoxlike and more elliptical. Additional information is given separately for each species.

Tragelaphus buxtoni.   Head and body length is 1,900 to 2,600 mm, shoulder height is 900 to 1,350 mm, weight is 200 to 225 kg, and record horn length is 1,187 mm. According to Dorst and Dandelot (1969), the coat is rather shaggy, the general color is grayish chestnut, there is a white chevron between the eyes and two white spots on the cheek, there are poorly defined white stripes on the back and upper flanks, and there is a short brown mane on the neck.

Dorst and Dandelot (1969) wrote that the mountain nyala inhabits forests and heathland at elevations of 2,900 to 3,800 meters. It normally feeds in the evening and early morning, and is mainly a browser. It occurs alone or in groups of up to 15 individuals. Old males are usually solitary. *T. buxtoni* was not discovered until 1908, and its numbers are now much reduced in some areas.

Tragelaphus spekei.   Head and body length is 1,150 to 1,700 mm, tail length is about 200 to 260 mm, shoulder height is 750 to 1,250 mm, weight is about 125 kg, and horn length in males is about 508 to 924 mm. The general coloration is dull brown. There are two whitish areas on the throat, one near the head and one near the chest. There are also whitish marks on the face, ears, cheeks, body, legs, and feet. The females are slightly browner than the males, generally chestnut. The young are the color of the mother. The hair is much shaggier than in other antelopes. The great elongation of the hooves and the peculiar flexibility of the joints at the feet (pasterns), which are bare and rest on the ground, are striking structural characters adapted to walking on boggy and marshy ground.

The sitatunga is semiaquatic, spending most of its life in dense beds of *Papyrus* and *Phragmites* within swamps (Smithers 1971). It is both diurnal and nocturnal, and may move onto marshy land at night. It swims well and often submerges entirely when feeding in the water. If pursued, it may hide under the water with only the nostrils exposed. The sitatunga is a grazer of reeds, sedges, and grasses. Breeding apparently occurs throughout the year, females produce a single young at an average interval of 11.6 months, the mean gestation period is 247 days, and sexual maturity is attained at approximately 1 year of age by females and at 1½ years by males (Densmore 1980). Potential longevity is about 20 years.

Tragelaphus angasi.   Head and body length is 1,350 to 1,550 mm, shoulder height is 800 to 1,150 mm, weight is 112 to 127 kg, and record horn length is 835 mm. *T. angasi* differs from other species of *Tragelaphus* in having a fringe of long pendant hairs that forms a line around the lower neck, lower shoulders, sides of the belly, lower thighs, and backs of the thighs. The color is grayer than in most of the other species.

The nyala is found on plains and mountains, usually in areas near water and with dense cover (Grzimek 1975). In a study in Zululand, Anderson (1980) found feeding activity mainly in the early morning and late afternoon. Average individual home range was 0.65 sq km in males and 0.83 sq km in females. There was extensive overlap both between and within sexes, and no evidence of territoriality. Although herds of up to 30 nyala have been reported, Anderson observed an average of 2.38 individuals per sighting. Young males typically occurred in loose associations of about 3 animals, and females in groups of about 6, sometimes with their young. Older males tended to live alone for most of the year, but attempted to join groups that contained females in estrus. Several males would compete for the females by postural displays and sparring with the horns; serious fights were apparently rare. The alarm call of the nyala is a staccato, doglike bark.

The gestation period is a little over 7 months, and the single young is usually born in August (Grzimek 1975). A female reached sexual maturity at about 14½ months of age and bore its first litter when nearly 22 months old (Dittrich 1972).

Tragelaphus scriptus.   Head and body length is 1,050 to 1,500 mm, tail length is about 200 to 275 mm, shoulder height is 650 to 1,100 mm, weight is about 50 kg, and horn length is 350 to 635 mm. The horns have pronounced keels in front and back and are twisted into spirals. The females are smaller than the males and are generally without horns. The color of the back and sides ranges from light tawny or reddish in females to dark brown or almost black in males. The underparts are usually slightly darker than the sides or back. There are a variety of white markings on the throat and lower neck, a line down the middle of the back, and stripes or rows of dots vertically on the sides. In some forms there is a pronounced mane the full length of the back.

The bushbuck inhabits the edges of swamps and other areas of dense vegetation near water. It is adept at following small, overgrown tunnels through tangles of interwoven vines and shrubbery. It is either diurnal or nocturnal. It usually keeps to a restricted area, sometimes only a few hundred meters across, but may undertake seasonal movements in search of food. It is mainly a browser of the leaves and twigs of trees and shrubs, but it also grazes.

Odendaal and Bigalke (1979) reported that individual home range evidently decreases as population density increases. Studies in savannah areas revealed densities of over 25 individuals per sq km and ranges of 0.2 sq km or less. Studies in forests indicated densities of only 4 per sq km, but ranges of around 1 sq km. All available information (Allsopp 1978; Estes 1974; Jacobsen 1974; Odendaal and Bigalke 1979; Smithers 1971) indicates that *T. scriptus* is usually found alone. The only regular exceptions are mothers with calves, and courting pairs. Males may compete vigorously for estrous females, but are not territorial. The bushbuck gives a call that is said to resemble closely the barking of a dog.

Breeding occurs throughout the year, though there are reproductive peaks in some areas. The gestation period is about 6 months, and the mean interval between births is about 8 months. The single young lies in concealment away from the mother for the first few weeks of life. Sexual maturity is attained after 11 or 12 months of life (Allsopp 1978; Dittrich 1972; Jacobsen 1974; Von Ketelhodt 1976; Morris and Hanks 1974; Odendaal and Bigalke 1979).

Tragelaphus strepsiceros.   Head and body length is 1,950 mm to 2,450 mm, tail length is 400 to 470 mm, shoulder height is 1,200 to 1,500 mm, and weight is about 200 to 270 kg. The horns of males usually measure about 1,016 mm in a straight line and 1,320 mm along the curves. There is one record of 1,816 mm in a straight line. The hair of the neck is short and scant. The general coloration ranges from reddish to pale slaty with blue gray and white markings.

The greater kudu is found mainly in woodland and thickets, and requires adequate cover for concealment (Smithers 1971). It may be active by day or night. Its hearing is acute, and all observers agree that its extreme wariness makes it difficult to approach. It has the often fatal habit, however, of stopping after a short run to look back. Its ability to leap is marvelous; there is one report of its clearing bushes 2.5

A. Nyala (*Tragelaphus angasi*); B. Greater kudu (*T. strepsiceros*); photos from San Diego Zoological Garden.

Lesser kudu (*Tragelaphus imberbis*), photo by Leonard Lee Rue III.

meters high with ease. It feeds mainly by browsing (Smithers 1971).

This species is gregarious, often occurring in herds of 6 to 20 individuals. Such groups include mostly adult females and young. Mature males live alone or in groups of 2 to 4, and compete with one another for possession of estrous females. Breeding occurs throughout the year in at least some areas. Gestation periods of about 7 to 9 months have been reported, and a single calf is born (Smithers 1971; Grzimek 1975; Dittrich 1972). A captive kudu lived for 20 years and 9 months (Marvin L. Jones, Zoological Society of San Diego, pers. comm.).

Tragelaphus imberbis. Head and body length is 1,100 to 1,400 mm, shoulder height is 900 to 1,050 mm, weight is about 60 to 100 kg, and horn length is about 600 to 900 mm. The general coloration of males is usually deep yellowish gray, but some individuals are darker. The general coloration of females is fawn. There are white patches on the throat, like in *T. scriptus,* and 11 to 14 vertical stripes on the body. There is no throat mane like that of *T. strepsiceros.* The young are redder than the adult females and are strongly marked with white.

The lesser kudu inhabits areas of dense vegetation. It hides by day in secluded places, and grazes or browses in the late evening and early morning. An individual reportedly made a horizontal leap of 9.2 meters, in which it went over the top of a bush 1.5 meters high and 1.8 meters in diameter.

Leuthold (1974) reported that some individuals remain throughout the year in one area, while others make seasonal movements in search of food. Individual home range varies from 0.4 to 6.3 sq km, and averages 2.2 sq km in males and 1.8 sq km in females. Territorial behavior is not apparent.

Most groups contain 4 or fewer individuals, but aggregations of up to 24 occasionally form. A group may be either unisexual or bisexual, but an adult male associates only temporarily with females. A young male remains with its mother's group for 1½ to 2 years, and then becomes solitary or joins a group of males. A female may temporarily separate from her group to bear her single calf. Dittrich (1972) reported a gestation period of 222 days, and that a male attained sexual maturity at about 29 months of age.

Tragelaphus euryceros. Head and body length is 1,700 to 2,500 mm, tail length is 450 to 650 mm, shoulder height is 1,100 to 1,400 mm, and weight is 150 to 220 kg. The bongo has short hair, an erect mane from the shoulders to the rump, and a long, tufted tail. It is among the most beautiful of bovids. The back and sides are bright chestnut red, and the belly is black. A pure white chevron crosses the forehead, and other white patches are located on the sides of the head. The breast bears a large white crescent. There are 11 to 12 narrow, vertical stripes on the sides of the body. Apparently, the number of stripes on each side is rarely the same. Females are usually brighter colored, and old males may become much darker, the chestnut red turning dark mahogany brown and the body stripes becoming buffy. A dark dorsal stripe is present and the tail tuft is dark maroon or black. The outsides of the legs are dark or blackish, with a black chevron above white knees and a white patch above the hooves. The insides of the legs are white. The horns have yellow or buffy tips.

Both sexes have horns that spiral in one complete twist. The longest on record measured 1,002 mm along the front curve and were borne by a male from Kenya. The average horn length is about 835 mm, and is smaller in females.

Greater Kudu (*Tragelaphus strepsiceros*), photo by Lothar Schlawe.

The bongo inhabits lowland forests in most of its range, but in Kenya occurs in montane forests at elevations of 2,000 to 3,000 meters (Ralls 1978). It lives in the densest, most tangled parts of the forest. It is apparently diurnal and most active in the morning and afternoon. At midday it rests in heavy cover, while ruminating. It depends more on its sense of hearing than on sight or smell. It is very shy and swift, and can quickly disappear when startled. It runs gracefully, and at full speed, through even the thickest tangles of lianas, laying its heavy, spiraled horns on its back so that the brush cannot impede its flight. So common is this habit that most older animals have bare, rubbed patches on their backs, where the horn tips rest. Also, the front surfaces of the horns of older animals are often much worn and frayed from the friction caused by their rapid passage through the heavy undergrowth. The bongo prefers to go under or around obstacles rather than over them. It likes to wallow in mud puddles and then rub the mud against a tree, at the same time polishing its horns.

The diet is varied, but in general the bongo is a browser. It eats the tips, shoots, and trailers of many plants, and, like the bushbuck and nyala, shows a marked preference for the tender bush herbage that grows around the bases of trees. Roots, bamboo leaves, cassava, and sweet potato leaves are also preferred items. The bongo sometimes raids coco yam farms for the tender leaves. It often uproots saplings with its horns in order to eat the roots. Like many other browsing animals, it can rear up on its hind legs, bracing the forelegs against a tree trunk. It can reach leaves and twigs as high as 2.5 meters above the ground. The bongo is said to eat earth at times, and also to chew and swallow pieces of burned wood from lightning-killed forest trees, apparently to obtain salt.

The bongo does not congregate in large herds, but sometimes travels in parties of up to 20 individuals, including cows, calves, and young males. Old bulls are usually solitary. Births are said to occur in December or January in the wild, and have taken place in December, April, and August in captivity. According to Ralls (1978), females have an estrous cycle of 21–22 days, estrus is about 3 days long, two gestation periods lasted 282 and 285 days, all 11 births in captivity yielded a single offspring, and average weight at birth is 19.5 kg. The young have the same color pattern as

Bongo (*Tragelaphus euryceros*), female, photo from New York Zoological Society.

adults, but are a rich tawny shade and much lighter. Ralls (1978) wrote that two captive-born females first conceived at the ages of 27 and 31 months, and that a captive female lived for about 19 years and 5 months.

ARTIODACTYLA; BOVIDAE; **Genus TAUROTRAGUS**
Wagner, 1855

Elands

There are two species (Meester and Setzer 1971):

T. oryx (common eland), open country from Ethiopia and southern Zaire to South Africa;

T. derbianus (Derby or giant eland), savannah zone from Senegal to southern Sudan.

Some authorities, such as Van Gelder (1977*b*), include *Taurotragus* within the genus *Tragelaphus*.

Head and body length is 1,800 to 3,450 mm, tail length is 500 to 900 mm, shoulder height is 1,000 to 1,800 mm, and weight is 400 to 1,000 kg. Males average larger than females. The color of *T. oryx* is a fairly uniform grayish fawn. The general coloration of *T. derbianus* is rich fawn, its head is lighter than its body, and its neck is black with a white band at the base. Both species have whitish or creamy verti-

Eland (*Taurotragus oryx*), photo by C. A. Spinage through East African Wildlife Society.

cal stripes on the upper parts. There is a short mane on the nape and longer hairs on the throat.

Taurotragus is characterized by oxlike massiveness, rounded hooves, a hump on the withers, a dewlap between the throat and chest, and heavy spiral horns that are carried by both sexes. The horns of *T. derbianus* are longer and more massive, the record length being about 1,200 mm.

The preferred habitat of *Taurotragus* is plains or moderately rolling country with brush and scattered trees. Elands are alert, keen of sense, and thus difficult to approach. They are sometimes said to be slow and easily caught, but Schaller (1972) reported maximum speed to be at least 70 km per hr. Despite their size, elands are good jumpers and can take a 1.5-meter fence with ease. They prefer to lie in some shelter during the heat of the day. In the morning and evening they seek more open areas, where they browse upon leaves and succulent fruits.

Schaller (1972) estimated that there were 7,000 *T. oryx* in the 25,000-sq-km Serengeti ecosystem. Herds there usually consisted of fewer than 25 individuals, including a number of cows and subadults, and one or more large bulls. According to Leuthold (1977a), the home ranges of *T. oryx* vary considerably by sex. Those of males have been found to be only 6 to 71 sq km in size and to be located mainly in wooded areas. Those of females are 34 to 360 sq km and are generally on the open plains. There is much overlap of home range, and evidently no exclusive use of space. Males occur alone or in small groups, whereas females and young may form aggregations of 200 or more individuals. Several adult males may be present in a female herd at a given time, but they have a strict dominance hierarchy that governs access to estrous females. Young elands have a strong tendency to associate in peer groups of their own.

There are distinct breeding seasons in some areas. In Zambia the calves are born in July and August (Wilson 1969). Gestation periods of 254 to 277 days have been recorded (Dittrich 1972). There is normally a single calf. Sexual maturity is usually attained at about 3 years of age by females and 4 years by males. According to Treus and Lobanov (1971), captive female *T. oryx* have given birth when as young as 22 months and as old as 19 years, and record longevity is 23 years and 6 months.

Elands have disappeared from large parts of their range, mainly because of excessive hunting by people. The western giant eland (*T. derbianus derbianus*), originally found in Senegal, Gambia, Mali, Guinea, and Guinea-Bissau, is classified as endangered by the IUCN (1976) and the USDI (1980). Elands yield a large quantity of tender meat, and the quality of the thick hide is excellent. These animals are docile and easily tamed, and efforts have been made in Africa and the Soviet Union to domesticate them for meat and milk production. Eland milk has about triple the fat content and twice the protein of the milk from a dairy cow (Treus and Lobanov 1971).

ARTIODACTYLA; BOVIDAE; *Genus BOSELAPHUS*
De Blainville, 1816

Nilgai, Bluebuck

The single species, *B. tragocamelus*, is found in eastern Pakistan and India (Ellerman and Morrison-Scott 1966; Roberts 1977).

Head and body length is 1,800 to 2,100 mm, tail length is 456 to 535 mm, shoulder height is 1,200 to 1,500 mm, and weight is up to 270 kg. Males average about one-fifth larger and heavier than females (Grzimek 1975). The hair on the body is short and wiry. Although in both sexes the neck is ornamented with a mane, only the bulls develop a tuft of hair on the throat. The upper parts of males are generally iron gray, but the lower surface of the tail, stripes inside the ears, rings on the fetlocks, and underparts are white. The head and limbs are tawny, and the throat tuft and tip of the tail are black. The females are more lightly colored. The forelegs are somewhat longer than the hind ones, and the head is long and pointed. The horns are short and carried only by males.

The nilgai frequents forests, low jungles, and, occasionally, open plains. It is diurnal, but also rests during the day. It is both a browser and a grazer, and is fond of fruit and sugar cane.

In a study in central India, Schaller (1967) found a population density of about 0.07 individuals per sq km. Males were observed to establish territories during the breeding season, and to attempt to gather groups of 2 to 10 females thereon. The males fought one another, for possession of the territories and the females, by dropping to their knees and lunging with the horns. Nonterritorial males, in groups of up to 18 animals, were also present. Some breeding occurred throughout the year, but most calves were born from June to October. Mating takes place immediately after the cow has given birth. Gestation periods of 245 to 277 days have been recorded (Dittrich 1972). The number of young per litter is one, or more commonly two. A captive bluebuck lived for 21 years and 8 months (Marvin L. Jones, Zoological Society of San Diego, pers. comm.).

Because its horns are not impressive and its meat poor tasting, *Boselaphus* is not much sought after by modern hunters. Moreover, it is regarded as a close relative of the sacred cow by the Hindu religion. Consequently, the nilgai enjoys immunity from molestation and sometimes displays remarkably little concern in the presence of people. It is rather docile for so large an animal, is said to tame easily, and thrives in captivity.

ARTIODACTYLA; BOVIDAE; *Genus TETRACERUS*
Leach, 1825

Four-horned Antelope, Chousingha

The single species, *T. quadricornis*, occurs in India and Nepal (Ellerman and Morrison-Scott 1966; Mitchell 1977).

Head and body length is about 800 to 1,000 mm, tail length is about 126 mm, shoulder height is about 600 mm, and weight is 17 to 21 kg. The short, thin, coarse hair is a uniform brownish bay above, lighter on the lower sides, and white on the insides of the legs and middle of the belly. The

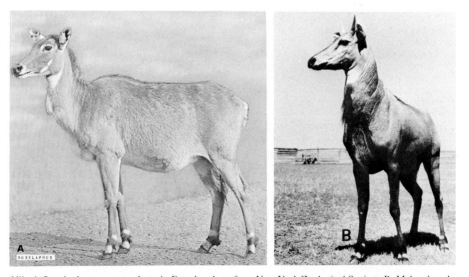

Nilgai (*Boselaphus tragocamelus*): A. Female, photo from New York Zoological Society; B. Male, photo by Bernhard Grzimek.

Four-horned antelope (*Tetracerus quadricornis*), photo from Field Museum of Natural History. Inset: photo from *Guide to Great Game Animals*, British Museum (Natural History).

muzzle, outer surface of the ears, and a line down the front of each leg are blackish brown. The horns, borne only by males, are short, conical, smooth, and usually four in number. The posterior two horns are 80 to 100 mm long. The front two are often small, about 25 to 38 mm long, and are sometimes represented by only a slightly raised area of black hairless skin. In having four horns, *Tetracerus* is unique among the Bovidae. The hooves are small and are rounded in front.

The chousingha is found most frequently in open forests. It is shy and swift, dashing into dense cover at the first sign of danger. It is sometimes confused in the field with the hog deer *(Axis porcinus)*, but can be distinguished by its peculiar jerky manner of walking or running. It is a grazer and drinks regularly, being seldom found far from water.

Tetracerus is not gregarious, and rarely are more than two individuals found together. Mating takes place during the rainy season from July to September. The gestation period is 7½ to 8 months (Grzimek 1975). There are one to three young per litter. A captive lived for 10 years (Marvin L. Jones, Zoological Society of San Diego, pers. comm.).

Despite the small size of the horns, the presence of two pairs makes the chousingha much sought after by trophy hunters. Most persons agree that the meat is not as good as that of some of the other antelopes. When captured young, the chousingha is easily tamed, but it is apparently quite delicate.

ARTIODACTYLA; BOVIDAE; *Genus BUBALUS*
Hamilton Smith, 1827

Asian Water Buffaloes, Anoas

There are two subgenera and four species (Groves 1969*a*; IUCN 1972, 1976):

subgenus *Bubalus*

B. bubalis (Asian water buffalo, carabao), originally found from at least Nepal and India to Viet Nam and Malaysia;
B. mindorensis (tamaraw), Mindoro Island (Philippines);

subgenus *Anoa*

B. depressicornis (lowland anoa), lowlands of Celebes;
B. quarlesi (mountain anoa), highlands of Celebes.

Anoa Hamilton Smith, 1827 has sometimes been used as a separate genus, and *B. mindorensis* has sometimes been placed in *Anoa*. *B. bubalis* has been widely domesticated and introduced by people, and certain wild-living populations outside of the region indicated above are sometimes thought to represent the natural distribution of the species. Some authorities, including Groves (1969*a*), use the name *B. arnee* in place of *B. bubalis*. The tamaraw, or tamarao (*B. min-*

Four-horned antelope (*Tetracerus quadricornis*), photo by N. Das.

Asian water buffaloes (*Bubalus bubalis*), photo by R. Van Nostrand through San Diego Zoological Garden.

A. Tamaraws (*Bubalus mindorensis*), photo from "Saugethiere vom Celebes- und Philippinen- Archipel," Adolph B. Meyer, *Abh. Zool. anthrop.-ethn. Mus. Dresden*. B. Anoas (*B. depressicornis*), photo from San Diego Zoological Garden.

dorensis), should not be confused with the carabao, a small domestic form of *B. bubalis* used in the Philippines.

In *B. bubalis*, head and body length is 2,400 to 3,000 mm, tail length is 600 to 1,000 mm, shoulder height is 1,500 to 1,900 mm, and weight is 700 to 1,200 kg. The hair is moder-ately long, coarse, and sparse, and is directed forward from the haunches to the head. There is a tuft on the forehead, and the tip of the tail is bushy. The general coloration is ash gray to black. Some domestic breeds are black and white, brown and white, or entirely white (Grzimek 1975). The face is long and narrow, the hooves are large and splayed, and the ears are comparatively small. The horns, carried by both sexes, are heavy at the base, normally curve backward and inward, are somewhat triangular in cross section, and are conspicuously marked with wrinkles. The spread of the horns, up to 1,200 mm along the outer edge, exceeds that of any other living bovid. Lekagul and McNeely (1977) stated that a wild *B. bubalis* is larger, quicker, and more aggressive, and has much more widely spreading horns, than a domestic buffalo.

Bubalus mindorensis is much smaller than *B. bubalis*, being only about 1,000 mm high at the shoulders. It has more hair on the body than does *B. bubalis*, and is dark brown to grayish black. Its horns are stout and short, only about 355 to 510 mm long.

In the subgenus *Anoa*, head and body length is about 1,600 to 1,720 mm, tail length is 180 to 310 mm, and shoulder height is 690 to 1,060 mm. Grzimek (1975) listed weight as 150 to 300 kg. Although the young are thickly covered with yellowish brown, woolly hair, the skin of old individuals is almost bare. The general color of adults varies from dark brown to blackish, and there are often blotches of white on the face, nape, throat, and lower limbs. The underparts are usually light brown. Males are generally darker than females. The hide is of exceptional thickness, the limbs are rather short, the body is plump, and the neck is thick. According to Groves (1969a), *B. depressicornis* has white forelegs, a long tail, and horns that are triangular in section, flattened, wrinkled, and 183 to 373 mm long; and *B. quarlesi* has legs that are generally the same color as the body, more hair than does *B. depressicornis*, a short tail, and horns that are rounded in section, nonwrinkled, and only 146 to 199 mm long.

The anoas and tamaraw are basically forest animals. The tamaraw requires dense vegetation for resting, water for drinking and wallowing, and open grazing land (IUCN 1976). Anoas feed in the morning and rest in the shade during the afternoon. Their gait is a trot, but at times they make clumsy leaps. It is said that they feed chiefly on young cane shoots and various water plants.

The water buffalo is associated with wet grasslands, swamps, and densely vegetated river valleys. It is pestered by insects, and as a means of protection it wallows in the water and soil. It is thus often completely caked by a layer of mud, through which insects cannot penetrate. It also escapes from insects by submerging in the water with only the nostrils exposed. The time of activity apparently varies, but animals of the introduced population in northern Australia leave the resting area at sunrise, graze for three or four hours, drink and wallow in the middle of the day, graze again in the late afternoon, and return to the resting area at sunset (Tulloch 1978). The favorite foods are lush grass and vegetation growing in or beside rivers and lakes.

The anoas and tamaraw are rare and wary, and inhabit remote areas; they apparently associate in pairs, rather than herds, except when the cows are about to give birth. The water buffalo is much more common, at least in the domestic form, and is gregarious. According to Tulloch (1978), about 100,000 to 200,000 feral buffalo inhabit 100,000 sq km of the Northern Territory of Australia. Adult females and their young form stable "clans" of up to 30 individuals. Each such clan has a home range of 170 to 1,000 ha., containing areas for resting, grazing, wallowing, and drinking. There is a dominance hierarchy, and the leader is always an old cow, even when bulls accompany the group. Young females remain in their mother's group, but young males are driven off when they are two or three years old. Several clans of females form a herd of 30 to 500 animals, which gathers at the same nightly resting area. There are also bachelor groups, usually containing about 10 males, which spend the dry season (May–September) apart from the females. Their home

ranges are larger than those of the females, and overlap one another. During the wet season (October–April) the bulls move into the areas used by the cows. Dominant males then mate with, but do not control, the females of a clan, and are subsequently driven off. Very old males are usually solitary.

Anoas have no specific breeding season, a gestation period of 275 to 315 days, generally a single offspring, and a maximum known longevity in captivity of 28 years. In *B. bubalis*, breeding is seasonal in some areas, the gestation period is 300 to 340 days, there is normally one calf, nursing lasts for 6 to 9 months, and known longevity is up to 25 years in the wild and 29 years in captivity (Grzimek 1975).

The species *B. mindorensis*, *B. depressicornis*, and *B. quarlesi* are classified as endangered by the IUCN (1976) and the USDI (1980), and are on appendix 1 of the CITES. All have very restricted ranges and have declined through excessive hunting and loss of habitat. The tamaraw, with an estimated 150 to 200 individuals in existence, is one of the world's rarest mammals. *B. bubalis* is designated as vulnerable by the IUCN (1972). This species has disappeared from most of its original range, because of usurpation of habitat by agriculture, hunting by people, and competition from and diseases transmitted by domestic livestock. A few herds, thought to be descended from original native stock, are still scattered from India to Indochina.

Domestication of *B. bubalis* may have begun in the third millenium B.C. in southern Asia. There are now an estimated 75,000,000 domestic buffalo in the world, mostly in India, Southeast Asia, and the East Indies. The species also has been brought to southern Europe, Asia Minor, northern and eastern Africa, Madagascar, Mauritius, Australia, Japan, Hawaii, and Central and South America. In some of these areas, large feral populations have become established (Grzimek 1975). Domestic buffalo are especially suitable for tilling rice fields, and are also used as beasts of burden. Their milk is richer in fat and protein than that of the dairy cow, and leather made from their skin is of superior quality. They are docile and tractable with persons whom they know, and can even be controlled by young children.

ARTIODACTYLA; BOVIDAE; ***Genus SYNCERUS***
Hodgson, 1847

African Buffalo

The single species, *S. caffer*, originally occurred in most of Africa south of the Sahara (Meester and Setzer 1971).

Head and body length is 2,100 to 3,000 mm, tail length is 750 to 1,100 mm, shoulder height is 1,000 to 1,700 mm, and weight is 500 to 900 kg. There is much variation in size and other characters, and the subspecies *S. c. caffer* of the eastern savannahs may be twice as large as the subspecies *S. c. nana* of the equatorial forests (Sinclair 1977a; Grubb 1977). There is a thick covering of hair on young individuals, a sparse covering on adults, and little hair on old animals. Coloration is brownish to black. The head and limbs are massive, the breadth of the chest is great, and the ears are relatively large, drooping, and fringed by soft hairs. The horns spread outward and downward, and then upward in some animals, and out and back in others. In males the two horns are joined by a boss, a large shield covering the whole top of the head.

The African buffalo occurs in a great variety of habitats, but prefers areas with grass, water, and some dense cover. It seems to enjoy splashing about in water and wallowing in mud. It is a powerful and deadly fighter, and can run at 57 km per hr. It usually goes to drink in the evening and morning, and feeds during the early part of the night. Later in the night it generally rests and chews its cud. It usually retires to a

African buffaloes (*Syncerus caffer*): Top, photo from San Diego Zoological Garden; Inset: photo from U.S. National Zoological Park; Bottom, photo from East Berlin Zoo.

shady area during the heat of the day. The African buffalo is mainly a grazer, but occasionally browses on leaves.

Sinclair (1977*a*) reported population densities of about 3 to 18 individuals per sq km, and herd ranges of about 10 to 300 sq km. Generally, density is greatest, and range smallest, in areas with the highest rainfall. Mean herd size is only about 20 individuals, probably all closely related, in the forests of Zaire. On the Serengeti Plains, however, herd size ranged from about 50 to 1,500 and averaged 350. In the latter

area, Sinclair found herds to be fairly stable and each to have a largely separate range. Herds were permanently composed of units consisting of a female and her young of the previous two birth seasons. There seemed to be no overall social hierarchy. Most mature males lived with the herds for the rainy half of the year, but split off during the dry season, then forming bachelor groups of usually 3 or 4 animals. Some males, especially very old ones, lived permanently apart from the herds. Male groups had a dominance hierarchy, established by agonistic displays and fighting, and individuals competed for estrous females.

Reproduction occurs throughout the year in some areas, but there are seasonal peaks associated with rainfall. In the Serengeti, the heavy rains extend from about February to July. According to Sinclair (1977a), most conceptions there occur at the end of this season, and births take place mainly in the second half of the following wet season. Females have an estrous cycle of about 23 days, an estrus of 5 or 6 days, and a mean gestation period of 340 days. The single reddish or blackish brown calf has an average weight of 40 kg at birth. Young males leave their mother and join a bachelor group when about 2 years old, but females remain until they produce their own young, or even longer. Sexual maturity seems to come mainly between 3½ and 5 years of age. Wild individuals up to 18 years old were found in the Serengeti. A captive specimen lived for 29 years and 6 months (Marvin L. Jones, Zoological Society of San Diego, pers. comm.).

Syncerus is often considered the most dangerous big game animal of Africa. Old bulls are said to stalk human victims and attack without provocation. Sinclair (1977a) stated that the buffalo's reputation is unjustified and based mainly on the tales of hunters, especially those that tried to follow a wounded animal. The buffalo has been heavily hunted for sport and food. It has been exterminated in most of South Africa, and its range is becoming fragmentary in other regions, especially West Africa, as people occupy its range. It apparently became scarce in East Africa by the late 19th century, but numbers recovered from 1900 to 1920, when an epidemic of rinderpest prevented human populations from expanding into buffalo habitat. Schaller (1972) estimated that there were 50,000 buffalo in the Serengeti ecosystem.

ARTIODACTYLA; BOVIDAE; **Genus BOS** Linnaeus, 1758

Oxen

There are four subgenera and five species (Lekagul and McNeely 1977; Ellerman and Morrison-Scott 1966; Grzimek 1975; Corbet 1978; Medway 1977):

subgenus *Bos*

B. taurus (aurochs, domestic cattle), originally found throughout Europe, southern Asia, and northern Africa;
B. javanicus (banteng), Burma, Thailand, Indochina, Malay Peninsula, Java, Borneo;

subgenus *Bibos*

B. gaurus (gaur, seladang), Nepal and India to Indochina and Malay Peninsula;

subgenus *Novibos*

B. sauveli (kouprey), northern Cambodia and adjacent parts of Thailand, Laos, and Viet Nam;

subgenus *Poephagus*

B. grunniens (yak), Tibet and adjacent highland regions.

Each subgenus has sometimes been used as a full genus. Some authorities use the names *B. primigenius* and *B. mutus*, respectively, in place of *B. taurus* and *B. grunniens*. The names *B. sondaicus* and *B. banteng* have sometimes been used in place of *B. javanicus*. The zebu, a domestic form of cattle from India, and the gayal, a domestic form of gaur from eastern India and Burma, have sometimes been given

Domestic bull, near aurochs (*Bos taurus*), photo from West Berlin Zoo.

Scotch cattle (*Bos taurus*), photo by Ernest P. Walker.

British park cattle (*Bos taurus*), photo by Ernest P. Walker.

A. Zebu (*Bos taurus*), photo by R. Pucholt. B. Ankole cattle (*B. taurus*), photo by C. A. Spinage.

the respective scientific names *B. indicus* and *B. frontalis*.

According to Lekagul and McNeely (1977), *Bos* is characterized by large size, a massive body, stout limbs, and a tail that is long and usually tufted at the tip. There are no suborbital, inguinal, or interdigital glands. Horns are borne by both sexes, are larger in males, are inserted rather far apart on each extremity of the top of the skull, and are approximately oval in cross section. The occipital area of the skull forms an acute angle with the face, not a right angle as in *Bubalus*. Additional information is provided separately for each species.

Bos taurus. According to Grzimek (1975), the wild aurochs, ancestor of domestic cattle, had a head and body length of up to 3,100 mm, a shoulder height of 1,750 to 1,850 mm in males, and a weight of 800 to 1,000 kg. Males were about one-fourth larger and heavier than females. The anterior part of the body was more massive than the hindquarters, the legs were rather long, the back was straight, and the head was slender. The horns were pointed, sturdy, and up to 800 mm long. The coat was short and smooth, and was denser in winter. The general coloration was some shade of brown. The original habitat was open forest and meadows. Activity was probably mainly diurnal. The diet included grass, leaves, and acorns. Herds consisted of a bull, several females, and their young. During the mating season in August and September, males had severe fights. The young were born in May and June after a gestation period of about nine months. The aurochs apparently disappeared in southern Asia early in historical times, but survived in western and central Europe until the Middle Ages. The decline was associated with habitat alteration and hunting by people. Efforts were made to protect the last known herd, which lived in Poland, but all the animals died by 1627.

The aurochs was domesticated by people about 8,000 years ago, and gave rise to the numerous breeds of cattle that are now used for the production of milk, meat, leather, fertilizer, and other items. In Europe there are still some primitive breeds that resemble the aurochs in certain external characters. These forms include the Spanish fighting oxen, Corsican country cattle, Scottish highland cattle, and English park cattle (Grzimek 1975). Some of these kinds have been used experimentally in efforts to breed animals that look like the original aurochs. The English park cattle of the Chillingham and Chartley estates are generally white in color, but are like the aurochs in build and behavior. Their precise origin is unknown. The Chillingham herd has been kept fenced, but in a semiwild state, since about 1220 A.D. The only change in this herd over the centuries, at least in external appearance, has been a slight decrease in size. The herd is an absolute monarchy, the "king" bull being the only male that mates with the females. He usually reigns for two or three years, or until he is deposed by another bull.

Modern domestic cattle usually have a shoulder height of 900 to 1,100 mm and a weight of 450 to 1,000 kg. The body is usually covered with short hair. Coloration depends on breed, ranges from white to black, and is often marked with spots and blotches. Cattle graze approximately 8 hours a day, during which time they consume about 70 kg of grass. The remaining time is spent resting or chewing the cud. Mating takes place throughout the year, and a single offspring (occasionally twins) is born after a gestation of 277 to 290 days. Females attain sexual maturity at 18 months of age and remain fertile for about 12 years. The life span may be more than 20 years.

The zebus, the sacred cattle of India, are forms of *Bos taurus*. Of the 30 or more breeds, each of which originated in a province of India, 4 main strains have been introduced to the United States. Zebus have a characteristic hump over the shoulder, drooping ears, and a large dewlap. The coloration is pale fawn, bay, gray, or black. In India, they are allowed to roam the streets and villages without molestation. They interbreed readily with other kinds of cattle, and both they and their hybrids are valued for the ability to resist heat, ticks, and insects. A breed known as the Santa Gertrudis, recognized in 1940, was developed on the King Ranch of Texas by crossing zebu bulls with stock that originated from Texas longhorns, herefords, and shorthorns.

Bos javanicus. Head and body length is 1,800 to 2,250 mm, tail length is 650 to 700 mm, shoulder height is 1,300 to 1,650 mm, and weight is 500 to 900 kg (Lekagul and McNeely 1977; Grzimek 1975). The coloration varies, but tends to be more blue black than that of *B. gaurus*. There are white stockings and a white rump patch.

The banteng is usually found in drier, more open areas than *B. gaurus*. Nonetheless, it depends on dense thickets and forest for shelter. It may be active at any hour, but has become nocturnal in areas where molested by people (Lekagul and McNeely 1977). It sometimes feeds continuously through the night, with many pauses for rest and chewing. It is extremely wary and shy. During the monsoon season, herds may leave the lowlands and drift up into the hill forests, where they feed on tender new herbage, including bamboo shoots. In the dry season, they return to the valleys and more open wooded districts, where they feed on grass. *B. javanicus* apparently does not have to drink as often as *B. gaurus*.

The banteng is usually found in groups of 2 to 40 animals. Generally, there is only a single fully adult bull per herd. Other males either are solitary or live in unisexual groups. Mating may occur at any time of year in captivity, but takes place in May and June in the wild in Thailand. Females are capable of giving birth each year. The gestation period is 9½ to 10 months, there are one or two offspring per birth, nursing lasts up to 9 months, and females attain sexual maturity at 2 years of age (Lekagul and McNeely 1977; Van Bemel 1967).

The banteng is classified as vulnerable by the IUCN (1978) and as endangered by the USDI (1980). It has disappeared in many areas through loss of habitat to the expanding human population, hunting pressure, and hybridization with domestic cattle. Several thousand individuals are still thought to survive from Burma to Borneo, but numbers are decreasing. The banteng has long been domesticated on Java and Bali, and is used as a work animal and for meat and milk production (Grzimek 1975).

Bos gaurus. Head and body length is 2,500 to 3,300 mm, tail length is 700 to 1,050 mm, shoulder height is 1,650 to 2,200 mm, weight is 650 to 1,000 kg, and horn length is 600 to 1,150 mm. Males are approximately one-fourth larger and heavier than females (Grzimek 1975). The general coloration is dark reddish brown to almost blackish brown, and there are white stockings. In adult males there is a large hump over the shoulders.

The gaur inhabits forested hills and associated grassy clearings up to elevations of 1,800 meters. It requires water for drinking and bathing, but does not seem to wallow (Grzimek 1975). It is not excessively wary, but, when startled, it crashes off through the jungle at high speed. In some areas it is basically diurnal, feeding in the morning and late afternoon and resting during the hot hours of the day. In other areas it has become largely nocturnal, because of molestation by people (Lekagul and McNeely 1977). In a study in central India, Schaller (1967) found most activity at night; the animals filed from the meadows to the forest at dawn and were rarely seen in the open after 0800 hrs. Most herds remained in a relatively small area until the beginning of the monsoon,

Banteng (*Bos javanicus*), photo from West Berlin Zoo.

Gaur (*Bos gaurus*), photo from West Berlin Zoo.

Kouprey (*Bos sauveli*), young male with horns not fully developed, photo from Paris Zoo.

and then dispersed into the hills. The gaur was found to be both a grazer and a browser, preferring green grass when it was available, but otherwise eating coarse, dry grasses, and forbs and leaves.

In Schaller's study area, population density was about 0.6 individuals per sq km, and overall home range of a herd was around 78 sq km. Groups contained 2 to 40 individuals, usually about 8 to 11. There was generally not more than a single fully mature bull in a herd. Some males lived alone or in bachelor groups. At the peak of the mating season, the bulls wandered widely in search of estrous females and did not spend more than a few days in any one herd. They competed and sparred with one another, but serious fighting was not observed. Dominance seemed to be based primarily on size. Schaller described the mating call of the bull as a pleasant song, composed of clear resonant tones, each successively lower, that carries about 1.6 km. The gaur has also been reported to have a whistling snort as an alarm call, and to bellow and "moo" like domestic cattle.

Reproduction takes place throughout the year, but in central India mating seems to occur mainly from December to June, and the peak calving season is in the cool months of December and January. The gestation period is 270 to 280 days. There is normally a single offspring, which is nursed for 9 months. Females attain sexual maturity in their second or third year of life (Schaller 1967; Grzimek 1975; Medway 1978). A captive gaur lived for 26 years and 2 months (Marvin L. Jones, Zoological Society of San Diego, pers. comm.).

The gaur is classified as vulnerable by the IUCN (1976) and as endangered by the USDI (1980), and is on appendix 1 of the CITES. It is still found in scattered areas from India to Indochina and the Malay Peninsula, but has declined dras-

tically through hunting and habitat alteration by people, and exposure to the diseases of domestic cattle. The gaur has been reported occasionally to ambush and kill persons that pursue it (Lekagul and McNeely 1977). The gayal, a domestic form of gaur, is smaller than the wild animal, and has shorter legs and a less prominent dorsal hump. It may live in a semitame or feral state in eastern India and Burma, and is used for work and meat production (Grzimek 1975). A gestation period of 293 to 303 days has been reported for the gayal (Scheurmann 1975).

Bos sauveli. Except as noted, the information for the account of this species was taken from Lekagul and McNeely (1977). Head and body length is 2,100 to 2,225 mm, tail length is 1,000 to 1,100 mm, shoulder height is 1,710 to 1,900 mm, and weight is 700 to 900 kg. In females and young animals the general coloration is gray, the underparts are lighter, and the chest, neck, and forelegs are darker. Old males are a very rich, dark brown. The lower legs are white or grayish. From the neck hangs a very long dewlap, which in old males almost reaches the ground. The tail is longer than in *B. gaurus* and *B. javanicus,* and has a bushier tip. Suggestions that the kouprey is only a hybrid between *B. javanicus* and either *B. gaurus, B. taurus,* or *Bubalus bubalis* do not appear valid.

There is considerable difference between the horns of the sexes. In females the horns are lyre shaped and corkscrew upward in a manner resembling those of *Tragelaphus imberbis* of Africa. In males the horns are much larger, spread outward and downward from the base, and then curve forward and upward. When the bulls are about three years old, the horns become split at the tips, and, as the horns grow larger, the split pieces also continue to grow. This fraying of

Yak (*Bos grunniens*), photo from U.S. National Zoological Park.

the horn tips is said to be caused largely by their being used for digging into the earth or thrusting into tree stumps.

The kouprey inhabits low, rolling hills, covered by open country interrupted with patches of forest. It generally grazes in the open areas and enters the forest for shelter from the sun, for refuge from predators, and to seek food when the grasslands are dry. The kouprey moves considerably more than does *B. javanicus*, covering up to 15 km in a day's undisturbed feeding. Herds often contain more than 20 individuals, including several adult males. The herds are not particularly cohesive. Males are sometimes found alone or in groups of their own. Mating takes place in April, and births in December and January. A mother and young may stay away from the herd for about a month.

The kouprey did not become known to science until 1937. Since its discovery, its limited range has been the scene of almost constant human warfare. Little information about the species has emerged, except that it is one of the world's most critically endangered mammals. Its total numbers were estimated at 1,000 in 1940, 500 in 1951, and 100 in 1969. At times it has been feared extinct, but recent observations indicate that a few herds survive. The species is classified as endangered by the IUCN (1976) and the USDI (1980).

Bos grunniens.   In wild males, head and body length is up to 3,250 mm, shoulder height may be over 2,000 mm, and weight is about 1,000 kg. Females weigh only one-third as much (Grzimek 1975). Long hairs that reach almost to the ground form a fringe around the lower part of the shoulders, the sides of the body, and the flanks and thighs. The tail also has long hairs. The general coloration is blackish brown. The large black horns curve upward and forward in the males. The domestic yak is much smaller than the wild form, has weaker horns, and has more varied coloration, being red, mottled, brown, or black.

The yak inhabits desolate steppes at elevations of up to 6,100 meters. In spite of its awkward appearance, it is an expert climber, sure footed, and sturdy. According to Grzimek (1975), the yak stays in high areas with permanent snow during the relatively warm months of August and September, and spends the rest of the year at lower elevations, feeding on grass, herbs, and lichens. Females and young congregate in large herds, formerly reported sometimes to contain thousands of individuals, and adult males spend most of the year alone or in groups of up to 12. The mating season begins in September, and for the next several weeks the bulls join the herds and fight one another for the females. At this time, the wild yak utters a strange grunting sound, but domestic animals emit this call throughout the year. In the wild, births occur in June, the gestation period is 9 months, females have 1 calf every other year, independence is attained after 1 year of life, full size is reached at 6 to 8 years of age, and maximum longevity is 25 years. Domestic females have a less regular reproductive cycle and may give birth every year.

The wild yak is classified as endangered by the IUCN (1974) and the USDI (1980). Although it is officially protected in China, numbers and distribution have declined drastically through uncontrolled hunting by native tribes. The yak was probably domesticated in Tibet during the first millenium B.C., and it now occurs throughout the high plateaus and mountains of Central Asia, in association with people. It is docile but powerful, and is the most useful of domestic mammals at elevations above 2,000 meters. It serves as a mount and beast of burden, is used for milk and meat production, and is sheared for its wool (Grzimek 1975).

European bison (*Bison bonasus*), photo from East Berlin Zoo.

ARTIODACTYLA; BOVIDAE; *Genus BISON*
Hamilton Smith, 1827

Bison

There are two species (Hall 1981; Corbet 1978; Grzimek 1975):

B. bonasus (European bison, wisent), found in historical time in much of Europe and possibly Asia;

B. bison (American bison), found in historical time in western Canada, most of the conterminous United States, northeastern Mexico, and possibly Alaska.

Some authorities, including Corbet (1978), consider *B. bonasus* conspecific with *B. bison*. Some authorities, including Van Gelder (1977*b*), consider *Bison* not more than subgenerically distinct from *Bos*. Groves (1981), who adopted both of these views, suggested that bison are actually less distinct members of *Bos* than are *B. gaurus* and *B. javanicus*.

Head and body length is 2,100 to 3,500 mm, tail length is 500 to 600 mm, shoulder height is 1,500 to 2,000 mm, and weight is 350 to 1,000 kg. Males average larger than females. The pelage of the head, neck, shoulders, and forelegs is long, shaggy, and brownish black. There is usually a beard on the chin. The remainder of the body is covered with short hairs of a lighter color. Young calves are reddish brown. The forehead is short and broad, the head is heavy, the neck is short, and the shoulders have a high hump. The horns, borne by both sexes, are short, upcurving, and sharp. Females have smaller humps, thinner necks, and thinner horns than males.

Of the two species, *B. bison* has longer and more luxuriant hair on the neck, head, and forequarters, giving this species the appearance of larger size. Its body is lower, its pelvis smaller, and its hindquarters less powerful than in *B. bonasus*, though its body on the whole is more massively built. The horns of *B. bison* are shorter and more curved, the front of its head more convex, and its tail shorter and less bushy.

The American bison is traditionally associated with the prairies, but it also occurred extensively in mountainous areas and open forests. The European bison inhabited both woodland and grassland. Bison frequently wallow in dust or mud, and then rub against boulders, tree trunks, and other objects to rid themselves of parasites. Activity may occur at any hour. Banfield (1974) stated that *B. bison* is mostly diurnal and has an average daily movement of about 3 km. Populations of this species on the Great Plains formerly made migrations of several hundred kilometers, generally in a southward direction, to seek better feeding grounds for the winter. The existing population in northern Alberta still migrates up to 250 km each November and May, between the wooded hills and the Peace River Valley. The Yellowstone Park herd also shifts to the high country in the spring and to low valleys in the fall (Meagher 1973). *B. bison* feeds mostly on grass, while *B. bonasus* takes mainly the leaves, twigs, and bark of trees.

Population density of *B. bonasus* has been calculated at about 12 individuals per 1,000 ha. in Poland, and 3 to 4 bison per 1,000 ha. in the Caucasus (Krasinski 1978). A small herd of *B. bison* has been estimated to use a range of about 30 sq km in summer and 100 sq km in winter (Banfield 1974). Both species seem to occur basically in small groups of probably related individuals, but sometimes, as during migrations or on favorable feeding grounds, to associate in herds of hundreds of animals. Formerly, such aggregations of *B. bison* contained many thousands and perhaps millions of individuals. In a study of *B. bison* in Montana, Lott (1974) found adult females, and their young up to three years of age, to exist in groups with an average membership of 57 animals.

American bison (*Bison bison*), photo from U.S. National Zoological Park.

Mature males moved about alone or in small groups for most of the year, but joined the female groups during the mating season. At this time the males fought fiercely for the females, mainly by head-to-head ramming. A successful male would tend a female for several days and prevent other males from approaching within eight meters of her. The voice of *Bison* is a bellow.

Mating occurs mainly from July to September in both North America and Europe, and births take place in the spring. Females are seasonally polyestrous and may produce young every 1 or 2 years. The gestation period averages 285 days in *B. bison* and is 260–70 days in *B. bonasus*. There is almost always a single calf. It weighs about 30 kg at birth, can run after 3 hours, and is weaned at about 7 months of age. The mother guards it closely and will charge intruders. Both sexes seem to attain sexual maturity at between 2 and 4 years of age. Wild individuals are known to have lived about 20 years, and maximum potential longevity may be as much as 40 years (Banfield 1974; Grzimek 1975; Haugen 1974; Krasinski 1978; Meagher 1973).

Mainly because of the spread of human settlements and agriculture, by the early 20th century *B. bonasus* became restricted to the Caucasus and the Bialowieza Forest on what is now the border between Poland and the Soviet Union. The population of the Caucasus became extinct by 1925. The Bialowieza population, which had numbered 727 individuals and been well protected prior to World War I, was completely destroyed by 1919. Some animals had previously been distributed to zoos, however, and some of these bison were eventually reintroduced to Bialowieza and other areas. There are now about 250 *B. bonasus* in the Bialowieza Forest, and about 1,500 in the world. A herd has been established in the Caucasus, but it contains a small admixture of genes from *B. bison* (Krasinski 1978; Grzimek 1975). The IUCN (1972) classifies *B. bonasus* as out of danger.

There were once an estimated 50,000,000 *B. bison* in North America. Certain Indian tribes depended largely on the hunting of these animals. The decline of the great bison herds began almost as soon as European explorers arrived on the continent. The bison were hunted commercially and for subsistence in order to obtain both meat and skins. They were also shot to protect agricultural interests and to help subdue the Indians of the Great Plains. *B. bison* disappeared from east of the Mississippi early in the 20th century. By 1890, fewer than 1,000 individuals survived on the continent, mostly in Canada. Subsequent private and governmental conservation efforts have allowed bison numbers to increase to about 50,000. Many of these animals are captive or are descended from captive stock, and almost all represent the plains subspecies, *B. bison bison*. The only place in the United States where a wild bison herd has been maintained continually is Yellowstone National Park. The herd there originally represented the mountain or woodland subspecies, *B. bison athabascae*, but plains bison were introduced to the area (Meagher 1973). The wood bison survived in larger numbers in northern Alberta and adjacent parts of the Northwest Territories, and the Wood Buffalo National Park was established to protect them, but many plains bison were also released there. A small, apparently pure herd of wood bison was discovered in 1957 in Wood Buffalo Park. *B. bison athabascae* is listed as endangered by the USDI (1980) and is on appendix 1 of the CITES.

ARTIODACTYLA; BOVIDAE; *Genus CEPHALOPHUS*
Hamilton Smith, 1827

Duikers

There are 16 species (Meester and Setzer 1971; Groves and Grubb 1974):

C. monticola, southeastern Nigeria to Kenya, and south to South Africa;

C. maxwelli, Senegal to southwestern Nigeria;

C. nigrifrons, southern Cameroon to western Kenya and northern Angola;

C. rufilatus, Senegal to southwestern Sudan;

C. jentinki, Liberia, probably Ivory Coast and Sierra Leone;

C. sylvicultor, Gambia to Kenya, and south to northern Angola and Zambia;

C. spadix, northeastern and southern highlands of Tanzania;

C. zebra, western Sierra Leone to central Ivory Coast;

C. niger, Guinea to southwestern Nigeria;

C. adersi, east coast of Kenya, Zanzibar;

C. natalensis, Zaire and southern Somalia to South Africa;

C. weynsi, Zaire, Uganda, Rwanda, western Kenya;

C. callipygus, southern Cameroon, Gabon, Congo;

C. dorsalis, Guinea-Bissau or possibly Gambia to Zaire and northern Angola;

C. leucogaster, southern Cameroon to Zaire;

C. ogilbyi, Sierra Leone to western Cameroon, Fernando Poo.

The species *C. monticola* and *C. maxwelli* have sometimes been placed in a separate genus or subgenus, *Philantomba* Blyth, 1840 (a synonym is *Guevei* Gray, 1852).

Size varies considerably, ranging from that of a hare to that of a deer (Grzimek 1975). Head and body length is 550 to 1,450 mm, tail length is 70 to 175 mm, shoulder height is 300 to 800 mm, and weight is about 5 to 65 kg. The upper parts vary in color from almost buffy through shades of brown to almost black. Most forms have a stripe along the middle of the back, which differs in color from the remainder of the back. One species, *C. zebra,* has a bright orange coat marked with dark vertical stripes. The underparts range from white to almost as dark as the back. All species have relatively short legs, pointed hooves, and an arched back. The horns, possessed by both sexes, are short, project backward from the skull, and are frequently hidden by a long tuft of hair on the crown.

Some species prefer open country with scattered trees and brush, whereas others are partial to the dense growth of the jungle. Although these antelope are quite numerous, they are seldom observed, owing to their shyness, largely nocturnal habits, and inaccessible habitats. When alarmed, they dart away with great speed into the protection of dense vegetation; hence the origin of the common name, duiker, which means diving buck. They are sturdy, active, and fleet footed. The horns serve several purposes, such as defense against enemies and combat with others of their own kind. They feed on grass and leaves, and at times scramble upon logs or climb vine-entangled shrubs to obtain choice leaves and fruit. They also have been observed to eat insects and carrion. According to Farst et al. (1980), they kill and eat small animals, especially birds, and in captivity are given dog food in addition to vegetable matter.

Duikers are not gregarious and usually move about alone or in pairs. According to Ralls (1973), *C. maxwelli* is probably territorial and scent marks an area with secretions from maxillary glands. Individuals are highly intolerant of others of the same sex, and fights are frequent. In the Ivory Coast, the young of this species are produced mainly during the two dry seasons, January to March and August to September. A female can give birth each year, the gestation period is 120 days, and there is a single offspring. It weighs 710 to 954 grams at birth, resembles the adults in color pattern, nibbles leaves after 14 days, and is completely weaned at 2 months of age. Observations of captives indicate that estrus occurs about once a month throughout the year, that gestation varies from 126 days in *C. niger* to as long as 245 days in *C. rufilatus,* and that females of the latter species reach sexual maturity at around 9 months of age (Farst et al. 1980; Dittrich

A. Black-fronted duiker (*Cephalophus niger*), photo by Ernest P. Walker. B. Striped-backed duikers (*C. zebra*), photo by Bernhard Grzimek. C. Light-backed duiker (*C. sylvicultor*), photo by Bernhard Grzimek.

1972). A captive *C. nigrifrons* lived for 19 years and 8 months (Marvin L. Jones, Zoological Society of San Diego, pers. comm.).

Duikers are hunted for food by native people, but are generally not sought by sportsmen, because their head does not make a showy trophy. *C. jentinki* is jeopardized by habitat destruction and probably numbers only a few hundred individuals; it is classified as endangered by the IUCN (1976) and the USDI (1980). *C. monticola* is on appendix 2 of the CITES.

ARTIODACTYLA; BOVIDAE; Genus *SYLVICAPRA* Ogilby, 1837

Gray or Common Duiker

The single species, *S. grimmia,* occurs in the savannah zones and mountainous areas from Senegal to Ethiopia, and

Common duikers (*Sylvicapra grimmia*): Top, photo from New York Zoological Society; Bottom, photo by John Visser.

south to South Africa (Meester and Setzer 1971). Van Gelder (1977*b*) considered *Sylvicapra* a synonym of *Cephalophus*.

Head and body length is 850 to 1,150 mm, tail length is 80 to 180 mm, shoulder height is 450 to 700 mm, and weight is 12 to 25 kg. Females are usually larger than males. Coloration is grayish, yellowish, or reddish yellow. *Sylvicapra* is distinguished from *Cephalophus* in having ears that are pointed, rather than rounded, and horns that are sharply pointed, rise above the plane of the face, and are usually present only in males. Females occasionally carry small, stunted horns, and such females seem to be more common in some localities than in others.

The gray duiker avoids dense forests, but usually is found in places with adequate vegetative cover. This hardy and adaptable antelope lives at higher elevations than any other hoofed mammal in Africa. It reportedly is common in alpine meadows in equatorial regions. It has great speed and

stamina, and is usually able to outdistance dogs. It rests during the day in scrub or grass, often in favorite hiding places, and comes out at night to feed. It is predominantly a browser, and the diet includes a great variety of leaves, stems, flowers, fruit, bulbs, tubers, and cultivated crops. It occasionally takes insects and birds (Smithers 1971; Wilson 1966*a*).

Sylvicapra usually travels alone, but is occasionally seen in pairs. Studies by Dunbar and Dunbar (1979) in Ethiopia indicate that there is little overlap between the home ranges of individuals of the same sex, but more overlap between those of opposite sexes. Males are apparently territorial, scent marking their areas with preorbital glands and fighting other males that intrude. Pregnant and lactating females have been taken throughout the year (Smithers 1971). A captive pair produced offspring at intervals of 232 to 298 days, and the actual gestation period seems to be about 210 days (Von Ketelhodt 1977). There is usually a single young per birth, occasionally two. A captive lived for 14 years and 4 months (Marvin L. Jones, Zoological Society of San Diego, pers. comm.).

ARTIODACTYLA; BOVIDAE; *Genus KOBUS* A. Smith, *1840*

Waterbuck, Lechwes, Kob, and Puku

There are five species (Meester and Setzer 1971):

K. ellipsiprymnus (waterbuck), savannah zones of Africa south of the Sahara;

K. megaceros (Nile lechwe), swamps of southern Sudan and western Ethiopia;

K. leche (lechwe), wetlands of southern Zaire, eastern Angola, the Caprivi Strip of Namibia, Zambia, and northern Botswana;

K. kob (kob), savannah zone from Senegal to western Kenya;

K. vardoni (puku), savannah zone of southern Zaire, eastern Angola, Zambia, southwestern Tanzania, Malawi, and northern Botswana.

The name *Adenota* Gray, 1872 has sometimes been used for a genus or subgenus comprising *K. kob* and *K. vardoni*.

Lechwe (*Kobus leche*), photo from New York Zoological Society.

The name *Onotragus* Gray, 1872 has sometimes been used for a genus or subgenus comprising *K. megaceros* and *K. leche*. The northern and central populations of *K. ellipsiprymnus* have often been considered a distinct species, *K. defassa*.

Head and body length is 1,250 to 2,200 mm, tail length is 180 to 450 mm, shoulder height is 700 to 1,340 mm, and weight is 50 to 300 kg. Males average larger than females. The hair is long and coarse, being longest on the throat and tip of the tail. There is considerable color variation among the species; some are yellowish brown and others range to almost black. Some species have white markings; *K. ellipsiprymnus*, for example, has a large white patch or an elliptical white ring on the buttocks.

The horns are long, usually measuring 510 to 1,020 mm along the front from base to tip. They are normally carried only by males, curve forward at the tips, lie at an angle slightly above the level of the face, and have transverse corrugations. The hooves vary in size and shape, but usually the false hooves (the second and fifth digits) are conspicuous. In *K. megaceros* and *K. leche* the backs of the pasterns are bare, an adaptation to wetland habitat.

These animals are seldom found far from water, and often remain in swampy tracts, where they frequent reedbeds and shrubby growth. *K. megaceros* and *K. leche* are entirely restricted to flood plains and adjacent ground. *K. kob* and *K. vardoni* inhabit moist savannah, flood plains, and the margins of adjacent light woodlands. Despite the common name "waterbuck," *K. ellipsiprymnus* is less confined to wet areas than are the other species. Although it is always within

Nile lechwe (*Kobus megaceros*), photo from San Diego Zoological Society.

reach of water, it ranges farther into woodlands than do *K. kob* and *K. vardoni*.

All species are graceful in their movements and run swiftly. *K. megaceros* and *K. leche* are particularly adept at wading and swimming; they travel by a series of leaps in water that is too shallow for swimming. Like most antelopes, *Kobus* is active mainly in the morning and evening. There are no extensive migrations, but there may be local seasonal movements. Detailed studies of *K. leche* on the Kafue Flats of Zambia show that the species is highly specialized for life on flood plains (Sayer and Van Lavieren 1975; Schuster 1980). It often grazes in water up to shoulder height. Populations move back and forth in relation to the rising and falling of the annual floods, feeding on the grasses that are exposed at various times. Because of the abundance of forage produced by the floods, population density of the Kafue lechwe may be as great as 200 individuals per sq km. In contrast, densities of *K. kob* in two areas of Uganda were 8.6 and 45 per sq km (Modha and Eltringham 1976), and reported densities of *K. ellipsiprymnus* are about 0.15 to 10.9 per sq km (Herbert 1972; Spinage 1974).

These antelopes are gregarious, sometimes being found in aggregations of many thousands. Systematic observations, however, have revealed an intricate social structure that differs between species. In *K. ellipsiprymnus*, males leave their mother when about 8 or 9 months old, and join a bachelor group with a mean membership of 5.3 animals. These groups have a dominance hierarchy based on size, strength, and horn length, and fighting is common. When a male reaches an age of about 6 years, he separates from his group and attempts to establish an individual territory. Such an area covers about 1 or 2 sq km, and is associated with a water supply and other

resources needed for the year-round, permanent support of the owner. Bachelor groups are tolerated and may wander over several territories, but other fully mature males are excluded through displays and combat. By the time a male is about 10 years old, he is ousted from his territory by a younger animal, and may then become solitary or rejoin a bachelor group. Females may wander in groups of about 5 individuals, and have home ranges that cover several male territories (Spinage 1974; Herbert 1972).

Perhaps because of the limited availability and seasonal nature of their habitat, the males of *K. kob* and *K. leche* do not maintain large, permanent territories. Instead, the adult males in given areas of high population density periodically gather on a cluster of tiny territories called a lek. In *K. kob* the leks are in an elevated area and seem to be used for many years. In *K. leche*, leks may form in different spots, depending on the seasonal movements of the herd. Within a lek, from 20 to 200 adult males defend areas ranging from about 15 to 200 meters in diameter. The smaller areas are in the middle of the lek, and there is such intensive competition for these central territories that they are seldom held by one individual for more than a day or two. Each group of lekking males is associated with a herd of up to about 1,000 females, most of which seek to mate with those males on the central territories. There is also an associated herd of young males, some of which may split off and attempt to gain a territory (Buechner 1974; Floody and Arnold 1975; E. Joubert 1972; Schuster 1980).

Breeding may occur at any time of year, though there are peaks in some areas. Leks seem to form continuously in some populations of *K. kob*, but form mainly from November to January in the population of *K. leche* on the Kafue Flats. In

Kob (*Kobus kob*), photo by Leonard Lee Rue III.

the latter area, most births occur from July to September after a gestation period of about 8 months. Gestation is around 9 months in *K. ellipsiprymnus* and *K. kob*. There is usually a single offspring. Weaning occurs at 6 to 8 months of age in *K. ellipsiprymnus*. Females begin to mate when about 1 year old, but males normally must wait several more years (Buechner 1974; Dittrich 1972; Sayer and Van Lavieren 1975; Schuster 1980; Spinage 1974). A captive *K. ellipsiprymnus* lived for 18 years and 8 months (Marvin L. Jones, Zoological Society of San Diego, pers. comm.).

The lechwe *(K. leche)* is classified as vulnerable by the IUCN (1974) and as threatened by the USDI (1980), and is on appendix 2 of the CITES. Most populations have declined drastically through uncontrolled hunting and habitat loss. The greatest remaining concentration, that on the Kafue Flats of Zambia, is jeopardized by the recent construction of two huge dams for hydroelectric power. The dams will interfere with the natural fluctuation of water levels, and thereby prevent maximum food production and perhaps disrupt the lek breeding system (Schuster 1980).

ARTIODACTYLA; BOVIDAE; **Genus REDUNCA**
Hamilton Smith, 1827

Reedbucks

There are three species (Meester and Setzer 1971):

R. arundinum, savannah zone from Gabon and Tanzania to South Africa;

R. redunca, savannah zone from Senegal to Ethiopia and Tanzania;

R. fulvorufula, known from northern Cameroon, southeastern Sudan, southwestern Ethiopia, northeastern Uganda, western Kenya, northeastern Tanzania, southern Botswana, southern Mozambique, and South Africa.

Head and body length is 1,100 to 1,600 mm, tail length is 150 to 445 mm, shoulder height is 600 to 1,050 mm, and weight is 20 to 95 kg (Grzimek 1975). The upper parts vary in color from bright fawn to grayish, the underparts are light, and the underside of the tail is white. The body hair is short and stiff, but the tail is bushy. The young have a woolly coat. The neck is thin, and the legs are long and slender. Horns, carried only by the males, are large at the base, prominently ringed, and usually 203 to 254 mm in length. They point backward, diverge in a graceful upward curve, and curve sharply forward at the tips. There is a conspicuous bare glandular patch below each ear. Females have four mammae.

Reedbucks frequent rolling grasslands, mountain plateaus, open forests, and reedbeds. They usually live near water, but do not enter it freely. They are perhaps the least wary of riverside antelopes, and, once set to flight, they seldom go far before stopping to glance back. If suddenly alarmed, they crouch close to the ground. When fleeing, they hold the tail upright, exposing the white undersurface. They graze early in the morning and evening, and frequently throughout the night, eating grass and tender shoots of reeds. Jungius (1971*a*, 1971*b*) reported that *R. arundinum* is largely nocturnal, but is also active by day during the dry season, and that grass makes up the bulk of its diet throughout the year.

In a study of *R. redunca* along the Seronera River in the

Reedbuck (*Redunca fulvorufula*), photo from New York Zoological Society.

Serengeti, Hendrichs (1975*a*) found a population density of 14 to 21 individuals per sq km. Females lived alone with their fawns of one or two years in overlapping home ranges of 15 to 40 ha. Adult males defended territories that comprised the ranges of 1 to 5 females. *R. fulvorufula* also lives alone or in small groups, with males defending a territory of 15 to 48 ha. throughout the year (Irby 1979; Smithers 1971). In Kruger National Park, Jungius (1971*b*) found *R. arundinum* sometimes to form temporary aggregations of up to 20 animals around limited winter water sources. Otherwise, the animals lived in groups consisting of a mated pair and their young. The male defended a territory of 35 to 60 ha., with a core area containing favorite resting and grazing sites, and a water supply. Other males were excluded, but fighting was not serious enough to cause injury. Vocalizations included clicks, pops, and whistles, the last being used for alarm, territorial expression, and contact between sexes.

According to Jungius (1970), *R. arundinum* breeds throughout the year, but most births occur from December to May in Kruger National Park. The female separates from the male for 3 or 4 months to give birth and rear the newborn. The calf remains hidden for about 2 months, during which time the mother visits it for 10 to 30 minutes a day. The maternal bond is broken shortly before the birth of the next calf. Irby (1979) stated that *R. fulvorufula* is also capable of reproduction throughout the year, but that births peak from March to June in East Africa and in the wet summer in South Africa. Females have an interbirth interval of 9 to 14 months, an estrous cycle of 2 to 4 weeks, and a gestation period of about 8 months. Females reach sexual maturity between 9 and 24 months of age. A captive *R. redunca* lived for 18 years (Marvin L. Jones, Zoological Society of San Diego, pers. comm.).

Most hunters agree that reedbucks are among the easiest of antelopes to approach and kill. *R. arundinum* was formerly common in South Africa, but is now rare in that country, except in Natal and Transvaal (Jungius 1971*b*).

ARTIODACTYLA; BOVIDAE; **Genus PELEA** Gray, 1851

Rhebok

The single species, *P. capreolus*, is known to occur only in South Africa (Meester and Setzer 1971).

Reedbuck (*Redunca* sp.), photo by E. L. Button.

Head and body length is 1,150 to 1,250 mm, tail length is 150 to 300 mm, shoulder height is 710 to 787 mm, and weight is 20 to 30 kg. The body is covered with hair that is woollier and curlier than that of other antelopes. The upper parts are brownish gray, the face and legs are yellowish, and the underparts of the body and tail are white. *Pelea* may be readily distinguished by its long, pointed, erect ears, and by the absence of a bare patch below the ear. The horns, borne only by males, are straight, upright, and 200 to 250 mm in length. A naked area around the nostrils extends to the top of the nose and is swollen; it becomes studded with moisture when the animal is excited. Females have four mammae.

The rhebok lives among rocks and tangled growth on mountain sides and plateaus, but, where protected, will venture to grassy valleys. It probably frequented such valleys regularly before being driven out by human activity. It has a jerky gait, resembling that of a rocking horse, and is agile, fleet, and a good jumper. The tail is carried erect, so the white undersurface is conspicuous. The rhebok seeks water and grass during the night and early morning.

This antelope consorts in family parties and small groups that occasionally combine to form herds containing as many as 30 individuals. A herd is usually led by a mature buck. Very old males are frequently solitary. Generally one of the group acts as a sentinel while the others feed or rest. If danger appears, the sentinel gives a warning grunt or cough, and leads the herd to more rugged country. Males are extraordinarily aggressive, often killing others of their sex during the rutting season, and even attacking and killing sheep and goats. One or two young are usually born in November or December. They are hidden by the mother for the first few days. Longevity in the wild has been reported as 8 to 10 years (Grzimek 1975). *Pelea* does not thrive in captivity and is seldom found in zoological parks.

ARTIODACTYLA; BOVIDAE; Genus **HIPPOTRAGUS**
Sundevall, 1846

Roan and Sable Antelopes

There are three species (Meester and Setzer 1971):

H. leucophaeus (blue buck), formerly found in southwestern South Africa;

H. equinus (roan antelope), savannah zones from Senegal to western Ethiopia, and south to South Africa;

H. niger (sable antelope), savannah zone from southeastern Kenya to Angola and eastern South Africa.

Head and body length is 1,880 to 2,670 mm, tail length is 370 to 760 mm, shoulder height is 1,000 to 1,600 mm, and

Rhebok (*Pelea capreolus*), photo by Herbert Lang through J. Meester.

weight is 150 to 300 kg. Males average approximately one-fifth larger and heavier than females (Grzimek 1975). The general coloration is bluish gray in *H. leucophaeus,* pale reddish brown in *H. equinus,* and rich chestnut to black in *H. niger* (Meester and Setzer 1971). The underparts are white. *Hippotragus* is characterized by thick, tough skin; a well-developed and often upright mane on the nape; a short mane on the throat; a moderately long tail with a tufted tip; large, long, and pointed ears; and long white hairs below the eyes. The horns, borne by both sexes, are stout and heavily ringed, and rise at an obtuse angle from the plane of the face. When fully developed, the horns are usually 510 to 1,020 mm long, and they rarely attain a length of 1,520 mm.

These antelopes occur in a variety of habitats, but generally *H. equinus* is found in more open grassland, while *H. niger* requires some wooded country. Neither species moves more than 2 to 4 km from drinking water (Child and Wilson 1964; Wilson and Hirst 1977; Estes and Estes 1974). Where not persecuted, these antelopes are not extremely wary, often running a short distance and then stopping to look back. When closely pursued, however, they can run as fast as 57 km per hr with great endurance. When wounded or cornered, they become savage, charging and using their horns with amazing speed and dexterity. The movements of *H. equinus* are fairly localized, but *H. niger* may change its range several times in the course of a year, sometimes covering 40 km or more. Both species feed primarily on grass, but the sable may also browse extensively in the dry season (Child and Wilson 1964; Estes 1974; Estes and Estes 1974).

Population density of *H. niger* has varied from 0.4 to 9.2 individuals per sq km, but Wilson and Hirst (1977) calculated that maximum sustainable density is not over 4 per sq

A. Roan antelope (*Hippotragus equinus*), photo from New York Zoological Society. B. Sable antelope and young (*H. niger*), photo from San Diego Zoological Garden.

km, and that the minimum area needed to support a healthy population of 40 to 50 animals is 1,200 to 1,500 ha. There seems to be great variation in home range. In Rhodesia, Grobler (1974) found herds of female and young *H. niger* to use ranges of 240 to 280 ha., while mature males had individual ranges of 25 to 40 ha. According to Estes and Estes (1974), however, home range of *H. niger* varies from about 1,000 to 32,000 ha. In a study of *H. equinus* in Kruger National Park, Joubert (1974) found herds generally to use 6,400 to l0,400 ha. per year. The area regularly used at any given time was about 200 to 400 ha., and there was seldom overlap between the ranges of two herds.

Both species are gregarious, but herds of *H. equinus* commonly do not exceed 12 to 15 animals, while those of *H. niger* may be several times as large and occasionally contain over 100 antelope. The herds consist mostly of females and young, and are dominated by a single adult male (Wilson and Hirst 1977). Joubert (1974) found herds of *H. equinus* to contain 6 to 12 individuals. The dominant male drove other males out when they reached 2½ years of age. The latter then formed bachelor groups of 2 to 5 animals, which established a dominance hierarchy through much fighting. At about 5 or 6 years of age, the males became solitary and sought to take over a herd of females. Victorious males defended an area extending outward about 300 to 500 meters from their herd. The mature females of a herd established a dominance hierarchy among themselves through extensive fighting, and the highest-ranking individual initiated most herd movements, even in the presence of the adult male.

Studies of *H. niger* in Angola and Rhodesia (Estes and Estes 1974; Grobler 1974) indicate that there is seasonal variation in the structure of herds of females and young. During the dry season the entire herd concentrates. When the rains begin, there is dispersal into small groups, and toward the end of the rainy season, when the calves are about to be born, there is still more fragmentation. The main herds form again after the calves leave concealment. Despite such fluctuations, Estes and Estes found herds to be stable from year to year and to have little interchange of members. The average number of individuals per herd was 25.6, and herds were generally spaced 5 to 10 km apart. In addition to the main herds of females and young, populations of *H. niger* may have bachelor groups of 2 to 12 males. Young males are driven from the main herds when about three or four years old. At the age of five or six years they become solitary and attempt to establish a territory. The size of such territories was reported by Grobler as 25 to 40 ha., but by Estes and Estes as 1,000 to 1,400 ha. Territories are marked by defecation and vigorously defended against other fully mature males. A territory may be overlapped by the ranges of several female herds. The strongest males obtain the most centrally located and desirable territories. A male takes control of any female herd that enters his territory.

In *H. equinus,* there seems to be no specific breeding season. Females of this species enter estrus 2 or 3 weeks after giving birth and have the potential of producing a calf every 10 to 10½ months. *H. niger* is a seasonal breeder; the time and duration of calving varies across the range and seems to coincide with the height of the growing season. Births occur mainly from January to March in the Transvaal, May to July in Angola, and June to September in Zambia. The gestation period has been variously reported as 268 to 280 days in *H. equinus* and 240 to 281 days in *H. niger*. The single calf weighs 13 to 18 kg at birth and remains hidden at least 10 days. Females attain sexual maturity at about 2 years of age (Wilson and Hirst 1977; Estes and Estes 1974; Dittrich 1972; Grobler 1974). A captive *H. niger* lived for 19 years and 6 months (Marvin L. Jones, Zoological Society of San Diego, pers. comm.).

The blue buck (*H. leucophaeus*) was one of the first mammals to become extinct in modern times. It apparently had been declining from natural causes since the Pleistocene, and was restricted to the southern coastal portion of South Africa, when that area was being settled by Europeans in the 18th century (Klein 1974). It was quickly eliminated by hunting, the last individual being killed about 1800. The other two species still occur in large areas, but have declined drastically in recent decades through habitat deterioration, agricultural encroachment, illegal hunting, and deliberate slaughter for tsetse fly control (Wilson and Hirst 1977). *H. equinus* is on appendix 2 of the CITES. The subspecies *H. niger variani,* the giant sable antelope, is on appendix 1, and is classified as endangered by the IUCN (1976) and the USDI (1980). This subspecies is restricted to an isolated part of central Angola, and an estimated 2,000 to 3,000 individuals were thought to survive in 1970. Although it is totally protected by law and occurs largely in government reserves, conservation efforts in the area have been disrupted by a recent civil war.

ARTIODACTYLA; BOVIDAE; *Genus ORYX De Blainville, 1816*

Oryx, Gemsbok

There are three species (Meester and Setzer 1971; Corbet 1978):

O. dammah (scimitar oryx), originally found in the semi-desert zones from Morocco and Senegal to Egypt and the Sudan;

O. leucoryx (Arabian oryx), originally found in Syria, Iraq, Israel, Jordan, Sinai, and the Arabian Peninsula;

O. gazella (gemsbok), arid country from Ethiopia and Somalia to Namibia and eastern South Africa.

Head and body length is 1,600 to 2,350 mm, tail length is 450 to 900 mm, shoulder height is 900 to 1,400 mm, and weight is 100 to 210 kg (Grzimek 1975). The general adult coloration varies from cream to grays and browns, and there are striking markings of black and brown. The young are brownish and have markings only on the tail and knees. A mane extends from the head to the shoulders, the tail is tufted, and males have a tuft of hair on the throat. The ears are fairly short, broad, and rounded at the tips. Both sexes have horns ranging from 600 to 1,500 mm in length. Those of females are usually longer and more slender than those of males. The horns of *O. leucoryx* and *O. gazella* are fairly

straight and directed backward from the eyes, but those of *O. dammah* curve back in a large arc.

These animals usually live on arid plains and deserts, but in some areas they also inhabit rocky hillsides and thick brush. According to Jungius (1978), the habitat of *O. leucoryx* consists of flat and undulating gravel plains intersected by shallow wadis and depressions, and the dunes edging sand deserts, with a diverse vegetation of trees, shrubs, herbs, and grasses. It apparently digs shallow depressions in soft ground under trees and shrubs for resting. It is able to detect rainfall over great distances, and moves accordingly in the direction of fresh plant growth. Since rainfall is irregular, it must travel over hundreds of square kilometers in no fixed pattern. One individual is known to have moved over 90 km in 18 hours. In the Serengeti, Walther (1978) observed that *O. gazella* became active just after dawn, grazed until about 1000 hrs, rested until 1400–1500 hrs, grazed again, and began moving toward the sleeping area around sunset. Oryx have been described as alert, wary, and keen sighted. They defend themselves by lowering the head so that the sharp horns point forward. They feed on grasses and shrubs, and go to streams and waterholes to drink. When free water is not available, they can obtain sufficient moisture for lengthy periods from such sources as melons and succulent bulbs.

Oryx are gregarious. *O. dammah* is generally found in groups of 20 to 40 individuals, but at certain times of the year, in areas of fresh pasture or surface water after rainfall, or during the wet season migrations, herds may contain 1,000 animals or more (Newby 1978). The normal group size of *O. leucoryx* is 10 animals or fewer, but herds of up to 100 have been reported. Captive *O. leucoryx* have been maintained in herds consisting of a single dominant adult male and several adult females and young. Groups of bachelor males are kept separately, and these animals establish a dominance hierarchy through nonsevere fighting and chases (Jungius 1978). In a study of *O. gazella* in the Serengeti, Walther (1978) found groups of up to 22 individuals, as well as lone males. Some herds contained only males and others were harems controlled by a single dominant male. There was

Gemsbok (*Oryx gazella*), photo by Karl H. Maslowski.

Scimitar oryx (*Oryx dammah*), photo from Paris Zoo.

considerable competition between males, but it involved mainly ritualized, horn-to-horn sparring, with no serious injury.

In *O. gazella* the young are born from September to January, the gestation period is 260 to 300 days, and a single offspring is normal. In *O. leucoryx*, reproductive timing varies, but under favorable conditions a female can produce a calf once a year and during any month. The gestation period in this species is apparently 9 months, the young are weaned by 4½ months of age, and captive females initially give birth when 2½ to 3½ years old (Jungius 1978). The potential longevity of *Oryx* appears to be about 20 years.

The Arabian oryx *(O. leucoryx)* is classified as endangered by the IUCN (1976) and the USDI (1980), and is on appendix 1 of the CITES. By the early 20th century this species survived only in the Arabian Peninsula. It continued to decline as the expansion of the oil industry led to hunting from motor vehicles with modern firearms. Its meat is greatly esteemed, its hide is valued as leather, other parts have alleged medicinal uses, and the head makes a choice trophy. The last known individuals in the wild were killed in 1972. In 1962, however, some Arabian oryx had been brought into captivity and a breeding program begun (Dolan 1976). According to the "Census of Rare Animals in Captivity" in the 1980 *International Zoo Yearbook,* there are 150 *O. leucoryx* in captivity, the majority at the Phoenix and San Diego zoos. Plans are now under way to reintroduce the species in Jordan and Oman, the governments of which have been staunch advocates of wildlife conservation (Strassburger 1978; Jungius 1978).

The scimitar oryx *(O. dammah)* is classified as vulnerable by the IUCN (1976) and is on appendix 2 of the CITES. It has disappeared from much of its former range through uncontrolled hunting and excessive grazing of the limited vegetation by domestic livestock. The gemsbok *(O. gazella)* is still common in some parts of Africa. The natives have sometimes used the sharp tips of its horns for spear points, and its thick, tough skin for shield coverings.

ARTIODACTYLA; BOVIDAE; **Genus ADDAX** Rafinesque, *1815*

Addax

The single species, *A. nasomaculatus,* originally occurred in desert and semidesert areas from Western Sahara and Mauritania to Egypt and Sudan. There are old, inconclusive reports from Palestine and the Arabian Peninsula (Meester and Setzer 1971).

Head and body length is 1,500 to 1,700 mm, tail length is 250 to 350 mm, shoulder height is 950 to 1,150 mm, and weight is 60 to 125 kg (Grzimek 1975). The body and neck are grayish brown in winter, but in summer the body coloration becomes sandy to almost white. The legs, hips, belly, ears, and facial markings are white, and the tuft on the forehead is black. Both sexes have horns that have some resemblance to those of *Oryx,* but have a spiral twist of 1½ to

Addaxes (*Addax nasomaculatus*), photo from New York Zoological Society.

almost 3 turns. The horns measure 762 to 890 mm along the curves. The widely splayed hooves are adapted for traveling on desert sands.

The addax is the most desert adapted of antelopes (Corbet 1978). It lives most of its life without drinking, deriving sufficient moisture from the plants upon which it feeds. It travels great distances to search for the scant vegetation of the Sahara, and, in spite of the seeming adversity of its habitat, is said always to appear in good condition. It seems to move about in herds of 5 to 20 individuals, led by an old male. A single offspring is born at a time, generally in the winter or early spring. Dittrich (1972) reported gestation periods of 257 to 264 days, and that sexual maturity was reached at about 30 months of age by a female and about 24 months by a male. A captive addax lived for 25 years and 4 months (Marvin L. Jones, Zoological Society of San Diego, pers. comm.).

This antelope is heavily built and is not capable of great speed, and thus is easy prey to people with camels, horses, dogs, and modern weapons. Both the meat and the skin are prized by the natives, the latter being used for shoe and sandal soles. Ruthless hunting has eliminated resident populations in Egypt, Tunisia, Western Sahara, and many other parts of the original range. The addax is classified as vulnerable by the IUCN (1976) and is on appendix 2 of the CITES. During World War II, American soldiers in North Africa observed the addax and described it in letters to their families, who then contacted zoos to learn where such a white antelope lived, in order that they might know where the soldiers were.

ARTIODACTYLA; BOVIDAE; **Genus DAMALISCUS**
Sclater and Thomas, 1894

Sassabies

There are two subgenera and three species (Meester and Setzer 1971):

subgenus *Beatragus*

D. *hunteri* (Hunter's hartebeest), eastern Kenya, extreme southwestern Somalia;

subgenus *Damaliscus*

D. *dorcas* (bontebok, blesbok), South Africa.
D. *lunatus* (topi), open country from Senegal to Ethiopia, and south to South Africa.

Beatragus Heller, 1912 has sometimes been given full generic rank. In contrast, Grzimek (1975) treated D. *hunteri* as a subspecies of D. *lunatus*. Van Gelder (1977b) considered *Damaliscus* a synonym of *Alcelaphus*.

Head and body length is 1,300 to 2,050 mm, tail length is 177 to 600 mm, shoulder height is 880 to 1,200 mm, and weight is 68 to 136 kg. The hair is generally soft and has an iridescent sheen, but some animals have a mixture of soft and coarse hairs. The general coloration ranges from grays and reds to rich brown and almost black. The underparts are lighter, and there may be various white markings on the face, legs, and hips. D. *dorcas* has a large white area on the face and D. *hunteri* has a white line passing from one eye to the

Bontebok (*Damaliscus dorcas*), photo by Karl H. Maslowski.

other across the forehead. *Damaliscus* is characterized by its slightly convex forehead and its somewhat elongated muzzle. The horns, which are well developed in both sexes, are angular and curved, are ringed for most of their length, lack pedicels, and are up to about 700 mm long.

These antelopes inhabit grassland and sparsely timbered regions. If suddenly alarmed, they may jump over each other's backs in their haste to escape. The normal gait is a lumbering canter, but maximum speed is at least 70 km per hr (Schaller 1972). Most activity occurs in the morning and early evening. There are seasonal migrations in some areas. The diet consists mainly of grass.

At one time, these antelopes gathered in the thousands, at least in certain seasons. Recent studies show that populations have an intricate but highly variable structure (David 1973, 1975; DuPlessis 1972; Jewell 1972; S.C.J. Joubert 1972; Joubert and Bronkhorst 1977; Monfort-Braham 1975; Novellie 1979). In general, fully mature males are solitary and compete with one another, and with younger males, for possession of females. Most competition is through postural displays and ritualistic sparring with the horns, rather than serious fighting. In Kruger National Park, a male *D. lunatus* establishes a territory of 200 to 400 ha., which he marks by urine, feces, secretions from preorbital glands, and horning of the soil. Within his territory, he permanently maintains a harem of about seven or eight adult females and associated offspring. Young males are ejected from the area before they are one year old, and form bachelor groups of up to six individuals. When the young males are three or four years old, they begin to challenge the older bulls.

Males of *D. lunatus* in southern Rwanda also have large territories, up to 100 ha., and a permanent harem of up to 20 females. Males of *D. dorcas* in southern South Africa have territories that are small, 10 to 40 ha., but possibly occupied for life. Groups of adult females there, with an average of 3 individuals, are not permanently associated with any one male, but wander from one territory to another. There are also large herds containing the young animals of both sexes. Some males of *D. dorcas* in northern South Africa, and of *D. lunatus* in Uganda and northern Rwanda, have small territories of about 1 to 3 ha., which are occupied only temporarily during the mating season. There are large associated herds of females, which spend only a brief time in any one territory. Other adult males of *D. dorcas* in northern South Africa have no territories, but move with and defend a herd of females.

There are generally well-defined seasons of birth: July–September for *D. lunatus* in Uganda, September–December for *D. lunatus* in Kruger National Park, September–October for *D. dorcas* in southern South Africa, and November–December for *D. dorcas* in northern South Africa. The gestation period is 7–8 months long, and normally a single calf is born. Females attain sexual maturity in their second or third year of life. Males do not usually mate until they are old enough to compete successfully with their elders (David 1973, 1975; DuPlessis 1972; Jewell 1972; S.C.J. Joubert 1972). A captive *D. dorcas* lived for 21 years and 5 months (Marvin L. Jones, Zoological Society of San Diego, pers. comm.).

These antelopes have been greatly reduced in numbers and distribution through excessive hunting and the encroachment

Hunter's hartebeests (*Damaliscus hunteri*): Top, photo from Gladys Porter Zoo; Left, photo from San Diego Zoological Society.

of agriculture. The subspecies *D. dorcas dorcas* was nearly exterminated by 1830, but several herds have been maintained on government and private preserves, and about 1,000 individuals now exist. The subspecies is considered out of danger by the IUCN (1972), but is still listed as endangered by the USDI (1980), and is on appendix 1 of the CITES. *D. hunteri* is classified as rare by the IUCN (1976), but recent estimates put the total number at around 14,000 (Bunderson 1977).

ARTIODACTYLA; BOVIDAE; **Genus ALCELAPHUS**
De Blainville, 1816

Hartebeests

There are two species (Meester and Setzer 1971):

A. lichtensteini, savannah zone of Tanzania, southeastern Zaire, northeastern Angola, Zambia, eastern Rhodesia, and Mozambique;

A. buselaphus, originally found in open country from Morocco to Egypt, from Senegal to Somalia and northeastern Tanzania, and from southern Angola and western Rhodesia to South Africa.

Some authorities, including Vrba (1979), place *A. lichtensteini* in a separate genus, *Sigmoceros* Heller, 1912. In contrast, other authorities, including Grzimek (1975), consider *A. lichtensteini* only a subspecies of *A. buselaphus*.

Hartebeests (*Alcelaphus buselaphus*), photo by C. A. Spinage.

Head and body length is 1,500 to 2,450 mm, tail length is 300 to 700 mm, shoulder height is 1,100 to 1,500 mm, and weight is 100 to 225 kg. The body hair is about 25 mm in length and is particularly fine textured. Coloration ranges from fawn through brownish gray to chestnut. *A. buselaphus* has prominent white patches on the hips, and black on the forehead, muzzle, shoulders, and thighs. *A. lichtensteini* has a chestnut area along the back. The genus is characterized by the rump being lower than the shoulders, a long head, slender legs, a tufted tail, and large, conspicuous glands below the eyes. The horns, borne by both sexes, unite or rise from a single pedicel, have heavy encircling rings, and are usually 255 to 380 mm long. Females have two mammae.

Hartebeests inhabit dry savannahs and grasslands. Owing to the difference in height between the forequarters and hindquarters, the gait may appear clumsy, but Schaller (1972) reported maximum speed to be 70 to 80 km per hr. Hartebeests are diurnal, feeding in the early morning and late afternoon and resting in the shade during the heat of the day. The diet consists mainly of grass.

Schaller (1972) estimated that there were 18,000 *A. buselaphus* in the 25,500-sq-km Serengeti ecosystem, and Laws, Parker, and Johnstone (1975) reported population density in Uganda to be 1.4 individuals per sq km. Hartebeests are gregarious, living in organized herds of up to 300 animals and sometimes uniting in aggregations of up to 10,000 (Smithers 1971). In Nairobi National Park, Gosling (1974) found *A. buselaphus* to have four social classes: territorial adult males, nonterritorial adult males, groups of up to 100 young males, and groups of females and young. Young males left the females when about 20 months old and joined the male groups. At 3 or 4 years of age they began to attempt to take over a territory, and hence control of the females thereon. Males generally lost their territory at 7 or 8 years of age. Animals holding a territory represented 38 percent of the adult male population, and defended an average area of 31 ha. In territorial struggles, the males leapt forward in a kneeling position, with their horns lowered, but serious fights were rare.

In southern Ethiopia, according to Lewis and Wilson (1979), *A. buselaphus* occurs in groups of up to 180 animals, but breaks into smaller units prior to the calving season. In Zambia, groups of *A. lichtensteini* contain up to 5 individuals, which may be led by a single large male (Wilson 1966b). A herd of hartebeests may have a sentinel posted on a summit to warn the others of danger. When alarmed, the animals gallop off in a single file. These antelope are usually nonagressive and silent.

Reproduction may be markedly seasonal. The young of *A. buselaphus* are born from October to November in South Africa and from mid-December to mid-February in southern Ethiopia, and those of *A. lichtensteini* are born from July to August in Zambia (Lewis and Wilson 1979; Skinner, Van Zyl, and Van Heerden 1973; Wilson 1966b). In Nairobi National Park, births of *A. buselaphus* occur throughout the year, but there is a major peak in February and March and a lesser peak in July (Gosling 1974). The gestation period is 214 to 242 days, and normally a single calf is produced. Potential longevity is 11 to 20 years.

Hartebeests were once abundant, but have been greatly reduced in numbers through excessive hunting and habitat destruction by people. The subspecies *A. buselaphus buselaphus* of North Africa apparently became extinct by 1923 (Goodwin and Goodwin 1973). The subspecies *A. b. tora* and *A. b. swaynei*, which formerly occurred from southern Egypt to Somalia, are classified as endangered by the IUCN (1972, 1978) and the USDI (1980). *A. buselaphus* has also disappeared from most of South Africa, but both that species and *A. lichtensteini* are still locally common in certain areas from Kenya to Botswana.

ARTIODACTYLA; BOVIDAE; *Genus* **CONNOCHAETES**
Lichtenstein, 1814

Wildebeests, Gnus

There are two subgenera and two species (Meester and Setzer 1971):

subgenus *Connochaetes*

C. gnou (black wildebeest, white-tailed wildebeest), central and eastern South Africa;

subgenus *Gorgon*

C. taurinus (blue wildebeest, brindled gnu, white-bearded wildebeest), southern Kenya and southern Angola to northern South Africa.

Head and body length is 1,500 to 2,400 mm, tail length is 350 to 550 mm (not including the tuft), shoulder height is 1,000 to 1,450 mm, and weight is 140 to 275 kg. In *C. gnou* the general coloration is buffy brown to black; tufts of long, black hair protrude from the muzzle, throat, and chest; the mane is upright; and the tail is black at the base, but otherwise white. In *C. taurinus* the general coloration is grayish silver; there are brownish bands on the neck, shoulders, and forepart of the body; and the face, mane, beard, and tail are black, except in the subspecies *C. t. albojubatus* and *C. t. mearnsi* of Kenya and Tanzania, in which the beard is white. Both sexes of *Connochaetes* possess horns, which arise separately and are usually heavy and recurved. The oxlike head and horns, the bristly facial hair, and the general body form give *Connochaetes* a ferocious appearance, which is misleading. Females have four mammae.

Wildebeests prefer open grassy plains with a nearby source of water. When disturbed, they begin curious antics, prancing about, pawing the ground, probing the earth with their horns, and thrashing their tails. If a person approaches to within 500 meters, they snort and dash off a short distance, wheel about to face the intruder, and then resume the ritual. They are active mainly in the early morning and late afternoon, resting in the hot hours of the day. Some herds are migratory, constantly moving about in search of food, but others remain in one area throughout the year. Schaller (1972) estimated that there were 400,000 migratory and 10,000 nonmigratory *C. taurinus* in the 25,500-sq-km Serengeti ecosystem. Wildebeests are primarily grazers, though they also feed on succulent plants and karroo bushes.

In some areas, during the dry season, wildebeests form vast aggregations of all age and sex classes. Usually, however, females and young are found in discrete herds of 10 to over 1,000 individuals. Such herds may be nomadic or have a regular home range of perhaps 1 sq km. Males separate from these herds as yearlings, and join a group of bachelors. When a male is about three or four years old, it becomes solitary and attempts to win a territory. Some males establish territories only temporarily, when the associated herd of females is not moving about. Other males may continuously occupy the same area for years. The distance between territorial males is usually about 100 to 400 meters, but varies from as little as 9 meters in favorable areas to about 1,600 meters in poor habitat. Territorial competition between males involves ritualized displays, loud calls ("ge-nu"), and pushing with the horns, but rarely severe fighting. Only males with territories can mate, and they attempt to control all females that enter their areas (Estes 1969; Schaller 1972; Von Richter 1972, 1974).

Black wildebeest (*Connochaetes gnou*), photo from East Berlin Zoo.

Blue wildebeests (*Connochaetes taurinus*), photo by Lothar Schlawe.

In a given area, reproduction is highly seasonal and births occur mainly within a period of 2 or 3 weeks at the beginning of the rains. This period falls from November to January in South Africa, and in January or February in the Serengeti (Estes 1976; Skinner, Van Zyl, and Van Heerden 1973; Von Richter 1974). Sinclair (1977b) suggested that the lunar cycle is the external timer for synchronization of mating in the Serengeti. The gestation period is 8 to 9 months, and a single offspring is normal. It can stand within 15 minutes of birth, begins to follow its mother shortly thereafter, and is weaned by 9 months of age. Females attain sexual maturity in their second or third year of life. A captive *C. taurinus* lived for 21 years and 5 months (Marvin L. Jones, Zoological Society of San Diego, pers. comm.).

The flesh of *Connochaetes* is coarse, dry, and hard; the skin makes good leather; and the silky tail is used for fly whisks called "chowries." Wildebeest populations have been reduced through hunting and are also subject to decline periodically from drought, but there are still vast herds of *C. taurinus* in East Africa. *C. gnou* was nearly exterminated by hunting during the settlement of South Africa in the 19th century, but was maintained on a few government and private preserves. Numbers increased to about 3,100 individuals by 1970, and the IUCN (1976) considers the species out of danger.

ARTIODACTYLA; BOVIDAE; **Genus OREOTRAGUS**
A. Smith, 1834

Klipspringer

The single species, *O. oreotragus*, occurs in rocky hills and mountains in northern Nigeria, probably in the eastern Central African Republic, in northeastern and extreme south-eastern Sudan, and from Ethiopia and Angola to South Africa (Meester and Setzer 1971).

Head and body length is 770 to 1,150 mm, tail length is 50 to 130 mm, shoulder height is 450 to 600 mm, and weight is 11 to 16 kg. The horns, ringed at the base and usually borne only by males, are 75 to 153 mm in length. The pelage has a yellowish olive gloss and is speckled with yellow and brown or orange, shading to white below. The coat harmonizes with the background of rocks. The texture of the pelage has been variously described as mosslike and grasslike. The thick, pithy hair lies like a mat on the body to cushion the animal against bumps and bruises in its rocky environment. Females have four mammae (Grzimek 1975).

This little "cliff-springing" antelope might be likened to the chamois *(Rupicapra rupicapra)* or the mountain goat *(Oreamnos americanus)* in its mode of life among the rocks, where it makes rapid progress without apparent suitable footholds. It walks and stands on the tips of the relatively rounded hooves, and can jump onto a rocky projection the size of a silver dollar, landing on it with all four feet. It is active mainly in the morning and late afternoon, sheltering at other times among rocks and under overhanging cliffs. It is mainly a browser, but also eats some grass, and apparently does not drink water regularly.

In three parts of Ethiopia, Dunbar and Dunbar (1974b) found population densities of 13.4, 19.1, and 46.7 individuals per ha. Groups had exclusive home ranges averaging 8.1 ha., which they defended against conspecifics and rarely left. In both this region and Namibia (Tilson 1980), groups were found to consist basically of a mated, monogamous pair and their offspring of one or two years. The adult female initiates most group movements, but the male seems mainly responsible for defense. Generally, one member of the group stands at a point above the others and constantly watches for danger. When alarmed, it gives a loud shrill whistle and the animals ascend higher into the rocks.

Klipspringer (*Oreotragus oreotragus*), photo from San Diego Zoological Garden.

Reproduction seems to be nonseasonal in south-central Africa, but in Ethiopia mating apparently occurs in August and September (Dunbar and Dunbar 1974b; Smithers 1971; Wilson and Child 1965). The gestation period is about 214 days. Young males separate from the group before they are 1 year old, but females remain longer and may eventually mate with their father (Dunbar and Dunbar 1974b). A captive klipspringer lived for 12 years and 1 month (Marvin L. Jones, Zoological Society of San Diego, pers. comm.).

The skins of *Oreotragus* are used by the hill tribes to make bags for carrying bread. The meat is said to be excellent. Because of its elasticity and lightness, klipspringer hair was once in demand for stuffing saddlebags.

ARTIODACTYLA; BOVIDAE; **Genus OUREBIA**
Laurillard, 1842

Oribi

The single species, *O. ourebi,* is found in the savannah zones of Africa south of the Sahara (Meester and Setzer 1971).

Head and body length is 920 to 1,100 mm, tail length is 60 to 150 mm, shoulder height is 500 to 700 mm, and weight is 14 to 21 kg. The hair is fine and silky, and the coloration of the upper parts is sandy rufous to tawny. The underparts and chin are white. The tail is black above and white below. Females usually have a dark crown patch. Both sexes have tufts of long hair on the knees. Beneath the ears is a bare glandular area, which is usually dark and conspicuous. The ears are large, the legs are slender, and the tail is short and bushy. The horns, borne only by males, are ringed at the base, set at an angle of about 45° to the plane of the face, and 75 to 125 mm long. Females have four mammae.

The oribi inhabits areas that are generally open but that have a sufficient growth of vegetation for protection. It customarily shelters in tall grass and does not attempt to escape until an intruder is within a few meters. When flushed, it bounds along, leaping above the grass. When fleeing, it conspicuously displays the tail, which serves as a warning signal to conspecifics. It is capable of traveling at considerable speed. It is active both day and night, lying in thickets during the heat of the day and emerging in the morning and evening to graze and browse.

Oribi (*Ourebia ourebi*), photo by Lothar Schlawe.

Ourebia is commonly found in groups of up to five individuals. Such groups do not contain more than one adult male, but often have more than one adult female. Groups are territorial and may be separated from each other by one to several kilometers. Both sexes defend the territory and mark it with urine, feces, and secretions from interdigital and ant-

orbital glands (Monfort and Monfort 1974; Leuthold 1977*b*). A loud, shrill whistle may be uttered when an animal is alarmed. The single offspring is generally born from September to December following a gestation period of about 210 days. One specimen was still living after 12 years in captivity (Marvin L. Jones, Zoological Society of San Diego, pers. comm.).

The flesh of *Ourebia* is said to be excellent, and it is extensively hunted. It is highly vulnerable to such hunting and to habitat alteration, and has been greatly reduced in numbers and distribution (Meester and Setzer 1971; Skinner, Fairall, and Bothma 1977). The oribi is relatively easy to raise and is said to make a delightful pet.

ARTIODACTYLA; BOVIDAE; **Genus RAPHICERUS**
Hamilton Smith, 1827

Steenbok, Grysboks

There are three species (Meester and Setzer 1971):

R. campestris (steenbok), Kenya and southern Angola to South Africa;
R. melanotis (grysbok), southern South Africa;
R. sharpei (Sharpe's grysbok), Tanzania to northeastern South Africa.

Head and body length is 630 to 900 mm, tail length is 40 to 80 mm, shoulder height is about 500 mm, and weight is 7 to 14 kg. The hair is rather coarse. Grysboks are tawny rufous to chocolate red, speckled with white, and have dark markings on the crown. The steenbok is reddish fawn to brown or gray.

Steenbok (*Raphicerus campestris*), photo by John Hanks.

Steenbok (*Raphicerus campestris*): Left, photo by John Visser; Right, photo by Lothar Schlawe.

The underside of the body and tail of all three species is white. In contrast to *Ourebia*, there are no bare patches below the ears and no tufts on the knees, and the straight horns are not conspicuously ridged, but are rather smooth. The horns are borne only by males and are 25 to 127 mm long. *R. campestris* and *R. sharpei* do not have false hooves, but small ones are present in *R. melanotis*. The ears are long and narrow, the tail is short, and the legs are slender. Females have four mammae.

These antelopes inhabit grassland with scattered scrub, or stony country. Instead of running when danger threatens, they hide in the grass, lying flat against the ground with the neck outstretched; they do not run away until nearly trodden upon. Their gait is unlike that of *Ourebia*, being a swift, less bounding run. It is said that when hard pressed they will seek refuge in burrows of the aardvark *(Orycteropus)*. Activity is mainly in the early morning and late afternoon, but the animals may become nocturnal in areas where molested by people. They are both browsers and grazers (Smithers 1971). Although they will drink water, if available, they do not seem to require free water.

Raphicerus is usually solitary, but occasionally two or three individuals will live together. Breeding seems to occur throughout all or most of the year, the gestation period is apparently 168 to 177 days, and a single offspring is normal (Smithers 1971; Kerr and Wilson 1967; Wilson and Kerr 1969).

The flesh of these antelopes is palatable but somewhat dry. It has been claimed that they sometimes damage crops. They have disappeared or become rare in some parts of their original range.

ARTIODACTYLA; BOVIDAE; **Genus NEOTRAGUS**
Hamilton Smith, 1827

Dwarf Antelopes

There are two subgenera and three species (Meester and Setzer 1971):

subgenus *Neotragus*

N. pygmaeus (royal antelope), lowland forest zone from Sierra Leone to Ghana;

subgenus *Nesotragus*

N. batesi (Bates's dwarf antelope), lowland forest zone from southeastern Nigeria to western Uganda;
N. moschatus (suni), dry country from Kenya to eastern South Africa, Zanzibar, Mafia Island.

Suni (*Neotragus moschatus*), top, photo from Los Angeles Zoo. Royal antelope (*N. pygmaeus*), bottom, photo from *Proc. Zool. Soc. London.*

Nesotragus Hamilton Smith, 1827 has sometimes been treated as a separate genus.

With the exception of some individuals of the genus *Tragulus*, *N. pygmaeus* is the smallest hoofed mammal. It has a head and body length of about 500 mm, a tail length of about 75 mm, and a shoulder height of 250 to 305 mm. In *N. batesi*, shoulder height reaches about 355 mm. In *N. moschatus*, head and body length is 580 to 620 mm, tail length is 115 to

130 mm, shoulder height is about 300 mm, and weight is 8 to 9 kg. The general coloration of *N. pygmaeus* and *N. batesi* is cinnamon to russet above and white below. The tuft and underside of the tail are white in *N. pygmaeus* and usually dusky in *N. batesi*. *N. moschatus* has brownish gray to chestnut upper parts, rufous sides, and white underparts.

The genus *Neotragus* is characterized by a slender build, relatively high hindquarters, no tufts on the head and knees, and no false hooves. The horns, borne only by males, slope backward at about the same angle as that of the plane of the face. In *N. moschatus* the horns are 65 to 90 mm long and are heavily ringed for at least three-fourths of their length. In *N. pygmaeus* the horns are 12 to 25 mm long, smooth, and black in color. In *N. batesi* the horns are 38 to 50 mm long, smooth, and usually brown and/or fawn in color. Female *N. pygmaeus* have four mammae.

The species *N. pygmaeus* and *N. batesi* dwell in forests and clearings therein, whereas *N. moschatus* is found mainly in dry country with tangled underbrush. All species are shy and secretive. Their coloration harmonizes with their surroundings and they are generally not seen until almost underfoot, when they bound away with considerable speed, dodging and twisting about the underbrush and then quickly disappearing. *N. pygmaeus* has been reported to make leaps of 2.8 meters. *N. moschatus* sleeps in the shade during the heat of the day; as evening approaches it moves toward the glades, where it feeds on grass and shrubs. It is almost independent of free water, apparently deriving sufficient moisture from vegetation. *N. batesi* often eats the tops of peanut plants and is caught in snares set around peanut patches.

The young of *N. moschatus* are usually born in the wild from mid-November to mid-December and are somewhat darker than the adults. According to Izard and Umfleet (1971), births of this species have occurred at the Dallas Zoo throughout the year, the gestation period is thought to be 180 days, and sexual maturity is attained at about 6 months of age. One captive *N. moschatus* lived for 10 years and 2 months (Marvin L. Jones, Zoological Society of San Diego, pers. comm.).

In the folklore of Liberia, *N. pygmaeus* has a position comparable to that of our "Br'er Rabbit," being renowned for its speed and sagacity. Although the meat of *N. moschatus* is of mediocre quality and the species is protected in some preserves, it has declined drastically through illegal hunting and is considered endangered in South Africa (Skinner, Fairall, and Bothma 1977). The subspecies *N. moschatus moschatus* of Zanzibar is classified as endangered by the IUCN (1974) and the USDI (1980).

ARTIODACTYLA; BOVIDAE; **Genus MADOQUA** Ogilby, *1837*

Dik-diks

There are two subgenera and four species (Meester and Setzer 1971; Yalden 1978):

Dik-dik (*Madoqua saltiana*), photo by J. Meester. Inset: *M. kirki*, photo by C. A. Spinage.

subgenus *Madoqua*

M. saltiana, northeastern Sudan, northern and eastern
  Ethiopia, Somalia;
M. piacentinii, coast of southeastern Somalia;

subgenus *Rhynchotragus*

M. guentheri, extreme southeastern Sudan, northeastern
  Uganda, southern Ethiopia, northern Kenya, Somalia;
M. kirki, extreme southeastern Somalia, central and
  southern Kenya, northern and central Tanzania,
  southwestern Angola, Namibia.

Head and body length is 520 to 670 mm, tail length is 35 to
55 mm, shoulder height is 305 to 405 mm, and weight is 3 to
6 kg. The pelage is soft and lax. The coloration varies from
yellowish gray to reddish brown above and from grayish to
white below. Horns, possessed only by the males, are ringed,
stout at the base, somewhat longitudinally grooved, and,
occasionally, partially concealed by a tuft of hair on the
forehead. A characteristic feature of the genus is the elongated snout. In the subgenus *Rhynchotragus* the snout is
especially long and can be turned in all directions (Grzimek
1975). The accessory hooves are small and the tail is not
conspicuous. Females have four mammae.

Dik-diks usually frequent areas that are rather dry but have

Beira (*Dorcatragus megalotis*), photo from Zoological Garden of Naples, Italy. Inset: photo from *The Book of Antelopes,* P. L. Sclater and Oldfield Thomas.

considerable brush for concealment. When startled, they
may dash off in a series of erratic, zigzag leaps, uttering a call
resembling ''zik-zik'' or ''dik-dik''—hence the origin of the
common name. These animals are shy and elusive, and usually stay in dense vegetation, but they utilize definite paths.
They are active mainly in the morning and late afternoon, and
occasionally at night (Grzimek 1975). Although at least one
species regularly occurs along streambanks, others seem to
live for months without moisture except that obtained from
dew and the vegetation they consume. For food, dik-diks
browse on such shrubs as acacias.

In a study of *M. kirki* in the Serengeti, Hendrichs (1975b)
found a population density of 24 adults per sq km. The animals lived in permanently mated pairs, with their young, on
territories of 5 to 20 ha. All members of the family marked
the territory at regular points with urine and dung, but only
the adult male actively defended the area. Young individuals
were ejected at about seven or eight months of age.

Observations of *M. kirki* indicate that females are polyestrous and usually bear young twice a year. Births in the
Serengeti area peak at the beginning and toward the end of
the rainy season, November–December and April–May.
The gestation period is 169 to 174 days and there is normally
a single offspring that weighs about 600 grams at birth.
Females produce their own first young when about 15 to 18
months of age. Wild individuals are known to have lived at
least 10 years (Hendrichs 1975b; Kellas 1955; Dittrich 1967,
1972).

Some hunters dislike these little antelopes, because they
flush and warn the larger game. Dik-diks are extensively
hunted in some areas, and their skins used in the manufacture
of gloves (Grzimek 1975).

ARTIODACTYLA; BOVIDAE; **Genus DORCATRAGUS**
Noack, 1894

Beira

The single species, *D. megalotis,* occurs in Afars and
Issas, eastern Ethiopia, and northern Somalia (Meester and
Setzer 1971).

Head and body length is 762 to 865 mm, tail length is 51 to
76 mm, shoulder height is 500 to 760 mm, and weight is 9 to
11 kg. The hair is thick and coarse. The upper parts are
reddish gray, the underparts are white, the head is yellowish
red, and the legs are fawn colored. There are white areas
around the eyes, and a distinct dark line occurs on each side
from the lower shoulder to the flank. The horns, borne only
by males, are upright, straight, spiked, spaced widely on top
of the head, and 76 to 102 mm long. The ears, borne by both
sexes, are large, about 152 mm long and 76 mm wide, and
curiously marked inside with white hairs. *Dorcatragus* is
also characterized by a relatively short body, long legs, a
short and bushy tail, and short, padded hooves.

The beira inhabits stony hills and hot, dry plateaus. It
appears to be goatlike in habit, leaping from rock to rock with
great agility. Hunters report that it is extremely difficult to
obtain, because of its protective coloration, extreme wariness, and great speed when alarmed. *Dorcatragus* feeds on
coarse grass and the leaves of mimosas, and from these it
seems to derive enough moisture to live without additional
water (Dorst and Dandelot 1969). It associates in herds of
four to seven individuals, which include one or two males.
The beira apparently has always been uncommon and its
range is thought to have declined in recent decades. It is
classified as vulnerable by the IUCN (1972).

Blackbucks (*Antilope cervicapra*), photos from New York Zoological Society and San Diego Zoological Garden.

ARTIODACTYLA; BOVIDAE; *Genus ANTILOPE Pallas, 1766*

Blackbuck

The single species, *A. cervicapra,* is found in Pakistan and India, from the Punjab and Sind to Bengal and Cape Comorin (Ellerman and Morrison-Scott 1966).

Head and body length is about 1,200 mm, tail length is about 178 mm, shoulder height is about 812 mm, and average weight is 37 kg. The blackbuck is one of the few antelopes whose coloration differs between male and female. The buck is rich dark brown above, on the sides, and on the outsides of the legs, whereas the doe is yellowish fawn on the head and back. In both sexes the underparts, insides of the legs, and an area encircling the eyes are white. The males gradually become darker with age. The build is graceful and slender. The horns, borne only by males, are 456 to 685 mm long, ringed at the base, and twisted spirally up to five turns. The narrow muzzle is sheeplike, the tail is short, and the hooves are delicate and sharply pointed.

The blackbuck lives on generally open plains, though according to Schaller (1967) it reaches greatest abundance in areas with thorn and dry deciduous forest. It is very fast and can usually outrun even a greyhound, but the cheetah (*Acinonyx*) has been trained to capture it. When danger approaches, the alert females are usually the first to warn the herd. When alarmed, a single animal will bound into the air to a surprising height and is soon followed by others until the whole herd is in motion. These bounds are continued for a few strides, after which the herd settles down to a regular gallop. Schaller (1967) reported *Antilope* to be diurnal during the cool season, with intermittent activity all day. In the hot season it is active in the very early morning and late afternoon, and rests in the shade at other times. It is almost exclusively a grazer, feeding on short grass and various cultivated cereals.

At one time, *Antilope* reportedly formed aggregations of thousands of individuals. Most recent observations are of herds of 5 to 50 animals, though in some areas several groups may join during December and January (Schaller 1967). In a study in southeastern India, Nair (1976) reported the number of individuals per herd to average 23 and vary from 2 to 129. Most herds were essentially harems, with a single adult male and a number of adult females and young. Some herds had 1 or 2 peripheral adult males, and there were also small separate groups of bachelors. Observations by both Schaller and Nair indicate that adult males are territorial, at least during the breeding season. The size of territories apparently varies from about 25 to 100 ha. The adult male drives younger males from the herd, and competes with other adults for possession of territories and the females thereon. Dominance is achieved mainly by display of the horns and threatening gestures. The horns are potentially dangerous weapons, but serious fighting is rare.

Mating occurs throughout the year in central India, but the peak periods are March-April and August-October (Schaller 1967). The gestation period is 6 months and there seems to be almost always a single young. A captive blackbuck lived for 16 years and 9 months (Marvin L. Jones, Zoological Society of San Diego, pers. comm.).

The blackbuck is prized for its meat and as a sporting trophy. It was probably once the most abundant hoofed mammal in India, but has been greatly reduced in numbers and distribution through excessive hunting and loss of habitat to agricultural development (Schaller 1967).

ARTIODACTYLA; BOVIDAE; *Genus AEPYCEROS Sundevall, 1847*

Impala

The single species, *A. melampus,* occurs in the open country from Kenya and southern Angola to northern South Africa (Meester and Setzer 1971).

Head and body length is 1,100 to 1,500 mm, tail length is 250 to 400 mm, and shoulder height is 775 to 1,000 mm. The

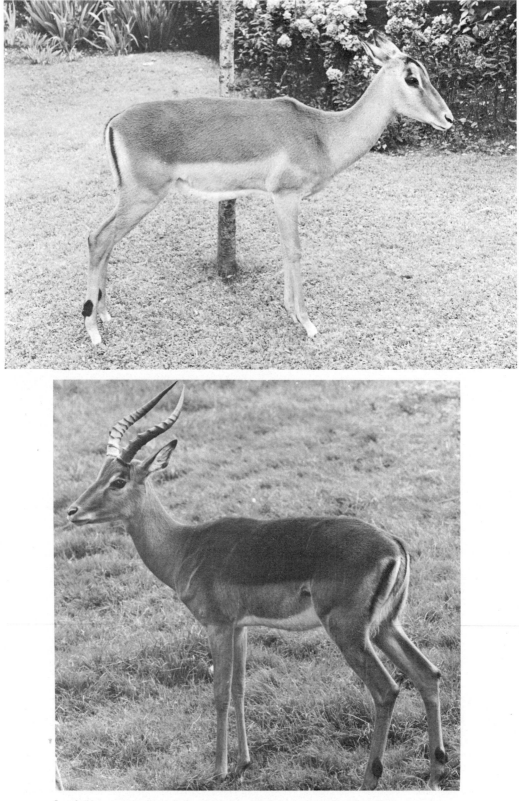

Impala (*Aepyceros melampus*), female above, male below, photos from Paris Zoo.

usual weight is 40 to 45 kg in females and 60 to 65 kg in males (Jarman and Jarman 1974). The hair is sleek and glossy. The coloration is dark fawn or reddish above, and is lighter on the thighs and legs. A distinct vertical black streak lies on each side of the hindquarters. The underside of the body and tail, the inside of the upper foreleg, the upper lip, and the chin are white. The horns, borne only by males, are lyrate, 500 to 750 mm long, and ridged only on the front surface. The hooves lack clefts, and there are glandular tufts of black hair on the hind feet.

The impala inhabits open woodlands, sandy bush country, and acacia savannahs. When alarmed, it makes prodigious leaps, seemingly without effort, the animals often jumping over bushes and over each other, and even springing high into the air in places with no obstacle to clear. The speed of the impala is considerable; when running, it has been known to make successive leaps of eight, five, and nine meters. Although the gazelles *(Gazella)* and hartebeests *(Alcelaphus)*, with which *Aepyceros* often grazes, run away on the open plains when disturbed, the impala seeks shelter in dense vegetation. It is active both day and night, alternately feeding and resting. It grazes on grass and browses on the leaves of bushes and trees, and drinks at least once a day. Unlike certain other antelopes, the impala apparently requires a source of free water.

Schaller (1972) estimated that there were 65,000 impala in the 25,500-sq-km Serengeti ecosystem. During the dry season, all age and sex classes may form large aggregations, sometimes numbering hundreds of individuals, that move about in search of green vegetation. At other times of the year there is a more intricate social structure, as well as territorial activity (Fairall 1972; Jarman and Jarman 1973, 1974; Leuthold 1970). In parts of East Africa, where there are two wet seasons, territoriality may extend over most of the year, while in South Africa it may last only about three months. At such times the sexes segregate. Females and young form herds of usually 10 to 100 individuals, and males form groups of up to 60 bachelors. Such herds have regular home ranges of about 2 to 6 sq km. The male groups establish a dominance hierarchy through displays and fighting, and the highest-ranking animals attempt to take over a territory. At any given time, about one-third of the adult males in a particular population hold a territory. Males that lose their territory return to the bachelor herds. Territories are about 0.2 to 0.9 sq km in size and are marked by urination and defecation. The resident male attempts to control any herds of females and young that enter his area. From such groups, he ejects the males that have reached an age of about six to nine months, and the latter subsequently join bachelor herds. During the mating season, males frequently utter loud, hoarse grunts.

In East Africa the two mating seasons extend from February to April and from August to October (Grzimek 1975). In South Africa there is a major peak in mating from April to June and a lesser peak in September and October (Anderson 1975). The gestation period is about 6 to 7 months, normally a single offspring is produced, weaning occurs at 5 to 7 months of age, males are physiologically capable of reproduction at 13 months of age, and individuals up to 13 years old have been found in the wild (Jarman and Jarman 1973; Kerr 1965). One impala was still living after 17 years and 5 months in captivity (Marvin L. Jones, Zoological Society of San Diego, pers. comm.).

Aepyceros is vulnerable to overhunting and has disappeared in much of southern Africa, but has been introduced in some areas beyond the original range (Meester and Setzer 1971). The subspecies *A. m. petersi* of Angola and Namibia is classified as endangered by the IUCN (1972) and the USDI (1980).

ARTIODACTYLA; BOVIDAE; Genus **AMMODORCAS**
Thomas, 1891

Dibatag

The single species, *A. clarkei,* is found in Somalia and extreme eastern Ethiopia (Meester and Setzer 1971).

Head and body length is about 1,170 mm, tail length is about 355 mm, shoulder height is 762 to 890 mm, and weight is 27 to 34 kg. The dark upper parts are purplish rufous, the underparts are white, and the buttocks are light and lack a dark band on the flank. The streak down the middle of the face is rich chestnut, and the tail is black. The peculiar purplish tint of the rufous coat blends so well with the surroundings that the dibatag is difficult to see.

Ammodorcas is slender and resembles *Gazella* in body form and *Redunca* in horn structure. The neck and tail are long, the skull is flat, and the hooves are small. The horns, borne only by males, are 150 to 250 mm long; the basal half is ringed and curved backward, and the terminal half is smooth and curved forward. The ears of *Ammodorcas* are rounded, whereas those of *Litocranius* are pointed.

The dibatag frequents sandy areas with a scattered growth of thorn scrub and grass. It is not readily seen in its natural habitat, as it conceals its body behind vegetation and peers over the top. The neck is so slender, the head so pointed, and the coloration so much like that of the natural cover that the animal is practically invisible. It remains motionless until discovered. *Ammodorcas* bounds away with the head arched back and the tail thrown forward, whereas *Litocranius* runs with the head and tail outstretched in line with the body. The long neck of the dibatag allows it to reach a considerable height when browsing on shrubs, and is reminiscent of the giraffe. *Ammodorcas* often stands on its hind legs, with its forefeet in a tree, to reach as high as possible. The long upper lip also facilitates browsing. The dibatag probably can exist without free water.

Ammodorcas generally travels alone or in family parties of three to five individuals. Females are thought to give birth in October and November. A gestation period of 204 days has been recorded (Dittrich 1972).

The dibatag is classified as vulnerable by the IUCN (1974) and as endangered by the USDI (1980). Its numbers appear to have dropped because of heavy poaching, the effects of drought, and competition from domestic livestock.

ARTIODACTYLA; BOVIDAE; Genus **LITOCRANIUS**
Kohl, 1886

Gerenuk

The single species, *L. walleri,* is found in the arid parts of central and southern Ethiopia, Afars and Issas, Somalia, Kenya, and northeastern Tanzania. In early historical time this species also occurred in northeastern Sudan and eastern Egypt (Meester and Setzer 1971).

Head and body length is 1,400 to 1,600 mm, tail length is 250 to 350 mm, shoulder height is 900 to 1,050 mm, and weight is 35 to 52 kg (Grzimek 1975). The general coloration is reddish fawn. A broad, dark brown band runs down the back and along the upper third of the sides. The underparts and the front of the neck are white. The giraffelike neck is only 180 to 255 mm in circumference. The legs are long and slender, the muzzle elongate, the skull wide and flat, and the head wedge shaped. The horns, developed only in males, are

Dibatags (*Ammodorcas clarkei*), photo of mounted group in Field Museum of Natural History.

about 355 mm long and are comparatively massive. They curve backward and upward, and finally hook forward near the tips.

The gerenuk inhabits dry country with a light covering of brush and thorn scrub. Upon seeing a strange object, it usually stands motionless, hides behind a bush, and then looks over or around the cover by means of its long neck. When frightened, it usually leaves in a stealthy, crouched trot, with the neck and tail carried horizontally. It is not speedy compared to other antelopes. It is active throughout the day.

Litocranius is rather sedentary, but may make seasonal shifts within its home range. It is highly adapted for an arid

Gerenuks (*Litocranius walleri*): A. Photo by Hans-Jurg Kuhn; B. Photo by Leonard Lee Rue III.

Gerenuks (*Litocranius walleri*): Left, male; Right, female and young; photos by W. Leuthold.

environment and seems independent of free water, though individuals living near water may occasionally drink. It is exclusively a browser, taking the leaves and shoots of a wide variety of trees and shrubs (Leuthold 1978a). Its eating habits are somewhat like those of *Giraffa* in that it plucks acacia leaves with its long upper lip and long tongue. To reach the food, it often stands upright on its hind legs, with the body practically vertical. It then places its forelegs against a tree to reach for branches, from which it takes the highest and most succulent leaves.

In the Tsavo National Park of Kenya, Leuthold (1978a, 1978b) found population density to average 0.6 individuals per sq km, and home range size to vary from 1.5 to 3.5 sq km. Ranges overlapped only slightly and seemed to constitute fairly stable territories. Males began to compete for territories when they matured at around 3 years of age. Apparently, all adult males obtained an individual territory. Younger males wandered from place to place, alone or in small groups. Territories were marked with urine, feces, and secretions from antorbital glands. Adult females and their offspring occurred in groups of two to five individuals. Territorial males associated with the females and kept younger males away. Reproduction was nonseasonal. According to Leuthold, the gestation period is 6½ to 7 months, a female may mate again within 1 month of giving birth, a single offspring is usual, young females attain sexual maturity at about 1 year of age, and maximum known longevity is about 8 years in the wild and well over 13 years in captivity.

ARTIODACTYLA; BOVIDAE; **Genus GAZELLA**
De Blainville, 1816

Gazelles

There are 3 subgenera and 12 species (Meester and Setzer 1971; Corbet 1978; Groves 1969b):

subgenus *Trachelocele*

G. subgutturosa, desert and subdesert steppes from Palestine and the Arabian Peninsula to the Gobi Desert and northern China;

subgenus *Gazella*

G. dorcas, desert and subdesert zones from Morocco and Senegal to central India and Somalia;

G. gazella, Palestine, Sinai, Arabian Peninsula;

G. spekei, eastern Ethiopia, Somalia;

G. rufina, known only by specimens purchased in markets in northern Algeria during the late 19th century;

G. leptoceros, desert zone from central Algeria to Egypt;

G. cuvieri, mountains of Morocco, Algeria, and Tunisia;

G. rufifrons, subdesert and savannah zones from Senegal to Ethiopia;

G. thomsoni, arid parts of southeastern Sudan, Kenya, and Tanzania;

subgenus *Nanger*

Gazelles: Top, *Gazella thomsoni;* Bottom, *G. subgutturosa;* photos from Paris Zoo.

G. dama, southern Morocco and Senegal to Sudan;

G. soemmerringi, eastern Sudan, Ethiopia, Somalia;

G. granti, southeastern Sudan, southern Ethiopia, south-western Somalia, northeastern Uganda, Kenya, northern Tanzania.

Head and body length is 850 to 1,700 mm, tail length is 150 to 300 mm, shoulder height is 500 to 1,100 mm, and weight is 12 to 85 kg (Grzimek 1975). The back and sides vary in color from rich brown through fawn and gray to white. In many species a dark band extends along the sides, immediately above the light area of the belly, and frequently there is a light band above the dark one. Most of the species have white areas around the base of the tail and on the back of the thighs.

The muzzle is normal and is neither expanded, as in *Pantholops,* nor elongated, as in *Saiga.* The neck is not long, as in *Litocranius,* and the back lacks evertible folds, as in *Antidorcas.* Both sexes of most species possess horns, which are smaller and more slender in females. One exception is *G. subgutturosa,* in which horns usually occur only in the males. Well-developed horns measure from 150 to 760 mm, but generally average 255 to 355 mm, and all are strongly ringed. Many gazelles have lyre-shaped horns, but there is considerable variation among the species.

Gazelles usually occur in dry open country or brushland. The elevational range of the genus is sea level to 5,750 meters. The name "gazelle" has come to suggest grace and beauty, for all species are dainty, alert, and graceful. Maximum speed of *G. thomsoni* is 80 km per hr (Schaller 1972). In some areas there are large-scale seasonal migrations to different elevations or new feeding grounds. The populations of *G. dama* to the south of the Sahara move northward into the desert during the rainy season. Most *G. thomsoni* in the Serengeti spend the rainy season in the grasslands and the dry season in bush country (Grzimek 1975). The diet varies; *G. thomsoni,* for example, is a grazer, whereas *G. granti* is a browser (Estes 1967).

Schaller (1972) estimated that there were 180,000 *G. thomsoni* in the 25,500-sq-km Serengeti ecosystem. This species, and some others, may form temporary aggregations of hundreds or thousands of individuals, but herd size is usually much smaller. Generally, groups contain about 10 to 30 females and young. There are also small groups of males and lone adult males. The latter establish territories, and temporarily win control of the females that enter. Female herds of *G. granti* wander over an annual home range of about 300 sq km, while males of this species have territories 500 to 2,000 meters in diameter. Defense against other territorial males and young males is mainly ritualistic and involves gestures and display of the horns. The territories of male *G. thomsoni* are smaller, the animals being spaced only about 200 to 300 meters apart, but defense is more vigorous (Estes 1967; Grzimek 1975; Walther 1972).

Reproduction occurs throughout the year in some areas, though there may be peaks of breeding in the wet seasons. A female may experience a postpartum estrus and give birth more than once a year. Reported gestation periods vary from 156 to 203 days. There is normally a single offspring. Average weight of newborn *G. spekei* is 1,250 grams. Females may attain sexual maturity in their first year of life (Dittrich 1972; Estes 1967; Read and Frueh 1980). A captive *G. dorcas* lived for 17 years and 1 month (Marvin L. Jones, Zoological Society of San Diego, pers. comm.).

Gazelles have long been hunted by people for use as food. The flesh of some species is considered excellent, while that of others is strong, coarse, and dry. Some hunters use falcons to harass a gazelle so that it can be readily overtaken by dogs. In Israel are the remains of large stone corrals into which

Gazelles (*Gazella dorcas*), photo by Lothar Schlawe.

gazelles were driven in prehistoric times and which were still in use in the early 20th century (Mendelssohn 1974). Excessive hunting by people and excessive grazing by domestic livestock have adversely affected a number of species and subspecies of *Gazella* in the Middle East and North Africa. *G. rufina* has apparently long been extinct. Both the IUCN (1972, 1974) and the USDI (1980) classify the following as endangered: *G. subgutturosa marica* (Jordan, Arabian Peninsula), *G. dorcas massaesyla* (Morocco to Tunisia), *G. dorcas pelzelni* (Somalia), *G. dorcas saudiya* (Syria, Iraq, Israel, Jordan, Arabian Peninsula), *G. gazella, G. leptoceros, G. cuvieri, G. dama lozanoi* (Western Sahara), and *G. dama mhorr* (Morocco). In addition, the IUCN (1974) classifies *G. spekei* as indeterminate.

ARTIODACTYLA; BOVIDAE; **Genus ANTIDORCAS**
Sundevall, 1847

Springbuck, Springbok

The single species, *A. marsupialis,* occurs in Angola, Namibia, Botswana, and South Africa (Meester and Setzer 1971).

Head and body length is 1,200 to 1,400 mm, tail length is 190 to 275 mm, shoulder height is 730 to 870 mm, and weight is about 32 to 36 kg. The springbok is cinnamon fawn above, with a dark reddish brown horizontal band extending from the upper foreleg to the edge of the hip, separating the upper color from the white underside. The inside of the legs, the back of the thighs, the tail, and a patch extending onto the rump are also white. Both sexes have black, ringed horns. False hooves are present and there are no tufts of hair on the knees.

The general external appearance is very much like that of *Gazella. Antidorcas* was separated generically from *Gazella* because of its teeth; in the springbok there are five pairs of grinding teeth in the lower jaw, whereas there are six pairs in the gazelles. The one peculiar and striking external difference of *Antidorcas* is the fold of skin extending along the middle of the back to the base of the tail. This fold is covered

Springbuck (*Antidorcas marsupialis*): Left, photo from San Diego Zoological Garden; Right, photo from Paris Zoo.

with hair, much lighter in color than the rest of the back. When the animal becomes alarmed, it opens and raises this fold so the white hair shows as a conspicuous crest along the back. At the same time the white hairs on the rump are erected.

The springbok lives on open, dry savannahs and grassland. Its common name is based on its habit of leaping up to 3½ meters into the air when startled or at play. In springing, the body is curved, the legs are held stiff and close together, and the head is lowered. As soon as the animal hits the ground, it rebounds with no apparent effort. It is suspicious of roads and wagon paths, and clears these obstacles at a bound. It may be active at any hour, but sometimes rests during the hot part of the day (Smithers 1971). Drought occasionally forces the springbok to undertake large-scale migrations in search of new pastures. *Antidorcas* is both a grazer and a browser. It thrives on karroo shrubs and grass, and is able to get along without water, though it will drink if water is available.

The springbok is highly gregarious, and migratory herds formerly contained over 1 million individuals. The animals were so numerous that it would take days for the herd to pass a given point. Numbers have now been greatly reduced, but groups still occasionally have up to 1,500 individuals. Herds may contain both sexes or only males, and there are also lone males. The main reproductive season varies from place to place and from year to year, but most births take place from about August to January. The gestation period lasts about 171 days, a single offspring is normal, and some females become pregnant before they are 1 year old (Bigalke 1970; Smithers 1971). A captive springbok lived for 19 years (Marvin L. Jones, Zoological Society of San Diego, pers. comm.).

Antidorcas is easily tamed and thrives in captivity. It has been extensively hunted by people for its excellent meat, and because its mass migrations were ruinous to crops. As a result, it disappeared from most of South Africa, but there are about 20,000 in the Kalahari Gemsbok National Park and reintroduced populations in several other parks (De Graaf and Penzhorn 1976). The springbok is the national emblem of the Republic of South Africa. To the north, it still occurs in most of its original range, but is much less common than formerly (Meester and Setzer 1971; Smithers 1971).

ARTIODACTYLA; BOVIDAE; **Genus PROCAPRA**
Hodgson, 1846

Central Asian Gazelles

There are two subgenera and three species (Groves 1967*a*; Corbet 1978):

subgenus *Procapra*

P. picticaudata (goa), Himalayan region, Tibet, highlands of central China;
P. przewalskii, subdesert steppes of north-central China;

subgenus *Prodorcas*

P. gutturosa (zeren), dry steppe and subdesert of Mongolia and adjacent parts of southern Siberia and northern China.

Head and body length is 950 to 1,480 mm, tail length is 20 to 120 mm, shoulder height is 540 to 840 mm, and weight is 20 to 40 kg (Grzimek 1975). *P. gutturosa* is the largest species; it is orange buff above in the summer, with pinkish cinnamon sides, but is paler in the winter. The other species are generally brownish gray in summer and paler in winter. Both have a white rump patch, but that of *P. picticaudata* is continuous, whereas that of *P. przewalskii* is divided by a median line of darker color. All species have white underparts. The horns, present only in males, are about 200 to 250 mm long. The backward deflection of the horns is not as conspicuous in *P. gutturosa* as in the other species. The horns of *P. przewalskii* are curved in two places, with the points turned inward, while the horns of *P. picticaudata* are curved in only one plane.

These gazelles are found in dry grassland at elevations of up to 5,750 meters. *P. picticaudata* is a wary and swift animal that scrapes out a bedding place in the ground and feeds in areas of scant vegetation. Its mating season begins in December and lasts for about one month; the young are born the following May. A single offspring is normal.

The mating season of *P. gutturosa* is in the fall. At this time the males have swollen throats. In the spring there is a northward migration, and herds of 6,000 to 8,000 individuals

Zerens (*Procapra gutturosa*):Left, two views of male;Lower right,female and young; photos from Osaka Municipal Tennoji Zoological Garden. Upper right, *P.* sp., photo by Howard E. Uible of skull in U.S. National Museum.

Tibetan antelope or chirus (*Pantholops hodgsoni*); Left, photo from *Mammals of the Second Yarkland Mission,* Blandford; Right, photo by Feng Zuo-Jian.

then form. Upon reaching the summer pastures in June, the sexes separate and the young are born. Twins are commonly produced (Grzimek 1975). Within a few days the young can keep up with the mother. A captive *P. gutturosa* lived for seven years (Marvin L. Jones, Zoological Society of San Diego, pers. comm.). This species is hunted by people for its flesh and skin.

ARTIODACTYLA; BOVIDAE; **Genus PANTHOLOPS**
Hodgson, 1834

Chiru, Tibetan Antelope

The single species, *P. hodgsoni*, occurs in the highlands of Kashmir, Tibet, and adjacent parts of north-central China (Corbet 1978).

Head and body length is 1,300 to 1,400 mm, tail length is about 100 mm, shoulder height is 790 to 940 mm, and weight is 25 to 50 kg. The hair is short, dense, and woolly. The back and sides are pale fawn with a pinkish suffusion, the face and fronts of the legs are dark, and the underparts are white. The horns, carried only by males, are slender, black, ridged in front, almost vertical in relation to the head, and 510 to 710 mm long. The nose is swollen at the tip, the legs are slender, and the tail is relatively short.

The chiru inhabits plateau steppes at elevations of 3,700 to 5,500 meters. It is quite wary, and the gait is a trot that can be rapid enough to outdistance dogs and wolves. When a male is in motion, the horns are held high, greatly enhancing the appearance. When at rest, *Pantholops* often lies in a shallow depression that it has excavated to a depth of about 300 mm. Thus sheltered from the wind and partly concealed, it watches for enemies. In the morning and evening it grazes and perhaps browses along glacial streams.

During the mating season in November and December, adult males eat little and are in a state of great excitement. Each attempts to form a harem of 10 to 20 females, which are jealously guarded. As soon as a male spots a rival, he lowers his saberlike horns and rushes to attack. The battles are fierce. The long, sharp horns can inflict terrible wounds, and sometimes both contestants die as a result. If one doe attempts to leave a harem, the male tries to drive her back. Meanwhile, the other does may use the opportunity to desert, as there is apparently no lasting bond between the sexes. The young are born in May.

ARTIODACTYLA; BOVIDAE; **Genus SAIGA** *Gray, 1843*

Saiga

The single species, *S. tatarica*, was found in historical time in the steppe zone from western Ukraine to western Mongolia. In the Pleistocene this species evidently occurred from England to Alaska (Sokolov 1974).

Head and body length is 1,000 to 1,400 mm, tail length is 60 to 120 mm, shoulder height is 600 to 800 mm, and weight is 26 to 69 kg. The coat is heavy and woollike, with a fringe of long hairs extending from the chin to the chest. In summer the upper parts are cinnamon buff, the nose and sides of the face are dark, the crown is grizzled, and the rump patch, underparts, and tail are white. In winter the coat is longer and thicker, and is uniformly whitish. The horns, possessed only by males, are 203 to 255 mm in length, irregularly lyrate, heavily ridged, and pale amber in color.

A remarkable character of *Saiga* is the inflated and pro-boscislike nose. The nostril openings point downward, and there are unusual internal structures. The bones of the nose are greatly developed and convoluted, and the nasal openings are lined with hairs, glands, and mucous tracts. In each nostril is a sac lined with mucous membranes, a feature found in no other mammal except whales. The inflated nose and associated structures may be adaptations for warming and moistening inhaled air. The saiga's exceptionally keen sense of smell may also be related to the nasal development.

Except as noted, the information for the remainder of this account was taken from Bannikov et al. (1967) and Sokolov (1974). The saiga is found mainly on grassy plains, often in arid areas. It avoids broken country and dense vegetation, but may enter the forest steppe zone in the summer. It can run at speeds of up to 80 km per hr. The saiga is active throughout the day for most of the year. During the summer it grazes in the early morning and evening, and rests at midday. It usually has no fixed home range and commonly wanders several dozen kilometers per day. Some populations undertake extensive seasonal migrations. The animals on the west side of the Caspian Sea concentrate to the south in the winter, move northward in April and May, and return to the south in the fall. They may cover 80 to 120 km per day during such movements. The diet consists mainly of grass, and also includes various herbs and shrubs. During the summer, under favorable conditions, the saiga visits waterholes twice daily.

In 1958, an estimated 2,000,000 saiga inhabited a total range of 2,500,000 sq km, giving a mean density of 0.8 individuals per sq km. In local areas, however, density has reached 14 per sq km during winter concentration and up to 40 per sq km in concentrations around lakes during drought. On the wintering grounds to the west of the Caspian Sea, both sexes winter together. At the beginning of the spring migration the males form herds of 10 to 2,000 individuals, and push out ahead of the females. The latter form vast aggregations and move off in search of a place to give birth. As soon as their young are able to travel, the females follow the males. In 1957, a continuous stream of females and young, estimated to contain 150,000 to 200,000 animals, was observed. In summer the saiga is generally found in groups of 30 to 40 individuals. Large migratory herds again form in the autumn. During the mating season in early winter, adult males become territorial and each attempts to gather a harem, generally consisting of 5 to 15 females. The male constantly herds the females and challenges other males that approach his area. There are fierce fights, often resulting in death. At the end of the rut, the exhausted males perish in large numbers.

Mating extends from December to January, and births occur in late April and May. The gestation period is 139 to 152 days. About two-thirds of the females have twins and the rest produce a single calf. The newborn weighs an average of 3.5 kg and can outrun a human by its second day of life. It begins to graze at 4 to 8 days of age, but is not completely weaned for 4 months. Females usually attain sexual maturity before their first birthday and continue to grow until they are 20 months old. Males can mate at 19–20 months of age and grow until they are 24 months old. Known maximum longevity in the wild is 10 to 12 years.

The saiga was exterminated in the Crimea by the 13th century A.D., but survived elsewhere in the southern Ukraine until the 18th century. Subsequent uncontrolled hunting led to drastic declines in numbers and distribution, and by the early 20th century fewer than 1,000 saiga were thought to survive. Many animals were killed merely to obtain the horns, which could be sold in the Orient for their alleged medicinal value. Total protection was established in Europe in 1919 and in Soviet Central Asia in 1923. The saiga made a remarkable comeback in the Soviet Union and by

Saigas (*Saiga tatarica*), photo from Dierenpark "Wassenaar," Wassenaar, Holland.

1958 an estimated 2,000,000 individuals were present. Sport and commercial hunting is now allowed in some areas. Saiga products include muttonlike meat, hides, fat, and, still, horns for the pharmaceutical trade. The Mongolian subspecies, *S. t. mongolica,* has not shared in the general recovery of the species, and now numbers only about 200 individuals (Dash et al. 1978). It is listed as endangered by the USDI (1980).

ARTIODACTYLA; BOVIDAE; **Genus NEMORHAEDUS**
Hamilton Smith, 1827

Gorals

There are two species (Ellerman and Morrison-Scott 1966; Corbet 1978; Lekagul and McNeely 1977; Hayman 1961):

N. goral, extreme southeastern Siberia to northern Burma and northern Thailand, Himalayan region;
N. cranbrooki, Tibet, Assam, northern Burma.

Volf (1976) suggested that *N. caudatus* of northern China is a species distinct from *N. goral.*

Head and body length is 820 to 1,300 mm, tail length is 76 to 203 mm, shoulder height is 584 to 711 mm, and weight is 22 to 35 kg. The body is covered by a short, woolly undercoat, which is covered by long, coarse guard hairs. The male has a short, semierect mane. Coloration in *N. goral* is buffy gray to dark brown above and paler below. There is a black stripe on the foreleg, a white patch on the throat, and a dark stripe down the middle of the back. In *N. cranbrooki* the coloration is bright foxy red to tawny buff, with a blackish dorsal stripe from the back of the neck nearly to the base of the tail.

Nemorhaedus resembles *Capricornis* in general appearance, but is distinguished by smaller size, absence of facial glands, and shorter horns. The horns, which are carried by both sexes, are conical in form, are 127 to 178 mm in length, curve backward, and are marked by small irregular ridges. The facial profile is concave, the back is somewhat arched, and the limbs are stout and long, well adapted to climbing and jumping. Females have four mammae.

Gorals are found on rugged, wooded mountains, generally at elevations of 1,000 to 4,000 meters. They inhabit slopes even more precipitous than those used by *Capricornis,* seeming to prefer the most difficult terrain possible (Lekagul and McNeely 1977). They are most active during the early morning and late evening, but on cloudy days they roam throughout the day. After eating in the morning, they usually drink water, and then they retire to a sunny rock ledge to stretch out and rest until evening. They are quite difficult to recognize, even though they are in full view, for they lie motionless and their color blends with that of the rocks. The diet consists of twigs, low shrubs, grass, and nuts (Lekagul and McNeely 1977).

Nemorhaedus usually occurs in groups of 4 to 12 individuals, though old males commonly live alone most of the year. When frightened, they emit a hissing or sneezing sound. Mating takes place from late September to November in Siberia, and in November and December farther south. Gestation periods of 6 to 8 months have been reported. There are 1 or 2 offspring per birth. Sexual maturity is attained in the third year of life (Dobroruka 1968; Lekagul and McNeely 1977; Schaller 1977). A captive lived for 17 years and 7 months (Marvin L. Jones, Zoological Society of San Diego, pers. comm.).

Although the heads of gorals do not make showy trophies, they are frequently hunted for sport. *N. goral* is listed as

Goral (*Nemorhaedus goral*), photo by R. Pucholt.

endangered by the USDI (1980) and is on appendix 1 of the CITES.

ARTIODACTYLA; BOVIDAE; *Genus CAPRICORNIS*
Ogilby, 1837

Serows

There are two species (Ellerman and Morrison-Scott 1966; Corbet 1978; Lekagul and McNeely 1977):

C. sumatrensis, central and southern China, Himalayan region, Assam, Burma, Thailand, Indochina, Malay Peninsula, Sumatra;
C. crispus, Honshu, Shikoku, Kyushu, Taiwan.

Head and body length is 1,400 to 1,800 mm, tail length is 80 to 160 mm, shoulder height is 850 to 940 mm, and weight is 50 to 140 kg. The upper parts are generally gray or black, the mane ranges in color from white to black, and the underparts are whitish. *Capricornis* resembles *Nemorhaedus,* but has large preorbital glands and a straighter facial profile. The rhinarium is naked, and the ears are long, narrow, and pointed. The horns, carried by both sexes, are slightly curved, marked with narrow transverse ridges on the basal three-fourths, and 152 to 255 mm long. The hooves are short and the tail is moderately bushy.

Serows inhabit rugged mountains or ridges, covered with thick brush or forest, at elevations of up to 2,700 meters. Their gait is clumsy and not particularly rapid, but they are sure footed in descending steep, rocky slopes. They are commonly hunted with dogs, and, when brought to bay, defend themselves with their horns in a deadly manner. According to Lekagul and McNeely (1977), they have acute senses of smell, vision, and hearing; they may have well-defined runways along steep rock faces; and they are known to swim

between the small islands near the Malay Peninsula. They feed during the early morning and late evening, and shelter at other times in favorite resting places, often in caves or under overhanging rocks and cliffs. The diet consists of grass, shoots, and leaves.

Capricornis is usually solitary, but sometimes occurs in groups of up to seven members. According to Schaller (1977), lone males, pairs, and small family groups of *C. crispus* tend to occupy small, discrete home ranges for most of the year. Home range size is 1.3 to 4.4 ha. for solitary individuals and 9.7 to 21.7 ha. for family units. In the summer and autumn, males roam beyond their home ranges and grown young leave their mothers' ranges. These home ranges may constitute exclusive territories, as males have been seen to chase other males away, and both sexes mark the areas with secretions from their preorbital glands. Mating takes place in October and November, the gestation period lasts about 7 months, usually a single offspring is produced, and captives have lived for over 10 years (Lekagul and McNeely 1977; Yamamoto 1967).

The meat of serows is of mediocre quality, but some people believe that various parts of the animals have medicinal value. The subspecies on Sumatra, *C. sumatrensis sumatrensis,* has been severely reduced in numbers and distribution through excessive hunting and habitat loss. It is classified as endangered by the IUCN (1972) and the USDI (1980), and is on appendix 1 of the CITES.

ARTIODACTYLA; BOVIDAE; *Genus OREAMNOS*
Rafinesque, 1817

Mountain Goat

The single species, *O. americanus,* originally occurred in the mountainous region from southeastern Alaska and south-

Formosa serow (*Capricornis crispus*), photo by Constance P. Warner.

Serow (*Capricornis sumatrensis*), photo by Wang Sung.

North American mountain goat (*Oreamnos americanus*), photo by Leonard Lee Rue III.

western Northwest Territories to north-central Oregon and western Montana (Hall 1981). *Oreamnos* is a member of the goat-antelope tribe (Rupicaprini) of the subfamily Caprinae, and is not a true goat (tribe Caprini).

Head and body length is 1,200 to 1,600 mm, tail length is 100 to 200 mm, and weight is 46 to 140 kg. Males generally exceed females by 10 to 30 percent in linear dimensions. The pelage is white or yellowish white, and the underfur is thick and woolly. The hair is long and soft along the midline of the neck and shoulders, forming a ridge or hump. A beard is also present. The hooves have a hard, sharp rim, enclosing a soft inner pad, and are well suited to climbing over rocks and ice. Females have four mammae.

Except as noted, the information for the remainder of this account was taken from the review papers by Rideout (1978) and Rideout and Hoffmann (1975). The mountain goat is usually found among steep slopes and cliffs in alpine tundra or subalpine areas, associated with low temperature and heavy snowfall. It is renowned for its ability to climb and jump easily through rugged terrain, and has been known to gain 460 meters in elevation within 20 minutes. Activity generally occurs in the early morning and late afternoon, and frequently goes on through the night. Daily movements generally cover several hundred meters. For bedding, shallow depressions may be excavated with the forefeet. In the fall there is a general downward movement to south- and west-facing slopes that are often free of snow. The distance between summer and winter centers of activity was found to range from 1.7 to 11.1 km in Montana. The diet is broad, varies from place to place and from season to season, and may involve either browsing or grazing. Grass, mosses, lichens, woody plants, and herbs are all significant foods. Salt

licks are important to the mountain goat in the spring and summer, and it may travel several miles to reach them.

The average annual home range in Montana was found to be 21.5 sq km for adult males and 24 sq km for adult females. In winter, individual ranges are as small as 81 ha., and large groups of animals may form at that time. In other seasons, groups usually do not contain more than 4 animals, except around salt licks, and adult males are frequently seen alone. Social structure appears to be variable, with some reports indicating that males are dominant and others suggesting that males are subordinant to both females and juveniles. Except during the breeding season, the sexes seem indifferent to one another, and sometimes clash over limited food resources. In agonistic interaction there is no butting of the heads, as in *Ovis*, but the sharp horns may be thrust at the opponent's flanks and rump, sometimes causing serious injury.

Mating occurs from November to early January, and births take place in late May and early June. Gestation periods of 147 and 178 days have been reported. There is usually a single kid, but twins are not uncommon and triplets are rare. One newborn weighed 2,950 grams. The young is able to follow its mother within a week, and is completely weaned by September. It is driven away by the aggression of the female, when she again gives birth. Sexual maturity is attained by both sexes at around 30 months of age. Maximum known longevity in the wild is 14 years for males and 18 years for females.

Because of its generally inaccessible habitat, the mountain goat has been affected less by human activity than has any other big game animal in North America. Nonetheless, some populations have declined seriously through excessive hunting, which has resulted in part from improved access by

Chamois (*Rupicapra rupicapra*), photo by Fritz Grögl.

people on newly constructed roads. Introduced populations of *Oreamnos* have been successfully established in Colorado, central Montana, the Black Hills of South Dakota, northeastern Oregon, and Olympic National Park, and on the Alaskan islands of Kodiak, Baranof, and Chichagof. The mountain goat is subject to limited sport hunting in most parts of its range. Various population estimates suggest that the total number of individuals in North America is around 100,000 (Samuel and Macgregor 1977; Rideout 1978).

ARTIODACTYLA; BOVIDAE; *Genus* **RUPICAPRA**
De Blainville, 1816

Chamois

The single species, *R. rupicapra*, occurs in the mountains of central and southern Europe, Asia Minor, and the Caucasus (Corbet 1978). A possibly related genus and species, *Myotragus balearicus*, lived on the Balearic Islands until about 4,000 years ago, when it apparently was exterminated by neolithic human invaders (Burleigh and Clutton-Brock 1980; Kurten 1968).

Head and body length is 900 to 1,300 mm, tail length is 30 to 40 mm, shoulder height is 760 to 810 mm, and weight is 24 to 50 kg. The pelage is stiff and coarse. The hairs of the summer coat are only about 40 mm long and are tawny brown in color. The winter coat is blackish brown, and is composed of guard hairs 100 to 200 mm long, and a thick, woolly underfur. The underparts are pale and the throat patch is white. The horns, borne by both sexes, are slender, black, and 152 to 203 mm long. They are set closely together, rise

almost vertically, and then bend abruptly backward to form hooks. The pad of the hoof is slightly depressed and somewhat elastic, providing a sure foothold on uneven, slippery terrain.

This beautiful animal is nimble, agile, daring, and graceful. Its senses are acute. When alarmed, it flees to the most inaccessible places, often making prodigious leaps. It may shift to ranges at a relatively low elevation in the fall, and return to alpine areas in the spring. During the summer months the diet consists chiefly of herbs and flowers, but in winter the chamois eats lichens, mosses, and young pine shoots. It has been known to fast for two weeks and survive, when the snow was so deep that food could not be secured.

Females and young commonly form herds of 15 to 30 individuals. A herd is said to post a sentinel that warns the other animals of approaching danger by stamping its feet and uttering a sharp, high-pitched, whistling call. Old males generally live alone for most of the year, but begin to join the herds in the late summer. During the autumn rut the old males drive the younger males from the herds, and occasionally kill them. The young are born in a shelter of grass and lichens after a gestation period of 153 to 210 days. The usual number of young is one, but twins and triplets sometimes occur. Kids are able to follow their mother almost immediately after being born, and rapidly improve their leaping ability within the first few days of life. If a mother is killed, other chamois take care of its young. Potential longevity is 22 years.

The flesh of *Rupicapra* is prized as food by some people, the skin is made into "shammy" leather for cleaning glass and polishing automobiles, and the winter hair from the back is used to make the "gamsbart," the brush of Tyrolean hats. Excessive hunting has greatly reduced the number of cham-

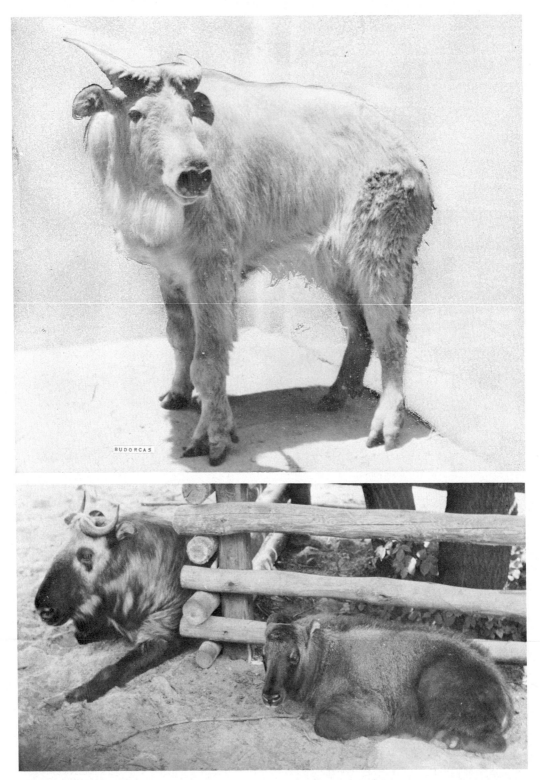

Takins (*Budorcas taxicolor*): Top, photo from New York Zoological Society; Bottom, photo by André Verstraete through Lothar Schlawe.

ois in some areas. The subspecies *R. r. ornata* of the Appenines is now restricted to Abruzzo National Park in central Italy, where about 300 to 400 individuals survive (Lovari 1977). This subspecies is listed as endangered by the USDI (1980) and is on appendix 1 of the CITES.

ARTIODACTYLA; BOVIDAE; **Genus BUDORCAS**
Hodgson, 1850

Takin

The single species, *B. taxicolor,* occurs in Bhutan, southeastern Tibet, central China, and possibly northern Burma and adjacent parts of Yunnan (Ellerman and Morrison-Scott 1966).

Head and body length is about 1,200 mm, tail length is about 100 mm, shoulder height is 760 to 1,070 mm, and weight is usually 230 to 275 kg. Coloration of the shaggy fur varies from yellowish white through straw brown to blackish brown. There is a dark stripe along the back. The build is heavy and oxlike, the front limbs are stout, the lateral hooves (dew claws) are large, the profile is convex, and the muzzle is hairy. The horns, carried by both sexes, are fairly massive, transversely ribbed at the bases, and up to 635 mm long. They arise near the midline of the head, abruptly turn outward, and then sweep backward and upward.

The takin prefers dense thickets near the upper limit of tree growth at an elevation of around 2,400 to 4,250 meters. It makes narrow paths through this thick growth, which it uses regularly in passing to and from grazing areas and salt licks. It commonly spends the day in dense vegetation and emerges in the evening to feed. On cloudy and foggy days, it may remain active throughout the day. In summer, it eats grasses and other tender mountain vegetation, but the winter diet consists principally of bamboo and willow shoots.

Budorcas gathers in herds of considerable size near and above the tree line during the summer, but in winter there is dispersion into smaller bands and migration to grassy valleys at somewhat lower elevations. The old bulls are usually solitary, but during August and September the sexes are often seen together. When alarmed, an individual gives a warning cough to alert the others of its herd, and the animals then quickly dash for the safety of dense underbrush. Mating occurs in July and August, and births take place in March or April. A gestation period of 200 to 220 days has been recorded in captivity (Aung 1968). A single kid is normal and within 3 days it is able to accompany its mother almost anywhere. One captive takin lived for 15 years and 10 months (Marvin L. Jones, Zoological Society of San Diego, pers. comm.).

Native peoples highly esteem the flesh of *Budorcas,* and capture it by means of deadfalls, spear traps, and snares. The subspecies of central China, *B. t. tibetana* and *B. t. bedfordi,* are classified by the IUCN (1972), respectively as indeterminate and rare.

ARTIODACTYLA; BOVIDAE; **Genus OVIBOS**
De Blainville, 1816

Muskox

The single living species, *O. moschatus,* occurred in historical time from northern Alaska to Hudson Bay, on the northern and western islands of the Canadian Arctic, and in Greenland (Hall 1981). The same species, or a close relative,

Four-year-old muskox bull (*Ovibos moschatus*), photo from U.S. Fish & Wildlife Service.

Muskox (*Ovibos moschatus*), photo by Lothar Schlawe.

was present in northern Eurasia during the late Pleistocene, and may have survived in Siberia until about 2,000 years ago (Corbet 1978; Tener 1965).

Head and body length is 1,900 to 2,300 mm, tail length is 90 to 100 mm, shoulder height is 1,200 to 1,510 mm, and weight is generally 200 to 410 kg. Captive males have attained weights of up to 650 kg (Lent 1978). Males are larger,

on the average, than females. The dark brown, coarse guard hairs reach nearly to the ground, shed rain and snow, and take the wear. An inner coat of fine, soft, light brown hair is so dense that neither cold nor frost can penetrate it. The legs and middle of the back are pale. The general form is massive, there is a slight hump at the shoulders, and the neck, legs, and tail are short. The horns, borne by both sexes, are broad,

Muskoxen (*Ovibos moschatus*) in defensive position, photo from U.S. Fish & Wildlife Service.

curve down and outward, and nearly meet in the midline of the skull to form a large boss. The horns of the males are much more massive at the base than are those of the females. The common name refers to a characteristic odor that emanates from males during the rut. Females have four mammae.

The information for the remainder of this account was taken largely from Banfield (1974), Lent (1978), and Tener (1965). The muskox dwells exclusively on the arctic tundra. In the summer it prefers moist habitats, such as river valleys, lake shores, and seepage meadows. In the winter it may shift to hilltops, slopes, and plateaus, where prevailing winds keep snow levels to a minimum. Although the muskox appears to be clumsy, and is usually slow and stolid, it moves with surprising agility and can run swiftly if necessary. It seems to have acute vision and hearing. A herd usually does not remain in one place for more than a day during the summer, but may feed in a favorable spot for several days during the winter. Movements between winter and summer ranges do not cover more than 80 km. The summer food includes mainly grasses and sedges, and the winter diet consists primarily of browse, such as crowberry, cowberry, and willow.

Ovibos is gregarious, occuring in herds with up to 100 individuals. The usual number of animals per group is 15 to 20 in winter and about 10 in summer. When approached by an enemy, the members of a herd bunch together, often in the form of a circle or semicircle with the calves inside. Such a formation provides effective defense against wolves, but has allowed entire herds to be easily killed by human hunters with modern firearms.

Some questions remain regarding social structure, but main herds generally seem to contain adults and young of both sexes. During the summer rut, however, a dominant bull, commonly 6 to 10 years of age, drives the other males away and takes charge of the herd. Dominance is achieved by displays, threats, and serious fights. The latter involve repeated head-on charges at 40 km per hr, with the horns clashing together. The males that are forced out, mainly young animals and some that are very old, either become solitary or form small groups of their own. They usually rejoin the main herd in the autumn, but some may wander widely and attempt to acquire a herd of their own.

Mating takes place from July to September, and the young are born from late April to mid-June. A female gives birth every 1 or 2 years. The gestation period is about 8 to 9 months. A single offspring is normal, but twins have been reported on rare occasion. The calf weighs 10 to 14 kg at birth and is highly precocious. It begins to nibble some vegetation within 1 week, but may not be completely weaned for over 1 year. Mating under fully natural conditions does not usually occur until females are over 2 and males over 5 years of age, but captive animals have bred as yearlings. Wild individuals have been known to live up to 23 years.

The muskox has been exploited by people for its meat and hide, and the Eskimos have used the horns to fashion bows. As human activity increased in the Arctic during the 19th and early 20th centuries, the muskox was exterminated in Alaska and nearly wiped out on the mainland of Canada. Subsequent conservation efforts allowed a limited recovery in some areas, and there are now an estimated 25,000 individuals in the world. There are small introduced populations in various parts of Alaska, the Soviet Union, Svalbard, Norway, and northern Quebec.

ARTIODACTYLA; BOVIDAE; *Genus* **HEMITRAGUS**
Hodgson, 1841

Tahrs

Tahrs (*Hemitragus jemlahicus*), photo from New York Zoological Society.

There are three species (Ellerman and Morrison-Scott 1966):

H. jemlahicus, the Himalayan region from Kashmir to Sikkim;
H. jayakari, Oman;
H. hylocrius, Nilgiri Hills and adjacent ranges in southern India.

Head and body length is about 900 to 1,400 mm, tail length is about 90 to 120 mm, shoulder height is 610 to 1,060 mm, and weight is about 50 to 100 kg. The pelage varies between species. *H. jemlahicus* is reddish to dark brown in color, and has a shaggy mane around the neck and shoulders and extending to the knees. *H. jayakari* is covered with shaggy, brittle, short hairs, which are grayish to tawny brown. *H. hylocrius* is dark yellowish to brownish and has a grizzled area across the back. Although *Hemitragus* bears a general resemblance to *Capra,* it differs in that the males lack the beard, the muzzle is naked, the feet have glands, and the horns are not twisted but are somewhat laterally flattened. Female *H. hylocrius* have two mammae, whereas female *H. jemlahicus* and *H. jayakari* have four.

All species occur on rugged hills and mountain slopes. *H. jayakari* occupies areas of sparse scrub vegetation, *H. jemlahicus* occurs mainly on wooded slopes, and *H. hylocrius* prefers grass-covered hills. These animals are wary and difficult to approach, especially from below. It is surprising with what ease and confidence they scamper about their uneven, rocky haunts. *H. hylocrius* is active intermittently from dawn to late evening, and is mainly a grazer (Davidar 1978). *H. jayakari* is evidently crepuscular and nocturnal in the warm summer months, and eats a variety of vegetation, but seems to prefer such parts as seeds, fruits, and new growth (IUCN 1978).

In New Zealand, where *H. jemlahicus* has been introduced, population density was found to be 4.5 to 6.8 individuals per sq km. In its natural habitat this species was observed to travel in herds of 2 to 23 animals. The average membership of all-female groups was 4.0, while that of herds with adults of both sexes was 10.2. Herd size in *H. hylocrius* was 6 to 104, with means of 9.3 for all-female groups and 27.4 for mixed herds (Schaller 1977). *H. jayakari* has been seen alone, in pairs, and in trios of a male, a female, and a kid. Males or pairs of this species are thought to be territorial (IUCN 1978).

Reproduction occurs throughout the year in *H. hylocrius,* with a birth peak in winter (Davidar 1978). In the Himalayas, *H. jemlahicus* mates from mid-October to mid-January (Schaller 1977). A newborn *H. jayakari* was found in April (Harrison 1968). Gestation periods of 180 to 242 days have been reported. There is usually a single offspring, occasionally two. A captive *H. jemlahicus* lived for 21 years and 9 months (Marvin L. Jones, Zoological Society of San Diego, pers. comm.).

The Arabian tahr *(H. jayakari)* is classified as endangered by the IUCN (1978) and the USDI (1980). Although protected by law, it has declined through competition with domestic goats and illegal hunting. Fewer than 2,000 individuals are thought to survive in the limited available habitat. The Nilgiri tahr *(H. hylocrius)* is classified as vulnerable by the IUCN (1976). Its range has declined drastically through poaching and habitat destruction. Its current total population has been estimated at 2,230 (Davidar 1978). *H. jemlahicus* has been introduced to New Zealand; a population of 20,000 to 30,000 individuals is now established there and is subject to sport and commercial hunting (Schaller 1977).

ARTIODACTYLA; BOVIDAE; *Genus CAPRA* Linnaeus, *1758*

Goats

There are eight species (Corbet 1978; Meester and Setzer 1971):

C. aegagrus (wild goat), mountains from Asia Minor to Afghanistan and Pakistan, Oman, Crete, Aegean Islands;
C. hircus (domestic goat), world-wide in association with people, feral in many areas;
C. ibex (ibex), European Alps, Palestine, Sinai, Arabian Peninsula, mountainous region from Afghanistan and northern India to Lake Baikal, Egypt and Sudan east of Nile, northern Ethiopia;
C. walie (Walia ibex), Simien Mountains of north-central Ethiopia;
C. caucasica (west Caucasian tur), western Caucasus;
C. cylindricornis (east Caucasian tur), eastern Caucasus;
C. pyrenaica (Spanish ibex), Spain, Portugal;
C. falconeri (markhor), mountains of southern Uzbek, Tadzhik, Afghanistan, northern and central Pakistan, and Kashmir.

There is much controversy regarding the systematics of this genus. Both Corbet (1978) and Meester and Setzer (1971) included *Ammotragus* in *Capra.* Van Gelder (1977*b*) considered *Ammotragus* and *Ovis* to be synonyms of *Capra.* Ellerman and Morrison-Scott (1966) placed *C. falconeri* in a separate subgenus, *Orthaegoceros* Trouessart, 1905, but Schaller (1977) did not think this designation warranted. Corbet (1978), Grzimek (1975), and Schaller (1977) listed *C. walie* as a subspecies of *C. ibex.* Grzimek (1975) and Schaller (1977) included *C. caucasica* in *C. ibex,* and Grzimek (1975) also treated *C. cylindricornis* as a subspecies of *C. ibex.* Some authorities, such as Corbet (1978), prefer not to use the name *C. hircus;* others, such as Ellerman and Morrison-Scott (1966), include both *C. aegagrus* and the domestic goat in *C. hircus;* and others, such as Grzimek (1975), treat *C. hircus* as a subspecies of *C. aegagrus.* American authorities, including Schaller (1977) and Lowery (1974), have generally recognized *C. hircus* as a species distinct from *C. aegagrus.* Schaller (1977) indicated that *C. aegagrus* may have lived in the Balkan Peninsula until 1891. The IUCN (1976) stated that *C. pyrenaica* occurred on the French side of the Pyrenees in the 14th century A.D.

Head and body length is 1,150 to 1,700 mm, tail length is 100 to 200 mm, shoulder height is 650 to 1,050 mm, and weight is 18 to 150 kg (Grzimek 1975; Schaller 1977). The general coloration is some shade of brown or gray. The underparts may be almost white or the same color as the rest of the coat. *Capra* bears some resemblance to *Ovis,* but differs in that males are odorous, there is a beard, the feet lack scent glands, and the forehead is convex, not concave. Females have two mammae (Grzimek 1975).

In wild species the horns of males are about 500 to 1,650 mm long, and those of females are 150 to 380 mm long. There is much structural variation (Schaller 1977; Grzimek 1975). In *C. aegagrus* the horns are scimitar shaped and have a generally sharp anterior keel. In *C. ibex, C. walie,* and *C. caucasica* the horns are also scimitar shaped, but the anterior surface is relatively flat and is broken by prominent transverse ridges. *C. walie* is distinguished by a bony boss on the forehead (Meester and Setzer 1971). In *C. falconeri* the horns are sharp keeled and twisted into a tight or open spiral. In *C. pyrenaica* the horns curve out and up, and then back,

Cretian wild goat (*Capra aegagrus*), photo by Ernest P. Walker.

inward, and up again, and they have a sharp posterior keel. In *C. cylindricornis* the horns curve somewhat like those of *C. pyrenaica*, but are heavy and almost round in cross section. The horns of the domestic goat *(C. hircus)* may be either scimitar shaped or spiraling, the former apparently being the original condition.

Wild species of *Capra* are usually found in rugged mountain country, rocky crags, and alpine meadows. They may become active in the afternoon and feed throughout the night. Some populations have rather definite seasonal migrations; in spring they move upward in the mountains to new areas for feeding, and in the fall or winter they move to lower levels to avoid the deep snow. These animals are both browsers and grazers.

Population density varies widely, being around 1 to 8 individuals per sq km for several wild species in Pakistan, 9 per sq km for *C. ibex* in the Alps, 143 per sq km for *C. aegagrus* on Theodoru Island in the Aegean, and 1,000 per sq km for feral *C. hircus* on Macauley Island in New Zealand (Schaller 1977). *Capra* usually occurs in groups of 5 to 20 individuals, though some herds contain over 100 members and there are also solitary animals. Groups may consist of both sexes, of only males, or of only females and young. In some populations, such as that of *C. ibex* in the Alps, adult males join the females only in the breeding season, while in other populations, such as that of *C. falconeri* in northern Pakistan, some adult males remain with the females throughout the year. Groups establish a dominance hierarchy through threats and fighting, and males become especially aggressive toward each other during the rut. Fights involve shoving and clashing with the horns, sometimes after the animals rear onto their hind legs (Schaller 1977; Grzimek 1975).

In most wild populations, mating occurs during some period from late summer to midwinter (Schaller 1977; Grzimek 1975). *C. walie* of Ethiopia, however, shows rutting behavior throughout the year, with a peak from March to May (Nievergelt 1974). Gestation periods of 147 to 180 days have been reported. There are usually one or two precocious young. The mother protects her kids by fighting with the horns or by decoying the intruder. Sexual maturity may be attained around 1 year of age in captivity, but a wild female *C. ibex* does not bear young until she is 3 to 6 years old (Grzimek 1975). A captive of this species lived for 22 years and 3 months (Marvin L. Jones, Zoological Society of San Diego, pers. comm.).

Present evidence indicates that the earliest domestication of *Capra* took place in southwestern Asia between 8,000 and 9,000 years ago. *C. aegagrus* is generally considered to be directly ancestral to, if not conspecific with, *C. hircus*. There have been suggestions that *C. falconeri* also contributed to the ancestry of the domestic goat, but Schaller (1977) consid-

Markhors (*Capra falconeri*), photo from San Diego Zoological Society.

ered such a view unwarranted. There are numerous breeds of *C. hircus,* with a multitude of forms and colors, and some are hornless. They are used for the production of milk, meat, and wool (Grzimek 1975). Goats are very nimble and aggressive, and can live and obtain food in places not generally accessible to other domestic mammals. Herds of goats have been highly destructive to natural vegetation, especially in the Mediterranean region and Middle East, thereby contributing to erosion, the spread of deserts, and the disappearance of native wildlife. *C. hircus* has become feral on Hawaii and many other islands, and has been partly responsible for the endangerment and extinction of several species of forest birds.

The domestic goat also has been a factor in the decline of its wild relatives by competing with them for available food. In addition, wild species of *Capra* have been adversely affected by pressure from other domestic livestock, loss of habitat to agriculture, and excessive hunting by people for trophies and meat. Both the IUCN (1976, 1978) and the USDI (1980) classify the following as endangered: *C. walie* of Ethiopia, with an estimated total population of 300 individuals; *C. pyrenaica pyrenaica* of the Pyrenees, thought to number 20 or fewer; and *C. falconeri megaceros* of central Pakistan and adjacent Afghanistan, perhaps numbering fewer than 2,000. The species *C. falconeri* is generally clas-

sified as vulnerable by the IUCN (1976) and is on appendix 2 of the CITES. The subspecies *C. falconeri jerdoni* and *C. f. chialtanensis* of Pakistan and Afghanistan are officially listed as endangered by the USDI and are on appendix 1 of the CITES, but the former is probably a synonym of *C. f. megaceros* and the latter is actually a form of *C. aegagrus* (Schaller 1977). *C. aegagrus* itself has declined in many areas and is classified as rare by the Union of Soviet Socialist Republics Ministry of Agriculture (1978). It survives in small numbers on Crete and has been introduced on Theodoru Island just to the north, but populations on other Aegean islands have hybridized with *C. hircus*. Because of sport hunting, the European population of *C. ibex* survived only in the Gran Paradiso of Italy by the early 19th century. Animals from that area, however, were later used to restore the species to other parts of the Alps (Grzimek 1975).

ARTIODACTYLA; BOVIDAE; **Genus PSEUDOIS**
Hodgson, 1846

Bharals, Blue Sheep

There are two species (Corbet 1978; Groves 1978*b*):

Blue sheep (*Pseudois nayaur*): Left, male; Right, female and young; photos by Lothar Schlawe.

P. nayaur, alpine zone from Himalayan region to Inner Mongolia;

P. schaeferi, southeastern Tibet.

Head and body length is 1,150 to 1,650 mm, tail length is 100 to 200 mm, shoulder height is 750 to 910 mm, and weight is 25 to 80 kg (Grzimek 1975; Schaller 1977). In *P. nayaur* the head and upper parts are brownish gray with a tinge of slaty blue, and the underparts and insides of the legs are white. This coloration blends well with the blue shale, rocks, and brown grasses of the open hillsides. *P. schaeferi* is generally smaller than *P. nayaur*, and has more drab coloration with a silvery sheen (Groves 1978b).

In structure and habits, *Pseudois* is intermediate to *Capra* and *Ovis*, but is apparently more closely related to *Capra*. Schaller (1977) described bharals as "aberrant goats with sheep-like affinities." He noted that *Pseudois* resembles *Capra* in having a broad flat tail with a bare central surface, large dew claws, no inguinal glands, no preorbital glands, and usually no pedal glands. The horns of *Pseudois* are rounded and smooth, and curve backward over the neck. They are borne by both sexes, but are much larger in males, reaching a length of 820 mm.

Bharals occur on open slopes and plateaus, with abundant grass, at elevations of 2,500 to 5,500 meters. Because of their protective coloration, and in the absence of any brush in which to hide, these animals remain motionless when approached. When they discover that they have been seen, however, they take to the precipitous cliffs, ascending to the most difficult and inaccessible places. They feed and rest alternately throughout the day on the grassy slopes of mountains. The diet consists of grasses, herbs, and lichens (Grzimek 1975).

Population density of *P. nayaur* in Nepal was found to be 0.9 to 1.3 individuals per sq km, but to increase to 8.8 to 10.0 per sq km during winter concentrations in valleys (Schaller 1977). This species may form herds of up to 400 individuals, but the largest herd seen by Schaller contained 61, and mean size in different study areas ranged from 4.8 to 18.4. Males tend to separate from the females after the mating season, and either to become solitary or to form bachelor herds. A few males, however, associate with the females throughout the year. During the mating season, Schaller observed consider-

able aggression, usually directed against individuals of the same sex. In most fights, one combatant rears up and then lunges down at its opponent, which catches the blow between its horns. In another study of *P. nayaur*, Wegge (1979) determined average group size to be 11.1 individuals, and observed no sexual segregation. *P. schaeferi* reportedly occurs in considerably smaller groups than does *P. nayaur* (Grzimek 1975).

Mating occurs from October to December, and the young are born in May or June. The gestation period is 160 days, a single offspring is normal, lactation lasts 6 months, and sexual maturity is attained at 1½ years of age (Grzimek 1975; Schaller 1977). Wegge (1979) observed that males do not reach trophy size until they are 7 years old, and it is unlikely that they are able to mate under natural conditions prior to that age. Blue sheep do well in captivity, and one specimen lived for 20 years and 3 months (Marvin L. Jones, Zoological Society of San Diego, pers. comm.).

ARTIODACTYLA; BOVIDAE; Genus **AMMOTRAGUS**
Blyth, 1840

Aoudad, Barbary Sheep

The single species, *A. lervia*, occurs in highlands within the desert and subdesert zones from Morocco and Western Sahara to Egypt and Sudan. Corbet (1978), Meester and Setzer (1971), and Van Gelder (1977b) all included *Ammotragus* within the genus *Capra*, mainly on the basis of the production of fertile offspring from a cross between an aoudad and a domestic goat. In an assessment of the biochemical characters of *Ammotragus*, however, Manwell and Baker (1977) found 5 characters to be sheeplike, 5 to be goatlike, 7 to be common to both groups, and 10 to be unique. A recent review (Gray and Simpson 1980) maintained *Ammotragus* as a full genus.

Head and body length is 1,300 to 1,650 mm, tail length is 150 to 250 mm, shoulder height is 750 to 1,120 mm, and weight is 40 to 55 kg in females and 100 to 145 kg in males (Grzimek 1975; Gray and Simpson 1980). The general coloration is rufous tawny. The insides of the ears, the chin, a line

Barbary sheep (*Ammotragus lervia*): Top, photo by Lothar Schlawe; Bottom, photo by Ernest P. Walker.

on the underparts, and the insides of the legs are whitish. There is no beard, but there is a ventral mane of long, soft hairs on the throat, chest, and upper part of the forelegs. The horns of males sweep outward, backward, and then inward; they are rather heavy and wrinkled, and measure up to 840 mm in length. Females also have prominent horns.

The aoudad is found mainly in rough, rocky, arid country. The almost total lack of vegetation tall enough to conceal this animal has seemingly resulted in its developing exceptional ability to hide by remaining motionless whenever danger threatens. Within its range, sources of water are few and far between, but it is usually able to obtain sufficient moisture from green vegetation and the dew that condenses on leaves during the cold desert nights. Nonetheless, populations decrease sharply during periods of drought. In captivity, the aoudad seems to like water and to enjoy taking a bath. The diet consists of grass, herbaceous plants, and stunted bushes.

Reported home range size for individuals introduced in western Texas is 1 to 5 sq km in winter and 13 to 31 sq km in summer (Gray and Simpson 1980). *Ammotragus* generally occurs alone or in small groups under natural conditions. A stable linear dominance hierarchy is established in captive herds, with adult males ranking highest but an adult female actually leading group movements (Gray and Simpson 1980). In the wild, adult males may fight savagely for control of a group of females (Grzimek 1975). Two males frequently stand about 10 to 15 meters apart, then walk rapidly toward each other, gradually gaining speed, and finally break into a run shortly before they collide. Just prior to impact, the heads are lowered. A male will not attack, however, if the other is off balance or unprepared.

Schaller (1977) stated that mating takes place mainly in July and August in Niger. According to Gray and Simpson (1980), reproductive activity goes on throughout the year, but mating occurs largely from September to November and births from March to May. Females are known to have given birth twice in 1 year. The gestation period is 154 to 161 days. A single offspring is usual, twins are common, and triplets are rare. The young weighs an average of 4.5 kg at birth, and is able to negotiate moderately rugged terrain almost immediately thereafter. Sexual maturity is reached at around 1½ years of age. A captive aoudad lived for 20 years and 11 months (Marvin L. Jones, Zoological Society of San Diego, pers. comm.).

The native peoples of the Sahara kill many *Ammotragus*, as this animal is an important source of meat, hides, hair, and sinews, all of which are valuable in the desert economy. The genus has thus declined alarmingly in much of its natural range, and has disappeared entirely in some areas (Meester and Setzer 1971; Grzimek 1975). The aoudad, however, was introduced in the United States for purposes of sport hunting, and large populations are now established in California, New Mexico, and Texas. There is concern that *Ammotragus* will spread into the range of the native desert bighorn sbeep *(Ovis canadensis)*, and that the two animals will compete for limited resources, to the probable detriment of the bighorn (Seegmiller and Simpson 1980).

ARTIODACTYLA; BOVIDAE; *Genus OVIS Linnaeus, 1758*

Sheep

There are two subgenera and eight species (Nadler et al. 1973b; Korobitsyna et al. 1974; Ellerman and Morrison-Scott 1966; Hall 1981):

subgenus *Ovis*

O. vignei (urial), southern Soviet Central Asia, northeastern Iran, Afghanistan, Pakistan, Kashmir, northwestern India, Oman;

O. ammon (argali), southern Siberia, eastern Soviet Central Asia, Mongolia, northern China, Sinkiang, Tibet, Himalayan region;

O. orientalis (Asiatic mouflon), Asia Minor, western Iran;

O. musimon (mouflon), Corsica, Sardinia, Cyprus;

O. aries (domestic sheep), world-wide in association with people;

subgenus *Pachyceros*

O. canadensis (bighorn sheep), southwestern Canada, western conterminous United States, northern Mexico, Baja California;

O. dalli (Dall's sheep), Alaska, northwestern Canada;

O. nivicola (snow sheep), northeastern Siberia.

There is much controversy regarding the systematics of sheep. Van Gelder (1977b) considered *Ovis* a synonym of *Capra*, on the basis of the production of fertile offspring through hybridization between *C. hircus* and *O. aries*. Some authorities, including Corbet (1978) and Grzimek (1975), include all members of the subgenus *Ovis* in the species *O. ammon*, and all members of *Pachyceros* in *O. canadensis*. Schaller (1977) did not agree with the specific separation of *O. vignei* from *O. ammon* or of *O. musimon* from *O. orientalis*. Valdez, Nadler, and Bunch (1978) described a zone of hybridization between *O. vignei* and *O. orientalis* in northern Iran.

Head and body length is 1,200 to 1,800 mm, tail length is 70 to 150 mm, shoulder height is 650 to 1,270 mm, and weight is 20 to 200 kg. There is much variation in size, the average weight being about 32 kg for male *O. musimon* on Cyprus, around 60 kg for male *O. vignei* in southern Central Asia, and 180 kg for male *O. ammon* in the Altai Mountains (Schaller 1977). In any given population, males average larger than females. In *O. canadensis* of the northern Rocky Mountains, males weigh 73 to 143 kg, and females 53 to 91 kg. In *O. canadensis* of the southwestern deserts, males weigh 58 to 86 kg, and females 34 to 52 kg (Wishart 1978). Coloration varies from creamy white to dark gray and brown. Some males possess a fringe of long hair down the front of the neck, but they do not have beards such as are present in *Capra*. Also unlike all goats, there is no gland at the base of the tail. *Ovis* has a narrow nose and pointed ears. Males have massive spiral horns that have an average length of 1,106 mm in *O. canadensis* of the northern Rockies (Wishart 1978) and a record length of 1,690 mm in *O. ammon* (Schaller 1977). The horns of females are much smaller and only slightly curved. Some domestic breeds are hornless.

Wild sheep are usually found in upland areas. Banfield (1974) stated that *O. canadensis* inhabits alpine meadows, grassy mountain slopes, and foothill country in proximity to rugged, rocky cliffs and bluffs. This species is very alert, has remarkable eyesight, is an excellent climber, and swims freely. When alarmed, it bounds away over jagged rocks with surprising speed and agility. In contrast, Schaller (1977) reported that Eurasian wild sheep occur mainly in gentle to steeply rolling, but nonprecipitous terrain, at elevations from sea level to 5,000 meters. Geist (1971) noted that *O. ammon* actually seems to be a poor jumper and avoids cliffs, when escaping. Sheep may feed intermittently throughout the day or rest during the hottest hours, and are sometimes active at night. Most populations undergo seasonal movements, generally dispersing upward and over a larger area in the summer, and concentrating in sheltered valleys during the winter. The diet consists largely of grasses, sedges, and forbs, and also includes some browse.

A. Bighorn sheep (*Ovis canadensis*), photo of mounted group in Denver Museum of Natural History. B. Mouflons (*O. musimon*); C. Barbados sheep (*O. aries*); D. Young Dall's sheep (*O. dalli*); photos by Ernest P. Walker.

In the Canadian Rockies, Geist (1971) found that males use up to 7 and females up to 4 separate ranges in the course of a year, the different areas being about 1 to 32 km apart. The midwinter ranges average about 0.6 km across, while those of spring and fall have a mean diameter of 6 km. In a radio-tracking study of the desert bighorn in Nevada, Leslie and Douglas (1979) found 10 individual home ranges to average 16.9 sq km. Some reported population densities are about 0.2 individuals per sq km for *O. dalli*, 2 per sq km for *O. canadensis*, 11 to 13 per sq km for *O. vignei*, and 1 to 1.2 per sq km for *O. ammon* (Geist 1971; Schaller 1977).

Sheep are gregarious, sometimes gathering in herds of over 100 individuals. In some species, mature males stay apart from the females and young for most of the year. In Nevada, from January to May 1976, Leslie and Douglas (1979) found average group size of the desert bighorn to be 4.0 for rams and 5.8 for ewes. In Yellowstone National Park, Woolf, O'Shea, and Gilbert (1970) obtained respective figures of 5.7 and 8.0. Observations in Central Asia indicate that groups are about the same size as in North America, but that a few males may associate with the female herds throughout the year (Schaller 1977).

The most detailed studies of the social life of wild *Ovis* were made by Geist (1971) in the Canadian Rockies. He found that a young female remains in its mother's group, but that a male departs when two to four years old and joins a group of rams. Young are tolerated by the adults, seem to receive guidance from them, and eventually inherit their home range and migratory pattern. In male groups there is a strict dominance hierarchy based on age and size of the horns; the animal with the biggest horns is the leader. In the ritualized struggle for dominance, two individuals push and shove one another, and finally back away some distance, rear up on their hind legs, and lunge forward and down, crashing their horns together. Young animals are the most aggressive, but the older rams easily neutralize the attacks with their massive horns, and thus injuries are rare. The most spectacular battles occur on the rutting grounds. A dominant ram is not territorial, but courts one estrous female at a time, and drives other males away. For their part, ewes prefer mates with large horns and thwart the advances of others. At other times of the year there are also fights for dominance within groups of males. Homosexual activity then may occur, with the dominant animal behaving like a courting male and the subordinate playing the role of an estrous female.

In both the Old and New worlds, the rut occurs mainly in the autumn and early winter, and births take place in the spring. The desert bighorn of southwestern North America, however, has an unusually long mating season for a ruminant, lasting from July to December (Wishart 1978). Gestation periods of about 150 to 180 days have been reported, and litters contain one to four young. A single lamb seems to be the usual case in the subgenus *Pachyceros,* but multiple births are very common in the subgenus *Ovis,* especially in the domestic sheep (Geist 1971; Grzimek 1975). The newborn weighs about 3 to 5 kg and is precocious. Within a few weeks of birth the lambs form bands of their own and seek their mothers only periodically to suckle. They are weaned by 4 to 6 months of age (Geist 1971). A captive female *O. canadensis* successfully mated when only 10–11 months old (McCutchen 1977), but ewes generally do not breed until their second or third year of life. Because of social factors, males are usually not able to mate until they are 7 years old, but they will do so sooner if the dominant rams of their group are killed (Nichols 1978).

According to Geist (1971), longevity depends on population status. In a declining or stable population, with low recruitment, most sheep live over 10 years, with a maximum of 20 for rams and 20–24 for ewes. In an expanding popula-

tion, with heavy reproduction, the average life span is only 6 or 7 years. The quality of an individual and its longevity are inversely related. Large, vigorous animals that reproduce well quickly reach advanced physiological age and die young, but a social and reproductive failure, especially if in a group characterized by lethargic behavior and slow reproduction, may attain a venerable age.

Available chromosomal and archeological evidence indicates that the domestic sheep is descended from a mouflonlike animal, and that domestication occurred about 10,000–11,000 years ago in the eastern Mediterranean region (Nadler et al. 1973b). No wild sheep has a woolly coat comparable to that of *O. aries. O. musimon* has a woolly underfur in winter, but this is well hidden by the coarse heavy coat. It has been reported that when domestic sheep become feral, they gradually lose much of their woolly pelage and develop a coat of coarse hairs, approaching the kind found in the wild species. There are now more than 800 breeds of domestic sheep (Valdez, Nadler, and Bunch 1978), and over 800,000,000 individuals (Grzimek 1975). These animals are valued for the production of wool and meat, but also are sometimes considered detrimental to natural vegetation and native wildlife.

The domestic sheep has adversely affected its wild relatives in many areas by competing with them for forage and spreading disease. The wild species have also declined through indiscriminate hunting by people for meat and tro-

Dall's sheep (*Ovis dalli*), photo by Leonard Lee Rue III.

phies. Several Central Asian subspecies have become very rare and may not be able to survive without increased protection (Schaller 1977). The Union of Soviet Socialist Republics Ministry of Agriculture (1978) considers a majority of the subspecies of *O. vignei* and *O. ammon* that occur in the Soviet Union to be rare or endangered. *O. ammon hodgsoni* of Tibet and the Himalayas, *O. vignei vignei* of Kashmir, and *O. musimon ophion* of Cyprus are listed as endangered by the USDI (1980) and are on appendix 1 of the CITES. The other subspecies of *O. ammon* and the species *O. canadensis* are on appendix 2 of the CITES. The IUCN (1976) considers the populations of *O. musimon* (mouflon) on Cyprus, Corsica, and Sardinia to be endangered. The mouflon was present on the mainland of Europe through the end of the Pleistocene, but has apparently been restricted to the Mediterranean islands in historical time. Numbers in the natural habitat fell drastically after 1900, there now being about 700 survivors. Since the mid-19th century, however, large introduced mouflon populations have become established in much of Europe and in various other parts of the world (Grzimek 1975).

In northwestern North America there are now about 65,000 to 95,000 Dall's sheep *(O. dalli)*. The largest numbers are in Alaska, where populations appear to have recovered from earlier declines (Nichols 1978). To the south, the bighorn sheep *(O. canadensis)* suffered severe losses in numbers and distribution during the 19th century, mainly through excessive hunting, competition from domestic livestock, and the introduction of disease from domestic sheep. The subspecies *O. c. auduboni* of the Black Hills and adjacent areas became extinct, and other subspecies underwent increasing fragmentation as settlement proceeded. Although hunting has been legally prohibited or controlled since the early 1900s, and while some bighorn populations appear to be doing well, there has been no general recovery comparable to that of certain other North American ungulates during the same period. There are now about 35,000 bighorns, fewer than at the turn of the century (Buechner 1960; Boone and Crockett Club 1975).

Geist (1971, 1974) has suggested that one reason for the failure of the bighorn to rebuild its numbers is the discontinuous nature of its habitat. Only certain terrain is suitable, and animals will not readily cross intervening areas to places formerly occupied by other populations. A bighorn establishes its home range not by venturing out on its own but by following and learning from the others of its group. A young animal depends on the knowledge and guidance of its elders, with respect to becoming familiar with the migratory paths and the rather complex system of ranges that are used in the course of a year. Moreover, the reproductive potential of a bighorn population seems to adjust itself to the limitations of traditionally available habitat, and there are no surplus animals to spread elsewhere. Human hunting of a restricted population may remove a disproportionately large number of dominant males, thereby upsetting the cohesiveness of groups and preventing the young from properly adapting to the intricacies of their social system and habitat. These problems demonstrate the value of behavioral and ecological information to people as they work to save their fellow mammals from extinction.

LITERATURE CITED

A.

Ables, E. D. 1975. Ecology of the red fox in North America. *In* Fox (1975), pp. 216–236.

Abravaya, J. P., and J. O. Matson. 1975. Notes on a Brazilian mouse, *Blarinomys breviceps* (Winge). Los Angeles Co. Mus. Contrib. Sci., no. 270, 8 pp.

Adams, C. E. 1976. Measurements and characteristics of fox squirrel, *Sciurus niger rufiventer*, home ranges. Amer. Midl. Nat. 95:211–215.

Adams, D. B. 1979. The cheetah: native American. Science 205:1155–1158.

Adams, W. 1958. A flying squirrel that really flew. Malayan Nat. J. 13:31.

Adler, H. E., and L. L. Adler. 1978. What can dolphins (*Tursiops truncatus*) learn by observation? Cetology, no. 30, 10 pp.

Aellen, V. 1973. Un *Rhinolophus* nouveau d'Afrique Centrale. Period. Biol. 75:101–105.

Aggundey, I. 1977. First record of *Potamogale velox* in Kenya. Mammalia 41:368.

Agrawal, V. C., and S. Chakraborty. 1969. Occurrence of the woolly flying squirrel, *Eupetaurus cinereus* Thomas (Mammalia: Rodentia: Sciuridae) in north Sikkim. J. Bombay Nat. Hist. Soc. 66:615–616.

———. 1971. Notes on a collection of small mammals from Nepal, with the description of a new mouse-hare (Lagomorpha: Ochotonidae). Proc. Zool. Soc. Calcutta 24:41–46.

Agrawal, V. C., and D. K. Ghosal. 1969. A new field-rat (Mammalia: Rodentia: Muridae) from Kerala, India. Proc. Zool. Soc. Calcutta 22:41–45.

Ahnlund, H. 1980. Sexual maturity and breeding season of the badger, *Meles meles* in Sweden. J. Zool. 190:77–95.

Ahroon, J. K., and F. G. Fidura. 1976. The influence of the male on maternal behaviour in the Mongolian gerbil (*Meriones unguiculatus*). Anim. Behav. 24:372–375.

Aitken, P. F. 1971a. The distribution of the hairy-nosed wombat [*Lasiorhinus latifrons* (Owen)]. Part 1: Yorke Peninsula, Eyre Peninsula, the Gawler Ranges and Lake Harris. S. Austral. Nat. 45:93–104.

———. 1971b. *Planigale tenuirostris* Troughton, the narrow-nosed planigale, an addition to the mammal fauna of South Australia. S. Austral. Nat. 46:18.

———. 1971c. Rediscovery of the large desert sminthopsis (*Sminthopsis psammophilus* Spencer) on Eyre Peninsula, South Australia. Victorian Nat. 88:103–111.

———. 1972. *Planigale gilesi* (Marsupialia, Dasyuridae); a new species from the interior of south eastern Australia. Rec. S. Austral. Mus. 16:1–14.

———. 1977a. The little pigmy possum [*Cercartetus lepidus* (Thomas)] found living on the Australian mainland. S. Austral. Nat. 51:63–66.

———. 1977b. Rediscovery of swamp antechinus in South Australia after 37 years. S. Austral. Nat. 52:28–30.

———. 1979. The status of endangered Australian wombats, bandicoots and the marsupial mole. *In* Tyler (1979), pp. 61–65.

Ajayi, S. S. 1975. Observations on the biology, domestication and reproductive performance of the African giant rat *Cricetomys gambianus* Waterhouse in Nigeria. Mammalia 39:344–364.

Akersten, W. A. 1973. Upper incisor grooves in the Geomyinae. J. Mamm. 54:349–355.

Albignac, R. 1969. Notes ethologiques sur quelques carnivores Malgaches: le *Galidia elegans* I. Geoffroy. Terre Vie 23:202–215.

———. 1970a. Notes ethologiques sur quelques carnivores Malgaches: le *Fossa fossa* (Schreber). Terre Vie 24:383–394.

———. 1970b. Notes ethologiques sur quelques carnivores Malgaches: le *Cryptoprocta ferox* (Bennett). Terre Vie 24:395–402.

———. 1971. Notes ethologiques sur quelques carnivores Malgaches: le *Mungotictis lineata* Pocock. Terre Vie 25:328–343.

———. 1972. The Carnivora of Madagascar. *In* Battistini and Richard-Vindard (1972), pp. 667–682.

———. 1974. Observations éco-ethologiques sur le genre *Eupleres*, viverridé de Madagascar. Terre Vie 28:321–351.

———. 1975. Breeding the fossa *Cryptoprocta ferox* at Montpelier Zoo. Int. Zoo Yearbook 15:147–150.

———. 1976. L'écologie de *Mungotictis decemlineata* dans les forêts décidues de l'ouest de Madagascar. Terre Vie 30:347–376.

Alcasid, G. L. Undated. Checklist of Philippine mammals. Natl. Mus. Philippines, Manila, 51 pp.

Allen, G. M. 1939a. Bats. Harvard Univ. Press, Cambridge, 368 pp.

———. 1939b. A checklist of African mammals. Bull. Mus. Comp. Zool. 83:1–763.

———. 1940. The mammals of China and Mongolia. Amer. Mus. Nat. Hist., New York, 2 vols.

———. 1942. Extinct and vanishing mammals of the Western Hemisphere with the marine species of all the oceans. Spec. Publ. Amer. Comm. Int. Wildl. Protection, no. 11, xv + 620 pp.

Allen, J. A. 1905. Mammalia of southern Patagonia. Princeton Univ. Exped. Patagonia 3:1–210.

———. 1911. Mammals from Venezuela collected by Mr. M. A. Carriker, Jr., 1909–1911. Bull. Amer. Mus. Nat. Hist. 30:239–273.

———. 1915. Review of the South American Sciuridae. Bull. Amer. Mus. Nat. Hist. 34:147–309.

———. 1917. The skeletal characters of *Scutisorex* Thomas. Bull. Amer. Mus. Nat. Hist. 37:769–784.

Allsopp, R. 1978. Social biology of bushbuck (*Tragelaphus scriptus* Pallas 1776) in the Nairobi National Park, Kenya. E. Afr. Wildl. J. 16:153–165.

Al-Robaae, K. 1977. Distribution of *Nesokia indica* (Gray and Hardwicke, 1830) in Basrah Iiwa, south Iraq; with some biological notes. Saugetierk. Mitt. 25:194–197.

Altmann, S. A., ed. 1967. Social communication among primates. Univ. Chicago Press, xiv + 392 pp.

Altmann, S. A., and J. Altmann. 1970. Baboon ecology. Univ. Chicago Press, vii + 220 pp.

Alvarez del Toro, M. 1967. A note on the breeding of the Mexican tree porcupine. Int. Zoo Yearbook 7:118.

Ambrose, H. W., III. 1969. A comparison of *Microtus pennsylvanicus* home ranges as determined by isotope and live trap methods. Amer. Midl. Nat. 81:535–555.

Amstrup, S. C., and J. Beecham. 1976. Activity patterns of radio-collared black bears in Idaho. J. Wildl. Mgmt. 40:340–348.

Anadu, P. A. 1979a. The occurrence of *Steatomys jacksoni* Hayman in south-western Nigeria. Acta Theriol. 24:513–517.

———. 1979b. Gestation period and early development in *Myomys daltoni* (Rodentia: Muridae). Terre Vie 33:59–69.

Anand Kumar, T. C. 1965. Reproduction in the rat-tailed bat *Rhinopoma kinneari*. J. Zool. 147:147–155.

Andersen, D. C. 1978. Observations on reproduction, growth, and behavior of the northern pocket gopher (*Thomomys talpoides*). J. Mamm. 59:418–422.

Andersen, H. T., ed. 1969. The biology of marine mammals. Academic Press, New York, 511 pp.

Andersen, K. 1912. Catalogue of the Chiroptera in the collection of the British Museum. I. Megachiroptera. British Mus. (Nat. Hist.), London, ci + 854 pp.

Andersen, S. 1976. The taming and training of the harbor porpoise *Phocoena phocoena*. Cetology, no. 24, 9 pp.

Anderson, E. 1970. Quaternary evolution of the genus *Martes* (Carnivora, Mustelidae). Acta Zool. Fennica, no. 130, 132 pp.

_____. 1973. Ferret from the Pleistocene of central Alaska. J. Mamm. 54:778–779.

Anderson, J. L. 1975. The occurrence of a secondary breeding peak in the southern impala. E. Afr. Wildl. J. 13:149–151.

_____. 1980. The social organisation and aspects of behaviour of the nyala *Tragelaphus angasi* Gray, 1849. Z. Saugetierk. 45:90–123.

Anderson, J. W., and W. A. Wimsatt. 1963. Placentation and fetal membranes of the Central American noctilionid bat, *Noctilio labialis minor*. Amer. J. Anat. 112:181–201.

Anderson, P. K. 1979. Dugong behavior: on being a marine mammalian grazer. Biologist 61:113–144.

Anderson, R. R., and K. N. Sinha. 1972. Number of mammary glands and litter size of the golden hamster. J. Mamm. 53:382–384.

Anderson, S. 1967a. Primates. *In* Anderson and Jones (1967), pp. 151–177.

_____. 1967b. Introduction to the rodents. *In* Anderson and Jones (1967), pp. 206–209.

_____. 1969. *Macrotus waterhousii*. Mammalian Species, no. 1, 4 pp.

Anderson, S., and J. K. Jones, Jr., eds. 1967. Recent mammals of the world. A synopsis of families. Ronald Press, New York, viii + 453 pp.

Anderson, S. S.; R. W. Burton; and C. F. Summers. 1975. Behaviour of gray seals (*Halichoerus grypus*) during a breeding season at North Rona. J. Zool. 177:179–195.

Andreev, V. N. 1978. State of world reindeer breeding and its classification. Soviet J. Ecol. 8:291–295.

Andrews, P. 1978. Taxonomy and relationships of fossil apes. *In* Chivers and Joysey (1978), pp. 43–56.

Andrews, P., and C. P. Groves. 1976. Gibbons and brachiation. *In* Rumbaugh, D. M., ed., Gibbon and Siamang, IV, S. Karger, Basel, pp. 167–218.

Andrews, R. D., and E. K. Boggess. 1978. Ecology of coyotes in Iowa. *In* Bekoff (1978), pp. 249–265.

Angst, W. 1975. Basic data and concepts on the social organization of *Macaca fascicularis*. Primate Behav. 4:325–388.

Ansell, W. F. H. 1960. Mammals of Northern Rhodesia. Government Printer, Lusaka, xxxi + 155 + 24 pp.

_____. 1974. Some mammals from Zambia and adjacent countries. Occas. Pap. Natl. Parks and Wildl. Serv. Zambia, Suppl., no. 1, 48 pp.

_____. 1978. The mammals of Zambia. Natl. Parks and Wildl. Serv., Chilanga, Zambia, ii + 126 pp.

Ansell, W. F. H., and P. D. H. Ansell. 1973. Mammals of the north-eastern montane areas of Zambia. Puku 7:21–69.

Anthony, H. E. 1929. Two new genera of rodents from South America. Amer. Mus. Novit., no. 383, 6 pp.

Aranson, U. 1974. Comparative chromosome studies in Pinnipedia. Hereditas 76:179–225.

Arata, A. A. 1967. Muroid, gliroid, and dipodoid rodents. *In* Anderson and Jones (1967), pp. 226–253.

Arata, A. A., and J. B. Vaughan. 1970. Analyses of the relative abundance and reproductive activity of bats in southwestern Colombia. Caldasia 10:517–528.

Archer, M. 1974a. New information about the Quaternary distribution of the thylacine (Marsupialia, Thylacinidae) in Australia. J. Roy. Soc. W. Austral. 57:43–50.

_____. 1974b. Regurgitation or mercyism in the western native cat, *Dasyurus geoffroii*, and the red-tailed wambenger, *Phascogale calura* (Marsupialia, Dasyuridae). J. Mamm. 55:448–452.

_____. 1975. *Ningaui*, a new genus of tiny dasyurids (Marsupialia) and two species, *N. timealeyi* and *N. ridei*, from arid western Australia. Mem. Queensland Mus. 17:237–249.

_____. 1976. Revision of the marsupial genus *Planigale* Troughton (Dasyuridae). Mem. Queensland Mus. 17:341–365.

_____. 1977. Revision of the dasyurid marsupial genus *Antechinomys* Krefft. Mem. Queensland Mus. 18:17–29.

_____. 1979a. Two new species of *Sminthopsis* Thomas (Dasyuridae: Marsupialia) from northern Australia, *S. butleri* and *S. douglasi*. Austral. Zool. 20:327–345.

_____. 1979b. The status of Australian dasyurids, thylacinids and myrmecobiids. *In* Tyler (1979), pp. 29–43.

_____. 1981. Results of the Archbold Expeditions. No. 104. Systematic revision of the marsupial dasyurid genus *Sminthopsis* Thomas. Bull. Amer. Mus. Nat. Hist. 168:61–224.

Archer, M., and J. A. W. Kirsch. 1977. The case for the Thylacomyidae and Myrmecobiidae Gill, 1872, or why are marsupial families so extended? Proc. Linnean Soc. New South Wales 102:18–25.

Archer, M.; M. D. Plane; and N. S. Pledge. 1978. Additional evidence for interpreting the Miocene *Obdurodon insignis* Woodburne and Tedford, 1975, to be a fossil platypus (Ornithorhynchidae: Monotremata) and a reconsideration of the status of *Ornithorhynchus agilis* De Vis, 1885. Austral. Zool. 20:9–27.

Armitage, K. B. 1973. Population changes and social behavior following colonization by the yellow-bellied marmot. J. Mamm. 54:842–854.

_____. 1975. Social behavior and population dynamics of marmots. Oikos 26:341–354.

_____. 1977. Social variety in the yellow-bellied marmot: a population-behavioural system. Anim. Behav. 25:585–593.

Armitage, K. B., and J. F. Downhower. 1974. Demography of yellow-bellied marmot populations. Ecology 55:1233–1245.

Armstrong, D. M. 1972. Distribution of mammals in Colorado. Monogr. Mus. Nat. Hist. Univ. Kansas, no. 3, x + 415 pp.

Armstrong, D. M., and J. K. Jones, Jr. 1971. Mammals from the Mexican state of Sinaloa. I. Marsupialia, Insectivora, Edentata, Lagomorpha. J. Mamm. 52:747–757.

_____. 1972a. *Megasorex gigas*. Mammalian Species, no. 16, 2 pp.

_____. 1972b. *Notiosorex crawfordi*. Mammalian Species, no. 17, 5 pp.

Armstrong, D. M.; J. K. Jones, Jr.; and E. C. Birney. 1972. Mammals from the Mexican state of Sinaloa. III. Carnivora and Artiodactyla. J. Mamm. 53:48–61.

Armstrong, J. J. 1979. The California sea otter: emerging conflicts in resource management. San Diego Law Rev. 16:249–285.

Arndt, R. G.; F. C. Rohde; and J. A. Bosworth. 1978. Additional records of the rice rat, *Oryzomys palustris* (Harlan), from New Jersey and Delaware. Bull. New Jersey Acad. Sci. 23:65–72.

Arundel, J. H.; I. K. Barker; and I. Beveridge. 1977. Diseases of marsupials. *In* Stonehouse and Gilmore (1977), pp. 141–154.

Asdell, S. A. 1964. Patterns of mammalian reproduction. Cornell Univ. Press, Ithaca, viii + 670 pp.

Ashby, K. R. 1967. Studies on the ecology of field mice and voles (*Apodemus sylvaticus*, *Clethrionomys glareolus* and *Microtus agrestis*) in Houghall Wood, Durham. J. Zool. 152:389–513.

Asibey, E. O. A. 1974a. The grasscutter, *Thryonomys swinderianus* Temminck, Ghana. Symp. Zool. Soc. London 34:161–170.

_____. 1974b. Reproduction in the grasscutter (*Thryonomys swinderianus* Temminck) in Ghana. Symp. Zool. Soc. London 34:251–263.

Aslin, H. J. 1974. The behaviour of *Dasyuroides byrnei* (Marsupialia) in captivity. Z. Tierpsychol. 35:187–208.

_____. 1975. Reproduction in *Antechinus maculatus* Gould (Dasyuridae). Austral. Wildl. Res. 2:77–80.

_____. 1977. New records of *Sminthopsis ooldea* Troughton from South Australia. S. Austral. Nat. 52:9–12.

Aslin, H. J., and C. H. S. Watts. 1980. Breeding of a captive colony of *Notomys fuscus* Wood Jones (Rodentia: Muridae). Austral. Wildl. Res. 7:379–383.

Atallah, S. I. 1977. Mammals of the eastern Mediterranean region; their ecology, systematics and zoogeographical relationships. Saugetierk. Mitt. 25:241–320.

Augee, M. L. 1978. Monotremes and the evolution of homeothermy. Austral. Zool. 20:111–119.

Augee, M. L.; T. J. Bergin; and C. Morris. 1978. Observations on behaviour of echidnas at Taronga Zoo. Austral. Zool. 20:121–129.

Augee, M. L.; E. H. M. Ealey; and I. P. Price. 1975. Movements of echidnas, *Tachyglossus aculeatus*, determined by marking-recapture and radio-tracking. Austral. Wildl. Res. 2:93–101.

Aung, H. 1968. A note on the birth of a Mishmi takin *Budorcas t. taxicolor* at Rangoon Zoo. Int. Zoo Yearbook 8:145.

Australian National Parks and Wildlife Service. 1977, 1978. Australian endangered species. Mammals. Nos. 1–23.

Autenrieth, R. E., and E. Fichter. 1975. On the behavior and socialization of pronghorn fawns. Wildl. Monogr., no. 42, 111 pp.

Avila-Pires, F. D. de. 1960. Um novo gênero de roedor Sul-Americano (Rodentia-Cricetidae). Bol. Mus. Nac. Rio de Janeiro (Zool.), n.s., no. 220, 6 pp.

————. 1967. The type-locality of "*Chaetomys subspinosus*" (Olfers, 1818) (Rodentia, Caviomorpha). Rev. Brasil. Biol. 27:177–179.

————. 1972. A new subspecies of *Kunsia fronto* (Winge, 1888) from Brazil (Rodentia, Cricetidae). Rev. Brasil. Biol. 32:419–422.

Avila-Pires, F. D. de, and M. R. Caldatto Wutke. 1981. Taxonómia e evolução de *Clyomys* Thomas, 1916 (Rodentia, Echimyidae). Rev. Brasil. Biol. 41:529–534.

Ayala, S. C., and A. D'Alessandro. 1973. Insect feeding of some Colombian fruit-eating bats. J. Mamm. 54:266–267.

B.

Baar, S. L.; E. D. Fleharty; and M. F. Artman. 1975. Utilization of deep burrows and nests by cotton rats in west-central Kansas. Southwestern Nat. 19:440–444.

Backhaus, D. 1964. Zum Verhalten des nördlichen Breitmaulnashornes (*Diceros simus cottoni* Lydekker, 1908). Zool. Garten 29:93–107.

Bailey, T. N. 1974. Social organization in a bobcat population. J. Wildl. Mgmt. 38:435–446.

Bailey, V. 1930. Animal life of Yellowstone National Park. Charles C Thomas, Springfield, Illinois, 241 pp.

Bain, J. R. 1978. The breeding system of *Nycticeius humeralis* in Florida. Amer. Soc. Mamm., Abstr. Tech. Pap., 58th Ann. Mtg., p. 29.

Baker, A. N. 1977. Spectacled porpoise, *Phocoena dioptrica*, new to the subantarctic Pacific Ocean. New Zealand J. Mar. Freshwater Res. 11:401–406.

————. 1978. The status of Hector's dolphin, *Cephalorhynchus hectori* (Van Beneden), in New Zealand waters. Rept. Int. Whaling Comm. 28:331–334.

Baker, H. G., and B. J. Harris. 1957. The pollination of *Parkia* by bats and its attendant evolutionary problems. Evolution 11:449–460.

Baker, R. C.; F. Wilke; and C. H. Baltzo. 1970. The northern fur seal. U.S. Dept. Interior Circ., no. 336, iii + 19 pp.

Baker, R. H. 1968. Habitats and distribution. *In* King (1968), pp. 98–121.

————. 1974. Records of mammals from Ecuador. Michigan State Univ. Mus. Publ., Biol. Ser., 5:129–146.

————. 1977. Mammals of the Chihuahuan Desert region—future prospects. *In* Wauer, R. H., and D. H. Riskind, eds., Transactions of the symposium on the biological resources of the Chihuahuan Desert region, United States and Mexico, U.S. Natl. Park Serv. Trans. Proc. Ser., no. 3, pp. 221–225.

Baker, R. H., and J. K. Greer. 1962. Mammals of the Mexican state of Durango. Michigan State Univ. Mus. Publ., Biol. Ser., 2:25–154.

Baker, R. H., and M. K. Petersen. 1965. Notes on a climbing rat, *Tylomys*, from Oaxaca, Mexico. J. Mamm. 46:694–695.

Baker, R. J.; P. V. August; and A. A. Steuter. 1978. *Erophylla sezekorni*. Mammalian Species, no. 115, 5 pp.

Baker, R. J.; H. H. Genoways; and J. C. Patton. 1978. Bats of Guadeloupe. Occas. Pap. Mus. Texas Tech Univ., no. 50, 16 pp.

Baker, R. J., and J. K. Jones, Jr. 1975. Additional records of bats from Nicaragua, with a revised checklist of Chiroptera. Occas. Pap. Mus. Texas Tech Univ., no. 32, 13 pp.

Baker, R. J.; J. K. Jones, Jr.; and D. C. Carter, eds. 1976. Biology of bats of the New World family Phyllostomatidae. Part I. Spec. Publ. Mus. Texas Tech Univ., no. 10, 218 pp.

————. 1977. Biology of bats of the New World family Phyllostomatidae. Part II. Spec. Publ. Mus. Texas Tech Univ., no. 13, 364 pp.

————. 1979. Biology of bats of the New World family Phyllostomatidae. Part III. Spec. Publ. Mus. Texas Tech Univ., no. 16, 441 pp.

Baker, R. J., and S. L. Williams. 1974. *Geomys tropicalis*. Mammalian Species, no. 35, 4 pp.

Baldwin, J. D. 1970. Reproductive synchronization in squirrel monkeys (*Saimiri*). Primates 11:317–326.

Baldwin, J. D., and J. I. Baldwin. 1971. Squirrel monkeys (*Saimiri*) in natural habitats in Panama, Colombia, Brazil, and Peru. Primates 12:45–61.

————. 1972. Population density and use of space in howling monkeys (*Alouatta villosa*) in south-western Panama. Primates 13:371–379.

————. 1976. Primate populations in Chiriqui, Panama. *In* Thorington and Heltne (1976), pp. 20–31.

Banfield, A. W. F. 1961. A revision of the reindeer and caribou, genus *Rangifer*. Natl. Mus. Canada Bull., no. 177, vi + 137 pp.

————. 1974. The mammals of Canada. Univ. Toronto Press, xxv + 438 pp.

Banholzer, U. 1976. Water balance, metabolism, and heart rate in the fennec. Naturwissenschaften 63:202.

Banks, E. 1978. Mammals from Borneo. Brunei Mus. J. 4:165–227.

Banks, E. M.; R. J. Brooks; and J. Schnell. 1975. A radiotracking study of home range and activity of the brown lemming (*Lemmus trimucronatus*). J. Mamm. 56:888–901.

Banks, R. C., and R. L. Brownell. 1969. Taxonomy of the common dolphins of the eastern Pacific Ocean. J. Mamm. 50:262–271.

Bannikov, A. G.; L. V. Zhirnov; L. S. Lebedeva; and A. A. Fandeev. 1967. Biology of the saiga. Israel Progr. Sci. Transl., Jerusalem, iv + 252 pp.

Barash, D. P. 1973a. Social variety in the yellow-bellied marmot (*Marmota flaviventris*). Anim. Behav. 21:579–584.

————. 1973b. The social biology of the Olympic marmot. Anim. Behav. Monogr. 6:173–244.

————. 1973c. Territorial and foraging behavior of pika (*Ochotona princeps*) in Montana. Amer. Midl. Nat. 89:202–207.

————. 1974a. The social behaviour of the hoary marmot (*Marmota caligata*). Anim. Behav. 22:256–261.

————. 1974b. Mother-infant relations in captive woodchucks (*Marmota monax*). Anim. Behav. 22:446–448.

————. 1976. Social behaviour and individual differences in free-living Alpine marmots (*Marmota marmota*). Anim. Behav. 24:27–35.

Barbehenn, K. R.; J. P. Sumangil; and J. L. Libay. 1972–1973. Rodents of the Philippine croplands. Philippine Agriculturist 56:217–242.

Barbour, R. W., and W. H. Davis. 1969. Bats of America. University Press of Kentucky, Lexington, 286 pp.

————. 1974. Mammals of Kentucky. University Press of Kentucky, Lexington, xii + 322 pp.

Barkalow, F. S., Jr., and R. F. Soots, Jr. 1975. Life span and reproductive longevity of the gray squirrel, *Sciurus c. carolinensis* Gmelin. J. Mamm. 56:522–524.

Barker, I. K.; I. Beveridge; A. J. Bradley; and A. K. Lee. 1978. Observations on spontaneous stress-related mortality among males of the dasyurid marsupial *Antechinus stuartii* Macleay. Austral. J. Zool. 26:435–447.

Barlow, J. C. 1967. Edentates and pholidotes. *In* Anderson and Jones (1967), pp. 178–191.

————. 1969. Observations on the biology of rodents in Uruguay. Roy. Ontario Mus. Life Sci. Contrib., no. 75, 59 pp.

Barnett, J. L.; A. A. How; and W. F. Humphreys. 1977. Small mammal populations in pine and native forests in north-eastern New South Wales. Austral. Wildl. Res. 4:233–240.

Barquez, R. M., and R. A. Ojeda. 1979. Nueva subespecie de *Phylloderma* (Chiroptera Phyllostomidae). Neotrópica 25:83–89.

Barquez, R. M., and C. C. Olrog. 1980. Tres nuevas especies de *Vampyrops* para Bolivia. Neotrópica 26:53–56.

Barrette, C. 1977. Some aspects of the behaviour of muntjacs in Wilpattu National Park. Mammalia 41:1–34.

Barritt, M. K. 1978. Two further specimens of the little pigmy possum [*Cercartetus lepidus* (Thomas)] from the Australian mainland. S. Austral. Nat. 53:12–13.

Barzdo, J. 1980. International trade in harp and hooded seals. Oryx 15:275–279.

Bateman, G. C., and T. A. Vaughan. 1974. Nightly activities of mormoopid bats. J. Mamm. 55:45–65.

Batten, P., and A. Batten. 1966. Notes on breeding the small-toothed palm civet *Arctogalidia trivirgata* at Santa Cruz Zoo. Int. Zoo Yearbook 6:172–173.

Battistini, R., and G. Richard-Vindard, eds. 1972. Biogeography and ecology in Madagascar. Dr. W. Junk B. V., The Hague. xv + 765 pp.

Baud, F. J. 1979. *Myotis aelleni*, nov. spec., chauve-souris nouvelle d'Argentine (Chiroptera: Vespertilionidae). Rev. Suisse Zool. 86:267–278.

Baumgardner, G. D., and D. J. Schmidly. 1981. Systematics of the southern races of two species of kangaroo rats (*Dipodomys compactus* and *D. ordii*). Occas. Pap. Mus. Texas Tech Univ., no. 73, 27 pp.

Baverstock, P. R.; C. H. S. Watts; M. Adams; and M. Gelder. 1980. Chromosomal and electrophoretic studies of Australian *Melomys* (Rodentia: Muridae). Austral. J. Zool. 28:553–574.

Baverstock, P. R.; C. H. S. Watts; and S. R. Cole. 1977. Electrophoretic comparisons between allopatric populations of five Australian pseudomyine rodents (Muridae). Austral. J. Biol. Sci. 30:471–485.

Baverstock, P. R.; C. H. S. Watts; J. T. Hogarth; A. C. Robinson; and J. F. Robinson. 1977. Chromosome evolution in Australian rodents. II. The *Rattus* group. Chromosoma 61:227–241.

Beamish, P. 1978. Evidence that a captive humpback whale (*Megaptera novaengliae*) does not use sonar. Deep-Sea Res. 25:469–472.

Beatley, J. 1976. Environments of kangaroo rats (*Dipodomys*) and effects of environmental change on populations in southern Nevada. J. Mamm. 57:67–93.

Beck, A. J., and Lim Boo Liat. 1973. Reproductive biology of *Eonycteris spelaea*, Dobson (*Megachiroptera*) in West Malaysia. Acta Tropica 30:251–260.

Beck, A. M. 1973. The ecology of stray dogs. A study of free-ranging urban animals. York Press, Baltimore. xiv + 98 pp.

———. 1975. The ecology of "feral" and free-roving dogs in Baltimore. *In* Fox (1975), pp. 380–390.

Beck, B.; T. G. Smith; and A. W. Mansfield. 1970. Occurrence of the harbour seal, *Phoca vitulina*, Linnaeus in the Thlewiaza River, N. W. T. Can. Field-Nat. 84:297–300.

Bee, J. W., and E. R. Hall. 1956. Mammals of northern Alaska on the Arctic Slope. Univ. Kansas Mus. Nat. Hist. Misc. Publ., no. 8, 309 pp.

Beebe, B. F. 1978. Two new Pleistocene

mammal species from Beringia. Amer. Quaternary Assoc. Abstr., 5th Bien. Mtg., p. 159.

Beg, M. A., and R. S. Hoffmann. 1977. Age determination and variation in the red-tailed chipmunk, *Eutamias ruficaudus*. Murrelet 58:26–36.

Begg, R. J., and C. R. Dunlop. 1980. Security eating, and diet in the large rock-rat, *Zyzomys woodwardi* (Rodentia: Muridae). Austral. Wildl. Res. 7:63–70.

Bekenov, A., and J. Mirzabekov. 1977. Reproduction of *Allactaga elater* in north Kyzylkum and Ustyurt. Zool. Zhur. 56:769–778.

Bekoff, M. 1975. Social behavior and ecology of the African Canidae: a review. *In* Fox (1975), pp. 120–142.

———. 1977. *Canis latrans*. Mammalian Species, no. 79, 9 pp.

———, ed. 1978. Coyotes. Biology, behavior, and management. Academic Press, New York, xx + 384 pp.

Bekoff, M., and M. C. Wells. 1980. The social ecology of coyotes. Sci. Amer. 242(4):130–148.

Belden, R. C., and D. J. Forrester. 1980. A specimen of *Felis concolor coryi* from Florida. J. Mamm. 61:160–161.

Bengtson, J. L., and D. B. Siniff. 1981. Reproductive aspects of female crabeater seals (*Lobodon carcinophagus*) along the Antarctic Peninsula. Can. J. Zool. 59:92–102.

Benjaminsen, T., and I. Christensen. 1979. The natural history of the bottlenose whale, *Hyperoodon ampullatus* (Forster). *In* Winn and Olla (1979), pp. 143–164.

Bennett, D. K. 1980. Stripes do not a zebra make, part I: a cladistic analysis of *Equus*. Syst. Zool. 29:272–287.

Benson, D. L., and F. R. Gehlbach. 1979. Ecological and taxonomic notes on the rice rat (*Oryzomys couesi*) in Texas. J. Mamm. 60:225–228.

Bere, R. M. 1962. The wild mammals of Uganda and neighbouring regions of East Africa. Longmans, Green, & Co., London, xii + 148 pp.

Berg, W. E., and R. A. Chesness. 1978. Ecology of coyotes in northern Minnesota. *In* Bekoff (1978), pp. 229–247.

Bergerud, A. T. 1967. The distribution and abundance of arctic hares in Newfoundland. Can. Field-Nat. 81:242–248.

———. 1974. Decline of caribou in North America following settlement. J. Wildl. Mgmt. 38:757–770.

———. 1978. Caribou. *In* Schmidt and Gilbert (1978), pp. 83–101.

Bergmans, W. 1973. New data on the rare African fruit bat *Scotonycteris ophiodon* Pohle, 1943. Z. Saugetierk. 38:285–289.

———. 1975a. A new species of *Dobsonia* Palmer, 1898 (Mammalia, Megachiroptera) from Waigeo, with notes on other members of the genus. Beaufortia 23:1–13.

———. 1975b. On the differences between sympatric *Epomops franqueti* (Tomes, 1860) and *Epomops buettikoferi* (Matschie, 1899), with additional notes on the latter species (Mammalia, Megachiroptera). Beaufortia 23:141–152.

———. 1976. A revision of the African genus *Myonycteris* Matschie, 1899 (Mammalia, Megachiroptera). Beaufortia 24:189–216.

———. 1977. Notes on new material of *Rousettus madagascariensis* Grandidier, 1929 (Mammalia, Megachiroptera). Mammalia 41:67–74.

———. 1978a. On *Dobsonia* Palmer 1898 from the Lesser Sunda Islands (Mammalia: Megachiroptera). Senckenbergiana Biol. 59:1–18.

———. 1978b. Rediscovery of *Epomophorus pousarguesi* Trouessart, 1904 in the Central African Empire (Mammalia: Megachiroptera). J. Nat. Hist. 12:681–687.

———. 1979a. Taxonomy and zoogeography of the fruit bats of the People's Republic of Congo, with notes on their reproductive biology (Mammalia: Megachiroptera). Bijdragen Tot de Dierkunde 48:161–186.

———. 1979b. Taxonomy and zoogeography of *Dobsonia* Palmer, 1898, from the Louisiade Archipelago, the D'Entrecasteaux Group, Trobriand Island and Woodlark Island (Mammalia: Megachiroptera). Beaufortia 29:199–214.

———. 1979c. First records of *Epomops dobsonii* (Bocage, 1889) from Tanzania and Rwanda, with a note on its size range (Mammalia: Megachiroptera). Z. Saugetierk. 44:240–241.

———. 1980. A new fruit bat of the genus *Myonycteris* Matschie, 1899, from eastern Kenya and Tanzania (Mammalia, Megachiroptera). Zool. Meded. 55:171–181.

Bergmans, W.; L. Bellier; and J. Vissault. 1974. A taxonomical report on a collection of Megachiroptera (Mammalia) from the Ivory Coast. Rev. Zool. Afr. 88:18–48.

Bergmans, W., and J. E. Hill. 1980. On a new species of *Rousettus* Gray, 1821, from Sumatra and Borneo (Mammalia: Megachiroptera). Bull. British Mus. (Nat. Hist.), Zool. 38:95–104.

Bergmans, W., and P. J. H. Van Bree. 1972. The taxonomy of the African bat *Megaloglossus woermanni* Pagenstecher, 1885 (Megachiroptera: Macroglossinae). Biol. Gabonica 3–4:291–299.

Berkes, F. 1977. Turkish dolphin fisheries. Oryx 14:163–167.

Berland, B. 1966. The hood and its extrusible balloon in the hooded seal—*Cystophora cristata* ERXL. Norsk Polarinst. Arbok, 1965, pp. 95–102.

Berry, P. S. M. 1973. The Luangwa Valley giraffe. Puku 7:71–92.

———. 1978. Range movements of giraffe in the Luangwa Valley, Zambia. E. Afr. Wildl. J. 16:77–83.

Berry, R. J. 1970. The natural history of the house mouse. Field Studies 3:219–262.

————, ed. 1981. Biology of the house mouse. Symp. Zool. Soc. London, no. 47, xxx + 715 pp.

Bertram, B. C. R. 1975. Social factors influencing reproduction in wild lions. J. Zool. 177:463–482.

Bertram, G. C. L. 1940. The biology of the Weddell and crabeater seal, with a study of the comparative behaviour of the Pinnipedia. British Graham Land Exped., 1934–1937, Sci. Rept., British Mus. (Nat. Hist.), 1:1–139.

Berzin, A. A. 1972. The sperm whale. Israel Progr. Sci. Transl., Jerusalem, v + 394 pp.

————. 1978. Whale distribution in tropical eastern Pacific waters. Rept. Int. Whaling Comm. 28:173–177.

Best, P. B. 1968. The sperm whale (Physeter catodon) off the west coast of South Africa. 2. Reproduction in the female. S. Afr. Div. Sea Fish. Investig. Rept., no. 66, 32 pp.

————. 1970a. The sperm whale (Physeter catodon) off the west coast of South Africa. 5. Age, growth and mortality. S. Afr. Div. Sea Fish. Investig. Rept., no. 79, 27 pp.

————. 1970b. Exploitation and recovery of right whales Eubalaena australis off the Cape Province. S. Afr. Div. Sea Fish. Investig. Rept., no. 80, 20 pp.

————. 1974. The biology of the sperm whale as it relates to stock management. In Schevill (1974), pp. 257–293.

————. 1979. Social organization in sperm whales, Physeter macrocephalus. In Winn and Olla (1979), pp. 227–289.

Best, P. B., and P. D. Shaughnessy. 1981. First record of the melon-headed whale Peponocephala electra from South Africa. Ann. S. Afr. Mus. 83:33–47.

Best, R. C., and H. D. Fisher. 1974. Seasonal breeding of the narwhal (Monodon monoceros L.). Can. J. Zool. 52:429–431.

Best, T. L. 1972. Mound development by a pioneer population of the banner-tailed kangaroo rat, Dipodomys spectabilis baileyi Goldman in eastern New Mexico. Amer. Midl. Nat. 87:201–206.

————. 1978. Variation in kangaroo rats (genus Dipodomys) of the heermanni group in Baja California, Mexico. J. Mamm. 59:160–175.

Best, T. L., and E. B. Hart. 1976. Swimming ability of pocket gophers (Geomyidae). Texas J. Sci. 27:361–366.

Betts, B. J. 1976. Behaviour in a population of Columbian ground squirrels, Spermophilus columbianus columbianus. Anim. Behav. 24:652–680.

Bianchi, N. O.; O. A. Reig; O. J. Molina; and F. N. Dulout. 1971. Cytogenetics of the South American akodont rodents (Cricetidae). I. A progress report of Argentinian and Venezuelan forms. Evolution 25:724–736.

Bibikov, D. I. 1980. Wolves in the USSR. Nat. Hist. 89(6):58–63.

Bigalke, R. C. 1970. Observations on springbok populations. Zool. Afr. 5:59–70.

Bigg, M. A. 1969. The harbour seal in British Columbia. Fish. Res. Bd. Can. Bull., no. 172, vii + 33 pp.

————. 1973. Census of California sea lions on southern Vancouver Island, British Columbia. J. Mamm. 54:285–287.

Bigg, M. A., and A. A. Wolman. 1975. Live-capture killer whale (Orcinus orca) fishery, British Columbia and Washington, 1962–73. J. Fish. Res. Bd. Can. 32:1213–1221.

Bigler, W. J. 1974. Seasonal movements and activity patterns of the collared peccary. J. Mamm. 55:851–855.

Bindra, O. S., and P. Sagar. 1968. Breeding habits of the field rat Millardia meltada. J. Bombay Nat. Hist. Soc. 65:477–481.

Birkenholz, D. E. 1963. A study of the life history and ecology of the round-tailed muskrat (Neofiber alleni True) in north-central Florida. Ecol. Monogr. 33:255–280.

————. 1972. Neofiber alleni. Mammalian Species, no. 15, 4 pp.

Birkenholz, D. E., and W. O. Wirtz, II. 1965. Laboratory observations on the vesper rat. J. Mamm. 46:181–189.

Birkenstock, P. J., and J. A. J. Nel. 1977. Laboratory and field observations on Zelotomys woosnami (Rodentia: Muridae). Zool. Afr. 12:429–443.

Birney, E. C. 1973. Systematics of three species of woodrats (genus Neotoma) in central North America. Univ. Kansas Mus. Nat. Hist. Misc. Publ., no. 58, 1973 pp.

————. 1974. Twentieth century records of wolverine in Minnesota. Loon 46:78–81.

Birney, E. C.; J. B. Bowles; R. M. Timm; and S. L. Williams. 1974. Mammalian distributional records in Yucatan and Quintana Roo, with comments on reproduction, structure, and status of peninsular populations. Occas. Pap. Bell Mus. Nat. Hist., no. 13, 25 pp.

Bishop, I. R. 1979. Notes on Praomys (Hylomyscus) in eastern Africa. Mammalia 43:519–530.

Bjorge, R. R.; J. R. Gunson; and W. M. Samuel. 1981. Population characteristics and movements of striped skunks (Mephitis mephitis) in central Alberta. Can. Field-Nat. 95:149–155.

Blanchard, H. M. 1974. The marten makes a comeback. Maine Fish and Game 16(4):21.

Bland, K. P. 1973. Reproduction in the female African tree rat (Grammomys surdaster). J. Zool. 171:167–175.

Bleich, V. C. 1977. Dipodomys stephensi. Mammalian Species, no. 73, 3 pp.

————. 1979. Microtus californicus scirpensis not extinct. J. Mamm. 60:851–852.

Bleich, V. C., and O. A. Schwartz. 1975. Observations on the home range of the desert woodrat, Neotoma lepida intermedia. J. Mamm. 56:518–519.

Bloxam, Q. 1977. Breeding the spectacled bear Tremarctos ornatus at Jersey Zoo. Int. Zoo Yearbook 17:158–161.

Blus, L. J. 1971. Reproduction and survival

of short-tailed shrews (Blarina brevicauda) in captivity. Lab. Anim. Sci. 21:884–891.

Bockstoce, J. 1980. Battle of the bowheads. Nat. Hist. 89(5):52–61.

Bogan, M. A. 1972. Observations on parturition and development in the hoary bat, Lasiurus cinereus. J. Mamm. 53:611–614.

————. 1978. A new species of Myotis from the Islas Tres Marias, Nayarit, Mexico, with comments on variation in Myotis nigricans. J. Mamm. 59:519–530.

Bogart, M. H.; R. W. Cooper; and K. Benirschke. 1977. Reproductive studies of black and ruffed lemurs Lemur macaco macaco and L. variegatus. Int. Zoo Yearbook 17:177–182.

Boggess, E. K.; F. R. Henderson; and J. R. Choate. 1980. A black-footed ferret from Kansas. J. Mamm. 61:571.

Bollinger, A., and T. C. Backhouse. 1960. Blood studies on the echidna Tachyglossus aculeatus. Proc. Zool. Soc. London 135:91–97.

Bolton, B. L., and P. K. Latz. 1978. The western hare-wallaby, Lagorchestes hirsutus (Gould) (Macropodidae), in the Tanami Desert. Austral. Wildl. Res. 5:285–293.

Bonaccorso, F. J., and J. H. Brown. 1972. House construction of the desert woodrat, Neotoma lepida lepida. J. Mamm. 53:283–288.

Bonner, W. N. 1968. The fur seal of South Georgia. British Antarctic Surv. Sci. Rept., no. 56, 81 pp.

————. 1972. The grey seal and common seal in European waters. Oceanogr. Mar. Biol. Ann. Rev. 10:461–507.

Boone and Crockett Club, ed. 1975. The wild sheep in modern North America. Winchester Press, New York, xv + 302 pp.

Boonstra, R., and C. J. Krebs. 1978. Pitfall trapping of Microtus townsendii. J. Mamm. 59:136–148.

Borsboom, A. C. 1975. Pseudomys gracilicaudatus (Gould): range extension and notes on a little-known Australian murid rodent. Austral. Wildl. Res. 2:81–84.

Boskoff, K. J. 1977. Aspects of reproduction in ruffed lemurs (Lemur variegatus). Folia Primatol. 28:241–250.

Bothma, J. D. P. 1966. Food of the silver fox Vulpes chama. Zool. Afr. 2:205–210.

————. 1971a. Control and ecology of the black-backed jackal Canis mesomelas in the Transvaal. Zool. Afr. 6:187–193.

————. 1971b. Food of Canis mesomelas in South Africa. Zool. Afr. 6:195–203.

Bothma, J. D. P., and J. A. J. Nel. 1980. Winter food and foraging behaviour of the aardwolf Proteles cristatus in the Namib-Naukluft Park. Madoqua 12:141–149.

Bouchard, R., and G. Moisan. 1974. Chasse contrôlée á l'original dans les parcs et réserves du Québec (1962–1972). Naturaliste Canadien 101:689–704.

Boulenge, E. L., and E. R. Fuentes. 1978. Preliminary population dynamics of Octodon *degus*. Abstr. 2nd Congr. Theriol. Int., Brno, p. 180.

Boulva, J., and I. A. McLaren. 1979. Biology of the harbor seal, *Phoca vitulina*, in eastern Canada. Fish. Res. Bd. Can. Bull., no. 200, viii + 24 pp.

Bourliere, F.; C. Hunkeler; and M. Bertrand. 1970. Ecology and behavior of Lowe's guenon (*Cercopithecus campbelli lowei*) in the Ivory Coast. *In* Napier and Napier (1970), pp. 297–350.

Bowen, D. W., and R. J. Brooks. 1978. Social organization of confined male collared lemmings (*Dicrostonyx groenlandicus* Traill). Anim. Behav. 26:1126–1135.

Bowen, W. W. 1968. Variation and evolution of Gulf Coast populations of beach mice, *Peromyscus polionotus*. Bull. Florida State Mus. 12:1–91.

Bower, J. N. 1979. Père David's deer: the trek from extinction. Natl. Parks and Conserv. Mag. 53(4):20–23.

Bowles, J. B. 1972. Notes on reproduction in four species of bats from Yucatan, Mexico. Trans. Kansas Acad. Sci. 75:271–272.

Boyd, R. J. 1978. American elk. *In* Schmidt and Gilbert (1978), pp. 11–29.

Bradbury, J. W. 1977. Lek mating behavior in the hammer-headed bat. Z. Tierpsychol. 45:225–255.

Bradbury, J. W., and L. H. Emmons. 1974. Social organization of some Trinidad bats. I. Emballonuridae. Z. Tierpsychol. 36:137–183.

Bradbury, J. W., and S. L. Vehrencamp. 1977. Social organization and foraging in emballonurid bats. I. Field studies. Behav. Ecol. Sociobiol. 1:337–381.

Bradley, W. G., and R. A. Mauer. 1971. Reproduction and food habits of Merriam's kangaroo rat, *Dipodomys merriami*. J. Mamm. 52:497–507.

Brady, C. A. 1978. Reproduction, growth and parental care in crab-eating foxes *Cerdocyon thous*. Int. Zoo Yearbook 18:130–134.

⸻. 1979. Observations on the behavior and ecology of the crab-eating fox (*Cerdocyon thous*). *In* Eisenberg (1979), pp. 161–171.

Brady, C. A., and M. K. Ditton. 1979. Management and breeding of maned wolves *Chrysocyon brachyurus* at the National Zoological Park, Washington. Int. Zoo Yearbook 19:171–176.

Braham, H. W.; R. D. Everitt; and D. J. Rugh. 1980. Northern sea lion population decline in the eastern Aleutian Islands. J. Wildl. Mgmt. 44:25–33.

Braithwaite, R. W. 1980. The ecology of *Rattus lutreolus*. III. The rise and fall of a commensal population. Austral. Wildl. Res. 7:199–215.

Bramblett, C. A.; L. D. Pejaver; and D. J. Drickman. 1975. Reproduction in captive vervet and Sykes' monkeys. J. Mamm. 56:940–946.

Brand, C. J., and L. B. Keith. 1979. Lynx demography during a snowshoe hare decline in Alberta. J. Wildl. Mgmt. 43:827–849.

Brand, C. J.; L. B. Keith; and C. A. Fischer. 1976. Lynx responses to changing showshoe hare densities in central Alberta. J. Wildl. Mgmt. 40:416–428.

Brand, C. J.; R. H. Vowles; and L. B. Keith. 1975. Snowshoe hare mortality monitored by telemetry. J. Wildl. Mgmt. 39:741–747.

Brand, L. R. 1974. Tree nests of California chipmunks (*Eutamias*). Amer. Midl. Nat. 91:489–491.

⸻. 1976. The vocal repertoire of chipmunks (genus *Eutamias*) in California. Anim. Behav. 24:319–335.

Brander, R. B. 1971. Longevity of wild porcupines. J. Mamm. 52:835.

⸻. 1973. Life-history notes on the porcupine in a hardwood-hemlock forest in upper Michigan. Michigan Academician 5:425–433.

Brattstrom, B. H. 1973. Social and maintenance behavior of the echidna, *Tachyglossus aculeatus*. J. Mamm. 54:50–70.

Briggs, K. T., and C. W. Davis. 1972. A study of predation by sea lions on salmon in Monterey Bay. California Fish and Game 58:37–43.

Broadbooks, H. E. 1965. Ecology and distribution of the pikas of Washington and Alaska. Amer. Midl. Nat. 73:299–335.

⸻. 1970*a*. Home ranges and territorial behavior of the yellow-pine chipmunk, *Eutamias amoenus*. J. Mamm. 51:310–326.

⸻. 1970*b*. Populations of the yellow-pine chipmunk, *Eutamias amoenus*. Amer. Midl. Nat. 83:472–488.

⸻. 1974. Tree nests of chipmunks with comments on associated behavior and ecology. J. Mamm. 54:630–639.

Brockie, R. 1974. Self-annointing by wild hedgehogs, *Erinaceus europaeus*, in New Zealand. Anim. Behav. 24:68–71.

Brodie, E. D., Jr. 1977. Hedgehogs use toad venom in their own defense. Nature 268:627–628.

Brodie, P. F. 1971. A reconsideration of aspects of growth, reproduction, and behavior of the white whale (*Delphinapterus leucas*), with reference to the Cumberland Sound, Baffin Island, population. J. Fish. Res. Bd. Can. 28:1309–1318.

Bronson, F. H. 1979. The reproductive ecology of the house mouse. Quart. Rev. Biol. 54:265–299.

Brooks, P. M. 1972. Post-natal development of the African bush rat. Zool. Afr. 7:85–102.

Brooks, R. J., and E. M. Banks. 1971. Radio-tracking study of lemming home range. Commun. Behav. Biol. 6:1–5.

⸻. 1973. Behavioural biology of the collared lemming [*Dicrostonyx groenlandicus* (Traill)]: an analysis of acoustic communication. Anim. Behav. Monogr. 6:1–83.

Brosset, A. 1962. La reproduction des chiroptères de l'ouest et du centre de l'Inde. Mammalia 26:176–213.

⸻. 1976. Social organization in the African bat, *Myotis boccagei*. Z. Tierpsychol. 42:50–56.

Brosset, A.; G. Dubost; and H. Heim de Balsac. 1965. Mammifères inédits récoltés au Gabon. Biol. Gabonica 1:147–174.

Brothwell, D. 1981. The Pleistocene and Holocene archaeology of the house mouse and related species. *In* Berry (1981), pp. 1–13.

Brown, D. H.; D. K. Caldwell; and M. C. Caldwell. 1966. Observations on the behavior of wild and captive false killer whales, with notes on associated behavior of other genera of captive delphinids. Los Angeles Co. Mus. Contrib. Sci., no. 95, 32 pp.

Brown, J. C. 1964. Observations on the elephant shrews (Macroscelididae) of equatorial Africa. Proc. Zool. Soc. London 143:103–119.

Brown, J. H., and R. C. Lasiewski. 1972. Metabolism of weasels: the cost of being long and thin. Ecology 53:939–943.

Brown, L. N. 1970. Population dynamics of the western jumping mouse (*Zapus princeps*) during a four-year study. J. Mamm. 51:651–658.

⸻. 1975. Ecological relationships and breeding of the nutria (*Myocastor coypus*) in the Tampa, Florida area. J. Mamm. 56:928–930.

⸻. 1977. Litter size and notes on reproduction in the giant water vole (*Arvicola richardsoni*). Southwestern Nat. 22:281–282.

Brown, M. K. 1979. Two old beavers from the Adirondacks. New York Fish and Game J. 26:92.

Brown, P. 1976. Vocal communication in the pallid bat, *Antrozous pallidus*. Z. Tierpsychol. 41:34–54.

Brown, R. 1973. Has the thylacine really vanished? Animals 15:416–419.

Brown, R. E. 1978. First Iranian record of *Dipus sagitta* (Pallas, 1773). Mammalia 42:257.

Brownell, R. L., Jr. 1975*a*. Progress report on the biology of the Franciscana dolphin, *Pontoporia blainvillei*, in Uruguayan waters. J. Fish. Res. Bd. Can. 32:1073–1078.

⸻. 1975*b*. *Phocoena dioptrica*. Mammalian Species, no. 66, 3 pp.

Brownell, R. L., Jr.; A. Aguayo L.; and D. Torres N. 1976. A Shepherd's beaked whale, *Tasmacetus shepherdi*, from the eastern South Pacific. Sci. Rept. Whales Res. Inst. 28:127–128.

Brownell, R. L., Jr., and C. Chun. 1977. Probable existence of the Korean stock of the gray whale (*Eschrichtius robustus*). J. Mamm. 58:237–239.

Brownell, R. L., Jr., and E. S. Herald. 1972. *Lipotes vexillifer*. Mammalian Species, no. 10, 4 pp.

Brownell, R. L., Jr., and R. Praderi. 1976. Records of the delphinid genus *Stenella* in western South Atlantic waters. Sci. Rept. Whales Res. Inst. 28:129–135.

————. In Press. *Phocoena spinipinnis*. Mammalian Species.

Brumback, R. A. 1974. A third species of the owl monkey (*Aotus*). J. Hered. 65:321–323.

Bruner, P. L., and H. D. Pratt. 1979. Notes on the status and natural history of Micronesian bats. Elepaio 40:1–4.

Bryden, M. M. 1969. Growth of the southern elephant seal, *Mirounga leonina* (Linn.), Growth 33:69–82.

Bryden, M. M.; R. J. Harrison; and R. J. Lear. 1977. Some aspects of the biology of *Peponocephala electra* (Cetacea: Delphinidae). I. General and reproductive biology. Austral. J. Mar. Freshwater Res. 28:703–715.

Bucher, J. E., and H. I. Fritz. 1977. Behavior and maintenance of the woolly opposum (*Caluromys*) in captivity. Lab. Anim. Sci. 27:1007–1012.

Buchler, E. R. 1976. The use of echolocation by the wandering shrew (*Sorex vagrans*). Anim. Behav. 24:858–873.

Buchmann, O. L. K., and E. R. Guiler. 1974. Locomotion in the potoroo. J. Mamm. 55:203–206.

————. 1977. Behavior and ecology of the Tasmanian devil, *Sarcophilus harrisii*. *In* Stonehouse and Gilmore (1977), pp. 155–168.

Buckner, C. H. 1969. Some aspects of the population ecology of the common shrew, *Sorex araneus,* near Oxford, England. J. Mamm. 50:326–332.

Budd, G. M. 1970. Rapid population increase in the Kerguelen fur seal, *Arctocephalus tropicalis gazella,* at Heard Island. Mammalia 34:410–414.

————. 1972. Breeding of the fur seal at McDonald Islands, and further population growth at Heard Island. Mammalia 36:423–427.

Buden, D. W. 1976. A review of the endemic West Indian genus *Erophylla*. Proc. Biol. Soc. Washington 89:1–16.

————. 1977. First records of bats of the genus *Brachyphylla* from the Caicos Islands, with notes on geographic variation. J. Mamm. 58:221–225.

Budnitz, N., and K. Dainis. 1975. *Lemur catta:* ecology and behavior. *In* Tattersall and Sussman (1975), pp. 219–235.

Buechner, H. K. 1960. The bighorn sheep in the United States, its past, present, and future. Wildl. Monogr., no. 4, 174 pp.

————. 1974. Implications of social behavior in the management of the Uganda Kob. *In* Geist and Walther (1974), pp. 853–870.

Bueler, L. E. 1973. Wild dogs of the world. Stein and Day, New York, 274 pp.

Buettner-Janusch, J. 1966. Origins of man; physical anthropology. John Wiley and Sons, New York, 674 pp.

Bulir, L. 1972. Breeding binturongs *Arctictis binturong* at Liberec Zoo. Int. Zoo Yearbook 12:11–12.

Bunderson, W. T. 1977. Hunter's antelope. Oryx 14:174–175.

Burleigh, R., and J. Clutton-Brock. 1980. The survival of *Myotragus balearicus* Bate, 1909 into the neolithic on Mallorca. J. Archaeol. Sci. 7:385–388.

Burns, J. J., and F. H. Fay. 1970. Comparative morphology of the skull of the ribbon seal, *Histriophoca fasciata,* with remarks on systematics of the Phocidae. J. Zool. 161:363–394.

Burns, J. J., and A. Gavin. 1980. Recent records of hooded seals, *Cystophora cristata* Erxleben, from the western Beaufort Sea. Arctic 33:326–329.

Burt, W. H., and R. P. Grossenheider. 1976. A field guide to the mammals. Houghton Mifflin, Boston, xxv + 289 pp.

Burton, M. 1955. Bulldog bats. Ill. London News 226:28.

Butynski, T. M. 1979. Reproductive ecology of the springhaas *Pedetes capensis* in Botswana. J. Zool. 189:221–232.

Butynski, T. M., and R. Mattingly. 1979. Burrow structure and fossorial ecology of the springhare *Pedetes capensis* in Botswana. Afr. J. Ecol. 17:205–215.

C.

Cabrera, A. 1957, 1961. Catálogo de los mamíferos de América del Sur. Rev. Mus. Argentino Cien. Nat. "Bernardo Rivadavia," 4:1–732.

Caire, W.; J. E. Vaughan; and V. E. Diersing. 1978. First record of *Sorex arizonae* (Insectivora: Soricidae) from Mexico. Southwestern Nat. 23:532–533.

Calaby, J. H. 1966. Mammals of the upper Richmond and Clarence rivers, New South Wales. CSIRO Div. Wildl. Res. Tech. Pap., no. 10, 55 pp.

————. 1971. The current status of Australian Macropodidae. Austral. Zool. 16:17–31.

Calaby, J. H.; L. K. Corbett; G. B. Sharman; and P. G. Johnston. 1974. The chromosomes and systematic position of the marsupial mole, *Notoryctes typhlops*. Austral. J. Biol. Sci. 27:529–532.

Calaby, J. H.; H. Dimpel; and I. M. Cowan. 1971. The mountain pygmy possum, *Burramys parvus* Broom (Marsupialia) in the Kosciusko National Park, New South Wales. CSIRO Div. Wildl. Res. Tech. Pap., no. 23, 11 pp.

Calaby, J. H., and J. M. Taylor. 1980. Reevaluation of the holotype of *Mus ruber* Jentink, 1880 (Rodentia: Muridae) from western New Guinea (Irian Jaya). Zool. Meded. 55:215–219.

Calaby, J. H., and C. White. 1967. The Tasmanian devil (*Sarcophilus harrisi*) in northern Australia in Recent times. Austral. J. Sci. 29:473–475.

Calder, W. A. 1969. Temperature relations and underwater endurance of the smallest homeothermic diver, the water shrew. Comp. Biochem. Physiol. 30:1075–1082.

Caldwell, D. K., and M. C. Caldwell. 1970.

Echolocation-type signals by two dolphins, genus *Sotalia*. Quart. J. Florida Acad. Sci. 33:124–131.

————. 1971*a*. Underwater pulsed sounds produced by captive spotted dolphins, *Stenella plagiodon*. Cetology, no. 1, 7 pp.

————. 1971*b*. The pygmy killer whale, *Feresa attenuata,* in the western Atlantic, with a summary of world records. J. Mamm. 52:206–209.

————. 1977. Cetaceans. *In* Sebeok, T. A., ed., How animals communicate, Indiana Univ. Press, pp. 794–807.

Caldwell, D. K.; M. C. Caldwell; and D. W. Rice. 1966. Behavior of the sperm whale, *Physeter catodon* L. *In* Norris (1966), pp. 678–717.

Caldwell, D. K.; M. C. Caldwell; and R. V. Walker. 1976. First records for Fraser's dolphin (*Lagenodelphis hosei*) in the Atlantic and the melon-headed whale (*Peponocephala electra*) in the western Atlantic. Cetology, no. 25, 4 pp.

Caldwell, M. C., and D. K. Caldwell. 1969. The ugly dolphin. Sea Frontiers 15(5):1–7.

————. 1979. The whistle of the Atlantic bottlenosed dolphin (*Tursiops truncatus*)—ontogeny. *In* Winn and Olla (1979), pp. 369–401.

Caldwell, M. C.; D. K. Caldwell; and W. E. Evans. 1966. Sounds and behavior of captive Amazon freshwater dolphins, *Inia geoffrensis*. Los Angeles Co. Mus. Contrib. Sci., no. 108, 24 pp.

Caldwell, M. C.; D. K. Caldwell; and J. F. Miller. 1973. Statistical evidence for individual whistles in the spotted dolphin, *Stenella plagiodon*. Cetology, no. 16, 21 pp.

Calhoun, J. B. 1963. The ecology and sociology of the Norway rat. U.S. Pub. Health Serv., Baltimore, viii + 288 pp.

California Department of Fish and Game. 1978. At the Crossroads. Sacramento, 103 pp.

Callahan, J. R. 1980. Taxonomic status of *Eutamias bulleri*. Southwestern Nat. 25:1–8.

Camenzind, F. J. 1978. Behavioral ecology of coyotes on the National Elk Refuge, Jackson, Wyoming. *In* Bekoff (1978), pp. 267–294.

Cameron, G. N., and D. G. Rainey. 1972. Habitat utilization by *Neotoma lepida* in the Mohave Desert. J. Mamm. 53:251–266.

Campbell, C. B. G. 1974. On the phyletic relationships of the tree shrews. Mamm. Rev. 4:125–143.

Canivenc, R., and M. Bonnin. 1979. Delayed implantation is under environmental control in the badger (*Meles meles* L.). Nature 278:849–850.

Cant, J. G. H. 1977. A census of the agouti (*Dasyprocta punctata*) in seasonally dry forest at Tikal, Guatemala, with some comments on strip censusing. J. Mamm. 58:688–690.

Cao Kequing. 1978. On the time of extinction of the wild mi-deer in China. Acta Zool. Sinica 24:289–291.

Carl, E. A. 1971. Population control in arctic ground squirrels. Ecology 52:395–413.

Carleton, M. D. 1973. A survey of gross stomach morphology in the New World Cricetinae (Rodentia, Muroidea), with comments on functional interpretations. Misc. Publ. Mus. Zool. Univ. Michigan, no. 146, 43 pp.

————. 1977. Interrelationships of populations of the *Peromyscus boylii* species group (Rodentia, Muridae) in western Mexico. Occas. Pap. Mus. Zool. Univ. Michigan, no. 675, 47 pp.

————. 1979. Taxonomic status and relationships of *Peromyscus boylii* from El Salvador. J. Mamm. 60:280–296.

————. 1980. Phylogenetic relationships in neotomine-peromyscine rodents (Muroidea) and a reappraisal of the dichotomy within New World Cricetinae. Misc. Publ. Mus. Zool. Univ. Michigan, no. 157, vii + 146 pp.

Carleton, M. D., and P. Myers. 1979. Karyotypes of some harvest mice, genus *Reithrodontomys*. J. Mamm. 60:307–313.

Carley, C. J. 1979. Status summary: the red wolf (*Canis rufus*). U.S. Fish and Wildl. Serv., Albuquerque, Endangered Species Rept., no. 7, iv + 36 pp.

Carlisle, J. G., Jr. 1973. The census of northern elephant seals on San Miguel Island, 1965–1973. California Fish and Game 59:311–313.

Carman, M. 1979. The gestation and rearing periods of the mandrill *Mandrillus sphinx*. Int. Zoo Yearbook 19:159–160.

Caro, T. M. 1976. Observations on the ranging behaviour and daily activity of lone silverback mountain gorillas (*Gorilla gorilla beringei*). Anim. Behav. 24:889–897.

Carpenter, C. R. 1965. The howlers of Barro Colorado Island. *In* De Vore (1965), pp. 250–291.

Carrick, R.; S. E. Csordas; S. E. Ingham; and K. Keith. 1962. Studies on the southern elephant seal, *Mirounga leonina* (L.). III. The annual cycle in relation to age and sex. CSIRO Wildl. Res. 7:119–160.

Carroll, L. E., and H. H. Genoways. 1980. *Lagurus curtatus*. Mammalian Species, no. 124, 6 pp.

Carter, C. H., and H. H. Genoways. 1978. *Liomys salvini*. Mammalian Species, no. 84, 5 pp.

Carter, D. C., and J. K. Jones, Jr. 1978. Bats from the Mexican state of Hidalgo. Occas. Pap. Mus. Texas Tech Univ., no. 54, 12 pp.

Cary, J. R., and L. B. Keith. 1979. Reproductive change in the 10-year cycle of snowshoe hares. Can. J. Zool. 57:375–390.

Casimir, M. J. 1975. Some data on the systematic position of the eastern gorilla population of the Mt. Kahuzi region (République du Zaire). Z. Morph. Anthrop. 66:188–201.

Casimir, M. J., and E. Butenandt. 1973. Migration and core area shifting in relation to some ecological factors in a mountain gorilla

group (*Gorilla gorilla beringei*) in the Mt. Kahuzi region (République du Zaire). Z. Tierpsychol. 33:514–522.

Castro, R., and P. Soini. 1977. Field studies on *Saguinus mystax* and other callitrichids in Amazonian Peru. *In* Kleiman (1977a), pp. 73–78.

Caughley, G.; R. G. Sinclair; and G. R. Wilson. 1977. Numbers, distribution and harvesting rate of kangaroos on the inland plains of New South Wales. Austral. Wildl. Res. 4:99–108.

Chakraborty, S., and V. C. Agrawal. 1977. A melanistic example of woolly flying squirrel, *Eupetaurus cinereus* Thomas (Rodentia: Sciuridae). J. Bombay Nat. Hist. Soc. 74:346–347.

Chaline, J.; P. Mein; and F. Petter. 1977. Les grandes lignes d'une classification évolutive des Muroidea. Mammalia 41:245–252.

Chalmers, N. R. 1968a. Group composition, ecology, and daily activities of free living mangabeys in Uganda. Folia Primatol. 8:247–262.

————. 1968b. The social behaviour of free living mangabeys in Uganda. Folia Primatol. 8:263–281.

————. 1973. Differences in behavior between some arboreal and terrestrial species of African monkeys. *In* Michael and Crook (1973), pp. 69–100.

Chan, K. L.; S. S. Dhaliwal; and Yong Hoi-Sen. 1978. Protein variation and systematics in Malayan rats of the subgenus *Lenothrix* (Rodentia: Muridae, genus *Rattus* Fischer). Comp. Biochem. Physiol. 59B:345–351.

Chanin, P. R. F., and D. J. Jefferies. 1978. The decline of the otter *Lutra lutra* L. in Britain: an analysis of hunting records and discussion of causes. Biol. J. Linnean Soc. 10:305–328.

Channing, A., and D. T. Rowe-Rowe. 1977. Vocalizations of South African mustelines. Z. Tierpsychol. 44:283–293.

Chaplin, R. E., and G. Dangerfield. 1973. Breeding records of muntjac deer (*Muntiacus reevesi*) in captivity. J. Zool. 170:150–151.

Chapman, B. R. 1972. Food habits of Loring's kangaroo rat, *Dipodomys elator*. J. Mamm. 53:877–880.

Chapman, B. R., and R. L. Packard. 1974. An ecological study of Merriam's pocket mouse in southeastern Texas. Southwestern Nat. 19:281–291.

Chapman, D., and N. Chapman. 1975. Fallow deer. Their history, distribution and biology. Terence Dalton, Ltd., Lavenham, United Kingdom, 271 pp.

Chapman, D. G. 1974. Status of Antarctic rorqual stocks. *In* Schevill (1974), pp. 218–238.

Chapman, J. A. 1974. *Sylvilagus bachmani*. Mammalian Species, no. 34, 4 pp.

————. 1975a. *Sylvilagus transitionalis*. Mammalian Species, no. 55, 4 pp.

————. 1975b. *Sylvilagus nuttalii*. Mammalian Species, no. 56, 3 pp.

Chapman, J. A., and G. A. Feldhamer. 1981. *Sylvilagus aquaticus*. Mammalian Species, no. 151, 4 pp.

Chapman, J. A.; A. L. Harman; and D. E. Samuel. 1977. Reproductive and physiological cycles in the cottontail complex in western Maryland and nearby West Virginia. Wildl. Monogr., no. 56, 73 pp.

Chapman, J. A.; J. G. Hockman; and M. M. Ojeda C. 1980. *Sylvilagus floridanus*. Mammalian Species, no. 136, 8 pp.

Chapman, J. A., and G. R. Willner. 1978. *Sylvilagus audubonii*. Mammalian Species, no. 106, 4 pp.

————. 1981. *Sylvilagus palustris*. Mammalian Species, no. 153, 3 pp.

Chapman, N. G., and D. I. Chapman. 1980. The distribution of fallow deer: a worldwide review. Mamm. Rev. 10:61–138.

Charles-Dominique, P. 1974. Aggression and territoriality in nocturnal prosimians. *In* Holloway (1974), pp. 31–48.

————. 1977. Ecology and behaviour of nocturnal primates. Prosimians of Equatorial West Africa. Duckworth, London, x + 277 pp.

————. 1978. Écologie et vie sociale de *Nandinia binotata* (Carnivores, Viverrides): comparaison avec les prosimiens sympatriques du Gabon. Terre Vie 32:477–528.

Charles-Dominique, P., and S. K. Bearder. 1979. Field studies of lorisid behavior: methodological aspects. *In* Doyle and Martin (1979), pp. 567–629.

Chasen, F. N. 1940. A handlist of Malaysian mammals. Bull. Raffles Mus., Singapore, no. 15, xx + 209 pp.

Chaudhry, M. A., and M. A. Beg. 1977. Reproductive cycle and population structure of the northern palm squirrel, *Funambulus pennanti*. Pakistan J. Zool. 9:183–189.

Cheatheam, L. K. 1977. Density and distribution of the black-tailed prairie dog in Texas. Texas J. Sci. 29:33–40.

Chen Peixun; Liu Peilin; Liu Renjun; Lin Kejie; and G. Pilleri. 1979. Distribution, ecology, behaviour and conservation of the dolphins of the middle reaches of Changjiang (Yangtze) River (Wuhan-Yueyang). Investig. Cetacea 10:87–103.

Chesemore, D. L. 1972. History and economic importance of the white fox, *Alopex*, fur trade in northern Alaska 1798–1963. Can. Field-Nat. 86:259–267.

————. 1975. Ecology of the Arctic fox (*Alopex lagopus*) in North America—a review. *In* Fox (1975), pp. 143–163.

Chiarelli, A. B. 1972. Taxonomic atlas of living primates. Academic Press, London, vii + 363 pp.

Chidumayo, E. N. 1977. The ecology of the single striped grass mouse, *Lemniscomys griselda*, in Zambia. Mammalia 41:411–418.

Child, G., and V. J. Wilson. 1964. Observations on ecology and behaviour of roan and sable in three tsetse control areas. Arnoldia 1(16):1–8.

Chivers, D. J. 1972. The siamang and the

gibbon in the Malay Peninsula. *In* Rumbaugh (1972a), pp. 103–135.

————. 1973. An introduction to the socio-ecology of Malayan forest primates. *In* Michael and Crook (1973), pp. 101–146.

————. 1974. The siamang in Malaya. Contrib. Primatol. 4:i–xiii + 1–335.

————. 1977. The lesser apes. *In* Rainier III and Bourne (1977), pp. 539–598.

Chivers, D. J., and S. P. Gittins. 1978. Diagnostic features of gibbon species. Int. Zoo Yearbook 18:157–164.

Chivers, D. J., and J. Herbert, eds. 1978. Recent advances in primatology. I. Behaviour. Academic Press, London, xxiv + 980 pp.

Chivers, D. J., and K. A. Joysey, eds. 1978. Recent advances in primatology. III. Evolution. Academic Press, London, xvii + 509 pp.

Chivers, D. J., and W. Lane-Petter, eds. 1978. Recent advances in primatology. II. Conservation. Academic Press, London, xiii + 312 pp.

Choate, J. R. 1970. Systematics and zoogeography of Middle American shrews of the genus *Cryptotis*. Univ. Kansas Publ. Mus. Nat. Hist. 19:195–317.

Choate, J. R., and E. D. Fleharty. 1973. Habitat preference and spatial relations of shrews in a mixed grassland in Kansas. Southwestern Nat. 18:110–112.

Choate, J. R., and J. K. Jones, Jr. 1970. Additional notes on reproduction in the Mexican vole, *Microtus mexicanus*. Southwestern Nat. 14:356–358.

Choate, T. S. 1972. Behavioural studies on some Rhodesian rodents. Zool. Afr. 7:103–118.

Chorazyna, H., and G. U. Kurup. 1975. Observations on the ecology and behaviour of *Anathana ellioti* in the wild. Proc. 5th Int. Congr. Primatol., pp. 342–344.

Chorn, J., and R. S. Hoffmann. 1978. *Ailuropoda melanoleuca*. Mammalian Species, no. 110, 6 pp.

Christensen, P. 1975. The breeding burrow of the banded ant-eater or numbat (*Myrmecobius fasciatus*). W. Austral. Nat. 13:32–34.

Christenson, T. E., and B. J. LeBoeuf. 1978. Aggression in the female northern elephant seal, *Mirounga angustirostris*. Behaviour 64:158–172.

Christian, D. P. 1977. Diurnal activity of the four-striped mouse, *Rhabdomys pumilio*. Zool. Afr. 12:238–239.

Clark, M. J., and W. E. Poole. 1967. The reproductive system and embryonic diapause in the female grey kangaroo, *Macropus giganteus*. Austral. J. Zool. 15:441–459.

Clark, T. W. 1976. The black-footed ferret. Oryx 13:275–280.

————. 1977. Ecology and ethology of the white-tailed prairie dog (*Cynomys leucurus*). Milwaukee Pub. Mus. Publ. Biol. Geol., no. 3, vi + 97 pp.

————. 1978. Current status of the black-footed ferret in Wyoming. J. Wildl. Mgmt. 42:128–134.

Clarke, M. R. 1979. The head of the sperm whale. Sci. Amer. 240(1):128–141.

Cleveland, A. G. 1970. The current geographic distribution of the armadillo in the United States. Texas J. Sci. 22:87–92.

Clough, G. C. 1972. Biology of the Bahaman hutia, *Geocapromys ingrahami*. J. Mamm. 53:807–823.

————. 1973. A most peaceable rodent. Nat. Hist. 82(6):66–74.

————. 1974. Additional notes on the biology of the Bahamian hutia, *Geocapromys ingrahami*. J. Mamm. 55:670–672.

————. 1976. Current status of two endangered Caribbean rodents. Biol. Conserv. 10:43–47.

Clutton-Brock, J.; G. B. Corbet; and M. Hills. 1976. A review of the family Canidae, with a classification by numerical methods. Bull. British Mus. (Nat. Hist.), Zool. 29:117–199.

Clutton-Brock, T. H., ed. 1977. Primate ecology: studies of feeding and ranging behaviour in lemurs, monkeys and apes. Academic Press, London, xxii + 631 pp.

Cockburn, A. 1978. The distribution of *Pseudomys shortridgei* (Muridae: Rodentia) and its relevance to that of other heathland *Pseudomys*. Austral. Wildl. Res. 5:213–219.

Cockrum, E. L. 1969. Migration in the guano bat, *Tadarida brasiliensis*. Univ. Kansas Mus. Nat. Hist. Misc. Publ. 51:303–336.

————. 1973. Additional longevity records for American bats. J. Arizona Acad. Sci. 8:108–110.

————. 1977. Status of the hairy footed gerbil, *Gerbillus latastei* Thomas and Trouessart. Mammalia 41:75–80.

Cockrum, E. L.; T. C. Vaughan; and P. J. Vaughan. 1976a. A review of North African short-tailed gerbils (*Dipodillus*) with description of a new taxon from Tunisia. Mammalia 40:313–326.

————. 1976b. *Gerbillus andersoni* de Winton, a species new to Tunisia. Mammalia 40:467–473.

————. 1977. Status of the pale sand rat, *Psammomys vexillaris* Thomas, 1925. Mammalia 41:321–326.

Cohen, J. A. 1977. A review of the biology of the dhole or Asiatic wild dog (*Cuon alpinus* Pallas). Anim. Reg. Studies 1:141–158.

————. 1978. *Cuon alpinus*. Mammalian Species, no. 100, 3 pp.

Cohen, J. A.; M. W. Fox; A. J. T. Johnsingh; and B. D. Barnett. 1978. Food habits of the dhole in south India. J. Wildl. Mgmt. 42:933–936.

Coimbra-Filho, A. F. 1972. Mamíferos ameaçados de extinção no Brasil. *In* Academia Brasileira de Ciencias, espécies da fauna Brasileira ameaçados de extinção, Rio de Janeiro, pp. 13–98.

Coimbra-Filho, A. F., and R. A. Mitter-

meier. 1973a. Distribution and ecology of the genus *Leontopithecus* Lesson, 1840 in Brazil. Primates 14:47–66.

————. 1973b. New data on the taxonomy of the Brazilian marmosets of the genus *Callithrix* Erxleben, 1777. Folia Primatol. 20:241–264.

————. 1977a. Tree-gouging, exudate eating and the "short-tusked" condition in *Callithrix* and *Cebuella*. *In* Kleiman (1977a), pp. 105–115.

————. 1977b. Conservation of the Brazilian lion tamarins (*Leontopithecus rosalia*). *In* Rainier III and Bourne (1977), pp. 59–94.

Cole, G. F. 1976. Management involving grizzly and black bears in Yellowstone National Park 1970–75. U.S. Natl. Park Serv. Nat. Res. Rept., no. 9, 26 pp.

Cole, L. R. 1972. A comparison of *Malacomys longipes* and *Malacomys edwardsi* (Rodentia: Muridae) from a single locality in Ghana. J. Mamm. 53:616–619.

————. 1975. Foods and foraging places of rats (Rodentia: Muridae) in the lowland evergreen forest of Ghana. J. Zool. 175:453–471.

Collier, C., and S. Emerson. 1973. Handraising bush dogs *Speothos venaticus* at the Los Angeles Zoo. Int. Zoo Yearbook 13:139–140.

Collier, G. D., and J. J. Spillett. 1973. The Utah prairie dog—decline of a legend. Utah Sci. 34:83–87.

Collins, L. R. 1973. Monotremes and marsupials. A reference for zoological institutions. Smithson. Inst. Press, Washington, D.C., v + 323 pp.

Collins, L. R., and J. F. Eisenberg. 1972. Notes on the behaviour and breeding of pacaranas *Dinomys branickii* in captivity. Int. Zoo Yearbook 12:108–114.

Committee for Whaling Statistics. 1980. International Whaling Statistics, vols. 85 and 86. Sandefjord, Norway.

Condy, P. R. 1978. Annual cycle of the southern elephant seal *Mirounga leonina* (Linn.) at Marion Island. S. Afr. J. Zool. 14:95–102.

Contreras, J. R. 1968. *Akodon molinae* una nueva especie de ratón de campo del sur de la Provincia de Buenos Aires. Zool. Platense 1:9–12.

Contreras, J. R., and E. R. Justo. 1974. Aportes a la mastozoología pampeana. I. Nuevas localidades para roedores Cricetidae. Neotrópica 20:91–96.

Contreras, J. R.; V. G. Roig; and C. M. Suzarte. 1977. *Ctenomys validus*, una nueva especie de "tunduque" de la Provincia de Mendoza (Rodentia: Octodontidae). Physis, sec. C, 36:159–162.

Conway, M. C., and C. G. Schmitt. 1978. Record of the Arizona shrew (*Sorex arizonae*) from New Mexico. J. Mamm. 59:631.

Cooper, J. E.; S. S. Robinson; and J. B. Funderburg, eds. 1977. Endangered and threatened plants and animals of North Carolina. North Carolina State Mus. Nat. Hist., Raleigh, xvi + 444 pp.

Corbet, G. B. 1978. The mammals of the Palaearctic Region: a taxonomic review. British Mus. (Nat. Hist.), London, 314 pp.

_____. 1979. The taxonomy of *Procavia capensis* in Ethiopia, with special reference to the aberrant tusks of *P. c. capillosa* Brauer (Mammalia, Hyracoidea). Bull. British Mus. (Nat. Hist.), Zool. 36:251–259.

Corbet, G. B., and J. Hanks. 1968. A revision of the elephant-shrews, family Macroscelididae. Bull. British Mus. (Nat. Hist.), Zool. 16:1–111.

Corbet, G. B., and J. E. Hill. 1980. A world list of mammalian species. British Mus. (Nat. Hist.), London, viii + 226 pp.

Corbet, G. B., and H. N. Southern. 1977. The handbook of British mammals. Blackwell Scientific Publications, London, xxxii + 520 pp.

Corbett, L. K. 1975. Geographical distribution and habitat of the marsupial mole, *Notoryctes typhlops*. Austral. Mamm. 1:375–378.

Corbett, L. K., and A. Newsome. 1975. Dingo society and its maintenance: a preliminary analysis. *In* Fox (1975), pp. 369–379.

Costa, W. R.; K. A. Nagy; and V. H. Shoemaker. 1976. Observations of the behavior of jackrabbits (*Lepus californicus*) in the Mojave Desert. J. Mamm. 57:399–402.

Cottrell, W. 1978. The fisher (*Martes pennanti*) in Maryland. J. Mamm. 59:886.

Coulter, M. W. 1974. Maine's "black cat." Maine Fish and Game 16(3):23–26.

Cousins, D. 1976. The breeding of gorillas, *Gorilla gorilla*, in zoological collections. Zool. Garten 46:215–236.

Cousteau, J. Y., and P. Diole. 1972. Killer whales have fearsome teeth and a strange gentleness to man. Smithsonian 3(3):66–73.

Cowan, I. M. 1971. Summary of the symposium on the native cats of North America. *In* Jorgenson, S. E., and L. D. Mech, eds., Proceedings of a symposium on the native cats of North America, U.S. Bur. Sport Fish. and Wildl., pp. 1–8.

_____. 1972. The status and conservation of bears (Ursidae) of the world—1970. *In* Herrero (1972), pp. 343–367.

Craighead, F. C., Jr. 1976. Grizzly bear ranges and movement as determined by radiotracking. *In* Pelton, Lentfer, and Folk (1976), pp. 97–109.

Craighead, F. C., Jr., and J. J. Craighead. 1972. Grizzly bear prehibernation and denning activities as determined by radiotracking. Wildl. Monogr., no. 32, 35 pp.

Craighead, J. J.; F. C. Craighead, Jr., and J. Sumner. 1976. Reproductive cycles and rates in the grizzly bear, *Ursus arctos horribilis*, of the Yellowstone ecosystem. *In* Pelton, Lentfer, and Folk (1976), pp. 337–356.

Craighead, J. J.; J. R. Varney; and F. C. Craighead, Jr. 1974. A population analysis of the Yellowstone grizzly bears. Montana Cooperative Wildl. Res. Unit, 20 pp.

Crandall, L. S. 1964. The management of wild animals in captivity. Univ. Chicago Press, xv + 761 pp.

Cranford, J. A. 1977. Home range and habitat utilization by *Neotoma fuscipes* as determined by radiotelemetry. J. Mamm. 58:165–172.

Crawford-Cabral, J. 1981. A new classification of genets. Afr. Small Mamm. Newsl., no. 6, pp. 8–10.

Crawley, M. C. 1969. Movements and home ranges of *Clethrionomys glareolus* Schreber and *Apodemus sylvaticus* L. in north-east England. Oikos 20:310–319.

_____. 1973. A live-trapping study of Australian brush-tailed possums, *Trichosurus vulpecula* (Kerr), in the Orongorongo Valley, Wellington, New Zealand. Austral. J. Zool. 21:75–90.

Crawley, M. C., and D. B. Cameron. 1972. New Zealand sea lions, *Phocarctos hookeri*, on the Snares Islands. New Zealand J. Mar. Freshwater Res. 6:127–132.

Crawley, M. C., and G. J. Wilson. 1976. The natural history and behaviour of the New Zealand fur seal (*Arctocephalus forsteri*). Tuatara 22:1–29.

Crespo, J. A. 1975. Ecology of the pampas gray fox and the large fox (culpeo). *In* Fox (1975), pp. 179–191.

Crichton, E. G. 1969. Reproduction in the pseudomyine rodent *Mesembriomys gouldii* (Gray) (Muridae). Austral. J. Zool. 17:785–797.

_____. 1974. Aspects of reproduction in the genus *Notomys* (Muridae). Austral. J. Zool. 22:439–447.

Crocker-Bedford, D. C., and J. J. Spillett. 1977. Home ranges of Utah prairie dogs. J. Mamm. 58:672–673.

Crowcroft, P. 1977. Breeding of wombats (*Lasiorhinus latifrons*) in captivity. Zool. Garten 47:313–322.

Crowe, D. M. 1975. Aspects of ageing, growth, and reproduction of bobcats from Wyoming. J. Mamm. 56:177–198.

Cumming, D. H. M. 1975. A field study of the ecology and behaviour of warthog. Trustees Natl. Museums and Monuments Rhodesia Mus. Mem., no. 7, 179 pp.

Cumming, H. G. 1974. Moose management in Ontario from 1948 to 1973. Naturaliste Canadien 101:673–687.

Cunha, F. L. D. S., and J. F. Cruz. 1979. Novo gênero de Cricetidae (Rodentia) de Castelo, Espirito Santo, Brasil. Bol. Mus. Biol. Prof. Mello Leitao, Santa Teresa, E. E. Santo, Brasil, ser. Zool., no. 96, 5 pp.

Czekala, N. M., and K. Benirschke. 1974. Observations on a twin pregnancy in the African long-tongued fruit bat (*Megaloglossus woermanni*). Bonner Zool. Beitr. 25:220–230.

D.

Dagg, A. I. 1971. *Giraffa camelopardalis*. Mammalian Species, no. 5, 8 pp.

Dagg, A. I., and J. B. Foster. 1976. The giraffe. Its biology, behavior, and ecology. Van Nostrand Reinhold, New York, xiii + 210 pp.

Daggett, P. M., and D. R. Henning. 1974.

The jaguar in North America. Amer. Antiquity 39:465–469.

Dalby, P. L. 1975. Biology of pampa rodents. Publ. Mus. Michigan State Univ., Biol. Ser., 5:149–272.

Dalby, P. L., and M. A. Mares. 1974. Notes on the distribution of the coney rat, *Reithrodon auritus*, in northwestern Argentina. Amer. Midl. Nat. 92:205–206.

Dalquest, W. W. 1953. Mammals of the Mexican state of San Luis Potosi. Louisiana State Univ. Studies, Biol. Ser., no. 1, 229 pp.

_____. 1957. Observations on the sharp-nosed bat, *Rhynchiscus naso* (Maximilian). Texas J. Sci. 9:219–226.

Daly, M. 1979. Of Libyan jirds and fat sand rats. Nat. Hist. 88(2):64–71.

Daly, M., and S. Daly. 1973. On the feeding ecology of *Psammomys obesus* (Rodentia, Gerbillidae) in the Wadi Saoura, Algeria. Mammalia 37:545–561.

_____. 1975a. Socio-ecology of Saharan gerbils, especially *Meriones libycus*. Mammalia 39:289–311.

_____. 1975b. Behavior of *Psammomys obesus* (Rodentia: Gerbillinae) in the Algerian Sahara. Z. Tierpsychol. 37:289–321.

Danell, K. 1977. Dispersal and distribution of the muskrat (*Ondatra zibethica* (L.)) in Sweden. Viltrevy 10:1–26.

Daniel, M. J. 1975. First record of an Australian fruit bat (Megachiroptera: Pteropodidae) reaching New Zealand. New Zealand J. Zool. 2:227–231.

_____. 1976. Feeding by the short-tailed bat (*Mystacina tuberculata*) on fruit and possibly nectar. New Zealand J. Zool. 3:391–398.

_____. 1979. The New Zealand short-tailed bat, *Mystacina tuberculata;* a review of present knowledge. New Zealand J. Zool. 6:357–370.

Dapson, R. W. 1968. Reproduction and age structure in a population of short-tailed shrews, *Blarina brevicauda*. J. Mamm. 49:205–214.

Dash, Y.; A. Szaniawski; G. S. Child; and P. Hunkeler. 1978. Observations on some large mammals of the Transaltai, Djungarian and Shargin Gobi, Mongolia. Tigerpaper 5(2):5–10.

Da Silveira, E.K.P. 1968. Notes on the care and breeding of the maned wolf *Chrysocyon brachyurus* at Brasilia Zoo. Int. Zoo Yearbook 8:21–23.

Dathe, H. 1970. A second generation birth of captive sun bears *Helarctos malayanus* at East Berlin Zoo. Int. Zoo Yearbook 10:79.

David, J. H. M. 1973. The behaviour of the bontebok, *Damaliscus dorcas dorcas* (Pallas 1766), with special reference to territorial behaviour. Z. Tierpsychol. 33:38–107.

_____. 1975. Observations on mating behaviour, parturition, suckling and the mother-young bond in the bontebok (*Damaliscus dorcas dorcas*). J. Zool. 177:203–223.

Davidar, E. R. C. 1975. Ecology and behavior of the dhole or Indian wild dog *Cuon alpinus* (Pallas). *In* Fox (1975), pp. 109–119.

————. 1978. Distribution and status of the Nilgiri tahr (*Hemitragus hylocrius*)—1975–1978. J. Bombay Nat. Hist. Soc. 75:815–844.

Davidge, C. 1978. Ecology of baboons (*Papio ursinus*) at Cape Point. Zool. Afr. 13:329–350.

Davis, B. L., and R. J. Baker. 1974. Morphometrics, evolution, and cytotaxonomy of mainland bats of the genus *Macrotus* (Chiroptera: Phyllostomatidae). Syst. Zool. 23:26–39.

Davis, D. D. 1962. Mammals of the lowland rain-forest of north Borneo. Bull. Singapore Natl. Mus., no. 31, 129 pp.

Davis, J. L.; P. Valkenburg; and H. V. Reynolds. 1980. Population dynamics of Alaska's western Arctic caribou herd. *In* Reimers, E.; E. Gaare; and S. Skjenneberg, eds., Proc. 2nd Int. Reindeer/caribou Symp., Direktoratet for vilt og ferskvannsfisk, Trondheim, Norway, pp. 595–604.

Davis, R. 1970. Carrying of young by flying female American bats. Amer. Midl. Nat. 83:186–196.

Davis, R. A.; W. J. Richardson; S. R. Johnson; and W. E. Renaud. 1978. Status of Lancaster Sound narwhal population in 1976. Rept. Int. Whaling Comm. 28:209–215.

Davis, R. M. 1972. Behaviour of the vlei rat, *Otomys irroratus* (Brants, 1827). Zool. Afr. 7:119–140.

Davis, W. B. 1966*a*. Review of South American bats of the genus *Eptesicus*. Southwestern Nat. 11:245–274.

————. 1966*b*. The mammals of Texas. Texas Parks and Wildl. Dept. Bull., no. 41, 267 pp.

————. 1970*a*. *Tomopeas ravus* Miller (Chiroptera). J. Mamm. 51:244–247.

————. 1970*b*. A review of the small fruit bats (genus *Artibeus*) of Middle America. Part II. Southwestern Nat. 14:389–402.

————. 1976*a*. Notes on the bats *Saccopteryx canescens* Thomas and *Micronycteris hirsuta* (Peters). J. Mamm. 57:604–607.

————. 1976*b*. Geographic variation in the lesser noctilio, *Noctilio albiventris* (Chiroptera). J. Mamm. 57:687–707.

————. 1980. New *Sturnira* (Chiroptera: Phyllostomidae) from Central and South America, with key to currently recognized species. Occas. Pap. Mus. Texas Tech Univ., no. 70, 5 pp.

Davis, W. B., and D. C. Carter. 1978. A review of the round-eared bats of the *Tonatia silvicola* complex, with descriptions of three new taxa. Occas. Pap. Mus. Texas Tech Univ., no. 53, 12 pp.

Davis, W. B., and J. R. Dixon. 1976. Activity of bats in a small village clearing near Iquitos, Peru. J. Mamm. 57:747–749.

Davis, W. B., and L. A. Follansbee. 1945. The Mexican volcano mouse, *Neotomodon*. J. Mamm. 26:401–411.

Davis, W. H., and J. W. Hardin. 1967. New records of mammals from Mesa Verde Na-tional Park, Colorado. J. Mamm. 48:322–323.

Dawbin, W. H. 1966. The seasonal migratory cycle of humpback whales. *In* Norris (1966), pp. 145–170.

Dawson, G. A. 1977. Composition and stability of social groups of the tamarin, *Saguinus oedipus geoffroyi*, in Panama: ecological and behavioral considerations. *In* Kleiman (1977*a*), pp. 23–37.

Dawson, T. J.; D. Fanning; and T. J. Bergin. 1978. Metabolism and temperature regulation in the New Guinea monotreme *Zaglossus bruijni*. Austral. Zool. 20:99–103.

Deag, J. M. 1977. The status of the Barbary macaque *Macaca sylvanus* in captivity and factors influencing its distribution in the wild. *In* Rainier III and Bourne (1977), pp. 267–287.

Dean, F. C. 1976. Aspects of grizzly bear population ecology in Mount McKinley National Park. *In* Pelton, Lentfer, and Folk (1976), pp. 111–119.

Dean, W. R. J. 1978. Conservation of the white-tailed rat in South Africa. Biol. Conserv. 13:133–140.

Deems, E. F., Jr., and D. Pursley, eds. 1978. North American furbearers. Int. Assoc. Fish and Wildl. Agencies, Univ. Maryland Press, College Park, Maryland, x + 165 pp.

Deerson, D.; R. Dunn; D. Spittall; and P. Williams. 1975. Mammals of the upper Lerderberg Valley. Victorian Nat. 92:28–43.

Defler, T. R. 1979*a*. On the ecology and behavior of *Cebus albifrons* in eastern Colombia: I. Ecology. Primates 20:475–490.

————. 1979*b*. On the ecology and behavior of *Cebus albifrons* in eastern Colombia: II. Behavior. Primates 20:491–502.

De Graaf, G. 1972. On the mole-rat (*Cryptomys hottentotus damarensis*) (Rodentia) in the Kalahari Gemsbok National Park. Koedoe 15:25–35.

De Graff, G., and B. L. Penzhorn. 1976. The re-introduction of springbok *Antidorcas marsupialis* into South African national parks—a documentation. Koedoe 19:75–82.

Delany, M. J. 1969. The ecological distribution of small mammals on Bugala Island, Lake Victoria. Zool. Afr. 4:129–133.

————. 1971. The biology of small rodents in Mayanja Forest, Uganda. J. Zool. 165:85–129.

————. 1975. The rodents of Uganda. British Mus. (Nat. Hist.), London, vii + 165 pp.

DeLong, R. L. 1978. Northern elephant seal. *In* Haley (1978), pp. 206–211.

DeMaster, D. P., and I. Stirling. 1981. *Ursus maritimus*. Mammalian Species, no. 145, 7 pp.

Dene, H.; M. Goodman; and W. Prychodko. 1978. An immunological examination of the systematics of Tupaioidea. J. Mamm. 59:697–706.

Dennis, E., and J. I. Menzies. 1978. Systematics and chromosomes of New Guinea *Rattus*. Austral. J. Zool. 26:197–206.

————. 1979. A chromosomal and morphometric study of Papuan tree rats *Pogonomys* and *Chiruromys* (Rodentia, Muridae). J. Zool. 189:315–332.

Densmore, M. A. 1980. Reproduction of sitatunga *Tragelaphus spekei* in captivity. Int. Zoo Yearbook 20:227–229.

De Santis, L. J. M., and E. R. Justo. 1980. *Akodon (Abrothrix) mansoensis* sp. nov. un nuevo "Raton Lanoso" de la provincia de Rio Negro, Argentina. Neotrópica 26:121–127.

Desy, E. A., and J. D. Druecker. 1979. The estrous cycle of the plains pocket gopher, *Geomys bursarius*, in the laboratory. J. Mamm. 60:235–236.

De Vore, I., ed. 1965. Primate behavior. Field studies of monkeys and apes. Holt, Rinehart and Winston, New York, xiv + 654 pp.

De Vore, I., and K. R. L. Hall. 1965. Baboon ecology. *In* De Vore (1965), pp. 20–52.

De Vos, A., and A. Omar. 1971. Territories and movements of Sykes monkeys (*Cercopithecus mitis kolbi* Neuman) in Kenya. Folia Primatol. 16:196–205.

De Vree, F. 1972. Description of a new form of *Pipistrellus* from Ivory Coast. Rev. Zool. Bot. Afr. 85:412–416.

————. 1973. New data on *Scotophilus gigas* Dobson, 1875 (Microchiroptera—Vespertilionidae). Z. Saugetierk. 38:189–196.

De Winton, W. E. 1898. On a small collection of mammals made by Mr. C. V. A. Peel in Somaliland. Ann. Mag. Nat. Hist., ser. 7, 1:247–251.

D'Huart, J.-P. 1976. Actogramme journalier de l'hylochère (*Hylochoerus meinertzhageni* Thomas) au Parc National des Virunga, Zaire. Terre Vie 30:165–180.

Diersing, V. E. 1980. Systematics and evolution of the pygmy shrews (subgenus *Microsorex*) of North America. J. Mamm. 61:76–101.

————. 1981. Systematic status of *Sylvilagus brasiliensis* and *S. insonus* from North America. J. Mamm. 62:539–556.

Diersing, V. E., and D. F. Hoffmeister. 1977. Revision of the shrews *Sorex merriami* and a description of a new species of the subgenus *Sorex*. J. Mamm. 58:321–333.

Dieterlen, F. 1969. *Dendromus kahuziensis* (Dendromurinae; Cricetidae; Rodentia)—eine neue Art aus Zentralafrika. Z. Saugetierk. 34:348–353.

————. 1974. Bemerkungen zur Systematik der Gattung *Pelomys* (Muridae; Rodentia) in Athiopien. Z. Saugetierk. 39:229–231.

————. 1976. Die afrikanische Muridengattung *Lophuromys* Peters, 1874. Stuttgarter Beitr. Naturkunde, ser. A, no. 285, 96 pp.

Dieterlen, F., and H. Rupp. 1976. Die Rotnasenratte *Oenomys hypoxanthus* (Pucheran, 1855) (Muriden, Rodentia)—Erstnachweis fur Athiopien und dritter Fund aus Tansania. Saugetierk. Mitt. 24:229–235.

————. 1978. *Megadendromus nikolausi*, gen. nov., sp. nov. (Dendromurinae; Roden-

tia), ein neuer Nager aus Athiopien. Z. Saugetierk. 43:129–143.

Dimpel, H., and J. H. Calaby. 1972. Further
observations on the mountain pigmy possum
(*Burramys parvus*). Victorian Nat. 89:101–
106.

Dippenaar, N. J. 1980. New species of *Cro-
cidura* from Ethiopia and northern Tanzania
(Mammalia: Soricidae). Ann. Transvaal
Mus. 32:125–154.

Dittrich, L. 1966. Breeding Indian elephants
Elephas maximus at Hanover Zoo. Int. Zoo
Yearbook 6:193–196.

————. 1967. Breeding Kirk's dik-dik
Madoqua kirki thomasi at Hanover Zoo. Int.
Zoo Yearbook 7:171–173.

————. 1972. Gestation periods and age
of sexual maturity of some African antelopes. Int. Zoo Yearbook 12:184–187.

————. 1976. Age of sexual maturity in
the hippopotamus *Hippopotamus amphibius*.
Int. Zoo Yearbook 16:171–173.

Dittus, W. 1977. The socioecological basis
for the conservation of the toque monkey
(*Macaca sinica*) of Sri Lanka (Ceylon). *In*
Rainier III and Bourne (1977), pp. 237–265.

Dixon, J. M. 1971. *Burramys parvus* Broom
(Marsupialia) from Falls Creek area of the
Bogong high plains, Victoria. Victorian Nat.
88:133–138.

————. 1978. The first Victorian and
other records of the little pigmy possum *Cercartetus lepidus* (Thomas). Victorian Nat.
95:4–7.

Dixson, A. F. 1977. Observations on the
displays, menstrual cycles and sexual behaviour of the "black ape" of Celebes
(*Macaca nigra*). J. Zool. 182:63–84.

Dixson, A. F.; D. M. Scruton; and J. Herbert. 1975. Behaviour of the talapoin
monkey (*Miopithecus talapoin*) studied in
groups, in the laboratory. J. Zool.
176:177–210.

Dobroruka, L. J. 1968. Breeding group of
gorals *Nemorhaedus goral* at Prague Zoo.
Int. Zoo Yearbook 8:143–145.

————. 1970a. Fecundity of the Chinese
water deer, *Hydropotes inermis* Swinhoe,
1870. Mammalia 34:161–162.

————. 1970b. To the supposed former
occurrence of the David's deer, *Elaphurus
davidianus* Milne Edwards, 1866, in Hainan. Mammalia 34:162–164.

————. 1973. Yellow-spotted dassie *Heterohyrax brucei* (Gray, 1868) feeding on a
poisonous plant. Saugetierk. Mitt. 21:365.

Doebl, J. H., and B. S. McGinnes. 1974.
Home range and activity of a gray squirrel
population. J. Wildl. Mgmt. 38:860–867.

Dolan, J. M. 1976. The Arabian oryx *Oryx
leucoryx*. Its destruction, captive history and
propagation. Int. Zoo Yearbook 16:230–
239.

Dolan, P. G., and D. C. Carter. 1979. Distributional notes and records for Middle
American Chiroptera. J. Mamm. 60:644–
649.

Dolan, P. G., and T. L. Yates. 1981. Interspecific variation in *Apodemus* from the

northern Adriatic islands of Yugoslavia. Z.
Saugetierk. 46:151–161.

Dolgov, V. A., and R. S. Hoffmann. 1977.
The Tibetan shrew. *Sorex thibetanus* Kast-
schenko, 1905 (Soricidae, Mammalia).
Zool. Zhur. 56:1687–1692.

Domning, D. P. 1978. Sirenian evolution in
the North Pacific Ocean. Univ. California
Publ. Geol. Sci. 118:1–176.

Donaldson, S. L.; T. B. Wirtz; and A. E.
Hite. 1975. The social behaviour of capybaras *Hydrochoerus hydrochaeris* at Evansville Zoo. Int. Zoo Yearbook 15:201–206.

Donnelly, B. G., and J. H. Grobler. 1976.
Notes on food and anvil using behaviour by
the Cape clawless otter, *Aonyx capensis*, in
the Rhodes Matopos National Park, Rhodesia. Arnoldia 7(37):1–7.

Dorst, J., and P. Dandelot. 1969. A field
guide to the larger mammals of Africa.
Houghton Mifflin, Boston, 287 pp.

Douglas, A. M. 1967. The natural history of
the ghost bat, *Macroderma gigas* (Microchiroptera, Megadermatidae), in Western
Australia. W. Austral. Nat. 10:125–137.

Douglas-Hamilton, I. 1973. On the ecology
and behavior of the Lake Manyara elephants.
E. Afr. Wildl. J. 11:401–403.

Douglass, R. J. 1977. Population dynamics,
home ranges, and habitat associations of the
yellow-cheeked vole, *Microtus xanthognathus*, in the Northwest Territories. Can.
Field-Nat. 91:237–247.

Dowler, R. C., and H. H. Genoways. 1978.
Liomys irroratus. Mammalian Species, no.
82, 6 pp.

Doyle, G. A. 1979. Development of behavior in prosimians with special reference to
the lesser bushbaby, *Galago senegalensis
moholi*. *In* Doyle and Martin (1979), pp.
157–206.

Doyle, G. A., and S. K. Bearder. 1977. The
galagines of South Africa. *In* Rainier III and
Bourne (1977), pp. 1–35.

Doyle, G. A., and R. D. Martin, eds. 1979.
The study of prosimian behavior. Academic
Press, New York, xvii + 696 pp.

Drabek, C. M. 1973. Home range and daily
activity of the round-tailed ground squirrel,
Spermophilus tereticaudus neglectus. Amer.
Midl. Nat. 89:287–293.

Drickamer, L. C.; J. G. Vandenbergh; and
D. R. Colby. 1973. Predictors of dominance
in the male golden hamster (*Mesocricetus
auratus*). Anim. Behav. 21:557–563.

Drickamer, L. C., and B. M. Vestal. 1973.
Patterns of reproduction in a laboratory colony of *Peromyscus*. J. Mamm. 54:523–528.

Dubost, G. 1978. Un aperçu sur l'écologie
du chevrotain africain *Hyemoschus aquaticus* Ogilby, Artiodactyle Tragulide. Mammalia 42:1–62.

Dubost, G., and H. Genest. 1974. Le com-
portement social d'une colonie de maras
Dolichotis patagonum Z. dans le Parc de
Branféré. Z. Tierpsychol. 35:225–302.

Dubost, G., and F. Petter. 1978. Une espèce
nouvelle de "rat-pecheur" de Guyane française: Daptomys oyapocki sp. nov.
(Rongeurs, Cricetidae). Mammalia
42:435–439.

Du Mond, F. V. 1968. The squirrel monkey
in a seminatural environment. *In* Rosenblum
and Cooper (1968), pp. 87–145.

Dunaway, P. B. 1968. Life history and population aspects of the eastern harvest mouse.
Amer. Midl. Nat. 79:48–67.

Dunbar, R. I. M. 1974. Observations on the
ecology and social organization of the green
monkey, *Cercopithecus sabaeus*, in Senegal. Primates 15:341–350.

————. 1977a. The gelada baboon: status
and conservation. *In* Rainier III and Bourne
(1977), pp. 363–383.

————. 1977b. Feeding ecology of gelada
baboons: a preliminary report. *In* Clutton-
Brock (1977), pp. 251–273.

Dunbar, R. I. M., and P. Dunbar. 1974a.
On hybridization between *Theropithecus
gelada* and *Papio anubis* in the wild. J.
Human Evol. 3:187–192.

————. 1974b. Social organization and
ecology of the klipspringer (*Oreotragus
oreotragus*) in Ethiopia. Z. Tierpsychol.
35:481–493.

————. 1975. Social dynamics of gelada
baboons. Contrib. Primatol. 6:i–vii +
1–157.

————. 1979. Observations on the social
organization of common duiker in Ethiopia.
Afr. J. Ecol. 17:249–252.

Duncan, P., and R. W. Wrangham. 1970.
On the ecology and distribution of subterranean insectivores in Kenya. J. Zool.
164:149–163.

Dunford, C. 1972. Summer activity of eastern chipmunks. J. Mamm. 53:176–180.

————. 1974. Annual cycle of cliff chipmunks in the Santa Catalina Mountains, Arizona. J. Mamm. 55:401–416.

————. 1977. Social system of round-
tailed ground squirrels. Anim. Behav.
25:885–906.

Duplaix, N. 1980. Observations on the ecol-
ogy and behavior of the giant river otter
Pteronura brasiliensis in Suriname. Rev.
Ecol. (Terre Vie) 34:496–620.

Duplaix-Hall, N. 1975. River otters in captivity: a review. *In* Martin (1975b), pp.
315–327.

DuPlessis, S. S. 1972. Ecology of blesbok
with special reference to productivity. Wildl.
Monogr., no. 30, 70 pp.

Durham, N. M. 1971. Effects of altitude differences on group organization of wild black
spider monkeys (*Ateles paniscus*). Proc. 3rd
Int. Congr. Primatol. 3:32–40.

Durrell, G., and J. Mallinson. 1970. The
volcano rabbit *Romerolagus diazi* in the wild
and at Jersey Zoo. Int. Zoo Yearbook
10:118–122.

Dusi, J. L. 1976. Mammals. *In* Boschung,
H., ed., Endangered and threatened plants
and animals of Alabama. Bull. Alabama
Mus. Nat. Hist., no. 2, pp. 88–92.

Dwyer, P. D. 1966. Observations on *Chalinolobus dwyeri* (Chiroptera: Vespertilionidae) in Australia. J. Mamm. 47:716–
718.

————. 1968. The biology, origin, and adaption of *Miniopterus australis* (Chiroptera) in New South Wales. Austral. J. Zool. 16:49–68.

————. 1970. Latitude and breeding season in a polyestrous species of *Myotis*. J. Mamm. 51:405–410.

————. 1975a. Notes on *Dobsonia moluccensis* (Chiroptera) in the New Guinea highlands. Mammalia 39:113–118.

————. 1975b. Observations on the breeding biology of some New Guinea murid rodents. Austral. Wildl. Res. 2:33–45.

————. 1977. Notes on *Antechinus* and *Cercartetus* (Marsupialia) in the New Guinea highlands. Proc. Roy. Soc. Queensland 88:69–73.

Dzneladze, A. M. 1974. Distribution of the long-clawed mole-vole (*Prometheomys schaposchnikovi* Satun. 1901) in the Georgian SSR. Vestn. Zool. 6:80–82.

E.

Easterla, D. A. 1973. Ecology of the 18 species of Chiroptera at Big Bend National Park, Texas. Part II. Northwest Missouri State Univ. Studies. 34:54–165.

————. 1976. Notes on the second and third newborn of the spotted bat, *Euderma maculatum*, and comments on the species in Texas. Amer. Midl. Nat. 96:499–501.

Easterla, D. A., and L. C. Watkins. 1970. Breeding of *Lasionycteris noctivagans* and *Nycticeius humeralis* in southwestern Iowa. Amer. Midl. Nat. 84:254–255.

Eaton, R. L. 1974. The cheetah. Van Nostrand Reinhold, New York, xii + 178 pp.

————. 1976. The brown hyena: a review of biology, status and conservation. Mammalia 40:377–399.

Eberhard, I. H. 1978. Ecology of the koala, *Phascolarctos cinereus* (Goldfuss) Marsupialia: Phascolarctidae, in Australia. *In* Montgomery (1978), pp. 315–327.

Eberhardt, L. E., and W. C. Hanson. 1978. Long-distance movements of Arctic foxes tagged in northern Alaska. Can. Field-Nat. 92:386–389.

Edmunds, R. M.; J. W. Goertz; and G. Linscombe. 1978. Age ratios, weights, and reproduction of the Virginia opossum in northern Louisiana. J. Mamm. 59:884–885.

Edwards, G. P., and E. H. M. Ealey. 1975. Aspects of the ecology of the swamp wallaby *Wallabia bicolor* (Marsupialia: Macropodidae). Austral. Mamm. 1:307–317.

Egbert, A. L., and A. W. Stokes. 1976. The social behavior of brown bears on an Alaskan salmon stream. *In* Pelton, Lentfer, and Folk (1976), pp. 41–56.

Eger, J. L. 1977. Systematics of the genus *Eumops* (Chiroptera: Molossidae). Roy. Ontario Mus. Life Sci. Contrib., no. 110, 69 pp.

Egoscue, H. J. 1970. A laboratory colony of the Polynesian rat, *Rattus exulans*. J. Mamm. 51:261–268.

————. 1972. Breeding the long-tailed pouched rat, *Beamys hindei*, in captivity. J. Mamm. 53:296–302.

————. 1979. *Vulpes velox*. Mammalian Species, no. 122, 5 pp.

Egoscue, H. J.; J. G. Bittmenn; and J. A. Petrovich. 1970. Some fecundity and longevity records for captive small mammals. J. Mamm. 51:622–623.

Eibatov, T. M. 1976. Natural life span in *Phoca caspica*. Zool. Zhur. 55:1893–1896.

Eisenberg, J. F. 1963. The behavior of heteromyid rodents. Univ. California Publ. Zool. 69:i–iv + 1–100.

————. 1974. The function and motivational basis of hystricomorph vocalizations. Symp. Zool. Soc. London, 34:211–247.

————. 1975. Tenrecs and solenodons in captivity. Int. Zoo Yearbook 15:6–12.

————. 1977. Comparative ecology and reproduction of New World monkeys. *In* Kleiman (1977a), pp. 13–22.

————, ed. 1979. Vertebrate ecology in the northern neotropics. Smithson. Inst. Press, Washington, D.C., 271 pp.

Eisenberg, J. F., and E. Gould. 1966. The behavior of *Solenodon paradoxus* in captivity with comments on the behavior of other Insectivora. Zoologica 51:49–58.

————. 1967. The maintenance of tenrecoid insectivores in captivity. Int. Zoo Yearbook 7:194–196.

————. 1970. The tenrecs: a study in mammalian behavior and evolution. Smithson. Contrib. Zool., no. 27, v + 138 pp.

Eisenberg, J. F., and M. C. Lockhart. 1972. An ecological reconnaissance of Wilpattu National Park, Ceylon. Smithson. Contrib. Zool., no. 101, 118 pp.

Eisenberg, J. F., and E. Maliniak. 1973. Breeding and captive maintenance of the lesser bamboo rat *Cannomys badius*. Int. Zoo Yearbook 13:204–207.

————. 1974. The reproduction of the genus *Microgale* in captivity. Int. Zoo Yearbook 14:108–110.

————. 1978. Reproduction by the two-toed sloth, *Choloepus hoffmanni*, in captivity. Amer. Soc. Mamm. Abstr. Tech. Pap., 58th Ann. Mtg., pp. 41–42.

Eisenberg, J. F., and G. M. McKay. 1970. An annotated checklist of the Recent mammals of Ceylon. Ceylon J. Sci., Biol. Sci. 8:69–99.

Eisenberg, J. F.; G. M. McKay; and M. R. Jainudeen. 1971. Reproductive behavior of the Asiatic elephant (*Elephas maximus maximus* L.). Behaviour 38:193–225.

Eisentraut, M. 1970. Beitrag zur Fortpflanzungsbiologie der Zwergbeutelratte *Marmosa murina* (Didelphidae, Marsupialia). Z. Saugetierk. 35:159–172.

Elbl, A.; U. H. Rahm; and G. Mathys. 1966. Les mammifères et leurs tiques dans la forêt du Rueggege (République Rwandaise). Acta Tropica 23:223–263.

Elgmork, K. 1978. Human impact on a brown bear population (*Ursus arctos* L.). Biol. Conserv. 13:81–103.

El Hilali, M., and J.-P. Veillat. 1975. *Jaculus orientalis*: a true hibernator. Mammalia 39:401–404.

Ellefson, J. O. 1974. A natural history of white-handed gibbons in the Malayan Peninsula. *In* Rumbaugh, D. M., ed., Gibbon and siamang, III, S. Karger, Basel, pp. 1–136.

Ellerman, J. R. 1941. The families and genera of living rodents. II. Family Muridae. British Mus. (Nat. Hist.), London, xii + 690 pp.

————. 1946. Further notes on two little-known murine genera, and preliminary diagnosis of a new species of *Rattus* (subgenus *Cremnomys*) from the eastern Ghats. Ann. Mag. Nat. Hist., ser. 11, 13:204–208.

Ellerman, J. R., and T. C. S. Morrison-Scott. 1966. Checklist of Palaearctic and Indian mammals. British Mus. (Nat. Hist.), London, 810 pp.

Elliott, J. P., and I. M. Cowan. 1978. Territoriality, density, and prey of the lion in Ngorongoro Crater, Tanzania. Can. J. Zool. 56:1726–1734.

Elliott, L. 1978. Social behavior and foraging ecology of the eastern chipmunk (*Tamias striatus*) in the Adirondack Mountains. Smithson. Contrib. Zool., no. 265, vi + 107 pp.

Ellis, L. S., and L. R. Maxson. 1979. Evolution of the chipmunk genera *Eutamias* and *Tamias*. J. Mamm. 60:331–334.

Ellis, L. S.; V. E. Diersing; and D. F. Hoffmeister. 1978. Taxonomic status of short-tailed shrews (*Blarina*) in Illinois. J. Mamm. 59:305–311.

Ellis, R. 1980. The book of whales. Alfred A. Knopf, New York, xvii + 202 pp.

Elwood, R. W. 1975. Paternal and maternal behavior in the Mongolian gerbil. Anim. Behav. 23:766–772.

Emison, W. B.; J. W. Porter; K. C. Norris; and G. J. Apps. 1975. Ecological distribution of the vertebrate animals of the volcanic plains-Otway Range area of Victoria. Victoria Fish. and Wildl. Pap., no. 6, 93 pp.

————. 1978. Survey of the vertebrate fauna in the Grampians-Edenhope area of southwestern Victoria Mem. Natl. Mus. Victoria 39:281–363.

Emmons, L. H. 1978. Sound communication among African rainforest squirrels. Z. Tierpsychol. 47:1–49.

————. 1979. A note on the forefoot of *Myosciurus pumilio*. J. Mamm. 60:431–432.

Enders, R. K. 1966. Attachment, nursing and survival of young in some didelphids. Symp. Zool. Soc. London 15:195–203.

Engstrom, M. D., and D. E. Wilson. 1981. Systematics of *Antrozous dubiaquercus* (Chiroptera: Vespertilionidae), with comments on the status of *Bauerus* Van Gelder. Ann. Carnegie Mus. 50:371–383.

Erdbrink, D. P. 1953. A review of fossil and Recent bears of the Old World with remarks on their phylogeny based upon their dentition. Deventer, Netherlands, 597 pp.

Erlinge, S. 1967. Home range of the otter *Lutra lutra* L. in southern Sweden. Oikos 18:186–209.

————. 1968. Territoriality of the otter *Lutra lutra* L. Oikos 19:81–98.

————. 1977. Spacing strategy in stoat *Mustela erminea*. Oikos 28:32–42.

Eshelkin, I. I. 1976. On the reproduction of *Alticola strelzovi* in the south-east Altai. Zool. Zhur. 55:437–442.

Esher, R. J.; J. L. Wolfe; and J. N. Layne. 1978. Swimming behavior of rice rats (*Oryzomys palustris*) and cotton rats (*Sigmodon hispidus*). J. Mamm. 59:551–558.

Estes, J. A. 1980. *Enhydra lutris*. Mammalian Species, no. 133, 8 pp.

Estes, R. D. 1967. The comparative behavior of Grant's and Thomson's gazelles. J. Mamm. 48:189–209.

————. 1969. Territorial behavior of the wildebeest (*Connochaetes taurinus* Burchell, 1823). Z. Tierpsychol. 26:284–370.

————. 1974. Social organization of the African Bovidae. *In* Geist and Walther (1974), pp. 166–205.

————. 1976. The significance of breeding synchrony in the wildebeest. E. Afr. Wildl. J. 14:135–152.

Estes, R. D., and R. K. Estes. 1974. The biology and conservation of the giant sable antelope, *Hippotragus niger variani* Thomas, 1916. Proc. Acad. Nat. Sci. Philadelphia 126:73–104.

Evans, C. D., and L. S. Underwood. 1978. How many bowheads? Oceanus 21(2):17–23.

Evans, G. D., and E. W. Pearson. 1980. Federal coyote control methods used in the western United States, 1971–77. Wildl. Soc. Bull. 8:34–39.

Everard, C. O. R., and E. S. Tikasingh. 1973. Ecology of the rodents, *Proechimys guyannensis trinitatis* and *Oryzomys capito velutinus*, on Trinidad. J. Mamm. 54:875–886.

Ewer, R. F. 1967. The behaviour of the African giant rat (*Cricetomys gambianus* Waterhouse). Z. Tierpsychol. 24:6–79.

————. 1968. A preliminary survey of the behaviour in captivity of the dasyurid marsupial, *Sminthopsis crassicaudata* (Gould). Z. Tierpsychol. 25:319–365.

————. 1971. The biology and behaviour of a free-living population of black rats (*Rattus rattus*). Anim. Behav. Monogr. 4:127–174.

————. 1973. The carnivores. Cornell Univ. Press, Ithaca, xv + 494 pp.

Ewer, R. F., and C. Wemmer. 1974. The behaviour in captivity of the African civet, *Civettictis civetta* (Schreber). Z. Tierpsychol. 34:359–394.

F.

Fairall, N. 1972. Behavioural aspects of the reproductive physiology of the impala, *Aepyceros melampus* (Licht.). Zool. Afr. 7:167–174.

Fairley, J. S. 1971. The present distribution of the bank vole *Clethrionomys glareolus* Schreber in Ireland. Proc. Roy. Irish Acad., sec. B, 71:183–188.

Fairley, J. S., and J. M. Jones. 1976. A woodland population of small rodents (*Apodemus sylvaticus* (L.) and *Clethrionomys glareolus* Schreber) at Adare, Co. Limerick. Proc. Roy. Irish Acad., sec. B, 76:323–336.

Farentinos, R. C. 1972. Observations on the ecology of the tassel-eared squirrel. J. Wildl. Mgmt. 36:1234–1239.

Farst, D. D.; D. P. Thompson; G. A. Stones; P. M. Burchfield; and M. L. Hughes. 1980. Maintenance and breeding of duikers *Cephalophus* spp. at Gladys Porter Zoo, Brownsville. Int. Zoo Yearbook 20:93–99.

Faust, R., and C. Scherpner. 1967. A note on the breeding of the maned wolf. Int. Zoo Yearbook 7:119.

Fay, F. H. 1978. Belukha whale. *In* Haley (1978), pp. 132–137.

Fayenuwo, J. O., and L. B. Halstead. 1974. Breeding cycle of straw-colored fruit bat, *Eidolon helvum*, Ile-Ife, Nigeria. J. Mamm. 55:453–454.

Feiler, A. 1978a. Bemerkungen über *Phalanger* der "orientalis-Gruppe" nach Tate (1945). Zool. Abhandlungen (Dresden) 34:385–395.

————. 1978b. Über artliche Abgrenzung und innerartliche Ausformung bei *Phalanger maculatus* (Mammalia, Marsupialia, Phalangeridae). Zool. Abhandlungen (Dresden) 35:1–30.

Feist, J. D., and D. R. McCullough. 1976. Behavior patterns and communication in feral horses. Z. Tierpsychol. 41:337–371.

Feldhamer, G. A. 1980. *Cervus nippon*. Mammalian Species, no. 128, 7 pp.

Fellner, K. 1965. Natural rearing of clouded leopards *Neofelis nebulosa* at Frankfurt Zoo. Int. Zoo Yearbook 5:111–113.

Felten, H. 1964. Flughunde der Gattung *Pteropus* von Neukaledonien und den Loyalty-Inseln (Mammalia, Chiroptera). Senckenbergiana Biol. 45:671–683.

Felten, H., and D. Kock. 1972. Weitere Flughunde der Gattung *Pteropus* von den Neuen Hebriden, sowie den Banks- und Torres- Inseln, Pazifischer Ozean. Senckenbergiana Biol. 53:179–188.

Feng Tso-Chien. 1973. A new species of *Ochotona* (Ochotonidae, Mammalia) from Mount Jolmo-Lungma area. Acta Zool. Sinica 19:69–75.

Fenton, M. B. 1975. Observations on the biology of some Rhodesian bats, including a key to the Chiroptera of Rhodesia. Roy. Ontario Mus. Life Sci. Contrib., no. 104, 27 pp.

Fenton, M. B., and T. H. Kunz. 1977. Movements and behavior. *In* Baker, Jones, and Carter (1977), pp. 351–364.

Fenton, M. B., and R. L. Peterson. 1972. Further notes on *Tadarida aloysiisabaudiae* and *Tadarida russata* (Chiroptera: Molossidae—Africa). Can. J. Zool. 50:19–24.

Ferron, J. 1976. Cycle annuel d'activité de l'écureuil roux (*Tamiasciurus hudsonicus*), adultes et jeunes en semi-liberté au Québec. Naturaliste Canadien 103:1–10.

————. 1977. Le comportement de marquage chez le spermophile à mante dorée (*Spermophilus lateralis*). Naturaliste Canadien 104:407–418.

Ferron, J., and J. Prescott. 1977. Gestation, litter size, and number of litters of the red squirrel (*Tamiasciurus hudsonicus*) in Quebec. Can. Field-Nat. 91:83–85.

Field, R. J., and G. Feltner. 1974. Wolverine. Colorado Outdoors 23(2):1–6.

Figala, J.; K. Hoffmann; and G. Goldau. 1973. Zur Jahresperiodik beim Dsungarischen Zwerghamster *Phodopus sungorus* Pallas. Oecologia 12:89–118.

Filonov, C. P., and C. D. Zykov. 1974. Dynamics of moose populations in the forest zone of the European part of the USSR and in the Urals. Naturaliste Canadien 101:605–613.

Findley, J. S. 1967. Insectivores and dermopterans. *In* Anderson and Jones (1967), pp. 87–108.

————. 1972. Phenetic relationships among bats of the genus *Myotis*. Syst. Zool. 21:31–52.

Findley, J. S., and D. E. Wilson. 1974. Observations on the neotropical disk-winged bat, *Thyroptera tricolor* Spix. J. Mamm. 55:562–571.

Finerty, J. P. 1979. Cycles in Canadian lynx. Amer. Nat. 114:453–455.

Fish, J. F.; J. L. Sumich; and G. L. Lingle. 1974. Sounds produced by the gray whale, *Eschrichtius robustus*. Mar. Fish. Rev. 36(4):38–45.

Fisher, J. 1978. Tiger! Tiger! Int. Wildl. 8(3):4–12.

Fisher, J.; N. Simon; and J. Vincent. 1969. Wildlife in danger. Viking Press, New York, 368 pp.

Fisler, G. F. 1971. Age structure and sex ratio in populations of *Reithrodontomys*. J. Mamm. 52:653–662.

————. 1977. Potential behavioral dominance by the Mongolian gerbil. Amer. Midl. Nat. 97:33–41.

Fitch, H. S.; P. Goodrum; and C. Newman. 1952. The armadillo in the southeastern United States. J. Mamm. 33:21–37.

Fitch, J. E., and R. L. Brownell, Jr. 1968. Fish otoliths in cetacean stomachs and their importance in interpreting feeding habits. J. Fish. Res. Bd. Can. 25:2561–2574.

————. 1971. Food habits of the franciscana *Pontoporia blainvillei* (Cetacea: Platanistidae) from South America. Bull. Mar. Sci. 21:626–636.

Fitch, J. H., and K. A. Shump, Jr. 1979. *Myotis keenii*. Mammalian Species, no. 121, 3 pp.

Fitzgerald, A. E. 1976. Diet of the opossum, *Trichosurus vulpecula* (Kerr) in the Orongorongo Valley, Wellington, New Zealand, in relation to food-plant availability. New Zealand J. Zool. 3:399–419.

————. 1978. Aspects of the food and nutrition of the brush-tailed opossum, *Trichosurus vulpecula* (Kerr, 1792), Marsupialia: Phalangeridae, in New Zealand. *In* Montgomery (1978), pp. 289–303.

Fitzgerald, J. P., and R. R. Lechleitner. 1974. Observations on the biology of Gunnison's prairie dog in central Colorado. Amer. Midl. Nat. 92:146–163.

Fleharty, E. D., and M. A. Mares. 1973. Habitat preference and spatial relations of *Sigmodon hispidus* on a remnant prairie in west-central Kansas. Southwestern Nat. 18:21–29.

Fleming, M. R., and A. Cockburn. 1979. *Ningaui:* a new genus of dasyurid for Victoria. Victorian Nat. 96:142–145.

Fleming, T. H. 1971. Population ecology of three species of neotropical rodents. Misc. Publ. Mus. Zool. Univ. Michigan, no. 143, 77 pp.

———. 1972. Aspects of the population dynamics of three species of opossums in the Panama Canal Zone. J. Mamm. 53:619–623.

———. 1973. The reproductive cycles of three species of opossums and other mammals in the Panama Canal Zone. J. Mamm. 54:439–455.

———. 1974. The population ecology of two species of Costa Rican heteromyid rodents. Ecology 55:493–510.

———. 1977. Growth and development of two species of tropical heteromyid rodents. Amer. Midl. Nat. 98:109–123.

Fleming, T. H.; E. T. Hooper; and D. E. Wilson. 1972. Three Central American bat communities: structure, reproductive cycles, and movement patterns. Ecology 53:555–569.

Floody, O. R., and A. P. Arnold. 1975. Uganda kob (*Adenota kob thomasi*): territoriality and the spatial distribution of sexual and agnostic behaviors at a territorial ground. Z. Tierpsychol. 37:192–212.

Floyd, B. L., and M. R. Stromberg. 1981. New Records of the swift fox (*Vulpes velox*) in Wyoming. J. Mamm. 62:650–651.

Flux, J. E. C. 1970. Life history of the mountain hare (*Lepus timidus scoticus*) in north-east Scotland. J. Zool. 161:75–123.

Fogden, M. P. L. 1974. A preliminary field study of the western tarsier, *Tarsius bancanus* Horsfield. *In* Martin, Doyle, and Walker (1974), pp. 151–165.

Folk, G. E., Jr.; A. Larson; and M. A. Folk. 1976. Physiology of hibernating bears. *In* Pelton, Lentfer, and Folk (1976), pp. 373–380.

Fons, R. 1974. Le répertoire comportemental de la pachyure Etrusque, *Suncus etruscus* (Savi, 1822). Terre Vie 28:131–157.

Fontaine, P. A. 1965. Breeding clouded leopards *Neofelis nebulosa* at Dallas Zoo. Int. Zoo Yearbook 5:113–114.

Fontaine, R., and F. V. Du Mond. 1977. The red ouakari in a seminatural environment: potentials for propagation and study. *In* Rainier III and Bourne (1977), pp. 167–236.

Fooden, J. 1963. A revision of the woolly monkeys (genus *Lagothrix*). J. Mamm. 44:213–247.

———. 1969. Taxonomy and evolution of the monkeys of Celebes (Primates: Cercopithecidae). Bibl. Primatol. 10:1–148.

———. 1971. Report on primates collected in western Thailand January–April, 1967. Fieldiana Zool. 59:1–62.

———. 1975. Taxonomy and evolution of liontail and pigtail macaques (Primates: Cercopithecidae). Fieldiana Zool. 67:1–169.

———. 1976a. Primates obtained in peninsular Thailand June–July, 1973, with notes on the distribution of continental Southeast Asian leaf-monkeys (*Presbytis*). Primates 17:95–118.

———. 1976b. Provisional classification and key to living species of macaques (Primates: *Macaca*). Folia Primatol. 25:225–236.

———. 1979. Taxonomy and evolution of the *sinica* group of macaques: I. Species and subspecies accounts of *Macaca sinica*. Primates 10:109–140.

Ford, J. K. B., and H. D. Fisher. 1978. Underwater acoustic signals of the narwhal (*Monodon monoceros*). Can. J. Zool. 56:552–560.

Ford, S. D. 1977. Range, distribution and habitat of the western harvest mouse, *Reithrodontomys megalotis*, in Indiana. Amer. Midl. Nat. 98:422–432.

Foresman, K. R., and R. A. Mead. 1973. Duration of post-implantation in a western subspecies of the spotted skunk (*Spilogale putorius*). J. Mamm. 54:521–523.

Fossey, D. 1972. Vocalizations of the mountain gorilla (*Gorilla gorilla beringei*). Anim. Behav. 20:36–53.

———. 1974. Observations on the home range of one group of mountain gorillas (*Gorilla gorilla beringei*). Anim. Behav. 22:568–581.

Fossey, D., and A. H. Harcourt. 1977. Feeding ecology of free-ranging mountain gorilla (*Gorilla gorilla beringei*). *In* Clutton-Brock (1977), pp. 415–447.

Foster, M. S., and R. M. Timm. 1976. Tent-making by *Artibeus jamaicensis* (Chiroptera: Phyllostomatidae) with comments on plants used by bats for tents. Biotropica 8:265–269.

Fourie, P. B. 1977. Acoustic communication in the rock hyrax. Z. Tierpsychol. 44:194–219.

Fox, B. J., and D. A. Briscoe. 1980. *Pseudomys pilligaensis*, a new species of murid rodent from the Pilliga Scrub, northern New South Wales. Austral. Mamm. 3:109–126.

Fox, M. W. 1975. The wild canids. Van Nostrand Reinhold, New York, xvi + 508 pp.

Fradrich, H. 1974. A comparison of behaviour in the Suidae. *In* Geist and Walther (1974), pp. 133–143.

Frame, L. H., and G. W. Frame. 1976. Female African wild dogs emigrate. Nature 263:227–229.

Frame, L. H.; J. R. Malcolm; G. W. Frame; and H. Van Lawick. 1979. Social organization of African wild dogs (*Lycaon pictus*) on the Serengeti Plains, Tanzania (1967–1978). Z. Tierpsychol. 50:225–249.

Franklin, W. L. 1974. The social behavior of the vicuna. *In* Geist and Walther (1974), pp. 477–487.

Franzmann, A. W. 1978. Moose. *In* Schmidt and Gilbert (1978), pp. 67–81.

———. 1981. *Alces alces*. Mammalian Species, no. 154, 7 pp.

Fraser, J. F. D. 1973. Specific foetal growth rates of cetaceans. J. Zool. 169:111–126.

Frazer, F. C. 1970. An early 17th century record of the Californian grey whale in Icelandic waters. Investig. Cetacea 2:13–20.

Freese, C. 1976. Censusing *Alouatta palliata, Ateles geoffroyi,* and *Cebus capucinus* in the Costa Rican dry forest. *In* Thorington and Heltne (1976), pp. 4–19.

Freese, C. H.; M. A. Freese; and N. Castro R. 1977. The status of callitrichids in Peru. *In* Kleiman (1977a), pp. 121–130.

French, N. R.; D. M. Stoddart; and B. Bobek. 1975. Patterns of demography in small mammal populations. *In* Golley, F. B.; K. Petrusewicz; and L. Ryszkowski, eds., Small mammals: their productivity and population dynamics, Cambridge Press, London, pp. 73–102.

French, T. W. 1981. Notes on the distribution and taxonomy of short-tailed shrews (genus *Blarina*) in the southeast. Brimleyana 6:101–110.

Fritts, S. H., and L. D. Mech. 1981. Dynamics, movements, and feeding ecology of a newly protected wolf population in northwestern Minnesota. Wildl. Monogr., no. 80, 79 pp.

Fritts, S. H., and J. A. Sealander. 1978. Reproductive biology and population characteristics of bobcats (*Lynx rufus*) in Arkansas. J. Mamm. 59:347–353.

Fritzell, E. K. 1978. Aspects of raccoon (*Procyon lotor*) social organization. Can. J. Zool. 56:260–271.

Fry, E. 1971. The scaly-tailed possum *Wyulda squamicaudata* in captivity. Int. Zoo Yearbook 11:44–45.

Fulk, G. W. 1975. Population ecology of rodents in the semiarid shrublands of Chile. Occas. Pap. Mus. Texas Tech Univ., no. 33, 40 pp.

———. 1976. Notes on the activity, reproduction, and social behavior of *Octodon degus*. J. Mamm. 57:495–505.

Fulk, G. W., and A. R. Khokhar. 1980. Observations on the natural history of pika (*Ochotona rufescens*) from Pakistan. Mammalia 44:51–58.

Fuller, T. K. 1978. Variable home-range sizes of female gray foxes. J. Mamm. 59:446–449.

Fuller, T. K., and L. B. Keith. 1980. Wolf population dynamics and prey relationships in northwestern Alberta. J. Wildl. Mgmt. 44:583–601.

Funaioli, U. 1971. Guida breve dei mammiferi della Somalia. Istituto Agronomico per l'Oltremare, Firenze, 232 pp.

Funmilayo, O. 1977. Distribution and abundance of moles (*Talpa europaea* L.) in relation to physical habitat and food supply. Oecologia 30:277–283.

G.

Gabow, S. A. 1975. Behavioral stabilization of a baboon hybrid zone. Amer. Nat. 109:701–712.

Gade, D. W. 1967. The guinea pig in Andean folk culture. Geogr. Rev. 57:213–224.

Gaines, M. S., and R. K. Rose. 1976. Population dynamics of *Microtus ochrogaster* in eastern Kansas. Ecology 57:1145–1161.

Gaisler, J.; V. Holas; and M. Homolka. 1976. Ecology and reproduction of Gliridae (Mammalia) in northern Moravia. Folia Zool. 26:213–228.

Gaisler, J.; G. Madkour; and J. Pelikan. 1972. On the bats (Chiroptera) of Egypt. Acta Sci. Nat. Brno 6:1–40.

Galat, G., and A. Galat-Luong. 1976. La colonisation de la mangrove par *Cercopithecus aethiops sabaeus* au Sénégal. Terre Vie 30:3–30.

————. 1977. Démographie et régime alimentaire d'une troupe de *Cercopithecus aethiops sabaeus* en habitat marginal au nord Sénégal. Terre Vie 31:557–577.

————. 1978. Diet of green monkeys in Senegal. *In* Chivers and Herbert (1978), pp. 257–258.

Galat-Luong, A. 1975. Notes préliminaires sur l'écologie de *Cercopithecus ascanius schmidti* dans les environs de Bangui (R. C. A.). Terre Vie 29:288–297.

Gallagher, M. D., and D. L. Harrison. 1977. Report on the bats (Chiroptera) obtained by the Zaire River Expedition. Bonner Zool. Beitr. 28:19–32.

Gangloff, B. 1975. Beitrag zur Ethologie der Schleichkatzen (Banderlinsang, *Prionodon linsang* [Hardw.] und Banderpalmenroller, *Hemigalus derbyanus* [Gray]). Zool. Garten 45:329–376.

Gardner, A. L. 1971a. Karyotypes of two rodents from Peru, with a description of the highest diploid number recorded from a mammal. Experientia 27:1088–1089.

————. 1971b. Postpartum estrus in a red brocket deer, *Mazama americana*, from Peru. J. Mamm. 52:623–624.

————. 1973. The systematics of the genus *Didelphis* (Marsupialia: Didelphidae) in North and Middle America. Spec. Publ. Mus. Texas Tech Univ., no. 4, 81 pp.

————. 1976. The distributional status of some Peruvian mammals. Occas. Pap. Mus. Zool. Louisiana State Univ., no. 48, 18 pp.

————. 1977. Feeding habits. *In* Baker, Jones, and Carter (1977), pp. 293–350.

Gardner, A. L.; R. K. LaVal; and D. E. Wilson. 1970. The distributional status of some Costa Rican bats. J. Mamm. 51:712–729.

Gardner, A. L., and J. P. O'Neill. 1969. The taxonomic status of *Sturnira bidens* (Chiroptera: Phyllostomatidae) with notes on its karyotype and life history. Occas. Pap. Mus. Zool. Louisiana State Univ., no. 38, 8 pp.

————. 1971. A new species of *Sturnira* (Chiroptera: Phyllostomatidae) from Peru. Occas. Pap. Mus. Zool. Louisiana State Univ., no. 42, 7 pp.

Gardner, A. L., and J. L. Patton. 1972. New species of *Philander* (Marsupialia: Didelphidae) and *Mimon* (Chiroptera: Phyllostomatidae) from Peru. Occas. Pap. Mus. Zool. Louisiana State Univ., no. 43, 12 pp.

————. 1976. Karyotypic variation in oryzomyine rodents (Cricetinae) with comments on chromosomal evolution in the neotropical cricetine complex. Occas. Pap. Mus. Zool. Louisiana State Univ., no. 49, 48 pp.

Garrott, R. A., and D. A. Jenni. 1978. Arboreal behavior of yellow-bellied marmots. J. Mamm. 59:433–434.

Gartlan, J. S. 1970. Preliminary notes on the ecology and behavior of the drill, *Mandrillus leucophaeus* Ritgen, 1824. *In* Napier and Napier (1970), pp. 445–480.

Gartlan, J. S., and C. K. Brain. 1968. Ecology and social variability in *Cercopithecus aethiops* and *C. mitis*. *In* Jay (1968), pp. 253–292.

Gartlan, J. S., and T. T. Struhsaker. 1972. Polyspecific associations and niche separation of rain-forest anthropoids in Cameroon, West Africa. J. Zool. 168:221–266.

Gashwiler, J. S. 1972. Life history notes on the Oregon vole, *Microtus oregoni*. J. Mamm. 53:558–569.

————. 1976. Notes on the reproduction of Trowbridge shrews in western Oregon. Murrelet 57:58–62.

Gaskin, D. E. 1968. The New Zealand Cetacea. New Zealand Mar. Dept. Fish. Res. Bull., no. 1, n.s., 92 pp.

————. 1970. Composition of schools of sperm whales *Physeter catodon* Linn. east of New Zealand. New Zealand J. Mar. Freshwater Res. 4:456–471.

————. 1976. The evolution, zoogeography and ecology of Cetacea. Oceanogr. Mar. Biol. Ann. Rev. 14:247–346.

————. 1977. Harbour porpoise *Phocoena phocoena* (L.) in the western approaches to the Bay of Fundy 1969–75. Rept. Int. Whal. Comm. 27:487–492.

Gaskin, D. E.; P. W. Arnold; and B. A. Blair. 1974. *Phocoena phocoena*. Mammalian Species, no. 42, 8 pp.

Gaskin, D. E., and B. A. Blair. 1977. Age determination of harbour porpoise, *Phocoena phocoena* (L.), in the western North Atlantic. Can. J. Zool. 55:18–30.

Gates, G. R. 1978. Vision in the monotreme echidna (*Tachyglossus aculeatus*). Austral. Zool. 20:147–169.

Gauckler, A., and M. Kraus. 1970. Kennzeichen und Verbreitung von *Myotis brandti* (Eversman, 1845). Z. Saugetierk. 35:113–124.

Gauthier-Pilters, H. 1974. The behaviour and ecology of camels in the Sahara, with special reference to nomadism and water management. *In* Geist and Walther (1974), pp. 542–551.

Gautier-Hion, A. 1971. L'écologie du talapoin du Gabon. Terre Vie 25:427–490.

————. 1973. Social and ecological features of talapoin monkey—comparisons with sympatric cercopithecines. *In* Michael and Crook (1973), pp. 147–170.

Gautun, J. 1975. Périodicité de la reproduction de quelques rongeurs d'une savane préforestière du centre de la Côte-D'Ivoire. Terre Vie 29:265–287.

Gee, C. K.; R. Magleby; W. R. Bailey; R. L. Gum; and L. M. Arthur. 1977. Sheep and lamb losses to predators and other causes in the western United States. U.S. Dept. Agric., Agric. Econ. Rept., no. 369, 41 pp.

Geist, V. 1971. Mountain sheep. A study in behavior and evolution. Univ. Chicago Press, xv + 383 pp.

————. 1974. Bound by tradition. Anim. Kingdom 77(6):2–9.

Geist, V., and F. Walther. 1974. The behaviour of ungulates and its relation to management. Int. Union Conserv. Nat. Publ., n.s., no. 24, 2 vols.

Geluso, K. N.; J. S. Altenbach; and D. E. Wilson. 1976. Bat mortality: pesticide poisoning and migratory stress. Science 194:184–186.

Genest, H., and G. Dubost. 1974. Pair-living in the mara (*Dolichotis patagonum* Z.). Mammalia 38:155–162.

Genest-Villard, H. 1978a. Radio-tracking of a small rodent, *Hybomys univittatus*, in an African equatorial forest. Bull. Carnegie Mus. Nat. Hist., no. 6, pp. 92–96.

————. 1978b. Révision systématique du genre *Graphiurus* (Rongeurs, Gliridae). Mammalia 42:391–426.

————. 1979. Écologie de *Steatomys opimus* Pousargues, 1894 (Rongeurs, Dendromurides) en Afrique centrale. Mammalia 43:275–294.

Gengozian, N.; J. S. Batson; and T. A. Smith. 1977. Breeding of tamarins (*Saguinus* spp.) in the laboratory. *In* Kleiman (1977a), pp. 207–213.

Genoways, H. H., and R. J. Baker. 1972. *Stenoderma rufum*. Mammalian Species, no. 18, 4 pp.

Genoways, H. H., and E. C. Birney. 1974. *Neotoma alleni*. Mammalian Species, no. 41, 4 pp.

Genoways, H. H., and J. R. Choate. 1972. A multivariate analysis of systematic relationships among populations of the short-tailed shrew (genus *Blarina*) in Nebraska. Syst. Zool. 21:106–116.

Genoways, H. H., and J. K. Jones, Jr. 1972. Variation and ecology in a local population of the vesper mouse (*Nyctomys sumichrasti*). Occas. Pap. Mus. Texas Tech Univ., no. 3, 22 pp.

Genoways, H. H., and S. L. Williams. 1979a. Notes on bats (Mammalia: Chiroptera) from Bonaire and Curacao, Dutch West Indies. Ann. Carnegie Mus. 48:311–321.

————. 1979b. Records of bats (Mammalia: Chiroptera) from Suriname. Ann. Carnegie Mus. 48:323–335.

————. 1980. Results of the Alcoa Foundation-Suriname Expeditions. I. A new species of bat of the genus *Tonatia* (Mammalia: Phyllostomatidae). Ann. Carnegie Mus. 49:203–211.

Gensch, W. 1963a. Breeding tupaias. Int. Zoo Yearbook 4:75–76.

————. 1963b. Successful rearing of the binturong Arctictis binturong. Int. Zoo Yearbook 4:79–80.

————. 1965. Birth and rearing of a spectacled bear Tremarctos ornatus at Dresden Zoo. Int. Zoo Yearbook 5:11.

George, G. G. 1979. The status of endangered Papua New Guinea mammals. In Tyler (1979), pp. 93–100.

George, G. G., and U. Schurer. 1978. Some notes on macropods commonly misidentified in zoos. Int. Zoo Yearbook 18:152–156.

George, S. B.; J. R. Choate; and H. H. Genoways. 1981. Distribution and taxonomic status of Blarina hylophaga Elliot (Insectivora: Soricidae). Ann. Carnegie Mus. 50:493–513.

George, W. 1974. Notes on the ecology of gundis (F. Ctenodactylidae). Symp. Zool. Soc. London 34:143–160.

————. 1978. Reproduction in female gundis (Rodentia: Ctenodactylidae). J. Zool. 185:57–71.

————. 1979. The chromosomes of the hystricomorphous family Ctenodactylidae (Rodentia: ?Sciuromorpha) and their bearing on the relationships of the four living genera. Zool. J. Linnean Soc. 65:261–280.

Germain, M., and F. Petter. 1973. Addition à la liste des rongeurs myomorphes de la République Centrafricaine: Zelotomys instans Thomas, 1915. Mammalia 37:683.

Getz, L. L. 1972. Social structure and aggressive behavior in a population of Microtus pennsylvanicus. J. Mamm. 53:310–317.

Gewalt, W. 1979a. Eine "neue" Delphinart in Delphinarien—der Karib-Delphin Sotalia guianensis (Van Beneden, 1864) (?). Saugetierk. Mitt. 27:288–291.

————. 1979b. The Commerson's dolphin (Cephalorhynchus commersonii)—capture and first experiences. Aquatic Mammals 7:37–40.

Ghose, R. K. 1965. A new species of mongoose (Mammalia: Carnivora: Viverridae) from West Bengal, India. Proc. Zool. Soc. Calcutta 18:173–178.

————. 1978. Observations on the ecology and status of the hispid hare in Rajagarh Forest, Darrang District, Assam, in 1975 and 1976. J. Bombay Nat. Hist. Soc. 75:206–209.

Ghose, R. K., and U. Chaturvedi. 1972. Extension of range of the mongoose Herpestes palustris Ghose (Mammalia: Carnivora: Viverridae), with a note on its endoparasitic nematode. J. Bombay Nat. Hist. Soc. 69:412–413.

Gier, H. T. 1975. Ecology and behavior of the coyote (Canis latrans). In Fox (1975), pp. 247–262.

Giger, R. D. 1973. Movements and homing in Townsend's mole near Tillamook, Oregon. J. Mamm. 54:648–659.

Gijzen, A., and S. Smet. 1974. Seventy years okapi. Acta Zool. Pathol., no. 59, 111 pp.

Gilbert, B. K., and L. D. Roy. 1977. Prevention of black bear damage to beeyards using aversive conditioning. In Phillips and Jonkel (1977), pp. 93–102.

Gilmore, D. P. 1977. The success of marsupials as introduced species. In Stonehouse and Gilmore (1977), pp. 169–178.

Gilmore, R. M. 1978. Right whale. In Haley (1978), pp. 62–69.

Gipson, P. S., and J. A. Sealander. 1977. Ecological relationships of white-tailed deer and dogs in Arkansas. In Phillips and Jonkel (1977), pp. 3–16.

Glander, K. E. 1978. Howling monkey feeding behavior and plant secondary compounds: a study of strategies. In Montgomery (1978), pp. 561–573.

Glenn, L. P.; J. W. Lentfer; J. B. Faro; and L. H. Miller. 1976. Reproductive biology of female brown bears (Ursus arctos), McNeil River, Alaska. In Pelton, Lentfer, and Folk (1976), pp. 381–390.

Gliwicz, J. 1973. Characteristics of a population of Proechimys semispinosus (Tomes, 1860)—a rodent species of the tropical rain forest. Bull. Acad. Polonaise Sci., Ser. Sci. Biol. 21:413–418.

Godfrey, G. K. 1969. Reproduction in a laboratory colony of the marsupial mouse Smithopsis larapinta (Marsupialia: Dasyuridae). Austral. J. Zool. 17:637–654.

————. 1975. A study of oestrus and fecundity in a laboratory colony of mouse opossums (Marmosa robinsoni). J. Zool. 175:541–555.

————. 1979. Gestation period in the common shrew, Sorex coronatus (araneus) fretalis. J. Zool. 189:548–555.

Godfrey, G. K., and P. Crowcroft. 1971. Breeding the fat-tailed marsupial mouse Sminthopsis crassicaudata in captivity. Int. Zoo Yearbook 11:33–38.

Godin, A. J. 1977. Wild mammals of New England. Johns Hopkins Univ. Press, Baltimore, xii + 304 pp.

Goehring, H. H. 1972. Twenty-year study of Eptesicus fuscus in Minnesota. J. Mamm. 53:201–207.

Goertz, J. W. 1970. An ecological study of Neotoma floridana in Oklahoma. J. Mamm. 51:91–104.

————. 1971. An ecological study of Microtus pinetorum in Oklahoma. Amer. Midl. Nat. 86:1–12.

Goertz, J. W.; R. M. Dawson; and E. E. Mowbray. 1975. Response to nest boxes and reproduction by Glaucomys volans in northern Louisiana. J. Mamm. 56:933–939.

Goodall, A. G. 1977. Feeding and ranging behaviour of a mountain gorilla group (Gorilla gorilla beringei) in the Tshibinda-Kahuzi region (Zaire). In Clutton-Brock (1977), pp. 449–479.

Goodall, A. G., and C. P. Groves. 1977. The conservation of eastern gorillas. In Rainier III and Bourne (1977), pp. 599–637.

Goodall, J. 1977. Infant killing and cannibalism in free-living chimpanzees. Folia Primatol. 28:259–282.

Goodall, R. N. P. 1978. Report on the small cetaceans stranded on the coasts of Tierra del Fuego. Sci. Rept. Whales Res. Inst. 30:197–230.

Goodman, M. 1975. Protein sequence and immunological specificity. Their role in phylogenetic studies of primates. In Luckett and Szalay (1975), pp. 219–248.

Goodwin, G. G., and A. M. Greenhall. 1961. A review of the bats of Trinidad and Tobago. Bull. Amer. Mus. Nat. Hist. 122:187–302.

Goodwin, H. A., and J. M. Goodwin. 1973. List of mammals which have become extinct or are possibly extinct since 1600. Int. Union Conserv. Nat. Occas. Pap., no. 8, 20 pp.

Goodwin, M. K. 1979. Notes on caravan and play behavior in young captive Sorex cinereus. J. Mamm. 60:411–413.

Goodwin, R. E. 1970. The ecology of Jamaican bats. J. Mamm. 51:571–579.

————. 1979. The bats of Timor: systematics and ecology. Bull. Amer. Mus. Nat. Hist. 163:75–122.

Gopalakrishna, A. 1955. Observations on the breeding habits and ovarian cycle in the Indian sheath-tailed bat, Taphozous longimanus (Hardwicke). Proc. Natl. Inst. Sci. India 21B:29–41.

Gopalakrishna, A., and P. N. Choudhari. 1977. Breeding habits and associated phenomena in some Indian bats. Part I—Rousettus leschenaulti (Desmarest)—Megachiroptera. J. Bombay Nat. Hist. Soc. 74:1–16.

Gordon, G., and P. M. Johnson. 1973. Rock rat in north Queensland. Queensland Div. Plant Industry Adv. Leaf., no. 1192, 3 pp.

Gordon, G., and B. C. Lawrie. 1977. The rufescent bandicoot, Echymipera rufescens (Peters & Doria) on Cape York Peninsula. Austral. Wildl. Res. 5:41–45.

Gordon, G.; D. G. McGreevy; and B. C. Lawrie. 1978. The yellow-footed rock-wallaby, Petrogale xanthopus Gray (Macropodidae), in Queensland. Austral. Wildl. Res. 5:295–297.

Gordon, G. H. 1978. Distribution of sibling species of the Praomys (Mastomys) natalensis group in Rhodesia (Mammalia: Rodentia). J. Zool. 186:397–401.

Gorman, M. L. 1976. Seasonal changes in the reproductive pattern of feral Herpestes auropunctatus (Carnivora: Viverridae), in the Fijian Islands. J. Zool. 178:237–246.

Gosling, L. M. 1974. The social behavior of Coke's hartebeest (Alcelaphus buselaphus cokei). In Geist and Walther (1974), pp. 488–511.

Gosnell, M. 1977. Carlsbad's famous bats are dying off. Natl. Wildl. 15(4):28–33.

Goss, C. M.; L. T. Popejoy, II; J. L. Fusiler; and T. M. Smith. 1968. Observations on the relationship between embryological development, time of conception, and gestation. In Rosenblum and Cooper (1968), pp. 171–191.

Gould, E. 1965. Evidence for echolocation in the Tenrecidae of Madagascar. Proc. Amer. Philos. Soc. 109:352–360.

————. 1969. Communication in three genera of shrews (Soricidae): Suncus, Blarina, & Cryptotis. In Communications in Behavioral Biology, Academic Press, part A, 3:11–31.

————. 1977. Foraging behavior of

Pteropus vampyrus on the flowers of *Durio zibethinus*. Malayan Nat. J. 30:53–57.

―――. 1978a. Rediscovery of *Hipposideros ridleyi* and seasonal reproduction in Malaysian bats. Biotropica 10:30–32.

―――. 1978b. The behavior of the moonrat, *Echinosorex gymnurus* (Erinaceidae) and the pentail shrew, *Ptilocercus lowi* (Tupaiidae) with comments on the behavior of other Insectivora. Z. Tierpsychol. 48:1–27.

Gould, E., and J. F. Eisenberg. 1966. Notes on the biology of the Tenrecidae. J. Mamm. 47:660–686.

Gould, E.; N. C. Negus; and A. Novick. 1964. Evidence for echolocation in shrews. J. Exp. Zool. 156:19–38.

Goulden, E. A., and J. Meester. 1978. Notes on the behaviour of *Crocidura* and *Myosorex* (Mammalia: Soricidae) in captivity. Mammalia 42:197–207.

Grant, T. R. 1973. Dominance and association among members of a captive and free-ranging group of grey kangaroos (*Macropus giganteus*). Anim. Behav. 21:449–456.

Grant, T. R., and F. N. Carrick. 1978. Some aspects of the ecology of the platypus, *Ornithorhynchus anatinus*, in the upper Shoalhaven River, New South Wales. Austral. Zool. 20:181–199.

Gray, G. G., and C. D. Simpson. 1980. *Ammotragus lervia*. Mammalian Species, no. 144, 7 pp.

Greegor, D. H., Jr. 1980. Diet of the little hairy armadillo, *Chaetophractus vellerosus*, of northwestern Argentina. J. Mamm. 61:331–334.

Green, C. A.; H. Keogh; D. H. Gordon; M. Pinto; and E. K. Hartwig. 1980. The distribution, identification, and naming of the *Mastomys natalensis* species complex in southern Africa (Rodentia: Muridae). J. Zool. 192:17–23.

Green, C. P. 1979. People: an endangered species? Natl. Wildl. Fed., Washington, D.C. 14 pp.

Green, J. S., and J. T. Flinders. 1980. *Brachylagus idahoensis*. Mammalian Species, no. 125, 4 pp.

Green, R. H. 1967. Notes on the devil (*Sarcophilus harrisii*) and the quoll (*Dasyurus viverrinus*) in north-eastern Tasmania. Rec. Queen Victoria Mus., no. 27, 13 pp.

―――. 1968. The murids and small dasyurids in Tasmania. Parts 3 and 4. Rec. Queen Victoria Mus., no. 32, 19 pp.

―――. 1972. The murids and small dasyurids in Tasmania. Parts 5, 6 and 7. Rec. Queen Victoria Mus., no. 46, 34 pp.

Green, S., and K. Minkowski. 1977. The lion-tailed monkey and its South Indian forest habitat. *In* Rainier III and Bourne (1977), pp. 289–337.

Greenbaum, I. F., and J. K. Jones, Jr. 1978. Noteworthy records of bats from El Salvador, Honduras, and Nicaragua. Occas. Pap. Mus. Texas Tech Univ., no. 55, 7 pp.

Greenhall, A. M. 1968. Notes on the behavior of the false vampire bat. J. Mamm. 49:337–340.

―――. 1972. The biting and feeding habits of the vampire bat, *Desmodus rotundus*. J. Zool. 168:451–461.

Greer, J. K. 1965. Mammals of Malleco Province Chile. Publ. Mus. Michigan State Univ., Biol. Ser., 3:49–152.

Griffiths, M. 1968. Echidnas. Pergamon Press, Oxford, ix + 282 pp.

―――. 1978. The biology of monotremes. Academic Press, New York, 367 pp.

Grimwood, I. R. 1969. Notes on the distribution and status of some Peruvian mammals. Spec. Publ. Amer. Comm. Int. Wildl. Protection, no. 21, v + 86 pp.

―――. 1976. The Palawan stink badger. Oryx 13:297.

Grobler, J. H. 1974. Aspects of the biology, population ecology and behaviour of the sable *Hippotragus niger niger* (Harris, 1838) in the Rhodes Matopos National Park, Rhodesia. Arnoldia 7(6):1–36.

Gross, J. E.; L. C. Stoddart; and F. H. Wagner. 1974. Demographic analysis of a northern Utah jackrabbit population. Wildl. Monogr., no. 40, 68 pp.

Groves, C. P. 1967a. On the gazelles of the genus *Procapra* Hodgson, 1864. Z. Saugetierk. 32:144–149.

―――. 1967b. On the rhinoceroses of south-east Asia. Saugetierk. Mitt. 15:221–237.

―――. 1969a. Systematics of the anoa (Mammalia, Bovidae). Beaufortia 17:1–12.

―――. 1969b. On the smaller gazelles of the genus *Gazella* de Blainville, 1916. Z. Saugetierk. 34:38–60.

―――. 1970a. The forgotten leaf-eaters, and the phylogeny of the Colobinae. *In* Napier and Napier (1970), pp. 555–587.

―――. 1970b. Population systematics of the gorilla. J. Zool. 161:287–300.

―――. 1971a. Systematics of the genus *Nycticebus*. Proc. 3rd Int. Congr. Primatol. 1:44–53.

―――. 1971b. Distribution and place of origin of the gorilla. Man 6:44–51.

―――. 1972a. Systematics and phylogeny of gibbons. *In* Rumbaugh (1972a), pp. 1–89.

―――. 1972b. *Ceratotherium simum*. Mammalian Species, no. 8, 6 pp.

―――. 1974. Taxonomy and phylogeny of prosimians. *In* Martin, Doyle, and Walker (1974), pp. 449–473.

―――. 1975a. The taxonomy of *Moschus* (Mammalia, Artiodactyla), with particular reference to the Indian region. J. Bombay Nat. Hist. Soc. 72:662–676.

―――. 1975b. Taxonomic notes on the white rhinoceros *Ceratotherium simum* (Burchell, 1817). Saugetierk. Mitt. 23:200–212.

―――. 1976. The origin of the mammalian fauna of Sulawesi (Celebes). Z. Saugetierk. 41:201–216.

―――. 1978a. Phylogenetic and population systematics of the mangabeys (Primates: Cercopithecoidea). Primates 19:1–34.

―――. 1978b. The taxonomic status of the dwarf blue sheep (Artiodactyla: Bovidae). Saugetierk. Mitt. 26:177–183.

―――. 1980a. Notes on the systematics of *Babyrousa* (Artiodactyla, Suidae). Zool. Meded. 55:29–46.

―――. 1980b. A further note on *Moschus*. J. Bombay Nat. Hist. Soc. 77:130–132.

―――. 1981. Systematic relationships in the Bovini (Artiodactyla, Bovidae). Z. Saugetierk. 19:264–278.

Groves, C. P., and P. Grubb. 1974. A new duiker from Rwanda. Rev. Zool. Afr. 88:189–196.

Groves, C. P., and F. Kurt. 1972. *Dicerorhinus sumatrensis*. Mammalian Species, no. 21, 6 pp.

Groves, C. P., and D. P. Willoughby. 1981. Studies on the taxonomy and phylogeny of the genus *Equus*. I. Subgeneric classification of the Recent species. Mammalia 45:321–354.

Grubb, P. 1973. Distribution, divergence and speciation of the drill and mandrill. Folia Primatol. 20:161–177.

―――. 1977. Variation and incipient speciation in the African buffalo. Z. Saugetierk. 37:121–144.

―――. 1981. *Equus burchelli*. Mammalian Species, no. 157, 9 pp.

Grzimek, B., ed. 1975. Grzimek's animal life encyclopedia. Mammals, I–IV. Van Nostrand Reinhold, New York, vols. 10–13.

Gucwinska, H. 1971. Development of six-banded armadillos *Euphractus sexcinctus* at Wroclaw Zoo. Int. Zoo Yearbook 11:88–89.

Guggisberg, C.A.W. 1975. Wild cats of the world. Taplinger, New York, 328 pp.

Guiler, E. R. 1961. Breeding season of the thylacine. J. Mamm. 42:396–397.

―――. 1970a. Observations on the Tasmanian devil, *Sarcophilus harrisii* (Marsupialia: Dasyuridae). I. Numbers, home range, movements, and food in two populations. Austral. J. Zool. 18:49–62.

―――. 1970b. Observations on the Tasmanian devil, *Sarcophilus harrisii* (Marsupialia: Dasyuridae). II. Reproduction, breeding, and growth of pouch young. Austral. J. Zool. 18:63–70.

―――. 1971a. Food of the potoroo (Marsupialia, Macropodidae). J. Mamm. 52:232–234.

―――. 1971b. The husbandry of the potoroo *Potorous tridactylus*. Int. Zoo Yearbook 11:21–22.

―――. 1971c. The Tasmanian devil *Sarcophilus harrisii* in captivity. Int. Zoo Yearbook 11:32–33.

Guiler, E. R., and D. A. Kitchener. 1967. Further observations on longevity in the wild potoroo, *Potorous tridactylus*. Austral. J. Sci. 30:105–106.

Gulland, J. A. 1974. Distribution and abundance of whales in relation to basic productivity. *In* Schevill (1974), pp. 27–52.

―――. 1976. Antarctic baleen whales:

history and prospects. Polar Record 18:5–13.

Gulotta, E. F. 1971. *Meriones unguiculatus.* Mammalian Species, no. 3, 5 pp.

Gum, R. L.; L. M. Arthur; and R. S. Magleby. 1978. Coyote control: a simulation evaluation of alternative strategies. U.S. Dept. Agric., Agric. Econ. Rept., no. 408, 49 pp.

Gunderson, H. L. 1978. A mid-continent irruption of Canada lynx, 1962–1963. Prairie Nat. 10:71–80.

Gunn, A.; F. L. Miller; and D. C. Thomas. 1981. The current status and future of Peary caribou *Rangifer tarandus pearyi* on the Arctic islands of Canada. Biol. Conserv. 19:283–296.

Gunson, J. R., and R. R. Bjorge. 1979. Winter denning of the striped skunk in Alberta. Can. Field-Nat. 93:252–258.

H.

Hadidian, J., and I. S. Bernstein. 1979. Female reproductive cycles and birth data from an Old World monkey colony. Primates 20:429–442.

Hafner, D. J.; J. C. Hafner; and M. S. Hafner. 1979. Systematic status of kangaroo mice, genus *Microdipodops:* morphometric, chromosomal, and protein analyses. J. Mamm. 60:1–10.

Hafner, D. J.; K. E. Petersen; and T. L. Yates. 1981. Evolutionary relationships of jumping mice (genus *Zapus*) of the southwestern United States. J. Mamm. 62:501–512.

Hafner, M. S., and D. J. Hafner. 1979. Vocalizations of grasshopper mice (genus *Onychomys*). J. Mamm. 60:85–94.

Haiduk, M. W.; J. W. Bickham; and D. J. Schmidly. 1979. Karyotypes of six species of *Oryzomys* from Mexico and Central America. J. Mamm. 60:610–615.

Haley, D., ed. 1978. Marine mammals of eastern North Pacific and Arctic waters. Pacific Search Press, Seattle, 256 pp.

Hall, E. R. 1946. Mammals of Nevada. Univ. California Press, Berkeley, xi + 710 pp.

———. 1951. American weasels. Univ. Kansas Publ. Mus. Nat. Hist. 4:1–466.

———. 1955. Handbook of mammals of Kansas. Univ. Kansas Mus. Nat. Hist. Misc. Publ., no. 7, 303 pp.

———. 1981. The mammals of North America. John Wiley & Sons, New York, 2 vols.

Hall, E. R., and W. W. Dalquest. 1963. The mammals of Veracruz. Univ. Kansas Publ. Mus. Nat. Hist. 14:165–362.

Hall, E. S., Jr. 1973. Archaeological and recent evidence for expansion of moose range in northern Alaska. J. Mamm. 54:294–295.

Hall, H. T., and J. D. Newsom. 1978. Summer home ranges and movements of bobcats in bottomland hardwoods of southern Louisiana. Proc. Ann. Conf. S. E. Assoc. Fish and Wildl. Agencies 30:427–436.

Hall, K.R.L. 1967. Social interactions of the adult male and adult females of a patas monkey group. *In* Altmann (1967), pp. 261–280.

———. 1968. Behaviour and ecology of the wild patas monkey, *Erythrocebus patas,* in Uganda. *In* Jay (1968), pp. 32–119 (reprinted from J. Zool. 148:15–87).

Hall, K.R.L., and I. De Vore. 1965. Baboon social behavior. *In* De Vore (1965), pp. 53–110.

Hallett, A. F., and J. Meester. 1971. Early postnatal development of the South African hamster *Mystromys albicaudatus.* Zool. Afr. 6:221–228.

Hallett, J. G. 1978. *Parascalops breweri.* Mammalian Species, no. 98, 4 pp.

Hall-Martin, A. J., and J. D. Skinner. 1978. Observations on puberty and pregnancy in female giraffe (*Giraffa camelopardalis*). S. Afr. J. Wildl. Res. 8:91–94.

Halls, L. K. 1978. White-tailed deer. *In* Schmidt and Gilbert (1978), pp. 43–65.

Haltenorth, T., and H. H. Roth. 1968. Short review of the biology and ecology of the red fox. Saugetierk. Mitt. 16:339–352.

Hamilton, R. B., and D. T. Stalling. 1972. *Lasiurus borealis* with five young. J. Mamm. 53:190.

Hamilton, W. J., III. 1962. Reproductive adaptations of the red tree mouse. J. Mamm. 43:486–504.

Hamilton, W. J., III; R. E. Buskirk; and W. H. Buskirk. 1978. Omnivory and utilization of food resources by chacma baboons, *Papio ursinus.* Amer. Nat. 112:911–924.

Hamilton-Smith, E. 1979. Endangered and threatened Chiroptera of Australia and the Pacific region. *In* Tyler (1979), pp. 85–91.

———. 1980. The status of Australian Chiroptera. Proc. 5th Int. Bat Res. Conf., pp. 199–205.

Handley, C. O., Jr. 1959. A review of the genus *Hoplomys* (thick-spined rats), with description of a new form from Isla Escudo de Veraguas, Panama. Smithson. Misc. Coll. 139:1–10.

———. 1976. Mammals of the Smithsonian Venezuelan Project. Brigham Young Univ. Sci. Bull., Biol. Ser., 20(5):1–89.

———. 1980a. Inconsistencies in formation of family-group and subfamily-group names in Chiroptera. Proc. 5th Int. Bat Res. Conf., pp. 9–13.

———. 1980b. Mammals. *In* Linzey, D. W., ed., Endangered and threatened plants and animals of Virginia, Virginia Polytechnic Inst., Blacksburg, pp. 483–621.

Handley, C. O., Jr., and L. K. Gordon. 1979. New species of mammals from northern South America: mouse possums, genus *Marmosa* Gray. *In* Eisenberg (1979), pp. 65–72.

Handley, C. O., Jr., and E. Mondolfi. 1963. A new species of fish-eating rat, *Ichthyomys,* from Venezuela (Rodentia, Cricetidae). Acta Biol. Venezuelica 3:417–419.

Hanks, J. 1979. The struggle for survival. The elephant problem. Mayflower Books, New York, 176 pp.

Hanselka, C. W.; J. M. Inglis; and H. G. Applegate. 1971. Reproduction in the black-tailed jackrabbit in southwestern Texas. Southwestern Nat. 16:214–217.

Hansen, C., and E. D. Fleharty. 1974. Structural ecological parameters of a population of *Peromyscus maniculatus* in west-central Kansas. Southwestern Nat. 19:293–303.

Hansson, L. 1968. Population densities of small rodents in open field habitats in south Sweden in 1964–1967. Oikos 19:53–60.

Happold, D.C.D. 1977. A population study on small rodents in the tropical rain forest of Nigeria. Terre Vie 31:385–457.

Happold, D.C.D., and M. Happold. 1978a. The fruit bats of western Nigeria. Nigerian Field 43:30–37.

———. 1978b. The fruit bats of western Nigeria. Part 2. Nigerian Field 43:72–77.

Happold, M. 1976a. Social behavior of the conilurine rodents (Muridae) of Australia. Z. Tierpsychol. 40:113–182.

———. 1976b. Reproductive biology and developments in the conilurine rodents (Muridae) of Australia. Austral. J. Zool. 24:19–26.

Harcourt, A. H. 1978. Strategies of emigration and transfer by primates, with particular reference to gorillas. Z. Tierpsychol. 48:401–420.

———. 1979. Social relationships between adult male and female mountain gorillas in the wild. Anim. Behav. 27:325–342.

Harcourt, A. H., and Kai Curry-Lindhal. 1978. The FPS mountain gorilla project—a report from Rwanda. Oryx 14:316–324.

Harcourt, A. H.; K. J. Stewart; and D. Fossey. 1976. Male emigration and female transfer in wild mountain gorilla. Nature 263:226–227.

Hardin, J. W.; R. W. Barbour; and W. H. Davis. 1970. Observations on the home range of a yellow-nosed cotton rat, *Sigmodon ochrognathus.* Southwestern Nat. 14:353–355.

Harding, L. E. 1976. Den-site characteristics of arctic coastal grizzly bears (*Ursus arctos* L.) on Richards Island, Northwest Territories, Canada. Can. J. Zool. 54:1357–1363.

Harington, C. R. 1968. Denning habits of the polar bear (*Ursus maritimus* Phipps). Can. Wildl. Serv. Rept. Ser., no. 5, 30 pp.

Harland, R. M.; P. J. Blancher; and J. S. Millar. 1979. Demography of a population of *Peromyscus leucopus.* Can. J. Zool. 57:323–328.

Harper, F. 1945. Extinct and vanishing mammals of the Old World. Spec. Publ. Amer. Comm. Int. Wildl. Protection, no. 12, xv + 850 pp.

Harper, R. J. 1977. "Caravanning" in *Sorex* species. J. Zool. 183:541.

Harrington, F. H., and L. D. Mech. 1979. Wolf howling and its role in territory maintenance. Behaviour 68:207–249.

Harrington, J. E. 1975. Field observations of social behavior of *Lemur fulvus fulvus* E. Geoffroy 1812. *In* Tattersall and Sussman (1975), pp. 259–279.

————. 1978. Diurnal behavior of *Lemur mongoz* at Ampijora, Madagascar. Folia Primatol. 29:291–302.

Harrison, C. S. 1979. Sighting of Cuvier's beaked whale (*Ziphius cavirostris*) in the Gulf of Alaska. Murrelet 60:35–36.

Harrison, C. S., and J. D. Hall. 1978. Alaskan distribution of the beluga whale, *Delphinapterus leucas*. Can. Field-Nat. 92:235–241.

Harrison, D. L. 1964, 1968, 1972. The mammals of Arabia. Ernest Benn Ltd., London, 3 vols.

————. 1967. Observations on some rodents from Tunisia, with the description of a new gerbil (Gerbillinae: Rodentia). Mammalia 31:381–389.

————. 1975a. *Macrophyllum macrophyllum*. Mammalian Species, no. 62, 3 pp.

————. 1975b. A new species of African free-tailed bat (Chiroptera: Molossidae) obtained by the Zaire River Expedition. Mammalia 39:313–318.

————. 1978. A critical examination of alleged sibling species in the lesser three-toed jerboas (subgenus *Jaculus*) of the North African and Arabian deserts. Bull. Carnegie Mus. Nat. Hist., no. 6, pp. 77–80.

————. 1979. A new species of pipistrelle bat (*Pipistrellus:* Vespertilionidae) from Oman, Arabia. Mammalia 43:573–576.

Harrison, R. J. 1969. Reproduction and reproductive organs. *In* Andersen (1969), pp. 296–322.

Harrison, R. J., and R. L. Brownell, Jr. 1971. The gonads of the South American dolphins, *Inia geoffrensis, Pontoporia blainvillei*, and *Sotalia fluviatilis*. J. Mamm. 52:413–419.

Hart, J. A., and R. M. Timm. 1978. Observations on the aquatic genet in Zaire. Carnivore 1:130–131.

Hartman, D. S. 1979. Ecology and behavior of the manatee (*Trichechus manatus*) in Florida. Amer. Soc. Mamm. Spec. Publ., no. 5, viii + 153 pp.

Harvey, M. J. 1976. Home range, movements, and diel activity of the eastern mole, *Scalopus aquaticus*. Amer. Midl. Nat. 95:436–445.

Harvie, A. E. 1973. Diet of the opossum (*Trichosurus vulpecula* Kerr) on farmland northeast of Waverly, New Zealand. Proc. New Zealand Ecol. Soc. 20:48–52.

Hasler, J. F. 1975. A review of reproduction and sexual maturation in the microtine rodents. The Biologist 57:52–86.

Hasler, J. F., and E. M. Banks. 1975. Reproductive performance and growth in captive collared lemmings (*Dicrostonyx groenlandicus*). Can. J. Zool. 53:777–787.

Hasler, J. F., and M. W. Sorenson. 1974. Behavior of the tree shrew, *Tupaia chinensis,* in captivity. Amer. Midl. Nat. 91:294–314.

Hasler, M. J.; J. F. Hasler; and A. V. Nalbandov. 1977. Comparative breeding biology of musk shrews (*Suncus murinus*) from Guam and Madagascar. J. Mamm. 58:285–290.

Hassinger, J. D. 1973. A survey of the mammals of Afghanistan resulting from the 1965 Street Expedition (excluding bats). Fieldiana Zool. 60:i–xi + 1–195.

Hatt, R. T. 1959. The mammals of Iraq. Misc. Publ. Mus. Zool. Univ. Michigan, no. 106, 113 pp.

Haugen, A. O. 1974. Reproduction in the plains bison. Iowa State J. Res. 49:1–8.

————. 1975. Reproductive performance of white-tailed deer in Iowa. J. Mamm. 56:151–159.

Hausfater, G. 1974. History of three little-known New World populations of macaques. Lab. Primate Newsl. 13(1):16–18.

Hawes, M. L. 1977. Home range, territoriality, and ecological separation in sympatric shrews, *Sorex vagrans* and *Sorex obscurus*. J. Mamm. 58:354–367.

Hawkins, R. E., and W. D. Klimstra. 1970. A preliminary study of the social organization of white-tailed deer. J. Wildl. Mgmt. 34:407–419.

Hayden, P., and R. G. Lindberg. 1976. Survival of laboratory-reared pocket mice, *Perognathus longimembris*. J. Mamm. 57:266–272.

Hayes, S. R. 1977. Home range of *Marmota monax* (Sciuridae) in Arkansas. Southwestern Nat. 22:547–550.

Hayman, R. W. 1946. A new genus of fruit-bat and a new squirrel, from Celebes. Ann. Mag. Nat. Hist., ser. 11, 12:766–775.

————. 1961. The red goral of the northeast frontier region. Proc. Zool. Soc. London, 136:317–324.

Hayward, B. J., and E. L. Cockrum. 1971. The natural history of the western long-nosed bat *Leptonycteris sanborni*. Western New Mexico State Univ. Res. Sci. 1:75–123.

Heaney, L. R. 1979. A new species of tree squirrel (*Sundasciurus*) from Palawan Island, Philippines (Mammalia: Sciuridae). Proc. Biol. Soc. Washington 92:280–286.

Heaney, L. R., and R. S. Hoffmann. 1978. A second specimen of the neotropical montane squirrel, *Syntheosciurus poasensis*. J. Mamm. 59:854–855.

Heaney, L. R., and R. W. Thorington. 1978. Ecology of neotropical red-tailed squirrels, *Sciurus granatensis,* in the Panama Canal Zone. J. Mamm. 59:846–851.

Heim de Balsac, H. 1972. Insectivores. *In* Battistini and Richard-Vindard (1972), pp. 629–660.

Heithaus, E. R., and T. H. Fleming. 1978. Foraging movements of a frugivorous bat, *Carollia perspicillata* (Phyllostomatidae). Ecol. Monogr. 48:127–143.

Hellwing, S. 1970. Reproduction in the white-toothed shrew *Crocidura russula monacha* Thomas in captivity. Israel J. Zool. 19:177–178.

————. 1973. Husbandry and breeding of white-toothed shrews (Crocidurinae) in the research zoo of the Tel-Aviv University. Int. Zoo Yearbook 13:127–134.

Helm, J. D., III. 1975. Reproductive biology of *Ototylomys* (Cricetidae). J. Mamm. 56:575–590.

Heltne, P. G.; D. C. Turner; and N. J. Scott, Jr. 1976. Comparison of census data on *Alouatta palliata* from Costa Rica and Panama. *In* Thorington and Heltne (1976), pp. 10–19.

Hemmer, H. 1972. *Uncia uncia*. Mammalian Species, no. 20, 5 pp.

————. 1974. Untersuchungen zur Stammesgeschichte der Pantherkatzen (Pantherinae). III. Zur Artgeschichte des Löwen *Panthera (Panthera) leo* (Linnaeus 1758). Veroff. Zool. Staatssaml. München 17:167–280.

————. 1976. Gestation period and postnatal development in felids. *In* Eaton, R. L., ed., Proc. 3rd Int. Symp. World's Cats, vol. 3, Carnivore Res. Inst., Univ. Washington, pp. 143–165.

————. 1978. The evolutionary systematics of living Felidae: present status and current problems. Carnivore 1:71–79.

Henderson, F. R.; P. F. Springer; and R. Adrian. 1969. The black-footed ferret in South Dakota. South Dakota Dept. Game, Fish, and Parks Tech. Bull., no. 4, 37 pp.

Hendey, Q. B. 1977. Fossil bear from South Africa. S. Afr. J. Sci. 73:112–116.

Hendrichs, H. 1975a. Observations on a population of Bohor reedbuck, *Redunca redunca* (Pallas 1767). Z. Tierpsychol. 38:44–54.

————. 1975b. Changes in a population of dikdik, *Madoqua (Rhynchotragus) kirki* (Günther 1880). Z. Tierpsychol. 38:55–69.

————. 1975c. The status of the tiger *Panthera tigris* (Linne, 1758) in the Sundarbans Mangrove Forest (Bay of Bengal). Saugetierk. Mitt. 23:161–199.

Hennings, D., and R. S. Hoffmann. 1977. A review of the taxonomy of the *Sorex vagrans* species complex from western North America. Occas. Pap. Mus. Nat. Hist. Univ. Kansas, no. 68, 35 pp.

Henry, V. G. 1968. Length of estrous cycle and gestation in European wild hogs. J. Wildl. Mgmt. 32:406–408.

Henson, O. W., Jr., and A. Novick. 1966. An additional record of the bat *Phyllonycteris aphylla*. J. Mamm. 47:351–352.

Heptner, W. G. 1975. Uber einige Besonderheiten der Formbildung und der geographischen Verbreitung der Rennmaus, *Meriones (Pallasiomys) meridianus* Pallas, 1773, in den Wusten Mittelasiens. Z. Saugetierk 40:261–269.

Herald, E. S.; R. L. Brownell, Jr.; F. L. Frye; E. J. Morris; W. E. Evans; and A. B. Scott. 1969. Blind river dolphin: first side-swimming cetacean. Science 166:1408–1410.

Herbert, H. J. 1972. The population dynamics of the waterbuck *Kobus ellipsiprymnus* (Ogilby, 1833) in the Sabi-Sand Wildtuin. Mammalia Depicta, Verlag Paul Parey, Hamburg, 68 pp.

Herman, L. M. 1980. Humpback whales in Hawaiian waters: a study in historical ecology. Pacific Sci. 33:1–15.

Herman, L. M., and R. C. Antinoja. 1977.

Humpback whales in the Hawaiian breeding waters: population and pod characteristics. Sci. Rept. Whales Res. Inst. 29:59–85.

Herman, L. M.; C. S. Baker; P. H. Forestell; and R. C. Antinoja. 1980. Right whale *Balaena glacialis* sightings near Hawaii: a clue to the wintering grounds? Mar. Ecol. Prog. Ser., 2:271–275.

Herman, T., and K. Fuller. 1974. Observations of the marten, *Martes americana*, in the Mackenzie District, Northwest Territories. Can. Field-Nat. 88:501–503.

Hernandez-Camacho, J. 1960. Primitiae mastozoologicae Colombianae—I. Status taxonómico de *Sciurus pucheranii santanderensis*. Caldasia 8:359–368.

Hernandez-Camacho, J., and A. Cadena. 1978. Notas para la revisión del genero *Lonchorhina* (Chiroptera, Phyllostomidae). Caldasia 12:199–251.

Hernandez-Camacho, J., and R. W. Cooper. 1976. The nonhuman primates of Colombia. *In* Thorington and Heltne (1976), pp. 35–69.

Herrero, S. 1970. Human injury inflicted by grizzly bears. Science 170:593–598.

———, ed. 1972. Bears—their biology and management. Int. Union Conserv. Nat. Publ., n.s., no. 23, 371 pp.

———. 1976. Conflicts between man and grizzly bears in the national parks of North America. *In* Pelton, Lentfer, and Folk (1976), pp. 121–145.

Hershkovitz, P. 1959a. Nomenclature and taxonomy of the neotropical mammals described by Olfers, 1818. J. Mamm. 40:337–353.

———. 1959b. Two new genera of South American rodents (Cricetinae). Proc. Biol. Soc. Washington 72:5–10.

———. 1960. Mammals of northern Colombia, preliminary report no. 8: arboreal rice rats, a systematic revision of the subgenus *Oecomys*, genus *Oryzomys*. Proc. U.S. Natl. Mus. 110:513–568.

———. 1962. Evolution of neotropical cricetine rodents (Muridae) with special reference to the phyllotine group. Fieldiana Zool. 46:1–524.

———. 1963. A systematic and zoogeographic account of the monkeys of the genus *Callicebus* (Cebidae) of the Amazonas and Orinoco river basins. Mammalia 27:1–79.

———. 1966a. South American swamp and fossorial rats of the scapteromyine group (Cricetinae, Muridae) with comments on the glans penis in murid taxonomy. Z. Saugetierk. 31:81–149.

———. 1966b. Mice, land bridges and Latin American faunal exchange. *In* Wenzel, R. L., and V. J. Tipton, eds., Ectoparasites of Panama, Field Mus. Nat. Hist., Chicago, pp. 725–751.

———. 1970. Supplementary notes on neotropical *Oryzomys dimidiatus* and *Oryzomys hammondi* (Cricetinae). J. Mamm. 51:789–794.

———. 1971. A new rice rat of the *Oryzomys palustris* group (Cricetinae, Murinae) from northwestern Colombia, with remarks on distribution. J. Mamm. 52:700–709.

———. 1972. Notes on New World Monkeys. Int. Zoo Yearbook 12:3–12.

———. 1975a. The scientific name of the lesser *Noctilio* (Chiroptera), with notes on the chauve-souris de la Vallee d'Ylo (Peru). J. Mamm. 56:242–247.

———. 1975b. Comments on the taxonomy of Brazilian marmosets (Callithrix, Callithricidae). Folia Primatol. 24:137–172.

———. 1976. Comments on generic names of four-eyed opossums (family Didelphidae). Proc. Biol. Soc. Washington 89:295–304.

———. 1977. Living New World monkeys (Platyrrhini). Volume I. Univ. Chicago Press, xiv + 1117 pp.

———. 1979. The species of sakis, genus *Pithecia* (Cebidae, Primates), with notes on sexual dichromatism. Folia Primatol. 31:1–22.

Herzig-Straschil, B. 1977. Notes on the feeding habits of the yellow mongoose *Cynictis penicillata*. Zool. Afr. 12:225–226.

———. 1978. On the biology of *Xerus inaurius* (Zimmermann, 1780) (Rodentia, Sciuridae). Z. Saugetierk. 43:262–278.

Heuvelmans, B. 1958. On the track of unknown animals. Hill and Wang, New York, 558 pp.

Hewson, R., and M. Taylor. 1975. Embryo counts and length of the breeding season in European hares in Scotland from 1960–1972. Acta Theriol. 20:247–254.

Hick, U. 1975. Breeding and maintenance of douc langurs at Cologne Zoo. *In* Martin (1975b), pp. 223–233.

Hickman, D. L., and E. M. Grigsby. 1978. Comparison of signature whistles in *Tursiops truncatus*. Cetology, no. 31, 10 pp.

Hickman, G. C. 1977a. Burrow system structure of *Pappogeomys castanops* (Geomyidae) in Lubbock County, Texas. Amer. Midl. Nat. 97:50–58.

———. 1977b. Swimming behavior in representative species of the three genera of North American geomyids. Southwestern Nat. 21:531–538.

———. 1978. Reactions of *Cryptomys hottentotus* to water (Rodentia: Bathyergidae). Zool. Afr. 13:319–328.

———. 1979. Burrow system structure of the bathyergid *Cryptomys hottentotus* in Natal, South Africa. Z. Saugetierk. 44:153–162.

Hickman, V. V., and J. L. Hickman. 1960. Notes on the habits of the Tasmanian dormouse phalangers *Cercartetus nanus* (Desmarest) and *Eudromicia lepida* (Thomas). Proc. Zool. Soc. London 135:365–374.

Hill, C. 1975. The longevity record for *Colobus*. Primates 16:235.

Hill, C. A. 1967. A note on the gestation period of the siamang *Hylobates syndactylus*. Int. Zoo Yearbook 7:93–94.

Hill, D. O. 1975. Vanishing giants. Audubon 77(1):56–107.

Hill, J. E. 1956. The mammals of Rennell Island. *In* Wolff, T., The natural history of Rennell Island, British Solomon Islands, vol. 1, Univ. Copenhagen, pp. 73–84.

———. 1958. Some observations on the fauna of the Maldive Islands. II. Mammals. J. Bombay Nat. Hist. Soc. 55:3–10.

———. 1961a. Fruit bats from the Federation of Malaya. Proc. Zool. Soc. London 136:629–642.

———. 1961b. Indo-Australian bats of the genus *Tadarida*. Mammalia 25:29–56.

———. 1962. Notes on some insectivores and bats from upper Burma. Proc. Zool. Soc. London 139:119–137.

———. 1963a. A revision of the genus *Hipposideros*. Bull. British Mus. (Nat. Hist.), Zool. 11:1–129.

———. 1963b. Notes on some tube-nosed bats, genus *Murina*, from southeastern Asia, with descriptions of a new species and a new subspecies. Fedn. Mus. J. (n.s.) 8:48–59.

———. 1965. Asiatic bats of the genera *Kerivoula* and *Phoniscus* (Vespertilionidae), with a note on *Kerivoula aerosa* Tomes. Mammalia 29:524–556.

———. 1966. A review of the genus *Philetor* (Chiroptera: Vespertilionidae). Bull. British Mus. (Nat. Hist.), Zool. 14:371–387.

———. 1971a. A note on *Pteropus* (Chiroptera: Pteropodidae) from the Andaman Islands. J. Bombay Nat. Hist. Soc. 68:1–8.

———. 1971b. The status of *Vespertilio brachypterus* Temminck, 1840 (Chiroptera: Vespertilionidae). Zool. Meded. 45:139–146.

———. 1971c. Bats from the Solomon Islands. J. Nat. Hist. 5:573–581.

———. 1972a. A note on *Rhinolophus rex* Allen, 1923 and *Rhinomegalophus paradoxolophus* Bourret, 1951 (Chiroptera: Rhinolophidae). Mammalia 36:428–434.

———. 1972b. The Gunong Benom Expedition 1967. 4. New records of Malayan bats, with taxonomic notes and the description of a new *Pipistrellus*. Bull. British Mus. (Nat. Hist.), Zool. 23:21–42.

———. 1974a. A review of *Laephotis* Thomas, 1901 (Chiroptera: Vespertilionidae). Bull. British Mus. (Nat. Hist.), Zool. 27:73–82.

———. 1974b. New records of bats from southeastern Asia, with taxonomic notes. Bull. British Mus. (Nat. Hist.), Zool. 27:127–138.

———. 1974c. A review of *Scotoecus* Thomas, 1901 (Chiroptera: Vespertilionidae). Bull. British Mus. (Nat. Hist.), Zool. 27:169–188.

———. 1974d. A new family, genus and species of bat (Mammalia: Chiroptera) from Thailand. Bull. British Mus. (Nat. Hist.), Zool. 27:301–336.

———. 1976a. A note on *Pipstrellus rusticus* (Tomes, 1861) (Chiroptera: Vespertilionidae). Rev. Zool. Afr. 90:626–633.

———. 1976b. Bats referred to *Hesperoptenus* Peters, 1869 (Chiroptera: Vespertilionidae) with the description of a new subgenus. Bull. British Mus. (Nat. Hist.), Zool. 30:1–28.

————. 1977a. A review of the Rhinopomatidae (Mammalia: Chiroptera). Bull. British Mus. (Nat. Hist.), Zool. 32:29–43.

————. 1977b. African bats allied to *Kerivoula lanosa* (A. Smith, 1847). Rev. Zool. Afr. 91:623–633.

————. 1979. The flying fox *Pteropus tonganus* in the Cook Islands and on Niue Island, Pacific Ocean. Acta Theriol. 24:115–117.

————. 1980. A note on *Lonchophylla* (Chiroptera: Phyllostomatidae) from Ecuador and Peru, with the description of a new species. Bull. British Mus. (Nat. Hist.), Zool. 38:233–236.

Hill, J. E., and W. N. Beckon. 1978. A new species of *Pteralopex* Thomas, 1888 (Chiroptera: Pteropodidae) from the Fiji Islands. Bull. British Mus. (Nat. Hist.), Zool. 34:65–82.

Hill, J. E., and B. Boeadi. 1978. A new species of *Megaerops* from Java (Chiroptera: Pteropodidae). Mammalia 42:427–434.

Hill, J. E., and P. Morris. 1971. Bats from Ethiopia collected by the Great Abbai Expedition, 1968. Bull. British Mus. (Nat. Hist.), Zool. 21:25–49.

Hill, J. E., and K. Thonglongya. 1972. Bats from Thailand and Cambodia. Bull. British Mus. (Nat. Hist.), Zool. 22:171–196.

Hill, J. E., and G. Topal. 1973. The affinities of *Pipistrellus ridleyi* Thomas, 1898 and *Glischropus rosseti* Oey, 1951 (Chiroptera: Vespertilionidae). Bull. British Mus. (Nat. Hist.), Zool. 24:447–454.

Hill, J. E., and M. Yoshiyuki. 1980. A new species of *Rhinolophus* (Chiroptera, Rhinolophidae) from Iriomote Island, Ryukyu Islands, with notes on the Asiatic members of the *Rhinolophus pusillus* group. Bull. Natl. Sci. Mus. (Tokyo), ser. A, 6:179–189.

Hill, R. W., and E. T. Hooper. 1971. Temperature regulation in mice of the genus *Scotinomys*. J. Mamm. 52:806–816.

Hill, W.C.O. 1970. Primates. Comparative anatomy and taxonomy. VIII. Cynopithecinae. *Papio, Mandrillus, Theropithecus*. Wiley—Interscience, New York, xix + 680 pp.

Hillman, C. N., and J. W. Carpenter. 1980. Masked mustelid. Nature Conservancy News, 30(2):20–23.

Hillman, C. N., and T. W. Clark. 1980. *Mustela nigripes*. Mammalian Species, no. 126, 3 pp.

Hillman, C. N.; R. L. Linder; and R. B. Dahlgren. 1979. Prairie dog distribution in areas inhabited by black-footed ferrets. Amer. Midl. Nat. 102:185–187.

Hills, D. M., and R.H.N. Smithers. 1980. The "king cheetah." A historical review. Arnoldia 9(1):1–22.

Hilton, H. 1978. Systematics and ecology of the eastern coyote. *In* Bekoff (1978), pp. 209–228.

Hinesley, L. L. 1979. Systematics and distribution of two chromosome forms in the southern grasshopper mouse, genus *Onychomys*. J. Mamm. 60:117–128.

Hirth, D. H. 1977. Social behavior of white-tailed deer in relation to habitat. Wildl. Monogr., no. 53, 55 pp.

Hitchcock, H. B. 1965. Twenty-three years of bat banding in Ontario and Quebec. Can. Field-Nat. 79:4–14.

Hladik, C. M. 1977. Chimpanzees of Gabon and Chimpanzees of Gombe: some comparative data on the diet. *In* Clutton-Brock (1977), pp. 481–501.

————. 1978. Adaptive strategies of primates in relation to leaf-eating. *In* Montgomery (1978), pp. 373–395.

————. 1979. Diet and ecology of prosimians. *In* Doyle and Martin (1979), pp. 307–357.

Hladik, C. M., and P. Charles-Dominique. 1974. The behaviour and ecology of the sportive lemur (*Lepilemur mustelinus*) in relation to its dietary peculiarities. *In* Martin, Doyle, and Walker (1974), pp. 23–37.

Hocking, G. J. 1980. The occurrence of the New Holland mouse, *Pseudomys novaehollandiae* (Waterhouse), in Tasmania. Austral. Wildl. Res. 7:71–77.

Hoeck, H. N. 1975. Differential feeding behaviour of the sympatric hyrax *Procavia johnstoni* and *Heterohyrax brucei*. Oecologia 22:15–47.

————. 1978. Systematics of the Hyracoidea: toward a clarification. Bull. Carnegie Mus. Nat. Hist., no. 6, pp. 146–151.

Hoese, H. D. 1971. Dolphin feeding out of water in a salt marsh. J. Mamm. 52:222–223.

Hoffmann, R. S., and R. D. Fisher. 1978. Additional distributional records of Preble's shrew (*Sorex preblei*). J. Mamm. 59:893–894.

Hoffmann, R. S.; J. W. Koeppl; and C. F. Nadler. 1979. The relationships of the amphiberingian marmots (Mammalia: Sciuridae). Occas. Pap. Mus. Nat. Hist. Univ. Kansas, no. 83, 56 pp.

Hoffmeister, D. F., and V. E. Diersing. 1978. Review of the tassel-eared squirrels of the subgenus *Otosciurus*. J. Mamm. 59:402–413.

Hoffmeister, D. F., and W. W. Goodpaster. 1962. Life history of the desert shrew *Notiosorex crawfordi*. Southwestern Nat. 7:236–252.

Holisova, V.; J. Pelikan; and J. Zejoa. 1962. Ecology and population dynamics in *Apodemus microps* Krat. & Ros. (Mamm.: Muridae). Acta. Acad. Sci. Cechoslovenciae, Brno 34:493–540.

Hollien, H.; P. Hollien; D. K. Caldwell; and M. C. Caldwell. 1976. Sound production by the Atlantic bottlenosed dolphin *Tursiops truncatus*. Cetology, no. 26, 8 pp.

Holloway, R. L., ed. 1974. Primate aggression, territoriality, and xenophobia. A comparative perspective. Academic Press, New York, xiv + 513 pp.

Homan, J. A., and J. K. Jones, Jr. 1975a. *Monophyllus redmani*. Mammalian Species, no. 57, 3 pp.

————. 1975b. *Monophyllus plethodon*. Mammalian Species, no. 58, 2 pp.

Homewood, K. 1975. Can the Tana mangabey survive? Oryx 13:53–59.

Hoogstraal, H. 1962. A brief review of contemporary land mammals of Egypt (including Sinai). 1. Insectivora and Chiroptera. J. Egyptian Pub. Health Assoc. 37:143–162.

Hooper, E. T. 1968. Classification. *In* King (1968), pp. 27–74.

————. 1972. A synopsis of the rodent genus *Scotinomys*. Occas. Pap. Mus. Zool. Univ. Michigan, no. 665, 32 pp.

Hooper, E. T., and M. D. Carleton. 1976. Reproduction, growth and development in two contiguously allopatric rodent species, genus *Scotinomys*. Misc. Publ. Mus. Zool. Univ. Michigan, no. 151, 52 pp.

Horacek, I. 1975. Notes on the ecology of the bats of the genus *Plecotus* Geoffroy, 1818 (Mammalia: Chiroptera). Vest. Cesk. Spol. Zool. 39:195–210.

Horn, A. D. 1979. The taxonomic status of the bonobo chimpanzee. Amer. J. Phys. Anthropol. 51:273–282.

Hornocker, M. G. 1969. Winter territoriality in mountain lions. J. Wildl. Mgmt. 33:457–464.

————. 1970. An analysis of mountain lion predation upon mule deer and elk in the Idaho Primitive Area. Wildl. Monogr., no. 21, 39 pp.

Horr, D. A. 1975. The Borneo orang-utan: population structure and dynamics in relationship to ecology and reproductive strategy. Primate Behavior 4:307–323.

Houseknecht, C. R., and J. R. Tester. 1978. Denning habits of striped skunks (*Mephitis mephitis*). Amer. Midl. Nat. 100:424–430.

How, R. A. 1976. Reproduction, growth and survival of young in the mountain possum, *Trichosurus caninus* (Marsupialia). Austral. J. Zool. 24:189–199.

————. 1978. Population strategies of four species of Australian "possums." *In* Montgomery (1978), pp. 305–313.

Howard, W. E., and J. N. Amaya. 1975. European rabbit invades western Argentina. J. Wildl. Mgmt. 39:757–761.

Howe, D. 1975. Observations on a captive marsupial mole, *Notoryctes typhlops*. Austral. Mamm. 1:361–365.

Howe, H. F. 1974. Additional records of *Phyllonycteris aphylla* and *Ariteus flavescens* from Jamaica. J. Mamm. 55:662–663.

Howe, R. J. 1974. Marking behaviour of the Bahaman hutia (*Geocapromys ingrahami*). Anim. Behav. 22:645–649.

————. 1976. Social behavior of Bahamian hutias in captivity. Florida Sci. 39:8–14.

Hoyt, E. 1977. *Orcinus orca*. Separating facts from fantasies. Oceans 10(4):22–26.

Hubbard, C. A. 1970. A new species of *Tatera* from Tanzania with a description of its life history and habits studied in captivity. Zool. Afr. 5:237–247.

————. 1972. Observations on the life histories and behaviour of some small rodents from Tanzania. Zool. Afr. 7:419–449.

Hubbard, W. P., and S. Harris. 1960. Notorious grizzly bears. Sage Books, Denver, 205 pp.

Hubbs, C. L., and K. S. Norris. 1971. Original teeming abundance, supposed extinction, and survival of the Juan Fernandez fur seal. *In* Burt, W. H., ed., Antarctic Pinnipedia, Amer. Geophys. Union, Washington, D.C., pp. 35–51.

Hubbs, C. L.; W. F. Perrin; and K. C. Balcomb. 1973. *Stenella coeruleoalba* in the eastern and central tropical Pacific. J. Mamm. 54:549–552.

Hubert, B. 1978a. Revision of the genus *Saccostomus* (Rodentia, Cricetomyinae), with new morphological and chromosomal data from specimens from the lower Omo Valley, Ethiopia. Bull. Carnegie Mus. Nat. Hist., no. 6, pp. 48–52.

————. 1978b. Modern rodent fauna of the lower Omo Valley, Ethiopia. Bull. Carnegie Mus. Nat. Hist., no. 6, pp. 109–112.

Hubert, B.; F. Adam; and A. Poulet. 1973. Liste préliminaire des rongeurs du Sénégal. Mammalia 37:76–87.

Huckaby, D. G. 1980. Species limits in the *Peromyscus mexicanus* group (Mammalia: Rodentia: Muroidea). Los Angeles Co. Mus. Contrib. Sci., no. 326, 24 pp.

Hudson, J. W. 1974. The estrous cycle, reproduction, growth, and development of temperature regulation in the pygmy mouse, *Baiomys taylori.* J. Mamm. 55:572–588.

Hulbert, A. J.; G. Gordon; and T. J. Dawson. 1971. Rediscovery of the marsupial *Echymipera rufescens* in Australia. Nature 231:330–331.

Humphrey, S. R. 1975. Nursery roosts and community diversity of Nearctic bats. J. Mamm. 56:321–346.

————. 1978. Status, winter habitat, and management of the endangered Indiana bat, *Myotis sodalis.* Florida Sci. 41:65–76.

Humphrey, S. R., and T. H. Kunz. 1976. Ecology of a Pleistocene relict, the western big-eared bat (*Plecotus townsendii*), in the southern Great Plains. J. Mamm. 57:470–494.

Humphrey, S. R.; A. R. Richter; and J. B. Cope. 1977. Summer habitat and ecology of the endangered Indiana bat, *Myotis sodalis.* J. Mamm. 58:334–346.

Hunsaker, D., II, ed. 1977a. The biology of marsupials. Academic Press, New York, xv + 537 pp.

————. 1977b. Ecology of New World marsupials. *In* Hunsaker (1977a), pp. 95–156.

Hunsaker, D., II, and D. Shupe. 1977. Behavior of New World marsupials. *In* Hunsaker (1977a), pp. 279–347.

Hunt, J. H. 1974. The little-known lynx. Maine Fish & Game 16(2):28.

Husar, S. L. 1977. *Trichechus inunguis.* Mammalian Species, no. 72, 4 pp.

————. 1978a. *Dugong dugon.* Mammalian Species, no. 88, 7 pp.

————. 1978b. *Trichechus senegalensis.* Mammalian Species, no. 89, 3 pp.

————. 1978c. *Trichechus manatus.* Mammalian Species, no. 93, 5 pp.

Husson, A. M. 1978. The mammals of Suriname. E. J. Brill, Leiden, xxxiv + 569 pp.

Husson, A. M., and L. B. Holthuis. 1974. *Physeter macrocephalus* Linnaeus, 1758, the valid name for the sperm whale. Zool. Meded. 48:205–217.

Hutterer, R. 1978. Verbreitung und Systematik von *Sorex minutus* Linnaeus, 1766 (Insectivora; Soricidae) im Nepal-Himalaya und angrenzenden Gebieten. Z. Saugetierk. 44:65–80.

————. 1980. A record of Goodwin's shrew, *Cryptotis goodwini,* from Mexico. Mammalia 44:413.

Hutterer, R., and U. Hirsch. 1979. Ein neuer *Nesoryzomys* von der Insel Fernandina, Galapagos (Mammalia, Rodentia). Bonner Zool. Beitr. 30:276–283.

Hutterer, R., and P. D. Jenkins. 1980. A new species of *Crocidura* from Nigeria (Mammalia: Insectivora). Bull. British Mus. (Nat. Hist.), Zool. 39:305–310.

Hyndman, D., and J. I. Menzies. 1980. *Aproteles bulmerae* (Chiroptera: Pteropodidae) of New Guinea is not extinct. J. Mamm. 61:159–160.

I.

Ichihara, T. 1966. The pygmy blue whale *Balaenoptera musculus brevicauda,* a new subspecies from the Antarctic. *In* Norris (1966), pp. 79–113.

Ikeda, H.; K. Eguchi; and Y. Ono. 1979. Home range utilization of a raccoon dog, *Nyctereutes procyonides viverrinus,* Temminck, in a small islet in western Kyushu. Japanese J. Ecol. 29:35–48.

Ilmen, M., and S. Lahti. 1968. Reproduction, growth and behaviour in the captive wood lemming, *Myopus schisticolor* (Lilljeb). Ann. Zool. Fennici 5:207–219.

Ingles, J. M.; P. N. Newton; M.R.W. Rands; and G. R. Bowden. 1980. The first record of a rare murine rodent *Diomys* and further records of three shrew species from Nepal. Bull. British Mus. (Nat. Hist.), Zool. 39:205–211.

Ingles, L. G. 1961. Home range and habitats of the wandering shrew. J. Mamm. 42:455–462.

Innes, D.G.L. 1978. A reexamination of litter size in some North American microtines. Can. J. Zool. 56:1488–1496.

Insley, H. 1977. An estimate of the population density of the red fox (*Vulpes vulpes*) in the New Forest, Hampshire. J. Zool. 183:549–553.

Irby, L. R. 1979. Reproduction in mountain reedbuck (*Redunca fulvorufula*). Mammalia 43:191–213.

IUCN (International Union for Conservation of Nature and Natural Resources). 1972–1978. Red data book. I. Mammalia. Morges, Switzerland.

Ivanter, E. V. 1972. Contribution to the ecology of *Sicista betulina* Pall. Aquilo Ser. Zool. 13:103–108.

Iverson, S. L., and B. N. Turner. 1972. Natural history of a Manitoba population of Franklin's ground squirrels. Can. Field-Nat. 86:145–149.

Iwano, T. 1975. Distribution of Japanese monkey (*Macaca fuscata*). Proc. 5th Int. Congr. Primatol., pp. 389–391.

Izard, J., and K. Umfleet. 1971. Notes on the care and breeding of the suni *Nesotragus moschatus* at Dallas Zoo. Int. Zoo Yearbook 11:129.

Izawa, K. 1970. Unit groups of chimpanzees and their nomadism in the savanna woodland. Primates 11:1–46.

————. 1976. Group sizes and compositions of monkeys in the upper Amazon basin. Primates 17:367–399.

————. 1978. A field study of the ecology and behavior of the black-mantle tamarin (*Saguinus nigricollis*). Primates 19:241–274.

Izor, R. J. 1979. Winter range of the silverhaired bat. J. Mamm. 60:641–643.

Izor, R. J., and L. de la Torre. 1978. A new species of weasel (*Mustela*) from the highlands of Colombia, with comments on the evolution and distribution of South American weasels. J. Mamm. 59:92–102.

J.

Jachowski, R. L. 1981. Proposal to remove the bobcat from Appendix II of the Convention on International Trade in Endangered Species of Wild Fauna and Flora. Federal Register 46:45652–45656.

Jackson, H.H.T. 1961. Mammals of Wisconsin. Univ. Wisconsin Press, Madison, xii + 504 pp.

Jackson, P.F.R. 1978. Scientists hunt the Bengal tiger—but only in order to trace and save it. Smithsonian 9(5):28–37.

Jackson, R. 1979. Snow leopards in Nepal. Oryx 15:191–195.

Jacobsen, N.H.G. 1974. Distribution, home ranges, and behaviour patterns of bushbuck in the Lutope and Sengwa valleys, Rhodesia. J. S. Afr. Wildl. Mgmt. Assoc. 4:75–93.

Jacobsen, N.H.G., and E. Du Plessis. 1976. Observations on the ecology and biology of the Cape fruit bat *Rousettus aegyptiacus leachi* in the eastern Transvaal. S. Afr. J. Sci. 72:270–273.

Jameson, E. W., Jr., and G. S. Jones. 1978. The Soricidae of Taiwan. Proc. Biol. Soc. Washington 90:459–482.

Jannett, F. J., and J. Z. Jannett. 1974. Drum-marking by *Arvicola richardsoni* and its taxonomic significance. Amer. Midl. Nat. 92:230–234.

Jantschke, F. 1973. On the breeding and rearing of bush dogs *Speothos venaticus* at Frankfurt Zoo. Int. Zoo Yearbook 13:141–143.

Jarman, P. J., and M. V. Jarman. 1973. Social behaviour, population structure and reproductive potential in impala. E. Afr. Wildl. J. 11:329–338.

———. 1974. Impala behaviour and its relevance to management. *In* Geist and Walther (1974), pp. 871–881.

Jarvis, C. 1967. Tabulated data on the breeding biology of the black rhinoceros *Diceros bicornis* compiled from reports in the yearbook. Int. Zoo Yearbook 7:166.

Jarvis, J.U.M. 1969. The breeding season and litter size of African mole-rats. J. Reprod. Fert., Suppl. 6:237–248.

———. 1973a. Activity patterns in the mole-rats *Tachyoryctes splendens* and *Heliophobius argenteocinereus*. Zool. Afr. 8:101–119.

———. 1973b. The structure of a population of mole-rats, *Tachyoryctes splendens* (Rodentia: Rhizomyidae). J. Zool. 171:1–14.

———. 1978. Energetics of survival in *Heterocephalus glaber* (Rüppell), the naked mole-rat (Rodentia: Bathyergidae). Bull. Carnegie Mus. Nat. Hist., no. 6, pp. 81–87.

Jarvis, J.U.M., and J. B. Sale. 1971. Burrowing and burrow patterns of East African mole-rats *Tachyoryctes, Heliophobius* and *Heterocephalus*. J. Zool. 163:451–479.

Jay, P. C., ed. 1968. Primates. Studies in adaptation and variability. Holt, Rinehart, and Winston, New York, xii + 529 pp.

Jeanne, R. L. 1970. Note on a bat (*Phylloderma stenops*) preying upon the brood of a social wasp. J. Mamm. 51:624–625.

Jenkins, P. D. 1976. Variation in Eurasian shrews of the genus *Crocidura* (Insectivora: Soricidae). Bull. British Mus. (Nat. Hist.), Zool. 30:269–309.

Jenkins, S. H., and P. E. Busher. 1979. *Castor canadensis*. Mammalian Species, no. 120, 8 pp.

Jennings, T. J. 1975. Notes on the burrow systems of woodmice (*Apodemus sylvaticus*). J. Zool. 177:500–504.

Jewell, P. A. 1972. Social organisation and movements of topi (*Damaliscus korrigum*) during the rut, at Ishasha, Queen Elizabeth Park, Uganda. Zool. Afr. 7:233–255.

Johanson, D. C., and T. D. White. 1979. A systematic assessment of early African hominids. Science 203:321–330.

Johnson, A. S. 1970. Biology of the raccoon (*Procyon lotor varius* Nelson and Goldman) in Alabama. Auburn Univ. Agric. Exp. Sta. Bull., no. 402, vi + 148 pp.

Johnson, D. H. 1962. Two new murine rodents. Proc. Biol. Soc. Washington, 75:317–319.

Johnson, G. L., and R. L. Packard. 1974. Electrophoretic analysis of *Peromyscus comanche* Blair, with comments on its systematic status. Occas. Pap. Mus. Texas Tech Univ., no. 24, 16 pp.

Johnson, M. 1973. Characters of the heather vole, *Phenacomys*, and the red tree vole, *Arborimus*. J. Mamm. 54:239–244.

Johnson, M. L.; R. H. Taylor; and N. W. Winnick. 1975. The breeding and exhibition of capromyid rodents at Tacoma Zoo. Int. Zoo Yearbook 15:53–56.

Johnson, P. M. 1978. Husbandry of the rufous rat-kangaroo *Aeprymnus rufescens* and brush-tailed rock wallaby *Petrogale penicillata* in captivity. Int. Zoo Yearbook 18:156–157.

———. 1979. Reproduction in the plain rock-wallaby, *Petrogale penicillata inornata* Gould, in captivity, with age estimation of the pouch young. Austral. Wildl. Res. 6:1–4.

Johnson, R. E. 1977. An historical analysis of wolverine abundance and distribution in Washington. Murrelet 58:13–16.

Johnston, P. G., and G. B. Sharman. 1976. Studies on populations of *Potorous* Desmarest (Marsupialia). I. Morphological variation. Austral. J. Zool. 24:573–588.

———. 1977. Studies on populations of *Potorous* Desmarest (Marsupialia). II. Electrophoretic, chromosomal and breeding studies. Austral. J. Zool. 25:733–747.

Jolly, A. 1966. Lemur behavior. A Madagascar field study. Univ. Chicago Press, xiv + 187 pp.

———. 1972. Troop continuity and troop spacing in *Propithecus verreauxi* and *Lemur catta* at Berenty (Madagascar). Folia Primatol. 17:335–362.

Jones, C. 1971a. The bats of Rio Muni, West Africa. J. Mamm. 52:121–140.

———. 1971b. Notes on the anomalurids of Rio Muni and adjacent areas. J. Mamm. 52:568–572.

———. 1972. Comparative ecology of three pteropid bats in Rio Muni, West Africa. J. Zool. 167:353–370.

———. 1977. *Plecotus rafinesquii*. Mammalian Species, no. 69, 4 pp.

———. 1978. *Dendrohyrax dorsalis*. Mammalian Species, no. 113, 4 pp.

Jones, C., and S. Anderson. 1978. *Callicebus moloch*. Mammalian Species, no. 112, 5 pp.

Jones, C., and J. Sabater Pi. 1968. Comparative ecology of *Cercocebus albigena* (Gray) and *Cercocebus torquatus* (Kerr) in Rio Muni, West Africa. Folia Primatol. 9:99–113.

———. 1971. Comparative ecology of *Gorilla gorilla* (Savage and Wyman) and *Pan troglodytes* (Blumenbach) in Rio Muni, West Africa. Bibl. Primatol., no. 13, 96 pp.

Jones, C., and H. W. Setzer. 1970. Comments on *Myosciurus pumilio*. J. Mamm. 51:813–814.

———. 1971. The designation of a holotype of the West African pygmy squirrel, *Myosciurus pumilio* (LeConte, 1857) (Mammalia: Rodentia). Proc. Biol. Soc. Washington, 84:59–64.

Jones, E. 1977. Ecology of the feral cat, *Felis catus* (L.), (Carnivora: Felidae) on Macquarie Island. Austral. Wildl. Res. 4:249–262.

Jones, F. W. 1923–1925. The mammals of South Australia. A. B. Jones, Government Printer, Adelaide, 458 pp.

Jones, G. S., and R. E. Mumford. 1971. *Chimarrogale* from Taiwan. J. Mamm. 52:228–232.

Jones, J. K., Jr. 1964. Distribution and taxonomy of mammals of Nebraska. Univ. Kansas Publ. Mus. Nat. Hist. 16:1–356.

———. 1966. Bats from Guatemala. Univ. Kansas Publ. Mus. Nat. Hist. 16:439–472.

———. 1977. *Rhogeessa gracilis*. Mammalian Species, no. 76, 2 pp.

Jones, J. K., Jr., and R. J. Baker. 1979. Notes on a collection of bats from Montserrat, Lesser Antilles. Occas. Pap. Mus. Texas Tech Univ., no. 60, 6 pp.

Jones, J. K., Jr., and D. C. Carter. 1976. Annotated checklist, with keys to subfamilies and genera. *In* Baker, Jones, and Carter (1976), pp. 7–38.

———. 1979. Systematic and distributional notes. *In* Baker, Jones, and Carter (1979), pp. 7–11.

Jones, J. K., Jr.; D. C. Carter; and H. H. Genoways. 1979. Revised checklist of North American mammals north of Mexico, 1979. Occas. Pap. Mus. Texas Tech Univ., no. 62, 17 pp.

Jones, J. K., Jr.; J. R. Choate; and A. Cadena. 1972. Mammals from the Mexican State of Sinaloa. II. Chiroptera. Occas. Pap. Mus. Nat. Hist. Univ. Kansas, no. 6, 29 pp.

Jones, J. K., Jr., and H. H. Genoways. 1970. Harvest mice (genus *Reithrodontomys*) of Nicaragua. Occas. Pap. Western Foundation Vert. Zool., no. 2, 16 pp.

———. 1971. Notes on the biology of the Central American squirrel, *Sciurus richmondi*. Amer. Midl. Nat. 86:242–247.

———. 1973. *Ardops nichollsi*. Mammalian Species, no. 24, 2 pp.

Jones, J. K., Jr.; H. H. Genoways; and R. J. Baker. 1971. Morphological variation in *Stenoderma rufum*. J. Mamm. 52:244–247.

Jones, J. K., Jr.; H. H. Genoways; and T. E. Lawlor. 1974. Annotated checklist of mammals of the Yucatan Peninsula, Mexico. II. Rodentia. Occas. Pap. Mus. Texas Tech Univ., no. 22, 24 pp.

Jones, J. K., Jr.; H. H. Genoways; and J. D. Smith. 1974. Annotated checklist of mammals of the Yucatan Peninsula, Mexico. III. Marsupialia, Insectivora, Primates, Edentata, Lagomorpha. Occas. Pap. Mus. Texas Tech Univ., no. 23, 12 pp.

Jones, J. K., Jr., and J. A. Homan. 1974. *Hylonycteris underwoodi*. Mammalian Species, no. 32, 2 pp.

Jones, J. K., Jr., and R. R. Johnson. 1967. Sirenians. *In* Anderson and Jones (1967), pp. 366–373.

Jones, J. K., Jr.; J. D. Smith; and H. H. Genoways. 1973. Annotated checklist of mammals of the Yucatan Peninsula, Mexico. I. Chiroptera. Occas. Pap. Mus. Texas Tech Univ., no. 13, 31 pp.

Jones, J. K., Jr.; J. D. Smith; and R. W. Turner. 1971. Noteworthy records of bats from Nicaragua, with a checklist of the chiropteran fauna of the country. Occas. Pap. Mus. Nat. Hist. Univ. Kansas, no. 2, 35 pp.

Jones, J. K., Jr.; P. Swanepoel; and D. C. Carter. 1977. Annotated checklist of the bats of Mexico and Central America. Occas. Pap. Mus. Texas Tech Univ., no. 47, 35 pp.

Jones, R. E. 1967. A *Hydrodamalis* skull fragment from Monterey Bay, California. J. Mamm. 48:143–144.

Jonkel, C. J. 1978. Black, brown (grizzly), and polar bears. *In* Schmidt and Gilbert (1978), pp. 227–248.

Jonkel, C. J., and I. M. Cowan. 1971. The black bear in the spruce-fir forest. Wildl. Monogr., no. 27, 57 pp.

Jonkel, C. J., and F. L. Miller. 1970. Recent records of black bears (*Ursus americanus*) on the barren grounds of Canada. J. Mamm. 51:826–828.

Jonkel, C. J.; P. Smith; I. Stirling; and G. B. Kolenosky. 1976. The present status of the polar bear in the James Bay and Belcher Islands area. Can. Wildl. Serv. Occas. Pap., no. 26, 42 pp.

Jorge, W.; D. A. Meritt, Jr.; and K. Benirschke. 1977. Chromosome studies in Edentata. Cytobios 18:157–172.

Joubert, E. 1972. A note on the challenge rituals of territorial male lechwe. Madoqua, ser. 1, no. 5, pp. 63–67.

Joubert, E., and F. C. Eloff. 1971. Notes on the ecology and behaviour of the black rhinoceros *Diceros bicornis* Linn. 1758 in South West Africa. Madoqua, ser. 1, no. 3, pp. 5–53.

Joubert, S.C.J. 1972. Territorial behavoiur of the tsessebe (*Damaliscus lunatus lunatus* Burchell) in the Kruger National Park. Zool. Afr. 7:141–156.

———. 1974. The social organization of the roan antelope *Hippotragus equinus* and its influence on the special distribution of herds in the Kruger National Park. *In* Geist and Walther (1974), pp. 661–675.

Joubert, S.C.J., and P.J.L. Bronkhorst. 1977. Some aspects of the history and population ecology of the tsessebe *Damaliscus lunatus lunatus* in the Kruger National Park. Koedoe 20:125–145.

Jouventin, P. 1975a. Les roles des colorations du mandrill (*Mandrillus sphinx*). Z. Tierpsychol. 39:455–462.

———. 1975b. Observations sur la socioécologie du mandrill. Terre Vie 29:493–532.

Jungius, H. 1970. Studies on the breeding biology of the reedbuck (*Redunca arundinum* Boddaert, 1785) in the Kruger National Park. Z. Saugetierk. 35:129–146.

———. 1971a. Studies on the food and feeding behavior of the reedbuck *Redunca arundinum* Boddaert, 1785 in the Kruger National Park. Koedoe 14:65–97.

———. 1971b. The biology and behaviour of the reedbuck (*Redunca arundinum* Boddaert 1785) in the Kruger National Park. Mammalia Depicta, Paul Parey, Hamburg, 106 pp.

———. 1978. Plan to restore Arabian oryx in Oman. Oryx 14:328–336.

K.

Kahlke, H. D. 1973. A review of the Pleistocene history of the orang-utan (*Pongo* Lacépède 1799). Asian Perspectives 15:5–14.

Kaiser, H. E. 1974. Morphology of the Sirenia. S. Karger, Basel, 76 pp.

Kale, H. W., II. 1972. A high concentration of *Cryptotis parva* in a forest in Florida. J. Mamm. 53:216–218.

Kamiya, T., and F. Yamasaki. 1974. Organ weights of *Pontoporia blainvillei* and *Platanista gangetica*. Sci. Rept. Whales Res. Inst. 26:265–270.

Kaplan, H., and S. O. Hyland. 1972. Behavioural development in the Mongolian gerbil (*Meriones unguiculatus*). Anim. Behav. 20:147–154.

Kasuya, T. 1972a. Growth and reproduction of *Stenella caeruleoalba* based on the age determination by means of dentinal growth layers. Sci. Rept. Whales Res. Inst. 24:57–79.

———. 1972b. Some information on the growth of the Ganges dolphin with a comment on the Indus dolphin. Sci. Rept. Whales Res. Inst. 24:87–108.

———. 1973. Systematic consideration of the Recent toothed whales based on the morphology of tympano-periotic bone. Sci. Rept. Whales Res. Inst. 25:1–103.

———. 1975. Past occurrence of *Globicephala melaena* in the western North Pacific. Sci. Rept. Whales Res. Inst. 27:95–110.

———. 1976. Reconsideration of life history parameters of the spotted and striped dolphins based on cemental layers. Sci. Rept. Whales Res. Inst. 28:73–106.

———. 1977. Age determination and growth of the Baird's beaked whale with a comment on the fetal growth rate. Sci. Rept. Whales Res. Inst. 29:1–20.

———. 1978. The life history of Dall's porpoise with special reference to the stock off the Pacific coast of Japan. Sci. Rept. Whales Res. Inst. 30:1–63.

Kasuya, T., and R. L. Brownell, Jr. 1979. Age determination, reproduction, and growth of Franciscana dolphin *Pontoporia blainvillei*. Sci. Rept. Whales Res. Inst. 31:45–67.

Kasuya, T., and A.K.M. Aminul Haque. 1972. Some informations on distribution and seasonal movement of the Ganges dolphin. Sci. Rept. Whales Res. Inst. 24:109–115.

Kasuya, T., and K. Kureha. 1979. The population of finless porpoise in the Inland Sea of Japan. Sci. Rept. Whales Res. Inst. 31:1–44.

Kasuya, T.; N. Miyazaki; and W. H. Dawbin. 1974. Growth and reproduction of *Stenella attenuata* in the Pacific Coast of Japan. Sci. Rept. Whales Res. Inst. 26:157–226.

Kasuya, T., and M. Nishiwaki. 1975. Recent status of the population of Indus dolphin. Sci. Rept. Whales Res. Inst. 27:81–94.

Kaufmann, J. H. 1962. Ecology and social behavior of the coati, *Nasua narica*, on Barro Colorado Island, Panama. Univ. California Publ. Zool. 60:95–222.

———. 1974. Social ethology of the whiptail wallaby, *Macropus parryi*, in northeastern New South Wales. Anim. Behav. 22:281–369.

———. 1975. Field observations of the social behaviour of the eastern grey kan-

garoo, *Macropus giganteus*. Anim. Behav. 23:214–221.

Kaufmann, J. H.; D. V. Lanning; and S. E. Poole. 1976. Current status and distribution of the coati in the United States. J. Mamm. 57:621–637.

Kawabe, M., and T. Mano. 1972. Ecology and behavior of the wild proboscis monkey, *Nasalis larvatus* (Wurmb), in Sabah, Malaysia. Primates 13:213–228.

Kawai, M., ed. 1979. Ecological and sociological studies of gelada baboons. Contrib. Primatol. 16:i–xxiv + 1–344.

Kawamichi, T. 1976. Hay territory and dominance rank of pikas (*Ochotona princeps*). J. Mamm. 57:133–148.

Kawamichi, T., and M. Kawamichi. 1979. Spatial organization and territory of tree shrews (*Tupaia glis*). Anim. Behav. 27:381–393.

Kawamura, A. 1973. Food and feeding of sei whale caught in the waters south of 40°N in the North Pacific. Sci. Rept. Whales Res. Inst. 25:219–236.

———. 1974. Food and feeding ecology in the southern sei whale. Sci. Rept. Whales Res. Inst. 26:25–144.

Keen, R., and H. B. Hitchcock. 1980. Survival and longevity of the little brown bat (*Myotis lucifugus*) in southeastern Ontario. J. Mamm. 61:1–7.

Keith, K., and J. H. Calaby. 1968. The New Holland mouse, *Pseudomys novaehollandiae* (Waterhouse), in the Port Stephens district, New South Wales. CSIRO Wildl. Res. 13:45–58.

Keith, L. B., and L. A. Windberg. 1978. A demographic analysis of the snowshoe hare cycle. Wildl. Monogr., no. 58, 70 pp.

Kellas, L. M. 1955. Observations on the reproductive activities, measurements, and growth rate of the dikdik (*Rhynchotragus kirkii thomasi* Neumann). Proc. Zool. Soc. London, 124:751–784.

Kellogg, R. 1940. Whales, giants of the sea. Natl. Geogr. 77:35–90.

Kelsall, J. P. 1972. The northern limits of moose (*Alces alces*) in western Canada. J. Mamm. 53:129–138.

Kemp, G. A. 1976. The dynamics and regulation of black bear *Ursus americanus* populations in northern Alberta. *In* Pelton, Lentfer, and Folk (1976), pp. 191–197.

Kemp, G. A., and L. B. Keith. 1970. Dynamics and regulation of red squirrel (*Tamiasciurus hudsonicus*) populations. Ecology 51:763–779.

Kemper, C. M. 1976a. Growth and development of the Australian murid *Pseudomys novaehollandiae*. Austral. J. Zool. 24:27–37.

———. 1976b. Reproduction of *Pseudomys novaehollandiae* (Muridae) in the laboratory. Austral. J. Zool. 24:159–167.

Kenagy, G. J. 1976. Field observations of male fighting, drumming, and copulation in the Great Basin kangaroo rat, *Dipodomys microps*. J. Mamm. 57:781–785.

Kennelly, J. J. 1978. Coyote reproduction. *In* Bekoff (1978), pp. 73–93.

Kenyon, K. W. 1969. The sea otter in the eastern Pacific Ocean. N. Amer. Fauna, no. 68, ix + 352 pp.

———. 1972. Man versus the monk seal. J. Mamm. 53:687–696.

Kenyon, K. W., and C. H. Fiscus. 1963. Age determination in the Hawaiian monk seal. J. Mamm. 44:280–281.

Kenyon, K. W., and D. W. Rice. 1961. Abundance and distribution of the Steller sea lion. J. Mamm. 42:223–234.

Keogh, H. J. 1973. Behaviour and breeding in captivity of the Namaqua gerbil *Desmodillus auricularis* (Cricetidae: Gerbillinae). Zool. Afr. 8:231–240.

Kern, J. A. 1964. Observations on the habits of the proboscis monkey, *Nasalis larvatus* (Wurmb), made in the Brunei Bay area, Borneo. Zoologica 49:183–192.

Kerr, M. A. 1965. The age at sexual maturity in male impala. Arnoldia 1(24):1–6.

Kerr, M. A., and V. J. Wilson. 1967. Notes on reproduction in Sharpe's grysbok. Arnoldia, 3(17):1–4.

Kerridge, D. C., and R. J. Baker. 1978. *Natalus micropus*. Mammalian Species, no. 114, 3 pp.

Khan, Mohd Khan Bin Momin. 1978. Man's impact on the primates of peninsular Malaysia. *In* Chivers and Lane-Petter (1978), pp. 41–46.

Kilgore, D. L., Jr. 1969. An ecological study of the swift fox (*Vulpes velox*) in the Oklahoma Panhandle. Amer. Midl. Nat. 81:512–534.

King, C. 1975. The home range of the weasel (*Mustela nivalis*) in an English woodland. J. Anim. Ecol. 44:639–669.

King, J. A., ed. 1968. Biology of *Peromyscus* (Rodentia). Amer. Soc. Mamm. Spec. Publ., no. 2, xiii + 593 pp.

King, J. E. 1964. Seals of the world. British Mus. (Nat. Hist.), London, 154 pp.

———. 1978. On the specific name of the southern sea lion (Pinnipedia, Otariidae). J. Mamm. 59:861–863.

Kingdon, J. 1971. East African mammals. An atlas of evolution in Africa. I. Academic Press, London, ix + 446 pp.

———. 1974a. East African mammals. An atlas of evolution in Africa. II(A). Insectivores and bats. Academic Press, London, xi + 341 + 1 pp.

———. 1974b. East African mammals. An atlas of evolution in Africa. II(B). Hares and rodents. Academic Press, London, ix + 362 + lvii pp.

———. 1977. East African mammals. An atlas of evolution in Africa. III(A). Carnivores. Academic Press, London, viii + 475 pp.

Kinsey, K. P. 1976. Social behaviour in confined populations of the Allegheny woodrat, *Neotoma floridana magister*. Anim. Behav. 24:181–187.

Kinzey, W. G. 1977. Diet and feeding behaviour of *Callicebus torquatus*. *In* Clutton-Brock (1977), pp. 127–151.

Kinzey, W. G.; A. L. Rosenberger; P. S. Heisler; D. L. Prowse; and J. S. Trilling. 1977. A preliminary field investigation of the yellow handed titi monkey, *Callicebus torquatus*, in northern Peru. Primates 18:159–181.

Kipp, H. 1965. Beitrag zur Kenntnis der Gattung *Conepatus* Molina, 1782. Z. Saugetierk. 30:193–232.

Kirkby, R. J. 1977. Learning and problem-solving in marsupials. *In* Stonehouse and Gilmore (1977), pp. 193–208.

Kirkland, G. L., Jr.; D. F. Schmidt; and C. J. Kirkland. 1979. First record of the long-tailed shrew (*Sorex dispar*) in New Brunswick. Can. Field-Nat. 93:195–198.

Kirkland, G. L., and H. M. Van Deusen. 1979. The shrews of the *Sorex dispar* group: *Sorex dispar* Batchelder and *Sorex gaspensis* Anthony and Goodwin. Amer. Mus. Novit., no. 2675, 21 pp.

Kirkpatrick, R. L., and G. L. Valentine. 1970. Reproduction in captive pine voles, *Microtus pinetorum*. J. Mamm. 51:779–785.

Kirkpatrick, T. H. 1965. Studies of Macropodidae in Queensland. 2. Age estimation in the grey kangaroo, the red kangaroo, the eastern wallaroo and the red-necked wallaby, with notes on dental abnormalities. Queensland J. Agric. Anim. Sci. 22:301–317.

———. 1967. The grey kangaroo in Queensland. Queensland Agric. J. 93:550–552.

———. 1968. Studies on the wallaroo. Queensland Agric. J. 94:362–365.

———. 1970a. The agile wallaby in Queensland. Queensland Agric. J. 96:169–170.

———. 1970b. The swamp wallaby in Queensland. Queensland Agric. J. 96:335–336.

Kirsch, J.A.W. 1968. Burrowing by the quenda, *Isoodon obesulus*. W. Austral. Nat. 10:178–180.

———. 1977a. The six-percent solution: second thoughts on the adaptedness of the Marsupialia. Amer. Sci. 65:276–288.

———. 1977b. The classification of marsupials. *In* Hunsaker (1977a), pp. 1–50.

———. 1977c. The comparative serology of Marsupialia, and a classification of marsupials. Austral. J. Zool., suppl. ser., no. 52, 152 pp.

Kirsch, J.A.W., and J. H. Calaby. 1977. The species of living marsupials: an annotated list. *In* Stonehouse and Gilmore (1977), pp. 9–26.

Kirsch, J.A.W., and W. E. Poole. 1972. Taxonomy and distribution of the grey kangaroos, *Macropus giganteus* Shaw and *Macropus fuliginosus* (Desmarest), and their subspecies (Marsupialia: Macropodidae). Austral. J. Zool. 20:315–339.

Kirsch, J.A.W., and P. F. Waller. 1979. Notes on the trapping and behavior of the Caenolestidae (Marsupialia). J. Mamm. 60:390–395.

Kistchinski, A. A. 1972. Life history of the

brown bear (*Ursus arctos* L.) in north-east Siberia. *In* Herrero (1972), pp. 67–73.

Kitchen, D. W. 1974. Social behavior and ecology of the pronghorn. Wildl. Monogr., no. 38, 96 pp.

Kitchener, D. J. 1972. The importance of shelter to the quokka, *Setonix brachyurus* (Marsupialia), on Rottnest Island. Austral. J. Zool. 20:281–299.

———. 1973. Reproduction in the common sheath-tailed bat, *Taphozous georgianus* (Thomas) (Microchiroptera: Emballonuridae), in Western Australia. Austral. J. Zool. 21:375–389.

———. 1975. Reproduction in female Gould's wattled bat, *Chalinolobus gouldii* (Gray) (Vespertilionidae), in Western Australia. Austral. J. Zool. 23:29–42.

———. 1976. Further observations on reproduction in the common sheath-tailed bat, *Taphozous georgianus* Thomas, 1915 in Western Australia, with notes on the gular pouch. Rec. W. Austral. Mus. 4:335–347.

———. 1980a. *Taphozous hilli* sp. nov. (Chiroptera: Emballonuridae), a new sheath-tailed bat from Western Australia and Northern Territory. Rec. W. Austral. Mus. 8:161–169.

———. 1980b. A new species of *Pseudomys* (Rodentia: Muridae) from Western Australia. Rec. W. Austral. Mus. 8:405–414.

Kitchener, D. J., and G. Sanson. 1978. *Petrogale burbidgei* (Marsupialia, Macropodidae), a new rock wallaby from Kimberley, Western Australia. Rec. W. Austral. Mus. 6:269–285.

Kitchener, S. L. 1971. Observations on the breeding of the bush dog *Speothos venaticus* at Lincoln Park Zoo, Chicago. Int. Zoo Yearbook 11:99–101.

Kleiman, D. G. 1970. Reproduction in the female green acouchi, *Myoprocta pratti* Pocock. J. Reprod. Fert. 23:55–65.

———. 1971. The courtship and copulatory behaviour of the green acouchi (*Myoprocta pratti*). Z. Tierpsychol. 29:259–278.

———. 1972. Social behavior of the maned wolf (*Chrysocyon brachyurus*) and bush dog (*Speothos venaticus*): a study in contrast. J. Mamm. 53:791–806.

———, ed. 1977a. The biology and conservation of the Callitrichidae. Smithson. Inst. Press, Washington, D.C., 354 pp.

———. 1977b. Characteristics of reproduction and sociosexual interactions in pairs of lion tamarins (*Leontopithecus rosalia*) during the reproductive cycle. *In* Kleiman (1977a), pp. 181–192.

———. 1980. 1979 International studbook. Golden lion tamarin, *Leontopithecus rosalia rosalia*. Natl. Zool. Park, Washington, D.C.

Kleiman, D. G.; J. F. Eisenberg; and E. Maliniak. 1979. Reproductive parameters and productivity of caviomorph rodents. *In* Eisenberg (1979), pp. 173–183.

Kleiman, D. G., and P. A. Racey. 1969. Observations on noctule bats (*Nyctalus noctula*) breeding in captivity. Lynx 10:65–77.

Klein, L. L., and D. J. Klein. 1976. Neotropical primates: aspects of habitat usage, population density, and regional distribution in La Macarena, Colombia. *In* Thorington and Heltne (1976), pp. 70–78.

———. 1977. Feeding behaviour of the Colombian spider monkey. *In* Clutton-Brock (1977), pp. 153–181.

Klein, R. G. 1974. On the taxonomic status, distribution and ecology of the blue antelope, *Hippotragus leucophaeus* (Pallas, 1766). Ann. S. Afr. Mus. 65:99–143.

Kleinenberg, S. E.; A. V. Yablokov; B. M. Bel'Kovich; and M. N. Tarasevich. 1969. Beluga (*Delphinapterus leucas*). Investigation of the species. Israel Progr. Sci. Transl., Jerusalem, vi + 376 pp.

Klimchenko, I. Z., et al. 1975. Change of the population structure of Vinogradov's gerbil and red-tailed Libyan jird in Azerbaidzhan after their extermination by the bait method. Soviet J. Ecol. 6:436–439.

Klingel, H. 1974. A comparison of the social behaviour of the Equidae. *In* Geist and Walther (1974), pp. 124–132.

———. 1977. Observations on social organization and behaviour of African and Asiatic wild asses (*Equus africanus* and *E. hemionus*). Z. Tierpsychol. 44:323–331.

Klingener, D. 1968. Anatomy. *In* King (1968), pp. 127–147.

Klingener, D.; H. H. Genoways; and R. J. Baker. 1978. Bats from southern Haiti. Ann. Carnegie Mus. 47:81–97.

Klopfer, P. H., and K. J. Boskoff. 1979. Maternal behavior in prosimians. *In* Doyle and Martin (1979), pp. 123–156.

Knap, J. J. 1975. Martens on the move. Int. Wildl. 5(5):32–35.

Knight, R. R.; B. M. Blanchard; and K. C. Kendall. 1981. Yellowstone grizzly bear investigations. Report of the Interagency Study Team. U.S. Natl. Park Serv., v + 55 pp.

Knopf, F. L., and D. F. Balph. 1977. Annual periodicity of Uinta ground squirrels. Southwestern Nat. 22:213–224.

Knorre, E. P. 1974. Changes in the behavior of moose with age and during the process of domestication. Naturaliste Canadien 101:371–377.

Knowlton, F. F. 1972. Preliminary interpretations of coyote population mechanics with some management implications. J. Wildl. Mgmt. 36:369–382.

Knox, E. 1978. A note on the identification of *Melomys* species (Rodentia: Muridae) in Australia. J. Zool. 185:276–277.

Knudsen, B. 1978. Time budgets of polar bears (*Ursus maritimus*) on North Twin Island, James Bay, during summer. Can. J. Zool. 56:1627–1628.

Kock, D. 1969a. *Dyacopterus spadiceus* (Thomas 1890) auf den Philippinen (Mammalia, Chiroptera). Senckenbergiana Biol. 50:1–7.

———. 1969b. Eine neue Gattung und Art cynopteriner Flughunde von Mindanao, Philippinen (Mammalia, Chiroptera). Senckenbergiana Biol. 50:319–327.

———. 1969c. Eine bemerkenswerte neue Gattung und Art Flughunde von Luzon, Philippinen. Senckenbergiana Biol. 50:329–338.

———. 1974a. Eine neue *Suncus*-Art von Flores, Kleine Sunda-Inseln (Mammalia: Insectivora). Senckenbergiana Biol. 55:197–203.

———. 1974b. Egyptian tomb bat *Taphozous perforatus* E. Geoffroy 1818. First record from Uganda. E. Afr. Nat. Hist. Soc. Bull., July, p. 130.

———. 1975. Ein originalexemplar von *Nyctinomus ventralis* Heuglin 1861 (Mammalia: Chiroptera: Molossidae). Stuttgarter Beitr. Nat., ser. A, no. 272, 9 pp.

———. 1978a. A new fruit bat of the genus *Rousettus* Gray 1821, from the Comoro Islands, western Indian Ocean (Mammalia: Chiroptera). Proc. 4th Int. Bat Conf., Nairobi, pp. 205–216.

———. 1978b. The identity of *Gerbillus bottai* Lataste, 1882 (Mammalia: Rodentia), from Sennar, Sudan. Bull. Carnegie Mus. Nat. Hist., no. 6, pp. 31–37.

Koenig, L. 1970. Zur Fortpflanzung und Jugendentwicklung des Wüstenfuchses (*Fennecus zerda* Zimm. 1780). Z. Tierpsychol. 27:205–246.

Koffler, B. R. 1972. *Meriones crassus*. Mammalian Species, no. 9, 4 pp.

Koopman, K. F. 1967. Artiodactyls. *In* Anderson and Jones (1967), pp. 385–406.

———. 1971. Taxonomic notes on *Chalinolobus* and *Glauconycteris* (Chiroptera, Vespertilionidae). Amer. Mus. Novit., no. 2451, 10 pp.

———. 1972. *Eudiscopus denticulus*. Mammalian Species, no. 19, 2 pp.

———. 1973. Systematics of Indo-Australian *Pipistrellus*. Period. Biol. 75:113–116.

———. 1975. Bats of the Sudan. Bull. Amer. Mus. Nat. Hist. 154:353–444.

———. 1978a. Zoogeography of Peruvian bats with special emphasis on the role of the Andes. Amer. Mus. Novit., no. 2651, 33 pp.

———. 1978b. The genus *Nycticeius* (Vespertilionidae), with special reference to tropical Australia. Proc. 4th Int. Bat Conf., Nairobi, pp. 165–171.

———. 1979. Zoogeography of mammals from islands off the northeastern coast of New Guinea. Amer. Mus. Novit., no. 2690, 17 pp.

Koopman, K. F., and F. Gudmundsson. 1966. Bats in Iceland. Amer. Mus. Novit., no. 2262, 6 pp.

Koopman, K. F., and J. K. Jones, Jr. 1970. Classification of bats. *In* Slaughter, B. H., and D. W. Walton, eds., About bats, S. Methodist Univ. Press, Dallas, pp. 22–28.

Koopman, K. F.; R. E. Mumford; and J. F. Heisterberg. 1978. Bat records from Upper Volta, West Africa. Amer. Mus. Novit., no. 2643, 6 pp.

Kooyman, G. L. 1968. An analysis of some behavioral and physiological characteristics related to diving in the Weddell seal. Antarctic Res. Ser., Amer. Geophys. Union, 11:227–261.

Kooyman, G. L.; R. L. Gentry; and D. L. Urquhart. 1976. Northern fur seal diving behavior: a new approach to its study. Science 193:411–412.

Korobitsyna, K. V.; C. F. Nadler; N. N. Vorontsov; and R. S. Hoffmann. 1974. Chromosomes of the Siberian snow sheep, *Ovis nivicola*, and implications concerning the origin of amphiberingian wild sheep (subgenus *Pachyceros*). Quaternary Res. 4:235–245.

Kortlucke, S. M. 1973. Morphological variation in the kinkajou, *Potos flavus* (Mammalia: Procyonidae), in Middle America. Occas. Pap. Mus. Nat. Hist. Univ. Kansas, no. 17, 36 pp.

Kovalskaya, J. M., and V. E. Sokolov. 1980. *Microtus evoronensis* sp. n. (Rodentia, Cricetidae) from the Lower Amur territory. Zool. Zhur. 59:1409–1416.

Kowalski, K. 1955. Our bats and their protection. Polish Acad. Sci. Nat. Protection Res. Cent. Publ., no. 11, 110 pp.

Koyama, N.; K. Norikoshi; and T. Mano. 1975. Population dynamics of Japanese monkeys at Arashiyama. Proc. 5th Int. Congr. Primatol., pp. 411–417.

Krasinski, Z. A. 1978. Dynamics and structure of the European bison population in the Bialowieza Primeval Forest. Acta Theriol. 23:3–48.

Kratochvil, J.; L. Rodriguez; and V. Barus. 1978. Capromyinae (Rodentia) of Cuba. I. Prirod. Prace Ustavu Cesk. Akad. Ved v Brne, Acta Sci. Nat. Brno 12(11):1–60.

Krefting, L. W. 1974. Moose distribution and habitat selection in north central North America. Naturaliste Canadien, 101:81–100.

Kruuk, H. 1972. The spotted hyena. A study of predation and social behavior. Univ. Chicago Press, xvi + 335 pp.

———. 1976. Feeding and social behaviour of the striped hyaena (*Hyaena vulgaris* Desmarest). E. Afr. Wildl. J. 14:91–111.

———. 1978a. Spatial organization and territorial behaviour of the European badger *Meles meles*. J. Zool. 184:1–19.

———. 1978b. Foraging and spatial organization of the European badger, *Meles meles* L. Behav. Ecol. Sociobiol. 4:75–89.

Krzanowski, A. 1977. Contribution to the history of bats on Iceland. Acta Theriol. 22:272–273.

Kubiak, H. 1965. The appearance of a raccoon dog, *Nyctereutes procyonides* (Gray, 1834) in Cracow District (Poland). Prezelgl. Zool. 9:417–422.

Kucera, T. E. 1978. Social behavior and breeding system of the desert mule deer. J. Mamm. 59:463–476.

Kucherenko, S. P. 1976. The common otter (*Lutra lutra*) in the Amur-Ussury district. Zool. Zhur. 55:904–911.

Kucherenko, S. P., and V. G. Yudin. 1973. Distribution, population density and economical importance of the raccoon-dog

(*Nyctereutes procyonides*) in the Amur-Ussury district. Zool. Zhur. 52:1039–1045.

Kuhn, H. 1971. An adult female *Micropotamogale lamottei*. J. Mamm. 52:477–478.

Kummer, H. 1968. Social organization of hamadryas baboons. A field study. Univ. Chicago Press, vi + 189 pp.

Kummer, H.; W. Goetz; and W. Angst. 1970. Cross-species modifications of social behavior in baboons. *In* Napier and Napier (1970), pp. 351–363.

Kuntz, R. E., and B. J. Myers. 1969. A checklist of parasites and commensals reported for the Taiwan macaque. Primates 10:71–80.

Kunz, T. H. 1973a. Population studies of the cave bat (*Myotis velifer*): reproduction, growth, and development. Occas. Pap. Mus. Nat. Hist. Univ. Kansas, no. 15, 43 pp.

———. 1973b. Resource utilizations: temporal and spatial components of bat activity in central Iowa. J. Mamm. 54:14–32.

Kurland, J. A. 1973. A natural history of Kra macaques (*Macaca fascicularis* Raffles, 1821) at the Kutai Reserve, Kalimantan Timur, Indonesia. Primates 14:245–262.

———. 1977. Kin selection in the Japanese monkey. Contrib. Primatol. 12:i–x + 1–145 pp.

Kurt, F. 1974. Remarks on the social structure and ecology of the Ceylon elephant in the Yala National Park. *In* Geist and Walther (1974), pp. 618–634.

Kurten, B. 1968. Pleistocene mammals of Europe. Aldine, Chicago, viii + 317 pp.

———. 1973. Transberingean relationships of *Ursus arctos* Linne (brown and grizzly bears). Commentat. Biol. 65:1–10.

Kurz, J. C., and R. L. Marchinton. 1972. Radiotelemetry studies of feral hogs in South Carolina. J. Wildl. Mgmt. 36:1240–1248.

Kuschinski, L. 1974. Breeding binturongs *Arctictis binturong* at Glasgow Zoo. Int. Zoo Yearbook 14:124–126.

Kuyt, E. 1972. Food habits and ecology of wolves on barren-ground caribou range in the Northwest Territories. Can. Wildl. Serv. Rept. Ser., no. 21, 36 pp.

L.

Lackey, J. A. 1976. Reproduction, growth, and development in the Yucatan deer mouse, *Peromyscus yucatanicus*. J. Mamm. 57:638–655.

Laerm, J. 1981. Systematic status of the Cumberland Island pocket gopher, *Geomys cumberlandius*. Brimleyana 6:141–151.

Lamprecht, J. 1979. Field observations on the behaviour and social system of the bat-eared fox *Otocyon megalotis* Desmarest. Z. Tierpsychol. 49:260–284.

Lang, H., and J. P. Chapin. 1917. Notes on the distribution and ecology of Central African Chiroptera. Bull. Amer. Mus. Nat. Hist. 37:479–563.

Langguth, A. 1975a. La identidad de *Mus lasiotis* Lund y el status del gênero *Thalpomys* Thomas (Mammalia, Cricetidae). Papeis Avulsos Zool. (São Paulo) 29:45–54.

———. 1975b. Ecology and evolution in the South American canids. *In* Fox (1975), pp. 192–206.

Langman, V. A. 1973. Radio-tracking giraffe for ecological studies. J. S. Afr. Wildl. Mgmt. Assoc. 3:75–78.

———. 1977. Cow-calf relationships in giraffe (*Giraffa camelopardalis giraffa*). Z. Tierpsychol. 43:264–286.

Lanning, D. V. 1976. Density and movements of the coati in Arizona. J. Mamm. 57:609–611.

Largen, M. J.; D. Kock; and D. W. Yalden. 1974. Catalogue of the mammals of Ethiopia. 1. Chiroptera. Italian J. Zool., suppl., n.s., 5:221–298.

Larkin, P., and M. Roberts. 1979. Reproduction in the ring-tailed mongoose *Galidia elegans* at the National Zoological Park, Washington, Int. Zoo Yearbook 19:189–193.

Larsen, T. 1975. Polar bear den surveys in Svalbard in 1973. Norsk Polarinst. Arbok, 1973, pp. 101–112.

Larsson, T., L. Hansson, and E. Nyholm. 1973. Winter reproduction in small rodents in Sweden. Oikos 24:475–476.

Laurie, A., and J. Seidensticker. 1977. Behavioural ecology of the sloth bear (*Melursus ursinus*). J. Zool. 182:187–204.

Laurie, E.M.O., and J. E. Hill. 1954. List of land mammals of New Guinea, Celebes and adjacent islands 1758–1952. British Mus. (Nat. Hist.), London, 175 pp.

Laursen, L., and M. Bekoff. 1978. *Loxodonta africana*. Mammalian Species, no. 92, 8 pp.

LaVal, R. K. 1973a. A revision of the neotropical bats of the genus *Myotis*. Los Angeles Co. Nat. Hist. Mus. Sci. Bull., no. 15, 54 pp.

———. 1973b. Systematics of the genus *Rhogeessa* (Chiroptera: Vespertilionidae). Occas. Pap. Mus. Nat. Hist. Univ. Kansas, no. 19, 47 pp.

———. 1973c. Observations on the biology of *Tadarida brasiliensis cynocephala* in southeastern Louisiana. Amer. Midl. Nat. 89:112–120.

———. 1976. Voice and habitat of *Dactylomys dactylinus* (Rodentia: Echimyidae) in Ecuador. J. Mamm. 57:402–404.

———. 1977. Notes on some Costa Rican bats. Brenesia 10–11:77–83.

LaVal, R. K., and H. S. Fitch. 1977. Structure, movements and reproduction in three Costa Rican bat communities. Occas. Pap. Mus. Nat. Hist. Univ. Kansas, no. 69, 28 pp.

LaVal, R. K., and M. L. LaVal. 1977. Reproduction and behavior of the African banana bat, *Pipistrellus nanus*. J. Mamm. 58:403–410.

Lawlor, T. E. 1969. A systematic study of the rodent genus *Ototylomys*. J. Mamm. 50:28–42.

Lawrence, B. 1939. Collections from the Philippine Islands. Mammals. Bull. Mus. Comp. Zool., 86:28–73.

Lawrence, B., and W. H. Bossert. 1967. Multiple character analysis of *Canis lupus*, *latrans*, and *familiaris*, with a discussion of the relationships of *Canis niger*. Amer. Zool. 7:223–232.

———. 1975. Relationships of North American *Canis* shown by a multiple character analysis of selected populations. *In* Fox (1975), pp. 73–86.

Laws, R. M.; I.S.C. Parker; and R.C.B. Johnstone. 1975. Elephants and their habitats/the ecology of elephants in north Bunyoro, Uganda. Oxford Univ. Press, London, xii + 376 pp.

Lay, D. M. 1967. A study of the mammals of Iran. Fieldiana Zool. 54:1–282.

———. 1975. Notes on rodents of the genus *Gerbillus* (Mammalia: Muridae: Gerbillinae) from Morocco. Fieldiana Zool. 65:89–101.

———. 1978. Observations on reproduction in a population of pocket gophers, *Thomomys bottae*, from Nevada. Southwestern Nat. 23:375–380.

Lay, D. M., and C. F. Nadler. 1975. A study of *Gerbillus* (Rodentia: Muridae) east of the Euphrates River. Mammalia 39:423–445.

Layne, J. N. 1967. Lagomorphs. *In* Anderson and Jones (1967), pp. 192–205.

———. 1968. Ontogeny. *In* King (1968), pp. 148–253.

———. 1972. Tail autotomy in the Florida mouse, *Peromyscus floridanus*. J. Mamm. 53:62–71.

———, ed. 1978. Rare and endangered biota of Florida. I. Mammals. University Presses of Florida, Gainesville, xx + 52 pp.

Layne, J. N., and D. Glover. 1977. Home range of the armadillo in Florida. J. Mamm. 58:411–413.

Leakey, L.S.B.; P. V. Tobias; and J. R. Napier. 1964. A new species of the genus *Homo* from Olduvai Gorge. Nature 202:7–9.

Leatherwood, J. S. 1974. Aerial observations of migrating gray whales, *Eschrichtius robustus*, off southern California, 1969–72. Mar. Fish. Rev. 36(4):45–49.

Leatherwood, S.; D. K. Caldwell; and H. E. Winn. 1976. Whales, dolphins, and porpoises of the western North Atlantic. A guide to their identification. U.S. Natl. Mar. Fish. Serv., NOAA Tech. Rept. NMFS CIRC–396, iv + 176 pp.

Leatherwood, S.; L. J. Harrington-Coulombe; and C. L. Hubbs. 1978. Relict survival of the sea otter in central California and evidence of its recent dispersal south of Point Conception. Bull. S. California Acad. Sci. 77:109–115.

Leatherwood, S., and R. R. Reeves. 1978. Porpoises and dolphins. *In* Haley (1978), pp. 97–111.

Leatherwood, S., and W. A. Walker. 1979. The northern right whale dolphin *Lissodelphis borealis* Peale in the eastern North Pacific. *In* Winn and Olla (1979), pp. 85–141.

Le Boeuf, B. J.; D. G. Ainley; and T. J. Lewis. 1974. Elephant seals on the Farallones: population structure of an incipient breeding colony. J. Mamm. 55:370–385.

Le Boeuf, B. J., and K. J. Panken. 1977. Elephant seals breeding on the mainland in California. Proc. California Acad. Sci., ser. 4, 41:267–280.

Le Boeuf, B. J., and R. S. Peterson. 1969. Social status and mating activity in elephant seals. Science 163:91–93.

Le Boeuf, B. J., and L. F. Petrinovich. 1974. Elephant seals: interspecific comparisons of vocal and reproductive behavior. Mammalia 38:16–32.

Lee, A. K.; A. J. Bradley; and R. W. Braithwaite. 1977. Corticosteroid levels and male mortality in Antechinus stuartii. In Stonehouse and Gilmore (1977), pp. 209–220.

Lee, M. R., and D. J. Schmidly. 1977. A new species of Peromyscus (Rodentia: Muridae) from Coahuila, Mexico. J. Mamm. 58:263–268.

Lee, M. R.; D. J. Schmidly; and C. C. Huheey. 1972. Chromosomal variation in certain populations of Peromyscus boylii and its systematic implications. J. Mamm. 53:697–707.

Le Gros Clark, W. E. 1955. The fossil evidence for human evolution. Univ. Chicago Press, 180 pp.

Lehner, P. N. 1978. Coyote communication. In Bekoff (1978), pp. 128–162.

Lekagul, B., and J. A. McNeely. 1977. Mammals of Thailand. Sahakarnbhat, Bangkok, li + 758 pp.

Lemke, T. O., and J. R. Tamsitt. 1979. Anoura cultrata (Chiroptera: Phyllostomatidae) from Colombia. Mammalia 43:579–581.

Lenglet, G., and G. Coppois. 1979. Description du crâne et de quelques ossements d'un genre nouveau éteint de Cricetidae (Mammalia-Rodentia) géant des Galapagos: Megaoryzomys (gen. nov.). Bull. Acad. Roy. Belgique, Classe des Sciences, 55:632–648.

Lensing, J. E., and E. Joubert. 1972. Intensity distribution patterns for five species of problem animals in South West Africa. Madoqua 10:131–141.

Lent, P. C. 1974. A review of rutting behavior in moose. Naturaliste Canadien 101:307–323.

———. 1978. Musk-ox. In Schmidt and Gilbert (1978), pp. 135–147.

Leopold, A. S. 1959. Wildlife of Mexico. Univ. California Press, Berkeley, xiii + 568 pp.

LeResche, R. E. 1974. Moose migrations in North America. Naturaliste Canadien 101: 393–415.

Leslie, D. M., Jr., and C. L. Douglas. 1979. Desert bighorn sheep of the River Mountains, Nevada. Wildl. Monogr., no. 66, 56 pp.

Leslie, G. 1971. Further observations on the oriental short-clawed otter Amblonyx cinerea at Aberdeen Zoo. Int. Zoo Yearbook 11:112–113.

Leuthold, B. M. 1979. Social organization and behaviour of giraffe in Tsavo East National Park. Afr. J. Ecol. 17:19–34.

Leuthold, B. M., and W. Leuthold. 1978. Ecology of the giraffe in Tsavo East National Park, Kenya. E. Afr. Wildl. J. 16:1–20.

Leuthold, W. 1970. Observations on the social organization of impala (Aepyceros melampus). Z. Tierpsychol. 27:693–721.

———. 1974. Observations on home range and social organization of lesser kudu, Tragelaphus imberbis (Blyth, 1869). In Geist and Walther (1974), pp. 206–234.

———. 1977a. African ungulates. Springer-Verlag, Berlin, xiii + 307 pp.

———. 1977b. A note on group size and composition in the oribi Ourebia ourebi (Zimmerman, 1783) (Bovidae). Saugetierk. Mitt. 25:233–235.

———. 1978a. On the ecology of the gerenuk Litocranius walleri. J. Anim. Ecol. 47:561–580.

———. 1978b. On social organization and behaviour of the gerenuk Litocranius walleri (Brooke 1878). Z. Tierpsychol. 47:194–216.

Leuthold, W., and J. B. Sale. 1973. Movements and patterns of habitat utilization of elephants in Tsavo National Park, Kenya. E. Afr. Wildl. J. 11:369–384.

Lewis, J. G., and R. T. Wilson. 1979. The ecology of Swayne's hartebeest. Biol. Conserv. 15:1–12.

Leyhausen, P. 1979. Cat behavior. Garland STPM Press, New York, 340 pp.

Liberg, O. 1980. Spacing patterns in a population of rural free roaming domestic cats. Oikos 35:336–349.

Lidicker, W. Z., Jr. 1976. Social behaviour and density regulation in house mice living in large enclosures. J. Anim. Ecol. 45:677–697.

Lidicker, W. Z., Jr., and B. J. Marlow. 1970. A review of the dasyurid marsupial genus Antechinomys Krefft. Mammalia 34:212–227.

Lidicker, W. Z., Jr., and A. C. Ziegler. 1968. Report on a collection of mammals from eastern New Guinea including species keys for fourteen genera. Univ. California Publ. Zool. 87:i–v + 1–64.

Liers, E. E. 1966. Notes on breeding the Canadian otter Lutra canadensis in captivity and longevity records of beavers Castor canadensis. Int. Zoo Yearbook 6:171–172.

Lilly, J. C. 1977. The cetacean brain. Oceans 10(4):4–7.

Lim Boo Liat. 1967. Note on the food habits of Ptilocercus lowii Gray (pentail treeshrew) and Echinosorex gymnurus (Raffles) (moonrat) in Malaya with remarks on "ecological labelling" by parasite patterns. J. Zool. 152:375–379.

———. 1970. Distribution, relative abundance, food habits, and parasite patterns of giant rats (Rattus) in West Malaysia. J. Mamm. 51:730–740.

Lim Boo Liat, and I. Muul. 1975. Notes on a rare species of arboreal rat, Pithecheir parvus Kloss. Malayan Nat. J. 28:181–185.

Lim Boo Liat; Chai Koh Shin; and I. Muul. 1972. Notes on the food habit of bats from the fourth division, Sarawak, with special reference to a new record of Bornean bat. Sarawak Mus. J. 20:351–357.

Linares, O. J. 1969. Notas acerca de la captura de una rata acuática (Nectomys squamipes) en la Cueva del Agua (AN–1), Anzoátegui, Venezuela. Bol. Soc. Venezolana Espeleol. 2:31–34.

———. 1973. Présence de l'oreillard d'Amerique du Sud dans les Andes Vénézuéliennes (Chiroptères, Vespertilionidae). Mammalia 37:433–438.

Lindburg, D. G. 1971. The rhesus monkey in north India: an ecological and behavioral study. Primate Behav. 2:1–106.

Linder, R. L., and C. N. Hillman, eds. 1973. Proceedings of the black-footed ferret and prairie dog workshop. Dept. Wildl. and Fish. Sci., S. Dakota State Univ., v + 208 pp.

Lindsay, S. L. 1981. Taxonomic and biogeographic relationships of Baja California chickarees (Tamiasciurus). J. Mamm. 62:673–682.

Lindzey, F. G., and E. C. Meslow. 1976. Winter dormancy in black bears in southwestern Washington. J. Wildl. Mgmt. 40:408–415.

———. 1977a. Population characteristics of black bears on an island in Washington. J. Wildl. Mgmt. 41:408–412.

———. 1977b. Home range and habitat use by black bears in southwestern Washington. J. Wildl. Mgmt. 41:413–425.

Ling, J. K. 1977. Children of the bight. Oceans (Australia) 1:116–121.

———. 1979. The status of endangered Australian marine mammals. In Tyler (1979), pp. 67–74.

Ling, J. K., and G. E. Walker. 1976. Seal studies in South Australia: progress report for the year 1975. S. Austral. Nat. 50:59–68, 72.

———. 1977. Seal studies in South Australia: progress report for the period January 1976 to March 1977. S. Austral. Nat. 52:18–27, 30.

Linzey, D. W., and R. L. Packard. 1977. Ochrotomys nuttalli. Mammalian Species, no. 75, 6 pp.

Lippold, L. K. 1974. The duoc langur: a time for conservation. In Rainier III and Bourne (1977), pp. 513–538.

Lloyd, H. G. 1975. The red fox in Britain. In Fox (1975), pp. 207–215.

Lloyd, J. A. 1975. Social structure and reproduction in two freely-growing populations of house mice (Mus musculus L.). Anim. Behav. 23:413–424.

Lock, J. M. 1972. The effects of hippopotamus grazing on grasslands. J. Ecol. 60:445–467.

Lockard, R. B. 1978. Seasonal change in the activity pattern of Dipodomys spectabilis. J. Mamm. 59:563–568.

Lockard, R. B., and D. H. Owings. 1974. Moon-related surface activity of bannertail (*Dipodomys spectabilis*) and Fresno (*D. nitratoides*) kangaroo rats. Anim. Behav. 22:262–273.

Long, C. A. 1972a. Taxonomic revision of the mammalian genus *Microsorex* Coues. Trans. Kansas Acad. Sci. 74:181–196.

———. 1972b. Notes on habitat preference and reproduction in pigmy shrews, *Microsorex*. Can. Field-Nat. 86:155–160.

———. 1972c. Taxonomic revision of the North American badger, *Taxidea taxus*. J. Mamm. 53:725–759.

———. 1973a. Reproduction in the white-footed mouse at the northern limits of its geographical range. Southwestern Nat. 18:11–20.

———. 1973b. *Taxidea taxus*. Mammalian Species, no. 26, 4 pp.

———. 1974. *Microsorex hoyi* and *Microsorex thompsoni*. Mammalian Species, no. 33, 4 pp.

———. 1978. A listing of Recent badgers of the world, with remarks on taxonomic problems in *Mydaus* and *Melogale*. Rept. Fauna and Flora Wisconsin, Univ. Wisconsin Mus. Nat. Hist., 14:1–6.

Lopez-Forment, W. 1980. Longevity of wild *Desmodus rotundus* in Mexico. Proc. 5th Int. Bat. Res. Conf., Texas Tech Press, pp. 143–144.

Lorenz, R.; C. O. Anderson; and W. A. Mason. 1973. Notes on reproduction in captive squirrel monkeys (*Saimiri sciureus*). Folia Primatol. 19:286–292.

Lott, D. F. 1974. Sexual and aggressive behaviour of adult male American bison (*Bison bison*). *In* Geist and Walther (1974), pp. 382–394.

Lotze, J. 1979. The raccoon (*Procyon lotor*) on St. Catherines Island, Georgia. 4. Comparisons of home ranges determined by live-trapping and radiotracking. Amer. Mus. Novit., no. 2664, 25 pp.

Lotze, J., and S. Anderson. 1979. *Procyon lotor*. Mammalian Species, no. 119, 8 pp.

Loughlin, T. R. 1980. Home range and territoriality of sea otters near Monterey, California. J. Wildl. Mgmt. 44:576–582.

Loukashkin, A. S. 1940. On the pikas of north Manchuria. J. Mamm. 21:402–405.

Louwman, J.W.W. 1973. Breeding the tailless tenrec *Tenrec ecaudatus* at Wassenaar Zoo. Int. Zoo Yearbook 13:125–126.

Lovari, S. 1977. The Abruzzo chamois. Oryx 14:47–50.

Lovejoy, B. P., and H. C. Black. 1974. Growth and weight of the mountain beaver, *Aplodontia rufa pacifica*. J. Mamm. 55:364–369.

———. 1979a. Movements and home range of the Pacific mountain beaver *Aplodontia rufa pacifica*. Amer. Midl. Nat. 101:393–402.

———. 1979b. Population analysis of the mountain beaver, *Aplodontia rufa pacifica*, in western Oregon. Northwest Sci. 53:82–89.

Lovejoy, B. P.; H. C. Black; and E. F. Hooven. 1978. Reproduction, growth, and development of the mountain beaver (*Aplodontia rufa pacifica*). Northwest Sci. 52:323–328.

Lowery, G. H., Jr. 1974. The mammals of Louisiana and its adjacent waters. Louisiana State Univ. Press, xxiii + 565 pp.

Luckett, W. P., and F. S. Szalay, eds. 1975. Phylogeny of the primates. A multidisciplinary approach. Plenum Press, New York, xiv + 483 pp.

Lutton, L. M. 1975. On the territorial behavior and response to predators of the pika, *Ochotona princeps*. J. Mamm. 56:231–234.

Lyman, C. P., and R. C. O'Brien. 1977. A laboratory study of the Turkish hamster *Mesocricetus brandti*. Breviora, no. 442, 27 pp.

Lynch, C. D. 1981. The status of the Cape grey mongoose, *Herpestes pulverulentus* Wagner, 1839 (Mammalia: Viverridae). Navorsinge Nasionale Mus., Bloemfontein 4:121–168.

Lynch, G. R.; F. D. Vogt; and H. R. Smith. 1978. Seasonal study of spontaneous daily torpor in the white-footed mouse, *Peromyscus leucopus*. Physiol. Zool. 51:289–299.

Lyne, A. G. 1974. Gestation period and birth in the marsupial *Isoodon macrourus*. Austral. J. Zool. 22:303–309.

———. 1976. Observations on oestrus and the oestrous cycle in the marsupials *Isoodon macrourus* and *Perameles nasuta*. Austral. J. Zool. 24:513–521.

M.

MacArthur, R. A. 1978. Winter movements and home range of the muskrat. Can. Field-Nat. 92:345–349.

Macdonald, D. W. 1978. Observations on the behaviour and ecology of the striped hyaena, *Hyaena hyaena*, in Israel. Israel J. Zool. 27:189–198.

MacKinnon, J. 1971. The orang-utan in Sabah today. Oryx 11:141–191.

———. 1973. Orang-utans in Sumatra. Oryx 12:234–242.

———. 1974. The behaviour and ecology of wild orang-utans (*Pongo pygmaeus*). Anim. Behav. 22:3–74.

———. 1981. The structure and function of the tusks of babirusa. Mamm. Rev. 11:37–40.

Macintosh, N.W.G. 1975. The origin of the dingo: an enigma. *In* Fox (1975), pp. 87–106.

Mackintosh, J. H. 1973. Factors affecting the recognition of territory boundaries by mice (*Mus musculus*). Anim. Behav. 21:464–470.

Mackintosh, N. A. 1966. The distribution of southern blue and fin whales. *In* Norris (1966), pp. 125–144.

Macpherson, A. H. 1965. The barren-ground grizzly bear and its survival in northern Canada. Can. Audubon 27(1):2–8.

———. 1969. The dynamics of Canadian arctic fox populations. Can. Wildl. Serv. Rept. Ser., no. 8, 52 pp.

Madden, J. R. 1974. Female territoriality in a Suffolk County, Long Island, population of *Glaucomys volans*. J. Mamm. 55:647–652.

Maddock, T. H., and A. McLeod. 1974. Polyoestry in the little brown bat, *Eptesicus pumilus*, in central Australia. S. Austral. Nat. 48:50, 63.

Madhavan, A. 1971. Breeding habits in the Indian vespertilionid bat, *Pipistrellus ceylonicus chrysothrix* (Wroughton). Mammalia 25:283–306.

Madhavan, A.; D. R. Patil; and A. Gopalakrishna. 1978. Breeding habits and associated phenomena in some Indian bats. Part IV—*Hipposideros fulvus fulvus* (Gray)—Hipposideridae. J. Bombay Nat. Hist. Soc. 75:96–103.

Madison, D. M. 1977. Movements and habitat use among interacting *Peromyscus leucopus* as revealed by radiotelemetry. Can. Field-Nat. 91:273–281.

Maeda, K. 1980. Review on the classification of little tube-nosed bats, *Murina aurata* group. Mammalia 44:531–551.

Magnusson, W. E.; G.J.W. Webb; and J. A. Taylor. 1976. Two new locality records, a new habitat and a nest description for *Xeromys myoides* Thomas (Rodentia: Muridae). Austral. Wildl. Res. 3:153–157.

Mahe, J. 1976. Crâniométrie des lémuriens analyses multivariables-phylogénie. Mem. Mus. Natl. Hist. Nat., ser. C, 32:1–342.

Mahoney, J. A. 1968. *Baiyankamys* Hinton 1943 (Muridae, Hydromyinae) a New Guinea rodent genus named for an incorrectly associated skin and skull (Hydromyinae, *Hydromys*) and mandible (Murinae, *Rattus*). Mammalia 32:64–71.

———. 1975. *Notomys macrotis* Thomas, 1921, a poorly known Australian hopping mouse [Rodentia: Muridae]. Austral. Mamm. 1:367–374.

———. 1977. Skull characters and relationships of *Notomys mordax* Thomas (Rodentia: Muridae), a poorly known Queensland hopping-mouse. Austral. J. Zool. 25:749–754.

Mahoney, J. A., and B. J. Marlow. 1968. The rediscovery of the New Holland mouse. Austral. J. Sci. 31:221–223.

Mahoney, J. A., and H. Posamentier. 1975. The occurrence of the native rodent, *Pseudomys gracilicaudatus* (Gould, 1845) [Rodentia: Muridae], in New South Wales. Austral. Mamm. 1:333–346.

Main, A. R., and M. Yadav. 1971. Conservation of macropods in reserves in Western Australia. Biol. Conserv. 3:123–133.

Mainstone, B. J. 1974. Tigers reduce honey production. Malayan Nat. J. 28:36.

Malagnoux, M., and J. C. Gautun. 1976. Un ennemi des plantations d'araucaria en Côte d'Ivoire. Bois et Forêts Tropiques 165:35–38.

Malcolm, J. R. 1980. African wild dogs play every game by their own rules. Smithsonian 11(8):62–71.

Malcolm, J. R., and H. Van Lawick. 1975. Notes on wild dogs (*Lycaon pictus*) hunting zebras. Mammalia 39:231–240.

Maliniak, E., and J. F. Eisenberg. 1971. Breeding spiny rats *Proechimys semispinosus* in captivity. Int. Zoo Yearbook 11:93–98.

Mallinson, J.J.C. 1973. The reproduction of the African civet *Viverra civetta* at Jersey Zoo. Int. Zoo Yearbook 13:147–150.

————. 1974. Establishing mammal gestation at the Jersey Zoological Park. Int. Zoo Yearbook 14:184–187.

————. 1977. Maintenance of marmosets and tamarins at Jersey Zoological Park with special reference to the design of the new marmoset complex. *In* Kleiman (1977*a*) pp. 323–329.

Mallory, F. F., and R. J. Brooks. 1978. Infanticide and other reproductive strategies in the collared lemming, *Dicrostonyx groenlandicus*. Nature 273:144–146.

Mansfield, A. W. 1966. The grey seal in eastern Canadian waters. Can. Audubon 28:160–166.

————. 1967. Distribution of the harbor seal, *Phoca vitalina* Linnaeus, in Canadian arctic waters. J. Mamm. 48:249–257.

Mansfield, A. W.; T. G. Smith; and B. Beck. 1975. The narwhal, *Monodon monoceros*, in eastern Canadian waters. J. Fish. Res. Bd. Can. 32:1041–1046.

Manville, R. H. 1966. The extinct sea mink, with taxonomic notes. Proc. U.S. Natl. Mus. 122:1–12.

Manwell, C., and C.M.A. Baker. 1977. *Ammotragus lervia:* Barbary sheep or Barbary goat. Comp. Biochem. Physiol. 58:267–271.

Maples, W. R. 1972. Systematic reconsideration and a revision of the nomenclature of Kenya baboons. Amer. J. Phys. Anthropol. 36:9–19.

Mares, M. A. 1977. Water balance and other ecological observations on three species of *Phyllotis* in northwestern Argentina. J. Mamm. 58:514–520.

Mares, M. A.; M. D. Watson; and T. E. Lacher. 1976. Home range perturbations in *Tamias striatus*. Oecologia 25:1–12.

Marinina, L. S. 1971. Vital activity of the *Rhombomys opimus* population in the central Karakum after the winter of 1968/69. Soviet J. Ecol. 2:77–80.

Marinkelle, C. J., and A. Cadena. 1972. Notes on bats new to the fauna of Colombia. Mammalia 36:50–58.

Markgren, G. 1974. The moose in Fennoscandia. Naturaliste Canadien 101:185–194.

Markham, D. D., and F. W. Whicker. 1973. Seasonal data on reproduction and body weights of pikas (*Ochotona princeps*). J. Mamm. 54:496–498.

Marlow, B. J. 1961. Reproductive behavior of the marsupial mouse, *Antechinus flavipes* (Waterhouse) (Marsupialia) and the development of pouch young. Austral. J. Zool. 9:203–218.

————. 1975. The comparative behavior

of the Australian sea lions *Neophoca cinerea* and *Phocarctos hookeri* (Pinnipedia: Otariidae). Mammalia 39:159–230.

Marlow, B. J., and J. E. King. 1974. Sea lions and fur seals of Australia and New Zealand—the growth of knowledge. Austral. Mamm. 1:117–135.

Marples, T. G. 1973. Studies on the marsupial glider, *Schoinobates volans* (Kerr). IV. Feeding biology. Austral. J. Zool. 21:213–216.

Marquette, W. M. 1978. Bowhead whale. *In* Haley (1978), pp. 70–81.

————. 1979. The 1977 catch of bowhead whales (*Balaena mysticetus*) by Alaskan Eskimos. Rept. Int. Whaling Comm. 29:281–289.

Marshall, J. T., Jr. 1977. A synopsis of Asian species of *Mus* (Rodentia, Muridae). Bull. Amer. Mus. Nat. Hist. 158:173–220.

————. 1979. Classification of genus *Mus*. *In* Altman, P. C., and D. D. Katz, eds., Inbred and genetically defined strains of laboratory animals, part I, Mouse and rat, Fed. Amer. Soc. Exp. Biol., Bethesda, Maryland, pp. 212–220.

Marshall, J. T., Jr., and R. O. Sage. 1981. Taxonomy of the house mouse. *In* Berry (1981), pp. 15–25.

Marshall, L. G. 1977. *Lestodelphys halli*. Mammalian Species, no. 81, 3 pp.

————. 1978*a*. *Lutreolina crassicaudata*. Mammalian Species, no. 91, 4 pp.

————. 1978*b*. *Dromiciops australis*. Mammalian Species, no. 99, 5 pp.

————. 1978*c*. *Glironia venusta*. Mammalian Species, no. 107, 3 pp.

————. 1978*d*. *Chironectes minimus*. Mammalian Species, no. 109, 6 pp.

Martin, P. 1971. Movements and activities of the mountain beaver (*Aplodontia rufa*). J. Mamm. 52:717–723.

Martin, P. G. 1977. Marsupial biogeography and plate tectonics. *In* Stonehouse and Gilmore (1977), pp. 97–115.

Martin, P. S., and H. E. Wright, eds. 1967. Pleistocene extinctions. Yale Univ. Press, New Haven, x + 453 pp.

Martin, R. A. 1972. Synopsis of late Pliocene and Pleistocene bats of North America and the Antilles. Amer. Midl. Nat. 87:326–335.

Martin, R. D. 1968. Reproduction and ontogeny in tree-shrews (*Tupaia belangeri*), with reference to their general behaviour and taxonomic relationships. Z. Tierpsychol. 25:409–495, 25:505–532.

————. 1973. A review of the behaviour and ecology of the lesser mouse lemur (*Microcebus murinus* J. F. Miller 1777). *In* Michael and Crook (1973), pp. 1–68.

————. 1975*a*. The bearing of reproductive behavior and ontogeny on strepsirhine phylogeny. *In* Luckett and Szalay (1975), pp. 265–297.

————, ed. 1975*b*. Breeding endangered species in captivity. Academic Press, London, xxv + 420 pp.

Martin, R. D.; G. A. Doyle; and A. C. Walker, eds. 1974. Prosimian biology. Duckworth, London, xxi + 983 pp.

Martin, R. E. 1970. Cranial and bacular variation in populations of spiny rats of the genus *Proechimys* (Rodentia: Echimyidae) from South America. Smithson. Contrib. Zool., no. 35, 19 pp.

Martin, R. E., and K. G. Matocha. 1972. Distributional status of the kangaroo rat, *Dipodomys elator*. J. Mamm. 53:873–877.

Martinka, C. J. 1974. Population characteristics of grizzly bears in Glacier National Park, Montana. J. Mamm. 55:21–29.

————. 1976. Ecological role and management of grizzly bears in Glacier National Park, Montana. *In* Pelton, Lentfer, and Folk (1976), pp. 147–156.

Masaki, Y. 1976. Biological studies on the North Pacific sei whale. Bull. Far Seas Fish. Res. Lab. 14:1–104.

————. 1978. Yearly change in the biological parameters of the Antarctic sei whale. Rept. Int. Whaling Comm. 28:421–429.

Maser, C., ed. 1974. The sage vole, *Lagurus curtatus* (Cope, 1868), in the Crooked River National Grassland, Jefferson County, Oregon. A contribution to its life history and ecology. Saugetierk. Mitt. 22:193–222.

Mason, W. A. 1968. Use of space by *Callicebus* groups. *In* Jay (1968), pp. 200–216.

————. 1971. Field and laboratory studies of social organization in *Saimiri* and *Callicebus*. Primate Behav. 2:107–137.

Massoia, E. 1963. *Oxymycterus iheringi* (Rodentia-Cricetidae) nueva especie para la Argentina. Physis 24:129–136.

————. 1973. Descripción de *Oryzomys fornesi*, nueva especie y nuevos datos sobrealgunas especies y subespecies argentinas del subgénero *Oryzomys* (*Oligoryzomys*) (Mammalia—Rodentia—Cricetidae). Rev. Investig. Agropec., Buenos Aires, ser. 1, 10:21–37.

————. 1977. Sobre la identidad del holotipo de *Graomys hypogaeus* Cabrera, 1934 (Mammalia—Rodentia—Cricetidae). Rev. Investig. Agropec., Buenos Aires, ser. 5, 13:15–20.

————. 1979. Descripción de un género y especie nuevos: *Bibimys torresi* (Mammalia—Rodentia—Cricetidae—Sigmodontinae—Scapteromyini). Physis, sec. C, 38:1–7.

————. 1980. El estado sistemático de cuatro especies de cricétidos Sudamericanos y comentarios sobre otras especies congenéricas. Ameghiniana, Rev. Asoc. Paleontol. Argentina 17:280–287.

————. 1981. El estado sistemático y zoogeografía de *Mus brasiliensis* Desmarest y *Holochilus sciureus* Wagner. Physis, sec. C, 39:31–34.

Massoia, E., and A. Fornes. 1967. El estado sistemático, distribución geográfica y datos etoecológicos de algunos mamíferos neotropicales (Marsupialia y Rodentia) con la descripción de *Cabreramys*, género nuevo (Cricetidae). Acta Zool. Lilloana 23:407–430.

Masui, K.; Y. Suigiyama; A. Nishimura; and H. Ohsawa. 1975. The life table of Japanese monkeys at Takasakiyama. A preliminary report. Proc. 5th Int. Congr. Primatol., pp. 401–406.

Matocha, K. G. 1977. The vocal repertoire of *Spermophilus tridecemlineatus*. Amer. Midl. Nat. 98:482–487.

Matson, J. O., and J. P. Abravaya. 1977. *Blarinomys breviceps*. Mammalian Species, no. 74, 3 pp.

Mayer, J. J., and P. N. Brandt. 1978. Pelage description and reproductive data of the Chacoan peccary, *Catagonus wagneri*. American Soc. Mamm., Abstr. Tech. Pap., 58th Ann. Mtg., p. 44.

Maynes, G. M. 1973. Reproduction in the parma wallaby, *Macropus parma* Waterhouse. Austral. J. Zool. 21:331–351.

————. 1974. Occurrence and field recognition of *Macropus parma*. Austral. Zool. 18:72–87.

————. 1977a. Distribution and aspects of the biology of the Parma wallaby, *Macropus parma*, in New South Wales. Austral. Wildl. Res. 4:109–125.

————. 1977b. Breeding and age structure of the population of *Macropus parma* on Kawau Island, New Zealand. Austral. J. Ecol. 2:207–214.

Mayr, E. 1951. Taxonomic categories in fossil hominids. Cold Spring Harbor Symp. Quant. Biol. 15:109–118.

————. 1963. The taxonomic evaluation of fossil hominids. *In* Washburn, S. L., ed., Classification and human evolution, Aldine, Chicago, pp. 332–345.

Maza, B. G.; N. R. French; and A. P. Aschwanden. 1973. Home range dynamics in a population of heteromyid rodents. J. Mamm. 54:405–425.

Mazak, V. 1981. *Panthera tigris*. Mammalian Species, no. 152, 8 pp.

Mazin, V. N. 1975. Data on reproduction of jerboas of the genus *Allactaga* in southeast Kazakhstan. Soviet J. Ecol. 6:334–339.

McBride, R. T. 1980. The Mexican wolf (*Canis lupus baileyi*): a historical review and observations on its status and distribution. U.S. Fish and Wildl. Serv., Albuquerque, Endangered Species Rept., no. 8, vi + 38 pp.

McCann, C. 1975. A study of the genus *Berardius* Duvernoy. Sci. Rept. Whales Res. Inst. 27:111–137.

McCann, T. S. 1980. Population structure and social organization of southern elephant seals, *Mirounga leonina* (L.). Biol. J. Linnean Soc. 14:133–150.

McCarley, H., and C. J. Carley. 1979. Recent changes in distribution and status of wild red wolves (*Canis rufus*). U.S. Fish and Wildl. Serv., Albuquerque, Endangered Species Rept., no. 4, v + 38 pp.

McCarty, R. 1975. *Onychomys torridus*. Mammalian Species, no. 59, 5 pp.

————. 1978. *Onychomys leucogaster*. Mammalian Species, no. 87, 6 pp.

McClenaghan, L. R., and M. S. Gaines.

1978. Reproduction in marginal populations of the hispid cotton rat (*Sigmodon hispidus*) in northwestern Kansas. Occas. Pap. Mus. Nat. Hist. Univ. Kansas, no. 74, 16 pp.

McCord, C. 1974. Selection of winter habitat by bobcats (*Lynx rufus*) on the Quabbin Reservoir, Massachusetts. J. Mamm. 55:428–437.

McCrane, M. P. 1966. Birth, behaviour and development of a hand-reared two-toed sloth. Int. Zoo Yearbook 6:153–163.

McCully, H. 1967. The broad-handed mole, *Scapanus latimanus*, in a marine littoral environment. J. Mamm. 48:480–482.

McCusker, J. S. 1974. Breeding Malayan sun bears *Helarctos malayanus* at Fort Worth Zoo. Int. Zoo Yearbook 14:118–119.

McCutchen, H. E. 1977. A minimum breeding age for a desert bighorn ewe. Southwestern Nat. 22:153.

McDowell, S. B., Jr. 1958. The Greater Antillean insectivores. Bull. Amer. Mus. Nat. Hist. 115:113–214.

McEvoy, J. S. 1970. Red-necked wallaby in Queensland. Queensland Div. Plant Industry Advis. Leaf., no. 1050, 4 pp.

McGhee, M. E., and H. H. Genoways. 1978. *Liomys pictus*. Mammalian Species, no. 83, 5 pp.

McGrew, J. C. 1979. *Vulpes macrotis*. Mammalian Species, no. 123, 6 pp.

McGuire, M. T., et al. 1974. The St. Kitts vervet. Contrib. Primatol. 1:i–x + 1–199.

McHugh, J. L. 1974. The role and history of the International Whaling Commission. *In* Schevill (1974), pp. 305–335.

McKay, G. M. 1973. Behavior and ecology of the Asiatic elephant in southeastern Ceylon. Smithson. Contrib. Zool., no. 125, iv + 113 pp.

McKean, J. L. 1972. Notes on some collections of bats (order Chiroptera) from Papua-New Guinea and Bougainville Island. CSIRO Div. Wildl. Res. Tech. Pap., no. 26, 35 pp.

————. 1975. The bats of Lord Howe Island with the description of a new nyctophiline bat. Austral. Mamm. 1:329–332.

McKean, J. L., and J. H. Calaby. 1968. A new genus and two new species of bats from New Guinea. Mammalia 32:372–378.

McKean, J. L., and L. S. Hall. 1964. Notes on microchiropteran bats. Victorian Nat. 81:36–37.

McKean, J. L., and W. J. Price. 1967. Notes on some Chiroptera from Queensland, Australia. Mammalia 31:101–119.

McKean, J. L.; G. C. Richards; and W. J. Price. 1978. A taxonomic appraisal of *Eptesicus* (Chiroptera: Mammalia) in Australia. Austral. J. Zool. 26:529–537.

McKenna, M. C. 1975. Toward a phylogenetic classification of the Mammalia. *In* Luckett and Szalay (1975), pp. 21–46.

McLanahan, E. B., and K. M. Green. 1977. The vocal repertoire and an analysis of the contexts of vocalizations in *Leontopithecus rosalia*. *In* Kleiman (1977a), pp. 251–269.

McLaren, I. A. 1960. Are the Pinnipedia biphyletic? Syst. Zool. 9:18–28.

McLaughlin, C. A. 1967. Aplodontoid, Sciuroid, Geomyoid, Castoroid, and Anomaluroid rodents. *In* Anderson and Jones (1967), pp. 210–225.

McManus, J. J. 1970. Behavior of captive opossums, *Didelphis marsupialis virginiana*. Amer. Midl. Nat. 84:144–169.

————. 1974. *Didelphis virginiana*. Mammalian Species, no. 40, 6 pp.

McNally, R. 1977. Echolocation. Cetaceans' sixth sense. Oceans 10(4):27–33.

McWhinnie, M., and C. J. Denys. 1980. The high importance of the lowly krill. Nat. Hist. 89(3):66–73.

Mead, J. G. 1975. Preliminary report on the former net fisheries for *Tursiops truncatus* in the western North Atlantic. J. Fish. Res. Bd. Can. 32:1155–1162.

————. 1981. First records of *Mesoplodon hectori* (Ziphiidae) from the Northern Hemisphere and a description of the adult male. J. Mamm. 62:430–432.

Mead, J. G.; D. K. Odell; R. S. Wells; and M. D. Scott. 1980. Observations on a mass stranding of spinner dolphin, *Stenella longirostris*, from the west coast of Florida. Fishery Bull. 78:353–360.

Mead, J. G., and R. S. Payne. 1975. A specimen of the Tasman beaked whale, *Tasmacetus shepherdi*, from Argentina. J. Mamm. 56:213–218.

Mead, R. A. 1968a. Reproduction in eastern forms of the spotted skunk (genus *Spilogale*). J. Zool. 156:119–136.

————. 1968b. Reproduction in western forms of the spotted skunk (genus *Spilogale*). J. Mamm. 49:373–390.

Meagher, M. M. 1973. The bison of Yellowstone National Park. U.S. Natl. Park Serv. Sci. Monogr. Ser., no. 1, xv + 161 pp.

Mech, L. D. 1961. The marten. Symbol of wilderness. Anim. Kingdom 44(5):133–137.

————. 1970. The wolf: the ecology and behavior of an endangered species. Natural History Press, Garden City, New York, xx + 384 pp.

————. 1974a. A new profile for the wolf. Nat. Hist. 83(4):26–31.

————. 1974b. *Canis lupus*. Mammalian Species, no. 37, 6 pp.

————. 1977a. A recovery plan for the eastern timber wolf. Natl. Parks and Conserv. Mag. 50(1):17–21.

————. 1977b. Wolf-pack buffer zones as prey reservoirs. Science 198:320–321.

————. 1977c. Productivity, mortality, and population trends of wolves in northeastern Minnesota. J. Mamm. 58:559–574.

————. 1977d. Population trend and winter deer consumption in a Minnesota wolf pack. *In* Phillips and Jonkel (1977), pp. 55–83.

————. 1977e. Record movement of a Canadian lynx. J. Mamm. 58:676–677.

————. 1979. Why some deer are safe from wolves. Nat. Hist. 88(1):70–77.

————. 1980. Age, sex, reproduction, and spatial organization of lynxes colonizing northeastern Minnesota. J. Mamm. 61:261–267.

Mech, L. D.; L. D. Frenzel, Jr.; R. R. Ream; and J. W. Winship. 1971. Movements, behavior, and ecology of timber wolves in northeastern Minnesota. In Mech, L. D., and L. D. Frenzel, Jr., eds., Ecological studies of the timber wolf in northeastern Minnesota, U.S. Forest Serv. Res. Pap., no. NC-52, pp. 1–35.

Mech, L. D., and P. D. Karns. 1977. Role of the wolf in a deer decline in the Superior National Forest. U.S. Forest Serv. Res. Pap., no. NC-148, 23 pp.

Mech, L. D., and M. Korb. 1978. An unusually long pursuit of a deer by a wolf. J. Mamm. 59:860–861.

Mech, L. D., and R. M. Nowak. 1981. Return of the gray wolf to Wisconsin. Amer. Midl. Nat. 105:408–409.

Mech, L. D., and L. L. Rogers. 1977. Status, distribution, and movements of martens in northeastern Minnesota. U.S. Forest Serv. Res. Pap., no. NC-143, 7 pp.

Medjo, D. C., and L. D. Mech. 1976. Reproductive activity in nine- and ten-month-old wolves. J. Mamm. 57:406–408.

Medway, Lord. 1964. The marmoset rat, Hapalomys longicaudatus Blyth. Malayan Nat. J. 18:104–110.

————. 1970. The monkeys of Sundaland. In Napier and Napier (1970), pp. 513–553.

————. 1972. Reproductive cycles of the flat-headed bats Tylonycteris pachypus and T. robustula (Chiroptera: Vespertilioninae) in a humid equatorial environment. Zool. J. Linnean Soc. 51:33–61.

————. 1973. The taxonomic status of Tylonycteris malayana Chasen 1940 (Chiroptera). J. Nat. Hist. 7:125–131.

————. 1977. Mammals of Borneo. Monogr. Malaysian Branch Roy. Asiatic Soc., no. 7, xii + 172 pp.

————. 1978. The wild mammals of Malaya (peninsular Malaysia) and Singapore. Oxford Univ. Press, Kuala Lumpur, xxii + 128 pp.

Medway, Lord, and A. G. Marshall. 1972. Roosting associations of flat-headed bats, Tylonycteris species (Chiroptera: Vespertilionidae) in Malaysia. J. Zool. 168:463–482.

Medway, Lord, and Yong Hoi-Sen. 1976. Problems in the systematics of the rats (Muridae) of peninsular Malaysia. Malaysian J. Sci. 4:43–53.

Meester, J. 1972. A new golden mole from the Transvaal (Mammalia: Chrysochloridae). Ann. Transvaal Mus. 28:37–46.

————. 1973. Mammals collected during the Bernard Carp Expedition to the Western Province of Zambia. Puku, 7:137–149.

————. 1976. South African red data book—small mammals. S. Afr. Natl. Sci. Programmes Rept., no. 11, vi + 59 pp.

Meester, J., and N. J. Dippenaar. 1978. A new species of Myosorex from Knysna, South Africa (Mammalia: Soricidae). Ann. Transvaal Mus. 31:29–42.

Meester, J., and A. F. Hallett. 1970. Notes on early postnatal development in certain southern African Muridae and Cricetidae. J. Mamm. 51:703–711.

Meester, J., and H. W. Setzer. 1971 (as revised 1977). The mammals of Africa. An identification manual. Smithson. Inst. Press, Washington, D.C.

Mehrer, C. F. 1976. Gestation period in the wolverine, Gulo gulo. J. Mamm. 57:570.

Meier, M. N., and V. N. Yatsenko. 1980. On taxonomic status and distribution of the common (Microtus arvalis Pallas 1778) and Kirghiz (M. kirgisorum Ognev 1950) voles in south-east Kazakhstan. Zool. Zhur. 59:283–288.

Mein, P., and Y. Tupinier. 1977. Formule dentaire et position systématique du Minioptere (Mammalia, Chiroptera). Mammalia 41:207–211.

Melchior, H. R. 1971. Characteristics of arctic ground squirrel alarm calls. Oecologia 7:184–190.

Mendelssohn, H. 1965. Breeding the Syrian hyrax Procavia capensis syriaca Schreber, 1784. Int. Zoo Yearbook 5:116–125.

————. 1974. The development of the populations of gazelles in Israel and their behavioural adaptions. In Geist and Walther (1974), pp. 722–743.

Menzies, J. I. 1971. The lobe-lipped bat (Chalinolobus nigrogriseus Gould) in New Guinea. Rec. Papua and New Guinea Mus. 1:6–8.

————. 1973. A study of leaf-nosed bats (Hipposideros caffer and Rhinolophus landeri) in a cave in northern Nigeria. J. Mamm. 54:930–945.

————. 1977. Fossil and subfossil fruit bats from the mountains of New Guinea. Austral. J. Zool. 25:329–336.

Menzies, J. I., and E. Dennis. 1979. Handbook of New Guinea rodents. Wau Ecology Inst., Wau, Papua New Guinea, v + 68 pp.

Mercer, W. E., and F. Manuel. 1974. Some aspects of moose management in Newfoundland. Naturaliste Canadien 101:657–671.

Merchant, J. C. 1976. Breeding biology of the agile wallaby, Macropus agilis (Gould) (Marsupialia: Macropodidae), in captivity. Austral. Wildl. Res. 3:93–103.

Meritt, D. A., Jr. 1973. Some observations on the maned wolf, Chrysocyon brachyurus, in Paraguay. Zoologica 58:53.

————. 1975. The lesser anteater Tamandua tetradactyla in captivity. Int. Zoo Yearbook 15:41–44.

————. 1976a. Sex ratios of Hoffmann's sloth, Choloepus hoffmanni Peters and three-toed sloth (Bradypus infuscatus Wagler in Panama. Amer. Midl. Nat. 96:472–473.

————. 1976b. The La Plata three-banded armadillo Tolypeutes matacus in captivity. Int. Zoo Yearbook 16:153–155.

————. 1977. Second-generation owl monkey birth. Amer. Assoc. Zool. Parks Aquar. Newsl. 18(3):12.

Merritt, J. F. 1978. Peromyscus californicus. Mammalian Species, no. 85, 6 pp.

Merritt, J. F., and J. M. Merritt. 1978. Population ecology and energy of Clethrionomys gapperi in a Colorado subalpine forest. J. Mamm. 59:576–598.

Meserve, P. L. 1971. Population ecology of the prairie vole, Microtus ochrogaster, in the western mixed prairie of Nebraska. Amer. Midl. Nat. 86:417–433.

Meserve, P. L.; J. Rodriguez-M; and R. E. Martin. 1978. Demography and community ecology of the caviomorph rodent, Octodon degus, in central and northern Chile. Amer. Soc. Mamm., Abstr. Tech. Pap., 58th Ann. Mtg., p. 69.

Messick, J. P., and M. G. Hornocker. 1981. Ecology of the badger in southwestern Idaho. Wildl. Monogr., no. 76, 53 pp.

Meyer-Holzapfel, M. 1968. Breeding the European wild cat Felis s. silvestris at Berne Zoo. Int. Zoo Yearbook 8:31–38.

Meylan, A. 1977. Fossorial forms of the water vole, Arvicola terrestris (L.), in Europe. EPPO Bull. 7:209–221.

Michael, R. P., and J. H. Crook, eds. 1973. Comparative ecology and behaviour of primates. Academic Press, London, 847 pp.

Mihok, S. 1976. Behaviour of subarctic red-backed voles (Clethrionomys gapperi athabascae). Can. J. Zool. 54:1932–1945.

Mikkola, H. 1974. The raccoon dog spreads to western Europe. Wildlife 16:344–345.

Mikuriya, M. 1976. Notes on the Japanese otter, Lutra lutra whiteyi (Gray). J. Mamm. Soc. Japan 6:214–217.

Millar, J. S. 1972. Timing of breeding of pikas in southwestern Alberta. Can. J. Zool. 50:665–669.

————. 1973. Evolution of litter-size in the pika, Ochotona princeps (Richardson). Evolution 27:134–143.

————. 1974. Success of reproduction in pikas, Ochotona princeps (Richardson). J. Mamm. 55:527–542.

Millar, J. S., and F. C. Zwickel. 1972a. Determination of age, age structure, and mortality of the pika, Ochotona princeps (Richardson). Can. J. Zool. 50:229–232.

————. 1972b. Characteristics and ecological significance of hay piles of pikas. Mammalia 36:657–667.

Millar, R. P. 1971. Reproduction in the rock hyrax (Procavia capensis). Zool. Afr. 6:243–261.

Miller, D. H., and L. L. Getz. 1977. Comparisons of population dynamics of Peromyscus and Clethrionomys in New England. J. Mamm. 58:1–16.

Miller, E. H. 1975a. Body and organ measurements of fur seals, Arctocephalus forsteri (Lesson), from New Zealand. J. Mamm. 56:511–513.

————. 1975b. Walrus ethology. I. The social role of tusks and applications of multidimensional scaling. Can. J. Zool. 53:590–613.

Miller, E. H., and D. J. Boness. 1979. Remarks on display functions of the snout of the grey seal, *Halichoerus grypus* (Fab.), with comparative notes. Can. J. Zool. 57:140–148.

Miller, F. L.; R. H. Russell; and A. Gunn. 1977. Interisland movements of Peary caribou (*Rangifer tarandus pearyi*) on western Queen Elizabeth Islands, arctic Canada. Can. J. Zool. 55:1029–1037.

Miller, G. S., Jr. 1907. The families and genera of bats. Bull. U.S. Natl. Mus. 57:i–xvii + 1–282.

Mills, M.G.L. 1978. Foraging behaviour of the brown hyaena (*Hyaena brunnea* Thunberg, 1820) in the southern Kalahari. Z. Tierpsychol. 48:113–141.

Mills, M.G.L., and M.E.J. Mills. 1978. The diet of the brown hyaena *Hyaena brunnea* in the southern Kalahari. Koedoe 21:125–149.

Mills, R. S.; G. W. Barrett; and M. P. Farrell. 1975. Population dynamics of the big brown bat (*Eptesicus fuscus*) in southwestern Ohio. J. Mamm. 56:591–604.

Mineau, P., and D. Madison. 1977. Radiotracking of *Peromyscus leucopus*. Can. J. Zool. 55:465–468.

Mishra, A. C., and V. Dhanda. 1975. Review of the genus *Millardia* (Rodentia: Muridae), with description of a new species. J. Mamm. 56:76–80.

Misonne, X. 1969. African and Indo-Australian Muridae. Evolutionary trends. Ann. Mus. Roy. Afr. Cent., Terurven, Belgium, ser. IN–8°, no. 172, 219 pp.

―――. 1974. Une nouvelle gerbille de Libye (Mammalia, Rodentia). Bull. Inst. Roy. Sci. Nat. Belgium 50(6):1–6.

―――. 1979. Muridae collected in Irian Jaya, Indonesia. Bull. Inst. Roy. Sci. Nat. Belgium 51(8):1–16.

Misonne, X., and J. Verschuren. 1976. Les rongeurs du Nimba Liberien. Acta Zool. Pathol. 66:199–220.

Mitchell, E. D. 1975a. Report of the meeting on smaller Cetaceans, Montreal, April 1–11, 1974. J. Fish. Res. Bd. Can. 32:889–983.

―――. 1975b. Porpoise, dolphin and small whale fisheries of the world. Int. Union Conserv. Nat. Monogr., no. 3, 129 pp.

―――. 1978. Finner whales. *In* Haley (1978), pp. 36–45.

Mitchell, E. D., and V. M. Kozicki. 1975. Supplementary information on minke whale (*Balaenoptera acutorostrata*) from Newfoundland fishery. J. Fish. Res. Bd. Can. 32:985–994.

Mitchell, E. D., and J. G. Mead. 1977. History of the gray whale in the Atlantic Ocean (abstr.). Proc. 2nd Conf. Biol. Mar. Mamm., San Diego.

Mitchell, R. M. 1975. A checklist of Nepalese mammals (excluding bats). Saugetierk. Mitt. 23:152–157.

―――. 1977. Accounts of Nepalese

mammals and analysis of the host-ectoparasite data by computer techniques. Unpubl. Ph.D. Diss., Iowa State Univ., 557 pp.

―――. 1979. The sciurid rodents (Rodentia: Sciuridae) of Nepal. J. Asian Ecol. 1:21–28.

Mitchell, R. M., and F. Punzo. 1975. *Ochotona lama* sp. n. (Lagomorpha: Ochotonidae) a new pika from the Tibetan highlands of Nepal. Mammalia 39:419–422.

Mittermeier, R. A. 1973. Group activity and population dynamics of the howler monkey on Barro Colorado Island. Primates 14:1–19.

Mittermeier, R. A., and A. Coimbra-Filho. 1977. Primate conservation in Brazilian Amazonia. *In* Rainier III and Bourne (1977), pp. 117–166.

Mittermeier, R. A.; H. de Macedo-Ruiz; B. A. Luscombe; and J. Cassidy. 1977. Rediscovery and conservation of the Peruvian yellow-tailed woolly monkey (*Lagothrix flavicauda*). *In* Rainier III and Bourne (1977), pp. 95–115.

Mizue, K.; M. Nishiwaki; and A. Takemura. 1971. The underwater sounds of Ganges River dolphins (*Platanista gangetica*). Sci. Rept. Whales Res. Inst. 23:123–128.

Mock, O. B., and C. H. Conaway. 1975. Reproduction of the least shrew (*Cryptotis parva*) in captivity. *In* Antikatzides, T.; S. Erichsen; and A. Spiegel, eds., The laboratory animal in the study of reproduction, Gustav Fischer Verlag, Stuttgart, pp. 59–74.

Modha, K. L., and S. K. Eltringham. 1976. Population ecology of the Uganda kob [*Adenota kob thomasi* (Neumann)] in relation to the territorial system in the Rwenzori National Park, Uganda. J. Appl. Ecol. 13:453–473.

Moehlman, P. D. 1978. Jackals of the Serengeti. Wildl. News, African Wildl. Leadership Foundation 13(3):2–6.

Mohler, L. L. 1974. Threatened wildlife of Idaho. Idaho Wildl. Rev. 26(5):3–5.

Mohnot, S. M. 1978. The conservation of non-human primates in India. *In* Chivers and Lane-Petter (1978), pp. 47–53.

Mohr, C. E. 1972. The status of threatened species of cave-dwelling bats. Bull. Natl. Speleol. Soc. 34:33–47.

Mones, A. 1973. Estudios sobre la familia Hydrochoeridae (Rodentia). I.—Introducción e historia taxonómica. Rev. Brasil. Biol. 33:277–283.

―――. 1975. Estudios sobre la familia Hydrochoeridae (Rodentia). VI. Catálogo anotado de los ejemplares tipo. Comun. Paleontol. Mus. Hist. Nat. Montevideo 1:99–128.

Monfort, A., and N. Monfort. 1974. Notes sur l'écologie et le comportement des oribis (*Ourebia ourebi*, Zimmerman, 1783). Terre Vie 28:169–208.

Monfort-Braham, N. 1975. Variations dans la structure sociale du topi, *Damaliscus korrigum* Ogilby, au Parc National de l'Akagera, Rwanda. Z. Tierpsychol. 39:332–364.

Montgomery, G. G., ed. 1978. The ecology of arboreal folivores. Smithson. Inst. Press, Washington, D.C., 573 pp.

Montgomery, G. G., and Y. D. Lubin. 1977. Prey influences on movements of neotropical anteaters. *In* Phillips and Jonkel (1977), pp. 103–131.

―――. 1978. Movements of *Coendou prehensilis* in the Venezuelan llanos. J. Mamm. 59:887–888.

Montgomery, G. G., and M. E. Sunquist. 1978. Habitat selection and use by two-toed and three-toed sloths. *In* Montgomery (1978), pp. 329–359.

Montgomery, W. I. 1976. On the relationship between yellownecked mouse (*Apodemus flavicollis*) and woodmouse (*A. sylvaticus*) in a Cotswold valley. J. Zool. 179:229–233.

Moore, J. C. 1958a. A new species and a redefinition of the squirrel genus *Prosciurillus* of Celebes. Amer. Mus. Novit., no. 1890, 5 pp.

―――. 1958b. New genera of East Indian squirrels. Amer. Mus. Novit., no. 1914, 5 pp.

―――. 1959. Relationships among living squirrels of the Sciurinae. Bull. Amer. Mus. Nat. Hist. 118:153–206.

―――. 1968. Relationships among the living genera of beaked whales with classifications, diagnoses and keys. Fieldiana Zool. 53:209–298.

―――. 1972. More skull characters of the beaked whale *Indopacetus pacificus* and comparative measurements of austral relatives. Fieldiana Zool. 62:1–19.

Moore, J. C., and G.H.H. Tate. 1965. A study of the diurnal squirrels, Sciurinae, of the Indian and Indochinese subregions. Fieldiana Zool. 48:1–351.

Moore, N. W. 1975. The diurnal flight of the Azorean bat (*Nyctalus azoreum*) and the avifauna of the Azores. J. Zool. 177:483–486.

Moore, R. E., and N. S. Martin. 1980. A recent record of the swift fox (*Vulpes velox*) in Montana. J. Mamm. 61:161.

Moors, P. J. 1975. The urogenital system and notes on the reproductive biology of the female rufous rat-kangaroo, *Aepyprymnus rufescens* (Gray) (Macropodidae). Austral. J. Zool. 23:355–361.

Morcombe, M. K. 1967. The rediscovery after 83 years of the dibbler *Antechinus apicalis* (Marsupialia, Dasyuridae). W. Austral. Nat. 10:103–111.

Morejohn, G. V. 1979. The natural history of Dall's porpoise in the North Pacific Ocean. *In* Winn and Olla (1979), pp. 45–83.

Morgan, G. S.; C. E. Ray; and O. Arredondo. 1980. A giant extinct insectivore from Cuba (Mammalia: Insectivora: Solenodontidae). Proc. Biol. Soc. Washington 93:597–608.

Morlok, W. F. 1978. Nagetiere aus der Turkei. Senckenbergiana Biol. 59:155–162.

Morrell, S. 1972. Life history of the San Joaquin kit fox. California Fish and Game 53:162–174.

Morris, N. E., and J. Hanks. 1974. Reproduction in the bushbuck *Tragelaphus scriptus ornatus*. Arnoldia 1(7):1–8.

Morris, P. A., and J. R. Malcolm. 1977. The Simien fox in the Bale Mountains. Oryx 14:151–160.

Morrison, D. W. 1978. Influence of habitat on the foraging distance of the fruit bat, *Artibeus jamaicensis*. J. Mamm. 59:622–624.

———. 1979. Apparent male defense of tree hollows in the fruit bat, *Artibeus jamaicensis*. J. Mamm. 60:11–15.

Morton, S. R. 1974. First record of Forrest's mouse *Leggadina forresti* (Thomas, 1906) in N. S. W. Victorian Nat. 91:91–94.

———. 1978*a*. An ecological study of *Sminthopsis crassicaudata* (Marsupialia: Dasyuridae). I. Distribution, study areas and methods. Austral. Wildl. Res. 5:151–162.

———. 1978*b*. An ecological study of *Sminthopsis crassicaudata* (Marsupialia: Dasyuridae). II. Behaviour and social organization. Austral. Wildl. Res. 5:163–182.

———. 1978*c*. An ecological study of *Sminthopsis crassicaudata* (Marsupialia: Dasyuridae). III. Reproduction and life history. Austral. Wildl. Res. 5:183–211.

———. 1978*d*. Torpor and nest-sharing in free-living *Sminthopsis crassicaudata* (Marsupialia) and *Mus musculus* (Rodentia). J. Mamm. 59:569–575.

Morton, S. R., and T. C. Burton. 1973. Observations on the behaviour of the macropodid marsupial *Thylogale billardieri* (Desmarest) in captivity. Austral. Zool. 18:1–14.

Morton, S. R., and A. K. Lee. 1978. Thermoregulation and metabolism in *Planigale maculata* (Marsupialia: Dasyuridae). J. Thermal Biol. 3:117–120.

Moynihan, M. 1976. Notes on the ecology and behavior of the pygmy marmoset (*Cebuella pygmaea*) in Amazonian Colombia. *In* Thorington and Heltne (1976), pp. 79–84.

Muckenhirn, N. A., and J. F. Eisenberg. 1973. Home ranges and predation of the Ceylon leopard. *In* Eaton, R. L., ed., The world's cats, vol. I, World Wildlife Safari, Winston, Oregon, pp. 142–175.

Mullen, D. A. 1977. The striped dolphin, *Stenella coeruleoalba*, in the Gulf of California. Bull. S. California Acad. Sci. 76:131–132.

Müller, J. P. 1977. Populationsökologie von *Arvicanthis abyssinicus* in der Grassteppe des Semien Mountains National Park (Äthiopien). Z. Saugetierk. 42:145–172.

Mumford, R. E. 1973. Natural history of the red bat (*Lasiurus borealis*) in Indiana. Period. Biol. 75:155–158.

Mumford, R. E., and D. M. Knudson. 1978. Ecology of bats at Vicosa, Brazil. Proc. 4th Int. Bat Res. Conf., pp. 287–295.

Munro, W. T. 1979. Vancouver Island marmot. Committee on the status of Endangered Wildlife in Canada, 2 pp.

Munthe, K., and J. H. Hutchison. 1978. A wolf-human encounter on Ellesmere Island, Canada. J. Mamm. 59:876–878.

Murie, A. 1981. The grizzlies of Mount McKinley. U.S. Natl. Park Serv. Sci. Monogr., no. 14, xvi + 251 pp.

Murie, J. O. 1973. Population characteristics and phenology of a Franklin ground squirrel (*Spermophilus franklinii*) colony in central Alberta. Amer. Midl. Nat. 90:334–340.

Murie, J. O., and M. A. Harris. 1978. Territoriality and dominance in male Columbian ground squirrels (*Spermophilus columbianus*). Can. J. Zool. 56:2402–2412.

Murie, O. J. 1959. Fauna of the Aleutian Islands and Alaska Peninsula. N. Amer. Fauna, no. 61, xiv + 406 pp.

Murphy, E. T. 1976. Breeding the clouded leopard *Neofelis nebulosa* at Dublin Zoo. Int. Zoo Yearbook 16:122–124.

Mussehl, T. W., and F. W. Howell. 1971. Game management in Montana. Montana Fish and Game Dept., Helena, xi + 238 pp.

Musser, G. G. 1969*a*. Results of the Archbold Expeditions. No. 91. A new genus and species of murid rodent from Celebes, with a discussion of its relationships. Amer. Mus. Novit., no. 2384, 41 pp.

———. 1969*b*. Results of the Archbold Expeditions. No. 92. Taxonomic notes on *Rattus dollmani* and *Rattus hellwaldi* (Rodentia, Muridae) of Celebes. Amer. Mus. Novit., no. 2386, 24 pp.

———. 1970*a*. Species-limits of *Rattus brahma*, a murid rodent of northeastern India and northern Burma. Amer. Mus. Novit., no. 2406, 27 pp.

———. 1970*b*. *Rattus masaretes*: a synonym of *Rattus rattus moluccarius*. J. Mamm. 51:606–609.

———. 1970*c*. Results of the Archbold Expeditions. No. 93. Reidentification and reallocation of *Mus callitrichus* and allocations of *Rattus maculipilis*, *R. m. jentinki*, and *R. microbullatus* (Rodentia, Muridae). Amer. Mus. Novit., no. 2440, 35 pp.

———. 1971*a*. The taxonomic association of *Mus faberi* Jentink with *Rattus xanthurus* (Gray), a species known only from Celebes (Rodentia: Muridae). Zool. Meded. 45:107–118.

———. 1971*b*. The identities and allocations of *Taeromys paraxanthus* and *T. tatei*, two taxa based on composite holotypes (Rodentia, Muridae). Zool. Meded. 45:127–138.

———. 1971*c*. Results of the Archbold Expeditions. No. 94. Taxonomic status of *Rattus tatei* and *Rattus frosti*, two taxa of murid rodents known from middle Celebes. Amer. Mus. Novit., no. 2454, 19 pp.

———. 1971*d*. The taxonomic status of *Rattus tondanus* Sody and notes on the holotypes of *R. beccarii* (Jentink) and *R. thysanurus* Sody (Rodentia: Muridae). Zool. Meded. 45:147–157.

———. 1971*e*. The taxonomic status of *Rattus dammermani* Thomas and *Rattus toxi* (Rodentia, Muridae) of Celebes. Beaufortia 18:205–216.

———. 1972*a*. The species of *Hapalomys* (Rodentia, Muridae). Amer. Mus. Novit., no. 2503, 27 pp.

———. 1972*b*. Identities of taxa associated with *Rattus rattus* (Rodentia, Muridae) of Sumba Island, Indonesia. J. Mamm. 53:861–865.

———. 1973*a*. Zoogeographical significance of the ricefield rat, *Rattus argentiventer*, on Celebes and New Guinea and the identity of *Rattus pesticulus*. Amer. Mus. Novit., no. 2511, 30 pp.

———. 1973*b*. Notes on additional specimens of *Rattus brahma*. J. Mamm. 54:267–270.

———. 1973*c*. Species-limits of *Rattus cremoriventer* and *Rattus langibanis*, murid rodents of Southeast Asia and the Greater Sunda Islands. Amer. Mus. Novit., no. 2525, 65 pp.

———. 1977*a*. *Epimys benguetensis*, a composite, and one zoogeographical view of rat and mouse faunas in the Philippines and Celebes. Amer. Mus. Novit., no. 2624, 15 pp.

———. 1977*b*. Results of the Archbold Expeditions. No. 100. Notes on the Philippine rat, *Limnomys*, and the identity of *Limnomys picinus*, a composite. Amer. Mus. Novit., no. 2636, 14 pp.

———. 1979. Results of the Archbold Expeditions. No. 102. The species of *Chiropodomys*, arboreal mice of Indochina and the Malay Archipelago. Bull. Amer. Mus. Nat. Hist. 162:377–445.

———. 1981*a*. Results of the Archbold Expeditions. No. 105. Notes on systematics of Indo-Malayan murid rodents, and descriptions of new genera and species from Ceylon, Sulawesi, and the Philippines. Bull. Amer. Mus. Nat. Hist. 168:225–334.

———. 1981*b*. The giant rat of Flores and its relatives east of Borneo and Bali. Bull. Amer. Mus. Nat. Hist. 169:67–176.

———. 1981*c*. A new genus of arboreal rat from West Java, Indonesia. Zool. Verhandelingen, no. 189, 35 pp.

Musser, G. G., and Boeadi. 1980. A new genus of murid rodent from the Komodo Islands in Nusatenggara, Indonesia. J. Mamm. 61:395–413.

Musser, G. G., and S. Chiu. 1979. Notes on taxonomy of *Rattus andersoni* and *R. excelsior*, murids endemic to western China. J. Mamm. 60:581–592.

Musser, G. G., and P. W. Freeman. 1981. A new species of *Rhynchomys* (Muridae) from the Philippines. J. Mamm. 62:154–159.

Musser, G. G., and A. L. Gardner. 1974. A new species of the Ichthyomine *Daptomys* from Peru. Amer. Mus. Novit., no. 2537, 23 pp.

Musser, G. G., and L. K. Gordon. 1981. A new species of *Crateromys* (Muridae) from the Philippines. J. Mamm. 62:513–525.

Musser, G. G.; J. T. Marshall, Jr.; and Boeadi. 1979. Definition and contents of the Sundaic genus *Maxomys* (Rodentia, Muridae). J. Mamm. 60:592–606.

Mutere, F. A. 1967. The breeding biology of equatorial vertebrates: reproduction in the fruit bat, *Eidolon helvum*, at latitude 0°20′ N. J. Zool. 153:153–161.

———. 1970. The breeding biology of

equatorial vertebrates: reproduction in the insectivorous bat, *Hipposideros caffer*, living at 0°27′ N. Bijdragen Tot De Dierkunde 40:56–58.

_____. 1973*a*. Reproduction in two species of equatorial free-tailed bats (Molossidae). E. Afr. Wildl. J. 11:271–280.

_____. 1973*b*. A comparative study of reproduction in two populations of the insectivorous bats, *Otomops martiensseni*, at latitudes 1°5′ S and 2°30′ S. J. Zool. 171:79–92.

Muul, I., and Lim Boo Liat. 1970. Ecological and morphological observations of *Felis planiceps*. J. Mamm. 51:806–808.

_____. 1971. New locality records for some mammals of West Malaysia. J. Mamm. 52:430–437.

_____. 1974. Reproductive frequency in Malaysian flying squirrels, *Hylopetes* and *Pteromyscus*. J. Mamm. 55:393–400.

_____. 1978. Comparative morphology, food habits, and ecology of some Malaysian arboreal rodents. *In* Montgomery (1978), pp. 361–368.

Muul, I., and K. Thonglongya. 1971. Taxonomic status of *Petinomys morrisi* (Carter) and its relationship to *Petinomys setosus* (Temminck and Schlegel). J. Mamm. 52:362–369.

Myers, L. G. 1969. Home range and longevity in *Zapus princeps* in Colorado. Amer. Midl. Nat. 82:628–629.

Myers, N. 1975*a*. The silent savannahs. Int. Wildl. 5(5):4–11.

_____. 1975*b*. The cheetah *Acinonyx jubatus* in Africa. Int. Union Conserv. Nat. Monogr., no. 4, 88 pp.

_____. 1976. The leopard *Panthera pardus* in Africa. Int. Union Conserv. Nat. Monogr., no. 5, 79 pp.

Myers, P. 1977*a*. Patterns of reproduction of four species of vespertilionid bats in Paraguay. Univ. California Publ. Zool. 107:1–41.

_____. 1977*b*. A new phyllotine rodent (genus *Graomys*) from Paraguay. Occas. Pap. Mus. Zool. Univ. Michigan, no. 676, 5 pp.

Myers, P., and M. D. Carleton. 1981. The species of *Oryzomys* (*Oligoryzomys*) in Paraguay and the identity of Azara's "Rat sixième ou Rat à Tarse noir." Misc. Publ. Mus. Zool. Univ. Michigan, no. 161, iii + 41 pp.

Myers, P., and R. M. Wetzel. 1979. New records of mammals from Paraguay. J. Mamm. 60:638–641.

Myllymäki, A. 1977*a*. Demographic mechanisms in the fluctuating populations of the field vole *Microtus agrestis*. Oikos 29:468–493.

_____. 1977*b*. Intraspecific competition and home range dynamics in the field vole *Microtus agrestis*. Oikos 29:553–569.

Mysterud, I.; J. Viitala; and S. Lahti. 1972. On winter breeding of the wood lemming (*Myopus schisticolor*). Norwegian J. Zool. 20:91–92.

N.

Nader, I. A. 1974. A new record, bushy-tailed jird, *Sekeetamys calurus calurus* (Thomas, 1892) from Saudi Arabia. Mammalia 38:347–349.

_____. 1979. The present status of the viverrids of the Arabian Peninsula (Mammalia: Carnivora: Viverridae). Senckenbergiana Biol. 59:311–316.

Nadler, C. F., et al. 1973*a*. Zoogeography of transferins in arctic and long-tailed ground squirrel populations. Comp. Biochem. Physiol., 44B:33–40.

_____. 1973*b*. Cytogenetic differentiation, geographic distribution, and domestication in Palearctic sheep (*Ovis*). Z. Saugetierk. 38:109–125.

_____. 1974. Evolution in ground squirrels—I. Transferins in Holarctic populations of *Spermophilus*. Comp. Biochem. Physiol. 47A:663–681.

_____. 1975. Chromosomal evolution in holarctic ground squirrels (*Spermophilus*). I. Giemsa-band homologies in *Spermophilus columbianus* and *S. undulatus*. Z. Saugetierk. 40:1–7.

_____. 1977. Chromosomal evolution in chipmunks, with special emphasis on A and B karyotypes of the subgenus *Neotamias*. Amer. Midl. Nat. 98:343–353.

_____. 1978. Biochemical relationships of the holarctic vole genera [*Clethrionomys*, *Microtus*, and *Arvicola* (Rodentia: Arvicolinae)]. Can. J. Zool. 56:1564–1575.

Nagel, U. 1971. Social organization in a baboon hybrid zone. Proc. 3rd Int. Congr. Primatol. 3:48–57.

_____. 1973. A comparison of anubis baboons, hamadryas baboons and their hybrids at a species border in Ethiopia. Folia Primatol. 19:104–165.

Nagorsen, D., and J. R. Tamsitt. 1981. Systematics of *Anoura cultrata*, *A. brevirostrum*, and *A. werckleae*. J. Mamm. 62:82–100.

Nagy, K. A.; R. S. Seymour; A. K. Lee; and R. Braithwaite. 1978. Energy and water budgets in free-living *Antechinus stuartii* (Marsupialia: Dasyuridae). J. Mamm. 59:60–68.

Nair, S. S. 1976. A population survey and observations on the behaviour of the blackbuck in the Point Calimere Sanctuary, Tamil Nadu. J. Bombay Nat. Hist. Soc. 73:304–310.

Naito, Y., and M. Nishiwaki. 1975. Ecology and morphology of *Phoca vitulina largha* and *Phoca kurilensis* in the southern Sea of Okhotsk and northeast of Hokkaido. Rapp. P.-v. Reun. Cons. Int. Explor. Mer. 169:379–386.

Napier, J. R., and P. H. Napier. 1967. A handbook of living primates. Academic Press, New York, xiv + 456 pp.

_____, eds. 1970. Old World monkeys. Evolution, systematics, and behavior. Academic Press, New York, xiv + 660 pp.

Nass, R. D. 1977. Movements and home ranges of Polynesian rats in Hawaiian sugarcane. Pacific Sci. 31:135–142.

National Geographic Society. 1981. Book of mammals. Spec. Publ. Div., Natl. Geogr. Soc., Washington, D.C., 2 vols.

Naumov, N. P., and V. S. Lobachev. 1975. Ecology of desert rodents of the U.S.S.R. *In* Prakash and Ghosh (1975), pp. 465–598.

Neal, B. R. 1977*a*. Reproduction of the punctated grass-mouse, *Lemniscomys striatus* in the Ruwenzori National Park, Uganda (Rodentia: Muridae). Zool. Afr. 12:419–428.

_____. 1977*b*. Reproduction of the multimammate rat, *Praomys* (*Mastomys*) *natalensis* (Smith), in Uganda. Z. Saugetierk. 42:221–231.

Neal, E. 1970. The banded mongoose, *Mungos mungo* Gmelin. E. Afr. Wildl. J. 8:63–71.

Nel, J.A.J. 1978. Notes on the food and foraging behavior of the bat-eared fox, *Otocyon megalotis*. Bull. Carnegie Mus. Nat. Hist., no. 6, pp. 132–137.

Nel, J.A.J., and C. J. Stutterheim. 1973. Notes on early post-natal development of the Namaqua gerbil *Desmodillus auricularis*. Koedoe 16:117–125.

Nellis, C. H., and L. B. Keith. 1976. Population dynamics of coyotes in central Alberta, 1964–1968. J. Wildl. Mgmt. 40:389–399.

Nellis, C. H.; S. P. Wetmore; and L. B. Keith. 1972. Lynx-prey interactions in central Alberta. J. Wildl. Mgmt. 36:320–329.

Nellis, D. W., and C. P. Ehle. 1977. Observations on the behavior of *Brachyphylla cavernarum* (Chiroptera) in Virgin Islands. Mammalia 41:403–409.

Nellis, D. W.; N. F. Eichholz; T. W. Regan; and C. Feinstein. 1978. Mongoose in Florida. Wildl. Soc. Bull. 6:249–250.

Nesbitt, W. H. 1975. Ecology of a feral dog pack on a wildlife refuge. *In* Fox (1975), pp. 391–395.

Neuhauser, H. N., and A. F. DeBlase. 1974. Notes on bats (Chiroptera: Vespertilionidae) new to the faunal lists of Afghanistan and Iran. Fieldiana Zool. 62:85–96.

Neuweiler, G. 1969. Verhaltensbeobachtungen an einer indischen Flughundkolonie (*Pteropus g. giganteus* Brünn). Z. Tierpsychol. 26:166–199.

Neville, M.; N. Castro; A. Marmol; and J. Revilla. 1976. Censusing primate populations in the reserved area of the Pacaya and Samiria rivers, Department Loreto, Peru. Primates 17:151–181.

Nevo, E. 1961. Observations of Israeli populations of the mole-rat *Spalax ehrenbergi* Nehring 1898. Mammalia 25:127–144.

Nevo, E., and H. Bar-El. 1976. Hybridization and speciation in fossorial mole rats. Evolution 30:831–840.

Newby, J. E. 1978. Scimitar-horned oryx—the end of the line? Oryx 14:219–221.

Newman, M. A. 1978. Narwhal. *In* Haley (1978), pp. 138–144.

Newsome, A. E. 1971*a*. The ecology of red kangaroos. Austral. Zool. 16:32–50.

_____. 1971*b*. Competition between wildlife and domestic livestock. Austral. Vet. J. 47:577–586.

_____. 1975. An ecological comparison of the two arid-zone kangaroos of Australia, and their anomalous prosperity since the introduction of ruminant stock to their environment. Quart. Rev. Biol. 50:389–424.

Newsome, A. E.; L. K. Corbett; and S. M. Carpenter. 1980. The identity of the dingo. I. Morphological discriminants of dingo and dog skulls. Austral. J. Zool. 28:615–625.

Neyman, P. F. 1977. Aspects of the ecology and social organization of free-ranging cotton-top tamarins (*Saguinus oedipus*) and the conservation status of the species. *In* Kleiman (1977*a*), pp. 39–71.

Nicholls, D. G. 1971. Daily and seasonal movements of the quokka, *Setonix brachyurus* (Marsupialia), on Rottnest Island. Austral. J. Zool. 19:215–226.

Nichols, L., Jr. 1978. Dall's sheep. *In* Schmidt and Gilbert (1978), pp. 173–189.

Niebauer, T. J., and O. J. Rongstad. 1977. Coyote food habits in northwestern Wisconsin. *In* Phillips and Jonkel (1977), pp. 237–251.

Niemitz, C. 1973. Field research on the Horsfield's tarsier (*Tarsius bancanus*) at Sarawak Museum. Borneo Res. Bull. 5:61–63.

Niethammer, G. 1970. Beobachtungen am Pyrenaen-Desman, *Galemys pyrenaica*. Bonner Zool. Beitr. 21:157–182.

Nievergelt, B. 1974. A comparison of rutting behaviour and grouping in the Ethiopian and Alpine ibex. *In* Geist and Walther (1974), pp. 324–340.

Nishida, T. 1972*a*. A note on the ecology of the red-colobus monkeys (*Colobus badius tephrosceles*) living in the Mahali Mountains. Primates 13:57–64.

_____. 1972*b*. Preliminary information of the pygmy chimpanzees (*Pan paniscus*) of the Congo Basin. Primates 13:415–425.

Nishida, T.; S. Uehara; and R. Nyundo. 1979. Predatory behavior among wild chimpanzees of the Mahale Mountains. Primates 20:1–20.

Nishimura, A., and K. Izawa. 1975. The group characteristics of woolly monkeys (*Lagothrix lagothrica*) in the upper Amazonian Basin. Proc. 5th Int. Congr. Primatol., pp. 351–357.

Nishiwaki, M. 1966. A discussion of rarities among the smaller cetaceans caught in Japanese waters. *In* Norris (1966) pp. 192–204.

_____. 1975. Ecological aspects of smaller cetaceans, with emphasis on the striped dolphin (*Stenella coeruleoalba*). J. Fish. Res. Bd. Can. 32:1069–1072.

Nishiwaki, M., and T. Kasuya. 1970. A Greenland right whale caught at Osaka Bay. Sci. Rept. Whales Res. Inst. 22:45–62.

Nishiwaki, M.; T. Kasuya; N. Miyazoki; T. Tobayama; and T. Kataoka. 1979. Present distribution of the dugong in the world. Sci. Rept. Whales Res. Inst. 31:133–141.

Nishiwaki, M., and K. S. Norris. 1966. A new genus, *Peponocephala*, for the odontocete cetacean species *Electra electra*. Sci. Rept. Whales Res. Inst. 20:95–100.

Nishiwaki, M., and N. Oguro. 1971. Baird's beaked whales caught on the coast of Japan in recent 10 years. Sci. Rept. Whales Res. Inst. 23:111–122.

Norris, K. S. 1966. Whales, dolphins, and porpoises. Univ. California Press, Berkeley, xv + 789 pp.

Norris, K. S., and G. W. Harvey. 1972. A theory for the function of the spermaceti organ of the sperm whale (*Physeter catodon* L.). *In* Galler, S. R.; K. Schmidt-Koenig; G. J. Jacobs; and R. E. Belleville, eds., Animal orientation and navigation, U.S. Natl. Aeronautics and Space Admin., Washington, D.C., pp. 397–417.

Norris, M. L., and C. E. Adams. 1972. The growth of the Mongolian gerbil, *Meriones unguiculatus*, from birth to maturity. J. Zool. 166:277–282.

Nosek, J.; O. Kozuch; and J. Chmela. 1972. Contribution to the knowledge of home range in common shrew *Sorex araneus* L. Oecologia 9:59–63.

Novellie, P. A. 1979. Courtship behaviour of the blesbok (*Damaliscus dorcas phillipsi*). Mammalia 43:263–274.

Novick, A., and B. A. Dale. 1971. Foraging behavior in fishing bats and their insectivorous relatives. J. Mamm. 52:817–818.

Novikov, G. A. 1962. Carnivorous mammals of the fauna of the U.S.S.R. Israel Progr. Sci. Transl., Jerusalem, 284 pp.

Nowak, R. M. 1972. The mysterious wolf of the south. Nat. Hist. 81(1):50–53, 74–77.

_____. 1973. Return of the wolverine. Natl. Parks and Conserv. Mag. 47(2):20–23.

_____. 1974. Red wolf. Our most endangered mammal. Natl. Parks and Conserv. Mag. 48(8):9–12.

_____. 1975*a*. The cosmopolitan wolf. Natl. Rifle Assoc. Conserv. Yearbook, pp. 76–82.

_____. 1975*b*. Retreat of the jaguar. Natl. Parks and Conserv. Mag. 49(12):10–13.

_____. 1976. The cougar in the United States and Canada. Unpubl. Rept. to U.S. Fish and Wildl. Serv., 190 pp.

_____. 1978. Evolution and taxonomy of coyotes and related *Canis*. *In* Bekoff (1978), pp. 3–16.

_____. 1979. North American Quaternary *Canis*. Monogr. Mus. Nat. Hist. Univ. Kansas, no. 6, 154 pp.

O.

Oates, J. F. 1977*a*. The guereza and man. *In* Rainier III and Bourne (1977), pp. 419–467.

_____. 1977*b*. The guereza and its food. *In* Clutton-Brock (1977), pp. 276–321.

_____. 1977*c*. The social life of a black-and-white colobus monkey, *Colobus guereza*. Z. Tierpsychol. 45:1–60.

Obidina, V. A. 1972. Winter breeding of *Alticola argentatus* under natural conditions. Soviet J. Ecol. 3:567–568.

Odell, D. K. 1971. Censuses of pinnipeds breeding on the California Channel Islands. J. Mamm. 52:187–190.

_____. 1975. Status and aspects of the life history of the bottlenose dolphin, *Tursiops truncatus*, in Florida. J. Fish. Res. Bd. Can. 32:1055–1058.

Odell, D. K.; D. Forrester; and E. Asper. 1978. Growth and sexual maturation in the West Indian manatee. American Soc. Mamm. Abstr. Tech. Pap., 58th Ann. Mtg., pp. 7–8.

Odendaal, P. B., and R. C. Bigalke. 1979. Home range and groupings of bushbuck in the southern Cape. S. Afr. J. Wildl. Res. 9:96–101.

O'Farrell, M. J. 1978. Home range dynamics of rodents in a sagebrush community. J. Mamm. 59:657–668.

O'Farrell, M. J., and A. R. Blaustein. 1974*a*. *Microdipodops megacephalus*. Mammalian Species, no. 46, 3 pp.

_____. 1974*b*. *Microdipodops pallidus*. Mammalian Species, no. 47, 2 pp.

O'Farrell, M. J., and E. H. Studier. 1973. Reproduction, growth, and development in *Myotis thysanodes* and *M. lucifugus* (Chiroptera: Vespertilionidae). Ecology 54:18–30.

O'Farrell, T. P.; R. J. Olson; R. O. Gilbert; and J. D. Hedlund. 1975. A population of Great Basin pocket mice, *Perognathus parvus*, in the shrub-steppe of south-central Washington. Ecol. Monogr. 45:1–28.

O'Gara, B. W. 1978. *Antilocapra americana*. Mammalian Species, no. 90, 7 pp.

O'Gara, B. W., and G. Matson. 1975. Growth and casting of horns by pronghorns and exfoliation of horns by bovids. J. Mamm. 56:829–846.

Ognev, S. I. 1962–1963. Mammals of eastern Europe and northern Asia. Israel Progr. Sci. Transl., 8 vols.

Ohsawa, H., and M. Kawai. 1975. Social structure of gelada baboons. Studies of the gelada society (I). Proc. 5th Int. Congr. Primatol., pp. 464–469.

Ohsumi, S. 1965. Reproduction of the sperm whale in the north-west Pacific. Sci. Rept. Whales Res. Inst. 19:1–35.

_____. 1966. Sexual segregation of the sperm whale in the North Pacific. Sci. Rept. Whales Res. Inst. 20:1–16.

_____. 1971. Some investigations on the school structure of sperm whale. Sci. Rept. Whales Res. Inst. 23:1–25.

_____. 1979. Population assessment of the Antarctic minke whale. Rept. Int. Whaling Comm. 29:407–420.

Ohsumi, S., and Y. Masaki. 1975. Biological parameters of the Antarctic minke whale at the virginal population level. J. Fish. Res. Bd. Can. 32:995–1004.

Ohsumi, S.; Y. Masaki; and A. Kawamura. 1970. Stock of the Antarctic minke whale. Sci. Rept. Whales Res. Inst. 22:75–110.

Ojasti, J. 1972. Revisión preliminar de los pícures o aguites de Venezuela (Rodentia, Dasyproctidae). Mem. Soc. Ciencias Nat. La Salle 32:159–204.

Ojasti, J., and O. J. Linares. 1971. Adiciones a la fauna de murciélagos de Venezuela con notas sobre las especies del género Diclidurus (Chiroptera). Acta Biol. Venez. 7:421–441.

Ojasti, J., and C. J. Naranjo. 1974. First record of Tonatia nicaraguae in Venezuela. J. Mamm. 55:248–249.

Ojasti, J., and G. M. Padilla. 1972. The management of capybara in Venezuela. Trans. N. Amer. Wildl. Conf. 33:268–277.

Okia, N. O. 1974a. Breeding in Franquet's bat, Epomops franqueti (Tomes), in Uganda. J. Mamm. 55:462–465.

———. 1974b. The breeding pattern of the eastern epauletted bat, Epomophorus anurus Heuglin, in Uganda. J. Reprod. Fert. 37:27–31.

———. 1976. The biology of the bush rat, Aethomys hindei Thomas in southern Uganda. J. Zool. 180:41–56.

Olds, T. J., and L. R. Collins. 1973. Breeding Matschie's tree kangaroo Dendrolagus matschiei in captivity. Int. Zoo Yearbook 13:123–125.

Olert, J.; F. Dieterlen; and H. Rupp. 1978. Eine neue Muriden—Art aus Südäthiopien. Z. Zool. Syst. Evolutionforsch. 16:297–308.

Oliver, W.L.R. 1977. The hutias of the West Indies. Int. Zoo Yearbook 17:14–20.

———. 1978. The doubtful future of the pigmy hog and the hispid hare. J. Bombay Nat. Hist. Soc. 75:341–372.

Olivier, R. 1978. Distribution and status of the Asian elephant. Oryx, 14:380–424.

Olsen, S. J., and J. W. Olsen. 1977. The Chinese wolf, ancestor of New World dogs. Science 197:533–535.

Olterman, J. H., and B. J. Verts. 1972. Endangered plants and animals of Oregon. IV. Mammals. Oregon State Univ. Agric. Exp. Sta. Spec. Rept., no. 364, 47 pp.

Omar, A., and A. DeVos. 1971. The annual reproductive cycle of an African monkey (Cercopithecus mitis kolbi Neuman). Folia Primatol. 16:206–215.

Omura, H. 1974. Possible migration route of the gray whale on the coast of Japan. Sci. Rept. Whales Res. Inst. 26:1–14.

Oppenheimer, J. R. 1969. Changes in forehead patterns and group composition of the white-faced monkey (Cebus capucinus). Proc. 2nd Int. Congr. Primatol. 1:36–42.

———. 1977. Presbytis entellus, the hanuman langur. In Rainier III and Bourne (1977), pp. 469–512.

Oppenheimer, J. R., and E. C. Oppenheimer. 1973. Preliminary observations of Cebus nigrivittatus (Primates: Cebidae) on the Venezuelan Llanos. Folia Primatol. 19:409–436.

Øritsland, T. 1975. Sexual maturity and reproductive performance of female hooded seals at Newfoundland. Int. Comm. Northwest Atlantic Fish. Res. Bull. 11:37–41.

Orlov, V. N., and Yu. M. Kovalskaya. 1978. Microtus mujanensis sp. n. from the Vitim River Basin. Zool. Zhur. 57:1224–1232.

Osborn, D. J., and I. Helmy. 1980. The contemporary land mammals of Egypt (including Sinai). Fieldiana Zool., n.s., no. 5, xix + 579 pp.

Osgood, W. H. 1932. Mammals of the Kelley-Roosevelts and Delacour Asiatic Expeditions. Field Mus. Nat. Hist. Publ., Zool. Ser., 18:193–339.

———. 1943. The mammals of Chile. Field Mus. Nat. Hist. Zool. Ser., 30:1–268.

———. 1947. Cricetine rodents allied to Phyllotis. J. Mamm. 28:165–174.

O'Shea, T. J. 1976. Home range, social behavior, and dominance relationships in the African unstriped ground squirrel, Xerus rutilus. J. Mamm. 57:450–460.

O'Shea, T. J., and T. A. Vaughan. 1977. Nocturnal and seasonal activities of the pallid bat, Antrozous pallidus. J. Mamm. 58:269–284.

Owens, D. D., and M. J. Owens. 1979a. Communal denning and clan associations in brown hyenas (Hyaena brunnea, Thunberg) of the central Kalahari Desert. Afr. J. Ecol. 17:35–44.

———. 1979b. Notes on social organization and behavior in brown hyenas (Hyaena brunnea). J. Mamm. 60:405–408.

Owens, M. J., and D. D. Owens. 1978. Feeding ecology and its influence on social organization in brown hyenas (Hyaena brunnea, Thunberg) of the central Kalahari Desert. E. Afr. Wildl. J. 16:113–135.

Owen-Smith, N. 1972. Territoriality: the example of the white rhinoceros. Zool. Afr. 7:273–280.

———. 1974. The social system of the white rhinoceros. In Geist and Walther (1974), pp. 341–351.

———. 1975. The social ethology of the white rhinoceros Ceratotherium simum (Burchell 1817). Z. Tierpsychol. 38:337–384.

Owings, D. H.; M. Borchert; and R. A. Virginia. 1977. The behaviour of California ground squirrels. Anim. Behav. 25:221–230.

Owings, D. H., and R. A. Virginia. 1978. Alarm calls of California squirrels (Spermophilus beecheyi). Z. Tierpsychol. 46:58–70.

P.

Packard, R. L. 1967. Octodontoid, bathyergoid, and ctenodactyloid rodents. In Anderson and Jones (1967), pp. 273–290.

———. 1968. An ecological study of the fulvous harvest mouse in eastern Texas. Amer. Midl. Nat. 79:68–88.

Packard, R. L., and J. B. Montgomery, Jr. 1978. Baiomys musculus. Mammalian Species, no. 102, 3 pp.

Pagels, J. F., and C. Jones. 1974. Growth and development of the free-tailed bat, Tadarida brasiliensis cynocephala (Le Conte). Southwestern Nat. 19:267–276.

Pages, E. 1978. Home range, behaviour and tactile communication in a nocturnal Malagasy lemur Microcebus coquereli. In Chivers and Joysey (1978), pp. 171–177.

Palacios, F. 1976. Descripción de una nueva especie de liebre (Lepus castroviejoi), endémica de la Cordillera Cantábrica. Doñana Acta Vert. 3:205–223.

Paradiso, J. L. 1971. A new subspecies of Cynopterus sphinx (Chiroptera: Pteropodidae) from Serasan (South Natuna) Island, Indonesia. Proc. Biol. Soc. Washington 84:293–300.

———. 1972. Status report on cats (Felidae) of the world, 1971. U.S. Bur. Sport Fish. and Wildl. Spec. Sci. Rept.—Wildl., no. 157, iv + 43 pp.

Paradiso, J. L., and A. M. Greenhall. 1967. Longevity records for American bats. Amer. Midl. Nat. 78:251–252.

Paradiso, J. L., and R. M. Nowak. 1972. Canis rufus. Mammalian Species, no. 22, 4 pp.

Parker, C. 1979. Birth, care and development of Chinese hog badgers Arctonyx collaris albogularis at Metro Toronto Zoo. Int. Zoo Yearbook 19:182–185.

Parker, G. R. 1973. Distribution and densities of wolves within barren-ground caribou range in northern mainland Canada. J. Mamm. 54:341–348.

———. 1977. Morphology, reproduction, diet, and behavior of the arctic hare (Lepus arcticus monstrabilis) on Axel Heiberg Island, Northwest Territories. Can. Field-Nat. 91:8–18.

Parker, P. 1977. An ecological comparison of marsupial and placental patterns of reproduction. In Stonehouse and Gilmore (1977), pp. 273–286.

Parker, S. A. 1971. Notes on the small black wallaroo Macropus bernardus (Rothschild, 1904) of Arnhem Land. Victorian Nat. 88:41–43.

Partridge, J. 1967. A 3,300 year old thylacine (Marsupialia: Thylacinidae) from the Nullarbor Plain, Western Australia. J. Roy. Soc. W. Austral. 50:57–59.

Patten, D. R., and L. T. Findley. 1970. Observations and records of Myotis (Pizonyx) vivesi Menegaux (Chiroptera: Vespertilionidae). Los Angeles Co. Mus. Contrib. Sci., no. 183, 9 pp.

Patterson, B. D. 1980. A new subspecies of Eutamias quadrivittatus (Rodentia: Sciuridae) from the Organ Mountains, New Mexico. J. Mamm. 61:455–464.

Pattie, D. 1973. Sorex bendirii. Mammalian Species, no. 27, 2 pp.

Patton, J. L. 1973. An analysis of natural hybridization between the pocket gophers, Thomomys bottae and Thomomys umbrinus, in Arizona. J. Mamm. 54:561–584.

Patton, J. L., and A. L. Gardner. 1972. Notes on the systematics of Proechimys (Rodentia: Echimyidae), with emphasis on Peruvian forms. Occas. Pap. Mus. Zool. Louisiana State Univ., no. 44, 30 pp.

Patton, J. L.; H. MacArthur; and S. Y. Yang. 1976. Systematic relationships of the four-toed populations of *Dipodomys heermanni*. J. Mamm. 57:159–163.

Patton, J. L.; S. W. Sherwood; and S. Y. Yang. 1981. Biochemical systematics of chaetodipine pocket mice, genus *Perognathus*. J. Mamm. 62:477–492.

Paul, J. R. 1968. Risso's dolphin, *Grampus griseus*, in the Gulf of Mexico. J. Mamm. 49:746–748.

Paula Couto, C. 1979. Tratado de Paleomastozoologia. Acad. Brasil. Cien., Rio de Janeiro, 590 pp.

Pavlenko, T. A., and A. G. Daveletshina. 1971. Nutrition of lesser jerboa in the Fergana Valley. Soviet J. Ecol. 2:69–71.

Pavlinov, I. Ya. 1980. Superspecies groupings Cardiocraniinae Satunin (Mammalia, Dipodidae). Vestn. Zool. 1980(2):47–50.

Payne, M. R. 1978. Population size and age determination in the Antarctic fur seal *Arctocephalus gazella*. Mamm. Rev. 8:67–73.

Payne, R. 1976. At home with right whales. Natl. Geogr. 149:322–339.

Pearson, A. M. 1975. The northern interior grizzly bear *Ursus arctos* L. Can. Wildl. Serv. Rept. Ser., no. 34, 86 pp.

———. 1976. Population characteristics of the arctic mountain grizzly bear. *In* Pelton, Lentfer, and Folk (1976), pp. 247–260.

Pearson, E. W. 1978. A 1974 coyote harvest estimate for 17 western states. Wildl. Soc. Bull. 6:25–32.

Pearson, O. P. 1948. Life history of mountain viscachas in Peru. J. Mamm. 29:345–374.

———. 1972. New information on ranges and relationships within the rodent genus *Phyllotis* in Peru and Ecuador. J. Mamm. 53:677–686.

Pearson, O. P., and J. L. Patton. 1976. Relationships among South American phyllotine rodents based on chromosome analysis. J. Mamm. 57:339–350.

Pearson, O. P., et al. 1968. Estructura social, distribución espacial y composición por edades de una población de tuco-tucos (*Ctenomys talarum*). Inv. Zool. Chileñas 13:47–80.

Pedersen, A. 1962. The walrus. Neue Brehm-Bucherei 306. Wildl. Serv. Transl., no. 125, 60 pp.

Peek, J. M.; R. E. LeResche; and D. R. Stevens. 1974. Dynamics of moose aggregations in Alaska, Minnesota, and Montana. J. Mamm. 55:126–137.

Peek, J. M.; D. L. Urich; and R. J. Mackie. 1976. Moose habitat selection and relationships to forest management in northeastern Minnesota. Wildl Monogr., no. 48, 65 pp.

Pefaur, J.; W. Hermosilla; F. Di Castri; R. Gonzalez; and F. Salinas. 1968. Estudio preliminar de mamiferos silvestres Chilenos: su distribución, valor económico e importancia

zoonótica. Rev. Soc. Med. Vet. (Chile) 18(1–4):3–15.

Pelikan, J., and V. Holisova. 1969. Movements and home ranges of *Arvicola terrestris* on a brook. Zool. Listy 18:207–224.

Pelton, M. R.; J. W. Lentfer; and G. E. Folk. 1976. Bears—their biology and management. Int. Union Conserv. Nat. Publ., n.s., no. 40, 467 pp.

Pelton, M. R.; C. D. Scott; and G. M. Burghardt. 1976. Attitudes and opinions of persons experiencing property damage and/or injury by black bears in the Great Smoky Mountains National Park. *In* Pelton, Lentfer, and Folk (1976), pp. 157–167.

Pembleton, E. F., and S. L. Williams. 1978. *Geomys pinetis*. Mammalian Species, no. 86, 3 pp.

Penrod, B. 1976. Fisher in New York. Conservationist, 31(2):20.

Penzhorn, B. L. 1979. Social organisation of the Cape mountain zebra *Equus z. zebra* in the Mountain Zebra National Park. Koedoe 22:115–156.

Peracchi, A. L. 1968. Sobre os hábitos de "*Histiotus velatus*" (Geoffroy, 1824) (Chiroptera, Vespertilionidae). Rev. Brasil. Biol. 28:469–473.

Perez, G.S.A. 1972. Observations of Guam bats. Micronesia 8:141–149.

———. 1973. Notes on the ecology and life history of Pteropodidae on Guam. Period. Biol. 75:163–168.

Perkins, J. E. 1945. Biology at Little America III, the west base of the United States Antarctic Service Expedition 1939–1941. Proc. Amer. Phil. Soc. 89:270–284.

Pernetta, J. C. 1977. Population ecology of British shrews in grassland. Acta Theriol. 22:279–296.

Perrers, C. 1965. Notes on a pigmy possum, *Cercartetus nanus* Desmarest. Austral. Zool. 13:126.

Perrin, W. F. 1975a. Distribution and differentiation of populations of dolphins of the genus *Stenella* in the eastern tropical Pacific. J. Fish. Res. Bd. Can. 32:1059–1067.

———. 1975b. Variation of spotted and spinner porpoise (genus *Stenella*) in the eastern tropical Pacific and Hawaii. Bull. Scripps Inst. Oceanogr. Univ. California 21:1–206.

———. 1976. First record of the melonheaded whale, *Peponocephala electra*, in the eastern Pacific, with a summary of world distribution. Fishery Bull. 74:457–458.

Perrin, W. F.; P. B. Best; W. H. Dawbin; K. C. Balcomb; R. Gambell; and G.J.B. Ross. 1973. Rediscovery of Fraser's dolphin *Lagenodelphis hosei*. Nature 241:345–350.

Perrin, W. F.; J. M. Coe; and J. R. Zweifel. 1976. Growth and reproduction of the spotted porpoise, *Stenella attenuata*, in the offshore eastern tropical Pacific. Fishery Bull. 74:229–269.

Perrin, W. F.; D. B. Holts; and R. B. Miller. 1977. Growth and reproduction of the eastern spinner dolphin, a geographical form of *Stenella longirostris* in the eastern tropical Pacific. Fishery Bull. 75:725–750.

Perrin, W. F.; R. B. Miller; and P. A. Sloan. 1977. Reproductive parameters of the offshore spotted dolphin, a form of *Stenella attenuata*, in the eastern tropical Pacific, 1973–75. Fishery Bull. 75:629–633.

Perrin, W. F.; E. D. Mitchell; J. G. Mead; D. K. Caldwell; and P.J.H. Van Bree. 1981. *Stenella clymene*, a rediscovered tropical dolphin of the Atlantic. J. Mamm. 62:583–598.

Perrin, W. F., and W. A. Walker. 1975. The rough-toothed porpoise, *Steno bredanensis*, in the eastern tropical Pacific. J. Mamm. 56:905–909.

Peters, R. P., and L. D. Mech. 1975. Scentmarking in wolves. Amer. Sci. 63:628–637.

Petersen, L. R.; M. A. Martin; and C. M. Pils. 1977. Status of fishers in Wisconsin, 1975. Wisconsin Dept. Nat. Resources Res. Rept., no. 92, 9 pp.

Peterson, R. L. 1965a. A review of the flatheaded bats of the family Molossidae from South America and Africa. Roy. Ontario Mus. Life Sci. Contrib., no. 64, 32 pp.

———. 1965b. A review of the bats of the genus *Ametrida*, family Phyllostomidae. Roy. Ontario Mus. Life Sci. Contrib., no. 65, 13 pp.

———. 1969. Notes on the Malaysian fruit bats of the genus *Dyacopterus*. Roy. Ontario Mus. Life Sci. Occas. Pap., no. 13, 4 pp.

———. 1971a. The African molossid bat *Tadarida russata*. Can. J. Zool. 49:297–301.

———. 1971b. Notes on the African long-eared bats of the genus *Laephotis* (family Vespertilionidae). Can. J. Zool. 49:885–888.

———. 1971c. The systematic status of the African molossid bats *Tadarida bemmeleni* and *Tadarida cistura*. Can. J. Zool. 49:1347–1354.

———. 1972. Systematic status of the African molossid bats *Tadarida congica, T. niangarae* and *T. trevori*. Roy. Ontario Mus. Life Sci. Contrib., no. 85, 32 pp.

———. 1973. The first known female of the African long-eared bat *Laephotis wintoni* (Vespertilionidae: Chiroptera). Can. J. Zool. 51:601–603.

———. 1974a. Variation in the African bat, *Tadarida lobata*, with notes on habitat and habits (Chiroptera: Molossidae). Roy. Ontario Mus. Life Sci. Occas. Pap., no. 24, 8 pp.

———. 1974b. A review of the general life history of the moose. Naturaliste Canadien 101:9–21.

Peterson, R. L., and M. B. Fenton. 1970. Variation in the bats of the genus *Harpyionycteris*, with the description of a new race. Roy. Ontario Mus. Life Sci. Occas. Pap., no. 17, 15 pp.

Peterson, R. L., and D. L. Harrison. 1970. The second and third known specimens of the African molossid bat, *Tadarida lobata*. Roy. Ontario Mus. Life Sci. Occas. Pap., no. 16, 6 pp.

Peterson, R. L., and D. A. Smith. 1973. A new species of *Glauconycteris* (Vespertilionidae, Chiroptera). Roy. Ontario Mus. Life Sci. Occas. Pap., no. 22, 9 pp.

Peterson, R. O. 1977. Wolf ecology and prey relationships on Isle Royale. U.S. Natl. Park Serv. Sci. Monogr. Ser., no. 11, xx + 210 pp.

Peterson, R. S., and G. A. Bartholomew. 1967. The natural history and behavior of the California sea lion. Amer. Soc. Mamm. Spec. Publ., no. 1, xi + 79 pp.

Peterson, R. S.; C. L. Hubbs; R. L. Gentry; and R. L. DeLong. 1968. The Guadalupe fur seal: habitat, behavior, population size, and field identification. J. Mamm. 49:665–675.

Petter, F. 1972. The rodents of Madagascar: the seven genera of Malagasy rodents. *In* Battistini and Richard-Vindard (1972), pp. 661–665.

———. 1973. Les noms de genre *Cercomys, Trichomys* et *Proechimys* (Rongeurs, Echimyides). Mammalia 37:422–426.

———. 1975. Les *Praomys* de République Centrafricaine (Rongeurs, Murides). Mammalia 39:51–56.

———. 1977. Les rats à mamelles multiples d'Afrique occidentale et centrale: *Mastomys erythroleucus* (Temminck, 1853) et *M. huberti* (Wroughton, 1908). Mammalia 41:441–444.

———. 1978a. Une souris nouvelle du sud de l'Afrique: *Mus setzeri* sp. nov. Mammalia 42:377–379.

———. 1978b. Epidémiologie de la leishmaniose en Guyane française, en relation avec l'existence d'une espèce nouvelle de rongeurs echimyides, *Proechimys cuvieri* sp. n. Compt. Rend. Acad. Sci. Paris, ser. D, 287:261–264.

———. 1979. Une nouvelle espèce de rat d'eau de Guyane française, *Nectomys parvipes* sp. nov. (Rongeurs, Cricetidae). Mammalia 43:507–510.

Petter, J.-J. 1965. The lemurs of Madagascar. *In* De Vore (1965), pp. 292–319.

———. 1975. Breeding of Malagasy lemurs in captivity. *In* Martin (1975b), pp. 187–202.

———. 1977. The aye-aye. *In* Rainier III and Bourne (1977), pp. 37–57.

———. 1978. Ecological and physiological adaptations of five sympatric nocturnal lemurs to seasonal variations in food production. *In* Chivers and Herbert (1978), pp. 211–223.

Petter, J.-J., and P. Charles-Dominique. 1979. Vocal communication in prosimians. *In* Doyle and Martin (1979), pp. 247–305.

Petter, J.-J., and A. Petter. 1967. The aye-aye of Madagascar. *In* Altmann (1967), pp. 195–205.

Petter, J.-J., and A. Petter-Rousseaux. 1979. Classification of the prosimians. *In* Doyle and Martin (1979), pp. 1–44.

Petter, J.-J., and A. Peyrieras. 1975. Preliminary notes on the behavior and ecology of *Hapalemur griseus*. *In* Tattersall and Sussman (1975), pp. 281–286.

Petter, J.-J.; A. Schilling; and G. Pariente. 1975. Observations on behavior and ecology

of *Phaner furcifer*. *In* Tattersall and Sussman (1975), pp. 209–218.

Pettifer, H. L., and J.A.J. Nel. 1977. Hoarding in four southern African rodent species. Zool. Afr. 12:409–418.

Peyton, B. 1980. Ecology, distribution, and food habits of spectacled bears, *Tremarctos ornatus*, in Peru. J. Mamm. 61:639–652.

Phillips, C. J. 1966. A new species of bat of the genus *Melonycteris* from the Solomon Islands. J. Mamm. 47:23–27.

———. 1967. A collection of bats from Laos. J. Mamm. 48:633–636.

———. 1968. Systematics of megachiropteran bats in the Solomon Islands. Univ. Kansas Publ. Mus. Nat. Hist. 16:777–837.

Phillips, C. J., and E. C. Birney. 1968. Taxonomic status of the vespertilionid genus *Anamygdon* (Mammalia; Chiroptera). Proc. Biol. Soc. Washington 81:491–498.

Phillips, C. J., and J. K. Jones, Jr. 1968. Additional comments on reproduction in the woolly opossum (*Caluromys derbianus*) in Nicaragua. J. Mamm. 49:320–321.

———. 1969. Notes on reproduction and development in the four-eyed opossum, *Philander opossum*, in Nicaragua. J. Mamm. 50:345–348.

———. 1971. A new subspecies of the long-nosed bat, *Hylonycteris underwoodi*, from Mexico. J. Mamm. 52:77–80.

Phillips, R. L., and C. Jonkel, eds. 1977. Proceedings of the 1975 predator symposium. Montana Forest and Conserv. Exp. Sta., Univ. Montana, Missoula, ix + 268 pp.

Pidduck, E. R., and J. B. Falls. 1973. Reproduction and emergence of juveniles in *Tamias striatus* (Rodentia: Sciuridae) at two localities in Ontario, Canada. J. Mamm. 54:693–707.

Pidoplichko, I. G. 1973. On time of *Lagurus lagurus* Pall. extinction on the Ukraine right bank area. Vestn. Zool. 1973(5):35–41.

Piechocki, R. 1966. Uber die Nachweise der Langohr-Fledermause, *Plecotus auritus* L. und *Plecotus austriacus* Fischer im mitteldeutschen Raum. Hercynia 3·407–415.

Piekielek, W., and T. S. Burton. 1975. A black bear population study in northern California. California Fish and Game 61:4–25.

Pielowski, Z. 1972. Home range and degree of residence of the European hare. Acta Theriol. 17:93–103.

———. 1976. On the present state and perspectives of the European hare breeding in Poland. *In* Ecology and management of European hare populations, Proc. Int. Symp., 23–24 December 1974, Poznan, Poland, pp. 25–27.

Pienaar, U.D.V. 1970. A note on the occurrence of bat-eared fox *Otocyon megalotis megalotis* (Desmarest) in the Kruger National Park. Koedoe 13:23–27.

———. 1972. A new bat record for the Kruger National Park. Koedoe 15:91–93.

Pieper, H. 1978. Eine neue *Crocidura*-Art (Mammalia: Soricidae) von der Insel Kreta. Bonner Zool. Beitr. 4:281–286.

Pilgrim, W. 1980. Fisher, *Martes pennanti* (Carnivora: Mustelidae) in Labrador. Can. Field-Nat. 94:468.

Pilleri, G. 1971. On the La Plata dolphin *Pontoporia blainvillei* off the Uruguayan coast. Investig. Cetacea 3:59–67.

———. 1979. Observations on the ecology of *Inia geoffrensis* from the Rio Apure, Venezuela. Investig. Cetacea 10:137–143.

Pilleri, G., and Chen Peixun. 1979. How the finless porpoise (*Neophocaena asiaeorientalis*) carries its calves on its back, and the function of the denticulated area of skin, as observed in the Changjiang River, China. Investig. Cetacea 10:105–108.

Pilleri, G., and M. Gihr. 1971. Zur Systematik der Gattung *Platanista* (Cetacea). Rev. Suisse Zool. 78:746–759.

———. 1975. On the taxonomy and ecology of the finless black porpoise, *Neophocaena* (Cetacea, Delphinidae). Mammalia 39:657–673.

———. 1977a. Observations on the Bolivian (*Inia boliviensis* d'Orbigny, 1834) and the Amazonian bufeo (*Inia geoffrensis* de Blainville, 1817) with description of a new subspecies (*Inia geoffrensis humboldtiana*). Investig. Cetacea 8:11–76.

———. 1977b. Neotype for *Platanista indi* Blyth, 1859. Investig. Cetacea 8:77–81.

———. 1977c. Radical extermination of the South American sea lion *Otaria byronia* (Pinnipedia, Otariidae) from Isla Verde, Uruguay. Brain Anat. Inst., Univ. Berne, Switzerland, 15 pp.

Pilleri, G.; M. Gihr; P. E. Purves; K. Zbinden; and C. Kraus. 1976. On the behaviour, bioacoustics and functional morphology of the Indus River dolphin (*Platanista indi* Blyth, 1859). Investig. Cetacea 6:1–151.

Pilleri, G., and O. Pilleri. 1979a. Precarious situation of the dolphin population (*Platanista indi* Blyth, 1859) in the Punjab, upstream from the Taunsa Barrage, Indus River. Investig. Cetacea 10:121–127.

———. 1979b. Observations on the dolphins in the Indus Delta (*Sousa plumbea* and *Neophocaena phocaenoides*) in winter 1978–1979. Investig. Cetacea 10:129–135.

Pilleri, G.; K. Zbinden; and M. Gihr. 1976. The 'black finless porpoise' (*Neophocaena phocaenoides* Cuvier, 1829) is not black. Investig. Cetacea 7:161–164.

Pils, C. M., and M. A. Martin. 1978. Population dynamics, predator-prey relationships and management of the red fox in Wisconsin. Wisconsin Dept. Nat. Res. Tech. Bull., no. 105, 56 pp.

Pimlott, D. H., ed. 1975. Wolves. Int. Union Conserv. Nat. Suppl. Pap., n.s., no. 43, 144 pp.

Pimlott, D. H.; J. A. Shannon; and G. B. Kolenosky. 1969. The ecology of the timber wolf in Algonquin Provincial Park. Ontario Dept. Lands and Forests, 92 pp.

Pine, R. H. 1971. A review of the long-whiskered rice rat, *Oryzomys bombycinus* Goldman. J. Mamm. 52:590–596.

———. 1972a. A new subgenus and species of murine opossum (genus *Marmosa*) from Peru. J. Mamm. 53:279–282.

————. 1972*b*. The bats of the genus *Carollia*. Texas Agric. Exp. Sta. Tech. Monogr., no. 8, 125 pp.

————. 1973*a*. Anatomical and nomenclatural notes on opossums. Proc. Biol. Soc. Washington 86:391–402.

————. 1973*b*. Una nueva especie de *Akodon* (Mammalia: Rodentia: Muridae) de la Isla de Wellington, Magallanes, Chile. An. Inst. Patagonia 4:423–426.

————. 1975. A new species of *Monodelphis* (Mammalia: Didelphidae) from Bolivia. Mammalia 39:320–322.

————. 1976*a*. A new species of *Akodon* (Mammalia: Rodentia: Muridae: Cricetinae) from Isla de los Estados, Argentina. Mammalia 40:63–68.

————. 1976*b*. *Monodelphis umbristriata* (A. de Miranda-Ribeiro) is a distinct species of opossum. J. Mamm. 57:785–787.

————. 1977. *Monodelphis iheringi* (Thomas) is a recognizable species of Brazilian opossum (Mammalia: Marsupialia: Didelphidae). Mammalia 41:235–237.

————. 1979. Taxonomic notes on "*Monodelphis dimidiata itatiayae* (Miranda-Ribeiro)," *Monodelphis domestica* (Wagner) and *Monodelphis maraxina* Thomas (Mammalia: Marsupialia: Didelphidae). Mammalia 43:495–499.

————. 1980. Notes on rodents of the genera *Wiedomys* and *Thomasomys* (including *Wilfredomys*). Mammalia 44:195–202.

Pine, R. H., and J. P. Abravaya. 1978. Notes on the Brazilian opossum *Monodelphis scalops* (Thomas) (Mammalia: Marsupialia: Didelphidae). Mammalia 42:379–382.

Pine, R. H.; D. C. Carter; and R. K. LaVal. 1971. Status of *Bauerus* Van Gelder and its relationships to other nyctophiline bats. J. Mamm. 52:663–669.

Pine, R. H.; S. D. Miller; and M. L. Schamberger. 1979. Contributions to the mammalogy of Chile. Mammalia 43:339–376.

Pine, R. H., and R. M. Wetzel. 1975. A new subspecies of *Pseudoryzomys wavrini* (Mammalia: Rodentia: Muridae: Cricetinae) from Bolivia. Mammalia 39:649–655.

Pinter, A. J. 1970. Reproduction and growth for two species of grasshopper mice (*Onychomys*) in the laboratory. J. Mamm. 51:236–243.

Pinto da Silveira, E. K. 1968. Notas sôbre a historia natural do tamanduá mirim (*Tamandua tetradactyla chiriquensis* J. A. Allen 1904, Myrmecophagidae), com referências à fauna do Istmo do Panamá. Vellozia, Rio de Janeiro, no. 6, pp. 9–31.

————. 1969. História natural do tamanduá-bandeira *Myrmecophaga tridactyla* Linn. 1758, Myrmecophagidae. Vellozia, Rio de Janeiro, no. 7, 20 pp.

Pitcher, K. W., and D. G. Calkins. 1981. Reproductive biology of Steller sea lions in the Gulf of Alaska. J. Mamm. 62:599–605.

Pizzimenti, J. J. 1975. Evolution of the prairie dog genus *Cynomys*. Occas. Pap. Mus. Nat. Hist. Univ. Kansas, no. 39, 73 pp.

Pizzimenti, J. J., and L. R. McClenaghan, Jr. 1974. Reproduction, growth and development, and behavior in the Mexican prairie dog *Cynomys mexicanus* Merriam. Amer. Midl. Nat. 92:130–145.

Platt, W. J. 1976. The social organization and territoriality of short-tailed shrew (*Blarina brevicauda*) populations in old-field habitats. Anim. Behav. 24:305–318.

Poche, R. M. 1975. Notes on reproduction in *Funisciurus anerythrus* from Niger, Africa. J. Mamm. 56:700–701.

Poche, R. M., and G. L. Bailie. 1974. Notes on the spotted bat (*Euderma maculatum*) from southwest Utah. Great Basin Nat. 34:254–256.

Poelker, R. J., and H. D. Hartwell. 1973. Black bear of Washington. Washington State Game Dept. Biol. Bull., no. 14, viii + 180 pp.

Poglayen-Neuwall, I. 1966. Notes on care, display and breeding of olingos *Bassaricyon*. Int. Zoo Yearbook 6:169–171.

————. 1973. Preliminary notes on maintenance and behaviour of the Central American cacomistle *Bassariscus sumichrasti*. Int. Zoo Yearbook 13:207–211.

————. 1975. Copulatory behavior, gestation and parturition of the tayra (*Eira barbara* L., 1758). Z. Saugetierk. 40:176–189.

Poglayen-Neuwall, I., and I. Poglayen-Neuwall. 1980. Gestation period and parturition of the ringtail *Bassariscus astutus* (Liechtenstein, 1830). Z. Saugetierk. 45:73–81.

Poirier, F. E. 1970. The Nilgiri langur (*Presbytis johnii*) of south India. Primate Behav. 1:254–383.

————. 1972. The St. Kitts green monkey (*Cercopithecus aethiops sabaeus*): ecology, population dynamics, and selected behavior traits. Folia Primatol. 17:20–55.

Pola, Y., and C. T. Snowdon. 1975. The vocalizations of pygmy marmosets (*Cebuella pygmaea*). Anim. Behav. 23:826–842.

Pollock, J. I. 1975. Field observations on *Indri indri*: a preliminary report. *In* Tattersall and Sussman (1975), pp. 287–311.

————. 1977. The ecology and sociobiology of feeding in *Indri indri*. *In* Clutton-Brock (1977), pp. 37–69.

————. 1979. Spatial distribution and ranging behavior in lemurs. *In* Doyle and Martin (1979), pp. 359–409.

Poole, T. B., and H.D.R. Morgan. 1973. Differences in aggressive behaviour between male mice (*Mus musculus* L.) in colonies of different sizes. Anim. Behav. 21:788–795.

————. 1976. Social and territorial behaviour of laboratory mice (*Mus musculus* L.) in small complex areas. Anim. Behav. 24:476–480.

Poole, W. E. 1973. A study of breeding in grey kangaroos, *Macropus giganteus* Shaw and *M. fuliginosus* (Desmarest), in central New South Wales. Austral. J. Zool. 21:183–212.

————. 1975. Reproduction in the two species of grey kangaroos, *Macropus giganteus* Shaw and *M. fuliginosus* (Desmarest). II. Gestation, parturition and pouch life. Austral. J. Zool. 23:333–353.

————. 1976. Breeding biology and current status of the grey kangaroo, *Macropus fuliginosus fuliginosus*, of Kangaroo Island, South Australia. Austral. J. Zool. 24:169–187.

————. 1977. The eastern grey kangaroo, *Macropus giganteus*, in south-east South Australia: its limited distribution and need of conservation. CSIRO Div. Wildl. Res. Tech. Pap., no. 31, 15 pp.

————. 1978. Management of kangaroo harvesting in Australia. Austral. Natl. Parks and Wildl. Serv. Occas. Pap., no. 2, 28 pp.

————. 1979. The status of the Australian Macropodidae. *In* Tyler (1979), pp. 13–27.

Poole, W. E., and P. C. Catling. 1974. Reproduction in the two species of grey kangaroos, *Macropus giganteus* Shaw and *M. fuliginosus* (Desmarest). I. Sexual maturity and oestrus. Austral. J. Zool. 22:277–302.

Porter, F. L. 1978. Roosting patterns and social behavior in captive *Carollia perspicillata*. J. Mamm. 59:627–630.

————. 1979*a*. Social behavior in the leaf-nosed bat, *Carollia perspicillata*. I. Social organization. Z. Tierpsychol. 49:406–417.

————. 1979*b*. Social behavior in the leaf-nosed bat, *Carollia perspicillata*. II. Social communication. Z. Tierpsychol. 50:1–8.

Posamentier, H., and H. F. Recher. 1974. The status of *Pseudomys novaehollandiae* (the New Holland mouse). Austral. Zool. 18:66–71.

Poulet, A. R. 1972. Recherches écologiques sur une savane Sahélienne du Ferlo Septentrional Sénégal: les mammifères. Terre Vie 26:440–472.

————. 1978. Evolution of the rodent population of a dry bush savanna in the Senegalese Sahel from 1969 to 1977. Bull. Carnegie Mus. Nat. Hist., no. 6, pp. 113–117.

Poulet, A. R., and H. Poupon. 1978. L'invasion d'*Arvicanthis niloticus* dans le Sahél Sénégalais en 1975–1976 et ses conséquences pour la strate ligneuse. Terre Vie 32:161–194.

Powell, J. A., Jr. 1978. Evidence of carnivory in manatees (*Trichechus manatus*). J. Mamm. 59:442.

Powell, R. A. 1981. *Martes pennanti*. Mammalian Species, no. 156, 6 pp.

Powell, R. A., and R. B. Brander. 1977. Adaptations of fishers and porcupines to their predator prey system. *In* Phillips and Jonkel (1977), pp. 45–53.

Prakash, I. 1971. Breeding season and litter size of Indian desert rodents. Z. Angewandte Zool. 58:441–454.

————. 1975. The population ecology of the rodents of the Rajasthan Desert, India. *In* Prakash and Ghosh (1975), pp. 75–116.

Prakash, I., and P. K. Ghosh, eds. 1975.

Rodents in desert environments. W. Junk, The Hague, xv + 624 pp.

Prakash, I.; A. P. Jain; and B. D. Rana. 1975. A study of field populations of rodents in the Indian Desert. IV. Ruderal habitat. Z. Angewandte Zool. 62:339–348.

Prakash, I., and B. D. Rana. 1972. A study of field populations of rodents in the Indian Desert. II. Rocky and piedmont zones. Z. Angewandte Zool. 59:129–139.

Pratt, H. D.; B. F. Bjornson; and K. S. Littig. 1977. Control of domestic rats and mice. U.S. Pub. Health Serv., Center for Disease Control, Atlanta, 47 pp.

Prescott, J., and J. Ferron. 1978. Breeding and behaviour development of the American red squirrel *Tamiasciurus hudsonicus*. Int. Zoo Yearbook 18:125–130.

Preston, E. M. 1975. Home range defense in the red fox *Vulpes vulpes* L. J. Mamm. 56:645–652.

Preuschoft, H. 1971. Mode of locomotion in subfossil giant lemuroids from Madagascar. Proc. 3rd Int. Congr. Primatol. 1:79–90.

Prynn, D. 1980. Tigers and leopards in Russia's Far East. Oryx 15:496–503.

Pulliainen, E. 1965. On the distribution and migrations of the arctic fox (*Alopex lagopus* L.) in Finland. Aquilo, Ser. Zool., 2:25–40.

————. 1968. Breeding biology of the wolverine (*Gulo gulo* L.) in Finland. Ann. Zool. Fennici 5:338–344.

————. 1974. Seasonal movements of moose in Europe. Naturaliste Canadien 101:379–392.

————. 1975. Wolf ecology in northern Europe. *In* Fox (1975), pp. 292–299.

————. 1980. The status, structure and behaviour of populations of the wolf (*Canis l. lupus* L.) along the Fenno-Soviet border. Ann. Zool. Fennici 17:107–112.

Pulliainen, E., and P. Ovaskainen. 1975. Territory marking by a wolverine (*Gulo gulo*) in northeastern Lapland. Ann. Zool. Fennici 12:268–270.

Purves, P. E., and G. Pilleri. 1974–1975. Observations on the ear, nose, throat and eye of *Platanista indi*. Investig. Cetacea 5:13–57.

————. 1978. The functional anatomy and general biology of *Pseudorca crassidens* (Owen) with a review of the hydrodynamics and acoustics in Cetacea. Investig. Cetacea 9:68–227.

Q.

Quadagno, D. M.; J. T. Allin; R. J. Brooks; R. D. St. John; and E. M. Banks. 1970. Some aspects of the reproductive biology of *Baiomys taylori ater*. Amer. Midl. Nat. 84:550–551.

Quris, R. 1975. Ecologie et organisation sociale de *Cercocebus galeritus agilis* dans le nord-est du Gabon. Terre Vie 29:337–398.

R.

Rabor, D. S. 1939. *Sciuropterus mindanensis* sp. nov., a new species of flying squirrel

from Mindanao. Philippine J. Sci. 69:389–393.

————. 1952. Two new mammals from Negros Island, Philippines. Nat. Hist. Misc. (Chicago Acad. Sci.), no. 96, 7 pp.

Racey, P. A. 1973. The time of onset of hibernation in pipistrelle bats, *Pipistrellus pipistrellus*. J. Zool. 171:465–467.

Rageot, R. 1978. Observaciones sobre el monito del monte. Chile Min. Agric. Corp. Nac. For. Dept. Tec. IX—Reg. Interp.—V. Silvestre, 16 pp.

Rahm, U. 1962. L'élevage et la reproduction en captivité de l'*Atherurus africanus* (Rongeurs, Hystricidae). Mammalia 26:1–9.

————. 1966. Les mammifères de la forêt équatoriale de l'Est du Congo. Ann. Mus. Roy. Afr. Cent. (Tervuren), ser. 8, 149:39–121.

————. 1969a. Dokumente über *Anomalurus* and *Idiurus* des ostlichen Kongo. Z. Saugetierk. 34:75–84.

————. 1969b. Zur Fortpflanzungsbiologie von *Tachyoryctes ruandae* (Rodentia, Rhizomyidae). Rev. Suisse Zool. 76:695–702.

————. 1970a. Note sur la reproduction des sciurides et murides dans la forêt équatoriale au Congo. Rev. Suisse Zool. 77:635–646.

————. 1970b. Ecology, zoogeography, and systematics of some African forest monkeys. *In* Napier and Napier (1970), pp. 589–626.

————. 1972. Zur Oekologie der Muriden im Regenwaldgebiet des ostlichen Kongo (Zaire). Rev. Suisse Zool. 79:1121–1130.

Rainier III, Prince of Monaco, and G. H. Bourne. 1977. Primate Conservation. Academic Press, New York, xviii + 658 pp.

Rajagopalan, P. K. 1968. Notes on the Malabar spiny dormouse, *Platacanthomys lasiurus* Blyth, 1859, with new distribution record. J. Bombay Nat. Hist. Soc. 65:214–215.

Ralls, K. 1973. *Cephalophus maxwelli*. Mammalian Species, no. 31, 4 pp.

————. 1976. Mammals in which females are larger than males. Quart. Rev. Biol. 51:245–276.

————. 1978. *Tragelaphus euryceros*. Mammalian Species, no. 111, 4 pp.

Ramirez, M. F.; C. H. Freese; and J. Revilla C. 1977. Feeding ecology of the pygmy marmoset, *Cebuella pygmaea*, in northeastern Peru. *In* Kleiman (1977a), pp. 91–104.

Ramirez-Pulido, J., and W. Lopez-Forment. 1979. Additional records of some Mexican bats. Southwestern Nat. 24:541–544.

Ranck, G. L. 1968. The rodents of Libya. Bull. U.S. Natl. Mus. 275:1–264.

Rasa, O.A.E. 1972. Aspects of social organization in captive dwarf mongooses. J. Mamm. 53:181–185.

————. 1973. Marking behaviour and its social significance in the African dwarf mongoose, *Helogale undulata rufula*. Z. Tierpsychol. 32:293–318.

————. 1975. Mongoose sociology and behaviour as related to zoo exhibition. Int. Zoo Yearbook 15:65–73.

————. 1976. Invalid care in the dwarf mongoose (*Helogale undulata rufula*). Z. Tierpsychol. 42:337–342.

————. 1977. The ethology and sociology of the dwarf mongoose (*Helogale undulata rufula*). Z. Tierpsychol. 43:337–406.

Rasweiler, J. J., IV. 1973. Care and management of the long-tongued bat, *Glossophaga soricina* (Chiroptera: Phyllostomatidae), in the laboratory, with observations on estivation induced by food deprivation. J. Mamm. 54:391–404.

Rathbun, G. B. 1973. Territoriality in the golden-rumped elephant shrew. E. Afr. Wildl. J. 11:405.

————. 1979. *Rhynchocyon chrysopygus*. Mammalian Species, no. 117, 4 pp.

Rathbun, G. B., and M. Gache. 1980. Ecological survey of the night monkey, *Aotus trivirgatus*, in Formosa Province, Argentina. Primates 21:211–219.

Rau, R. 1978. The end of the hunt? Natl. Wildl. 16(3):4–11.

Rau, R. E. 1978. Additions to the revised list of preserved material of the extinct Cape Colony quagga and notes on the relationship and distribution of southern plains zebras. Ann. S. Afr. Mus. 77:27–45.

Rausch, R. A., and A. M. Pearson. 1972. Notes on the wolverine in Alaska and the Yukon Territory. J. Wildl. Mgmt. 36:249–268.

Rausch, R. A.; R. J. Somerville; and R. H. Bishop. 1974. Moose management in Alaska. Naturaliste Canadien 101:705–721.

Rausch, R. L. 1953. On the status of some arctic mammals. Arctic 6:91–148.

————. 1963. Geographic variation in size in North American brown bears, *Ursus arctos* L., as indicated by condylobasal length. Can. J. Zool. 41:33–45.

Rausch, R. L., and V. R. Rausch. 1975. Taxonomy and zoogeography of *Lemmus* sp. (Rodentia: Arvicolinae), with notes on laboratory-reared lemmings. Z. Saugetierk. 40:8–34.

Rautenbach, I. L., and J.A.J. Nel. 1975. Further records of smaller mammals from the Kalahari Gemsbok National Park. Koedoe 18:195–198.

Rautenbach, I. L., and D. A. Schlitter. 1978. Revision of genus *Malacomys* of Africa (Mammalia: Muridae). Ann. Carnegie Mus. 47:385–422.

Ray, C. E. 1964. The taxonomic status of *Heptaxodon* and dental ontogeny in *Elasmodontomys* and *Amblyrhiza*. Bull. Mus. Comp. Zool. 131:107–127.

Ray, G. C., and W. E. Schevill. 1974. Feeding of a captive gray whale, *Eschrichtius robustus*. Mar. Fish. Rev. 36(4):31–38.

Ray, G. C.; W. A. Watkins; and J. J. Burns. 1969. The underwater song of *Erignathus* (bearded seal). Zoologica 54:79–83.

Read, B., and R. J. Frueh. 1980. Management and breeding of Speke's gazelle

Gazella spekei at the St. Louis Zoo, with a note on artificial insemination. Int. Zoo Yearbook 20:99–104.

Ream, R. R. 1980. Wolf ecology project. Annual report. Wilderness Inst., Univ. Montana, 58 pp.

Redding, R. W., and D. M. Lay. 1978. Description of a new species of shrew of the genus *Crocidura* (Mammalia: Insectivora: Soricidae) from southwestern Iran. Z. Saugetierk. 43:306–310.

Redhead, T. D., and J. L. McKean. 1975. A new record of the false water-rat, *Xeromys myoides*, from the Northern Territory of Australia. Austral. Mamm. 1:347–354.

Reeves, R. R. 1977. Hunt for the narwhal. Oceans 10(4):50–57.

———. 1978. Atlantic walrus (*Odobenus rosmarus rosmarus*): a literature survey and status report. U.S. Fish and Wildl. Serv. Wildl. Res. Rept., no. 10, 41 pp.

Reeves, R. R.; J. G. Mead; and S. Katona. 1978. The right whale, *Eubalaena glacialis*, in the western North Atlantic. Rept. Int. Whaling Comm. 28:303–312.

Reeves, R. R., and E. Mitchell. 1981. The whale behind the tusk. Nat. Hist. 90(8):50–57.

Reeves, R. R., and S. Tracey. 1980. *Monodon monoceros*. Mammalian Species, no. 127, 7 pp.

Reichman, O. J. 1975. Relation of desert rodent diets to available resources. J. Mamm. 56:731–751.

Reichman, O. J., and R. J. Baker. 1972. Distribution and movements of two species of pocket gophers (Geomyidae) in an area of sympatry in the Davis Mountains, Texas. J. Mamm. 53:21–33.

Reig, O. A. 1970. Ecological notes on the fossorial octodont rodent *Spalacopus cyanus* (Molina). J. Mamm. 51:592–601.

———. 1978. Roedores cricetidos del Plioceno Superior de la Provincia de Buenos Aires (Argentina). Publ. Mus. Munic. Cien. Nat. Mar del Plata "Lorenzo Scaglia," 2:164–190.

———. 1980. A new fossil genus of South American cricetid rodents allied to *Wiedomys*, with an assessment of the Sigmodontinae. J. Zool. 192:257–281.

Reilly, S. B. 1978. Pilot Whale. *In* Haley (1978), pp. 112–119.

Reinhardt, J. 1852. Description of *Carterodon sulcidens*, Lund. Ann. Mag. Nat. Hist., ser. 2, 10:420.

Renfree, M. B. 1980. Embryonic diapause in the honey possum *Tarsipes spencerae*. Search 11(3):81.

Repenning, C. A. 1967. Subfamilies and genera of the Soricidae. U.S. Geol. Surv. Prof. Pap., no. 565, iv + 74 pp.

Repenning, C. A., and C. E. Ray. 1977. The origin of the Hawaiian monk seal. Proc. Biol. Soc. Washington 89:667–688.

Reynolds, H. G., and F. Turkowski. 1972. Reproductive variations in the round-tailed ground squirrel as related to winter rainfall. J. Mamm. 53:893–898.

Reynolds, V., and F. Reynolds. 1965. Chimpanzees of the Budongo Forest. *In* De Vore (1965), pp. 368–424.

Ricciuti, E. R. 1978. Dogs of war. Int. Wildl. 8(5):36–40.

———. 1980. The ivory wars. Anim. Kingdom 83(1):6–58.

Rice, D. W. 1967. Cetaceans. *In* Anderson and Jones (1967), pp. 291–324.

———. 1977. A list of the marine mammals of the world (third edition). U.S. Natl. Mar. Fish. Serv., NOAA Tech. Rept. NMFS SSRF–711, iii + 15 pp.

———. 1978a. Blue whale. *In* Haley (1978), pp. 30–35.

———. 1978b. Gray whale. *In* Haley (1978), pp. 54–61.

———. 1978c. Sperm whales. *In* Haley (1978), pp. 82–87.

———. 1978d. Beaked whales. *In* Haley (1978), pp. 88–95.

Rice, D. W., and A. A. Wolman. 1971. The life history and ecology of the gray whale (*Eschrichtius robustus*). Amer. Soc. Mamm. Spec. Publ., no. 3, viii + 142 pp.

Richard, A. 1974. Patterns of mating in *Propithecus verreauxi verreauxi*. *In* Martin, Doyle, and Walker (1974), pp. 49–74.

———. 1977. The feeding behaviour of *Propithecus verreauxi*. *In* Clutton-Brock (1977), pp. 72–96.

———. 1978. Variability in the feeding behavior of a Malagasy prosimian, *Propithecus verreauxi*: Lemuriformes. *In* Montgomery (1978), pp. 519–533.

Richard, A. F., and R. W. Sussman. 1975. Future of the Malagasy lemurs: conservation or extinction? *In* Tattersall and Sussman (1975), pp. 335–350.

Richard, P. B. 1973. Capture, transport and husbandry of the Pyrenean desman *Galemys pyrenaicus*. Int. Zoo Yearbook 13:175–177.

———. 1976. Détermination de l'age et de la longévité chez le desman des Pyrénées (*Galemys pyrenaicus*). Terre Vie 30:181–192.

Richard, P. B., and A. V. Viallard. 1969. Le desman des Pyrénées (*Galemys pyrenaicus*): premières notes sur sa biologie. Terre Vie 23:225–245.

Richardson, B. J., and G. B. Sharman. 1976. Biochemical and morphological observations on the wallaroos (Macropodidae: Marsupialia) with a suggested new taxonomy. J. Zool. 179:499–513.

Richardson, E. G. 1977. The biology and evolution of the reproductive cycle of *Miniopterus schreibersii* and *M. australis* (Chiroptera: Vespertilionidae). J. Zool. 183:353–375.

Rick, A. M. 1968. Notes on bats from Tikal, Guatemala. J. Mamm. 49:516–520.

Rickart, E. A. 1977. Reproduction, growth and development in two species of cloud forest *Peromyscus* from southern Mexico. Occas. Pap. Mus. Nat. Hist. Univ. Kansas, no. 67, 22 pp.

Ride, W.D.L. 1964. A review of Australian fossil marsupials. J. Proc. Roy. Soc. W. Austral. 47:97–131.

———. 1970. A guide to the native mammals of Australia. Oxford Univ. Press, Melbourne, xiv + 249 pp.

Rideout, C. B. 1978. Mountain goat. *In* Schmidt and Gilbert (1978), pp. 149–159.

Rideout, C. B., and R. S. Hoffmann. 1975. *Oreamnos americanus*. Mammalian Species, no. 63, 6 pp.

Rieger, I. 1979. A review of the biology of striped hyaenas, *Hyaena hyaena* (Linne, 1758). Saugetierk. Mitt. 27:81–95.

———. 1981. *Hyaena hyaena*. Mammalian Species, no. 150, 5 pp.

Rigby, R. G. 1972. A study of the behaviour of caged *Antechinus stuartii*. Z. Tierpsychol. 31:15–25.

Rijksen, H. D. 1978. A field study on Sumatran orang utans (*Pongo pygmaeus abelli* Lesson 1827). Meded. Landbouwhogeschool, Wageningen, Netherlands, 78–2, 420 pp.

Riley, G. A., and R. T. McBride. 1975. A survey of the red wolf (*Canis rufus*). *In* Fox (1975), pp. 263–277.

Ritcey, R. W. 1974. Moose harvesting programs in Canada. Naturaliste Canadien 101:631–642.

Robbins, C. B. 1973. Nongeographic variation in *Taterillus gracilis* (Thomas) (Rodentia: Cricetidae). J. Mamm. 54:222–238.

———. 1974. Comments on the taxonomy of the West African *Taterillus* (Rodentia: Cricetidae) with the description of a new species. Proc. Biol. Soc. Washington 87:395–404.

———. 1977. A review of the taxonomy of the African gerbils, *Taterillus* (Rodentia: Cricetidae). *In* Sokolov, V. E., ed., Advances in modern theriology, Publishing House "Nauka," Moscow, pp. 178–194.

———. 1978. Taxonomic identification and history of *Scotophilus nigrita* (Schreber) (Chiroptera: Vespertilionidae). J. Mamm. 59:212–213.

———. 1980. Small mammals of Togo and Benin. I. Chiroptera. Mammalia 44:83–88.

Robbins, L. W.; J. R. Choate; and R. L. Robbins. 1980. Nongeographic and interspecific variation in four species of *Hylomyscus* (Rodentia: Muridae) in southern Cameroon. Ann. Carnegie Mus. Nat. Hist. 49:31–48.

Robbins, L. W., and D. A. Schlitter. 1981. Systematic status of dormice (Rodentia: Gliridae) from southern Cameroon, Africa. Ann. Carnegie Mus. 50:271–288.

Robbins, L. W., and H. W. Setzer. 1979. Additional records of *Hylomyscus baeri* Heim de Balsac and Aellen (Rodentia, Muridae) from western Africa. J. Mamm. 60:649–650.

Roben, P. 1974. Zum vorkommen des Otters, *Lutra lutra* (Linne, 1758) in der Bündesrepublik Deutschland. Saugetierk. Mitt. 22:29–36.

————. 1975. Zur ausbreitung des Waschbaren, *Procyon lotor* (Linne, 1758), und der Marderhundes, *Nyctereutes procyonides* (Gray, 1834), in der Bündesrepublik Deutschland. Saugetierk. Mitt. 23:93–101.

Roberts, A. 1951. Mammals of South Africa. Central News Agency, Johannesburg, xlviii + 700 pp.

Roberts, M. S. 1975. Growth and development of mother-reared red pandas *Ailurus fulgens*. Int. Zoo Yearbook 15:57–63.

Roberts, T. J. 1977. The mammals of Pakistan. Ernest Benn Ltd., London, xxvi + 361 pp.

Robinson, A. C. 1975. The sticknest rat, *Leporillus conditor*, on Franklin Island, Nuyts Archipelago, South Australia. Austral. Mamm. 1:319–327.

Robinson, F. J.; A. C. Robinson; C.H.S. Watts; and P. R. Baverstock. 1978. Notes on rodents and marsupials and their ectoparasites collected in Australia in 1974–75. Trans. Roy. Soc. S. Austral. 102:59–70.

Robinson, J. W., and R. S. Hoffmann. 1975. Geographical and interspecific cranial variation in big-eared ground squirrels (*Spermophilus*): a multivariate study. Syst. Zool. 24:79–88.

Robinson, W. L., and G. J. Smith. 1977. Observations on recently killed wolves in upper Michigan. Wildl. Soc. Bull. 5:25–26.

Roche, J. 1972a. Systématique du genre *Procavia* et des damans en général. Mammalia 36:22–49.

————. 1972b. Capture de *Zenkerella insignis* (Rongeurs, Anomalurides) en République Centrafricaine. Mammalia 36:305–306.

Rodgers, D. H., and W. H. Elder. 1977. Movements of elephants in Luangwa Valley, Zambia. J. Wildl. Mgmt. 41:56–62.

Rodman, P. S. 1973. Population composition and adaptive organisation among orangutans of the Kutai Reserve. *In* Michael and Crook (1973), pp. 171–209.

————. 1977. Feeding behaviour of orang-utans of the Kutai Nature Reserve, East Kalimantan. *In* Clutton-Brock (1977), pp. 384–413.

————. 1978. Diets, densities, and distributions of Bornean Primates. *In* Montgomery (1978), pp. 465–478.

Roe, N. A., and W. E. Rees. 1976. Preliminary observations of the taruca (*Hippocamelus antisensis*: Cervidae) in southern Peru. J. Mamm. 57:722–730.

Rogers, L. L.; L. D. Mech; D. K. Dawson; J. M. Peek; and M. Korb. 1980. Deer distribution in relation to wolf pack territory edges. J. Wildl. Mgmt. 44:253–258.

Rohwer, S. A., and D. L. Kilgore, Jr. 1973. Interbreeding in the arid-land foxes, *Vulpes velox* and *V. macrotis*. Syst. Zool. 22:157–165.

Roig, V. G., and O. A. Reig. 1969. Precipitin test relationships among Argentinian species of the genus *Ctenomys* (Rodentia, Octodontidae). Comp. Biochem. Physiol. 30:665–672.

Romer, A. S. 1968. Notes and comments on vertebrate paleontology. Univ. Chicago Press, viii + 304 pp.

Romer, J. D. 1974. Notes on the care and breeding of tree squirrels *Callosciurus* spp. Int. Zoo Yearbook 14:115–116.

Rongstad, O. J., and J. R. Tester. 1971. Behavior and maternal relations of young snowshoe hares. J. Wildl. Mgmt. 35:338–346.

Rood, J. P. 1970a. Notes on the behavior of the pygmy armadillo. J. Mamm. 51:179.

————. 1970b. Ecology and social behavior of the desert cavy (*Microcavia australis*). Amer. Midl. Nat. 83:415–454.

————. 1972. Ecological and behavioural comparisons of three genera of Argentine cavies. Anim. Behav. Monogr. 5:1–83.

————. 1974. Banded mongoose males guard young. Nature 248:176.

————. 1975. Population dynamics and food habits of the banded mongoose. E. Afr. Wildl. J. 13:89–111.

————. 1978. Dwarf mongoose helpers at the den. Z. Tierpsychol. 48:277–287.

————. 1980. Mating relationships and breeding suppression in the dwarf mongoose. Anim. Behav. 28:143–150.

Rood, J. P., and F. H. Test. 1968. Ecology of the spiny rat, *Heteromys anomalus*, at Rancho Grande, Venezuela. Amer. Midl. Nat. 79:89–102.

Rookmaaker, L. C. 1977. The distribution and status of the rhinoceros, *Dicerorhinus sumatrensis*, in Borneo—a review. Bijdragen Tot de Dierkunde 47:197–204.

————. 1980. The distribution of the rhinoceros in eastern India, Bangladesh, China, and the Indo-Chinese region. Zool. Anz., Jena 205:253–268.

Roonwal, M. L. 1949. Systematics, ecology and bionomics of mammals studied in connection with tsutsugamushi disease (scrub typhus) in the Assam-Burma war theatre during 1945. Trans. Natl. Inst. Sci. India 3:67–122.

Roonwal, M. L., and S. M. Mohnot. 1977. Primates of south Asia. Harvard Univ. Press, Cambridge, xviii + 421 pp.

Roots, C. G. 1966. Notes on the breeding of white-lipped peccaries *Tayassu albirostris* at Dudley Zoo. Int. Zoo Yearbook 6:198–199.

Rose, K. D. 1975. *Elpidophorus*, the earliest dermopteran (Dermoptera, Plagiomenidae). J. Mamm. 56:675–679.

Rose, R. K. 1981. *Synaptomys* not extinct in the Dismal Swamp. J. Mamm. 62:844–845.

Rose, R. K., and M. S. Gaines. 1976. Levels of aggression in fluctuating populations of the prairie vole, *Microtus ochrogaster*, in eastern Kansas. J. Mamm. 57:43–57.

————. 1978. The reproductive cycle of *Microtus ochrogaster* in eastern Kansas. Ecol. Monogr. 48:21–42.

Rose, R. W. 1978. Reproduction and evolution in female Macropodidae. Austral. Mamm. 2:65–72.

Rosenberg, H. 1971. Breeding the bat-eared fox *Otocyon megalotis* at Utica Zoo. Int. Zoo Yearbook 11:101–102.

Rosenblum, L. A. 1968. Some aspects of female reproductive physiology in the squirrel monkey. *In* Rosenblum and Cooper (1968), pp. 147–169.

Rosenblum, L. A., and R. W. Cooper, eds. 1968. The squirrel monkey. Academic Press, New York, xii + 451 pp.

Rosenthal, M. A. 1975a. Observations on the water opossum or yapok *Chironectes minimus* in captivity. Int. Zoo Yearbook 15:4–6.

————. 1975b. The management, behavior and reproduction of the short-eared elephant shrew *Macroscelides proboscideus* (Shaw). Unpubl. M.A. thesis, Northeastern Illinois Univ., 61 pp.

Rosenthal, M. A., and D. A. Meritt, Jr. 1973. Hand-rearing springhaas *Pedetes capensis* at Lincoln Park Zoo, Chicago. Int. Zoo Yearbook 13:135–137.

Rosevear, D. R. 1965. The bats of West Africa. British Mus. (Nat. Hist.), London, xvii + 418 pp.

————. 1969. The rodents of West Africa. British Mus. (Nat. Hist.), London, xii + 604 pp.

————. 1974. The carnivores of West Africa. British Mus. (Nat. Hist.), London, xii + 548 pp.

Ross, G.J.B. 1977. The taxonomy of bottlenosed dolphins *Tursiops* species in South African waters, with notes on their biology. Ann. Cape Prov. Mus. (Nat. Hist.) 11:135–194.

————. 1979. Records of pygmy and dwarf sperm whales, genus *Kogia*, from southern Africa, with some biological notes and some comparisons. Ann. Cape Prov. Mus. (Nat. Hist.) 11:259–327.

Ross, G.J.B.; P. B. Best; and B. G. Donnelly. 1975. New records of the pygmy right whale (*Caperea marginata*) from South Africa, with comments on distribution, migration, appearance, and behavior. J. Fish. Res. Bd. Can. 32:1005–1017.

Ross, W. G. 1974. Distribution, migration, and depletion of bowhead whales in Hudson Bay, 1860 to 1915. Arctic and Alpine Res., 6:85–98.

————. 1979. The annual catch of Greenland (bowhead) whales in waters north of Canada 1719–1915: a preliminary compilation. Arctic 32:91–121.

Rossolimo, O. L. 1976a. Taxonomic status of the mouse-like dormouse *Myomys* (Mammalia, Myoxidae) from Bulgaria. Zool. Zhur. 55:1515–1524.

————. 1976b. *Myomimus setzeri* (Mammalia, Myoxidae), a new species of mouse-like dormouse from Iran. Vestn. Zool., 1976(4):51–53.

Rothman, R. J., and L. D. Mech. 1979. Scent-marking in lone wolves and newly formed pairs. Anim. Behav. 27:750–760.

Rowell, T. E. 1971. Organization of caged groups of *Cercopithecus* monkeys. Anim. Behav. 19:625–645.

————. 1977a. Variation in age at puberty in monkeys. Folia Primatol. 27:284–296.

————. 1977b. Reproductive cycles of the talapoin monkey (Miopithecus talapoin). Folia Primatol. 28:188–202.

Rowe-Rowe, D. T. 1977. Prey capture and feeding behaviour of South African otters. Lammergeyer, no. 23, pp. 13–21.

————. 1978a. Reproduction and post-natal development of South African mustelines (Carnivora: Mustelidae). Zool. Afr. 13:103–114.

————. 1978b. The small carnivores of Natal. Lammergeyer, no. 25, 48 pp.

Rowlands, I. W. 1974. Mountain viscacha. Symp. Zool. Soc. London 34:131–141.

Rudnai, J. A. 1973. The social life of the lion. Washington Square East, Publishers, Wallingford, Pennsylvania, 122 pp.

Rudran, R. 1973. The reproductive cycles of two subspecies of purple-faced langurs (Presbytis senex) with relation to environmental factors. Folia Primatol. 19:41–60.

————. 1978. Socioecology of the blue monkeys (Cercopithecus mitis stuhlmani) of the Kibale Forest, Uganda. Smithson. Contrib. Zool., no. 249, iv + 88 pp.

Rumbaugh, D. M. 1970. Learning skills of anthropoids. Primate Behav. 1:1–70.

————, ed. 1972a. Gibbon and siamang. I. Evolution, ecology, behavior, and captive maintenance. S. Karger, Basel, x + 263 pp.

————. 1972b. Preface. In Rumbaugh (1972a), pp. ix–x.

Rumpler, Y. 1974. Cytogenetic contributions to a new classification of lemurs. In Martin, Doyle, and Walker (1974), pp. 865–869.

————. 1975. The significance of chromosomal studies in the systematics of the Malagasy lemurs. In Tattersall and Sussman (1975), pp. 25–40.

Rumpler, Y., and R. Albignac. 1975. Intraspecific chromosome variability in a lemur from the north of Madagascar: Lepilemur septentrionalis, species nova. Amer. J. Phys. Anthropol. 42:425–430.

————. 1977. Chromosome studies of the Lepilemur, an endemic Malagasy genus of lemurs: contribution of the cytogenetics to their taxonomy. J. Human Evol. 7:191–196.

Rumpler, Y., and B. Dutrillaux. 1978. Chromosomal evolution in Malagasy lemurs. III. Chromosome banding studies in the genus Hapalemur and the species Lemur catta. Cytogenet. Cell Genet. 21:201–211.

Rusch, D. A., and W. G. Reeder. 1978. Population ecology of Alberta red squirrels. Ecology 59:400–420.

Russell, E. M. 1974a. The biology of kangaroos (Marsupialia-Macropodidae). Mamm. Rev. 4:1–59.

————. 1974b. Recent ecological studies on Australian marsupials. Austral. Mamm. 1:189–211.

————. 1979. The size and composition of groups in the red kangaroo, Macropus rufus. Austral. Wildl. Res. 6:237–244.

Russell, J. K. 1981. Exclusion of adult male coatis from social groups: protection from predation. J. Mamm. 62:206–208.

Russell, K. R. 1978. Mountain lion. In Schmidt and Gilbert (1978), pp. 207–227.

Russell, R. H. 1975. The food habits of polar bears of James Bay and southwest Hudson Bay in summer and autumn. Arctic 28:117–129.

Russell, R. J. 1968. Evolution and classification of the pocket gophers of the subfamily Geomyinae. Univ. Kansas Publ. Mus. Nat. Hist. 16:473–579.

Ryan, R. M. 1966. A new and some imperfectly known Australian Chalinolobus and the taxonomic status of African Glauconycteris. J. Mamm. 47:86–91.

Ryberg, O. 1947. Studies on bats and bat parasites. Univ. Lund and Zool. Lab. Agric., Dairy, and Hort. Inst. Alnarp, Stockholm, 330 pp.

S.

Saayman, G. S., and C. K. Tayler. 1973. Social organisation of inshore dolphins (Tursiops aduncus and Sousa) in the Indian Ocean. J. Mamm. 54:993–996.

————. 1979. The socioecology of humpback dolphins (Sousa sp.). In Winn and Olla (1979), pp. 165–226.

Sabater Pi, J. 1972. Contribution to the ecology of Mandrillus sphinx Linnaeus 1758 of Rio Muni (Republic of Equatorial Guinea). Folia Primatol. 17:304–319.

————. 1973. Contribution to the ecology of Colobus polykomos satanas (Waterhouse, 1838) of Rio Muni, Republic of Equatorial Guinea. Folia Primatol. 19:193–207.

Sade, D. S. 1967. Determinants of dominance in a group of free-ranging rhesus monkeys. In Altmann (1967), pp. 99–114.

Sadleir, R.M.F.S. 1970. The establishment of a dominance rank order in male Peromyscus maniculatus and its stability with time. Anim. Behav. 18:55–59.

Sahu, A., and B. R. Maiti. 1978. Estrous cycle of the bandicoot rat—a rodent pest. Zool. J. Linnean Soc. 63:309–314.

Sailler, H., and U. Schmidt. 1978. Die sozialen Laute der Gemeinen vampirfledermaus Desmodus rotundus bei Konfrontation am Futterplatz unter experimentellen Bedingungen. Z. Saugetierk. 43:249–261.

Samaras, W. F. 1974. Reproductive behavior of the gray whale Eschrichtius robustus, in Baja California. Bull. S. California Acad. Sci. 73:57–64.

Samuel, W., and W. G. Macgregor, eds. 1977. Proceedings of the First International Mountain Goat Symposium. British Columbia Fish and Wildlife Branch, iii + 243 pp.

Sanborn, C. C. 1931. Bats from Polynesia, Melanesia, and Micronesia. Field Mus. Nat. Hist., Zool. Ser., 18:7–29.

————. 1950. New Philippine fruit bats. Proc. Biol. Soc. Washington 63:189–190.

————. 1952. Philippine Zoological Expedition 1946–1947. Mammals. Fieldiana Zool. 33:89–158.

————. 1953. Mammals from Mindanao, Philippine Islands collected by the Danish Philippine Expedition 1951–1952. Vidensk. Medd. fra Dansk Naturhist Foren. 115:283–288.

Sanborn, C. C., and A. J. Nicholson. 1950. Bats from New Caledonia, the Solomon Islands, and New Hebrides. Fieldiana Zool. 31:313–338.

Sandegren, F. E. 1976. Courtship display, agonistic behavior and social dynamics in the Steller sea lion (Eumetopias jubatus). Behaviour 57:159–172.

Sanderson, G. C., and A. V. Nalbandov. 1973. The reproductive cycle of the raccoon in Illinois. Illinois Nat. Hist. Surv. Bull. 31:29–85.

Sargeant, A. B., and D. W. Warner. 1972. Movements and denning habits of a badger. J. Mamm. 53:207–210.

Sarich, V. M. 1969. Pinniped phylogeny. Syst. Zool. 18:416–422.

Sauer, E.G.F. 1973. Zum Sozialverhalten der kurzohrigen Elefantenspitzmaus, Macroscelides proboscidens. Z. Saugetierk. 38:65–97.

Savic, I. R. 1973. Ecology of the species Spalax leucodon Nordm. in Yugoslavia. Zbornik za prir. nauke Matice srpske 44:66–70.

Sayer, J. A., and L. P. Van Lavieren. 1975. The ecology of the Kafue lechwe population of Zambia before the operation of hydroelectric dams on the Kafue River. E. Afr. Wildl. J. 13:9–37.

Sazima, I. 1976. Observations on the feeding habits of phyllostomatid bats (Carollia, Anoura, and Vampyrops) in southeastern Brazil. J. Mamm. 57:381–382.

————. 1978. Vertebrates as food items of the woolly false vampire, Chrotopterus auritus. J. Mamm. 59:617–618.

Sazima, I., and V. A. Taddei. 1976. A second Brazilian record of the South American flat-headed bat, Neoplatymops mattogrossensis. J. Mamm. 57:757–758.

Sazima, I., and W. Uieda. 1977. O morcego Promops nasutus no sudeste Brasileiro (Chiroptera, Molossidae). Ciencia e Cultura 29:312–314.

Sazima, I.; L. D. Vizotto; and V. A. Taddei. 1978. Uma nova espécie de Lonchophylla da serra do Cipo, Minas Gerais, Brasil (Mammalia, Chiroptera, Phyllostomatidae). Rev. Brasil. Biol. 38:81–89.

Scarlett, N. 1969. The bilby, Thylacomys lagotis in Victoria. Victorian Nat. 86:292–294.

Schaller, G. B. 1963. The mountain gorilla. Univ. Chicago Press, xvii + 431 pp.

————. 1965. The behavior of the mountain gorilla. In De Vore (1965), pp. 324–367.

————. 1967. The deer and the tiger. Univ. Chicago Press, 370 pp.

_____. 1972. The Serengeti lion. A study of predator-prey relations. Univ. Chicago Press, xiii + 480 pp.

_____. 1976. The mouse that barks. Int. Wildl. 6(5):12–16.

_____. 1977. Mountain monarchs. Wild sheep and goats of the Himalaya. Univ. Chicago Press, xviii + 425 pp.

_____. 1980a. Movement patterns of jaguar. Biotrópica 12:161–168.

_____. 1980b. Epitaph for a jaguar. Anim. Kingdom 83(2):4–11.

_____. 1981. Pandas in the wild. Natl. Geogr. 160:735–749.

Schaller, G. B., and J.M.C. Vasconcelos. 1978. A marsh deer census in Brazil. Oryx 14:345–351.

Schauenberg, P. 1969. Contribution à l'étude du tapir pinchaque Tapirus pinchaque Roulin 1829. Rev. Suisse Zool. 76:211–256.

_____. 1978. Note sur le rat de Cuming Phloeomys cumingi Waterhouse 1839 (Rodentia, Phloeomyidae). Rev. Suisse Zool. 85:341–347.

Scheffer, V. B. 1978a. Killer whale. In Haley (1978), pp. 120–127.

_____. 1978b. False killer whale. In Haley (1978), pp. 128–131.

Schempf, P. F., and M. White. 1977. Status of six furbearer populations in the mountains of northern California. U.S. Forest Serv., iv + 51 pp.

Schenkel, R., and L. Schenkel-Hullinger. 1969. Ecology and behaviour of the black rhinoceros (Diceros bicornis L.). Mammalia Depicta, Paul Parey, Hamburg, 100 pp.

Scheurmann, E. 1975. Beobachtungen zur Fortflanzung des Gayal, Bibos frontalis Lambert, 1837. Z. Saugetierk. 40:113–127.

Schevill, W. E. 1974. The whale problem. A status report. Harvard Univ. Press, Cambridge, x + 419 pp.

Schlawe, L. 1980. Zur geographischen Verbreitung der Ginsterkatzen, Gattung Genetta Cuvier, 1816. Faun. Abh. Mus. Tierk. Dresden 7:147–161.

Schliemann, H., and B. Maas. 1978. Myzopoda aurita. Mammalian Species, no. 116, 2 pp.

Schlitter, D. A. 1973. A new species of gerbil from South West Africa with remarks on Gerbillus tytonis Bauer and Niethammer, 1959 (Rodentia: Gerbillinae). Bull. S. California Acad. Sci. 72:13–18.

_____. 1974. Notes on the Liberian mongoose, Liberiictis kuhni Hayman, 1958. J. Mamm. 55:438–442.

_____. 1976. Taxonomy, zoogeography and evolutionary relationships of hairy-footed gerbils, Gerbillus (Gerbillus), of Africa and Asia (Mammalia: Rodentia). Unpubl. Ph.D. Diss., Univ. Maryland, xvi + 558 pp.

Schlitter, D. A., and S. B. McLaren. 1981. An additional record of Myonycteris relicta Bergmans, 1980, from Tanzania (Mammalia: Chiroptera). Ann. Carnegie Mus. 50:385–389.

Schlitter, D. A.; J. Phillips; and G. E. Kemp. 1973. The distribution of the white-collared mangabey, Cercocebus torquatus in Nigeria. Folia Primatol. 19:380–383.

Schlitter, D. A., and H. W. Setzer. 1972. A new species of short-tailed gerbil (Dipodillus) from Morocco. Proc. Biol. Soc. Washington 84:385–392.

Schlitter, D. A., and K. Thonglongya. 1971. Rattus turkestanicus (Satunin, 1903), the valid name for Rattus rattoides Hodgson, 1845 (Mammalia: Rodentia). Proc. Biol. Soc. Washington 84:171–174.

Schmidly, D. J. 1973. The systematic status of Peromyscus comanche. Southwestern Nat. 18:269–278.

_____. 1974a. Peromyscus attwateri. Mammalian Species, no. 48, 3 pp.

_____. 1974b. Peromyscus pectoralis. Mammalian Species, no. 49, 3 pp.

Schmidly, D. J.; M. H. Beleau; and H. Hildebran. 1972. First record of Cuvier's dolphin from the Gulf of Mexico with comments on the taxonomic status of Stenella frontalis. J. Mamm. 53:625–628.

Schmidly, D. J., and W. A. Brown. 1979. Systematics of short-tailed shrews (genus Blarina) in Texas. Southwestern Nat. 24:39–48.

Schmidly, D. J., and F. S. Hendricks. 1976. Systematics of the southern races of Ord's kangaroo rat, Dipodomys ordii. Bull. S. California Acad. Sci. 75:225–237.

Schmidt, C. R. 1973. Breeding season and notes on some aspects of reproduction in captive camelids. Int. Zoo Yearbook 13:387–390.

_____. 1975. Captive breeding of the vicuña. In Martin (1975b), pp. 271–283.

Schmidt, J. L., and D. L. Gilbert, eds. 1978. Big game of North America. Stackpole Books, Harrisburg, Pennsylvania, xv + 494 pp.

Schmidt, U. 1974. Die Tragzeit der Vampirfledermaus (Desmodus rotundus). Z. Saugetierk. 39:129–132.

Schmidt, U.; C. Schmidt; W. Lopez-Forment; and R. F. Crespo. 1978. Rückfunde beringter Vampirfledermause Desmodus rotundus in Mexiko. Z. Saugetierk. 43:65–70.

Schneider, D. G.; L. D. Mech; and J. R. Tester. 1971. Movements of female raccoons and their young as determined by radio-tracking. Anim. Behav. Monogr. 4:1–43.

Schowalter, D. B.; J. R. Gunson; and L. D. Harder. 1979. Life history characteristics of little brown bats (Myotis lucifugus) in Alberta. Can. Field-Nat. 93:243–251.

Schowalter, D. B.; L. D. Harder; and B. H. Treichel. 1978. Age composition of some vespertilionid bats as determined by dental annuli. Can. J. Zool. 56:355–358.

Schroder, G. D. 1979. Foraging behavior of the bannertail kangaroo rat (Dipodomys spectabilis). Ecology 60:657–665.

Schultz, A. H. 1942. Growth and development of the proboscis monkey. Bull. Mus. Comp. Zool. 89:279–314.

Schuster, R. 1980. Will the Kafue lechwe survive the Kafue dams? Oryx 15:476–489.

Schwartz, C. W., and E. R. Schwartz. 1959. The wild mammals of Missouri. Univ. Missouri Press, vi + 341 pp.

Schweinsburg, R. E. 1971. Home range, movements, and herd integrity of the collared peccary. J. Wildl. Mgmt. 35:455–460.

Sclater, W. L. 1900. The mammals of South Africa. R. H. Porter, London, 2 vols.

Scott, M. D., and K. Causey. 1973. Ecology of feral dogs in Alabama. J. Wildl. Mgmt. 37:253–265.

Seal, U. S., and D. G. Makey. 1974. ISIS mammalian taxonomic directory. International Species Inventory System, Minnesota Zoological Garden, St. Paul, viii + 645 pp.

Sealander, J. A. 1979. A guide to Arkansas mammals. River Road Press, Conway, Arkansas, x + 313 pp.

Seebeck, J. H., and P. G. Johnston. 1980. Potorous longipes (Marsupialia: Macropodidae); a new species from eastern Victoria. Austral. J. Zool. 28:119–134.

Seegmiller, R. F., and C. D. Simpson. 1980. The Barbary sheep: some conceptual implications of competition with desert bighorn. Desert Bighorn Council Trans., 1979, pp. 47–49.

Seidensticker, J. C., IV; M. G. Hornocker; W. V. Wiles; and J. P. Messick. 1973. Mountain lion social organization in the Idaho Primitive Area. Wildl. Monogr., no. 35, 60 pp.

Seidensticker, J.; R. K. Lahiri; K. C. Das; and A. Wright. 1976. Problem tiger in the Sundarbans. Oryx 13:267–273.

Sergeant, D. E. 1973. Biology of white whales (Delphinapterus leucas) in western Hudson Bay. J. Fish. Res. Bd. Can. 30:1065–1090.

_____. 1974. A rediscovered whelping population of hooded seal Cystophora cristata Erxleben and its possible relationship to other populations. Polarforschung 44:1–7.

_____. 1976. History and present status of populations of harp and hooded seals. Biol. Conserv. 10:96–117.

Sergeant, D. E., and P. F. Brodie. 1975. Identity, abundance, and present status of populations of white whales, Delphinapterus leucas, in North America. J. Fish. Res. Bd. Can. 32:1047–1054.

Sergeant, D. E.; D. K. Caldwell; and M. C. Caldwell. 1973. Age, growth, and maturity of bottlenosed dolphin (Tursiops truncatus) from northeast Florida. J. Fish. Res. Bd. Can. 30:1009–1011.

Sergeant, D. E.; K. Ronald; J. Boulva; and F. Berkes. 1978. The recent status of Monachus monachus, the Mediterranean monk seal. Biol. Conserv. 14:259–287.

Sergeant, D. E.; D. J. St. Aubin; and J. R. Geraci. 1980. Life history and northwest Atlantic status of the Atlantic white-sided dolphin, *Lagenorhynchus acutus*. Cetology, no. 37, 12 pp.

Setzer, H. W. 1971. New bats of the genus *Laephotis* from Africa (Mammalia: Chiroptera). Proc. Biol. Soc. Washington 84:259–264.

Shapiro, A. E. 1981. Florida panther population studies. Endangered Species Tech. Bull. 6(7):1, 3.

Sharman, G. B. 1970. Reproductive physiology of marsupials. Science 167:1221–1228.

Sharman, G. B., and P. E. Pilton. 1964. The life history and reproduction of the red kangaroo (*Megaleia rufa*). Proc. Zool. Soc. London 142:29–48.

Shaughnessy, P. D., and F. H. Fay. 1977. A review of the taxonomy and nomenclature of North Pacific harbour seals. J. Zool. 182:385–419.

Shaw, J. F. 1979. Kangaroo shooter at work! Int. Wildl. 9(3):4–11.

Shaw, J. H., and P. A. Jordan. 1977. The wolf that lost its genes. Nat. Hist. 86(10):80–88.

Shaw, T. H., and S. Wong. 1959. A new insectivore from Hainan. Acta Zool. Sinica 11:422–429.

Sheets, R. G.; R. L. Linder; and R. B. Dahlgren. 1971. Burrow systems of prairie dogs in South Dakota. J. Mamm. 52:451–453.

Sheng Helin and Lu Hogee. 1980. Current studies on the rare Chinese black muntjac. J. Nat. Hist. 14:803–807.

Sheppe, W. A. 1972. The annual cycle of small mammal populations in a Zambian floodplain. J. Mamm. 53:445–460.

———. 1973. Notes on Zambian rodents and shrews. Puku 7:167–190.

Shield, J. 1968. Reproduction of the quokka, *Setonix brachyurus*, in captivity. J. Zool. 155:427–444.

Short, H. L. 1961. Age at sexual maturity of Mexican free-tailed bats. J. Mamm. 42:533–536.

Shubin, I. G. 1972. Reproduction and numbers of steppe lemming in northern Balkhash area. Soviet J. Ecol. 3:450–452.

———. 1974. Ecology of *Lagurus luteus* in the Zaisan Hollow. Zool. Zhur. 53:272–277.

Shubin, I. G., and N. G. Suchkova. 1975. The biology of the Altai birch mouse (*Sicista napaea*). Zool. Zhur. 54:475–479.

Sikes, S. K. 1971. The natural history of the African elephant. American Elsevier, New York, xxv + 397 pp.

Silva Taboada, G., and R. H. Pine. 1969. Morphological and behavioral evidence for the relationship between the bat genus *Brachyphylla* and the Phyllonycterinae. Biotrópica 1:10–19.

Silverman, H. B., and M. J. Dunbar. 1980. Aggressive tusk use by the narwhal (*Monodon monoceros* L.). Nature 284:57–58.

Simonds, P. E. 1965. The bonnet macaque in south India. *In* De Vore (1965), pp. 175–196.

Simonetta, A. M. 1968. A new golden mole from Somalia with an appendix on the taxonomy of the family Chrysochloridae (Mammalia, Insectivora). Italian J. Zool., suppl., n.s., 2:27–55.

———. 1979. First record of *Caluromysiops* from Colombia. Mammalia 43:247–248.

Simonetti, J., and A. Spotorno. 1980. Posición taxonómica de *Phyllotis micropus* (Rodentia: Cricetidae). An. Mus. Hist. Nat. Valparaiso 13:285–297.

Simpson, C. D. 1964. Notes on the banded mongoose, *Mungos mungo* (Gmelin), Arnoldia, 1(19):1–8.

Simpson, G. G. 1945. The principles of classification and a classification of the mammals. Bull. Amer. Mus. Nat. Hist. 85:i–xvi + 1–350.

Sinclair, A.R.E. 1977a. The African buffalo. Univ. Chicago Press, xii + 355 pp.

———. 1977b. Lunar cycle and timing of mating season in Serengeti wildebeest. Nature 267:832–833.

Sinha, Y. P., and S. Chakraborty. 1971. Taxonomic status of the vespertilionid bat, *Nycticejus emarginatus* Dobson. Proc. Zool. Soc. Calcutta 24:53–59.

Siniff, D. B., and J. L. Bengtson. 1977. Observations and hypotheses concerning the interactions among crabeater seals, leopard seals, and killer whales. J. Mamm. 58:414–416.

Siniff, D. B.; I. Stirling; J. L. Bengtson; and R. A. Reichle. 1979. Social and reproductive behavior of crabeater seals (*Lobodon carcinophagus*) during the austral spring. Can. J. Zool. 57:2243–2255.

Skinner, J. D. 1976. Ecology of the brown hyaena *Hyaena brunnea* in the Transvaal with a distribution map for southern Africa. S. Afr. J. Sci. 72:262–269.

Skinner, J. D.; G. J. Breytenbach; and C.T.A. Maberly. 1976. Observations on the ecology and biology of the bushpig *Potamochoerus porcus* Linn. in the northern Transvaal. S. Afr. J. Wildl. Res. 6:123–128.

Skinner, J. D.; S. Davis; and G. Ilani. 1980. Bone collecting by striped hyaenas, *Hyaena hyaena*, in Israel. Paleontol. Afr. 23:99–104.

Skinner, J. D.; N. Fairall; and J.D.P. Bothma. 1977. South African red data book—large mammals. S. Afr. Natl. Sci. Programmes Rept., no. 18, v + 29 pp.

Skinner, J. D., and A. J. Hall-Martin. 1975. A note on foetal growth and development of the giraffe *Giraffa camelopardalis giraffa*. J. Zool. 177:73–79.

Skinner, J. D., and G. Ilani. 1979. The striped hyaena *Hyaena hyaena* of the Judean and Negev Deserts and a comparison with the brown hyaena *H. brunnea*. Israel J. Zool. 28:229–232.

Skinner, J. D.; J.H.M. Van Zyl; and J.A.H. Van Heerden. 1973. The effect of season on

reproduction in the black wildebeest and red hartebeest in South Africa. J. Reprod. Fert., suppl., 19:101–110.

Skirrow, M. H., and M. Rysan. 1976. Observations on the social behaviour of the Chinese hamster, *Cricetulus griseus*. Can. J. Zool. 54:361–368.

Skoog, P. 1970. The food of the Swedish badger, *Meles meles* L. Viltrevy 7:1–120.

Slade, N. A., and D. F. Balph. 1974. Population ecology of Uinta ground squirrels. Ecology 55:989–1003.

Slobodyan, A. A. 1976. The European brown bear in the Carpathians. *In* Pelton, Lentfer, and Folk (1976), pp. 313–319.

Sly, G. R. 1975. Second record of the bronzed tube-nosed bat (*Murina aenea*) in peninsular Malaysia. Malayan Nat. J. 28:217.

Small, G. L. 1971. The blue whale. Columbia Univ. Press, New York, xiii + 248 pp.

Smit, C. J., and A. Van Wijngaarden. 1976. Threatened mammals of Europe. European Committee for the Conservation of Nature and Natural Resources, 189 pp.

Smith, A. T. In Press *a*. Territoriality and social behavior of *Ochotona princeps*. Proc. World Lagomorph Conf., Guelph, Ontario.

———. In Press *b*. Population dynamics of pikas (genus *Ochotona*). Proc. World Lagomorph Conf., Guelph, Ontario.

Smith, C. C. 1977. Feeding behaviour and social organization in howling monkeys. *In* Clutton-Brock (1977), pp. 96–126.

———. 1978. Structure and function of the vocalizations of tree squirrels (*Tamiasciurus*). J. Mamm. 59:793–808.

Smith, D. A., and L. C. Smith. 1975. Oestrus, copulation, and related aspects of reproduction in female eastern chipmunks, *Tamias striatus* (Rodentia: Sciuridae). Can. J. Zool. 53:756–767.

Smith, G. W. 1977. Population characteristics of the porcupine in northeastern Oregon. J. Mamm. 58:674–676.

Smith, H. D., and C. D. Jorgensen. 1975. Reproductive biology of North American desert rodents. *In* Prakash and Ghosh (1975), pp. 305–330.

Smith, J. D. 1972. Systematics of the chiropteran family Mormoopidae. Univ. Kansas Mus. Nat. Hist. Misc. Publ., no. 56, 132 pp.

———. 1977. On the nomenclatorial status of *Chilonycteris gymnonotus* Natterer, 1843. J. Mamm. 58:245–246.

Smith, J. D., and H. H. Genoways. 1974. Bats of Margarita Island, Venezuela, with zoogeographic comments. Bull. S. California Acad. Sci. 73:64–79.

Smith, J. D., and J. E. Hill. 1981. A new species and subspecies of bat of the *Hipposideros bicolor*-group from Papua New Guinea, and the systematic status of *Hipposideros calcaratus* and *Hipposideros cupidus* (Mammalia: Chiroptera: Hipposideridae). Los Angeles Co. Mus. Nat. Hist. Contrib. Sci., no. 331, 19 pp.

Smith, J. R.; C.H.S. Watts; and E. G. Crichton. 1972. Reproduction in the Australian desert rodents *Notomys alexis* and *Pseudomys australis* (Muridae). Austral. Mamm. 1:1–7.

Smith, L. C., and D. A. Smith. 1972. Reproductive biology, breeding seasons, and growth of eastern chipmunks, *Tamias striatus* (Rodentia: Sciuridae) in Canada. Can. J. Zool. 50:1069–1085.

Smith, M. J. 1971. Breeding the sugar-glider *Petaurus breviceps* in captivity; and growth of pouch young. Int. Zoo Yearbook 11:26–28.

————. 1973. *Petaurus breviceps*. Mammalian Species, no. 30, 5 pp.

Smith, M. J.; B. K. Brown; and H. J. Frith. 1969. Breeding of the brush-tailed possum, *Trichosurus vulpecula* (Kerr), in New South Wales. CSIRO Wildl. Res. 14:181–193.

Smith, M. J., and R. A. How. 1973. Reproduction in the mountain possum, *Trichosurus caninus* (Ogilby), in captivity. Austral. J. Zool. 21:321–329.

Smith, N. 1974. Agouti and babassu. Oryx 12:581–582.

Smith, N. S., and I. O. Buss. 1973. Reproductive ecology of the female African elephant. J. Wildl. Mgmt. 37:524–534.

Smith, P. A., and C. J. Jonkel. 1975. Résumé of the trade in polar bear hides in Canada, 1973–74. Can. Wildl. Serv. Progress Notes, no. 48, 5 pp.

Smith, R.F.C. 1969. Studies on the marsupial glider, *Schoinobates volans* (Kerr). Austral. J. Zool. 17:625–636.

Smith, R. M. 1977. Movement patterns and feeding behaviour of leopard in the Rhodes Matopos National Park, Rhodesia. Arnoldia 8(13):1–16.

Smith, T. G. 1973. Population dynamics of the ringed seal in the Canadian eastern Arctic. Fish. Res. Bd. Can. Bull., no. 181, viii + 55 pp.

————. 1976. Predation of ringed seal pups (*Phoca hispida*) by the arctic fox (*Alopex lagopus*). Can. J. Zool. 54:1610–1616.

Smith, T. G., and I. Stirling. 1975. The breeding habitat of the ringed seal (*Phoca hispida*). The birth lair and associated structures. Can. J. Zool. 53:1297–1305.

Smith, W. J.; S. L. Smith; E. C. Oppenheimer; and J. G. Devilla. 1977. Vocalizations of the black-tailed prairie dog, *Cynomys ludovicianus*. Anim. Behav. 25:152–164.

Smithers, R.H.N. 1971. The mammals of Botswana. Trustees Natl. Mus. Rhodesia Mus. Mem., no. 4, 340 pp.

————. 1978. The serval *Felis serval* Schreber, 1776. S. Afr. J. Wildl. Res. 8:29–37.

Smolen, M. J.; H. H. Genoways; and R. J. Baker. 1980. Demographic and reproductive parameters of the yellow-cheeked pocket gopher (*Pappogeomys castanops*). J. Mamm. 61:224–236.

Smythe, N. 1970. The adaptive value of the social organization of the coati (*Nasua narica*). J. Mamm. 51:818–820.

————. 1978. The natural history of the Central American agouti (*Dasyprocta punctata*). Smithson. Contrib. Zool., no. 257, iii + 52 pp.

Snyder, P. A. 1974. Behavior of *Leontopithecus rosalia* (golden-lion marmoset) and related species: a review. J. Human Evol. 3:109–122.

Snyder, R. L., and S. C. Moore. 1968. Longevity of captive mammals in Philadelphia Zoo. Int. Zoo Yearbook 8:175–183.

Sokolov, V. E. 1974. *Saiga tatarica*. Mammalian Species, no. 38, 4 pp.

————. 1981. A new species of five-toed jerboa *Allactaga nataliae* sp. n. (Rodentia, Dipodidae) from Mongolia. Zool. Zhur. 60:793–795.

Sokolov, V. E.; M. I. Baskevich; and Yu. M. Kovalskaya. 1981. Revision of birch mice of the Caucasus: sibling species *Sicista caucasica* Vinogradov, 1925 and *S. kluchorica* sp. n. (Rodentia, Dipodidae). Zool. Zhur. 60:1386–1393.

Soldatovic, B., and I. Savic. 1974. The karyotype forms of the genus *Spalax* Guld. in Yugoslavia and their areas of distribution. *In* Kratochvil, J., and R. Obrtel, eds., Proceedings of the International Symposium on Species and Zoogeography of European Mammals, Academia Publishing House, Prague, pp. 125–130.

Sorenson, M. W. 1970. Behavior of tree shrews. Primate Behav. 1:141–194.

————. 1974. A review of aggressive behavior in the tree shrews. *In* Holloway (1974), pp. 13–30.

Sorenson, M. W., and C. H. Conway. 1968. The social and reproductive behavior of *Tupaia montana* in captivity. J. Mamm. 49:502–512.

Sosnovskii, I. P. 1967. Breeding the red dog or dhole *Cuon alpinus* at Moscow Zoo. Int. Zoo Yearbook 7:120–122.

Southwick, C. H.; M. A. Beg; and M. R. Siddiqi. 1965. Rhesus monkeys in north India. *In* De Vore (1965), pp. 111–159.

Southwick, C. H., and M. F. Siddiqi. 1977. Population dynamics of rhesus monkeys in northern India. *In* Rainier III and Bourne (1977), pp. 339–362.

Southwick, C. H.; M. F. Siddiqi; M. Y. Farooqui; and B. C. Pal. 1974. Xenophobia among free-ranging rhesus groups in India. *In* Holloway (1974), pp. 185–209.

Soutiere, E. C. 1979. Effects of timber harvesting on marten in Maine. J. Wildl. Mgmt. 43:850–860.

Sowls, L. K. 1974. Social behaviour of the collared peccary *Dicotyles tajacu* (L.). *In* Geist and Walther (1974), pp. 144–165.

————. 1978. Collared peccary. *In* Schmidt and Gilbert (1978), pp. 191–205.

Spiess, A. 1976. Labrador grizzly (*Ursus arctos* L.): first skeletal evidence. J. Mamm. 57:787–790.

Spinage, C. A. 1974. Territoriality and population regulation in the Uganda Defassa waterbuck. *In* Geist and Walther (1974), pp. 635–643.

Spitzenberger, F. 1978. Die Stachelmaus von Kleinasien, *Acomys cilicicus* n. sp. Ann. Naturhist. Mus. Wien 81:443–446.

Spitzer, N. C., and J. D. Lazell, Jr. 1978. A new rice rat (genus *Oryzomys*) from Florida's lower Keys. J. Mamm. 59:787–792.

Spotte, S.; C. W. Radcliffe; and J. L. Dunn. 1979. Notes on Commerson's dolphin (*Cephalorhynchus commersonii*) in captivity. Cetology, no. 35, 9 pp.

Sreenivasan, M. A.; H. R. Bhat; and G. Geevarghese. 1973. Breeding cycle of *Rhinolophus rouxi* Temminck, 1835 (Chiroptera: Rhinolophidae), in India. J. Mamm. 54:1013–1017.

————. 1974. Observations on the reproductive cycle of *Cynopterus sphinx sphinx* Vahl, 1797 (Chiroptera: Pteropidae). J. Mamm. 55:200–202.

Stains, H. J. 1967. Carnivores and pinnipeds. *In* Anderson and Jones (1967), pp. 325–354.

Stanley, W. C. 1963. Habits of the red fox in northeastern Kansas. Univ. Kansas Mus. Nat. Hist. Misc. Publ., no. 34, 31 pp.

Starrett, A. 1967. Hystricoid, erethizontoid, cavioid, and chinchilloid rodents. *In* Anderson and Jones (1967), pp. 254–272.

————. 1972. *Cyttarops alecto*. Mammalian Species, no. 13, 2 pp.

Starrett, A., and R. S. Casebeer. 1968. Records of bats from Costa Rica. Los Angeles Co. Mus. Contrib. Sci., no. 148, 21 pp.

Starrett, A., and G. F. Fisler. 1970. Aquatic adaptations of the water mouse, *Rheomys underwoodi*. Los Angeles Co. Mus. Contrib. Sci., no. 182, 14 pp.

Start, A. N. 1972a. Notes on *Dyacopterus spadiceus* from Sarawak. Sarawak Mus. J. 20:367–369.

————. 1972b. Some bats of Bako National Park, Sarawak. Sarawak Mus. J. 20:371–376.

————. 1975. Another specimen of *Dyacopterus spadiceus* from Sarawak. Sarawak Mus. J. 23:267.

Statistics Canada. 1981. Fur production. Can. Min. Supply and Services, Ottawa.

Steadman, D. W., and C. E. Ray. In Press. The relationships of *Megaoryzomys curioi*, an extinct cricetine rodent (*Muroidea, Muridae*) from the Galapagos Islands, Ecuador. Smithson. Contrib. Paleobiol.

Stejneger, L. 1936. Georg Wilhelm Steller. Harvard Univ. Press, Cambridge, 623 pp.

Sterner, R. T., and S. A. Shumake. 1978. Coyote damage-control research: a review and analysis. *In* Bekoff (1978), pp. 297–325.

Stevenson, M. F. 1973a. Notes on pregnancy in the sooty mangabey *Cercocebus atys*. Int. Zoo Yearbook 13:134–135.

————. 1973b. Observations of maternal behaviour and infant development in the De Brazza monkey *Cercopithecus neglectus* in captivity. Int. Zoo Yearbook 13:179–184.

Stickel, L. F. 1968. Home range and travels. *In* King (1968), pp. 373–411.

Stiemie, S., and J.A.J. Nel. 1973. Nest-building behaviour in *Aethomys chrysophilus, Praomys (Mastomys) natalensis* and *Rhabdomys pumilio.* Zool. Afr. 8:91–100.

Stinson, N., Jr. 1977. Home range of the western jumping mouse, *Zapus princeps,* in the Colorado Rocky Mountains. Great Basin Nat. 37:87–90.

Stirling, I. 1971*a.* Studies on the behaviour of the South Australian fur seal, *Arctocephalus forsteri* (Lesson). I. Annual cycle, postures and calls, and adult males during the breeding season. Austral. J. Zool. 19:243–266.

———. 1971*b. Leptonychotes weddelli.* Mammalian Species, no. 6, 5 pp.

———. 1972*a.* Observations on the Australian sea lion, *Neophoca cinerea* (Peron). Austral. J. Zool. 20:271–279.

———. 1972*b.* Regulation of numbers of an apparently isolated population of Weddell seals (*Leptonychotes weddelli*). J. Mamm. 53:107–115.

———. 1972*c.* The economic value and management of seals in South Australia. Austral. Dept. Fish. Publ., no. 2, 11 pp.

———. 1973. Vocalization in the ringed seal (*Phoca hispida*). J. Fish. Res. Bd. Can. 30:1592–1594.

———. 1974. Midsummer observations on the behavior of wild polar bears (*Ursus maritimus*). Can. J. Zool. 52:1191–1198.

Stirling, I.; W. Calvert; and D. Andriashek. 1980. Population ecology studies of the polar bear in the area of southeastern Baffin Island. Can. Wildl. Serv. Occas. Pap., no. 44, 33 pp.

Stirling, I.; C. Jonkel; P. Smith; R. Robertson; and D. Cross. 1977. The ecology of the polar bear (*Ursus maritimus*) along the western coast of Hudson Bay. Can. Wildl. Serv. Occas. Pap., no. 33, 64 pp.

Stirling, I., and H.P.L. Kiliaan. 1980. Population ecology studies of the polar bear in northern Labrador. Can. Wildl. Serv. Occas. Pap., no. 42, 21 pp.

Stirling, I., and G. L. Kooyman. 1971. The crabeater seal (*Lobodon carcinophagus*) in McMurdo Sound, Antarctica, and the origin of mummified seals. J. Mamm. 52:175–180.

Stirling, I., and R. M. Warneke. 1971. Implications of a comparison of the airborne vocalizations and some aspects of the behaviour of the two Australian fur seals, *Arctocephalus* spp., on the evolution and present taxonomy of the genus. Austral. J. Zool. 19:227–241.

Stock, A. D. 1972. Swimming ability in kangaroo rats. Southwestern Nat. 17:98–99.

Stodart, E. 1977. Breeding and behaviour of Australian bandicoots. *In* Stonehouse and Gilmore (1977), pp. 179–191.

Stonehouse, B., and D. Gilmore, eds. 1977. The biology of marsupials. University Park Press, Baltimore, viii + 486 pp.

Stones, R. C., and C. L. Hayward. 1968. Natural history of the desert woodrat, *Neotoma lepida.* Amer. Midl. Nat. 80:458–476.

Storer, T. I., and L. P. Tevis, Jr. 1955. California grizzly. Univ. California Press, Berkeley, vii + 335 pp.

Storm, G. L.; R. D. Andrews; R. L. Phillips; R. A. Bishop; D. B. Siniff; and J. R. Tester. 1976. Morphology, reproduction, dispersal, and mortality of midwestern red fox populations. Wildl. Monogr., no. 49, 81 pp.

Storm, G. L., and G. G. Montgomery. 1975. Dispersal and social contact among red foxes: results from telemetry and computer simulation. *In* Fox (1975), pp. 237–246.

Storro-Patterson, R. 1977. Gray whale protection. How well is it working? Oceans 10(4):44–49.

Strahan, R. 1975. Status and husbandry of Australian monotremes and marsupials. *In* Martin (1975*b*), pp. 171–182.

Strassburger, S. 1978. The oryx express. Zoonooz 51(6):4–8.

Stringham, S. F. 1974. Mother-infant relations in moose. Naturaliste Canadien 101:325–369.

Stroganov, S. U. 1969. Carnivorous mammals of Siberia. Israel Progr. Sci. Transl., Jerusalem, x + 522 pp.

Stroman, H. R., and L. M. Slaughter. 1972. The care and breeding of the pygmy hippopotamus *Choeropsis liberiensis* in captivity. Int. Zoo Yearbook 12:126–131.

Struhsaker, T. T. 1967*a.* Ecology of vervet monkeys (*Cercopithecus aethiops*) in the Masai-Amboseli Game Reserve, Kenya. Ecology 48:891–904.

———. 1967*b.* Auditory communication among vervet monkeys (*Cercopithecus aethiops*). *In* Altmann (1967), pp. 281–324.

———. 1970. Phylogenetic implications of some vocalizations of *Cercopithecus* monkeys. *In* Napier and Napier (1970), pp. 365–444.

———. 1971. Notes on *Cercocebus a. atys* in Senegal, West Africa. Mammalia 35:343–344.

———. 1975. The red colobus monkey. Univ. Chicago Press, xiv + 311 pp.

Struhsaker, T. T., and J. S. Gartlan. 1970. Observations on the behaviour and ecology of the patas monkey (*Erythrocebus patas*) in the Waza Reserve, Cameroon. J. Zool. 161:49–63.

Stuart, C. T. 1980. The distribution and status of *Manis temmincki* Smuts, 1832 (Pholidota: Manidae). Saugetierk. Mitt. 28:123–129.

Subbaraj, R., and M. K. Chandrashekaran. 1977. 'Rigid' internal timing in the circadian rhythm of flight activity in a tropical bat. Oecologia 29:341–348.

Sugiyama, Y. 1973. The social structure of wild chimpanzees. A review of field studies. *In* Michael and Crook (1973), pp. 375–410.

Sung, C. V. 1976. New data on morphology and biology of some rare small mammals from North Vietnam. Zool. Zhur. 55:1880–1885.

Sunquist, M. E. 1974. Winter activity of striped skunks (*Mephitis mephitis*) in east-central Minnesota. Amer. Midl. Nat. 92:434–446.

———. 1981. The social organization of tigers (*Panthera tigris*) in Royal Chitawan National Park, Nepal. Smithson. Contrib. Zool., no. 336, vi + 98 pp.

Sunquist, M. E., and G. G. Montgomery. 1973. Activity patterns and rates of movement of two-toed and three-toed sloths (*Choloepus hoffmanni* and *Bradypus infuscatus*). J. Mamm. 54:946–954.

Sussman, R. W. 1975. A preliminary study of the behavior and ecology of *Lemur fulvus rufus* Audebert 1800. *In* Tattersall and Sussman (1975), pp. 237–258.

———. 1977. Feeding behaviour of *Lemur catta* and *Lemur fulvus. In* Clutton-Brock (1977), pp. 1–36.

Sussman, R. W., and A. Richard. 1974. The role of aggression among diurnal prosimians. *In* Holloway (1974), pp. 49–76.

Suthers, R. A., and J. M. Fattu. 1973. Fishing behaviour and acoustic orientation by the bat (*Noctilio labialis*). Anim. Behav. 21:61–66.

Suzuki, A. 1971. Carnivority and cannibalism observed among forest-living chimpanzees. J. Anthropol. Soc. Nippon 79:30–48.

Svendsen, G. E. 1970. Notes on the ecology of the harvest mouse, *Reithrodontomys megalotis,* in southwestern Wisconsin. Trans. Wisconsin Acad. Sci., Arts, and Letters 58:163–166.

———. 1974. Behavioral and environmental factors in the spatial distribution and population dynamics of a yellow-bellied marmot population. Ecology 55:760–771.

———. 1976. Vocalizations of the long-tailed weasel (*Mustela frenata*). J. Mamm. 57:398–399.

———. 1978. Castor and anal glands of the beaver (*Castor canadensis*). J. Mamm. 59:618–620.

———. 1979. Territoriality and behavior in a population of pikas (*Ochotona princeps*). J. Mamm. 60:324–330.

Swanepoel, P. 1975. Small mammals of the Addo Elephant National Park. Koedoe 18:103–130.

Swanepoel, P., and H. H. Genoways. 1978. Revision of the Antillean bats of the genus *Brachyphylla* (Mammalia: Phyllostomatidae). Bull. Carnegie Mus. Nat. Hist., no. 12, 53 pp.

Swanepoel, P., and D. A. Schlitter. 1978. Taxonomic review of the fat mice (genus *Steatomys*) of West Africa (Mammalia: Rodentia). Bull. Carnegie Mus. Nat. Hist., no. 6, pp. 53–76.

Swanson, H. H. 1974. Sex differences in behaviour of the Mongolian gerbil (*Meriones unguiculatus*) in encounters between pairs of same or opposite sex. Anim. Behav. 22:638–644.

Syroechkovskiy, E. E., and E. V. Rogacheva. 1974. Moose of the Asiatic part of the U.S.S.R. Naturaliste Canadien 101:595–604.

T.

Taber, R. D.; A. N. Sheri; and M. S. Ahmad. 1967. Mammals of the Lyallpur region, West Pakistan. J. Mamm. 48:392–407.

Taddei, V. A. 1976. The reproduction of some Phyllostomatidae (Chiroptera) from the northwestern region of the state of São Paulo. Bol. Zool. Univ. São Paulo 1:313–330.

————. 1979. Phyllostomidae (Chiroptera) do norte-ocidental do estado de São Paulo. III—Stenodermatinae. Ciencia e Cultura 31:900–914.

Taddei, V. A.; L. D. Vizotto; and S. M. Martins. 1976. Notas taxionómicas e biológicas sobre *Molossops brachymeles cerastes* (Thomas, 1901) (Chiroptera—Molossidae). Naturalia 2:61–69.

Tamarin, R. H., and S. R. Malecha. 1972. Reproductive parameters in *Rattus rattus* and *R. exulans* of Hawaii, 1968 to 1970. J. Mamm. 53:513–528.

Taruski, A. G. 1979. The whistle repertoire of the North Atlantic pilot whale (*Globicephala melaena*) and its relationship to behavior and environment. *In* Winn and Olla (1979), pp. 345–368.

Tate, C. M.; J. F. Pagels; and C. O. Handley, Jr. 1980. Distribution and systematic relationship of two kinds of short-tailed shrews (Soricidae: *Blarina*) in south-central Virginia. Proc. Biol. Soc. Washington 93:50–60.

Tate, G.H.H. 1931. Random observations on habits of South American mammals. J. Mamm. 12:248–256.

————. 1934. Bats from the Pacific Islands, including a new fruit bat from Guam. Amer. Mus. Novit., no. 713, 3 pp.

————. 1942. Results of the Archbold Expeditions. No. 47. Review of the vespertilionine bats, with special attention to genera and species of the Archbold Collections. Bull. Amer. Mus. Nat. Hist. 80:221–297.

————. 1951a. *Harpyionycteris*, a genus of rare fruit bats. Amer. Mus. Novit., no. 1522, 9 pp.

————. 1951b. Results of the Archbold Expeditions. No. 65. The rodents of Australia and New Guinea. Bull. Amer. Mus. Nat. Hist. 97:183–430.

Tate, G.H.H., and R. Archbold. 1939. A revision of the genus *Emballonura* (Chiroptera). Amer. Mus. Novit., no. 1035, 14 pp.

Tattersall, I. 1971. Revision of the subfossil Indriinae. Folia Primatol. 16:257–269.

————. 1977a. Ecology and behavior of *Lemur fulvus mayottensis* (Primates, Lemuriformes). Amer. Mus. Nat. Hist. Anthropol. Pap. 54:421–482.

————. 1977b. The lemurs of the Comoro Islands. Oryx 13:445–448.

————. 1978a. Behavioural variation in *Lemur mongoz* (=*L. m. mongoz*). *In* Chivers and Joysey (1978), pp. 127–132.

————. 1978b. Functional cranial anatomy of the subfossil Malagasy lemurs. Natl. Geogr. Soc. Res. Rept., 1969 Proj., pp. 559–568.

Tattersall, I., and J. H. Schwartz. 1975. Relationships among the Malagasy lemurs. The craniodental evidence. *In* Luckett and Szalay (1975), pp. 299–312.

Tattersall, I., and R. W. Sussman. 1975. Lemur biology. Plenum Press, New York, xiii + 365 pp.

Taub, D. M. 1977. Geographic distribution and habitat diversity of the Barbary macaque *Macaca sylvanus* L. Folia Primatol. 27:108–133.

Taylor, E. H. 1934. Philippine land mammals. Philippine Bur. Sci. Monogr., no. 30, 548 pp.

Taylor, J. M., and B. E. Horner. 1970a. Gonodal activity in the marsupial mouse, *Antechinus bellus*, with notes on other species of the genus (Marsupialia: Dasyuridae). J. Mamm. 51:659–668.

————. 1970b. Reproduction in the mosaic-tailed rat, *Melomys cervinipes* (Rodentia: Muridae). Austral. J. Zool. 18:171–184.

————. 1971. Reproduction in the Australian tree-rat *Conilurus penicillatus* (Rodentia: Muridae). CSIRO Wildl. Res. 16:1–9.

————. 1972. Observations on the reproductive biology of *Pseudomys* (Rodentia: Muridae). J. Mamm. 53:318–328.

————. 1973a. Results of the Archbold Expeditions. No. 98. Systematics of native Australian *Rattus* (Rodentia, Muridae). Bull. Amer. Mus. Nat. Hist. 150:1–130.

————. 1973b. Reproductive characteristics of wild native Australian *Rattus* (Rodentia: Muridae). Austral. J. Zool. 21:437–475.

Taylor, M. E. 1972. *Ichneumia albicauda*. Mammalian Species, no. 12, 4 pp.

————. 1975. *Herpestes sanguineus*. Mammalian Species, no. 65, 5 pp.

Tedford, R. H. 1976. Relationships of pinnipeds to other carnivores (Mammalia). Syst. Zool. 25:363–374.

Temme, M. 1974. Neue Belege der Philippinischen Streifenratte *Chrotomys whiteheadi* Thomas, 1895. Z. Saugetierk. 39:342–345.

Tenaza, R. R. 1975. Pangolins rolling away from predation risks. J. Mamm. 56:257.

Tener, J. S. 1965. Muskoxen in Canada. Can. Wildl. Serv., Ottawa, 166 pp.

Tesh, R. B. 1970a. Observations on the natural history of *Diplomys darlingi*. J. Mamm. 51:197–199.

————. 1970b. Notes on the reproduction, growth, and development of echimyid rodents in Panama. J. Mamm. 51:199–202.

Thaeler, C. S., Jr. 1968. An analysis of three hybrid populations of pocket gophers (genus *Thomomys*). Evolution 22:543–555.

————. 1972. Taxonomic status of the pocket gophers, *Thomomys idahoensis* and *Thomomys pygmaeus* (Rodentia, Geomyidae). J. Mamm. 53:417–428.

————. 1980. Chromosome numbers and systematic relations in the genus *Thomomys* (Rodentia: Geomyidae). J. Mamm. 61:414–422.

Thaeler, C. S., Jr., and L. L. Hinesley. 1979. *Thomomys clusius*, a rediscovered species of pocket gopher. J. Mamm. 60:480–488.

Thein, U T. 1977. The Burmese freshwater dolphin. Mammalia 41:233–234.

Thenius, E. 1979. Zur systematischen und phylogenetischen Stellung des Bambusbären: *Ailuropoda melanoleuca* David (Carnivora, Mammalia). Z. Saugetierk. 44:286–305.

Thomas, K. R. 1974. Burrow systems of the eastern chipmunk (*Tamias striatus pipilans* Lowery) in Louisiana. J. Mamm. 55:454–459.

Thomas, M. E., and D. N. McMurray. 1974. Observations on *Sturnira aratathomasi* from Colombia. J. Mamm. 55:834–836.

Thomas, O. 1906. On a second species of *Lenothrix* from the Liu Kiu Islands. Ann. Mag. Nat. Hist., ser. 7, 17:88–89.

————. 1910a. Further new African mammals. Ann. Mag. Nat. Hist., ser. 8, 5:191–202.

————. 1910b. A new genus of fruit-bats and two new shrews from Africa. Ann. Mag. Nat. Hist., ser. 8, 6:111–114.

————. 1916. On Muridae from Darjiling and the Chin Hills. J. Bombay Nat. Hist. Soc. 24:404–415.

————. 1921. A new genus of opossum from southern Patagonia. Ann. Mag. Nat. Hist., ser. 9, 8:136–139.

Thomas, W. D. 1975. Observations on captive brockets *Mazama americana* and *M. gouazoubira*. Int. Zoo Yearbook 15:77–78.

Thompson, D. C. 1977. Diurnal and seasonal activity of the grey squirrel (*Sciurus carolinensis*). Can. J. Zool. 55:1185–1189.

Thompson, T. J.; H. E. Winn; and P. J. Perkins. 1979. Mysticete sounds. *In* Winn and Olla (1979), pp. 403–431.

Thonglongya, K. 1973. First record of *Rhinolophus paradoxolophus* (Bourret, 1951) from Thailand, with the description of a new species of the *Rhinolophus philippinensis* group (Chiroptera, Rhinolophidae). Mammalia 37:587–597.

Thorington, R. W., Jr. 1968. Observations of squirrel monkeys in a Colombian forest. *In* Rosenblum and Cooper (1968), pp. 69–85.

————. 1978. Some problems relevant to the conservation of the Callitrichidae. *In* Goldsmith, E. I., and J. Moor-Jankowski, eds., Primates in medicine, S. Karger, Basel, Switzerland, pp. 1–11.

Thorington, R. W., Jr., and C. P. Groves. 1970. An annotated classification of the Cercopithecoidea. *In* Napier and Napier (1970), pp. 629–647.

Thorington, R. W., Jr., and P. G. Heltne, eds. 1976. Neotropical primates. Field studies and conservation. Natl. Acad. Sci., Washington, D.C., v + 135 pp.

Thorington, R. W., Jr.; N. A. Muckenhirn; and G. G. Montgomery. 1976. Movements of a wild night monkey (*Aotus trivirgatus*).

In Thorington and Heltne (1976), pp. 32–34.

Thorington, R. W., Jr., and R. E. Vorek. 1976. Observations on the geographic variation and skeletal development of *Aotus*. Lab. Anim. Sci. 26:1006–1021.

Thornton, W. A., and G. C. Creel. 1975. The taxonomic status of kit foxes. Texas J. Sci. 26:127–136.

Tien, D. V. 1972. Données écologiques sur l'écureuil géant de McClelland (*Ratufa bicolor gigantea*) (Rodentia, Sciuridae) au Vietnam. Zool. Garten 41:240–243.

Tien, D. V., and C. V. Sung. 1971. Données écologiques sur les rats de bambou (*Rhizomys pruinosus* Blyth et *Rhizomys sumatrensis cinereus* McClelland) au Vietnam. Zool. Garten 40:227–231.

Tilson, R. L. 1977. Social organization of simakobu monkeys (*Nasalis concolor*) in Siberut Island, Indonesia. J. Mamm. 58:202–212.

————. 1980. Klipspringer (*Oreotragus oreotragus*) social structure and predator avoidance in a desert canyon. Madoqua 11:303–314.

Timm, R. M.; L. R. Heaney; and D. D. Baird. 1977. Natural history of rock voles (*Microtus chrotorrhinus*) in Minnesota. Can. Field-Nat. 91:177–181.

Timm, R. M., and J. Mortimer. 1976. Selection of roost sites by Honduran white bats, *Ectophylla alba* (Chiroptera: Phyllostomatidae). Ecology 57:385–389.

Timmis, W. H. 1971. Observations on breeding the oriental short-clawed otter *Amblonyx cinerea* at Chester Zoo. Int. Zoo Yearbook 11:109–111.

Tiwari, K. K.; R. K. Ghose; and S. Chakraborty. 1971. Notes on a collection of small mammals from Western Ghats, with remarks on the status of *Rattus rufescens* (Gray) and *Bandicota indica malabarica* (Shaw). J. Bombay Nat. Hist. Soc. 68:378–384.

Tobias, P. V. 1978. The place of *Australopithecus africanus* in hominid evolution. *In* Chivers and Joysey (1978), pp. 373–394.

Topachevskii, V. A. 1976. Fauna of the USSR: Mammals. Mole rats, Spalacidae. Amerind Publishing Co., New Delhi, x + 308 pp.

Topal, G. 1970a. The first record of *Ia io* Thomas, 1902 in Vietnam and India, and some remarks on the taxonomic position of *Parascotomanes beaulieui* Bourret, 1942, *Ia longimana* Pen, 1962, and the genus *Ia* Thomas, 1902 (Chiroptera: Vespertilionidae). Opusc. Zool. Budapest 10:341–347.

————. 1970b. On the systematic status of *Pipistrellus annectans* Dobson,1871 and *Myotis primula* Thomas, 1920 (Mammalia). Ann. Hist.-Nat. Mus. Natl. Hung., Zool. 62:373–379.

Tranier, M. M. 1976. Nouvelles données sur l'évolution non parallèle du caryotype et de la morphologie chez les Phyllotines (Rongeurs, Cricetides). Compt. Rend. Acad. Sci. Paris, ser. D, 283:1201–1203.

Trapp, G. R. 1972. Some anatomical and behavioral adaptations of ringtails, *Bassariscus astutus*. J. Mamm. 53:549–557.

Trapp, G. R., and D. L. Hallberg. 1975. Ecology of the gray fox (*Urocyon cinereoargenteus*): a review. *In* Fox (1975), pp. 164–178.

Trebbau, P. 1975. Measurements and some observations on the freshwater dolphin, *Inia geoffrensis*, in the Apure River, Venezuela. Zool. Garten 45:153–167.

Trebbau, P., and P.J.H. Van Bree. 1974. Notes concerning the freshwater dolphin *Inia geoffrensis* (de Blainville, 1817) in Venezuela. Z. Saugetierk. 39:50–57.

Trent, T. T., and O. J. Rongstad. 1974. Home range and survival of cottontail rabbits in southwestern Wisconsin. J. Wildl. Mgmt. 38:459–472.

Treus, V. D., and N. V. Lobanov. 1971. Acclimatisation and domestication of the eland *Taurotragus oryx* at Askanya-Nova Zoo. Int. Zoo Yearbook 11:147–156.

Trimarchi, C. V. 1980. Rabies in bats in the United States. Natl. Speleol. Soc. Bull. 42:23.

Troughton, E. Le G. 1931. Three new bats of the genera *Pteropus, Nyctimene,* and *Chaerephon* from Melanesia. Proc. Linnaean Soc. New South Wales 56:204–209.

————. 1936. The mammalian fauna of Bougainville Island, Solomons Group. Rec. Austral. Mus. 19:341–353.

————. 1971. The early history and relationships of the New Guinea highland dog (*Canis hallstromi*). Proc. Linnean Soc. New South Wales 96:93–98.

Trout, R. C. 1978a. A review of studies on populations of wild harvest mice [*Micromys minutus* (Pallas)]. Mamm. Rev. 8:143–158.

————. 1978b. A review of studies on captive harvest mice [*Micromys minutus* (Pallas)]. Mamm. Rev. 8:159–175.

Tulloch, D. G. 1978. The water buffalo, *Bubalus bubalis*, in Australia: grouping and home range. Austral. Wildl. Res. 5:327–354.

Tupinier, Y. 1977. Description d'une chauve-souris nouvelle: *Myotis nathalinae* nov. sp. (Chiroptera—Vespertilionidae). Mammalia 41:327–340.

Turnbull, C. M. 1976. Man in Africa. Anchor Press/Doubleday, Garden City, New York, xx + 313 pp.

Turner, D. C. 1975. The vampire bat. A field study in behavior and ecology. Johns Hopkins Univ. Press, Baltimore, viii + 145 pp.

Turner, K. 1970. Breeding Tasmanian devils *Sarcophilus harrisii* at Westbury Zoo. Int. Zoo. Yearbook 10:65.

Tuttle, M. D. 1970. Distribution and zoogeography of Peruvian bats, with comments on natural history. Univ. Kansas Sci. Bull. 49:45–86.

————. 1976a. Population ecology of the gray bat (*Myotis grisescens*): philopatry, timing and patterns of movement, weight loss during migration, and seasonal adaptive strategies. Occas. Pap. Mus. Nat. Hist., Univ. Kansas, no. 54, 38 pp.

————. 1976b. Population ecology of the gray bat (*Myotis grisescens*): factors influencing growth and survival of newly volant young. Ecology 57:587–595.

————. 1979. Status, causes of decline, and management of endangered gray bats. J. Wildl. Mgmt. 43:1–17.

Twigg, G. I. 1965. Studies on *Holochilus sciureus berbicensis*, a cricetine rodent from the coastal region of British Guiana. Proc. Zool. Soc. London 145:263–283.

Tyler, M. J., ed. 1979. The status of endangered Australian wildlife. Proc. Cent. Symp. Roy. Zool. Soc. S. Australia, Adelaide, ix + 210 pp.

Tyndale-Biscoe, C. H. 1968. Reproduction and post-natal development in the marsupial *Bettongia lesueur* (Quoy & Gaimard). Austral. J. Zool. 16:577–602.

————. 1973. Life of marsupials. American Elsevier, New York, viii + 254 pp.

Tyndale-Biscoe, C. H., and R. B. MacKenzie. 1976. Reproduction in *Didelphis marsupialis* and *D. albiventris* in Colombia. J. Mamm. 57:249–265.

Tyndale-Biscoe, C. H., and R.F.C. Smith. 1969. Studies on the marsupial glider, *Schoinobates volans* (Kerr). II. Population structure and regulatory mechanisms. J. Anim. Ecol. 38:637–649.

U.

Ulmer, F. A., Jr. 1968. Breeding fishing cats *Felis viverrina* at Philadelphia Zoo. Int. Zoo Yearbook 8:49–55.

Union of Soviet Socialist Republics Ministry of Agriculture. 1978. Red data book of USSR. Lesnaya Promyshlenmost Publishers, Moscow, 460 pp.

USDI (United States Department of the Interior). 1976. Proposal to list 27 species of primates as endangered or threatened species. Federal Register 41:1646–1649.

————. 1980. Republication of the Lists of Endangered and Threatened Species and correction of technical errors in final rules. Federal Register 45:33768–33781.

USDI (United States Department of the Interior). 1982. Threatened status for the leopard in southern Africa. Federal Register 47:4204–4211.

United States Department of Health, Education and Welfare. 1975. Restrictions on importation of nonhuman primates. Federal Register 40:33659.

U.S. (United States) Fish and Wildlife Service. 1980. Administration of the Marine Mammal Protection Act of 1972. April 1, 1979 to March 31, 1980. Washington, D.C., v + 86 pp.

U.S. (United States) National Marine Fisheries Service. 1978. The Marine Mammal Protection Act of 1972. Annual Report 1977–78. Washington, D.C., v + 183 pp.

————. 1981. Marine Mammal Protection Act of 1972. Annual Report 1980/81. Washington, D.C., v + 143 pp.

Urban, D. 1970. Raccoon populations, movement patterns, and predation on a managed waterfowl marsh. J. Wildl. Mgmt. 34:372–382.

Uspensky, S. M., and S. E. Belikov. 1976. Research on the polar bear in the USSR. *In* Pelton, Lentfer, and Folk (1976), pp. 321–323.

Ustinov, S. K. 1976. The brown bear on Baikal: a few features of vital activity. *In* Pelton, Lentfer, and Folk (1976), pp. 325–326.

V.

Valdez, R., and R. K. LaVal. 1971. Records of bats from Honduras and Nicaragua. J. Mamm. 52:247–250.

Valdez, R.; C. F. Nadler; and T. D. Bunch. 1978. Evolution of wild sheep in Iran. Evolution 32:56–72.

Valtonen, M. H.; E. J. Rajakoski; and J. I. Mäkelä. 1977. Reproductive features in the female raccoon dog. J. Reprod. Fert. 51:517–518.

Van Aarde, R. J. 1980. Harem structure of the southern elephant seal *Mirounga leonina* at Kerguelen Island. Rev. Ecol. (Terre Vie) 34:31–44.

Van Ballenberghe, V., and A. W. Erickson. 1973. A wolf pack kills another wolf. Amer. Midl. Nat. 90:490–493.

Van Ballenberghe, V.; A. W. Erickson; and D. Byman. 1975. Ecology of the timber wolf in northeastern Minnesota. Wildl. Monogr., no. 43, 43 pp.

Van Ballenberghe, V., and L. D. Mech. 1975. Weights, growth, and survival of timber wolf pups in Minnesota. J. Mamm. 56:44–63.

Van Bemmel, A.C.V. 1967. The banteng *Bos javanicus* in captivity. Int. Zoo Yearbook 7:222–223.

Van Bree, P.J.H. 1971. On *Globicephala sieboldii* Gray, 1846, and other species of pilot whales (Notes on Cetacea, Delphinoidea III). Beaufortia 19:79–87.

———. 1976. On the correct Latin name of the Indus susu (Cetacea, Platanistoidea). Bull. Zool. Mus. Univ. Amsterdam 5:139–140.

Van Bree, P.J.H., and M. D. Gallagher. 1978. On the taxonomic status of *Delphinus tropicalis* Van Bree, 1971 (Notes on Cetacea Delphinoidea IX). Beaufortia 28:1–8.

Van Bree, P.J.H., and P. E. Purves. 1972. Remarks on the validity of *Delphinus bairdii* (Cetacea, Delphinidae). J. Mamm. 53:372–374.

Van Camp, J., and R. Gluckie. 1979. A record long-distance movement by a wolf (*Canis lupus*). J. Mamm. 60:236–237.

Van Den Brink, F. H. 1968. A field guide to the mammals of Britain and Europe. Houghton Mifflin, Boston, 221 pp.

Van der Horst, G. 1972. Seasonal effects on the anatomy and histology of the reproductive tract of the male rodent mole *Bathyergus suillus suillus* (Schreber). Zool. Afr. 7:491–520.

Van der Merwe, M. 1975. Preliminary study on the annual movements of the Natal clinging bat. S. Afr. J. Sci. 71:237–241.

———. 1978. Postnatal development and

mother-infant relationships in the Natal clinging bat *Miniopterus schreibersi natalensis* (A. Smith 1834). Proc. 4th Int. Bat Conf., Nairobi, pp. 309–322.

Van der Merwe, M.; J. D. Skinner; and R. P. Millar. 1980. Annual reproductive pattern in the springhaas, *Pedetes capensis*. J. Reprod. Fert. 58:259–266.

Van der Straeten, E. 1975. *Lemniscomys bellieri*, a new species of Muridae from the Ivory Coast. Rev. Zool. Afr. 89:906–908.

———. 1976. *Lemniscomys striatus dieterleni*, a new subspecies of Muridae from Zaire. Rev. Zool. Afr. 90:431–434.

———. 1980a. Etude biométrique de *Lemniscomys linulus* (Afrique Occidentale) (Mammalia, Muridae). Rev. Zool. Afr. 94:185–201.

———. 1980b. A new species of *Lemniscomys* (Muridae) from Zambia. Ann. Cape Prov. Mus. (Nat. Hist.) 13:55–62.

———. 1981. Some biometrical data for *Stenocephalemys albocaudata* and *Stenocephalemys griseicauda*. Afr. Small Mamm. Newsl., no. 6, pp. 11–14.

Van der Straeten, E., and W. N. Verheyen. 1978a. Karyological and morphological comparisons of *Lemniscomys striatus* (Linnaeus, 1758) and *Lemniscomys bellieri* Van der Straeten, 1975, from Ivory Coast (Mammalia: Muridae). Bull. Carnegie Mus. Nat. Hist., no. 6, pp. 41–47.

———. 1978b. Taxonomical notes on the West-African *Myomys* with the description of *Myomys derooi* (Mammalia—Muridae). Z. Saugetierk. 43:31–41.

———. 1979a. Notes taxonomiques sur les *Malacomys* de l'Ouest africain avec redescription du patron chromosomique de *Malacomys edwardsi* (Mammalia, Muridae). Rev. Zool. Afr. 93:10–35.

———. 1979b. Note sur la position systématique de *Lemniscomys macculus* (Thomas et Wroughton, 1910) (Mammalia, Muridae). Mammalia 43:377–389.

———. 1980. Relations biométriques dans le groupe spécifique *Lemniscomys striatus* (Mammalia, Muridae). Mammalia 44:73–82.

Van Deusen, H. M. 1968. Carnivorous habits of *Hypsignathus monstrosus*. J. Mamm. 49:335–336.

———. 1969. Results of the 1958–1959 Gilliard New Britain Expedition. 5. A new species of *Pteropus* (Mammalia, Pteropidae) from New Britain, Bismarck Archipelago. Amer. Mus. Novit., no. 2371, 16 pp.

Van Deusen, H. M., and J. K. Jones, Jr. 1967. Marsupials. *In* Anderson and Jones (1967), pp. 61–86.

Van Deusen, H. M., and K. Keith. 1966. Range and habitat of the bandicoot, *Echymipera clara*, in New Guinea. J. Mamm. 47:721–723.

Van Deusen, H. M., and K. F. Koopman. 1971. Results of the Archbold Expeditions. No. 95. The genus *Chalinolobus* (Chiroptera, Vespertilionidae) taxonomic review of *Chalinolobus picatus*, *C. nigrogriseus*, and *C. rogersi*. Amer. Mus. Novit., no. 2468, 30 pp.

Van Dyck, S. 1980. The cinnamon antechinus, *Antechinus leo* (Marsupialia: Dasyuridae), a new species from the vineforests of Cape York Peninsula. Austral. Mamm. 3:5–17.

Van Ee, C. A. 1966. A note on breeding the Cape pangolin *Manis temmincki* at Bloemfontein Zoo. Int. Zoo Yearbook 6:163–164.

Van Gelder, R. G. 1977a. An eland x kudu hybrid, and the content of the genus *Tragelaphus*. Lammergeyer, no. 23, 6 pp.

———. 1977b. Mammalian hybrids and generic limits. Amer. Mus. Novit., no. 2635, 25 pp.

———. 1978. A review of canid classification. Amer. Mus. Novit., no. 2646, 10 pp.

———. 1979. Mongooses on mainland North America. Wildl. Soc. Bull. 7:197–198.

Van Horn, R. N., and G. G. Eaton. 1979. Reproductive physiology and behavior in prosimians. *In* Doyle and Martin (1979), pp. 79–122.

Van Lawick, H., and J. Van Lawick-Goodall. 1971. Innocent killers. Houghton Mifflin, Boston, 222 pp.

Van Lawick-Goodall, J. 1968. The behaviour of free-living chimpanzees in the Gombe Stream Reserve. Anim. Behav. Monogr. 1:165–311.

———. 1973. Cultural elements in a chimpanzee community. Symp. 4th Int. Congr. Primatol. 1:144–184.

Van Peenen, P.F.D.; R. H. Light; F. J. Duncan; R. See; J. Sulianti Saroso; Boeadi; and W. P. Carney. 1974. Observations on *Rattus bartelsii* (Rodentia: Muridae). Treubia 28:83–117.

Van Strien, N. J. 1975. *Dicerorhinus sumatrensis* (Fischer) the Sumatran or two-horned Asiatic rhinoceros. A study of literature. Netherlands Comm. Int. Nat. Protection Meded., no. 22, 82 pp.

Van Valen, L. 1967. New Paleocene insectivores and insectivore classification. Bull. Amer. Mus. Nat. Hist. 135:217–284.

Van Weers, D. J. 1976. Notes on Southeast Asian porcupines (Hystricidae, Rodentia). I. On the taxonomy of the genus *Trichys* Günther, 1877. Beaufortia 25:15–31.

———. 1977. Notes on Southeast Asian porcupines (Hystricidae, Rodentia). II. On the taxonomy of the genus *Atherurus* F. Cuvier, 1829. Beaufortia 26:205–230.

———. 1978. Notes on Southeast Asian porcupines (Hystricidae, Rodentia). III. On the taxonomy of the subgenus *Thecurus* Lyon, 1907 (genus *Hystrix* Linnaeus, 1758). Beaufortia 28:17–33.

———. 1979. Notes on Southeast Asian porcupines (Hystricidae, Rodentia). IV. On the taxonomy of the subgenus *Acanthion* F. Cuvier, 1823 with notes on the other taxa of the family. Beaufortia 29:215–272.

Van Zyll de Jong, C. G. 1972. A systematic review of the nearctic and neotropical river otters (genus *Lutra*, Mustelidae, Carnivora). Roy. Ontario Mus. Life Sci. Contrib., no. 80, 104 pp.

———. 1975. The distribution and abundance of the wolverine (*Gulo gulo*) in Canada. Can. Field-Nat. 89:431–437.

————. 1976. Are there two species of pygmy shrews (*Microsorex*)? Can. Field-Nat. 90:485–487.

Varland, K. L.; A. L. Lovaas; and R. B. Dahlgren. 1978. Herd organization and movements of elk in Wind Cave National Park, South Dakota. U.S. Natl. Park Serv. Nat. Res. Rept., no. 13, 28 pp.

Varona, L. S. 1979. Subgenero y especie nuevos de *Capromys* (Rodentia: Caviomorpha) para Cuba. Poeyana, no. 194, 33 pp.

Varona, L. S., and O. Arredondo. 1979. Nuevos taxones fosiles de Capromyidae (Rodentia: Caviomorpha). Poeyana, no. 195, 51 pp.

Vasey, D. E. 1979. Capybara ranching for Amazonia? Oryx 15:47–49.

Vaughan, T. A. 1976. Nocturnal behavior of the African false vampire bat (*Cardioderma cor*). J. Mamm. 57:227–248.

Vaughan, T. A., and G. C. Bateman. 1970. Functional morphology of the forelimb of mormoopid bats. J. Mamm. 51:217–235.

Vaughan, T. A., and T. J. O'Shea. 1976. Roosting ecology of the pallid bat, *Antrozous pallidus*. J. Mamm. 57:19–41.

Vaughn, R. 1974. Breeding the tayra *Eira barbara* at Antelope Zoo, Lincoln. Int. Zoo Yearbook 14:120–122.

Veal, R., and W. Caire. 1979. *Peromyscus eremicus*. Mammalian Species, no. 118, 6 pp.

Vehrencamp, S. L.; F. G. Stiles; and J. W. Bradbury. 1977. Observations on the foraging behavior and avian prey of the neotropical carnivorous bat, *Vampyrum spectrum*. J. Mamm. 58:469–478.

Velte, F. F. 1978. Hand-rearing springhaas *Pedetes capensis* at Rochester Zoo. Int. Zoo Yearbook 18:206–208.

Vereschagin, N. K. 1976. The brown bear in Eurasia, particularly the Soviet Union. *In* Pelton, Lentfer, and Folk (1976), pp. 327–335.

Verheyen, W. N., and E. Van der Straeten. 1977. Description of *Malacomys verschureni*, a new murid-species from Central Africa (Mammalia-Muridae). Rev. Zool. Afr. 91:737–744.

Verme, L. J. 1970. Some characteristics of captive Michigan moose. J. Mamm. 51:403–405.

Vermeiren, L.J.P., and W. N. Verheyen. 1980. Notes sur les *Leggada* de Lamto, Côte d'Ivoire, avec la description de *Leggada baoulei* sp. n. Rev. Zool. Afr. 94:570–590.

Veselovsky, Z. 1966. A contribution to the knowledge of the reproduction and growth of the two-toed sloth *Choloepus didactylus* at Prague Zoo. Int. Zoo Yearbook 6:147–153.

Vesmanis, I. 1976. Beitrag zur Kenntnis der Crociduren-Fauna Siziliens (Mammalia: Insectivora). Z. Saugetierk. 41:257–273.

————. 1977. Eine neue *Crocidura*-Art aus der Cyrenaica, Libyen: *Crocidura alek-* *sandrisi* n. sp. (Mammalia: Insectivora: *Crocidura*). Bonner Zool. Beitr. 28:3–12.

Vessey, S. H.; B. K. Mortenson; and N. A. Muckenhirn. 1978. Size and characteristics of primate groups in Guyana. *In* Chivers and Herbert (1978), pp. 187–188.

Vestjens, W.J.M., and L. S. Hall. 1977. Stomach contents of forty-two species of bats from the Australian region. Austral. Wildl. Res. 4:25–35.

Viljoen, S. 1977. Behaviour of the bush squirrel, *Paraxerus cepapi cepapi* (A. Smith, 1836). Mammalia 41:119–166.

————. 1978. Notes on the western striped squirrel *Funisciurus congicus* (Kuhl 1820). Madoqua 11:119–128.

Villa, B. 1976. Report on the status of *Phocoena sinus*, Norris and McFarland 1958, in the Gulf of California. An. Inst. Biol. Univ. Nal. Auton. México, Ser. Zool. 47:203–208.

Villa, B., and J. R. Pulido. 1968. *Diclidurus virgo* Thomas, el murciélago blanco, en la costa de Nayarit, México. An. Inst. Biol. Univ. Nal. Auton. México, Ser. Zool. 39:155–158.

Vogel, P. 1970. Biologische Beobachtungen an Etruckerspitzmausen (*Suncus etruscus* Savi, 1832). Z. Saugetierk. 35:173–185.

Vogt, B. 1978. The big ones are back! Natl. Wildl. 16(6):4–13.

Voight, D. R.; G. B. Kolenosky; and D. H. Pimlott. 1976. Changes in summer foods of wolves in central Ontario. J. Wildl. Mgmt. 40:663–668.

Volf, J. 1976. Some remarks on the taxonomy of the genus *Nemorhaedus* H. Smith, 1827. Vest. Cesk. Spol. Zool. 40:75–80.

Von Ketelhodt, H. F. 1976. Observations on the lambing interval of the Cape bushbuck, *Tragelaphus scriptus sylvaticus*. Zool. Afr. 11:221–225.

————. 1977. Observations on the lambing interval of the grey duiker, *Sylvicapra grimmia grimmia*. Zool. Afr. 12:232–233.

Von Richter, W. 1972. Territorial behaviour of the black wildebeest *Connochaetes gnou*. Zool. Afr. 7:207–231.

————. 1974. *Connochaetes gnou*. Mammalian Species, no. 50, 6 pp.

Vorontsov, N. N.; I. V. Kartavtseva; and E. G. Potapova. 1979. Systematics of the genus *Calomyscus* (Cricetinae). Zool. Zhur. 59:283–288.

Vose, H. M. 1973. Feeding habits of the western Australian honey possum, *Tarsipes spenserae*. J. Mamm. 54:245–247.

Voss, R. S., and A. V. Linzey. 1981. Comparative gross morphology of male accessory glands among neotropical Muridae (Mammalia: Rodentia) with comments on systematic implications. Misc. Publ. Mus. Zool. Univ. Michigan, no. 159, 41 pp.

Vrba, E. S. 1979. Phylogenetic analysis and classification of fossil and Recent Alcelaphini Mammalia: Bovidae. Biol. J. Linnean Soc. 11:207–228.

W.

Wackernagel, H. 1968. A note on breeding the serval cat *Felis serval* at Basle Zoo. Int. Zoo Yearbook 8:46–47.

Wade, D. A. 1978. Coyote damage: a survey of its nature and scope, control measures and their application. *In* Bekoff (1978), pp. 347–368.

Wade-Smith, J., and M. E. Richmond. 1975. Care, management, and biology of captive striped skunks (*Mephitis mephitis*). Lab. Anim. Sci. 25:575–584.

————. 1978. Reproduction in captive striped skunks (*Mephitis mephitis*). Amer. Midl. Nat. 100:452–455.

Wadsworth, C. E. 1972. Observations of the Colorado chipmunk in southeastern Utah. Southwestern Nat. 16:451–454.

Waechter, A. 1975. Ecologie de la fouine en Alsace. Terre Vie 29:399–457.

Wainer, J. W. 1976. Studies of an island population of *Antechinus minimus* (Marsupialia, Dasyuridae). Austral. Zool. 19:1–7.

Waithman, J. 1979. A report on a collection of mammals from southwest Papua, 1972–1973. Austral. Zool. 20:313–326.

Wakefield, N. A. 1970a. Notes on Australian pigmy-possums (*Cercartetus*, Phalangeridae, Marsupialia). Victorian Nat. 87:11–18.

————. 1970b. Notes on the glider-possum, *Petaurus australis* (Phalangeridae, Marsupialia). Victorian Nat. 87:221–236.

————. 1971. The brush-tailed rock-wallaby (*Petrogale penicillata*) in western Victoria. Victorian Nat. 88:92–102.

————. 1972. Studies in Australian Muridae: review of *Mastacomys fuscus*, and description of a new subspecies of *Pseudomys higginsi*. Mem. Natl. Mus. Victoria 33:15–31.

Walker, A. 1970. Nuchal adaptations in *Perodicticus potto*. Primates 11:134–144.

————. 1979. Prosimian locomotor behavior. *In* Doyle and Martin (1979), pp. 543–565.

Walker, G. E., and J. K. Ling. 1980. Body weights and dimensions of adult female Australian sea lions, *Neophoca cinerea*. J. Mamm. 61:164–165.

Walker, P. L., and J.G.H. Cant. 1977. A population survey of kinkajous (*Potos flavus*) in a seasonally dry tropical forest. J. Mamm. 58:100–102.

Walley, H. D., and W. L. Jarvis. 1971. Longevity record for *Pipistrellus subflavus*. Trans. Illinois Acad. Sci. 64:305.

Wallin, L. 1969. The Japanese bat fauna. Zool. Bidrag Uppsala, 37:223–440.

Wallmo, O. C. 1978. Mule and black-tailed deer. *In* Schmidt and Gilbert (1978), pp. 31–41.

Walther, F. R. 1972. Social grouping in Grant's gazelle (*Gazella granti* Brooke 1827) in the Serengeti National Park. Z. Tierpsychol. 31:348–403.

———. 1978. Behavioral observations on oryx antelope (*Oryx beisa*) invading Serengeti National Park, Tanzania. J. Mamm. 59:243–260.

Walton, D. W.; J. E. Brooks; U M. M. Tun; and U H. Naing. 1978. Observations on reproductive activity among female *Bandicota bengalensis* in Rangoon. Acta Theriol. 23:489–501.

Walton, G. M., and D. W. Walton. 1973. Notes on hedgehogs of the lower Indus Valley. Korean J. Zool. 16:161–170.

Wang Tsiang-Ke. 1974. On the taxonomic status of species, geological distribution and evolutionary history of *Ailuropoda*. Acta Zool. Sinica 20:201.

Wang Youzhi; Hu Jinchu; and Chen Ke. 1980. A new species of Murinae—*Vernaya foramena* sp. nov. Acta Zool. Sinica 26:393–397.

Ward, G. D. 1978. Habitat use and home range of radio-tagged opossums *Trichosurus vulpecula* (Kerr) in New Zealand lowland forest. *In* Montgomery (1978), pp. 267–287.

Waring, G. H. 1970. Sound communications of black-tailed, white-tailed, and Gunnison's prairie dogs. Amer. Midl. Nat. 83:167–185.

Waser, P. 1977. Feeding, ranging and group size in the mangabey *Cercocebus albigena*. *In* Clutton-Brock (1977), pp. 183–222.

Waser, P. M., and O. Floody. 1974. Ranging patterns of the mangabey, *Cercocebus albigena*, in the Kibale Forest, Uganda. Z. Tierpsychol. 35:85–101.

Watkins, L. C. 1972. *Nycticeius humeralis*. Mammalian Species, no. 23, 4 pp.

———. 1977. *Euderma maculatum*. Mammalian Species, no. 77, 4 pp.

Watkins, L. C.; J. K. Jones, Jr.; and H. H. Genoways. 1972. Bats of Jalisco, Mexico. Spec. Publ. Mus. Texas Tech Univ., no. 1, 44 pp.

Watkins, W. A. 1976. A probable sighting of a live *Tasmacetus shepherdi* in New Zealand waters. J. Mamm. 57:415.

Watkins, W. A.; W. E. Schevill; and P. B. Best. 1977. Underwater sounds of *Cephalorhynchus heavisidii* (Mammalia: Cetacea). J. Mamm. 58:316–318.

Watts, C.H.S. 1974. The native rodents of Australia: a personal view. Austral. Mamm. 1:109–116.

———. 1975a. Vocalizations of Australian hopping-mice (Rodentia: *Notomys*). J. Zool. 177:247–263.

———. 1975b. The neck and chest glands of Australian hopping-mice, *Notomys*. Austral. J. Zool. 23:151–157.

———. 1976a. Vocalizations of the plains rat *Pseudomys australis* Gray (Rodentia: Muridae). Austral. J. Zool. 24:95–103.

———. 1976b. *Leggadina lakedownensis*, a new species of murid rodent from north Queensland. Trans. Roy. Soc. S. Austral. 100:105–108.

———. 1977. The foods eaten by some Australian rodents (Muridae). Austral. Wildl. Res. 4:151–157.

———. 1979. The status of endangered Australian rodents. *In* Tyler (1979), pp. 75–83.

Weaver, J. 1978. The wolves of Yellowstone. U.S. Natl. Park Serv. Nat. Res. Rept., no. 14, 38 pp.

Webster, W. D., and J. K. Jones, Jr. 1980. Taxonomic and nomenclatorial notes on bats of the genus *Glossophaga* in North America, with description of a new species. Occas. Pap. Mus. Texas Tech Univ., no. 71, 12 pp.

Wegge, P. 1979. Aspects of the population ecology of blue sheep in Nepal. J. Asian Ecol. 1:10–20.

Weir, B. J. 1974a. The tuco-tuco and plains viscacha. Symp. Zool. Soc. London 34:113–130.

———. 1974b. Reproductive characteristics of hystricomorph rodents. Symp. Zool. Soc. London 34:265–301.

———. 1974c. Notes on the origin of the domestic guinea-pig. Symp. Zool. Soc. London 34:437–446.

Weise, T. F.; W. L. Robinson; R. A. Hook; and L. D. Mech. 1975. An experimental translocation of the eastern timber wolf. Audubon Conserv. Rept., no. 5, 28 pp.

Wellington, G. M., and Tj. De Vries. 1976. The South American sea lion, *Otaria byronia*, in the Galapagos Islands. J. Mamm. 57:166–167.

Wells, R. T. 1978. Field observations of the hairy-nosed wombat, *Lasiorhinus latifrons* (Owen). Austral. Wildl. Res. 5:299–303.

Wemmer, C. M. 1977. Comparative ethology of the large-spotted genet *Genetta tigrina* and some related viverrids. Smithson. Contrib. Zool., no. 239, iii + 93 pp.

Wemmer, C. M., and J. Murtaugh. 1981. Copulatory behavior and reproduction in the binturong, *Arctictis binturong*. J. Mamm. 62:342–352.

Werdelin, L. 1981. The evolution of lynxes. Ann. Zool. Fennici 18:37–71.

West, S. D. 1977. Midwinter aggregation in the northern red-backed vole, *Clethrionomys rutilus*. Can. J. Zool. 55:1404–1409.

Wetzel, R. M. 1975. The species of *Tamandua* Gray (Edentata, Myrmecophagidae). Proc. Biol. Soc. Washington 88:95–112.

———. 1977a. The extinction of peccaries and a new case of survival. Ann. New York Acad. Sci. 288:538–544.

———. 1977b. The Chacoan peccary *Catagonus wagneri* (Rusconi). Bull. Carnegie Mus. Nat. Hist., no. 3, 36 pp.

———. 1980. Revision of the naked-tailed armadillos, genus *Cabassous* McMurtie. Ann. Carnegie Mus. 49:323–357.

———. 1981. The hidden Chacoan peccary. Carnegie Mag. 55(2):24–32.

———. In Press. The identification and distribution of Recent Xenarthra, and Taxonomy and distribution of armadillos (Dasypodidae). *In* Montgomery, G. G., ed., The evolution and ecology of sloths, anteaters, and armadillos (Mammalia, Xenarthra=Edentata), Smithson. Inst. Press, Washington, D.C.

Wetzel, R. M., and F. D. de Avila-Pires. 1980. Identification and distribution of the Recent sloths of Brazil (Edentata). Rev. Brasil. Biol. 40:831–836.

Wetzel, R. M.; R. E. Dubos; R. L. Martin; and P. Myers. 1975. *Catagonus*, an "extinct" peccary, alive in Paraguay. Science 189:379–381.

Wetzel, R. M., and D. Kock. 1973. The identity of *Bradypus variegatus* Schinz (Mammalia, Edentata). Proc. Biol. Soc. Washington 86:25–34.

Wetzel, R. M., and J. W. Lovett. 1974. A collection of mammals from the Chaco of Paraguay. Univ. Connecticut Occas. Pap., Biol. Sci. Ser., 2:203–216.

Wetzel, R. M., and E. Mondolfi. 1979. The subgenera and species of long-nosed armadillos, genus *Dasypus* L. *In* Eisenberg (1979), pp. 43–63.

Wheeler, M. E., and C. F. Aguon. 1978. The current status and distribution of the Marianas fruit bat on Guam. Guam Aquatic and Wildl. Res. Div. Tech. Rept., no. 1, 29 pp.

Whitaker, J. O., Jr. 1972. *Zapus hudsonius*. Mammalian Species, no. 11, 7 pp.

Whitaker, J. O., Jr., and H. Black. 1976. Food habits of cave bats from Zambia, Africa. J. Mamm. 57:199–204.

Whitaker, J. O., Jr., and R. E. Mumford. 1972. Ecological studies on *Reithrodontomys megalotis* in Indiana. J. Mamm. 53:850–860.

Whitaker, J. O., Jr., and R. E. Wrigley. 1972. *Napaeozapus insignis*. Mammalian Species, no. 14, 6 pp.

Whitehouse, S.J.O. 1977. The diet of the dingo in Western Australia. Austral. Wildl. Res. 4:145–150.

Whittow, G. C.; E. Gould; and D. Rand. 1977. Body temperature, oxygen consumption, and evaporative water loss in a primitive insectivore, the moon rat, *Echinosorex gymnurus*. J. Mamm. 58:233–235.

Wickler, W., and U. Seibt. 1976. Field studies of the African fruit bat *Epomophorus wahlbergi* (Sundevall), with special reference to male calling. Z. Tierpsychol. 40:345–376.

Widholzer, F. L., and W. A. Voss. 1978. Breeding the giant anteater *Myrmecophaga tridactyla* at São Leopoldo Zoo. Int. Zoo Yearbook 18:122–123.

Wiegl, P. D. 1974. Study of the northern flying squirrel, *Glaucomys sabrinus* by temperature telemetry. Amer. Midl. Nat. 92:482–486.

Wiley, R. W. 1980. *Neotoma floridana*. Mammalian Species, no. 139, 7 pp.

Williams, D. F. 1975. Distributional records for shrew-moles, *Neurotrichus gibbsii*. Murrelet 56:2–3.

———. 1978a. Karyological affinities of the species groups of silky pocket mice (Rodentia, Heteromyidae). J. Mamm. 59:599–612.

———. 1978b. Taxonomic and karyologic comments on small brown bats, genus *Eptesicus*, from South America. Ann. Carnegie Mus. 47:361–383.

Williams, D. F.; J. D. Druecker; and H. L. Black. 1970. The karyotype of *Euderma maculatum* and comments on the evolution of the plecotine bats. J. Mamm. 51:602–606.

Williams, D. F., and M. A. Mares. 1978. A new genus and species of phyllotine rodent (Mammalia: Muridae) from northwestern Argentina. Ann. Carnegie Mus. 47:193–221.

Williams, S. L., and R. J. Baker. 1974. *Geomys arenarius*. Mammalian Species, no. 36, 3 pp.

Williams, S. L., and H. H. Genoways. 1980*a*. Results of the Alcoa Foundation-Suriname Expeditions. II. Additional records of bats (Mammalia: Chiroptera) from Suriname. Ann. Carnegie Mus. 49:213–236.

————. 1980*b*. Results of the Alcoa Foundation-Suriname Expeditions. IV. A new species of bat of the genus *Molossops* (Mammalia: Molossidae). Ann. Carnegie Mus. 49:487–498.

————. 1980*c*. Morphological variation in the southeastern pocket gopher, *Geomys pinetis* (Mammalia: Rodentia). Ann. Carnegie Mus. 49:405–453.

Williams, T. C.; L. C. Ireland; and J. M. Williams. 1973. High altitude flights of the free-tailed bat, *Tadarida brasiliensis*, observed with radar. J. Mamm. 54:807–821.

Willner, G. R.; J. A. Chapman; and J. R. Goldsberry. 1975. A study and review of muskrat food habits with special reference to Maryland. Maryland Dept. Nat. Res. Publ. Wildl. Ecol., no. 1, 25 pp.

Willner, G. R.; J. A. Chapman; and D. Pursley. 1979. Reproduction, physiological responses, food habits, and abundance of nutria on Maryland marshes. Wildl. Monogr., no. 65, 43 pp.

Willner, G. R.; G. A. Feldhamer; E. E. Zucker; and J. A. Chapman. 1980. *Ondatra zibethicus*. Mammalian Species, no. 141, 8 pp.

Willoughby, D. P. 1966. The vanished quagga. Nat. Hist. 75(2):60–63.

Wilson, D(avid). E., and S. M. Hirst. 1977. Ecology and factors limiting roan and sable antelope populations in South Africa. Wildl. Monogr., no. 54, 111 pp.

Wilson, D(on). E. 1971. Ecology of *Myotis nigricans* (Mammalia: Chiroptera) on Barro Colorado Island, Panama Canal Zone. J. Zool. 163:1–13.

————. 1976. The subspecies of *Thyroptera discifera* (Lichtenstein and Peters). Proc. Biol. Soc. Washington, 89:305–312.

————. 1979. Reproductive patterns. *In* Baker, Jones, and Carter (1979), pp. 317–378.

Wilson, D(on). E., and E. L. Tyson. 1970. Longevity records for *Artibeus jamaicensis* and *Myotis nigricans*. J. Mamm. 51:203.

Wilson, R. T., and J. G. Lewis. 1979. The ecology of Swayne's hartebeest. Biol. Conserv. 15:1–12.

Wilson, V. J. 1966*a*. Notes on the food and feeding habits of the common duiker, *Sylvicapra grimmia* in eastern Zambia. Arnoldia 2(14):1–19.

————. 1966*b*. Observations on Lichtenstein's hartebeest, *Alcelaphus lichtensteini*, over a three-year period, and their response to various tsetse control measures in eastern Zambia. Arnoldia 2(15):1–14.

————. 1969. Eland, *Taurotragus oryx*, in eastern Zambia. Arnoldia 4(12):1–9.

————. 1975. Mammals of the Wankie National Park, Rhodesia Natl. Mus. Monuments Mus. Mem., no. 5, i + 147 pp.

Wilson, V. J., and G. Child. 1965. Notes on klipspringer from tsetse fly control areas in eastern Zambia. Arnoldia 1(35):1–9.

Wilson, V. J., and M. A. Kerr. 1969. Brief notes on reproduction in Steenbok, *Raphicerus campestris*, Thunberg. Arnoldia 4(23):1–5.

Wilsson, L. 1971. Observations and experiments on the ethology of the European beaver (*Castor fiber* L.). Viltrevy 8:115–266.

Winchester, B. H.; R. S. Delotelle; J. R. Newman; and J. T. McClave. 1978. Ecology and management of the colonial pocket gopher: a progress report. *In* Odom, R. R., and L. Landers, eds., Proceedings of the rare and endangered wildlife symposium, Georgia Dept. Nat. Resources, pp. 173–184.

Winn, H. E.; R. K. Edel; and A. G. Taruski. 1975. Population estimate of the humpback whale (*Megaptera novaeangliae*) in the West Indies by visual and acoustic techniques. J. Fish. Res. Bd. Can. 32:499–506.

Winn, H. E., and B. L. Olla, eds. 1979. Behavior of marine animals. Volume 3: Cetaceans. Plenum Press, New York, xix + 438 pp.

Winn, H. E.; P. J. Perkins; and L. Winn. 1970. Sounds and behavior of the northern bottle-nosed whale. Proc. Ann. Conf. Biol. Sonar and Diving Mammals 7:53–59.

Winn, H. E., and L. K. Winn. 1978. The song of the humpback whale (*Megaptera novaeangliae*) in the West Indies. Mar. Biol. 47:97–114.

Winter, J. W. 1979. The status of endangered Australian Phalangeridae, Petauridae, Burramyidae, Tarsipedidae and the Koala. *In* Tyler (1979), pp. 45–59.

Winter, P. 1968. Social communication in the squirrel monkey. *In* Rosenblum and Cooper (1968), pp. 235–253.

Wirtz, W. O., II. 1972. Population ecology of the Polynesian rat, *Rattus exulans*, on Kure Atoll, Hawaii. Pacific Sci. 26:433–464.

————. 1973. Growth and development of *Rattus exulans*. J. Mamm. 54:189–202.

Wishart, W. 1978. Bighorn sheep. *In* Schmidt and Gilbert (1978), pp. 161–171.

Wistrand, H. 1974. Individual, social, and seasonal behavior of the thirteen-lined ground squirrel (*Spermophilus tridecemlineatus*). J. Mamm. 55:329–347.

Witte, G. R. 1971. Jungentransport in den Backentaschen beim Syrischen Goldhamster (*Mesocricetus auratus* Waterhouse, 1839). Z. Saugetierk. 36:216–219.

Wodzicki, K., and H. Felten. 1975. The peka, or fruit bat (*Pteropus tonganus tonganus*) (Mammalia, Chiroptera), of Niue

Island, South Pacific. Pacific Sci. 29:131–138.

Wodzicki, K., and J.E.C. Flux. 1971. The parma wallaby and its future. Oryx 11:40–47.

Wolfe, J. L., and A. V. Linzey. 1977. *Peromyscus gossypinus*. Mammalian Species, no. 70, 5 pp.

Wolfe, M. L., and D. L. Allen. 1973. Continued studies of the status, socialization, and relationships of Isle Royale wolves, 1967 to 1970. J. Mamm. 54:611–633.

Wolff, J. O., and G. C. Bateman. 1978. Effects of food availability and ambient temperature on torpor cycles of *Perognathus flavus* (Heteromyidae). J. Mamm. 59:707–716.

Wolman, A. A. 1978. Humpback whale. *In* Haley (1978), pp. 46–53.

Wolman, A. A., and D. W. Rice. 1979. Current status of the gray whale. Rept. Int. Whaling Comm. 29:275–279.

Womochel, D. R. 1978. A new species of *Allactaga* (Rodentia: Dipodidae) from Iran. Fieldiana Zool. 72:65–73.

Won, Pyong-Oh. 1968. Notes on the first propagating record of the Palaearctic flying squirrel, *Pteromys volans aluco* (Thomas) from Korea. J. Mamm. Soc. Japan 4:40–43.

Wood, A. E. 1977. The evolution of the rodent family Ctenodactylidae. J. Paleontol. Soc. India 20:120–137.

Wood, D. H. 1970. An ecological study of *Antechinus stuartii* (Marsupialia) in a southeast Queensland rain forest. Austral. J. Zool. 18:185–207.

————. 1971. The ecology of *Rattus fuscipes* and *Melomys cervinipes* (Rodentia: Muridae) in a south-east Queensland rain forest. Austral. J. Zool. 19:371–392.

Wood, G. W., and R. H. Barrett. 1979. Status of wild pigs in the United States. Wildl. Soc. Bull. 7:237–246.

Wood, G. W., and R. E. Brenneman. 1980. Feral hog movements and habitat use in coastal South Carolina. J. Wildl. Mgmt. 44:420–427.

Woods, C. A. 1973. *Erethizon dorsatum*. Mammalian Species, no. 29, 6 pp.

Woods, C. A., and D. K. Boraker. 1975. *Octodon degus*. Mammalian Species, no. 67, 5 pp.

Woolard, P.; W. J. M. Vestjens; and L. MacLean. 1978. The ecology of the eastern water rat *Hydromys chrysogaster* at Griffith, N.S.W.: food and feeding habits. Austral. Wildl. Res. 5:59–73.

Woolf, A.; T. O'Shea; and D. L. Gilbert. 1970. Movements and behavior of bighorn sheep on summer ranges in Yellowstone National Park. J. Wildl. Mgmt. 34:446–450.

Woolley, P. 1971*a*. Observations on the reproductive biology of the dibbler, *Antechinus apicalis* (Marsupialia: Dasyuridae). J. Roy. Soc. W. Austral. 54:99–102.

————. 1971*b*. Maintenance and breeding of laboratory colonies of *Dasyuroides byrnei* and *Dasycercus cristicauda*. Int. Zoo Yearbook 11:351–354.

————. 1977. In search of the dibbler, *Antechinus apicalis* (Marsupialia: Dasyuridae). J. Roy. Soc. W. Austral. 59:111–117.

Wrangham, R. W. 1977. Feeding behavior of chimpanzees in Gombe National Park, Tanzania. *In* Clutton-Brock (1977), pp. 503–537.

Wrigley, R. E. 1972. Systematics and biology of the woodland jumping mouse, *Napaeozapus insignis*. Illinois Biol. Monogr., no. 47, 117 pp.

Wrigley, R. E.; J. E. Dubois; and H.W.R. Copland. 1979. Habitat, abundance, and distribution of six species of shrews in Manitoba. J. Mamm. 60:505–520.

Wrigley, R. E., and D.R.M. Hatch. 1976. Arctic fox migrations in Manitoba. Arctic 29:147–158.

Würsig, B., and M. Würsig. 1980. Behavior and ecology of the dusky dolphin, *Lagenorhynchus obscurus*, in the South Atlantic. Fishery Bull. 77:871–890.

X.

Xanten, W. A.; H. Kafka; and E. Olds. 1976. Breeding the binturong *Arctictis binturong* at the National Zoological Park, Washington. Int. Zoo Yearbook 16:117–119.

Ximenez, A. 1975. *Felis geoffroyi*. Mammalian Species, no. 54, 4 pp.

Ximenez, A., and A. Langguth. 1970. *Akodon cursor montensis* en el Uruguay (Mammalia—Cricetinae). Comun. Zool. Mus. Hist. Nat. Montevideo 10(128):1–7.

Ximenez, A.; A. Langguth; and R. Praderi. 1972. Lista sistemática de los mamíferos del Uruguay. An. Mus. Nac. Hist. Nat. Montevideo 2(5):1–49.

Y.

Yager, R. H., and C. B. Frank. 1972. The nine-banded armadillo for medical research. Inst. Lab. Anim. Res. News 15(2):4–5.

Yahner, R. H. 1978. Burrow system and home range use by eastern chipmunks, *Tamias striatus*: ecological and behavioral considerations. J. Mamm. 59:324–329.

————. 1980. Activity patterns of captive Reeve's muntjacs (*Muntiacus reevesi*). J. Mamm. 61:368–371.

Yahr, P. 1977. Social subordination and scent-marking in male Mongolian gerbils (*Meriones unguiculatus*). Anim. Behav. 25:292–297.

Yalden, D. W. 1975. Some observations on the giant mole-rat *Tachyoryctes macrocephalus* (Rüppell, 1842) (Mammalia Rhizomyidae) of Ethiopia. Italian J. Zool., suppl., n.s., 15:275–303.

————. 1978. A revision of the dik-diks of the subgenus *Madoqua (Madoqua)*. Italian J. Zool., suppl., n.s., 11:245–264.

Yalden, D. W.; M. J. Largen; and D. Kock. 1976. Catalogue of the mammals of Ethiopia. 2. Insectivora and Rodentia. Italian J. Zool., suppl., n.s., 8:1–118.

————. 1977. Catalogue of the mammals of Ethiopia. 3. Primates. Italian J. Zool., suppl., n.s., 9:1–52.

————. 1980. Catalogue of the mammals of Ethiopia. 4. Carnivora. Italian J. Zool., suppl., n.s., 13:169–272.

Yamamoto, S. 1967. Breeding Japanese serows *Capricornis crispus* in captivity. Int. Zoo Yearbook 7:174–175.

Yates, T. L.; R. J. Baker; and R. K. Barnett. 1979. Phylogenetic analysis of karyological variation in three genera of peromyscine rodents. Syst. Zool. 28:40–48.

Yates, T. L., and D. J. Schmidly. 1977. Systematics of *Scalopus aquaticus* (Linnaeus) in Texas and adjacent states. Occas. Pap. Mus. Texas Tech Univ., no. 45, 36 pp.

————. 1978. *Scalopus aquaticus*. Mammalian Species, no. 105, 4 pp.

Yates, T. L.; D. J. Schmidly; and K. L. Culbertson. 1976. Silver-haired bat in Mexico. J. Mamm. 57:205.

Yeaton, R. I. 1972. Social behavior and social organization in Richardson's ground squirrel (*Spermophilus richardsonii*) in Saskatchewan. J. Mamm. 53:139–147.

Yoakum, J. D. 1978. Pronghorn. *In* Schmidt and Gilbert (1978), pp. 103–121.

Yocom, C. F. 1974. Status of marten in northern California, Oregon and Washington. California Fish and Game 60:54–57.

Yocom, C. F., and M. T. McCollum. 1973. Status of the fisher in northern California, Oregon and Washington. California Fish and Game 59:305–309.

Yong Hoi-Sen. 1970. A Malayan view of *Rattus edwardsi* and *R. sabanus* (Rodentia: Muridae). Zool. J. Linnean Soc. 49:359–369.

Yoshiyuki, M. 1979. A new species of the genus *Ptenochirus* (Chiroptera, Pteropodidae) from the Philippine Islands. Bull. Natl. Sci. Mus. (Tokyo), ser. A (Zool.) 5:75–81.

Youngman, P. M. 1975. Mammals of the Yukon Territory. Natl. Mus. Can. Publ. Zool., no. 10, 192 pp.

Z.

Zannier, F. 1965. Verhaltensuntersuchungen an der Zwergmanguste *Helogale undulata rufula* in Zoologischen Garten Frankfurt am Main. Z. Tierpsychol. 22:672–695.

Zara, J. L. 1973. Breeding and husbandry of the capybara *Hydrochoerus hydrochaeris* at Evansville Zoo. Int. Zoo Yearbook 13:137–139.

Zhou Kaiya; G. Pilleri; and Li Yuemin. 1979. Observations on the baiji (*Lipotes vexillifer*) and the finless porpoise (*Neophocaena asiaeorientalis*) in the Changjiang (Yangtze) River between Nanjing and Taiyangzhou, with remarks on some physiological adaptations of the baiji to its environment. Investig. Cetacea 10:109–120.

Zhou Kaiya; Qian Weijuan; and Li Yuemin. 1977. Studies on the distribution of Baiji, *Lipotes vexillifer* Miller. Acta Zool. Sinica 23:72–79.

————. 1979. The osteology and the systematic position of the baiji, *Lipotes vexillifer*. Acta Zool. Sinica 25:58–74.

Ziegler, A. C. 1972. Additional specimens of *Planigale novaeguineae* (Dasyuridae: Marsupialia) from Territory of Papua. Austral. Mamm. 1:43–45.

————. 1977. Evolution of New Guinea's marsupial fauna in response to a forested environment. *In* Stonehouse and Gilmore (1977), pp. 117–138.

Zihlman, A., and J. M. Lowenstein. 1979. False start of the human parade. Nat. Hist. 88(7):86–91.

Zimen, E. 1975. Social dynamics of the wolf pack. *In* Fox (1975), pp. 336–362.

————. 1981. Italian wolves. Nat. Hist. 90(2):66–81.

Zimina, R. P., and I. P. Gerasimov. 1973. The periglacial expansion of marmots (*Marmota*) in middle Europe during late Pleistocene. J. Mamm. 54:327–340.

Zimina, R. P., and E. V. Yasny. 1977. Observations on the ecology of *Prometheomys schaposchnikovi* Sat. Bull. Moscow Soc. Nat., Biol. Ser., 82:24–30.

INDEX

The scientific names of orders, families, and genera, which have titled accounts in the text, are in boldface type. The page numbers on which such accounts *begin* are also in boldface type. Other scientific names, and vernacular names, appear in ordinary type. If the scientific and common names of a mammal are identical, they are usually indexed separately (for example, **Addax** and Addax). If, however, the scientific name does not qualify for boldface type, and the common name is used in the singular, the two names are indexed as one (for example, Anoa). If no illustration of the genus appears on the page indexed for that genus, the reader should look one or two pages either before or after the account. However, not every account has an accompanying illustration (see Preface).

about the authors of the fourth edition

Ronald M. Nowak and John L. Paradiso are mammalogists with the Office of Endangered Species, Fish and Wildlife Service, U.S. Department of the Interior. Nowak is a staff specialist and Paradiso is senior zoologist. Nowak served as editorial consultant for the National Geographic Society's *Wild Mammals of North America*, to which Paradiso also was a contributor. In addition, Nowak is the author of *North American Quaternary Canis*. Paradiso, the author of *Mammals of Maryland*, edited the second and third editions of Walker's *Mammals of the World*. Nowak and Paradiso have worked together for fourteen years; this is their eighth joint publication.

TABLES FOR CONVERSION

U.S. Customary to Metric   Metric to U.S. Customary

—— Length ——

To convert	Multiply by	To convert	Multiply by
in. to mm.	25.4	mm. to in.	0.039
in. to cm.	2.54	cm. to in.	0.394
ft. to m.	0.305	m. to ft.	3.281
yd. to m.	0.914	m. to yd.	1.094
mi. to km.	1.609	km. to mi.	0.621

—— Area ——

To convert	Multiply by	To convert	Multiply by
sq. in. to sq. cm.	6.452	sq. cm. to sq. in.	0.155
sq. ft. to sq. m.	0.093	sq. m. to sq. ft.	10.764
sq. yd. to sq. m.	0.836	sq. m. to sq. yd.	1.196
sq. mi. to ha.	258.999	ha. to sq. mi.	0.004

—— Volume ——

To convert	Multiply by	To convert	Multiply by
cu. in. to cc.	16.387	cc. to cu. in.	0.061
cu. ft. to cu. m.	0.028	cu. m. to cu. ft.	35.315
cu. yd. to cu. m.	0.765	cu. m. to cu. yd.	1.308

—— Capacity (liquid) ——

To convert	Multiply by	To convert	Multiply by
fl. oz. to liter	0.03	liter to fl. oz.	33.815
qt. to liter	0.946	liter to qt.	1.057
gal. to liter	3.785	liter to gal.	0.264

—— Mass (weight) ——

To convert	Multiply by	To convert	Multiply by
oz. avdp. to g.	28.35	g. to oz. avdp.	0.035
lb. avdp. to kg.	0.454	kg. to lb. avdp.	2.205
ton to t.	0.907	t. to ton	1.102
l. t. to t.	1.016	t. to l. t.	0.984

Abbreviations

U.S. Customary

avdp.—avoirdupois
ft.—foot, feet
gal.—gallon(s)
in.—inch(es)
lb.—pound(s)
l. t.—long ton(s)
mi.—mile(s)
oz.—ounce(s)
qt.—quart(s)
sq.—square
yd.—yard(s)

Metric

cc.—cubic centimeter(s)
cm.—centimeter(s)
cu.—cubic
g.—gram(s)
ha.—hectare(s)
kg.—kilogram(s)
m.—meter(s)
mm.—millimeter(s)
t.—metric ton(s)

SCALES FOR COMPARISON OF METRIC AND U.S. UNITS OF MEASUREMENT

WEIGHT

GRAMS AND OUNCES

GRAMS	OUNCES
453.59	16
450	
440	
430	15
420	
410	
400	14
390	
380	
370	13
360	
350	
340	12
330	
320	
310	11
300	
290	
280	10
270	
260	
250	9
240	
230	8
220	
210	
200	7
190	
180	
170	6
160	
150	
140	5
130	
120	
110	4
100	
90	3
80	
70	
60	2
50	
40	
30	1
20	
10	
0	0

KILOGRAMS AND POUNDS

KG.	LB.
45.36	100
45	
	95
	90
40	
	85
	80
35	75
	70
30	65
	60
25	55
	50
	45
20	40
	35
15	30
	25
10	20
	15
5	10
	5
0	0

KILOGRAMS	POUNDS

KG. / LB.

KG.	LB.
907.18	2000
900	
800	1750
700	1500
600	1250
500	1102.31
453.59	1000
450	
400	900
	800
350	700
300	
250	600
	500
200	400
150	300
100	200
50	100
0	0

KILOGRAMS	POUNDS

METRIC TONS AND U.S. SHORT TONS

MET.	U.S.
90	100
85	95
80	90
75	85
70	80
65	75
60	70
55	65
50	60
45	55
40	50
35	45
30	40
25	35
20	30
15	25
13.61	20
9.07	15
9	10
8	9
7	8
6	7
5	6
4	5
3	4
2	3
1	2
0	1
	0

METRIC TONS	U.S. TONS